Biology of
Disease Vectors

SECOND EDITION

Biology of Disease Vectors

SECOND EDITION

Edited by

William C. Marquardt

Department of Biology
Colorado State University
Fort Collins, Colorado

Section Editors

Boris C. Kondratieff

Department of BioAgricultural Sciences and Pest
Management
Colorado State University
Fort Collins, Colorado

Chester G. Moore

Environmental Health Advanced Systems Laboratory
Department of Environmental and Radiological
Health Sciences
Colorado State University
Fort Collins, Colorado

Jerome E. Freier

USDA, APHIS, VS Centers for Epidemiology and
Animal Health
Fort Collins, Colorado

Henry H. Hagedorn

Department of Entomology
University of Arizona
Tucson, Arizona

William C. Black IV

Department of Microbiology
Colorado State University
Fort Collins, Colorado

Anthony A. James

Department of Molecular Biology and Biochemistry
University of California
Irvine, California

Janet Hemingway

Liverpool School of Tropical Medicine
Liverpool, United Kingdom

Stephen Higgs

Department of Pathology, Center for Biodefense and
Emerging Infectious Diseases, Sealy Center for
Vaccine Development, and WHO Collaborating
Center for Tropical Diseases
University of Texas Medical Branch (UTMB)
Galveston, Texas

ELSEVIER
ACADEMIC
PRESS

Amsterdam · Boston · Heidelberg · London · New York · Oxford
Paris · San Diego · San Francisco · Singapore · Sydney · Tokyo

Publisher	Dana Dreibelbis
Associate Acquisitions Editor	Kelly D. Sonnack
Project Manager	Troy Lilly
Marketing Manager	Linda Beattie
Cover Design	Eric DeCicco
Composition	SNP Best-set Typesetter Ltd., Hong Kong
Cover Printer	C & C Offset Printing Company, Limited
Interior Printer	C & C Offset Printing Company, Limited

Elsevier Academic Press
30 Corporate Drive, Suite 400, Burlington, MA 01803, USA
525 B Street, Suite 1900, San Diego, California 92101-4495, USA
84 Theobald's Road, London WC1X 8RR, UK

This book is printed on acid-free paper. ♾

Library of Congress Cataloging-in-Publication Data
Biology of disease vectors / edited by William C. Marquardt . . . [et al. — 2nd ed.
    p. cm.
    Includes bibliographical references and index.
    ISBN 0-12-473276-3 (hardcover : alk. paper)
    1. Insects as carriers of disease. 2. Insects — Molecular aspects. I. Marquardt, William C.
    RA639.5.B56 2004
    614.4′32 — dc22

                                                    2004011925

British Library Cataloguing in Publication Data
A catalogue record for this book is available from the British Library

ISBN: 0-12-473276-3

For all information on all Elsevier Academic Press Publications
visit our Web site at www.books.elsevier.com

Printed in China
04 05 06 07 08 09   9 8 7 6 5 4 3 2 1

Gratis Book Program

In order that individual investigators in developing countries might have access to *Biology of Disease Vectors*, the following individuals and organizations have contributed funds to allow the distribution of books without charge. We hope that the second edition will be met with the same approval that was the first.

William C. Black IV

Carol D. Blair

Department of Biology, Colorado State University

Department of Microbiology, Immunology & Pathology, Colorado State University

Florida Medical Entomology Laboratory, University of Florida

Jerome E. Freier

Henry H. Hagedorn

Janet Hemingway

Stephen Higgs

Anthony A. James

Marc J. Klowden

Boris C. Kondratieff

Kitsos Louis and the Crete Mosquito Group

William C. Marquardt

Chester G. Moore

Diane McMahon-Pratt

Leonard E. Munstermann

Walter J. Tabachnick

Michael A. Wells

Contents

Addresses of Authors and Contributors

Dr. Peter Adler
Department of Entomology, Soils, and Plant Sciences
Clemson University
Box 340315, 114 Long Hall
Clemson, South Carolina 29634-0315
USA

Dr. Serap Aksoy
Yale University
60 College Street, 606 LEPH
New Haven, Connecticut 06510-3210
USA

Dr. Francisco Alarcon-Chaidez
Center for Microbial Pathogenesis
School of Medicine
University of Connecticut Health Center
262 Farmington Avenue, MC3710
Farmington, Connecticut 06030
USA

Jennifer Anderson
Johns Hopkins Bloomberg School of Public Health
W. Harry Feinstone Department of Molecular
    Microbiology and Immunology
615 N. Wolfe Street, Room E3402
Baltimore, Maryland 21205
USA

Dr. Peter Atkinson
Department of Entomology, Entomology 339
University of California-Riverside
Riverside, California 92521-0314
USA

Dr. Abdu F. Azad
Department of Microbiology and Immunology
School of Medicine
University of Maryland
655 West Baltimore Street
Bressler Research Building 13-009
Baltimore, Maryland 21201
USA

Dr. Stephen C. Barker
Department of Microbiology and Parasitology
Institute for Molecular Bioscience
The University of Queensland, Brisbane, 4072
Australia

Dr. Carolina Bariilas-Mury
Laboratory of Malaria and Vector Research
12735 Twinbrook Parkway
Room 2E32
National Institutes of Health, NIAID
Rockville, Maryland 20852
USA

Dr. C. Ben Beard
Chief, Bacterial Zoonosis Branch
Division of Vector-Borne Diseases
Centers for Disease Control and Prevention
Rampart Road
Fort Collins, Colorado 80525
USA

Dr. Barry J. Beaty
Department of Microbiology and Arethropiod-Bone
    and Infectious Diseases Laboratory
Colorado State University
Fort Collins, Colorado 80523
USA

Dr. Klaus Beyenbach
Department of Biomedical Sciences
VRT 8014
Cornell University
Ithaca, New York 14853
USA

Dr. William C. Black
Department of Microbiology
Colorado State University
Fort Collins, Colorado 80523
USA

Dr. Carol. D. Blair
Department of Microbiology
Colorado State University
Fort Collins, Colorado 80523
USA

Dr. Art Borkent
691-8th Ave. SE,
Salmon Arm, British Columbia, V1E 2C2
Canada

Dr. Keith R. Bouchard
Center for Microbial Pathogenesis School of Medicine
University of Connecticut Health Center
263 Farmington Avenue, MC3710
Farmington, Connecticut 06030
USA

Dr. Jonathon O. Carlson
Department of Microbiology
Colorado State University
Fort Collins, Colorado 80523
USA

Dr. Craig J. Coates
Assistant Professor
Center for Advanced Invertebrate Molecular Sciences
Department of Entomology
Texas A&M University
College Station, Texas 77843-2475
USA

Dr. Martin Devenport
Johns Hopkins University
Bloomberg School of Public Health
Department of Molecular Microbiology and
    Immunology
Malaria Research Institute
615 North Wolfe Street, W4109
Baltimore, Maryland 21205
USA

Dr. Uwe Müeller-Doblies
Center for Microbial Pathogenesis
School of Medicine
University of Connecticut Health Center
263 Farmington Avenue, MC3710
Farmington, Connecticut 06030
USA

Dr. Lance A. Durden
Dept. of Biology and Institute of Arthropodology and
    Parasitology
Georgia Southern University
P.O. Box 8042
Statesboro, Georgia 30460-8042
USA

Dr. Bruce Eldridge
Center for Vectorborne Diseases
University of California
Old Davis Road
Davis, California 95616
USA

Dr. Ann M. Fallon
Department of Entomology
University of Minnesota
1980 Folwell Avenue
Saint Paul, Minnesota 55108
USA

Pierre-Edouard Fournier, MD, PhD
Unité des Rickettsies
Faculté de Médecine
27 Blvd Jean Moulin
13385 Marseille cedex 5
France

Dr. Jerome E. Freier
USDA, APHIS, VS Centers for Epidemiology and
    Animal Health
555 South Howes Street
Suite 100
Fort Collins, Colorado 80521-2865
USA

Dr. Kenneth Gage
CDC
CSU Foothills Campus/Rampart Rd.
Fort Collins, Colorado 80521
USA

Dr. Ronald H. Gooding
Department of Biological Sciences
University of Alberta
Edmonton, Alberta T6G 2E9
Canada

Dr. Henry Hagedorn
Department of Entomology
410 Forbes Building
University of Arizona
Tucson, Arizona 85721
USA

Dr. Lee R.Haines
Department of Biochemistry and Microbiology
Petch Building, Ring Road
P.O. Box 3055
Victoria, British Columbia V8W 3P6
Canada

Dr. Janet Hemingway
Liverpool School of Tropical Medicine
Pembroke Place
Liverpool L3 5QA
United Kingdom

Stephen Higgs B.Sc., Ph.D., F.R.E.S.
2.138A Keiller Building
Department of Pathology, Center for Biodefense and
   Emerging Infectious Diseases, Sealy Center for
   Vaccine Development, and WHO Collaborating
   Center for Tropical Diseases
University of Texas Medical Branch (UTMB)
301 University Boulevard
Galveston, Texas 77555-0609
USA

Gregg J. Hunt, Director
Beaufort County Mosquito Control
84 Shanklin Road
Beaufort, South Carolina 29906
USA

Dr. Jun Isoe
Department of Biochemistry and Molecular
   Biophysics and Center for Insect Science
Biological Sciences West
University of Arizona
Tucson, Arizona 85721-0088
USA

Dr. Marcelo Jacobs-Lorena
Department of Genetics
School of Medicine
Case Western Reserve University
10900 Euclid Avenue
Cleveland, Ohio 44106-4955
USA

Dr. Anthony A. James, Ph.D.
Department of Molecular Biology and Biochemistry
3205 Bio Science II
University of California, Irvine
Irvine, California 92697-3900
USA

Dr. Deborah C. Jaworski
Department of Molecular Biology and Biochemistry
University of California, Irvine
McGaugh Hall
Irvine, California 92697
USA

Dr. Michael Kanost
Department of Biochemistry
449 Chem-Biochem Building
Kansas State University
Manhattan, Kansas 66506-3702
USA

Dr. Hans Klompen
Ohio State University
Museum of Biological Diversity
1315 Kinnear Road
Columbus, Ohio 43212-1192
USA

Dr, Marc Klowden
Division of Entomology
University of Idaho
Moscow, Idaho 83844-2339
USA

Dr. Vladimir Kokoza
Department of Entomology
University of California, Riverside
3401 Watkins Drive
Riverside, California 92521
USA

Dr. Boris Kondratieff
Colorado State University
Department of BioAgricultural Sciences and Pest
   Management
Fort Collins, Colorado 80523
USA

Timothy J. Kurtti, Ph.D.
Department of Entomology
University of Minnesota
219 Hodson Hall
1980 Folwell Avenue
St. Paul, Minnesota 55108-6125
USA

Dr. James B. Lok, Ph.D
Department of Pathobiology
School of Veterinary Medicine
University of Pennsylvania
3800 Spruce Street
Philadelphia, Pennsylvania 19104-6008
USA

Dr. Christos (Kitsos) Louis
IMBB-FORTH
Vassilika Vouton, P.O. Box 1527
GR-711 10 Heraklion, Crete
Greece

Dr. Kevin R. Macaluso
Department of Microbiology and Immunology
School of Medicine
University of Maryland, Baltimore
655 West Baltimore Street
Bressler Research Building 13-009
Baltimore, Maryland 21201
USA

Dr. William C. Marquardt
Department of Biology
Colorado State University
Fort Collins, Colorado 80523
USA

Dr. Uwe Müeller-Doblies
Center for Microbial Pathogenesis
School of Medicine
University of Connecticut Health Center
262 Farmington Avenue, MC3710
Farmington, Connecticut 06030
USA

Dr. Chester G. Moore
Environmental Health Advanced Systems Laboratory
Department of Environmental and Radiological
    Health Sciences
1681 Campus Delivery
Colorado State University
Fort Collins, Colorado 80523-1681
USA

Dr. Leonard E. Munstermann
Yale University School of Medicine
P.O. Box 208034
706 LEPH
60 College Street
New Haven, Connecticut 06510-8034
USA

Dr. Fernando G. Noriega
Department of Biological Sciences
Florida International University
Miami, Florida 33199
USA

Dr. Kenneth E. Olson
Foothills Research Campus AIDL
3107 Rampart Road
Department of Microbiology
Colorado State University
Fort Collins, Colorado 80523
USA

Dr. Susan Paskewitz
Department of Entomology
University of Wisconsin-Madison
1630 Linden Drive
Madison, Wisconsin 53706
USA

Dr. James E. Pennington
Department of Biochemistry and Molecular
    Biophysics and Center for Insect Science
Biological Sciences West
University of Arizona
Tucson, Arizona 85721-0088
USA

Pamela M. Pennington, Ph.D.
Center for Health Studies
Universidad del Valle de Guatemala
18 Ave. 11-95 Z. 15 V.H.III
Guatemala City
Guatemala

Dr. Terry W. Pearson
Department of Biochemistry and Microbiology
Petch Building
University of Victoria
P.O. Box 3055
Victoria, BC V8W3P6
Canada

Dr. Hilary Ranson
Liverpool School of Tropical Medicine
Pembroke Place
Liverpool L3 5QA
United Kingdom

Dr. Alexander S. Raikhel
Department of Entomology
3401 Watkins Dr.
University of California,
Riverside, California 92521
USA

Didier Raoult, MD, Ph.D.
Unité des Rickettsies
Faculté de Médecine
27 Blvd Jean Moulin
13385 Marseille cedex 5
France

Dr. Mowafak D. Salman
Department of Environmental Health
14B Environmental Health Building
Colorado State University
Fort Collins, Colorado 80523
USA

Dr. Patricia Y. Scaraffia
Department of Biochemistry and Molecular
  Biophysics and Center for Insect Science
Biological Sciences West
University of Arizona
Tucson, Arizona 85721-0088
USA

Dr. Edward T. Schmidtmann
Arthropod-Borne Animal Diseases Research
  Laboratory
USDA, ARS
P.O. Box 3965, University Station
Laramie, Wyoming 82071-3965
USA

Dr. Dave Severson
Center for Tropical Disease Research and Training
Department of Biological Sciences
University of Notre Dame
Notre Dame, Indiana 46556
USA

Dr. Graham Small, FRES
Insect Investigations Ltd. (12L)
Capital Business Park, Wentloog, Cardiff CF3 2PX
United Kingdom

Dr. Walter J. Tabachnick
Florida Medical Entomology Laboratory
200 9th Street SE
Vero Beach, FL 32962-4699
USA

Dr. Rex Thomas
Bioenvironmental Associates
4117 Sumter Street
Fort Collins, Colorado 80525
USA

Dr. Jesus Valenzuela
Room 126, Building 4
4 Center Drive
MSC 0425
Bethesda, Maryland 20892-0425
USA

Dr David Evans Walter
Department of Biological Sciences
The University of Alberta
Edmonton, Alberta T6G 2E9
Canada

Dr. Michael A. Wells
Department of Biochemistry
Biosciences West
Room 445A
University of Arizona
Tucson, Arizona 85721
USA

Dr. Stephen K. Wikel, Ph.D.
Center for Microbial Pathogenesis
School of Medicine
University of Connecticut Health Center
263 Farmington Avenue, MC3710
Farmington, Connecticut 06030
USA

Dr. Roger John Wood
School of Biological Sciences
3.614 Stopford Building
Oxford Road
Manchester M13 9PT
United Kingdom

Dr. Guoli Zhou
Department of Biochemistry and Molecular
  Biophysics and Center for Insect Science
Biological Sciences West
University of Arizona
Tucson, Arizona 85721-0088
USA

Dr. Jinson Zhu
Department of Entomology
University of California-Riverside
3401 Watkins Drive
Riverside, California 92521
USA

Dr. Laurence J. Zweibel
Departments of Biology & Molecular Biology
Box 82, Station B
Vanderbilt University
Nashville, Tennessee 37235
USA

Preface

In the relatively few years since Dr. Barry J. Beaty and I edited the first edition of *The Biology of Disease Vectors,* much has happened that stimulated us to produce this second edition. The first edition was written to accompany the *Biology of Disease Vectors* course given, during most of the years since 1990, at Colorado State University. Users and reviewers generally felt that the book was a good addition to the literature available in medical entomology. As time passed, faculty and course content changed somewhat, and significant advances were made in the field.

This edition, like the first, is directed toward graduate students, postdoctoral fellows, and independent investigators working in the broad areas of vector biology and vector-borne diseases. We have attempted to include all of those subject areas that may concern our readers. The 57 chapters are meant to be not technical reviews but rather fairly sophisticated presentations for students and workers in vector biology. The first edition was a direct outgrowth of the disease vectors course, but the section editors and I felt that we should serve a broader readership in this second edition.

We have divided the subject matter into seven parts, with a different editor responsible for the material that would appear within a part. Each part editor has functioned more or less autonomously while interacting with the others as needed. This has generally worked well, because the part editors selected their authors and suggested the content of the chapters.

The manuscript for the first edition was completed in 1995. Since that time, methods and concepts have changed considerably. For example, a major molecular advance was completing the nucleic acid sequences of the *Anopheles gambiae* genome, the principal vector of malaria in sub-Saharan Africa. As with the completion of the human genome, these data now open many areas for investigation on this vector that were inaccessible a few years ago. Genomics has advanced so that we can now determine which genes are up-regulated or down-regulated and under what conditions. An alternative approach to gene function, proteomics, now allows an investigator to isolate and characterize the proteins produced in specific organs under different conditions. By combining these powerful techniques with relevant biological questions we can anticipate rapid advances in vector biology.

In the broad areas of systematics and epidemiology, widely differing methods have had significant impacts. In molecular systematics, for example, an explosion in the availability of nucleic acid sequences in insect vectors has allowed for more in-depth analysis of phylogenetic relationships among species. This has also impacted studies of both gene flow and the relationships of closely related, sometimes cryptic, species. This area has developed as *molecular epidemiology.* Questions on evolutionary processes and the identity of species are now more accessible using newly available methods and the genomes of vectors.

Epidemiology has also benefited from the Global Information System (GIS) to show the impact of environmental changes on vector populations. The satellite data used in GIS studies have progressively shown greater resolution, so predictions can also be more refined.

Bioinformatics, the use of computers in biology, undergirds many recent advances. The first conference on bioinformatics was held in 1984. Since that time, computer power has increased dramatically and programming has become more sophisticated. The human element of bioinformatics has changed also, because students who come into graduate work are already computer literate and they can move the field ahead without the lag time of earlier cohorts of graduate students. It might be said that the computer has changed the way we do research, but it can also be said that the computer had to be invented just at the time that it was. Those of us who entered biology precomputer struggled to encompass and analyze even moderate amounts of data; that can now be done in a matter of minutes.

The authors of the chapters joined this project out of a commitment to their subject areas. Likewise, the part editors participated out of devotion, rather than for financial return. It has been an experience that exemplifies the best in science: knowledge and effort freely given while searching for advances in understanding and providing solutions to signfiicant problems.

I have included the preface to the first edition, because it gives a broad view of the evolution of knowledge about vectors and vector-borne diseases, and the observations are still valid. We have come a long way since mosquitoes were first implicated in transmitting filariasis in humans, and it is now exactly 100 years since Sir Ronald Ross won the Nobel Prize for his discovery that mosquitoes transmit the malarial parasite. But we still have a long way to go in ameliorating the impact of vector-borne diseases.

I believe this is the golden age of biology. With methods now available and with well-trained investigators, we can learn and we can apply knowledge in ways undreamed of a generation ago. Benefits are accruing to humankind's welfare not only in vector-borne diseases, but also in all other areas of biology.

William C. Marquardt

Preface to the First Edition

The 19th and 20th centuries were exciting times in investigations on transmissible diseases. Pasteur, Koch, Bruce, Grassi, Schaudinn, Ross, and scores of other individuals found and described many microbial and eukaryotic disease agents. Organisms were cultivated from diseased humans and animals, vaccines were developed, and immunity was studied for use in diagnosis and prevention of disease. Koch's postulates were promulgated and used to prove that various microorganisms were the causes of diseases.

As it became clear what kinds of agents decimated both human and animal health, effective control techniques were instituted by newly established public health entities. Water treatment and sewage treatment improved. Quarantine was practiced to prevent the spread of diseases that were transmitted directly from one person to another. Vaccination for smallpox, typhoid fever, and diphtheria reduced both morbidity and mortality. Antibiotics and other chemotherapeutic agents improved so that many serious diseases seemed to be a thing of the past. By the mid-1950s effective vaccines all but eliminated childhood scourges such as poliomyelitis. Smallpox was globally eradicated in 1977. Many predicted that infectious diseases had been conquered.

Despite these successes, failures lurked in the background. The dark shadow was cast by many vector-borne diseases, which had complex epidemiologies and reservoirs in various animals other than humans and flared up unpredictably. The vector-borne agents included the whole spectrum of infectious agents: viruses, rickettsiae, bacteria, protists, and helminths. Except for the blood flukes (*Schistosoma* spp.), most of the disease agents were found to be transmitted by arthropods: lice, bugs, mosquitoes, black flies, midges, sand flies, ticks, and mites.

The disease agents were studied and their life cycles were described. The vectors were classified, raised in the laboratory, and their intimate habits probed. Nevertheless, malaria, dengue, trypanosomiasis, filariasis, and the viral encephalitides simmered and surged periodically. Vaccines were developed for some, such as yellow fever. Drugs were developed for malaria and trypanosomiasis, among others. Source reduction for control of insect vectors was effective when continual effort could be given to maintaining the system, but outbreaks persisted.

A turning point came in the late 1930s with the development of DDT. An effective, broad spectrum insecticide, it had all of the qualities that a good insecticide was thought to have. DDT aborted a typhus outbreak in North Africa and Italy in WWII. It controlled mosquitoes and thus could control mosquito-borne diseases. In the 1950s investigators thought that malaria could be eradicated by killing mosquitoes with DDT and by administering chloroquine to humans to prevent and treat cases. Good insecticides, good drugs, and a concerted campaign meant that success was at hand. But it was not to be, and we are now resigned to containing malaria and keeping its depredations to an acceptable level.

Wherein lay the problem in eradicating malaria? Part of it was economics and politics; as soon as eradication was imminent, other diseases gained priority and money was shifted elsewhere. Evolution also intervened as the mosquitoes developed resistance to insecticides and the malaria organisms became resistant to chloroquine.

In other vector-borne diseases, reservoirs maintained disease agents in the wild, and the agents essentially disappeared between outbreaks in the human populations. Plague would break out in rodents and then disappear, perhaps after causing some human infections and deaths. Typhus would become epidemic in difficult times, disappear, and then reappear to visit another disaster on humanity. Dengue waxed and waned and then spread from the Pacific Rim throughout the tropical world. New diseases, such as the tick-borne Lyme disease, emerged.

By the 1980s it became clear that the campaigns against various vector-borne diseases were at a stalemate at best. The long-term support of vector-borne disease research by institutions such as the Rockefeller Foundation, the Walter Reed Army Institute of Research, the U.S. Army Medical Research Institute of Infectious Diseases, the National Institutes of Health in the United States, and other institutions in many parts of the world laid the groundwork for a shift in emphasis in research involving vector-borne diseases. New molecular techniques, genetic cloning, genomic mapping, and the ability to transform organisms permitted identifying and expressing genes in ways that would not have been believed a few years earlier. A molecular revolution was under way in laboratories across the globe.

In the early 1980s the John D. and Catherine T. MacArthur Foundation initiated and supported a project to increase knowledge at the molecular level of the main parasitic diseases of humans: malaria trypanosomiasis, leishmaniasis, and filariasis. The result was an influx of investigators with expertise in immunology, biochemistry, and molecular biology who focused on these disease agents. A new cadre of individuals was recruited into parasitology.

In 1989 a similar program was initiated by the MacArthur Foundation; the Network on the Biology of Parasitic Vectors was formed. Eight academic and research institutions were joined in the network. One component of the program involved training a new generation of vector biologists. To help accomplish this, an annual course called The Biology of Disease Vectors was initiated. The first effort was undertaken at Colorado State University in June 1990. The course was subsequently offered at Colorado State University through 1993 and at the Institute of Molecular Biology and Biotechnology in Crete in 1994.

The objectives of the course have been to introduce molecular biologists to medical entomology, to give experience to medical entomologists in molecular concepts and methods, and to develop a worldwide network of vector biologists. As the instructors and organizers gained experience, it became clear that the course was unique and that much of the information presented would be valuable to those who could not attend. Thus, in 1992, the decision was made to publish

a multiauthor textbook based on the course, *The Biology of Disease Vectors*. The volume follows much the same pattern of exposition as the lecture portion of the course, but considerably expanded. It includes basic knowledge of arthropods of medical and veterinary importance; epidemiology, development, physiology, feeding, and metabolism, especially as they pertain to vectors; population biology and genetics; and methods of surveillance and control. In each chapter the emphasis has been on cutting-edge molecular biological approaches that illuminate the difficult problems facing vector biologists. Because of the nature of the material covered, the book complements rather than replaces other medical entomology textbooks; for example, *Medical Insects and Arachnids* by Lane and Crosskey (Chapman and Hall, New York) and *Medical and Veterinary Entomology* by Kettle (C.A.B. International, Wallingford, U.K., 1990). *Entomology in Human and Animal Health* by Harwood and James (Macmillan, New York, 1979) is no longer in print.

The contributors are drawn mostly from the faculty of the course and the MacArthur Network, but a number of other investigators agreed to contribute chapters. Because the volume is so directly connected to the course, the organization of the text and the chapters is different from other medical entomology texts. Contributors were given considerable latitude in selecting and organizing the information presented in their respective chapters. Some contributors presented the information in conventional textbook form with minimal referencing; others chose to present their material more in the form of a review article. We are indebted to them all for providing their expertise and for completing their chapters in a timely fashion. We are also indebted to the MacArthur Foundation and to Dr. Denis Prager for support of this effort.

When we began editing this volume, we hoped to produce a landmark textbook in vector biology. We have tapped the talents of 45 investigators and leaders in vector biology (s.l.), and their chapters provide state-of-the-art introductions into their respective areas of research. The contributors all have active research programs, and it was only through a labor of love that they took their time to produce readable, current summaries of knowledge. As editors, we thank them for their efforts, and we hope that we have achieved our objectives.

Because of the support of the MacArthur Foundation for the Network on the Biology of Parasite Vectors, research and educational opportunities are influencing vector biology. More than 150 students from all over the world have attended the 2-week course. Their total immersion in the subject has introduced them to a number of areas of vector biology. They have also made friends and colleagues that, we hope, will last a lifetime. The network is established, the techniques are at hand, and the individuals have been trained by experts in the field. We expect and anticipate great things from both students and faculty who have participated in the disease vectors course.

Perhaps we are on the edge of what will become known as a new golden age of medical entomology. Let us hope that is true.

Barry J. Beaty
William C. Marquardt

INTRODUCTION AND VECTORS

BORIS C. KONDRATIEFF

1

The Arthropods

BORIS C. KONDRATIEFF AND WILLIAM C. BLACK IV

INTRODUCTION

Arthropods, the joint-footed animals, are a highly successful group of animals unrivaled by any others in diversity of structure and function. As a group they first appeared from the Precambrian to the Cambrian divisions of earth's history, and they have radiated into unsurpassed lineages, both extinct and extant. Arthropods occur practically everywhere and have adapted to nearly every imaginable source of food, including vertebrate blood.

Arthropods are notable for having a segmental body plan made up of chitinous exoskeletal plates separated by less hardened membranous areas. Paired segmental appendages are usually present on at least some body segments. Additionally, important distinguishing characters include a ventral chain of segmental ganglia, an open circulatory system, and a body cavity that is a hemocoel. Most arthropods ventilate by means of a tracheal system and have well-developed appendicular mouthparts.

Extant arthropods include the chelicerates, crustaceans, myriapods, and insects. The distinctions among these groups are based largely on the manner in which the different segments are grouped together to form compact and distinct parts of the body and on the number and position of the appendages (Fig. 1.1). Two of these groups, the chelicerates (represented by mites, ticks, scorpions, and relatives) and the insects, include many important vector taxa; these are discussed in detail.

THE CHELICERATA

This diverse group of arthropods is characterized by possessing chelicerae and pedipalpi, lacking antennae, and having four pairs of legs as adults and a cephalothorax and abdomen. The class Arachnida constitutes the largest and most important class of Chelicerata, with nearly 70,000 species worldwide. Most authorities recognize at least 11 major groups of arachnids, but with much disagreement on names for the groups. The most important arachnid groups that may affect human health directly or indirectly are the acarines and ticks (Chapters 3 and 4).

THE INSECTS (HEXAPODA)

The 31 extant hexapoda orders (Table 1.1) (following Borror et al. 1989) include some of the most unique and spectacular animals on earth. About 1 million species have been described, and there have been estimates of as high as 8 million total species existing on our planet. The impact of insects on the "speciescape" (Wheeler 1990) is immense. Their high species richness has been attributed to their small size, their short generation time, their rather sophisticated nervous system, their coevolution with plants, their ability to fly, and their different development strategies.

As a group, hexapods are recognized by three main body regions (head, thorax, abdomen), three pairs of legs (restricted to the thorax), and one pair of antennae (Fig. 1.2). Adult insects normally have wings, but

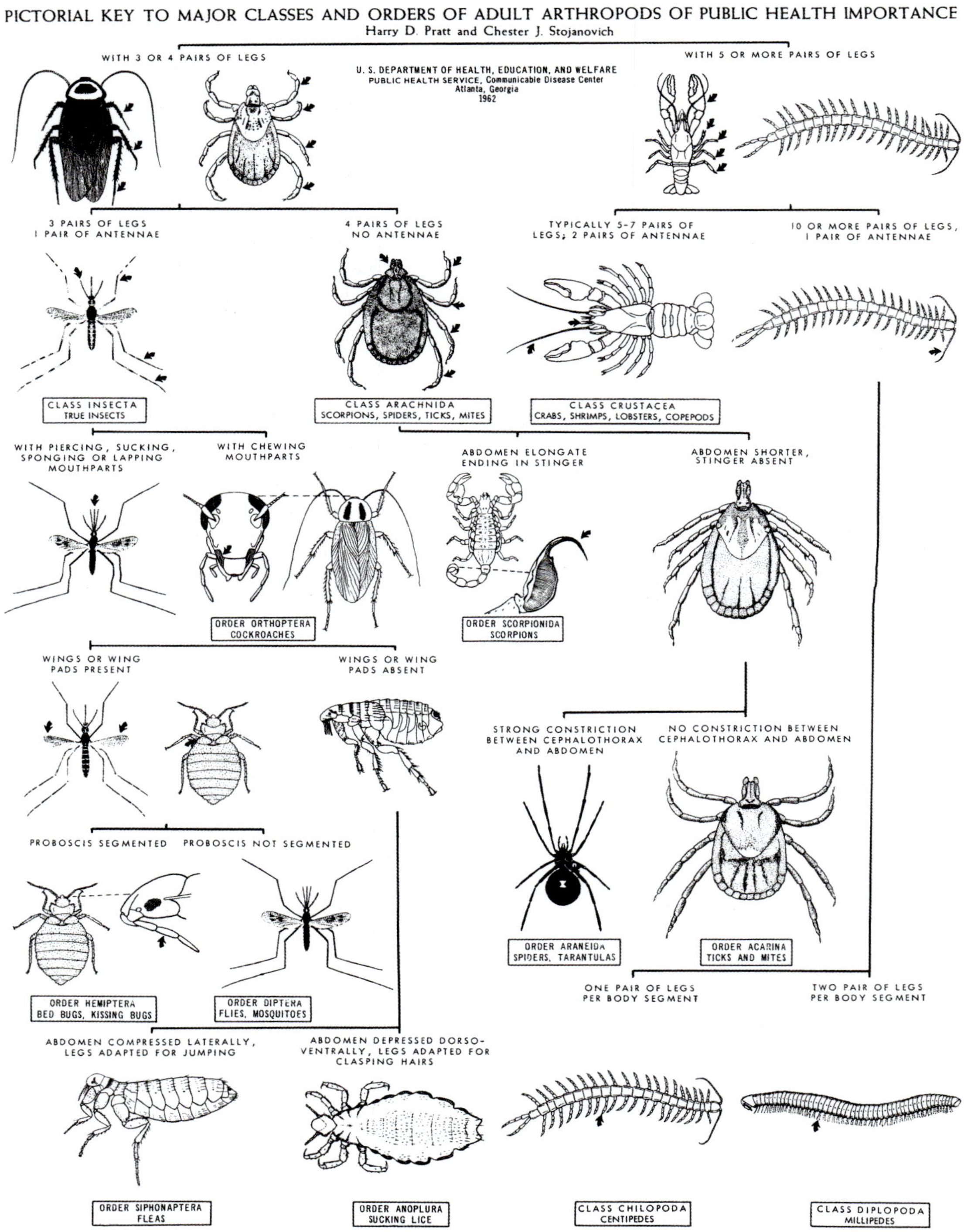

FIGURE 1.1   Representatives of the major groups of arthropods.

TABLE 1.1   Outline of the Hexapoda (following Borror et al. 1989)

Entognatha
1. Protura (Myrientomata) — proturans
2. Collembola (Oligentomata) — springtails
3. Diplura (Entognatha, Entotrophi, Aptera) — diplurans

Insecta
4. Microcoryphia (Archaeognatha; Thysanura, Ectognatha, and Ectotrophi in part) — bristletails
5. Thysanura (Ectognatha, Ectotrophi, Zygentoma) — silverfish, firebrats

Pterygota — winged and secondarily wingless insects
6. Ephemeroptera (Ephemerida, Plectoptera) — mayflies
7. Odonata — dragonflies and damselflies
8. Grylloblattaria (Grylloblattodea, Notoptera) — rock crawlers
9. Phasmida (Phasmatida, Phasmatoptera, Phasmatodea, Cheleutoptera; Orthoptera in part) — walkingsticks and timemas
10. Orthoptera (Saltatoria, including Grylloptera) — grasshoppers and crickets
11. Mantodea (Orthoptera, Dictyoptera, Dictuoptera in part) — mantids
12. Blattaria (Blattodea; Orthoptera, Dictyoptera, Dictuoptera in part) — cockroaches

13. Isoptera (Dictyoptera, Dictuoptera in part) — termites
14. Dermaptera (Euplexoptera) — earwigs
15. Embiidina (Embioptera) — webspinners
16. Plecoptera — stone flies
17. Zoraptera — zorapterans
18. Psocoptera (Corrodentia) — psocids
19. Phthiraptera (Mallophaga, Anoplura, Siphunculata) — lice
20. Heteroptera (Hemiptera) — bugs
21. Homoptera (Hemiptera in part) — cicadas, hoppers, psyllids, whiteflies, aphids, and scale insects
22. Thysanoptera (Physapoda) — thrips
23. Neuroptera (including Megaloptera and Raphidiodea) — alderflies, dobsonflies, fishflies, snakeflies, lacewings, antlions, and owlflies
24. Coleoptera — beetles
25. Strepsiptera (Coleoptera in part) — twisted-wing parasites
26. Mecoptera (including Neomecoptera) — scorpion flies
27. Siphonaptera — fleas
28. Diptera — flies
29. Trichoptera — caddisflies
30. Lepidoptera (including Zeugloptera) — butterflies and moths
31. Hymenoptera — sawflies, ichneumonids, chalcids, ants, wasps, and bees

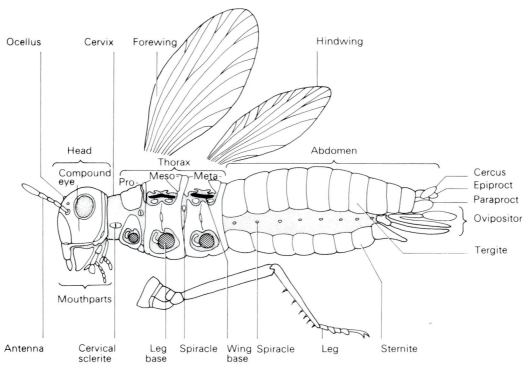

FIGURE 1.2   The external anatomy of a generalized insect. From William S. Romoser and John G. Stoffolano, Jr., *The Science of Entomology.* Copyright © 1994 Wm. C. Brown Communications, Inc., Dubuque, Iowa. All rights reserved. Reprinted by permission.

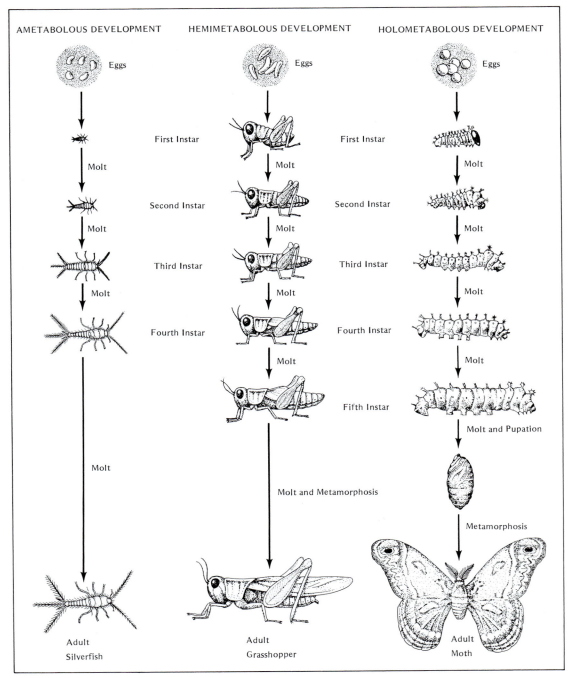

AMETABOLOUS DEVELOPMENT   HEMIMETABOLOUS DEVELOPMENT   HOLOMETABOLOUS DEVELOPMENT

Eggs

First Instar
Molt

Second Instar
Molt

Third Instar
Molt

Fourth Instar

Molt

Molt

Adult
Silverfish

Eggs

First Instar
Molt

Second Instar
Molt

Third Instar
Molt

Fourth Instar
Molt

Fifth Instar

Molt and Metamorphosis

Adult
Grasshopper

Eggs

First Instar
Molt

Second Instar
Molt

Third Instar
Molt

Fourth Instar
Molt

Molt and Pupation

Metamorphosis

Adult
Moth

FIGURE 1.3 Patterns of development in insects. From Clyde F. Herried, *Biology*. Copyright © 1977 Macmillan Publishing Company, Inc. Reprinted with permission of Simon & Schuster.

there are groups of primitively wingless insects and many groups that have secondarily lost wings. Within the hexapods, three major kinds of developmental patterns are recognized. The noninsect hexapods, or Entognatha (Protura, Diplura, Collembola), and the Microcoryphia and Thysanura (Table 1.1) develop to adulthood with little change in body form (ametaboly), except for sexual maturation (Fig. 1.3). All other insects have either gradual change in body form (hemimetaboly) or a pronounced change from a simplified wingless immature stage to usually a winged adult stage via a pupal stage (holometaboly) (Fig. 1.3).

Families of four orders of insects (Heteroptera, Phthiraptera, Siphonaptera, and Diptera) are covered in succeeding chapters (Chapters 5–13) that contain representatives affecting human health directly or indirectly.

Readings

Borror, D. J., Triplehorn, C. A., and Johnson, N. F. 1989. *An Introduction to the Study of the Insects*, 6th ed. Saunders, Philadelphia.

Wheeler, Q. D. 1990. Insect diversity and cladistic constraints. *Ann. Entomol. Soc. Amer.* 83: 1031–1047.

CHAPTER

2

Evolution of Arthropod
Disease Vectors

WILLIAM C. BLACK IV AND BORIS C. KONDRATIEFF

INTRODUCTION

Fossils resembling arthropods first appeared during the late Proterozoic era, about 600 to 540 million years ago. Today the phylum Arthropoda contains ~80% of all extant, metazoan animal species, and they have come to occupy virtually every marine, freshwater, terrestrial, and aerial habitat on earth. Arthropods are essential components of most of the major food chains. In many ecosystems arthropods consume and recycle detritus and the associated bacteria, algae, and fungi. Many species consume either the living or dead tissues of terrestrial and aquatic plants and animals, others are voracious predators or parasites. Given this huge taxonomic and ecological diversity, it is not surprising that some arthropods have evolved the ability to utilize a rich and abundant source of nutrients, the blood of vertebrates.

By plotting the occurrences of *hematophagy* (the habit of feeding on blood) (Fig. 2.1, solid circles) on the current phylogenetic hypothesis for arthropods (Wheeler et al. 2001), we can see that blood feeding has evolved independently at least 21 times in disparate arthropod taxa. Within the single insect Order Diptera, hematophagy has probably evolved independently in nine families. There is no evidence of monophyly among hematophagous fly species in the family Muscidae, and it is likely that blood feeding arose at least four additional times in this family.

Hematophagy was probably exploited quickly by viral, bacterial, protozoan, and helminth species, both as a means for increased mobility and, more impor-

tantly, as a means to find and occupy novel vertebrate hosts. Great benefits accrued to parasite and pathogen species or populations that possessed the correct combination of morphological, physiological, and biochemical characters necessary to survive in hematophagous hosts. Parasites and pathogens abruptly acquired many adaptations beneficial to their own survival and dispersal. These adaptations included physiological and behavioral mechanisms evolved for host location and precise morphological and neurosensory mechanisms for proximal location of blood within the vertebrate vascular system. This chapter reviews the evolution of disease vectors by examining the many ways that different arthropod lineages have independently overcome a common set of developmental, morphological, behavioral, physiological, and biochemical barriers to become hematophagous.

HOST LOCATION

Three different strategies have evolved in hematophagous arthropods to locate suitable vertebrate hosts. Some arthropods evolved highly sensitive neurosensory apparatus for long-distance, followed by proximal, host location. Other arthropods bypassed the need for long-distance host localization by adopting a *nidiculous* ("nest-dwelling") lifestyle in which they live on or near their hosts. Many free-living arthropod vectors have evolved an intermediate host-location strategy. These species maximized the

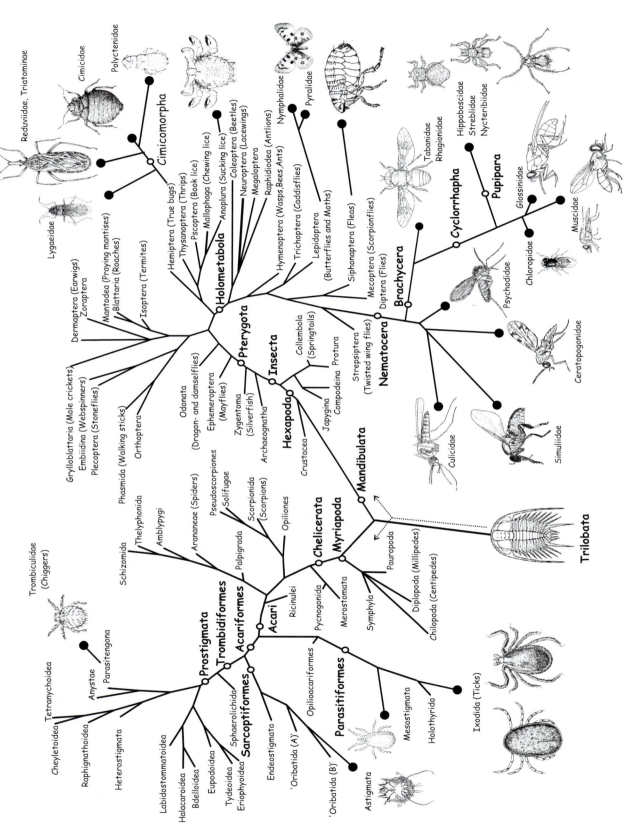

FIGURE 2.1 Current hypothesis of phylogenetic relationships among ancestral and extant arthropod groups. Phylogenetic relationships among hexapod orders is based upon Wheeler et al. (2001), relationships among Chelicerata is based upon Wheeler and Hayashi (1998) and Weygoldt (1998), and relationships among the Acari is based upon Lindquist (1984). The evolution of a hematophagous taxon is indicated with a solid circle. Representative pictures of hematophagous taxa are included.

opportunity for host contact by positioning themselves in the host's habitat without actually becoming ectoparasites or adopting a nidiculous habit.

Flies (Order Diptera) are undoubtedly the masters of long-distance host location. Adult mosquitoes (Culicidae), blackflies (Simuliidae), sand flies (Psychodidae), biting midges (Ceratopogonidae), horseflies/deerflies (Tabanidae), and the hematophagous cyclorrhaphous flies (e.g., Muscidae—horn flies and stable flies) associate with their vertebrate prey only when in need of a blood meal. Once physiological signals indicate the need for a blood meal, the flies actively locate and orient on suitable hosts through a number of cues.

Visual cues are used by flies that feed during the day (e.g., tabanids, tsetse, blackflies, and some mosquitoes). Blackflies and mosquitoes appear to prefer certain colors on which to land, and large moving objects elicit the attention of tsetse, which normally feed on large ungulates. Tabanids that normally live in flat habitats such as pastures are attracted to large contours that block the sky.

Odor also acts as a long-distance attractant for many hematophagous arthropods. Carbon dioxide (CO_2), water vapor (from breath), and lactic acid (or lactic acid oxidation products) are common components of sweat. Incredibly minute quantities of these substances elicit the attention and movement of hematophagous insects toward a host. The lactic acid receptor in the antennae of Aedes aegypti has been the object of elegant work. Butanol, a product of bacterial fermentation in ruminants, is excreted in their urine and is a potent attractive agent to tsetse. CO_2 in host breath is also a powerful cue for most hematophagous arthropods.

For insects feeding at night, when the environmental temperature is low and light is not available, orientation toward heat becomes an important cue. This may explain why nocturnal feeders are often attracted to a uniform, albeit dim, light source for navigational purposes. Normally these sources are distant and guide the arthropod along a linear path toward a potential host. Triatomine bugs are nocturnal feeders that sense irradiated heat as well as convection currents. Night-feeding mosquitoes and some species of blackflies also move toward a heat source that has a temperature compatible with that of a host.

Like other animals, bloodsucking arthropods have defined circadian cycles. Even closely related vector species frequently feed at different times of the day. For example, Anopheles gambiae and An. funestus, important malaria vectors in Africa, and Culex pipiens, the main vector of urban filariasis, are all night biters, their activity peaking between 22:00 and 2:00 h. Alternatively, many Aedes and Psorophora mosquitoes and

blackflies have crepuscular feeding habits; they seek hosts early in the morning and again late in the afternoon. In contrast, most tabanids and tsetse are strictly day feeders.

All of these factors interact to determine a unique feeding schedule for each vector species. For instance, Ae. aegypti is chiefly a crepuscular mosquito that also bites, albeit with less intensity, in the middle of the day. It quests for prey between 50 cm and 1 m from the ground. It senses CO_2, water vapor, and lactic acid, which are all heavier than air and thus close to the ground. The mosquito does not seek objects with temperatures higher than 37°C but is attracted to moving objects. In contrast, Culex mosquitoes search high in a room for ascending warm air currents given off by a warm body and then dive down into those currents.

Both chewing and sucking lice, the pupiparous Diptera and some flea, tick, and mite species spend little or no time finding a host because, being true ectoparasites, they are associated with their host's body most or all of their lives. Many flea and tick species and the bedbugs and triatomines adopt an intermediate approach, associating with their hosts' nests or dwellings. Consider An. gambiae in this regard. Adults rest in human dwellings most of the time (a habit known as endophagy), thus maximizing the chances of being close to a blood source.

PROXIMAL BLOOD LOCATION

Once an arthropod has landed or crawled onto a host's skin, a different set of proximal stimuli is used to identify optimal locations for blood meals. Chemoreceptors located either at the tip of the mouthparts or in the antennae inspect the skin surface for appropriate "flavors" (Fox et al. 2001; Merrill et al. 2002, Ch. 20). Mechanoreceptors are present in the tips of mouthparts and signal appropriate positioning of the feeding apparatus for penetration. Most hematophagous arthropods touch the host skin with the tips of their mouthparts, moving a few millimeters to either side of a potential feeding site before initiating penetration. Alternatively, some lice, tick, and mite species may spend a great deal of time searching the host body for an optimal feeding site.

Ribeiro et al. (1985) identified and quantified relevant parameters of Ae. aegypti probing behavior. Mosquitoes thrust their mouthparts repeatedly at 7-s intervals through the host's skin while searching for blood. If this search is successful, feeding ensues. If not, the mosquito withdraws the mouthpart stylets and attempts to feed at another site. Functions for the probability of feeding success and failure over time were derived using data from observations of 300 mosquitoes. The probability of feeding success was interpreted as being a function of the density of vessels in the skin, their geometric distribution, and the conditions locally affecting hemostasis. Mosquitoes in the genus *Culex* have a strong tendency to *ornithophagy* (feeding on birds) and appear to have only recently adapted to mammals. Thus, they may not yet have evolved efficient mechanisms to counteract mammalian platelet responses, while birds have only relatively inefficient thrombocytes. *Culex*, in fact, takes much more time to find blood on a mammalian host (human or mouse) than do other mosquito genera.

BLOOD ACQUISITION

The evolution of hematophagous mouthparts has been a subject of great interest. Snodgrass (Snodgrass 1943, 1959) was one of the earliest investigators in this area. Snodgrass (1943) performed an intensive comparative analysis of hematophagous mouthparts and concluded that arthropod lineages have modified the basic mouthpart blueprint in a wide variety of ways to enable hematophagy. He identified two general hematophagous strategies. The first, and probably the most primitive, he referred to as *telmophagy*, or "pool-feeding," in which the arthropod mouthparts slice through superficial capillary beds in the skin and then lap or suck up the blood that floats to the surface. Telmophagy is probably an evolutionary intermediate step toward a more derived form, called *solenophagy*, in which the mouthparts penetrate the vertebrate skin and either sever and rest within a capillary bed or actually cannulate individual blood vessels.

It easy to envision telmophagy arising from an ancestral species that occasionally fed on blood exposed through a superficial wound. The next step probably entailed using mouthparts to abrade and irritate the surface of wounds, mucous membranes, or skin to expose superficial blood vessels. This behavior is actually seen among cyclorrhaphous flies in the families Muscidae (*Musca*, *Hydrotaea*, and *Morellia* genera) and Chloropidae (*Hippelates* and *Siphulunculina*). Eventually, rasping mouthparts gave rise to cutting mouthparts and telmophagy. However, telmophagy is irritating to the vertebrate host and was probably limited to hosts without much defensive behavior or restricted to arthropods sturdy enough to withstand being beaten by a tail, chewed, scratched, or even being rolled upon. Alternatively, many telmophagous arthropods feed at a location on the host that cannot be groomed, for example, along the midline of the belly, around the anus, or in the ears.

Restricted host range, damage due to mauling, and preening all generated selective pressure for gradual modification and elongation of mouthparts toward solenophagy. Gradually hematophagy became unnoticed by the host because the mechanical damage was minor and did not elicit defensive behavior. However, there are many arthropod groups (Hemiptera, Lepidoptera, and the Acari) in which the ancestral mouthparts were preadapted for piercing and sucking and thus for solenophagy.

Insect mouthparts are subdivided into six operational structures (Fig. 2.2A). The labrum is usually a large chitinous flap anterior to the other mouthparts. The labrum acts as a "lid," preventing food from escaping anteriorly during the feeding process. It is frequently covered densely with chemo- and mechanoreceptors. The mandibles usually sit behind the labrum. These are highly sclerotized (hardened) opposable structures with sharp dentations for crushing and slicing food. The maxillae are posterior to the mandibles. The maxillae contain highly sclerotized opposable bladelike structures called the *galea* and *lacinia* for slicing and manipulating food. The maxillae also each have a large lateral palp covered with chemo-, thermo-, and mechanoreceptors. A hypo- or epipharynx lies dorsal to and midway between the mandibles and maxillae. This connects to the esophagus. Musculature surrounds the walls of the esophagus. When contracted these muscles create strong negative pressure, causing food to be drawn into the preoral cavity. This musculature can be highly developed in hematophagous arthropods, creating a powerful sucking apparatus. Lying posterior to the maxillae is the labium, which acts as a second "lid," preventing food from escaping posteriorly during feeding. The labium has a pair of large, laterally-produced palps that are covered with chemo- and mechanoreceptors.

The most primitive hematophagous insects are true bugs in the Order Heteroptera. All true bugs have a cylindrical labium (Fig. 2.2B) that forms a prominent 3- to 4-segmented beaklike proboscis. The proboscis encloses a fascicle comprising a pair of mandibles, often terminally barbed, and a pair of maxillae; all four structures act as piercing stylets. The maxillae operate as a unit, and the mandibles function separately. The maxillae are closely apposed, forming the food and salivary tubes; the mandibles add rigidity. The labrum is quite short and inconspicuous.

The subfamily Triatominae (Reduviidae—vectors of *Trypanosoma cruzi*, causative agent of Chagas disease), *Cimex* ("bedbugs"—Cimicidae), a few members of the Lygaeidae, and the Polyctenidae are hematophagous. All families arose from ancestors preadapted for solenophagy because most true bugs

A

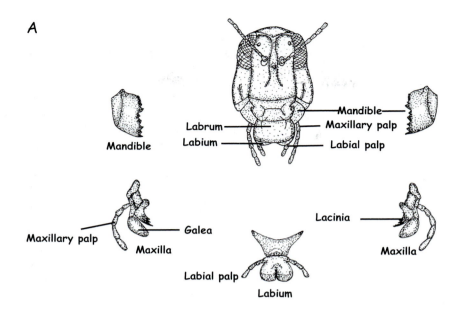

B

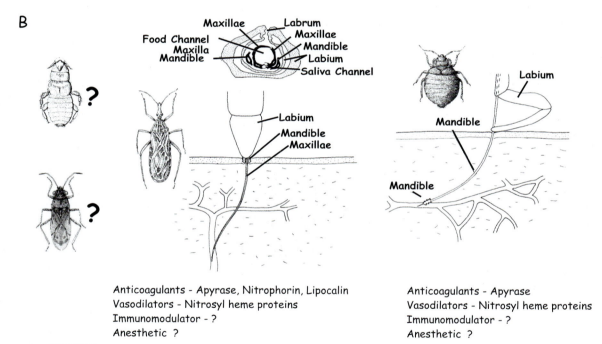

Anticoagulants - Apyrase, Nitrophorin, Lipocalin
Vasodilators - Nitrosyl heme proteins
Immunomodulator - ?
Anesthetic ?

Anticoagulants - Apyrase
Vasodilators - Nitrosyl heme proteins
Immunomodulator - ?
Anesthetic ?

FIGURE 2.2 (A) Generalized mouthpart morphology based upon an orthopteran type of design. (B) Mouthpart and feeding morphology in the heteropteran taxa Reduviidae subfamily Triatominae and Cimicidae.

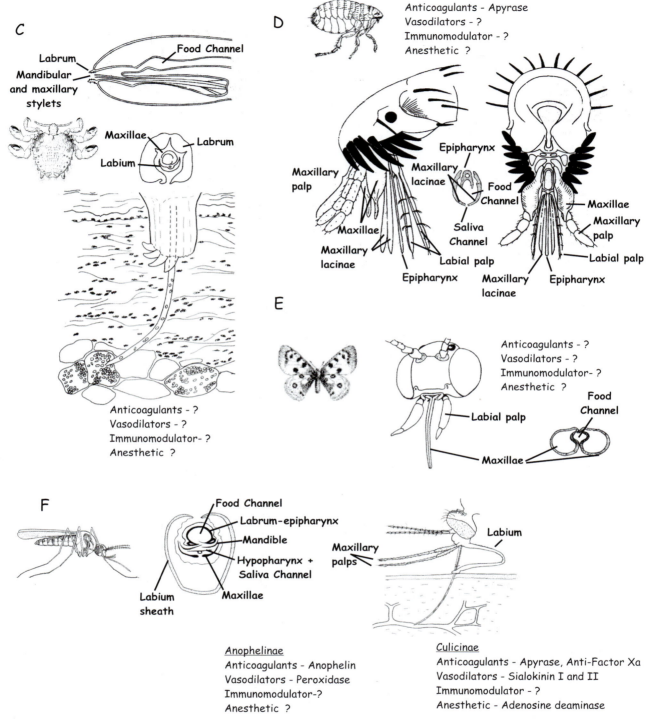

C

Labrum
Mandibular and maxillary stylets
Food Channel

Maxillae
Labrum
Labium

Anticoagulants - ?
Vasodilators - ?
Immunomodulator- ?
Anesthetic ?

D

Anticoagulants - Apyrase
Vasodilators - ?
Immunomodulator - ?
Anesthetic ?

Maxillary palp
Maxillae
Maxillary lacinae
Epipharynx

Epipharynx
Maxillary lacinae
Food Channel
Saliva Channel
Labial palp

Maxillae
Maxillary palp
Labial palp
Maxillary lacinae
Epipharynx

E

Anticoagulants - ?
Vasodilators - ?
Immunomodulator- ?
Anesthetic ?

Labial palp
Maxillae
Food Channel

F

Food Channel
Labrum-epipharynx
Mandible
Hypopharynx + Saliva Channel
Maxillae
Labium sheath

Maxillary palps
Labium

Anophelinae
Anticoagulants - Anophelin
Vasodilators - Peroxidase
Immunomodulator-?
Anesthetic ?

Culicinae
Anticoagulants - Apyrase, Anti-Factor Xa
Vasodilators - Sialokinin I and II
Immunomodulator - ?
Anesthetic - Adenosine deaminase

FIGURE 2.2 (*Continued*) (C) Mouthpart and feeding morphology in Anoplura. (D) Mouthparts in Siphonaptera. (E) Mouthparts in Lepidoptera. (F) Mouthpart and feeding morphology in Culicidae.

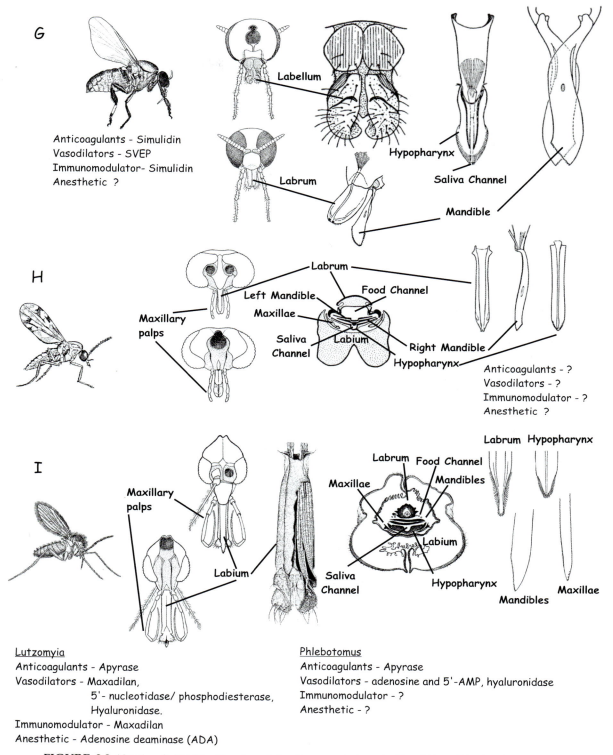

G

Anticoagulants - Simulidin
Vasodilators - SVEP
Immunomodulator- Simulidin
Anesthetic ?

Labellum

Labrum

Hypopharynx

Saliva Channel

Mandible

H

Maxillary palps

Labrum

Left Mandible

Food Channel

Maxillae

Saliva Channel

Labium

Right Mandible

Hypopharynx

Anticoagulants - ?
Vasodilators - ?
Immunomodulator - ?
Anesthetic ?

I

Labrum Hypopharynx

Maxillary palps

Labrum

Food Channel

Maxillae

Mandibles

Labium

Saliva Channel

Hypopharynx

Labium

Saliva Channel

Mandibles

Maxillae

Lutzomyia
Anticoagulants - Apyrase
Vasodilators - Maxadilan,
          5'- nucleotidase/ phosphodiesterase,
          Hyaluronidase.
Immunomodulator - Maxadilan
Anesthetic - Adenosine deaminase (ADA)

Phlebotomus
Anticoagulants - Apyrase
Vasodilators - adenosine and 5'-AMP, hyaluronidase
Immunomodulator - ?
Anesthetic - ?

FIGURE 2.2 (*Continued*) (G) Mouthparts and feeding morphology in Simuliidae. (H) Mouthparts and feeding morphology in Ceratopogonidae. (I) Mouthparts and feeding morphology in Psychodidae.

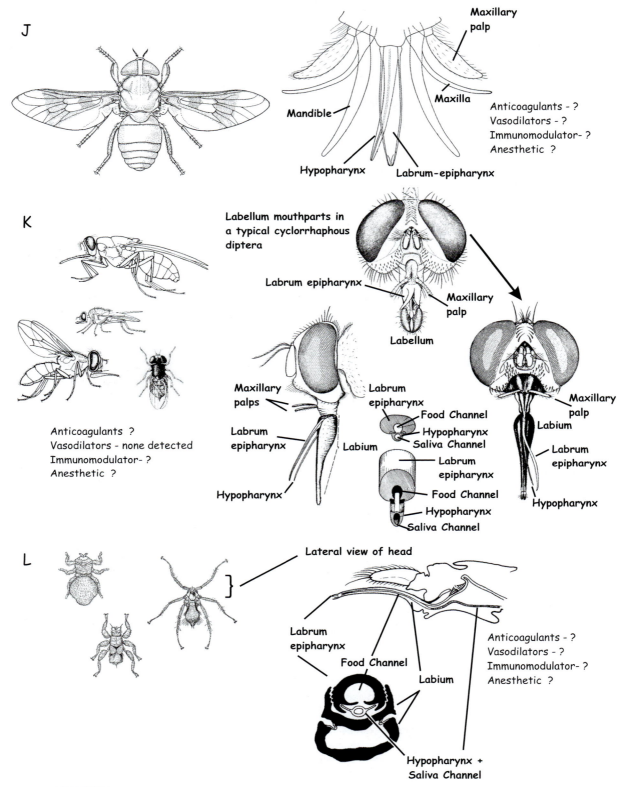

J

Maxillary
palp

Mandible

Maxilla

Anticoagulants - ?
Vasodilators - ?
Immunomodulator- ?
Anesthetic ?

Hypopharynx   Labrum-epipharynx

K

Labellum mouthparts in
a typical cyclorrhaphous
diptera

Labrum epipharynx

Maxillary
palp

Labellum

Anticoagulants ?
Vasodilators - none detected
Immunomodulator- ?
Anesthetic ?

Maxillary
palps

Labrum
epipharynx

Labrum
epipharynx

Food Channel

Hypopharynx

Saliva Channel

Labium

Labrum
epipharynx

Food Channel

Hypopharynx

Saliva Channel

Hypopharynx

Maxillary
palp

Labium

Labrum
epipharynx

Hypopharynx

L

Lateral view of head

Labrum
epipharynx

Food Channel

Labium

Anticoagulants - ?
Vasodilators - ?
Immunomodulator- ?
Anesthetic ?

Hypopharynx +
Saliva Channel

FIGURE 2.2 (*Continued*) (J) Mouthparts and feeding morphology in Tabanidae. (K) Mouthparts and feeding morphology in Muscidae and Glossinidae. (L) Mouthparts and feeding morphology in Pupipara.

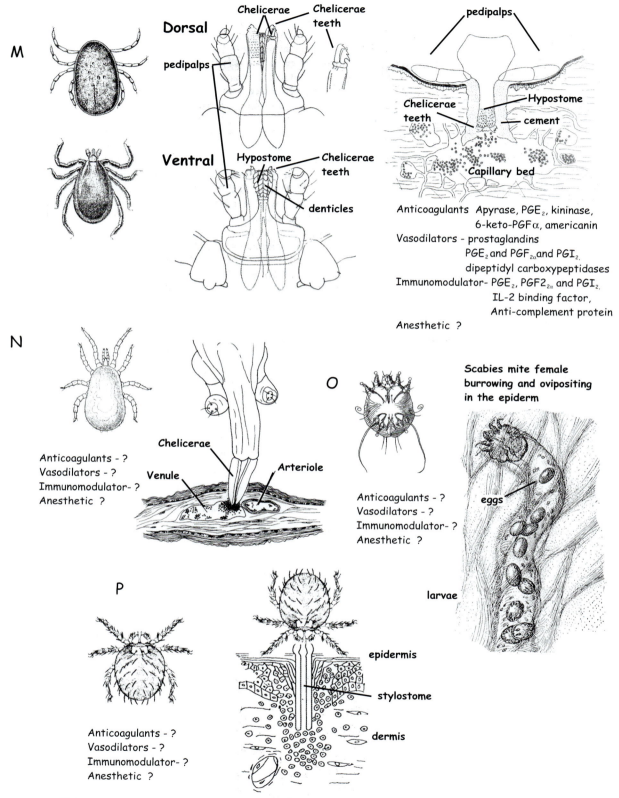

M

Dorsal

Chelicerae
Chelicerae teeth
pedipalps

Ventral

Hypostome
Chelicerae teeth
denticles

pedipalps

Chelicerae teeth
Hypostome
cement
Capillary bed

Anticoagulants Apyrase, PGE$_2$, kininase,
6-keto-PGFα, americanin
Vasodilators - prostaglandins
PGE$_2$ and PGF$_{2\alpha}$ and PGI$_2$,
dipeptidyl carboxypeptidases
Immunomodulator- PGE$_2$, PGF2$_{2\alpha}$ and PGI$_2$,
IL-2 binding factor,
Anti-complement protein
Anesthetic ?

N

Chelicerae
Venule
Arteriole

Anticoagulants - ?
Vasodilators - ?
Immunomodulator- ?
Anesthetic ?

O

Anticoagulants - ?
Vasodilators - ?
Immunomodulator- ?
Anesthetic ?

Scabies mite female
burrowing and ovipositing
in the epiderm

eggs

larvae

P

Anticoagulants - ?
Vasodilators - ?
Immunomodulator- ?
Anesthetic ?

epidermis
stylostome
dermis

FIGURE 2.2 (*Continued*) (M) Mouthpart and feeding morphology in Ixodidae. (N) Mouthparts and feeding morphology in Mesostigmatid mites. (O) *Sarcoptes scabei* feeding and burrowing in the host epidermis. (P) Trombiculidae larva feeding through the epidermis with the secreted stylostome.

feed on plants through sucking-piercing stylets or capture and suck nutrients from other invertebrate fauna; however, the four families arose independently in Heteroptera and therefore arrived at hematophagy via different routes. There are fundamental differences in the manner in which the fascicle operates. In Triatominae the barbed mandibles anchor into the superficial skin tissue, and the maxillary bundle cannulates a blood vessel (Fig. 2.2B). The tips of the maxillae differ: One is hooked, and the other is spiny so that they curve through the skin when sliding over one another. In *Cimex* the basic details of mandibular and maxillary construction are quite similar, but the mandibles penetrate deeply into peripheral blood vessels. In triatomines the labium is swung forward but does not bend during feeding. In *Cimex*, the labium folds back at the basal segments (Fig. 2.2B). Little is known of the blood-feeding morphology of the Lygaeidae and the Polyctenidae.

Hematophagy next arose in the sucking lice (Phthiraptera, suborder Anoplura) (Fig. 2.2C). It is generally agreed that the sucking lice arose from the chewing lice (Mallophaga; see Chapter 6). The chewing lice in the suborder Amblycera contain many genera (e.g., *Menacanthus*) that gnaw through skin and feed on blood and lymph and may thus represent the ancestor to Anoplura. Mouthparts of all extant Anoplura are distinctly adapted for solenophagy, but the stylets lie in a sac concealed within the head (Figure 2.2C). The opening to this sac is at the extreme anterior portion of a proboscis. The proboscis is eversible and is thought to consist of the ancestral labrum, the tip of which contains small, recurved hooks. These are pushed forward into the host skin by muscular action until firm attachment is achieved. The piercing fascicle consists of three stylets that lie within a long internal sac and are composed dorsally of united maxillae and the hypopharynx. The labium is attached posteriorly to the walls of the sac. The mandibles are vestigial. The apposed maxillae form the food duct, and the hypopharynx forms the salivary channel. Salivary secretion is poured into the wound, and the cibarial and pharyngeal pumps draw blood.

The mouthparts of fleas are typically solenophagous (Fig. 2.2D) but are unique in representing yet another model for hematophagous mouthpart anatomy. The mouth contains broad maxillary lobes and an elongate labium and palpi. The principal bladelike piercing organs are a pair of independently movable maxillary laciniae. The labrum, the mandibles, the hypopharynx, and the labium are all believed to be rudimentary. The epipharynx is a medial unpaired stylet that is closely embraced by the lacinial blades. The paired maxillary stylets and the epipharyngeal stylet form the fascicle that is held in a channel formed by grooves on the inner margin of the labial palps. The wound is made by the protraction and retraction of the laciniae, and as blood begins to flow it is drawn into the pharynx by the cibarial and pharyngeal pumps.

In butterflies and moths, mouthparts consist of a coiled sucking tube capable of very elongate protrusion (Fig. 2.2E). The labrum is greatly reduced, and mandibles are absent. The only obvious portions of the maxillae are the galea, which, by apposition of their inner grooved surfaces, form the long, coiled proboscis. The labium is represented by the labial palpi. Like the Hemiptera, extant hematophagous lepidopterans arose from ancestors preadapted for solenophagy. Banziger (1975, 1979) suggests that the hematophagous noctuids of Southeast Asia arose from fructivorous ancestors, with mouthparts adapted for burrowing deeply through fruit skins. Banziger (1970) showed that this skin-piercing group has a proboscis that can be stiffened by increased hemolymph pressure and that is stouter than that in nectar-feeding or fruit-piercing species. The proboscis may be sharp at the tip and bear erectile barbs.

Hematophagy arose independently many times during the evolution of the flies (Diptera). In the primitive suborder Nematocera, our current understanding of systematics (Yeates and Wiegmann, 1999) suggests that blackflies (Simuliidae—*Onchocerca* vectors), biting midges (Ceratopogonidae—arbovirus vectors), and midges (Chironomidae) arose from a common ancestor. However, it is likely that hematophagy arose independently in the Simuliidae and Ceratopogonidae. The mosquitoes (Culicidae) arose independently and are a sister group to the phantom midges (Chaoboridae), none of which are hematophagous. Interestingly, an entire subfamily of Culicidae (Toxorhynchitinae) subsequently lost the ability to blood feed. The other hematophagous Nematocera group contains the sand flies (Psychodidae—*Leishmania* vectors). This family also arose independently. Thus, hematophagy arose independently at least three times, and probably four, in the nematocerous Diptera.

Mosquito mouthparts contain many individual stylets that are loosely ensheathed within an elongated labium that forms a prominent proboscis (Fig. 2.2F). The six stylets consist of two mandibles, two maxillae, the hypopharynx, and the labrum-epipharynx. The food channel is formed by the labrum-epipharynx, and the mandibular stylets are positioned outside the maxillary stylets. The fascicle is highly flexible; and though blood vessels may be lacerated during penetration to form a blood pool, prolonged feeding takes place through cannulating of a blood vessel.

Simuliids, ceratopogonids, and psychodids are small insects—typically only 0.5–5mm in length. Mouthparts consist of mandibles with serrated teeth along both edges. Recurved teeth are found at the tip of the maxillary lacinia that anchors the mouthparts during feeding. These flies salivate onto the host skin and then lacerate superficial blood vessels. Blood and saliva are then drawn into the insect through a labrum-epipharyngeal structure (Figs. 2.2G–I).

Some nematocerous Diptera and most advanced brachyceran flies have a highly derived type of mouth consisting of a prominent fleshy proboscis comprising mainly the labium, which terminates in the pair of corrugated sponging organs, the labella, and is attached in an elbow-like form to the head. The structure is called the *labellum* (Fig. 2.2K). Mouthparts of blackflies and biting midges both contain a labellum-like structure that is used to ingest blood brought to the surface during telmophagy (Figs. 2.2G, H). The cutting mouthparts of the blackfly consist anteriorly of the labrum, behind which sit a pair of opposable bladelike mandibles that are loosely attached in the middle at a pivot. The blackfly produces saliva onto the surface of the bite site and then cuts with the mandibles in a scissor-like motion through the epidermis into shallow capillary beds. A similar ultrastructure exists in the mouthparts of biting midges (Fig. 2.2H). However, the labellum is less pronounced, and the labrum and hypopharynx are narrow, stylet-like structures with sharp tips that may actually be used to feed within shallow capillary beds exposed when the mandibles cut into the host epidermis.

Sand flies have a different type of mouthpart (Fig. 2.2I). The maxillae, mandibles, and labrum are all narrow, sharp-tipped, stylet-like structures contained within an involution of the labium. The sand fly places the tip of the labium onto the host skin surface, cuts into and exposes shallow capillary beds, and then ingests blood through a central food channel.

Tabanids and Rhagionidae are the most primitive hematophagous members of the Brachycera (Fig. 2.2J). Their mouthparts contain coarse, bladelike stylets that are clearly adapted for telmophagy. The mandibles move transversely and are distinctly flattened and saber-like. The bladelike maxillae are narrower. Both the hypopharynx and labrum-epipharynx are lancet-like. The labium is conspicuous at the median and loosely ensheathes the maxillary and mandibular stylet blades. The labium terminates in a pair of large lobes, the labella. Penetration of the vertebrate host is by means of a thrusting action, with the mandibles and maxillae lacerating tissues. The mandibles move with a scissor-like motion while the maxillae thrust and retract. This action ruptures both small and large blood vessels, and a large pool of blood is often formed. The blood is drawn into the insect via the labrum-epipharynx.

The mouthparts of most advanced brachyceran Diptera have evolved into a well-developed labellum. In the hematophagous advanced brachyceran Diptera, the labellum has become distinctly specialized for solenophagy (Fig. 2.2K). At rest the labium is oriented in front of and parallel to the body and joins the body at a prominent muscular basal joint. This proboscis terminates in a highly modified labellum that is armed with a dense set of rasping denticles. The denticles are derived from the same sponging structures on the labellum, but they are proportionally much smaller and more heavily sclerotized. The proboscis is strongly forced into the flesh of the host, and entry is probably facilitated by the rasping action of the labellar denticles. The labrum-epipharynx lies in the upper groove of the labium. It is uncertain how many times this highly sclerotized structure evolved within the advanced brachyceran Diptera. However, the structure appears in many unrelated species in the large family Muscidae. This mouthpart is most thoroughly characterized in the horn fly (*Haematobia irritans*), the stable fly (*Stomoxys calcitrans*), and tsetse flies (*Glossina* spp.).

The Hippoboscidae, Streblidae, and Nycteribiidae (the Pupipara) are all closely related, highly derived hematophagous advanced brachyceran families that have similar, solenophagous mouthparts. The proboscis is heavily sclerotized, with labial denticles for penetration into the skin of the host. The labrum epipharynx is stylet-shaped, its proximal portion is strongly sclerotized and rigid, and the distal end is membranous and flexible (Fig. 2.2L).

All evidence suggests that arachnid ancestors (subphylum Chelicerata—Figs. 2.2M–P) never possessed mouthparts for mastication and probably subsisted on liquids. All arachnids have a highly developed sucking apparatus and paired chelicerae that are used for grasping, holding, tearing, crushing, or piercing food items. The chelicerae also serve as cutting structures that penetrate into the vertebrate host. All hard and soft ticks have a highly sclerotized structure called the *hypostome* that bears many recurved denticles and anchors the mouthparts in place (Fig. 2.2M). According to Snodgrass (1943), the lobes or processes often associated with the distal part of the hypostome cannot be homologized with other structures present in ancestral Arachnida. Extensive laceration of blood vessels to locate chelicerae in a superficial capillary bed is characteristic of tick feeding. Ticks do not cannulate individual vessels but by the same token do not bring blood to the surface of the wound. Soft ticks (Argasidae) generally feed rapidly; hard ticks

(Ixodidae) characteristically remain attached for a period of 3–7 days and fix their mouthparts in place using a cement secreted from the salivary glands. Electrical recording of hard tick feeding has shown that initial feeding is characterized by short periods of sucking followed by bursts of salivation. As days of feeding progress, sucking, salivation, and intermittent resting intervals lengthen. A major blood pool in the host tissue does not form until after about $2^{1}/_{2}$ hours of feeding, and secretory activity reaches a peak after about 5 days.

The mouthparts of other mites lack the hypostome. The mesostigmatid mites have fine, hypodermic-like chelicerae that they insert into superficial capillary beds to withdraw blood (Fig. 2.2N). The mesostigmatid species *Dermanyssus gallinae*, the red chicken mite, infests the nests of poultry and other birds and feeds on them. The astigmatid mites are a large and diverse group, some of which have evolved to feed on lymph and tissues other than blood (e.g., species in the families Acaridae, Psoroptidae, Knemidokoptidae, and Analgidae). Species in these families are the major causes of mange in domesticated animals and diseased, weakened, or aged wildlife. The astigmatid mites also contain a highly derived family, Sarcoptidae, members of which have evolved the ability to burrow, mate, and oviposit within and beneath the host epidermis (Fig. 2.2O).

The final hematophagous family is Trombiculidae. Among trombidiform mites ("chiggers") the chelicerae became progressively adapted for piercing by a transformation of the movable digits into hooks or stylets. Larval trombiculid mites remain attached in the skin of the vertebrate host for some time and form a feeding tube called the *stylostome* around the inserted mouthparts (Fig. 2.2P).

THE EVOLUTION OF HEMATOPHAGOUS SALIVA

A number of adaptations in arthropod mouthparts enabled hematophagy; however, once an arthropod is physically capable of withdrawing blood, an array of additional problems present themselves. Vertebrate hemostasis is highly redundant, in that platelet aggregation, blood coagulation, and vasoconstriction prevent blood loss from injured tissue. Consequently, there has been selective pressure for the evolution of salivary antihemostatic components for maintaining flow during feeding (Ribeiro 1987). All hematophagous arthropods salivate during intradermal probing before ingesting blood (Chapter 28).

Antiplatelet, anticoagulant, and vasodilation agents have been found in the saliva of all bloodsucking arthropods. In addition to hemostasis, there must be also intense selection to reduce pain and inflammation at the feeding site.

The saliva of hematophagous arthropods also contains potent immunomodulators, and selection pressure must have been intense for the evolution of these compounds. Prior to the evolution of immunomodulators, vertebrates probably produced antibodies against salivary components. Thus, only immunologically naïve ("unbitten") vertebrates would have been suitable blood sources. Proteins that prevented inflammatory/immune responses from a host to the various salivary proteins would have had an immediate selective advantage both for individual insects and for a species. In a sense these substances maintained the population density of susceptible vertebrate hosts.

Physiological and biochemical adaptations in response to hemostasis, pain, and inflammatory/immune responses from a host have resulted in the evolution of a huge diversity of pharmacologically active substances in hematophagous arthropod saliva. Most of what we know of these substances comes from the pioneering work of Drs. Jose Ribeiro, Richard Titus (Titus and Ribeiro 1988), Stephen Wikel, Anthony James, and Jesus Valenzuela.

The probing of hosts is prolonged when the salivary ducts are severed, but this prolongation is absent when mosquitoes feed from an artificial membrane feeder. Little is known of the ways that the quantity of salivary secretion is regulated. There is a dense plexus of serotonin-producing axons surrounding the proximal medial lobe of the salivary gland of *Ae. aegypti*, which originates in the stomatogastric nervous system. Mosquitoes treated with the serotonin-depleting agent alpha-methyltryptophan (AMTP) and then allowed to feed on a rat exhibit a significantly longer mean probing period and a lower blood-feeding success rate. When mosquitoes are induced to salivate into mineral oil, AMTP-treated individuals secrete significantly less saliva and injection of serotonin increases in the volume of secreted saliva. Serotonin thus appears to play a critical role in controlling salivation in adult female mosquitoes.

As arthropods penetrate or lacerate their host's skin, they salivate and respond to whatever is available. Purine nucleotides provide a positive stimulus for most bloodsucking arthropods. Depending upon the arthropod group, adenosine, AMP, ADP, ATP, or 2,3-diphosphoglycerate stimulate the uptake of a warm saline meal through a membrane. The chemoreceptors for purine nucleotides are usually located in or near the cibarial pump. Because most purine nucleotides are inside red blood cells, it is presumed that shearing forces acting at the pump and the shredding action of the cibarial and pharyngeal armatures and other papillae and spines in the foregut release nucleotides from the erythrocytes.

The evolution of antihemostatic agents, vasodilators, immunomodulators, and proteins associated with other critical components the saliva of hematophagous arthropods is reviewed in Black (2002). A detailed discussion is also presented in Chapter 28.

WATER BALANCE

Once arthropods have taken an enormous amount of blood, locomotion or flying becomes difficult and many hematophagous insects rapidly eliminate water from the blood meal to reduce weight. For example, a 5th instar *Rhodnius* nymph, weighing 30 mg, can ingest 300 mg of blood in less than 15 minutes and urinate 150 mg (five times its body weight) in 6 hours (Ribeiro 1996).

The volume of blood ingestion is regulated by stretch receptors located in the gut or crop. Normally, as the blood meal distends the abdomen, a signal is sent to the brain that arrests the feeding reflex. When the abdominal nerves to stretch receptors are severed in mosquitoes and *Rhodnius*, the insects feed until the abdomen literally bursts. Once stretch receptors are stimulated, the insect ceases feeding, withdraws its mouthparts and flees the host. The engorged insect seeks a suitable resting place to initiate digestion and water loss.

Heteropterans, lice, and advanced brachyceran flies feed exclusively on blood, whereas mosquitoes, blackflies, and sand flies drink water and nectar and are more flexible in their ability to maintain water balance. Ticks have developed a unique method to capture water vapor from ambient air. They secrete hygroscopic saliva that is spread over their palps. This saliva captures moisture from the atmosphere and is then imbibed (Chapter 4). Soft ticks can remain alive without desiccating for more than 10 years.

Because most of the lipids and carbohydrates in bloodsucking insects must be made from amino acids, excess nitrogen has to be excreted. In all arthropods, uric acid is usually secreted by the excretory organs (Malpighian tubules) into the insect rectum. Diuretic hormones have been isolated from Malpighian tubules of *Rhodnius* that are present in the bug's hemolymph a few seconds after ingestion of the meal. The insect begins urinating on the host while still feeding.

Hormonal regulation of water transport from the midgut to the hemolymph and from the hemolymph to the Malpighian tubules is under the control of a number of neuropeptides such as serotonin.

Ticks eliminate excess water in a different way. Soft ticks, which feed for no longer than an hour, have evolved a unique filtration apparatus called the *coxal glands*. These glands open at the base of the legs. The tick integument is lined with smooth muscle that contracts to create filtration pressure. A dilute saline solution and some protein are eliminated from the coxal glands while the tick is feeding. Alternatively, the gut of hard ticks absorbs water from the blood meal into the hemolymph. Excess water is pumped back into the host in saliva.

DIGESTING THE BLOOD MEAL

As a blood meal is ingested in mosquitoes, sand flies, blackflies, and fleas, it travels directly to the gut. Tsetse flies instead store the blood in a diverticulum before gradually shunting the meal into the midgut. The anterior midgut of *Rhodnius* serves the same function. A protective chitin/protein web known as the *peritrophic matrix* is secreted around the blood meal in the midgut within a few hours of ingestion (Chapter 22).

The presence of a food bolus in the midgut causes digestive enzymes to be secreted into the lumen of the gut (Chapter 21). Within a few hours or days after taking blood, protease levels can rise 20-fold. In *Aedes* spp., a small amount of one type of early trypsin is produced following stretching of the gut, but the main trypsin is not produced if the meal does not contain protein. Early trypsin is apparently a part of a unique signal transduction system (Noriega et al. 1999). A large pool of transcribed message resides in the midgut of newly eclosed adults. Translation of early trypsin is induced by the presence of free amino acids, as might be found in blood but not by nectar feeding. Its function may be to "taste" the incoming meal to determine if there is sufficient protein to support a gonadotrophic cycle. If so, the signal transduction pathway activates transcription of late trypsin and other genes encoding other digestive enzymes to digest the blood meal.

Most hematophagous arthropods digest blood proteins using trypsins. Further digestion is accomplished by amino- and carboxypeptidases. These enzymes typically have optimal activity at neutral-to-alkaline pH. In contrast, triatomine midguts have sulfhydryl proteolytic enzymes with pH optima of around 5.0, similar to the lysosomal cathepsins. Ticks secrete a hemolysin and then intestinal cells slowly pinocytose the gut contents, resulting in intracellular digestion of the meal by lysosomal enzymes.

Blood is rich in proteins and essential amino acids but deficient in carbohydrate, fat, and adequate amounts of many B vitamins. Accordingly, all insects that feed exclusively on blood, such as the triatomines, lice, bedbugs, and advanced brachyceran Diptera, have bacterial symbionts that either supplement the blood with B vitamins or balance the diet by converting proteins to carbohydrates (glyconeogenesis). Without these endosymbionts the insects die or are infertile. These endosymbionts live in the insect gut, either free (as in triatomines) or, in mycetomes, special structures of the anterior midgut (as in tsetse or lice).

GROWTH AND EGG DEVELOPMENT

Arthropods that feed on large quantities of blood have to accommodate a large volume within a normally hard, rigid exoskeleton. Hematophagous arthropods overcome this problem by having an exoskeleton that is not entirely rigid but instead consists of hard plates (tergites and sternites) joined by less sclerotized lateral membranes (the pleura). In addition, plasticization of the exocuticle occurs in triatomine nymphs and hard ticks. A few minutes after initiating the blood meal, the cuticle becomes elasticized and can be pulled to a thin membrane. This can be accomplished in vitro by the addition of serotonin to the cuticle, and it involves water flow from the medium to the cuticle made by the epidermal cells underlying the tissue. The hydration of the cuticle allows the cuticular proteins to roll over one another. Little chitin is found in this elasticized cuticle.

The products of blood digestion accumulate in the arthropod's hemolymph. In females, products are rapidly sequestered into the tissues or cells known as the *fat body* (Chapter 25). After a period of time, a substance known as *juvenile hormone* (JH) stimulates the female fat body to start vitellogenin synthesis and prepares ovaries to receive vitellogenins. Vitellogenins are large glycolipophosphoproteins. After vitellogenin passes through spaces between follicular cells in the oocyst, it binds with receptor proteins on the oocyte surface membrane and is taken in by a process of receptor-mediated endocytosis. As vitellogenesis proceeds, the oocyte grows and activates stretch receptors that signal the arthropod brain to biosynthesize and release allatostatin. One of the activities of allatostatin is to block JH production, which then decreases vitellogenesis.

In mosquitoes, adult eclosion causes JH synthesis from a portion of the arthropod brain. JH then circulates to the ovary, where it induces growth, signals synthesis of vitellogenins, and initiates secretion of another hormone, known as *ecdysone*. JH presence also triggers feeding and mating behaviors. *Culex* mosquitoes do not bite in the absence of JH and have two egg production cycles. The first cycle is induced by release of JH following adult emergence and leads to primary growth of oocysts but discontinues until blood feeding. After a meal, vitellogenesis occurs in the oldest, primary oocysts and another hormone (20-hydroxyecdysone) stimulates the separation of secondary follicles from the germarium in preparation for the second gonadotrophic cycle. In the *Culex*, after mating, the female brain is stimulated to produce and release EDNH (egg development neurohormone), which circulates and activates the ovaries to begin biosynthesis and release of ecdysteroids, which in turn trigger vitellogenin synthesis by the fat body, which is quiescent from the earlier JH peak and is receptive to EDNH. The postvitellogenic ovary then resignals the brain to stop EDNH production and end the stimulation of the ovary. In Triatominae, an antigonadotrophic hormone is produced in the abdomen when mature eggs are present in the ovaries. This blocks the effect of JH on the follicle cells and prevents vitellogenesis.

GENERAL PATTERNS IN THE EVOLUTION OF ARTHROPOD HEMATOPHAGY

We have traced the adaptations associated with vertebrate host location, proximal localization of shallow capillary beds, blood acquisition through either telmophagy or solenophagy, the role of saliva in disrupting hemostasis and inflammation, and other functions at the feeding site that are subsequently involved in immunomodulation. We finish by examining the adaptations for processing the blood meal both through digestion and excretion and the use of the digested blood in maturing eggs for the subsequent generations.

All of these adaptations for hematophagy in arthropods have arisen on at least 20 independent occasions during arthropod evolution. In every case, the prehematophagous ancestral lineages faced a common set of problems. We have reviewed the ways that the six basic arthropod mouthparts have been modified to enable either telmophagy or solenophagy. In each case, it is clear that mouthpart and saliva evolution has followed different paths to derive a common set of phlebotomist tools, initially to permit telmophagy and eventually to enable solenophagy. The overriding message in considering all of these adaptations is that no general, consistent morphological, physiological, or biochemical adaptations for hematophagy are conserved among all of the hematophagous arthropod lineages. However, as a general theme, the arthropods, when faced with a common set of problems associated with gaining access to vertebrate blood, have taken up many independent but ultimately convergent paths.

Readings

Banziger, H. 1970. The piercing mechanism of the fruit-piercing moth *Calpe [calyptra] thalictri* bkh. (Noctuidae) with reference to the skin-piercing blood-sucking moth *C. eustrigata* hmps. *Acta Tropica* 27: 54–88.

Banziger, H. 1975. Skin-piercing blood-sucking moths I: Ecological and ethological studies on *Calpe eustrigata* (Lepid., Noctuidae). *Acta Tropica* 32: 125–144.

Banziger, H. 1979. Skin-piercing blood-sucking moths II: Studies on a further 3 adult *Calyptra [Calpe]* sp. (Lepid., Noctuidae). *Acta Tropica* 36: 23–37.

Black, W. C., IV. 2002. Evolution of arthropod vectors. In: *Emerging Pathogens: The Archaeology, Ecology, and Evolution of Infectious Disease* (C. L. Greenblatt, ed.). Oxford University Press, New York, pp. 49–64.

Edwards, J. F., Higgs, S., and Beaty, B. J. 1998. Mosquito feeding-induced enhancement of Cache Valley virus (Bunyaviridae) infection in mice. *J. Med. Entomol.* 35: 261–265.

Fox, A. N., Pitts, R. J., Robertson, H. M., Carlson, J. R., and Zwiebel, L. J. 2001. Candidate odorant receptors from the malaria vector mosquito *Anopheles gambiae* and evidence of down-regulation in response to blood feeding. *Proc. Natl. Acad. Sci. USA* 98: 14693–14697.

Gillespie, R. D., Mbow, M. L., and Titus, R. G. 2000. The immunomodulatory factors of bloodfeeding arthropod saliva. *Parasite Immunol.* 22: 319–331.

Gillespie, R. D., Dolan, M. C., Piesman, J., and Titus, R. G. 2001. Identification of an IL-2–binding protein in the saliva of the Lyme disease vector tick, *Ixodes scapularis. J. Immunol.* 166: 4319–4326.

Jones, L. D., Hodgson, E., and Nuttall, P. A. 1989. Enhancement of virus transmission by tick salivary glands. *J. Gen. Virol.* 70: 1895–1898.

Labuda, M., Jones, L. D., Williams, T., and Nuttall, P. A. 1993. Enhancement of tick-borne encephalitis virus transmission by tick salivary gland extracts. *Med. Vet. Entomol.* 7: 193–196.

Lindquist, E. E. 1984. Current theories on the evolution of major groups of Acari and on their relationships with other groups of Arachnida, with consequent implications for their classification. In: *Acarology VI*, Vol. 1 (D. A. Griffiths and C. E. Bowman, eds.). Wiley, New York, pp. 28–62.

Merrill, C. E., Riesgo-Escovar, J., Pitts, R. J., Kafatos, F. C., Carlson, J. R., and Zwiebel, L. J. 2002. Visual arrestins in olfactory pathways of *Drosophila* and the malaria vector mosquito *Anopheles gambiae. Proc. Natl. Acad. Sci. USA* 99: 1633–1638.

Noriega, F. G., Colonna, A. E., and Wells, M. A. 1999. Increase in the size of the amino acid pool is sufficient to activate translation of early trypsin mRNA in *Aedes aegypti* midgut. *Insect Biochem. Mol. Biol.* 29: 243–247.

Ribeiro, J. M. 1987. Vector salivation and parasite transmission. *Mem. Inst. Oswaldo Cruz* 82: 1–3.

Ribeiro, J. M. C. 1996. Common problems of arthropod vectors of disease. In *Biology of Disease Vectors* (W. C. Marquardt and B. J. Beaty, eds.). University Press of Colorado, Niwot, CO, pp. 393–416.

Ribeiro, J. M., Rossignol, P. A., and Spielman, A. 1985. *Aedes aegypti*: Model for blood-finding strategy and prediction of parasite manipulation. *Exp. Parasitol.* 60: 118–132.

Schoeler, G. B., Manweiler, S. A., and Wikel, S. K. 1999. *Ixodes scapularis*: Effects of repeated infestations with pathogen-free nymphs on macrophage and T lymphocyte cytokine responses of BALB/c and C3H/HeN mice. *Exp. Parasitol.* 92: 239–248.

Shultz, J. W. 1990. Evolutionary morphology and phylogeny of Arachnida. *Cladistics* 6: 1–38.

Snodgrass, R. E. 1943. The feeding apparatus of biting and disease-carrying flies: A wartime contribution to medical entomology. *Smithsonian Miscellaneous Collections* 104: 1–51.

Snodgrass, R. E. 1959. The Anatomical Life of the Mosquito. *Smithsonian Miscellaneous Collections* 139: 1–87.

Titus, R. G., and Ribeiro, J. M. 1988. Salivary gland lysates from the sandfly *Lutzomyia longipalpis* enhance *Leishmania* infectivity. *Science* 239: 1306–1308.

Weygoldt, P. 1998. Evolution and systematics of the Chelicerata. *Experimental Appl. Acarol.* 22: 63–79.

Wheeler, W. C., and C. Y. Hayashi. 1998. The phylogeny of the extant chelicerate orders. *Cladistics* 14: 173–192.

Wheeler, W. C., Whiting, M., Wheeler, Q. D., and Carpenter, J. M. 2001. The phylogeny of the extant hexapod orders. *Cladistics* 17: 113–169.

Yeates, D. K., and Wiegmann, B. M. 1999. Congruence and controversy: Toward a higher-level phylogeny of the Diptera. *Annu. Rev. Entomol.* 44: 397–428.

Zeidner, N., Dreitz, M., Belasco, D., and Fish, D. 1996. Suppression of acute *Ixodes scapularis*–induced *Borrelia burgdorferi* infection using tumor necrosis factor-alpha, interleukin-2, and interferon-gamma. *J. Infect. Dis.* 173: 187–195.

Zeidner, N., Mbow, M. L., Dolan, M., Massung, R., Baca, E., and Piesman, J. 1997. Effects of *Ixodes scapularis* and *Borrelia burgdorferi* on modulation of the host immune response: Induction of a TH2 cytokine response in Lyme disease–susceptible (C3H/HeJ) mice but not in disease-resistant (BALB/c) mice. *Infect. Immun.* 65: 3100–3106.

Zhu, K., Bowman, A. S., Brigham, D. L., Essenberg, R. C., Dillwith, J. W., and Sauer, J. R. 1997. Isolation and characterization of americanin, a specific inhibitor of thrombin, from the salivary glands of the lone star tick *Amblyomma americanum* (L.). *Exp. Parasitol.* 87: 30–38.

3

Mites and Disease

DAVID EVANS WALTER AND MATTHEW SHAW

INTRODUCTION

Mites (Acari or Acarina) comprise a large group of small to tiny arthropods that are closely related to spiders and their kin (the Arachnida). Three major lineages (superorders within the subclass Acari—see Table 3.1) are recognized, of which two, the Parasitiformes and the Acariformes, contain species of medical-veterinary importance. Mites both cause disease themselves and vector disease agent organisms. They are increasingly relevant to human welfare because of their importance in resurgent and emergent diseases (e.g., various rickettsial infections, delusional parasitosis); because of rampant parasitic mite infestations in immunocompromised patients (e.g., crustose scabies); because of rising levels of allergic reactions (e.g., to house-dust mites); and because of the global spread of generalist parasites that affect us, domestic animals, and wildlife (e.g., tropical rat mite; see Fig. 3.1A).

The most important acarine vectors of disease are ticks (Parasitiformes: Ixodida), an order of about 900 species of large, obligately blood-feeding mites that traditionally have been treated separately from other mites. This tradition is continued in this book (Chapter 4). When we use the word *mite* we are referring to nontick acarines; however, it is worth remembering that ticks are mites—an understanding that sheds light on many aspects of their life cycle, morphology, physiology, and potential to transmit disease. Excluding ticks, medically important mites are distributed across three large orders: Mesostigmata (Figs. 3.1A, B), Trombidiformes (Fig. 3.1C), and Sarcoptiformes (Figs. 3.1D, E). A fourth, small order, the Holothyrida,

is an anecdotal exception. These mites are large (2–7 mm in length) but rare, restricted to neotropical forests, Indo-Pacific Islands, and Australasia. They are slow-moving, heavily armored scavengers (Walter and Proctor 1998) that protect themselves with a bitter, burning secretion. Children who put holothyrans in their mouths have been discomforted, and chickens that have eaten the mites have died (Southcott 1976).

The remaining parasitiform order, Mesostigmata (also Gamasida), includes many significant vertebrate parasites in the superfamily Dermanyssoidea (Figs. 3.3A, B, 3.4). Adult dermanyssoids are mostly 0.6–1.2 mm in length and form at least two ecological groupings: nest-associated temporary ectoparasites and permanent ectoparasites. Some of the latter inhabit ears or respiratory systems (nasal passages, lungs, air sacs) but only penetrate the body cavity in unusual circumstances. No dermanyssoid parasite specific to humans is known, but those associated with synanthropic rodents, birds nesting around homes, and poultry are typically the most common mites biting man and causing transient dermatitis. Some of these mites are capable of transmitting viral, rickettsial, and filarial diseases of their hosts, and a few have been implicated in zoonotic transmission of disease to humans. Another, indirect, effect on human health by dermanyssoids is caused by honeybee parasites (*Varroa* spp., *Tropilaelaps* spp.) that can devastate hives, reducing pollination efficiency (and hence food resources) and honey production (an effective microbicide, especially in developing nations).

The superorder Acariformes contains about four times as many described species as the Parasitiformes and comprises two orders, both of which contain

---

TABLE 3.1   Systematic Synopsis and Distribution of Major Mites of Medical and Veterinary Importance

Class Arachnida, Subclass **Acari** (Acarina) — mites
  Superorder **Opilioacariformes** — Order Opilioacarida, Family Opilioacaridae
  Superorder **Parasitiformes** — ticks and ticklike mites
    Order **Ixodida (Metastigmata)** — ticks (Ixodidae, Argasidae)
    Order **Holothyrida** — rare, toxic reddish to brown mites
    Order **Mesostigmata**
        Superfamily Dermanyssoidea — poultry mites, nasal mites, rat mites
  Superorder **Acariformes** — mitelike mites
    Order **Trombidiformes**
      Suborder **Prostigmata**
        Superfamily Trombiculoidea — chiggers
        Superfamily Cheyletoidea — mange, skin, and follicle mites; cheletiellosis
        Family Pyemotidae — straw itch mites
    Order **Sarcoptiformes**
      Suborder **Astigmata**
       Infrasuborder Acaridida
        Family Acaridae — grocer's-itch, stored-product pests
        Family Glycyphagidae — furniture mite
       Infrasuborder Psoroptida
        Family Sarcoptidae — the human itch mite (scabies); sarcoptic mange mites
        Family Knemidokoptidae — scaly-leg and beak mites of birds
        Family Pyroglyphidae — house-dust mites
      Suborder **Oribatida** — oribatid, beetle, and moss mites

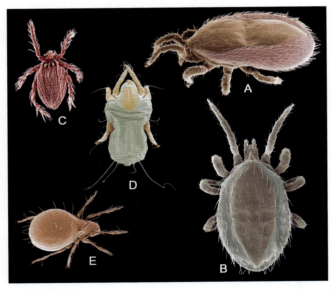

FIGURE 3.1   Scanning electron micrographs of some mites of medical importance: (A) *Ornithonyssus bacoti* (Hirst), the tropical rat mite; (B) *Ornithonyssus bursa* (Berlese), the tropical fowl mite; (C) *Guntheria* sp., a chigger; (D) *Dermatophagoides farinae* Hughes, a house-dust mite; (E) *Zygoribatula* sp., an oribatid mite.

medically significant species, and numerous parasites that cause mange in wild and domestic animals. The Trombidiformes includes the suborder Prostigmata, which contains chiggers (Figs. 3.1C, 3.6), biting mites that vector rickettsial disease, and numerous other transient and permanent ectoparasites, including human-follicle mites (Figs. 3.5B, C). Ectoparasitic prostigmatans are a frequent cause of transient dermatitis and, as in other biting or burrowing mites, contribute to the development of delusional parasitosis. The Sarcoptiformes includes the Oribatida (Fig. 3.1E) and the Astigmata. The most significant disease-causing acarines are found in the latter: the human-itch mite and house-dust mites, as well as stored-product pests and numerous permanent ectoparasites of domestic and wild animals. Fortunately, sarcoptiform mites do not appear to be significant vectors of human disease, although oribatid mites are intermediate hosts of tapeworms and are of some veterinary concern.

FORM, FUNCTION, AND LIFE CYCLES

Anatomy

Most mites are minute—usually less than 1 mm long—and easily overlooked. Like insects, mites are arthropods; but unlike insects, mites have only two body regions: an anterior headlike capitulum (gnathosoma) and a posterior saclike body (idiosoma) (Fig. 3.2). The capitulum is devoted to feeding and bears two pairs of appendages—chewing or stabbing chelicerae and sensory/feeding palps (pedipalps)—but may be vestigial in nonfeeding stages (Fig. 3.2B). The idiosoma bears the legs (three pairs in larvae, four pairs in other active stages), simple eyes (zero to two pairs) and numerous setae and other sensory struc-

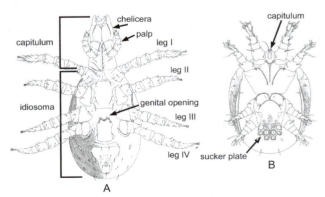

FIGURE 3.2   (A) Ventral view of mesostigmatan mite showing capitulum (gnathosoma), idiosoma, and appendages; (B) ventral view of astigmatan heteromorphic deutonymph.

tures (e.g., trichobothria) and is devoted to movement, most sensory inputs (including sight), digestion, excretion, and reproduction. Unlike insects, mites do not have antennae, mandibles, compound eyes, a true head, wings, a thorax, or a segmented abdomen.

Some mites (e.g., those that cause chelytiellosis) spin silk from the capitulum, but presumably the silk glands are located in the idiosoma, as is the brain and main nervous system. The distal segments of the legs often bear chemosensory setae and legs I may be antenniform. Colors range from shades of red and other primary colors to brown, beige, white, pale, or grayish. In females, mating and egg laying often occur through the same opening (usually located in the region of the hind leg bases); however, secondary copulatory pores have originated in various groups.

Feeding

Like most arachnids, many parasitic mites feed only on fluids; digestion is external and blood, lymph, or predigested skin is sucked into the esophagus by the action of a muscular pharynx. In dermanyssoid mesostigmatans, the entire cheliceral shafts are modified into slashing organs or into piercing-sucking structures that are often long enough to reach capillaries and allow rapid engorgement on blood. Extreme elongation of the chelicerae into a pair of stylets with concave paraxial surfaces has occurred independently in at least two of the most important families of biting mites: the Dermanyssidae and the Macronyssidae. The midgut usually bears a pair of saclike caecae that may fill much of the body cavity and leg bases.

In parasitic prostigmatans, the fixed digit of the chelicera is regressed and the movable digit is needle-like. Usually, these digits are used for puncturing the skin and injecting enzymes and other salivary fluids, but they are not long enough to reach capillaries for true blood feeding. Chiggers have an unusual feeding behavior. They attach to the surface of the skin with their needle-like movable digits and secrete salivary fluids into the wound. These fluids interact with the skin to form a ramifying system of semipermeable membranes with a hollow core (the *stylostome*—Fig. 3.6D) that collects digested skin and lymph and conducts it toward the mouth. Although chiggers feed only on the surface of the skin, the stylostome remains after the chigger has detached and may contribute to the illusion that they burrow into the skin.

In contrast to other mites, many sarcoptiform mites, including house-dust mites, bite off and swallow solid foods; the peritrophic matrix coating the resulting gut bolus/fecal pellet is a primary source of allergens. The mouthparts of particulate-feeding sarcoptiform mites are primitively chelate-dentate but are less robust in skin parasites that are evolving toward external digestion and fluid feeding. For example, sheep scab mites, *Psoroptes ovis* (Hering), use their mouthparts to suck or mop up an emulsion of lymph, skin cells, secretions, and microbes from the surface of the skin. The human itch mite, *Sarcoptes scabiei* (L.), secretes enzymes that weaken skin and then uses its spiny legs and chelicerae to break up the lysed skin and suck it in—creating the burrow in which it lays its eggs.

LIFE CYCLES AND LIFE HISTORY

Mite eggs may be laid singly or en masse, and on or off the host. Dermanyssoids develop and lay only a single large egg at a time (20–40% of the female body mass); however, several eggs may be laid each day. Maternal care is usually limited to hiding eggs in crevices or attaching them to hosts with silken webs or other secretions. Human itch mites and scaly leg mites of birds, however, lay their eggs in the burrow the females create in the skin. Some permanent vertebrate-parasitic Dermanyssoidea give birth to protonymphs, skipping an active larva entirely.

Like all arthropods, mites regularly shed their external skeletons during development and often add external structures as well as double in size with each molt. Cuticular growth within an instar (*neosomy*) is known for chiggers and some dermanyssoids, but, in general, intramolt increase in size is restricted to swelling or unfolding of previously deposited cuticle. For example, the unengorged cuticle of chiggers is usually folded in accordion-like pleats that allow the idiosoma to expand as the larva feeds and the protonymphal stage develops internally.

Acariform mites have what is considered to be the ancestral acarine developmental series: egg, prelarva, larva, protonymph, deutonymph, tritonymph, and adult (Fig. 3.3A-G). When present, the prelarva is inactive and usually poorly formed and retained within the egg. Prelarval and larval stages are hexapod, but larvae may be feeding or nonfeeding. Dermanyssoid mesostigmatans have a nonfeeding larva (Fig. 3.3C) and have suppressed the tritonymphal stage (as have chelytoid prostigmatans). In contrast, house-dust mites and astigmatan fur mites suppress the deutonymphal stage (a specialized dispersal morph in free-living Astigmata—Fig. 3.2E). Parasitengona (chiggers) have the most unusual acarine life cycle. All of the fundamental stages are present, but inactive regressive stages alternate with active stages. Thus, an inactive prelarva molts into an active parasitic

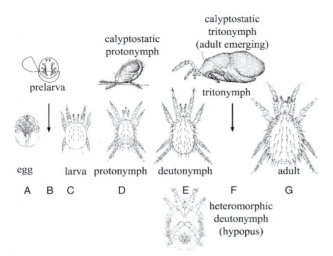

FIGURE 3.3 The fundamental ontogeny of mites includes 7 stages: egg (A), prelarva (B), larva (C), protonymph (D), deutonymph (E), tritonymnph (F), and sexually mature adult (G). In chiggers and their relatives (Parasitengona), the prelarva, protonymph, and tritonymph are inactive stages (calyptostases) hidden within the cuticle of the previous stage, so that only larvae (the parasitic stage), deutonymphs and adults (both predatory) are actively feeding. Prelarvae and tritonymphs are not present in the Dermanyssoidea. Astigmatan deutonymphs are absent in most parasites and house dust mites, but present as a specialized, non-feeding dispersal stage (hypopus) in many stored product mites.

larva. The protonymphal stage forms rudimentarily within the cuticle of the larva and eventually ecloses into an active, predatory deutonymph. This pattern is repeated with an active, predatory adult eventually emerging from the inactive tritonymph within the deutonymphal cuticle.

Dispersal

Migration from host to host typically occurs as mated adult females or wandering adult males in the Dermanyssoidea; however, when infection occurs through close contact (e.g., mother to offspring), then any active stage may change hosts. Some dermanyssoids mount their hosts only to feed, and eggs are laid in or around nests. The larvae are ephemeral and do not feed, so movement to the host is probably by the feeding nymph. Like most ticks, dermanyssoids that leave the host to molt must find another host. Unlike ticks, however, there are no data that indicate nymphs quest in surrounding habitats; most likely, the mites return to the same or a related host in the same nest, except perhaps in colonially nesting hosts. One interesting exception is that some parasites of flower-feeding birds enter flowers to await the next host. These mites can also be transported on nesting materials or commercial products associated with commercial production of the hosts; e.g., egg crates.

MITES AND HUMAN DISEASE

With the exception of chiggers and some dermanyssoid mesostigmatans, nontick acarines are generally considered to be minor vectors of disease. However, mites themselves cause diseases that range from annoying, but transient, dermatitis to severe infestations that may result in debilitating complications or death. Even nonparasitic mites, for example, commensal house-dust mites, contribute to the development of skin diseases, allergies and atopic asthma.

Parasites of Pets, Pests, and Synanthropic Wildlife That Bite People

Nests and dens of birds and mammals are usually home to a diverse assemblage of free-living mites (predators, scavengers, fungivores) and various facultative to obligate parasites. The latter include the most common mites that bite people, especially those ectoparasites that spend part (e.g., the egg stage) or most of their lives off the host. These mites are the most likely to wander from their obligate hosts and encounter people as transient pests. Human commerce has spread many of these pests around the world, resulting in numerous exotic occurrences.

One superfamily is responsible for the majority of the problems. The Dermanyssoidea (Mesostigmata) is a phylogenetically unresolved complex of families whose basal species are free-living predators in soil (Radovsky 1985, 1994). These basal lineages, however, have repeatedly adopted symbiotic associations with larger animals and have developed into a bewildering array of vertebrate and arthropod parasites. Many bees, ants, beetles, cockroaches, millipedes, mygalomorph spiders, reptiles, birds, and mammals have dermanyssoid parasites, some of which affect human health directly (e.g., bird and rat mites) or indirectly (e.g., varroa, poultry mites); others are major pests of pets, zoo animals, and wildlife. Adults typically range in size from 0.6 to 1.2 mm, or larger after a blood meal, so they are noticed by their victims more commonly than are other biting mites. Additionally, mange mites (Prostigmata, Cheyletoidea; Astigmata; Psoroptida) sometimes leave their hosts and bite people, especially if the host is handled, either as pets or in zoos.

Bird Mites

Ectoparasitic mites of birds that spill over and attack man belong primarily to one superfamily, the Dermanyssoidea (Mesostigmata). The starling mite, *Ornithonyssus sylviarum* (C and F) (Macronyssidae) (Figs. 3.4E, F) is well known because of its temperate

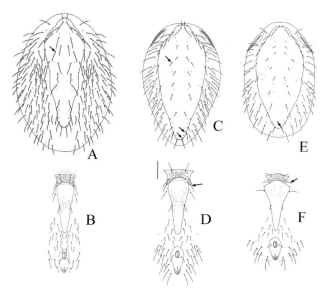

FIGURE 3.4 How to differentiate among the adult females of the common species of *Ornithonyssus* (Macronyssidae) biting man. *O. bacoti* (Hirst): (A) dorsum, (B) venter; *O. bursa* (Berlese): (C) dorsum, (D) venter; *O. sylviarum* (C and F): (E) dorsum, (F) venter. Scale bar = 0.1 mm.

distribution and broad host range, including poultry, widely distributed urban birds (e.g., house sparrows, starlings, pigeons, swallows), and many native song birds. Populations often reach extraordinary levels (tens of thousands per nest), and some spill over from nests into human habitations as birds rear their young. The real problem for people, however, occurs after a clutch has fledged or died; then thousands of starving mites may wander into homes or offices through windows, attics, or eaves and begin biting people. The bites are sharp and painful and develop into a substantial wheal. Scratching further inflames the bites and easily leads to secondary infections.

The tropical fowl mite, *Ornithonyssus bursa* (Berlese) (Figs. 3.1B, 3.4C, D), has a broader host range and geographical distribution than the starling mite, extending from warm temperate regions well into the tropics, and causes similar problems. The tropical fowl mite is less closely tied to its hosts (e.g., eggs are laid off the host) than the starling mite (which usually lay its eggs on the host), is more difficult to treat, and is a major pest of poultry, synanthropic birds, and native birds, even in remote protected areas. The spread of the tropical fowl mite appears to be one of those ecological disasters that we have caused but hardly noticed.

The pigeon, chicken, or poultry red mite, *Dermanyssus gallinae* (de Geer) (Dermanyssidae), is less invasive, but it is a significant economic pest of poultry. Those who rear or handle infested poultry are

often bitten, and the mite has been reported to take up residence in human ears (Rossiter 1997). Pigeons nesting on buildings can also provide a source of infestation (Regan et al. 1987). Usually, the poultry red mite feeds at night and hides in cracks and crevices during the day (where eggs are laid); however, in massive infestations mites may be almost continuously resident on birds. When large populations develop, they suppress egg production and cause death through blood loss. Under favorable conditions, generation time is about a week, and adult females can survive up to 9 months without feeding (Nordenfors et al. 1999). Several pathogens of birds have been found associated with the chicken mite, and it may be a vector of St. Louis encephalitis virus (Durden et al. 1993). It would be interesting to see if dermanyssoid mites are vectors of West Nile virus.

Rodent Mites

Ornithonyssus bacoti (Hirst), the tropical rat mite (Figs. 3.1A, 3.4A, B), infests rats, other rodents, and marsupials, and bites people living or working in rodent-infested buildings. Warehouses are a common source of contact with these annoying biters, as are laboratory rodent colonies and small mammal houses in zoos. As with many biting-mite problems, attacks by tropical rat mites are often confused with scabies, lice infestations, or the bites of insects.

The tropical rat mite occurs well outside the tropics and is strongly associated with synanthropic rodents, but it readily moves onto native species—and is widely reported from wild-caught small mammals in many parts of the world. Generation time is about two weeks, and large populations are capable of exsanguinating their hosts. It is a vector of the filarial parasite of rodents, *Litomosoides carinii*. In the laboratory, it has been demonstrated to transmit rickettsial pox (*Rickettsia akari*), encephalitis, and hemorrhagic fever viruses, the protozoans that cause Chagas disease, and the spirochetes that cause Lyme disease (Cortes et al. 1994; Durden and Turell 1993, Lopatina et al. 1999), but its importance as a natural vector is not clear. In the case of epidemic hemorrhagic fever, however, it is strongly suspected of being both vector and reservoir in urban China.

Other dermanyssoid parasites of rodents are problematic. *Liponyssoides sanguineus* (Hirst) (Dermanyssidae) is a parasite of the house mouse and a known vector of rickettsial pox—a disease that emerged in New York City after World War II. Spiny rat mites (*Laelaps echidninus* (Berlese) and their relatives (e.g., *L. nuttalli* Hirst), have also been introduced around the world, have spread to native mammals, and occasionally bite people. Junin virus (the causative agent for Argentine hemorrhagic fever), hantaan virus (Korean hemorrhagic fever), and Mossman virus—a paramyxovirus—have been isolated from *Laelaps* species. Mossman virus was isolated from introduced mites, *L. echidninus*, that were feeding on a native Queensland rat, *Rattus leucopus* (Campbell et al. 1977; Desch 1984). The rodent associates in the genus *Haemogamasus* have spread similarly, are sometimes associated with biting problems, and can carry hantavirus (Yakimenko et al. 2000).

Walking Dandruff and Cheyletiellosis

Cheyletiella spp. (Cheyletoidea, Cheyletidae) are mites that live on the epidermis of their hosts, sometimes cause a mange called *cheyletiellosis*, and may cause a pruritic dermatitis in pet owners. On mammalian hosts, silken webs are spun at the bases of hairs and eggs are attached to individual hairs.

Cheyletiella yasguri Smiley (Fig. 3.5A) is the species associated with dogs and can cause extensive irritation to their pet owners when lapdogs are infested. Similarly, *C. blakei* Smiley infests cats and those that pet them or let them sleep on their beds. Rabbits are host to *C. parasitivorax* (Mégnin). Infestations of pets can be without symptom, but in severe cases pruritis, hair

loss, and crustose lesions can develop. *Cheyletiella* also retain the ambulatory nature of their predatory ancestors, giving them their common name, *walking dandruff*, and allowing them to move easily from pets to pet owners, where they may cause a transient dermatitis characterized by raised red wheals with blister-like centers.

Mange Mites Biting People

The skin parasites that cause mange in us, our pets, livestock, and wildlife are treated in following sections; however, it is worth noting here that some of these mites bite people, cause transient dermatitis, and sometimes establish, usually self-limiting, infections. Except in immunocompromised patients, these infestations seem to be limited to bites, without the formation of burrows, or short burrows without successful reproduction.

Chiggers and Scrub Itch
(Black Soil Itch, Trombidiosis)

The diversity of common names for chiggers (Figs. 3.1C, 3.6A–D), including scrub itch mites, harvest mites, red bugs, *colorados*, *coloradillas*, *isangos*, *bêtes rouge*, *aoutat*, *lepte automal*, *mocuims*, etc., testify to the frequency with which we invade their habitats. These small (ca. 0.25 mm) red mites feed on their hosts only as protelean parasites; that is, only the first active stage, the six-legged larva, is parasitic. Eggs transform into an inactive prelarval stage before hatching into active larvae that seek out hosts using carbon dioxide,

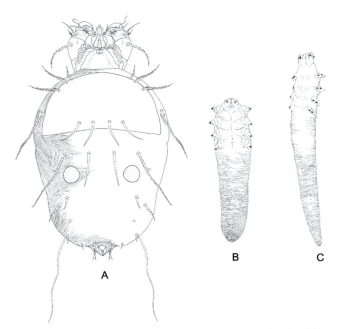

FIGURE 3.5 Some prostigmatan skin mites (Cheyletoidea): *Cheyletiella yasguri*, the dog cheyletiellosis mite (Cheyletidae): (A) dorsal view; *Demodex folliculorum*, the human follicle mite (Demodicidae): (B) active adult stage; (C) quiescent, pharate adult within nymphal cuticle.

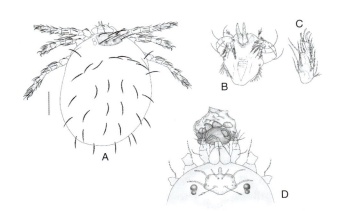

FIGURE 3.6 Chiggers (Acari: Prostigmata: Trombiculoidea): (A) engorged larva of *Leptotrobidium delinse* (Walch) (Trombiculidae), a vector of scrub typhus; (B) details of ventral capitulum; (C) details of legs I; (D) *Odontacarus* sp. (Leeuwenhoekiidae) attached to skin by stylostome.

body heat, or other host cues. After wandering on the host's body, the larva finds an acceptable spot and attaches, often around the eyes or genitals or in the ears of small mammals. The larvae usually accumulate in areas where the clothing is tight (e.g., tops of socks, belts, brassiere straps). The larva feeds externally, but the injected digestive salivary secretions and resulting stylostome may give a false impression that the mite is burrowing in the skin. Usually the first scratch crushes or dislodges the larva. On other hosts, and less commonly on people, the feeding larva begins to swell to many times its original size. A quiescent stage (the protonymph) is passed within the cuticle of the fully engorged larva, which eventually detaches and gives rise to the predatory deutonymph.

The postlarval stages (Fig. 3.3D) of this group are some of the smaller of the red velvet mites that are familiar to many. Deutonymphs feed on other arthropods or their eggs until they have greatly increased in size. A quiescent tritonymphal stage is passed within the deutonymphal cuticle. Adults emerge directly from the combined deutonymphal–tritonymphal cuticle and are often several millimeters long. Many species are bright red in color and covered with a dense pelage of plumose hairs (Fig. 3.1C). Similarly brightly colored water mites of streams and ponds are close relatives, but only a single species, *Thermacarus nevadensis* Marshall, is known to annoy people—in hot springs in the Sierras of California. The same species is also known from geyser fields in Chile, where the larval host is a toad (*Bufo spinulosus*) (Schwoerbel 1987).

Most of the chiggers that attack vertebrates belong to the Trombiculoidea (*trombicula* = little blood clot), which consists of two families: Leeuwenhoekiidae (Fig. 3.6D) and Trombiculidae (Figs. 3.1C, 3.6A). The Trombiculidae contains the chiggers of most medical concern. The normal hosts of these mites are reptiles (rarely amphibians), birds, and mammals (including pets and livestock). When people are bitten, it results in an itchy dermatitis (trombidiosis), commonly on the legs, but species vary in their preferences for attachment sites. Further complicating diagnosis, numerous pestiferous species co-occur in temperate to tropical regions of the world; e.g., *Eutrombicula sarcina* Womersley (the black soil itch mite), *E. hirsti* (Sambon), *E. samboni* (Womersley) (the ti-tree itch mite), *Odontacarus australiensis* (Hirst) (the Sydney grass itch mite), *O. hirsti* (Womersley), *Ascoschoengastia* spp., and others in Australia; *E. alfreddugesi* (Oudemans), *E. splendens* (Ewing), *E. batatas* (L.), and numerous others in the Americas. The harvest mite, *Neotrombicula autumnalis* (Shaw), appears to be the primary cause of human trombidiosis in western Europe.

Rodent chiggers are the primary reservoir and vector of a rickettsia (*Orientia* [formerly *Rickettsia*] *tsutsugamushi*) that causes the debilitating and sometimes fatal scrub typhus. The genus *Leptotrombidium* contains the best-known vectors; e.g., *L. akamushi* (Brumpt), distributed from Japan throughout southeast Asia, including many Pacific islands, and *L. deliense* (Walch), in Burma, Malaysia, Indonesia, the Philippines, and Australasia, including tropical Queensland, The Northern Territory, and Western Australia. *Blankaartia acustellaris* Walch and species of *Ascoschoengastia* also carry the rickettsia that causes scrub typhus (Tanskul et al. 1994), and one species in the latter genus is also a suspected vector of murine typhus (*Rickettsia typhi*) (Nadchatram et al. 1980).

Scrub Typhus (Chigger-Borne Rickettsiosis, Tsutsugamushi Disease)

Scrub typhus has long been a major disease in some rural areas of Japan and other regions in mainland (as far west as Pakistan and Tadzhikistan) and island southeast Asia and tropical Australasia. During World War II, almost 20,000 cases of scrub typhus were diagnosed in Allied troops, and about 1 in 10 American soldiers in Vietnam that were debilitated by fevers were victims of scrub typhus (Berman and Kundin 1973). Infections are still commonly reported from Japan, Vietnam, Thailand, New Guinea, Indonesia, and tropical Australia.

People contract scrub typhus when they enter a habitat with small mammals and chiggers infected with the rickettsia and are bitten by an infected mite. Chiggers often attach to people when they walk through local concentrations that result from mass hatchings (thousands of eggs can be laid by a single female). Disease foci are quite patchy, in time and space, and are often associated with grassy areas, especially ecotones or areas with disturbed vegetation (e.g., slash-burn agriculture, fields along rivers). Less disturbed habitats in central Asia (deserts, alpine areas), however, have had histories of disease (Varma 1993). Since potential vectors inhabit undisturbed forests and other pristine habitats (Domrow 1967), the chance for atypical disease foci (and the emergence of previously unrecognized rickettsial diseases) should be kept in mind.

Patients with scrub typhus exhibit acute fever with headache, profuse sweating, lethargy, and sometimes muscle pain, a rash, and pathologies of spleen, lymph system, and brain. Misdiagnosis appears to be a common problem, at least along the margins of the current distribution of scrub typhus. The incubation period is 7–14 days, and initial symptoms may be difficult to distinguish from other common infections. A primary *eschar* (ulcerous scab) 4–8 mm in diameter eventually forms at the site of attachment (commonly the genitals, buttocks, lower abdomen, arm, or armpit). Variable mortality is associated with different strains of *Orientia tsutsugamushi*, usually between 5 and 35%, but lethal strains have mortality as high as 60% (without treatment). Death can be due to heart failure, circulatory collapse, pneumonia, bleeding, or general organ failure. Mortality is highest in older patients, the chronically ill, and those delaying treatment (Currie et al. 1996; Kettle 1995; McBride et al. 1999; Quinlan et al. 1993).

The bites caused by *Leptotrombidium* species may be minor and hardly noticed by the patient, unless an eschar develops. Only a single eschar may be present, complicating diagnosis.

Straw Itch Mites

Pyemotes spp. (Prostigmata, Pyemotidae) are minute (0.22 mm) ectoparasites of insects. Each female mite attaches to an insect, injects a venom, and feeds on host fluids. The venom is capable of paralyzing the muscles of insects 150,000 times the size of the mite, and it is of current interest as a new class of insecticides (Tomalski and Miller 1991). The body of the female mite becomes grotesquely swollen as up to 300 new mites develop within the mother (a process called *physogastry*). Males mate with females within the mother or on emergence from the mother's ruptured body.

Members of the *Pyemotes ventricosis* group (Moser 1975) are parasites of insects; but if they wander onto larger animals (especially *P. tritici* LeGreze-Fossat and Montagne), they may bite (Yeruham et al. 1997). Those harvesting hay or lucerne or working in barns or in warehouses with stored grains or hay infested with beetles are most at risk; however, any accumulation of insect hosts can give rise to attacks from hay itch mites. The mites do not develop on people, and probably only bite when their normal hosts are depleted. However, severe dermatitis and anaphylactic shock may develop in some people.

Casual Biters

Many mites that are not parasitic can and do bite people on occasion. For example, *Androlaelaps casalis* (Berlese), a common inhabitant of hay, feed, and other stored grains, bites readily. The round, red, and fast-moving predatory whirligig mites, or footballers (Anystidae), wander from plants or outdoor furniture onto people and deliver a painful stab. Domestic species of Cheyletidae, close relatives of the mites that cause cheyletiellosis, usually feed on house-dust mites or stored-product mites, but at high population densities they can be annoying to people.

HUMAN SKIN PARASITES

Follicle Mites

Two species of demodecid mites (Prostigmata, Demodecidae) infect humans: *Demodex folliculorum* (Simon) (Figs. 3.5B, C), the hair-follicle mite, and *D. brevis* Akbulatova, the sebaceous-gland mite. Population levels increase with age, and the mites are especially common on the cheeks, eyelashes, and foreheads of adult men. The mites are parasites and feed on epithelial and sebaceous-gland cells, leading to some swelling and keratinization, but in most individuals they appear to be nonpathogenic. However, elevated populations of *Demodex* mites are found in severe cases of acne rosacea, and they may contribute to the inflammatory reaction (e.g., by blocking pores or inducing allergic reaction or by vectoring bacteria on their cuticles). Cases of pityriasis (scaly skin), chalazion, granuloma of the eyelid, and blepharitis also are associated with high populations of the mites (Bonnar et al. 1993; English and Nutting 1981; English et al. 1985; Roihu and Kariniemi 1998).

The Human Itch Mite and Scabies

Scabies, sometimes called the *seven-year itch*, has been known since ancient times. Its causative agent was identified by G. S. Bonomo and D. Cestoni in a letter to Francesco Redi (18 July 1687), although their hypothesis was ignored until much later (Arlian 1989a; Green 1989; Montesu and Cottoni 1991). Clinical scabies is characterized by lesions with burrowing adult female *Sarcoptes scabiei* (L.) (Astigmata, Sarcoptidae).

Adult female itch mites initiate burrows and become subspherical burrowing and egg-laying machines; they live 4–6 weeks and are capable of laying two to three eggs per day. Ovigerous females secrete lytic enzymes and dig their way into the horny layer of the skin using chelicerae and spines on legs in less than 30 minutes. Burrows extend into the stratum corneum and are extended from 0.5 to 5.0 mm/day by secreting enzymes that lyse skin and pushing through while engulfing lysed skin. Eggs are glued to the burrow walls and hatch in 50–53 h. The female larvae move to the skin surface and shelter in hair follicles for

77–101 h. The male protonymphal stage lasts 55–79 h, and the female 58–82 h. The deutonymph is not expressed. The tritonymphal male lasts 58–82 h, and the female 53–77 h. Development from egg to adult takes 10–13 days (Arlian and Vyszenski-Moher 1988). Transmission occurs when mites (presumably mated but unengorged females) move from person to person during close personal contact and from contact with fomites (Burkhart et al. 2000).

Most infestations, and most itch mites, are associated with the folded or wrinkled areas of forearms; e.g., hands, wrists, and elbows, or any susceptible areas of the body that the hands may touch. In infants, the mites often have an atypical distribution and can be found over most of body, including upper back, face, scalp, palms, and soles (all soft, wrinkled areas that may facilitate infection by the mites). Secondary infections are common, and scabies-initiated streptococcal infections in children may be associated with chronic renal disease later in life. People with diseases or disorders that reduce the scratching response develop larger populations of mites and may develop crusted scabies (see later).

Epidermal changes may result in an eczema-like reaction with hyperkeratosis, acanthosis, and spongiotic edema; microabcesses may form as the immune system responds, and ulcerations may develop from bacterial infection associated with scratching (Alexander 1984; Head et al. 1990). Misdiagnosis as eczema, psoriasis, impetigo, insect bites, or other skin diseases is common and leads to delayed treatment, recurrence of skin infections, and outbreaks and spread of scabies (Woltman 1994). Itching results from sensitization to mite cuticle, eggs, and fecal pellets (scybala) in the stratum corneum of epidermis; killing the mites does not immediately end itching. Nodular scabies may develop as an allergic reaction to mite bites but are usually free of mites.

In initial infestations by scabies in healthy people, surprisingly few mites are found. In the seminal studies of Bartley and Mellanby (1944) and Johnson and Mellanby (1942), the average number of mites found were 11 in men ($n = 886$), 12–13 in women ($n = 119$), and 19–20 in children ($n = 18$). Two-thirds of mites usually are found on the hands and wrists, but the feet and ankles also tend to be infested in children (Alexander 1984). Mellanby (1944) found that in newly-infested volunteers, itching developed after 1 month and other symptoms in 6–10 weeks; peak numbers were reached in 3–4 months. On reinfection (<1 yr), however, itching was induced almost immediately, and mite numbers remained low or infection failed. Protective immunity is known in humans, rabbits, and dogs, both cell-mediated and circulating antibody immune responses (Arlian et al. 1996). Scabies-specific IgE antibodies may be produced and cross-react with antigens of various house-dust mites (Arlian et al. 1991, 1994, 1995; Morgan et al. 1996).

In advanced scabies (so-called "Norwegian," crustose, or crusted scabies), the skin becomes scaly and hyperkeratotic; dense populations of mites (>1,400/cm^2) inhabit the thick crusts (2–20 mm) that develop, up to 2 million mites per person having been estimated (Mellanby 1942) and 4,700 mites/g of skin (Currie et al. 1995). Bed linen, curtains, furniture, and floors become infested with numerous mites, and epidemics are common (Carslaw et al. 1975).

Crusted scabies was first reported from lepers in Norway. It once was considered rare (Mellanby 1942; Espy and Holly 1976) but is now cosmopolitan and associated with HIV and other immune-system diseases, Down's syndrome, aged people in feeble condition, and those exhibiting dementia (Currie et al. 1995).

Various host-specific strains of *Sarcoptes scabiei* are known from a variety of domestic and wild animals, 40 known mammalian hosts in 17 families (Elgart 1990), but it is unclear if they are locally adapted populations of *S. scabiei* or true sibling species. The itch mite probably originated on people (related mites occur on the gorilla), and the known races appear to represent a single heterogeneous species based on morphological (Fain 1978; Estes and Estes 1993) and nuclear ribosomal (ITS2) studies (Zahler et al. 1999). However, Walton et al. (1999), using microsatellite studies on human and canine itch mites from Ohio, Panama, and Australia (NT), found that dog-derived and human itch mites are genetically distinct.

Animal-transmitted scabies is usually self-limiting in humans, burrows are rarely formed, and papules are present at sites of contact with infested animals. Removal or successful treatment of the animals usually results in spontaneous clearing of human infestations (Beek and Mellanby 1953; Emde 1961; Parish and Schwartzman 1993).

House-Dust Mites and Atopic Asthma

House-dust mites (Astigmata, Pyroglyphidae, Echymipodidae) live in our homes and feed on cast skin flakes, hair, and other detritus, and on the microbes that grow on this minute refuse. More than 20 species of mites inhabit house dust, with some unusual species being common in certain climatic zones; e.g., *Blomia tropicalis* Bronswijk et al. (Echymipodidae). The three most common species in temperate areas are *Dermatophagoides pteronyssinus* (Trouessart) (often 80% or more of the mites in homes),

Dermatophagoides farinae Hughes (Fig. 3.1D), and *Euroglyphus maynei* (Cooreman) (Colloff 1991). Pyroglyphid house-dust mites are the most important source of allergens in house dust. Severe allergies, rhinitis (hay fever–like symptoms), eczema (skin inflammation), and asthma (respiratory disorder) are induced by the allergens these mites produce and affect 50–100 million people worldwide (Fain et al. 1990; Tovey 1992). Contact dermatitis (e.g., "grocer's itch") and cross-reactivity with allergens of related astigmatans infesting straw or stored grains is often reported; e.g., with the brown-legged mite, *Aleuroglyphus ovatus* (Tropeau) (Acari: Acaridae) (Geary et al. 2000).

Mite feces are dry pellets 0.02–0.05 mm in diameter and covered with a peritrophic matrix. On average, about 20 are produced by each mite each day, and they readily become airborne (Arlian 1989b). Allergenic enzymes, e.g., cysteine proteases like *Der p I*, serine proteases, and amylase, are associated with mite feces. Other allergens present in the skin and exuviae or pieces of dead mite also cause reactions. The number of house-dust mites in the mattresses of asthmatic children is correlated with symptoms; in most asthmatic patients, exposure to house dust aggravates symptoms. Although dust and mites seem to be the most important factors causing allergies, it appears that there are both genetic and additional environmental factors that affect the expression of the disease (Tovey and Mahnic 1993).

Dust mites and other Astigmata have a small gland that opens at the base of the first pair of legs (supracoxal gland) and secretes a hyperosmotic solution of sodium chloride and potassium chloride into a gutter that runs to the buccal opening. The salt solution is hygroscopic and absorbs water from the atmosphere, down to a species-specific critical threshold of relative humidity (RH) (often 70–75% RH at 25°C). At lower humidities the solution crystallizes and blocks the gland opening, preventing water loss. Thus, dust mites obtain water from the atmosphere at relative humidities above the critical threshold, but they are unable to "drink" at lower relative humidities and eventually desiccate and die. At 25°C and 75–80% RH, development from egg to adult takes about 4 wk. Males mount the backs of newly emerged females and attach with suckers near the anus and on the hind legs. The aedeagus is extruded and inserted into a porelike opening on the female, copulation may take up to 48 h, and adult females live 4–5 wk and lay 50–80 eggs (Arlian 1989b; Colloff 1991a, 1991b, 1993; De Weck and Todt 1988; Fain et al. 1990).

Any conditions that help to maintain high humidity may encourage house-dust mite populations (Colloff 1994; P. J. Thompson and Stewart 1989; Tovey and Mahnic 1993). Poorly ventilated homes and those with furnishings that retain humidity (deep rugs, upholstered furniture) tend to have high average relative humidities, especially in the areas where the mites live. Also, the critical relative humidity decreases with temperature. The evaporative coolers used in some hot, dry areas will also lead to increased humidities within homes.

Within an hour of occupation, beds rise in temperature to 30–32°C and 95% RH as body heat and moisture are generated. Mattress-inhabiting dust mites will rise through the mattress toward people as the temperature and humidity rise and retreat back toward the center of the mattress as they drop. House-dust mite populations tend to be positively correlated with humidity in the home, and this relationship can be reflected in geographical population trends. Geographic areas that are humid and relatively cool tend to have more house-dust mites than those that are warm and dry. Allergy problems tend to be highest in homes located at low elevation in coastal regions, but even in dry regions homes with conditions favorable to mites can develop high populations (Kivity et al. 1993; Mumcuoglu et al. 1988; Veale et al. 1996; Tovey 1992). Additional resources and moisture added to beds during certain activities can contribute to high dust mite population levels (Colloff 1994).

Delusional and Illusional Parasitosis

An increasingly common psychological syndrome with which any acarologist working in a public institution is likely to have firsthand experience is delusional (delusory) parasitosis (DP); i.e., the mistaken belief that one is chronically infested by parasites (often mites). A recent overview of the problem is available in Hinkle (2000), and several Web sites maintain excellent treatments, with examples and critical literature (e.g., UC Davis 2002). Perhaps the most important things for the entomologist or acarologist to remember are: (1) the problems that victims of DP experience are very real to them and require appropriate medical treatment; (2) the role of the insect scientist is to rule out real infestations by parasites and to cooperate with the patient's health care professionals; and (3) DP is often curable.

Patients exhibiting DP mistakenly believe they are infested with or are being chronically bitten by minute animals (usually mites but also insects, nematodes, etc.), but they are not. Characteristically, a person with DP has visited several doctors and pest control operators in attempts to alleviate their problem, without success. Also, they tend to have treated themselves and pets with strong cleansing agents, pesticides, and disinfectants—sometimes causing serious damage to themselves and injury or death to their pets. Usually the delusional infestation is in or under the skin, and scratching at imagined burrows (or real bites) can cause extensive excoriation and secondary infection. In cases where deeper psychological problems seem to be indicated, the DP victim may believe they have mites living inside body openings (anus, mouth, or ears) or in the stomach or intestines.

Perhaps, not surprisingly, people exhibiting DP often react with anger and sometimes with hostility to yet another negative diagnosis, even when you show them under a microscope that their "mites" are bits of lint or scabs. More surprisingly, it is our experience that DP victims are often articulate and obviously intelligent people that have read widely on the subject of parasitism and can often describe their preferred parasite in extraordinary detail. Many are so convincing on the phone or in e-mail that it is difficult to believe they do not have a real problem with which you may be able to help. In our experience, the multiply shared delusions are usually most strongly held by only one of the individuals, and although the others may exhibit scratches and bites, they continually defer to the lead individual when it comes to detailed explanations.

Unlike delusions of parasitosis, *illusions of parasitosis* result from real irritations, but the cause is incorrectly interpreted and attributed to insects or other small organisms, which are then thought to be biting or infesting the person, the home, and/or the work environment. Itching and the sensations of things crawling on or under the skin may be associated with many seemingly unrelated medical conditions, including: allergic sensitivity to nonliving substances in the environment (e.g., office paper dust); use or abuse of drugs, especially alcohol, amphetamines, and cocaine; skin cancers; vitamin, protein, or other deficiencies; diabetes mellitus and other organic disease.

Mites of Domestic and Wild Animals

Given the ubiquity of mites on mammals and birds and their frequency on reptiles, reports of pathological effects on wild animals outside of zoos have not been areas of major discussion in the past. However, either significant parasitic mite diseases are emerging and/or our abilities to recognize such diseases have improved. In particular, skin parasites that cause severe mange or deformations (e.g., scaly leg of birds) in wildlife are now reported commonly, often in rare or endangered populations. Such outbreaks may be more likely to be reported for such animals, but isolated, stressed populations also may suffer mange outbreaks more

frequently than healthy ones. Mange can be so severe that local population extinction has been surmised (Martin et al. 1998) or even demonstrated in the well-documented example of the Spanish ibex (Leon-Vizcaino et al. 1999). Other factors also may be important. For instance, mountain brushtail possums (*Trichosurus caninus*) in peripheral habitats have greater numbers of mesostigmatans than those in contiguous optimal habitat (Presidente et al. 1982). Thus, fragmented habitats may be subject to "edge effects" and elevated mite burdens. Also, although the transmission or maintenance of disease-causing microbes by nontick acarines may be most important in restricted foci (e.g., underground burrows) (Pavlovsky 1964), the continued spread of synanthropic animals and their associated mites and microbes is undoubtedly affecting the epidemiology of mite-borne disease. Next, we briefly review the mites that attack reptiles, birds, and mammals. Mites affecting livestock and poultry are reviewed in other publications; our main focus here is on mites that cause pathology or transmit disease in wildlife or that have zoonotic potential.

Mites and Reptiles

The exotic snake mite (ESM) *Ophionyssus natricis* (Gervais) (Macronyssidae) is a cosmopolitan parasite of reptiles, especially snakes and large lizards. It is well known to reptile keepers, who implement stringent quarantine procedures to prevent its establishment in their collections. At 25°C the complete life cycle takes from 13 to 19 days (Camin 1953). Snakes can become anemic from high mite populations, and ESM also transmits *Pseudomonas hydrophilus*, causative agent of hemorrhagic septicemia. Like other macronyssids, *O. natricis* is a nest parasite, so it is vital that housing cages be treated along with the host itself for effective control. Other *Ophionyssus* species also transmit haemogregarine parasites, such as *Karyolysus lacertae* (Strandtmann and Wharton 1958; Shanavas and Ramachandra 1990). ESM has spread onto wild hosts in native habitats in Africa (Strandtmann & Wharton 1958) and Australia, including various snakes and blue-tongue skinks (Domrow 1987; Walter and Shaw 2002).

Pterygosomatidae are bright red mites found on lizards in Africa, Asia, Australia, and the Americas. They attach and feed in a similar manner to chiggers, with which they can be confused. Unlike chiggers, pterygosomatids undergo most if not all of their life cycle on the host and thrive in captive conditions. *Hirstiella* spp. are associated with disease and debilitation of captive reptiles, including anemia, skin lesions, and the transmission of parasitic blood diseases

(Walter and Shaw 2002). *Geckobia* spp. are found on geckoes in the Pacific, Australia, and Asia. Certain species may show preferences for particular microhabitats, such as around the eyes and on the feet. Others have become much broader than long as an adaptation to living under scales. Interestingly, parthenogenetic geckoes are especially susceptible and may endure heavy *Geckobia* infestations involving hundreds of mites (Moritz 1991).

Some of the most unusual associations include the Cloacaridae (Astigmata) inhabitants of cloacas of turtles (Camin et al. 1967) and the unusual nasal chiggers of sea snakes (*Vatacarus* spp.). Some debate has raged about whether certain lizards attempt to manipulate chigger attachment sites away from vital organs and into "mite pockets," but this is as yet unresolved (Arnold 1986, 1993; Bauer et al. 1990, 1993).

Mites and Birds

Birds maintained as pets in cages or aviaries can be plagued by a diversity of mites that are divided into three ecological groupings: external parasites and commensals, nest parasites, and internal parasites. Internal parasites such as the canary lung mite (*Sternostoma tracheacolum* Lawrence) can devastate caged and wild populations of Gouldian finches. Lung mites also infest the budgerigar (budgie, parakeet) *Melopsittacus undulatus*. Nest parasites such as the chicken mite (*Dermanyssus gallinae*) and the tropical fowl mite (*Ornithonyssus bursa*) can build up large populations around aviaries, chicken coops, and nests of introduced birds (e.g., pigeons and starlings). These blood-feeding parasites are generalists (and will bite humans) and can cause significant pathology in caged birds.

A small sampler of the different types of mite pathology is offered by the humble budgie. *Knemidokoptes pilae* (Lavoipierre and Griffiths) can cause lesions on the face and bill of caged budgies (Domrow 1992). Additionally, feathers are often infested with feather mites. Worldwide, only two species of feather mites have been reported from budgies: *Protolichus lunula* (Robin) (Pterolichidae) lives on the exposed surfaces of the wing and tail feathers, and *Dubininia melopsittaci* Atyeo and Gaud (Fig. 3.7F) (Xolalgidae) is found on the smaller feathers of the body. Black dots or a line of dots running parallel and close to the rachis may be seen with the naked eye, or the mites may readily be seen with magnification. Feather mites are usually nonpathogenic; however, heavy infestations of feather mites can cause weight loss, reduced egg production, depluming, and self-inflicted injuries (Atyeo and Gaud 1987).

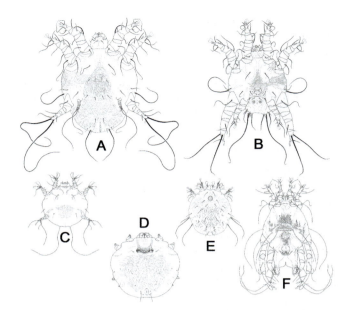

FIGURE 3.7 Some psoroptid astigmatans (Psoroptida). *Otodectes cynotis* (Hering) (Psoroptidae), the ear mite: (A) female venter, (B) male venter; *Knemidokoptes pilae* Lavoipierre and Griffiths (Knemidokoptidae), the scaly leg mite of parakeets: (C) larva venter, (D) female dorsum; *Trixacarus caviae* Fain et al. (Sarcoptidae), the guinea pig scab mite: (E) female dorsum; *Dubininia melopsittaci* Atyeo and Gaud (Xolalgidae), a parakeet feather mite: (F) male venter.

Feather Mites

With the exception of the Sphenisciformes (penguins), most birds have 2–25 species of astigmatic mites living on the contour on wing or tail feathers or within the quills of the flight feathers (Pterolichoidea, Analgoidea, Freyanoidea). Most birds become infected through contact with their parents in the nest, usually when flight feathers are 50–75% developed. Large populations may develop on unhealthy nestlings; but once a bird fledges, mite numbers drop drastically and individual species become highly restricted in their distribution on the birds. Feather mites live between barbules on the undersides of feathers. Although usually referred to as *parasites* in the literature, pathology associated with feather mites is rare.

Respiratory Mites of Birds

Perhaps the best-documented example of the destructive spread of a bird-parasitic mite is that of the canary "lung" mite (also air sac mite or tracheal mite), *Sternostoma tracheacolum* Lawrence (Rhinonyssidae), a parasite that infests the lungs and air sacs of many bird species. The lung mite appears to be a major factor in the decline of the Gouldian finch (*Erythrura gouldiae*) in the Northern Territory, especially in the early death of juvenile females during drought and reduced breeding success in survivors. Lung mites were found in 62% of Gouldian finches that died in mist nets (Tidemann 1996; Bell 1996). Other Rhinonyssidae, and the Speleognathinae (Ereynetidae) that inhabit the nasal mucosa of birds, appear to be less pathogenic.

Skin Parasites of Birds

Wild birds also are attacked by a variety of skin parasites, including relatives of the mites that cause cheyletiellosis in mammals. For example, *Bakericheyla chanayi* (Berlese and Trouessart) (Cheyletidae) feeds on the blood of several species of wild sparrows; in captivity, the infestations sometimes became acute (Furman and Sousa 1968). Two subspecies of white crowned sparrows (*Zonotrichia leucophrys*) that were migrating in the San Francisco area were found to have a higher infestation of *B. chanayi* (Cheyletidae), ca. 16%, compared to the resident subspecies, whose infestation rates averaged about 3%. Wild birds kept in captivity sometimes develop particularly dense infestations and acute symptoms that rarely are observed in the wild. These heavy infestations were characterized by large amounts of silk webbing coating the plumage. Birds became emaciated and less active and alert and exhibit ruffled plumage (Furman and Sousa 1968). Another cheyletid, *Ornithocheyletia* sp., can cause severe cases of hyperkeratinosis in the common pigeon (*Columba livia*) (Haarlov and Morch 1974). A more derived family of the Cheyletoidea, the Harpirhynchidae, consists entirely of bird parasites, some of which can form skin nodules or tumors (Filipich and Domrow 1985). North American passerine birds in four families are known to be parasitized by *Harpirhynchus novoplumaris* Moss et al. Most are found on the head and neck, where the female feeds at the base of a feather and forms a silken sheath that covers both her and her eggs. Reproduction is synchronized with the host, and females die after oviposition (Moss et al. 1968).

Scaly leg mites (*Knemidokoptes* spp.) are sarcoptoid mites of birds, with skin lesions, feather loss, and lesions around the beak and face (scaly face) and on the feet and legs (scaly leg) being the most commonly observed sign/symptom. Epizootics of scaly leg disease caused by *Knemidokoptes jamaicensis* Turk have been reported in migratory American robins (*Turdus migratorius*) in several places in the United States, with 60–80% of the birds showing lesions (Pence et al. 1999). This mite has also been found causing scaly leg in wild greenfinch (*Chloris chloris*) in South Australia (Domrow 1974).

Chiggers of Birds

Chiggers commonly attack birds, although the effects outside of the poultry industry are rarely assessed; however, the unusual bird-parasitizing genus *Womersia* (Leeuwenhoekiidae) can cause highly irritating skin inflammation and dermatitis. In the Chesapeake Bay, 10% of juvenile black ducks (*Anas rubripes*) and 8% of juvenile mallards (*Anas platyrhynchos*) had noticeable inflammation of the belly and vent caused by the presence of *Womersia strandtmanii* Wharton, located in tiny pits in the skin (G. M. Clark and Stotts 1960). In an extensive survey of 127 Laysan albatross chicks, several chicks had mild infections of chiggers *Womersia midwayensis* Goff et al., and three had dermatitis so severe that it was the apparent cause of death (Sileo et al. 1990). In another study from this same population, 24 out of 31 Laysan albatross fledglings were found to have mild to marked dermatitis due to unrelated mites, *Myialges nudus* (Epidermoptidae), most of which were found embedded in the stratum corneum of the skin (Gilardi et al. 2001).

Mites and Mammals

Almost any companion animal can have parasitic mites. Even the guinea pig has a scabies-like sarcoptic mange mite, *Trixacarus caviae* Fain et al. (Fig. 3.7D). Although infestations are often benign, or at least asymptomatic, dogs and cats may develop cheyletiellosis; demodectic mange, sometimes in puppies, from follicle mites (*Demodex* spp.); notoedric mange from *Notoedres cati* (Hering); or sarcoptic mange from an itch mite. The ear mite *Otodectes cynotis* (Hering) (Figs. 3.7A, B) infests the ears of most carnivore species commonly used as pets. Captive wild felids are also commonly affected and even free-ranging Eurasian lynx (*Lynx lynx*) (Degiorgis et al. 2001).

The mites mentioned cause disease, but others have the potential to transmit disease. This is true of any synanthropic parasite that moves into a wild population. For example, *Cheyletiella parasitivorax* has been shown capable of transmitting myxomatosis virus among wild rabbits in Australia (Mykytowycz 1958).

Mange Mites

Several kinds of mites cause mange, but the itch mite *Sarcoptes scabiei*, already mentioned in the context of human disease, is certainly the most dangerous, considering both its pathogenicity and the range of animals affected. The itch mite has a close association with humans, and consideration of the phylogeny of the Sarcoptidae strongly suggests that humans were the original host of *S. scabiei* (O'Connor 1984; Klompen 1992). Just as humans are labeled a superspecies, so, *S. scabiei* might be called a superparasite. Many case studies show how virulent this parasite can be to a range of mammal hosts; e.g., coyotes (Pence and Windberg 1994), foxes (Lindstrom et al. 1984), Spanish ibex (Leon-Vizcaino et al. 1999), wombats (Martin et al. 1998), and koalas (A. S. Brown et al. 1982).

A radiation of sarcoptid mites in the genus *Notoedres* is responsible for mange in a variety of wild mammals. Felids are the regular hosts for the mange mite *Notoedres cati* (Hering), which has caused severe mange in Florida panthers (Maehr et al. 1995), Queensland koalas (Domrow 1958), and Texan bobcats (Pence et al. 1982). *Notoedres muris* (Mégnin) (Sarcoptidae) is normally associated with synanthropic rodents, but infestations may flare up on other mammals. For example, *N. muris*–infested marsh rats from Argentina (*Holochilus brasiliensis*) had thick encrustations on the ears and nose, along with alopecia (Klompen and Nachman 1990). *Notoedres* outbreaks can have an intense but restricted focus, as has been demonstrated for the white-footed mouse, *Peromyscus leucopus* (Telford 1998). Other records of *N. muris* include collections from thick, crusty lesions on the ears, tail, and snout, and around the eyes of common brushtail possums (*Trichosurus vulpecula*) in Victoria, Australia (Booth 1994); from the ears and face and causing mange on fawn-footed mosaic-tailed rats (*Melomys cervinipes*) from palm forest in Australia; and from the forequarters and head and causing mange of a bandicoot (*Isoodon macrourus*) in Queensland, Australia (Domrow 1974). A mange outbreak by *Notoedres centrifera* Jansen devastated a population of western gray squirrels in Washington State (Cornish et al. 2001).

Demodex mites are minute, slender, spindle-shaped mites adapted to the skin glands of mammals, where they feed on dermal or epithelial cell contents. *Demodex* spp. are adapted to the Meibomian gland complex (Nutting et al. 1978). Pathology due to *Demodex* is limited in most wild animals (Nutting 1985), but massive infestations inducing hyperkeratosis and alopecia occur for reasons that are not understood. Captive wild animals (Nutting 1986; Holz 1998; Vogelnest et al. 2000) or domestic animals in poor physical condition are susceptible to such conditions. More rarely, wild animals such as black bears (*Ursus americanus*) may suffer, perhaps in restricted foci (Forrester et al. 1993). *Demodex antechini* Nutting and Sweatman on captive dasyurid marsupials (*Antechinus stuartii*) may cause large nodules 3–24 mm in diameter. However, most failed to cause gross pathology, and tumors occurred only in response to mites trapped in hair follicles (Nutting 1986).

Nest Associates

Nidicolous dermanyssoids are associated with disease-causing microbes. For example, a high incidence (ca. 40–60%) of infection by the haemogregarine blood parasite *Hepatozoon balfouri* is maintained in jerboas (*Jaculus jaculus*) in Egypt by the nest parasite *Haemolaelaps longipes* (Furman 1966). One infective mite ingested by *Jaculus jaculus* was shown to be capable of inducing a fatal infection. Another hepatozoon, *H. griseiscuri*, is found in, and develops in, *Haemogamasus ambulans* mites infesting gray squirrel (*Sciurus carolinensis*) nests. Similarly, three haemogamasine mite species inhabiting mole (*Talpa europea*) nests were shown to harbor a coccidian (Mohamed et al. 1987), and an extract made from several species of parasitic laelapid mites found in mole nests (*Talpa europea*) was found to contain tick-borne encephalitis virus (Kocianova and Kozuch 1988).

Nidicoles can both cause and transmit disease. For example, the skin-infesting hirstionyssine *Trichosurolaelaps crassipes* Womersley may spread wobbly possum disease virus in common brushtail possums (*Trichosurus vulpecula*) (Kocianova and Kozuch 1988). It also causes irritation and alopecia, especially on the lower back of common brushtail possums, a condition that may be synonymous with the widely recognized *rump wear*—the loss of extensive patches of hair on the lower body of common brushtail and mountain brushtail possums (*Trichosurus caninus*) (Booth 1994; Viggers and Spratt 1995).

Irritation by nest- or roost-associated mites can cause significant damage to populations. For example, a captive population of the endangered New Zealand bat, *Mystacina tuberculata*, was found to be infested by a previously undescribed laelapid mite, *Chirolaelaps mystacinae* Heath et al. The bats were extremely irritated by high numbers of these mites, and self-inflicted wounds, caused by scratching with their robust claws, resulted in secondary infections, septicemia, and death (Heath et al. 1987a, 1987b). Irritation caused by mite parasites and the associated costs due to increased grooming, decreased rest, and lost body weight can be substantial, as convincingly demonstrated for *Myotis* bats infested with experimentally manipulated loads of *Spinturnix* wing mites (Giorgi et al. 2001).

Endofollicular Hypopi

The deutonymph of many nest-based astigmatid mites of various families has a highly derived morphology adapted to developing endoparasitically within the follicles of a nesting host. This heteromorphic deutonymph, also known as a hypopus

(Fig. 3.2B), is mostly considered mildly or nonpathogenic. For instance, hypopi of *Aplodontopus sciuricola* were found in the skin of 95% of chipmunks (*Tamias striatus*) sampled, often in high numbers. Their presence caused minor pathology, including hyperkeratosis of the epidermis (Tadkowski and Hyland, 1974). Not all endofollicular hypopi are so benign; hypopi of *Marsupiopus zyzomys* Lukoschus et al. (Echymipodidae) cause the loss of tails in the Western Australian rock rats (*Zyzomys argurus*) (Lukoschus et al. 1979). This finding is curious because the conventional explanation for tail loss in rodents is autotomy to escape predators (Shargal et al. 1999), but these rodents store large amounts of fat in their tails, and the loss of this food store would be detrimental for their survival, particularly during the dry season.

Respiratory Mites of Mammals

At least five families of mites have invaded the nasal passages of mammals, and two of these (Halarachninae and Lemurnyssidae) have colonized the lower respiratory tract, including the lungs (Timm and Clauson 1985). Best known are the halarachnines (Dermanyssoidea), and these include most pathological examples. Halarachnines have invaded the respiratory passages of a diverse range of mammals, including rodents, dogs, pigs, seals, sea otters, hyraxes, phalangerid possums, and primates, including gorillas and chimpanzees. Heavy infestation by two species of *Orthohalarachne* larvae in the nasal turbinates of fur seals (*Callorhinus ursinus*) results in expiratory dyspnea as a result of chronic inflammation of the mucous membranes (Radovsky 1985). In one interesting case, a visitor to a marine park subsequently felt something moving under his eyelid. Three days after his visit, the pain was found to be due to the activities of a living adult seal mite, *Orthohalarachne attenuata* (Banks), which evidently had been snorted into his eye by an infected walrus (Webb et al. 1985). Species of *Halarachne* are found in the nares of earless seals and sea otters. Most infections are benign, but the fatality of a zoo sea otter is a notable exception (Kenyon and Yunker 1965). The nasal mite of dogs (*Pneumonyssoides caninum* Chandler and Ruhe) may induce profuse nasal secretion and hyperemia of the nasal mucosa.

Fur Mites of Mammals

The extraordinary diversity of fur mites of wild animals is largely ignored here, because they have rarely been associated with disease. For unknown reasons, massive infestations of the usually rare *Lynxacarus radovskyi* Tenorio occurred on feral and household cats in Florida, causing substantial irritation to these animals (Foley 1991). Also, fur mites of laboratory mice and rats often reach levels where treatment seems needed.

Oribatid Mites and Tapeworms

Oribatid mites (order Oribatida), also called *beetle mites* because the adults are heavily sclerotized (Fig. 3.1E), are usually brown to black and range in size from 0.22 mm to over 1 mm. Litter and soil have high abundances and diversities of these mites feeding on microbes, small worms, and detritus. Many species inhabit vegetation, where they feed on molds and lichens.

Like the related house-dust mites, oribatids feed on microbial foods, but they also readily eat small invertebrates and the eggs of anoplocephalan tapeworms. Cestode cysticercoids develop within the mites; when the infected mites climb onto vegetation, they in turn are eaten by herbivores. *Moniezia expansa*, a cattle tapeworm, and other economically important cestodes are spread by these mites. *Bertiella studeri* (Anoplocephalidae) is normally a parasite of New World monkeys but has been reported many times from humans in Asia. *B. mucronata* appears to infect New World monkeys and may also infect humans (Roberts and Janovy 1996). The intermediate hosts of the cestodes that infect koalas and other leaf-feeding marsupials are unknown, but arboreal oribatid mites are likely candidates. *Glycyphagus domesticus* (de Geer) (Astigmata, Glycyphagidae), a derived oribatid mite, is the intermediate host for the plerocercoid stage of *Catenotaenia pusilla*, a mouse cestode. The plerocercoid develops in the mite hemocoel over a period of 15 days, after which it is infective (Desch 1984).

Readings

Adjei, O., and Brenya, R. C. 1997. Secondary bacterial infection in Ghanaian patients with scabies. *East African Medical Journal* 74: 729–731.

Alexander, J. O. 1984. Scabies. In *Arthropods and Human Skin* (J. O. Alexander, ed.) Springer-Verlag, Berlin, pp. 227–292.

Allan, S. A., Simmons, L., and Burridge, M. J. 1998. Establishment of the tortoise tick *Amblyomma marmoreum* (Acari: Ixodidae) on a reptile-breeding facility in Florida. *J. Med. Entomol.* 35: 621–624.

Arends, J. J. 1991. External parasites and poultry pests. In B. W. Calnek, ed. *Diseases of Poultry*. Iowa State University Press, Ames, pp. 702–730.

Arlian, L. G. 1989a. Biology, host relations, and epidemiology of *Sarcoptes scabiei*. *Annu. Rev. Entomol.* 34: 139–161.

Arlian, L. G. 1989b. Biology and ecology of house dust mites, *Dermatophagoides* spp. and *Euroglyphus* spp. *Immunol. Allergy Clin. North America* 9: 2339–2335.

Arlian, L. G., and Vyszenski-Moher, D. L. 1988. Life cycle of *Sarcoptes scabiei* var *canis*. *J. Parasitol.* 74: 427–430.

Arlian, L. G., Vyszenski-Moher, D. L., Ahmed, S. G., and Estes, S. A. 1991. Cross-antigenicity between the scabies mite, *Sarcoptes scabiei*, and the house dust mite *Dermatophagoides pteronyssinus*. *J. Investigative Dermatol.* 96: 349–354.

Arlian, L. G., Vyszenski-Moher, D. L., and Fernandez-Caldas, E. 1993. Allergenicity of the mite *Blomia tropicalis*. *J. Allergy Clin. Immunol.* 91: 1042–1050.

Arlian, L. G., Morgan, M. S., Vyszenski-Moher, D. L., and Stemmer, B. L. 1994. *Sarcoptes scabiei*: The circulating antibody response and induced immunity to scabies. *Exp. Parasitol.* 78: 37–50.

Arlian, L. G., Rapp, C. M., and Morgan, M. S. 1995. Resistance and immune response in scabies-infested hosts immunized with *Dermatophagoides* mites. *Am. J. Trop. Med. Hygiene* 52: 539–545.

Arlian, L. G., Morgan, M. S., Rapp, C. M., and Vyszenski-Moher, D. L. 1996. The development of protective immunity in canine scabies. *Veterinary Parasitol.* 62: 133–142.

Arnold, E. N. 1986. Mite pockets of lizards, a possible means of reducing damage by ectoparasites. *Biolog. J. Linnaean Society* 29: 1–21.

Arnold, E. N. 1993. Comment—Function of the mite pockets of lizards: an assessment of a recent attempted test. *Can. J. Zool.* 71: 862–864.

Atyeo, W. T., and Gaud, J. 1987. Feather mites (Acarina) of the parakeet, *Melopsittacus undulatus* (Shaw) (Aves: Psittacidae). *J. Parasitol.* 73: 203–206.

Baker, A. S. 1998. A new species of *Hirstiella* Berlese (Acari: Pterygosomatidae) from captive rhinoceros iguanas *Cyclura cornuta* Bonnaterre (Reptilia: Iguanidae). *Systematic Appl. Acarol.* 3: 183–192.

Bartley, W. C., and Mellanby, K. 1944. The parasitology of human scabies (women and children). *Parasitology* 35: 207–208.

Bauer, A. M., Russell, A. P., and Dollahon, N. R. 1990. Skin folds in the gekkonid genus *Rhacodactylus*: A natural test of the damage limitation hypothesis of mite pocket function. *Can. J. Zool.* 68: 1196–1201.

Bauer, A. M., Russell, A. P., and Dollahon, N. R. 1993. Function of the mite pockets of lizards: A reply to E. N. Arnold. *Can. J. Zool.* 71: 865–868.

Beek, C. H., and Mellanby, K. 1953. Scabies. In *Handbook of Tropical Dermatology and Medical Mycology (II)*. Elsevier, Amsterdam, pp. 875–888.

Bell, P. J. 1996. The life history and transmission biology of *Sternostoma tracheacolum* Lawrence (Acari: Rhinonyssidae) associated with the Gouldian finch *Erythrura gouldiae*. *Exp. Appl. Acarol.* 20: 323–334.

Berman, S. J., and Kundin, W. D. 1973. Scrub typhus in South Vietnam. *Ann. Int. Med.* 79: 26–30.

Bonnar, E., Eustace, P., and Powell, F. C. 1993. The Demodex mite population in rosacea. *J. Am. Acad. Dermatol.* 28: 443–448.

Booth, R. 1994. *Medicine and Husbandry: Dasyurids, Possums and Bats, Wildlife*. The T. G. Hungerford Refresher Course for Veterinarians, Vol. 233. University of Sydney Post-Graduate Committee in Veterinary Science, Western Plains Zoo, Dubbo, Australia.

Bram, R. A., and George, J. E. 2000. Introduction of nonindigenous arthropod pests of animals. *J. Med. Entomol.* 37: 1–8.

Brown, A. S., Seawright, A. A., and Wilkinson, G. T. (1982). The use of amitraz in the control of an outbreak of sarcoptic mange in a colony of koalas. *Austral. Vet. J.* 58: 8–10.

Brown, S. K. 1994. Optimization of a screening procedure for house dust mite numbers in carpets and preliminary application to buildings. *Exp. Appl. Acarol.* 18: 423–434.

Bubenik, G. A. 1989. Can androgen deficiency promote an outbreak of psoroptic mange mites in male deer? *J. Wildlife Dis.* 25: 639–642.

Camin, J. H. 1953. Observations on the life history and sensory behavior of the snake mite, *Ophionyssus natricis* (Gervais) (Acarina: Macronyssidae). *Chicago Academy of Sciences Special Publication* 10: 1–75.

Camin, J. H., Moss, W. W., Oliver, Jr., J. H., and Singer, G. 1967. Cloacaridae, a new family of cheyletoid mites from the cloaca of aquatic turtles. *J. Med. Entomol.* 4: 261–272.

Campbell, R. W., Carley, J. G., Doherty, R. L., Domrow, R., Filippich, C., Gorman, B. M., and Karabatsos, N. 1977. Mossman virus, a paramyxovirus of rodents isolated in Queensland. *Search* 8: 435–436.

Carslaw, R. W., Dobson, R. M., Hood, A. J. K., and Taylor, R. N. 1975. Mites in the environment of cases of Norwegian scabies. *Br. J. Dermatol.* 92: 333–337.

Clark, G. M., and Stotts, V. D. 1960. Skin lesions on black ducks and mallards caused by chigger (*Womersia strandtmani* Wharton 1947). *J. Wildlife Management* 24: 106–108.

Clark, J. M. 1995. Morphological and life-cycle aspects of the parasitic mite *Trichosurolaelaps crassipes* Womersley, 1956 of trichosurid possums. *New Zealand Vet. J.* 4: 209–214.

Colloff, M. J. 1991a. Practical and theoretical aspects of the ecology of house dust mites (Acari: Pyroglyphidae) in relation to the study of mite-mediated allergy. *Rev. Med. Vet. Entomol.* 79: 611–630.

Colloff, M. J. 1991b. A review of biology and allergenicity of the house-dust mite *Euroglyphus maynei* (Acari: Pyroglyphidae). *Exp. Appl. Acarol.* 11: 177–198.

Colloff, M. J. 1993. House dust mites, asthma and allergy. *Biol. Sci. Rev.* 5: 38–40.

Colloff, M. J. 1994. Dust mite control and mechanical ventilation: When the climate is right. *Clin. Exp. Allergy*, 24: 94–96.

Colloff, M. J., Lever, R. S., and McSharry, C. 1989. A controlled trial of house dust mite eradication using natamycin in homes of patients with atopic dermatitis: Effects on clinical status and mite populations. *Brit. J. Dermatol.* 121: 199–208.

Connors, C. 1994. Scabies treatment. *Northern Territory Communicable Dis. Bull.* 2: 5–6.

Cornish, T. E., Linders, M. J., Little, S. E., and Haegen, W. M. V. 2001. Notoedric mange in western gray squirrels from Washington. *J. Wildlife Dis.* 37: 630–633.

Cortes, J. M., Torres, B. N., and Aguilar, R. A. 1994. Experimental transmission of *Trypanosoma cruzi* by *Ornithonyssus bacoti*. *Veterinaria* 25: 61–63.

Currie, B., Maguire, G. P., and Wood, Y. K. 1995. Ivermectin and crusted (Norwegian) scabies. *Med. J. Australia* 163: 559–560.

Currie, B., Lo, D., Mark, P., Lum, G., Whelan, P., and Krause, V. 1996. Fatal scrub typhus from Litchfield Park, Northern Territory. *Comm. Dis. Intell.* 20: 420–421.

Currie, B., Huffam, S., O'Brien, D., and Walton, S. 1997. Ivermectin for scabies. *Lancet* 350: 1551.

Degiorgis, M. P., Segerstad, C. H. A., Christensson, B., and Morner, T. 2001. Otodectic otoacariasis in free-ranging Eurasian lynx in Sweden. *J. Wildlife Dis.* 37: 626–629.

Derrick, E. H. 1961. The incidence and distribution of scrub typhus in north Queensland. *Aust. Ann. Med.* 10: 256–257.

Desch, C. E. 1984. Biology of biting mites (Mesostigmata). In *Mammalian Diseases and Arachnids*, Vol. 1 (W. B. Nutting, ed.), CRC Press, Boca Raton, FL, pp. 83–109.

De Weck, A., and Todt, A. (eds.) 1988. *Mite Allergy, a Worldwide Problem*. UCB Institute of Allergy, Brussels.

Domrow, R. 1958. A summary of the Atopomelinae (Acarina: Listrophoridae). *Proc. Linnean Soc. New South Wales* 83: 40–54.

Domrow, R. 1967. Mite parasites of small mammals from scrub typhus foci in Australia. *Austral. J. Zool.* 15: 759–798.

Domrow, R. 1974. Miscellaneous mites from Australian vertebrates. *Proc. Linnean Soc. New South Wales* 99: 15–35.

Domrow, R. 1985. Species of *Ophionyssus* Mégnin from Australian lizards and snakes (Acari: Dermanyssidae). *J. Austral. Entomological Soc.* 24: 149–153.

Domrow, R. 1988. Acari Mesostigmata parasitic on Australian vertebrates: An annotated checklist, keys and bibliography. *Invert. Taxon.* 1: 817–948.

Domrow, R. 1992. Acari Astigmata (excluding feather mites) parasitic on Australian vertebrates: An annotated checklist, keys and bibliography. *Invertebrate Taxonomy* 6: 1459–1606.

Durden, L. A., and Turell, M. J. 1993. Inefficient mechanical transmission of Langat (tick-borne encephalitis virus complex) by blood-feeding mites (Acari) to laboratory mice. *J. Med. Entomol.* 30: 639–641.

Durden, L. A., Linthicum, K. J., and Monath, T. P. 1993. Laboratory transmission of eastern equine encephalomyelitis virus to chickens by chicken mites (Acari: Dermanyssidae). *J. Med. Entomol.* 30: 281–285.

Elgart, M. L. 1990. Scabies. *Dermatologic Clinics* 8: 253–263.

Emde, R. N. 1961. Sarcoptic mange in humans. *Arch. Dermatol.* 84: 633–636.

English, F. P., and Nutting, W. B. 1981. Demodicosis of ophthalmic concern. *Am. J. Ophthalmol.* 91: 362–372.

Epsy, P. D., and Holly, H. W. 1976. Norwegian scabies: Occurrence in a patient undergoing immunosuppression. *Arch. Dermatol.* 112: 193–196.

Epton, M. J., Dilworth, R. J., Smith, W., and Thomas, W. R. 2001a. Sensitisation to the lipid-binding apolipophorin allergin *Der p 14* and the peptide Mag-1. *Int. Arch. Allergy Immunol.* 124: 57–60.

Epton, M. J., Malainual, N., Smith, W., and Thomas, W. R. 2001b. Vitellogenin-apolipophorin-like allergin *Der p 14* is a major specificity in house dust mite sensitization. *J. Allergy Clin. Immunol.* 107: S55.

Estes, A., and Estes, J. 1993. Scabies research: Another dimension. *Seminars Dermatol.* 12: 34–38.

Everitt, R., Price, M. A., and Kunz, S. E. 1973. Biology of the chigger *Neoschongastia americana* (Acarina: Trombiculidae). *Ann. Entomol. Soc. Amer.* 66: 429–435.

Fain, A. 1978. Epidemiological problems of scabies. *Int. J. Dermatol.* 17: 20–30.

Fain, A., and Hyland, K. E. 1962. The mites parasitic in the lungs of birds. The variability of *Sternostoma tracheacolum* Lawrence, 1948, in domestic and wild birds. *Parasitology* 52: 401–424.

Fain, A., and Laurence, B. R. 1975. *Satanicoptes armatus* n. gen., n. sp. (Astigmata: Sarcoptidae), a new mite producing mange in the Tasmanian devil (*Sarcophilus harrisii* Boitard). *J. Med. Entomol.* 12: 415–417.

Fain, A., and Lukoschus, F. S. 1979. Two new parasitic mites (Acari, Astigmata) from the skin of Australian vertebrates. *Mitteilungen aus dem Hamburgischen Zoologischen Museum und Institut* 76: 387–393.

Fain, A., Guerin, B., and Hart, B. J. 1990. *Mites and Allergenic Disease.* Allerbio, Varnenes-en-Argone, France.

Filippich, L. J., and Domrow, R. 1985. Harpyrhynchid mites in scaly-breasted lorikeet, *Trichoglossus chlorolepidotus* (Kuhl). *J. Wildlife Dis.* 21: 457–458.

Foreyt, W. J. 1993. Efficacy of in-feed formulation ivermectin against *Psoroptes* sp. in bighorn sheep. *J. Wildlife Dis.* 29: 85–89.

Foreyt, W. J., Coggins, V., and Parker, T. 1985. *Psoroptes ovis* (Acarina: Psoroptidae) in a Rocky Mountain bighorn sheep (*Ovis canadensis canadensis*) in Idaho. *J. Wildlife Dis.* 21: 456–457.

Forrester, D. J., Spalding, M. G., and Wooding, J. B. 1993. Demodicosis in black bears (*Ursus americanus*) from Florida. *J. Wildlife Dis.* 29: 136–138.

Frye, F. L. 1991. *Biomedical and Surgical Aspects of Captive Reptile Husbandry*, Vol. 1, 2nd ed. Krieger, Malabar, FL.

Fujita, T., Kumakiri, M., Ueda, K., and Takagi, H. 1997. A case of dermatitis due to *Notoedres cati* and scanning electron microscopic observation of mite. *Acta Dermatologica Kyoto* 92: 133–137.

Furman, D. P. 1966. *Hepatozoon balfouri* (Laveran, 1905): Sporogonic cycle, pathogenesis, and transmission by mites to jerboa hosts. *J. Parasitol.* 52: 373–382.

Furman, D. P., and Sousa, O. E. 1968. Morphology and biology of a nest-producing mite, *Bakericheyla chaneyi* (Acarina: Cheyletidae). *Ann. Entomological Soc. Amer.* 62: 858–863.

Gaud, J., and Atyeo, W. T. 1996. Feather mites of the world. *Ann. Mus. Roy. Afr. Cent., Sci. Zool.* 200: 1–193 (Part I), 1–436 (Part II).

Geary, M. J., Knihinicki, D. K., Halliday, R. B., and Russell, R. C. 2000. Contact dermatitis associated with the brown-legged mite, *Aleuroglyphus ovatus* (Tropeau) (Acari: Acaridae), in Australia. *Austral. J. Entomol.* 39: 351–352.

Gilardi, K. V. K., Gilardi, J. D., Frank, A., Goff, M. L., and Boyce, W. M. 2001. Epidermoptid mange in Laysan albatross fledglings in Hawaii. *J. Wildlife Dis.* 37: 185–188.

Giorgi, M. S., Arlettaz, R., Christe, P., and Vogel, P. 2001. The energetic grooming costs imposed by a parasitic mite (*Spinturnix myoti*) upon its bat host (*Myotis myotis*). *Proc. Royal Soc. London, B — Biological Sciences* 268: 2071–2075.

Goddard, J. 1996. *Physician's Guide to Arthropods of Medical Importance*, 2nd ed. CRC Press, Boca Raton, FL.

Goldberg, S. R., and Holshuh, H. J. 1993. Histopathology in a captive Yarrow's spiny lizard, *Sceloporus jarrovii* (Phrynosomatidae), attributed to the mite *Hirstiella* sp. (Pterygosomatidae). *Trans. Am. Microscopical Soc.* 112: 234–237.

Green, M. S. 1989. Epidemiology of scabies. *Epidemiologic Rev.* 11: 126–148.

Haarlov, N., and Morch, J. 1974. Interaction between *Ornithocheyletia hallae* Smiley, 1970 (Acarina: Cheyletiellidae) and *Micromonospora chalceae* (Faulerton, 1905) Orskov, 1923 (Streptomycetaceae, Ascomycetales) in the skin pigeon. *Proc. 4th Int. Cong. Acarol.*, pp. 309–314.

Halliday, B. 1998. *Mites of Australia, a Checklist and Bibliography.* Monographs on Invertebrate Taxonomy, Vol. 5. CSIRO, Collingwood, Victoria, Canada.

Hastriter, M. W., Kelly, D. J., Chan, T. C., Phang, O. W., and Lewis, Jr., G. E. 1987. Evaluation of *Leptotrombidium* (*Leptotrombidium*) *fletcheri* (Acari: Trombiculidae) as a potential vector of *Ehrlichia sennetsu*. *J. Med. Entomol.* 24: 542–546.

Head, E. S., Macdonald, E. M., Ewert, A., and Apisarnthanarax, P. 1990. *Sarcoptes scabiei* in histopathologic sections of skin in human scabies. *Arch. Dermatol.* 126: 1475–1477.

Heath, A. C. G., Bishop, D. M., and Daniel, M. J. 1987a. A new laelapine genus and species (Acari: Laelapidae) from the short-tailed bat *Mystacina tuberculata*, in New Zealand. *J. Royal Soc. New Zealand* 17: 31–39.

Heath, A. C. G., Julian, A. F., Daniel, M. J., and Bishop, D. M. 1987b. Mite infestation (Acari: Laelapidae) of New Zealand short-tailed bats, *Mystacina tuberculata*, in captivity. *J. Royal Soc. New Zealand* 17: 41–47.

Hinkle, N. 2000. Delusory Parasitosis. *Am. Entomologist* 46: 17–25

Hogue, C. L. 1983. *Eutrombicula* (Coloradillos, Chiggers). In *Costa Rican Natural History* (D. H. Janzen, ed.). University of Chicago Press, Chicago, pp. 723–724.

Hurwitz, S. 1985. Scabies in infants and children. In *Cutaneous Infestations and Insect Bites* (M. Orkin and H. I. Maibach, eds.). Marcel Dekker, New York, pp. 31–47.

Johnson, C. G., and Mellanby, K. 1942. The parasitology of human scabies. *Parasitology* 34: 285–290.

Kenyon, K. W., and Yunker, C. E. 1965. Nasal mites (Halarachnidae) in the sea otter. *J. Parasitol.* 51: 960.

Kettle, D. S. 1995. *Medical and Veterinary Entomology*, 2nd ed. CABI, Wallingford, VT.

Kivity, S., Solomon, A., Soferman, R., Schwarz, Y., Mumcuoglu, K. Y., and Topilsky, M. 1993. Mite asthma in childhood: A study of the relationship between exposure to house dust mites and disease activity. *J. Allergy Clin. Immunol.* 91: 844–849.

Klompen, J. S. H. 1992. Phylogenetic relationships in the mite family Sarcoptidae (Acari: Astigmata). Miscellaneous Publications of the Museum of Zoology, University of Michigan, No. 180, Ann Arbor.

Klompen, J. S. H., and Nachman, M. W. 1990. Occurrence and treatment of the mange mite *Notoedres muris* in marsh rats from South America. *J. Wildlife Dis.* 26: 135–136.

Kocianova, E., and Kozuch, O. 1988. A contribution to the parasite fauna in winter nests of the common mole (*Talpa europaea*) and incidence of its infection with tick-borne encephalitis virus (TBE) and rickettsia *Coxiella burnetti. Folia Parasitologica* 35: 175–180.

Lane, R. P., and Crosskey, R. W. (eds.) 1993. *Medical Insects and Arachnids.* Chapman and Hall, New York.

Leon-Vizcaino, L., Ruiz de Ybanez, M. R., Cubero, M. J., Ortiz, J. M., Espinosa, J., Pérez, L., Simon, M. A., and Alonso, F. 1999. Sarcoptic mange in Spanish Ibex from Spain. *J. Wildlife Dis.* 35: 647–659.

Lessof, M. H., Lee, T. H., and Kemeny, D. M. 1987. House dust mite allergens. pp. 94–95; House dust mites, pp. 415–416, In *Allergy: An International Textbook.* Wiley, London. Population growth of Dp and Df increase in presence of *Aspergillus amstelodami* and *niger.*

Levine, N. D. 1982. The genus *Atoxoplasma* (Protozoa, Apicomplexa). *J. Parasitol.* 68: 719–723.

Lindström, E., and Mörner, T. 1985. The spreading of sarcoptic mange among Swedish red foxes (*Vulpes vulpes* L.) in relation to fox population dynamics. *Rev. Ecol.* 40: 211–216.

Lindstrom, E. R., Andren, H., Angelstam, P., Cederlund, G., Hornfeldt, B., Jaderberg, L., Lemnell, P.-A., Martinsson, B., Skold, K., and Swenson, J. E. 1984. Disease reveals the predator: Sarcoptic mange, red fox predation, and prey populations. *Ecology* 75: 1042–1049.

Lombert, H. A. P. M., Gaud, J., and Lukoschus, F. S. 1984. *Apocnemidocoptes tragicola* gen. nov., spec. nov. (Acari: Astigmata: Knemidokoptidae) from the swift *Apus apus* (Aves: Apodiformes: Apodidae). *Acarologia* 25: 377–383.

Lopatina, Y. V., Vasilyeva, I. S., Gutova, V. P., Ershova, A. S., Burakova, O. V., Naumov, R. L., and Petrova, A. D. 1999. On *Ornithonyssus bacoti's* ability to perceive, support and transmit Lyme disease agent. *Meditsinskaya Parazitologiya I Parazitarnye Bolezni* April–June (2): 26–30 (in Russian).

Lounibos, L. P. 2002. Invasions by insect vectors of human disease. *Annu. Rev. Entomol.* 47: 233–266.

Lukoschus, F. S., Janssen Duijghuijsen, G. H. S., and Fain, A. 1979. Parasites of Western Australia IV: Observations on the genus *Marsupiopus* Fain, 1968 (Acaina: Astigmata: Glycyphagidae). *Records West Australian Museum* 7: 37–55.

MacDonald, J. W., and Gush, G. H. 1975. Knemidocoptic mange in chaffinches. *Br. Birds* 68: 103–107.

Maehr, D. S., Greiner, E. C., Lanier, J. E., and Murphy, D. 1995. Notoedric mange in the Florida panther (*Felis concolor coryi*). *J. Wildlife Dis.* 31: 251–254.

Martin, R. W., Handasyde, K. A., and Skerratt, L. F. 1998. Current distribution of sarcoptic mange in wombats. *Austral. Vet. J.* 76: 411–414.

Mason, R. W., and Fain, A. 1988. *Knemidocoptes intermedius* identified in forest ravens (*Corvus tasmanicus*). *Austral. Vet. J.* 65: 260.

McBride, W. J. H., Taylor, C. T., Pryor, J. A., and Simpson, J. D. 1999. Scrub typhus in north Queensland. *MJA* 170: 318–320.

Mellanby, K. 1942. Transmission and prevention of scabies. *Public Health* 55: 150–151.

Mellanby, K. 1944. The devlopment of symptoms, parasitic infection and immunity in human scabies. *Parasitology* 35: 197–226.

Michener, C. D. 1946. Observations on the habits and life history of a chigger mite, *Eutrombicula batatas* (Acarina, Trombiculinae). *Ann. Entomol. Soc. Amer.* 39: 101–118.

Mohamed, H. A., Molyneux, D. H., and Wallbanks, K. R. 1987. A coccidian in haemogamasid mites; possible vectors of *Elleipsisoma thomsoni* Franca, 1912. *Annales De Parasitologie Humaine Et Comparee* 62: 107–116.

Montesu, M. A., and Cottoni, F. 1991. G. C. Bonomo and D. Cestoni. Discoverers of the parasitic origin of scabies. *Am. J. Dermatopathol.* 13: 425–427.

Morgan, M. S., Arlian, L. G., and Fernandez-Caldas, E. 1996. Cross-allergenicity of the house-dust mites *Euroglyphus maynei* and *Blomia tropicalis. Ann. Allergy, Asthma Immunol.* 77: 386–392.

Moritz, C. 1991. Parasite loads in parthenogenetic and sexual lizards (*Heteronotia binoei*): Support for the Red Queen. *Proc. Royal Soc. London B* 244: 145–149.

Moser, J. C. 1975. Biosystematics of the straw itch mite with special reference to nomenclature and dermatology. *Trans. R. Entomol. Soc. London* 127: 185–191.

Moss, W. W., Oliver, J. H. Jr., and Nelson, B. C. 1968. Karyotypes and developmental stages of *Harpyrhynchus novoplumaris* sp. n. (Acari: Cheyletoidea: Harpyrhynchidae), a parasite of North American birds. *J. Parasitol.* 54: 377–392.

Mumcuoglu, K. Y., Zavaro, A., Samra, Z., and Lazarowitz, Z. 1988. House dust mites and vernal keratoconjunctivitis. *Ophthalmologica* 196: 175–181.

Mykytowycz, R. 1958. Contact transmission of infectious myxomatosis of the rabbit, *Oryctolagus cuniculus. CSIRO Wildlife Res.* 3: 1–6.

Nadchatram, M., Walton, D. W., and Telford, S. R. 1980. Species distribution of trombiculid mites on murine rodents in Rangoon, Burma. *Southeast Asian J. Trop. Med. Pub. Health* 11: 352–354.

Nordenfors, H., Hoglund, J., and Uggla, A. 1999. Effects of temperature and humidity on oviposition, molting, and longevity of *Dermanyssus gallinae* (Acari: Dermanyssidae). *J. Med. Entomol.* 36: 68–72.

Nutting, W. B. 1985. Prostigmata-Mammalia. In *Coevolution of Parasitic Arthropods and Mammals* (K. C. Kim, ed.). Wiley, New York.

Nutting, W. B. 1986. Histopathology of benign tumours due to *Demodex antechini* in *Antechinus stuartii.* In *Parasite Lives* (M. Cremin, C. Dobson, and D. E. Moorhouse, eds.). University of Queensland Press, Brisbane, pp. 95–108.

Nutting, W. B., Beekman, E. E., and Snyder, D. P. 1978. *Demodex gapperi* (Acari: Demodicidae) associated with eyelid sealing in the red-backed vole, *Clethrionomys gapperi. J. Med. Entomol.* 14: 646–648.

O'Callaghan, M. G., Carmichael, I. H., Finnie, J. W., and Conaghty, S. 1994. Lesions associated with infestation of a yellow-footed rock wallaby (*Petrogale xanthopus xanthopus*) with larvae of *Odontacarus* (*Legonius*) *adelaideae* (*Womersley*) (Acarina: Trombiculidae) in South Australia. *J. Wildlife Dis.* 30: 257–259.

O'Connor, B. M. 1984. Co-evolutionary patterns between astigmatid mites and primates. In *Acarology VI* (D. A., G. and C. E., B eds.). Ellis Horwood, Chichester.

Parish, L. C., and Schwartzman, R. M. 1993. Zoonoses of dermatological interest. *Sem. Dermatol.* 12: 57–64.

Pavlovsky, E. N. 1964. *Natural Nidality of Transmissive Diseases with Special Reference to the Landscape Epidemiology of Zooanthroponoses.* Nauka, Leningrad.

Pence, D. B., and Windberg, L. A. 1994. Impact of a sarcoptic mange epizootic on a coyote population. *J. Wildlife Management* 58: 624–633.

Pence, D. B., Matthews III, F. D., and Windberg, L. A. 1982. Notoedric mange in the bobcat, *Felis rufus*, from south Texas. *J. Wildlife Dis.* 18: 47–50.

Pence, D. B., Cole, R. A., Brugger, K. E., and Fischer, J. R. 1999. Epizootic podoknemidokoptiasis in American robins. *J. Wildlife Dis.* 35: 1–7.

Pérez Lozano, A. 1979. Environmental control in asthmatic homes. The role of *Cheyletus* mites. Preliminary report. *Allergol. et Immunopathol.* 7: 303–306.

Presidente, P. J. A., Barnett, J. L., How, R. A., and Humphreys, W. F. 1982. Effects of habitat, host sex and age on the parasites of *Trichosurus caninus* (Marsupialia: Phalangerida) in northeastern New South Wales. *Austral. J. Zool.* 30: 33–47.

Proctor, H. C., and Owens, I. 2000. Mites and birds. *Trends Ecol. Evol.* 15: 358–364.

Quinlan, M. L., Chappell, T., and Golledge, L. 1993. Scrub typhus in Western Australia. *Comm. Dis. Intell.* 17: 570–571.

Radovsky, F. J. 1985. Evolution of mammalian mesostigmate mites. In *Coevolution of Parasitic Arthropods and Mammals.* (K. C. Kim, ed.). Wiley, Brisbane, pp. 441–504.

Radovsky, F. J. 1994. The evolution of parasitism and the distribution of some dermanyssoid mites (Mesostigmata) on vertebrate hosts. In *Mites—Ecological and Evolutionary Analyses of Life-History Patterns* (M. A. Houck, ed.). Chapman and Hall, New York, pp. 186–217.

Regan, A. M., Metersky, M. L., and Craven, D. E. 1987. Nosocomial dermatitis and pruritis caused by pigeon mite infestation. *Arch. Int. Med.* 147: 2185–2187.

Roberts, L. S., and Janovy, J., Jr. 1996. *Gerald D. Schmidt and Larry S. Roberts' Foundations of Parasitology*, 5th ed. Wm. C. Brown, Dubuque, IA.

Roihu, T., and Kariniemi, A. L. 1998. Demodex mites in acne rosacea. *J. Cutaneous Pathol.* 25: 550–552.

Rosenberg, R. 1997. Drug-resistant scrub typhus: Paradigm and paradox. *Parasitol. Today* 13: 131–132.

Rossiter, A. 1997. Occupational otitis externa in chicken catchers. *J. Laryngol. Otol.* 111: 366–367.

Samuel, W. M. 1981. Attempted experimental transfer of sarcoptic mange (*Sarcoptes scabiei*: Acarina: Sarcoptidae) among red fox, coyote, wolf and dog. *J. Wildlife Dis.* 17: 343–347.

Sasa, M. 1961. Biology of chiggers. *Annu. Rev. Entomol.* 6: 221–244.

Schwoerbel, J. 1987. Rheophile water mites (Acari: Hydrachnellae) from Chile: 3. Species from hot springs. *Archiv für Hydrobiologie* 110: 399–408.

Service, M. W. 1980. *A Guide to Medical Entomology.* Macmillan, London.

Shanavas, K. R., and Ramachandra, N. P. 1990. Life-history of *Hepatozoon octosporei*, new species, a new haemogregarine from the skink, *Mabuya carinata* (Schneider), with notes on the in vitro excystment of its oocysts. *Archiv für Protistenkunde* 138: 127–137.

Shargal, E., Rath-Wolfson, L., Kronfeld, N., and Dayan, T. 1999. Ecological and histological aspects of tail loss in spiny mice (Rodentia: Muridae, *Acomys*) with a review of its occurrence in rodents. *J. Zoolog. Soc. London* 249: 187–193.

Sileo, L., Sievert, P. R., and Samuel, M. D. 1990. Causes of mortality of albatross chicks at Midway Atoll. *J. Wildlife Dis.* 26: 329–338.

Southcott, R. V. 1976. Arachnidism and allied syndromes in the Australian region. *Rec. Adelaide Children's Hospital* 1: 97–186.

Speare, R., Haffenden, A. T., Daniels, P. W., Thomas, A. D., and Seawright, C. D. 1984. Diseases of the Herbert River ringtail, *Pseudocheirus herbertensis*, and other north Queensland rainforest possums. In *Possums and Gliders* (A. P. Smith, and I. D. Hume, eds.). Surrey Beatty and Sons, Sydney.

Stone, W. B. Jr., Parks, E., Weber, B. L., and Parks, F. J. 1972. Experimental transfer of sarcoptic mange from red foxes and wild canids to captive wildlife and domestic animals. *N.Y. Fish Game J.* 19: 1–11.

Strandtmann, R. W., and Wharton, G. W. 1958. A manual of mesostigmatid mites parasitic on vertebrates. *Contributions of the Institute of Acarology, University of Maryland* 4: 1–330.

Tadkowski, T., and Hyland, K. E. 1974. The developmental stages of *Aplodontopus sciuricola* (Astigmata) from *Tamias striatus* L. (Sciuridae) in North America. *Proc. 4th Int. Cong. Acarol.*, pp. 321–326.

Tamara, A., Ohashi, N., Urakami, H., and Miyamura, S. 1995. Classification of *Rickettsia tsutsugamushi* in a new genus, *Orientia* gen. nov., as *Orientia tsutsugamushi* comb. nov. *Int. J. Syst. Bacteriol.* 45: 589–591.

Tanskul, P., Strickman, D., Eamsila, C., and Kelly, D. J. 1994. *Rickettsia tsutsugamushi* in chiggers (Acari: Trombiculidae) associated with rodents in central Thailand. *J. Med. Entomol.* 31: 225–230.

Telford, S. R., III. 1998. Focal epidemic of sarcoptid (Acarina: Sarcoptidae) mite infestation in an insular population of white-footed mice. *J. Med. Entomol.* 35: 538–542.

Thompson, C. F. 1999. Ectoparasite behavior and its effects on avian nest-site selection: Corrections and comment. *Ann. Entomological Soc. Amer.* 92: 108–109.

Thompson, P. J., and Stewart, G. A. 1989. House-dust mite reduction strategies in the treatment of asthma. *Med. J. Austr.* 151: 408, 411.

Tidemann, S. C. 1996. Causes of the decline of the Gouldian finch *Erythrura gouldiae*. *Bird Conservation Int.* 6: 49–61.

Tomalski, M. D., and Miller, L. K. 1991. Insect paralysis by baculvirus mediated expression of a mite neurotoxin gene. *Nature* 352: 82–85.

Tovey, E. R. 1992. Allergen exposure and control. *Exp. Appl. Acarol.* 16: 181–202.

Tovey, E. R., and Mahmic, A. (eds.) 1993. *Mites, Asthma and Domestic Design: Proceedings of a Conference Held at the Powerhouse Museum, 15th March, 1993.* University Printing Service, University of Sydney, pp. 23–25.

UC Davis. 2002. Delusional Parasitosis. http://delusion.ucdavis.edu/

Vargas, M., Bassols, B. I., Desch, C. E., Quintero, M. T., and Polaco, O. J. 1995. Description of two new species of the genus *Demodex* Owen, 1843 (Acari: Demodecidae) associated with Mexican bats. *Int. J. Acarol.* 21: 75–82.

Varma, M. G. R. 1993. Ticks and mites (Acari). In *Medical Insects and Arachnids* (R. P. Lane and R. W. Crosskey, eds.). Chapman and Hall, New York, pp. 597–658.

Veale, A. J., Peat, J. K., Tovey, E. R., Salome, C. M., Thompson, J. E., and Woolcock, A. J. 1996. Asthma and atopy in four rural Australian Aboriginal communities. *Med. J. Aus.* 165: 192–196.

Viggers, K. L., and Spratt, D. M. 1995. The parasites recorded from *Trichosurus* spp. (Marsupialia: Phalangeridae). *Wildlife Res.* 22: 311–332.

Vogelnest, L. J., Vogelnest, L., and Mueller, R. S. 2000. An undescribed *Demodex* sp. and demodicosis in a captive koala (*Phascolarctos cinereus*). *J. Zoo Wildlife Med.* 31: 100–106.

Walter, D. E., and Proctor, H. C. 1998. Feeding behaviour and phylogeny: Observations on early derivative Acari. *Exp. Appl. Acarol.* 22: 39–50.

Walter, D. E., and Proctor, H. C. 1999. *Mites: Ecology, Evolution and Behaviour.* University of NSW Press, Sydney.

Walter, D. E., and Shaw, M. D. 2002. First record of the mite *Hirstiella diolii* Baker (Prostigmata: Pterygosomatidae) from Australia, with a review of mites found on Australian lizards. *Austral. J. Entomol.* 41: 30–34.

Walton, S. F., Choy, J. L., Bonson, A., Valle, A., McBroom, J., Taplin, D., Arlian, L., Mathews, J. D., Currie, B., and Kemp, D. J. 1999a. Genetically distinct dog-derived and human-derived *Sarcoptes scabiei* in scabies-endemic communities in northern Australia. *Am. J. Trop. Med. Hygiene* 61: 542–547.

Walton, S. F., McBroom, J., Mathews, J. D., Kemp, D. J., and Currie, B. J. 1999b. Crusted scabies: A molecular analysis of *Sarcoptes scabiei* variety *hominis* populations from patients with repeated infestations *Clin. Infec. Dis.* 29: 1226–1230.

Webb, J. P., Furman, D. P., and Wang, S. 1985. A unique case of human opthalmic acariasis caused by *Orthohalarachne attenuata* (Banks, 1910) (Acari: Halarachnidae). *J. Parasitol.* 71: 388–389.

Woltman, L. 1994. Scabies: Treatment failures and hope for new success. *Austral. J. Rural Health* 2: 13–16.

Yakimenko, V. V., Dekonenko, A. Y., Malkova, M. G., Kuzmin, I., Tantsev, A. K., Dzagurova, T. K., and Tkachenko, Y. A. 2000. Spread of hantaviruses in West Siberia. *Meditsinskaya Parazitologiya i Parazitarnye Bolezni.* July–September (3): 21–28 (in Russian).

Yeruham, I., Rosen, S., and Braverman, Y. 1997. Dermatitis in horses and humans associated with straw itch mites (*Pyemotes tritici*) (Acarina: Pyemotidae). *Acarologia* 38: 161–164.

Zahler, M., Essig, A., Gothe, R., and Rinder, H. 1999. Molecular analyses suggest monospecificity of the genus *Sarcoptes* (Acari: Sarcoptidae). *Int. J. Parasitol.* 11: 8–15.

4

Ticks, the Ixodida

HANS KLOMPEN

INTRODUCTION

Ticks are chelicerates, bearing four pairs of walking legs, palps, and mouthparts in the form of chelicera. Specifically, they are mites, albeit quite large mites, with adults ranging in size from 0.5 to over 20 mm in some engorged females. Ticks are classified within the suborder Ixodida of the order Parasitiformes, the smaller of the two lineages of mites (Acari) (Norton et al. 1993). Also included in the Parasitiformes are the suborders Holothyrida, Opilioacarida, and Mesostigmata. The main diversity among Parasitiformes is concentrated in the suborder Mesostigmata (or Gamasida). Most Mesostigmata are free-living predators and/or phoretic associates of other arthropods, but several lineages have evolved into obligate, often permanent parasites of vertebrates (Krantz 1978; Walter and Proctor 1999) (Chapter 3). Like the Ixodida, the Holothyrida and Opilioacarida are moderate to large mites, but Holothyrida are very strongly armored, while Opilioacarida do not carry any shields at all. Often considered a bridge group between the Parasitiformes and Acariformes, Opilioacarida most likely belong in the Parasitiformes, based on molecular evidence. Both Holothyrida and Opilioacarida appear to be scavengers (Walter and Proctor 1998). Relationships among the four suborders are the focus of ongoing research, although it is most likely that the closest living relatives of ticks are the Holothyrida (Klompen et al. 2000; Lehtinen 1991).

Compared to other mite lineages, Ixodida is not species rich, with approximately 840 currently recognized species in 3 families: Ixodidae (hard ticks), Argasidae (soft ticks), and Nuttalliellidae (Keirans 1992; Keirans and Robbins 1999). The Nuttalliellidae are represented by a single species, *Nuttalliella namaqua*, collected from Namibia, South Africa, and Tanzania (Keirans et al. 1976). Little is known about its biology, and the species appears to have no medical-veterinary significance. The family Argasidae, the soft ticks, includes four genera: *Argas* (with five subgenera), *Ornithodoros*, *Otobius*, and *Carios*, and about 180 species (Klompen and Oliver 1993). The Ixodidae, or hard ticks, are more diverse, with approximately 660 recognized species. The Ixodidae are traditionally divided into Prostriata (one genus, *Ixodes*) and Metastriata (with 13 genera), based on the position of a ventral groove, which may pass anterior (Prostriata) or posterior (Metastriata) to the anus. The Metastriata have been arranged into four subfamilies: Amblyomminae (*Amblyomma*, *Aponomma*), Haemaphysalinae (*Haemaphysalis*), Hyalomminae (*Hyalomma*), and Rhipicephalinae (all other genera) (Hoogstraal and Aeschlimann 1982). This traditional ixodid classification is being revised based on recent phylogenetic analyses of tick relationships: (1) Monophyly of the Prostriata is uncertain (Klompen et al. 2000); (2) part of *Aponomma* (the "Australian endemic *Aponomma*") should be placed in a new genus and subfamily, while the remainder of *Aponomma* should be synonymized with *Amblyomma* (Dobson and Barker 1999); (3) the Hyalomminae have been synonymized with the Rhipicephalinae (Filippova 1997; Murrell et al. 1999, 2001); and (4) *Rhipicephalus* appears paraphyletic relative to *Boophilus* (and possibly *Margaropus*) (Murrell et al. 2000). These systematic problems are more than academic; they have an impact on how we perceive evolution in this group to have taken place and

45

how biological and epidemiological data are to be interpreted.

DEVELOPMENT AND MORPHOLOGY

Development

Ticks share the general acarine developmental pattern of a six-legged larva, followed by one or more eight-legged nymphal instars and eight-legged adults, but they differ from their close relatives in having only one (Ixodidae) or variable numbers (Argasidae) of nymphal instars. Ancestrally, Acari are assumed to have had three nymphal instars (Lindquist 1984), a condition retained by Opilioacarida and at least some Holothyrida. Reduction to a single active nymphal instar also occurs in other mite lineages, but the variable numbers observed in Argasidae, with the exact number depending on sex (females usually have one more) and feeding status (Pound et al. 1986), are unique.

External Structure

The body of arachnids is generally divided into a prosoma (cephalothorax), including the mouthparts and legs, and an opisthosoma (abdomen), containing the reproductive system and most of the digestive system. In Acari, including ticks, this division is completely obscured. The body is divided in two different functional units, the gnathosoma (or capitulum), which carries the chelicera, palps, and mouth; and the idiosoma, which includes the legs, brain, and digestive and reproductive structures.

Gnathosoma

In Ixodidae and larval Argasidae, the gnathosoma retains the anterior position most commonly found in other mites (Fig. 4.1), but in the postlarval instars of Argasidae the gnathosoma is inserted ventrally, and is not visible in dorsal view (Fig. 4.2). The chelicerae of ticks show some unique modifications; relative to other parasitiform mites, tick chelicera are small and slender. The movable digit of the chelicera no longer moves opposed to the fixed digit but has effectively rotated, with the cutting surface facing in the same direction as the fixed digit (Balashov 1972). Tick chelicerae do not function as pincers but as cutting organs, moving outward from the midline in cutting through the host skin. The hypostome is modified as an attachment organ, covered with rows of small retrorse spines, the denticles (Figs. 4.1C, D). The number of

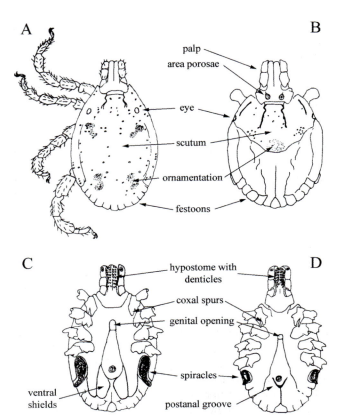

FIGURE 4.1 Generalized adult metastriate ixodid tick. Dorsal view of male (A) and female (B); ventral view of male (C) and female (D), with major morphological characters labeled. With minor adaptations from U.S.D.A. Agriculture Handbook No. 485 (1973).

denticles in each row is often used in keys of Ixodidae as a formula. For example, a formula of 3/3 indicates the presence of three denticles on each side of the hypostome over most of its length. Attachment to the host is achieved primarily by inserting part of the hypostome into the host's skin, with the hypostomal denticles preventing backsliding. In addition, many Ixodidae secrete "cement" around the bite site, which functions to glue them to the host. This "cement" is one of several substances produced by the salivary glands. In "long mouthpart" taxa, such as *Amblyomma* and *Hyalomma*, the hypostome is inserted deep into the host, with a core of cement around it. In "short mouthpart" taxa, such as *Dermacentor*, *Rhipicephalus*, and *Boophilus*, the cement forms a cone around the exposed mouthparts, a small mound. Detaching ticks can pull their mouthparts out of the cement cones without damage, but forcible removal of ticks commonly results in broken mouthparts. While the mouthparts are clearly the primary attachment organs for ticks, some Ixodidae also carry distinct spines on the palps and leg coxae (Figs. 4.1C, D). These may assist in anchoring of the tick by hooking around host hair. Argasid ticks lack palpal and coxal spines and do not produce cement.

The palps are leglike (Argasidae) or compressed (many Ixodidae) but never raptorial. They show the characteristic chelicerate segmentation of trochanter, femur, genu, and fused tibia and tarsus (tibiotarsus). In older tick literature these segments are often referred to as articles I–IV, respectively. The pretarsal claw is completely absent. Female ixodid ticks carry a pair of glandular fields, the areae porosae, dorsally on the subcapitulum (Fig. 4.1B). These glands function in coating the eggs with a waxy layer, thus waterproofing the eggs (Feldman-Muhsam 1963).

Idiosoma

Unlike most other Parasitiform mites, ticks have few idiosomal shields, and the cuticle of most of the body is only weakly sclerotized. Argasid nymphs and adults rarely have any shields at all (Fig. 4.2A), although most argasid larvae carry a mid-dorsal shield. Ixodid immatures and females always carry a small anterior dorsal shield, the scutum (Fig. 4.1B), which, among other functions, provides attachment points for the cheliceral musculature. The remaining parts of the idiosoma are weakly sclerotized and characterized by apparent striation resulting from cuticular folding. Immature and female hard ticks can take in enormous blood meals, with female ixodids swelling to 50–100 times their original weight (Balashov 1972). This is made possible by a combination of unfolding and stretching the unsclerotized cuticle and production of new cuticle (neosomy). Neosomy is restricted mostly to the early, slow feeding phase, while stretching appears restricted largely to the rapid feeding phase. Male ixodid ticks have more and larger shields. Their scutum covers the entire dorsum (holodorsal), and they often carry additional ventral shields (Figs. 4.1A, C). The added sclerotization offers added protection, but it does not allow significant engorgement. Not surprisingly, male ixodids do not feed much, and some male Ixodes do not feed at all. The scutum of adult metastriate ticks may show various forms of ornamentation, patches of color ranging from off-white to gold, green, and red (Figs. 4.1A, B). Some species of Amblyomma and Cosmiomma are particularly colorful. Ornamentation is nearly always restricted to the adults but may occur in the nymphs. The idiosoma of most Metastriata is characterized by the presence of festoons, uniform rectangular areas on the posterior margin, separated by distinct grooves (Fig. 4.1). These structures are best viewed in unengorged specimens and can be obscured in engorged ones. They are absent in Boophilus. The anus is generally ventral or ventroterminal, and the genital opening of both sexes is midsternal. Absence of a genital opening differentiates nymphs from adults, and the extent of coverage of the scutum (anterior dorsal only or holodorsal) differentiates females from males.

The idiosoma in Argasidae shows different sets of characteristics. The cuticle may be striate (larvae), wrinkled, granulate (most Argas and adult Otobius), or mammillate (most Ornithodoros and Carios), depending on the size of cuticular elevations. If present, cuticular discs, smooth areas forming attachment points for the dorsoventral muscles, may be arranged in radial (Fig. 4.2B) or dispersed patterns. Many Argasidae in the genus Argas are extremely flattened, with strong differentiation of the marginal cuticle (Fig. 4.2B). Overall, Argasidae show a greater diversity of cuticular structure and grooves than Ixodidae (Figs. 4.2C, D).

The idiosoma of all ticks carries a number of sensory and glandular structures. The pattern of these structures can be helpful in identification of larvae, but hypertrichy (the presence of "excessive" numbers of

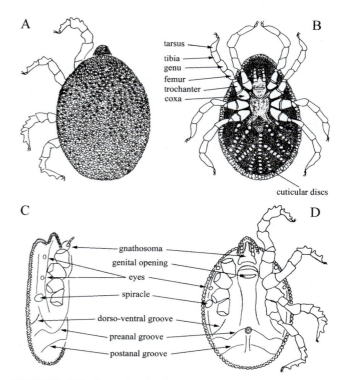

FIGURE 4.2 Generalized adult argasid tick. Dorsal view of Ornithodoros (A); ventral view of Argas (B); lateral (C) and ventral (D) views of Ornithodoros. With minor adaptations from U.S.D.A. Agriculture Handbook No. 485 (1973).

setae arranged in random patterns) usually makes those characters unworkable in the postlarval instars. In terms of function, the large wax glands (sensilla sagittiformia) in Metastriata have been shown to secrete defensive chemicals that are effective against ants (Yoder et al. 1993). This type of gland is absent in Prostriata and Argasidae.

Respiration is cuticular in the larva, but the post-larval instars have a tracheal system opening to the outside by means of a pair of spiracular plates. The spiracular plates in Ixodidae are located slightly posterior to coxae IV; they tend to be quite complex and are important in identification. Spiracles in Argasidae are simpler, and they are positioned lateral to coxae III–IV. Tracheal openings are closed by valves under nervous control, an important feature in limiting water loss. Overall respiratory rates can be low, ventilating only once every 4–6 hours.

Legs

The legs of ticks are made up of the standard complement of segments for Arachnida. The coxae are discrete but fused to the body. The trochanter, femur, genu (or patella), tibia, and tarsus (basal to distal) are all free (Fig. 4.2B). The coxae and trochanters in Ixodidae may bear spines (Figs. 4.1C, D). In most ticks, the tarsus bears a pair of distal claws and an ambulacrum, both assumed to be remnants of a pretarsal segment. Leg and palp setation patterns are useful in identification of larvae (Edwards and Evans 1967; Klompen 1992), but the patterns in the postlarval instars are once again too complex and erratic to be of much use. The tarsus I sensory complex provides another characteristic feature for all ticks, the *Haller's organ*. This structure consists of a capsule, an invagination of the cuticle, and an anterior pit. The capsule is often nearly closed off from the outside environment by a "roof." The structure of this roof can be genus—or even species—specific. The capsule contains three to seven chemosensitive "porose" sensilla. The anterior pit may also be delimited by cuticular ridges, but it is never covered. It usually contains seven to nine sensilla of multiple types. Based on light microscopic examination they are split into conical, fine, grooved, porose, and serrate sensilla. The function of each type of sensillum is still incompletely understood, but this grouping includes thermo-, hygro-, and chemoreceptors. Ixodidae can regenerate parts of their legs if lost in a previous instar. The regrown segments may not look perfectly normal, but they appear to be functional. In Argasidae, regeneration may take two molts, but it results in near-complete restoration of the amputated limb.

Sensory Systems

Many Metastriate Ixodidae and a few Argasidae have eyes, but sight is generally not well developed. Eyes are always simple, with a single pair inserted at the lateral margins of the scutum (Ixodidae) or two pairs inserted lateral (some Argasidae) (Figs. 4.1A, B, 4.2C, D, 4.3). Whether or not they have eyes with well-developed lenses, all ticks appear to be able to detect light–dark changes. The main sensory systems are geared toward chemical clues and temperature, especially heat. Most long-distance sensors are concentrated on the tips of the palps (the tibiotarsal region) and on the tarsi of legs I (Haller's organ).

NORTH AMERICAN TAXA

The North American fauna includes all four genera of Argasidae but only seven out of 14 genera of Ixodidae. Argasidae are rarely encountered, and adults especially are often hard to identify, but adults of the more familiar ixodid ticks can be differentiated to genus quite easily. *Ixodes* species have a distinct preanal groove and lack festoons, while all others have a postanal groove (Fig. 4.1D) and carry festoons (Fig. 4.1) (*Boophilus* carries no festoons). Other distinguishing characters are found in the gnathosoma, palps, and scutum (Fig. 4.3). *Ixodes* and *Haemaphysalis* lack eyes. *Ixodes, Amblyomma,* and *Aponomma* retain long palps in the adults, while members of the other genera have short, compressed palps. Among the latter genera, the palp femur in *Haemaphysalis* species is characteristically flared outwards (Fig. 4.3). The basis capitulum (Fig. 4.3) of *Rhipicephalus* and *Boophilus* is hexagonal in shape, while the remaining genera retain a rectangular or triangular shape. A good pictorial key to the genera and species east of the Mississippi is provided by Keirans and Litwak (1989).

BIOLOGY

Ticks are obligate parasites of vertebrates; that is, they can feed only on the blood. Unlike fleas or mosquitoes, all active instars (with the exception of males in some lineages) are parasitic. They are also nonpermanent parasites: Most ticks are not permanently attached to their hosts, but instead have to find a new host every time they feed. Beyond that point there are major differences between Ixodidae and Argasidae.

In Ixodidae, eggs are deposited in a large batch (from several hundred to 23,000 per female), after which the female dies. The eggs hatch, and the larva

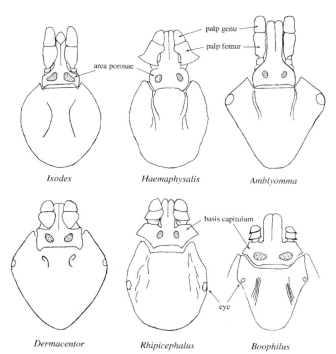

FIGURE 4.3 Comparative illustrations of the scutum and dorsal gnathosoma for the six most common genera of ixodid ticks in North America. With minor adaptations from U.S.D.A. Agriculture Handbook No. 485 (1973).

searches for a host (questing). Once a host is located, the larva climbs on, attaches, and feeds for 3–7 days. Feeding starts slowly but proceeds very rapidly in the final 12–36 hours. Often almost half of the total weight gain is achieved in that final period. This pattern of differential feeding rates is similar for the larva, nymph, and female. The engorged larva drops off and eventually molts to a nymph. This nymph will search for a new host. If successful, it attaches and feed for 3–8 days. In terms of percentage weight increase, larvae and nymphs feed less than females, but weight increases of, respectively, 7–20 and 9–80 times have been recorded (Balashov 1972). After engorgement, the nymph drops off and molts to an adult. At this point there are some differences among ixodid taxa.

In many *Ixodes*, mating takes place off the host, with the males never searching for a host. In other *Ixodes*, and in Metastriata, both sexes search for a host, and mating takes place on the host. In an interesting intermediate condition, males of the Australian *I. holocyclus* feed, but always on females of their own species. These differences are correlated with differences in sperm maturation.

In most Metastriata, males must feed to be able to produce viable sperm. But in many *Ixodes* (and in

Argasidae), males have completed sperm development at the adult molt. If males feed, weight increase tends to be limited to 0.7–3 times. Females of most Ixodidae need to be mated before fully engorging. Engorged females drop off and deposit their single batch of eggs, completing the cycle.

The foregoing life history pattern, where larvae, nymphs, and adults each have to find a host, is called a three-host life cycle (Fig. 4.4A). It does not characterize all Ixodidae. Some species of the genera *Hyalomma* and *Rhipicephalus* do not drop off after larval feeding; they molt on their host. These represent a two-host life cycle (Fig. 4.4B). Finally, species of *Boophilus*, *Margaropus*, some *Dermacentor*, and occasionally some *Hyalomma* remain on their hosts from first attachment to drop-off as mated females, a one-host life cycle (Fig. 4.4C).

The Argasidae follows an entirely different strategy. Larvae seek a host and feed for 3–8 days (just as in most Ixodidae) or, more rarely, for 1–2h (most members of the genus *Ornithodoros*), or not at all (some members of the *O. moubata* complex). They drop off and molt to the first of several nymphal instars, each of which will feed before molting to the next instar. Feeding in argasid nymphs and adults is short, with feeding bouts usually lasting 1 hour or less. Adults of the *Carios coprophilus* group (formerly *Antricola*) and *Otobius* do not feed at all. Adult Argasidae mate off the host, and females produce a small batch of eggs (50–600) after each feeding bout.

Nearly all Argasid ticks stay in the nests or burrows of their hosts, so even though they may feed on more occasions than Ixodidae, they often feed repeatedly on the same host species or even individual. Many Ixodidae are also nidicolous, but others, including most of the medically-veterinary important ones, are nonnidicolous (*field*) species, living and dropping off in open areas. These nonnidicolous species are especially prone to feeding on different species in different instars, a major factor in the epidemiology of tick-borne diseases.

What the foregoing summary does not convey is a sense of time. Many tick species live for long periods of time (2–4 years, over 10 years in some Argasidae), and unfed ticks may survive for more than a year off-host. These characteristics result in the uncommon situation that these parasites often outlive their hosts. The longevity of ticks, combined with the relatively short feeding periods on the host, also means that many ticks spend more than 95% of their life off-host (Needham and Teel 1991). Their ability to survive off of the host is truly astounding. Much of this time is spent in an inactive state resembling diapause. Diapause has been studied most often in temperate

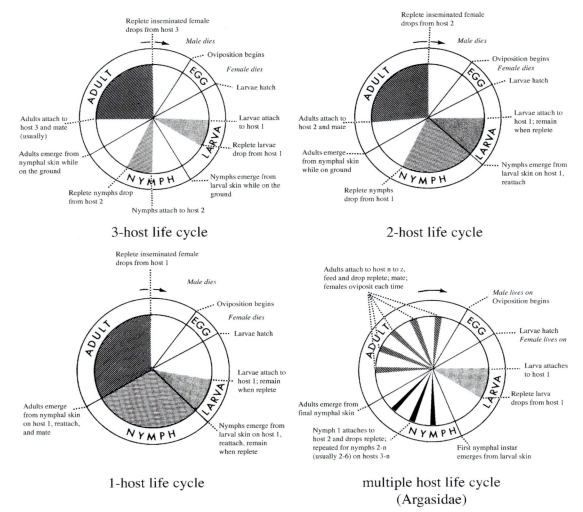

FIGURE 4.4 Comparative overviews of the various types of life cycle found in ticks. Please note that for purpose of illustration, the on-host periods are depicted as relatively longer than they really are (especially for the three-host and multiple-host life cycles). With minor adaptations from U.S.D.A. Agriculture Handbook No. 485 (1973).

species, but it may also occur in the tropics. Larvae, nymphs, and adults may diapause, and they may do so after their molt (*behavioral diapause*) or before (*developmental* or *ovipositional diapause*). Photoperiod and soil temperature are common clues for initiation and termination of diapause (Sonenshine 1994).

The main stress factor during the off-host period is desiccation, a factor that strongly influences questing behavior and thus is of some importance for assessing vector capabilities. Active, hungry ticks climb out of the relatively sheltered environment of the leaf-litter/soil layer into the vegetation layer to search for hosts. In this exposed habitat they start to lose body water. Once they start to lose excessive amounts of water they cease questing and retreat to the litter layer. In the

more humid litter layer they restore body water by sorption. The availability of cool, moist microclimates is therefore essential for survival, which may largely determine the range of a tick species. Overall, Ixodidae are more numerous and more diverse in areas with higher humidity, although quite a few have adapted to more open, savanna-like habitats. Differences in tolerance of desiccation among tick species are key in this respect. Larger-bodied species in the same lineage are often more resistant, but there are also differences in cuticle permeability. For example, *Amblyomma cajennense* is much more desiccation resistant than *A. americanum* (Needham and Teel, 1991). There may also be a connection between life history patterns and desiccation resistance. Two— and one-host life-cycles diminish the risk of desiccation, and it is not surprising that the one-host ticks tested for this characteristic appear to be relatively sensitive to desiccation.

In contrast to Ixodidae, Argasidae can be considered specialists for dry conditions, often occurring in habitats where Ixodidae would not survive. They are found in caves, near desert oases, in attics, etc., habitats that are often hot and dry. The characteristic leathery cuticle of postlarval Argasidae is effective at preventing water loss. Some argasids are capable of metabolizing body fat to generate body water.

Seasonality of nonnidicolous ticks in temperate areas is often pronounced, but this may vary over the entire geographic range. Thus *Dermacentor variabilis* has a 1-year life cycle in the southeastern United States, while having a 2-year cycle in Nova Scotia, Canada (Sonenshine 1994). Different instars may be active in different times of the year, or activity may overlap. In the case of *I. scapularis* in the northeastern United States, nymphs are often active before the larvae. Given that both instars tend to feed on the same small mammals, this sequence helps explain the often-extraordinary high infection rates of ticks with *Borellia* in that area.

HOSTS, HABITAT, AND GEOGRAPHIC RANGE

Geographic Range

Taken in a broad sense, ticks are cosmopolitan. The range of Ixodidae extends from the subarctic to the subantarctic. Argasidae have a more limited range, generally not extending beyond the warm temperate regions. Of course, this does not imply that ticks are equally common in all habitats. Microclimate and host availability (see earlier) are strong factors in determining tick distribution. Geographical distribution of the various tick genera throughout the world and in North America shows some interesting patterns. As might be expected, Argasidae are most common in drier parts of the western and especially the southwestern parts of North America. Notably, *Carios* and *Otobius* may well be New World genera, for this region contains the larger majority of species in those groups.

Within Ixodidae, *Ixodes* shows the broadest distribution, occurring on all major land masses except for the true arctic regions. Within North America, the genus is also widespread. But because most of its species are nidicolous, only a few taxa are generally encountered (see, next page box). Overall, the field species in this genus require relatively high humidity. Other genera show more distinct geographical con-

straints. For example, *Amblyomma* is most diverse and common in tropical and subtropical regions, and the North American species are most common in the humid parts of the south and southeast, moving along the coast up to Rhode Island. In contrast, *Dermacentor* is largely a holarctic genus, with a few representatives in Africa and the Oriental region. More drought and cold resistant than most *Amblyomma*, representatives of this genus occur commonly in the northern half of the United States, including the relatively dry Rocky Mountain region.

The remaining genera are not very diverse in North America. *Haemaphysalis* is a species rich in subtropical and tropical regions of the Old World and Australia, but only two species occur natively in North America, both associated with lagomorphs and birds. Finally, *Boophilus* and *Rhipicephalus* also have their main distribution in Africa and southern Asia. Their North American representatives are pest species associated with cattle and domestic dogs. All of these species are widespread, most probably transported around the world by human activity.

Host Associations

The overall host range of ticks is impressive, including the majority of terrestrial mammals and many bird, lizard, snake, and turtle species, and occasionally even amphibians. Larvae often feed on small hosts, with nymphs and adults feeding on progressively larger ones, although this pattern has many exceptions. Most of the literature on tick–host associations focuses on host specificity of the adults. In particular, that literature has been dominated by a view that ticks are generally host specific (Hoogstraal and Aeschlimann 1982). There is evidence for this: Some tick species; e.g., *Ixodes plumbeus* on sand martin (*Riparia riparia*), do not feed well on alternative host species (Balashov 1972). However, while there is no doubt about the close and prolonged contact of ticks with their hosts during feeding, some interpretations of the host association records may be biased. Species listed as being host specific are often those that have been collected only once or a few times, often from the same locality, while most widespread and frequently collected species are rarely host specific (Klompen et al. 1996). The exploited host range can perhaps be better understood by looking at the off-host habitat. There are strong indications that off-host habitat requirements may be among the most important factors determining tick distributions and host usage patterns (Klompen et al. 1996). This is certainly true within Argasidae, but it is also valid within Ixodidae. Extremely stated, ticks use all potential hosts that are available in their habitat. Under this

LYME DISEASE: AN EXAMPLE OF HOST, HABITAT, AND DISEASE INTERACTIONS

Lyme disease, or, more accurately, Lyme borrelliosis, is caused by *Borellia burgdorferi* and closely related species, and it is vectored by ixodid ticks in the *Ixodes ricinus* group (Keirans et al. 1992). The disease organism, vector ticks, and tick hosts are common in many parts of North America and Europe, but the disease is particularly prevalent in the northeastern part of the United States. Of the over 16,000 cases reported to the CDC in 1999, more than 83% were from New England and the North Atlantic states. Yet this disease was not common in this area earlier in the 20th century. The reasons why provide a fascinating tale in epidemiology. First, it is necessary to understand the ecology of Lyme disease in the northeast. The main vector is *Ixodes scapularis*, a species that requires relatively high humidity. In New England, the immatures generally feed on small rodents, such as white-footed mice, *Peromyscus leucopus*, while adults feed on large hosts, such as white-tailed deer, *Odocoileus virginianus*. Early in the 20th century most of the forest in New England was cut to make room for farms, and hunting had substantially reduced the deer herd relative to pre-Colonial times. In addition to reducing the number of hosts for the adults, the open fields formed a poor habitat for *I. scapularis*. Under these conditions, Lyme disease was uncommon. The upsurge in cases in the second half of the 20th century appears to be correlated with a number of developments: (1) The abandonment of farms, and the resurgence of the forest, created more habitat for the ticks; (2) more forest and diminished hunting pressure allowed the deer herds to grow to unprecedented levels; and (3) the increasing trend to develop housing in or near to the forest increased contact between deer, ticks, and humans. So why are there relatively few cases in the southeast? The area has large numbers of *I. scapularis* and deer, and, although perhaps to a lesser extent, houses are also built into wooded areas. Part of the answer may be in the far greater diversity of potential host species for immature ticks. Transmission of *Borellia* is transstadial but rarely transovarial. Therefore, for nymphs to become infected, the larval ticks have to feed on infected hosts. While the mice fed on in the northeast are excellent reservoirs for *Borellia*, easily infecting feeding ticks, many immature ticks in the southeast feed on lizards, a host group that is rare in the northeast. Lizards appear to have a low capacity to transmit the spirochetes, thus breaking the cycle that continues to infect ticks in the northeast. Presence of large populations of lizards may also depress tick infection levels on the Pacific coast, despite the presence of a competent vector (*I. pacificus*).

hypothesis it might be more profitable to investigate why a tick species does *not* use an available potential host species than to ask why it does. In this context it is also necessary to critically reexamine tick control programs based on a single, "key," host.

Approaches to studies of tick distributions that are based on off-host ecology can generate quite interesting results. Perry et al. (1991) used bioinformatics technology to combine data on tick off-host requirements with detailed information on local climate and vegetation conditions in order to predict the suitability of different areas for the African cattle pest *Rhipicephalus appendiculatus*.

Host Searching

The host-searching strategy of most nonnidicolous ticks is passive waiting, attaching when a host touches the tick or at least is very close. Questing ticks climb vegetation to a height conducible for encountering the appropriate host or host size class. Thus, *Ixodes scapularis* immatures tend to quest low to the ground, searching for lizards and rodents, while adults, looking for hosts the size of deer and horses, quest at heights often exceeding 50 cm. A few tick species, e.g., some *Hyalomma* species and *Argas cucumerinus*, are more aggressive and pursue potential hosts over distances of several meters. The principal limit on host-searching behavior appears to be water loss, and thus solar radiation and air temperature are the main factors determining tick activity (Balashov 1972).

MEDICAL AND VETERINARY SIGNIFICANCE

Direct Effects of Tick Feeding

Tick feeding can have direct negative effects on hosts. The allergenic nature of some materials in both salivary material and cement can cause considerable discomfort to humans and animals, and the bite of some species (e.g., *Ornithodoros coriaceus*) can be painful. But direct effects can be far more serious. Large numbers of ticks feeding on a single host can cause death through exsanguination, an effect commonly seen on nestlings in seabird colonies. Second, some ticks secrete powerful toxins that cause tick paralysis. Apart from general malaise, headaches, etc., this disease shows ascending paralytic symptoms starting in the limbs, which over a period of a few days may extend to the upper body and may soon lead to death from respiratory failure. All of this while the tick(s) are still attached. Removing the offending tick often results in dramatic resolution of the paralysis, but symptoms reach a peak 24–48 hours after removal in the case of *Ixodes holocyclus*. Tick paralysis can be caused by only a few ticks or even a single tick and can be deadly. Best known in this regard are the tick-paralysis tick, *I. holocyclus*, in Australia and *Dermacentor andersoni* and *D. variabilis* in North America, but tick paralysis can be caused by a variety of other ixodid species worldwide. Gothe and Neitz (1991) reviewed the literature on tick paralysis.

Ticks As Vectors

The ecology and physiology of ticks have combined to make these organisms second only to mosquitoes in the number of diseases vectored (Hoogstraal 1985). The use of different host individuals (and even species) in different instars of especially Ixodidae greatly enhances their vector capability. The fact that they can live for long periods, often longer than their vertebrate hosts, adds another component by allowing ticks themselves to become reservoir hosts. Intracellular digestion, a general characteristic of Chelicerata, may also make it easier for disease organisms to enter the tick tissue. The resulting range of disease organisms vectored is indeed impressive. An extensive review of tick-borne diseases is presented in Sonenshine (1994). The following is limited to pointing out the diversity of disease organisms vectored and the variation in ecology and transmission mode.

Several piroplasm protozoans (Phylum Apicomplexa) are specific to ticks. Sexual reproduction takes place in the tick, making those organisms the definitive host. Transmission is both transstadial and transovarial. *Babesia* species invade the vertebrate red blood cells, causing babesiosis in a wide range of animals and humans, while *Theileria* species invade T- and B-lymphocytes, causing several forms of theileriosis in Artiodactyla. These diseases cause considerable economic damage. For example, east coast fever, caused by *Theileria parva* and vectored by *Rhipicephalus appendiculatus*, strongly limits the range of cattle herding in eastern and southern Africa. Texas cattle fever, caused

by *Babesia bigemmina* and vectored by *Boophilus* spp., is another major cattle disease. It was eradicated from the United States after an extensive and continuing control program, but it remains a considerable problem in many other parts of the world. Most of the diseases caused by piroplasms are problems for domestic animals, but humans may be affected too: *Babesia microti*, vectored by *Ixodes scapularis*, may cause human babesiosis. The risk of developing this disease appears relatively small; most patients affected are splenectomized individuals. Symptoms resemble those of malaria but without the periodicity.

Ticks vector a considerable diversity of intracellular bacteria in the *Rickettsia* group. Many genera include tick-associated pathogens, e.g., *Rickettsia*, *Ehrlichia*, *Coxiella*, and *Anaplasma*, although others, such as *Wohlbachia*, do not appear to be pathogenic. Rocky Mountain spotted fever, caused by *Rickettsia rickettsii* and vectored mostly by *D. andersoni* and *D. variabilis*, was initially described from the northern Rocky Mountains area, specifically the Bitterroot Valley in Montana, but the disease is now diagnosed mostly in the eastern United States. Symptoms include fever with the distinct spotted rash that gave the disease its name. The pathogen infects all cells of the ticks and is transmitted transovarially. Rickettsiae in starving ticks appear to suspend multiplication, but they reactivate upon feeding. Virulence varies geographically, with western strains generally more virulent than eastern ones. Extremely virulent strains can kill their tick host.

Unlike *Rickettsia* species, which multiply freely in the cytoplasm, *Ehrlichia* species form colonies in vacuoles in the cell. Human granulocytic ehrlichiosis (HGE) was described in 1994 and is vectored by *I. scapularis*. Human ehrlichiosis, caused by *E. chaffeensis* and vectored by *A. americanum*, is another low-incidence disease of humans. Heartwater, caused by *E. ruminantium* (formerly *Cowdria ruminantium*) is one of the most important cattle diseases in sub-Saharan Africa and Madagascar, and it has now appeared on a few islands in the Caribbean. It is vectored by various species in the genus *Amblyomma*, specifically *A. variegatum* and *A. hebraeum*. *E. ruminantium* infects the cells of the midgut, salivary glands, and a few other organs but rarely the gonads. It is therefore rarely transmitted transovarially. Animal husbandry practices, such as concentrating cattle in enclosures for protection against predators, are likely to increase both tick populations and infestation rates. In this disease the presence of infected males, which remain on the host for long periods, may be important (Norval et al. 1992). Long-lived unfed ticks appear to be the principal reservoir of this disease. Interestingly, native cattle breeds in Africa appear to be largely asymptomatic.

Transmission can vary dramatically, even with related disease organisms. Among bacteria, the spirochaete genus *Borellia* includes species that appear limited largely to transstadial transmission; e.g., *B. burgdorferi*, and species that are quite frequently transmitted transovarially, such as the *Borellia* species in the tick-borne relapsing fever group. The most studied disease in this group in recent years is undoubtedly Lyme disease. It is caused by *B. burgdorferi* and perhaps some closely related species and is vectored by a number of species in the *Ixodes ricinus* complex. This species may provide one of the best examples of the interaction between ticks, disease organisms, host populations, and landscape ecology in disease transmission (see earlier box). Notably, *Ixodes scapularis* may be infected by three different human pathogens simultaneously: *Babesia*, *Ehrlichia*, and *Borellia* (see below). There is evidence that these disease organisms interact, influencing transmission rates.

Tick-borne relapsing fevers, vectored by argasid ticks in the genera *Ornithodoros* and *Carios*, are characterized by fever, headaches, and fatigue, usually lasting 3–5 days. After this, symptoms subside for about 7 days, before a slightly milder fever recurs. Recurring symptoms may be associated with antigenic variability in the spirochetes. Unlike many other tickborne diseases that are transmitted through salivary gland secretions, infection by tick-borne relapsing fever can also take place through coxal fluid (produced generously in these fast-feeding ticks). Many human cases in western North America are due to bites by *O. hermsi* after removal of squirrels in log cabins, attics etc., suggesting that humans are accidental hosts in a primarily zoonotic disease.

Arboviruses

The list of arboviruses vectored by ticks is long and continually growing. Some produce encephalitis-like symptoms, such as Russian spring–summer encephalitis, Louping ill, and Kyasanur Forest disease; others, such as Crimean Congo hemorrhagic fever (CCHF), are hemorrhagic fevers; and still others, such as Colorado tick fever, produce a generalized, systemic infection. Crimean Congo hemorrhagic fever is caused by a member of the Bunyaviridae, vectored most commonly by *Hyalomma marginatum*, *H. rufipes*, and *D. marginatus* (Hoogstraal 1979). Geographically extremely widespread, ranging from Asia to Africa, this virus is most common in steppe- and savanna-type environments, the natural habitat of *Hyalomma* species. Wild animals, especially hares and hedgehogs, appear to serve as reservoirs. While these hosts are generally asymptomatic, the virus causes disease in some rodents and humans. In humans, symptoms include fever, severe headache, and photophobia, often followed by rashes and bleeding. This disease may be lethal, with mortality levels of 10–50% during outbreaks (Hoogstraal 1979). Colorado tick fever is caused by a member of the Reoviridae, and it occurs mostly in the western United States and Canada. Confirmed vectors include *Dermacentor* species, especially *D. andersoni*, and *Haemaphysalis leporis-palustris*. This is another typical zoonotic disease, with the normal cycle involving small mammals. Symptoms are less severe than in CCHF but may include, among others, fever, headaches, vomiting, and occasionally meningitis.

Readings

Balashov, Y. S. 1972. Bloodsucking ticks (Ixodoidea) vectors of disease of man and animals. *Miscellaneous Publications Entomological Society America*, 8: 160–376.

Dobson, S. J., and Barker, S. C. 1999. Phylogeny of the hard ticks (Ixodidae) inferred from 18S RNA indicates the genus *Aponomma* in paraphyletic. *Molec. Phylogenet. Evol.* 11: 288–295.

Edwards, M. A., and Evans, G. O. 1967. Some observations on the chaetotaxy of the legs of larval Ixodidae (Acari: Metastigmata). *J. Nat. Hist. London* 4: 595–601.

Feldman-Muhsam, B. 1963. Function of the areae porosae of ixodid ticks. *Nature London* 197: 100.

Filippova, N. A. 1997. *Ixodid Ticks of Subfamily Amblyomminae*. Nauka Publishing House, St. Petersburg, FL.

Gothe, R., and Neitz, A. W. H. 1991. Tick paralysis: Pathogenesis and etiology. *Adv. Dis. Vector Res.* 8: 177–204.

Hoogstraal, H. 1979. The epidemiology of tick-borne Crimean-Congo hemorrhagic fever in Asia, Europe, and Africa. *J. Med. Entomol.* 15: 307–417.

Hoogstraal, H. 1985. Argasid and nuttalliellid ticks as parasites and vectors. *Adv. Parasitol.* 24: 135–238.

Hoogstraal, H., and Aeschlimann, A. 1982. Tick–host specificity. *Bull. Société Entomologique Suisse* 55: 5–32.

Keirans, J. E. 1992. Systematics of the Ixodida (Argasidae, Ixodidae, Nuttalliellidae): An overview and some problems. In *Tick Vector Biology. Medical and Veterinary Aspects* (B. H. Fivaz, T. N. Petney, and I. G. Horak, eds.). Springer-Verlag, Berlin, pp. 1–21.

Keirans, J. E., and Robbins, R. G. 1999. A world checklist of genera, subgenera, and species of ticks (Acari: Ixodida) published from 1973–1997. *J. Vector Ecol.* 24: 115–129.

Keirans, J. E., Clifford, C. M., Hoogstraal, H., and Easton, E. R. 1976. Discovery of *Nuttalliella namaqua* Bedford (Acarina: Ixodoidea: Nuttalliellidae) in Tanzania and redescription of the female based on scanning electron microscopy. *Ann. Entomological Soc. Amer.* 69: 926–932.

Keirans, J. E., and Litwak, T. R. 1989. Pictorial key to the adults of hard ticks, family Ixodidae (Ixodida: Ixodoidea), east of the Mississippi River. *J. Med. Entomol.* 26: 435–448.

Keirans, J. E., Oliver, J. H., Jr., and Needham, G. R. 1992. The *Ixodes ricinus/persulcatus* complex defined. In *First International Conference on Tick-Borne Pathogens at the Host–Vector Interface: An Agenda for Research* (U. G. Munderloh, and T. J. Kurtti, eds.). University of Minnesota Press, Saint Paul, p. 302.

Klompen, J. S. H. 1992. Comparative morphology of argasid larvae (Acari: Ixodida: Argasidae), with notes on phylogenetic relationships. *Ann. Entomological Soc. Amer.* 85: 541–560.

Klompen, J. S. H., and Oliver, J. H., Jr. 1993. Systematic relationships in the soft ticks (Acari: Ixodida: Argasidae). *System. Entomol.* 18: 313–331.

Klompen, J. S. H., Black, W. C., IV, Keirans, J. E., and Oliver, J. H., Jr. 1996. Evolution of ticks. *Annu. Rev. Entomol.* 41: 141–161.

Klompen, J. S. H., Black, W. C., IV, Keirans, J. E., and Norris, D. E. 2000. Systematics and biogeography of hard ticks, a total evidence approach. *Cladistics* 16: 79–102.

Krantz, G. W. 1978. *A Manual of Acarology*, 2nd ed. Oregon State University Book Stores, Corvallis, OR.

Lehtinen, P. T. 1991. Phylogeny and zoogeography of the Holothyrida. In *Modern Acarology*, Vol. II. (F. Dusbábek, and V. Bukva, eds.). SPB Academic, The Hague, pp. 101–113.

Lindquist, E. E. 1984. Current theories on the evolution of major groups of Acari and on their relationship with other groups of Arachnida, with consequent implications for their classification. In *Acarology VI*, Vol. 1. (D. A. Griffiths, and C. E. Bowman, eds.). Ellis Horwood, Chichester, UK, pp. 28–62.

Murrell, A., Campbell, N. J. H., and Barker, S. C. 1999. Re: Mitochondrial 12S rDNA indicates that the Rhipicephalinae (Acari: Ixodida: Ixodidae) is paraphyletic. *Molec. Phylogenet. Evol.* 12: 83–86.

Murrell, A., Campbell, N. J. H., and Barker, S. C. 2000. Phylogenetic analyses of the Rhipicephaline ticks indicate that the genus Rhipicephalus is paraphyletic. *Molec. Phylogenet. Evol.* 16: 1–7.

Murrell, A., Campbell, N. J. H., and Barker, S. C. 2001. A total-evidence phylogeny of ticks provides insights into the evolution of life cycles and biogeography. *Molec. Phylogenet. Evol.* 21: 244–258.

Needham, G. R., and Teel, P. D. 1991. Off-host physiological ecology of ixodid ticks. *Annu. Rev. Entomol.* 36: 659–681.

Norton, R. A., Kethley, J. B., Johnston, D. E., and O'Connor, B. M. 1993. Phylogenetic perspectives on genetic systems and reproductive modes of mites. In *Evolution and Diversity of Sex Ratio in Insects and Mites* (D. L. Wrench, and M. A. Ebbert, eds.). Chapman and Hall, New York, pp. 8–99.

Norval, R. A. I., Andrew, H. R., Yunker, C. E., and Burridge, M. J. 1992. Biological processes in the epidemiology of heartwater. In *Tick Vector Biology: Medical and Veterinary Aspects* (B. H. Fivaz, T. N. Petney, and I. G. Horak, eds.). Springer-Verlag, Berlin, pp. ••.

Perry, B. D., Kruska, R., Lessard, P., Norval, R. A. I., and Kundert, K. 1991. Estimating the distribution and abundance of *Rhipicephalus appendiculatus* in Africa. *Prevent. Vet. Med.* 11: 261–268.

Pound, J. M., Campbell, J. D., Andrews, R. H., and Oliver, J. H., Jr. 1986. The relationship between weights of nymphal stages and subsequent development of *Ornithodoros parkeri* (Acari: Argasidae). *J. Med. Entomol.* 23: 320–325.

Sonenshine, D. E. 1994. *Biology of Ticks*, Vol. 2. 1st ed. Oxford University Press, New York.

Walter, D. E., and Proctor, H. C. 1998. Feeding behavior and phylogeny: Observations on early derivative Acari. *Exp. Appl. Acarol.* 22: 39–50.

Walter, D. E., and Proctor, H. C. 1999. *Mites: Ecology, Evolution and Behavior*. CABI, New York.

Yoder, J. A., Pollack, R. J., and Spielman, A. 1993. An ant-diversionary secretion of ticks: First demonstration of an ant allomone. *J. Insect Physiol.* 39: 429–435.

5

Kissing Bugs and Bedbugs, the Heteroptera

C. BEN BEARD

GENERAL OVERVIEW

The Heteroptera, or true bugs, comprise a diverse group of hemimetabolous insects, most of which are phytophagous. A smaller number of species are predatory on other insects and arthropods, and an even smaller number feed on vertebrate blood. True bugs have piercing-sucking type of mouthparts that are used for ingesting fluids from plants or animals. Some authors prefer to call this order the Hemiptera, distinguishing it completely from its close relative the Homoptera. The name *Hemiptera* means literally "half wing" and is derived from the fact that the forewing is often a hemelytron, which has two distinct regions, a basal leathery portion and a clear or membranous apical region.

Two groups of Heteroptera contain members that are important from a public health standpoint. These are the triatomines, or kissing bugs, which are in the family Reduviidae, and the bedbugs, which are in the family Cimicidae. There are two other heteropteran families, the Polyctenidae (bat bugs) and the Lygaeidae (tribe Cleradini), that contain members ectoparasitic on vertebrates; however, they are not considered of public health importance. Consequently, this chapter is limited to triatomines and bedbugs and their importance as ectoparasites and vectors of diseases that affect man.

TRIATOMINE VECTORS OF CHAGAS DISEASE

General Identification

The insect vectors of Chagas disease are in the family Reduviidae, subfamily Triatominae (Fig. 5.1). This group comprises five tribes, 17 genera, and approximately 130 described species (Table 5.1). The common names for these insects vary by geographic region and include kissing bugs, conenoses, vinchucas, chinches, and barbeiros. These are large insects, averaging around 28 mm in length, but ranging from 5 mm (*Alberprosnia goyovargasi*) to 44 mm (*Dipetalogaster maxima*). The Triatominae demonstrate typical heteropteran morphology and bionomics. They generally have a subcylindrical head, large pronotum, and a distinctive triangular scutellum that divides at its base the two hemelytra. The adults and immature stages are similar in appearance; however, the adults can easily be distinguished, in most cases, by the presence of well-developed external genitalia, ocelli, and fully developed fore wings and hind wings. The females are typically larger than males and can generally be recognized by the presence of a pointed or truncated abdominal apex, which is rounded in males.

Biology and Life Cycle

Triatomines can require from 3 months up to 2 years to complete their life cycle in nature, depending on the

57

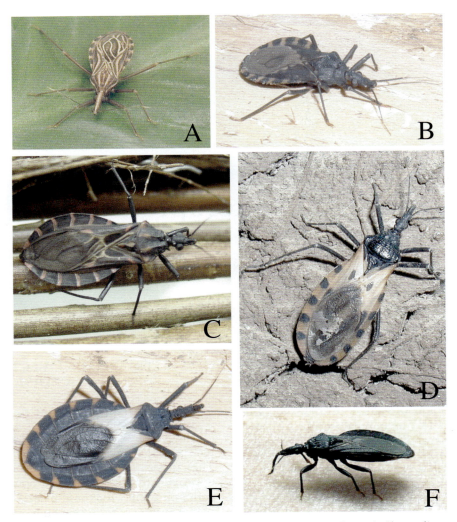

FIGURE 5.1 Various species of Triatominae: (A) *Rhodnius prolixus*, primary domestic Chagas disease vector in northern South America and Central America; (B) *Triatoma infestans*, primary Chagas disease vector in the southern cone region of South America; (C) *Panstrongylus megistus*, important historical vector of Chagas disease in Brazil; (D) *Triatoma dimidiata*, second most important domestic Chagas disease vector in Central America; (E) *Triatoma pallidipennis*, important Chagas disease vector in Mexico; (F) *Triatoma protracta*, North American triatomine species associated primarily with packrat nests but commonly encountered in California and responsible for allergic reactions in sensitized individuals. Images by C. B. Beard.

TABLE 5.1   Tribes and Genera of the Triatominae

Tribe Alberproseniini
    Genus: *Alberprosenia*

Tribe Bolboderini
    Genera: *Belminus, Bolbodera, Microtriatoma, Parabelminus*

Tribe Cavernicolini
    Genus: *Cavernicola, Torrealbai*

Tribe Rhodniini
    Genera: *Psammolestes, Rhodnius*

Tribe Triatominae
    Genera: *Dipetalogaster, Eratyrus, Hermanlentia, Linshcosteus, Mepraia, Panstrongylus, Paratriatoma, Triatoma*

species and on environmental conditions. They are oviparous, with a single female capable of producing up to 1,000 eggs in her lifetime. Both males and females are hematophagous, requiring blood for nutrition and for egg production. In *Rhodnius prolixus*, adult females produce eggs within approximately 1–2 weeks, following blood feeding. The fertilized eggs generally take 2–3 weeks to hatch, depending on temperature. Immature bugs go through five nymphal instars (Fig. 5.2), with at least one complete blood meal required for each successive molt. This species is capable of completing its entire life cycle, from egg

FIGURE 5.2 Triatomine life stages: illustrating egg mass, 1st, 3rd, 4th, and 5th instar nymphs and adult. Photograph from CDC Image Library.

to adult, in as little as 3 months, under optimum conditions.

Since all triatomines are obligate blood feeders throughout their entire developmental cycle, they maintain populations of bacterial symbionts in their intestinal tract, which are involved in providing essential nutrients that are lacking from their diet of vertebrate blood. The precise nature of the symbiotic association is unclear. Early studies suggested that the bacteria provided specific B vitamins that were lacking from the insect diet. More recent work, however, supports the view that the bacteria themselves are cultivated and digested by the insect, thus directly providing the required supplemental nutrients. Efforts aimed at controlling Chagas disease transmission through use of these bacteria are discussed in a later section in this chapter.

CLASSIFICATION AND SYSTEMATICS

Within the subfamily Triatominae are five tribes. Only two of these tribes, however, contain members that are of significant public health importance, the Triatomini and the Rhodniini. The differences in the two main tribes are striking, consisting of both physiological and biological distinctions. Within the Triatomini there are eight genera: *Triatoma, Panstrongylus, Eratyrus, Dipetalogaster, Linshcosteus, Mepraia, Hermanlentia*, and *Paratriatoma*. Of these eight genera, the important vectors to man of the Chagas disease agent, *Trypanosoma cruzi*, are contained within the genera *Triatoma* and *Panstrongylus*. Within the Rhodniini are only two genera: *Rhodnius* and *Psammolestes*. The genus

Psammolestes includes three species, none of which are important vectors of Chagas disease in man. Both *Rhodnius* and *Psammolestes* are strongly tied to arboreal habitats. They are frequently found in palm trees, where they are closely associated with mammals and birds that share the same niche. Recent molecular analyses indicate that the members of the genus *Psammolestes* represent a specialized lineage within the genus *Rhodnius* and are more closely related to the *Rhodnius prolixus* species group (i.e., *R. prolixus, R. robustus, R. neglectus*, and *R. nasutus*) than these species are to other members of their own genus (i.e., *R. pallescens* and *R. pictipes*). Consequently, in due time the *Psammolestes* genus undoubtedly will be collapsed into the genus *Rhodnius*.

Within the tribe Triatomini, the genus *Triatoma* is the most important, both in total numbers of species and in importance in disease transmission. With the single exception of *R. prolixus*, all of the major regional vectors of Chagas disease fall into the genus *Triatoma*. The genus *Panstrongylus* is also important. In former years, before the introduction and spread of *T. infestans* throughout Brazil and the southern cone region of South America, *Panstrongylus megistus* was the most important vector of Chagas disease in regions of central and eastern Brazil. The expansion of *T. infestans*, however, apparently resulted in displacement of *P. megistus* in domestic habitats in this region.

Recent DNA-based studies suggest the need for changes in the current classification scheme for the Triatominae. Clear support for the existence of several of the described genera cannot be provided based upon DNA sequence comparisons of both nuclear and mitochondrial gene sequences. For example, using comparisons at the mitochondrial large subunit ribosomal DNA and Cytochrome B loci, the monotypic genus, *Dipetalogaster*, is grouped together with the other *Triatoma* species from the Mexican–southwestern U.S. region (i.e., the protracta and phyllosoma complexes), sharing more similarity with these species than with the South American *Triatoma* species (e.g., *T. infestans* and *T. sordida*). Similar observations exist with respect to *Paratriatoma hirsuta*. In all likelihood, several of the small genera will eventually be absorbed into the larger genera, once more thorough analyses have been completed.

Important Vector Species, by Region

With a few key exceptions, the important vector species of Chagas disease are unique to a defined geographic region. *Rhodnius prolixus, T. infestans*, and *T. dimidiata* are key exceptions to this general rule. These species have broad geographic distributions, a fact that can be attributed directly to their ability to invade and colonize domestic habitats and their use of humans for dispersal. Domiciliation is also the single most important factor that determines the importance of the triatomine species as a vector of human disease. Many species of bugs fly into homes, but few species actually colonize homes, thereby greatly increasing the transmission risks.

In the southern cone region of South America, the most important vector species is *T. infestans*. The success of recent control measures (discussed later) has resulted in the near elimination of this species from significant geographic regions, including all or much of Paraguay, Uruguay, Argentina, Brazil, and Chile. Nevertheless, this species continues to be of great importance in other countries, such as Bolivia, where house infestation rates in some regions can be extremely high. In parts of Brazil where *T. infestans* has been successfully eliminated, *T. brasiliensis* has now replaced the former species as the most important human vector. Unlike *T. infestans*, *T. brasiliensis* is native to Brazil. It occurs in natural ecotopes, such as rocky outcroppings, and invades and colonizes homes. This species is reported in 12 states and can frequently be found inside homes in high population densities. Likewise, *T. sordida* has increased in its importance and has been reported to invade and colonize homes where *T. infestans* had previously been eliminated, specifically in the central part of Brazil. Two other species, found natively in palm trees, have focal epidemiologic importance in human disease transmission. These are *R. neglectus* in Goiás State and *R. nasutus* in Ceará and Rio Grande do Norte states. Both species can be found colonizing domiciles in areas now considered to be *T. infestans* free. In northern South America, *R. prolixus* is the most important domestic Chagas disease vector. This species is morphologically indistinguishable from the closely related sibling species *R. robustus*. DNA-based methods demonstrated clear differences in these species and showed that the taxon currently called *R. robustus* is, in fact, a complex of at least three different species, all of which are sylvatic. *Rhodnius prolixus*, on the other hand, appears to be exclusively domestic throughout all of its range, with the exception of possible sylvatic populations that have been reported from Venezuela, where the species probably first adapted to the domestic lifestyle that allowed it to be dispersed throughout its present range. *Rhodnius prolixus* is also the most important domestic Chagas disease vector throughout Central America. In this region, *T. dimidiata* is considered the most important secondary vector. One chief difference in these species, however, is the fact that *R. prolixus* in Central America has never been reported outside of homes. *Triatoma dimidiata*, however, is frequently found both in and around homes, and the peridomiciliary populations provide a constant risk for reinfestations of homes in the aftermath of insecticide treatment. In studies conducted in both Guatemala and Honduras, it is a general observation that infection rates in humans tend to be higher in regions where *R. prolixus* is found, as opposed to *T. dimidiata*. Consequently, the species that is easier to eliminate (*R. prolixus*) appears to be a more efficient vector, and the less efficient vector (*T. dimidiata*) is more difficult to eliminate. This observation suggests that insecticides should be effective in reducing overall transmission rates in Central America but virtually ineffective at long-term disease elimination efforts. A similar situation exists for *T. infestans* and *T. brasiliensis* in Brazil.

In addition to *R. prolixus* and *T. dimidiata*, there are other species that are important vectors in other Central American countries, such as *R. pallescens* in Panama. This species is found primarily in palms but frequently invades homes. It is not thought actually to colonize homes; however, it is nevertheless an important vector to humans, much more so than other palm tree–inhabiting species, such as *R. robustus* in Brazil.

In Mexico, there are some 39 species of triatomines that have been reported, many of which are found only there. The primary vector species are members of the phyllosoma and protracta species complexes and include *T. dimidiata*, *T. pallidipennis*, and *T. barberi*. Natural infection rates in these species have been reported to range from 16% to 92%, depending on the species and location. Members of the protracta species complex are common in Mexico and in the southwestern United States. These species are generally associated with rodent burrows, particularly of the genus *Neotoma*. Their importance as vectors of the *T. cruzi* to humans is probably minimal; however, they are important in maintaining zoonotic transmission cycles.

There are approximately 10 species of triatomines found in the United States. Of these, three are the most important in terms of pest species and potential vectors of Chagas disease. These include *T. sanguisuga* (southern United States), *T. gerstaeckeri* (Texas, New Mexico, and Mexico), and *T. protracta* (southwestern United States and Mexico). There have been only five reported cases of autochthonous vector-borne transmission of *T. cruzi* in the United States, the most recent being an 18-month-old infant in Tennessee who acquired the infection through exposure to an infected *T. sanguisuga*. Consequently, vector-transmitted Chagas disease is not considered a significant public health problem in the United States. There have been quite a number of cases in dogs, however, particularly in the southwestern United States. These cases have been associated primarily with *T. gerstaeckeri*. In California, the chief complaints related to triatomines involve problems associated with allergic reactions to the bite of *T. protracta*, which has been reported to cause life-threatening anaphylactic reactions in some people. All three of these species are primarily sylvatic. The adults occasionally fly into homes and feed readily on humans but do not, as a rule, colonize homes. This observation provides the greatest explanation for why Chagas disease exists almost exclusively as a zoonosis in the United States. While defecation patterns—specifically, the time required for defecation following a blood meal—may have an impact on vector competency with some triatomine species and/or populations, it is the fact that the bug species found in the United States do not colonize and infest homes that has the greatest impact on the risk of disease transmission.

CHAGAS DISEASE

Chagas disease was first described by Carlos Chagas in 1909 (see Fig. 5.3). He determined the etiologic agent, the vector, and the transmission cycle. This disease, also known as *American trypanosomiasis*, affects an estimated 12 million people throughout the Americas. Currently, there are approximately 100 million persons in 21 countries who are living in areas of risk for acquiring the disease. The disease is also common as a zoonosis throughout much of the Americas and has frequently been reported from opossums and raccoons as well as in stray dogs in the United States.

Approximately 80% of transmission in man is vector borne; the other significant transmission routes include blood transfusions and congenital transmission. Oral infection in humans has been reported in association with food sources that have become contaminated with infected bugs. Oral infection is probably the most frequent route of transmission in dogs and in insectivorous mammals. Transmission to humans occurs via infectious metacyclic-stage trypomastigotes that are shed in feces and deposited on the skin of the person while the bug is feeding. Infection takes place when the infective parasite is rubbed into the bite wound or a nearby membrane, such as the conjunctiva of the eye.

Acute Chagas disease can range in severity from a mild flu-like or even asymptomatic illness to acute, severe myocarditis and death. Periorbital edema, referred to as *Romaña's sign*, is considered a hallmark of acute Chagas disease, although it occurs in fewer

FIGURE 5.3 Brazilian currency: (A) 10,000 Cruzado note, honoring Carlos Chagas; (B) magnified image of Cruzado note demonstrating a triatomine that is feeding on an infected individual, the trypanosome in the feces, penetration of the trypanosome through the skin, where it is engulfed by a phagocytic cell and passes into the vasculature and on to muscle tissue. The backdrop of the illustration is an adobe wall. Image by C. B. Beard.

than 50% of all cases and can also result from allergic reactions to the bites of uninfected bugs. Following initial infection, there is a subclinical period that can last from a few years to several decades. The majority of patients remain asymptomatic for life, but up to 30% of patients develop symptoms of severe chronic cardiac or digestive system disease.

Chronic Chagas cardiomyopathy is characterized by cardiomegaly, ECG abnormalities, pronounced fibrosis, apical ventricular aneurysms and heart failure, or rhythm disturbances, which can produce sudden death. Chronic digestive system disorders involve enlargement of the esophagus (*megaesophagus*) or colon (*megacolon*), due to destruction of the autonomic nerves and subsequent discordant motility. Patients can present with cardiac or digestive system disorders, or both, depending largely on the geographic region. The immunopathologic basis for this phenomenon is unclear, but it is probably related to differences either in host immunologic response or in parasite strain virulence. Once chronic disease has developed, chemotherapeutic treatment cannot reverse established lesions. No vaccine is available for disease prevention.

Control

Since there is neither a vaccine for prevention nor an effective chemotherapy for reversal of damage in chronic symptomatic patients, disease prevention measures rely on two primary activities: (1) spraying homes, primarily with residual pyrethroid insecticides (Fig. 5.4), for elimination of domiciliary triatomine populations, and (2) screening of blood donors to prevent transmission via contaminated blood. Currently there are three multinational control programs for Chagas disease. These include the Southern Cone, Andean Pact, and Central American control initiatives. The goal is the elimination of new cases of Chagas disease through the aforementioned approaches.

The primary focus of vector control activities is domestic populations of *T. infestans* throughout its range in the southern cone countries of South America and *R. prolixus* in northern South America and Central America. These campaigns have been highly successful in Uruguay, Brazil, and Chile, where new cases of Chagas disease in humans have been virtually eliminated. Some significant challenges persist, however, namely, reinfestation of treated homes by either (1) residual populations of insects that were not completely eliminated by spraying or (2) insects from peridomestic and/or sylvatic populations that enter and recolonize homes following successful insecticide applications. These problems have been largely associated with *T. dimidiata* in Central America and *T. brasiliensis* in Brazil. In Guatemala, insecticide-treated homes have been found to be reinfested within as little as 6 months following insecticide application. The emergence of low-level insecticide resistance has also been reported for *T. infestans* populations from Brazil and *R. prolixus* populations from Venezuela. One additional limitation of wide-scale insecticide use for Chagas disease elimination is implicit in the current theory of the distribution of *R. prolixus* in Central America. This species is originally indigenous to northern South America and considered to have been introduced into Central America, presumably through the escape of a laboratory colony in El Salvador in the early 1900s. From this single release point, the species is thought to have spread over the last 90 years to all of the Central American countries, with the exception of Belize and Panama. If in such a short period of time this species has become so broadly distributed, valid concerns must be raised over the likelihood that insecticide-based control efforts aimed at eradicating this species would be successful, since residual popula-

tions can lead to widespread reinfestations in areas thought to be under control.

These and other limitations have resulted in the recognition of the need for additional vector control approaches. One such approach has been the evaluation of genetically modified bacterial symbionts for use in generating paratransgenic triatomines that are incapable of transmitting Chagas disease. As mentioned earlier, because triatomines feed exclusively throughout their entire developmental cycle on vertebrate blood, they maintain populations of symbiotic microorganisms within their intestinal tracts, which provide nutrients lacking in the insect's restricted diet. Bait formulations of these bacteria have been produced and used for introducing genetically modified symbiont strains, producing a gene product that can kill the trypanosomes or otherwise interfere with the parasite–host relationship, thus blocking transmission. This technology has the potential for use as a part of an integrated pest management program against Chagas disease, which could provide specific benefits in cases where reinfestation of insecticide-treated homes is a problem for control.

BEDBUGS

General Identification

The family Cimicidae includes bedbugs, bat bugs, swallow bugs, and other related taxa. The species that is of greatest importance to humans is the common bedbug, *Cimex lectularius* (Fig. 5.5). The adults are small (5–7 mm), reddish brown, dorsoventrally flattened insects, with highly reduced nonfunctional wings. This species is ectoparasitic, primarily on mammals, having a particularly strong association with humans. Bedbugs are distributed worldwide, where they are reported inconsistently from diverse settings, ranging from low-income housing to luxury hotels. They are nocturnal in their feeding behavior, hiding in cracks and crevices during the day and creeping out at night to feed on their host. They particularly prefer to hide in the corners of wooden bed frames and in the seams of mattresses and box springs.

Biology and Life Cycle

Bedbugs are obligatorily hematophagous, feeding on blood throughout their entire developmental cycle. They are intermittent feeders, residing not on the host but in the host's nest or immediate surroundings. As with triatomines, they maintain a rich flora of symbiotic microorganisms that are involved in nutritional

FIGURE 5.4 Chagas disease epidemiology and control: (A) a royal palm in Panama, the primary habitat of *Rhodnius pallescens*; (B) *Didelphis marsupialis* (opossum), an important zoonotic reservoir host for the Chagas disease agent *Trypanosoma cruzi*; (C and D) domestic dwellings in Central America, commonly infested with either *Triatoma dimidiata* (C) or *Rhodnius prolixus* (D); (E) spraying of home for control of *T. dimidiata* in Guatemala. Images by C. B. Beard.

FIGURE 5.5 The bedbug, *Cimex lectularius*. Photograph from the CDC image library.

mutualism. These endosymbionts reside primarily in specialized organs called *mycetomes* but have also been reported in various other tissues, including the midgut and ovaries. At least two different, phylogenetically distinct populations of microorganisms have been described.

Following a blood meal, adult female bedbugs typically lay eggs in small batches of 10–15, up to a total of 200–500. The eggs require 10 days to hatch on the average, and the developmental time from egg to adult can range from a little over a month to 4 months. Most species molt five times and ordinarily require one complete blood meal between molts. An average of 8 days is required between molts; however, the exact time can vary widely, due to temperature and other biotic and abiotic factors. All instars, especially the later ones, can withstand prolonged periods without feeding (i.e., over 4 months).

Classification and Systematics

Ryckman et al. (1981) lists six subfamilies, 23 genera, and 91 species. Twelve of these genera are restricted to the New World, eight only in the Old World, and two, including *Cimex*, that are global in their distribution. Thirteen genera are associated with bats and/or other mammals, nine genera are parasitic exclusively on birds, and the host for one genus, *Bertilia*, is unknown. The genus *Cimex* contains members that parasitize both mammals and birds. *Cimex lectularius* is the most important species. It has a global distribution and parasitizes birds, bats, and domestic animals, in addition to humans. *Cimex hemipterous* is similarly found in both the New and Old Worlds, primarily on humans and chickens, occasionally being found on bats, but is limited in its distribution to more tropical regions. Additionally, *Cimex*

pilosellus is found in the western United States, where it is known to be an important species.

Public Health Importance

Bedbugs demonstrate many biological characteristics that suggest that they could serve as vectors of human disease agents. These have been summarized by Ryckman et al. (1981) and include the following: (1) They are obligate blood feeders throughout their entire developmental cycle; (2) they are frequently found in hotels and in other shared residences, where they may feed transiently on multiple human hosts throughout their entire life; and (3) they have been readily infected in laboratory studies with various pathogenic agents. Bedbugs have been suspected, at one time or another, in the transmission of 37 different human disease agents. Their importance, however, in the natural transmission of any known human pathogen has never been definitively proven. Most recently, their potential role in the transmission of hepatitis B virus (HBV) has been examined. These studies demonstrated that HBV could persist up to 35 days in laboratory-infected bugs following feeding on an infected individual. During this time, virus passed transtadially during molts and was shed continuously in bug feces. These studies suggest the possibility of mechanical transmission of HBV by bedbugs; however, the public health importance of this phenomenon remains unclear. In an earlier study, during a 2-year intervention program where domiciliary insecticides were used for bedbug control, while successful control was apparently achieved, the incidence of HBV infection remained unaffected. Additional studies will be required to clarify the potential role, if any, of bedbugs in the natural transmission and epidemiology of HBV.

Control

Successful control of bedbug infestations can be achieved by a combination of three intervention methods: (1) laundering of clothing and bedding that may contain insects or their eggs; (2) elimination and/or disinfection of resting sites, such as cracks, crevices, or seams in and around beds; and (3) insecticide treatment of infested furniture and room surfaces. A number of different insecticides have been shown to be effective. Care should be taken to follow applicable labeling instructions.

Readings

Beard, C. B., Cordon-Rosales, C., and Durvasula, R. V. 2002. Bacterial symbionts of the Triatominae and their potential use in control of Chagas disease transmission. *Ann. Rev. Entomol.* 47: 123–141.

Blow, J. A., Turell, M. J., Silverman, A. L., and Walker, E. D. 2001. Stercorial shedding and transtadial transmission of hepatitis B virus by common bedbugs (Hemiptera: Cimicidae). *J. Med. Entomol.* 38: 694–700.

Lent, H., and Wygodzinski, P. 1979. Revision of the Triatominae (Hemiptera: Reduviidae) and their significance as vectors of Chagas disease. *Bull. Am. Mus. Nat. His.* 163: 123–520.

Magile, A. J., and Reed, S. G. 2000. American Trypanosomiasis. In *Hunter's Tropical Medicine and Emerging Infectious Diseases* (G. T. Stickland, ed.). Saunders, Philadelphia, pp. 653–664.

Monteiro, F. A., Wesson, D. M., Dotson, E. M., Schofield, C. J., and Beard, C. B. 2000. Phylogeny and molecular taxonomy of the Rhodniini derived from mitochondrial and nuclear DNA sequences. *Am. J. Trop. Med. Hyg.* 62: 460–465.

Monteiro, F. A., Escalante, A. A., and Beard, C. B. 2001. The application of molecular tools in triatomine systematics: A public health perspective. *Trends Parasitol.* 17: 344–447.

Ryckman, R. R., Bentley, D. G., and Archbold, E. F. 1981. The Cimicidae of the Americas and Oceanic Islands, a checklist and bibliography. *Bull. Soc. Vect. Ecol.* 6: 93–142.

Schofield, C. J. 1988. Biosystematics of the Triatominae. In *Biosystematics of Haematophagous Insects* (M. W. Service, ed.). Clarendon Press, Oxford, UK, pp. 285–312.

Schofield, C. J. 1994. *Triatominae—Biology and Control.* Eurocommunica Publications, West Sussex, UK.

Usinger, R. L. 1966. Monograph of Cimidicae (Hemiptera-Heteroptera). Vol 7. Thomas Say Foundation, College Park, MD.

6

Lice, the Phthiraptera

LANCE A. DURDEN

INTRODUCTION

Lice are wingless, ectoparasitic insects that parasitize birds or mammals. Many species are host specific and feed on a single host species; some are even more specialized, in that they normally occur only on certain body regions of their hosts. Based on the morphology of their mouthparts, lice can be divided into chewing lice (biting lice of some works) and sucking lice. Chewing lice feed mainly on feathers, fur, skin debris, or (rarely) blood of birds or mammals, whereas sucking lice feed exclusively on the blood of eutherian (placental) mammals. Because of their blood-feeding habits, sucking lice are much more important than chewing lice as vectors of pathogens, especially with respect to human diseases.

Infestation by lice is called *pediculosis*. Throughout human history, lice have been a major scourge, and the body louse has played an important role in shaping human civilization through its role as a vector of the agents of epidemic typhus, trench fever, and louse-borne relapsing fever. Today, because of louse control programs and improved hygienic standards, these louse-borne diseases are much less common than in former decades and centuries, although they persist in a few parts of the world. Also, any of these diseases, but especially epidemic typhus, can reemerge under certain conditions, such as during wars or famine or in crowded refugee camps. Also, trench fever is reemerging in some inner cities. Although they do not appear to transmit pathogens in nature, both head and crab lice are common ectoparasites of humans throughout the world, with head lice often infesting many preteenage children. Sucking lice of mammals,

especially rodents, may be important enzootic vectors of several zoonotic pathogens and serve to maintain these infections in nature. Although chewing lice have little involvement in human disease, one species serves as an intermediate host of a tapeworm that can be transmitted to humans.

GENERAL IDENTIFICATION

Adults

Adult lice are dorsoventrally flattened, small (0.4–10 mm) insects with chewing mandibles (chewing lice) or sucking stylet-like mouthparts that are withdrawn into a stylet sac in the head except during feeding (sucking lice). They are usually elongate (the crab louse is an exception), with the body distinctly divisible into a head, thorax, and abdomen; although the three thoracic segments are often wholly or partly fused. Antennae are well developed but are concealed in cephalic grooves in members of the Amblycera. Each of the three thoracic segments bears a pair of thickened legs that each terminate in a claw. These claws are relatively small in chewing lice but are large and highly adapted in sucking lice, where they are developed into tibiotarsal claws that tightly grasp the host hair and aid in host attachment. The body is covered with setae to varying degrees, depending on the species, and is usually leathery except for sclerotized thoracic and abdominal plates in many species that confer some degree of rigidity. Laterally, many lice have paired paratergal plates that curve around the anterior abdominal segments from dorsal to ventral on each side. Most of these plates typically enclose a spiracle that, in addition to a pair of thoracic spiracles, aids in respiration. Additional sclerotized plates are present in both sexes posteroventrally on the abdomen, where they serve to support and protect the genitalia. Also, tiny finger-like female appendages, called *gonopods*, facilitate egg positioning during oviposition. With one exception, louse eggs are glued onto the host fur or feathers close to the skin; the body louse differs by ovipositing on human clothing, especially along seams.

Internally, most blood-feeding lice have mycetomes situated next to the gut; these are colonized by symbiotic microorganisms that aid in blood digestion. The genitalia are conspicuous, especially in males, where the sclerotized aedeagus and associated structures can occupy up to half the length of the abdomen. As in other arthropods, the body cavity is a hemocoel that is bathed with hemolymph.

Immatures

Louse eggs ("nits") are subcylindical in shape and glued basally to the host hair, feathers, or clothing; they have an anterior operculum with respiratory pores (aeropyles) that is pushed off by the hatching first instar nymph. For medically important lice, care should be taken to distinguish louse eggs on hair samples from hair casts that are harmless accumulations of skin and scalp secretions. Immature lice (nymphs) closely resemble adult lice but are smaller, lack external genitalic openings, and have progressively more setae (first instar nymphs have fewer setae than second instars, which have fewer than third instars).

CLASSIFICATION

Worldwide, about 3,200 species of lice have been described, although additional, undescribed species exist, especially in association with birds and rodents. Both chewing and sucking lice are currently grouped into a single insect order, the Phthiraptera, by most workers. However, the ordinal names Anoplura (sucking lice) and Mallophaga (chewing lice) have been used in the past and are still used by some researchers. Currently, the most widely accepted higher classification of lice dispenses with the name Mallophaga, and treats the Anoplura and each of three groups of chewing lice (Amblycera, Ischnocera, and Rynchophthirina) as suborders of equal rank within the Phthiraptera.

As a group, the chewing lice are usually divided into 11 or 12 families, only one of which, the Trichodectidae, includes a species (the dog-biting louse, *Trichodectes canis*) with medical importance. However, some chewing lice are known to transmit pathogens to birds, and others may transmit pathogens to wild mammals.

The sucking lice are divided into 15 families, four of which include species that are of direct or indirect importance to humans. Of direct importance are the body louse (*Pediculus humanus humanus*) and head louse (*Pediculus humanus capitis*), both of which are assigned to the family Pediculidae, and the crab louse (*Pthirus pubis*), also referred to as the *pubic louse*, which belongs to the family Pthiridae. Of indirect importance to humans are certain rodent-infesting sucking lice belonging to the genera *Hoplopleura* (family Hoplopleuridae), *Neohaematopinus*, and *Polyplax* (both belonging to the family Polyplacidae). These rodent-infesting sucking lice are enzootic vectors of zoonotic pathogens. Some lagomorph-infesting lice in the polyplacid genus *Haemodipsus* may also be enzootic vectors of the agent of tularemia between their hosts.

IMPORTANT SPECIES

Body Louse (*Pediculus humanus humanus*)
(Fig. 6.1A)

Infestation by body lice is called *pediculosis corporis*. Although once a common, intimate companion of humans throughout the world, the body louse is now

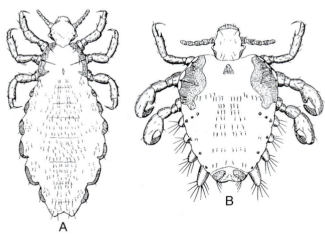

FIGURE 6.1 Lice of medical importance: (A) body/head louse (*Pediculus humanus*), female, dorsal view, (B) crab louse (*Pthirus pubis*), female, dorsal view. Reproduced from CDC, 1966.

rare in developed nations except on some homeless persons or others without access to a change of uninfested clothing. Nevertheless, it persists in many countries, especially in parts of Africa, Asia, and Central and South America. The reduced incidence of people infested with body lice has been achieved mainly through insecticidal intervention and increased hygienic standards, predictably accompanied by a global concomitant reduction in the prevalence of louse-borne diseases.

Body louse bites often cause intense irritation for a few days, and each bite-site develops into a small red papule. However, as an infested person is subjected to more bites over a prolonged period, desensitization occurs and little or no reaction at the bite site occurs. Persons with chronic body louse infestations often develop a generalized skin discoloration and thickening known as *vagabond disease* or *hobo disease*. Persons with chronic body louse infestations may also develop swollen lymph nodes, edema, elevated body temperature, headache, joint and muscle pain, and a diffuse rash. Occasionally, patients may become allergic to body louse bites and develop a generalized dermatitis or (rarely) a form of asthmatic bronchitis.

Head Louse (*Pediculus humanus capitis*)
(Fig. 6.1A)

Infestation by head lice is called p*ediculosis capitis*. As noted later, the head louse is morphologically almost indistinguishable from the body louse. However, it has a clear predilection for head hair and is still common throughout the world, including the United States, where 6–12 million people, primary children, are infested each year. Most children become more aware of their personal appearance and hygiene as they approach their teenage years, and this is thought to partly explain the lower incidence of head lice with increasing age. Female head lice glue their eggs to hair bases next to the scalp. As the hair grows, the nymph later hatches, but the empty egg case remains attached to the hair shaft. Based on hair growth rates and the distance of empty nits from the scalp surface, the duration of a head louse infestation can be estimated. In nature, head lice are not directly involved in pathogen transmission, but heavy infestations cause significant irritation, and the resultant scratching can lead to secondary infections such as impetigo, blood poisoning, or pyoderma. Also, swollen cervical lymph nodes may accompany severe head louse infestations. In these cases, a scabby crust may form on the scalp, with large numbers of head lice typically living beneath it.

Crab Louse (*Pthirus pubis*) (Fig. 6.1B)

Infestation by crab lice is called *pthiriasis* or *pediculosis inguinalis*. This louse, also referred to as the *pubic louse* in English or as *papillons d'amour* ("buttterflies of love") in French is a squat louse (1.1–1.8 mm long) with robust claws used for gripping the thicker pubic hairs. However, it can also grasp other thick body hairs, such as eyelashes, eyebrows, and those in the armpit of both sexes, as well as beard, moustache, and chest hairs of men; consequently it can also infest these body regions. Crab lice are common worldwide and are often diagnosed by health care workers at STD (sexually transmitted disease) clinics. Purplish lesions frequently develop at the intensely itchy bite sites, and small blood stains usually are present on the underwear from louse feces or squashed lice. Like the head louse, the crab louse is not a vector of pathogens in nature, although secondary infections may occur at bite sites.

Flying Squirrel Louse
(*Neohaematopinus sciuropteri*) (Fig. 6.2A)

In North America, this louse is an enzootic vector of a zoonotic strain of *Rickettsia prowazekii* between its flying squirrel hosts. However, the route by which humans can be infected has not been determined.

Spined Rat Louse (*Polyplax spinulosa*)
(Fig. 6.2B)

This louse of domestic rats is an enzootic vector of *R. typhi* and possibly also of other zoonotic pathogens. This louse occurs throughout much of the world in both warm and cool climates, including all of North America.

Tropical Rat Louse (*Hoplopleura pacifica*)
(Fig. 6.2C)

This louse is widely distributed in warmer climates throughout the world and parasitizes domestic rats in the southern United States. It is an enzootic vector of *Rickettsia typhi*, the agent of murine typhus, and possibly also of other zoonotic pathogens.

Rabbit Louse (*Haemodipsus setoni*)

Under laboratory conditions, this louse is capable of transmitting *Francisella tularensis*, the bacterial agent that causes tularemia (rabbit fever), suggesting that rabbit lice may be important enzootic vectors of this zoonotic agent in nature.

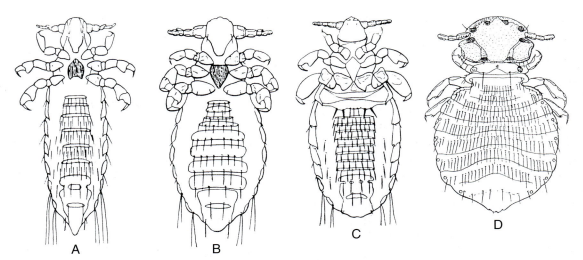

FIGURE 6.2 Lice of indirect medical importance: (A) Flying squirrel louse (*Neohaematopinus sciuropteri*), male, ventral view; (B) spined rat louse (*Polyplax spinulosa*), male, ventral view; (C) tropical rat louse (*Hoplopleura pacifica*), male, ventral view; (D) dog-biting louse (*Trichodectes canis*), female, dorsal view. Redrawn from CDC, 1966.

Dog-Biting Louse (*Trichodectes canis*)
(Fig. 6.2D)

Although this louse feeds exclusively on dogs and some other canids, it is an intermediate host of a zoonotic tapeworm, as discussed later.

CRYPTIC SPECIES

The head louse and body louse are morphologically similar, and, in the absence of data on the collection site on the body, they can be difficult to distinguish, especially if only a single specimen is available. Some workers treat these two lice as separate species (*Pediculus humanus*—body louse; and *Pediculus capitis*—head louse), whereas others treat them as separate subspecies (*Pediculus humanus humanus*—body louse; *Pediculus humanus capitis*—head louse). Here, they are tentatively treated as separate subspecies. Some other names for body lice that appear in the literature, such as *Pediculus corporis*, *Pediculus vestimenti*, and *Pediculus humanus corporis*, are junior synonyms and should therefore *not* be used. Various researchers have argued for decades over the correct taxonomic status of head and body lice and will, undoubtedly, continue to do so. However, there are valid arguments against both the specific and subspecific status of these lice. Although these lice virtually never interbreed in nature, even in dual infestations on the same person, they interbreed to produce viable progeny under laboratory conditions; this suggests that they are not good species. However, both head and body lice are

distributed worldwide, showing that there is no geographical separation between them; this suggests they are not good subspecies either. Perhaps both forms of lice started to evolve from a common ancestor (possibly from *Pediculus schaeffi*, which parasitizes chimpanzees, or the progenitor of this louse) on humans in different parts of the world, but, before speciation could occur, worldwide remixing of head and body louse–infested human populations occurred. New molecular techniques applied to human head and body louse populations from different parts of the world may provide a more definitive answer to this question in the future.

Although head and body lice are similar, morphologically extreme individuals, or specimens from a large sample, can often be identified as one species or the other. Body lice tend to be longer (males, 2.3–3.0 mm; females, 2.4–3.6 mm) than head lice (males, 2.1–2.6 mm; females, 2.4–3.3 mm) and are lighter in color, with a longer third antennal segment, less prominent indentations between abdominal segments, and shorter lobes on their paratergal plates. Obviously, these are subtle differences, and many persons would be unable to distinguish these lice without very careful microscopical examination.

BIOLOGY/LIFE CYCLE

Lice have a hemimetabolous ("simple") life cycle, with three nymphal stages superficially resembling small adults. Mated females glue 0.2–10 eggs per day, one egg at a time, onto a hair, feather, or clothing, depending on the species. Most louse eggs hatch in 4–15 days, with each of the nymphal instars typically lasting 3–8 days before the molt to the next stage occurs, and adults live up to 35 days. Specific periods for head and body lice are 4–5 eggs laid per day by fertile females, eggs hatching after about 8 days, each nymphal instar lasting 3–5 days, and adults surviving up to 30 days. Similar data for the crab louse include an average of three eggs laid per day by gravid females, egg hatch after 7–8 days, each nymphal instar lasting 3–6 days, and adults surviving up to 25 days.

Unlike louse infestations of some livestock animals, such as cattle, sheep, and horses, which peak in numbers during the winter and early spring, there are typically no distinct seasonal trends in the population densities of head, body, or crab lice associated with humans. In some regions, however, lice may become more prevalent during cooler seasons, when more clothing is required and washing may be less frequent.

Mating

Mating is initiated in lice when the smaller male pushes himself beneath the female from behind while both sexes are on the host. Once he is completely underneath the female, he curls the tip of his abdomen upwards and partly extrudes his aedeagus into the female opening to initiate mating and sperm transfer, which typically lasts for several minutes. Mating body lice assume a vertical position along a hair shaft, with the female supporting the weight of the male, whereas mating crab louse partners hold onto a hair rather than to each other.

Biting Behavior

Chewing lice feed by chewing skin, fur, or feathers on their hosts. The few hematophagous species typically chew the host skin until it bleeds and then imbibe blood from the wound site. Biting and feeding by sucking lice is more refined. Prior to feeding, the three sharp stylets that are used to penetrate the host to initiate blood feeding are withdrawn into a stylet sac inside the head. Externally, the labrum is modified into a broad, partly flattened *haustellum* with tiny *haustellar teeth* that latch onto the skin surface. Once the haustellum is in place, the stylet bundle is pushed through the skin until a host blood capillary is penetrated. The stylets consist of the fused maxillae, which form a blood canal, the smaller hypopharynx, which forms a salivary canal, and the much larger, more robust labium, in which both the blood and salivary canals are supported. A cocktail of enzymes, anticoagulants, and other compounds is secreted in the saliva, and some of these are recognized as antigens by the host, which can result in local or (rarely) systemic host reactions (see Chapter 28).

Dispersal

Most lice transfer during close physical contact between their hosts. Some mammal lice transfer from infested mothers to their offspring during suckling or when the offspring sleep or rest next to the mother. Others transfer between partners during host mating. Some chewing lice attach to a larger winged louse-fly (family Hippoboscidae), which then transports the louse as it flies to a new host; this is called *phoresy*. A few sucking lice have been found attached to house-flies and livestock-associated flies and may occasionally also transfer between hosts in this way.

Body lice or their eggs are often transferred in the clothing of an infested (lousy) individual. However, body lice will also crawl from one person to another under crowded conditions, and they tend to leave human hosts with elevated body temperatures to search for a new host. The latter behavior has epidemiological significance, because individuals with louse-borne diseases often have elevated body temperatures and their associated lice are often infected with the causative agent(s).

Although head lice are also exchanged during physical contact, as often occurs in daycare institutions or schools for children, these lice can also transfer on shared objects, such as combs, brushes, headphones, caps, hats, and scarfs. They can also be contracted by occupying a seat and headrest recently vacated by an infested person on trains and buses and in hair salons, etc.

Crab lice are usually exchanged during sexual intimacy. However, as with other human lice, transfer sometimes also occurs from inanimate objects such as shared bedsheets, clothing, or toilet seats.

PATHOGENS TRANSMITTED

Three significant pathogens are transmitted by the body louse. These are the causative agents of epidemic typhus, trench fever, and louse-borne relapsing fever, respectively. Although head lice and crab lice can also transmit some of these pathogens under optimal laboratory conditions, in nature only the body louse is involved in disease outbreaks caused by these agents. Under certain conditions, body lice can also transmit *Salmonella* spp., which cause food poisoning (salmonellosis, typhoid, etc.), but these bacteria are more efficiently transmitted by other means. Although several additional pathogens have been detected in human lice, available evidence suggests that these were imbibed from an infected host during the blood meal and are not actually transmitted by lice. Other lice are enzootic vectors of human pathogens or (in one case) serve as intermediate hosts of a tapeworm that can parasitize humans.

Epidemic Typhus

Epidemic typhus, also known as *jail fever*, *louse-borne fever*, and *exanthematic typhus*, is caused by infection with the rickettsial bacterium *Rickettsia prowazekii*. Body lice become infected after feeding on an infectious (rickettsemic) person. Rickettsiae ingested by the louse colonize the cells that line its gut, where they replicate and later burst free into the gut lumen. Some of these infectious rickettsiae are then voided in the louse feces, which are typically deposited on the host while the louse is feeding. When the host later scratches the louse bite area, rickettsiae are abraded into the skin to initiate an infection. Although this posterior station route is the typical mode of transmission, infectious rickettsiae can remain viable in louse feces for about 30 days, suggesting that aerosol transmission may also be possible.

About 10–14 days after the initial exposure, early clinical signs of epidemic typhus usually appear and include malaise, muscle aches, headache, coughing, rapid onset of fever, and a blotchy rash on the chest or abdomen. In severe cases, the rash can eventually cover most of the body. Later-stage symptoms in untreated cases include delirium, prostration, low blood pressure, and coma, which may culminate in death. Case fatality rates of 10–20% are typical, but figures as high as 50% have been recorded in untreated outbreaks; however, prompt administration of an antibiotic such as doxycycline, tetracycline, or (formerly) chloramphenical is usually curative. Historically, epidemic typhus has significantly shaped human history, especially during military campaigns. For example, Napolean's great army of 1812 was defeated more by epidemic typhus than by other factors during the attempted invasion of Russia.

Several different forms of epidemic typhus can be recognized. Classic epidemic typhus involves direct transmission of *R. prowazekii* by body lice. Another form of epidemic typhus, called *recrudescent typhus* or *Brill–Zinsser disease*, does not involve transmission by lice. Instead, it is a recurrence of the disease in individuals who were infected months or years previously. After the patient has recovered from the initial bout of epidemic typhus, infectious rickettsiae can remain dormant in human tissues and later cause a second bout of disease in the presence or absence of body lice. Intervals as great as 30 years have been recorded between disease bouts in some individuals. Most of the larger cities in the northeastern United States experienced cases of recrudescent typhus in the 18th, 19th, and early 20th centuries because some immigrants from Europe, central Asia, or other regions had previously been exposed to classic epidemic typhus in their countries of origin. If a patient experiencing a bout of recrudescent typhus is also infested with body lice, then the lice could become infected during blood feeding and then transfer to other persons to initiate an outbreak of classic epidemic typhus. The last recorded North American outbreak of epidemic typhus, which occurred in Philadelphia in 1877, may have started in this way.

Curiously, North American flying squirrels (*Glaucomys* spp.) also harbor a zoonotic strain of *R. prowazekii*, which can be molecularly distinguished from strains isolated from body lice or humans. Although this strain is infectious to people, the exact mode of transmission remains unknown. The flying squirrel louse, *Neohaematopinus sciuropteri* (Fig. 6.2A), and, perhaps, fleas appear to be the principal enzootic vectors, serving to maintain the infection in flying squirrel populations. However, *N. sciuropteri* does not bite humans. Because flying squirrels often colonize attics or eaves of houses, it has been suggested that humans might become infected when frequenting these areas by inhaling aerosolized rickettsiae from infectious louse or flea feces. Flying squirrel–associated epidemic typhus is often referred to as *sporadic epidemic typhus* or *sylvatic epidemic typhus*. To date, no human deaths have been recorded from this infection, which appears to be uncommon; typically, fewer than 10 cases are diagnosed in the United States each year.

Although epidemic typhus is relatively rare today, it persists in several parts of the world, and recent outbreaks have been recorded in Algeria, Burundi, China, Ethiopia, Peru, Russia, and various other parts of Asia, Africa, South America, and Central America. An outbreak associated with refugee camps in Burundi in 1996–1997 represented the largest epidemic of this disease since World War II and may have involved as many as 500,000 people. This amply demonstrates that, although epidemic typhus is certainly a disease that is less prevalent now than it has been throughout human history, it has the potential to rapidly reemerge under certain conditions.

Trench Fever

This disease, also known as *5-day fever* or *wolhynia*, is caused by infection with the bacterium *Bartonella* (formerly *Rochalimaea*) *quintana.* Human infection ranges from asymptomatic through mild to severe, but death is a rare outcome. The disease was unknown prior to World War I, when, in 1916, European troops who were engaged in trench warfare presented with symptoms of headache, muscle aches, fever, and nausea, with disease episodes often alternating with afebrile periods. More than 200,000 cases were recorded among British troops alone. Later, the causative agent was discovered, the disease was named, and body lice were implicated as vectors. Trench fever virtually disappeared after World War I, only to resurface again under similar conditions in World War II. Since that time, the disease has been recorded from time to time in various parts of the world, but, with one exception (see below), it is generally considered to be rare today. Nevertheless, serosurveys suggest that asymptomatic exposure is fairly widespread in many human populations. Antibiotics recommended for treating trench fever are doxycycline, erythromycin, and azithromycin.

Body lice become infected with *B. quintana* when feeding on the blood of an infectious person, who may or may not show clinical symptoms. Like *R. prowazekii,* the bacteria then invade the louse midgut, where they replicate in the lumen and epithelial cells, eventually being voided in louse feces. As in epidemic typhus, *B. quintana* is transmitted by the posterior station route from infectious louse feces scratched into the skin.

Recently, infection with *B. quintana* has been recorded as an opportunistic infection in some homeless or chronic alcoholic persons living in inner cities in North America, Europe and Asia. Some patients have also been immunocompromised, principally through HIV infection. Under these circumstances, the disease does not manifest as typical trench fever but, instead, mainly as vascular tissue lesions (bacillary angiomatosis, bacillary peliosis, etc.), chronically swollen lymph nodes, and endocarditis (inflammation of the heart valves). This disease has been called *urban trench fever.* Although body lice may be associated with these outbreaks, lice were not recorded on some of the patients, suggesting that an alternate transmission route could be implicated for some cases.

Louse-Borne Relapsing Fever

Also known as *epidemic relapsing fever,* this disease is caused by infection with the spirochete bacterium *Borrelia recurrentis.* Clinical symptoms include head and muscle aches, nausea, anorexia, dizziness, coughing, vomiting, thrombocytopenia (a decrease in blood platelets), and abrupt onset of fever. However, the most characteristic symptom is the presence of afebrile

periods followed by periods of fever. These relapses usually occur two to five times before the disease dissipates and reflect changes in the bacterial antigens in response to host antibody responses. However, in severe infections the liver and spleen become swollen, breathing becomes painful, and the patient typically lies prostrate, shaking and taking shallow breaths. Mortality rates for untreated cases range from 5% to 40%. Antibiotics commonly used to combat this disease are penicillin and tetracycline.

As with the two previously discussed diseases, body lice are the vectors of *B. recurrentis,* and they become infected after feeding on an infectious person. However, after ingestion by the louse, some spirochetes pass through the gut wall and colonize the hemocoel, where they multiply into huge populations. Because these spirochetes are effectively trapped inside the louse and are not secreted or excreted, the only way they can be transmitted to a person is by crushing lice on the skin and causing a small abrasion through which the spirochetes can then enter the body.

As with epidemic typhus and trench fever, outbreaks of louse-borne relapsing fever have affected human history. For example, outbreaks of this disease swept through several English towns and villages during the 18th century, sometimes killing all of the inhabitants. An epidemic in eastern Europe and Russia from 1919 to 1923 resulted in 5 million deaths, and millions of people were infected during an epidemic in North Africa in the 1920s. Today, most cases of louse-borne relapsing fever are recorded in Ethiopia, where 1000–5000 cases are typically recorded annually. However, recent outbreaks have also occurred in other countries, including Burundi, China, Peru, Russia, Rwanda, Sudan, and Uganda. As with the previously discussed louse-borne diseases, reemergence of this disease is an ominous possibility under conditions such as war and famine.

Lice of Wild Animals as Enzootic Vectors

Although little is currently known on this subject, sucking lice associated with various rodents and lagomorphs can transmit zoonotic pathogens between their hosts and thereby serve as enzootic or maintenance vectors for these agents. Included in this category are lice of domestic rats, which can transmit the agent of murine typhus, and almost certainly other agents also, to their hosts. The flying squirrel louse also functions as an enzootic vector through its role of transmitting zoonotic strains of the agent of epidemic typhus to its hosts. Also, rabbit lice, as well as some other ectoparasites of rabbits, can transmit the agent of tularemia between their lagomorph hosts.

Double-Pored Tapeworm

Adults of the double-pored tapeworm (*Dipylidium caninum*) typically parasitize carnivores, but humans can also become parasitized under certain conditions. Eggs of this tapeworm are voided in the feces of the definitive host, which is often a domestic dog or cat. As the host feces dry or dissipate, chewing lice or flea larvae, both of which serve as intermediate hosts, can ingest them during feeding. After ingestion by a dog-biting louse, each tapeworm egg hatches and develops into a cysticercoid stage inside the hemocoel, where it remains quiescent, unless the louse is eaten by a definitive host. Dogs may ingest chewing lice as they groom themselves with their teeth. Although it seems unlikely that a human would consume a tapeworm-infected louse to initiate an infection, this can happen when, for example, children with wet or sticky fingers are playing with a dog and then place their fingers inside their mouth. Various anthelmintics can be administered to kill these tapeworms in humans or in pets.

CONTROL EFFORTS

Body louse infestations can be eliminated if the infested person and the clothing are both carefully treated with approved insecticides; agents used to kill lice are called *pediculicides*. However, louse removal from an infested person will almost invariably be futile if the same, unwashed clothes are reworn, because of the presence of lice and nits on these. Humans can be "deloused" with insecticidal sprays or lotions, and clothes should be either burned and discarded or washed in very hot (not warm or cold) water. The hot temperatures kill the lice and nits, as does subsequent careful ironing, especially along clothing seams. Body lice are difficult to control on some infested homeless persons who do not have a change of clean (uninfested) clothes.

Head lice can be difficult to control, because some louse populations have developed resistance to currently approved pesticides. Also, most of the pediculicidal shampoos currently on the market are not completely efficacious against the nits, so the treatment needs to be repeated after a new batch of nymphs has hatched about a week later. Because of insecticide resistance, the age-old technique of carefully and repetitiously pulling a "louse comb" through the hair of infested individuals once or twice a day until the infestation has been eliminated is still a useful control technique. Louse combs have very small spaces between the teeth and are designed to pull lice and nits from the hair.

Crab lice can be controlled using techniques similar to those for body lice, taking care to treat all areas with thick hairs, not just the pubic region. These include especially the armpits, eyelashes, and eyebrows of both sexes and beard, moustache, and chest hairs of men.

Rodent and lagomorph ectoparasites, including lice that may be enzootic vectors of zoonotic pathogens, can be controlled using a variety of techniques, including dusting of burrows with insecticides, bait stations equipped with insecticides, and supplying insecticide-impregnated food or nesting material. These same techniques are also used to control flea and tick vectors associated with these mammals. Louse-infested mammals or birds in residences, laboratories, breeding facilities, pet stores, zoos, etc. can be spot-treated or individually treated with any of a number of contact pediculicides. Insecticidal strips placed inside or adjacent to cages also can control lice on caged birds and small mammals.

Readings

Burgess, I. F. 1995. Human lice and their management. *Adv. Parasitol.* 36: 271–342.

CDC. 1966. Pictorial keys: Arthropods, reptiles, birds and mammals of public health significance. U.S. Department of Health, Education, and Welfare, Public Health Service, Communicable Disease Center, Atlanta.

Durden, L. A. 2001. Lice. pp. 3–17, In: *Parasitic diseases of wild mammals*, 2nd. ed. (W. M. Samuel, M. J. Pybus, and A. A. Kocan, eds.). Iowa State University Press, Ames.

Durden, L. A., and Musser, G. G. 1994. The sucking lice (Insecta, Anoplura) of the world: A taxonomic checklist with records of mammalian hosts and geographical distributions. *Bull. Am. Mus. Nat. Hist.* 218: 1–90.

Ferris, G. F. 1919–1935. Contributions toward a monograph of the sucking lice. Parts I–VIII. *Stanford Univ. Publ. Univ. Ser., Biol. Sci.* 2: 1–634.

Ferris, G. F. 1951. The sucking lice. *Mem. Pac. Coast Entomol. Soc.* 1: 1–320.

Gratz, N. G. 1997. Human lice: Their prevalence, control and resistance to insecticides: A review 1985–1997. World Health Organization/CTD/WHOPES/97.8, Geneva.

Kim, K. C., Pratt, H. D., and Stojanovich, C. J. 1986. *The Sucking Lice of North America: An Illustrated Manual for Identification.* Pennsylvania State University Press, University Park.

McDade, J. E. 1987. Flying squirrels and their ectoparasites: Disseminators of epidemic typhus. *Parasitol. Today* 3: 85–87.

Mumcuoglu, K. Y. 1996. Control of human lice (Anoplura: Pediculidae) infestations: Past and present. *Am. Entomol.* 42: 175–178.

Ohl, M. E., and Spach, D. H. 2000. *Bartonella quintana* and urban trench fever. *Clin. Inf. Dis.* 31: 131–135.

Orkin, M., and Maibach, H. I. (eds.) 1985. Cutaneous infestations and insect bites. *Dermatology*, Vol. 4, Marcel Dekker, New York.

Orkin, M., Maibach, H. I., Parish, L. C., and Schwartzman, R. M. (eds.) 1977. *Scabies and Pediculosis.* Lippincott, Philadelphia.

Price, M. A., and Graham, O. H. 1997. Chewing and sucking lice as parasites of mammals and birds. *Tech. Bull. No. 1849, USDA.*

Raoult, D., Ndihokubwayo, J. B., Tissot-Dupont, H., Roux, V., Faugere, B., Abegbinni, R., and Birtles, R. J. 1998. Outbreak of epidemic typhus associated with trench fever in Burundi. *Lancet* 352: 353–358.

Raoult, D., Roux, V., Ndihokubwayo, J. B., Bise, G., Baudon, D., Martet, G., and Birtles, R. 1997. Jail fever (epidemic typhus) outbreak in Burundi. *Emerg. Inf. Dis.* 3: 357–360.

Roux, V., and Raoult, D. 1999. Body lice as tools for diagnosis and surveillance of reemerging diseases. *J. Clin. Microbiol.* 37: 596–599.

Tarasevich, I., Rydkina, E., and Raoult, D. 1998. Outbreak of epidemic typhus in Russia. *Lancet* 352: 1151.

Zinsser, H. 1935. *Rats, Lice, and History.* Bantam, New York.

7

Fleas, the Siphonaptera

KENNETH L. GAGE

INTRODUCTION

The Siphonaptera is a relatively small order, with fewer than 2000 described species, a majority of which are found in northern temperate regions. The adults are quite distinctive in appearance (Fig. 7.1) and live as blood-feeding ectoparasites on a wide variety of mammals and birds. Nearly three-fourths of all known species of fleas infest rodents and insectivores.

Fossil fleas, which are known from only a few specimens identified in Baltic or Dominican amber, provide few clues to the origins of the Siphonaptera. Morphological studies and molecular investigations suggest that the Mecoptera (scorpion flies) are the closest relatives of fleas. Other evidence suggests that fleas first arose on mammals 125 or more million years ago, while all avian-specific fleas evolved from mammalian fleas at a later date. These conclusions are based on a number of observations. The diversity of flea families, genera, and species is much higher on mammals than on birds, suggesting a longer association with the former class. Mammals also host certain types of primitive fleas not found on birds. In many instances these fleas occur on primitive host groups, and occasionally both flea and host groups occur in distinct biogeographic regions that have been separated for a long period. This is well illustrated by the presence of primitive stephanocircid fleas on marsupials in South America and Australia, suggesting that these insects were well established on marsupials prior to the breakup of Gondwanaland. By contrast, true bird fleas generally belong to more advanced flea taxa, such as the Ceratophyllidae, and in each instance these groups also have many members that parasitize mammals.

IDENTIFICATION AND MORPHOLOGY

Adults

Morphologically, adult fleas are unique and unlikely to be confused with other types of insects (Fig. 7.1). Many of the most distinctive features of fleas are related to their ectoparasitic lifestyle and blood-feeding habits, including the lack of wings, laterally compressed body, and modifications of the hind legs and metathorax for jumping.

Head and Mouthparts

The head varies considerably, depending on the degree of development of the eyes, setae, and combs (ctenidia) on the genal or frons regions. The numbers and placement of setae on the head often differ among species, as do the presence of frontal tubercles, the shape of the antennae, and the appearance of the mouthparts. The antennae lie in grooves on either side of the head and bear sensory structures that are believed to be sensitive to touch, smell, heat, humidity, and vibrations. Male antennae are typically somewhat longer than those of females and bear copulatory suckers on their inner surfaces that are used for clasping and restraining females during mating. The eyes lie directly beneath or slightly forward of the base of the antennae. The eyes are simple ocelli-like structures (Fig. 7.2A) rather than compound eyes as seen in most other insects. Eyes are absent in some burrowing species (Fig. 7.2B), while other fleas, such as *Malaraeus* spp., have a distinctive clearing that can be seen in the

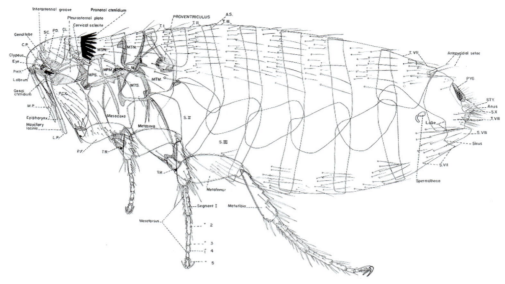

FIGURE 7.1   General anatomy, female *Phalacropsylla allos*. After Stark (1958).

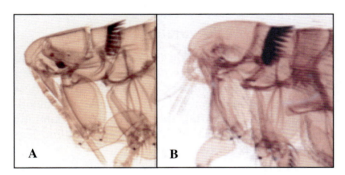

FIGURE 7.2   Flea eyes: (A) *Oropsylla hirsuta* (well-developed eye); (B) *Foxella ignota* (lacks eyes). CDC photograph.

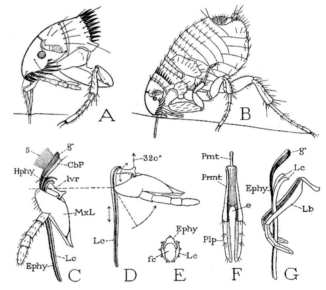

FIGURE 7.3   Feeding attitudes and mouthparts of the cat flea (*Ctenocephalides felis*). After Snodgrass (1946).

middle of the eye. Within the head are rodlike sclerites, including the trabecula centralis and tentorium, that strengthen the head capsule and have taxonomic value.

The adult mouthparts are specialized for piercing the host's skin and sucking blood (Figs. 7.3 and 2.2D). The labium has a channel running along its anterior length and bears a pair of four- or five-segmented labial palps that can have taxonomic significance. Lying inside the labial palps are three stylets. One of these stylets represents the median epipharynx, while the other two, which are often highly serrated or denticulate, are modified maxillary lacinae. The maxillae and the epipharynx join to form a tubelike structure through which the host's blood is imbibed. The maxillary palps are four-segmented and vary little in relative length between flea species. The labrum

consists of a rudimentary sclerite located just in front of the epipharynx.

Thorax

The adults of many species have a row of heavy black spines, termed the *pronotal comb*, that projects from the posterior edge of the pronotum. The presence or absence of this comb and the number and shape of its spines are often taxonomically significant. Certain bird fleas, such as *Ceratophyllus* spp., typically have combs with unusually high numbers of spines (24 or more). One or more rows of setae are usually found on both the dorsal and pleural sclerites of the thorax. The pattern of these setae is often consistent within a genus. Distinctive features of the metanotum include an anterior intercostal sclerotization and a posterior notal ridge that terminates at the upper edge of another sclerotized structure, the pleural ridge. The junction of these last two structures forms a joint that contains a mass of elastic resilin. The metanotum of some species bears distinctive spines along its posterior margin that have taxonomic significance. Two pairs of thoracic spiracles are also located on the thorax, one between the prothorax and mesothorax and another immediately below the lateral metanotal area.

Legs and Jumping Ability

The coxa of each leg is relatively large and often bears setae or, occasionally, small spiniforms that can be taxonomically significant. The trochanter is relatively small and short, while the femur is broader than the tibia but of similar length. The tarsus is divided into five tarsomeres. The relative length of the first tarsomere and the positioning of plantar bristles on tarsomere V can have taxonomic significance.

Surprisingly, the flea's jumping ability is not due directly to its leg muscles but rather to the presence of resilin, a highly elastic protein that is secreted by specialized cells within the developing pleural arch of the metathorax and presumed to be homologous to the resilin in the wing hinge ligament of flying insects. In order to jump, the flea uses muscles to compress the resilin, which is then held in place by a catch mechanism. When the flea is ready to jump the catch is undone and the muscles are relaxed, resulting in the sudden release of the elastic energy stored in the resilin. This causes the femur to rapidly rotate downward, which, in turn, results in the upward thrusting of the tibia and the distinctive jumping movement of the flea. It has been estimated that fleas develop an incredible 149 g's of force during their jumps, enabling some species to leap farther in relation to their body length than any other animal.

Certain species have lost the ability to jump. These fleas typically lack the resilin mass, and many show severe reduction or loss of the pleural ridge. Loss of jumping ability is especially common in species that infest certain birds, bats, or gliding mammals. Fleas that spend the majority of their time in their hosts' nests or roosts are also likely to exhibit reduced jumping ability. By contrast, good jumping ability seems to occur most commonly among species, such as *Ctenocephalides* spp. and *Pulex* spp., that infest relatively large and mobile hosts. It is appropriate to note here that the leaping abilities of fleas are often greatly overestimated. Even good jumpers, such as *Xenopsylla cheopis* and *Pulex irritans*, are incapable of jumping more than about 20–30 cm.

Abdomen

The abdomen is divided into pregenital and genital segments. The seven pregenital segments usually show relatively few specializations, but the presence and pattern of setae or spines on the tergites or pleural regions can have taxonomic significance. Sternites are typically present on pregenital segments II–VII but absent on segment I. Segments II–VI are generally similar in size and appearance. The tergites and sternites of segments II–VII can be retracted, telescope fashion, over or under those of the previous segment, a state that is observed most often in starved fleas. When fleas take blood meals, the abdomen expands lengthwise and the segments no longer overlap. The posterior edge of abdominal segment VII in female fleas often has a distinctive contour that can be used for separating closely related species. The tergite of segment VII also bears one to five obvious setae that can have taxonomic significance. Tergites on the pregenital abdominal segments also bear laterally placed spiracles that are connected to the tracheal system.

Abdominal segments VIII–X are referred to as the genital segments. In general, sternum VIII is quite small, while tergum VIII is enlarged and obscures the rudimentary segment IX. Segment X is represented by dorsal and ventral plates located near the anus. The sensilium (pygidium), which might also comprise part of segment X, lies on the dorsal surface of the abdomen posterior to tergite VIII and is covered with a number of easily discernible circular depressions (trichobothria) that each contain a long seta. It has been proposed that this structure might be sensitive to ultrasonic waves or play a sensory role in mating.

The genital segments and genitalia of male fleas are highly modified structures that bear many useful taxonomic features (Fig. 7.4A). The tergum and sternum of segment VIII are sometimes modified to form a presumably protective shield around the external genitalia. Arising from the posterior edge of the clasper are a fixed and a movable process, the latter of which often is referred to as the *movable finger* or *digitoid*. The fixed process contains a depression that can interlock with a toothlike structure on the movable process, thereby allowing the male to grasp the

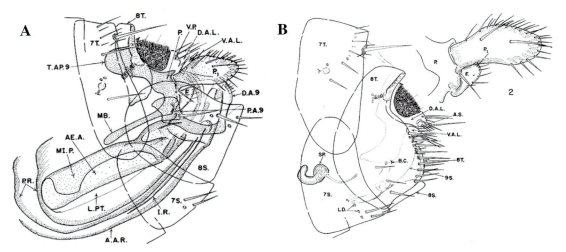

FIGURE 7.4 Genital segments of *Ctenocephalides felis*: (A) Male *C. felis*; (B) female *C. felis*. After Johnson (1957).

female's seventh sternite during copulation. The size, shape, and arrangement of bristles on the movable process are of great taxonomic importance, occasionally representing the only morphological means for separating species that have nearly identical females. Other structures of the male genitalia are located posteriorly in the abdominal cavity, including the aedeagal apodeme (plate of the penis) and extendable penis rods, which are coiled and normally retracted within a structure termed the *endophallic sac*.

The female genital segments are not as highly modified as those of the male flea (Fig. 7.4B). The most obvious female reproductive structure is the spermatheca, which is clearly visible through the side of the abdomen in cleared and mounted specimens. This structure has a head (*bulga*) and tail arrangement (*hilla*) that vary greatly among different taxa of fleas. Some fleas, such as those in the genera *Atyphloceras* or *Hystrichopsylla*, have paired spermathecae, a presumably primitive condition. A sclerotized tube connects the spermatheca with the bursa copulatrix.

Setae, Spines, and Combs

The various setae, spines, and combs found on fleas are thought to help fleas move efficiently through the host's pelage or plumage, prevent dislodgement from the host, steady the flea during feeding, and perhaps provide some protection against being crushed. Setae and spines appear on many regions of the flea's body, while combs are usually located on the first thoracic segment (pronotal combs) and/or the side of the head (genal combs). Combs occur less often on the front of the head, as in the helmet fleas (Stephanocircidae), or on the first abdominal segment, as in the genus

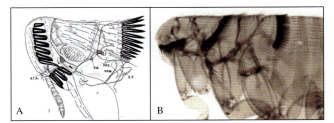

FIGURE 7.5 (A) Helmet comb of *Cleopsylla townsendi*. After Johnson (1957). (B) Abdominal comb of *Stenoponia americana*. CDC photograph.

Stenoponia, which also has pronotal and genal combs (Figs. 7.5A, B).

The arrangement and characteristics of the setae, spines, and combs often correspond to the characteristics of their hosts' pelage or plumage. For example, spiny hosts, such as hedgehogs, porcupines, and echidnas, are hosts to a few highly specialized fleas that exhibit heavy, widely spaced comb spines. The fleas found on each of these hosts are not closely related to those found on the other two types of mammals, and it appears that the similar appearance of their comb spines represents a case of convergent evolution. Other hosts, such as rodents, have less coarse pelages, and their fleas typically have comb spines that are finer and more numerous. Even relatively minor differences in host pelage appear to affect the characteristics of the comb spines. For example, shrew fleas belonging to the genera *Ctenophthalmus*, *Palaeopsylla*, *Corypsylla*, and *Nearctopsylla* all have spines that are up-curved with ovate tips. By comparison, *Palaeopsylla setzeri*, which infests voles rather than shrews, has pronotal comb spines that are stilletto shaped. Setae, spines, and combs are often most well developed on those fleas that actively course over a host's body. These same features are likely to be much reduced in other fleas, such as tungids and sticktights, that attach more firmly to the host and feed for long periods at a particular site.

Selected Internal Structures

Four pear-shaped salivary glands lie in the abdomen and are connected to the foregut by long ducts. These glands do not fully develop in adult females until blood feeding and egg development begins, at which time the epithelial cells increase greatly in size and exude large droplets of saliva into the gland lumen. Muscles lying along the foregut contract and provide suction for the cibarial and pharyngeal "pumps," thereby allowing blood to be drawn through the pharynx, esophagus, and proventriculus and passed into the midgut, where blood-meal digestion occurs. The globular proventriculus contains numerous spinelike setae that can vary greatly in appearance among species. Muscles associated with the proventriculus allow it to act as a valve, thus preventing regurgitation of the midgut contents into the esophagus. Digestion occurs in the distensible saclike midgut (Chapter 21). Posterior to the midgut is a short hindgut that ends in a rectal ampule and the anus. The Malpighian tubules attach near the midgut–hindgut junction and excrete nitrogenous waste into the gut.

Eggs

Eggs are round to oval and often white. Adult females typically lay eggs singly rather than as a large batch, usually in the somewhat protected environment of the host's nest or burrow but occasionally on the host itself. In the latter situation, the eggs fall from the host into its nest or other areas visited by the host.

Larvae

The legless larva has a wormlike appearance with a well-developed eyeless head capsule bearing two antennae. The last abdominal segment bears a pair of projections referred to as *anal struts*. Unlike the ectoparasitic adults, the typically free-living larvae possess chewing mouthparts. Larval labial glands secrete silk for construction of the pupal cocoon.

Pupae

The quiescent pupae typically develop within silken cocoons. In general, a flea pupa resembles a pale adult that has its legs folded up next to the body.

CLASSIFICATION

According to Lewis (1998), only 1957 species of fleas, belonging to 220 genera and 15 families, have been described (Table 7.1). The currently accepted grouping of fleas into families, genera, and species is based almost entirely on morphological characters; molecular genetic studies might shed light on previously unrecognized groups and their phylogenetic relationships. The geographic distributions and major host associations of the 15 flea families are given in Table 7.1.

BIONOMICS AND SOME IMPORTANT SPECIES

Basic Life Cycle

Fleas undergo complete metamorphosis, in nearly all cases passing through three larval instars. The time for the life cycle to go from egg to adult is affected by temperature and relative humidity (RH). Most fleas complete the life cycle in 30–75 days, but some may require months to do so. Those with extended life cycles are usually associated with migratory birds or burrow-dwelling mammals.

Bionomics

The larval habitat is determined by the female's oviposition behavior. Eggs are typically laid singly, often on the host, from which they then fall off. Suitable habitat includes acceptable ranges of RH and temperature for the species. Larval food consists of organic debris, often including the dried blood in the feces of the adult fleas. Larvae of the northern rat flea, *Nosopsyllus fasciatus*, pinch the abdomens of adults fleas with their mandibles to cause them to defecate droplets of blood, which they ingest directly. Two species of larvae are phoretic on their mammalian hosts, *Uropsylla tasmanica* on dasyurid marsupials and *Hoplopsyllus glacialis* on the artic hare. Pupation occurs in a silken cocoon that is often covered with debris from the host nest.

Host–Parasite Interactions

Adults may be categorized by the amount of time they spend on their hosts. Some species remain on their hosts and feed almost continuously, while others, usually in burrows or certain types of nests, feed for a short period and then leave the host.

Host specificity is important from the standpoint of transmission of disease agents, especially plague. In general, hosts that are taxonomically related or are

TABLE 7.1 Siphonapteran Families (From Lewis 1998)

Family	Distribution	Genera	Species	Major Hosts
Ancistropsyllidae	Oriental	1	3	Ungulates
Ceratophyllidae	Cosmopolitan but predominantly Holarctic	44	403	Primarily rodents, occasionally viverrids, mustelids, birds, and a single species on an insectivore (Siberian mole)
Chimaeropsyllidae	Ethiopian	8	26	Rodents, insectivores, elephant shrews
Coptopsyllidae	Palearctic	1	19	Rodents (gerbils and their allies)
Ctenophthalmidae	Primarily Holarctic, some in southern hemisphere	42	548	Rodents, occasionally pikas, insectivores (shrews and moles), marsupials, and a single species on mustelids
Hystrichopsyllidae	Nearctic, Palaearctic, Neotropical, Australian	6	36	Rodents, insectivores
Ischnopsyllidae	Cosmopolitan	20	122	Bats
Leptopsyllidae	Palaearctic, Nearctic, Oriental, a few species in Australian or Ethiopian realms (Madagascar)	29	230	Rodents, lagomorphs (hares, rabbits, pikas), insectivores, and rarely elephant shrews and foxes
Malacopsyllidae	Neotropical	2	2	Edentates (armadillos)
Pulicidae (includes tungid fleas)	Cosmopolitan	27	182	Very broad host range, including carnivores, ungulates, bats, edentates (armadillos), and occasionally birds (*Cariama* spp.)
Pygiopsyllidae	Ethiopian, Oriental, Australian, and one Neotropical genus	37	166	Rodents, marsupials, insectivores, and occasionally monotremes, birds, or tree shrews
Rhopalopsyllidae	Neotropical, southern Nearctic, Oceanic	10	122	Primarily rodents, some on oceanic seabirds
Stephanocircidae	Primarily Neotropical, two Australian species	9	51	Rodents, a few species on marsupials
Vermipsyllidae	Holarctic	3	39	Carnivores and ungulates
Xiphiopsyllidae	Ethiopian	1	8	Rodents

similar in their ecologies are likely to share flea species. For example, of the half dozen species of wood rats, *Neotoma* spp., in the United States, each is infested by one or more subspecies of *Orchopeas sexdentatus*. A high degree of specificity is uncommon; of 1,667 species of fleas of mammals, only 14% are found on one species of host. About 30% of fleas inhabit a single genus of host, and only 1% are limited to a single family of hosts.

Members of the bird flea genus, *Ceratophyllus*, can serve as an example of mating in adults. The male moves behind the female and clasps the female's second abdominal sternite with his antennae. The male then telescopes his abdomnal terga and arches the abdomen against the underside of the female's abdomen, where he clasps sternum VII with his raised terminal segments. He then inserts his aedeagus into the genital chamber, and one of the two penis rods is inserted into the spermathecal duct as far as the

spermatheca itself, where he deposits sperm. Copulation may last 3 hours or more. In species that are sessile, such as *Tunga penetrans* and *Echniophaga gallinaceae*, copulation takes place as the female continues to feed. In other variations, copulation may precede feeding; in still others it may be delayed until after one or two blood meals have been taken.

The timing of flea reproduction is often influenced by the seasonal availability of hosts or the timing of host reproduction. For example, reproduction in the European rabbit flea, *Spilopsyllus cuniculi,* is timed to coincide with the hormonal cycles of its host (*Oryctolagus cuniculi*) and ovary maturation does not begin in this flea until it feeds on pregnant rabbits. Immediately after the rabbit gives birth, the female fleas leave her body but remain in the nest, where they feed on the young rabbits, copulate, and lay eggs. Rabbit fleas fail to copulate in the absence of newborn rabbits and it has been suggested that copulation is stimulated by an airborne kairomone released by newborn rabbits. The onset of maturation in adult female fleas also corresponds with increased feeding activity and deposition of blood-containing feces in the rabbits' nests. The deposition of adult flea feces peaks about a week before parturition, which is when the pregnant rabbit is actively making her nest. The deposition by female fleas of large amounts of blood-containing feces provides an important nutritional boost for developing larvae, which hatch soon after the rabbits give birth. Under such favorable circumstances the larvae mature, pupate and emerge as unfed adults ready to infest young rabbits as they disperse from the nest.

Although it seems logical that other species would utilize similar cues for ovarial maturation, this has not been easy to demonstrate. The onset of reproduction in *Cediopsylla simplex*, another rabbit flea, is reported to be influenced by host hormones but this does not appear to be the case for the Spanish rabbit flea (*Xenopsylla cunicularis*), which, like *S. cuniculi*, infests European rabbits. Ovarian maturation in ground squirrel fleas (*Oropsylla bruneri*) in Manitoba also occurred independently of the estrous cycle of its host, *Spermophilus franklini*. *O. bruneri* populations in this study were reported to be bivoltine while their ground squirrel hosts produced only a single litter per year. Experimental studies also indicate that *X. cheopis*, *X. astia*, *Nosopsyllus fasciatus*, and *Leptopsylla segnis* are able to mature and reproduce when raised on hypophysectomized or castrated hosts.

The life cycles of certain bird fleas, including *Ceratophyllus celsus* on swallows and *C. galllinae* on tits, are timed so that adult fleas are ready to emerge from their pupal cocoons when their avian hosts return each spring to nesting sites. In other instances, climatic factors clearly influence the timing of life cycles. For example, the abundance and activity of *Oropsylla montana* adults increase during periods of relatively high humidity and moderate temperatures but decrease markedly when conditions are excessively hot and dry. Another ground squirrel flea, *Hoplopsyllus anomalus*, is more tolerant of hot, dry conditions and can replace *O. montana* as the dominant flea on rock squirrels during mid-summer in the American Southwest. The factors influencing the timing of life cycles of other flea species are often unknown. For example, on an annual basis prairie dogs (*Cynomys spp.*) are frequently infested by two species of *Oropsylla* fleas, *O. hirsuta* and *O. tuberculata cynomuris*. For unknown reasons, *O. hirsuta* predominates in the warmer months of the year, while *O. tuberculata cynomuris* peaks during the cooler months.

Adult fleas are often inactive unless stimulated to move by the presence of a potential host. Some remain in the protective cocoon until they sense a host by vibrations. In additiion to vibrations, fleas are stimulated by a passing shadow, CO_2, or certain other chemicals. For example, the rabbit flea is attracted by rabbit urine. Most fleas have the ability to survive for extended periods when fed regularly (Table 7.2)

Dispersal of fleas is primarily passive. Although fleas can crawl or jump over short distances, transport of these insects from one potential habitat to another relies mainly on the movements of their primary hosts. Temporary hosts, especially predators, also can transport adult fleas from site to site.

Some Representative Fleas

The following provide brief descriptions of certain important flea species. The list chosen is heavily biased toward members of the family Pulicidae (*Ctenocephalides felis*, *Xenopsylla cheopis*, *Pulex irritans*, *Echidnophaga gallinaceae*, *Tunga penetrans*), which contains a disproportionate number of species important to man and, therefore, well studied. The nonpulicid fleas included in this section also are important to humans as either disease vectors or pests of livestock.

TABLE 7.2 Median and Maximum Durations of Survival of Adult Female Fleas Fed Daily on Rats and Held in Pill Boxes at 20–21°C and 92–94% RH. From Burroughs (1953)

Species	Normal Hosts	Median (days)	Maximum (days)
Xenopsylla cheopis	Various commensal rats (*Rattus* spp.)	14	158
Nosopsyllus fasciatus	Norway rats (*Rattus norvegicus*)	20	281
Oropsylla montana	Ground squirrels (*Spermophilus* spp.)	62	331
Oropsylla idahoensis	Ground squirrels (*Spermophilus* spp.)	45	326
Opisodasys nesiotus	Mice and voles (*Peromyscus* spp. and *Microtus* spp.)	2	43
Malaraeus telchinum	Mice and voles (*Peromyscus* spp. and *Microtus* spp.)	21	182
Orchopeas sexdentatus	Wood rats (*Neotoma* spp.)	4	319
Megabothris abantis	Mice and voles (*Peromyscus* spp. and *Microtus* spp.)	21	291

Ctenocephalides felis

The cat flea (Figs. 7.6A, B) is extremely common on cats and dogs in many temperate and tropical regions, but it also infests other animals, including opossums, raccoons, and commensal rats.

Development is strongly influenced by temperature and relative humidity (RH), which is especially significant because *C. felis* lays its eggs on the host. These eggs later fall from the host onto sleeping areas or other sites frequented by the host. Optimal hatching of eggs occurs when temperatures are between 16 and 27°C (70% hatch success) and RH is greater than 50%. Eggs exposed to less than 50% RH fail to hatch, and 65% of eggs die following a 1-day exposure at 3°C. Hatching times are also greatly influenced by temperature, requiring as little as 1.5 days at 32°C.

Larval development is optimal at 20–30°C and greater than 70% RH. About 80% of larvae held at 32°C and 75% RH completed development within 8 days, while those held at 13°C and the same RH required 34 days. Larvae are more susceptible to low RH than eggs, and mortality increases dramatically when the former are held at temperatures above or below the optimal range for larval development. Successful development of larvae also is influenced by the availability of adequate nutrients, including the blood-containing feces of adult cat fleas. Upon completion of the third instar, larvae void any remaining gut contents and begin to construct cocoons for pupation.

Pupation requires a little less than two days to complete at 27°C but can take considerably longer at lower temperatures, with mean female and male pupation times reported to be 19.5 and 23.5 days, respectively, at 15°C. Pupae are quite resistant to desiccation, requiring as little as 2% RH for 80% emergence success at 27°C.

Typically, females emerge as adults within 5–8 days after pupation, with males taking somewhat longer (7–10 days). The stimuli for adult emergence appears to be mechanical pressure and heat, factors that can indicate the presence of a mammalian host. If members

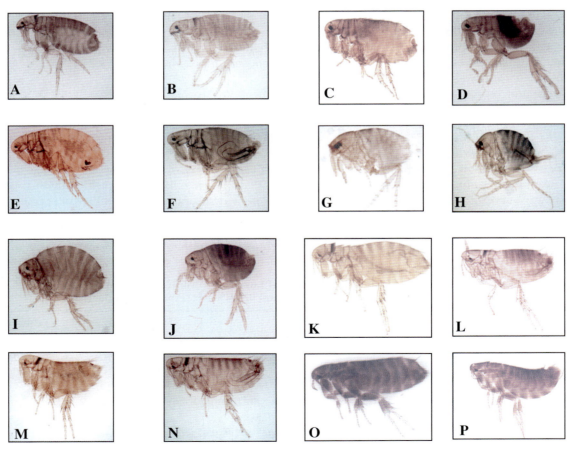

FIGURE 7.6   Common fleas: *Ctenocephalides felis* female (A) and male (B); *Pulex irritans* female (C) and male (D); *Xenopsylla cheopis* female (E) and male (F); *Tunga penetrans* male (G) and female (H); *Echidnophaga gallinacea* female (I) and male (J); *Oropsylla montana* female (K) and male (L); *Nosopsyllus fasciatus* female (M) and male (N); *Ceratophyllus gallinae* female (O) and male (P).

of the opposite sex are available, newly emerged females usually mate within 8–24 hours after infesting a host. Egg laying begins 24–36 hours after these fleas take their first blood meal. Females reach a reproductive peak 4–9 days after the onset of egg development, at which time they can produce 40–50 eggs per day in the absence of host grooming, a factor that can increase flea mortality and lower egg production. If protected from host grooming, most adult cat fleas can survive on their hosts for 50 or more days, although reproductive output continues to diminish after the initial peak.

Cat fleas use a number of visual and thermal cues to locate their mammalian hosts. Sensitivity to light is greatest at wavelengths of 510–550 nm, and traps that utilize yellow-green light filters are twice as attractive as those that use white light. Cat fleas are also attracted by the contrast of a dark object moving across a light background.

The roles of cat fleas as vectors of pathogens and hosts of parasites are discussed later in the chapter. Cat fleas also are exceptionally important pests of dogs and cats, and these infestations often mean that the owners of these pets receive numerous annoying flea bites. The most significant problem caused by these infestations is flea allergy dermatitis, which is caused by hypersensitization of hosts to certain antigens found in the cat flea's saliva.

Ctenocephalides canis

The dog flea is closely related to the cat flea and is very similar to it in appearance and biology. Despite its name it is actually less common than cat fleas on dogs in the United States. The dog flea and the cat flea can be distinguished by the characteristics of the teeth in their genal combs.

Pulex irritans

This flea has a nearly cosmopolitan distribution. It is often mistakenly called the human flea, although it attacks a wide range of mammals, including pigs, goats, wild carnivores, and domestic dogs and cats. Infestations can reach tremendous levels in the dirt-floored huts of persons living in some developing countries, particularly when farmers share their dwellings with their livestock or hold these animals in corrals or buildings immediately adjacent to their homes. *Pulex irritans* lacks combs and has a body that is relatively free of bristles and spines (Figs. 7.6C, D).

Xenopsylla cheopis

Various species of *Xenopsylla* are found throughout much of Africa and central and southern Asia, particularly in regions where gerbils or certain species of

wild and commensal rats can be found. *Xenopsylla cheopis* probably first arose in northeastern Africa from an ancestor that parasitized gerbils or grass rats. Regardless of its origin, the so-called Oriental rat flea now occurs primarily on commensal rats (*Rattus* spp.), although it readily attacks humans and other animals when its natural hosts are absent. Despite this, *X. cheopis* clearly prefers rat hosts. In one study *X. cheopis* responded positively to the odor of *R. norvegicus* but were unresponsive to the odors of three other murine hosts not normally parasitized.

Xenopsylla cheopis is currently common in many rat-infested tropical and warm temperate environments around the world. Although rat-infested ships have widely disseminated this flea, its major rat hosts (*R. rattus* and *R. norvegicus*) are actually more widely distributed than *X. cheopis*, perhaps because these rats often occupy habitats that are unsuitable for the survival and development of the immature stages of *X. cheopis*.

Xenopsylla cheopis lays 300–400 eggs during a lifetime, typically at a rate of two to six eggs per day. As with other fleas, egg development is dependent on the female acquiring a suitable blood meal. Larval *X. cheopis* usually live on dirt floors or in the nests and burrows of their hosts, where they feed on organic debris, including spilled cereal grains. Unlike some species, successful development of *X. cheopis* larvae does not appear to depend on the the shedding of blood-containing feces by adult fleas. *Xenopsylla cheopis* cocoons typically contain dirt, sand and other bits of debris. The duration of the pupal stage is influenced by temperature, with adult emergence times being markedly delayed at temperatures of less than 18°C. Emergence from the cocoon usually occurs 2–9 weeks after the onset of pupation, depending on temperature.

Although adult *X. cheopis* spend most of their time on their hosts, they can survive for relatively long periods in off-host environments (Table 7.2). In one study, unfed adult *X. cheopis* survived for 38 days after emergence, while those given blood meals lived for 100 days. Survival of *X. cheopis* is known to depend heavily on climatic conditions, and adult populations often decrease quickly during hot, dry weather.

Xenopsylla cheopis lacks combs and is relatively devoid of bristles and spines (Figs. 7.6E, F). Females of the different species of *Xenopsylla*, including *X. cheopis*, often can be recognized by the pattern of pigmentation in their spermathecae. As discussed later, the Oriental rat flea is an important disease vector and intermediate host for parasitic worms.

Tunga penetrans

Tunga penetrans, commonly called the *chigoe* or *jigger*, often attacks humans but also has been recovered from other animals, including pigs, dogs, cats, guinea pigs, and chickens. Originally found in the neotropics, it has since been introduced into Africa. Prior to feeding, both sexes of *T. penetrans* are small but relatively normal-appearing fleas (Figs. 7.6G, H). Unlike most fleas, however, the females use their serrated mouthparts to firmly attach to the host and establish a permanent feeding site. Within 7–10 days after attaching, the impregnated female's abdomen grows rapidly, through a process known as *neosomy*, until it swells to as much as 50–75 mm in diameter. As the female feeds, the host's immune response causes the skin to expand upward, forming a globular nodule that surrounds the flea and leaves only the tip of its abdomen exposed. Breathing is accomplished through large spiracles (sclerotized openings of the respiratory system), which are crowded together at the end of the abdomen. For the rest of its life, the female remains attached to its host, expelling feces and thousands of eggs from the exposed tip of its abdomen. Following oviposition, these eggs fall to the ground and hatch into typical detritivorous larvae. Development from egg to adult can require as little as 3 weeks under optimum conditions. Male feeding activities do not induce the capsular swellings characteristic of the females. Male *T. penetrans* also maintain the appearance of a typical flea as well as the ability to detach after feeding in order to search for mates.

Infestation with female *T. penetrans* fleas is termed *tungiasis*. Penetration of *T. penetrans* under toenails, between toes, or into soles of feet causes great irritation, painful swellings, and often pus-filled lesions and secondary infections involving *Clostridium tetani* or other bacteria. Persons with exceptionally severe cases of tungiasis have reportedly amputated their own toes to relieve the pain or to remove toes that were badly ulcerated or gangrenous. *Tunga penetrans* also is a significant pest of swine, attacking the feet, snout, scrotums of males, and teats of sows. Teat infestations can result in decreased milk production and starvation of piglets.

Echidnophaga gallinacea

This small, angular-headed species (Figs. 7.6I, J) is widely distributed in tropical and semitropical environments. It belongs to a group of fleas called *sticktights*, because of the females' habit of using their serrated laciniae to anchor themselves to their hosts. Males and females move actively about the host prior to mating, but females soon anchor themselves to their hosts and begin feeding while they wait for a male to locate them and initiate copulation. Once a female begins to feed, she typically remains attached for many days while acquiring the necessary blood for egg development. Eggs are deposited in the hosts' nests or in the ulcers caused by heavy infestations of these fleas. Larvae feed on droppings or other debris.

Echidnophaga gallinacea are also called *poultry fleas*, because they can become severe pests on these birds, requiring the use of insecticidal dusts or other means to rid poultry houses of infestations. These fleas are by no means restricted to poultry, however, because they also infest a wide variety of mammals, including dogs, cats, rabbits, and horses. In the United States it also is commonly encountered on certain ground squirrels, such as the rock squirrel (*Spermophilus variegatus*). Sticktights are relatively insignificant as vectors of disease, probably in part because of their tendency to anchor themselves in one place and remain on the same host for long periods.

Oropsylla montana (syn. Diamanus montanus)

This flea (Figs. 7.6K, L) is common on the rock squirrel, *Spermophilus variegatus*, and the California ground squirrel, *S. beecheyi*, in the western United States. It is also frequently encountered on golden-mantled ground squirrels in the Sierra Nevada Mountains but is absent on these animals in the Rocky Mountains. Many other hosts are attacked on an occasional basis, especially when plague epizootics cause high mortality among their normal hosts. Although *O. montana* is often found on hosts away from their burrows, it also spends considerable time in the nest. It is the primary vector of plague to humans in the United States.

Nosopsyllus fasciatus

The northern rat flea is common on commensal rats, especially *Rattus norvegicus*, in many temperate environments. It spends more time in the host's nest than does the Oriental rat flea (*X. cheopis*) and is more likely to occur on rats that have underground harborages than on ones that live in above-ground sites. *Nosopsyllus fasciatus* occasionally infests other mammals, including mice, voles, ground squirrels, carnivores, and even occasionally humans. Although considered to be a relatively poor vector of plague, it has been reported to transmit *Salmonella enteriditis* and is a vector of *Trypanosoma lewisi*, a blood protozoan of rats. These fleas can easily be distinguished from *X. cheopis* by the presence of a pronotal comb and the coiled penis rods present in the males (Figs. 7.6M, N).

Ceratophyllus gallinae

This common bird flea (Figs. 7.6O, P) is found on a wide variety of birds in Europe, including the domestic chicken. The main hosts are blue tits (*Parus caeruleus*) and great tits (*P. major*). It also has been introduced into other parts of the world and is now found in eastern North America. It does best in relatively dry nests in shrubs, trees, and nesting boxes constructed for songbirds or domestic poultry operations, where this flea can become a great pest. Reproduction occurs during the breeding season of its bird hosts. Adults spend most of their time in the nest feeding on both adults and nestlings. Larvae live within the nest material, where they feed on debris and the partially digested blood in flea feces. When development is complete, the larva spins a cocoon, pupates, and then molts to the adult stage.

Populations on *Parus* spp. can produce two generations per year when temperature and other environmental conditions are suitable. If pupation and development of the newly formed adult occurs after the nesting season, the latter will remain quiescent within the cocoon until the following spring, when the flea's avian hosts return to their nests. Reproduction in poultry houses is less restricted by seasonal factors, allowing populations of *C. gallinae* to increase quickly over a short period.

Experimental evidence indicates that the amount of nest cleaning undertaken by nesting female blue tits is proportional to the levels of *Ceratophyllus gallinae* infestation in their nests, suggesting that these fleas have a negative impact on their natural hosts. *Ceratophyllus gallinae* is known to bite humans, particularly farmers working in infested henhouses.

FLEAS AS VECTORS OF PATHOGENS

Fleas transmit microbial pathogens, serve as hosts for parasitic worms, and cause allergic dermatitis or other conditions as a result of their feeding activities. This section briefly describes the major diseases transmitted by fleas.

Plague

Plague is an often-fatal flea-borne zoonosis that is best known as the cause of the Black Death of the Middle Ages and other historically significant outbreaks. This disease also can cause high mortality among rodents and certain other mammalian species, including domestic cats, lagomorphs, and some wild carnivores, such as bobcats, lynx, and black-footed ferrets. In some regions of the United States plague threatens efforts to reestablish prairie dogs and black-footed ferrets, both of which are highly susceptible to

plague. When a plague epizootic sweeps through a region, black-footed ferrets are at a greater-than-expected disadvantage because they are likely to succumb to plague and also lose their preferred prey (prairie dogs) source to the disease. In humans, plague is an acute febrile illness often accompanied by swollen lymph glands or buboes, which are the hallmark of bubonic plague, the most common form of the disease. In the absence of appropriate antibiotic therapy, mortality among persons with bubonic plague can approach 60%, while cases of untreated septicemic and pneumonic plague, which are more severe forms of the disease, are almost invariably fatal.

The causative agent of plague is *Yersinia pestis*, a gram-negative bacterium that is maintained in nature through transmission between certain rodent species and their fleas (Fig. 7.7). Humans usually become infected with *Y. pestis* as a result of being bitten by infectious fleas, but they also can acquire plague through handling infected animals or by inhaling infectious respiratory droplets or other infectious materials.

Plague exists in many foci around the world, where it is usually maintained in cycles involving wild rodents and their fleas, although a few foci, such as those in Madagascar and Vietnam, appear to persist primarily through transmission between commensal rats and their fleas (Fig. 7.8). Usually only sporadic human cases or small clusters of cases occur in those areas where plague is maintained in wild rodent–associated cycles. Most of these cases occur after widespread epizootics have caused high mortality among susceptible rodents, placing humans at increased risk of acquiring plague through the bites of infectious wild rodent fleas or as a result of handling infected animals. In some poverty-stricken, rat-infested regions, plague can spread to commensal rats, causing epizootics among these animals (Fig. 7.7). Epizootics in rats are most likely to occur when they are heavily infested with *Xenopsylla cheopis*, an exceptionally good vector of plague and one that readily feeds on humans when its normal hosts are absent. Rat-associated plague outbreaks can spread rapidly and remain sources of infection for thousands of human cases each year. Another rat-associated species, *X. brasiliensis*, is an important vector of plague in some regions of Africa where *X. cheopis* is absent or rare.

The ability of fleas to transmit plague varies greatly among species. The most efficient vectors are those that become blocked by the massive proliferation of *Y. pestis* in the proventriculus. The development of plague bacteria in *Xenopsylla cheopis* is used as a model for describing blocking. Under natural conditions *X. cheopis* becomes infected with *Y. pestis* while feeding on

Plague Cycle

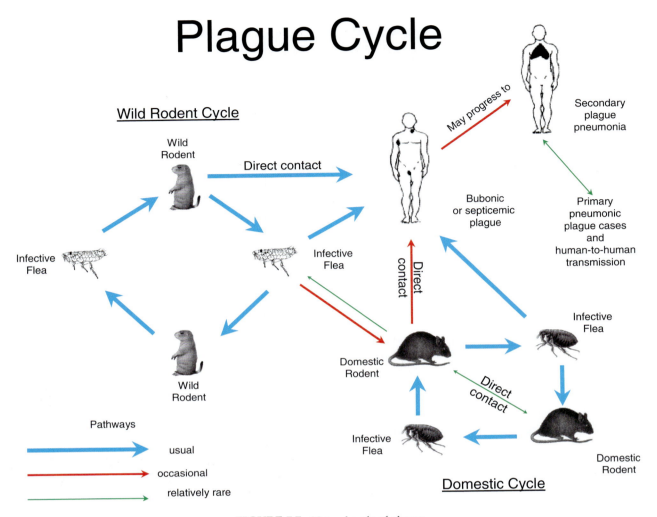

FIGURE 7.7 Natural cycle of plague.

a bacteremic rat. Once ingested, the bacteria multiply in the midgut and proventriculus, eventually causing occlusion of the proventriculus and blockage of the gut.

The block prevents the flow of host blood from the foregut into the midgut, and the flea starves despite its attempt to feed. While trying to obtain a blood meal, a blocked *X. cheopis* regurgitates repeatedly in an apparent effort to disrupt the proventricular blockage, and some of the *Y. pestis* are dislodged and injected into the wound. Blockage of *X. cheopis* by *Y. pestis* can occur in as few as 5 days.

The factors that influence blocking of fleas by *Y. pestis* have been described only recently. The characteristics and arrangement of the proventricular spines have been proposed to be important determinants of whether fleas will become blocked, but this has yet to be proved. Previously it had been suggested that blockage was promoted by a *Y. pestis* factor that had

weak coagulase activity at temperatures below about 28°C. The same authors proposed that *Y. pestis* exhibited fibrinolytic activity at higher temperatures (>28°C), resulting in breakdown of the blockage. Somewhat paradoxically, the fibrinolytic and weak coagulase activities are due to the actions of the same plasminogen activator (Pla) of *Y. pestis* that is encoded by a gene on a small plasmid.

While this study was widely cited as an explanation for the blockage, more recent experiments demonstrated that block formation can occur in the absence of Pla. In these experiments fleas were infected with a *Y. pestis* strain that had been cured of the plasmid-bearing the *pla* gene. The fact that these fleas still became blocked in the absence of *pla* expression provides evidence that Pla is not required for blockage. Breakdown of blockages at temperatures above 28°C also did not depend on Pla activity, because fleas infected with the same *Y. pestis* strain that had been

Global Distribution of Plague

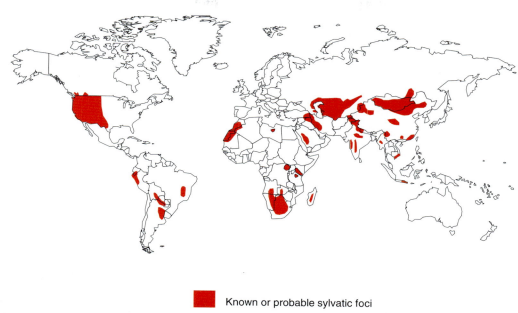

Known or probable sylvatic foci

Compiled from WHO, CDC, and country sources

FIGURE 7.8 World plague foci.

cured of the *pla*-bearing plasmid experienced loss of blockage at elevated temperatures. More recent work has demonstrated that plague bacteria have the means to induce blockages in *X. cheopis* and that blockage depends on the presence of a functional hemin storage (*hms*) locus, which lies on the chromosomal DNA of *Y. pestis*. When *hms* is expressed normally, plague bacteria aggregate, forming clumps that presumably promote the formation of a block in the proventriculus. If *hms* is not expressed normally, the proventriculus remains unblocked and the flea is unlikely to transmit plague. Another gene (*ymt*) of *Y. pestis*, which encodes for a phospolipase D, has recently been found to be important for survival of plague bacteria in the midgut of fleas.

It is tempting to consider the blocking process as observed in *X. cheopis* to be representative of what occurs when other potential flea vectors are infected with *Y. pestis*. In reality, many of the fleas that are most common on the wild rodent hosts of plague do not become blocked under laboratory conditions or have done so only after extrinsic incubation periods of many weeks. Other wild rodent fleas appear to transmit plague only when their mouthparts are contaminated with viable *Y. pestis* (mechanical transmission).

Despite experimental data indicating that many rodent fleas are relatively inefficient plague vectors, other epidemiological and ecological evidence suggest that some of these fleas play significant roles in the transmission of plague to humans or other animals, especially during severe epizootics. Clearly, when compared to *X. cheopis*, certain commensal rat fleas, including *N. fasciatus* and *X. astia*, and some common pest species, such as *C. felis*, *C. canis*, *E. gallinacaea*, and *P. irritans*, appear to be relatively poor vectors of plague. Cat and dog fleas probably play little or no role in plague transmission, but some of the other species listed earlier may have limited roles. The role of *P. irritans* as a plague vector is particularly intriguing because epidemiological studies have suggested that it is an important vector in poverty-stricken villages in developing countries. Others have suggested it was a significant vector during the Black Death, especially in those areas where *X. cheopis* probably did not exist. Most evidence suggests that *P. irritans* rarely, if ever, becomes blocked with *Y. pestis*, and transmission by this flea probably occurs only through mechanical means. Because mechanical transmission is relatively inefficient, this flea is likely to be an important vector of plague only when it is abundant and host densities are high, thus facilitating the rapid transfer of *P. irritans* from host to host and transmission of *Y. pestis* before they die on the exposed mouthparts of the flea. Such conditions indeed seem to exist in some villages in developing countries and might well have occurred in many medieval villages affected by the Black Death.

Investigations of *Oropsylla montana* provide excellent examples of the difficulties encountered in trying to assess what role a particular flea plays in the transmission of plague. This flea species is considered to be the primary vector of plague to humans in the United States as well as an important vector of plague between ground squirrels and certain other rodents during epizootics. Its importance as a vector is probably explained in part by its willingness to feed on humans and other animals in the absence of its normal hosts, the fact that it is common and lives in close association with humans, and perhaps by its ability to transmit *Y. pestis* while being only partially blocked or through mechanical means. Most laboratory studies indicate that *O. montana* is a relatively inefficient vector of plague, typically requiring many weeks to become blocked and transmitting plague at a much lower rate than *X. cheopis*. In one study the development of *Y. pestis* in *X. cheopis* and *O. montana* fleas was compared using a strain of *Y. pestis* that had been transformed with the gene for green fluorescent protein. When fleas were examined at various intervals after ingesting an infectious blood meal, it was found that colonies of the transformed *Y. pestis* strain became established within just a few days in both the proventriculus and midgut of *X. cheopis*, but only in the midgut of *O. montana*. The lack of early proventricular infections in *O. montana* caused these fleas to become blocked less often and much more slowly than *X. cheopis*. Transmission of *Y. pestis* also occurred at much lower rates for *O. montana* than for *X. cheopis*. At least one *O. montana* transmitted plague without being blocked although it is uncertain whether this occurred as a result of transmission by a partially blocked flea or through mechanical transmission.

Flea-Borne Bacteria and Viruses

Murine typhus is a flea-borne rickettsial disease caused by infection with *Rickettsia typhi* (syn. *R. mooseri*) (Fig. 7.9). Typical symptoms include fever, macular rash, headache, chills, prostration, and generalized achiness. Case fatality rates are generally low (1–5%; highest among elderly persons), and the disease can be treated with appropriate antibiotics. Infection occurs when infectious flea feces are rubbed into the flea bite wound or other breaks in the skin. *Xenopsylla cheopis* is the primary vector of murine

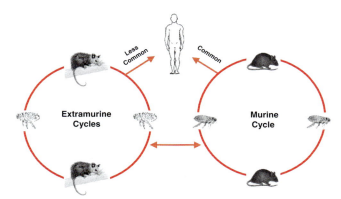

FIGURE 7.9 Natural cycle of murine typhus.

typhus and transmits *R. typhi* from rat to rat (murine cycle) and to humans. Extramurine cycles also have been described, including one involving opossums and cat fleas (*Ctenocephalides felis*) (Fig. 7.9). Other potential vectors include *Echidnophaga gallinaceae*, *Leptopsylla segnis*, *Nosopsyllus fasciatus*, and *Pulex irritans*. Additional mammalian hosts include other rodents as well as skunks, shrews, and cats.

Within the past decade, another flea-borne rickettsia, *R. felis*, has been identified. Originally identified in a colony of cat fleas, *R. felis* also has been found in wild-caught cat fleas and opossums. This rickettsia typically causes heavy infections in the ovaries of infected fleas and can be maintained in these insects through transovarial transmission. It has been suggested to cause a mild rickettsial illness in humans.

Bartonella

The gram-negative *Bartonella* spp. parasitize red blood cells. They are the etiologic agents of Carrión's disease (*B. bacilliformis*), trench fever (*B. quintana*), and cat-scratch disease (*B. henselae*). The first of these diseases is transmitted by sand flies, while *B. quintana* can be transmitted by lice. Cat-scratch disease is normally transmitted through the scratches or bites of *B. henselae*-infected cats, but recent evidence indicates that cat fleas transmit this agent from cat to cat, apparently as a result of the contamination of flea feeding sites or other skin lesions with flea feces. Human cases arising from flea bite have yet to be described. Cat-scratch disease occurs most often among children and young adults, often causing a lesion at the site of initial infection and lymph node swelling that gradually resolves without treatment. Limited laboratory evidence indicates that some rodent-associated bartonellae can be transmitted by fleas, but the importance of fleas as vectors of these agents needs to be further investigated.

Tularemia

Tularemia is a bacterial zoonosis caused by infection with *Francisella tularensis*. Humans and animals acquire infection through exposure to infectious vectors, contact with infected animals, or ingestion of contaminated food or water. Symptoms in humans include fever, swollen and painful lymph glands, and skin ulcers, the last of which are often located at the site of an arthropod bite. Fatality rates are generally lower than 5%. The most important vectors are ticks and deerflies, although fleas might play a limited role as mechanical vectors. One study reported positive results in only 6 of 116 experiments performed with fleas. In another investigation transmission occurred only within the first 5 days after fleas had fed on infected hosts, which is suggestive of a mechanical rather than biological mechanism of transmission.

Myxomatosis

Myxoma virus was originally found in South America, where it causes a mild disease in local rabbit populations, *Sylvilagus braziliensis*. Domestic European rabbits, *Oryctolagus cuniculi*, in that continent were eventually exposed to the virus, causing infections that were almost always fatal. The most characteristic signs of the illness are skin lesions and edematous swellings of the eyelids, lips, nose, ears, and genitalia. Because myxoma virus caused such a high fatality rate in *O. cuniculi*, it was imported into Australia in the 1950s as a means for controlling these animals, which had proliferated out of control after being introduced from Europe as a game animal. Although initially quite effective in controlling Australia's rabbit populations, they evolved a level of resistance, and the virus became less pathogenic.

The virus was accidentally introduced into Europe, where it also devastated wild *O. cuniculi* populations. A number of arthropods have been implicated as vectors, but the most important are mosquitoes and fleas. Both of these insects appear to transmit the virus mechanically through contaminated mouthparts. Flea-borne transmission appears to be relatively unimportant in Australia but is significant in Europe, especially in Great Britain.

Fleas As Intermediate Hosts for Tapeworms (Subclass Cestoda)

Fleas serve as intermediate hosts for cyclophyllidean tapeworms, including *Hymenolepis nana*, *H. diminuta*, and *Dipylidium caninum*. Among the potential flea hosts of the *Hymenolepis* spp. are *C. felis*, *C. canis*, *P. irritans*, and *X. cheopis*. Cat fleas and dog fleas are important hosts for *D. caninum*. The definitive hosts are normally rats (*H. nana*, *H. diminuta*), mice (*H. nana*), or dogs, cats and other carnivores (*D. caninum*), but humans also can become infested through ingestion of fleas (or crushed flea material) that contain the cysticercoids. The adults of these species mature in the vertebrate gut and shed eggs (*H. nana* and *H. diminuta*) or egg-bearing proglottids (*D. caninum*). Fleas and certain other insects become infested through ingesting tapeworm eggs containing the oncosphere. An arthropod intermediate host is obligatory for *H. diminuta* and *D. caninum*, but *H. nana* is unusual in that cysticercoids can develop either in an arthropod or directly in the vertebrate host's gut. In the latter situation, the *H. nana* eggs are ingested, invade the intestinal villi, complete development of the cysticercoid stage, and then shed the mature cysticercoid into the gut lumen, where it attaches to the gut and matures to an adult worm. In most instances persons can be infected with these worms and suffer no obvious adverse effects, although occasionally gastrointestinal discomfort, diarrhea, and unrest occur. Preventive measures include reducing rodent access to houses and food supplies, avoiding insect-contaminated foods, and, for *H. nana*, personal hygiene measures, including hand washing and avoidance of materials contaminated by feces.

CONTROL

Flea control is typically undertaken for one of two reasons, (1) to reduce the risks of disease transmission, or (2) to address a pest problem or economic losses associated with parasitization of domestic animals by fleas. The strategies used for each situation are often different, and the best results are achieved when the biology and behavior of the host are taken into account.

Flea control is an effective means of reducing the risks of flea-borne plague and murine (flea-borne) typhus. Because both diseases can be transmitted by rat fleas, the same control techniques can be used to control both plague and murine typhus. Typically, this involves using insecticidal dusts to treat rat runways and burrows. In emergencies, liquid spray formulations of insecticides can be applied to runways and burrow entrances, but these should be used only when dust formulations are unavailable. In some situations insecticide can be applied to hosts through the use of bait stations that contain food or some other attractant, along with an appropriate insecticide that is placed on the floor of the station or applied to the host's body by forcing it to brush against an applicator as it enters the station. Limited attempts have been made to use insecticide-treated cotton or other material that rodents take back to their nests. The advantage of this last method is that fleas are controlled not only on the animal but also in the nest.

Controlling fleas on pets and domestic animals is of great concern, as suggested by the fact that more than $1 billion U.S. is spent annually in attempts to control cat flea infestations and the problems they cause. Flea control on pets and other domestic animals can take many forms, including the use of insecticidal dust formulations, granules, sprays, flea collars, topically applied liquids, shampoos, and oral systemics. Insect growth regulators (IGRs) that act as chitin synthesis inhibitors or mimic insect juvenile hormones are also popular (Chapter 41), especially for controlling fleas on pets. Control measures do not always have to involve the use of insecticides. Vacuuming, when properly done with a suitable vacuum, has been shown to remove about 90% of cat flea eggs and 50% of larvae from carpets. Cleaning or removal of bedding or nests and other environmental modifications also can have favorable effects.

Readings

Azad, A. F. 1990. Epidemiology of murine typhus. *Annu. Rev. Entomol.* 35: 553–569.

Azad, A. F., and Beard, C. B. 1998. Rickettsial pathogens and their arthropod vectors. *Emer. Infect. Dis.* 4: 179–186.

Bacot, A. W. 1914. A study of the bionomics of the common rat fleas and other species associated with human habitations, with special reference to the influence of temperature and humidity at various periods of the life history of the insect. *J. Hyg.* Plague Supplement III. 13: 447–654.

Burroughs, A. L. 1953. Sylvatic plague studies. X. Survival of rodent fleas in the laboratory. *Parasitology.* 43: 36–47.

Chomel, B. B., Kasten, R. W., Floyd-Hawkins, K., et al. 1996. Experimental transmission of *Bartonella henselae* by the cat flea. *J. Clin. Microbiol.* 34: 1952–1956.

Engelthaler, D. M., Hinnebusch, B. J., Rittner, C. M., and Gage, K. L. 2000. Quantitative Competitive PCR as a method for exploring flea-*Yersinia pestis* dynamics. *Am. J. Trop. Med. Hyg.* 62: 552–560.

Foil, L., Andress, E., Freeland, R. L., Roy, A. F., Rutledge, R., Triche, P. C., and O'Reilly, K. L. 1998. Experimental infection of domestic cats with *Bartonella henselae* by innoculation of *Ctenocephalides felis* (Siphonaptera: Pulicidae) feces. *J. Med. Entomol.* 35: 625–628.

Gage, K. L. 1998. Plague. In *Topley and Wilson's Microbiology and Microbial Infections*, 9th ed. (L. Collier, A. Balows, and M. Sussman, general eds.). Vol. 3, *Bacterial Infections*. (W. J. Hausler and M. Sussman, vol. eds.). Edward Arnold, London, pp. 885–904.

Gratz, N. G. 1999. Control of Plague Transmission. In *Plague Manual — Epidemiology, Distribution, Surveillance and Control* (D. T. Dennis, K. L. Gage, N. Gratz, J. D. Poland, and E. Tikhomirov, principal authors). World Health Organization, Geneva, pp. 97–134.

Hinnebusch, B. J., Perry, R. D., and Schwan, T. G. 1996. Role of the *Yersinia pestis* hemin storage (*hms*) locus in the transmission of plague by fleas. *Science* 273: 367–370.

Hinnebusch, B. J., Rudolph, A. E., Cherepenov, P., Dixon, J. E., Schwan, T. G., and Forsberg, A. 2002. Role of *Yersinia* murine toxin in survival of *Yersinia pestis* in the midgut of the flea vector. *Science* 296: 733–735.

Holland, G. P. 1984. *The Fleas of Canada, Alaska and Greenland (Siphonaptera).* Mem. Entomol. Soc. Can. No. 130.

Hopla, C. E., and Loye, J. E. 1983. The ectoparasites and microorganisms associated with cliff swallows in west-central Oklahoma. I. Ticks and fleas. *Bull. Soc. Vector Ecol.* 8: 111–121.

Lewis, R. E. 1972. Notes on the geographical distribution and host preferences in the order Siphonaptera. Part 1. Pulicidae. *J. Med. Entomol.* 6: 511–520.

Lewis, R. E. 1973. Notes on the geographical distribution and host preferences in the order Siphonaptera. Part 2. Rhopalopsyllidae, Malacopsyllidae, and Vermipsyllidae. *J. Med. Entomol.* 10: 255–260.

Lewis, R. E. 1974a. Notes on the geographical distribution and host preferences in the order Siphonaptera. Part 3. Hystrichopsyllidae. *J. Med. Entomol.* 11: 147–167.

Lewis, R. E. 1974b. Notes on the geographical distribution and host preferences in the order Siphonaptera. Part 4. Coptopsyllidae, Pygiopsyllidae, Stephanocircidae, and Xiphiopsyllidae. *J. Med. Entomol.* 11: 403–413.

Lewis, R. E. 1974c. Notes on the geographical distribution and host preferences in the order Siphonaptera. Part 5. Ancistropsyllidae, Chimaeropsyllidae, Ischnopsyllidae, Leptopsyllidae, and Macropsyllidae. *J. Med. Entomol.* 11: 525–540.

Lewis, R. E. 1975. Notes on the geographical distribution and host preferences in the order Siphonaptera. Part 6. Ceratophyllidae. *J. Med. Entomol.* 11: 658–676.

Lewis, R. E. 1985. Notes on the geographical distribution and host preferences in the order Siphonaptera. Part 7. New taxa described between 1972 and 1983, with a superspecific classification of the order. *J. Med. Entomol.* 22: 134–152.

Lewis, R. E. 1998. Resume of the Siphonaptera (Insecta) of the world. *J. Med. Entomol.* 35: 377–389.

Lewis, R. E., Lewis, J. H., and Maser, C. 1988. *The Fleas of the Pacific Northwest.* Oregon State University Press. Corvallis, OR.

Loye, J. E., and Hopla, C. E. 1983. Ectoparasites and microorganisms associated with the cliff swallow in west-central Oklahoma. II. Life history patterns. *Bull. Soc. Vector Ecol.* 8: 79–84.

Pratt, H. D., Maupin, G. O., and Gage, K. L. 1994. Fleas of public health importance and their control. In *Vector-Borne Disease Control* (2nd ed.). U.S. Dept. HHS/PHS/CDC, Atlanta.

Rothschild, M. 1975. Recent advances in our knowledge of the order Siphonaptera. *Annu. Rev. Entomol.* 20: 241–257.

Rothschild, M., Schlein, Y., and Ito, S. 1986. *A Colour Atlas of Insect Tissues via the Flea.* Wolfe, London.

Rust, M. K., and Dryden, M. W. 1997. The biology, ecology, and management of the cat flea. *Annu. Rev. Entomol.* 42: 451–473.

Traub, R. 1985. Coevolution of fleas and mammals. In *Coevolution of Parasitic Arthropods and Mammals* (K. C. Kim, ed.). Wiley Interscience, New York, pp. 295–437.

Tripet, F., and Richner, H. 1999. Dynamics of hen flea *Ceratophyllus gallinae* subpopulations in blue tit nests. *J. Insect Behavior* 12: 159–174.

Whiting, M. F., Carpenter, J. C., Wheeler, Q. D., and Wheeler, W. C. 1997. The Strepsiptera problem: Phylogeny of the holometabolous insect orders inferred from 18S and 28S ribosomal DNA sequences and morphology. *Syst. Biol.* 46: 1–68.

8

Introduction to the Diptera

BORIS C. KONDRATIEFF

The Diptera are an order that includes more than 250,000 species in approximately 120 families. Several thousand species are known to be of medical and veterinary importance, the majority of them belonging to the suborder Nematocera. Treated in the following chapter are the families Culicidae (mosquitoes), Ceratopogonidae (biting midges), Psychodidae (Phebotominae, sand flies), and Simuliidae (the blackflies). The larvae of these families are aquatic to semiaquatic. They form a monophyletic group, the infraorder Culicomorpha (Hennig 1973; Wood and Borkent 1989; Miller et al. 1997). The biting and sucking type of proboscis in the adults are considered a primitive feature; apparently it has evolved and often been modified in several lineages. The blood-feeding females of the Culicidae, Ceratopogonidae, Psychodidae (Phebotominae), and Simuliidae generally have similar mouthparts, but they differ in depth of penetration and probing.

The remaining family, the Glossinidae (tsetse flies), is included in the infraorder Muscomorpha of the suborder Brachycera. Most authors have previously treated these flies as members of the suborder Cyclorrhapha. McAlpine (1989) provides a thorough discussion of the preferred usage of the subordinal category Brachycera. The reader is referred to McAlpine (1989) for the delimitation and evaluation of the main groups of cyclorrhaphous Brachycera.

Service (2001) treats the diseases and other infections these five families of flies are known to transmit. The authors of the following chapters review the taxonomy, the biology, the role of the flies in transmission, and, where appropriate, control.

Readings

Hennig, W. 1973. Ordnung Diptera (Zweiflugler). *Handb. Zool.* 4(2) 2/31 (Lfg. 20): 1–337.

McAlpine, J. F. 1989. Phylogeny and classification of the Muscomorpha. In *Manual of Nearctic Diptera*, Vol. 3. (J. F. McAlpine and D. M. Wood, eds.). *Res. Br. Agri. Canada. Monogr.* 32: 1397–1518.

Miller, B. R., Crabtree, M. B., and Savage, H. M. 1997. Phylogenetic relationships of the Culicomorpha inferred from 18S and 5.8S ribosomal DNA sequence (Diptera: Nematocera). *Insect Mol. Biol.* 6: 105–114.

Service, M. W. (ed.) 2001. *Encyclopedia of Arthropod-Transmitted Infections of Man and Domesticated Animals.* CABI, New York.

Wood, D. M., and Borkent, A. 1989. Phylogeny and classification of the Nematocera. In *Manual of Nearctic Diptera*, Vol. 3 (J. F. McAlpine and D. M. Wood, eds.). *Res. Board Agri. Canada. Monogr.* 32: 1333–1370.

CHAPTER

9

Mosquitoes, the Culicidae

BRUCE F. ELDRIDGE

INTRODUCTION

Mosquitoes are the most important group of arthropods of medical and veterinary importance. There are over 3,500 species and subspecies of mosquitoes in the world. Mosquitoes are classified in the family Culicidae of the order Diptera, the two-winged, or true, flies. They fall within the suborder Nematocera, one of two suborders of Diptera, and thus are related to biting midges (Ceratopogonidae, Chapter 10), (Simuliidae, Chapter 11), and sand flies (Psychodidae, Chapter 12).

The majority of mosquito species fall into three groups, commonly referred to as the anophelines, the culicines, and the aedines. These names derive from the names of the major taxonomic groups within Culicidae, discussed later. Mosquitoes are found everywhere in the world where standing water occurs, which is needed for the development of their aquatic immature stages. They occur on every continent except Antarctica and from below sea level to above tree line at elevations of 3,000 meters or more.

Adult mosquitoes share the characteristic of most Diptera in having a single pair of wings, and they are relatively strong fliers. Further, as is the case in several other dipteran groups, adult female mosquitoes have piercing-sucking mouthparts adapted for sucking the blood of vertebrate animals. It is this bloodsucking habit that accounts for their importance as disease vectors. Among the disease pathogens transmitted by mosquitoes are those causing the diseases malaria, yellow fever, and dengue.

All immature stages of mosquitoes are aquatic. Some mosquito species have larvae that are found only in a narrow range of habitats, while others have larvae that develop in a wide range of situations. Some of the habitats that have been exploited by mosquitoes are stream edges, ponds, ditches, rainwater puddles, marshes, swamps, tree holes, rock pools, snow pools, and various kinds of water-filled artificial containers, such as tires and tin cans.

Mosquitoes probably first appeared, with other Diptera, no later than the Triassic period, about 200–245 million years ago. However, the first fossils recognizable as mosquitoes are associated with the Eocene epoch of the Quaternary Period (37–58 million years ago), and most of these specimens are very similar in appearance to modern *Culex* species. Aedine mosquito fossils appear first in sediments from the middle Oligocene Epoch, about 30 million years ago (Edwards 1932).

Bloodsucking in mosquitoes probably had already developed by the time birds and mammals arose late in the Mesozoic Era (about 100 million years ago). This notion is supported by the fact that certain species of mosquitoes feed on cold-blooded animals (e.g., reptiles) and lack certain sensory receptors found in species that feed on birds and mammals.

In view of the apparently long association of humans and mosquitoes, it is remarkable that our understanding of the role mosquitoes play in the transmission of human pathogens was developed only in the latter part of the 19th century. The first pathogen discovered to be transmitted by mosquitoes was the filarial nematode causing human lymphatic filariasis. This discovery, made by Sir Patrick Manson in 1876, earned him, in many people's eyes, the title "father of medical entomology" (Eldridge 1993). This discovery was followed by discoveries of other mosquito-borne disease relationships: malaria in 1898, yellow fever in 1900, dengue in 1902, and western equine encephalomyelitis in 1933 (Philip and Rozeboom 1973).

The literature on mosquitoes is voluminous, and several excellent books are devoted to these insects. A book written nearly 50 years ago by Bates (1954) is still timely and informative. A more recent general book on mosquitoes is that of Gillett (1971). A wealth of information on the ecology of mosquitoes can be found in Service (1993). Books by Clements (1963, 1992) cover the biology and physiology of mosquitoes. A book by Spielman and D'Antonio (2001) is an entertaining and insightful account of the importance of mosquitoes in today's world.

GENERAL IDENTIFICATION

Immatures

Mosquitoes, as is the case with all Diptera, have complete metamorphosis and an egg stage, four larval stages, a single pupal stage, and the adult. Mosquito eggs are of three basic types. Culicine mosquitoes deposit batches of 100 or more eggs on the surface of the water. The eggs in each batch are held together by surface tension to form a concave "raft" (Figs. 9.1a, b). Anophelines deposit individual eggs that float on the surface of the water until they hatch. Each egg has a pair of floats along the sides (Figs. 9.1c, d). The shape and structure of these floats may be used to identify some species. Aedine mosquitoes lay individual eggs, but often on moist substrates (e.g., mud, moist tree tissue) subject to later flooding (Fig. 9.1e).

All four of the larval stages of culicine and aedine mosquitoes are easy to recognize because of the presence of an elongate air tube used for breathing (Fig. 9.2a). However, anopheline larvae lack this air tube and lie parallel to, and just beneath, the surface of the water to maintain contact with the air–water interface for breathing (Fig. 9.2b). Anophelines are able to maintain this orientation with the aid of pairs of palmate hairs that line the body and act as floats (Fig. 9.2c). Anophelines feed at the water surface by swiveling their head 180° so that the mouthparts face up toward the water surface.

One group of culicine mosquitoes (in the genera *Mansonia* and *Coquillettidia*) have short, sharpened air tubes adapted for piercing roots of aquatic plants for the purpose of obtaining oxygen (Fig. 9.3).

First-stage mosquito larvae are barely visible to the naked eye. Second-stage larvae are larger, but their presence is still not obvious without careful inspection. Third- and fourth-stage larvae are readily observable. Fourth-stage larvae of some species are a centimeter or more in length.

Mosquito larvae have characteristic patterns of hairs or setae used for identification, with the number

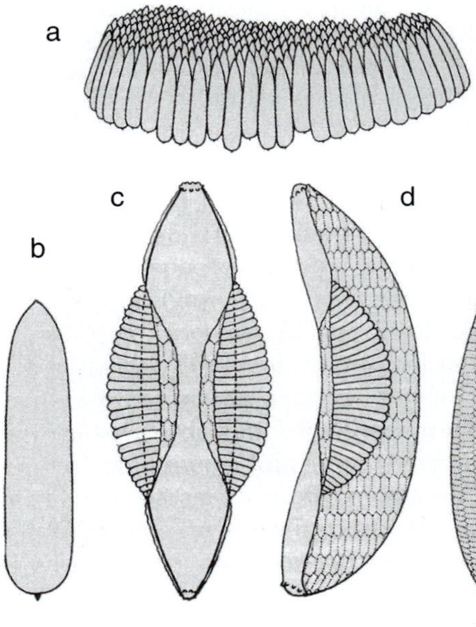

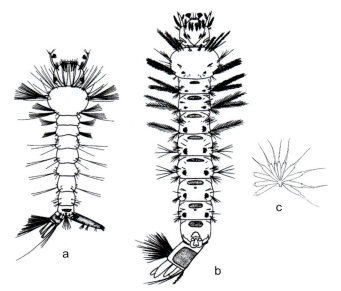

FIGURE 9.1 The eggs of mosquitoes: (a) Egg raft of *Culex*; (b) single egg of *Culex*; (c) egg of *Anopheles* (dorsal view); (d) egg of *Anopheles* (lateral view); (e) egg of *Aedes aegypti*. From Carpenter and LaCasse (1955).

FIGURE 9.2 Larvae of mosquitoes: (a) Culicine; (b) anopheline; (c) palmate hair of anopheline. (a, b) From Harry Pratt, Centers for Disease Control and Prevention; (c) from Darsie and Ward (1981).

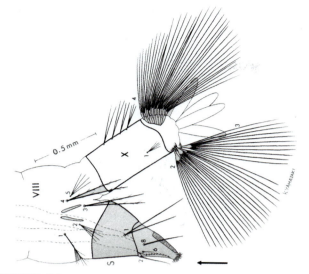

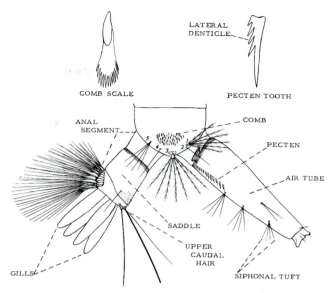

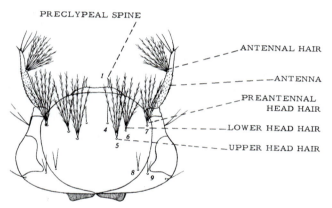

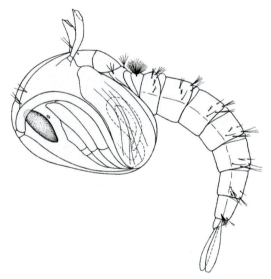

FIGURE 9.3 Modified air tube of *Mansonia uniformis* used to pierce aquatic plants (arrow). From LaCasse and Yamaguti (1950).

FIGURE 9.5 Terminal segments of fourth-stage showing structures commonly used for identification. From Stojanovich and Scott (1965).

FIGURE 9.4 Head of fourth-stage culicine larva showing structures commonly used for identification. From Stojanovich and Scott (1965).

FIGURE 9.6 Mosquito pupa. From Carpenter and LaCasse (1955).

and complexity of the setal patterns increasing from the first to the last stages. Because of this, first-stage larvae are very difficult to identify to species; fourth-stage larvae are the easiest stage to identify. Usually, larvae are better for identification than any other stage. Hair patterns on the head of culicine and aedine larvae are used commonly for identification (Fig. 9.4). Also, structures on the siphon are useful for identification, including hairs, pecten teeth, and comb scales (Fig. 9.5).

The appearance of all mosquito pupae is remarkably similar (Fig. 9.6). Mosquito pupae can be recognized among all other aquatic insects by their general appearance (C-shaped) and by a characteristic pair

of structures called *respiratory trumpets* or just *horns*. Although mosquito taxonomists have studied the pupae of many species in great detail, the pupal stage remains the least valuable stage for mosquito identification.

Adults

Mosquito adults are small flying midgelike insects with long, slender wings (Fig. 9.7). Most female mos-

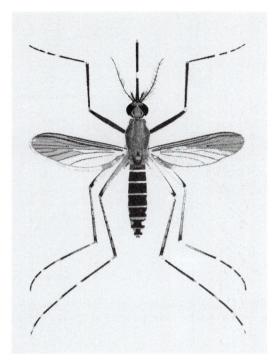

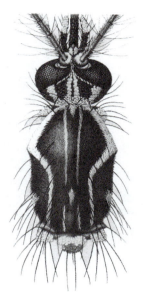

FIGURE 9.9 *Aedes aegypti* thorax. From Tanaka et al. (1979).

FIGURE 9.7 Adult female *Aedes taeniorhynchus.* From Carpenter and LaCasse (1955).

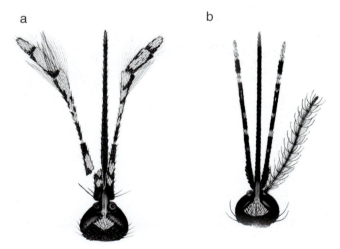

FIGURE 9.8 Head of adult *Anopheles stephensi*: (a) Male; (b) female. Courtesy of Southeast Asia Mosquito Project, U.S. Army, Washington, DC.

quitoes have a long, slender proboscis (Fig. 9.8a) that is adapted for piercing skin and sucking blood. Male mosquitoes also have a proboscis, but it is adapted for sucking plant juices and other sources of sugars (Fig. 9.8b). Most male mosquitoes are generally smaller than females of the same species and have much longer and hairier maxillary palps.

The bodies of mosquito adults are covered with small scales, often with patterns of contrasting colors (e.g., black and white, brown and white). These patterns are used frequently to identify mosquitoes to species. The yellow fever mosquito, *Aedes aegypti*, has a characteristic patch of white scales on the dorsal surface of the thorax in the shape of a lyre against a background of black scales. This feature (Fig. 9.9) permits easy recognition of this important species in North American specimens, but a number of Old World *Aedes* species have very similar markings.

CLASSIFICATION

The complete classification of the Culicidae is contained in a world catalog (Knight and Stone 1977), plus three supplements (Knight 1978; Ward 1984, 1992). A Web-accessible electronic version of the catalog is maintained by the Walter Reed Biosystematics Unit in Washington, DC (URL: http://wrbu.si.edu).

There are three subfamilies recognized within the family Culicidae (Table 9.1). The species of importance from the standpoint of public health are contained in the subfamilies Anophelinae (i.e., the anophelines) and Culicinae. The latter subfamily contains the tribes Culicini (culicines) and Aedini (aedines) plus several other tribes. Females of species in a third subfamily, Toxorhynchitinae, lack mouthparts adapted for

TABLE 9.1 Classification of the Family Culicidae

Subfamily	Tribe	Representative genera
Anophelinae	Anophelini	*Anopheles* *Bironella* *Chagasia*
Culicinae	Aedini	*Aedes* *Haemagogus* *Ochlerotatus* *Psorophora*
	Culicini	*Culex* *Deinocerites*
	Culisetini	*Culiseta*
	Mansoniini	*Coquillettidia* *Mansonia*
	Orthopodomyiini	*Orthopodomyia*
	Sabethini	*Sabethes* *Trichoprosopon* *Wyeomyia*
	Uranotaeniini	*Uranotaenia*
Toxorhynchitinae		*Toxorhynchites*

After Knight and Stone (1977).

FIGURE 9.10 *Aedes aegypti*. Photograph by Leonard E. Munstermann.

sucking blood from vertebrates. The larvae of this subfamily are predaceous on other aquatic organisms and have been proposed as biological control agents of mosquito larvae. The tribe Sabethini of the subgenus Culicinae contains many genera and species of mosquitoes that occur in the Old and New World tropics. It is relatively difficult to identify larvae and adults of sabethines; in some cases, immature stages have never been found.

Important Species

Several mosquito species stand out because of their medical importance. *Aedes aegypti* is one of the most medically important species in the world (Fig. 9.10). Its common name is yellow fever mosquito. But in addition to its role as the principal urban vector of the yellow fever virus, it is the primary vector of the dengue viruses. *Culex quinquefasciatus* is a vector of the nematode worms causing lymphatic filariasis and of several arboviruses. There are many important vectors of malarial parasites in the world. Examples of malaria vectors include *Anopheles gambiae* and *Anopheles funestus* in Africa, *Anopheles albimanus* and *Anopheles darlingi* in the New World tropics, and *Anopheles stephensi* and *Anopheles culicifacies* in Asia. Most medical entomologists consider about 40 species of *Anopheles* as important malaria vectors in some part of the world.

Many species of *Aedes* and *Culex* are vectors of arboviruses that infect various vertebrates, including humans. *Culex pipiens* is the vector of Tahyna virus in Europe, St. Louis encephalitis virus (SLE) in North America, and West Nile virus on several continents. *Culex tritaeniorhynchus* is the primary vector of Japanese encephalitis virus in Asia, and *Culex tarsalis* is the primary vector of western equine encephalomyelitis and SLE in western North America. *Aedes (Neomelaniconion) macintoshi* and related species are important enzootic vectors of Rift Valley fever virus in Africa.

There are some species of mosquitoes that, although they may be vectors of pathogens in certain situations, owe their importance to their roles as severe pests of humans and livestock. *Aedes vexans* is a notorious human biter worldwide.

Cryptic Species

In the 1930s, it was discovered that various populations of the species *Anopheles maculipennis* varied in their importance as malaria vectors in Europe. This led to the expression "anophelism without malaria." Further investigation led to the discovery that these populations differed somewhat in the shape of their eggs. It was eventually recognized that what had previously been considered a single species, *Anopheles maculipennis*, was actually a complex of species nearly indistinguishable from one another. Such species now are known as *cryptic* or *sibling* species. With the advent of modern biochemical tools permitting the analysis of insect populations in detail came the realization that many species complexes exist in nature (Chapter 32), including several groups of medically important mosquitoes. Examples of mosquito species complexes include *Anopheles maculipennis*, already mentioned, *Anopheles gambiae*, *Anopheles quadrimaculatus*, *Anopheles nuneztovari*, and *Anopheles albimanus*. Cryptic species satisfy the biological definition of a species in that various barriers exist to prevent breeding with closely related cryptic species. However, such species are difficult or even impossible to differentiate on the basis of conventional morphological characters.

LIFE CYCLE

Life Cycle Stages

Egg Stage

The egg-laying habits of female mosquitoes vary widely from species to species. Some female mosquitoes lay eggs on water surfaces; others lay single eggs on moist soil where later flooding is likely. Eggs deposited on water surfaces usually hatch within a day or so, but eggs laid on soil surfaces do not hatch until flooded, which may occur months or even years later. Oviposition site selection by female mosquitoes is the primary determining factor in the habitat distribution of mosquito larvae, but the various clues used for oviposition remain only partially known (Bentley and Day 1989). The various combinations of chemical and physical factors involved have been reviewed by Maire (1983). Color, moisture, and volatile chemical stimulants appear to play a role in most species. An oviposition hormone has been isolated from droplets that appear at the apices of eggs of *Culex quinquefasciatus*. This pheromone is attractive to gravid females of this species (Pickett and Woodcock 1996). However, in general, attempts to explain the occurrence of various mosquito species in different aquatic habitats based strictly on oviposition cues have been unsuccessful.

Embryogenesis occurs after oviposition, as does hardening and darkening of the chorion (egg shell).

Larval Stages

Small larvae that are nearly invisible to the naked eye hatch from eggs. Larvae molt three times to become a fourth-stage larva. Several days later, this larval form molts again to become a pupa. The time required for development of the larval stages depends on several factors, the most important of which is water temperature. Availability of food and larval density are also factors. Water temperature and food are inversely related to time of development; larval density is directly related.

Adult Stage

Adult mosquitoes emerge from pupae 1–2 days after the appearance of pupae, with males emerging first. In warm summer temperatures, the entire development cycle, from egg to adult, may be completed in 10 days or less. Adult female mosquitoes can fly long distances in search of blood meals or oviposition sites, and studies commonly have demonstrated flights of several kilometers. Mating probably occurs close to larval sources, and male mosquitoes probably do not fly as far as do females. After completion of a blood meal by the female, mating usually occurs, and nutrients from the blood meal are used to develop a batch of eggs (some females can develop eggs without a blood meal, as described later). If a suitable oviposition site is found, the fully formed eggs are fertilized as they are deposited, embryogenesis begins, and the life cycle continues.

Phenology

At any one geographic location, species of mosquito are *univoltine* (a single generation per year) or *multivoltine* (many generations per year). A single species may be univoltine in one part of its range and multivoltine in another. Within a given species this is dependent upon the length of the season favoring the activity of the adult stages. To avoid seasons of the year not favorable to adult activity (usually the winter), mosquitoes may undergo some kind of diapause. Diapause is a physiological adaptation of insects whereby certain reproductive and behavior processes are altered to allow survival of the insect under adverse conditions, such as freezing temperatures. In aedine mosquito populations living in temperate or subarctic areas, winter usually is spent in the egg stage. These eggs are desiccation resistant and freeze resistant. To protect against premature hatching of the eggs that could occur as a result of a late-winter flooding, the eggs are in diapause and will not hatch even if flooded. Often, diapause will not be broken until after exposure to several months of low temperatures. In the older literature, this necessary exposure to cold temperatures was called *conditioning*. Diapause in aedines is triggered in one of two ways: Females deposit diapausing eggs as a result of exposure of the female to short day lengths, or the eggs enter diapause as a result of exposure of the eggs themselves to short day lengths. A brief example of aedine diapause is *Ochlerotatus communis*, a univoltine snow-pool species. In late spring or summer, gravid females deposit eggs on moist borders of snow pools that are shrinking from evaporation.

These eggs remain in place through the fall and, after snowfall, through the winter. The following spring melting snow floods the eggs. After development of the immature stages, adults emerge and mate and the females take a vertebrate blood meal. After ovarian development is complete, eggs are deposited on the margins of the snow pool, and the annual cycle is complete. Floodwater species such as *Aedes vexans* have a similar life cycle, except that the larvae develop in seasonally flooded sites at lower elevations, such as river floodplains and irrigated pastures.

Members of the genera *Anopheles, Culex,* and *Culiseta* usually survive unfavorable conditions as diapausing or quiescent adult females. Male mosquitoes do not survive unfavorable periods, so insemination must occur before the onset of diapause. The life cycle of *Culex pipiens* exemplifies this type of diapause mechanism. This species is multivoltine and continues to produce new generations of mosquitoes until immature stages (usually late-stage larvae and pupae) develop under day lengths that fall below some critical value. After this, adult females from these larvae are altered physiologically in several ways. They cease blood feeding and build up extensive reserves of fat. Development of eggs within the ovaries is markedly slowed; if a prehibernating female takes a rare blood meal, resources from the blood are used to produce fat rather than eggs. When temperatures fall, the females seek out protected sites (called *hibernacula*), such as animal burrows, caves, and underground concrete structures. Many females do not survive the winter in these hibernacula because of predation (e.g., by spiders), starvation, or freezing. Surviving females terminate diapause in the spring, most likely in response to a combination of warmer temperatures and longer days, venture out of the hibernacula, and take a blood meal, providing new generations for the coming season.

Still other mosquito species survive unfavorable periods as diapausing larvae (e.g., certain species of *Aedes, Anopheles,* and *Culiseta*). Diapause is a variable phenomenon in some of these species, with populations from warmer latitudes having a larval diapause and those from cooler latitudes showing diapause as eggs.

In mosquitoes, diapause is controlled by JH (juvenile hormone), produced in the corpora allata, a pair of neurosecretory structures located in the thorax (Chapter 24). This was demonstrated by inducing diapause in laboratory-reared females by exposure to JH (Meola and Readio 1988).

Most tropical and subtropical mosquitoes do not undergo a true diapause. However, some tropical species have mechanisms for avoidance of hot, dry seasons, but these mechanisms have been little studied.

BIOLOGY

Larvae

Common Larval Habitats

Mosquitoes exploit a wide variety of aquatic habitats for larval development (Fig. 9.11). The vast majority of mosquito species have larvae that are restricted to freshwater. However, a few species can tolerate highly polluted aquatic environments (e.g., septic tanks), and others are adapted to highly saline environments. Even saline water–adapted species do not occur in the open ocean. Rather, they occur in brackish water in tidal marshes and swamps, and some species have larvae that can tolerate salinity higher than seawater's, occurring in rock pools in the intertidal regions along ocean shores.

Only a relatively few mosquito species have larvae that develop in permanent bodies of water (e.g., lakes, reservoirs). Permanent bodies of water usually are deep and open and provide little protection from mosquito predators, such as certain fishes and larval insects that commonly are present. Rather, most species inhabit temporary collections of water, such as river floodplains, livestock troughs, tin cans, discarded tires, mud puddles, irrigation ponds, sewage treatment ponds, ponds of melted snow, vernal pools, rice fields, and seasonally flooded ponds used to attract water fowl.

Mosquito larvae rarely are found in moving water. Some species occur in slowly moving water, such as might be found at the margins of streams, seepage areas, and slowly moving melted snow, but they do not occur in free-running rivers and streams. An interesting larval habitat may be found in the tropics, where larvae of certain species can develop in extremely small quantities of water held between layers of plant tissues.

Larvae of freshwater-adapted species have the problem of excess water uptake and loss of ions (Chapter 26). To minimize these adverse effects, they maintain osmotic balance by restricting water intake, by producing dilute urine, and by uptake of ions from water through four (rarely two) large bladder-like structures on the tip of the abdomen called *anal papillae* (Bradley 1987). The anal papillae of freshwater species tend to be very large, and their size can vary according to water quality (Clements 1992) (Fig. 9.12).

FIGURE 9.11 Some representative larval mosquito habitats:
(a) A rot hole in a tree; (b) a tire and a bedpan, typical discarded
artificial containers; (c) a pasture flooded from pivot irrigation
in Oregon; (d) a pool of melted snow in the Oregon Cascades;
(e) a degraded salt marsh in California; (f) an Oregon log pond.
(a) Courtesy of University of California, Berkeley; (b–f) photographs by Bruce Eldridge.

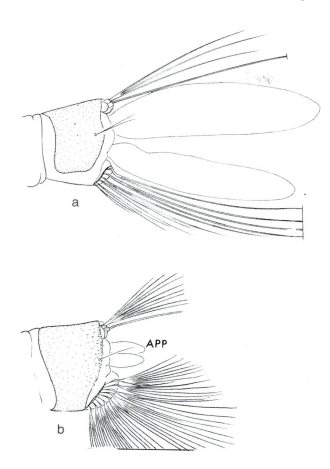

FIGURE 9.12 Anal segments of (a) *Aedes aegypti*, a freshwater species, and (b) *Aedes taeniorhynchus*, a saline species, showing difference in size of anal papillae. From Darsie and Ward (1981).

habitats for mosquitoes. Degree of shade, presence or absence of emergent vegetation, and degree of water motion may also play a part in determining which species is present at a particular site. All of these factors probably influence the attractiveness of sites for ovipositing females, but survival factors also come into play.

Larval Feeding and Drinking

As already discussed, all mosquito larvae drink water, and freshwater species can obtain ions through their anal papillae. However, most larval food is taken in through the mouth. Except for species that have predaceous larvae (e.g., *Toxorhynchites* spp.), mosquitoes feed by creating a current of water via the movement of brushlike structures on their mouthparts and filtering out particulate matter such as plant and animal matter (diatoms, microcrustaceans, etc.) that are then taken into the digestive tract. Nutritional requirements of mosquito larvae have been studied extensively, and most studies have shown that protein and amino acids are required for development to the adult stage (Clements 1963). The amount of food available to larvae is one of the factors that determine the time required to complete development. When nutritional resources are scarce, the time to complete development is extended.

Adults

Nutrition

Ordinarily, both male and female mosquitoes require carbohydrate, usually in the form of nectar from flowers or fruit, but they can utilize other sugar sources as well, such as honeydew. Both sexes utilize the proboscis, an organ greatly modified from ancestral chewing mouthparts for feeding. Female mouthparts are further modified for piercing of vertebrate skin and penetration of blood vessels (Fig. 9.13). Nectar and other sugars are stored in a structure called the *crop* or *ventral diverticulum* (Chapter 21), which is a blind sac that branches from the alimentary canal in the thorax. In time, the contents of the crop are moved to the midgut. Control of the destination of imbibed fluids is through the action of sensory receptors and valves at the entrance to the crop, whereby fluids with high concentrations of carbohydrates are shunted to the crop, but blood and water go directly to the midgut (Day 1954). Blood ordinarily is used to support ovarian development, but female mosquitoes can also use it as an energy source under certain circumstances. Edman et al. (1992) report that *Aedes aegypti* females in Thailand rarely feed on sugar in the tropics but instead

Species with larvae adapted to saline habitats face the challenge of maintaining water and ion balance in varying situations, ranging from fresh to highly saline environments (Chapter 26). In freshwater conditions they osmoregulate in the same manner as do strictly freshwater species. In saline conditions, where excess water loss and ion uptake is a problem, they regulate osmotic pressure within their bodies by drinking substantial amounts of water and by selective removal of ions from their hemolymph through their Malpighian tubules and rectum (Bradley 1987).

As a general rule, saline species can also develop in freshwater, but they do not compete well with freshwater species in freshwater habitats; consequently, saline-adapted species are found only rarely in freshwater habitats.

Besides water quality, there are other factors that determine suitability of various types of aquatic

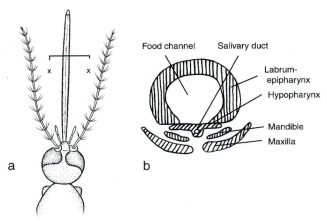

Food channel   Salivary duct

Labrum-epipharynx

Hypopharynx

Mandible

Maxilla

a   b

FIGURE 9.13 Female mosquito mouthparts: (a) Head and appendages of a female; (b) cross section of structures of mouthparts at point shown by cross-section label ("x") of proboscis. Head and appendages from Darsie and Ward (1981). Drawing of mouthparts based on various authors.

rely on vertebrate blood for nutrition. Mosquitoes that survive winters as diapausing adult females can convert nutritional substances in blood to fat in the fall of the year, although the blood-feeding drive is usually suppressed at this time.

Blood Feeding

Blood feeding in insects is believed to have evolved several times independently from ancestral forms adapted for sucking plant juices or for preying on other insects. The Anophelinae are considered to be the most primitive among the major groups of mosquitoes, and the Toxorhynchitinae, then the Culicinae, are believed to have branched off later. This phylogeny suggests that blood feeding is ancestral in mosquitoes and that lack of this function among Toxorhynchitinae evolved later (Wood and Borkent 1989).

The means by which female mosquitoes locate suitable blood-feeding hosts has been studied for many years, but the overall process is still poorly understood. The best explanation is that females are attracted by warmth, moisture, and carbon dioxide from hosts, but other factors are involved. Other substances, such as lactic acid, a component of human sweat, may serve as attractants (see Chapter 20).

Some mosquitoes take blood only from certain groups of vertebrate animals. For example, *Culex pipiens*, the southern house mosquito, feeds almost entirely on birds. *Ochlerotatus sierrensis*, the western

tree hole mosquito, feeds only on mammals. *Culex tarsalis*, the western encephalitis mosquito, feeds on birds and mammals (this is one characteristic of an effective vector of disease pathogens). In the past, this relative host specificity has been called *host preference*. However, this is not an appropriate term, because it ignores factors such as availability of hosts, host defensive behavior, and other factors unrelated to the mosquitoes themselves.

Female mosquitoes feed on hosts for blood meals only at certain phases of their life cycles, and the drive to seek blood meals is regulated by hormones (Chapters 20 and 25). Mosquitoes also have definite daily cycles, called *circadian rhythms*, in which various behavioral activities, such as blood feeding, oviposition, and flying, take place at only certain times of the day. The timing of these cycles varies from species to species, which helps to explain why certain mosquitoes are considered "day-biters" and others "night-biters."

Freshly emerged females, diapausing females, and recently blood-fed females usually do not seek hosts. After a full blood meal, stretch receptors in the abdomen signal a full midgut, and the blood-feeding drive is suppressed (Gwadz 1969). However, multiple blood meals in mosquitoes within a single gonotrophic cycle are relatively common (Edman and Downe 1964). Presumably, many such cases result from an interrupted blood meal in which the midgut is not full and the blood-feeding drive has not been suppressed by a signal from the midgut receptors. However, mosquitoes such as *Aedes aegypti* have been shown to take multiple blood meals routinely (Scott et al. 1993), and other factors, such as nutritional needs for fecundity, may be involved in anopheline mosquitoes (Briegel and Horler 1993).

Blood feeding by mosquitoes is a complex process. It can be divided into various stimulus–response phases, beginning with host seeking to landing, probing, and biting (Bowen 1991). Host-seeking flights may be simply random activity of females that are physiologically primed for blood feeding. When in the proximity of a potential host, movement to the host is probably by both olfactory and visual cues. Both carbon dioxide and lactic acid are considered important volatile substances released by vertebrate hosts, and both have been shown to attract female mosquitoes in laboratory experiments. Mosquitoes have also been shown to follow plumes of these substances upwind in wind tunnel studies (Cardé 1996). The final phases of blood feeding involve the landing of mosquitoes on the skin of the host, probing, actual penetration by the stylets of the mouthparts, and blood feeding. Chemical mosquito repellents such as *N,N*-diethyl-*m*-toluamide (DEET) are considered spatial repellents and probably interfere with the ability of mosquitoes to detect volatile substances such as CO_2 and lactic acid.

Blood feeding is facilitated by the injection of saliva into the feeding wound of the vertebrate host. Saliva comes from organs in the thorax of mosquitoes called *salivary glands*. Saliva contains a variety of substances, including numerous polypeptides (Chapter 28). Saliva has numerous functions in blood feeding, including vasodilation, bacteriolytic activity, and anticoagulant activity. Female mosquitoes produce the substance apyrase in their salivary glands. Apyrase is believed to affect probing behavior and to aid in the ability of probing female mosquitoes to find blood vessels (Ribeiro et al. 1985; Ribeiro 1987). Salivary glands assume great importance in the transmission of disease pathogens by insects because most of these pathogens are injected into a vertebrate host with saliva during blood feeding. This type of pathogen transmission thus is called *salivarian transmission*.

Uptake of blood is accomplished by the action of two muscular pumps: the cibarial pump, located in the head, and the pharyngeal pump, located in the thorax. Blood is forced through the alimentary canal of the mosquito through the esophagus into the midgut. Droplets of urine or blood may be released through the anus at the time of blood feeding.

Within a few hours after a blood meal, the midgut cells of the female mosquito begin secretion of a chitinous material that will eventually become a fine network called the *peritrophic matrix* (Chapter 22). By about 12 hours, this membrane largely surrounds the blood meal. The peritrophic *matrix* is of significance in medical entomology because pathogens or parasites ingested along with the blood meal must find a way to escape from the matrix to complete their development in the mosquito host. This may be by penetration, by passing through openings that occur before formation of the membrane is complete, or through openings that remain after formation. Digestion of the blood takes place in the midgut, within the peritrophic matrix, and is completed within 2–3 days, depending upon temperature.

Reproduction

Reproduction in mosquitoes is sexual. Fertilization of female eggs by male sperm occurs not at the time of mating, but at the time of oviposition by the female. During mating, sperm are injected into the female and stored in spherical structures called *spermathecae*

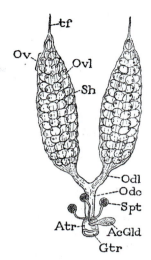

FIGURE 9.14 Reproductive structures of adult female culicine mosquito: AcGld, accessory gland; Atr, atrium; Gtr, gonotreme; Odc, common oviduct; Odl, lateral oviduct; Ov, ovary; Ovl, ovariole; Sh, ovarian sheath; Spt, spermatheca; tf, terminal filament. From Snodgrass (1959).

(Fig. 9.14). The number of spermathecae varies from one to three, depending on the species of mosquito. As eggs pass down the common oviduct during oviposition, sperm enter each egg through the micropyle. A single female may deposit 100–150 eggs at a time; over the course of her life she may lay three to four batches of eggs. Because of mortality factors, the average number of egg batches deposited by female mosquitoes is probably less than this.

Mating

Male and female mosquitoes mate in midair mating swarms, often at dusk. Copulation occurs by juxtaposition of complex genitalic structures located at the tip of the male and female abdomen. Male genitalia are especially complex, and their structure is often used for species identification (Fig. 9.15). Once a female has mated, she remains inseminated for life and her spermathecae remain filled. There is some evidence that females are monogamous because of the introduction of a substance from the male accessory gland (Craig 1967).

Some female mosquitoes can develop an initial batch of eggs without a blood meal. This is called *autogeny*, and its expression in nature is complex and highly variable. Autogeny is a genetic characteristic of certain subspecies or biotypes of a given species (e.g., *Culex pipiens*). The expression of autogeny may be

FIGURE 9.15 Male genitalia of *Anopheles stephensi*. Courtesy of Southeast Asia Mosquito Project, U.S. Army, Washington, DC.

obligatory, in which case the first ovarian cycle is always completed without a prior blood meal. In other species, autogeny may be facultative, and its expression may occur as a result of environmental influences of some type, such as day length. *Culex tarsalis* has facultative autogeny over much of its geographic range. In most cases, the nutrients required for ovarian development are carried over from the immature stages, and blood is required to complete the second and subsequent ovarian cycles. However, some species, such as *Wyeomyia smithii*, the pitcher plant mosquito, can complete several ovarian cycles without a blood meal. As was mentioned earlier, mosquitoes in the genus *Toxorhynchites* do not take blood meals from vertebrates at all. A complete discussion of this subject can be found in Spielman (1971) and Clements (1992).

ECOLOGY OF MOSQUITOES

The ecology of mosquitoes has been studied intensively, especially concerning species of public health importance. Many of these studies have extended classical approaches in general ecology to mosquitoes. An excellent source of information on the origin and development of these approaches is contained in the compendium of classical scientific papers edited by Real and Brown (1991). This book provides a good backdrop for many important concepts in mosquito ecology, such as predator–prey trophic relationships and life table analyses. Service (1993) provides comprehensive information on specific methods used in mosquito ecology.

Estimating Abundance

A basic attribute of mosquito populations is abundance; consequently, its estimation is one of the most important activities of mosquito ecologists. Abundance is a key factor in various types of studies, including life table analyses, assessment of control strategies, and estimation of vectorial capacity. Larval density usually is estimated based on sampling with a dipper of some kind; the most commonly used device is called simply the *one-pint dipper*. Estimates of larval density are usually very challenging, because their distribution in most aquatic habitats is highly clumped. Estimates of adult density usually involve traps of some kind, typically using an attractant such as light or CO_2 or both. By far the two most commonly used devices are the Standard New Jersey Light Trap (Fig. 9.16a) and some form of the CDC-style light trap (Fig. 9.16b). Complete descriptions of equipment used for sampling larval and adult mosquitoes as well as methods of data interpretation can be found in Service (1993).

Natural Mortality Factors

Only a small percentage of mosquito eggs deposited by females survives to the adult stage. Unsuccessful individuals are eliminated by a variety of natural mortality factors, including predation, starvation, desiccation, and exposure to temperatures above or below thermal death points. Mortality factors are considered to be either density dependent, in which mortality of a population is proportional to its density, or density independent, in which mortality effects are not dependent upon population density. Parasites and predators are usually density-dependent mortality factors; insecticides and various physical factors such as temperature are usually density independent. Density-dependent factors may depress natural larval mosquito populations, but usually to stable levels buffered by population density. Density-independent factors, on the other hand, may be associated with sharp rises and declines of population levels.

Studies of natural mortality factors are necessary for an understanding of the growth of mosquito populations. Such studies also form the basis for strategies for

FIGURE 9.16 Traps used to sample adult mosquitoes: (a) Standard New Jersey light trap; (b) CO_2-baited CDC-style trap. (a) Courtesy of Sacramento-Yolo Mosquito and Vector Control District; (b) photograph by Bruce Eldridge.

the biological control of mosquitoes. The study of mortality factors of populations is often done using an approach known as *life table analysis*. Southwood et al. (1972) provides an example of this approach applied to mosquitoes.

Adult mosquitoes also are subjected to natural mortality factors, and mortality can be expressed as daily survival rates. Data used to estimate survival rates can be obtained through a method known as *mark–release–recapture*, in which a known number of mosquitoes are marked with a fluorescent dust, released, and then recaptured at intervals, along with unmarked individuals. The change in the ratio of marked to unmarked individuals over time is used to estimate daily survival. This same approach also can be used to estimate flight range of adults and adult population density.

Parasites and predators of larval mosquitoes are among the most important mortality factors. Immature stages of other insects, such as dragonflies, consume larval mosquitoes, as do various species of fishes. Viruses, bacteria, protozoa, fungi, and nematodes also exact their toll. The mosquitofish, *Gambusia affinis*, is used in many biological control programs for mosquitoes (Fig. 9.17).

FIGURE 9.17 Mosquitofish, *Gambusia affinis*, female. Drawing by Deborah Dritz.

PUBLIC HEALTH AND VETERINARY IMPORTANCE

Information on disease agents transmitted by mosquitoes is presented in Table 9.2.

CONTROL OF MOSQUITOES AND MOSQUITO-BORNE DISEASES

A detailed description of all the ways used to control mosquitoes is beyond the scope of this chapter. Entire books have been written on the subject, such

TABLE 9.2   Some Disease Agents Transmitted by Mosquitoes

Disease agent and disease	Hosts		Important vectors
	Natural	Tangential	
Virus			
eastern equine encephalitis	Bird	Human	*Coquilletidia perturbans*
Venezuelan encephalomyelitis	Small mammal	Human, horse	*Cx. pipiens*, etc.
Western equine encephalomyelitis	Bird	Human, horse	*Cx. tarsalis*
Dengue	Human		*Ae. aegypti, Ae. albopictus*
Japanese encephalitis	Swine	Human	*Cx. tritaeniorhynchus*
St. Louis encephalitis	Bird	Human	*Cx. pipiens, Cx. nigripalpus*
Yellow fever	Primate	Human	*Ae. aegypti, Ae. africanus*
La Crosse encephalitis	Rodent	Human	*Ae. triseriatus*
Apicomplexa			
Malaria	Human		*Anopheles* spp.
Malaria	Bird		*Culex* spp.
Filarioid Nematode			
Wuchereria bancrofti	Human		*Culex, Mansonia*
Brugia malayi	Cat	Human	*Culex, Mansonia*
Dirofilaria immitis	Canid	Human	*Culex, Aedes* spp.

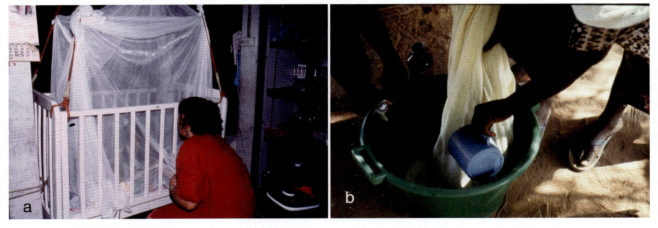

FIGURE 9.18 The use of insecticide-treated bed nets: (a) A mother tucks her young child under an insecticide-treated bed net in Thailand; (b) a woman treats bed nets with permethrin in Gambia. Photograph (a) courtesy of WHO/TDR/Crump; photograph (b) courtesy of WHO/TDR/S. Lindsay.

as the classic work by Herms and Gray (1944). There are two general approaches to reduction of risk to mosquito-borne infections. First, there are personal protective measures employed on an individual basis. These are provided either by direct purchase by individuals or through government programs. Mosquito repellents applied to exposed skin surfaces, bed nets, or window curtains are the most widely used and successful of these measures. Nearly all repellents applied to skin contain DEET; bed nets and curtains are usually treated with a synthetic pyrethroid, such as permethrin or deltamethrin (Fig. 9.18). Bed nets without the use of repellents can be effective if kept in good repair and used properly. Vaccines and prophylactic drugs also fall under this category. Unfortunately, there are few effective vaccines available for human use, an exception being the highly effective 17-D yellow fever vaccine. Prophylactic drugs are available that both

FIGURE 9.19 Power sprayer used for ultra-low-volume application of insecticides for adult mosquito control. Courtesy of Clark Equipment Company.

prevent and cure malaria infections, but widespread resistance by parasites to many of the drugs in various parts of the world continues to hamper their use.

The second type of mosquito control or abatement is that used area-wide. These are usually local or regional government-funded programs. The most widely used approach for mosquito abatement and prevention of mosquito-borne diseases worldwide continues to be the application of chemical pesticides to aquatic larval sources (*larvicides*) and into the air for adult mosquito control (*adulticides*). Such applications are done with aircraft-mounted or truck-mounted equipment. Two popular methods for pesticide application are spreading of granular formulations and fogging with small amounts of concentrated insecticides broken into very small particles (~10μ) by special nozzles called ultra-low volume, or ULV (Fig. 9.19).

In the years following World War II, highly effective and persistent insecticides such as DDT, an organochlorine, were used to control both mosquito larvae and adults. The worldwide malaria eradication effort of the 1950s and 1960s was based on treatment of interior walls of houses to kill indoor resting female anopheline mosquitoes. This program was responsible for the complete disappearance of malaria in some countries and the reduction of human cases in others. Unfortunately, a combination of factors, including resistance to DDT, increasing costs, and political instability, eventually doomed the program, and malaria has returned to nearly all the formerly malaria-free areas at incidences as high as or higher than before.

Organochlorines were phased out in most areas of the world and replaced by newer classes of conventional insecticides, such as organophosphates (e.g., malathion), carbamates (e.g., carbaryl), and synthetic pyrethroids (e.g., resmethrin). Some of the same problems that arose with DDT (resistance, environmental safety) have occurred with nearly all the classes of synthetic organic insecticides, and few chemical companies are developing new products for mosquito control. This has led to the use of pesticides known collectively as third-generation insecticides. These include synthetic materials that affect mosquito development (insect growth regulators), microbial insecticides such as *Bacillus thuringiensis israelensis*, or BTI, and chitin inhibitors, such as diflubenzuron. The use of oils to kill mosquito larvae predates the use of synthetic organic chemicals, and such use continues. Third-generation pesticides are more expensive than conventional pesticides but generally are less toxic to humans and other vertebrates. Because many are highly specific for mosquitoes, they are less disruptive to the environment.

The future of insecticides for mosquito abatement is uncertain, especially in the United States. Physiological resistance to pesticides in general has been a problem since the introduction of DDT, and resistance is now beginning to show up even among third-generation products, including *Bacillus sphaericus* (Bs), a microbial insecticide, and altosid, an insect growth regulator (Cornel et al. 2000). The greatest threat to the continued use of pesticides for mosquito control is economics. The costs involved in conducting vertebrate and environmental safety tests on new pesticides are rarely justified on the basis of a relatively small market for public health pesticides. Consequently, few products based on new active ingredients have become available over the past 10–15 years.

Source reduction, e.g., the management of standing water (Chapter 42) to avoid mosquito development, is an important tool in mosquito control (Fig. 9.20). In the early days of mosquito control, source reduction usually meant draining of swamps and marshes, and vast areas of wetlands were permanently lost. As appreciation of the value of wetlands increased in the latter part of the 20th century, a more balanced approach to source reduction was adopted by mosquito abatement agencies. Research has shown that mosquito breeding can be minimized by the timing of flooding in artificial freshwater wetlands and by restructuring of water channels in salt marshes in a way that restores natural tidal action (Collins and Resh 1989). Such approaches are desirable because they

FIGURE 9.20 Ditching of salt marsh to improve natural tidal flushing action for reduction of mosquito populations. Courtesy of Alameda County (California) Mosquito Control District.

proven practical, but their theoretical promise suggests that this kind of research should continue. Because of the continuing reduction in available mosquito pesticides and the fundamental discoveries that have been made over the last decade or so in the area of molecular genetics of mosquitoes, their application for vector mosquito control seems inevitable. For the near future, the most promising approach for area-wide control seems to be the selective use of combinations of modern, environmentally safe pesticides, source reduction, and biological control. For personal protection, locally implemented programs for repellent-treated bed nets and window curtains will continue to be important.

actually improve aquatic habitats while minimizing mosquito problems.

Biological control (BC) (Chapter 43) of mosquitoes is a persistent long-term goal in mosquito control. Biological control is the depression of population levels of mosquitoes using biological organisms or their products. In practice, BC involves the introduction of biological agents into mosquito habitats or the management of habitats to optimize the effect of such agents that occur there naturally. Most BC agents are parasites or predators of mosquitoes; many different organisms have been tested for their effectiveness. However, only a few organisms can be considered to be effective as BC agents. Two successful examples are mosquitofish (*Gambusia affinis*) and microbial toxins (Bti and Bs). Some investigators do not consider toxins to be BC agents, but rather a type of insecticide.

Several factors have prevented BC from being a viable alternative to insecticides for control of larval mosquitoes. Cost, difficulty in mass production of agents, and density-dependent effects all present serious challenges to effective mosquito control using this approach. Chapman (1974, 1985) presented reviews of BC agents for mosquito larvae. Service (1983) gave a critical, and somewhat pessimistic, assessment of the long-term promise of BC for mosquito control.

Considerable research emphasis has been placed recently on the use of molecular approaches (Chapter 44) to produce genetically altered strains of mosquitoes as a means of introducing lethal genes into natural populations or genes that may confer inability to transmit human pathogens. These schemes have yet to be

Readings

Bates, M. 1954. *The natural history of mosquitoes.* Macmillan, New York.

Bentley, M. D., and Day, J. F. 1989. Chemical ecology and behavioral aspects of mosquito oviposition. *Annu. Rev. Entomol.* 34: 401–422.

Bowen, M. F. 1991. The sensory physiology of host-seeking behavior in mosquitoes. *Annu. Rev. Entomol.* 36: 139–158.

Bradley, T. J. 1987. Physiology of osmoregulation in mosquitoes. *Annu. Rev. Entomol.* 32: 439–462.

Briegel, H., and Horler, E. 1993. Multiple blood meals as a reproductive strategy in *Anopheles* (Diptera: Culicidae). *J. Med. Entomol.* 30: 975–985.

Cardé, R. T. 1996. Plumes. Pages 32–33. In *Olfaction in Mosquito–Host Interactions* (G. R. Bock and G. Cardew, eds.). Wiley, Chichester, England.

Carpenter, S. J., and LaCasse, W. J. 1955. *Mosquitoes of North America.* University of California Press, Berkeley.

Chapman, H. C. 1974. Biological control of mosquito larvae. *Annu. Rev. Entomol.* 19: 33–59.

Chapman H. C. (ed.) 1985. *Biological Control of Mosquitoes.* American Mosquito Control Association, Fresno, CA.

Clements, A. N. 1963. *The Physiology of Mosquitoes.* Macmillan, New York.

Clements, A. N. 1992. *The Biology of Mosquitoes.* Chapman and Hall, London.

Collins, J. N., and Resh, V. H. 1989. *Guidelines for the Ecological Control of Mosquitoes in Nontidal Wetlands of the San Francisco Bay Area.* California Mosquito and Vector Control Assoc., Elk Grove, CA.

Cornel, A. J., Stanich, M. A., Farley, D., Mulligan, III, F. S., and Byde, G. 2000. Methoprene tolerance in *Aedes nigromaculis* in Fresno County in California. *J. Am. Mosquito Control Assoc.* 16: 223–228.

Craig, G. B., Jr. 1967. Mosquitoes: Female monogamy induced by male accessory gland substance. *Science* 156: 1499–1501.

Darsie, R. F., Jr., and Ward, R. A. 1981. Identification and geographical distribution of the mosquitoes of North America north of Mexico. *Mosquito Systematics* 1(Suppl.): 1–313.

Day, M. F. 1954. The mechanism of food distribution to midgut or diverticulum in the mosquito. *Austral. J. Med. Sci.* 7: 515–524.

Edman, J. D., and Downe, A. E. R. 1964. Host-blood sources and multiple feeding habits of mosquitoes in Kansas. *Mosquito News* 24: 154–160.

Edman, J. D., Strickman, D., Kittayapong, P., and Scott, T. W. 1992. Female *Aedes aegypti* (Diptera: Culicidae) in Thailand rarely feed on sugar. *J. Med. Entomol.* 29: 1035–1038.

Edwards, F. W. 1932. Diptera, Family Culicidae. Fasc. 194. In *Genera insectorum* (P. Wytsman, ed.). Desmet Verteneuil, Brussels, pp. 1–258.

Eldridge, B. F. 1993. Patrick Manson and the discovery age of vector biology. *J. Am. Mosquito Control Assoc.* 8: 215–218.

Gillett, J. D. 1971. *Mosquitos.* Weidenfeld and Nicolson, London.

Gwadz, R. W. 1969. Regulation of blood meal size in the mosquito. *J. Insect Physiol.* 15: 2039–2044.

Herms, W. B., and Gray, H. F. 1944. *Mosquito Control: Practical Methods for Abatement of Disease Vectors*, 2nd ed. Commonwealth Fund, New York.

Knight, K. L. 1978. *Supplement to a Catalog of the Mosquitoes of the World.* Thomas Say Foundation, College Park, MD.

Knight, K. L., and Stone, A. 1977. *A Catalog of Mosquitoes of the World*, 2nd ed. Thomas Say Foundation, College Park, MD.

LaCasse, W. J., and Yamaguti, S. 1950. *Mosquito Fauna of Japan and Korea.* Office of the Surgeon General, 8th U.S. Army, Japan.

Maire, A. 1983. Sélectivité des femelles de mostiques (Culicidae) pour leurs sites d'oviposition: État de la question. *Rev. Canadian Biol. Experimental* 42: 235–241.

Meola, R., and Readio, J. 1988. Juvenile hormone regulation of biting behavior and egg development in mosquitoes. *Adv. Dis. Vector Res.* 5: 1–24.

Philip, C. B., and Rozeboom, L. E. 1973. Medico-veterinary entomology: a generation of progress. In *History of Entomology* (R. F. Smith and T. E. Mittler, eds.). Annual Reviews, Palo Alto, CA, pp. 333–359.

Pickett, J. A., and Woodcock, C. M. 1996. The role of mosquito olfaction in oviposition site location and in the avoidance of unsuitable hosts. In *Olfaction in Mosquito–Host Interactions* (G. R. Bock and G. Cardew, eds.). Wiley, Chichester, England, pp. 109–123.

Real, L. A., and Brown, J. H. 1991. Foundations of ecology: Classic papers with commentaries.

Ribeiro, J. M. C. 1987. Role of saliva in blood feeding in arthropods. *Annu. Rev. Entomol.* 32: 463–478.

Ribeiro, J. M. C., Rossignol, P. A., and Spielman, A. 1985. Salivary gland apyrase determines probing time in anopheline mosquitoes. *J. Insect Physiol.* 31: 689–692.

Scott, T. W., Clark, G. G., Lorenz, L. H., Amerasinghe, P. H., Reiter, P., and Edman, J. D. 1993. Detection of multiple feeding in *Aedes aegypti* (Diptera: Culicidae) during a single gonotrophic cycle using a histologic technique. *J. Med. Entomol.* 30: 94–99.

Service, M. W. 1983. Biological control of mosquitoes — Has it a future? *Mosquito News* 43: 113–120.

Service, M. W. 1993. Mosquito ecology. *Field Sampling Methods*, 2nd ed. Chapman and Hall, London.

Snodgrass, R. E. 1959. *The Anatomical Life of the Mosquito* (Publication 4388). Smithsonian Institution, Washington, DC.

Southwood, T. R. E., Murdie, G., Yasuno, M., Tonn, R. J., and Reader, P. M. 1972. Studies on the life budget of *Aedes aegypti* in Wat Samphaya, Bangkok, Thailand. *Bull. World Health Org.* 46: 211–226.

Spielman, A. 1971. Bionomics of autogenous mosquitoes. *Annu. Rev. Entomol.* 16: 231–248.

Spielman, A., and D'Antonio, M. 2001. *Mosquito: A Natural History of Our Most Persistent and Deadly Foe.* Hyperion, New York.

Stojanovich, C. J., and Scott, H. G. 1965. *Illustrated Key to Culex Mosquitoes of Vietnam.* U.S. Public Health Service, Atlanta.

Tanaka, K., Mizusawa, K., and Saugstad, E. S. 1979. A revision of the adult and larval mosquitoes of Japan (including the Ryukyu Archipelago and the Ogasawara Islands) and Korea (Diptera: Culicidae). *Cont. Am. Entomological Inst.* 16: 1–987.

Ward, R. A. 1984. Second supplement to "A Catalog of the Mosquitoes of the World" (Diptera: Culicidae). *Mosquito Systematics* 16: 227–270.

Ward, R. A. 1992. Third supplement to "A Catalog of the Mosquitoes of the World" (Diptera: Culicidae). *Mosquito Systematics* 24: 177–230.

Wood, D. M., and Borkent, A. 1989. Phylogeny and classification of the Nematocera. In *Manual of Nearctic Diptera.* Agriculture Canada, Ottawa, Ontario, pp. 1333–1370.

10

The Biting Midges, the Ceratopogonidae (Diptera)

ART BORKENT

INTRODUCTION

The Ceratopogonidae, commonly known as biting midges, no-see-'ems, or sand flies (a name also applied to some Psychodidae), are a remarkably diverse and interesting family of biting flies. Species in this family occur in virtually all terrestrial areas of the planet, from coastal areas to high mountain peaks (at least 4,200 m), and from the tropics to the high arctic (within 150 km of permanent polar ice) and subantarctic islands. Immatures may be found in nearly any habitat with even a moderate amount of moisture, from rotting cacti in desert regions, epiphytes, and tiny rock pools to the benthic regions of large rivers and lakes.

The biting midges have a bad reputation as nasty biters that pester humans and domestic animals and as vectors of harmful diseases. Because of their small size, many species can pass through screens and mesh that keeps other biting pests outdoors, and these can make life insufferable. The female adults can occur in such huge numbers that in some areas people and live-stock suffer from numerous itchy welts or are driven indoors. Some species vector a variety of viruses, pro-tozoa, and nematodes, resulting in such important diseases as bluetongue, African horse sickness, and oropouche (Table 10.1).

Although known primarily as a significant pest group, biting midges, as a group, are a vital and impor-tant part of ecosystems; the vast majority of species are not pests at all and play important roles in a wide array of habitats. Some species are important pollinators of such plants as cacao (without them we wouldn't enjoy chocolate!) and rubber trees, and the larvae of many are important predators of other organisms in aquatic systems. The adults of most biting midges actually suck blood from other insects and may possibly be important vectors of viruses and spiroplasmas that can kill caterpillars and other insects. Finally, it is worth pointing out that even the pest species that attack humans and livestock play an important role by restricting human development in certain regions, thereby helping to preserve some habitats during a time when ecological destruction is rampant and many species are under great threat or have already become extinct.

The Ceratopogonidae, in nearly every regard, is by far the most poorly understood of all arthropod vector groups. Many species remain unnamed, identification of pest species is difficult or impossible in most regions of the planet, and our understanding of the role of many species as vectors is often, at best, superficial. It is highly likely that there are many diseases of wild life that are transmitted by biting midges, but we lack critical information concerning these. There have been difficulties in colonizing some species, and their small size has limited some aspects of field studies. The family is certainly in urgent need of further research.

The biting midges are currently placed in 103 genera worldwide, but vertebrate blood feeders are restricted to four of these. The remaining genera include species that obtain their blood meal from other insects or do not feed on blood at all (Fig. 10.1). In this chapter, infor-mation is presented primarily for those genera that

TABLE 10.1   Organisms Transmitted by Ceratopogonidae (Vectors in the Genera
Leptoconops, Forcipomyia, **and** *Culicoides*)

Taxon	Vector	Vertebrate hosts	Location
VIRUSES			
Alphavirus			
Barmah Forest	C. marxi	Birds, monkeys, humans	Australia
Eastern Equine encephalomyelitis	C. spp.	Birds, rodents, horses, dogs, humans	New World, Philippines
Bunyavirus			
Ananindeua	C. paraensis	Birds, marsupials, rodents, primates	Brazil
Aino	C. brevitarsis, C. spp.	Unknown; antibodies in mammals	Asia, Australia
Akabane	C. brevitarsis, C. wadai, C. imicola, C. milnei, C. oxystoma, C. spp.	Domestic ruminants	Africa, Asia, Cyprus, Australia
Belmont	C. bundyensis, C. marksi	Probably marsupials; antibodies in other mammals	Australia
Buttonwillow	C. variipennis, C. spp.	Lagomorphs; antibodies in other mammals	North America
Cul. 1/70	C. spp.	Cattle	Africa
Douglas	C. brevitarsis, C. austropalpalis, C. spp.	Cattle; antibodies in other mammals	Australia
Keterah	C. schultzei	Birds, bats, humans	Central Asia, Malaysia
Lokern	C. variipennis, C. spp.	Hares, rabbits; antibodies in other mammals	North America
Main Drain	C. variipennis, C. spp.	Lagomorphs	North America
Oropouche	C. paraensis	Sloths, humans; antibodies in other mammals	Trinidad, Panama, Brazil, Peru
Peaton	C. brevitarsis	Cattle; antibodies in other mammals	Australia
Rift Valley fever	C. spp.	Rodents, bats, various ungulates, humans	Africa, Egypt
Sabo	C. imicola, C. spp.	Cattle, goats; antibodies in other mammals	Nigeria
Sango	C. spp.	Cattle; antibodies in other mammals	Africa
Sathuperi	C. spp.	Cattle	India, Nigeria
Shamonda	C. imicola, C. spp.	Cattle	Nigeria
Shuni	C. spp.	Sheep, cattle, humans	Africa
Tahyna	C. spp.	Lagomorphs; antibodies in other mammals	Europe, Asia, Africa
Thimiri	C. histrio	Birds	Africa, India, Australia
Tinaroo	C. brevitarsis	Probably cattle; antibodies in other mammals	Australia
Utinga	Ceratopogonidae	Sloths; antibodies in other mammals	South and Central America
Weldona	C. spp.	Birds	North America
Unidentified Bunyavirus	C. variipennis	Unknown	North America
Flavivirus			
Israel turkey meningoencephalitis	C. spp.	Domestic turkeys	Asia, Africa
Nairovirus			
Congo-Crimean haemorrhagic fever	C. spp.	Various mammals including humans	Europe, Africa, Asia
Dugbe	C. spp.	Rodents, cattle, humans	Africa
Nairobi sheep disease	C. tororoensis, C. spp.	Sheep, goats, humans	Africa, India
Orbivirus			
African horse sickness	C. imicola, C. pulicaris, C. bolitinos, C. gulbenkiani, C. spp.	Equids, dogs	Southern Europe, Africa, Asia
Bluetongue	C. imicola, C. milnei, C. tororoensis, C. wadai, C. sonorensis, C. cockerellii, C. insignis, C. obsoletus, C. brevitarsis, C. filarifer, C. fulvus, C. pusillus, C. fulvus + orientalis, C. pycnostictus, C. bolitinos, C. gulbenkiani, C. spp.	Wild and domestic ruminants	Southern Europe, Africa, Asia New World, Australia
Bunyip Creek	C. brevitarsis, C. schultzei	Cattle, water buffalo; antibodies in other mammals	Australia
CSIRO Village	C. brevitarsis, C. spp.	Cattle; antibodies in other mammals	Australia
Cul. 1/69	C. spp.	Cattle	Africa
Cul. 2/69	C. spp.	Cattle	Africa

(continues)

TABLE 10.1 (*continued*)

Taxon	Vector	Vertebrate hosts	Location
Cul. 3/69	C. spp.	Unknown	Africa
D'Aguilar	C. brevitarsis, C. schultzei C. spp.	Cattle; antibodies in other mammals	Australia
Equine encephalosis	C. spp.	Equidae	Africa
Epizootic haemorrhagic disease	C. variipennis, C. brevitarsis, C. kingi, C. schultzei, C. spp.	Deer, antelopes; antibodies in other mammals	North America, Nigeria, Australia
Eubenangee	C. marksi	Antibodies in various mammals	Australia
Gweru	C. spp.	Antibodies in various mammals	South Africa
Kasba	C. schultzei, C. oxystoma C. spp.	Goats, cattle	Africa, Japan
*Letsitele	C. imicola, C. bolitinos, C. zuluensis, C. magnus	Unknown	Africa
Marrakai	C. schultzei, C. spp.	Cattle	Australia
Mitchell River	C. spp.	Antibodies in various mammals	Australia
Mudjinbarry	C. marksi	Antibodies in various mammals	Australia
Nyabira	C. imicola, C. spp.	Unknown	South Africa
Wallal	C. dycei, C. marksi, C. brevitarsis, C. spp.	Antibodies in various mammals	Australia
Warrego	C. marksi, C. dycei, C. actoni, C. spp.	Antibodies in various mammals	Australia
Wongorr	C. pallidothorax	Antibodies in various mammals	Australia
Vesiculovirus			
Vesicular stomatitis — New Jersey	C. variipennis, C. stellifer C. spp.	Horses, cattle, pigs, humans	New World
Lyssavirus			
Barur	C. punctatus	Antibodies in cattle	Japan, India
Bivens Arm	C. insignis	Antibodies in various mammals	North America
Bovine ephemeral fever	C. schultzei, C. coarctatus C. imicola, C. brevitarsis C. algecirensis, C. spp.	Cattle; antibodies in other mammals	Africa, Asia, Australia
Charieville	F. sp.	Lizards	Australia
Humpty Doo	C. marksi, F. sp.	Unknown	Australia
Kimberley	C. brevitarsis	Cattle	Australia, New Guinea
Kotonkan	C. spp.	Cattle; antibodies in other mammals	Africa
Kununurra	C. austropalpalis	Unknown	Australia
Ngaingan	C. brevitarsis, C. spp.	Antibodies in various mammals	Australia
Tibrogargan	C. brevitarsis	Antibodies in various mammals	Australia
Unclassified Arboviruses			
Beatrice Hill	C. peregrinus	Unknown	Australia
Buritirana	C. spp.	Unknown	Brazil
Itacaiunas	C. marksi	Unknown	Australia
Leanyer	C. marksi	Unknown	Australia
Parker's Farm	C. marksi	Unknown	Australia
Unidentified Virus Isolates			
Cul. 5/69	C. spp.	Cattle	Africa
	C. brevitarsis	Unknown	Australia
	C. brevitarsis + C. schultzei	Unknown	Australia
	C. kingi	Unknown	Africa
	C. marksi	Unknown	Australia
	C. imicola	Unknown	Africa
	C. imicola + C. schultzei	Unknown	Africa
	C. peregrinus + C. schultzei	Unknown	Australia
	C. zuluensis	Unknown	Africa
	C. zuluensis + C. imicola	Unknown	Africa
PROTOZOA			
Haemoproteus canachites	C. sphagnumensis	Birds (Tetraonidae)	North America
Haemoproteus danilewskyi	C. sphagnumensis, C. crepuscularis	Birds (Corvidae)	North America
Haemoproteus fringillae	C. crepuscularis, C. stilobezzioides	Birds (Fringillidae)	North America

(*continues*)

TABLE 10.1 (*continued*)

Taxon	Vector	Vertebrate hosts	Location
Haemoproteus meleagridis	*C. edeni, C. hinmani,* *C. arboricola*	Birds (gallinaceous)	North America
Haemoproteus nettionis	*C. downesi*	Birds (waterfowl)	North America
Haemoproteus velans	*C. sphagnumensis,* *C. crepuscularis*	Birds (Picidae)	North America
Hepatocystis kochi	*C. adersi, C. fulvithorax*	Monkeys	Africa
Hepatocystis levinei	*C.* sp.	Bats	Australia
Plasmodium malariae	*C.* sp.	Unknown	India
Leucocytozoon caulleryi	*C. arakawae* *C. circumscriptus*	Birds (chickens)	Asia
Leucocytozoon neavei	*C.* sp.	Birds	Africa
Trypanosoma avium	*C. sphagnumensis* *C. stilobezzioides*	Birds	North America
Trypanosoma bakeri	*C.* sp.	Birds	Thailand
Trypanosoma spp.	*C. variipennis*	Unknown	North America
Leishmania donovani	*C.* sp.	Unknown	India
Vavraia	*C. edeni*	Unknown	North America
FILARIOIDEA			
Chandlerella chitwoodae	*C. stilobezzioides, C. travisi*	Birds (Ploceidae)	Indonesia
Chandlerella quiscali	*C. crepuscularis* *C. haematopotus*	Birds (Icteridae)	North America
Chandlerella striatospicula	*C. haematopotus*	Birds (Corvidae)	North America
Dipetalonema caudispina	*C. hollensis*	Monkeys (Cebidae)	South America
Dipetalonema gracile	*C. hollensis*	Monkeys (Cebidae, Callithricidae)	Mexico, Central and South America
Dipetalonema llewellyni	*C. hollensis*	Raccoons	North America
Dipetalonema marmosetae	*C. hollensis, C. furens*	Monkeys (Cebidae, Callithricidae)	South America
Dipetalonema ozzardi	*L. becquaerti, C. furens* *C. phlebotomus*	Humans, other primates?	Central and South America
Dipetalonema perstans	*C. austeni, C. grahami* *C. inornatipennis*	Humans, other primates	Africa, Central and South America
Dipetalonema streptocerca	*C. austeni*	Humans, other primates	Africa
Eufilaria kalifai	*C. nubeculosus*	Birds (Corvidae)	Europe
Eufilaria longicaudata	*C. crepuscularis,* *C. haematopotus*	Birds (Corvidae)	North America
Eufilariella bartlettae	*C. nubeculosus*	Birds (Turdidae)	Europe
Eufilariella delicata	*C. nubeculosus*	Birds (Turdidae)	Europe
Filaria sp.	*L. laurae*	Donkey	Africa
Filaria sp.	*C. crepuscularis*	Birds (Sturnidae)	North America
Filaria sp.	*C. haematopotus*	Birds (Corvidae)	North America
Icosiella neglecta	*F. velox*	Frogs	Europe, Asia, Africa, South America
Macacanema formosana	*C. amamiensis*	Monkeys (Macaca)	China, Taiwan
Onchocerca cebei	*C. bundyensis*	Water buffalo	Australia
Onchocerca cervicalis	*C. nubeculosus, C. parroti* *C. variipennis, C. obsoletus*	Equidae	Worldwide
Onchocerca flexuosa	*C. nubeculosus*	Cervidae	Europe, Asia
Onchocerca gibsoni	*C. oxystoma, C. orientalis,* *C. pungens, C. shortii,* *C. nubeculosus, C. marksi,* *C.* spp., *F. townsvillensis*	Bovidae	Asia, India, Africa, Australia
Onchocerca gutturosa	*C. arakawae, C. kingi,* *C. krameri, C. fulvithorax,* *C. nubeculosus,* *C. trifasciellus*	Bovidae	Worldwide
Onchocerca reticulata	*C. nubeculosus, C.* spp.	Equidae	Europe, Asia, Africa, India, Australia
Onchocerca sp.	*C. arakawae*	Cattle	Japan
Ornithofilaria fallisensis	*C.* spp.	Birds (Anatidae)	North America, Asia
Splendidofilaria californiensis	*C. multidentatus*	Birds (Tetronidae)	North America
Splendidofilaria picardina	*C. crepuscularis*	Birds (Corvidae)	North America

* Perhaps misassociated, with external viral contamination of vector.

Disease vectors known		+		+			+			
Host type	Vertebrates	+	+	+			+			
	Large invertebrates			+	+					
	Other midges						+		+?	+
	Ephemeroptera									+
	Non-blood feeding			+	+	+	+	+		+
Number of extant species		135	1	1025 (159)	455	493	1270	9	1	2151

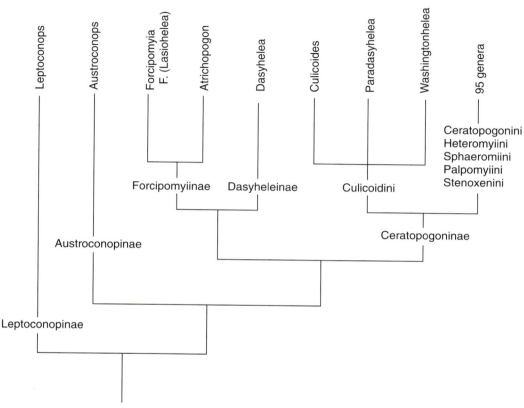

FIGURE 10.1 Phylogeny of the basal groups of Ceratopogonidae showing their classification, records of species acting as vectors, their hosts, and the number of included species.

feed on vertebrates, with additional information on other groups when this provides further context.

TAXONOMY AND IDENTIFICATION

The Ceratopogonidae are members of the superfamily Chironomoidea, which includes the closely related Chironomidae (nonbiting midges) and Simuliidae (blackflies). These families are placed together with the Culicidae (mosquitoes) and a few other small, nonpest families in the infraorder Culicomorpha. The presence of the biting habit in some of these families is clearly homologous, and studies in one family are therefore often applicable to the others. Ceratopogonidae are divided into five subfamilies (Fig. 10.1) and

the first four, representing the earlier lineages, include only one or two genera each. The subfamily Ceratopogoninae includes the bulk of the generic and nearly half of the species' diversity and is divided into six tribes, one of which includes the notorious genus *Culicoides*. The four genera that include vertebrate-biting species (Fig. 10.1) are *Leptoconops* (worldwide), *Austroconops* (from southwestern Australia), *Forcipomyia* (vertebrate feeders are restricted to the subgenus *Lasiohelea* (worldwide)), and *Culicoides* (worldwide); other genera either feed on other insects or do not bite at all. Some workers consider *Lasiohelea* to be a distinct genus, but there is no phylogenetic justification for this.

Biting midges have an excellent fossil record, primarily in amber, showing they were abundant and diverse by 120 million years ago. Species of both *Lep-*

toconops and *Austroconops* were present at that time, the earliest fossils of *Culicoides* are from about 90 million years ago, and fossil *Forcipomyia* (*Lasiohelea*) are known from 55-million-year-old amber. Some of these fossils have swollen abdomens and mouthpart structures that indicate they were feeding on vertebrate blood. Clearly, the relationship between biting midges and their vertebrate hosts is an ancient one!

Although 5540 species of Ceratopogonidae are now described, it would be reasonable to estimate that there are at least 15,000 morphologically distinct species on the planet; many species remain to be described and understood. Taxonomic studies are needed for most genera, including those of the vertebrate-feeding groups. Large regions of the globe have been only superficially surveyed. For example, the Andes of South America, a region known for its high diversity of other organisms, has been surveyed in only a few very local areas, and a concerted collecting effort is required; even so, museums contain many unnamed species. Even in some very accessible regions, including North America and Europe, there are numerous taxonomic problems to be solved. Ceratopogonidae is the only family of biting flies for which distribution maps and keys to all the North American pest species do not exist; the best we have are keys to *Leptoconops* and the species of *Culicoides* in a handful of states in the USA. Furthermore, there are species complexes that require resolution to determine the identity of pest species. It was only in 2000 that a taxonomic paper appeared resolving a North American species group that includes the primary vector of bluetongue (*C. sonorensis*); the three species in this complex had been previously recognized as five subspecies. Another problem is a reflection of the regional emphasis by some taxonomists: Comparisons of specimens from different regions will certainly reveal that some independently named taxa actually belong to the same species. For example, *C. obsoletus* has been separately named 14 times based on material from eight different countries!

These estimates reflect our understanding of morphospecies (species that can be distinguished by significant differences with laboratory microscopes). In the closely related Chironomidae and Simuliidae, it has become clear that many species are distinguishable only through study of chromosomes and molecular investigation, and it would be unbelievable if the same did not apply to the Ceratopogonidae. Chromosome study will likely be very limited within the Ceratopogonidae because they do not appear to have polytene chromosomes (very broad chromosomes with distinct banding patterns), at least in the salivary glands (other possible locations have not been systematically studied). Barely a handful of elec-

trophoretic studies have been employed within the Ceratopogonidae and only one analysis of DNA, used to separate the very similar species of the *Culicoides imicola* species group in Africa.

There is also a great need for comprehensive phylogenetic studies that would provide a predictive framework to better interpret the biology of the biting midges and their role as vectors. Although some applied scientists believe it is sufficient to restrict studies to pest species, it is clear that an understanding of nonpest relatives provides important comparative information that allows for a better interpretation of virtually every feature of pest species. Phylogenetic studies, which interpret the genealogical relationships of all members of a given group, provide this important reference system to interpret the behavior, population dynamics, phenology, vector capacity, host types and other characteristics of pest species. For example, understanding that vertebrate biters in the genera *Leptoconops*, *Austroconops*, *Forcipomyia* (*Lasiohelea*), and *Culicoides* are all early lineages within the family and that their biting habit is homologous to each other and to those of Simuliidae (blackflies) and Culicidae (mosquitoes) strongly suggests that many of their feeding strategies and behaviors are also homologous. In such instances, we can reasonably apply many of the conclusions of studies of one species to the others. A huge gap in our phylogenetic understanding, in particular, is in the genus *Culicoides*, with 1270 described species and including the great majority of pest species. This group requires a reappraisal of subgeneric relationships and the incorporation of a large number of unplaced "species groups" and miscellaneous species into a comprehensive phylogenetic framework.

MORPHOLOGY

Biting midge adults are relatively easy to recognize to family. Members have the following features that distinguish them from all other Diptera (Figs. 10.3A, 10.4A, 10.5A): ocelli absent, antennae generally with 6–13 flagellomeres (vertebrate feeders have either 11–12 flagellomeres (*Leptoconops*) or 13 flagellomeres (those in the other genera)), most males with plumose antenna (Fig. 10.5B), most females with a serrate mandible, postnotum lacking a medial longitudinal groove, wing length = 0.4–7 mm, wing with 1–3 radial veins and two median vein branches (posterior one may be weak) reaching the wing margin, wings overlapping each other over the abdomen for resting live individuals (except for a few *Stilobezzia*, which have wing somewhat spread apart when at rest), front legs shorter than the hind legs, and first tarsomere of each leg equal to or longer than the second tarsomere. The

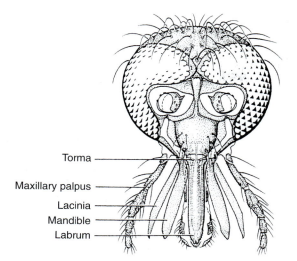

FIGURE 10.2 Mouthparts of female adult *Culicoides* from anterior view. From Downes and Wirth (1981).

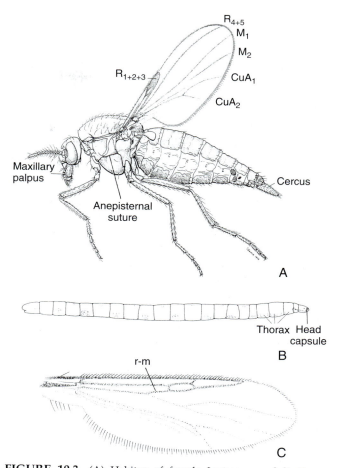

FIGURE 10.3 (A) Habitus of female *Leptoconops* adult. From Downes and Wirth (1981). (B) Habitus of *Leptoconops* larva. From Laurence and Mathias (1972). (C) Wing of *Austroconops mcmillani*.

mouthparts of female biting midges are composed of the same basic units as other biting Nematocera (Fig. 10.2): a well-developed anterior labrum with basal torma, a pair of apically serrate mandibles, a pair of long, slender laciniae with or without retrorse teeth, and a well-developed labium terminating in two lobes called the *labellum*. Vertebrate feeders have finely-toothed mandibles (to cut into the host) and laciniae with retrorse teeth (to hold the mouthparts in place), and virtually all biting midges that feed on invertebrates have coarsely toothed mandibles and simple laciniae. Within the vertebrate feeders there is a general correlation between the number of capitate sensilla on the third maxillary segment and the size of the host. These sensilla detect concentrations of CO_2, and the smaller the host, the more sensilla are present. Species that feed on larger mammals have 9–24 capitate sensilla, while those that feed on small mammals and birds have 29–75.

Pupae have well-developed, undivided respiratory organs with a series of small spiracles, the third leg is curled under the wing sheath, the apex of the abdomen is not curled under the thorax (as in Culicidae), and they have two pointed anal processes. Larvae (Figs. 10.3B, 10.4B, 10.5C) have a well-developed head capsule, a well-developed pharyngeal complex (a large "mortar and pestle" sclerotization in the center of the head capsule for sucking up and, in some taxa, grinding food), and no open spiracles. Eggs are either elongate and slender (some *Leptoconops*, *Austroconops*, and all Ceratopogoninae), short and somewhat oval (some *Leptoconops* and all Forcipomyiinae), or strongly C-shaped (*Dasyhelea*); they tend to be white when laid, and most turn dark brown or black after a short time.

Many eggs have distinguishing rows of bumps or tubercles.

There are significant differences between the different subfamilies of biting midges, and the following concentrates on the four that include pest species. Female adult Leptoconopinae, with its single genus *Leptoconops* (Fig. 10.3A), have anterior radial cells compacted into a thickened unit, an elongate but faint R_{4+5}, 10–12 antennal flagellomeres with the terminal flagellomere rounded or tapering apically, and an elongate anepisternal suture, and many species (in the subgenera *Leptoconops sensu stricto*, *Holoconops*, *Megaconops*, and *Proleptoconops*) have unique, very elongate cerci for laying eggs in sand and crevices. Larvae (Fig. 10.3B) are very peculiar, with a rather pale head capsule and secondarily divided abdominal segments.

The genus *Austroconops* is known from only one species in southwestern Australia. Adults have a distinctive wing venation (Fig. 10.3C) with a longitudinal r-m crossvein. The immatures are undescribed but are currently under study.

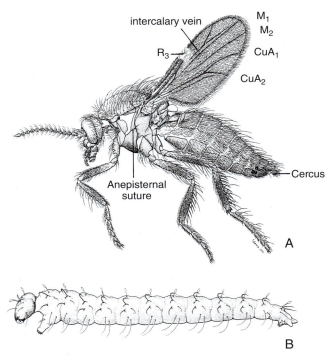

FIGURE 10.4 Habitus of *Forcipomyia*: (A) Female adult; (B) larva. From Downes and Wirth (1981).

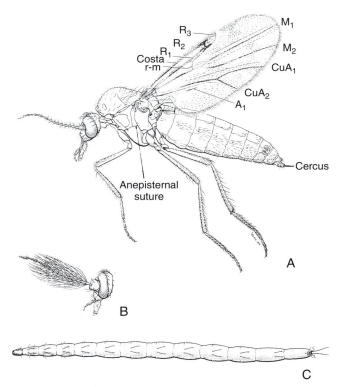

FIGURE 10.5 Habitus of *Culicoides*: (A) Female adult; (B) male head; (C) larva. From Downes and Wirth (1981).

Female adults of *Forcipomyia* (*Lasiohelea*) (Fig. 10.4A) have well-developed radial cells (although they may be narrow); they lack an R_{4+5} (although a peculiar forked intercalary vein may be present), have 13 antennal flagellomeres, with the terminal flagellomere bearing a basally constricted nipple apically, a short anepisternal suture, and all have short cerci. *Forcipomyia* larvae (Fig. 10.4B) are also rather modified, with a stout, hypognathous head capsule and with thick hairs and tubercles on the head and body. These features are shared only with members of *Atrichopogon* (a genus that lacks pest species).

Female adults of *Culicoides* (Fig. 10.5A) have well-developed radial cells; they lack an R_{4+5} (although a forked intercalary vein may be present), have 13 antennal flagellomeres, with the terminal flagellomere rounded or tapering apically, a short anepisternal suture, and all have short cerci. Many species have distinctively patterned wings. Larvae (Fig. 10.5C) have a prognathous, well-developed head capsule, with a long, slender body that may or may not terminate in a group of elongate setae. These adult and larval features are shared with most other members of the subfamily Ceratopogoninae, and students of the group need to use generic keys to be certain they have a *Culicoides* at hand.

Identification guides to genera and species are listed by Borkent and Spinelli (2000; area south of the United States); Borkent and Grogan (In press; Nearctic); Boorman (1997; Palaearctic); de Meillon and Wirth (1991; Afrotropical); Wirth (1973; Oriental); and Bugledich (1999; Australia).

BIOLOGY

Life Cycle

Species of Ceratopogonidae all share a basic life cycle, including an egg stage that lasts a few days, four larval instars, a short-lived pupal stage, and male and female adults (only two species of *Culicoides* are known to be parthenogenetic). In northern and southern temperate regions, it is nearly always the third or fourth instar larvae that diapause and overwinter, although a few species have been recorded as overwintering as eggs. Therefore, emerging adults are generally more diverse in spring and early summer than later in the season. In subtropical or warm temperate areas, some species complete two or more generations before going into diapause as larvae. No adults are known to overwinter in cold regions, but at least some species do so in warm temperate and subtropical regions (e.g., Spain, Florida). Our understanding of the life cycle of tropical species is poor, but some species are clearly more abundant and diverse during the wet season in areas with a distinct dry–wet season. In tropical regions with a less distinctive seasonality, it appears that some species may be found as adults throughout the year.

Eggs are laid directly on or very close to the microhabitat in which the larvae later develop, either separately in loose groups, in gelatinous masses (*Dasyhelea*), or arranged in closely packed strings (a few genera of Ceratopogoninae). Eggs are not dessication resistant and need to remain at least moist to survive. Larvae hatch by using a small tooth on the dorsum of the head capsule to break open the egg shell. They begin feeding almost immediately on microorganisms or, in the case of some Ceratopogoninae, on small invertebrates. As larvae mature, some groups switch from feeding on microorganisms to becoming predators of other invertebrates (or a mix of the two). Larvae of the different subfamilies have different behaviors and modes of feeding, as follows. Larvae of *Leptoconops* are elongate and rather sluggish as they move through wet sand or soil, sweeping their heads as they browse on microorganisms throughout all the instars. They use their powerful mandibles (with very elongate apodemes) to help move themselves through the substrate. The immatures of *Austroconops* are undescribed but are currently under study; their larvae feed on microorganisms and some small invertebrates. Members of the Forcipomyiinae have a unique mode of feeding in which the larvae graze moist or submerged surfaces for microorganisms or algae, sweeping their heads back and forth with a rapid motion as they move along using their anterior and posterior prolegs. Larvae of semiterrestrial species secrete a sticky substance from some of their dorsal hairs to protect themselves against ant predation. *Dasyhelea* larvae feed on microorganisms or are sometimes scavengers, and they propel themselves rather sluggishly along the substrate in small aquatic habitats using their mouthparts and posterior proleg. Members of Ceratopogoninae have a distinctive larval motion when they are in substrate, very reminiscent of a gliding snake, seeking areas in which they may browse for microorganisms or to find their invertebrate prey. Ceratopogoninae larvae use a distinctive, very rapid undulating motion to swim through water. Within *Culicoides*, the later instars of some species feed entirely on microorganisms, while others feed primarily on other invertebrates, such as nematodes. Earlier lineages within the Ceratopogoninae that are predaceous eat their prey whole. Members of the Sphaeromiini and Palpomyiini attack large larvae of Chironomidae and other Ceratopogonidae, latching onto and working their head into the body cavity of their prey and sucking the contents out.

Ceratopogonidae pupae have a pair of respiratory organs on their thorax that they use to obtain air. The larvae of most species pupate near the surface of the substrate in which the larvae are found, with the anterior portion of their bodies sticking up into the air or, in the case of aquatic species, either floating at the surface with the apices of the respiratory organs in contact with the air or, for most species, at least partially emergent above the water level. The pupae of lake species may often be found in beach drift, where they are deposited by wave action; some of these species have peculiar adhesive abdominal sternites that allow them to climb up and stick to emergent substrate. The pupae of many semiterrestrial Forcipomyiinae pupate communally, with their heads abutting or closely associated and their abdomens pointing outward (a likely defense against ant predation). Ceratopogonidae pupae are rather lethargic and can, at most, only move their abdomens in slow circular motions.

Adults have a variable life span, likely depending on their feeding behavior (described later) and other factors. Some species with reduced mouthparts are almost certainly quite short lived, while at least the females of other species, going through two or more gonotrophic cycles, may live for up to 90 days. The adults of most species probably live for 1–3 weeks. Study of some pest species indicates that temperature, humidity, and rainfall affect adult survivorship and can increase the number of ovarian cycles in a population (and hence increase the potential of disease transmission).

The total life cycle in many species takes 1 year but under favorable, warm conditions, may take as little as 2 weeks from egg to adult (for multovoltine species). Some arctic species likely take at least 2 years to complete their life cycle. A variety of environmental factors have been studied for a few pest species, and these indicate that temperature, moisture levels, chemical composition of the substrate, and population density affect the rate of development of immatures.

Seasonality of adult midges affects the time of disease outbreaks. Except in areas where some adults are present year-round, organisms transmitted by Ceratopogonidae must find a reservoir in resident hosts. In some areas, the disease must be reintroduced either by dispersing, infected biting midges, or the introduction of infected hosts.

Immature Habitats

All biting midges are associated with aquatic, semi-aquatic, or merely very moist habitats, and the immatures require moisture for their continual survival. Even those larvae of species that live in temporary habitats will burrow down into the mud to remain wet or will, in one species of *Dasyhelea*, build a protective moist cavity.

The immatures of Ceratopogonidae are found in an amazing diversity of habitats, from rotting vegetation and manure to rivers and lakes. The earlier lineages of the family are restricted to smaller habitats, and the more recently evolved lineages of the nonpest tribes Heteromyiini, Sphaeromiini, Palpomyiini, and Stenoxenini are those that inhabit the larger aquatic habitats, such as rivers and lakes (although some members of these tribes may also be in smaller habitats). When in such substrate as wet soil, mud, or sand, larvae are rarely deeper than 5cm from the surface. Larvae of some species of biting midges are quite rare and difficult to find, but some species of *Culicoides* may occur in huge numbers in wet soils or manure, with abundances of over 10,000 per square meter recorded.

The immatures of *Leptoconops* are generally found in wet or damp sand or sandy soil near freshwater or marine beaches, but some species are known from seepage areas in desert oases, the wet margins of salt flats, the margin of vernal ponds in desert areas, in halomorphic, calcareous soil, or in cracked clay soils. Initial observations of the immatures of *Austroconops* suggest that these also live in wet soils. The larvae and pupae of only 13 species of *Forcipomyia* (*Lasiohelea*) have been described, and these are generally associated with mosses and algae growing on wood or soil.

At least one immature stage has been described for only 238 of the 1270 known species of *Culicoides*, so there is a huge gap in our understanding of where most species of this important genus live. Those that have been reared are known from many types of environments, but all of these are small or restricted aquatic or subaquatic habitats and include the following: damp or wet decomposing vegetation, wet leaf packs, manure, fungi, many different types of phytotelmata, tree holes, seed husks, springs, seeps, bogs, fens, swamps, pond and lake margins, mangrove swamps, salt marshes, stream and river margins; a few species are known from tree sap, and one species is recorded from crab holes. Although there are many types of habitat represented within *Culicoides*, it is clear that individual species are nearly always quite restricted in where they may be found, and their habitat can often be described with precision.

Adult Habits

Upon emergence, adults share a behavior common to many other nematoceran Diptera: The soft emerging adults escape from the pupal exuvia, rest briefly, and then fly to a nearby spot to complete the hardening of their cuticle. On average, males emerge slightly earlier than females, but there is a broadly overlapping emergence period for the two sexes. The males of most species form swarms of varying size near to the habitat from which they emerged, while others appear to fly singly, apparently searching for females. The swarming sites are often species specific and are generally associated with a particular marker, such as above a dark rock or next to a tree branch at a certain height above the ground. Females fly through these swarms and the male, attracted by the wing beat frequency of the female and secreted female pheromones, grabs the passing female. Then they either mate during flight or land on the substrate. Males are able to twist their genitalia 180° so that the flying or resting couple face away from each other. Copulation generally takes only a few minutes. Males produce and insert a two-chambered spermatophore into the female genitalia, and these sperm are then released and find their way to the female spermathecae. After mating, the females of species that require a blood meal disperse in search of a suitable host, generally in the near vicinity but in some species up to 4km in distance (wind-blown dispersal is discussed later). Therefore, adult sex ratios in Ceratopogonidae tend to be close to 1:1 near the site of emergence, with the number of males rapidly decreasing with distance from the site of emergence.

The females of some species of Ceratopogonidae are able to develop their eggs without any further food (*autogenous*), but most require a blood meal to do so (*anautogenous*). Vertebrate biters are able to find their host initially by tracking the downwind plumes of carbon dioxide produced by their intended hosts. As the biting midges get closer, further cues, such as the size, shape, and color of the host, movement, heat, and skin secretions, direct them to specific locations on their host. Most species appear to prefer specific sites on the host, often at a certain height off the ground. For example, *Culicoides phlebotomus*, a neotropical coastal species and an important vector of *Dipetalonema ozzardi* (a filarial worm) in humans, sucks blood from humans on beaches from about the knees down; it also feeds on leatherback turtles as these are on the beach laying their eggs. Virtually nothing is known about how invertebrate biters find their hosts other than the attractiveness of cantharidin for some species of *Atrichopogon* that feed on blister beetles and the likely use of vision by some female Ceratopogoninae to find swarms of midges upon which they subsequently feed.

Females of most *Leptoconops* species require up to 5–12 minutes to complete feeding, a significantly longer period of time to draw the blood meal than the other vertebrate feeders, which often take 2–5 minutes. Fully fed biting midges may take more blood than their total body weight.

Female *Leptoconops*, *Austroconops*, and *Forcipomyia* (*Lasiohelea*) are daytime biters, while most *Culicoides* are crepuscular or nocturnal feeders (a few are also diurnal). Some nocturnal species of *Culicoides* are more active when the moon is shining, and some coastal species respond to ocean tides. Specific environmental conditions are required before females can begin to fly and feed. Obviously, daytime temperatures have an important effect (no flight if it is too cold and, for many, none if it is too hot), and higher winds, generally those more than 8 km/h, will severely restrict or eliminate flight. Other factors have also been discovered that affect rate of attack: movement of host, time of day, humidity, type of surrounding habitat (open or heavily wooded), whether it is raining or not, and number of closely congregated hosts.

The females of biting midges feed on a truly remarkable variety of hosts (Fig. 10.1). Of the vertebrate feeders, *Leptoconops* has been recorded from lizards, birds, and mammals, *Austroconops* from kangaroos and humans, and *Forcipomyia* (*Lasiohelea*) from lizards, frogs, toads, and mammals. Species of the diverse genus *Culicoides* have been recorded from turtles (marine and freshwater), frogs, birds, and mammals, and one species even feeds on emergent mud skipper fish in southeast Asia. It seems likely that most birds and mammal species are bitten by at least one species of Ceratopogonidae. Some vertebrate biters are very host specific, while others incorporate a range of hosts.

The remaining species of biting midges attack invertebrates, and these fall into three phylogenetic groups. First, most members of the Forcipomyiinae [those other than the vertebrate-feeding *F.* (*Lasiohelea*)] attack insects much larger than themselves, essentially replacing their ancestral vertebrate hosts with such large insect hosts as katydids, stick insects, damselflies, crane flies, caterpillars, the wing veins of lacewings, dragonflies, and butterflies, and blister beetles. Second, three *Culicoides* species in the Old World feed on the insect haemolymph of engorged mosquitoes and *Phlebotomus* sand flies. Third, the females of the majority of Ceratopogoninae (other than *Culicoides*) fly into the swarms of other small Diptera (generally Chironomidae) and capture another midge of ap-

proximately similar size. A few derived species are known to attack larger individuals of swarming Ephemeroptera. Often the female and prey fall to the ground or settle on nearby vegetation, where the female injects a proteolytic enzyme into the body cavity. This dissolves the body contents of the prey, and the female then proceeds to suck out the contents. In some members of the Heteromyiini, Sphaeromiini, and Palpomyiini, the female enters a male swarm of her own species and, while mating occurs, pierces the head capsule of the male and sucks up the contents of her mate! Such females may later be found with the dried male genitalia still attached to their own abdomen, after the bulk of the dried male has broken off. Female *Culicoides* develop a distinctive brown to burgundy-red pigment in the abdomen after feeding on blood, and at least some species also develop further pigmentation of the abdominal sternites and tergites 2–3. Because biting midges become infected only after feeding on blood, these useful features allow for the identification of individuals that may be vectoring organisms. There is no evidence that viruses are transmitted transovarially in Ceratopogonidae, and early emerging individuals are therefore free of viral infection.

After obtaining a blood meal, females require a couple of days to develop their eggs, during which time some females seek out suitable habitat (depending how far they are from the site of emergence). The females of some *Leptoconops* species have a peculiar behavior in which they rest by burying themselves just under the sand surface. Other adult Ceratopogonidae rest on surrounding vegetation. Females, depending on the species, lay 30–450 eggs.

Both sexes of longer-lived Ceratopogonidae require fuel for continued flight, and this is obtained as nectar from flowers or honeydew (the excretions of feeding aphids). The males and females of the earlier lineages of Ceratopogonidae, which include those of the vertebrate feeders, are particularly abundant on many flowers, especially those of smaller size, and are likely important pollinators of some of these.

In general, species of Ceratopogonidae that are associated with temporary habitats tend to disperse more broadly, while those that are associated with permanent habitats stay closer to their original larval habitat. For example, because many species of *Culicoides* occupy such temporary habitats as manure, fungi, rotting vegetation, and small pools, adults of many of these species are also good dispersers. There are two aspects to dispersal: One is the wing-propelled flight that females use to find relatively nearby hosts. The second is wind-borne dispersal, in which adults are caught in wind streams, generally at heights of 0.5–2 km, 10–40 km/h, and at temperatures of 12–35°C. Adult *Culicoides* (and some other genera) have been collected as aerial plankton with planes and are quite able to get to distant islands. There is reasonable indirect evidence that wind-dispersed *Culicoides* spread diseases into new regions 130–200 (and perhaps 700!) km away from their sites of emergence.

MEDICAL AND VETERINARY IMPORTANCE

The most obvious effect that biting midges have is the immediate impact of their nasty biting habit. Some species occur in such high populations that outdoor living becomes intolerable and livestock are traumatized. Their bite, both in terms of pain and the resulting welt, often seems entirely out of proportion to their small size. Biting midges can interfere with tourism, forestry, and, in one instance in Russia, the building of a dam. A large number of bites may lead to skin lesions and subsequent infections due to active scratching by the afflicted host. Heavy biting leads to human dermatitis in some regions with large populations of *Leptoconops* or *Culicoides*. High biting frequency of horses, which may involve thousands of bites in 1 day, leads to hypersensitivity in many regions worldwide (also called *sweet itch*, *summer eczema*, or *Queensland itch*), with symptomatic frequencies of 20–75% commonly reported. In British Columbia, one survey reported 6% of horses were destroyed because of this allergenic reaction. The condition also affects some sheep.

It is worth noting that most species of vertebrate-feeding biting midges are of no or little medical or veterinary significance. In reality, we have no idea of what the natural host might be for the majority of species (although we can see from the mouthparts that they must feed on vertebrates). The true culprits, as biting pests and/or as vectors of diseases, are restricted to a relative handful of species. For example, in Costa Rica there are at least 148 species of *Culicoides*, but only 12 have been recorded biting humans. And of these, just three occur in large enough numbers to be considered serious pests of humans.

Biting midges are known to transmit three types of organisms, including 66 viruses, 15 species of protozoa, and 26 species of filarial nematodes, to a diversity of hosts (Table 10.1), but it is only a few viruses that seriously affect humans or livestock. The others organisms produce no or relatively minor symptoms. Of the viruses, about 50% are known only from species of *Culicoides*. The others are also recorded in other biting fly families or ticks (for some of these viruses *Culicoides* probably play a minor or no significant role as vectors in nature). It is striking that many viruses are known from Africa ($n = 28$) and Australia ($n = 31$). This may be a reflection of greater diversity in these regions, but it seems more likely that the relatively low numbers elsewhere are a consequence of the scarcity of thorough surveys.

Oropouche virus is one of the most important arboviral diseases of humans in tropical America. It produces severe flulike symptoms, with fever and vomiting and, although generally not fatal, can produce severe symptoms for up to 2 weeks. It is widespread, and surveys suggest that about 500,000 people have been infected since the early 1960s in Brazil alone; there have been at least 27 epidemics attributed to this disease. The most common vector is *C. paraensis*, although the virus is occasionally also transmitted by mosquitoes.

African horse sickness is a miserable viral infection of equids that can produce mortality rates of over 90%. Characterized by internal hemorrhage, it is the most serious infectious disease of horses. Although originally known only from Africa, the disease has incursions into India and, more recently, into southern Europe; in all areas, the distribution of the disease is restricted by the distribution of the species of *Culicoides* that act as vectors, primarily *C. imicola*.

Bluetongue is a widespread disease of cattle and sheep and is known worldwide between latitudes 35°S to about 50°N. The virus affects mouth and nasal tissues and produces fever, muscle weakness, and often death. The impact can be huge, with high mortality rates (e.g., in 1956–1960 nearly 180,000 sheep died in Spain and Portugal). Although it can be transmitted through semen, the primary mode of transmission between hosts is through the bite of species of *Culicoides*, with each region having its own species or group of species acting as vectors.

Epizootic hemorrhagic disease attacks wild and domestic ruminants, and although it is often undetected in most species, it can be quite deadly in some species of deer. In North America, where it is vectored by *C. sonorensis* and likely at least one other species, there have been repeated reports of high mortality in white-tailed deer, particularly in western regions.

Akabane virus also attacks ruminants, but its presence is usually not evident. However, the virus can cross the placental barrier in pregnant females, causing abortions and congenital abnormalities in the offspring. The disease is widespread in the Old World, including Australia, and is vectored by a variety of species of *Culicoides*. Other significant viral human diseases transmitted by biting midges are eastern encephalitis in the New World and Rift Valley fever and Congo viruses in Africa. Domestic animals are further affected by equine encephalosis and bovine ephemeral fever.

The most important protozoan disease transmitted by biting midges is *Leucocytozoon caulleryi*, which attacks poultry in southeastern Asia. This acute hemorrhagic disease especially affects young birds, with mortality rates sometimes over 20%. In some locations over 92% of individual birds carry antibodies to the virus. The vector, *C. arakawae*, is a common pest of poultry from Japan to southeast Asia.

Filarial worms are transmitted by biting midges to a number of vertebrate hosts, from frogs to mammals, but only a few cause serious harm. Three species of these nematode parasites are present in humans, but only *Dipetalonema ozzardi* in Central and South America appears to be mildly pathogenic. Infection rates of over 96% have been recorded in some coastal areas. Several species of *Onchocera* are present in domestic ruminants and are the cause of a number of mild pathological conditions in cattle and horses. Some of these are worldwide and are transmitted by a variety of species of *Culicoides* and, in Australia, additionally by *Forcipomyia townvillensis*.

CONTROL EFFORTS

The early 1900s saw the beginning of the systematic effort to control numbers of biting midges. Early methods of massive drainage, filling, and flooding schemes and the use of crude oil had limited impact on populations of pest species. The years following the Second World War saw the indiscriminate use of chlorinated hydrocarbons, which provided relief in some areas but caused the eventual development of resistance by biting midges. Because of the severe impact on nontarget organisms (especially birds), stability in natural environments, and biological magnification, these chemicals were replaced with organophosphates and carbamates, again with limited success, but followed by subsequent resistance in the midges.

Today there are few effective controls of most biting Ceratopogonidae. Nearly all pest species occur in such high numbers and many in such broadly distributed habitats (e.g., bogs, wet soil, manure, mangrove swamps) that it is impractical to eliminate immatures in their breeding habitats or to control the flying adults. To avoid large numbers of bites, humans may effectively use repellents (DEET is by far the most effective). Outdoor experiences are enhanced by choosing windy areas, wearing light-colored clothing, and restricting movement (or moving quickly!). Live-stock will experience fewer bites if stabled during peak biting periods, and using blowing fans can further reduce biting midge activity. Finely-screened windows or regularly applying repellents to screens reduces numbers as well.

Control of some pest species in more specialized habitats is possible, especially in more arid regions, where moist or aquatic habitats are very localized and hence vulnerable to manipulation. For example, reducing standing or slow-moving water that is contaminated with manure will decrease populations of *C. sonorensis* in some localities in North America. There have been some attempts at modification of natural habitats to destroy the immatures of other pest species, with very mixed results and considerable alteration of natural environments. At a time when so many ecosystems are already under great duress, this can hardly be a recommended control measure!

There has been some study of the natural predators and parasites of Ceratopogonidae in the hopes of discovering biological control agents. In spite of a variety of infesting organisms, including viruses, spiroplasmas, protozoa, fungi, and nematodes, none has yet been found to be highly effective in reducing most populations of Ceratopogonidae. *Bacillus thuringiensis*, so effective against a number of other insect pests, appears to have little affect on *Culicoides* and *Leptoconops*.

The vaccination of livestock seems to hold some promise of limiting some diseases, but further research is required to determine this.

CONCLUSION

It has become clear in recent years that researchers have seriously underestimated the number of organisms, particularly viruses, that are likely vectored by biting midges, and there are several areas in need of urgent study. As noted earlier, a huge amount of taxonomic work remains to be completed before we will have a reasonable understanding of the diversity of species present in most regions. Phylogenetic studies need to be completed to provide a logical framework to interpret studies of biting midges. The immatures of most Ceratopogonidae cannot be identified with confidence, so the larval habitat remains unknown for these species. It is amazing that, although bluetongue costs the U.S. livestock industry over $120 million a year, no one has yet systematically studied the immatures of the species complex that vectors the disease!

Our understanding of the native hosts of most biting midges is nonexistent. For example, in the New World, where humans have been present for only about the last 12,000 years, it is nearly certain that all the biting species in this region recorded from humans and domestic animals have alternate native hosts. Many of these wild animals in North America and elsewhere are possible reservoirs of diseases (as is known to be the case for a few *Culicoides* species and the diseases they vector). African arthropod–borne viruses, in particular, are very common and widespread, and it seems certain that a number of biting midge species will be found to vector these. Although much progress has been made in predicting where and when outbreaks of some diseases occur, these models need to be applied more broadly to Ceratopogonidae.

A further serious difficulty in studying the vector capabilities of species of biting midges is the failure to successfully colonize most species. Studies of larval behavior, life cycle, natural habitat, and other bionomic features need to be undertaken to provide the groundwork allowing for further laboratory study.

There are presently only a few people worldwide studying Ceratopogonidae in any regard, and this is the most serious barrier to further progress in understanding the role of biting midges as disease vectors. A new generation of students is urgently needed to tackle the large gaps in our understanding of this family of biting flies. With further study it is certain that significant and fascinating discoveries will be made!

Readings

Blanton, F. S., and Wirth, W. W. 1979. The sand flies (*Culicoides*) of Florida (Diptera: Ceratopogonidae). In *Arthropods of Florida and Neighboring Land Areas* 10, XV+ 204 pp.

Boorman, J. 1993. Biting midges (Ceratopogonidae). In *Medical Insects and Arachnids* (R. P. Lane and R. W. Crosskey, eds.). Chapman and Hall, pp. 288–309.

Boorman, J. 1997. 2.21 Family Ceratopogonidae. In *Contributions to a Manual of Palaearctic Diptera (with special reference to flies of economic importance)*. Science Herald, Budapest, Hungary, pp. 349–368.

Borkent, A., and Grogan, W. L. (In press) *Catalog of the New Myia World Biting Midges North of Mexico (Ceratopogonidae: Diptera)*.

Borkent, A., and Spinelli, G. R. 2000. Catalog of New World Biting Midges South of the United States (Diptera: Ceratopogonidae). *Contributions Entomol. Int.* 4: 1–107.

Borkent, A., and Wirth, W. W. 1997. World species of biting midges (Diptera: Ceratopogonidae). *Bull. Am. Mus. Nat. His.* 233.

Bugledich, E.-M. A. 1999. Diptera: Nematocera. In *Zoological Catalogue of Australia*, Vol. 30.1 (A. Wells and W. W. K. Houston, eds.). CSIRO, Melbourne.

Clastrier, J., and Wirth, W. W. 1978. The *Leptoconops kerteszi* complex in North America (Diptera: Ceratopogonidae). *U.S. Dept. Agric. Tech. Bull. No. 1573*.

Downes, J. A., and Wirth, W. W. 1981. Ceratopogonidae. pp. 393–421 In *Manual of Nearctic Diptera*, Vol. 1. *Agric. Canada Monogr.* 27.

Halouzka, J., and Hubalek, Z. 1996. Biting midges (Ceratopogonidae) of medical and veterinary importance (a review). *Acta Scientifiarum Naturalium Academiae Scientiarum Bohemicae Brno* 30(2): 1–56.

Linley, J. R., Hoch, A. L., and Pinheiro, F. P. 1983. Biting midges (Diptera: Ceratopogonidae) and human health. *J. Med. Entomol.* 20: 347–364.

de Meillon, B., and Wirth, W. W. 1991. The genera and subgenera (excluding *Culicoides*) of the Afrotropical biting midges (Diptera: Ceratopogonidae). *Ann. Natal Mus.* 32: 27–147.

Meiswinkel, R., Nevill, E. M., and Venter, G. J. 1994. Vectors: *Culicoides* spp. In *Infectious Diseases of Livestock with Species Reference to Southern Africa* (J. A. W. Coetzer, G. R. Thomson, and R. C. Tustin, eds.). Oxford University Press, New York, pp. 68–89.

Mellor, P. S., Boorman, J., and Baylis, M. 2000. *Culicoides* biting midges: Their role as arbovirus vectors. *Annu. Rev. Entomol.* 45: 307–340.

Wirth, W. W. 1973. Family Ceratopogonidae. In *A Catalog of the Oriental Region*, Vol. 1. Suborder Nematocera. University of Hawaii, Honolulu, pp. 346–388.

Wirth, W. W. 1977. A review of the pathogens and parasites of the biting midges (Diptera: Ceratopogonidae). *J. Washington Acad. Sci.* 67: 60–75.

Wirth, W. W., and Atchley, W. R. 1973. A review of the North American *Leptoconops* (Diptera: Ceratopogonidae). *Graduate Studies Texas Tech University* 5.

Wirth, W. W., and Grogan, W. L. 1988. The predaceous midges of the world (Diptera: Ceratopogonidae; Tribe Ceratopogonini). *Flora and Fauna Handbook* 4. E. J. Brill, XV+ 160 pp.

Wirth, W. W., and Hubert, A. A. 1989. The *Culicoides* of Southeast Asia (Diptera: Ceratopogonidae). *Memoirs of the American Entomological Institute* 44, 508 pp.

CHAPTER

11

Black Flies, the Simuliidae

PETER H. ADLER

INTRODUCTION

Black flies, though tiny, are one of the more recognizable families of flies, not so much because they are black—some are yellow or orange, others are striped—but because of their robust build, humpbacked thorax, cigar-shaped antennae, and prominent venation at the anterior margin of each wing. Adding to this distinct appearance is the habit of some species to fly into the face and onto the skin of those who enter their domain. The stealthier approach of other species often belies their ability to transfer agents of disease to humans and animals.

Often ranked second, after the mosquitoes, as the most medically and economically important group of bloodsucking insects, black flies live a double life. The females, because they require a blood meal to mature their eggs, can be miserable pests and vectors of disease organisms, but their larvae and pupae—among the most ubiquitous and abundant aquatic insects in the world—play a significant role in the trophic webs of rivers and streams. Thus, the black flies that create horrific pest problems and transmit agents of disease are often the same flies that comprise a large portion of the diet of fish used for food and sport.

Black fly is the most universal name for these creatures, having originated in the northeastern United States in the 1780s, but other names are often used: *buffalo gnat* in the Mississippi River Valley, *jejen* in Argentina, *Kriebelmücke* in Germany, *moshka* in Russia, *mosca del cafe* in Costa Rica, *pium* in Brazil, and *sandfly* in New Zealand. The diversity of names reflects the worldwide distribution of black flies, which are absent only from Antarctica and some islands of the high Arctic and open ocean, such as Hawaii.

The female's quest for blood drives problems ranging from a few annoying flies darting around the heads of guests at a barbecue to vast clouds of flies attacking cattle and inflicting so many bites that the animals die from toxic shock and exsanguination. Equally insidious are the disease organisms transmitted by black flies. Among the blood-borne parasites most commonly transmitted are the protozoa that cause avian leucocytozoonosis; the filarial worms that cause human onchocerciasis (river blindness), mansonellosis, and bovine onchocerciasis; and the virus that causes vesicular stomatitis in livestock. Some of these diseases create enormous economic and sociological burdens.

The study of black flies has advanced markedly in recent years. Significant advances have been made in the systematics of the family, including cytotaxonomy, molecular taxonomy, and phylogenetic reconstruction. Insights into the salivary secretions of black flies, the cues used for acquiring a blood meal, and the interactions with disease organisms have broadened our understanding of vector physiology while opening new avenues of research. Increased rigor in ecological studies has enhanced our ability to predict species distributions and to understand the dynamics of larval filter feeding. Improvements in the application of biological control and implementation of drug therapy for human disease have greatly reduced the threat of some black flies as pests and vectors of disease agents. The discovery that the visual problems of human onchocerciasis involve not only black flies and filarial worms, but bacteria as well, promises to spawn new methods of combating the disease.

GENERAL IDENTIFICATION AND MORPHOLOGY

The shape of the eggs is characteristic for the family: elliptical or ovoid, somewhat triangular in lateral view, and rounded at both ends. Externally, the egg has few conspicuous details. Its surface is remarkably smooth and covered by a sticky, gelatinous matrix. Probably the majority of species have a *micropyle*, a minuscule opening at one end through which the spermatozoa enter to effect fertilization. More useful in taxonomy is the size of the eggs—0.15–0.55 mm in length—and their arrangement after oviposition, whether deposited singly, in sheets, or in strings and whether laid on their sides or upright. The eggs are cream colored when first deposited but darken as the embryo matures.

The larval body format is uniform throughout the family: basically elongate, with a well-sclerotized head capsule, typically bearing a pair of labral fans, and two prolegs, one located ventrally on the prothorax and the other at the end of the body continuous with the abdomen. Overlain on this general plan are many features of taxonomic importance.

The head has the greatest number of parts and taxonomic characters (Fig. 11.1). Anteriorly, a pair of labral fans adorns the head. Each actually consists of three sets of fans on a stalk, with the primary fan being the most conspicuous. The rays of the primary fan are lined on one side with microtrichia that aid in straining particulate matter from the current. Food is acquired by holding each fan open in the current and periodically flicking it or folding it toward the mouth, where the brushes of the mandibles scrape the adherent matter into the mouth. The larvae of only about 25 species completely lack labral fans. These larvae live in habitats that are virtually free of particulate organic matter, such as pristine springs and glacial meltwaters. Antennae arise near the base of the fan stalk; each consists of three articles and a distal, conelike sensory structure.

The head capsule usually bears spots corresponding to the sites of muscle attachment. The spots vary in intensity and arrangement, providing reliable identification aids. A pair of black eye spots, the ocelli, is located on each side of the head. The ventral portion of the head bears two important taxonomic features: the hypostoma, a somewhat trapezoidal plate with an anterior row of teeth, and the postgenal cleft, an area of thin, clear cuticle that gives the impression that the head capsule is incised to various depths. The hypostoma scrapes food from the substrate and, in conjunction with the mandibles, cuts the lines of silk spun from the salivary glands.

The body is characteristically pigmented, with the color and pattern useful in identification. Each proleg of the body is fitted with a circlet of tiny hooks used in pulling the silk and anchoring the larva to silken pads spun onto objects in flowing water. Minute hairs, tubercles, and bulges also can be found on the abdomen. The final-instar larva is recognized by a dark spot, the gill histoblast, on each side of its thorax indicating that pupation is imminent.

The body of the pupa reflects the external features of the adult, such as the head, legs, and wings, although these structures are closely appressed to the body. The two most characteristic attributes of the pupa are its silken cocoon and a pair of respiratory organs, or gills, that arise from the thorax. The cocoon varies among species from a small, loosely spun bit of silk covering only part of the abdomen to an elaborate slipper-shaped or boot-shaped shroud, often with various loops, anterior projections, and windows. Rows of tiny hooks and spines on the pupa secure it within the cocoon. The gill is probably the most diverse structure in the entire family, consisting of one to more than 150 filaments, some slender and others variously inflated or even balloon-like in the most extreme forms.

The adult black fly is robust and compact with rather short appendages, somewhat resembling the American bison or buffalo, the inspiration for the common name *buffalo gnat*. The wings typically span 5–10 mm. The head of most black flies differs markedly between males and females. That of the female is

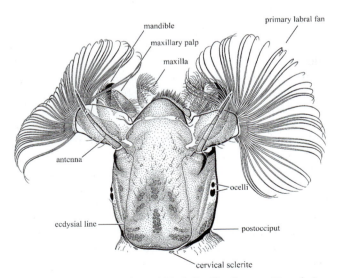

FIGURE 11.1 Head of larval black fly, *Simulium* sp.: Dorsal view, tilted slightly to left. Original illustration by R. M. Idema.

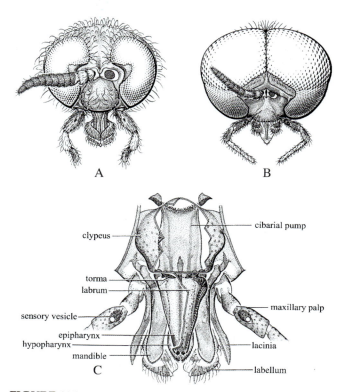

FIGURE 11.2 Head and mouthparts of adult black flies, anterior view: (A) Female of *Cnephia dacotensis*, left antenna omitted; (B) male of *Simulium decorum*, left antenna omitted; (C) mouthparts of female of *Simulium vittatum*, cutaway view. Reproduced from *Manual of Nearctic Diptera*, Monograph No. 27, Agriculture and Agri-Food Canada, with the permission of the Minister of Public Works and Government Services, 2002.

mouthparts are much like those of the female, but they lack the toothed edges of the mandibles and laciniae and the teeth at the apex of the labrum. The females of some species (e.g., the *Simulium ochraceum* complex) that transmit filarial nematodes to mammals have sharp cibarial teeth at the junction with the pharynx. These teeth damage and kill some of the microfilariae that are ingested with a blood meal, affording the fly some protection from the parasites.

The arched thorax is well developed to house the powerful indirect flight muscles of the forewings. The wing venation is not only characteristic of the family, but also useful in distinguishing genera. The legs feature a goldmine of taxonomic information as well as myriad sensory structures. The claws of the male are fitted with a large basal hood of grooved cuticle designed to grasp the hairs of the female during mating. Females that feed on birds typically have slender, curved claws with a conspicuous basal, thumblike lobe to aid in grasping feather barbules. Mammal-feeding females have similar claws but lack any kind of basal projection or have at most a tiny, basal tooth.

The weakly sclerotized abdomen allows for flexibility in mating and, in females, expansion to accommodate egg development. Situated at the end of the abdomen are the terminalia, consisting of the genitalia and associated structures, such as the cerci. Familiarity with the parts that comprise the genitalia and the nuances associated with ventral, terminal, and lateral views is essential for species-level identification of black flies.

dichoptic, the eyes being separated by the frons (Fig. 11.2A), whereas the head of the male is holoptic, the eyes meeting along the midline (Fig. 11.2B). The compound eyes of the male are composed of large upper facets and small lower ones. The eyes of the female are composed entirely of small facets, like the ventral ones of the male. Arising between the eyes are the characteristic antennae, consisting of a basal scape, a pedicel, and a terminal flagellum of seven to nine flagellomeres. An equally conspicuous pair of five-segmented maxillary palps arises at the base of the proboscis. They are sensory in nature and house the Lutz's organ, which detects carbon dioxide emitted by avian and mammalian hosts.

The mouthparts—the root of all pest and disease problems—are arranged in a compact, downward-projected proboscis (Fig. 11.2C). They are partially enveloped from behind by two large, fleshy lobes, the labella. The labrum, which has a food channel on its inner surface, forms the front of the proboscis. Male

CLASSIFICATION

The entire family Simuliidae includes about 1,800 formally named species arranged in two subfamilies: the Parasimuliinae, consisting of five evolutionarily basal, nonbloodsucking black flies endemic to the forests of North America's Pacific Northwest, and the Simuliinae, accounting for the vast remainder of the world's species. Within the Simuliinae, two tribes, the Prosimuliini and Simuliini, are recognized. The Prosimuliini occupy the northern hemisphere, whereas the Simuliini are worldwide and include 18 of the 23 simuliine genera and 93% of the species. The largest and most widely distributed genus is *Simulium*, which includes about 40 subgenera, 80% of all known species, and more than 90% of the major pests, including all but a few of the known vectors of disease agents. The major identification keys for taxonomic groups and geographic faunas of the world's black flies have been tabulated by Crosskey and Howard (1997).

The simuliid species of the world are divided rather unequally among the six principal zoogeographic regions. Unlike most insect groups, the number of species of black flies is greater in the temperate areas than in the tropics. Those in the Palearctic Region outnumber all others, with more than 600 known species, followed by the Neotropical Region with nearly 400, the Nearctic Region with more than 250, the Afrotropical and Oriental Regions, each with more than 200, and the Australasian Region with about 150.

The number of genera in the world is expected to remain rather constant, but many more species await discovery. These new species will be discovered by prospecting in poorly known areas, such as the Himalayas, and through chromosomal and molecular studies. Both chromosomal study (i.e., cytotaxonomy) and molecular investigation will reveal sibling, or cryptic, species (i.e., reproductively isolated, biologically unique, yet morphologically similar species).

Sibling species are routinely discovered through detailed study of certain large banded chromosomes, known as polytene, or giant, chromosomes, which are best developed in the larval silk glands. The paired silk glands extend the length of the body, enlarge in the abdomen, and double back on themselves. In the enlarged portion of the glands, these giant chromosomes reach their greatest expression. When stained and squashed on a microscope slide, the chromosomes show patterns of light and dark bands that are typically species specific and that can be used to demonstrate a lack of hybridization.

Sibling species are known from every continent inhabited by black flies. In North America, about one-quarter of the known species were discovered chromosomally. Perhaps the most famous example of sibling species involves the African black fly, *Simulium damnosum*, a complex (i.e., group) of as many as 40 sibling species, of which at least a dozen transmit the filarial worm (*Onchocerca volvulus*) that causes human onchocerciasis. In Latin America, *Simulium metallicum*, another vector of this parasite, is actually a complex of more than seven species. In North America, the infamous white-stockinged fly, *Simulium venustum*, is a complex of 10 species, all of which are unforgiving pests of humans and domestic animals and vectors of various parasites to mammals and birds. While simuliid vectors clearly include sibling species, the possibility must be recognized that the parasites they transmit also could consist of sibling species.

An even greater level of taxonomic complexity involves homosequential sibling species, those species that are virtually identical both morphologically and in the banding sequence of their chromosomes yet are reproductively isolated—the taxonomist's nightmare.

The new era of molecular taxonomy should not only aid the resolution of homosequential sibling species, but also reveal more of these hidden species.

BIOLOGY

Life History and Development

Black flies are holometabolous insects that pass through egg, larval, and pupal stages before becoming an adult (Fig. 11.3). The life cycle can be repeated from 1 to 20 or more times a year. Northern and high-altitude species are typically univoltine, completing only one generation per year. Nearly all species of the tribe Prosimuliini are univoltine. Many temperate and most tropical species are multivoltine, producing wave after wave of flies throughout the year. Some of the African vectors of the filarial worms that cause human onchocerciasis might complete more than 20 generations per year, while the multivoltine North America species complete about seven.

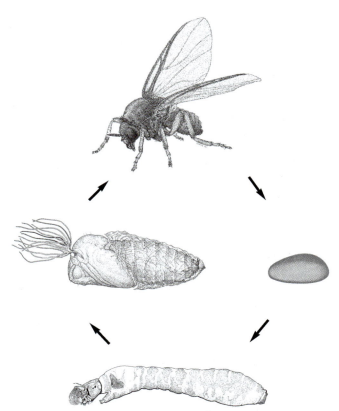

FIGURE 11.3 Life cycle of the black fly. Female, pupa, and larva reproduced from *Manual of Nearctic Diptera*, Monograph No. 27, Agriculture and Agri-Food Canada, with the permission of the Minister of Public Works and Government Services, 2002.

All preadult life stages develop in flowing water. Eggs are deposited in or alongside flowing water. They are not resistant to desiccation and can succumb to drying in a few hours, but they are able to remain hydrated in moist cracks and sediments of temporary flows. Hatching begins a few days to a year or more after the eggs are laid. Within the egg, the prolarva presses against its shell with a chitinized tubercle, the egg burster, located on its head. After rupturing the egg, the newly hatched larva spins a silken pad on the substrate and secures itself to the pad with the hooks of its posterior proleg.

Larvae can be found on practically any object in the current: sticks, stones, fallen leaves, trailing vegetation, and a variety of human debris, particularly smooth objects such as glass, aluminum, and plastic, which also can be exploited as artificial substrates for collecting larvae and pupae. The most unusual objects of attachment are mayfly larvae and river crabs and prawns in the streams of tropical Africa; mayfly larvae also are used as attachment sites in Central Asia. These carriers are sought by the larvae, for the eggs apparently are not laid on them. About 30 species of black flies obligatorily live as larvae and pupae on these hosts, a condition known as *phoresy*. Among phoretic black flies are at least two members of the *Simulium neavei* species group, which are vectors of *Onchocerca volvulus*. Phoretic larvae have a number of morphological adaptations, such as flattening of the labral fans. The nature of the relationship and its influence on the fitness of the black fly and the carrier are poorly understood.

Firmly secured to an object in the water, the larva leans in the direction of the current to filter its food (Fig. 11.4), a behavior that occupies the vast majority of larval life. Filter feeding is generally passive, much like room service, with the current bringing particulate matter to the labral fans. Particles smaller than 350 μm in diameter are consumed, as is dissolved organic matter that has flocculated. Larvae also graze adherent material such as diatoms and bacteria from the surface around them. The species that lack labral fans are obliged to obtain all of their food by engulfing smaller organisms or by grazing, using their mandibles, hypostomal teeth, and modified labrum.

Besides feeding, larvae also defend themselves against predators, jostle for position among themselves, and molt. Defensive and agonistic behaviors often involve curling into a tight C shape or drifting downstream on a silk lifeline and settling at a site remote from the disturbance. The molting process occurs five to eight times in the life of the larva, resulting in six to nine (typically seven) instars. Parasitism and poor nutrition can produce up to 11 instars. The

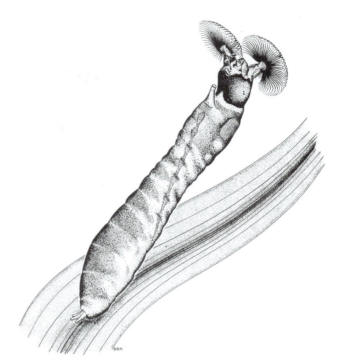

FIGURE 11.4 Larva of *Simulium vittatum* in typical filter-feeding position. By permission from D. C. Currie.

entire larval stage lasts a week to 6 months or more, largely depending on water temperature. Species at northern latitudes spend 6–9 months in the larval stage, while species in the baking sun of the tropics can race through the larval stage in less than 1 week, perhaps in as little as 4 days in some members of the *Simulium damnosum* complex.

When the larva reaches maturity, a silken cocoon is spun on an object in the water and pupation occurs as the larval skin is shed, allowing the gill histoblasts to uncurl and become the breathing organs (gills) of the pupa. These gills are specially adapted to work in water as well as out of water should the stationary pupa find itself high and dry as the water level drops. The pupal stage lasts less than a month, again depending on water temperature. The adult, fully formed within the pupal skin, begins to expel air through the spiracles of its respiratory system, eventually splitting its casing and shooting to the surface, aided by a bubble of air held in place by the nonwettable hairs on its body. Emergence is a diurnal phenomenon, often occurring in the morning. Most black flies are protandrous, with males emerging a day or two before females.

Natural Enemies

Black flies in all their life stages suffer the burden of natural enemies—parasites and predators (Chapter 43). Predators such as birds, fish, and arthropods, while diverse in kind, are typically generalists that prey on many kinds of organisms. Parasites come in the form of bacteria, fungi, helicosporidia, ichthyosporeans, mermithid and filarial nematodes, microsporidia, protists, stramenopiles, water mites, and viruses, nearly all of which are specialized to attack only black flies. Most parasites manifest themselves in the larval stage and are readily seen in living and fixed material. Larvae with patent microsporidian infections, for example, have large, dense, white or red cysts in the abdomen. Those infected with mermithid nematodes often sacrifice the entire body cavity to the worms. An infection can remain minimal in the larval stage and carry through to the adult. Mermithid nematodes, for example, can either exit the larva, killing it, or escape from the adult, rendering it sterile.

Habitats of the Immatures

The habitats of immature black flies are exceptionally diverse across the family. The females of each species, however, are quite specific where they choose to lay their eggs, and the larval habitats of individual species are, therefore, quite predictable. Permanent rocky streams come to mind as the typical habitat for black flies, but virtually no natural, flowing-water habitat remains uncolonized by black flies at some point during the year. Watercourses of all sizes, whether permanent or intermittent, are colonized.

Enormous rivers several kilometers or more in width sometimes support such colossal populations that they yield more than a billion flies per day per kilometer of river. One of the most notorious residents of the eastern United States, *Simulium jenningsi*, is a typical big-river species. So, too, are some of the vectors of the causal agent of human onchocerciasis: *Simulium albivirgulatum* and the *Simulium damnosum* complex in Africa and *Simulium oyapockense* in Brazil. Large rivers of North America's coastal plain accommodate sizable populations of black flies, including the continent's all-time villain, *Cnephia pecuarum* (southern buffalo gnat), as well as *Cnephia ornithophilia*, a vector of the agents (*Leucocytozoon* sp.) of avian leucocytozoonosis. In the summer these rivers appear slow and placid, but with the winter rains they swell into their mighty flood plains, becoming bitterly cold and dangerously swift while providing enormous breeding grounds for the flies.

At the other size extreme are the trickles and minute seepages, often so small and inconspicuous that they appear barely to move and are likely to be overlooked. Larvae of numerous species, such as members of the genus *Greniera* and the subgenera *Hellichiella* and *Nevermannia*, that live in these slow-flowing, often nutrient-poor trickles typically have long antennae and large labral fans to capture their food. The adults of these small-stream species often feed on birds and transmit the agents of leucocytozoonosis.

Highly productive waterways, such as the outflows of impoundments (e.g., beaver ponds and artificial lakes) are commonly colonized by black flies. Some species, such as *Simulium noelleri* in Europe, are outlet specialists and can achieve larval densities of more than 1 million larvae per square meter. The proliferation of artificial impoundments has greatly benefited species such as *Simulium decorum* and *Simulium vittatum* while destroying the habitats of other species. The alteration of habitat by impoundment or artificial warming can create pest and disease problems by providing ideal breeding grounds for species specifically adapted to these habitats. Larvae of some species that live in productive rivers, such as the *Leucocytozoon* vector *Simulium meridionale* of North America, have small labral fans, perhaps to cope with the heavy particulate loads that could overwhelm the filtering apparatus.

Thundering waterfalls and other torrential habitats test the mettle of any aquatic insect. Larval black flies must be able to cling to a substrate while the current moves at velocities that can approach the top speeds at which water flows in nature. Larvae that live in these watercourses, such as species of the subgenera *Hemicnetha* in North and South America and *Daviesellum* in Thailand and Malaysia, have an elongate, streamlined body, many hooks on their posterior proleg, short antennae, stout labral fans, and remarkably adhesive silk.

Black flies have colonized some of the most severe aquatic habitats on Earth: glacial meltwaters, hot springs, and subterranean flows. Glacial meltwater streams of northern North America contain almost no suspended food, only pulverized stone that is released into the current as the glaciers melt. Larvae in these habitats lack labral fans and must graze their food from the rocks. Females of these species are rarely pests or vectors; most do not have biting mouthparts. Opposite the glacial meltwaters in temperature are hot springs and desert streams. Larvae of North American species such as *Simulium argus* and *Simulium tescorum* have been found at temperatures above 30°C. Icelandic streams formed from condensed volcanic steam support populations of *Simulium vittatum*. Among the most unusual black flies are the five subterranean species of the genus *Parasimulium* in the Pacific

Northwest of North America. The immature stages live beneath the surface, where the poorly sorted basaltic rubble of past volcanic activity offers a wealth of space through which water can flow. The larvae resemble cave-dwelling organisms, lacking pigmentation and ocelli. Adults lack biting mouthparts.

Despite the remarkable adaptability of black flies to the entire panoply of flowing waters, no species has successfully exploited standing water or even the wave-washed shores of large lakes. Nonetheless, larvae that become stranded in pools as the streams dry up are sometimes capable of pupating and producing adults. A few unusual records exist for *Simulium adersi*, a vector of *Leucocytozoon* and *Trypanosoma* spp., developing—probably opportunistically—on the windy shore of Lake Victoria in Uganda. No black fly develops exclusively in brackish waters, although many species live in streams and rivers at their confluence with the ocean, thus experiencing saltwater and tidal wash.

Adult Habits

Black flies are almost exclusively diurnal, so all activities must be accomplished between dawn and dusk. The freshly emerged adult is soft and usually seeks a resting place to harden and tan its cuticle. Mating occurs shortly after emergence, and some females are probably intercepted and inseminated by males before they reach a resting site. The most well known, although rarely observed, mating behavior is swarming. Males aggregate in a loose swarm of a few to many individuals above or beside a landmark such as a waterfall, a host of the female, or a tree branch. Females that enter the swarm are readily detected by the large upper facets and corresponding retinular cells of the male eye, which is specifically designed to notice small, moving objects against the sky. The females are quickly seized and mated, the actual coupling typically lasting a few seconds to several minutes, long enough to pass the spermatophore containing the sperm bundle. A tiny fraction (<1%) of the world's species, primarily those at high altitudes and latitudes, meet and couple on the ground at the emergence site. About 10 species are parthenogenetic; males do not exist. For many years, the reluctance of black flies to mate in the laboratory hindered experimental investigations. In the past 25 years, a number of species have been colonized, including some of the African vectors of *Onchocerca volvulus*, but colony maintenance is labor intensive (Chapter 55).

Both sexes of nearly all species probably drink water, and most also take a sugar meal, obtained opportunistically from floral nectar, homopteran honeydew, or plant sap. The sugar meal is stored in the crop and used as energy, primarily to sustain flight and, in the female, to initiate oogenesis. The strong flight muscles of most species can take the female flies great distances in search of blood. While most species probably disperse no more than about 15 km, some species, especially the large-river species, can travel great distances, often aided by the wind: at least 55 km for *Simulium jenningsi*, up to 225 km for a member of the *Simulium arcticum* complex in Canada, and more than 500 km for some members of the *Simulium damnosum* complex in West Africa. On the other hand, *Simulium neavei*, the East African vector of *Onchocerca volvulus*, typically disperses no more than 4 km, and *Simulium slossonae*, a North American vector of *Leucocytozoon*, no more than 7 km.

Blood is required by the females of most species, although fewer than 10% (mostly in far-northern lands) do not have biting mouthparts and cannot cut vertebrate skin. The nonbiting species are obligatorily autogenous, maturing their eggs without blood but with the reserves carried over from the larval stage. Some blood-feeding species are facultatively autogenous, maturing their first batch of eggs without blood. These facultatively autogenous species develop under optimal larval food regimes and cooler temperatures. The maturation of subsequent egg lots, however, requires a blood meal. Anautogenous species need blood for each ovarian cycle (i.e., maturation of each egg batch) and are the most competent vectors of disease agents because they are more likely to acquire a parasite in one blood meal and transmit it during a subsequent meal.

All hosts of black flies are mammals and birds, with no known exceptions. The hosts are located by a series of cues, primarily visual (e.g., color and shape) and chemical (e.g., carbon dioxide). The oft-repeated observation that some people are more prone than others to attract black flies is based on greater production of exhaled carbon dioxide. Some of the forest-inhabiting members of the *Simulium damnosum* complex are attracted to human sweat, especially on the lower body.

Black flies tend to be either ornithophilic or mammalophilic. Some species are highly host specific. The most extreme example is that of *Simulium euryadminiculum*, which is attracted to the secretions of the uropygial gland (i.e., oil gland) of its only known host, the common loon (*Gavia immer*). Some species feed on a certain taxonomic group of hosts (e.g., ducks), whereas others feed on taxonomically unrelated hosts in a particular habitat (e.g., various birds in the tree canopy) or on hosts of a particular size (e.g., large mammals). Still other species (e.g., *Simulium venustum*) are highly catholic in their choice of hosts, taking blood from birds and mammals of all sizes in a wide variety of habitats. At least some members of the *Simulium damnosum* complex are also indiscriminate feeders, taking blood from humans and other mammals as well as wild and domestic birds. No black fly is specific to humans; fewer than 30 species regularly feed on humans. For many species of black flies, the hosts are simply unknown.

When a fly lands on a host, cues such as heat initiate probing. Once a female begins to bite, it retracts its labella and pushes against the skin, which is stretched tightly by tiny teeth and spines at the tips of the hypopharynx and labrum (Sutcliffe and McIver 1984). The minutely serrated mandibles then snip the host flesh, while the laciniae, with their backward-directed teeth, anchor the mouthparts in the wound (Chapter 2). The fly's decision to feed involves cues, or phagostimulants, such as adenosines (e.g., ATP). All the while, salivary compounds—anesthetics, vasodilators, and anticlotting factors—are pumped onto the skin and into the bite (Chapter 28). Capillaries are severed, and the pool of blood is imbibed via the food channel, which runs along the labral food canal beneath the overlapping mandibles. A tight seal is formed around the bite by the membranous cuticle of the proboscis. Uptake of the blood is facilitated by muscular pumps in the cibarium and pharynx. About 3–6 min are required to feed to repletion. The blood meal is then directed to the midgut, where it is digested.

The replete female flies from the host or drops to the ground or into the nest of its host to rest and convert the blood into a future generation of black flies. From 200–800 eggs are produced by anautogenous black flies in the first ovarian cycle. A fly that has deposited its eggs is referred to as *parous*, whereas one that has never laid eggs is *nulliparous*. Flies can be physiologically aged (i.e., parity can be determined) by examining the condition of the ovaries, fat body, and other structures, such as the Malpighian tubules. Host-seeking behaviors, such as the distance dispersed to find a host, can vary between parous and nulliparous flies. Most females probably live no more than a few weeks, completing two to four ovarian cycles during this time. The longevity record in nature is about 12 weeks.

Having matured a batch of eggs, the female returns to an appropriate habitat, sometimes the natal site, to deposit its eggs. The females of some species deposit their eggs in masses while stationed on a moist object in or beside a watercourse. Others scatter their eggs in the water a few at a time while dipping to the surface. Still others dab them onto moist surfaces while repeatedly descending from a hovering flight. And some

practice all three oviposition techniques. The species that lay their eggs in masses often do so communally, millions of eggs accumulating on objects such as a dam face or a stone. Communal oviposition is facilitated, at least in species of the *Simulium damnosum* complex, by pheromones that emanate from the eggs.

MEDICAL AND VETERINARY IMPORTANCE

The problems caused by black flies fall roughly into three categories: (1) nuisance, caused by swarming, crawling, and biting; (2) trauma, caused by loss of blood and injection of salivary toxins; and (3) disease, caused by transmission of parasites. Each of these broadly defined problems is inflicted by particular species, and each affects humans, domesticated animals, and wildlife. Remarkably, the sum total of significant pest and vector species is no more than about 5% of all species worldwide. Nonetheless, many species transmit disease organisms that go largely unnoticed in wildlife.

Nuisance and Biting Problems

Nuisance species are often a problem, not so much for their biting but for their persistent habit of flying into the eyes, ears, nose, and mouth, and crawling through the hair and on the skin. A classic example is *Simulium jenningsi* in eastern North America, for which one of the world's largest-ever control programs against black flies has been focused since the early 1980s. Nuisance thresholds vary from one or two flies per person on a golf course to much higher levels for communities of seasoned veterans. Some species, such as those of the *Simulium venustum* complex in Canada and New England, add vicious biting to their nuisance repertoire. Severe infestations of nuisance and biting black flies can negatively impact tourism, agriculture, forestry, and recreation. The effects of nuisance black flies on poultry and livestock are manifested through weight loss, stress-related illnesses, and reduced egg, milk, and meat production, all of which cost producers dearly.

Trauma

Nuisance problems associated with occasional biting grade into more serious attacks that can cause psychological stress and physical trauma. Assaults by black flies have caused serious health problems and death. Substantial carnage in the livestock industry has been inflicted by about seven legendary species, most of them emerging in astronomical numbers from large rivers: *Austrosimulium pestilens* in Queensland (Australia), *Cnephia pecuarum* in the lower Mississippi River Valley, *Simulium colombaschense* along central Europe's Danube River, the *Simulium arcticum* complex and *Simulium luggeri* on the Canadian prairies, and several species such as *Simulium reptans* in central Europe. Livestock deaths have been frequent—as many as 22,000 along the Danube in 1923. Death is caused by a number of factors, including exsanguination from withdrawal of excessive blood, suffocation from clogged respiratory passages, and trampling and crushing from stampedes. But probably the major mortality factor is toxic shock from the injection of salivary secretions, a condition known as *simuliotoxicosis*. Poultry also have been killed by some of these species, especially by the turkey gnat, *Simulium meridionale*, in midwestern North America. The massive onslaughts are now largely of historical interest as a result of habitat alteration and management programs, although a few deaths still occur in some areas.

Human deaths from black fly attacks have never been documented unequivocally, but reactions to the bites come in various forms. Most people experience a small, itching welt or hematoma (Fig. 11.5). Secondary infections can result from repeated scratching, and strategically placed bites can cause the eyes to swell shut. More severe reactions, often associated with more bites, can occur. *Black fly fever* is a generalized condition familiar to citizens of New England, involving fever, nausea, headaches, and swollen lymph nodes. In some children and adults, allergic reactions can necessitate hospitalization.

Diseases

During the process of feeding on avian and mammalian hosts, blood-borne parasites often are transmitted. Most of these parasites are specific to black flies, completing part of the life cycle in the flies. During a subsequent blood meal, the parasites can be transmitted to another host, precipitating disease. The estimated time between successive blood meals is 2–7 days in the *Simulium damnosum* complex and about 3–4 days in the Latin American vectors of *Onchocerca volvulus*. At least 15 protozoans, 12 filarial nematodes, and 1 virus are transmitted by black flies (Table 11.1). The bacterium responsible for tularemia can be mechanically transmitted, though rarely, by black flies. The protozoa are transmitted by ornithophilic black flies, whereas the filarial nematodes are transmitted by both mammalophilic and ornithophilic species. Probably many more parasite–simuliid relationships remain to be documented among wildlife.

The protozoa transmitted by black flies are in two genera. The most malignant of them are species of *Leucocytozoon*, which cause a malaria-like disease, leucocytozoonosis, in birds. The species of *Trypanosoma* transmitted by black flies are less insidious. Surveys in many areas of the world have shown that probably most, if not all, species of birds harbor *Leucocytozoon* spp. These taxonomically difficult protozoa are generally viewed as host specific at the avian family level. About 60 species of *Leucocytozoon* are considered to be valid, but the vectors of only about 11 are known and all but 1 are black flies.

The *Leucocytozoon* life cycle (Fig. 11.6) involves the acquisition of gametocytes from infected birds, asexual and sexual development of the parasite in the fly, and transmission of sporozoites to another bird in which asexual development occurs. The effects of most species of *Leucocytozoon* are not overt; however, domestic birds in North America, especially turkeys, ducks, and geese, have suffered. Turkeys are affected by *Leucocytozoon smithi*, whose primary vectors are *Simulium meridionale*, and *S. slossonae*. Ducks and geese are parasitized by *Leucocytozoon simondi*, whose principal vectors are *Simulium anatinum* and *S. rugglesi*. Birds with chronic infections have depressed immune systems and decreased reproductive capacity. Mortality can range from 5% to 100%. In wild birds, the young can be quite vulnerable in years of low food supply or inclement weather.

The relationship between black flies and avian trypanosomes has been little studied, and the number of

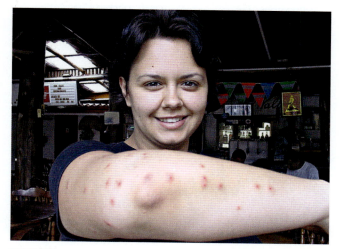

FIGURE 11.5 Bites on the arm caused by a member of the *Simulium metallicum* complex, one of the vectors of *Onchocerca volvulus*; the bites of these tropical species bleed little. Photograph by D. C. Currie.

TABLE 11.1   Disease Organisms Transmitted by Blackflies

Disease organism	Vector species	Host	Range
Filarial nematodes			
Dirofilaria ursi	*Simulium venustum* complex	Bears	North America
Mansonella ozzardi	*S. amazonicum* complex, *S. argentiscutum*, *S. exiguum* complex, *S. oyapockense* complex, *S. sanguineum*	Humans	Latin America
Onchocerca cervipedis	*Prosimulium impostor*, *S. decorum*, *S. venustum* complex	Deer, moose	North America
Onchocerca dukei	*S. bovis*	Cattle	Africa (Cameroon)
Onchocerca gutturosa	*S. bidentatum*, *S. erythrocephalum*	Cattle	Japan, Ukraine
Onchocerca lienalis	*S. erythrocephalum*, *S. jenningsi*, *S. ornatum* complex, *S. reptans*, *S. arakawae*, *S. daisense*, *S. kyushuense*	Cattle	North America, Russia, western Europe, Japan
Onchocerca ochengi	*S. damnosum* complex, *S. hargreavesi*	Cattle	West Africa
Onchocerca ramachandrini	*S. damnosum* complex	Warthogs	West Africa
Onchocerca possibly *skrjabini*	*S. aokii*, *S. arakawae*, *S. bidentatum*, *S. daisense*	Deer	Japan
Onchocerca tarsicola	*P. tomosvaryi*, *S. ornatum* complex	Deer	Western Europe
Onchocerca volvulus	Africa: *S. albivirgulatum*, *S. damnosum*, *S. dieguerense*, *S. ethiopiense*, *S. kilibanum*, *S. konkourense*, *S. leonense*, *S. mengense*, *S. neavei*, *S. rasyani*, *S. sanctipauli*, *S. sirbanum*, *S. soubrense*, *S. squamosum*, *S. woodi*, *S. yahense* Latin America: *S. callidum*, *S. exiguum* complex, *S. guianense* complex, *S. incrustatum*, *S. limbatum*, *S. metallicum* complex, *S. ochraceum* complex, *S. oyapockense*, *S. quadrivittatum*, *S. roraimense*	Humans	Africa, Latin America
Splendidofilaria fallisensis	*S. anatinum*, *S. rugglesi*	Ducks	North America
Protozoa			
Leucocytozoon cambournaci	*P. decemarticulatum*, *Cnephia ornithophilia*, *S. aureum* complex, *S. vernum* group	Sparrows	North America
Leucocytozoon dubreuili	*P. decemarticulatum*, *C. ornithophilia*, *S. aureum* complex, *S. vernum* group	Thrushes	North America
Leucocytozoon icteris	*P. decemarticulatum*, *C. ornithophilia*, *S. anatinum*, *S. aureum* complex, *S. vernum* group, *S. venustum* complex	Blackbirds	North America
Leucocytozoon lovati	*S. aureum* complex, *S. vernum* group	Grouse	North America
Leucocytozoon neavei	*S. adersi*	Guinea fowl	Eastern Africa
Leucocytozoon sakharoffi	*P. decemarticulatum*, *S. aureum* complex, *S. angustitarse*	Corvids	North America, Europe
Leucocytozoon schoutedeni	*S. adersi*	Chickens	Eastern Africa
Leucocytozoon simondi	*C. ornithophilia*, *S. anatinum*, *S. fallisi*, *S. rendalense*, *S. rugglesi*, *S. venustum* complex	Ducks, geese	North America, Norway
Leucocytozoon smithi	*S. aureum* complex, *S. congareenarum*, *S. jenningsi* group, *S. meridionale*, *S. slossonae*, possibly *S. ruficorne* group	Turkeys	North America, introduced to Africa
Leucocytozoon tawaki	*Austrosimulium ungulatum*	Penguins	New Zealand
Leucocytozoon toddi	*P. decemarticulatum*, *S. aureum* complex, *S. vernum* group	Hawks	North America
Leucocytozoon ziemanni	*S. aureum* complex, *S. vernum* group	Owls	North America
Trypanosoma confusum	*P. decemarticulatum*, *Simulium* spp.	Birds	North America
Trypanosoma corvi	*S. latipes*	Kestrels	Europe
Trypanosoma numidae	*S. adersi*	Chickens, guinea fowl	Eastern Africa
Viruses			
Vesicular stomatitis virus	*S. notatum*, *S. vittatum*	Horses, cattle	North America

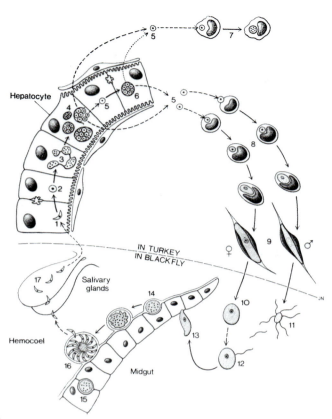

FIGURE 11.6 Life cycle of *Leucocytozoon smithi*: A sporozoite (1) inoculated by an infected female black fly into the blood of a turkey enters a hepatocyte and becomes an intracellular trophozoite (2). The immature first-generation meront (3) matures as a primary meront (4) and gives rise to daughter merozoites (5) that begin a second generation of merogony (6). An abortive meront in a macrophage host cell (7) develops no further. Immature gamonts (8) in a mononuclear phagocyte or leukocyte elongate to form mature microgametocytes and macrogametocytes (9) that will be ingested by a black fly. An extracellular macrogamete (10) and a microgamete from an exflagellating microgametocyte (11) in the lumen of the black fly midgut fuse (12) to form a zygote or migrating ookinete (13) that penetrates the midgut epithelium and develops extracellularly into a spherical oocyst (14–15). Asexual reproduction by sporogony occurs with the formation of sporozoites (16) that migrate through the hemocoel to the salivary glands of the female black fly. Reproduced by permission from Steele and Noblet (2001, *J. Eukaryotic Microbiol.* 48: 118–125).

species that transmit these parasites and the extent to which the host's fitness is impaired remain unknown. The best-studied trypanosome transmitted by black flies is *Trypanosoma confusum*, which infects many North American species of birds and has numerous simuliid vector species. The flies infect the birds when their fecal droplets or crushed bodies contaminate the bite wound. In Europe, the avian hosts probably acquire *Trypanosoma corvi* when they eat infected black flies.

Four genera of filarial nematodes are transmitted by black flies, and all except the single species of *Splendidofilaria* (*S. fallisensis*), which infects ducks, are parasites of mammals. *Dirofilaria ursi* is a parasite of bears in North America and Japan, although transmission by black flies has been demonstrated only in Canada. *Mansonella* and *Onchocerca* are parasites of humans, and the latter genus also includes at least eight simuliid-transmitted species that infect bovids, cervids, and warthogs. All have basically the same life cycle. The microfilariae circulate in the blood or diffuse in the skin of the vertebrate host, where they are acquired by a female black fly. Development occurs in various tissues of the fly, depending on the genus of filarial worms. The worms of *Mansonella* and *Onchocerca* are found in the thoracic flight muscles, where development to the infective third-stage larva typically requires 6–12 days, depending on temperature. The infective worms migrate to the vector's head and mouthparts. When the fly takes a blood meal from a vertebrate host, the worms break through the proboscis and move into the bite wound. The worms mature and mate in the vertebrate, and the female filarial worms produce microfilariae, restarting the life cycle.

Mansonella ozzardi is the causal agent of mansonellosis, a disease of humans on some Caribbean Islands and from the Yucatan Peninsula of Mexico to northern Argentina. At least five species of black flies serve as vectors, but only in northwestern Argentina and the rainforests of northern South America and Panama. Elsewhere, ceratopogonids (Chapter 10) serve as vectors. Humans are the only significant hosts, and in some areas, such as foci in Colombia, up to 70% of the human population can be infected. The disease is considered to have little overt pathology in most people.

At least eight species of *Onchocerca* that cause onchocerciasis in cattle and other large mammals are transmitted by black flies. Although the worms produce disease (e.g., bovine onchocerciasis), the impact, both clinical and economic, is typically superficial and minimal. The most widespread species, *Onchocerca lienalis*, is transmitted by species such as *Simulium jenningsi* in the United States and the *Simulium ornatum* complex in the Palearctic Region. The microfilariae of *Onchocerca lienalis* occur chiefly in the host's umbilical region, whereas those of *Onchocerca dukei* and *Onchocerca ochengi* are concentrated in the inguinal region. Species of *Onchocerca* that infect deer and moose typically occur in the subcutaneous connective tissues, especially in the legs, giving rise to the common name *leg worms*.

The most infamous of the diseases associated with black flies is human onchocerciasis, caused by

Onchocerca volvulus. Human onchocerciasis, or river blindness, is specific to humans. It is endemic in 28 tropical African countries, with scattered foci in southern Yemen, northern Brazil, Colombia, Ecuador, Guatemala, Mexico, and Venezuela. The disease in Latin America presumably was introduced by the slave trade. The World Health Organization estimates that nearly 18 million people are infected globally, with about 40,000 new cases annually. About one-third of infected individuals manifest skin disease, while half a million have impaired vision and another 350,000 are blind. The World Health Organization's Onchocerciasis Control Programme (OCP) in West Africa, initiated in 1974, eliminated onchocerciasis as a public health problem in 11 of the West African countries, but it has yet to be followed in the underpopulated valleys of the savanna lands by the economic development on which the scheme originally was predicated.

The adult filarial worms are found in humans, where they typically are encapsulated in nodules beneath the skin or in muscular and connective tissues; although cosmetically undesirable (Fig. 11.7), the nodules with their entombed worms cause no major discomfort. The worms mate in the nodules and the females, which can live up to 14 years, begin producing microfilariae, the demons of the host, which migrate to the skin and other organs, such as the eyes. The microfilariae can live for 2 years and are acquired from the skin by the simuliid vectors. To diagnose infections in humans, skin snips can be taken and examined for microfilariae or DNA probes can be used.

Heavy loads of microfilariae produce the most offensive aspects of the disease: severe itching, dermal lesions, depigmentation (Fig. 11.8), loss of skin elasticity coupled with lymphatic disturbances (e.g., "hanging groin"), and ocular problems. The skin problems, especially the itching, can be intolerable, leading to suicide in severe cases, but it is the eye problems that are so devastating. Visual impairment can include reduced peripheral vision, night blindness, and complete loss of all light perception. Blindness is most prevalent in the savannas of West Africa, where 10–15% of the people in some villages are blind. Recent research demonstrates that the major inflammatory response in the cornea is caused by a bacterium (*Wolbachia* sp.) that lives in *Onchocerca volvulus*, suggesting that antibiotic treatments might prevent ocular damage (Saint André et al. 2002). The socioeconomic effects of the disease are often far-reaching, destabilizing communities and causing abandonment of entire villages.

At least five viruses have been isolated from wild-caught female black flies. Although transmission of these viruses from simuliids to vertebrates might be possible, nonmechanical transmission has been documented only for vesicular stomatitis virus (VSV), including both the Indiana and New Jersey serotypes. The virus causes lesions in epithelial tissues of livestock, with subsequent weight loss, suppressed milk production, and government-imposed restrictions on the movement of animals from one area to another. Economic losses during outbreaks of VSV can soar into the millions of dollars.

FIGURE 11.7 Nodule containing adult filarial worms over the iliac crest of an early middle-aged man in Nigeria. The position of the nodule is characteristic of onchocerciasis in Africa. Photograph by R. W. Crosskey.

FIGURE 11.8 Man catching a biting black fly in Nigeria. The effects of advanced onchocerciasis can be seen in the thick, striated skin (*lichenification*) and irregular depigmentation characteristic of hyperendemic areas of the disease in Africa. Photograph by R. W. Crosskey.

CONTROL

The control of black flies is a relatively recent practice. No serious strides could be made to control black flies until the advent of chemical pesticides and the practical means of targeting the flies and economically manufacturing and dispensing the chemicals. DDT afforded this opportunity. But chemical control, chiefly using DDT and its replacement compounds, was short-lived, spanning the mid-1940s to late 1970s (Chapter 41). Falling into disfavor because of resistance problems and deleterious effects on nontarget organisms, it has been largely replaced with biological control, using *Bacillus thuringiensis* var. *israelensis* (Bti) (Chapter 43). Chemical control is now used in only a few programs, such as the OCP (Onchocerciasis Control Programme) in West Africa, which uses a rotation of multiple insecticides, including Bti. The last use of chemicals (methoxychlor) against larval black flies in North America was in 1998.

The study of natural enemies of black flies experienced a bustle of activity in the 1970s and early 1980s (Chapter 43). However, the natural enemies either were too poorly understood biologically and taxonomically or, except for Bti, were economically prohibitive to use as biological control agents. The 1978 discovery that Bti, originally isolated from mosquitoes, is a powerful killing agent of larval black flies initiated a new era of pest management and vector control. Its specificity for black flies and mosquitoes, high levels of mortality, and environmental safety have made it the management tool of choice. The toxins contained in the parasporal inclusions of the bacterium disrupt the cells of the alkaline larval midgut, causing massive trauma and death.

Management of black flies, including vector control, has been directed principally at the larvae because they tend to be localized in specific breeding areas. Treatments are applied from the ground or the air, depending on the extent of the problem and the size of the watercourses. Adulticiding has been implemented on military bases and in small communities, but flies quickly reinvade the treated areas, requiring sustained spraying. It rarely has been used for control of simuliid vectors.

Unlike control, the use of personal protection, especially repellents, has a long history. Traditional substances used on both people and animals include mud, grease, and herbal oils such as citronella. Synthetic repellents were introduced in World War II and have been used since, most notably those that include *N,N*-diethyl-3-methyl-benzamide (DEET). Dusts, ear tags, pour-ons, and sprays that incorporate pyrethroids and other synthetic compounds are used with livestock and poultry. Other means of protection include the age-old use of smoldering fires, known as *smudges*, which produce a dense smoke. Providing shelters for livestock and poultry also is effective because black flies typically do not enter buildings. When the turkey industry in the southeastern United States began housing their birds in shelters rather than in open arenas, leucocytozoonosis was sharply reduced.

A chemotherapy program was integrated with vector control in the early 1990s as a means of breaking the cycle of human onchocerciasis in Africa and was later implemented as a means of battling onchocerciasis in all endemic areas in Latin America. The program involves the use of ivermectin, also known as Mectizan, the formulation of the drug for human use. Taken orally about once every 6–12 months, ivermectin kills the microfilarial worms in the skin, easing the itching and reducing the number that can be ingested by the simuliid vectors.

SUMMARY

Sound taxonomy and predictive phylogenies will continue to provide the foundation for all future research on black flies. Current and emerging molecular techniques must be integrated with chromosomal and morphological approaches to solve the taxonomic problems of vector complexes (Chapter 29). The ability to identify females of isomorphic species, currently identifiable only by chromosomes of their larvae, is needed; resolution might come with new molecular techniques. Information on the frequency of homosequential sibling species and differences in their bionomics, such as host preferences and vector capabilities, must be resolved.

Many aspects of the basic behavior (Chapter 20) of black flies remain a mystery. Although females choose the breeding sites, ultimately determining the extent of disease problems, virtually nothing is known about how these decisions are made. Ovipositional cues used by females require experimental investigation. The recent demonstration that the eggs of some black flies emit pheromones that stimulate communal oviposition (McCall et al. 1997) suggests chemical cues might be more important in oviposition than previously appreciated. Similarly, the "invitation effect," whereby biting by one female stimulates biting by others, further hints at the importance of chemistry in mediating behavior (McCall and Lemoh 1997). Additional information on the hosts of black flies and the transmission of disease agents among wildlife is much needed, with special regard given to possible sibling species of both the hosts and disease organisms. Even among domestic animals, recent research has documented the ability of black flies to transmit viruses such as VSV (Mead et al. 2000), suggesting that the full extent of black flies as vectors of disease agents is not yet fully recognized.

As human development encroaches on vector habitats and as populations of urban wildlife increase, simuliid-borne diseases are likely to spread and, therefore, need to be monitored carefully. The possibility that long-term vector-management programs and habitat disturbance can cause increased rates of hybridization in simuliid vectors (Boakye et al. 2000) illustrates the need for long-term research programs that document baseline genetic information and that conduct frequent monitoring.

New techniques for fighting disease are on the horizon. For example, research is currently under way to find a vaccine for human onchocerciasis (Cook et al. 2001). At the same time, the potential for resistance to current and future control measures and drug therapy will require vigilance.

Finally, more students interested in the study of the organisms themselves—the black flies and the parasites they transmit—are essential if future progress is to be made in battling simuliid-borne diseases. The success of Bti has lulled some into believing that black flies no longer pose a threat. The reality is otherwise.

Acknowledgments

I thank C. R. L. Adler and R. W. Crosskey for providing insightful comments on the original manuscript and R. W. Crosskey, D. C. Currie, R. M. Idema, and G. P. Noblet for providing illustrations.

Readings

Adler, P. H., and McCreadie, J. W. 1997. The hidden ecology of blackflies: Sibling species and ecological scale. *Am. Entomol.* 43: 153–161.

Adler, P. H., Currie, D. C., and Wood, D. M. 2004. *The Black Flies (Simuliidae) of North America.* Cornell University Press, Ithaca, NY.

Basáñez, M. G., Yarzábal, L., Frontado, H. L., and Villamizar, N. J. 2000. Onchocerca–Simulium complexes in Venezuela: Can human onchocerciasis spread outside its present endemic areas? *Parasitology* 120: 143–160.

Boakye, D. A., Back, C., and Brakefield, P. M. 2000. Evidence of multiple mating and hybridization in *Simulium damnosum* s. 1. (Diptera: Simuliidae) in nature. *J. Med. Entomol.* 37: 29–34.

Cook, J. A., Steel, C., and Ottesen, E. A. 2001. Towards a vaccine for onchocerciasis. *Trends Parasit.* 17: 555–558.

Crosskey, R. W. 1990. *The Natural History of Blackflies.* Wiley, Chichester, England.

Crosskey, R. W. 1993. Blackflies (Simuliidae). In *Medical Insects and Arachnids* (R. P. Lane and R. W. Crosskey, eds.). Chapman & Hall, London, pp. 241–287.

Crosskey, R. W., and Howard, T. M. 1997. A new taxonomic and geographical inventory of world blackflies (Diptera: Simuliidae). *The Natural History Museum, London.*

Cupp, E. W., and Cupp, M. S. 1997. Blackfly (Diptera: Simuliidae) salivary secretions: Importance in vector competence and disease. *J. Med. Entomol.* 34: 87–94.

Cupp, E. W., and Ramberg, F. B. 1997. Care and maintenance of blackfly colonies. In *The Molecular Biology of Insect Disease Vectors: A Methods Manual* (J. M. Crampton, C. B. Beard, and C. Louis, eds.). Chapman and Hall, London. pp. 31–40.

Davies, J. B. 1994. Sixty years of onchocerciasis vector control: A chronological summary with comments on eradication, reinvasion, and insecticide resistance. *Annu. Rev. Entomol.* 39: 23–45.

Hunter, D. B., Rohner, C., and Currie, D. C. 1997. Mortality in fledgling great horned owls from blackfly hematophaga and leucocytozoonosis. *J. Wildl. Dis.* 33: 486–491.

Laird, M. (ed.) 1981. *Blackflies: The future for biological methods in integrated control.* Academic Press, New York.

McCall, P. J., and Lemoh, P. A. 1997. Evidence for the "invitation effect" during bloodfeeding by blackflies of the *Simulium damnosum* complex (Diptera: Simuliidae). *J. Insect Behav.* 10: 299–303.

McCall, P. J., Heath, R. R., Dueben, B. D., and Wilson, M. D. 1997. Oviposition pheromone in the *Simulium damnosum* complex: Biological activity of chemical fractions from gravid ovaries. *Physiol. Entomol.* 22: 224–230.

Mead, D. G., Ramberg, F. B., and Maré, C. J. 2000. Laboratory vector competence of blackflies (Diptera: Simuliidae) for the Indiana serotype of vesicular stomatitis virus. *Ann. New York Acad. Sci.* 916: 437–443.

Rothfels, K. H. 1988. Cytological approaches to blackfly taxonomy. In *Blackflies: Ecology, Population Management, and Annotated World List* (K. C. Kim and R. W. Merritt, eds.). Pennsylvania State University, University Park, PA, pp. 39–52.

Saint André, A., Blackwell, N. M., Hall, L. R., Hoerauf, A., Brattig, N. W., Volkmann, L., Taylor, M. J., Ford, L., Hise, A. G., Lass, J. H., Diaconu, E., and Pearlman, E. 2002. The role of endosymbiotic *Wolbachia* bacteria in the pathogenesis of river blindness. *Science* 295: 1892–1895.

Schofield, S. W., and Sutcliffe, J. F. 1996. Human individuals vary in attractiveness for host-seeking blackflies (Diptera: Simuliidae) based on exhaled carbon dioxide. *J. Med. Entomol.* 33: 102–108.

Shelley, A. J., and Coscarón, S. 2001. Simuliid blackflies (Diptera: Simuliidae) and ceratopogonid midges (Diptera: Ceratopogonidae) as vectors of *Mansonella ozzardi* (Nematoda: Onchocercidae) in northern Argentina. *Mem. Inst. Oswaldo Cruz* 96: 451–458.

Sutcliffe, J. F. 1986. Blackfly host location: A review. *Can. J. Zool.* 64: 1041–1053.

Sutcliffe, J. F., and McIver, S. B. 1984. Mechanics of blood feeding in blackflies (Diptera, Simuliidae). *J. Morphol.* 180: 125–144.

World Health Organization. 1995. Onchocerciasis and its control: Report of a WHO Expert Committee on Onchocerciasis Control. *WHO Technical Report Series* 852: 1–103.

12

Phlebotomine Sand Flies, the Psychodidae

LEONARD E. MUNSTERMANN

INTRODUCTION

Knowledge of phlebotomine biology suffers as a consequence of several characteristics: (1) They are relatively inconspicuous in size and behavior, (2) they are speciose but morphologically similar, (3) their larval stages are rarely found in the field setting, and (4) they are not readily reared. Phlebotomines serve as sole vectors of leishmaniases and other diseases in tropical and semitropical zones around the world. Only east Asia and sub-Saharan Africa are relatively free of leishmaniasis, although phlebotomine sand flies may be common. Hot zones for phlebotomine diseases are Central and South America and a broad band including the Mediterranean region and stretching across the Middle East to eastern India.

Phlebotomines are commonly known as *sand flies*, a reference arising from the phlebotomine–leishmaniasis associations studied extensively in the drier regions of the Mediterranean and Middle East. Unfortunately, this name is often confused with sand flies of the family Ceratopogonidae (Chapter 10), a family with very different behaviors and vector–disease associations. Second, phlebotomines in the western hemisphere have little association with sand. They are, instead, most commonly distributed in forests from southern United States to northern Argentina.

Of the nearly 1,000 species described, most adults are within a size range of 2.5–3.5 mm. Adult color varies from silvery gray to nearly black, and, characteristic of the family Psychodidae, adults are covered with narrow, erect scales, giving the impression of furriness. A characteristic peculiar to resting phlebotomine adults is the position of the wings, held erect in a V position. Flight is feeble and movement is by a peculiar hopping motion. Eggs and larvae have rarely been recovered in the field. Based on laboratory rearing studies, immatures apparently prefer organically rich, relatively dry soils.

The fly bites are typically painful—only the female requires blood—and often recognized by the small pool of blood at the bite site. Some species leave red swellings that persist for several hours, depending on the sensitivity of the individual. Generally, phlebotomines are not strongly anthropophilic; the preferred hosts seem to be those most readily available. This opportunistic tendency makes humans an auxiliary host for sand flies that have developed a capacity for surviving in peridomestic or urban environments. Development of vector–parasite model systems has been hampered by the lack of species specificity in hosts. A single species of cutaneous leishmanias may have a dozen or more incriminated or suspected vectors associated with it. As a consequence, epidemiologic models appropriate for one vector species–parasite pair may be irrelevant for others.

Many aspects of phlebotomine biology have not been explored to the extent seen in other vector systems. Several factors are responsible. The small size and rearing difficulty mentioned earlier have discouraged phlebotomines as a study organism in the classroom or in laboratories lacking specialized rearing facilities with skilled personnel. Second, because the larval stages have rarely been recovered from the field,

141

environmental predictors of density and field estimates of larval longevity are not available. Finally, the multiplicity of incriminated vector species and the several species with variable clinical manifestations have not permitted research focused on a representative vector–parasite model system. Four primary areas of research endeavor are necessary to better understand the vector biology and its relationship to leishmaniasis epidemiology: (1) detailed documentation of species distributions based on careful transect studies; (2) application of molecular methods to phlebotomine systematics to develop higher-level phylogenies and to produce better definitions of cryptic species; (3) host preference determination for suspected vector species to evaluate their roles in disease transmission; and (4) characterization of larval habitats and factors affecting larval development in the field.

GENERAL IDENTIFICATION

The family Psychodidae, with which the phlebotomines are classified (Fig. 12.1), is very old and maintains some of the most ancient of dipteran characters. Members of the family are distinguished by a dense covering of narrow scales on head, thorax, legs, and wing veins. Of the five psychodid subfamilies, only the Phlebotominae have piercing mouthparts capable of taking blood. The phlebotomines furthermore tend to have an elongate and more fragile structure, in contrast to a squatter and more robust appearance of the other psychodid flies.

Phlebotomine taxonomists have separated the subfamily into Old World (Europe, Africa, and Asia) and New World (North, Central, and South America) genera. Lane (1993) has provided taxonomic keys to the genera and subgenera of Old World phlebotomines, consisting mostly of *Phlebotomus* and *Sergentomyia* and their subgenera. *Phlebotomus*, a genus with representatives ranging from north and east Africa to eastern India, can be identified to species by the Lewis (1982) keys. Members of the second major genus, *Sergentomyia*, are found mainly in sub-Saharan Africa and southeast Asia—its members feed largely on reptiles and are not regarded as vectors of human disease. Taxonomic revision by Davidson (1990) provides keys to 27 sub-Saharan species of the dominant *Sergentomyia* subgenus *Sergentomyia*.

The New World genus *Lutzomyia* encompasses approximately 379 species. Young and Duncan (1994) have compiled distribution maps, illustrations, and keys to 361 of these in a volume that has served as the seminal *Lutzomyia* treatment since its publication. The other two New World genera are *Brumptomyia* (23 species) and *Warileya* (7 species). Each of the preceding citations contains reference listings for species descriptions, additional identification keys, and phlebotomine biology. Note that many species have been described from a single locality and even from a single specimen. For example, in the genus *Lutzomyia*, more than half the species have been recovered from five or fewer localities, and 91 species have been reported from only a single site (Fig. 12.2).

Egg and larval characters have played little role in the systematics of phlebotomines. In contrast, the adult flies have provided numerous reference points, most

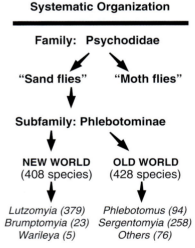

FIGURE 12.1 Systematic position of phlebotomine sand flies and their major genera.

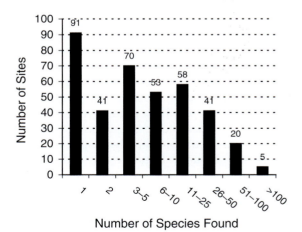

FIGURE 12.2 Number of sites associated with each of the 379 species of phlebotomine sand flies in the genus *Lutzomyia*. Note that each of 91 species has only the type locality associated with it.

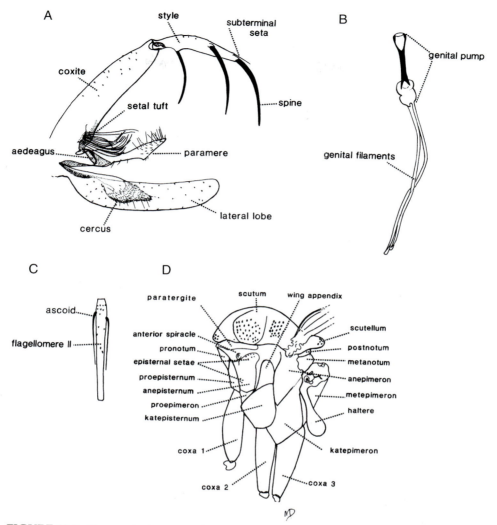

FIGURE 12.3 Taxonomic characters of Phlebotominae I: (A) External male genitalia; (B) internal male genitalic apparatus; (C) antennal segment, showing ascoid; (D) thoracic sclerites. From Young and Duncan (1994).

notably the male genitalia and the female spermathecae. Morphology of the ascoids of the antennae and numbers of cibarial teeth frequently differentiate species groups. Length ratios are used in combinations of wing veins, antennal and leg segments, or mouthparts. Figures 12.3 and 12.4 illustrate each of these characters. Approximately 90 characters have been demonstrated as effective species descriptors and have been recommended for inclusion in every new species description (Bermudez et al. 1991). Recognition and measurement of all character elements require mounting individual specimens on microscope slides and observation magnifications of 100–400×.

Long-term preservation of specimens is best done by clearing in potassium hydroxide and mounting in euparal. Although some taxonomists prefer Hoyer's, Berlese's, or other media, these preparations generally do not survive long-term museum storage. In preparing the slides, careful clearing, dissection, and arrangement of each specimen are essential to eventual identification to species. Folded wings, rotated heads, collapsed spermathecae, or lost antennae can prevent a secure final identification.

The taxonomic nomenclature of phlebotomines is undergoing substantial change at the genus and subgenus levels. Efforts are under way to revise the nomenclature, generally consisting of the elevation of currently recognized subgenera or groups to the genus level. Based on morphological characters, Artemiev (1991) produced a classification scheme of 24 genera

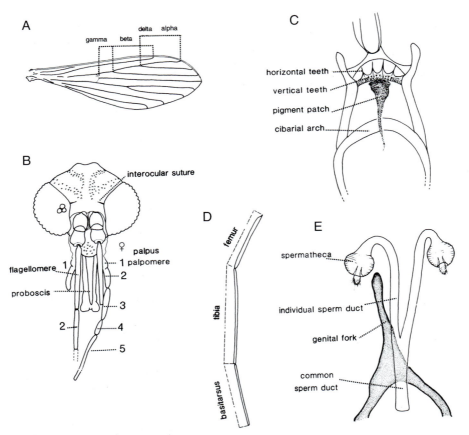

FIGURE 12.4 Taxonomic characters of Phlebotominae II: (A) Wing, with landmarks for length measurements; (B) head and mouthparts; (C) female cibarium; (D) hind leg; (E) sperm storage apparatus. After Young and Duncan (1994).

and 40 subgenera. These designations were supported and amplified by application of cladistic analysis (88 characters) to the New World phlebotomines (Galati 1995). Recently, her taxonomic reorganization of *Lutzomyia* was strongly supported by a molecular phylogeny based on ribosomal DNA sequences (Beati et al. 2004). As molecular methods become developed further for phlebotomine species, they will lend objective support to taxonomic reorganization among the phlebotomines. Together with biometric methods, the establishment of comprehensive phylogenetic relationships will be possible for many species whose current taxonomic position is not well defined.

LIFE CYCLE

The basic stages of the life cycle are typical of the Nematocera, but they require a longer period for completion, up to 6 weeks or more (Fig. 12.5). After mating and feeding on blood, the adult females develop eggs and oviposit them singly on a crevassed substrate. In 4–10 days the eggs hatch; the larvae grow through four instars over a period of 4–8 weeks. After pupation, the adults emerge in 4–6 days. The adult stage is relatively short, seldom lasting more than 3 weeks, during which mating and host-seeking occur and, for the female, blood feeding, search for oviposition substrate, and oviposition. Because many species serve as viral, bacterial, and protozoan vectors, the females must survive sufficiently long to take a second blood meal in order to infect a second host. The immature stages of egg, larva, and pupa are described entirely by observations in a laboratory setting. Attempts to locate adult emergence points by placing emergence traps at the most probable locations has been only moderately successful for the domestic species *L. longipalpis* (Fig. 12.6).

Egg

Retrieval of eggs from the field is nearly impossible, and information about them comes entirely from

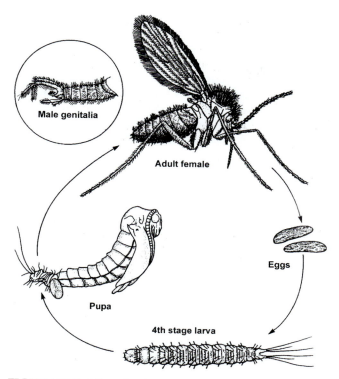

FIGURE 12.5 Life cycle of a phlebotomine sand fly. Approximate scale: egg, 0.25×0.8 mm; larva, 8–12 mm; pupal weight 0.4–0.65 mg; female adult length, 3–4 mm (after Marquardt et al. 2000).

FIGURE 12.6 Sampling potential larval habitats by capturing emerging adults, Magdalena Valley, Colombia.

laboratory observations. The female oviposits 20–50 eggs on average, preferring a rough, humid substrate. In laboratory life table studies, egg retention and oviposition without fertilization occurs with high frequency. Solidified, but damp, plaster of Paris provides an acceptable texture for many species that have been successfully colonized. The egg chorion has a distinctive structure of ridges, cells, and microvilli. These show species-specific patterns, using scanning electron microscopy (Fausto et al. 2001).

Larva

The newly hatched larva of 2 mm progresses in size through four larval instars, with the last instar reaching approximately 10 mm in length. Because of its terrestrial habitat, particularly in arid zones, the larval cuticle is thick and sclerotized to prevent water loss and desiccation. The body surface is spinose, with dark brown to black sclerites. The cuticular interstices are a translucent grey, a cryptic coloration suited for blending with its presumed organic topsoil environment. Respiration occurs through abdominal and thoracic spiracles in larval instars 2–4; the first-stage larva has only a pair of postabdominal openings (Fausto et al. 1999). Nutriment for larval growth can be deduced

only from laboratory rearing experience. The success of the cured feces-rabbit chow mixture indicates that at least some species prefer highly organic soils where substantial fecal deposition and decomposition is occurring. For only about 5% of species is the complete life cycle known. The period of larval development can be as short as a month at high temperatures and as long as 3 months at lower temperatures (Tesh and Guzman 1996). Seasonal cycles in peridomestic fly densities point to a capacity for larval estivation (or diapause) during inhospitable seasons (Morrison et al. 1995).

Pupa

Before pupation, the larva becomes sluggish and, in laboratory containers, prefers to move to higher, perhaps drier, vertical substrate. The fourth instar exuvium is collapsed and glued to the posterior segments of the pupa. Initially, the pupa is pale and cream-colored, but it becomes progressively darkened as the cuticle tans and hardens. The pupa remains relatively immobile, and, although it wriggles in response to probes, it does not demonstrate directed movement. After 4–6 days, the adult emerges.

Adult

The adults move about in a characteristic hopping motion, with the wings held over the abdomen in an erect V shape. The overall dimensions of male and female are similar, but the two are readily distinguished. The abdomen of the female is round and robust, whereas the male is slender and the terminalia claspettes in side view form a C shape. These differences are reflected in body weights: The female (approx. 0.6 mg) is approximately 50% heavier than the male (approx. 0.4 mg).

BIONOMICS

Habitat Diversity

Phlebotomines can be grouped as peridomestic, with close human associations, or feral, with associations independent of human activity. Peridomestic adult flies are found resting on walls of human and domestic animal domiciles, away from light and with high humidity. During preferred times of blood feeding, usually in early evening and throughout the night, adults can be collected directly from domestic animals, dogs, cattle, horses, chickens, and pigs. Pigsty crevices, undersides of chicken coops, dog kennels, and bathhouses are additional likely sites. Flies are found in highest densities near anticipated larval habitats, where soil and feces mix to form a conditioned, organically enriched environment.

Feral environments are less readily predictable. In moist, forested areas, protected cavities in trees and buttress roots are a common refuge. Common species, such as *L. shannoni*, can be found on the lower trunks of large trees in evenings. Organic litter on the soil surface consisting of fallen leaves, twigs, and dead grasses also yields adult flies. Dry and moist zones with rocky relief have multiple refugia on the undersides of rocks and in cracks and crevices. Caves are a particularly rich source of flies, furnishing a humid habitat protected from the environment and common predators. In arid areas, rodent burrows are often most productive because they afford the advantages of caves in addition to a convenient source of blood and larval food.

Dispersal

Adult phlebotomines are not strong fliers and therefore do not disperse far from the emergence site. Mark-recapture studies have indicated an average flight range of <60 m, with only 3% moving beyond 300 m. These results are typical of peridomestic species (Morrison et al. 1993), although forest species appear to be much less active (Alexander 1987). Low dispersal offers important insights into larval habitat, species distributions and diversity, speciation, and disease epidemiology. Field capture of any species as an adult permits inferring a nearby larval habitat. In a transect study, density peaks, particularly of nulliparous females, can point even more accurately to the larval areas. Alternatively, prediction of the presence of a given species is more difficult, particularly if the preferred environment is patchy or transient. The peridomestic fly that abounds near concentrations of domestic animals may not be able to colonize remote locations or may require lengthy periods to recolonize farmsteads that have been abandoned or treated with insecticides. Similarly, as forests become more fragmented, fly populations become disjunct. In the event of a local extinction, recolonization of forest patches becomes more difficult.

Sampling

Because of the low vagility, many of the inferences about microhabitats occupied by sand fly species come from typical retrieval locations of the adult flies. Phlebotomine diversity requires a variety of techniques for retrieving specimens from the field. Alexander (2000) has summarized sampling methods applicable for a spectrum of species and habitats. The most productive and widely used collecting device is the carbon dioxide–baited, CDC light trap (Fig. 12.7). The trap is suspended at the test site; it is turned on before dusk and retrieved after sunrise. Carbon dioxide produced from a dry ice source and light from a small flashlight bulb provide the attraction. Four 1.5 V dry cells drive a small fan that directs the incoming flies to a mesh chamber in which they are trapped.

A second sampling device is the hand aspirator, in association with a flashlight. Its use is labor intensive and time consuming and locating fly resting sites is difficult. Even when found, the resting sites may be difficult to access. This sampling method has the distinct advantages of portability, ease of use, and rapid deployment. In peridomestic habitats (Figs. 12.7, 12.8),

FIGURE 12.7 CDC light trap sampling in a peridomestic habitat, Corumba, Brazil.

the phlebotomines are more concentrated in predictable locations. In the forest, finding concentrations of flies may be optimized by searching large tree buttress roots (Fig. 12.9) or rock crevasses. More importantly, a direct association can be made between a presumed larval habitat and resting sites of sand fly species.

A third method is the Shannon trap (Fig. 12.10), most effective in the forest environment, where specific resting sites are not readily apparent. It consists of a tent of white sheeting with a strong interior light source, such as a pressure lantern or 100-watt incandescent bulb. It is typically used at early evening and into the night. The flies are attracted to the light and walk up the tent sides, from which they are hand aspirated.

Each of these methods has advantages and disadvantages. Light traps (including the Shannon trap) have the advantage of sampling in a standardized manner. Unfortunately, statically positioned traps collect only those flies within the sphere of their attractive powers; for those species not attracted at all, reliance on traps alone severely skews the perception of species composition of a locality. The hand aspirator has the advantage of being mobile, with the experience of the searcher employed to seek most likely resting spots. This is a highly stochastic process, particularly in a forest environment, and reliance solely on this method will severely skew estimates of species composition. Use of multiple methods, applied consistently through an entire year, is essential to begin an approximation of species diversity and densities for a particular area.

FIGURE 12.9 Aspiration from tree buttresses in a tropical forest, Canal Zone, Panama.

FIGURE 12.8 Adult capture by manual aspiration in a peridomestic habitat, Melgar, Colombia.

FIGURE 12.10 Shannon trap in a tropical forest habitat, Leticia, Colombia.

FIGURE 12.11 Feral blood sources in tropical forests. (A) gecko in buttress root, Canal Zone, Panama. (B) sloth near scrub forest, Puerto Suarez, Bolivia.

SAND FLY GENETICS

Length of life cycle, difficulty of colonization, and species diversity have hindered genetic investigations with sand flies. Even species for which longer-term colonies are available have provided limited genetic models. Formal genetics for the purpose of creating linkage maps that correspond to the four pairs of chromosomes has not been established for any phlebotomine species. One naturally occurring morphological polymorphism, two abdominal pale spots in the males, has been used in genetic crosses. It occurs naturally in some populations of *L. longipalpis* and has been shown to occur as a dominant phenotype in Hardy–Weinburg equilibrium (Chapter 33) in field populations (Mukhopadhyay et al. 1998). But it has not been linked with other genetic markers in laboratory crosses. Genetics of phlebotomines has been concentrated in the areas of cytogenetics and population genetics, with an eye to its taxonomic ramifications.

Cytogenetics

Attempts to extract polytene chromosomes have met with failure for the few species examined. Kary-

otypes based on mitoses in larval brain squashes have been established for nine species of the genus *Lutzomyia* and four *Phlebotomus* species (Kreutzer et al. 1987, 1988). The chromosome number ranges from three to five pairs in the latter genus and three to four pairs in *Lutzomyia*. Distribution of heteromorphism in sex chromosomes, chromosome number, and arm lengths can yield additional clues for subgeneric species relationships (Kreutzer et al. 1988). Intraspecific heteromorphism was described in geographic strains for *L. longipalpis*, carrying the potential of karyotypy a step forward (Yin et al. 1999). Meiotic events remain to be described. Male testes are small, and the meiotic chromosomes are so small that little can be distinguished at the light-microscope level. In *L. longipalpis*, meiosis and spermatogenesis commences within a few hours after the adult male emerges (Munstermann, unpublished observations).

Population Genetics

Three phlebotomine species have absorbed most of the work on sand fly population genetic structure. The best-known Old World vector species, *Phlebotomus papatasi*, has been investigated at a level of local populations and across the distribution range. The early indicators are that this species shows typical genetic variability across its range from northwestern Africa to eastern India, and without strong evidence of local differentiation (Essighir et al. 1997; Ghosh et al. 1999). This contrasts with the New World *L. longipalpis*, in which varying degrees of allopatric differentiation is strongly exhibited. Numerous comparisons of pheromones, allozymes, maxadilan genes, song genes, strain hybridization, and mitochondrial DNA segments have indicated "diagnostic" characters separating Central American, northern-tier South American and Brazilian populations, as well as several isolated populations within Brazil. The comparisons do not always corroborate one another very well, in part because sampling has been opportunistic rather than premeditated. To reassort the current, seemingly contradictory sets of evidence, field studies must involve a careful sampling design and multiple analytic methods. To date, only the Lampo et al. (1999) study has recovered samples of sympatric, but genetically distinct, sibling species of *L. longipalpis*, in east central Venezuela. Another important vector species, *L. whitmani*, did not show similar tendencies toward genetic compartmentalization, in spite of a similar broad distribution (Ishikawa et al. 1999).

The three preceding species are considered peridomestic, found in close association with humans and their domestic animals. A third species, *L. shannoni*, contrasts with them in its habitation within tropical and subtropical forests; however, it also has a broad geographic distribution, extending from Argentina to the southeastern United States. The U.S. population appears to be isolated from its congeners by several hundreds of kilometers. A long absence of gene flow between it and the nearest Latin American populations is reflected by a high degree of differentiation in allozyme constitution but without morphological divergence. In South America, the Andes Mountains appear to have provided a formidable barrier to gene flow, separating populations in the northern and eastern river basins from those of the Orinoco and Amazon basins (Mukhopadhyay et al. 2001). In the Andean region, the *L. shannoni* habitats are quite disjunct, and evidence from allozymes (Cárdenas et al. 2001) and endosymbiont distribution (Ono et al. 2001) indicate restricted levels of gene flow throughout this region.

HUMAN DISEASES, PARASITES, AND SYMBIONTS

The host–parasite/symbiont associations in phlebotomines have been reviewed comprehensively in several books and publications (Killick-Kendrick 1990; Lane 1994), ranging from the obligate heritable endosymbionts to human disease organisms for which phlebotomines serve as intermediaries.

Leishmania

The human diseases most commonly associated with phlebotomines are the leishmaniases, for which phlebotomines are the sole vector. This association has been dramatized by recent outbreaks as a consequence of war (Afghanistan, Iraq, Sudan), population expansion (Brazil), or even decreased use of insecticides (for malaria control in India). These phenomena have resulted in tens of thousands of new cases per year and thousands of deaths. Four categories of leishmaniasis have been described on the basis of clinical symptoms: visceral, cutaneous, mucocutaneous, and diffuse cutaneous. A system of parasite identification based on isozyme profiles has become standardized, with 30 or more recognized *Leishmania* spp. Certain of the species correspond with one of the four clinical categories, but others can produce symptoms characteristic of two or more.

Compounded with the complexity of the parasite identifications are the difficulties in resolving the vector(s) for each. Killick-Kendrick (1990) compiled a composite listing of phlebotomines associated with each of 29 putative *Leishmania* taxa. Eleven leishmanias were associated with their own vector species. However, the major *Leishmania* spp. (e.g., *Le. donovani, infantum, major,* and *braziliensis*) each has 11–16 phlebotomine species incriminated or suspected as vectors. Incrimination of a species proceeds in four steps, analogous to Koch's postulates: (a) the phlebotomine feeds on humans, (b) the phlebotomine occurs in sufficient densities to sustain cyclic transmission, (c) the phlebotomine is permissive to development and multiplication of the parasite in its midgut, (d) the phlebotomine transmits the infectious parasite back to the human or animal model. In many cases, however, a phlebotomine species is guilty by association— presence in high densities during a transmission episode. In urban or peridomestic situations, where phlebotomine species diversity is low, this association may readily indicate the vector. However, particularly in South America, species diversity is high even in the periurban settings. In the forest, where much of the transmission of cutaneous leishmaniases occurs, 40 or more species of phlebotomines may be present. Distinguishing vectors from nonvectors in this setting is problematic without careful identification of sand fly and parasite, characterization of densities and affinity for human feeding, and, finally, laboratory transmission studies.

Bacteria and Viruses

Other human disease agents transmitted are relatively local or have low human pathogenesis. The major bacterial agent, *Bartonella bacilliformis*, is the agent of bartonellosis, and it is localized in the Andean highland areas of Ecuador and Peru. Associated with bartonellosis is an 80% mortality rate in a virulent form or a disfiguring warty appearance in the more benign form. Sand fly species of the *L. verrucarum* group have been associated with disease outbreaks and endemic transmission; however, phlebotomines have not been satisfactorily incriminated as vectors by the principles stated earlier. Furthermore, the disease has occurred in areas where the *L. verrucarum* group vectors are apparently absent (Karem et al. 2000).

The best-studied New World viruses transmitted by phlebotomines are variants of the vesicular stomatitis virus (VSV) of the family Rhabdoviridae. In a series of studies spearheaded by Comer (summarized in Comer et al. 1995) on Ossabaw Island, Georgia, the VSV is described as an important pathogen of veterinary animals (bovines, equines, and swine), a zoonosis affecting feral pigs and deer, and transmitted by the phlebotomine *L. shannoni*. The increasing diagnosis of VSV in Colombian cattle has led to suspicions that

L. shannoni may have a VSV vector role in the tropics as well (Cárdenas et al. 1999).

A second viral group closely associated with phlebotomines are the Bunyaviridae, placed in the genus *Phlebovirus*. Tesh (1988) lists 23 *Phlebovirus* species isolated from Old and New World phlebotomines—8 from *Phlebotomus* species and 15 from *Lutzomyia* species. Only five *Phlebovirus* species are identified as agents of human disease, however, and only two "sand fly fevers" (Naples, Sicilian) have attained some level of notoriety. The latter occur mostly in the Mediterranean basin, where the two known vectors, *Phlebotomus papatasi* and *P. perfiliewi*, reside.

Symbiotic Bacteria

Recently, the widely distributed nature of endosymbiotic bacteria of the genus *Wolbachia* has been elucidated from a broad variety of invertebrates. A survey by PCR for its presence in 15 phlebotomine species, using primers for a segment coding for a surface protein, located it in two *Phlebotomus* species (*papatasi, perniciosis*) and two *Lutzomyia* species (*shannoni, whitmani*) (Ono et al. 2001). Isolates from the *Lutzomyia* were related to one form of hymenopteran *Wolbachia*, the *P. papatasi* to drosophilan *Wolbachia*, and the *P. perniciosus* to a coleopteran form. Notably, of five field samples of *L. shannoni* examined, only one was infected; apparently this endosymbiont is not strongly "driven" to infect the entire species.

CONCLUSIONS

The epidemiology of diseases transmitted by sand flies will prove to be a rich experimental field for years to come. New phlebotomine species continue to be described, including morphological species in newly explored regions and discovery of the occurrence of species complexes for previously described species. The number and diversity of genetic variants among the New World leishmanias are only beginning to be appreciated. As a consequence, rational control measures may be difficult to formulate in the face of an extremely complex parasite and vector interaction.

Nonetheless, new technologies will advance the understanding of the web of vector, parasite, and host associations. Molecular tools are beginning to quantify variability and assess relationships among vectors and the parasites in a rapid and objective format. Rapid and accurate detection and identification of the parasite by PCR in host and vector is now possible. Charting the distributions of vector species remains a monumental task, but field exploration and monitoring can now be coupled with accurate computerized mapping (GPS) and analysis (GIS) (Chapter 16). A combination of these methodologies will provide an initial assessment of risk that may be very broadly scaled, such as general topographic or vegetation zone delimitations. As the distribution data matrix becomes denser, the accuracy of risk assessment and prediction will greatly increase.

Acknowledgments

This work was supported by NIH grants AI-34521, "Genetics and biogeography of sand fly disease vectors", and AI-562554, "Phylogeography of Verrucarum sand fly disease vectors", to LEM.

Readings

Alexander, B. 2000. Sampling methods for phlebotomine sand flies. *Med. Veterin. Entomol.* 14: 109–122.

Alexander, J. B. 1987. Dispersal of phlebotomine sand flies (Diptera: Psychodidae) in a Colombian coffee plantation. *J. Med. Entomol.* 24: 552–558.

Artemiev, M. M. 1991. A classification of the subfamily Phlebotominae. *Parassitologia* 33 (suppl. 1): 69–77.

Beati, L., Cáceres, A. G., Lee, J. A., and Munstermann, L. E. 2004. Systematic relationships among Lutzomyia sand flies (Diptera: Psychodidae) of Peru and Colombia based on the analysis of 12S and 28S ribosomal DNA sequences. *Internat. J. Parasitol.* 34: 225–234.

Bermudez, H., Dedet, J. P., Falcao, A. L., Feliciangeli, D., Rangel, E. F., Ferro, C., et al. 1991. Proposition of a standard description for phlebotomine sand flies. *Parassitologia* 33 (suppl. 1): 127–133.

Cárdenas, E., Corredor, D., Munstermann, L. E., and Ferro, C. 1999. Reproductive biology of *Lutzomyia shannoni* (Dyar) (Diptera: Psychodidae) under experimental conditions. *J. Vector. Ecol.* 24: 158–170.

Cárdenas, E., Munstermann, L. E., Martínez, O., Corredor, D., and Ferro, C. 2001. Genetic variability among populations of *Lutzomyia (Psathyromyia) shannoni* (Dyar 1929) (Diptera: Psychodidae: Phlebotominae) in Colombia. *Mem. Inst. Oswaldo Cruz* 96(2): 189–196.

Davidson, I. H. 1990. *Sand Flies of Africa South of the Sahara. Taxonomy and Systematics of the Genus Sergentomyia*. South African Institute for Medical Research, Johannesburg.

Depaquit, J., Perrotey, S., Lecointre, G., Tillier, A., Tillier, S., Ferté, H., et al. 1998. Systématique moléculaire des Phlebotominae: étude pilote. Paraphylie du genre *Phlebotomus*. *C. R. Acad. Sci. Paris, Life Sci.* 321: 849–855.

Esseghir, S., Ready, P. D., Killick-Kendrick, R., and Ben-Ismail, R. 1997. Mitochondrial haplotypes and phylogeography of *Phlebotomus* vectors of *Leishmania major*. *Insect Mol. Biol.* 6: 211–225.

Fausto, A. M., Taddei, A. R., Mazzini, M., and Maroli, M. 1999. Morphology and ultrastructure of spiracles in phlebotomine sandfly larvae. *Med. Veterin. Entomol.* 13: 101–109.

Fausto, A. M., Feliciangeli, M. D., Maroli, M., and Mazzini, M. 2001. Ootaxonomic investigation of five *Lutzomyia* species (Diptera, Psychodidae) from Venezuela. *Mem. Inst. Oswaldo Cruz* 96: 197–204.

Galati, E. A. B. 1995. Phylogenetic systematics of Phlebotominae (Diptera, Psychodidae) with emphasis on American groups. *Bol. Dir. Malariol. San. Amb.* 35 (suppl. 1): 133–142.

Ghosh, K. N., Mukhopadhyay, J. M., Guzman, H., Tesh, R. B., and Munstermann, L. E. 1999. Interspecific hybridization and genetic variability in *Phlebotomus* sandflies. *Med. Veterin. Entomol.* 13: 78–88.

Guzman, H., and Tesh, R. 2000. Effects of temperature and diet on the growth and longevity of phlebotomine sand flies (Diptera: Psychodidae). *Biomédica* 20: 190–199.

Ishikawa, E. A. Y., Ready, P. D., Souza, A. A., Day, J. C., Rangel, E. F., Davies, C. R., and Shaw, J. J. 1999. A mitochondrial DNA phylogeny indicates close relationships between populations of *Lutzomyia whitmani* (Diptera: Psychodidae, Phlebotominae) from the rain-forest regions of Amazonia and northeast Brazil. *Mem. Inst. Oswaldo Cruz* 94: 339–345.

Karem, K. L., Paddock, C. D., and Regnery, R. L. 2000. *Bartonella henselae*, *B. quintana*, and *B. bacilliformis*: Historical pathogens of emerging significance. *Microbes Infection* 2: 1193–1205.

Killick-Kendrick, R. 1990. Phlebotomine vectors of the leishmaniases: A review. *Med. Veterin. Entomol.* 4: 1–24.

Kreutzer, R. D., Modi, G. B., Tesh, R. B., and Young, D. G. 1987. Brain cell karyotypes of six species of New and Old World sand flies. *J. Med. Entomol.* 24: 609–612.

Kreutzer, R. D., Morales, A., Cura, A., Ferro, C., and Young, D. G. 1988. Brain cell karyotypes of six New World sand flies (Diptera: Psychodidae). *J. Am. Mosq. Contr. Assoc.* 4: 453–455.

Lampo, M., Torgerson, D., Marquez, L. M., Rinaldi, M., Garcia, C. Z., and Arab, A. 1999. Occurrence of sibling species of *Lutzomyia longipalpis* (Diptera: Psychodidae) in Venezuela: First evidence from reproductively isolated sympatric populations. *Am. J. Trop. Med. Hyg.* 61: 1004–1009.

Lane, R. P. 1993. Sand flies (Phlebotominae). In *Medical Insects and Arachnids* (R. P. Lane and R. W. Crosskey, eds.). Chapman and Hall, London, Ch. 4. 723 p.

Lewis, D. J. 1982. A taxonomic review of the genus *Phlebotomus* (Diptera: Psychodidae). *Bull. Brit. Mus. Nat. Hist.* 45(2): 121–209.

Marquardt, W. C., Demaree, R. S., and Grieve, R. B. 2000. *Parasitology and Vector Biology.* 2nd ed. Harcourt/Academic Press. Burlington, MA.

Morrison, A. C., Ferro, C., Pardo, R., Torres, M., Devlin, B., Wilson, M. L., and Tesh, R. B. 1995. Seasonal abundance of *Lutzomyia longipalpis* (Diptera: Culicidae) at an endemic focus of visceral leishmaniasis in Colombia. *J. Med. Entomol.* 32: 538–548.

Mukhopadhyay, J., Ghosh, K., Azevedo, A. C. R., Rangel, E. F., and Munstermann, L. E. 1998. Genetic polymorphism of morphological and biochemical characters in a Natal, Brazil, population of *Lutzomyia longipalpis* (Diptera: Psychodidae). *J. Am. Mosq. Contr. Assoc.* 14: 277–282.

Mukhopadhyay, J., Ghosh, K. N., Ferro, C., and Munstermann, L. E. 2001. Distribution of phlebotomine sand fly genotypes (*Lutzomyia shannoni*, Diptera: Psychodidae) across a highly heterogeneous landscape. *J. Med. Entomol.* 38: 260–267.

Ono, M., Braig, H. R., Munstermann, L. E., Ferro, C., and O'Neill, S. L. 2001. *Wolbachia* infections of phlebotomine sand flies. *J. Med. Entomol.* 38: 237–241.

Santamaría, E., Munstermann, L. E., and Ferro, C. 2002. Estimating carrying capacity in a newly colonized sand fly, *Lutzomyia serrana* (Diptera: Psychodidae). *J. Econ. Entomol.* 95: 149–154.

Tesh, R. B., and Guzman, H. 1996. Sand flies and the agents they transmit. In *The Biology of Disease Vectors* (B. Beaty and W. Marquardt, eds.). pp. 117–127.

Yin, H., Mutebi, J. P., Marriott, S., and Lanzaro, G. C. 1999. Metaphase karyotypes and G-banding in sand flies of the *Lutzomyia longipalpis* complex. *Med. Veterin. Entomol.* 13: 72–77.

Young, D. G., and Duncan, M. A. 1994. Guide to the identification and geographic distribution of *Lutzomyia* sand flies in Mexico, the West Indies, Central and South America (Diptera: Psychodidae). *Mem. Am. Entomol. Inst.* No. 54, Associated Publ., Gainesville, FL.

Tsetse Flies, the Glossinidae, and Transmission of African Trypanosomes

R. H. GOODING, LEE R. HAINES, AND TERRY W. PEARSON

INTRODUCTION

Tsetse are higher flies that are strictly hematophagous and confined almost exclusively to sub-Saharan Africa (Mulligan 1970; Jordan 1993; Leak 1999); the exceptions are two species that have also been recorded from one location in southwestern Saudi Arabia (Elsen et al. 1990). Although the biology of tsetse flies and their relationships with other organisms make them inherently interesting, it is their role as vectors of pathogenic trypanosomes, which has been known for over 100 years, that has provided the main impetus for research on tsetse. Much of the voluminous literature on tsetse has been summarized in an excellent book by Stephen Leak (1999) and on informative Web sites developed by Claude Laveissière and Laurent Penchenier (www.sleeping-sickness.com; www.sleeping-sickness.org; www.trypano-humaine.com; www.trypano-humaine.net; www.trypano-humaine.org).

GENERAL IDENTIFICATION

Adult tsetses are most often encountered during their host-seeking activities. Adults of most species have a drab, dark appearance (Fig. 13.1), although some, such as *Glossina austeni*, are light brown. When at rest they hold their wings "closed," like the blades of a pair of scissors, over the abdomen. Males and females have a prominent haustellum (comprised of the labium, hypopharynx, and labrum) that projects forward from the head, between the palps, when the fly is not feeding. Adult tsetse are distinguished from other dipterans by the prominent "hatchet" cell, which has the general appearance of a meat cleaver and is located near the center of the wing (Fig. 13.2), and by the branched chetae on the arista of the antennae (Fig. 13.3). Details of body pigmentation, presence of chetae, and structure of the genitalia are used to identify members of various taxa. Jordan (1993) provides keys to identification of adults and puparia.

Because of their life cycle, described later, immature flies are rarely encountered. Third-instar larvae (Fig. 13.4) are characterized by the presence of large polypneustic lobes that carry the posterior spiracles. The fully formed puparium is black and is distinguished from other dipteran puparia by prominent polypneustic lobes (Fig. 13.5).

CLASSIFICATION

Tsetse flies are members of the genus *Glossina*, the sole genus in the family Glossinidae. Major taxonomic works have been published by Machado (1954, 1959, 1970) and Dias (1987). Morphology (McAlpine 1989) and molecular data (Nirmala et al. 2001) link the families Glossinidae, Hippoboscidae, Strebleidae, and Nycteribiidae as a monophyletic superfamily,

FIGURE 13.1 *G. m. centralis* about to feed on a human hand. Photo by Jack Scott, DBS, University of Alberta.

FIGURE 13.3 Head of a *G. m. centralis*. Note the antenna showing arista with branched hairs. Photo by Randy Mandryk, DBS, University of Alberta.

FIGURE 13.2 Pinned *G. m. centralis*. Note the hatchet cell near center of wing. Photo by Randy Mandryk, DBS, University of Alberta.

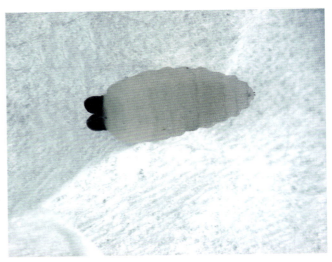

FIGURE 13.4 Third-instar larva crawling on tissue paper. Note the prominent polypneustic lobes at the posterior end. Photo by Randy Mandryk, DBS, University of Alberta.

Hipposcoidea (also referred to as the *Pupipara*), within the monophyletic Section Calyptratae of the suborder Brachycera in the Order Diptera. Most tsetse biologists treat *Glossina* as three subgenera (*Austenina*, *Nemorrhina*, and *Glossina* s. str.) that are also referred to as the species groups *fusca*, *palpalis*, and *morsitans*, respectively (Jordan 1993; Leak 1999). One problem with this arrangement is that one species usually considered to be in the *morsitans* group, *Glossina austeni*, is aberrant and has some morphological features and habitat preferences that place it closer to the *palpalis* group than to other species in the *morsitans* group. Dias (1987) considered these differences to be sufficient to warrant establishment of a fourth subgenus,

Machadomyia. The division of *Glossina* into four subgenera is supported by genetic evidence (Gooding et al. 1991; Chen et al. 1999).

Tsetse are important in medical and veterinary entomology, not because of their nuisance, but because of their ability to transmit pathogenic trypanosomes in Africa. Although all species of tsetse are believed to be capable of transmitting trypanosomes, only a few species are important vectors (Table 13.1). Human sleeping sickness is not distributed throughout the tsetse belt, but located in foci that are relatively well

FIGURE 13.5 Tsetse puparium. Note the prominent polypneustic lobes at the posterior end. Photo by Randy Mandryk, DBS, University of Alberta.

defined. The distribution of trypanosomiasis in livestock closely approximates the distribution of the vector species. Distribution maps for tsetse have been prepared by Ford and Katondo (1977) and are also available at the Web sites mentioned earlier. Major vectors of trypanosomes (Table 13.1) have also been tabulated by country of origin (Moloo 1993). The transmission of trypanosomes by tsetse is discussed in more detail later in this chapter.

Experiments in which *G. swynnertoni* was hybridized with *G. morsitans* from two localities led to the discovery that the two populations of *G. morsitans* were genetically different and ultimately led to the formal division of what had been referred to as *G. morsitans morsitans* into two subspecies (Vanderplank 1949). However, hybridization of flies from colonies that were established from widely separated populations of *G. m. morsitans* (Jordan et al. 1977), *G. m. centralis* (Gooding 1990a), and *G. pallidipes* (Langley et al. 1984) failed to find evidence of reproductive barriers, and hence of cryptic species, within these taxa. Recently, sterility among F_1 and F_2 males during hybridization of flies from inbred colonies of *G. p. palpalis* from Nigeria and Zaire raised the possibility that there are cryptic species within the nominal taxon *G. palpalis palpalis* (Gooding unpublished). Furthermore, there is genetic substructuring in natural populations of *G. p. gambiensis* (Solano et al. 1999), *G. pallidipes* (Krafsur and Wohlford 1999; Krafsur et al. 1997), *G. m. morsitans* (Wohlford et al. 1999), and *G. m. submorsitans* (Krafsur et al. 2000). However, such substructuring may be explicable by extinction of local populations during recent times; for example, as a result of a Rinderpest epizootic (Ford 1971) or other disasters that affected the local availability of normal hosts of tsetse. The possibility of cryptic species of tsetse remains an open question, the answer to which could help explain aspects of geographic variation in trypanosomiasis distribution and epidemiology.

TABLE 13.1   Important Vector Species of *Glossina*

Subgenus (species group)	Species name	Trypanosomiasis of	
		Humans	Livestock
Austenina (fusca)	*G. nashi*		•
	G. tabaniformis		•
	G. medicorum		•
	G. brevipalpis		•
	G. longipennis		•
Nemorhina (palpalis)	*G. tachinoides*	•	•
	G. fuscipes fuscipes	•	•
	G. palpalis palpalis	•	•
	G. p. gambiensis		•
Machadomyia (austeni)	*G. austeni austeni*		•
Glossina s. str. (morsitans)	*G. longipalpis*		•
	G. pallidipes	•	•
	G. swynnertoni		•
	G. morsitans morsitans	•	•
	G. m. centralis	•	•
	G. m. submorsitans		•

This table is based on the occurrence of trypanosomes in natural populations of tsetse, as cited in the abstracting journal *Tsetse and Trypanosomiasis Information Quarterly*.

BIOLOGY AND LIFE CYCLE

Adult males and females are strictly haematophagous, probably feeding every other day if hosts are readily available, and are relatively long lived. Females of most species mate when only a few days old. Eggs mature one at a time, ovulation occurs when females are about 1 week old, and the fertilized egg embryonates in the "uterus" (which is an expanded part of the bursa copulatrix). The egg hatches with the assistance of specialized structures in the female reproductive tract. All three larval instars are retained within the uterus and are nourished by a protein- and lipid-rich secretion ("milk") produced by the female's accessory glands ("milk glands"). Embryonation and larval development take approximately 9 days.

Deposition of the mature third-instar larva is delayed by an antiparturition hormone, until the larva has reached an appropriate size, at which time the antiparturition hormone is no longer released and release of a parturition hormone occurs. Both the pregnant female and the larva play an active role in parturition; both are involved in determining the time of parturition (generally in late afternoon). The female seeks a suitable larviposition site and relaxes the vaginal sphincter, and the mature larva undergoes waves of contractions to back out of the female's reproductive tract. Ovulation of a single mature egg then occurs, and the next (nine-day long) larviposition cycle begins.

The main requirements for a suitable larviposition site are that it be well enough shaded to prevent overheating and desiccation of the puparia and that the soil be friable enough to permit larvae to burrow to a suitable depth. Larviposition sites are relatively easy to find for savanna species and are often located in well-shaded soil below objects such as logs and leaning trees or at the base of trees, etc. For forest and riverine species, larviposition sites are more widely dispersed, less easily defined, and more difficult to locate.

After being deposited, the mature third-instar larva does not feed, but quickly burrows into the friable soil and pupariates. The puparial case is thick and hard and turns black within a few hours (Fig. 13.5). Development of the adult takes approximately 1 month. The adult emerges from the puparium and forces its way to the surface using its legs and ptilinum (an eversible saclike structure in the head). The wings are quickly expanded, and the adult flies to a suitable resting area. Adults seek their first blood meal within a day of eclosion.

Phenology

Tsetses have no diapause, hibernation, or aestivation mechanisms, nor do they have any mechanism for "reproductive diapause." Therefore, in order for a population to survive, the flies must reproduce continuously; generations overlap and all life stages are present at any given time. The abundance of tsetse in any locality may fluctuate, on an annual basis, due to changes in the birth, death, immigration, and emigration rates. The death rate, particularly of puparia, is significantly influenced by the evaporation rate, which is determined by temperature and humidity. High evaporation rates contribute significantly to high mortality; thus, apparent densities of populations may vary inversely with evaporation rates. Dry and wet seasons may, depending upon species, also affect availability of hosts and thus influence birth rates in tsetse. Annual effects can be cumulative if there are several consecutive years of higher or lower than normal rainfall; this may also affect the local distribution of tsetse (Leak 1999).

Common Adult Habitats

Adults of all species require suitable resting sites within wooded areas, and they may feed in these areas or in nearby open areas. Habitat choice is markedly influenced by the humidity (and thus may vary with the season) and on the availability of hosts, especially the preferred hosts. *Austenina* spp. are found primarily in forests, ranging from tropical rainforests to relatively dry forest islands. *Nemorrhina* spp. are in and near the riverine and lacustrine forests. *Machadomyia* sspp. are confined to relatively dense forests in coastal regions with high rainfall. Species of *Glossina* s. str. are found in wooded savanna, and some species are found in or near thickets. Within each subgenus there are species differences in the preferred habitats, and there are changes in resting sites within these habitats in different seasons (see Leak 1999 for details). Some species, most notably *G. tachinoides*, show signs of adapting to disturbed habitats; others are becoming locally rare or extinct as a result of human activities such as deforestation.

Mating

Laboratory studies have established that females of most species mate soon after their first meal and that, although enough sperm are received during one mating to fertilize all the eggs that a female can produce, females may mate more than once. Field observations are consistent with these findings. Since tsetse flies may occur at relatively low population densities, and since their muscle physiology and energy reserves limit them to flying less than 30 minutes per day, mate location may be a significant challenge. Male and female tsetse are haematophagous and use similar host-finding mechanisms; therefore hosts probably serve as locations for male/female interactions. For savanna species, such as *G. m. morsitans*, males that are not hungry are observed following hosts and other slow-moving objects, and this is presumed to be the site at which males locate virgin females. It is not known whether forest-inhabiting tsetse use hosts as a site for mating encounters.

Tsetses have good vision, and males in "following swarms" probably recognize potential mates by a combination of size and flight speed or pattern. Identification of females is by a species-specific contact sex pheromone (a branched, long-chain hydrocarbon) on her cuticle; there is no evidence for volatile sex pheromones in tsetse. The female sex pheromone induces copulatory behavior in the male, and copulation occurs immediately if the female is receptive. Mating may occur between members of closely related taxa, producing sterile male offspring and females with reduced fecundity (Gooding 1990b). The duration of copulation probably varies among tsetse species. Under laboratory conditions it commonly lasts an hour or more. During the last few minutes of mating, the male places accessory gland secretions in the female's reproductive tract to form a spermatophore, into which he ejaculates sperm. Shortly after mating, sperm begin to be transferred to the spermathecae.

Biting Behavior

Host seeking by tsetse is determined by endogenous factors (circadian rhythm, physiological state including available reserves, and metabolic demands such as stage of the pregnancy cycle) and by environmental conditions. Willemse and Takken (1994) divided host seeking into four phases: *ranging* (random flight prior to detecting stimuli from the host); *activation* (a behavioral change following detection of host stimuli); *orientation* (upwind flight stimulated by CO_2 and several organic chemicals emitted by the host and modified by visual stimuli); and *landing*. Wind tunnel experiments have recently demonstrated that *G. m. morsitans* fly upwind when exposed to windborne plumes of heat that fluctuate only fractions of a degree above ambient temperature. Structurally similar plumes were recorded 18 m downwind from a steer, suggesting that thermal stimuli are part of the repertoire of signals evoking host-locating behavior in tsetse. This raises the possibility that temperature fluctuation may be used by tsetse during host seeking and may help to explain why there is little long-range selectivity in attraction to hosts (Evans and Gooding, 2002). There is, however, short-range discrimination, and some tsetse species are not strongly attracted to

the vertical forms of humans and are repelled by lactic acid secreted onto human skin.

Relative abundance of vertebrates is generally not a good indicator of whether tsetse use them as hosts. Tsetses bite vertebrates that are in, or near, the tsetses' preferred habitat (wooded areas) at the time of day that the flies are most active (i.e., before or after the hottest part of the day). The vertebrates must also be relatively tolerant of tsetse attacks (i.e., minimal self-grooming, tail flicking, or skin rippling). Feeding on various vertebrate species by a particular species of tsetse shifts from one locality to another, and may undergo seasonal fluctuations. Some tsetse species take a significant portion of their blood meals from reptiles and birds. However, wild suids and bovids are the source of most tsetse blood meals; man and domestic pigs are also significant hosts for at least some members of the *palpalis* group (see summary by Moloo 1993).

When a tsetse lands on a host, its proboscis is held between the palps. In response to temperature (detected by receptors on the antennae and tarsi), the haustellum is positioned approximately perpendicular to the feeding site (Fig. 13.1). Short, stout chetae on the labellum penetrate the host's skin and draw the haustellum into the skin. Tsetse flies feed from a pool of blood that forms when blood vessels are damaged. The stimulus to ingest blood probably is ATP released from ruptured platelets; the signal to stop feeding likely originates in abdominal stretch receptors.

Tsetse saliva contains a powerful antithrombin (Parker and Mant 1979) that prevents blood from clotting and two platelet-aggregation inhibitors (Mant and Parker 1981). The meal, which varies in size from about 20 mg to more than 100 mg, depending upon species and physiological state, is taken in about 1 minute. The meal is passed first to the diverticulum or crop and then to the midgut. Immediately after feeding, most of the water is removed from the meal as it passes through the anterior part of the midgut. Digestion takes place in the posterior part of the midgut, by the action of a hemolytic agent (Gooding 1977) and several proteolytic enzymes (Cheeseman and Gooding 1985).

Dispersal

The dispersal of tsetse flies has implications for invasion of tsetse-free areas, disease transmission, and tsetse control by traps or insecticide-impregnated targets. Tsetse dispersal distances are also important, because the optimal resting and breeding sites are often different from the sites at which hosts are located. Using mark–release–recapture techniques, the mean displacement of tsetse, within their normal habitat, ranges from about 150 to 540 m/day, with a linear advance of populations being about 200–330 m/w (summarized by Leak 1999). There is, however, considerable variation within populations and between populations and species.

Tsetse dispersal is strongly influenced by both physiological and ecological factors. Incomplete oxidation of proline is the main source of energy for flight by tsetse (Chapter 23). In this process, proline is metabolized to alanine in flight muscles, and alanine is transported to the fat body, where it is converted back to proline; the carbon and energy for the latter process comes from fatty acids. Although this is an efficient system, the rate at which proline is used by the flight muscle exceeds the rate at which it can be restored, and tsetse flies cease flight when the proline concentration is significantly depressed. Although tsetses fly fast (ca. 5 m/sec), their flights are less than 5 minutes in duration (probably averaging 1–2 min), and the total time spent flying is probably 15–30 min/day. Thus, tsetse probably fly 4.5–9 km per day in a series of short, randomly oriented flights.

Tsetse need to avoid exposure to excessively high temperature and evaporation rates and thus normally remain in or close to wooded areas. However, when very hungry or when in or on vehicles, they do move relatively long distances from their normal habitat in search of hosts. The reinvasion of areas from which tsetse have been cleared indicates that they are capable of crossing clear-cut barriers of 1–5 km. This is accomplished by their efforts or with the aid of human activities, such as herding animals across the "barriers."

TRANSMISSION OF AFRICAN TRYPANOSOMES

Tsetse are infamous for their role as the primary vectors for pathogenic trypanosomes, the causative agents of African sleeping sickness in humans and of a variety of diseases in domestic livestock. Since antigenic variation in bloodstream form parasites (reviewed in Vanhamme et al. 2001) has presented a formidable barrier to immunization. In the past few years attention has shifted to the study of the forms of trypanosomes found in tsetse and, more recently still, to parasite–vector interactions. Such studies are deemed to be important for epidemiology and for devising methods for intervention in disease transmission. With the persistence of the diseases in domestic animals and an ongoing serious epidemic in humans (Smith et al. 1998), studies on the molecular entomology of tsetse are justified as perhaps the best

hope for control of these diseases that socially and economically devastate sub-Saharan Africa. Readers are referred to comprehensive reviews of the African trypanosomes and their tsetse vectors by Mulligan (1970) and by Leak (1999), although these books do not contain the most recent information on molecules of parasite surfaces or of the vectors or on molecules involved in trypanosome–tsetse interactions.

Surface Molecules of Insect Form Trypanosomes

During development in tsetse, African trypanosomes differentiate from anaerobic bloodstream forms (ingested in the blood meal) into aerobic forms that respire using the Krebs cycle. This transformation has been studied primarily in trypanosomes of the subgenus *Trypanozoon* (*Trypanosoma brucei* spp.), in the subgenus *Nannomonas* (*T. congolense* and *T. simiae*), and, to a lesser extent, in the subgenus *Duttonella* (*T. vivax*). All of these trypanosomes are pathogenic: *Trypanosoma brucei* spp. in humans domestic animals including: cattle, goats, horses, and camels; *T. congolense* in cattle; and *T. simiae* in pigs; and *T. vivax* in cattle. Normally these parasites are cyclically transmitted by tsetse, with only *T. vivax* exhibiting the ability to be occasionally mechanically transmitted by biting insects other than tsetse. Differentiation of all of these species is accompanied by extensive morphological changes, by loss of the variant surface glycoprotein surface coat, and by acquisition of a new set of coat molecules, the procyclins, although these molecules have not yet been described in *T. vivax* (reviewed in Roditi et al. 1998). Knockout mutants that do not express procyclins did not establish heavy tsetse infections (Ruepp et al. 1997), and thus the procyclins affect transit of parasites through tsetse. However, a puzzling observation using trypanosome knockout mutants is that those lacking glycosylphosphatidylinositol-anchored proteins (and thus surface expression of procyclins) did not show drastic changes in establishment, in comparison to wild-type *T. brucei* (Nagamune et al. 2000). It has been suggested that perhaps these parasites compensated by secretion of the procyclins (which could no longer be anchored to the membrane) or by expression of other membrane molecules that allowed survival in the midgut (Roditi and Liniger 2002).

Analysis of parasites taken directly from tsetse revealed that during the first few days in the midgut, trypanosomes expressed isoforms of both of the two major forms of procyclins. After 4–7 days, a subset of isoforms of only one of the major procyclins was displayed (Acosta-Serrano et al. 2001). The repression of the other major form of procyclin temporally paralleled the migration of the parasites from the endo- to the ectoperitrophic space and subsequent establishment of infection, suggesting a central role in interactions with tsetse. The dynamic surface mosaic presented by the procyclin molecules could influence differentiation and survival of the parasites by mechanisms where the organisms are selected for either differentiation or cell death (Pearson 2001). Alternative hypotheses are that the procyclins may be involved in protection against tsetse proteases or that they influence transit through the different compartments of the fly and tropism within tsetse (Roditi and Pearson 1990). For example, procyclins may prevent parasite adherence to tsetse midgut epithelium or allow adherence to salivary gland epithelium or mouthparts, depending on the trypanosome species. Some clues come from comparative analyses of the surface coat molecules of insect forms of different trypanosome species. First, *T. congolense* and *T. brucei* spp. display biochemically and immunochemically different procyclins (reviewed in Stebeck and Pearson 1994). Since epimastigotes of *T. congolense* develop in the tsetse proboscis, whereas epimastigotes of *T. brucei* develop in the salivary glands, it is tempting to think that the structurally distinct coat molecules are involved in this localization. Second is a comparative analysis of the procyclins from two species of the same subgenus, *T. congolense* and *T. simiae*. Although epimastigotes of both of these parasites develop in the proboscis, they express structurally different procyclins (Mookherjee and Pearson 2001). Even though this at first seems to contradict our hypothesis, it is intriguing that these molecules share surface-displayed, highly immunogenic carbohydrate epitopes (Mookherjee and Pearson 2002). Perhaps these shared carbohydrate structures are involved in interactions with tsetse molecules, thus anchoring the trypanosomes where the final differentiation to mammal-infective metacyclic forms occurs. It certainly puts carbohydrate structures under suspicion as potential ligands for possible interactions with tsetse molecules.

Molecules of Tsetse Origin

In contrast to studies with trypanosomes, molecular analyses of tsetse are less advanced. After tsetse ingest a trypanosome-infected blood meal, the establishment of midgut-adapted procyclic trypanosomes must involve midgut factors that mediate parasite transformation, growth, and survival. For example, Nguu et al. (1996) demonstrated that the mammalian bloodstream form trypanosomes transformed into procyclic forms *in vitro* when incubated with midgut homogenates from nonteneral tsetse. The relevant

molecules in the midgut homogenates were not characterized. Midgut proteins, including lectins (Welburn et al. 1989; Lehane and Msangi 1991), trypanolysins, and proteolytic enzymes (Endege et al. 1989; Imbuga et al. 1992; Yan et al. 2001), have been hypothesized to influence trypanosomes. Only a few of these kinds of molecules have been described. These include six proteolytic enzymes (Cheeseman and Gooding 1985), two genes encoding midgut serine proteases (Yan et al. 2001), and genes encoding a cathepsin and two zinc-metalloproteases (Yan et al. 2002).

That tsetse midgut molecules function during the developmental stages of trypanosomes in the tsetse vector is likely. This idea is supported by the observation, in an artificial membrane feeding system, that anti-*Anopheles* midgut monoclonal antibodies markedly reduced both vector competence and insect survival (Lal et al. 2001). These kinds of studies have not been performed in tsetse. However, recent protein microchemical studies of tsetse salivary glands (Haddow et al. 2002) and midguts (Haines et al. 2002) identified several of the major proteins present in these tissues and open the way to the determination of their functions in tsetse viability and parasite transmission. One intriguing finding is that the major midgut molecule identified is a molecular chaperone of microbial origin, a 60-kDa chaperone from *Wigglesworthia glossinidia*, one of several microbial symbionts that inhabit tsetse. Thus, tsetse midgut proteins influencing vector survival and competence may be of both eukaryotic and prokaryotic origin.

MICROBES INHABITING TSETSE

Wigglesworthia glossinidia

The obligate (primary) endosymbionts, *Wigglesworthia glossinidia*, are gram-negative, rod-shaped bacteria that are localized to the midgut of the tsetse. Their growth is restricted to the cytoplasm of specialized midgut epithelial cells termed *bacteriocytes* (previously called *mycetocytes*). In the tsetse, these bacteriocytes aggregate, forming a longitudinal organelle, the *bacteriome*, along the anterior region of the midgut (Aksoy 1995a, b; Aksoy et al. 1995). Often in the case of bacteriocyte symbiosis, the microbial partner supplements a nutritionally unbalanced host diet. By synthesizing cofactors and vitamins, in particular those of the B-complex, *W. glossinidia* is able to compensate for dietary deficiencies of its hematophagous host (Nogge 1981). Probing of heterologous *E. coli* gene arrays with *W. glossinidia* cDNAs revealed orthologous genes encoding amino acid biosynthetic enzymes and ABC-transporters (Akman and Aksoy 2001). This experiment did not detect genes unique to *W. glossinidia*, however; thus, genome sequencing is necessary to identify endosymbiont-specific genes. This has been accomplished (Akman et al. 2002). Since *in vitro* culture of *W. glossinidia* is difficult, experiments to determine the role of this symbiont have been limited to the use of symbiont-cured tsetse. The elimination of *W. glossinidia* often results in reduced fecundity and developmental abnormalities, but not a reduction in fly longevity (Nogge 1978). Larvae arising from aposymbiotic tsetse are also symbiont-free and sterile since larvae acquire symbionts from their mothers. The reduction in fertility may be an indirect result of insufficient essential vitamin biosynthesis required for normal larval development, since blood meals supplemented with B-complex vitamins can partially restore fly reproductive capabilities (Nogge 1981). The function of the aforementioned *W. glossinidia* chaperone, the most abundant protein in the midgut of teneral tsetse, is unknown.

Sodalis glossinidius

A secondary tsetse symbiont, *Sodalis glossinidius* (Fig. 13.6), (previously called a Rickettsia-like organism, RLO), was recently classified as a new bacterial species (Dale and Maudlin 1999). As with *W. glossini-*

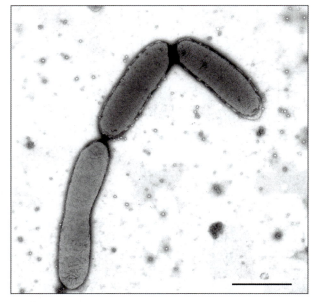

FIGURE 13.6 Transmission electron micrograph of *Sodalis glossinidius* isolated from a 5-day-old liquid culture. The scale bar represents 1.5 micrometers. Note pleomorphism and characteristic linear alignment of the cells. Photomicrograph by Lee Haines.

dia, this gram-negative, nonmotile member of the Enterobacteriaceae is transmitted through milk gland secretions. Initially *S. glossinidius* was observed both intracellularly in the midgut epithelium and as free-living forms within the gut lumen (Reinhardt et al. 1972). With the development of sensitive PCR assays, this symbiont was also detected in other tsetse tissues, such as the fat body, milk gland, hemolymph, muscle, and salivary glands. However, species-specific and age-specific population densities have been observed (Cheng and Aksoy 1999; Maudlin et al. 1990). Unlike *W. glossinidia*, *S. glossinidius* can be cultured *in vitro*, thus facilitating its characterization. The relationship between this bacterium and tsetse is unknown, although mutualism, perhaps involving the provision of metabolites, is possible. This idea is supported by evidence that genes encoding enzymes required for vitamin synthesis are found in *S. glossinidius* (Akman et al. 2001). Experimental elimination of *S. glossinidius* from tsetse using the antibiotic streptozotocin did not affect fly fecundity or pupal emergence; however, longevity was significantly reduced (Dale and Welburn 2001). Intriguingly, this observation is the mirror image of that seen with tsetse cured of *W. glossinidia*. Of perhaps greater interest is the hypothesis that *S. glossinidius* may modulate the transmission of African trypanosomes. Increased trypanosome infections in the tsetse midgut were reported in both lab-reared and field-caught tsetse possessing high numbers of *S. glossinidius* (Maudlin and Ellis 1985). Indeed, it was stated that a wild fly carrying this symbiont was six times more likely to be infected with trypanosomes than an *S. glossinidius*–free fly (Maudlin et al. 1990). Later studies on puparial development at lower temperatures (Welburn and Maudlin 1991) and the removal of *S. glossinidius* populations by antibiotic treatment (Dale and Welburn 2001) both resulted in a significant increase in refractoriness to trypanosome infection in the teneral tsetse midgut. Since *S. glossinidius* and trypanosomes coinhabit the tsetse midgut, it is plausible that this resident symbiont may influence parasite establishment in the teneral fly and, ultimately, transmission.

Wolbachia spp.

Some tsetse harbor a third symbiont, intracellular Rickettsia-like microbes belonging to the genus *Wolbachia* that are transovarially inherited through the female host. This bacterium is widespread among many insect species and is the cause of a variety of reproductive abnormalities, such as feminization of genetic males, parthenogenesis, male killing, and cytoplasmic incompatibility in which mating of *Wolbachia-*

infected males with uninfected females fails to produce viable progeny (Bandi et al. 2001). In tsetse, *Wolbachia* is found primarily in germ-line tissues, although in some flies it has also been localized in somatic tissues (O'Neill et al. 1993; Cheng et al. 2000). *Wolbachia* infections do not appear to benefit tsetse, and generation of *in vitro* cultures of *Wolbachia* or the maintenance of tsetse colonies with associated *Wolbachia*-mediated phenotypes will be necessary to fully understand and possibly exploit this symbiosis.

Although it is not known if *Wolbachia* is involved in cytoplasmic incompatibility in tsetse, it is assumed that it is capable of it (Cheng et al. 2000). Therefore, strategies have been proposed to use *Wolbachia* to drive maternally inherited symbionts, for example, *S. glossinidius* expressing antitrypanosomal factors, into a wild population (Aksoy et al. 2001). Experiments with antitrypanosome factors produced by symbionts have been successfully performed with triatomines in the laboratory (Beard et al. 1992). The potential risks and applications of this strategy are reviewed in Beard et al. (2001). The proposal to use *Wolbachia* as a drive mechanism is based upon the assumption that it causes cytoplasmic incompatibility in tsetse and thus females with *Wolbachia* have a fitness advantage over those that lack *Wolbachia*. Hybridization asymmetries occur between some closely related taxa of tsetse, and in the first generations of such crosses and back-crosses the hybridization patterns are similar to what would be expected with maternally inherited factors such as *Wolbachia* (Gooding 1990b). However, hybridization experiments, using lines that carried maternally inherited factors from one taxon and chromosomes from another, established that hybridization asymmetries in tsetse are more complicated than originally thought. Specifically, either hybridization asymmetries are determined by chromosomal factors, and not by maternally inherited factors, or there is paternal transmission of factors that are normally transmitted through the maternal lineage (Gooding 2000). Whether results with intertaxon hybridization apply to intrataxon hybridization remains to be determined.

LARGE-SCALE CONTROL EFFORTS

"Nearly 100 years of control efforts have failed to curb the distribution of tsetse fly infestations or the resulting incidence of trypanosomiasis in Africa" (Anonymous 2000). This brutally blunt assessment was the opening statement in a small glossy brochure outlining the *Programme Against African Trypanosomiasis*. The control programs to which the statement refers encompass the widespread use of insecticides (including ground spraying, aerial application, and application to "targets"), environmental modification (including habitat destruction and elimination of natural hosts), and, more recently, application of insecticide followed by eradication of the population by release of sterile males (Chapter 44). The latter procedure has recently been used to eradicate *G. austeni* from Zanzibar (Vreysen et al. 2000). A major problem with tsetse suppression or eradication has been sustainability of the efforts needed to eradicate a species or to maintain a "fly-proof barrier" between cleared and infested areas.

Tsetse control currently relies heavily upon insecticide-treated targets (i.e., insecticide-treated cloth screens to which tsetse are attracted) or insecticide-treated cattle (Allsopp 2001). However, the efficiency of insecticide-based control programs generally declines as the pest population declines, and release of sterile males is so far the method of choice for the final eradication of low-density tsetse populations (Feldmann and Hendrichs 2001).

Tsetse control operations are being carried out in about 128,000 km^2; i.e., in only about 1.5% of the 8.5 million km^2 of infested land (Allsopp 2001). In this context, the decision (OAU 2000) made by the heads of state and heads of government at an Organization of African Unity meeting in Lomé, Togo, in July 2000 to eradicate tsetse from Africa seems rather ambitious, particularly since the number of sterile males required to eradicate tsetse surviving at low levels is huge. One cannot help but recall the old saying that "the African landscape is littered with the graves of failed tsetse control projects." Something new must be tried.

Readings

Acosta-Serrano, A., Vassella, E., Liniger, M., Kunz Renggli, C., Brun, R., Roditi, I., and Englund, P. T. 2001. The surface coat of procyclic *Trypanosoma brucei*: Programmed expression and proteolytic cleavage of procyclin in the tsetse fly. *Proc. Natl. Acad. Sci. USA* 98: 1513–1518.

Akman, L., and Aksoy, S. 2001. A novel application of gene arrays: *Escherichia coli* array provides insight into the biology of the obligate endosymbiont of tsetse flies. *Proc. Natl. Acad. Sci. USA* 98: 7546–51.

Akman, L., Rio, R. V. M., Beard, C. B., and Aksoy, S. 2001. Genome size determination and coding capacity of *Sodalis glossinidius*, an enteric symbiont of tsetse flies, as revealed by hybridization of *Escherichia coli* gene array. *J. Bacteriol.* 183: 4517–4525.

Akman, L., Yamashita, A., Watanabe, H., Oshima, K., Shiba, T., Hattori, M., and Aksoy, S. 2002. Genome sequence of the endocellular obligate symbiont of tsetse flies, *Wigglesworthia glossinidia*. Nature Genetics 32: 402–407.

Aksoy, S. 1995a. Molecular analysis of the endosymbionts of tsetse flies: 16S rDNA locus and overexpression of a chaperonin. *Insect Mol. Biol.* 4: 23–29.

Aksoy, S. 1995b. *Wigglesworthia* gen. nov. and *Wigglesworthia glossinidia* sp. nov., taxa consisting of the mycetocyte-associated, primary endosymbionts of tsetse flies. *Int. J. Syst. Bacteriol.* 45: 848–851.

Aksoy, S., Pourhosseini, A., and Chow, A. 1995. Mycetome endosymbionts of tsetse flies constitute a distinct lineage related to Enterobacteriaceae. *Insect Mol. Biol.* 4: 15–22.

Aksoy, S., Maudlin, I., Dale, D., Robinson, A. S., and O'Neill, S. L. 2001. Prospects for control of African trypanosomiasis by tsetse vector manipulation. *Trends Parasitol.* 17: 29–35.

Allsopp, R. 2001. Options for vector control against trypanosomiasis in Aftrica. *Trends Parasitol.* 17: 15–19.

Anonymous. 2000. *Programme Against African Trypanosomiasis.* A single-fold glossy brochure (I/W8046E/1/5.98/2000), Publishing Management Group, FAO Information Division. PAAT Committee Chairman, Prof. Peter Holmes, (pholmes@mis.gla.ac.uk)

Bandi, C., Dunn, A. M., Hurst, G. D., and Rigaud, T. 2001. Inherited microorganisms, sex-specific virulence and reproductive parasitism. *Trends Parasitol.* 17: 88–94.

Beard, C. B., Mason, P. W., Aksoy, S., Tesh, R. B., and Richards, F. F. 1992. Transformation of an insect symbiont and expression of a foreign gene in the Chagas' disease vector *Rhodnius prolixus. Am. J. Trop. Med. Hyg.* 46: 195–200.

Beard, C. B., Dotson, E. M., Pennington, P. M., Eichler, S., Cordon-Rosales, C., and Durvasula, R. V. 2001. Bacterial symbiosis and paratransgenic control of vector-borne Chagas disease. *Int. J. Parasitol.* 31: 621–627.

Cheeseman, M. T., and Gooding, R. H. 1985. Proteolytic enzymes from tsetse flies, *Glossina morsitans* and *Glossina palpalis* (Diptera: Glossinidae). *Insect Biochem.* 15: 677–680.

Chen, S., Li, S., and Aksoy, S. 1999. Concordant evolution of a symbiont with its host insect species: Molecular phylogeny of genus *Glossina* and its bacteriome-associated endosymbiont, *Wigglesworthia glossinidia. J. Mol. Evol.* 48: 49–58.

Cheng, Q., and Aksoy, S. 1999. Tissue tropism, transmission and expression of foreign genes *in vivo* in midgut symbionts of tsetse flies. *Insect Mol. Biol.* 8: 125–132.

Cheng, Q., Ruel, T. D., Zhou, W., Moloo, S. K., Majiwa, P., O'Neill, S. L., and Aksoy, S. 2000. Tissue distribution and prevalence of *Wolbachia* infections in tsetse flies, *Glossina* spp. *Med. Vet. Entomol.* 14: 44–50.

Dale, C., and Maudlin, I. 1999. *Sodalis* gen. nov. and *Sodalis glossinidius* sp. nov., a microaerophilic secondary endosymbiont of the tsetse fly *Glossina morsitans morsitans. Int. J. Syst. Bacteriol.* 49: 267–275.

Dale, C., and Welburn, S. 2001. The endosymbionts of tsetse flies: Manipulating host–parasite interactions. *Int. J. Parasitol.* 31: 627–630.

Dias, J. A. Travassos Santos. 1987. Contribução para o estudo da sistemática do género *Glossina* Wiedemann, 1830 (Insecta, Brachycera, Cyclorrhapha, Glossinidae) Proposta para a criação de um novo subgénero. *Garcia de Orta, Série de Zoologia, Lisboa* 14: 67–78.

Elsen, P., Amoudi, M. A., and Leclercq, M. 1990. First record of *Glossina fuscipes fuscipes* Newstead, 1910 and *Glossina morsitans submorsitans* Newstead, 1910 in southwestern Saudi Arabia. *Ann. Soc. Belge Méd. Trop.* 70: 281–287.

Endege, W. O., Lonsdale-Eccles, J. D., Olembo, N. K., Moloo, S. K., and ole-MoiYoi, O. K. 1989. Purification and characterization of two fibrinolysins from the midgut of adult female *Glossina morsitans centralis. Comp. Biochem. Physiol. B* 92: 25–34.

Evans, W. G., and Gooding, R. H. 2002. Turbulent plumes of heat, moisture and carbon dioxide elicit upwind anemotaxis of tsetse flies *Glossina morsitans morsitans* Westwood (Diptera: Glossinidae). *Can. J. Zool.* 80: 1149–1155.

Feldmann, U., and Hendrichs, J. 2001. *Integrating the Sterile Insect Technique as a Key Component of Area-wide Tsetse and Trypanosomiasis Intervention.* PAAT Technical and Scientific Series, No. 3. Food and Agriculture Organization of the United Nations, Rome, 2001.

Ford, J. 1971. *The Role of the Trypanosomiases in African Ecology. A Study of the Tsetse Fly Problem.* Clarendon Press, Oxford, UK.

Ford, J., and Katondo, K. M. 1977. Maps of tsetse fly (*Glossina*) distribution in Africa 1973, according to subgeneric groups on scale 1:5,000,000. *Bull. Animal Health Production Africa* 15: 187–193.

Gooding, R. H. 1977. Digestive process of haematophagous insects. XIV. Haemolytic activity in the midgut of *Glossina morsitans morsitans* Westwood (Diptera: Glossinidae). *Can. J. Zool.* 55: 1899–1905.

Gooding, R. H. 1990a. Studies of *Glossina pallidipes* and *G. morsitans* subspecies related to the genetic control of tsetse flies. In *Sterile Insect Technique for Tsetse Control and Eradication.* Proceedings of the Final Research Coordination Meeting Organized by the Joint FAO/IAEA Division of Nuclear Techniques in Food and Agriculture, Vom, Plateau State, Nigeria, 6–10 June 1988. IAEA/RC/319.3/15, International Atomic Energy Agency, Vienna, 1990, pp. 213–226.

Gooding, R. H. 1990b. Postmating barriers to gene flow among species and subspecies of tsetse flies (Diptera: Glossinidae). *Can. J. Zool.* 68: 1727–1734.

Gooding, R. H. 2000. Hybridization asymmetries in tsetse (Diptera: Glossinidae): Role of maternally inherited factors and the tsetse genome. *J. Med. Entomol.* 37: 897–901.

Gooding, R. H., Moloo, S. K., and Rolseth, B. M. 1991. Genetic variation in *Glossina brevipalpis, G. longipennis*, and *G. pallidipes*, and the phenetic relationships of *Glossina* species. *Med. Vet. Entomol.* 5: 165–173.

Haddow, J. D., Poulis, B. A. D., Haines, L. R., Gooding, R. H., Aksoy, S., and Pearson, T. W. 2002. Identification of major soluble salivary gland proteins in teneral *Glossina morsitans morsitans. Insect Biochem. Mol. Biol.* 32: 1045–1053.

Haines, L. R., Haddow, J. D., Aksoy, S., Gooding, R. H., and Pearson, T. W. 2002. The major protein in the midgut of the tsetse fly, *Glossina morsitans morsitans*, is a molecular chaperone from the endosymbiotic bacterium *Wigglesworthia glossinidia. Insect Biochem. Mol. Biol.* 32: 1429–1438.

Imbuga, M. O., Osir, E. O., Labongo, V. L., Darji, N., and Otieno, L. H. 1992. Studies on tsetse midgut factors that induce differentiation of bloodstream *Trypanosoma brucei brucei in vitro. Parasitol. Res.* 78: 10–15.

Jordan, A. M. 1993. Tsetse flies (Glossinidae). In: *Medical Insects and Arachnids.* (R. P. Lane and R. W. Crosskey, eds.). Chapman and Hall, London, pp. 333–388.

Jordan, A. M., Nash, T. A. M., Southern, D. I., Pell, P. E., and Davies, E. D. G. 1977. Differences in laboratory performance between strains of *Glossina morsitans morsitans* Westwood from Rhodesia and Tanzania and associated chromosome diversity. *Bull. Entomol. Res.* 67: 35–48.

Krafsur, E. S., and Wohlford, D. L. 1999. Breeding structure of *Glossina pallidipes* populations evaluated by mitochondrial variation. *J. Heredity* 90: 635–642.

Krafsur, E. S., Griffiths, N. T., Brockhouse, C. L., and Brady, J. 1997. Breeding structure of *Glossina pallidipes* (Diptera: Glossinidae) populations in east and southern Africa. *Bull. Entomol. Res.* 87: 67–73.

Krafsur, E. S., Madsen, M., Wohlford, D. L., Mihok, S., and Griffiths, N. T. 2000. Population genetics of *Glossina morsitans submorsitans* (Diptera: Glossinidae). *Bull. Entomol. Res.* 90: 329–335.

Lal, A. A., Patterson, P. S., Sacci, J. B., Vaughan, J. A., Paul, C., Collins, W. E., Wirtz, R. A., and Azad, A. F. 2001. Anti-mosquito midgut antibodies block development of *Plasmodium falciparum* and *Plasmodium vivax* in multiple species of *Anopheles* mosquitoes and reduce vector fecundity and survivorship. *Proc. Natl. Acad. Sci. USA* 98: 5228–5233.

Langley, P. A., Maudlin, I., and Leedham, M. P. 1984. Genetic and behavioral differences between *Glossina pallidipes* from Uganda and Zimbabwe. *Ent. Exp. Appl.* 35: 55–60.

Leak, S. G. A. 1999. *Tsetse Biology and Ecology: Their Role in the Epidemiology and Control of Trypanosomosis*. CABI, Wallingford, VT, pp. 568.

Lehane, M. J., and Msangi, A. R. 1991. Lectin and peritrophic membrane development in the gut of *Glossina m. morsitans* and a discussion of their role in protecting the fly against trypanosome infection. *Med. Vet. Entomol.* 5: 495–501.

Machado, A. de Barros. 1954. Révision systématique des Glossines du groupe *palpalis* (Diptera). *Publicações Culturais da Companhia de Diamantes de Angola* 22: 1–189.

Machado, A. de Barros. 1959. Nouvelles contributions à l'étude systématique et biogéographique des Glossines (Diptera). *Publicações Culturais da Companhia de Diamantes de Angola* 46: 13–90.

Machado, A. de Barros. 1970. Les races géographiques de *Glossina morsitans*. In J. Fraga de Azevedo, ed., *Criaçõa da mosca tsé-tsé em laboratorio e sua aplicação prática (Tsetse Fly Breeding Under Laboratory Conditions and Its Practical Application)*. 1st International Symposium, 22 and 23 April 1969, Lisbon, pp. 471–486.

Mant, M. J., and Parker, K. R. 1981. Two platelet aggregation inhibitors in tsetse (*Glossina*) saliva with studies of roles of thrombin and citrate in *in vitro* platelet aggregation. *Brit. J. Haematol.* 48: 601–608.

Maudlin, I., and Ellis, D. S. 1985. Association between intracellular rickettsial-like infections of midgut cells and susceptibility to trypanosome infection in *Glossina* spp. *Zeitschrift für Parasitenkunde* 71: 683–687.

Maudlin, I., Welburn, S. C., and Mehlitz, D. 1990. The relationship between rickettsia-like organisms and trypanosome infections in natural populations of tsetse in Liberia. *Trop. Med. Parasitol.* 41: 265–267.

McAlpine, J. F. 1989. Phylogeny and classification of the Muscomorpha. In *Manual of Nearctic Diptera*, Vol. 3. (J. F. McAlpine, ed.). Research Branch Agriculture Canada (Monograph No. 32), Ottawa, Canada, pp. 1397–1518.

Moloo, S. K. 1993. The distribution of *Glossina* species in Africa and their natural hosts. *Insect Sci. Appl.* 14: 511–527.

Mookherjee, N., and Pearson, T. W. 2001. Surface molecules of procyclic forms of *Trypanosoma simiae* and *Trypanosoma congolense*, members of the subgenus *Nannomonas*, share immunodominant carbohydrate epitopes. *Mol. Biochem. Parasit.* 118: 123–126.

Mookherjee, N., and Pearson, T. W. 2002. *Trypanosoma simiae* and *Trypanosoma congolense*: Surface glycoconjugates of procyclic forms —the same coats on different hangers? *Exp. Parasitol.* 100: 257–268.

Mulligan, H. W. (ed.) 1970. *The African Trypanosomiases*. Allen and Unwin, Ministry of Overseas Development, London.

Nagamune, K., Nozaki, T., Maeda, Y., Ohishi, K., Fukuma, T., Hara, T., Schwarz, R. T., Sutterlin, C., Brun, R., Riezman, H., and Kinoshita, T. 2000. Critical roles of glycosylphosphidylinositol for *Trypanosoma brucei*. *Proc. Natl. Acad. Sci. USA* 97: 10336–10341.

Nguu, E. K., Osir, E. O., Imbuga, M. O., and Olembo, N. K. 1996. The effect of host blood in the in vitro transformation of bloodstream trypanosomes by tsetse midgut homogenates. *Med. Vet. Entomol.* 10: 317–322.

Nirmala, X., Hypsa, V., and Zurovec, M. 2001. Molecular phylogeny of Calyptratae (Diptera: Brachycera): The evolution of 18S and 16S ribosomal rDNAs in higher dipterans and their use in phylogenetic inference. *Insect Mol. Biol.* 10: 475–485.

Nogge, G. 1978. Aposymbiotic tsetse flies, *Glossina morsitans morsitans*, obtained by feeding on rabbits immunized specifically with symbionts. *J. Insect Physiol.* 24: 299–304.

Nogge, G. 1981. Significance of symbionts for the maintenance of an optimal nutritional state for successful reproduction in hematophagous arthropods. *Parasitology* 82: 299–304.

OAU. 2000. Assembly of Heads of State and Government, 36th Ordinary Session, 4th Ordinary Session of the African Economic Community, Decision of Proposal for Eradication of Tsetse Flies on the African Continent. Available at www.oau-oua.org/lome2000.

O'Neill, S. L., Gooding, R. H., and Aksoy, S. 1993. Phylogenetically distant symbiotic microorganisms reside in *Glossina* midgut and ovary tissues. *Med. Vet. Entomol.* 7: 377–383.

Parker, K. R., and Mant, M. J. 1979. Effects of tsetse (*Glossina morsitans morsitans* Westw.) (Diptera: Glossinidae) salivary gland homogenate on coagulation and fibrinolysis. *Thrombos. Haemostas.* 2: 743–751.

Pearson, T. W. 2001. Procyclins, proteases and proteomics: Dissecting trypanosomes in the tsetse fly. *Trends Microbiol.* 9: 299–301.

Reinhardt, C., Steiger, R., and Hecker, H. 1972. Ultrastructural study of the midgut mycetome-bacteroids of the tsetse flies *Glossina morsitans*, *G. fuscipes*, and *G. brevipalpis* (Diptera, Brachycera). *Acta Tropica* 29: 280–288.

Roditi, I., and Liniger, M. 2002. Dressed for success: The surface coats of insect-borne protozoan parasites. *Trends Microbiol.* 10: 128–134.

Roditi, I., and Pearson, T. W. 1990. The procyclin coat of African trypanosomes. *Parasitol. Today* 6: 79–81.

Roditi, I., Furger, A., Ruepp, S., Schurch, N., and Butikofer, P. 1998. Unraveling the procyclin coat of *Trypanosoma brucei*. *Mol. Biochem. Parasit.* 91: 117–130.

Ruepp, S., Furger, A., Kurath, U., Renggli, C. K., Hemphill, A., Brun, R., and Roditi, I. 1997. Survival of *Trypanosoma brucei* in the tsetse fly is enhanced by the expression of specific forms of procyclin. *J. Cell Biol.* 137: 1369–1379.

Smith, D. H., Pepin, J., and Stich, A. H. 1998. Human African trypanosomiasis: An emerging public health crisis. *Brit. Med. Bull.* 54: 341–355.

Solano, P., de la Rocque, S., Cuisance, D., Geoffroy, B., de Meeus, T., Cuny, G., and Duvalle, G. 1999. Intraspecific variability in natural populations of *Glossina palpalis gambiensis* from West Africa, revealed by genetic and morphometric analyses. *Med. Vet. Entomol.* 13: 401–407.

Stebeck, C. E., and Pearson, T. W. 1994. Major surface glycoproteins of procyclic stage African trypanosomes. *Exp. Parasitol.* 78: 432–436.

Vanderplank, F. L. 1949. The classification of *Glossina morsitans* Westwood Diptera, Muscidae, including a description of a new subspecies, varieties and hybrids. *Proc. R. Ent. Soc. Lond. (B)* 18: 56–64.

Vanhamme, L., Pays, E., McCulloch, R., and Barry, J. D. 2001. An update on antigenic variation in African trypanosomes. *Trends Parasitol.* 17: 338–343.

Vreysen, M. J. B., Saleh, K. M., Ali, M. Y., Abdulla, A. M., Zhu, Z.-R., Juma, K. G., Dyck, V. A., Msangi, A. R., Mkonyi, P. M., and Feldmann, U. 2000. The use of the sterile insect technique (SIT) for the eradication of the tsetse fly *Glossina austeni* (Diptera: Glossinidae) on the Island of Unguja (Zanzibar). *J. Econ. Entomol.* 93: 123–135.

Welburn, S. C., and Maudlin, I. 1991. Rickettsia-like organisms, puparial temperature and susceptibility to trypanosome infection in *Glossina morsitans*. *Parasitology* 102: 201–206.

Welburn, S. C., Maudlin, I., and Ellis, D. S. 1989. Rate of trypanosome killing by lectins in midguts of different species and strains of *Glossina*. *Med. Vet. Entomol.* 3: 77–82.

Willemse, L. P. M., and Takken, W. 1994. Odor-induced host location in tsetse flies (Diptera: Glossinidae). *J. Med. Entomol.* 31: 775–794.

Wohlford, D. L., Krafsur, E. S., Griffiths, N. T., Marquez, J. G., and Baker, M. D. 1999. Genetic differentiation of some *Glossina morsitans morsitans* populations. *Med. Vet. Entomol.* 13: 377–385.

Yan, J., Cheng, Q., Li, C. B., and Aksoy, S. 2001. Molecular characterization of two serine proteases expressed in gut tissue of the African trypanosome vector, *Glossina morsitans morsitans*. *Insect Mol. Biol.* 10: 47–56.

Yan, J., Cheng, Q., Li, C. B., and Aksoy, S. 2002. Molecular characterization of three gut genes from *Glossina morsitans*: Cathepsin B, zinc-metalloprotease and zinc-carboxypeptidase. *Insect Mol. Biol.* 11: 57–65.

EPIDEMIOLOGY AND SURVEILLANCE

CHESTER G. MOORE AND JEROME E. FREIER

14

Natural Cycles of
Vector-Borne Pathogens

STEPHEN HIGGS AND BARRY J. BEATY

INTRODUCTION

An astonishing variety and number of pathogens are maintained in nature by cycles involving vertebrate hosts and hematophagous arthropod vectors. The vector-borne disease cycle comprises a dynamic interaction between the pathogen, the vertebrate host(s), the vector(s), and the environment (Fig. 14.1). Vector-borne pathogens typically must infect and replicate and/or develop in both a vector and a vertebrate host. When feeding on blood to support oogenesis or to fulfill other nutritional requirements, the vector may ingest and be infected by pathogens in the vertebrate circulatory system or skin. In subsequent blood meals, the vector may transmit these disease agents to new, potentially susceptible, vertebrate hosts. Commonly, the vector-borne pathogen exerts little or no deleterious effect upon its arthropod vector, while infection of vertebrate hosts (especially tangential hosts) may result in significant morbidity and/or mortality.

At first glance, vector-borne cycles appear to be a tenuous mode of transmission and maintenance for pathogens; however, the multitude and diversity of pathogens involved suggests a highly successful strategy. Further, despite enormous effort expended toward controlling them, vector-borne diseases such as malaria, dengue, filariasis, and trypanosomiasis continue to cause overwhelming morbidity and mortality. The tenacity of these cycles in the face of sometimes-drastic human intervention attests to their ability to withstand perturbations. This chapter discusses the mechanisms by which vector-borne pathogens are transmitted and maintained in natural cycles. Understanding the interactions between pathogens, vectors, and vertebrates allows determination of the weak links of the respective cycles at which control efforts may be most efficaciously directed.

ARBOVIRUSES: CYCLES, VIRUSES, VECTORS, AND VERTEBRATE HOSTS

Arboviral Cycles

Arthropod-borne viruses (arboviruses) and arbovirus–vector interactions will be used to illustrate the basic mechanisms and generalities of vector-borne pathogen cycles in nature (Fig. 14.2). Arboviruses are naturally maintained in cycles by hematophagous arthropods that biologically transmit virus between vertebrate hosts. Vectors can become infected by a variety of mechanisms, depending upon the system, including ingesting viremic blood from a vertebrate host, transovarial transmission (virus crosses the ovaries to enter the follicles), venereal transmission during mating, and even vertical transmission of the virus to embryos during oviposition. Classically, an arbovirus does not exert a deleterious effect on its vector. Arboviruses do not include the vector-borne plant viruses or the invertebrate pathogenic viruses, such as baculoviruses, iridoviruses, entomopoxviruses, and cytoplasmic polyhedrosis viruses.

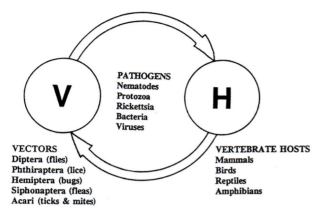

FIGURE 14.1 Principal pathogens, vectors, and vertebrates involved in vector-borne pathogen cycles. Vector-borne pathogens are transmitted between susceptible vertebrate hosts by blood-sucking arthropod vectors. Typically, each pathogen exploits only specific vectors and vertebrate hosts. Each vector cannot transmit every disease agent; nor is every vertebrate susceptible to all the vector-borne pathogens.

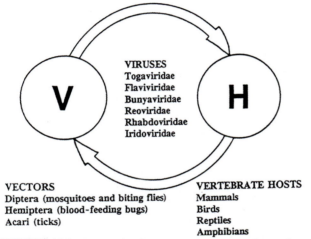

FIGURE 14.2 Principal arboviruses, vectors, and vertebrates involved in arbovirus cycles. Arboviruses are transmitted between susceptible vertebrate hosts by blood-sucking arthropod vectors. Some examples of arboviruses (and their respective families) include: Eastern and Western equine encephalitis (Togaviridae), yellow fever and dengue (Flaviviridae), La Crosse and Rift Valley fevers (Bunyaviridae), Colorado tick fever and bluetongue virus (Reoviridae), vesicular stomatitis virus (Rhabdoviridae), and the African swine fever virus (Iridoviridae).

Maintenance and Amplification Components

Arboviral cycles may be categorized into *maintenance* and *amplification* components. Maintenance is essentially the long-term survival of the virus. Arboviruses can be maintained by ongoing transmission between vertebrate hosts and vectors, by pro-

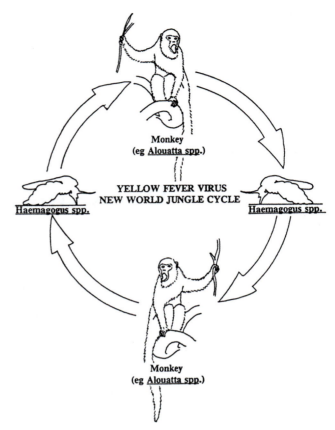

FIGURE 14.3   Simple arbovirus cycle: jungle yellow fever.

longed infections of a host or vector, and by other transeasonal survival mechanisms. During periods of reduced or no transmission, the number of infected hosts and vectors declines. Amplification results in increased prevalence of infected vertebrate hosts, vectors, and, ultimately, the amount of virus in nature. Amplification can occur with the onset of resumed and/or increased transmission and compensates for the loss of infected organisms during the maintenance component of the cycle. Thus, amplification is crucial for viral survival in nature. Amplification is also frequently responsible for epidemics of disease in vertebrates.

In another dimension, arboviral cycles are characterized by a dynamic interaction of the virus, the vector, and the vertebrate host (Fig. 14.3). At each step in a cycle, opportunities for, and barriers to, viral survival exist. The attributes of viruses, vectors, and vertebrate hosts that contribute to the integrity of arboviral cycles are examined in the following sections.

ARBOVIRUSES

Arboviral Biology

Viruses differ from other microorganisms in several ways. They are extremely small, ranging in size from 20 to 300 nm. The vast majority of arboviruses have small RNA genomes, which code for only a few proteins. Further, arboviruses possess neither an energy-generating system nor ribosomes. Thus, they are obligate intracellular parasites, completely dependent upon the host cell for their replication and protein synthesis. Viruses have an infectious stage, the virion, and an intracellular replication stage, known as the *eclipse phase*.

More than 500 arboviruses are recognized. Most have RNA genomes and enveloped virions. An arboviral virion is composed of protein and nucleic acid, associated to form the nucleocapsid, with smaller amounts of lipid and sugars found in the surrounding envelope. The virion obtains an envelope as it buds through a plasma or other cellular membrane containing viral-derived glycoproteins. Figure 14.4 illustrates the basic steps of viral replication in a cell.

In a *productive* infection, the virus infects a permissive cell, the sequence of events enumerated in Figure 14.4 occur, and large numbers of infectious virions result. In an *abortive* or *nonproductive* infection, nonpermissive cells permit viral entry but block replication. Infectious virions do not result, although transcription and translation of some viral genes and gene products may occur. Cells that are not susceptible to viral infection or that block infection and replication at an early stage are termed *resistant* or *refractory cells*. For example, resistant cells may not have appropriate receptors for viral infection.

Virus Considerations in Arboviral Cycles

To contribute to the integrity of their own cycles, arboviruses must infect both vectors and vertebrates in such a way as to be transmitted to the next host. Despite relatively small genomes of approximately 12,000 bases, arboviruses must be capable of infection and replication in two phylogenetically disparate systems, the invertebrate vector and the vertebrate host. Along with biochemical and physiological differences, significant temperature differences exist between these poikilothermic and homeothermic hosts. In light of these considerations, perhaps it is not surprising that an arbovirus may affect the vertebrate and invertebrate quite differently.

For the cycle to remain intact, arboviruses must cause a high-titered viremia (virus in the blood) in the

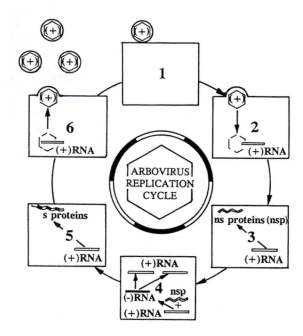

1 = **Reception**
2 = **Penetration and uncoating**
3 = **Translation**
4 = **Transcription and replication**
5 = **Translation**
6 = **Maturation and budding**

FIGURE 14.4 Arbovirus replication in host cells: a positive stranded RNA virus. (1) *reception*—Often the virion glycoprotein attaches to a host cell receptor, such as laminin; (2) *penetration*—viruses enter cells either by pinocytosis or by direct fusion with the plasma membrane, uncoating—the nucleocapsid enters the cytoplasm of the cell from the plasma membrane or from a lysosome and releases its nucleic acid; (3) *translation*—nonstructural proteins (polymerases) essential for transcription are produced; (4) *transcription*—viral RNA is transcribed and replicated; (5) *translation*—viral mRNA is translated on host ribosomes to produce structural glycoproteins; (6) *maturation/budding*—virus proteins and nucleic acids coalesce and mature by budding through host cell membranes. Steps 3–5 occur simultaneously and form the eclipse phase, when infectious virions cannot be detected.

natural vertebrate host. High-titered viremias of long duration maximize the opportunity for vector infection and for mechanical transmission of the virus. Viremias of short duration or low titers are less likely to infect vectors. Moreover, the virus should not be unduly virulent; e.g., killing the vertebrate host before new vectors can be infected. In a "good" host–parasite relationship, the parasite exerts little untoward effect on the usual host, because the virus and host have presumably evolved a benign relationship.

Arboviruses must also infect and replicate in the vector to be biologically transmitted to a new vertebrate host or to another vector. Alternatively, viruses that persist on vector body parts may be transmitted mechanically. Arboviruses, by definition, cause little or no detrimental effect on their natural or usual vectors. If the virus were to decrease the survival of the vector, the probability of transmission of the agent to a new vertebrate host would be reduced. Recent studies with alphaviruses suggest that some cytopathology may occur in natural vectors, although no effect on transmission could be demonstrated. Infection of *Ae. aegypti* with dengue-3 virus resulted in a significantly longer probing and engorgement time, perhaps due to viral infection of the sensory organs (ganglia, ommatidia, Johnston's organ). Feeding success was not affected, but one might presume that transmission efficiency could be enhanced through the prolonged feeding process. Mortality has been shown to be higher in mosquito eggs transovarially infected with La Crosse virus than in uninfected eggs, when the eggs are subjected to natural overwintering conditions and after they have emerged from diapause (McGaw et al. 1998).

Vectors of Arboviruses

Vector Transmission of Arboviruses

For our purposes, a vector is an arthropod, generally hematophagous, that can transmit pathogens from one vertebrate host to another. For some vector species, e.g., mosquitoes, only the females ingest blood and thus transmit viruses to vertebrate hosts. In other species, e.g., ticks, both females and males are hematophagous and can transmit the viruses.

Biological vs. Mechanical Transmission of Arboviruses

Vectors may transmit infectious agents biologically or mechanically. In biological transmission, the pathogens reproduce or develop in the arthropod vector before being transmitted to the next vertebrate host. Arboviruses undergo a propagative mode of development in vectors (Fig. 14.5); other types of vector-borne pathogens may undergo different modes of development.

In contrast, for mechanical transmission the pathogen does not reproduce or develop in the vector; the arthropod merely transmits the pathogen physically from one vertebrate host to another (Fig. 14.6). Most often, the mouthparts of the vector become contaminated with pathogens that are subsequently inoculated into another vertebrate during the vector's next feeding attempt. Transmission is possible only for a short period, until the contaminating virus becomes inactivated. Mechanical transmission has been com-

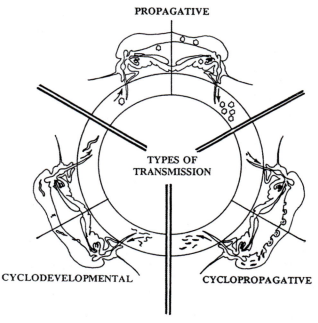

FIGURE 14.5 Three developmental modes of biologically transmitted pathogens. In propagative transmission, the pathogen multiplies an indeterminate number of times in the vector, remaining essentially in the same developmental form. Viruses, *rickettsia*, and bacteria usually propagate within the vector. Typically, the vector transmits many more infectious units than were ingested. With cyclopropagative transmission, the pathogen develops and multiplies in the vector. Protozoans are frequently transmitted in this fashion. For example, ingested malaria gametes form a zygote that develops into an oocyst on the vector's midgut wall (see Chapter 9). Infectious sporozoites emerge from the oocyst, migrate to the salivary gland ducts, and enter their next vertebrate host with the saliva. Again, the vector delivers more pathogenic units than originally consumed. Lastly, for cyclodevelopmental transmission, the pathogen develops but does not multiply in the vector. Filaria provide a good example of cyclodevelopmental transmission. Ingested microfilaria actively penetrate the vector's midgut, seek a species-specific site for growth and development, and then migrate to the head for subsequent transmission to a new vertebrate. Because of mortality in the vector, fewer pathogens normally leave the vector than entered it.

pared to a flying pin type of transmission. In a classic experiment, baby chicks were infected with Eastern equine encephalitis (EEE) and subsequently probed either by mosquitoes or by pins. At various time points before biological transmission was possible, noninfected chicks were exposed to the mosquitoes or pins, and mechanical transmission rates were determined (Table 14.1).

Biological and mechanical transmission are neither mutually exclusive nor mutually inclusive. By definition, arboviruses are vectored by propagative biological transmission; however, vectors that biologically transmit viruses may mechanically transmit for

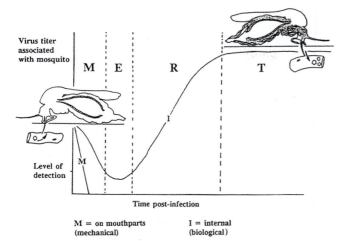

M = on mouthparts   I = internal
(mechanical)   (biological)

FIGURE 14.6 Viral titer in a vector over time as related to virus transmission. Curve *M* represents the virions on the mouthparts of the vector. Curve *I* shows the viral titer in a mosquito that will transmit propagatively. This curve initially drops due to viral inactivation in the midgut and entrance of infecting virions into the eclipse phase (*E*). The rise in the curve represents viral replication in the vector. During period *M* enough virions are present on the mouthparts for mechanical transmission (period *T*) to occur. During periods *E* and *R* the vector is infected but not yet capable of transmitting virus. From period *T* on, the vector can transmit the virus to a new host. The time from ingestion to *T* is the extrinsic incubation period (also see Fig. 14.13).

TABLE 14.1 Mechanical Transmission of EEE by Mosquitoes and Pins

Time postexposure (hours)	Percent transmission	
	Pins	Mosquitoes
0	100	100
1	100	90
4	100	70
20	20	60
70	5	10

Adapted from *Annu. Rev. Entomol.* 6: 371–390, 1961.

several hours after an infectious blood meal. Most hematophagous arthropods could theoretically transmit viruses mechanically, and some nonarboviruses are commonly transmitted in this manner (equine infectious anemia virus by stable flies) and others may be infrequently transmitted mechanically (hepatitis B virus by bedbugs). The ability to transmit a pathogen mechanically does not imply a corresponding ability for biological transmission. Usually only a single or a few species are competent for biological transmission,

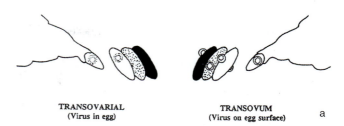

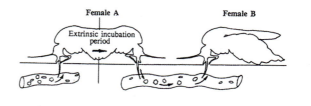

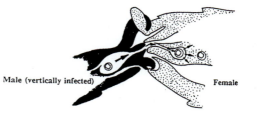

FIGURE 14.7 Vertical and horizontal transmission of arboviruses. Vertical transmission (a) can be through a transovarial (virus in the egg) or transovum (virus on the egg) route. Horizontal transmission may occur (b) from female A to female B via a viremia in the vertebrate host or (c) from an infected male to a female during copulation.

due to the multiple barriers the virus must cross. In explosive epidemics, undoubtedly both types of transmission occur. Whether acting alone or in concert, both biological and mechanical transmission can contribute to arboviral cycles. The hypothetical relationship between mechanical and biological transmission is shown in Figure 14.6.

Vertical vs. Horizontal Transmission of Arboviruses

Vertical transmission occurs with some arboviruses: The parent passes the pathogen to its progeny (Fig. 14.7). Vertical transmission occurs only with biologically transmitted viruses. Typically, a female arthropod infects her progeny through a transovarial or transovum route. California group viruses are efficiently transmitted transovarially, and the follicles are permissive to virus replication. In contrast, flaviviruses may be vertically transmitted during oviposition; apparently virus is transferred into the egg through the micropyle during fertilization as the egg passes through the oviduct. It may also be possible for infected males to vertically infect progeny via seminal fluid. Horizontal transmission (Fig. 14.7) is classically defined as transmission between vectors via viremia in the vertebrate host. Thus, both biological and mechanical transmission could be considered to be horizontal. In another form of horizontal transmission, arboviruses can be venereally transmitted from male to female mosquitoes. In vector species where only females are hematophagous, venereal transmission occurs only if vertical transmission to the male has first taken place and the virus has infected and replicated in the male reproductive organs.

Vectorial Capacity vs. Vector Competence

Vectorial capacity is the overall ability of a vector species in a given location at a specific time to transmit a pathogen. Quantitatively, vectorial capacity can be defined as the number of infectious bites a person (or other vertebrate of interest) receives daily. Vectorial capacity encompasses the vector interactions with the pathogen and the vertebrate host as well as innate vector characteristics not directly related to either the pathogen or the vertebrate host. Vector population size, longevity, length and number of gonadotrophic cycles, feeding behavior, and diel activity affect the vectorial capacity of a given arthropod population. These interactions of the vector, host, and pathogen are represented mathematically in the following model by Macdonald (1957):

$$\text{Vectorial capacity} = \frac{ma^2p^n}{-\log_n p}$$

where m = mean number of mosquitoes/individual
a = number of individuals bitten by a single mosquito/day
p = daily survival probability of a mosquito
n = length of pathogen life cycle in mosquito

Although this model is greatly simplified, it has been most helpful in understanding the critical components of the life cycle of pathogens and their relative contributions to their transmission and maintenance in nature.

In contrast to vectorial capacity, *vector competence* refers to the intrinsic ability of a vector (species, strain, individual, etc.) to biologically transmit a disease agent. Vector competence includes susceptibility to infection, permissiveness for pathogen reproduction/development, duration of extrinsic incubation

period, and transmission efficiency. Thus, we restrict vector competence to the defined vector–pathogen interaction.

Biological Transmission of Arboviruses

Viral Development in the Vector

In biological transmission, the period from ingestion of the infectious blood meal to transmission capability is designated the extrinsic incubation (EI) period (Fig. 14.8). During the extrinsic incubation period virus infects and replicates in the midgut cuboidal epithelial cells and then disseminates to infect secondary target organs. Recent evidence suggests that the tracheal system may provide one route by which virus disseminates from the midgut into the hemocoel. Virions then disperse in circulating hemolymph (insect blood). Once the salivary glands become infected and shed virions into the salivary ducts, the virus can be transmitted to vertebrates during a blood meal. In addition, if the developing follicles in the ovaries become infected, virus may be transmitted to the vector's progeny. Arthropod immunity differs from mammalian immunity; once infected with an arbovirus, a vector remains infected throughout its life span. After the EI period, the infective vector can potentially transmit virus to each new vertebrate host during feeding attempts or to new progeny with every bout of oviposition.

The duration of extrinsic incubation in a poikilothermic vector depends on the temperature. Within

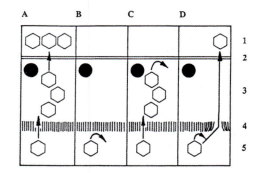

A = PERMISIVE
B = INFECTION BARRIER
C = DISSEMINATION BARRIER
D = LEAKY

MIDGUT	SALIVARY GLAND	OVARY
1 = Hemocoel	1 = Salivary duct	1 = Follicle
2 = Basement membrane	2 = Secretory surface	2 = Secretory surface
3 = Midgut epithelium	3 = Secretory cell	3 = Follicular epithelium
4 = Brush border	4 = Basement membrane	4 = Ovarian sheath
5 = Gut lumen	5 = Hemocoel	5 = Hemocoel

FIGURE 14.8 Barriers to biological transmission.

limits, higher temperatures shorten the EI period. The length of the EI also depends on the specific vector and virus involved. A typical EI period would be 10–14 days for a bunyavirus or flavivirus and perhaps 6–7 days for an alphavirus.

Barriers to Biological Transmission of Arboviruses

In any vector–arbovirus system, there are multiple barriers to productive vector infection (Fig. 14.8). The presence or absence of these barriers partially determines the vector competence of the arthropod. Several barriers to viral success in the vector have been shown or hypothesized at the midgut, salivary gland, and ovarian levels, and they are probably critical to arboviral cycle integrity.

The midgut barrier, a term used loosely to describe situations when the midgut is an impediment to productive viral infection, has long been recognized as a major determinant of vector competence. A midgut infection barrier, where the vector is actually refractory to infection, and a midgut escape barrier, where the midgut cells are infected but virus does not successfully disseminate to other organs, are both recognized. The midgut barrier can be bypassed experimentally or naturally. Experimentally, interspecific variability in infection rates disappears when mosquitoes are infected by intrahemocoelic inoculation of virus or when the midgut is penetrated physically with a needle or by microfilariae. A leaky-midgut phenomenon has also been described. In some populations, a certain percentage of the mosquitoes have virus in the hemocoel promptly after ingestion of the blood meal, before the virus could replicate in the midgut. The exact mechanism is unknown; however, the association of leaky midguts with having taken a previous blood meal suggests that damage to midgut integrity is likely to be the cause.

The midgut barrier is frequently not absolute. A species-specific dose–response phenomenon or infection threshold has been described for a variety of vector–virus systems. Below the threshold few of the vectors ingesting the blood meal become infected, whereas above the threshold significant numbers become infected. The infection threshold has been classically defined as the titer where 5% of the vectors become infected (Fig. 14.9). Most workers now quantify the threshold using an OID_{50} (oral infectious dose for 50% of the individuals tested). This statistical approach has the advantage of providing a more linear response between dose and infection.

A variety of hypotheses have been proposed to explain midgut infection barriers. Some workers suggest that the formation of the peritrophic matrix

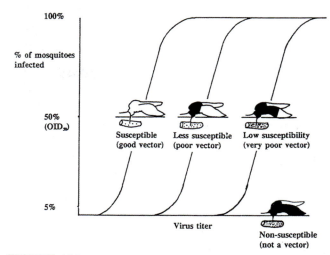

FIGURE 14.9 Infection thresholds for arthropods of differing susceptibilities to viral infection.

(PM) (Chapter 16) may prevent virus from contacting and infecting the midgut cells. However, PM formation probably occurs too late after the blood meal, to interfere with infection. It is also possible that refractory arthropods possess specific defensive mechanisms to protect against viral infections. Mucopolysaccharides or other antiviral agents could be secreted into the midgut. The explosion of information concerning the innate immune response of vectors to pathogens may be pertinent in this regard, especially the ability of vectors to respond by RNAi to viral infections (Adelman et al. 2001). Viral inactivation or inappropriate processing of viral surface glycoproteins by midgut enzymes is another potential explanation for the midgut infection barrier. Many now feel that arboviruses are truly enteric viruses of vectors, requiring proteolytic processing for infectivity. Infectivity of bluetongue virus (BTV) and La Crosse virus (LAC) for their respective vectors is enhanced by proteolytic processing of their virions. Alternatively, nonhomologous vectors might not possess specific receptors for viral attachment. This attractive hypothesis is the subject of a great deal of research, which will be discussed later.

Midgut escape barriers have been demonstrated for several viral families. In these nonproductive infections, the virus capably infects vector midgut cells and undergoes substantial transcription and translation but does not disseminate from the midgut cells to infect secondary target organs (Fig. 14.8). The virus could be trapped in midgut cells or, less likely, could reach the hemocoel but be unable to infect other organs. Some pathogens, for example, filarial nematodes, are certainly killed after entering the hemocoel due to vector immune factors, such as melanization (Chapter 27). Salivary gland infection and escape barriers have also been described (Fig. 14.8).

The infection barrier in salivary glands relates to those that are refractory to infection. With escape barriers, they become infected but the virus does not disseminate. For example, *Ae. hendersoni* is not normally an LAC vector, despite the replication of infectious virions in the salivary glands. Ovarian infection barriers also exist. When developing follicles do become infected, the virus must replicate and persist through embryogenesis, larval development, and metamorphosis for the adults of the successive generation to be infected. Transovarial infection is relatively rare for most arboviruses, and little is known concerning the molecular bases of ovarian barriers.

Genetic Determinants of Vector Competence

Both inter- and intraspecific variation in vector competence have been documented. In a classic study of interspecific variation, different species of mosquitoes were permitted to ingest blood meals from viremic chicks previously infected with EEE. Since all mosquitoes fed at approximately the same time and thus ingested the same dose of virus, the differences in infection and transmission rates reflect interspecific variation in barriers to viral infection and dissemination (Table 14.2).

Intraspecific variability in vector competence also exists (Chapter 30). Early studies demonstrated geographic and genetic bases to vector competence for dengue virus. Subsequent studies revealed a clear geographic distribution in the ability of *Ae. aegypti* to vector yellow fever (YF) virus. This intraspecific variation in vector competence may partially explain why the distribution of yellow fever does not mirror that of *Ae. aegypti*, the principal vector in many locations. Both the mosquito and the virus probably originated in

Africa, but *Ae. aegypti* is now pantropic in distribution. Yellow fever is established in the New World, probably introduced via slave trade with Africa, but has apparently not established in Asia, despite the longer history of trade between Africa and Asia and the seemingly favorable conditions in the Orient. *Ae. aegypti* is abundant in Asia, transmitting all four serotypes of dengue virus in a hyperendemic fashion. The Asian strain of *Ae. aegypti* proved to be a less competent vector for YF than the African and Caribbean strains.

Many studies have shown a genetic basis to vector competence in mosquitoes. *Ae. aegypti* competence for yellow fever virus has been studied extensively. Lines of *Ae. aegypti* selected for susceptibility and resistance to yellow fever virus react similarly to all flaviviruses. In resistant mosquitoes, viral antigen is restricted to the midgut; disseminated infections are not detected or are delayed. Viral titers are similar in susceptible and resistant lines early in the infection, and titers in susceptible mosquitoes increase presumably due to replication in the secondary target organs.

Quantitative genetic studies with *Ae. aegypti* and dengue-2 virus suggest that vector competence is determined by infection and dissemination barriers and that the latter is independent of viral titer. QTL analyses have revealed two loci that condition vector competence: one controlling the midgut infection barrier and another controlling the midgut escape barrier. Preliminary studies suggest that early trypsin, which maps to one QTL, may be a determinant of midgut infection. Interspecific and intraspecific variability in vector competence for western equine encephalitis (WEE) virus has been extensively studied. However, actual gene product(s) determining vector competence have not been revealed. Recent work suggests that cell receptors such as laminin may be utilized by viruses to infect midgut cells. The laminin receptor could bind with ligands on the envelope glycoproteins of the arbovirus, thereby permitting entry into the cell. There is little evidence to indicate that such generic receptors are determinants of vector specificity and thereby function to maintain arboviral cycle integrity. Nonetheless, such receptors could be critical determinants of viral infection of vectors and thus of the midgut barrier. Note that these hypotheses have all dealt with the vector; viral determinants of midgut infection will be discussed later.

Enhancement of Transmission

Many hematophagous vectors produce salivary substances that are secreted during feeding and promote feeding success. Some of these factors are well

TABLE 14.2 Interspecific Variation in Vector Competence

Mosquito	% infected	% transmission
Ochlerotatus triseriatus	100	56
Oc. sollicitans	100	75
Cx. restuans	45	33
Anopheles. quadrimaculatus	17	0
Cx. quinquefasciatus	5	0

Adapted from *Am. J. Hyg.* 60: 281.

characterized and include vasodilators and anticlotting factors (Chapter 28). In addition to being antihemostatic, some factors can have a rapid and sometimes prolonged effect on components of the vertebrate immune system. The effects are especially well documented for ticks, which feed for several days, but are also observed for diptera. For example, the expression levels of both Th-1 and Th-2 cytokines can be modulated even by the relatively brief feeding of mosquitoes (Zeidner et al. 1998). The host immune response may be perturbed; as a consequence, the infection efficiency of pathogens such as *Leishmania* spp. and arboviruses such as Cache Valley and vesicular stomatitis may be enhanced. Fewer pathogens may be required to produce a patent infection, an effect sometimes referred to as *salivary activated transmission* (SAT), and infections may result in higher viremias with subsequently greater host seroconversion rates/higher antibody titers. So-called nonviremic transmission, in which naive vectors become infected while cofeeding alongside infected ones despite the absence of a detectable viremia, has been observed for thogoto and vesicular stomatitis viruses, for example, and may be due to the effects of salivary factors, although the exact mechanism remains uncertain. These phenomena are further discussed in Chapters 2 and 28.

Vector Considerations in Arboviral Cycles

Vector Incrimination

Although many arthropods may be capable of transmitting a given etiological agent, only those invertebrates that are important in maintaining a natural transmission cycle are considered primary vectors. Four criteria of vector incrimination are used to identify primary vectors: (1) The putative vector must associate with and feed on the vertebrate host under field conditions. Moreover, the seasons and locations of arthropod activity should coincide with the incidence of vertebrate infection. (2) Naturally infected vectors must be consistently recovered from the field. (3) The presumed vector must be shown to become infected by feeding upon a viremic host. (4) The ability of an infected vector to transmit the pathogen to a new vertebrate host should be confirmed under controlled conditions.

Considerations for Biological Transmission

In vector-borne disease, the transmission rate is a function of the vector physiology, population dynamics, and behavior. Several attributes of the vector species and populations may aid in maintaining the integrity of arboviral cycles.

As previously defined, vector competence refers to the intrinsic ability of a vector to transmit a disease agent. What interactions make for a competent vector? A good vector species has a low infection threshold and is permissive for viral multiplication in at least the midgut and salivary glands (and ovaries in the case of transovarial transmission), and a high proportion of individuals transmit virus. Further, the vector should sustain little or no reduction in fitness or survival despite supporting viral replication. In addition, the shorter the extrinsic incubation period, generally the better the vector will be because of the potential to encounter more hosts after becoming infective.

There are many attributes of vectors in addition to vector competence that potentially increase the vectorial capacity of a population. During host-seeking and feeding activity, the vector must show a spatial and temporal coincidence with the vertebrate host. Vectors have characteristic diel activity (active during day, night, or crepuscular periods), feeding preferences (anthropophilic, zoophilic, etc.), and habitat activity (canopy vs. ground feeding, exophilic vs. endophilic). Moreover, the vector needs to feed on the vertebrate host. Host-feeding preferences may be a major determinant of specificity of arboviral cycles. Recent identification of odorant-binding proteins of mosquitoes (Fox et al. 2001) may provide insight into the vector determinants of host feeding and may permit development of new repellents and attractants for control of vectors. Some vectors feed on mammals, others on reptiles, and still others on birds, and some are diverse in their choice of hosts. Host preference may change during the transmission season. The ability to feed on alternate hosts may provide both a survival mechanism when the usual host is unavailable and a mechanism for transmitting virus to new, often tangential hosts. Vector mobility also affects arboviral cycles. A distance traveler may be more likely to spread a virus to new susceptibles than would a vector with low dispersive powers.

Vector longevity is a major factor in transmission potential and, hence, in vectorial capacity. The longer the life, the more feedings after the EI period and thus the more opportunities for transmission. Long-lived arthropods, such as ticks, may help maintain a pathogen for a long time in the environment. In addition, the shorter the gonadotrophic cycle, the more opportunity for transmission as blood is ingested to support production of the next batch of eggs. Arthropods that feed immediately after ovipositing, *Anopheles gambiae*, for example, are probably better vectors than those that take more postoviposition rest.

Seasonal and climatic effects on vectors cannot be ignored. As small invertebrates, vectors tend to be sensitive to temperature and humidity. Temperature affects the EI period, vector longevity, gonadotrophic cycle length, and activity. Humidity strongly impacts both vector longevity and mobility. Also, for water breeders, such as mosquitoes, standing water is required during a period when temperatures are permissive for reproduction. In many cases, cold winters and hot, dry summers are periods of greatly reduced mosquito breeding and feeding. Periods of vector inactivity may temporarily disrupt transmission cycles and require that an arbovirus possess a transeasonal survival mechanism.

Considerations for Mechanical Transmission

Feeding behavior and population density of the vector greatly influence the probability of mechanical transmission. Vectors that exhibit interrupted feeding behavior are likely to transmit mechanically. Interrupted feeding may occur when defensive behavior of the vertebrate host drives the vector away. Typically, the vector then promptly seeks to complete engorgement upon another host. Virus may be transmitted to this second host if the first host supported a viremia capable of contaminating the vector's mouthparts. As vector populations increase, so does the likelihood of mechanical transmission. This derives from two reasons. As more vectors feed, the odds that one will successfully transmit the virus increase. In addition, vertebrate populations usually become more defensive as vectors become more numerous, thus leading to more interrupted feeding.

Viral factors that facilitate mechanical transmission include viral stability, vertebrate infectivity, and ability to cause a high titered viremia in the vertebrate host. In general, the more resistant a virus is to environmental degradation, the more likely it is to be mechanically transmitted. Viruses are usually more labile at warm temperatures. Thus, cool temperatures are associated with viral stability and more efficient mechanical transmission. Viruses differ in their relative infectivity for vertebrate hosts. The more communicable the virus, the longer a sufficient infectious dose may remain on the vector and thus the greater the probability of mechanical transmission. Lastly, the potential for mechanical transmission is directly proportional to the titer of viremia in the vertebrate host. Obviously, the more virus present in the blood to contaminate the mouthparts of the vector, the greater the potential that virus will survive long enough to infect a new vertebrate. Note that all three of these factors are interdependent.

VERTEBRATE HOSTS OF ARBOVIRUSES

Vertebrate Hosts

Virtually any vertebrate capable of becoming infected with virus and developing sufficient viremia to infect subsequently feeding vectors may be considered a vertebrate host. Vertebrates may contribute to maintenance or amplification components of arboviral cycles. Different host names have classically been attached to these different roles, though it should be noted that maintenance and amplification roles are not necessarily mutually exclusive.

Classically, the reservoir host is involved in the maintenance of the virus over long periods of time and experiences no untoward effect from infection. The hallmark of a good reservoir host is a high-titered viremia of long duration, which can lead to infection of many vectors. Hosts that support recrudescent infections could also be reservoirs. A reservoir host population must usually generate sufficient new susceptible members for ongoing transmission to sustain the viral cycle. Species that breed year-round, thus providing a constant supply of new susceptibles, frequently serve in this capacity. In addition to cycle maintenance, migratory or highly agile reservoir hosts may disperse an arbovirus to new locations.

Vertebrates involved in amplifying the number of infected vectors (and consequently other vertebrates) are called *amplification hosts*. Amplification hosts reproduce annually or more frequently and provide large numbers of new susceptibles concurrent with vector activity. For example, each spring hatchling birds serve as a pool of new susceptibles. The first birds infected sustain viremias that lead to increasing waves of infection among the vectors and the other susceptibles. Susceptible amplification hosts may not be present year-round (most fledgling and adult birds will have developed immunity), but they provide an important opportunity for the quantity of virus in the environment to increase. For epidemiological purposes, vertebrates serving to bring an arbovirus in contact with vectors that may transmit the viral diseases to humans or livestock may also be amplification hosts.

A *tangential host* is not typically involved in amplification or maintenance of the virus. In contrast to the usual vertebrate hosts involved in the cycle, tangential hosts may support a viremia of relatively low titer. Because transmission to new vectors is reduced or prevented by the low-titered viremia, these hosts are also known as *dead-end hosts*. Tangential hosts are not important to cycle integrity (unless they distract vectors from the normal vertebrate hosts); however, humans and animals may suffer severe morbidity and mortality when they tangentially enter a disease cycle. Thus, for some serious arboviral diseases, humans and domestic animals are merely tangential hosts. By comparison, reservoir and amplification hosts usually do not develop any major untoward effects resulting from infection, although there are notable exceptions that cause some human and animal diseases (dengue and bluetongue, for example) of worldwide importance.

Pathogenesis in Vertebrate Hosts

Pathogenesis of many arboviruses is well understood. Infection probably occurs first in muscle or other cells near the site of the bite and is subsequently transmitted to regional lymph nodes. Viral replication in these tissues leads to a primary viremia (Fig. 14.10), resulting in infection of tissues associated with the vascular system. Viral replication in the vascular system produces a high-titered, long-duration secondary viremia. Infection of secondary target organs, such as the liver in yellow fever and the central nervous system in encephalitis, ensues. The period required from infection to disease in the vertebrate is termed the *intrinsic incubation period* (as opposed to the *extrinsic incubation period* in vectors).

Several potential syndromes can result from arboviral infection. Inapparent, asymptomatic infections frequently occur, especially in hosts normally involved in a particular cycle. Arboviral infection can also result in clinical manifestations ranging from fever or arthritis to hemorrhagic fever or encephalitis. Obviously, severe morbidity and mortality may result from these serious diseases. As noted previously, tangential hosts are probably more likely to experience severe syndromes, presumably due to lack of coadaptation between host and pathogen. Individuals may respond quite differently to infection with the same virus. Human infection with yellow fever, for example, may give rise to a spectrum of symptoms ranging from an inapparent infection to fatal hemorrhagic fever.

The vertebrate immune response is important in limiting arboviral infections. Unlike arthropods that remain infected for life with arboviruses, most vertebrates respond immunologically to infection (Fig. 14.10). Antibodies limit the duration of the viremia, thus halting further transmission to more vectors. Antibodies also render the individual immune to future infection by the particular virus.

Arthropod-borne pathogens other than viruses may develop differently in the vertebrate host, may cause their own distinctive syndromes, and may elicit less than full immunity. Details for these other systems may be found in the appropriate chapters.

Vertebrate Host Considerations in Arboviral Cycles

Several attributes of a vertebrate host impact its suitability for maintaining an arboviral cycle. Once fed upon by an infective vector, the vertebrate must be susceptible to viral infection and develop a viremia sufficient to infect or contaminate other arthropods, preferably a viremia of high titer and long duration. For biological transmission, a viremia surpassing the infection threshold of the vector species is critical. A high-titered viremia also aids mechanical transmission.

As noted previously, the vertebrate host and vector must share time and space when the vector is actively feeding. Diel activity patterns and accessibility to the vector must be considered. Vertebrates present on the forest floor encounter different vector species than those found primarily in the canopy. Stereotyped behavior of the vertebrate, such as returning to the same vector-infested water hole each evening, can serve to increase vertebrate–vector contact. Artifacts may also increase vertebrate–vector contact. For example, water jugs and other small containers around village homes provide excellent breeding sites for *Ae. aegypti* and bring the vector close to humans. Rodent burrows are environmentally suitable for both the rodents and their fleas, which may vector pathogens. Interestingly, some medical entomologists attribute air conditioning and prime time TV viewing to a decrease of arboviral transmission to humans in the United States. The air conditioning encourages people to limit vector access to their homes (by closing doors and windows), and TV keeps people in homes during

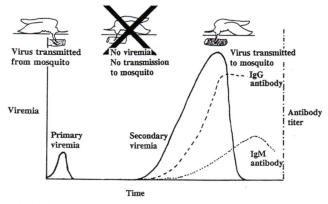

FIGURE 14.10 Viremia and antibody response in a vertebrate host infected by an arbovirus.

the twilight hours, when many vectors are actively host-seeking.

Good vertebrate hosts should not exhibit effective defensive behavior that could result in eluding or even killing the vector. Young vertebrates typically are less defensive than adults and may be more susceptible to both vector attack and to arboviral infection than adults. In some instances, defensive behaviors may increase viral transmission by promoting interrupted feeding, which may enhance mechanical pathogen transmission. In addition, an infected vector exhibiting interrupted feeding could potentially transmit virus to more than one host per gonotrophic cycle.

Vertebrate behavior may influence arboviral maintenance and transmission in yet another manner. Infected vertebrates that roam through vast territories or that migrate to distant locations may bring arboviruses into contact with new vector populations. Hence, disease cycles may begin in new regions or be reintroduced into areas where the cycle had been extinguished. Migratory birds are known to transfer some arboviruses in this manner, e.g., West Nile virus.

A vertebrate host should not exhibit undue mortality from an arboviral infection, especially a reservoir host. This is the general pattern observed. The exceptions can be fascinating. In its native home of Africa, YF is maintained in monkeys that typically show no clinical symptoms, as would be expected. In contrast, in South America, where the virus has probably been introduced, YF kills monkeys. YF persists because *Haemagogus* and other canopy mosquitoes transmit it between monkey troops. New susceptibles, born since the previous epizootic, become infected when troops come in close proximity and share vectors. Thus, the YF reservoir in the Amazon basin moves from region to region with approximately a 25-year periodicity in any given location.

Vertebrates typically develop a long-term immunity to viral pathogens; thus, the absolute size of a vertebrate host population may give little indication of its ability to support viral infections. Thus, the vertebrates should be fecund to provide for recruitment of new susceptibles to the population. Epidemic cycles often can be explained in terms of the time required for accumulation of sufficient susceptibles in the population. The Amazon monkeys and YF virus is a good example.

COMPLEXITY OF ARBOVIRAL CYCLES

The typical cycle described to this point has one or more principal vectors maintaining viral transmission between one or more species of vertebrate hosts. Many arboviral cycles do not normally involve humans as reservoir or amplifying hosts. Frequently, arboviruses are transmitted in two separate cycles: an urban cycle and a rural, jungle, or sylvatic cycle; also frequently called *epidemic* and *endemic cycles*, respectively. Humans impinge upon the sylvatic cycle and become infected; they can then transport the virus to urban areas (Fig. 14.11). In rural areas, wild vertebrates may normally be the amplification and maintenance hosts, with the relative sparse human populations only rarely impinging on the cycles. When humans do become infected, they probably do not contribute to the rural cycle. In contrast, human populations in cities are dense and are likely to include a high percentage of susceptible individuals. Under these conditions, urban vectors, usually well adapted to feeding on humans, can transmit virus in the susceptible population and cause epidemics.

More commonly, the rural or sylvatic cycle maintains the arbovirus. Humans or domestically important species become tangentially infected when they impinge upon this natural cycle or conditions bring the cycle in contact with them. In such situations, the vector is frequently catholic in its feeding preferences. The cycle of La Crosse virus (Fig. 14.12) is a good example. The cycles of WEE and rural Saint Louis encephalitis (SLE) in the American west are also good examples. *Cx. tarsalis*, the vector of WEE and rural SLE, prefers passerine hosts and transmits virus between birds. However, the mosquito will feed upon humans and mammals, and thus it can transmit the virus to tangential hosts. In other instances, the vector may change feeding preferences or lose access to its preferred host, thereby introducing the virus into different vertebrate hosts.

TRANSEASONALITY

Vectors are frequently inactive during parts of each year, whether winter in temperate regions or dry seasons in the tropics. During these periods, transmission of vector-borne pathogens ceases, a weak link in these disease cycles, and the cycle enters a transeasonal maintenance phase. Vectors in temperate regions must survive the cold season, frequently called *overwintering*, and may do so by going into diapause in one of the life stages. Vectors in tropical regions may have to survive the dry season, and may do so by going into estivation. Both diapause and estivation are states of dormancy with reduced metabolism and arrested development, which permits synchronization of the life cycle of the vector with the seasons.

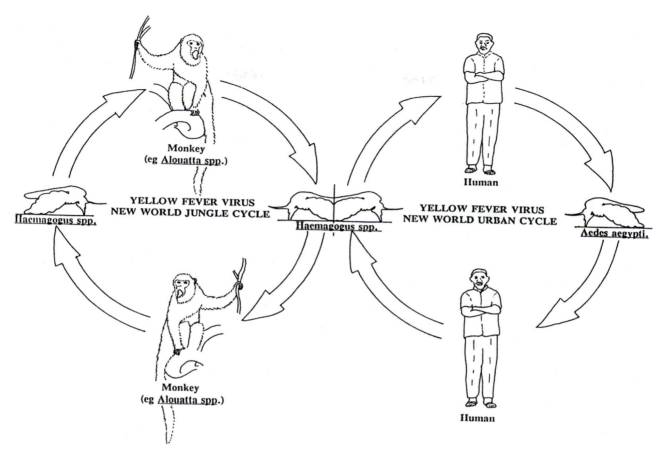

FIGURE 14.11 Jungle and urban yellow fever.

Vectors may contribute to viral persistence via three principal mechanisms. (1) The virus may persist in the adult vector when the adult is the stage that remains dormant during the unfavorable season. For example, SLE and West Nile viruses have been isolated from overwintering adult mosquitoes. (2) The egg or, more rarely, juvenile stages may serve as a reservoir for virus when vertical transmission occurs. In the family Bunyaviridae, transovarial transmission is clearly a major mechanism for viral persistence through harsh winters. (3) An undetected arthropod or other metazoan vector may harbor the virus during a season when the known vector is inactive. For example, a cimicid (bedbug) (Chapter 5) is the overwintering host of Fort Morgan virus, an alphavirus that infects swallows.

Vertebrate hosts may also contribute to viral persistence. One or more of the vertebrate hosts may act as a chronic reservoir, sustaining viremias on a very long-term basis. For example, Colorado tick fever virus replicates in developing erythrocytes. When mature erythrocytes enter the bloodstream, they protect the internal virus from circulating antibodies while causing an effective viremia. Recrudescent viremias may serve to reintroduce other arboviruses after a period of disrupted transmission. Hibernating reptiles and mammals with delayed viremias have been hypothesized to be involved with the overwintering of alphaviruses and flaviviruses. An arbovirus may actually disappear from an area when the vector(s) become inactive. Migrating species may then reintroduce arboviruses to the regions that are seasonally unfavorable to transmission. Some arboviruses, certain rhabdoviruses, for example, may actually persist in food chains when the vectors are inactive.

RESERVOIRS AND THE NIDUS OF INFECTION

Arboviral cycles may seem extremely tenuous, with many opportunities for their disruption. However, as

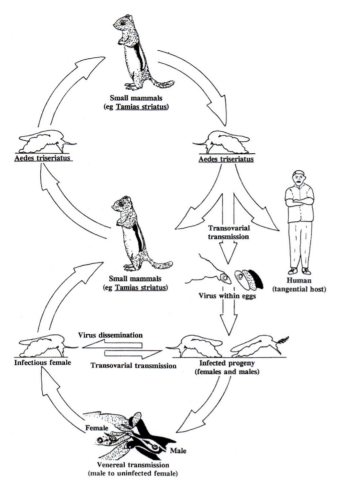

FIGURE 14.12   La Crosse virus cycle.

Two principal determinants of the nidus are: (1) the presence of competent vectors and (2) the availability of susceptible (nonimmune) vertebrate hosts. The nidus of infection of YF virus in Brazil may be as large as the Amazon basin forest. The nidus can be visualized as a moving wave of YF virus passing from infected to noninfected monkey troops in areas where territories overlap or are contested. In contrast, the nidus could be as small as a homesite or backyard, as for La Crosse virus, which is described in the following section.

TRANSMISSION CYCLE OF LA CROSSE VIRUS

A detailed description of the cycle of La Crosse (LAC) virus will be used to illustrate the principles of arboviral maintenance and amplification cycles enumerated in the preceding pages.

La Crosse Epidemiology

LAC virus is the leading cause of arboviral pediatric encephalitis in the upper midwestern and eastern United States. Humans are tangential hosts. Infections occur between June and September, with the majority reported in August and early September. Clinical cases are most often observed in children from 3 to 15 years of age. Boys are more frequently infected than girls, presumably due to more outdoor activity and thus greater exposure to the vector. Infection of older children or adults typically does not result in encephalitis; a headache and nuchal rigidity may be the only symptoms of infection. In the early 1960s, LAC virus was isolated postmortem from the brain of a child who died in La Crosse, Wisconsin. By the mid-1970s virtually the entire LAC cycle had been elucidated (Fig. 14.12), with chipmunks and tree squirrels identified as the primary natural vertebrate hosts and *Ochlerotatus triseriatus* (syn. *Aedes triseriatus*) incriminated as the principal vector. Importantly, the overwintering mechanism of the virus had also been determined; the virus is transovarially transmitted and overwinters in diapausing mosquito eggs. In the spring some adult females emerge immediately ready to transmit to newly born susceptible vertebrate hosts. Males may also contribute to the cycle by venereally infecting female mosquitoes during mating.

The natural cycle of LAC virus was elucidated in a rather short time. It is noteworthy that for other arboviral encephalitides, such as eastern and western equine encephalitis, overwintering mechanisms have

efforts to control vector-borne diseases have shown, these amazingly resilient cycles have proven to be virtually indestructible. The mechanisms that maintain arboviruses constitute the reservoir. In the early days of arbovirology, the vertebrate host was generally called the *viral reservoir*. With the realization that many arboviruses are transovarially transmitted, the definition of a *reservoir* has been expanded to include all of the underlying extrahuman mechanisms by which a specific pathogen population is maintained, including the specialized ecology necessary to support the biological relationships involved. The reservoir is then intimately associated with Pavlovsky's theory of nidality, or focus of endemicity. A nidus of infection must contain the components (vectors, vertebrate hosts, habitat) necessary to maintain the virus. Derived from the nidus concept is the field of landscape epidemiology, which focuses on habitats that are more likely to contain the components needed to sustain arboviral cycles.

yet to be discovered. Of course, the LAC cycle differs dramatically from the others due to transovarial and venereal transmission of the virus. Flaviviruses are inefficiently transmitted vertically; these viruses apparently infect the eggs during oviposition, not during follicle development as for transovarial transmission. Alphaviruses are not or very poorly vertically transmitted.

La Crosse Cycle Components

The Virus

LAC virus belongs to the family Bunyaviridae. It is a member of the California serogroup of the Bunyavirus genus. The virion measures approximately 90 nm in diameter and possesses an envelope containing two virally specified glycoproteins. The RNA genome is negative sense and tripartite, and the RNA segments are associated with nucleocapsid proteins. The large RNA segment (L) codes for the viral polymerase; the middle-sized RNA (M) codes for the two viral surface glycoproteins as well as a nonstructural protein; and the small RNA (S) codes for the nucleocapsid protein and a nonstructural protein.

Ochlerotatus triseriatus, the Principal Vector

Ochlerotatus triseriatus is the principal vector of the virus. This mosquito is common throughout the eastern portions of the United States; its distribution extends to the high plains. States bordering the Great Lakes and the Mississippi River historically record the highest incidence of LAC encephalitis. Few cases occur in the southern states, where *Oc. triseriatus* and vertebrate hosts are also relatively abundant. Thus, considerable intraspecific variability in vector competence may exist. The vector, principally a daytime feeder, becomes most active at crepuscular periods.

Those states with high LAC incidence contain oak-hickory or other climax forest areas that provide exceptional breeding opportunities for *Oc. triseriatus*. This vector oviposits in the tree holes that are especially abundant in oak forests. Other suitable oviposition sites, such as old tires and other water containers, often bring the vector out of the forested areas and into even closer contact with human hosts.

Recent isolations of LAC virus from field-collected *Oc. albopictus* mosquito larvae and association of this vector with LAC encephalitis cases in Tennessee and North Carolina are cause for concern (Gerhardt et al. 2001). *Ae. albopictus* transmission and maintenance of LAC virus could dramatically change the epidemiology of LAC encephaltis in the American south.

The Vertebrate Hosts

In the upper midwestern states, chipmunks and tree squirrels are the principal vertebrate hosts. Both vertebrate hosts are abundant, coexist with the mosquito vector, and have a relatively high reproductive rate. Active during the day, chipmunks and squirrels are available as hosts for the vectors. Both are permissive to viral infection, develop substantial viremias, and exhibit no apparent untoward effect from viral infection. In the forested areas, seroprevalence rates may be exceptionally high, approaching 100% in certain areas. Were it not for transovarial transmission, the arboviral cycles could probably not survive in locations with such high vertebrate host immunity. With efficient transovarial transmission, infected females can yield many infected progeny each gonadotrophic cycle, even when feeding on immune vertebrate hosts. Each spring, virus that overwintered in diapaused eggs is transmitted to newborn, susceptible hosts. The arboviral cycle is renewed, and virus prevalence is amplified in nature. Humans are tangential, dead-end hosts.

La Crosse/*Ochlerotatus triseriatus* Interactions

Productive Infection

In the LAC virus–*Oc. triseriatus* interaction, ingested virus first infects and replicates in cuboidal epithelial cells of the pyloric portion of the vector midgut; viral antigen is detected there between 3 and 6 days postingestion. Viral antigen is subsequently found in the foregut and cardiac portion of the midgut. After approximately 6 days, virus disseminates from the midgut and infects secondary target organs. Viral antigen becomes detectable at 8–9 days in fat body, heart, and pericardial cells and nervous and ovarian tissues. Simultaneous detection of viral antigen in so many organ systems suggests a hemolymph mode of spread of arboviruses in mosquitoes. LAC virus reaches *Oc. triseriatus* salivary gland cells approximately 10–14 days postinfection. Then the extrinsic incubation period ends, and the female can transmit the virus to new vertebrate hosts during subsequent blood meals (Fig. 14.13). The virus is virtually pantropic in the vector.

Barriers to Infection

Midgut infection and dissemination barriers as well as salivary gland and ovarian infection and dissemination barriers to LAC virus have been described in *Oc. triseriatus*. In addition, intraspecific variation in vector competence in oral susceptibility and transo-

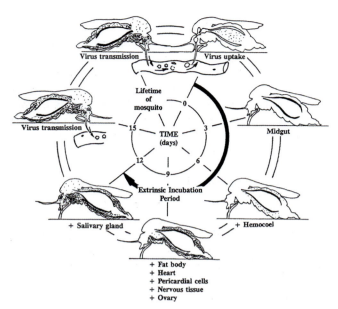

FIGURE 14.13 La Crosse virus dissemination in *Ochlerotatus triseriatus*.

TABLE 14.3 Infection, Dissemination, and Transmission of La Crosse, Snowshoe Hare, and Reassortant Viruses in *Oc. triseriatus* Mosquitoes

Viral genotypes	% infected	% disseminated	% transmitting
LAC/LAC/LAC SSH/LAC/SSH SSH/LAC/LAC LAC/LAC/SSH	98 (115/117)	98 (113/115)	9 (126/36)
SSH/SSH/LAC LAC/SSH/LAC LAC/SSH/SSH SSH/SSH/SSH	92 (92/100)	26 (24/92)	35 (36/104)

Adapted from *Virus Res.* 10: 289–302, 1988.

TABLE 14.4 Phenotypic Analysis of Parent LaCrosse, MARV 22, and Revertant Viruses

Virus	Infection Rate	Neutralization Titer	Fusion Index (pH 6.0)
Parent LaCrosse	74% (14/19)	160	>0.8
Variant V22	5% (4/47)	<10	0.2
Revertant Virus	85% (50/59)	320	>0.6

Adapted from *Virus Research,* 10: 289–302, 1988.

varial transmission have been demonstrated. For example, *Oc. triseriatus* from Connecticut are much less efficient than those from Wisconsin in transovarial transmission of the virus. *Oc. triseriatus* from southern Florida, where mosquitoes inefficiently diapause, exhibit lower filial infection rates than those from Wisconsin, which efficiently diapause. The genetic bases of these phenomena remain to be determined.

Due to the complexity of elucidating vector determinants of vector competence, considerable effort has been devoted to investigating viral genetic determinants of virus–vector interactions. Reassortant bunyaviruses were used to determine the molecular basis of virus–vector interactions (Table 14.3). When snowshoe hare (SSH) and LAC reassortant viruses (described according to their large/medium/small segment composition) infect *Oc. triseriatus* midgut cells, the parental origin of the medium RNA (M RNA) segment does not significantly affect infection rates. As quantified by detection of viral antigen, both viruses are equally capable of infecting midgut cells. Evidently, if efficient infection of midgut cells is mediated principally by specific receptors, then the receptor is not sufficiently discriminatory to distinguish between the LAC and SSH M RNA gene products (glycoproteins).

In contrast to midgut infections, there is a marked difference in the ability of the viruses to escape from the midgut cells or to infect secondary target organs (Table 14.3). Reassortant viruses containing the LAC M

RNA segment are significantly more efficient in disseminating than are those viruses containing the SSH M RNA segment. The molecular basis of this midgut escape barrier remains to be determined.

The presence of the LAC M RNA segment is also a major determinant of transmission of virus to mice. Viruses containing the LAC M RNA were transmitted by 93% of the mosquitoes; however, those containing the SSH M RNA were transmitted by only 35% of the mosquitoes with disseminated infections. Thus, the M RNA segment seems to be the major determinant of both LAC virus dissemination from the midgut and subsequent transmission by *Oc. triseriatus*. Unfortunately, genetic analysis of bunyaviruses is complicated by the fact that the M RNA segment codes for two glycoproteins (G1 and G2) and a nonstructural protein (N$_s$M). Studies with monoclonal antibody-resistant variant viruses revealed that a specific epitope on the G1 glycoprotein was a major determinant of midgut infection of *Oc. triseriatus*. A variant virus propagated in the presence of a specific monoclonal antibody was deficient in fusion function, was no longer neuroinvasive in mice, and exhibited greatly reduced infectivity for the mosquito midgut (Table 14.4). When the virus reverted at the specific epitope, infectivity for the mosquito was restored.

The midgut of mosquitoes is replete with proteolytic enzymes after engorging a blood meal (Chapter 21), and LAC virus glycoproteins are readily cleaved by such enzymes. Intact virions of LAC virus bind well to vertebrate cells but not to mosquito midguts; however, proteolytic processing of G1 apparently reveals sequences of the G2 glycoprotein, and the processed virions do bind to mosquito midguts. As noted previously, specific mosquito receptors utilized by viruses remain to be elucidated. Interestingly, the two glycoproteins of the virus seem to be differentially involved in the various transmission components of the arboviral cycle.

Vertical Transmission

LAC virus vertical transmission to progeny occurs through a transovarial, not a transovum, route. Transmission can be quite efficient; transovarial transmission rates (mothers transmitting virus to progeny) and filial infection rates (percent of progeny becoming infected) can exceed 80%. Reproductive tissues become infected concurrently with other tissues. LAC viral antigen can be demonstrated in virtually every embryonic cell of the follicle. Amazingly, there is no untoward effect or teratogenesis associated with LAC infection in laboratory studies, and filial infection rates of >80% may occur. Apparently coregulation of viral and vector transcription in the ovaries of LAC-infected mosquitoes modulates viral deleterious effects on the follicle. LAC virus is a cap-scavenger; that is, the virus requires ongoing host transcription to prime its own transcription. Thus, during ovarian quiescence, virus transcription is down-regulated. After ingestion of a blood meal and onset of oogenesis, virus transcription resumes. As noted previously, in laboratory conditions there is no untoward effect associated with LAC infection in terms of fecundity or fitness. In field conditions, however, overwintering LAC-infected larvae experience greater mortality rates than noninfected larvae, but only after the larvae have emerged from diapause. Interestingly, LAC virus apparently targets an inhibitor of apoptosis mRNA during persistent infections, perhaps thereby reducing virus virulence in vector cells by perturbing the host apoptotic capability. LAC infection of vertebrate cells (both neuronal and BHK-21 cells) can induce apoptosis, whereas infection of mosquito cells does not induce apoptosis. In addition, LAC virus is also apparently able to evade the vector RNAi response. The ability to avoid these two components of the vector host innate immune response is critical to establish long-term persistent infections and to be transovarially transmitted.

Transtadial Transmission and Metamorphosis

LAC virus is efficiently maintained transtadially (between molts) and through metamorphosis, when proteolytic enzymes and nucleases are abundant in the vector. Throughout transtadial passage, virus is detected virtually pantropically. During metamorphosis, virus is apparently preserved in tissues that are not histolyzed. LAC virions can then infect newly developed tissues after metamorphosis.

Horizontal Transmission

Virus is horizontally transmitted (1) to the vertebrate host from the adult female salivary glands or (2) to the adult female from vertically infected males. Females may become infected either horizontally or vertically; males become infected only vertically.

Venereal Transmission

Transovarially infected male mosquitoes can venereally transmit virus to females. Most of the male reproductive tract, including vas, seminal vesicles, and accessory sex glands, is permissive for viral replication. Accessory sex glands accumulate large amounts of virus that can be transmitted to females during mating. If females have not ingested a blood meal, less than 10% become infected venereally; if females have ingested a blood meal, up to 50% become infected. This latter observation suggests an interesting possible linkage between hormonal status and susceptibility to viral infection.

La Crosse Nidality

LAC virus may be maintained in small, forested areas, such as the islands of hardwood forests that are found throughout farming areas of the upper Midwest. Indeed, the nidus of infection may be as small as a backyard, if the yard contains at least moderate tree and shrub cover, tree holes, discarded tires or other breeding sites, and vertebrate hosts such as chipmunks or squirrels.

Viruses such as LAC that are efficiently transovarially transmitted may be maintained in rather small geographic regions because virus can be amplified by horizontal and transovarial transmission. LAC can be further amplified by venereal transmission from infected male mosquitoes to noninfected female mosquitoes. Note that infection of new vectors can occur even when the vertebrate host population displays high seroprevalence rates. In fact, the presence of antibodies to LAC virus has little effect upon transovarial transmission. Thus, a female could well oviposit 100 infected eggs, even after feeding upon an immune host. In contrast, arboviruses with no or inefficient transovarial transmission require a constant supply of susceptible vertebrate hosts for horizontal amplification of the virus.

Evolution of LAC Virus in the Vectors

General Principles of Arboviral Evolution

The Bunyaviridae, which includes LAC virus, is the largest family of viruses, with over 240 recognized members. Thus, the Bunyaviridae possesses considerable evolutionary prowess. Why is this group so successful? The evolutionary success of the family is probably attributable to its ability to evolve both by intramolecular mutations and by reassortment of the three RNA segments (the influenza viruses of arbovirology). Further, the long-term replication of bunyaviruses in their arthropod hosts provides many opportunities for these evolutionary events to occur. Understanding the evolutionary potential of bunyaviruses such as LAC virus in terms of spontaneous mutation rates and in terms of segment reassortment potential is not just an academic exercise. Evolutionary changes in the virus could result in more virulent viruses or viruses capable of being transmitted by different vectors that could introduce the virus to new susceptible vertebrate species.

Spontaneous Mutations

Spontaneous mutations provide evolutionary potential for all organisms. The spontaneous mutation rate for LAC virus is high; the base-substitution error frequency is probably in the order of 10^{-3}. Analyses of isolates of LAC virus from nature substantiates the genomic plasticity of LAC virus; no two isolates have identical genomes, as evidenced by oligonucleotide fingerprinting (ONF). This is true for viruses isolated from different geographic locales at the same or different times and for viruses isolated from the same locale simultaneously. Thus, the major evolutionary mechanism of bunyaviruses is genetic drift via the accumulation of point mutations, sequence deletions, and inversions.

Segment Reassortment

In addition to spontaneous mutations, the evolutionary potential of bunyaviruses is further enhanced by the segmented genome; segment reassortment has been documented to occur in vitro, in vivo, and in nature. For example, the Group C viruses in one little forest in Brazil seem to constitute a gene pool. The six viruses are related alternatively by hemagglutination-inhibiting antibody (HI)/neutralizing antibody (NT)

and complement fixation (CF) reactions that assayed M RNA and S RNA gene products, respectively. Reassortment of genomes could result in serious epidemiologic consequences.

Either the vertebrate host or the vector could serve as the site for reassortment in such circumstances. Studies done to demonstrate bunyavirus genome reassortment in vertebrates have been unsuccessful. Apparently the acute nature of infection in vertebrate hosts and the development of long-term immunity limits evolutionary opportunities in this host. In contrast, dual infection of Oc. triseriatus mosquitoes with two viruses (LAC and SSH) results in high-frequency reassortment of the viruses. Resulting reassortant viruses are efficiently horizontally transmitted to vertebrate hosts and vertically transmitted to progeny.

Clearly, the vector can serve as a site for viral segment reassortment, point mutations, sequence deletions, inversions, and so forth. The persistent infection of arthropods, the long life of many vectors (especially ticks), the habit of many vectors of taking multiple blood meals from different vertebrate hosts, and the lack of an effective immune response to viruses in arthropods may all contribute to arboviral evolution. As these evolutionary events occur, the vector conducts ongoing experimentation by introducing the virus into new vertebrates and potentially initiating an epidemic.

SUMMARY

The vector-borne disease cycle comprises a dynamic interaction between the pathogen, the vectors(s), the vertebrate host(s), and the environment. The pathogen must be capable of reproducing in two phylogenetically disparate organisms in such a way as to be biologically or mechanically transmitted to the next vector or vertebrate. Vectors contribute in multiple ways to cycles. Vector-borne cycles typically undergo maintenance and amplification phases. Several mechanisms may be employed by a pathogen to persist through seasons of reduced or no transmission. Vectors and usual vertebrate hosts frequently suffer little or no adverse effects from infection, though some exceptions cause disease of great medical or veterinary importance. Vertebrates respond immunologically to the pathogen, thereby limiting the duration or intensity of infection. In contrast, vectors are typically infected for life with no untoward effect. Pathogens can be amplified by horizontal and vertical transmission and have evolved mechanisms, which are in most cases poorly understood, for surviving adverse climatic conditions.

Vector-borne disease cycles seem to be a tenuous mode of transmission and maintenance for pathogens. However, the diversity of the pathogens involved and the tenacity of these cycles in the face of sometimes drastic human intervention attests to their ability to maintain and amplify the prevalence of pathogens in nature. Understanding the interactions between pathogens, vectors, and vertebrates may permit development of effective and novel control strategies for these important pathogens of humans and animals.

Readings

Adelman, Z. N., Blair, C. D., Carlson, J. O., Beaty, B. J., and Olson, K. E. 2001. Sindbis virus–induced silencing of dengue viruses in mosquitoes. *Insect Molec. Biol.* 10: 265–273.

Beaty, B. J., and Bishop, D. H. L. 1988. Bunyavirus-vector interactions. *Virus Res.* 10: 289–302.

Beaty, B. J., and Calisher, C. H. 1991. Bunyaviridae—Natural history. In *Bunyaviridae* (D. Kolakofsky, ed.). *Curr. Top. Micro. Immunol.* 169: 27–78.

Chamberlain, R. W., and Sudia, W. D. 1961. Mechanism of transmission of viruses by mosquitoes. *Annu. Rev. Entomol.* 6: 371–390.

Eldridge, B. F. 1990. Evolutionary relationships among California serogroup viruses (Bunyaviridae) and *Aedes* mosquitoes (Diptera: Culicidae). *J. Med. Entomol.* 27: 738–749.

Fox, A. N., Pitts, R. J., Robertson, H. M., Carlson, J. R., and Zwiebel, L. J. 2001. Candidate odorant receptors from the malaria vector mosquito *Anopheles gambiae* and evidence of down-regulation in response to blood feeding. *Proc. Natl. Acad. Sci. USA.* 98: 14693–14697.

Gerhardt, R. R., Gottfried, K. L., Apperson, C. S., Davis, B. S., Erwin, P. C., Smith, A. B., Panella, N. A., Powell, E. E., and Nasci, R. S. 2001. First isolation of La Crosse virus from naturally infected *Aedes albopictus. Emerg. Infect. Dis.* 7(5): 807–811.

Girard, Y. A., Klingler, K. A., and Higgs, S. in press. West Nile virus dissemination and tissue tropisms in orally-infected *Culex pipiens quinquefasciatus. Vector-borne and Zoonotic Diseases.*

Grimstad, P. R. 1983. Mosquitoes and the incidence of encephalitis. In *Advances in Virus Research*, (M. A. Lauffer and K. Maramorosch, eds.). Academic Press, New York, pp. 357–438.

Hardy, J. L. 1988. Susceptibility and resistance of vector mosquitoes. In *The Arboviruses. Epidemiology and Ecology*, Vol. I (T. P. Monath, ed.). CRC Press, Boca Raton, FL, pp. 87–126.

Higgs, S. 2004. How do mosquito vectors live with their viruses? In *Microbe-Vector Interactions in Vector-Borne Diseases* (Gillespie, S. H., Smith, G. L., and Osbourn, A., eds.). Cambridge University Press., pp. 103–137.

Karabatsos, N. (ed.) 1985. *International Catalogue of Arboviruses: Including Certain Other Viruses of Vertebrates*, 3rd ed. American Society of Tropical Medicine and Hygiene, San Antonio.

Kramer, L. D., and Ebel, G. D. 2003. Dynamics of flavivirus infection in mosquitoes. In *The Flaviviruses: Pathogenesis and Immunity.* 60: 187–232.

Ludwig, G. V., Israel, B. A., Christensen, B. M., Yuill, T. M., and Schultz, K. T. 1991. Role of LaCrosse virus glycoproteins in attachment of virus to host cells. *Virology* 181: 564–571.

Macdonald, G. 1957. *The Epidemiology and Control of Malaria.* Oxford University Press, London.

McGaw, M. M., Chandler, L. J., Wasieloski, L. P., Blair, C. D., and Beaty, B. J. 1998. Effect of La Crosse virus infection of overwintering of *Aedes triseriatus. Am. J. Trop. Med. Hyg.* 58: 168–175.

Miller, B. R., and Mitchell, C. J. 1991. Genetic selection of a flavivirus-refractory strain of the yellow fever mosquito, *Aedes aegypti. Am. J. Trop. Med. Hyg.* 45: 399–407.

Miller, B. R., Beaty, B. J., and Lorenz, L. 1982. Variation of LaCrosse virus filial infection rates in geographic strains of *Aedes triseriatus. J. Med. Entomol.* 19: 213–214.

Mitchell, C. J. 1983. Mosquito vector competence and arboviruses. In *Current Topics in Vector Research* (Harris, K. F., ed. Praeger Publishers, New York, pp. 63–92.

Monath, T. P. (ed.) *The Arboviruses: Epidemiology and Ecology.* Volumes 1–5. CRC Press, Boca Raton, FL.

Paulson, S., and Grimstad, P. R. 1989. Midgut and salivary gland barriers to LaCrosse virus dissemination in mosquitoes of the *Aedes triseriatus* group. *Med. Vet. Entomol.* 3: 113–123.

Romoser, W. S., Wasieloski, L. P., Pushko, P., Kondig, J. P., Lerdthusnee, K., Neira, M., and Ludwig, G. V. 2004. Evidence for arbovirus dissemination conduits from the mosquito (Diptera: Culicidae) midgut. *J. Med. Entomol.* 41(3): 467–475.

Tabachnick, W. J. 1991. Evolutionary genetics and arthropod-borne disease: The yellow fever mosquito. *Am. Entomol.* 37: 14–24.

Weaver, S. C., Scott, T. W., Lorenz, L. H., Lerdthusnee, K., and Romoser, W. S. 1988. Togavirus-associated pathologic changes in the midgut of a natural mosquito vector. *J. Virol.* 62: 2083–2090.

Zeidner, N. S., Higgs, S., Happ, C. M., Beaty, B. J., and Miller, B. R. 1998. Mosquito feeding modulates Th1 and Th2 cytokines in flavivirus susceptible mice: An effect mimicked by injection of sialokinins, but not demonstrated in flavivirus-resistant mice. *Parasite Immunol.* 21: 35–44.

15

Population Biology as a Tool to Study Vector-Borne Diseases

WILLIAM C. BLACK IV AND CHESTER G. MOORE

INTRODUCTION

If we wish to predict the course of an epidemic or, better yet, to predict its arrival, we must thoroughly understand the dynamics of disease transmission. In this chapter we describe the various biological components of vector populations that affect the rate of transmission of arthropod-borne diseases. The biotic factors to be discussed are chiefly birth rate, death rate, density, and age distribution. Migration rates, spatial distributions, and various genetic properties of populations also affect disease transmission; these are discussed in Chapter 32.

Most basic treatments of population biology explore each biotic component individually. Excellent references that adopt this approach are Anderson and May (1982) and Bailey (1982). Case (2000) published a superb textbook that illustrates, at an introductory level, the applications of linear algebra and provides source code for modeling and integrating individual biotic components.

We will take a slightly different approach in this chapter and instead explore many biotic factors *simultaneously* in the context of disease transmission models. These models are useful didactic tools that not only aid in learning about individual biotic components and how they interact but, in addition, illustrate their relative importance in the dynamics of disease transmission. In this regard, Excel files that we use for teaching various aspects of these chapters are available from W. C. Black at wcb4@lamar.colostate.edu.

The models presented in this chapter are also useful in designing optimal strategies for vector population suppression. The chapter ends with a discussion of the ways that the biotic components of these models are actually estimated in field populations using older, tried-and-true methods as well as some new molecular methods.

SIMPLE MODELS OF DISEASE TRANSMISSION

We generally turn to mathematical models to describe the series of events in disease transmission because of the large number of factors involved and the ease of notation. However, as we shall soon see, a mathematical model is really no more than a shorthand statement of what we know and how we think various components are related to each other. Epidemiologists and other health professionals have tried to formulate mathematical models of disease transmission since the early part of the 20th century. The early models are very much like those we use today. It is the data that we fit into the models that have changed; we now know in many cases that what was thought to be a simple constant might be a variable composed of several interacting components. Let's examine briefly what we know (or think we know) about the transmission of a simple contagious disease.

The Reed–Frost Equation

For any communicable disease, there is a probability that contact between an infected and an uninfected susceptible individual during a specified time period will result in transmission of the disease agent. For some diseases, the probability is high (e.g., influenza), while for others it is low (e.g., leprosy) or requires special forms of contact (e.g., AIDS, gonorrhea, kuru). This value is called the *probability of effective contact*. We represent this value by P. The probability that any two individuals will *not* have effective contact $= 1 - P$.

At any time during an epidemic there are "cases," individuals who may or may not be ill but are infected with the disease agent and are capable of transmission. Let's call the number of cases, at some time t, C_t. Then the probability that a given susceptible individual *does not* have contact with any of the cases in one time period is

$$(1-P)^{Ct} \tag{1}$$

By similar reasoning, the probability that a given susceptible *does* have contact with *at least one* of the cases during one time period is

$$(1-(1-P)^{Ct}) \tag{2}$$

If we know the number of susceptible individuals (let's denote them by S) at time t, then we can estimate the total number of cases we expect to occur at time $t + 1$ due to contacts with cases during time t. That value is given by

$$C_{t+1} = S_t(1-(1-P)^{Ct}) \tag{3}$$

Equation (3) is called the Reed–Frost equation. It is a *recurrence equation*, meaning that it describes what happens in one time period as a function of what has happened in the previous period. Note that all that we have done is to devise a shorthand statement of what was described in a somewhat "long-winded" fashion in the last few paragraphs. Cases at some future time depend on the current number of susceptible and diseased individuals, current cases, and the likelihood that they will come into "effective" contact.

Next, let's look at how the equation works through a simple simulation and see what it tells us. Assume that the probability of effective contact for a disease is .3. Assume also that the size of the susceptible population at time $t = 0$ is 10. At this time a single case enters the population. Other than this initial case the population is closed. There are no births, no deaths, and no immigration. Plugging these values into Eq. (3), the number of cases at time $t = 1$ is

$$C_1 = 10(1-(1-0.3)^1) = 3$$

There are now three cases. The diseased individual contacts everyone in the population but only three become infected. If we want to continue this simulation through more time periods, we next have to decide what to do with diseased individuals. If this is

a fatal disease, then we can simply drop them from the population. If this is a persistent infectious disease (e.g., schistosomiasis, giardiasis), then we should make the number of cases cumulative, so the number of diseased individuals in time $t = 1$ is 4 (the index case + 3 new cases). A third possibility is that the diseased individuals in a time period are infectious only within that time period and then become immune. We will apply this last assumption, because it best serves to illustrate components of the model; but you might want to explore the results of this model under the two earlier assumptions. The outcomes are very different.

Note that with an assumption of no mortality and acquired immunity, the population remains closed; people don't leave, even through death. Note also that we have created a third category of people: immunes. These don't enter the model; they just leave the susceptible population. At time $t = 2$,

$$C_2 = 7(1-(1-0.3)^3) = 4.6$$

We can repeat this process (called *iteration*) over 10 time periods to follow the course of the disease. This is shown in Figure 15.1. One case enters a population of 10 susceptible individuals and infects 3 people. Eventually all individuals are infected (i.e., become cases) and recover to become immune. The number of cases increases and then declines because all susceptible individuals are "used up." At the end of the simulation there are 11 immune people and no susceptible people left. Everyone lives happily ever after.

Now let's see what the model can teach us about the probability of effective contact. Figures 15.2A, B, and

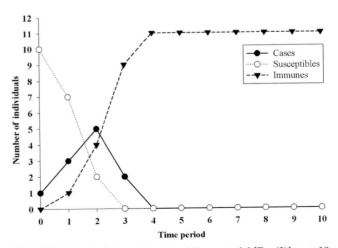

FIGURE 15.1 Iteration of the Reed–Frost model [Eq. (3)] over 10 time periods. There is one index case that enters a closed population of 10 susceptible individuals at time $t = 0$. The probability of effective contact (P) is .3.

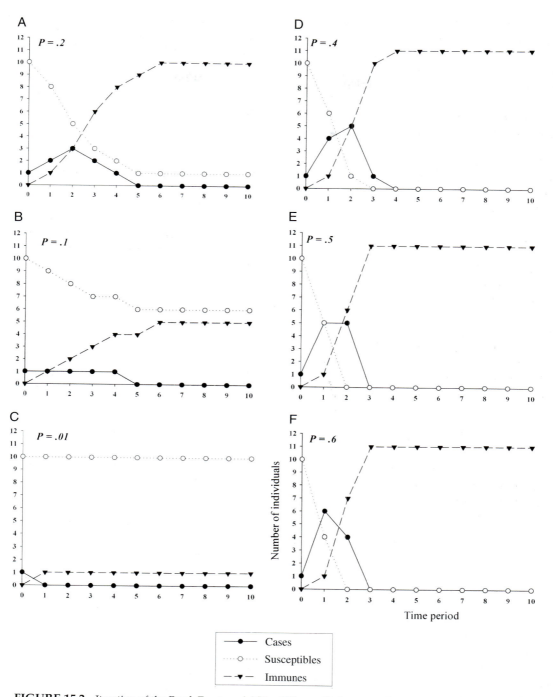

FIGURE 15.2 Iteration of the Reed–Frost model [Eq. (3)] over 10 time periods with different probabilities of effective contact (P). (A) P = .2; (B) P = 0.1; (C) P = .01; (D) P = .4; (E) P = .5; (F) P = .6. Otherwise, conditions are as in Figure 15.1.

C trace the progress of the disease for decreasing values of P = .2, .1, .01. As P decreases, the time required for transmission increases. The distribution of cases shifts to the right because it takes longer for transmission to occur. However, at P = .1, transmission levels off, with 5 immune and 6 susceptible people. No

further transmission can occur. At P = .01, the probability that a case will transmit the disease to a susceptible is 1/100. There are only 10 susceptible people in the population; no transmission occurs. The case enters and becomes immune without infecting anyone. Figures 15.2D, E, and F track disease transmission for increasing values of $P = .4, .5, .6$. As P increases, the time required for transmission decreases. The distribution of cases shifts to the left because it takes less time for transmission to occur.

Iteration of recurrence equations can provide many interesting, nonlinear trends in the long-term progress of disease. We could "open" this population by allowing susceptible individuals to enter through immigration or birth. Furthermore, we could allow people to leave the population through mortality or emigration. Figure 15.3 shows iterations of the Reed–Frost model with $P = .3$ through 50 time periods when the population receives one new susceptible individual every two time periods through immigration and 1 in every 10 cases leaves the population by dying from the disease. Now we see oscillations in the numbers of susceptible, diseased, and immune individuals. The trends in the initial time periods are as in Figure 15.1. However, note that the number of susceptible people now begins to rise again due to immigration and the number declines as transmission occurs. The number of immunes rises, levels off, and then increases again. Immunes initially level off because transmission temporarily ceases. The number of susceptible people is not large enough for transmission to occur. Oscillations in the numbers of susceptible and diseased individuals eventually dampen because the rate of transmission becomes constant. Increases in the susceptible population are offset by decreases due to disease (i.e., they enter the immune population). Once the susceptible population reaches equilibrium, the number of cases becomes constant.

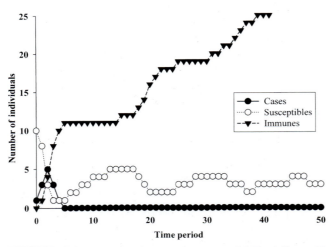

FIGURE 15.3 Iteration of the Reed–Frost model [Eq. (3)] over 50 time periods. Conditions as are as in Figure 15.1 except that the population is no longer closed. Susceptible individuals enter through immigration and some diseased individuals leave by dying.

Vectorial Capacity

The Reed–Frost equation can describe the dynamics of vector-borne diseases (Fine 1981) by slightly modifying the "effective contact" component of Eq. (3). The probability of effective contact will depend on the average number of *potentially infective bites* per individual in the host population. Following Fine (1981), we call that quantity V/T, where V is the total number of infective bites and T is the total size of the human population. Then Eq. (3) becomes

$$C_{t+1} = S_t \left[1 - (1 - V/T)^{C_t} \right] \qquad (4)$$

We can derive a simpler formulation for the Reed–Frost equation because V/T is usually very small. For example, for arboviruses such as western equine encephalitis, eastern equine encephalitis, and Saint Louis encephalitis, V/T is probably in the range of 10^{-4}–10^{-5}, while for dengue it is probably between 10^{-2} and 10^{-3}. Because

$$e^{-x} = 1 - x + (x^2/2) - (x^3/3) + \cdots$$

(where e is the base of the natural logarithm), we may write $e^{-x} = 1 - x$ if x^2 is small compared to x. If $(V/T)^2$ is small compared to (V/T), then Eq. (4) can be approximated as

$$C_{t+1} = S_t \left(1 - e^{-(V/T)C_t} \right) \qquad (5)$$

This is a common form of the Reed–Frost equation used for examining the epidemiology of vector-borne diseases.

Estimating Vectorial Capacity

The value V is called the *vectorial capacity*. It is defined as *the average number of potentially infective bites that will ultimately be delivered by all the vectors feeding on a single host in 1 day* (Fine 1981). Equation (5) is a fundamental theorem of vector-borne disease dynamics. It draws together all we know about the vector–host–pathogen relationship and expresses it in a single value, the number of infective bites. Equation (5) is deceptively simple. When we start examining each of the component parts of V we find more layers —like peeling an onion. Equation (5) suddenly becomes highly complex.

Basically, V is the product of three components: feeding, survival rate, and length of the extrinsic incubation period (EIP) and is represented by

V = [Number of vectors feeding on the host per unit time]
    × [Probability the vector survives the EIP]
    × [Number of blood meals on people after EIP]

(6)

EIP is the period of time, from ingestion of an infectious blood meal to the time of pathogen transmission capability. Let's now take each of the three main segments and expand them until we find things we can actually measure in the laboratory or in the field. The measurable values are denoted in italics in the following.

Vector Density on Hosts

The number of vectors feeding is the product of several factors:

[Number of vectors feeding on the host per unit time] =
[Vector density in relation to the host (m)] ×
[Probability a vector feeds on a host in one day (a)]

$$(7)$$

In turn:

a = [*Feeding frequency*] × [*Proportion of meals on this species* (Host index)]

$$(8)$$

Probability of Surviving the EIP

This is simply the probability of surviving 1 day raised to the power of the length of extrinsic incubation:

[Probability the vector survives EIP]
= [probability of living 1 day (p)]$^{[\text{length of EIP in days } (n)]}$
= p^n

$$(9)$$

Number of Blood Meals Taken After EIP

This is the product of the feeding probability [a in Eq. (8)] and the expected duration of the vector's life:

[Number of blood meals on people after EIP] =
[Duration of vector's life after surviving EIP
[$1/-\ln(p)$]] × [Probability a vector feeds on a host in 1 day (a)].

$$(10)$$

Macdonald's Equation

Now that we have exposed the layers to find things we can measure, we use our notational shorthand to put everything into perspective. Vectorial capacity consists of the following more or less readily measurable quantities.

$$V = [ma] \times [p^n] \times [(a/-\ln(p))] \qquad (11)$$

The equation is still a little clumsy, and we usually combine the two a values and rearrange the equation as follows:

$$V = \frac{m \times a^2 \times p^n}{-\ln p} \qquad (12)$$

A final quantity, b, is often incorporated into the equation. This is a measure of the proportion of vectors taking a meal from an infected host that actually become infective. It is a measure of the genetic and physiological "competence" of the vector. Vector competence is reviewed in Chapter 30. The full equation for V is

$$V = \frac{m \times a^2 \times p^n \times b}{-\ln p} \qquad (13)$$

where:

m = vector density in relation to the host
a = probability a vector feeds on a host in 1 day (= host preference index × feeding frequency)
b = "vector competence," the proportion of vectors ingesting an infective meal that successfully become infective
p = probability the vector will survive 1 day
n = duration of the extrinsic incubation period (in days)
$1/(-\ln p)$ = duration of the vector's life, in days, after surviving the extrinsic incubation period

This equation arose from Garrett-Jones' (1964a) modification of Macdonald's original equation (Macdonald 1957) for case reproduction number. Macdonald pioneered the use of the model to study the dynamics of malaria transmission in Africa.

Table 15.1 explores how each component of the model affects vector capacity. The results demonstrate that of the many variables defined earlier, there are three that strongly affect the magnitude of V: a, n, and p. a is important because it is squared; n is important because it is an exponent of p. p appears in both the numerator and denominator, and in both cases it has a nonlinear effect. Small changes in any one of these variables will cause large changes in V. But V is most sensitive to small changes in survival rate. This outcome led Macdonald to predict that adulticides rather than larvicides would be most effective in reducing malaria transmission rates.

The Vector Capacity of Populations Is Dynamic

Up to this point we have treated many of the biotic components of our models as static parameters; however, we need only consider the first component of Macdonald's equation, m (vector density in relation to the host), to sense the fallacy in this approach. It is well known that vector populations undergo large periodic fluctuations in size. We also know that daily survivorship and the age composition of vector populations

TABLE 15.1   Changes in Vector Capacity (*V*) in Macdonald's Equation as a Function of Changes in Vector-to-Host Ratio (*m*), Host Preference Index, Days Between Blood Meals, Infected Vectors That Become Infective (*b*), Daily Survivorship (*p*), and Extrinsic Incubation Period (*n*)

	Original value	Original value +10%	V	% ΔV	Original value −10%	V	% ΔV
1. Vector : host ratio (*m*)	1,000	1,100	0.646	**10**	900	0.529	**−10**
2. Probability that a vector feeds on a host in one day (*a*)	0.250		(0.227–0.275)			(0.225–0.278)	
a. Host preference index	(0.500)	0.55	0.711	**21**	0.45	0.476	**−19**
b. Days between blood meals	(2.000)	2.20	0.485	**−17**	1.80	0.725	**23**
3. Proportion of vectors ingesting an infective blood meal that successfully become infective (*b*)	0.010	0.011	0.646	**10**	0.009	0.529	**−10**
4. Daily survivorship (*p*)	0.800	0.88	1.998	**240**	0.72	0.191	**−68**
Vector's life in days after surviving through the EIP	4.500		7.8			3.0	
5. EIP (*n*)	7.000	7.70	0.502	**−14**	6.30	0.687	**17**
Vector capacity (*V*)	0.587						

Numbers in bold indicate the percentage change in *V* for a 10% change in the independent variable. Changes in *V* are linear and directly proportional to changes in *m* and *b*. A 10% increase or decrease in these parameters causes a corresponding 10% increase or decrease in *V*. *V* changes in a nonlinear fashion with changes in the host preference index, days between blood meals, and daily survivorship.

change both geographically and seasonally. The vector capacity of populations is therefore a dynamic, fluctuating process that should be estimated in the context of a particular population at a particular point in time. We need to understand the dynamics of vector density and age composition in order to completely appreciate the complexity of disease transmission.

Fluctuations in Vector Population Densities

Vector populations constantly fluctuate in density. The timing of these fluctuations is referred to as the seasonal *phenology* of a population. Density shifts can be a result of regular seasonal climatic changes in temperature, moisture, resources, or the emergence of new broods of adults. They also might arise through intraspecific and interspecific competition or predation by other species. Changes in density are important in disease epidemiology because, as we learned from Macdonald's equation, the vector-to-host ratio is one of the determinants of the vectorial capacity of a population.

There are two general elementary models for population growth: the exponential growth and logistic growth models. With the exponential growth model:

$$N_{t+1} = N_t + rN_t = N_t + N_t(b_0 - d_0) \qquad (14)$$

where:

N_t = number of vectors in a population at time t
r = *intrinsic rate of increase*

b_0 = *individual birth rate*, the number of offspring one individual will give birth to on average per unit of time when the population is small

d_0 = *individual death rate*, the average number of deaths per individual per unit of time when the population is small

This recurrence equation says simply that the change in population size at any point in time will be a function of the size of the population at the previous point in time multiplied by the difference between the rate that new individuals are born into the population and the rate that they leave the population through death. In "good" environments, $b_0 > d_0$, $r \gg 0$ and the population grows; but in "bad" environments, $d_0 > b_0$, $r \ll 0$ and the population declines. A model of exponential growth over 10 time periods (11 measurements) with $r = 2.7$ and starting from a single individual is shown in Figure 15.4. An *r* of 2.7 means that on average an individual produces 2.7 offspring per generation. This model explains the growth of a population over a short period of time and when a population is small; however, it is easy to see that a population would become infinitely large over only a few iterations. The exponential equation therefore has limited utility in modeling changes in vector density over time.

It is more likely that a population will either go extinct due to resource depletion or reach a density at which it no longer grows because the death rate and

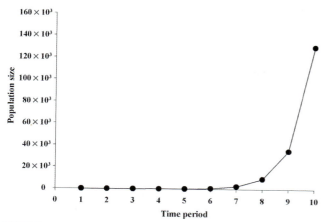

FIGURE 15.4 Plot of the exponential growth model [Eq. (14)]. The intrinsic rate of increase (r) was 2.7.

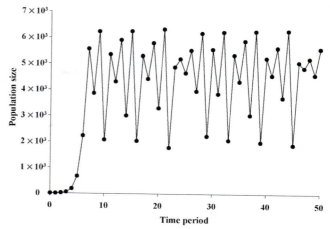

FIGURE 15.5 Plot of the logistic growth model [Eq. (15)]. Carrying capacity (K) was set to 5,000. The intrinsic rate of increase (r) was 2.7.

birth rate have become equal. A scenario commonly envisioned by population biologists is of a single pair of individuals or a single inseminated female arriving into a new region with abundant resources and few limitations to growth. Following oviposition, a new generation begins and reproduces; after a few generations, the population begins to grow exponentially as described in Eq. (14). However, over time, resources become limited and growth rate slows. Some time later, resources cannot continue to maintain growth and births are offset by deaths, so the population maintains itself in equilibrium. The density at which $N_{t+1} = N_t$ is called the *carrying capacity of the environment*. This scenario is modeled by this equation:

$$N_{t+1} = N_t + r[(K - N_t)/K]N_t \qquad (15)$$

where K is the carrying capacity of the environment. This *logistic growth model* is shown in Figure 15.5 over 50 time periods, with $r = 2.7$ and $K = 5,000$ and starting from a single individual. When N is low, $(K - N)/K \cong 1$ and Eq. (15) approximates Eq. (14). As N increases, $(K - N)/K < 1$ and the rate of population growth slows. As N approaches K, $(K - N)/K$ approaches 0 and $N_{t+1} = N_t$. If the population exceeds K due to a temporary increase in births or resources, then $(K - N)/K$ becomes negative and the population declines in size back to K. It is important to realize that K fluctuates around some mean value over time. Fluctuations in K may be regular due to seasonal changes in resource abundance (e.g., oviposition sites for mosquitoes during dry and wet seasons) or may be random and erratic when species are dependent on resources that are scarce in either time or space (e.g., head lice on schoolchildren).

The logistic growth model is a more realistic model of population growth than unlimited exponential

growth. The extent to which population births and deaths are conditioned by density is referred to as *density dependence*. The role of density dependence in natural populations is controversial. Some population biologists feel that species seldom reach the carrying capacity of the environment and that other factors, such as predation, interspecific competition, and disease, intervene to regulate population size at values well below K.

Density dependence has been identified in the artificial container habitat of some mosquitoes. Crowded larval habitats are often found in the field and produce smaller adult mosquitoes. Analysis of interspecific and intraspecific competition in vectors, especially in dipteran insect vectors, is an active area of research. The role of pathogens and predators in the regulation of vector population densities is the subject of Chapter 43.

There is also a wide variety of *density independent* processes that influence population growth. These include environmental factors such as food availability, adverse weather, and extremes in temperature and relative humidity. Random factors include catastrophic reductions in populations through habitat destruction, large-scale meteorological disturbances (e.g., droughts, floods), and massive eradication programs.

Because of the many factors that influence density, seasonal changes in phenology are currently impossible to predict with even the most sophisticated models. This stands as one of the most challenging areas in developing predictive models of vector-borne diseases. There are many models of density-dependent and density-independent growth that are far beyond the scope of this chapter. Interested readers are referred to Bellows (1981), Carey (1993), Charlesworth (1980), Goodenough and McKinion (1992), or Case (2000) for more detailed information on models and modeling. The review by Dye (1992) provides useful guidance for avoiding certain pitfalls in modeling vector-borne diseases.

Fluctuations in Age Structure

A seasonal change in density is only one factor that causes fluctuations in vector capacity. Fluctuations in the average age of individuals in a vector population are also important. The relative proportion of each age class in a population is referred to as the *age structure* of that population. Population biologists have shown that for many vector species, the proportions of a population consisting of eggs, immatures, and adults vary by time and location. Age structure is an important epidemiological factor because, as seen in our testing of Macdonald's equation, only older adults that have passed through the extrinsic incubation period are potentially infective.

To follow or predict the age structure of a population, we begin by collecting information on the daily survivorship and fecundity of a species. We might construct a frequency histogram of the age at which individuals in a cohort die to observe the general survivorship trend. This type of graph is called a *survivorship curve*, and there are three basic types of such curves (Fig. 15.6). Type I curves occur in species in which most deaths occur at an age of senescence. This

is the survivorship curve of humans in developed countries and is the survivorship curve for most of our domesticated plants and animals. Most of us live to an old age and then die. Domestic animals and plants grow to a certain age and are harvested or culled. Typically, organisms with Type I survivorship curves produce fewer offspring, and most of these survive to maturity. The tsetse fly is a good example of a vector species with a Type I curve. In Type II curves, the probability of daily survivorship remains constant throughout life. A constant fraction of offspring is removed in each time period by predators, accidents, or other natural sources of mortality. For example, if the probability of daily survivorship is .9, then the probability that an individual survives through 2 days is $.9 \times .9 = .81$, through 3 days is .73, and through 7 days is .48. This is the type of mortality that is most frequently assumed to occur in vector populations. Also implicit in this model is the constant rate of daily survivorship. Harrington et al. (2001) showed that this assumption is invalid for young vs. old adult *Aedes aegypti* in Puerto Rico and Thailand. Type III curves also describe a mortality pattern frequently found in nature. There is a high initial mortality among offspring (spores, seedlings, eggs), but the few that do survive stand a good chance of living to old age. Species with Type III curves tend to produce many offspring, and a small fraction of these survive to maturity. Many hard ticks produce 1,000–5,000 eggs, but very few survive through the larval period.

Knowledge of mortality patterns is insufficient to derive a complete life table. We also need to understand lifetime patterns in fecundity. Three such curves for vectors are shown in Figure 15.7. Figure 15.7A shows the fertility curve for a female mosquito. Once she has molted to an adult, she mates, blood feeds, and matures a batch of eggs. She oviposits and then must take another blood meal in order to produce another batch of eggs. She may do this several times during her life. A trend that is seen in many insects is that the average size of the oviposition declines on successive batches of eggs. This type of fertility curve is typical for most dipteran vectors, most cockroaches, and soft ticks. The fertility curve for body lice or bedbugs is shown in Figure 15.7B. These insects feed continually on the host and produce eggs continuously throughout their lives. Figure 15.7C shows the fertility curve for a hard tick female that matures a large single batch of eggs and then dies.

Once survivorship and fecundity curves are estimated, a *life table* can be derived. The interested reader should consult Carey (1993) or Case (2000) for a detailed treatment of life table construction. A life table for an imaginary mosquito is shown in Table 15.2. It

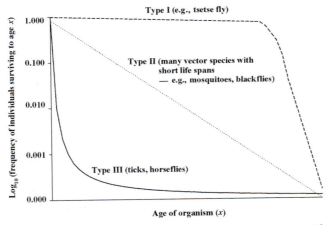

FIGURE 15.6 Three types of survivorship curves: Species with Type I curves have low mortality at an early age and usually survive to a late average age of senescence. Species with Type II curves have a constant rate of mortality. Type III curves describe survivorship in species with high rates of mortality among offspring.

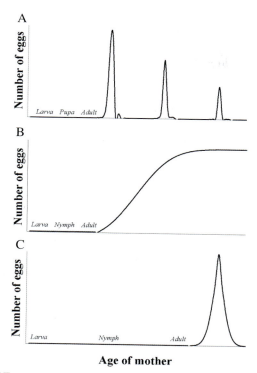

FIGURE 15.7 Three types of fertility schedules in vectors: (A) This is the schedule of reproduction in a species with multiple gonotrophic cycles. It is exhibited by most mosquito and muscid species. (B) This describes fertility rates in species that produce off-spring continuously throughout their life, such as a soft tick or a body louse. (C) This describes reproduction in species that produce one large batch of offspring, usually toward the end of their life, such as a hard tick.

TABLE 15.2 Life Table for an Imaginary Mosquito

Life stage	Chronological age (days i)	Probability of daily survival (p_i)	Fecundity (f_i)
Egg	0	.5	0
Larva	1	.6	0
	2	.6	0
	3	.6	0
	4	.6	0
	5	.6	0
	6	.6	0
Pupa	7	.6	0
Adult	8	.9	0
	9	.9	0
	10	.9	0
	11	.9	0
	12	.9	120
	13	.9	0
	14	.9	0
	15	.9	0
	16	.9	0
	17	.9	100
	18	.9	0
	19	.9	0
	20	.9	0
	21	.9	0
	22	.9	80
	23	.9	0
	24	.9	0
	25	.9	0
	26	.9	0
	27	.9	60
	28	.9	0

Each gonotrophic cycle requires 5 days. Each female oviposits 120 eggs during her first cycle and declining numbers in subsequent cycles. We assume that she passes through four gonotrophic cycles during her life. The egg stage lasts 1 day, larval development requires 6 days, and 1 day is required for adult development to occur in the pupal stage. Daily survivorship is 50% in eggs, 60% in larvae, and 90% in adults.

contains complete information on daily survivorship and fecundity. We assume that the maximum life span of this mosquito is 28 days, with 1 day spent as an egg, 7 days as a larva and pupa, and 20 days as an adult. The average probability of egg survivorship is .5. The probability that a larva survives through 1 day is .6. The probability that an adult lives through a day is .9. A female usually oviposits once every 5 days, and the average oviposition size in the first cycle is 120 eggs and declines by 20 eggs in the three subsequent cycles. This is therefore a type II survivorship curve with a fecundity schedule like that in Figure 15.7A. With this information we are prepared to build a model to predict fluctuations in age structure and density of populations for our imaginary mosquito.

The Lewis–Leslie Model

The Lewis–Leslie age structure model is often used to simulate changes in population density and age

structure in vector populations. The general form of the model in matrix notation is

$$\mathbf{n}_{t+1} = \mathbf{M} \times \mathbf{n}_t \qquad (16)$$

As with models discussed earlier, this is a system of recurrence equations. By iterating the system over several generations we can observe changes in density and the dynamics of age structure in a population.

$\mathbf{M}$ is a square matrix that contains information taken directly from life tables in the form of two elements, f_i and p_i, for each age class, where f_i is the fecundity of each age class i and p_i is the probability of survival through age class i. The $\mathbf{M}$ matrix appears in Figure 15.8. Comparison with the life table in Table

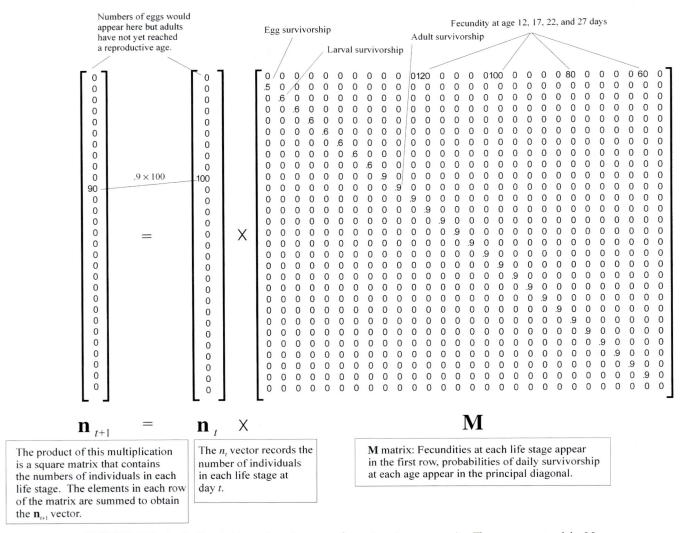

FIGURE 15.8 Leslie–Lewis **M** matrix and **n** vector for an imaginary mosquito. The components of the **M** matrix are as described in Table 15.2. The **n** vector indicates that the population is initiated with 100 newly emerged individuals.

15.2 indicates how the **M** matrix is derived. Fecundity values f_i are placed across the top row. The first value greater than zero appears in column 12. This corresponds with the first day of reproduction in the life table. Probability of daily survivorship (p_i) values appear below the principal diagonal of the **M** matrix. Egg survivorship appears in the first column on the second row, survivorship of first-day larvae appears in the second column on the third row. The $\mathbf{n}_t$ vector is the number of individuals in each age class at time t. This vector, therefore, records the age structure of the population at any point in time. By multiplying **M** by $\mathbf{n}_t$, two things happen: The number of individuals at time t is multiplied by the probability of daily survival

to give the number of individuals at $t+1$ in the $\mathbf{n}_{t+1}$ vector. In this way the model tracks survivorship at each life stage. Second, the number of individuals reproducing (days 12, 17, 22, and 27) is multiplied by the fecundity in each of these life stages. This product records the number of new individuals introduced into the system at that time. These appear as eggs at the top on the **n** vector. In the next iteration these will become 1-day-old larvae; in the next iteration, 2-day-old larvae, etc. The number passing from one day to the next is determined by the probability of survivorship through that period. What enters into each new day ($\mathbf{n}_{t+1}$ vector) is a function of: (1) what survived from the previous day and (2) new births from the reproductive

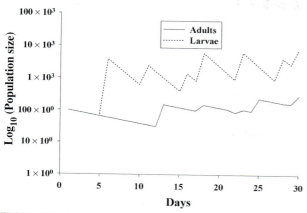

FIGURE 15.9 Iteration of the Leslie–Lewis matrix in Figure 15.8 over 30 days. Numbers of larvae and adults were obtained by summing over cells in the **n** vector for each time period.

age classes. This model could run indefinitely because new individuals are constantly regenerated through reproduction.

Iteration of the model over 30 days is shown in Figure 15.9. Note that adult density declines over the first 10 days as they die. Following day 5, the number of larvae increase because the eggs oviposited by the adults surviving to day 5 have hatched. These decline due to mortality over the next 5 days but increase again as the next set of ovipositions hatch. On day 12, larvae from the first oviposition emerge and the adult density increases. These fluctuations continue, and the census at 30 days is 8,000 larvae and 300 adults.

Note that larval densities are greater than adult densities. Even though survivorship is 30% lower in larvae than in adults, a large number of larvae are being produced and they experience fewer days of mortality.

Consideration of another model will help clarify this principle. In Table 15.3 the parameters of the model have been changed to match those of the tsetse fly. Recall that the tsetse fly doesn't oviposit but rather matures a single larva inside of its body and "lays" a fully developed third-instar larva that burrows into the soil and pupates. Egg, larval, and adult survivorship are therefore the same. The life table for this simulation is shown in Table 15.3. Figure 15.10 shows the population size of larvae and adults through 100 days. Notice that larval and adult fluctuations trace one another very closely, because, unlike the mosquito system (Fig. 15.9), larval survivorship is totally dependent on adult survivorship and because the fecundity of the tsetse fly is so low. What type of age distribution would you expect for the human population in a developed country?

TABLE 15.3   Life Table for a Tsetse Fly

Life stage	Chronological age (days i)	Probability of daily survival (p_i)	Fecundity (f_i)
Larva	1	.95	0
	2	.95	0
Pupa	3	.95	0
	4	.95	0
Adult	5	.95	0
	6	.95	1
	7	.95	0
	8	.95	1
	9	.95	0
	10	.95	1
	11	.95	0
	12	.95	1
	13	.95	0
	14	.95	1
	15	.95	0
	16	.95	1
	17	.95	0
	18	.95	1
	19	.95	0
	20	.95	1
	21	.95	0
	22	.95	1
	23	.95	0
	24	.95	1
	25	.95	0
	26	.95	1
	27	.95	0
	28	.95	1
	29	.95	0

The female tsetse gives birth to a fully grown larva. A larva is produced at intervals of 14 days. The pupal stage lasts 2 weeks. The adult tsetse produces 12 larvae during her life.

An important result of iterating the Lewis–Leslie matrix over several generations is that the population will come to contain a constant proportion of each age class. When this occurs the population is said to have reached a *stable age distribution (SAD)*. This is illustrated for our mosquito population in Figure 15.11, which shows the proportion of the population consisting of larvae and adults over a 200-day period. Notice that the magnitude of the oscillations dampens. Eventually, in the absence of any perturbations, they would become stable. The time required to reach an SAD is determined by the initial values of the model.

Figure 15.12 follows the mosquito population over 200 days. It is easy to see the similarity between Figures 15.4 (the exponential model), 15.10, and 15.12. Notice that in developing the Lewis–Leslie matrix, no adjustments were made for density dependence. The model as we have developed it is an exponential growth model. One could introduce density depend-

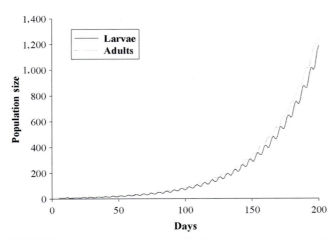

FIGURE 15.10 Iteration of the Leslie–Lewis matrix over 30 weeks. Components of the **M** matrix were as described in Table 15.3 for the tsetse fly.

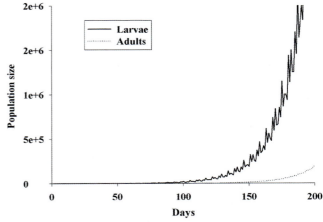

FIGURE 15.12 Exponential growth of larval and adult mosquito populations in an iteration of the Leslie–Lewis matrix in Figure 15.8 over 200 days.

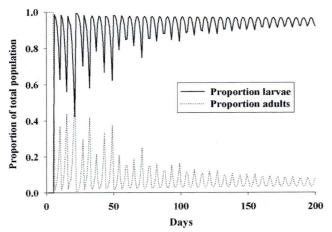

FIGURE 15.11 Approach to a stable age distribution: Plotted are the proportions of the total mosquito population, consisting of larvae and adults, in an iteration of the Leslie–Lewis matrix in Figure 15.8 over 200 days.

model. A model using days as age classes predicts trends in field populations. We might predict on what days, weeks, months, or years the vector capacity of a population would be at its greatest; however, even a scale based on days is inaccurate because insects are not homeothermic. The time spent in each stage is dependent on temperature, with accelerated development at warmer temperatures. We might therefore adjust our time units to be a product of time and temperature. A frequently used unit is *day-degrees (DD)*. Based on records of time and temperature, we could attempt to build a more predictive model. The nutritional resources of the larval habitat also will affect the duration of the larval stage. Similarly, ambient temperature and availability of hosts will affect the average length of the oviposition cycle.

Implications for Vector Control

It is important for people interested in suppressing vector populations to understand all of the factors we have discussed before initiating control practices. Many models have been developed to examine and predict the outcome of control programs [e.g., the Garki Project (Molineaux and Gramiccia 1980)]. Very sophisticated models employ factors such as density dependence and the presence of predators and parasites and will indicate to the control program managers what the outcome of insecticidal application is likely to be. Managers can use the programs to optimize the timing and placement of insecticidal control. An example of such a program that is easy to use and illustrates many of the principles discussed in this

ence by making daily survival and fecundity values dependent on population density. Then we would expect to see population densities approximate a logistic model.

A point that is frequently overlooked in developing the Lewis–Leslie model is that age classes must be of equal size in order for the model to be accurate. For example, we could have made our mosquito model a three-stage model with eggs, larvae, and adults and recorded survivorship and fecundity for these stages. But that model would be inaccurate because these stages vary in their lengths. Instead we have translated age classes to days. If a mosquito spends on average 7 days as a larva, we enter seven larval periods into the chapter is Fly Management Simulator (FMS) (Axtell 1992).

Dobson and Foufopoulos (2001) used "next generation" matrix techniques to study changes in the basic reproductive rate (R_0) for a variety of zoonotic diseases, including Lyme disease and West Nile virus. These models show that the presence of species that act as sinks for infection (dead-end hosts), especially when present in large numbers, can effectively buffer epidemic outbreaks. Thus, the dynamics of *Borrelia burgdorferi* in the species-rich southeastern United States show reduced rates of spread in comparison to the northeastern United States, where species diversity is lower. As expected, the models show that vertebrate host species that experience high mortality from infection may make a large contribution to the net force of infection, but they will not be effective reservoirs for the pathogen. Because the row and column totals of the matrices reflect the relative force of infection experienced by (columns) and exerted by (rows) each species, the efficacy of various control strategies, such as immunization and selective vector reduction, can be examined through the methodology of Dobson and Foufopoulos (2001).

Another application of R_0 in vector-borne disease epidemiology used estimates of R_0 for dengue virus transmission in Brazil to predict areas at risk for yellow fever (Massad et al. 2001). This application depends on the fact that the same vector, *Aedes aegypti*, transmits both dengue fever and yellow fever. Such techniques may be useful as an aid in assigning scarce resources to those regions that are at greatest risk.

Usually in vector control we must accept that a population cannot be reduced to zero or eradicated. The cost may be too high or the amount of damage to the environment too great. Instead we ask: What is an acceptable *economic threshold* for a vector population? This is the population density at which damage caused by the species is at or below acceptable levels of disease or economic damage. This threshold varies for each species and pathogen. In vector populations, economic thresholds may be dictated by the vectorial capacity of a population and the severity of diseases transmitted. Examination of the Lewis–Leslie model provides an important insight into this issue of how to most efficiently reduce the vector capacity of a population. Figure 15.13 shows, using the same mosquito model described earlier, the relationship between the size of the adult population after 100 days and survivorship in eggs, larvae, and adults. Survivorship was varied from .1 to 1.0 for each stage while holding survivorship at 1 for the other two stages. Let's say that we had set the economic threshold to 1×10^6 adult mosquitoes. To achieve this level through ovicidal

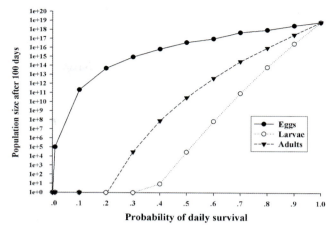

FIGURE 15.13 Effects of reducing the probability of daily survival (*p*) in eggs, larvae, and adults. Plotted are the final sizes of the egg, larvae, and adult populations in an iteration of the Leslie–Lewis matrix in Figure 15.8 over 100 days.

treatment would require reducing egg survivorship to .05. Adult survivorship would have to be reduced to approximately .35, but larval survivorship would only have to be reduced to .55 to achieve the same level of control. If we use insecticides to reduce the probability of daily survivorship and if the rate of insecticide use is proportional to the decrease in survivorship in each age class, then it is easy to see that application of insecticides to kill larvae is the most efficient and environmentally least destructive way to obtain the economic threshold. Fewer larvae must be killed to obtain the economic threshold.

This argument may be too simplistic if mortality is density dependent. For example, Agudelo-Silva and Spielman (1980) showed that, in food-limited environments, simulated larvicidal mortality (i.e., by removing a given percentage of the larval population) actually *increased* the number of adults that emerged. Hare and Nasci (1986) confirmed the foregoing observations using *Bacillus thuringiensis israeliensis* as a larvicide. Thus, the application of control measures in certain situations may eliminate naturally-occurring mortality, with a net *increase* in total numbers (compensatory mortality).

These examples illustrate the importance of understanding the degree of density dependence before initiating vector control. When population densities are low, populations often grow exponentially in size. Individuals are often larger and healthier and live longer (i.e., they have a larger vectorial capacity); however, eventually the carrying capacity of the environment is reached and individuals must compete for resources. The size, viability, and fecundity of individuals may be decreased. At this point the population either will stabilize at the carrying capacity of the environment or may undergo a drastic reduction in density. *The point is that **if** a population is at its carrying capacity and we reduce population densities, we may create a healthier population with a greater vectorial capacity than if it had been left alone.* Remember from our examination of Macdonald's equation that changes in vector-to-host densities had far less of an impact on vectorial capacity than changes in daily survivorship. Vectorial capacity of a population may be *increased* through control, or decreasing population densities may have little impact on vector capacity if it does not affect the probability of daily survivorship (Grimstad and Haramis 1984; Kay et al. 1989; Paulson and Hawley 1991; Grimstad and Walker 1991); however, this pattern does not always appear (Nasci and Mitchell 1994; Dye 1992).

In addition to larval nutrition, adult nutritional status can affect mortality and the potential for disease transmission. Females of some mosquitoes are less susceptible to insecticides if they have recently taken a blood meal (Moore et al. 1990). This could result in the preferential selection of recently fed (and possibly infected) females by using improperly timed applications of adulticides in a vector control program.

It is also important to know how parasites and predators regulate a population. If, in controlling a pest population, predators and parasites are destroyed, it has been commonly observed that the target population will rebound to an even greater level than before control began. Furthermore, other nonpest species that were regulated by predators and parasites can, all of a sudden, reach densities that make them pests or even vectors. Pathogens are spread most rapidly through dense populations, so a population at carrying may be more susceptible to decline due to a epizootic spread of a pathogen. A population with a significant parasite or pathogen load that is at carrying capacity may be at the point of collapse. Under these circumstances, expensive and environmentally destructive control operations are unnecessary and may, in fact, be wasteful and injurious.

MEASURING THE BIOTIC PARAMETERS THAT COMPRISE VECTOR CAPACITY

We have examined the biotic components that are important in regulating the vector capacity of populations. We have discussed how this knowledge can be used to design the most efficient and environmentally less destructive control strategies. We will finish this chapter by discussing methods to estimate the biotic components of vector capacity.

The Host-Biting Habit

Feeding Frequency (M)

In species that require a single blood meal for each batch of eggs, the feeding frequency is approximately equal to 1 divided by the length of one female reproductive cycle. This is either the time required for her to lay a batch of eggs following emergence as an adult or the time between ovipositions of separate egg batches. The cycle during which the ovaries mature to fully developed follicles is called the *gonotrophic cycle* (Detinova, 1962). Thus, for a species whose gonotrophic cycle is 4 days, the feeding frequency is 1/4, or 0.25. We have made four assumptions: (1) The gonotrophic cycle has constant length, (2) females lay their eggs as soon as ovarian development is complete, (3) females refeed on the same day that they oviposit (or there is a constant interval between oviposition and refeeding), and (4) females take only one blood meal per gonotrophic cycle.

The feeding frequency is best determined by mark–release–recapture studies, in which females coming to bite are marked and then released after feeding. Recaptures at subsequent biting catches provide exact measurements of the time between feeds. Another method uses sequential data on parous rates; this will be discussed in the section on daily survival rates.

Host Blood Index

Most vectors feed on more than one species of vertebrate. By collecting blood-filled females (and males in the case of some vectors), we can determine the relative importance of each host or the "preference" of the vector.

It is absolutely essential that a representative (i.e., random) sample of all blooded individuals be collected. Specimens collected inside houses or stables may represent only feeds on the predominant host. For mosquitoes, vacuum sweeper collections of resting females frequently provide the most uniform, random samples. Blooded specimens are usually uncommon, and very large samples must be collected to have sufficient data for statistical analysis.

The human blood index (HBI) (Garrett-Jones 1964b) is often used to make comparisons of host preference. This parameter is defined as the proportion of freshly engorged insects found to contain human blood. In studies estimating host preference indices, seldom are all potential vertebrate hosts present in equal numbers or are equally available to the vector. Therefore, *what appears to be a preference may really be only an indication of the most common or easily accessible host species.* The calculation of forage ratios (FRs) and feeding indices (FIs) can help to eliminate host density effects (Hess et al. 1978; Kay et al. 1979). FR adjusts the HBI by taking into account the number of available hosts. FI adjusts the HBI by taking into account the density of hosts, host size, age, and temporal and spatial concurrence between the host and vector. FR or FI values of 1 indicate no preference or avoidance of a host, while values greater than 1 indicate host preference.

Identification of the source of blood meals is usually conducted by immunological methods, such as microprecipitin, ELISA, soluble antigen–fluorescent antibody, agar-gel diffusion, or related tests (Tempelis 1975). The most difficult problem in blood meal testing is the preparation and standardization of antisera to the vertebrate host species under study (Washino and Tempelis 1983). Another method, hemoglobin crystallization, uses the morphology of crystals formed by the reaction of hemoglobin with different salt solutions to identify the host species. Precise conditions of temperature are required for crystal development, and large catalogs of crystal shapes must be maintained to permit identification.

Daily Survival (*p*)

The birth and death rates, discussed earlier in the context of the exponential and logistic models, are often immeasurable, and it is more feasible to estimate the probability of daily survival. A number of techniques have been developed to estimate daily survivorship (*p*) because it is such a critical component of vector capacity [Eq. (13)] (Gillies 1974). Four of the most promising techniques will be reviewed here.

Mark–Release–Recapture Methods

In mark–release–recapture studies, large numbers of individuals of known age are marked by some means and released into the study population. The population is then sampled on consecutive days following the release; hopefully, some of the marked individuals are recovered. If the logarithm of the number of marked individuals is plotted against time since release, a straight line should result. The slope of the fitted regression line is an estimate of log *p*, the survival rate.

There are several assumptions made in mark–recapture studies. First, neither the survival nor the behavior of the marked animals is affected by the marking method. Second, the animals become completely mixed in the population at large. Third, the probability of capturing a marked individual is the same as that of capturing an unmarked individual. Fourth, sampling is at discrete time intervals, and the time spent in sampling is short relative to total time. There are several variations on the mark–recapture theme; these are described in detail by Southwood (1978), Lounibos et al. (1985), and Service (1993). Conway et al. (1974) illustrated the use of mark–recapture in estimating *Aedes aegypti* densities in Tanzania.

Gonotrophic (Physiological) Age Grading

Much of the work on survival rates in mosquitoes and other biting flies has involved the use of changes in the ovarian status of the female (Detinova 1962). Early workers in this area were V. P. Polovodova and T. S. Detinova in Russia and D. S Bertram in England. As a female ages and passes through successive gonotrophic cycles, various changes occur in the ovaries and associated structures. If the time between various ovarian events can be determined (e.g., in laboratory experiments or by mark–release studies), then these physiologic "markers" can be used to tell the age of field-collected specimens. Two major events in the ovarian cycle are used: the unwinding of the ovarian tracheoles during the first gonotrophic cycle (Detinova method) (Fig. 15.14), and the formation of residual lumps or dilatations in the ovariole pedicel following each oviposition (Polovodova method).

The Detinova method is by far the easier of the two methods and for many species is the only reliable technique. The ovaries are removed from the abdomen and placed in a drop of water on a clean microscope slide. The water is allowed to evaporate, and the ovaries dry in place on the slide. As the ovary dries, air is drawn into the tracheoles so that they are easily seen under normal light microscopy at 250–400×. In the nulliparous female (no eggs developed), the ends of the tracheoles are coiled, looking like a skein of knitting yarn (Fig. 15.14A). In the parous female, the tracheoles are extended as a result of the expansion of the ovary during an earlier gonotrophic cycle (Fig. 15.14B). The problem with this method is that individuals can only be classified as parous or nulliparous, whereas with the Polovodova method several gonotrophic age classes are identifiable and the number of prior gonotrophic cycles can be determined. Details on classifying individual insects into gonotrophic age classes are too detailed for this chapter. Use of the Polovodova method in the examination of age structure of muscid fly species is clearly illustrated in Tyndale-Biscoe and Hughes (1969), Krafsur and Ernst (1983), and Krafsur et al. (1985).

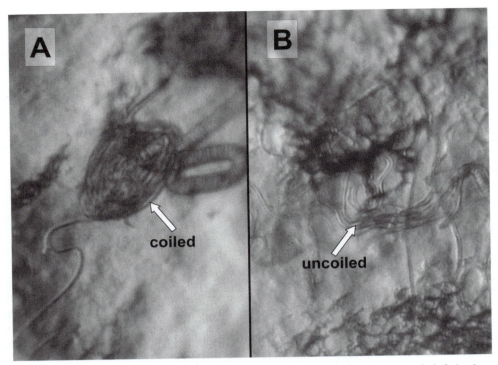

FIGURE 15.14 Tracheal skeins in the ovaries of *Culiseta melanura*, showing (A) coiled skeins in a nulliparous female and (B) uncoiled skeins in a parous female. Differential interference (Nomarski) microscopy. Dissections by T. Welch, photos by A. C. Brault and C. G. Moore, CDC.

With both methods, very large numbers of individuals are needed to get an accurate estimate of parous rates. The survival rate is estimated from the parous rate by the equation

$$m = p^d \qquad (17)$$

or

$$\log(p) = \log(m)/d \qquad (18)$$

where:
m = Proportion parous
p = Daily survival
d = Length of gonotrophic cycle

This method is widely used to estimate p. Not only is p critical in determining vector capacity, but it can also be useful in determining whether control efforts are having an impact on daily survival. Estimation of parous rates in the face fly, *Musca autumnalis*, was used to test the efficacy of insecticidal ear tags by comparing p in treated versus control populations (Krafsur 1984). However, there are several implicit assumptions made when using parous rates to estimate p: (1) We assume that the population has reached an SAD,

discussed earlier. (2) The sampling method provides an accurate cross section of the general population. (3) The survival rate is not age dependent. (4) The length of the gonotrophic cycle is known and is constant. Assumption 1 is almost never true in temperate regions, but it may hold for some tropical species. However, the parous rate will accurately estimate p in temperate regions if calculated from the overall parous rate measured throughout a breeding season (Birley et al. 1983). It is well known that different sampling methods or sites provide radically different parity estimates, so assumption 2 may also be violated. For example, Krafsur and Ernst (1983) showed that in horn fly (*Haematobia irritans*) populations, parous rates were greatest among flies sampled from cattle bellies and nulliparous males and females were most frequent on cattle backs. Survival may, in fact, be age dependent (Clements and Paterson 1981), thus invalidating assumption 3. It is often impossible to estimate the length of the gonotrophic cycle without conducting mark–recapture studies; and the cycle length may change in response to temperature or other environmental stimuli. A method developed by Birley and Rajagopalan (1981) overcomes this latter problem, but it requires daily catches for periods of 2–3 weeks or more. The method uses cross-correlation analysis to estimate the primary or dominant cycle length.

Chronological Age

Daily growth layers in the cuticle and daily bands on the apodemes were first observed by Neville (1963). Alternating light and dark bands are seen when sections through the cuticle are observed in polarized light microscopy. This banded pattern is caused by a circadian rhythm in the orientation of chitin rods in the protein matrix of the integument. Daytime deposition is unidirectional, while nighttime deposition rotates in a helicoidal fashion, giving rise to optically active material. By counting growth layers in the cuticle, the chronological age of an insect can be determined.

Yosef Schlein and Norman Gratz developed an alternative technique in 1972. They showed daily bands on the apodemes and apophyses of *Anopheles*, *Aedes*, and *Culex* and cyclorrhaphous dipterans *Glossina*, *Calliphora*, and *Sarcophaga* (Schlein 1979). *Apodemes* are infoldings of the integument that serve as muscle attachments, and they strengthen the skeleton. The muscle mass of an insect increases over the first few days of adult life; additional cuticle is laid down on the apodemes to handle the increase in muscle tissue. Electron microscope studies show that a daily variation in thickness of the apodemes gives rise to their banded appearance. The growth bands on apodemes probably result from differential stretching during the flight activity of the insect.

The Schlein–Gratz method requires a fairly elaborate and time-consuming staining process to make the bands visible. In addition, at least for *Culex pipiens*, the ages of known controls could not be determined with certainty. These factors are definite drawbacks for ecologically oriented work, in which we would like to process large numbers of specimens to obtain statistical reliability. Fortunately, another method has been found that is not only faster but provides more accurate results. Differential interference (Nomarski) microscopy causes structures of different composition, depth, or density to assume different shades or colors in a polarized light field. Since the bands have been shown by electron microscopy to vary in thickness, we can use Nomarski optics to see the bands. This method is faster than the Schlein–Gratz method because it requires little or no staining (Moore et al. 1986). It also offers greater resolution of the bands. Furthermore, Nomarski optics do not produce the halos that cause problems with techniques such as phase contrast microscopy. Figure 15.15 shows several apodemes in the thorax of *Culex pipiens*.

As with the other techniques, this one has its own particular problems. In order to achieve statistical reliability, a series of coded, known-age controls should be included with each batch of field material. By doing this, it can be determined if readings on a given day or by a particular worker are biased. However, it is difficult to prepare known-age controls in an insectary. Presumably, this is due to a lack of active flight during the normal activity period (and, therefore, the apodemes are not stretched and thickened at the appropriate times). Specimens with reasonably good band definition can be prepared if the insect can be reared outdoors in a large cage that permits normal daily flight activity.

Pteridine Quantification

Molecular biology has provided another tool for chronological age determination in insects. Pteridines are a major class of eye pigments in insects, and there are five major classes of pteridines: sepiapterin, biopterin, xanthopterin, isoxanthopterin, and pterine. Pteridines are synthesized in insect hemolymph, and while found throughout the body, they are highly concentrated in the eye. Each absorbs a different wavelength of light, and most of them fluoresce. They function presumably to control the amount and wavelengths of light striking the ommatidia, the light-sensitive structure in the insect eye. All insects accumulate pteridine compounds in their eyes at a continuous rate as they age. No one understands why pteridines accumulate in adult insect tissues rather than being excreted. Several workers have speculated that no mechanism for their excretion has evolved because the small amounts accumulated during the short life span of an adult are not deleterious. Pteridine accumulation is a natural process of aging in insects.

Estimation of pteridine concentration involves grinding an insect head in a homogenization buffer, spinning-down tissue debris in a microcentrifuge, and transferring the supernatant to a semimicro cuvette. A spectrofluorometer is then used to quantify pteridine concentrations. Mail et al. (1983) first reported on a high correlation between pteridine amounts and chronological age in the stable fly. They reported a regression coefficient (r) of 0.97 (on a scale of 0–1), a value that is never seen with gonotrophic age grading ($r = 0.5$–0.8, depending on the species). They demonstrated that the rate of accumulation was temperature dependent, that flies accumulated pteridines more rapidly at higher temperatures, and that the age of both sexes could be estimated. This allowed models to be built that incorporated temperature and pteridine amounts as independent variables to estimate chrono-

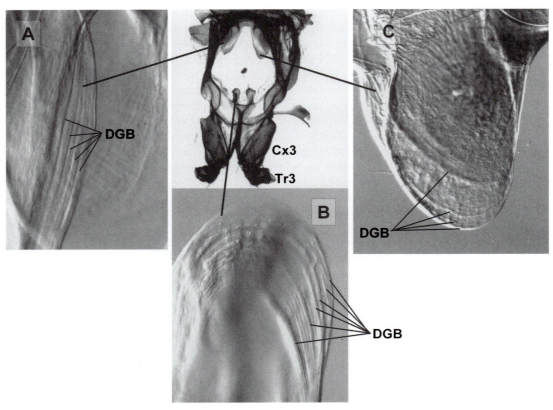

FIGURE 15.15 Metathorax of *Culex pipiens,* showing thoracic apodemes. (A) Lateral (metapleural) apodeme; (B) furcasternum-3; (C) second phragma. DGB—daily growth bands; Cx3—metacoxa; Tr3—metatrochanter. Cross section stained with haematoxylin; A, B, and C unstained; differential interference (Nomarski) microscopy. See Owen (1977) for a detailed discussion of the morphology of the thoracic skeleton and muscles of mosquitoes. Apodemes are from different specimens. Dissections and photos by C. G. Moore.

logical age. They tested their models in blind experiments with both males and females and were able to predict age to within 1.9 days in females and 1.4 days in males. This level of accuracy has never been achieved with any other age-grading technique, especially the gonotrophic age technique.

Lehane and Mail (1985) demonstrated that pteridine age grading was also accurate in the tsetse fly and showed that body size and temperature affected pteridine amounts. The size was easily adjusted for by measuring the width of the head. They also demonstrated that severed heads could be kept at room temperature in light-safe vials with desiccant for 8 days before an appreciable decline in pteridine could be detected. This meant that the time between which the head is severed and pteridines are measured is not critical and that the technique is therefore robust with regard to many laboratory and field applications. When the technique was used on recaptured insects that had been marked and released in the field, the regression coefficient declined to a range from 0.81 to

0.95 because the temperatures experienced by individual flies could not determined (Lehane and Hargrove 1988). Langley et al. (1988) demonstrated that this was not a major factor if flies were collected over finite intervals and the mean ambient temperatures were recorded. The use of pteridines in age grading has been expanded to other cyclorrhaphous flies [e.g., Krafsur et al. (1992) in *Haematobia irritans*; Krafsur et al. (1999) in *Musca autumnalis*]. Lardeux et al. (2000) attempted to use pteridine fluorescence levels to age grade individual *Aedes polynesiensis* and *Culex quinquefasciatus*. However, individual fluorescence measurements were mostly below the spectrofluorometer white noise level. With batches of mosquitoes of the same age, significant fluorescence levels were recorded but were not correlated with their calendar ages.

One of the major advantages of chronological methods is that they can be combined with gonotrophic data to establish the exact length of the gonotrophic cycle, time to the first blood meal, and many other characteristics of the population. A further advantage is that true horizontal (cohort) life tables can be constructed. This is particularly valuable in temperate regions, where rapid changes in recruitment or mortality may occur. It should be noted that mark–recapture methods provide chronologic data too. The advantage of techniques such as the apodeme or pteridine method is that essentially all specimens in a collection provide data.

Cuticular Hydrocarbon Quantification

Desena et al. (1999a) used gas chromatography with flame-ionization detection to measure quantitative age–related changes in the cuticular hydrocarbons of female *Aedes aegypti* (L.). The relative abundance of pentacosane decreased linearly from 0 to 132 day-degrees (DD), whereas nonacosane increased linearly over the same period. Crowding and starving of larval conditions that decreased size had negligible effect on the relative abundance of pentacosane and nonacosane. Additionally, the rate of change in the relative abundance of pentacosane and nonacosane did not vary between strains. The authors performed a blind study using laboratory-reared mosquitoes, and a mark–release–recapture experiment involving field mosquitoes validated these age-grading models and produced promising results for aging females up to 19, 12, and 10 calendar days at 24, 28, and 30°C, respectively.

Desena et al. (1999b) refined this method so that only the legs were used to age grade individual females. The use of leg cuticular hydrocarbons for estimating the age of female *Ae. aegypti* decreased variability associated with the relative abundance of pentacosane and the expanded range over which the models were able to predict age accurately by the addition of the relative abundance of octacosane.

CONCLUSION

We have described the various biological factors in vector populations that are thought to affect the rate of transmission of arthropod-borne diseases. We have presented a variety of models to explore interactions among these factors and explain key concepts in population biology. These models also provide useful ways to learn about the dynamics of populations and provide clues to general strategies for reducing vector densities. *However, we emphasize that these models are only didactic tools and are largely inaccurate in their predictions concerning natural populations.* For example, a significant part of this chapter centered on the use of Macdonald's equation. However, there are grave practical problems associated with measuring any and all of the components of the model. The interested reader is encouraged to consult references such as Dye (1992) to learn more about the problems associated with estimating the components of Macdonald's equation and ultimately the vector capacity of populations. Dye (1992) argues convincingly for taking a "comparative" rather than an "absolute" approach when estimating the importance of the biological factors affecting disease epidemiology. With the comparative approach, epidemiologists observe how disease prevalence and incidence change among populations that vary in the various factors we have discussed. Alternatively, observations can be made in a single population in which factors vary seasonally. The relative importance of each factor is judged by the degree to which variation in any one factor influences disease prevalence. Ultimately, this approach may be more useful and informative than trying to obtain "absolute" estimates of these factors that are then incorporated into predictive models. In a final analysis, there is no substitute for empirical observations in understanding the dynamics of vector populations.

References

Agudelo-Silva, F., and Spielman, A. 1980. Paradoxical effects of simulated larviciding on production of adult mosquitoes. *Am. J. Trop. Med. Hyg.* 33: 1267–1269.

Anderson, R. M., and May, R. M. 1982. *Population Biology of Infectious Disease.* Springer Verlag, New York.

Axtell, R. C. 1992. *Fly Management Simulator (FMS) User's Guide.* Department of Entomology, North Carolina State University, Raleigh, NC.

Bailey, N. T. J. 1982. *The Biomathematics of Malaria.* Chas. Griffin, London.

Bellows, T. S., Jr. 1981. The descriptive properties of some models for density dependence. *J. Animal Ecol.* 50: 139–156.

Birley, M. H., and Rajagopalan, P. K. 1981. Estimation of the survival and biting rates of *Culex quinquefasciatus* (Diptera: Culicidae). *J. Med. Entom.* 18: 181–186.

Birley, M. H., Walsh J. B., and Davies, J. B. 1983. Development of a model for *Simulium damnosus* s.l. recolonization dynamics at a breeding site in the OCP area when control is interrupted. *J. Appl. Ecol.* 20: 507–519.

Carey, J. R. 1993. *Applied Demography for Biologists.* Oxford University Press, New York.

Case, T. J. 2000. *An Illustrated Guide to Theoretical Ecology.* Oxford University Press, New York.

Charlesworth, B. 1980. *Evolution in Age-Structured Populations.* Cambridge University Press, Cambridge.

Clements, A. N., and Paterson, G. D. 1981. The analysis of mortality and survival rates in wild populations of mosquitoes. *J. Appl. Ecol.* 18: 373–399.

Conway, G. R., Trpis, M., and McClelland, G. A. H. 1974. Population parameters of the mosquito *Aedes aegypti* (L.) estimated by mark–release–recapture in a suburban habitat in Tanzania. *J. Anim. Ecol.* 43: 289–304.

Desena, M. L., Clark, J. M., Edman, J. D., Symington, S. B., Scott, T. W., Clark, G. G., and Peters, T. M. 1999a. Potential for aging female *Aedes aegypti* (Diptera: Culicidae) by gas chromatographic analysis of cuticular hydrocarbons, including a field evaluation. *J. Med. Entomol.* 36: 811–823.

Desena, M. L., Edman, J. D., Clark, J. M., Symington, S. B., and Scott, T. W. 1999b. *Aedes aegypti* (Diptera: Culicidae) age determination by cuticular hydrocarbon analysis of female legs. *J. Med. Entomol.* 36: 824–830.

Detinova, T. S. 1962. Age grouping methods in Diptera of medical importance. *Wld. Hlth. Org. Monogr. No.47.* WHO, Geneva.

Dobson, A., and Foufopoulos, J. 2001. Emerging infectious pathogens of wildlife. *Phil. Trans. Roy. Sot. Land. B* 356: 1001–1012.

Dye, C. 1992. The analysis of parasite transmission by bloodsucking insects. *Annu. Rev. Entomol.* 37: 1–19.

Fine, P. E. M. 1981. Epidemiological principles of vector-mediated transmission. In *Vectors of Disease Agents* (J. J. McKelvey, B. F. Eldridge, and K. Maramorosch, eds.). Praeger Scientific, New York.

Garrett-Jones, C. 1964a. Prognosis for the interruption of malaria transmission through assessment of a mosquito's vectorial capacity. *Nature* 204: 1173–1175.

Garrett-Jones, C. 1964b. The human blood index of malaria vectors in relation to epidemiological assessment. *Bull. WHO* 30: 241–261.

Gillies, M. T. 1974. Methods for assessing the density and survival of blood-sucking Diptera. *Annu. Rev. Entom.* 19: 345–362.

Goodenough, J. L., and McKinion, J. M. 1992. *Basics of Insect Modeling.* ASAE Monogr. No. 10. American Society of Agricultural Engineers, St. Joseph, MI.

Grimstad, P. R., and Haramis, L. D. 1984. *Aedes triseriatus* (Diptera: Culicidae) and La Crosse virus. III. Enhanced oral transmission by nutrient deprived mosquitoes. *J. Med. Entomol.* 21: 249–265.

Grimstad, P. R., and Walker, E. D. 1991. *Aedes triseriatus* (Diptera: Culicidae) and La Crosse virus. IV. Nutritional deprivation of larvae affects adult barriers to infection and transmission. *J. Med. Entomol.* 28: 378–386.

Hare, S. G. F., and Nasci, R. S. 1986. Effects of sublethal exposure to *Bacillus thuringiensis* var. *israelensis* on larval development and adult size in *Aedes aegypti*. *J. Am. Mosq. Control Assoc.* 2: 325–328.

Harrington, L. C., Buonaccorsi, J. P., Edman, J. D., Costero, A., Kittayapong, P., Clark, G. G., and Scott, T. W. 2001. Analysis of survival of young and old *Aedes aegypti* (Diptera: Culicidac) from Puerto Rico and Thailand. *J. Med. Entomol.* 38: 537–547.

Hess, A. D., Hayes R., and Tempelis, C. 1978. The use of forage ratio technique in mosquito host preference studies. *Mosq. News* 28: 386–389.

Kay, B. H., Boreham P., and Edman, J. D. 1979. Application of the "feeding index" concept to studies of mosquito host-feeding patterns. *Mosq. News* 39: 68–72.

Kay, B. H., Edman J. D., Fanning I. D., and Mottran, P. 1989. Larval diet and the vector competence of *Aedes aegypti*, *Culex annulirostris* and other mosquitoes (Diptera: Culicidae) for Murray Valley encephalitis virus. *J. Med. Entomol.* 26: 487–488.

Krafsur, E. S. 1984. Use of age structure to assess insecticidal treatments of face fly populations, *Musca autumnalis* DeGeer (Diptera: Muscidae). *J. Econ. Entomol.* 77: 1364–1366.

Krafsur, E. S., and Ernst, C. M. 1983. Physiological age composition and reproductive biology of horn fly populations, *Haematobia irritans irritans* (Diptera: Muscidae), in Iowa, USA. *J. Med. Entomol.* 20: 664–669.

Krafsur, E. S., Black IV, W. C., Church, C. J., and Barnes, D. A. 1985. Age structure and reproductive biology of a natural housefly (Diptera: Muscidae) population. *Environ. Entomol.* 14: 159–164.

Krafsur, E. S., Rosales, A. L., Robison-Cox, J. F., and Turner, J. P. 1992. Age structure of horn fly (Diptera: Muscidae) populations estimated by pterin concentrations. *J. Med. Entomol.* 29: 678–686.

Krafsur, E. S., Rosales, A. L., and Kim, Y. 1999. Age structure of overwintered face fly populations estimated by pteridine concentrations and ovarian dynamics. *Med. Vet. Entomol.* 13: 41–47.

Langley, P. A., Hall, M. J. R., and Felton, T. 1988. Determining the age of tsetse flies, *Glossina* spp. (Diptera: Glossinidae): An appraisal of the pteridine fluorescence technique. *Bull. Ent. Res.* 78: 387–395.

Lardeux, F., Ung, A., and Chebret, M. 2000. Spectrofluorometers are not adequate for aging *Aedes* and *Culex* (Diptera: Culicidae) using pteridine fluorescence. *J. Med. Entomol.* 37: 769–773.

Lehane, M. J., and Hargrove, J. 1988. Field experiments on a new method for determining age in tsetse flies (Diptera: Glossinidae). *Ecol. Ent.* 13: 319–322.

Lehane, P. A., and Mail, T. S. 1985. Determining the age of adult male and female *Glossina morsitans* using a new technique. *Ecol. Entomol.* 10: 219–224.

Lounibos, L. P., Rey, J. R., and Frank, J. H. 1985. *Ecology of Mosquitoes: Proceedings of a Workshop.* Florida Medical and Entomology Laboratory, Vero Beach.

Macdonald, G. 1957. *The Epidemiology and Control of Malaria.* Oxford University Press, London.

Mail, T. S., Chadwick, J., and Lehane, M. J. 1983. Determining the age of adults of *Stomoxys calcitrans* (L.) (Diptera: Muscidae). *Bull. Ent. Res.* 73: 501–525.

Massad, E. F., Coutinho, A. B., Burattini, M. N., and Lopez, L. F. 2001. The risk of yellow fever in a dengue-infested area. *Trans. Roy. Sot. Trop. Med. Hyg.* 95: 370–374.

Molineaux, L., and Gramiccia, G. 1980. *The Garki Project.* World Health Organization, Geneva.

Moore, C. G., Reiter P., and Xu, J.-J. 1986. Determination of chronological age in *Culex pipiens s.l. J. Am. Mosq. Control Assoc.* 2: 204–208.

Moore, C. G., Reiter, P., Eliason, D. A., Bailey, R. E., and Campos, E. G. 1990. Apparent influence of the stage of blood meal digestion on the efficacy of ground applied ULV aerosols for the control of urban *Culex* mosquitoes. III. Results of a computer simulation. *J. Am. Mosq. Control Assoc.* 6: 376–383.

Nasci, R. S., and Mitchell, C. J. 1994. Larval diet, adult size, and susceptibility of *Aedes aegypti* (Diptera: Culicidae) to infection with Ross River virus. *J. Med. Entom.* 31: 123–126.

Neville, A. C. 1963. Daily growth layers in locust rubber-like cuticle, influenced by an external rhythm. *J. Ins. Physiol.* 9: 177–186.

Owen, W. B. 1977. Morphology of the thoracic skeleton and muscles of the mosquito, *Culiseta inornata* (Williston) (Diptera: Culicidae). *J. Morphol.* 153: 427–460.

Paulson, S. L., and Hawley, W. A. 1991. Effect of body size on the vector competence of field and laboratory populations of *Aedes triseriatus* for La Crosse virus. *J. Am. Mosq. Control Assoc.* 7: 170–175.

Schlein, Y. 1979. Age grouping of anopheline malaria vectors (Diptera: Culicidae) by the cuticular growth lines. *J. Med. Entom.* 16: 502–506.

Service, M. W. 1993. *Mosquito Ecology: Field Sampling Methods*, 2nd ed. Elsevier Applied Science, New York.

Southwood, T. R. E. 1978. *Ecological Methods with Particular Reference to the Study of Insect Populations*, 2nd ed. Chapman and Hall, New York.

Tempelis, C. H. 1975. Host-feeding patterns of mosquitoes with a review of advances in analysis of blood meals by serology. *J. Med. Entom.* 11: 635–653.

Tyndale-Biscoe, M., and Hughes, R. D. 1969. Changes in female reproductive system as age indicators in the bushfly *Musca vetustissima* Wlk. *Bull. Entomol. Res.* 59: 129–141.

Washino, R. K., and Tempelis, C. H. 1983. Mosquito host bloodmeal identification: Methodology and data analysis. *Annu. Rev. Entomol.* 28: 179–201.

16

Use of Geographic Information System Methods in the Study of Vector-Borne Diseases

CHESTER G. MOORE AND JEROME E. FREIER

INTRODUCTION

In the study of vector-borne diseases, maps have played an important role in orienting investigators to local conditions and in guiding epidemiological activities within a study area. Today we are fortunate to have specialized methods available that extend our ability to work with spatial data well beyond paper maps. The digital world has opened new avenues in the way we look at spatial data. Geographic information systems (GIS) are a combination of computer technologies that integrate graphic elements with database information and enable the computation of spatial relationships. Use of GIS makes it possible to collect, manage, analyze, and report spatial information about vector-borne diseases. With this information, spatial relationships associated with vectors, hosts, pathogens, and the environment can be compared to learn more about the complex nature of these interactions.

Specialists in many fields use GIS, including emergency services, environmental monitoring agencies, and human health services. GIS methods are multidisciplinary, involving knowledge of physical geography, cartography, remote sensing, global positioning systems (GPS), surveying, and geodesy. GIS is a paradigm of science and technology combined into one approach. Although GIS methods have many scientific applications, the data-processing and analysis tools provided by this approach are scientifically neutral.

Therefore, it is essential that a study's design, observations, data analysis, and hypothesis formation be made with strict adherence to scientific method.

GIS methods used in the study of vector-borne diseases are aimed at identifying environmental factors responsible for pathogen transmission and survival. With GIS, the environment is dissected into a series of simplified themes, or thematic data layers. The information contained in each layer can be compared with event data, such as the location of infected animals, trap collections, and the application of control measures.

BASIC CONCEPTS OF GEOGRAPHIC INFORMATION SYSTEMS

In studying vector-borne diseases, investigators often use an ecological approach, in which the interaction between human, cultural, and natural environments is analyzed to identify factors associated with the survival of infectious agents. *Landscape ecology*, a concept first described by Carl Troll (Forman and Godron 1986), refers to the science that investigates the interactions among the biosphere, the anthroposphere, and the abiotic components of the earth's surface. The concepts and methods of landscape ecology have provided the scientific framework for the relatively new field of spatial epidemiology. Additional information on the methods and applications of spatial epidemiology can be found in Elliott et al. (2000).

With GIS methods, data models of the real world are created in two- or three-dimensional space. In constructing these models, landscape features are reduced to the simplest common unit, or *theme*, to characterize a feature. Examples of thematic data layers are terrain elevation, soil type, vegetation class, hydrologic feature, meteorological condition, population size, transportation feature, and the distribution of disease vectors over a geographic area. It is important that each data layer be aligned accurately with other data layers so that the real-world coordinates of a geographic location in one layer will be in exactly the same location in all other layers. The process of attaching geographic coordinates to spatial objects in each data layer is referred to as *georeferencing*; the process of aligning a data layer with other layers in a data set is called *registration*.

In modeling landscape features, vector and raster data formats are the principal data types used. In the case of vector data, locations are stored as pairs of X-Y coordinates, and each feature is linked to a row in a database table. Vector data use point, line, and polygon objects to model landscape features. The accompanying database contains basic information about each object. Points represent specific locations, lines have length and direction, and polygons have an area and a perimeter. With raster data, features are represented as a matrix of cells in continuous space. Because each cell contains a value, the database consists of the cell matrix itself. Each data format has advantages and disadvantages. For example, vector data appear more maplike and can be used where a high level of accuracy is needed. However, vector data are computationally more rigorous and require more specialized software tools to manage. In contrast, raster data have a simple structure, which is less intense computationally, and data can easily be generalized and classified for analysis purposes. A disadvantage to raster data is that it is generally lower in resolution and can introduce spatial inaccuracies. The decision to use one data format over the other is reversible; that is, once created, vector data can be converted to raster, and vice versa.

Every spatial object (point, line, or polygon) displayed in a GIS view is linked to an attribute database that contains an information record about each feature. Additional qualitative or quantitative data can be added to the attribute table, either directly as new fields or through attachment to other databases. Because each graphical feature represents one record in an attribute database, it is essential that one or more fields contain a unique identifier that distinguishes one record from another. The identifier is used to join database tables together, thus extending the information available from basic attributes to specialized data that can be used in complex analyses.

When a new GIS database table is created, a spatial reference is added for each record. This spatial reference may be a pair of X-Y coordinates, such as longitude-latitude, or the name of a specific geographic area. Areas commonly used for spatial referencing are state and county boundaries, census tracts, census blocks, zip code areas, and the U.S. Public Land Survey System grid. Once spatially referenced, or georeferenced, data are added to a GIS as a series of thematic layers, and information can be searched to identify patterns and trends. A collection of georeferenced themes covering a specific geographic area is referred to as a *project*. A single study may contain several projects, with each project focused on a different analysis objective. In any GIS study, it is important to determine the organizational structure of each data layer in every project and to plan how projects will be assembled for analysis.

An important aspect of using vector data is the ability to create topology. This is the establishment of spatial relationships among points, lines, and polygons. Although each object type possesses inherent information about location, length, perimeter, and area, depending on the type of object, an object's position and association with other objects are determined when topology is created. For example, when topology is created, polygons that are adjacent or lines that are connected or points contained within a polygon can be determined as spatial information. The ability to incorporate topology into spatial processes is important in searching, aggregating, splitting, and feature analysis operations.

An important early consideration in working with geographic data is determining the appropriate map scale. Map *scale* refers to the ratio of the distance between two points on a map and the earth distance between the same two points. This ratio is given as a *representative fraction* (RF). For example, a 1:25,000 scale map means that 1.0 cm on the map represents 25,000 cm on the earth's surface. Map scales are designated by the size of the resulting value when the RF is divided out. In this case, a 1:25,000-scale map results in a value of 1.0×10^{-4}. A 1:1,000,000 scale would give a value of 1.0×10^{-6}. Because 1.0×10^{-4} is a larger value than 1.0×10^{-6}, the 1:25,000 map is referred to as large scale, and a 1:1,000,000-scale map is considered small scale. Map scale is important when determining the level of detail and accuracy needed for each thematic data layer to provide information sufficient to resolve the biological issue in question. As map scales become larger, the detail and accuracy of the information increase. Consequently, in designing a study using GIS methods, map scale is an important consideration when selecting data sources. The scale of available data used in an analysis must be appropriate for the biological question under investigation.

Spatial data stored in a GIS may be output as views in computer displays, map layouts, spatial models, data reports, information for Internet map servers, spatial analysis summaries, and input into other systems. The type of output used depends on the nature of the specific information needed at a particular time. For example, a vector-borne disease study might involve a sequence of outputs designed to reveal possible spatial associations not detected by other methods. Spatial data may be output first as visual displays to generate hypotheses, leading to model building and the application of spatial statistical analyses to test hypotheses. Spatial-data summaries, analysis maps, and reports may then be exported to a spatial-data server for Internet distribution of information to users or imported into computerized decision-support systems.

GIS applications work best when data are used in an interactive way. That is, thematic data layers should be selectable and the view updated as new information is added. It should be possible to query either spatial features or the database while at the same time creating graphs, displaying images, and creating map layouts. GIS used in an interactive way allows the user to obtain and interpret the underlying numbers associated with spatial features. Because data layers are in either a vector or raster format, it is important to be able to use information in both forms simultaneously when conducting a spatial analysis.

SPATIAL DATA TYPES

Types and sources of spatial data frequently used as GIS data layers are listed in Table 16.1.

REMOTE SENSING

Remote sensing is the process of acquiring information about an object, area, or phenomenon from a distance. This process is an important source of data for GIS applications. Advantages to using remotely sensed data are that large geographic areas may be studied, including harsh environments and inaccessible areas. Because of the high calibration standards used with sensor systems, data comparisons can be made reliably, in both space and time. Advancements in sensor systems have given us the ability to observe

the earth in greater detail, and enhanced detector arrays provide increased spectral capabilities. Remotely sensed imagery provides a permanent record of landscape features in a given area at a specific time.

Remote sensors record electromagnetic radiation (EMR) that is reflected or reradiated from an object to the sensor. Most applications in vector-borne disease studies use data from sensors detecting EMR wavelengths in the ultraviolet, visible blue, visible green, visible red, near infrared, and middle infrared. Applications using longer wavelengths, such as thermal infrared, radar, and microwaves, are being used more frequently to characterize the landscape features. Radar is especially useful in penetrating cloud cover, smoke, and other atmospheric disturbances. In the characterization of vector habitats, it is best to work with data derived from sensors detecting various wavelengths.

Energy emitted from ERM sources is reflected, absorbed, and/or transmitted. The proportion of energy incident on an object that is reflected, absorbed, or transmitted depends on the nature of the object and the characteristics of energy at various wavelengths. Because objects vary substantially in the type of spectral response pattern produced, this information can be used to characterize landscape features. The energy response patterns are often referred to as a *spectral signature*. Once the spectral signature of a landscape object is determined, this information may be used to identify similar features in an area.

Remotely-sensed data can provide biophysical information directly, as in the case of location, elevation, color, chlorophyll absorption characteristics, vegetation biomass, vegetation moisture, temperature, and texture or surface roughness. This information can be applied to vector habitat studies involving vegetation, soils, rocks, water, atmosphere, and urban structure.

An important concept in remote sensing is *resolution*, which is the ability to distinguish between signals that are spatially or spectrally similar. Four types of resolution are generally considered in remote-sensing applications. (1) *Spatial resolution* is the shortest distance between two objects that still permits each object to be distinguished separately. (2) *Spectral resolution* refers to the number of specific wavelength intervals in the electromagnetic spectrum to which a sensor is sensitive. (3) *Temporal resolution* is the frequency of acquiring spectral data for a particular area. (4) *Radiometric resolution* refers to differences in the sensitivity of a detector in responding to the radiant flux reflected or emitted from an area. Although increasing resolution is generally desirable, it does increase

TABLE 16.1   Types and Sources of Spatial Data Frequently Used in GIS Data Layers

Name	Abbreviation	Data format	Source	Comments
Digital Line Graph	DLG	Vector	U.S. Geological Survey	
Digital Elevation Model	DEM	Raster	U.S. Geological Survey	
Digital Raster Graphics	DRG	Raster	U.S. Geological Survey	
Digital Orthophoto Quad	DOQ	Raster	U.S. Geological Survey	
Geographic Names Index	GNIS	Vector and text	U.S. Geological Survey	
National Global Elevation Data	GTOPO30	Raster	U.S. Geological Survey	
National Aerial Photography Program	NAPP	Raster	U.S. Geological Survey	Photos as paper prints or transparencies
Topologically Integrated Geographic Encoding and Referencing	TIGER	Vector	U.S. Census Bureau	
Natural Resources Inventory	NRI	ASCII text database	Natural Resource Conservation Service	
Geographic Approach to Planning	GAP	Raster and vector		
Advanced Very-High-Resolution Radiometer/Normalized-Difference Vegetation Index	AVHRR/NDVI	Raster	National Oceanic and Atmospheric Administration	
National Wetlands Inventory	NWI	Vector	U.S. Geological Survey	
	BASINS	Vector	Environmental Protection Agency	
Digital Chart of the World	DCW	Vector	National Imagery and Mapping Agency	
Digital Terrain Elevation Data	DTED	Raster	National Imagery and Mapping Agency	
State Soils Geographic	STATSGO	Vector	Natural Resources Conservation Service	

data-storage and processing requirements. When considering resolution, it is best to select the minimum resolution required to answer the biological question being investigated.

A practical aspect of resolution is related to the type and altitude of the sensor platform used to collect data. An effective approach is to use multistage remote sensing to acquire high-resolution data from ground and low-altitude aerial sensors. Medium- and low-resolution data are acquired from high-altitude and space-based sensors. In the multistage approach, satellite data may be analyzed in conjunction with high-altitude data, low-altitude data, and ground observations. Each successive data source might provide more detailed information over smaller geographic areas. Information extracted at any lower level of observation may be extrapolated to higher levels of observation. For example, high-resolution data may establish vegetation species found in aquatic habitats associated with a particular mosquito species; however, at medium resolution, this information may be generalized into establishing the community structure characteristic of this mosquito species.

Remotely-sensed data are well suited to incorporation into GIS as thematic layers that can be viewed and analyzed with other data layers for a specific area. However, before remotely-sensed data can be used, a series of preprocessing steps may be required to remove sensor errors and correct for terrain variation. Data are georeferenced from information obtained directly from field observations or indirectly from map sources. Contrast enhancement is often applied to raster imagery to help discern landscape features more clearly. Imagery that shows landscape features plainly can be used to make new themes specific for selected features. In addition, classification processes can be applied to the sensor values in each raster cell to organize data into classes. Sensor data that are classified using algorithms that require no additional input are referred to as *unsupervised* classifications. Those methods that use field observations to create classes from sensor values are called *supervised* classifications. Each approach to creating thematic information from remotely-sensed data is supported by a wide array of statistical methods. It is important to select methods that are appropriate for the biological issue under consideration.

SPATIAL ANALYSIS

Visualization

The process of spatial analysis begins when one looks at a map and examines the distribution of objects over a geographic area. Visualization is a useful first step in finding potential spatial patterns and trends among objects in one or more data layers. Unfortunately, spatial data displayed in map layouts can be misleading and biased; therefore, caution should be used in drawing conclusions from maps (Jacquez 1998). It is important to determine data sources and how the information was treated. Interpretations of point patterns and spatial associations should follow the same analytical procedures applied to other data; that is, testable hypotheses should be formulated that can be investigated more thoroughly using analytical procedures. GIS methods can be used to quantify observations relative to distance, direction, elevation, distribution, density, and pattern. In applying these quantitative methods, spatial relationships among landscape features can be used to characterize spatial neighborhoods, point patterns, object containment, line connectivity, and spatiotemporal changes. Typical steps in analyzing spatial data include: (1) viewing maps that show the spatial orientation and geographic distribution of selected objects, (2) stating hypotheses regarding potential spatial relationships, (3) selecting variables for analysis, (4) extracting values for specific objects or areas from each thematic layer, (5) organizing spatial data into a framework for statistical analysis, (6) applying statistical methods to the variables selected, and (7) using results of statistical analyses to accept or reject hypotheses.

Preparation of Data Layers for Analysis

Before spatial data can be analyzed, thematic layers must be prepared for analytical processing. Typical preparatory procedures include subdividing data, editing spatial features, recoding observations, and classifying variables. The process of subdividing spatial data is used to create smaller subsets that include the area of interest. Two types of processes are often used. *Splitting* is the process of extracting specific features, such as polygons containing administrative boundaries, from an area where a disease is occurring. Another method is the process of *tiling*, in which a large theme is subdivided into smaller segments called *tiles*. While tiling allows the use of smaller areas, it is a simple procedure to add tiles back to a theme as observations expand beyond their original geographic extent.

The ability to edit spatial features is valuable in analysis operations because extraneous features can be deleted from a dataset. For example, if a dataset contains polygons representing woodlots and it is known that only woodlots of a certain size are capable of supporting arbovirus activity, then a spatial delete process can be used to eliminate all woodlots from the dataset that are unlikely to maintain virus activity. Two other important spatial editing processes are *dissolve* and *merge*. In this dual operation, lines representing the border between two polygons are removed and the attributes of each area are merged into one record. An example would be a situation where arthropod vector collections are made within each county of a state. Afterwards, the data may be aggregated into northern, central, and southern regions by dissolving the polygon boundaries that divide counties in each region. The end result is a new dataset for the geographic distribution of arthropod vectors within a state based on region.

Recoding and classification are related processes that are useful in reducing data complexity, which may consist of simplifying large numbers of individual observations in preparation for analysis. *Recoding* is the process of renumbering values or renaming text, or the placing of numbers and text, into new groups or classes. An example of recoding and reclassifying involves the use population census data for a state. Human population estimates for all of counties within a state are likely to show a different value for each county. Displaying a disorganized series of large numbers makes it difficult to interpret this information. However, if three population ranges are established such that less than 50,000 equals 1, and 50,000–250,000 equals 2, and greater than 250,000 equals 3, the population values can now be recoded to new values. The recoded values can then be used to classify county populations as low, medium, or high.

Exploratory Data Analysis and Spatial Statistics

Exploratory data analysis (EDA) is applied to geographic data to look for "hot spots" in spatial patterns, create hypotheses about spatial associations, and establish zones of influence. In exploring spatial data, a variety of statistical methods are generally used; however, it is important to remember that many statistical methods are based on static situations and assume independence. Infectious diseases are not necessarily independent and move through space and time. Basic approaches to exploring spatial data are directed toward specific geographic features, continuous phenomena, numeric observations, data summaries by area, classification, and density. An investigation of *specific geographic features* refers to the exact location of spatial objects. At any given geographic location, a specific feature is either present or absent. Point, line, and polygon features can be described based on their location and proximity to other features. In analyzing discrete geographic features, attribute information about point, line, and polygon features are compared. For example, determining the number of small mammal traps placed within 500 meters of perennial streams is a comparison of discrete information contained in two different data layers. Another grouping of quantitative data includes counts, amounts, ratios, and ranks. Each type of information can be calculated from the spatial database and the results displayed.

Continuous phenomena are represented as raster data over an area. In the case of continuous phenomena, points of known information are interpolated to create a continuous grid of predicted values that cover a specific geographic area. An example is the process of collecting soil samples from a field to test for pesticide residue. The pesticide residue values from soil assays can be used to create a surface over the area in which each cell is an estimate of pesticide concentration based on the samples collected. This method can be used to estimate the location of specific sites where pesticide concentrations would be expected to be at a certain level.

Numeric observations are usually in the form of counts, ratios, and ranks. Observations in the form of counts associated with each record in a spatial database can be displayed directly or normalized relative to area or percent of total. Ratios can be derived from comparing data in two or more fields. For example, the amount of precipitation reported during the first week of June 2001 can be compared with the value for the same time period in 2000. The ratio of current year to previous year is the percentage change. This information can be used to show geographic areas that are either wetter or drier than in previous years.

In summarizing data by area, all of the features located within one or more polygons can be summarized, and then this information can be added to the attribute database. For example, if the point locations of each case are known during an outbreak, it is possible to use polygon boundaries, such as census block boundaries, to summarize the number of cases within each census block polygon. The number of cases found within each census block would appear as a new field in the census block attribute database. The census block attribute database could be merged with other census data for the blocks having one or more cases to determine specific demographic characteristics of the population at risk.

Before displaying *quantitative* information, it is important first to examine the distribution of values being included in a classification series. The most frequently used classifications are natural breaks, quantiles, equal intervals, and standard deviations (Mitchell 1999). To determine the most appropriate classification method, histogram techniques should be used to characterize the value range and frequency structure of quantitative data. Each classification method will have a significant effect in the display of the end results. Therefore, it is important that the type of classification method used be included in the description of any data analysis. An alternative to using quantitative classification methods is *qualitative* ranking, which is the process of converting observations to discrete groups, such as high, medium, and low.

When quantitative data are associated with a spatial object, GIS methods can be used to represent the quantities as different colors or as proportional symbols. In this process, the range of values is classified into a specified number of categories. Typical methods of classifying quantities are natural breaks, quantiles, equal intervals, equal area, and standard deviation. In a natural breaks classification, the class breaks are made where there is a gap between clusters of values. The number of classes chosen determines the point between values where the breaks will occur. This method is useful when values are not evenly distributed and clustered values are placed in the same class. However, this method makes it difficult to compare a distribution map with other maps derived from similar processes. In establishing a quantile classification, each class has an equal number of observations in it. This method is useful in comparing areas of similar size or when emphasizing the relative position of an observation relative to all other observations. In an *equal-interval* classification, each class has an equal range of values, which means that the difference between high and low values is the same for every class. Although this method is one of the easiest to interpret, it should be used only with data that are evenly distributed. If equal-interval methods are applied to clustered data, many aggregated observations may be placed into a few classes, while some classes will have no observations. In the *standard-deviation* classification, each class is based on the mean value of all the observations. This method is especially useful in finding observations with substantial differences from the mean and showing those observations with little change from an average value. The choice of which method to use depends on which is the most appropriate method relative to the distribution of the data.

Standard statistical methods can be applied directly to values in data layers, or to values derived from processing data layers, to yield measures that can be compared. The types of comparisons used may be as fundamental as 2×2 contingency tables, comparing presence or absence of disease with the presence or absence of a risk factor. Or the process may involve regression analysis of a series of variables.

In the analysis of point clusters, three general categories of spatial statistics are usually applied, referred to as global, local, and focused statistics. In the case of *global* statistics, one or more clusters may occur anywhere in the study area. *Local* statistics apply if a cluster occurs in specific locations, such as administrative subdivisions within a study area. If locations are prespecified, then *focused* statistics are used. Examples of global statistics include *K*-function tests, Grimson's method, Cuzick and Edwards tests, the join count, Moran's *I*, and Geary's *C*. Specific examples of local statistics include the spatial scan test, LISA statistics, and the geographical analysis machine tests. More information about these spatial statistical tests can be found in Bailey and Gatrell (1995) and Venables and Ripley (1999).

Spatial clusters are not always easy to identify, especially if the data are noisy or have a background population that makes it difficult to distinguish different groups. A major concern of point data is the possible influence of one point on other nearby points. When influence by nearby points is suspected, *spatial neighbor analysis* procedures should be applied. One method is to *tessellate* points to create polygon boundaries around each point. Polygons created this way can be used to calculate spatial autocorrelation indices, such as Moran's *I* and Geary's *C*. Indices of spatial autocorrelation can be used to estimate the likely influence neighboring points have on each other. Another method used to analyze the influence of spatial clusters is *intensity analysis*. This technique calculates the number of points per unit area, which is then expressed numerically in terms of intensity. The measure of intensity can be displayed on a map as a raster grid or as a series of contour lines.

In examining spatial point patterns, it is important to determine if an aggregation of points departs from complete spatial randomness (CSR). A frequently used collection of methods used to evaluate CSR are the *K*-function tests (Ripley 1976, 1981). The *F*-hat and *G*-hat tests indicate the probability of clustering when the distances relative to each statistic are small. Related tests, such as the *K*-hat and *L*-hat, also plot a calculated statistic relative to distance. However, these tests also include a plot of values expected from a process that simulates values if CSR is occurring. When observed *K*-hat and *L*-hat values fall outside of the confidence interval of the simulated values, then a point pattern can be considered to be clustered.

Overlay Analysis

Overlay analysis involves the use of information in two or more thematic data layers. Boolean methods are used to establish criteria for selecting information to be processed. These methods can be applied to both vector and raster data. An example would be when an aerial application of a mosquito larvicide was made over an area and the treated site was defined by a polygon. If potential larval breeding sites in a mosquito control district were previously mapped and stored in a GIS as polygons, then overlay analysis could be used to determine the percentage of the overall breeding area covered by the treatments and to calculate the amount of larvicide reaching the target areas.

Two principal overlay processes are intersect and union. *Intersect* is the process of comparing two or more data layers to determine spatial areas that share common space. For example, using two data layers, one being geographic boundaries of seven states in the southeastern United States and the other being a layer showing ecoregions in the Unites States, an intersection would result in a southeastern states data layer containing attribute records listing the ecoregions contained within each state's boundaries. In this case, the theme doing the intersecting, the Southeastern states, defines the geographic extent of the end result. However, in the *union* process, the full geographic extent of both input themes is maintained.

Some specialized GIS operations associated with intersect and union processes are clipping, masking and replacing, and database merging. In *clipping*, a selected part of one theme is used to remove part of another theme. Points, lines, and polygons can be used to clip parts of other spatial objects. *Masking and replacing* operations are similar to clipping, in that specific features in the masking theme contain information that can be applied to specific areas of another theme. In *database merging*, a common field is used to join two databases together.

Overlay analysis with raster data is somewhat different because cells are numerically coded for specific information. Mathematical operators are used to add, subtract, multiply, divide, or apply other operations to values in two or more data layers to create a new raster that contains a resulting value in each cell, based on the results of the algebraic equation applied.

Special Analysis Methods

Creating space around features is a fundamental technique in GIS, and the most common method used is creating a buffer. *Buffers* are zones around a point, line, or polygon feature that are set back a specified distance either inside or outside of a feature. In spatial analysis, buffers can be used to select objects within a given distance of a specific feature. For example, if the point location of a larval development site is known and the adult flight range of a mosquito species that develops in that site is known, a buffer can be created around the larval point that reflects the distance adult mosquitoes might travel from the larval breeding site.

Another way to create polygon space around a point object is the use *polygon-fitting* methods. Many of these methods, originally developed to estimate the home range of wildlife, are currently used to identify centers of activity within a cluster of points. In this approach, polygon-fitting algorithms are used to define space around observation points such that

groups of data points are organized into clusters with perimeters delineated by polygons. In epidemiological studies, polygon-fitting methods can be used to identify areas with the greatest frequency of disease occurrence. Although a variety of polygon-fitting methods have been developed, each technique uses different statistical methods and input parameters to determine the size and shape of the resulting polygons. Because each method gives a different result, it is important that the method selected be based on the type of data available. Those methods applicable to infectious diseases establish a series of vertical and horizontal grid lines over the data points and then a statistical surface is created using a density function. Polygon-fitting results are useful in identifying centers of activity, which may be significant in determining potential sources of infection in an outbreak.

Although buffer and polygon-fitting methods are useful in creating area data from spatial objects, such as points, polygon centroids can be used to create point data from geographic areas. *Centroids* are the center points calculated from a polygon's dimensions. All attributes in the polygon's database record are transferred to the centroid's attribute table. Points are the most common spatial objects used in disease investigations, and the creation of a buffer around a point is perhaps the simplest level of analysis that can be conducted. Buffers are based on a setback distance or radius and form a polygon boundary around a point. Buffers can be created at various distances from a point to form concentric rings. If buffers are created for several points and the rings overlap, the overlapping areas can be dissolved to form a single buffer polygon.

Proximity Analysis

Proximity analysis involves methods aimed at features that are close together. Typical considerations are distances from A to B, how many features are within a selected distance from a feature, and determining the distance between two or more features. Distance calculations are fundamental processes in GIS. Distances may be calculated either as straight-line (Euclidean) or as linear network distances, in which calculations consider distance along a linear feature, such as a road or stream. Using methods similar to buffering, it is possible to select all farms located within a specified radius. Point-to-point distances are commonly determined using techniques such as the spider diagram. With this technique, the distance to all points from one or more selected locations is determined. The straight-line distance is added as a database field, and a graphic is created showing point-to-point lines representing distance.

Although GIS methods work well with highly detailed data, this information can be generalized to provide a less detailed view of highly specific data. The processes of aggregation, generalization, and clustering are used to take detailed information and simplify the spatial organization of landscape features. For example, if a point data layer shows the location of all insect traps in an area, it is possible to aggregate traps according to the habitat being sampled. While the spatial database has records for all traps, aggregation procedures allow the separation of trap information into forest, marsh, fallow field, pasture, windbreak, and barnyard traps. Use of spatial aggregation as a generalization tool can provide organizational structure to the data being analyzed.

MODELING SPATIAL DATA

Spatial data models are based mostly on calculations derived from different data layers in an overlay analysis. Data processed in models are typically in a raster format. However, data in a vector format can be converted to a raster so that each cell representing an enclosed polygon (or cells shown along a line or cells representing points) has a specific numeric value associate with the object's attribute. Raster cells are used in modeling because each cell can be manipulated mathematically, especially through the use of algebraic functions. In addition, Boolean logic schemes can be applied to two or more data layers to determine how each layer will be used in an analysis.

Model building begins with the selection of a study area, followed by identifying those components necessary to produce the desired result. Each component of a model can be weighted according to its level of influence or the component's contribution to the probability of occurrence. When the model is run, data are automatically brought into the model, converted, edited, and manipulated mathematically to yield the end results. The results are in the form of a raster grid, and each cell reflects the outcome specified by the model.

Another type of modeling is the creation of surface models that are used to predict the occurrence of an event based on a series of observations. Surface models are functional, raster-based surfaces derived from estimates of three-dimensional data by combining geospatial information with a quantitative value (Z), such as elevation, population density, or trap capture numbers. Examples of surface modeling methods include minimum curvature, inverse distance weighted, contouring, profiling, polynomial, triangulation, meshing, and kriging. Each method is aimed at specific applications, and the method selected should be based on the nature of the data being modeling and the type of output needed.

DISTRIBUTIVE GIS

One of the most exciting new developments in GIS has been the creation of geodata servers that allow users to access specific information over the Internet. In the case of Internet map servers, data reside on a server and the client, using an interface viewer, selects specific data to be viewed in an Internet browser window. The map request can be from a list or from a more sophisticated query of several data layers. Once the map request is completed, it is then sent to the server, where the appropriate data are assembled into a view, an image file of that view is created, and the image is then sent back to the client. Depending on the nature of the client–server arrangement, actual data can also be distributed in this way. In some cases, the server-created view can be combined with thematic data layers that the user has on the computer client.

APPLICATIONS OF GEOSPATIAL METHODS TO RESEARCH IN VECTOR-BORNE DISEASES

The use of geospatial methods as a research tool has increased greatly during the past decade, and the role of these methods in studies on Lyme disease, human babesiosis, LaCrosse encephalitis, and malaria is summarized in a review by Albert (2000). After reviewing geospatial studies involving these vector-borne diseases, it was concluded that GIS and related geospatial tools are effective in population studies, risk mapping, predictive modeling, and the analysis of both social and environmental factors related to infectious diseases. The application of specific types of geospatial methods was reviewed in greater detail by Kitron (1998). Emphasized here is the importance of landscape ecology to epidemiologic studies in vector-borne diseases and the integration of geospatial tools to provide a flow of spatial data from a variety of sources for analysis purposes.

Examples of using satellite imagery to estimate the risk of malaria transmission are described by Beck et al. (1994, 1997). The authors applied discriminant analysis procedures to Landsat TM imagery to identify villages in southern Chiapas, Mexico, with a high risk of vector–human contact. Predictive models were developed to identify from satellite imagery those villages likely to have high mosquito abundance.

Similar studies on vector–environmental relationships using multispectral satellite data were reported by Roberts et al. (1996), leading to the development of predictive models of mosquito populations, particularly for *Anopheles pseudopunctipennis*. Hay et al. (1998) used multitemporal meteorological satellites to predict periods in Kenya when malaria transmission was likely to be greatest. These predictions were based mostly on correlations between mosquito abundance and an Advanced Very-High-Resolution Radiometer (AVHRR) satellite–derived index of vegetation biomass. Based on studies of malaria vectors and transmission dynamics, the application of geospatial methods to vector surveillance and control was reviewed by Washino and Wood (1994) and by Sharma and Srivastava (1997). Based on their observations, it was concluded that remotely-sensed data provides an important source of information to identify vector habitats and predict population abundance. It was also concluded that remotely-sensed data should be stored within a GIS to allow comparisons with environmental data and observations about vector populations.

The integration of vector habitat and population data with remotely-sensed data has been described in detail by Dale and Morris (1996) and Dale et al. (1998) for studies conducted in southeast Queensland, Australia. These authors used a combination of data sources to map breeding habitats for *Aedes vigilax* and *Culex annulirostris* to assess the risk of disease transmission by populations of these mosquito species. Meteorological data were used to establish critical times and locations when salt marsh flooding led to abundant populations of *Ae. vigilax*, and color infrared aerial photography was used to determine patterns of mangrove flooding that affected the abundance of *Cx. annulirostris*. The resulting spatial data models were used to plan effective strategies for vector management.

Data collected by the AVHRR on the National Oceanic and Atmospheric Administration's (NOAA) polar-orbiting meteorological satellites were used by Linthicum et al. (1987) to assess ecological observations associated with Rift Valley fever (RVF) viral activity in Kenya. A normalized-difference vegetation index (NDVI) was derived from the AVHRR data, and the resulting values showed that, in geomorphologic depressions called *dambos*, those areas remaining wetter and greener longer than normal were more likely to have abundant populations of mosquito species associated with epizootics of RVF virus. Subsequent studies by Pope et al. (1992) showed that the flooding status of dambos could be evaluated by airborne imaging radar. More recently, Linthicum et al. (1999) compared NDVI data with El Niño–Southern Oscillation (ENSO) indices to show that these indicators could be used to predict RVF outbreaks up to 5 months in advance of an outbreak in East Africa.

The use of geospatial methods to predict the size and distribution of tick populations has been the subject of many studies. To forecast the occurrence of *Ixodes ricinus*, a vector of tick-borne encephalitis virus in Europe, Daniel and Kolar (1990) and Daniel et al. (1998) used Landsat 5 multispectral and TM data to correlate landscape classes with tick presence and to assess the risk of human exposure. Peridomestic risk for Lyme disease was evaluated by Dister et al. (1997) using Landsat TM data to characterize landscape composition in two communities in suburban Westchester County, New York. Remote-sensing and GIS methods were used to characterize vegetation structure, moisture, and abundance–associated *Ixodes scapularis* ticks and their hosts. The authors concluded that a remote-sensing, GIS-based approach was an efficient way to characterize peridomestic risk of Lyme disease over large geographic areas. The application of geostatistics, especially cokriging, in modeling both vegetation and climate variables to predict the distribution of *I. scapularis* ticks was reported by Estrada-Peña (1998). Simultaneous modeling NDVI values and temperature variables produced distribution estimates with high levels of sensitivity and specificity.

A geographic information system was used by Glass et al. (1994, 1995) to predict *I. scapularis* abundance on white-tailed deer and to determine residential risk factors associated with Lyme disease in selected counties in Maryland. In these studies stepwise linear regression and logistic regression methods were used to identify environmental variables as risk factors associated with tick abundance and the presence of Lyme disease. Kitron and Kazmierczak (1997) used a combination of human infection data, tick survey data, and NOAA-AVHRR–derived NDVI data to analyze spatial patterns of Lyme disease risk in Wisconsin. Within a GIS, risk maps were created to show counties where risk of Lyme disease transmission was expected to be greatest. The use of GIS to investigate environmental determinants of tick distributions has been summarized by Randolph (2000). While this author advocates the application remotely-sensed data stored in a GIS for tick studies, the importance of using appropriate statistical tools is emphasized.

CONCLUSIONS

Use of spatial data in vector-borne disease investigations requires a suite of methods that may be collectively thought of as spatial data systems. Vector and raster data, including remotely-sensed data, are brought into a geographic information system's structure for data storage, maintenance, and retrieval. The key to GIS is in having one or more spatially enabled databases that link vector or raster objects with tabular information. Once information is stored in a GIS, spatial analysis and modeling can utilize this data to describe patterns, test hypotheses, and predict trends.

Geographic information systems provide an excellent means of managing and visualizing spatial data; however, displays of spatial data should be interpreted with caution and skepticism because of potential bias and interpretation problems. Appropriate analytical methods must be used to explore spatial data before drawing conclusions regarding potential associations. To evaluate the nature of spatial influences, it is best to build models that integrate data from a variety of sources. Finally, the Internet is an ideal way to provide spatial data users with rapid access to real-time information and to make interactive GIS a reality for everyone.

Readings

Albert, D. P. 2000. Infectious diseases and GIS. In *Spatial Analysis, GIS, and Remote Sensing Applications in the Health Sciences* (D. P. Albert, W. M. Gesler, and B. Levergood, eds.). Sleeping Bear Press, Ann Arbor, MI, pp. 111–127.

Bailey, T. C., and Gatrell, A. C. 1995. Interactive spatial data analysis. Addison Wesley Longman, Edinburgh.

Beck, L. R., Rodriguez, M. H., Dister, S. W., Rodriguez, A. D., Rejmankova, E., Ulloa, A., Meza, R. A., Roberts, D. R., Paris, J. F., and Spanner, M. A. 1994. Remote sensing as a landscape epidemiologic tool to identify villages at high risk of malaria transmission. *Am. J. Trop. Med. Hyg.* 51: 271–280.

Beck, L. R., Rodriguez, M. H., Dister, S. W., Rodriguez, A. D., Washino, R. K., Roberts, D. R., and Spanner, M. A. 1997. Assessment of a remote sensing–based model for predicting malaria transmission risk in villages of Chiapas, Mexico. *Am. J. Trop. Med. Hyg.* 51: 99–106.

Dale, P. E., and Morris, C. D. 1996. *Culex annulirostris* breeding sites in urban areas: Using remote sensing and digital image analysis to develop a rapid predictor of potential breeding areas. *J. Am. Mosq. Contr. Associ.* 12: 316–320.

Dale, P. E., Ritchie, S. A., Territo, B. M., Morris, C. D., Muhar, A., and Kay, B. H. 1998. An overview of remote sensing and GIS for surveillance of mosquito vector habitats and risk assessment. *J. Vector Ecol.* 23: 54–61.

Daniel, M., and Kolar, J. 1990. Using satellite data to forecast the occurrence of the common tick *Ixodes ricinus* (L.). *J. Hyg. Epidemiol. Microbiol. Immunol.* 34: 234–252.

Daniel, M., Kolar, J., Zeman, P., Pavelka, K., and Sadlo, J. 1998. Predictive map of *Ixodes ricinus* high-incidence habitats and a tick-borne encephalitis risk assessment using satellite data. *Exp. Appl. Acarol.* 22: 417–433.

Dister, S., Fish, D., Bros, S. M., Frank, D. H., and Wood, B. L. 1997. Landscape characterization of peridomestic risk for Lyme disease using satellite imagery. *Am. J. Trop. Med. Hyg.* 57: 687–692.

Elliott, P., Wakefield, J. C., Best, N. G., and Brigs, D. J. 2000. *Spatial Epidemiology: Methods and Applications.* Oxford University Press, Oxford.

Estrada-Peña, A. 1998. Geostatistics and remote sensing as predictive tools of tick distribution: A cokriging system to estimate *Ixodes scapularis* (Acari: Ixodidae) habitat suitability in the United States and Canada from advanced very-high-resolution radiometer satellite imagery. *J. Med. Entomol.* 35: 989–995.

Forman, R. T. T., and Godron, M. 1986. *Landscape Ecology.* Wiley, New York.

Glass, G. E., Amerasinghe, F. P., Morgan, J. M., and Scott, T. W. 1994. Predicting *Ixodes scapularis* abundance on white-tailed deer using geographic information systems. *Am. J. Trop. Med. Hyg.* 51: 538–544.

Glass, G. E., Schwartz, B. S., Morgan, J. M., Johnson, D. T., Noy, P. M., and Israel, E. 1995. Environmental risk factors for Lyme disease identified with geographic information systems. *Am. J. Pub. Health* 85: 944–948.

Haines-Young, R., Green, D. R., and Cousins, S. 1993. Landscape ecology and GIS. Taylor and Francis, London.

Hay, S. I., Snow, R. W., and Rogers, D. J. 1998. Prediction of malaria seasons in Kenya using multitemporal meteorological satellite sensor data. *Trans. Roy. Soc. Trop. Med. Hyg.* 92: 12–20.

Jacquez, G. M. 1998. GIS as an enabling technology. In *GIS and Health* (A. C. Gatrell and M. Loytonen, eds.). Taylor and Francis, London, pp. 17–28.

Kitron, U. 1998. Landscape ecology and epidemiology of vector-borne diseases: Tools for spatial analysis. *J. Med. Entomol.* 35: 435–445.

Kitron, U., and Kazmierczak, J. J. 1997. Spatial analysis of the distribution of Lyme disease in Wisconsin. *Am. J. Epidemiol.* 145: 558–566.

Linthicum, K. J., Bailey, C. L., Davies, F. G., and Tucker, C. J. 1987. Detection of Rift Valley fever viral activity in Kenya by satellite remote sensing imagery. *Science* 235: 1656–1659.

Linthicum, K. J., Anyamba, A., Tucker, C. J., Kelley, P. W., Myers, M. F., and Peters, C. J. 1999. Climate and satellite indicators to forecast Rift Valley fever epidemics in Kenya. *Science* 285: 397–400.

Limp, W. F. 2000. Put the "Fizz" into "Data Viz." *Geoworld* 13(9): 40–45.

Mitchell, A. 2000. *The ESRI Guide to GIS Analysis. Volume 1: Geographic Patterns and Relationships.* ESRI Press, Redlands, CAL.

Pope, K. O., Sheffner, E. J., Linthicum, K. J., Bailey, C. L., Logan, T. M., Kasischke, E. S., Birney, K., Njogu, A. R., and Roberts, C. R. 1992. Identification of central Kenyan Rift Valley fever virus vector habitats with Landsat TM and evaluation of their flooding status with airborne imaging radar. *Remote Sens. Environ.* 40: 185–196.

Randolf, S. E. 2000. Ticks and tick-borne disease systems in space and from space. In *Advances in Parasitology: Remote Sensing and Geographical Information Systems in Epidemiology*, vol. 47 (S. I. Hay, S. E. Randolph, and D. J. Rogers, ed.). Academic Press, London, pp. 217–243.

Ripley, B. D. 1976. The second-order analysis of stationary point processes. *J. Appl. Prob.* 13: 255–266.

Ripley, B. D. 1981. *Spatial Statistics.* Wiley, New York.

Roberts, D. R., Paris, J. F., Manguin, S., Harbach, R. E., Woodruff, R., Rejmankova, E., Polanco, J., Wullschleger, B., and Legters, L. J. 1996. Predictions of malaria vector distribution in Belize based on multispectral satellite data. *Am. J. Trop. Med. Hyg.* 54: 304–308.

Sharma, V., and Srivastava, A. 1997. Role of geographic information systems in malaria control. *Ind. J. Med. Res.* 106: 198–204.

Venables, W. N., and Ripley, B. D., 1999. *Modern Applied Statistics with S-Plus*, 3rd ed. Springer-Verlag, New York.

Vink, A. 1980. *Landschapsecologie en Landgebruik.* Bohn, Scheltema and Holkema, Utrecht.

Washino, R. K., and Wood, B. L. 1994. Application of remote sensing to arthropod vector surveillance and control. *Am. J. Trop. Med. Hyg.* 50: 134–144.

17

Invasive Species and Emerging Vector-Borne Infections

CHESTER G. MOORE AND JEROME E. FREIER

INTRODUCTION

A complete discussion of invasive species and emerging infections would be far too large to cover in a single chapter. Here we provide a brief historical perspective, some important definitions, and a general overview of some major issues, with examples from medical and veterinary entomology. Readers are referred to the Readings at the end of the chapter for additional information. In particular, there are several good reviews of invasive species issues (U.S. Congress Office of Technology Assessment 1993; Sakai et al. 2001; Lounibos 2002) and of emerging infections (Lederberg et al. 1992; Morse 1993, 1995; Wilson et al. 1994).

HISTORY OF INVASIONS AND EMERGENCE

From the beginning of recorded history, plagues of disease have affected human populations. No doubt many of these epidemics were the result of introductions of exotic pathogens, acquired as a result of contact with traders or other "outsiders." Beginning with the age of exploration in the 15th century, Europeans began a several-hundred-year process of spreading pathogens and disease vectors around the globe (see, e.g., Crosby 1994; Diamond 1997; Elton 1972). Today, with the advent of intensive international commerce and jet air transportation, people and goods

are distributed in increasing numbers at ever-faster rates. Human travel has increased both in speed and in volume. It is now possible to reach essentially any place on earth from any other place within 36 hours, well under the incubation period for the majority of infectious agents. Over 500 million travelers cross international borders annually by commercial aircraft alone. Mass migrations of refugees, workers, and displaced persons—especially into urban centers—increase the frequency of contact between infected and susceptible individuals. With this increasing movement comes an increase in the number and variety of pathogens and disease vectors that are introduced into new environments.

The movement of people and cargo within and between nations and continents serves as the primary means of moving species beyond their normal range. For example, Hughes (1961) and his colleagues in the U.S. Public Health Service Division of Quarantine recovered nearly a quarter of a million insects, including 20,692 mosquitoes from aircraft entering seven major airports in the United States. Since the time of Hughes' study, air traffic has increased greatly. For example, the total number of passenger enplanements in the United States grew from about 326 million in 1979 to over 710 million in 2000 (Fig. 17.1A). This includes passengers flying within the United States as well as those flying to foreign destinations. Shipments of cargo have grown in a similar fashion, increasing from 25.6 million tons in 1986 to 74.9 million tons in 2000 (Fig. 17.1B). Finally, the increasing speed of movement of air and sea transport makes it possible for

219

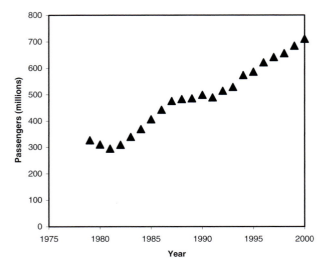

FIGURE 17.1A   Change in airline passenger volume from 1979 to 2000. *Source*: Federal Aviation Administration, *Historical Summary of Enplanement and All-cargo Data*.

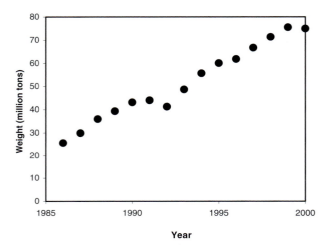

FIGURE 17.1B   Change in the amount of cargo moved by aircraft, 1986–2000. *Source*: Federal Aviation Administration, *Historical Summary of Enplanement and All-cargo Data*.

example, by commerce (shipping of goods by air, land or water), via tourism or business travel, and by military exercises (war, peacekeeping efforts, etc.).

Because cycles of vector-borne disease are complex processes that involve multiple species, it is relatively difficult for vector-borne pathogens to colonize new areas. Several conditions must be present to support the establishment of a vector-borne disease system. One or several underlying habitats or landscape elements (characterized by soil type, rainfall patterns, disturbance regimes, dominant plant species, and other characteristics) determine the ultimate potential range of the disease system. Suitable habitats may be patchily distributed within a surrounding matrix that is incompatible with one or more components of the disease system. Transmission of the pathogen is supported only where all components are present in time and space. Barriers such as rivers, oceans, deserts, forests, and grasslands can prevent the movement of vectors or vertebrate hosts between otherwise-suitable habitats. Human activities may eliminate barriers, creating opportunities for invasion of new localities.

Surveillance systems form a crucial part of disease-monitoring programs. Data must be shared internationally in near real time for proper international response. World Wide Web–based reporting systems will form an important link in this effort. Advances in molecular biology permit us rapidly to identify the source of new introductions. Techniques such as geographic information systems and remote sensing (Chapter 16) permit us to identify areas most at risk for the introduction and survival of a given vector species. Two recent introductions into the United States (West Nile virus, a human and animal disease agent, and *Aedes albopictus*, a mosquito capable of transmitting several diseases of humans) are used to illustrate the important features of detection, prevention, and control of exotic disease agents and disease vectors.

vectors, as well as infected humans or animals, to reach distant locations before they are detected.

Invading species, including disease vectors and human pathogens, can disperse in several ways: gradually from initial site of introduction to surrounding sites, or saltatorily; that is, to spatially disjunct areas in a patchy fashion. Plague and cholera provide examples of human pathogens that exhibit complex movement, as when they are transported by sea or air between continents and then spread locally in the usual manner. Humans can transfer vectors; or pathogens from point to point in several ways; for

DEFINITIONS

Invasive species: An invasive species is defined as a species that is (1) nonnative (or alien) to the ecosystem under consideration and (2) whose introduction causes or is likely to cause economic or environmental harm or harm to human health (Invasive Species Council 2001, Appendix 1). Organisms that have been moved from their native habitat to a new location are typically referred to as *nonnative*, *nonindigenous*, *exotic*, or *alien* to the new environment. Most U.S. food crops and domesticated animals are nonnative species that are not invasive. Many other nonnative species are simply benign. However, some nonnative species cause serious problems in their new environments and are collectively known as *invasive species*. Two invasive species of public health importance are the mosquitoes *Aedes albopictus* and *Ochlerotatus japonicus*.

Vectors and *pathways*: In some of the invasive species literature, the term *vector* is used to describe the conveyance (e.g., ship, plane) that moves an organism from one location to another. This has created confusion in the public health community, where the term vector has a different meaning. The U.S. Invasive Species Council (2001) has proposed the use of the term *pathway* to describe the means and routes by which invasive species are imported and introduced into new environments. *Vector* is reserved for organisms that transmit a disease agent from one host to another.

Emerging infection: Emerging infectious diseases are "clinically distinct conditions whose incidence in humans has increased ... within the past two decades" (Lederberg et al. 1992). Emergence can result from the introduction of a new agent (e.g., West Nile virus in the United States), recognition of an existing disease that has gone undetected (e.g., hantaviruses, La Crosse virus), or changes in the environment that provide increased contact between natural cycles and humans or domestic animals (e.g., Lyme disease, La Crosse virus).

CHARACTERISTICS OF INVASIVE SPECIES

Sakai et al. (2001) provide an extensive discussion of the population biology of invasive species. Successful introduction and invasion encompasses several stages: (1) introduction into a new habitat, (2) initial colonization and establishment, and (3) subsequent dispersal and secondary spread into new habitats.

While arthropod vectors can be dispersed to new habitats by natural means (e.g., ticks and mites on migrating birds), most, if not all, intercontinental introductions of arthropod vectors is the direct or indirect result of human activities. Thus, species that have close association with humans, either as ectoparasites of humans or their domestic animals (ticks, mites, lice) or as peridomestic or domestic species (e.g., some mosquitoes, triatomes, and cimicids), are the most likely candidates for dispersal to new regions of the globe. Additionally, species that deposit their eggs on items of commercial value may be dispersed through commercial activities. A particular example of this dispersal pathway is provided by the Asian "tiger mosquito," *Aedes albopictus*. This mosquito has been

collected at U.S. ports from scrap tires shipped from Asia (Craven et al. 1988) and in shipments of "lucky bamboo" (*Dracaena* spp.), also from Asia.

Once introduced, the colonizing species must become established. Good colonizers have several traits (Sakai et al. 2001). Arthropod vector species that (1) store sperm from an initial fertilization, (2) have multiple reproductive strategies (e.g., depositing small numbers of eggs in or on each of several habitats or hosts), (3) have high phenotypic plasticity (e.g., ability to respond to a wide range of photoperiods or ambient temperatures), and (4) have strong competitive ability are more likely to be successful colonists.

Once established, the invasion is accomplished by secondary spread to new locations. Secondary spread may result from (1) additional introductions from the original foreign source, (2) from long-distance transport (particularly by humans), or (3) by short-distance dispersal. Each of these pathways is documented for arthropod vectors. In molecular studies of the newly introduced mosquito *Ochlerotatus japonicus*, evidence indicates the presence of two distinct genetic populations, suggesting multiple introductions (Fonseca et al. 2001). The early distribution of *Ae. albopictus* in the United States was strongly associated with the interstate highway system (Moore and Mitchell 1997), suggesting human transport as a major means of dispersal. The original diffusion of *Ae. albopictus* within Harris County (Houston), Texas (Sprenger and Wuithiranyagool 1986), probably resulted from a combination of local dispersal by human activities and normal flight dispersal by the mosquito.

INVASIVE SPECIES OF MEDICAL AND VETERINARY IMPORTANCE

As already noted, invasions by insects and other organisms have occurred for centuries. Medically important vectors are no exception. One of the oldest invading vectors, the yellow fever mosquito, *Aedes aegypti*, probably entered the New World shortly after the Spanish monarchs granted permission to import African slaves to the New World in 1501 (Picó 1969). Two other exotic mosquitoes, *Culex pipiens* and *Cx. quinquefasciatus*, probably arrived aboard early sailing ships (Ross 1964). Recently, two mosquitoes, *Ae. albopictus* and *Oc. japonicus*, have been introduced into North America (Sprenger and Wuithiranyagool 1986; Peyton et al. 1999), with potential major public health implications.

From its initial appearance in Houston, Texas, in 1985, *Ae. albopictus* spread rapidly northward and eastward (Moore et al. 1988). By 1999, this species had been reported from 919 counties in 26 states (Moore 1999). The initial introduction probably was through shipments of used tires from Asia (Hawley et al. 1987; Craven et al. 1988), and the interstate movement of used tires probably played a role in the rapid dispersal of the species (Moore and Mitchell 1997). More recently, *Ae. albopictus* arrived at West Coast ports in shipments of "lucky bamboo" (*Dracaena* sp.) from Asia (Linthicum et al. 2002). The recent distribution of *Ae. albopictus* is shown in Figure 17.2A.

In 1998, another mosquito of Asian origin, *Ochlerotatus j. japonicus*, was collected by local mosquito control inspectors at locations in New Jersey and New York (Peyton et al. 1999). By 2003, populations had been reported in 144 counties in 12 states (Fig. 17.2B). The pathway of introduction and routes of dispersal are similar to those for *Ae. albopictus*.

Ticks are another group of arthropods that are easily transported from country to country. For example, studies in Florida (Burridge et al. 2000) found several exotic tick species on imported reptiles, including several vectors of heartwater (*Cowdria ruminantium*), a serious disease of cattle, sheep, goats, and deer. A review of the literature and unpublished records of the U.S. National Tick Collection (Keirans and Durden 2001) found 99 exotic tick species representing 11 genera. A significant number of those tick species are known or suspected vectors of human or animal pathogens.

EMERGING INFECTIONS OF HUMANS AND DOMESTIC ANIMALS

Some "emerging" diseases are best described as *resident* (also called *indigenous* or *autochthonous*) emerging diseases; that is, they probably have been established in a particular area for a long time. These agents suddenly become a problem because humans intrude into the natural focus of the zoonotic cycle, a new vector appears in the area that can serve as a bridge between the zoonotic cycle, the underlying ecology of the region changes in such a way that the zoonotic focus comes to overlap the human environment, or any of a number of similar changes. Good examples of resident emerging diseases are La Crosse (LAC) encephalitis and Lyme disease.

LAC virus was first identified in 1964 from a fatal case of encephalitis in a 4-year-old girl from Minnesota who died in 1960 from severe encephalitis in a La Crosse, Wisconsin, hospital (Henderson and Coleman 1971). Initially, nearly all cases of LAC encephalitis were reported from the upper midwest (Iowa, Minnesota, Wisconsin, Michigan, Illinois, Indiana, and Ohio). In recent years, increasing numbers of cases have been reported from the mid-Atlantic and southeastern states (West Virginia, Tennessee, Virginia, North Carolina). It is unclear if this represents dispersal of the LAC virus into new areas, the incursion of humans into areas where the virus has always been present, or the involvement of new vectors, such as *Aedes albopictus* (Nasci et al. 2000; Gerhardt et al. 2001).

Lyme disease (*Borrelia burgdorferi*) and human babesiosis (*Babesia microti*), two tick-borne diseases, have recently emerged as major public health problems in the United States. In this case, emergence appears to be a result of reforestation in the northeastern United States, coupled with a tremendous increase in the populations of white-tailed deer (*Odocoileus virginianus*) in the region. Rapid reforestation during the

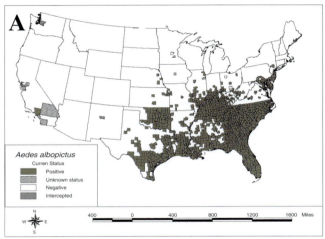

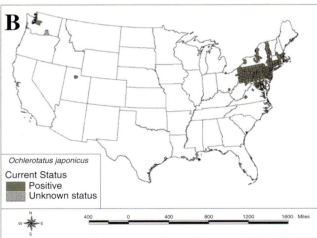

FIGURE 17.2 Distribution of two recently introduced mosquito species in the United States. (A) *Aedes albopictus*; (B) *Ochlerotatus j. japonicus*. Data from Centers for Disease Control and Prevention.

20th century created increased habitat for deer, and deer are a major host of adult *Ixodes scapularis* ticks, the vectors of both parasites. The relationship between deer abundance and Lyme disease has been demonstrated by recent exclusion experiments, in which deer are denied access to an area for extended periods of time (Fish 1995). Under these conditions, the incidence of Lyme disease decreased significantly.

Other examples of resident emerging infections transmitted by arthropods are Powassan virus, ehrlichiosis, and bartonellosis. Although not transmitted by arthropods, the hantaviruses are another important group of emerging pathogens.

The second group of emerging infections comprises those agents newly introduced into an area. West Nile virus is a good example. These agents are most likely moved from country to country by human activities, through transportation of infected vectors, livestock, or infected human hosts. Migrating birds offer an additional potential pathway of introduction.

West Nile virus (WNV), a member of the Japanese encephalitis serogroup, is closely related to St. Louis encephalitis (SLE) virus, which is widespread throughout the Americas. In fact, when it first appeared in the Borough of Queens in New York City in 1999, it was thought to be an outbreak of SLE. Since that original appearance, WNV has spread across the entire United States (Figs. 17.3A and B) and into Canada, Mexico, and the Caribbean. In 2002, over 4,000 human cases of WNV infection were reported to the CDC and over 14,700 equine cases were documented. Mortality in crows was widespread. Corvids and raptors appear to be particularly impacted by this virus, with potential severe impacts on endangered and threatened bird species.

Other examples of potential introduced exotic agents include Rift Valley fever, Japanese encephalitis, Ross River virus. Dengue, malaria, and yellow fever occasionally are brought into the United States by travelers from endemic areas. Suitable vectors for all of these agents are found in the United States, and there is a potential for sustained outbreaks of infection.

STRATEGIES AND CONSIDERATIONS IN SURVEILLANCE, PREVENTION, AND CONTROL

There are many ways in which countries can reduce the threat of importation of alien species or exotic pathogens. Several of the most important issues are described here.

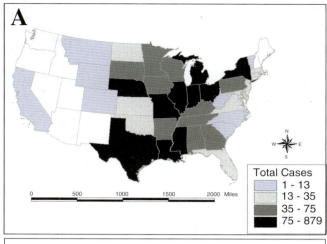

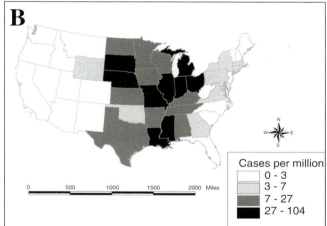

FIGURE 17.3 Distribution of laboratory-confirmed human West Nile infection in the United States, 2002. (A) Total cases by state; (B) Cases per million inhabitants (1999 census estimate). Case data from CDC (2003).

International Cooperation

It is in the interest of all nations to reduce the movement of vectors, either as free-living passengers in planes or ships or as eggs or larvae on or in agricultural or other commercial products. International programs, such as the Global Invasive Species Program (GISP) and the United Nations Convention on Biological Diversity, offer a forum for such cooperation.

Early Detection

It is extremely difficult to eradicate a vector or pathogen once it has become widely distributed. The cost of inaction during the earliest stages of introduction can be enormous. Ongoing active surveillance, both for vectors and for diseases, is crucial for early detection and response. Unfortunately, such programs are expensive, and many, such as the U.S. Public Health Service Quarantine program of the 1950s and '60s (e.g., see Hughes 1961), have been eliminated or drastically reduced in scope.

Communication and Data Sharing

The development and widespread availability of electronic communication via the World Wide Web provides an unparalleled means of rapid communication. Resources such as ProMED (http://www.promedmail.org/) provide rapid, worldwide sharing of information on diseases, vectors, and other relevant issues. International databases are being developed (e.g., the North American Non-Indigenous Arthropod Database—http://www. invasivespecies.org/NANIAD.html), but most databases tend to focus on agriculture, forestry, or other specific areas. This limits the utility of the data, since disease vectors, for example, may share an introduction pathway with an alien plant or animal species of concern to another regulatory agency. Since different databases may use different data formats or collect different information, it is not always possible to merge the information contained in different packages.

SUMMARY

Invasive vectors and emerging pathogens present a growing challenge to the vector-borne-disease community. Successful prevention and control of these problems is likely to result only by integrating knowledge over all scales, from molecular biology to landscape ecology, to gain a complete understanding of the system. Such an approach will identify the most sensitive points for intervention.

Readings

Burridge, M. J., Simmons, L. A., and Allan, S. A. 2000. Introduction of potential heartwater vectors and other exotic ticks into Florida on imported reptiles. *J. Parasitol*. 86: 700–704.

CDC. West Nile virus update, current case count (2002 data reported through March 12, 2003). Centers for Disease Control and Prevention, Division of Vector-Borne Infectious Diseases. Available at: http://www.cdc.gov/od/oc/media/wncount.htm (last accessed March 20, 2003).

Craven, R. B., Eliason, D. A., Francy, D. B., Reiter, P., Campos, E. G., Jakob, W. L., Smith, G. C., Bozzi, C. J., Moore, C. G., Maupin, G. O., et al. 1988. Importation of *Aedes albopictus* and other exotic mosquito species into the United States in used tires from Asia. *J. Am. Mosq. Control Assoc*. 4: 138–142.

Crosby, A. W. 1994. *Germs, Seeds, and Animals: Studies in Ecological History*. M.E. Sharpe, Armonk, NY.

Diamond, J. 1997. *Guns, Germs, and Steel: The Fates of Human Societies*. Norton, New York.

Elton, C. C. 1972. *The Ecology of Invasions by Animals and Plants*. Chapman and Hall, New York.

Fish, D. 1995. Environmental risk and prevention of Lyme disease. *Am. J. Med*. 98: 2S–8S [discussion 8S–9S].

Fonseca, D. M., Campbell, S., Crans, W. J., Mogi, M., Miyagi, I., Toma, T., Bullians, M., Andreadis, T., Berry, R. L., Pagac, B., Sardelis, M. R., and Wilkerson, R. C. 2001. *Aedes (Finlaya) japonicus* (Diptera: Culicidae), a newly recognized mosquito in the United States: Analyses of genetic variation in the United States and putative source populations. *J. Med. Entomol*. 38: 135–146.

Gerhardt, R. R., Gottfried, K. L., Apperson, C. S., Davis, B. S., Erwin, P. C., Smith, A. B., Panella, N. A., Powell, E. E., and Nasci, R. S. 2001. First isolation of La Crosse virus from naturally infected *Aedes albopictus*. *Emerg. Infect. Dis*. 7: 807–811.

Hawley, W. A., Reiter, P., Copeland, R. S., et al. 1987. *Aedes albopictus* in North America: Probable introduction in used tires from northern Asia. *Science* 236: 1114–1116.

Henderson, B. E., and Coleman, P. H. 1971. The growing importance of California arboviruses in the etiology of human disease. In *Progress in Medical Virology* (Melnick, J. L., ed.). S. Karger, Basel, pp. 404–461.

Hughes, J. H. 1961. Mosquito interceptions and related problems in aerial traffic arriving in the United States. *Mosq. News* 21: 93–100.

Invasive Species Council. 2001. *Meeting the Invasive Species Challenge: National Invasive Species Management Plan*. U.S. Government Printing Office, Washington, DC. Also available at: http://www.invasivespecies.gov/council/nmp.shtml.

Lederberg, J., Shope, R. E., and Oaks, Jr., S. C. (eds.) 1992. *Emerging Infections: Microbial Threats to Health in the United States*. National Academy Press, Washington, DC.

Linthicum, K. J., Kramer, V., and Surveillance Team. 2002. Update on *Aedes albopictus* infestations in California. *Vector Ecol. Newsl*. 33: 8–10.

Lounibos, L. P. 2002. Invasions by insect vectors of human disease. *Annu. Rev. Entomol*. 47: 233–266.

Moore, C. G. 1999. *Aedes albopictus* in the United States: Current status and prospects for further spread. *J. Am Mosq. Control Assoc*. 15: 221–227.

Moore, C. G., and Mitchell, C. J. 1997. *Aedes albopictus* in the United States: Ten-year presence and public health implications. *Emerg. Inf. Dis*. 3: 329–334.

Moore, C. G., Francy, D. B., Eliason, D. A., and Monath, T. P. 1988. *Aedes albopictus* in the United States: Rapid spread of a potential disease vector. *J. Am. Mosq. Control Assoc*. 4: 356–361.

Morse, S. S. (ed.) 1993. *Emerging Viruses*. Oxford University Press, New York.

Morse, S. S. 1995. Factors in the emergence of infectious diseases. *Emerg. Inf. Dis*. 1: 7–15.

Nasci, R. S., Moore, C. G., Biggerstaff, B., Liu, H. Q., Panella, N. A., Karabatsos, N., Davis, B., and Brannon, E. 2000. La Crosse encephalitis virus habitat associations in Nicholas County, West Virginia. *J. Med. Entom*. 37: 559–570.

Peyton, E. L., Campbell, S. R., Candeletti, T. M., Romanowski, M., and Crans, W. J. 1999. *Aedes (Finlaya) japonicus japonicus* (Theobald), a new introduction into the United States. *J. Am. Mosq. Control Assoc*. 15: 238–241.

Picó, R. 1969. *Nueva Geografía de Puerto Rico, Física, Económica y Social*. Editorial Universitaria, Rio Piedras, PR.

Ross, H. H. 1964. The colonization of temperate North America by mosquitoes and man. *Mosq. News* 24: 103–118.

Sakai A. K., Allendorf, F. W., Holt, J. S., Lodge, D. M., Molofsky, M., With, K. A., Baughman, S., Cabin, R. J., Cohen, J. E., Ellstrand, N. C., McCauley, D. E., O'Neil, P., Parker, I. M., Thompson, J. N., and Weller, S. G. 2001. Population biology of invasive species. *Annu. Rev. Ecol. Syst.* 32: 305–332.

Spielman, A. 1994. The emergence of Lyme disease and human babesiosis in a changing environment. In *Disease in Evolution: Global Changes and Emergence of Infectious Diseases* M. E. Wilson, R. Levins, and A. Spielman (eds.). Ann. NY Acad. Sci. 740: 146–156.

Sprenger, D., and Wuithiranyagool, T. 1986. The discovery and distribution of *Aedes albopictus* in Harris County, Texas. *J. Am. Mosq. Control Assoc.* 2: 217–219.

U.S. Congress Office of Technology Assessment. 1993. *Harmful Non-Indigenous Species in the United States.* OTA-F-565. U.S. Government Printing Office, Washington, DC.

Wilson, M. E., Levins, R., and Spielman, A. (eds.) 1994. *Disease in Evolution: Global Changes and Emergence of Infectious Diseases.* Ann. New York Acad. Sci. Vol. 740.

18

Molecular Techniques for Epidemiology and the Evolution of Arthropod-Borne Pathogens

WILLIAM C. BLACK IV AND M. D. SALMAN

INTRODUCTION

Epidemiology is the study of the patterns and the determinants of the spatial and temporal distributions of a disease. Epidemiology is essential in implementing control and preventive measures in a specific population. Furthermore, detection of or measuring disease and its etiological agents in a specific population is crucial in the process of understanding disease characteristics and behavior (Galen and Gambino 1975). *Molecular epidemiology* is the use of molecular genetic and biochemical markers to detect and identify pathogen species and to genetically characterize individual pathogen isolates and strains. Applied to vector biology, molecular epidemiology can rapidly identify arthropods that carry specific pathogen species and strains and thus estimate the rate of the spread (i.e., *incidence*) with which infected vectors occur in a particular location over a specified interval of time. In Chapter 15, we introduced McDonald's model as a general concept with which to understand the vectorial capacity of a population. The *basic reproductive ratio* (R_0) is a related, comprehensive measure of the transmissibility or spreading potential of an infectious agent in a population. It is defined as the average number of infected insects in a totally susceptible population during a defined period. Conceptually,

$$R_0 = b \times k \times D \tag{1}$$

where b is the infectivity rate, k is the number of potentially infectious contacts the average infected insect has per unit time, and D is the duration of infectivity

of an insect. An approximation of R_0 can be calculated as

$$R_0 = \frac{1 + \text{avg. life span of an arthropod vector in the population}}{\text{avg. age at infection}} \tag{2}$$

If $R_0 < 1$, then every new generation of infection will affect fewer individuals, and eventually the disease will die out. The value of R_0 and the percentage of the population that is refractory to infection ultimately determine whether the disease spreads or dies out. Patterns of change in incidence through time can provide a first approximation of R_0 and thus provide an epidemiologist data with which to predict if a disease is becoming established or will eventually become extinct in a population.

In recent years we have seen an explosion in the diversity of molecular genetic techniques for the detection of individual genotypes in a pathogen species, and this has, in turn, allowed epidemiologists to trace the origins of individual cases or outbreaks of arthropod-borne diseases. Ultimately, molecular epidemiology seeks to understand the emergence of pathogens and requires an understanding of how these pathogens have adapted to live in and be transmitted by the arthropod vector. Evolutionary relationships among pathogen species and isolates have been estimated, and we are beginning to understand the ways that arthropod-borne diseases have evolved and will continue to adapt to life in vertebrate and invertebrate hosts.

227

We begin this chapter with a discussion of epidemiological principles and tools and describe why an understanding of these principles is critical to the development of molecular epidemiological methods. A decade ago, there was a limited number of molecular genetic techniques available for the rapid detection and identification of arthropod-borne pathogens. The current molecular techniques used to identify species and differentiate strains of the most common arthropod-borne pathogens are listed in Table 18.1. We briefly define each of these techniques and then describe some recent applications of these tools to molecular epidemiological investigations of arthropod-borne diseases.

It is important to remember that arthropod-borne pathogens are not static entities, as shown by the recent emergence (or in many cases, reemergence) of known species and the appearance of altogether-new species. We, therefore, finish this chapter with a discussion of arthropod-borne pathogen evolution, focusing on adaptations that have enabled survival and transmission between alternating vertebrate and invertebrate hosts. We discuss briefly the concept of the *quasi-species* as a model for understanding the ways that arboviruses and possibly arthropod-borne prokaryotes have evolved to survive in alternating hosts. In the case of protozoan and helminth pathogens, we discuss models of evolution in sexually reproducing species. We finish the chapter with a description of very recent studies in which molecular epidemiology has been used to identify pathogen genes, or lesions within genes, associated with host specificity and virulence.

EPIDEMIOLOGICAL TOOLS TO ASSESS THE VALIDITY AND RELIABILITY OF MOLECULAR TECHNIQUES

Prior to utilizing a molecular technique for detection of a disease agent it is crucial to assess the validity and reliability of the technique, and classical epidemiological methodologies can be used for this type of assessment. When a molecular technique is applied toward detection of a disease agent, the outcome is either positive or negative. A positive result does not necessarily mean that the agent is present, only that there is the potential for the presence of the agent. For example, a positive result may be due to a cross-reaction with other agents similar to the disease agent. A test may also be positive at a given time due to improper laboratory sample handling or other errors. Similarly, a test may be negative when the

vector is actually infected. A test may not be sufficient to detect small quantities of the agent, particularly, for example, in the early stages of an infection. As with a positive test, a negative result could also be due to lab or sample-handling errors.

Screening tests involve the presumptive identification of an unrecognized disease agent by application of simple diagnostic tests to sort out infected vectors that may carry the disease agent from those that probably do not carry the agent. Therefore, screening tests are applied to apparently uninfected vectors in search of the disease agent. *Unlike prevalence surveys, done to measure the number of infected vectors, screening is mostly done with the objective of early detection of disease.* As a general rule, screening tests are applied to a large number of organisms and are often followed by a *diagnostic test (confirmatory test)* on those organisms that are found to be positive. When a screening test is applied to a "high-risk group" of organisms, it is defined as *case finding. High risk* means that organisms are suspected or known to have a higher prevalence of the disease as compared to the total population. They may be at a higher risk because of factors such as age, location, genetic components of vector competence (Chapter 30), or an increased exposure to disease agents. The aim of a screening test is usually for early detection, which leads to prevention, early treatment, and/or control.

Properties of Tests

Most tests are not generally 100% accurate in their ability to correctly identify infected or uninfected organisms. This is a problem of *misclassification.* The accuracy of a test can be measured and expressed by its ability to correctly classify organisms according to their disease status. These measures are termed *sensitivity* (Se) and *specificity* (Sp). *Sensitivity* is the probability that a test will correctly identify infected organisms. *Specificity* is the probability that a test will correctly identify uninfected organisms. To establish these two test attributes, the test must be conducted on organisms for which the disease status is known. The results can be tabulated in a 2-by-2 table from which Se and Sp can then be calculated.

Figure 18.1 is a 2-by-2 table for a generic disease. Viral or bacterial isolation is usually considered the "gold standard" for the definitive diagnosis of the presence of a disease agent (designated by D+) or for its absence (D–). The positive (T+) and negative (T–) test results are determined by one of the many screening or diagnostic procedures listed in the second part of this chapter.

TABLE 18.1 Current Molecular Techniques for Detecting Pathogens in Vectors

Taxonomic group	Method	Reference
Arboviruses		
Alphaviruses	Nested RT-PCR	Pfeffer et al. 1997
Chikungunya	RT-PCR	Pfeffer et al. 2002
Eastern equine encephalitis	RT-PCR	Huang et al. 2001
Western equine encephalitis	RT-PCR colorimetric	Linssen et al. 2000
Venezuelan equine encephalitis	RT-PCR	Linssen et al. 2000
Bunyaviruses	RT-PCR	Moreli et al. 2001
La Crosse virus	RT-PCR	Wasieloski et al. 1994
California serogroup	RT-PCR	Campbell and Huang 1996
Crimean-Congo hemorrhagic	RT-PCR	Burt et al. 1998
Rift Valley fever	RT-PCR, real-time PCR	S. Garcia et al. 2001
Flaviviruses	RT-PCR	Meiyu et al. 1997
Dengue	TaqMan PCR	Warrilow et al. 2002
Japanese encephalitis	RT-PCR	Paranjpe and Banerjee 1998
St. Louis encephalitis	RT-PCR	Kramer et al. 2001
West Nile	RT-PCR, real-time PCR	Shi et al. 2001
Yellow fever	Nested RT-PCR	Deubel et al. 1997
Orbiviruses		
African horsesickness	RT-PCR	Bremer and Viljoen 1998
Bluetongue	RT-PCR	Billinis et al. 2001
Bacteria		
Borrelia spp.	Real-time PCR, Reverse line blot	Rauter et al. 2002
Borrelia burgdorferi	PCR	Kahl et al. 1998
Borrelia recurrentis	PCR	Roux and Raoult 1999
Borrelia hermsii	PCR	Picken 1992
Yersinia pestis	PCR	Engelthaler et al. 1999
Francisella tularensis	PCR	Dolan et al. 1998
Rickettsia		
Anaplasma marginale	Nested PCR	Lew et al. 2002
Bartonella quintana	PCR immunoassay	La Scola et al. 2001
Cowdria ruminantium	PCR	Mahan et al. 1998
Ehrlichia equi	Nested PCR	Barlough et al. 1996
Human granulocytotropic ehrlichia	Nested PCR	Favia et al. 2001
Rickettsia tsutsugamushi	Nested PCR	Kelly et al. 1994
Rickettsia rickettsii	Nested PCR	Gage et al. 1992
Rickettsia prowazekii	PCR	Roux and Raoult 1999
Rickettsia typhi	PCR	Webb et al. 1990
Protozoa		
Babesia spp.	PCR, reverse line blot	Georges et al. 2001
Babesia bigemina	PCR	Figueroa et al. 1992
Babesia bovis	PCR	Fahrimal et al. 1992
Babesia equi, B. caballi	PCR	Bashiruddin et al. 1999
Babesia microti	PCR	Okabayashi et al. 2002
Theileria annulata	PCR	Kirvar et al. 2000
Theileria parva	PCR	Shayan et al. 1998
Theileria sergenti	PCR	Tanaka et al. 1993
Plasmodium falciparum	PCR	Moody 2002
Plasmodium malariae	PCR	Rubio et al. 1999b
Plasmodium vivax	PCR	Parkes et al. 2001
Plasmodium ovale	PCR	Rubio et al. 1999a
Trypanosoma cruzi	PCR	Hamano et al. 2001
Trypanosoma brucei	PCR	Masake et al. 2002
Trypanosoma evansi	PCR	Desquesnes et al. 2001
Trypanosoma rangeli	PCR	Machado et al. 2000
Trypanosoma vivax	PCR	Masake et al. 2002
Leishmania (Viannia) spp.	PCR	Weigle et al. 2002
Leishmania braziliensis	PCR	Aviles et al. 1999
Leishmania donovani	PCR	Salotra et al. 2001
Leishmania infantum	PCR-ELISA	Martin-Sanchez et al. 2001
Leishmania tropica	PCR	Chiurillo et al. 2001
Leishmania mexicana	PCR	Breniere et al. 1999
Nematodes		
Brugia malayi	PCR	Hoti et al. 2001
Dirofilaria immitis	PCR	Watts et al. 1999
Onchocerca volvulus	PCR	Zhang et al. 2000
Loa loa	PCR	Toure et al. 1998
Wuchereria bancrofti	Quantitative PCR	Fischer et al. 1999

This is not an exhaustive list of references for detection of any one pathogen species. Preference was given to the most recent references because these generally list the earlier references. Also, preference was given to techniques that identify the pathogen in the vector.

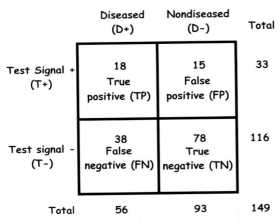

FIGURE 18.1 A 2-by-2 table for a diagnostic test of a generic disease. Viral or bacterial isolation was used to determine if the vector carried (D+) or did not carry the disease (D–). The positive (T+) and negative (T–) test results are determined by one of the many screening or diagnostic procedures listed in the second part of this chapter.

The following information is obtained from Figure 18.1. The total positive tests in the screening were 33 and the total negative tests were 116. The total number of diseased organisms was 56, and 93 nondiseased organisms were sampled. Those organisms that are misclassified by the screening test are either *false positives* (FP) (15 organisms who were not diseased but were positive for the screening test) or *false negatives* (FN) (38 vectors who were diseased but were negative in the screening test). Those organisms that were correctly identified (accuracy of the test) were *true positives* (TP) (18 diseased individuals were positive in the screening test) or were *true negatives* (TN) (78 nondiseased organisms were test negative).

Calculating Properties of Screening Tests

For most screening tests, Se and Sp are not known, and the consequences of misclassification must be understood. Interpretation of misclassified results depends upon the purpose of the test and upon the person who does the interpretation. *True prevalence* is the proportion of organisms tested that were diseased as assessed by the "gold standard":

$$\text{True prevalence} = \frac{D+}{N} \qquad (3)$$

where N is the total number of vectors tested. For our generic disease, true prevalence would be 56/149 = 37.6%. *Apparent prevalence* is the proportion of organisms that are positive in the screening test (also known as the test positive rate):

$$\text{Apparent prevalence} = \frac{T+}{N} \qquad (4)$$

For our generic disease, the apparent prevalence would be 33/149 = 22%.

We are also frequently interested in comparing *accuracy* and *misclassification*. *Accuracy* is the proportion of organisms correctly identified by a test:

$$\text{Accuracy} = \frac{TP + TN}{N} \qquad (5)$$

In our example the accuracy is 96/149 = 64.4%. *Misclassification* is the proportion of those organisms incorrectly identified as diseased by the test:

$$\text{Misclassification} = \frac{FP + FN}{N} \qquad (6)$$

In our example the misclassification is 53/149 = 35.6%. *Sensitivity* is the proportion of truly diseased organisms (D+) that the test correctly identifies (those among the diseased population that test positive). Sensitivity is calculated as

$$Se = \frac{TP}{TP + FN} \qquad (7)$$

In our example, Se = 18/56 = 32%. In other words, 32 out of 100 infected organisms will be positive on the screening test. *Specificity* is the proportion of the nondiseased organisms that the test correctly identifies (those among the nondiseased population that tested negative). Specificity is calculated as

$$Sp = \frac{TN}{TN + FP} \qquad (8)$$

In our example, Sp = 78/93 = 84%. This means that out of 100 nondiseased organisms, 84 will be negative on the screening test. Once Se and Sp are estimated, *true prevalence* can be recalculated as

$$\text{True prevalence} = \frac{\text{apparent prevalence} + Sp - 1}{Sp + Se - 1} \qquad (9)$$

In our example,

$$\text{True prevalence} = \frac{0.22 + 0.84 - 1.00}{0.84 + 0.32 - 1.00} = \frac{0.06}{0.16} = 37.5\%$$

This is very close to the true prevalence calculated in (3).

The *proportion of false positives* is the proportion of truly nondiseased vectors that the test misidentifies as positive:

$$\text{Proportion of false positive} = \frac{FP}{TN + FP} \qquad (10)$$

The proportion of false positives for the generic disease is 15/93 = 16%. This means that on the average 16 out of 100 truly nondiseased organisms will have a positive test. The *proportion of false negatives* is the proportion of diseased organisms that the test misidentifies as negative:

$$\text{Proportion of false negative} = \frac{FN}{TP + FN} \qquad (11)$$

The proportion of false negatives for the generic disease is $38/56 = 68\%$. On average 68 out of 100 infected organisms will have a negative test.

Predictive Values

The positive predictive value (PV+) is an important property of a test. It indicates the proportion of tested-positive organisms that are infected. It is the probability that a positive test result is correct.

$$PV+ = \frac{TP}{TP + FP} \qquad (12)$$

In Figure 18.1, $PV+ = 18/33 = 54\%$. There is a 54% chance that an individual is truly infected if it has a positive test result. The *negative predictive value (PV−)* is the probability that a negative test result is correct. It is calculated as

$$PV- = \frac{TN}{TN + FN} \qquad (13)$$

In our example, there is a 67% chance ($PV- = 78/116$) that an uninfected individual has a negative test result.

PV+ is closely related to specificity and PV− is closely related to sensitivity. PVs indicate test accuracy, given that the test result is known. Both PVs depend on the prevalence of the disease in the population and the Sp and Se of the test used. To see this, let us consider another way to calculate PVs:

$$PV = \frac{Se \times \text{True Prev.}}{(Se \times \text{True Prev.}) + (FP \times \text{Prev. nondiseased})} \qquad (14)$$

Thus, for example, if the true prevalence of an arbovirus in a vector population is 0.1% and we analyze 1,000 vectors, then 1 vector will be positive. If our test has Se = 0.95 and Sp = 0.95, then

$$PV+ = \frac{0.95 \times 1}{(0.95 \times 1) + (0.05 \times 999)} = 0.0187 = 1.87\%$$

However, if the true prevalence = 50%, then PV+ = 95%. Similarly:

$$PV- = \frac{Sp \times \text{True Prev. nondis.}}{(Sp \times \text{True Prev. nondis.}) + (FN \times \text{True Prev.})} \qquad (15)$$

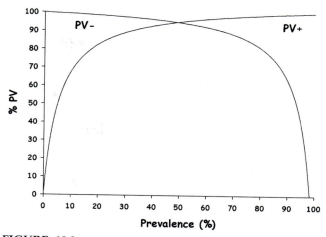

FIGURE 18.2 Predictive values of positive and negative test results vary directly with the prevalence of disease when Sp and Se are held constant.

If, again, the true prevalence of an arbovirus in a vector population is 0.1% and we analyze 1,000 vectors, then 999 vectors will not carry the virus. If our test has Se = 0.95 and Sp = 0.95, then

$$PV+ = \frac{0.95 \times 999}{(0.95 \times 999) + (0.05 \times 1)} = 0.999 = 99.9\%$$

However, if the true prevalence = 50%, then PV+ = 95%. Calculated in this way, the predictive values of positive and negative test results vary directly with the prevalence of disease when the Sp and Se are held constant (Fig. 18.2).

Evaluating the Usefulness of a Test

Figure 18.3 indicates the results of a screening test done by any of the techniques discussed in the second half of this chapter. Figure 18.3A shows the range of test readings (1–12 units) for vectors with and without the disease. Vectors without the disease produce normally distributed test results with a mean of 3 units and a range from 0 to 6 units, while disease-carrying vectors have normally distributed test results with a mean of 5 units and a range from 1 to 9 units. The epidemiologist chooses a criterion of 5 units for differentiating disease-carrying from disease-free organisms. Under these criteria, there are 92 true negatives, 1 false positive, 29 true positives, and 27 false negatives. Sensitivity is 51.3%, PV− = 77.2%, Sp = 98.9%, and PV+ is 96.6%. Se is low because there is too much overlap in the distributions of test results between diseased and nondiseased vectors. This is, then, a "poor" test.

Figure 18.3B shows the distributions of readings for a different test between vectors with and without the

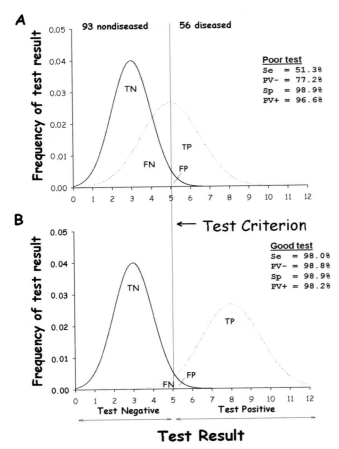

FIGURE 18.3   Results of a screening test done by any of the techniques discussed in the second half of this chapter: (A) Range of test readings for vectors with and without the disease with a "poor" test; (B) distributions of readings for a "good" test between vectors with and without the disease.

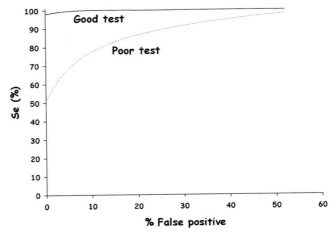

FIGURE 18.4   Sensitivity of a test is directly related to the number of false positives.

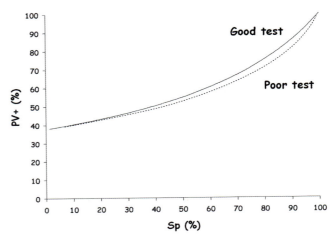

FIGURE 18.5   Relationship between specificity and PV+ of a test.

disease. The distribution for organisms without the disease is as in Figure 18.3A, while diseased vectors have normally distributed test results but with a mean of 8 units and a range between 4 and 12 units. The epidemiologist again chooses a criterion of 5 units for differentiating diseased from nondiseased vectors. Under these criteria, there are again 92 true negatives and 1 false positive but there are 55 true positives, and only 1 false negative. Sensitivity is 98.0%, PV− = 98.8%, Sp = 98.9%, and PV+ is 98.2%. Se is high because there is little overlap in the distributions of test results from diseased and nondiseased vectors. This is, then, a "good" test.

One can, of course, manipulate Se and Sp simply by selecting lower or higher criteria. For example, if we wanted to increase Sp for a screening test, the test criterion could be increased. This would decrease the false positives and increase the false negatives. The

sensitivity of a test is directly related to the number of false positives. This relationship is indicated for screening the poor and good tests in Figure 18.4. Conversely, the specificity is related to the number of false negatives.

In general, as the specificity of a test increases, PV+ increases, and therefore the probability that a positive result is true also increases (Fig. 18.5). Specificity allows more confidence in a positive test. Conversely, as the sensitivity of a test increases, PV− increases, and therefore the probability that a negative result is correct increases (Fig. 18.6). Sensitivity allows more confidence in a negative test.

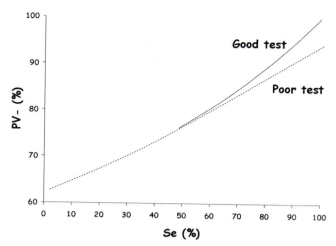

FIGURE 18.6   Relationship between sensitivity and PV– of a test.

Choosing Between Tests

It is generally recommended to use a test with high Se and PV– when it is advantageous to "rule out" a diagnosis in the early stages of a diagnostic workup to decrease the possible number of individuals to treat. Since there is more confidence in a negative test, this will allow confidence that those not treated (because they tested negative) will not spread the disease. High Se and high PV– are also recommended when FN is dangerous. For example, FN of a vector entering this country with an exotic disease would have serious consequences.

Alternatively, it is generally recommended to use a test with high Sp and PV+ when it is advantageous to confirm a diagnosis. Since there is more confidence in a positive test, this allows those that should be treated to be confidently determined. High Sp and high PV+ are again recommended when FP is dangerous or expensive. For example, if test and quarantine policy are the measures taken following a positive result, the cost of too many FPs could be quite high.

If screening tests are performed with the purpose of identifying cases for treatment, it is desirable for the test to have high PV+. Otherwise, a large proportion of individuals would be treated unnecessarily. Since screening aims to find potential cases of one specified disease (i.e., in the form of a diagnosis), the PV of a positive test can be termed its *diagnositability*. However, it is also very desirable for screening tests used in the early stages of a control program to be highly sensitive (so that there are few FNs) and the test used in the latter stages to be highly specific (to decrease FPs). This is especially true when prevalence is low (2%), when most of the organisms are free of the disease and the results of even a highly sensitive and specific test will include a large number of false positives.

Testing in Series and in Parallel

Testing in series means that the results of every test run must be positive or the vector is considered negative for disease. Therefore, only positive samples from a preceding test will be tested with the next test. For testing in series, it is desired to increase overall Sp and PV+ so that there is confidence in the final positive result. *Testing in parallel* means that the results of every test must be negative or the individual is considered positive for the disease. Therefore, only negative samples from a preceding test will be tested with the next test. It is desirable to increase overall Se and PV– so that there is confidence in the final negative result. *Test batteries* involve running all the available tests and panels for disease. The more tests administered, the greater the probability of detecting a false positive.

METHODS FOR DETECTION OF ARTHROPOD-BORNE PATHOGENS

Culturing of Arthropod-Borne Pathogens

Many virologists, microbiologists, and protozoologists consider isolation and culturing, or at a minimum visual identification, of a pathogen to represent the "gold standard" in assessing its presence in populations or in individual arthropods. Because of the time and expense involved in isolating a pathogen from single insects, from 25 to 50 arthropods are usually "pooled" and homogenized and the pathogen is then isolated from the homogenate to estimate the *minimum field infection rate*. This is the number of insects in which the pathogen can be isolated divided by the total number of insects that are exposed to the pathogens.

A valid concern with molecular genetic methods is that while they detect, with great sensitivity, the presence of pathogen nucleic acids, this may not correlate with the abundance of viable organisms. For example, an abundance of arboviral RNA usually occurs in the salivary gland of an infectious vector, but only a small proportion of this resides in viable infectious particles. Furthermore, because of issues of vector competence (discussed in Chapter 30), a vector may be infected but incapable of transmission. While culturing or dissection can detect the stage of pathogen development, molecular genetic methods only indicate the presence of pathogen nucleic acid.

Another issue concerns the pathogen gene used for detection. Often the probe or primers used for detection are specific for a genus of pathogen (e.g., *Borrelia* spp., rickettsia spp.) but are not species specific. Thus, molecular techniques might detect closely related but possibly nonpathogenic species. As sequence information becomes available for additional genes for more and more species, this concern will diminish.

For all of these reasons we agree that pathogens should be cultured from infected invertebrates and invertebrate hosts whenever possible. However, there are a number of practical issues associated with relying solely on isolation, visualization, and/or culturing for assessing pathogen presence. First, methods for culturing new pathogens may not be established. Sin nombre virus was detected with reverse transcriptase-PCR (RT-PCR) and sequence analysis months before techniques were established for culturing the virus (Nichol et al. 1993). Second, methods for culturing new pathogens may be slow and thus prevent rapid detection and intervention. Third, often it is difficult to maintain live pathogens in field-collected vectors because the pathogen either dies during the collection process or dies prior to the time it can be processed in the laboratory. In contrast, molecular genetic techniques can identify pathogen nucleic acids in ethanol-preserved and often dead, dry vectors. Fourth, a great deal of adaptation may be associated with the process of culturing a pathogen, and this may greatly underrepresent the amount of genetic variation present in nature. Norris et al. (1997) compared the frequencies of flagellin, 66-kDa, and outer surface protein A alleles among 71 *Borrelia burgdorferi* isolates in *Ixodes spinipalpis* ticks. He compared variability in *B. burgdorferi* DNA that had been cultured in BSK-II medium prior to DNA extraction with DNA that had been extracted directly from infected *I. spinipalpis* ticks. The frequencies of the *p66* and *ospA* alleles were significantly different between cultured and uncultured spirochetes, and the number of three-locus genotypes and the genetic diversity of alleles at all loci were consistently lower in cultured spirochetes, suggesting that culturing of *B. burgdorferi* in BSK-II medium selects for specific genotypes.

Multiple Locus Electrophoresis of Cultured Prokaryotic and Eukaryotic Pathogens

The earliest molecular epidemiological method that was applied to cultured prokaryotic and eukaryotic pathogens involved analysis of genetic variation at multiple genetic loci. Multiple locus electrophoresis

(MLE) typing of pathogens involves growing the isolate to high titer, centrifuging the organisms into a pellet, and lyzing the pellet to gently release soluble enzymes and/or to isolate nucleic acids. For isozyme electrophoresis, the homogenate is then subject to electrophoresis on several nondenaturing gels (Chapter 33). A different histochemical stain (Chapter 33) is applied to each gel, and the resultant band patterns are recorded. For restriction fragment length polymorphism (RFLP) analysis, the isolated DNA is digested with one or more restriction enzymes, Southern-blotted to a membrane, and probed with labeled insertion sequences or repetitive elements to genetically characterize cultured prokaryotic and eukaryotic pathogens (Bachellier et al. 1994). The investigator usually places homogenates or digested DNA of the same or closely related species that had previously been characterized alongside homogenates or DNA of the new isolate. This allows molecular epidemiologists to build large databases of information on genetic relatedness among isolates and closely related species and in many cases allows epidemiologists to trace the origins of individual cases or epidemics of arthropod-borne diseases.

The issues discussed earlier with culturing of arthropod-borne pathogens apply to MLE. In addition, MLE cannot be applied to arboviruses or rickettsia. Nevertheless, large and useful MLE databases have been developed for many arthropod-borne pathogens (e.g., Le Blancq et al. 1986), and MLE continues to be a popular method. Gallego et al. (2002) and Noyes et al. (2002) describe recent applications of MLE to examine new and emerging species of *Leishmania* spp. MLE has also been applied to *Plasmodium* species (Abderrazak et al. 1999). *Borrelia* spp. have been subject to MLE (Boerlin et al. 1992), as have *Yersinia* spp. (Dolina and Peduzzi 1993).

Detection of Pathogens with the Polymerase Chain Reaction (PCR)

The principles of PCR are described in Chapter 30. For detection of pathogen DNA, species-specific oligonucleotide primers are designed to amplify a region of the pathogen genome that is diagnostic, either in size or sequence, for the presence of the genus or species. Each of the PCR protocols listed in Table 18.1 targets different regions of the pathogen genome for amplification and diagnosis. PCR products are loaded and size-fractionated on an agarose gel to determine if the amplified product is of the anticipated size. Alternatively, biotinylated primers can be used and the PCR products can be bound to a nylon membrane. The presence of biotinylated PCR product is usually detected using streptavidin-horseradish peroxidase and a chromogenic, soluble substrate.

Because the amount of nucleic acids from a vector usually greatly exceeds that from the pathogen, there is a concern with sensitivity when using direct PCR amplification. In many cases this has been addressed using *nested PCR*. Following PCR with primers from conserved regions of the pathogen genome, a second PCR is performed using diluted PCR products from the first amplification as template. Species-specific primers "nested" between the more universal primers are used in the second PCR. As before, the products are then loaded and size-fractionated on an agarose gel to determine if the amplified product is of the anticipated size.

There is also concern over the specificity of PCR, especially when primers are used that are general for potentially many pathogen species within a family or genus. Rather than sequence individual PCR products, restriction enzymes are often used to digest the PCR products prior to analyzing them on an agarose gel to test for species-specific restriction patterns (Figure 33.6).

Reverse Transcriptase PCR (RT-PCR)

Conventional PCR is based upon the presence of a DNA template. At no point during its life cycle is arboviral DNA present. Thus, if PCR is to be used, the molecular epidemiologist must first make DNA from arboviral RNA transcripts. This is accomplished by isolating total RNA from the vector. A strand of virus-specific DNA is then generated using a virus-specific primer and one of several commercially available reverse transcriptases. Conventional PCR is then used to amplify virus-specific DNA. The issues of sensitivity and specificity in PCR discussed earlier also apply to RT-PCR. Nested PCR protocols have been developed; for some of the RT-PCR procedures listed in Table 18.1, restriction enzyme analyses are performed.

Earlier, we mentioned that molecular techniques generally only detect the presence of pathogen DNA without indicating if the pathogen is at an infectious stage of development in the vector. Eastern, western, and Venezuelan equine encephalitis viruses are alphaviruses (family Togaviridae) that have a single-stranded, positive-sense, nonsegmented RNA that is ~11.7 kb in size and resembles cellular mRNAs in having a 5′ cap and 3′ polyadenylation (Fig. 18.7A). Togavirus replication characteristically requires 2 rounds of translation. Replication occurs within a few hours in the cytoplasm. The positive-sense genomic RNA acts directly as mRNA, and the 5′ end is partially translated to produce nonstructural proteins. These proteins are responsible for replication, forming a complementary negative strand, the template for further positive-strand synthesis. Typically, two species of positive-sense RNA are synthesized: full-length genomic RNA and a subgenomic mRNA (Fig. 18.7A).

Dengue Fever virus, Japanese and St. Louis encephalitis viruses, and yellow fever virus are all flaviviruses. These also have a single-stranded, positive-sense RNA that is ~10.5 kb with a 5′ cap but without a polyadenylated 3′ end (Fig. 18.7B). Their genetic organization differs from togaviruses in having structural proteins at the 5′ end of the genome and nonstructural proteins at the 3′ end. The initial stages of flavivirus replication occur in the cytoplasm: The entire virus genome is translated as a single polyprotein that is then cleaved into mature nucleocapsid protein, matrix protein, and a glycoprotein. Complementary negative-strand RNA is synthesized by nonstructural proteins and is used as a template for genomic progeny RNA synthesis.

La Crosse and the California serogroup bunyaviruses contain three segments of antisense single-stranded RNA combined with nucleoprotein (Fig. 18.7C). Two external glycoproteins, G1, G2, form surface projections. A virus-encoded transcriptase is present in the virion. Bunyaviruses replicate in the cytoplasm. Their RNA genome is transcribed to mRNA. The host RNA sequence in some representative viruses primes viral mRNA synthesis. Bluetongue virus (BTV) is an orbivirus of the Reoviridae family. The viral double-stranded RNA genome consists of 10 segments that code for two outer capsid proteins, five core proteins, and three nonstructural proteins.

Primers specific for any of these positive or negative-sense strands can be used to indicate the stages and extent of alpha-, flavi-, bunya-, and orbiviral infection in a vector. Furthermore, as more is learned about stage-specific patterns of gene expression in prokaryotic and eukaryotic pathogens (or possibly even in vector tissue-specific expression patterns), RT-PCR may allow greater specificity by identifying the stage of arbovirus development.

Simultaneous Detection of Multiple Pathogen Species

Ticks and other vectors frequently harbor multiple pathogen species, and it is often of interest to simultaneously test for several species. *Multiplex PCR* with multiple primer pairs, each diagnostic for a different pathogen, can be used to simultaneously amplify markers from many pathogens. However, primer dimers often occur with multiplex PCR, and use of multiple primers increases the chances for nonspecific amplification.

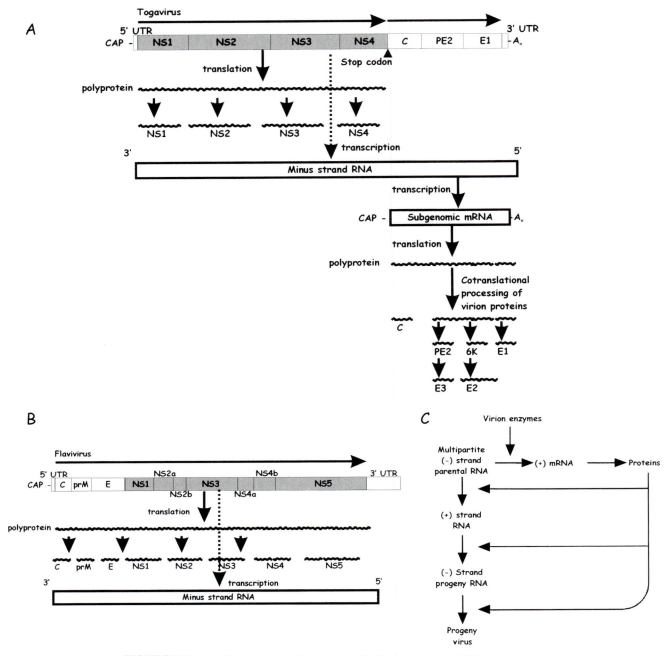

FIGURE 18.7 Replication of (A) alphaviruses, (B) flaviviruses, and (C) bunyaviruses.

The reverse line blot system was developed to simultaneously test DNA prepared from a single vector for the presence of DNA of potentially many different pathogens. The system utilizes pathogen species-specific oligonucleotide probes that are deposited in a parallel linear array along one axis of a nylon membrane (Fig. 18.8, Step 1). The oligonu-cleotide probes are covalently bound to the nylon membrane. DNA is isolated from a vector and PCR is performed using universal biotinylated primers for a common gene (e.g., bacterial 16s ribosomal rDNA). The PCR products from 40–50 organisms are then indi-vidually deposited in a parallel linear array perpendi-cular to the axis of the immobilized oligonucleotide

Pathogen species

1)

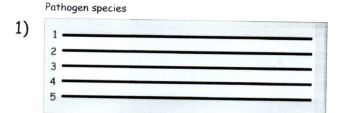

Pathogen species-specific oligonucleotide probes are deposited with a line blotter to a nylon membrane and then chemically bound to the membrane.

2)

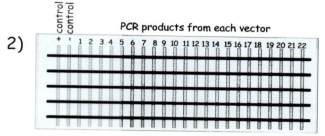

PCR products amplified with "universal" biotinylated primers for the 5 pathogens in each of 22 vectors are deposited with a line blotter in parallel linear array. PCR products are then hybridized to the immobilized oligonucleotide probes on the membrane.

3)

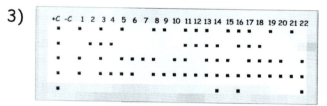

The nylon membrane is washed at high stringency. The presence of biotinylated PCR product bound to the species-specific probe is detected using streptavidin-horseradish peroxidase and a chromogenic, soluble substrate.

The positive control is positive for all 5 pathogens, no signal is detected in the negative control. Vector 1 is positive for DNA from pathogens 1, 3, and 4. Vector 9 is positive for DNA from pathogens 1 and 4, etc.

FIGURE 18.8   Reverse line blot procedure.

probes. The PCR products are then hybridized to the immobilized oligonucleotide probes on the membrane (thus the term *reverse*) (Fig. 18.8, Step 2). The nylon membrane is then washed at high stringency. The presence of biotinylated PCR product bound to the species-specific probe is detected using streptavidin-horseradish peroxidase and a chromogenic, soluble substrate to produce a hybridization signal at the position of the positive probe (Fig. 18.8, Step 3).

Real-Time PCR

Typically, molecular genetic methods detect only the presence, not the quantity, of pathogen nucleic acids. However, more expensive, recent advances in PCR allow for quantitation of the starting amount of template DNA. Figure 18.9 illustrates two current PCR-based techniques that can be used to quantify template DNA or RNA amounts. Both techniques require the use of a thermal cycler that is also able to quantify the increases in fluorescence as amplification of the target occurs. The prices of these machines average $100,000 at this time, but prices are declining rapidly as more companies release new products. Syber Green is a dye that only fluoresces when intercalated into double-stranded DNA. The amount of fluorescence, therefore, increases with rounds of PCR (Fig. 18.9A). There is legitimate concern with specificity when using Syber green because the dye intercalates in any DNA and is not necessarily specific to pathogen DNA. In contrast, Taqman technology analyzes the amount of a species- or strain-specific sequence contained within the amplicon (Fig. 18.9B). A species- or strain-specific oligonucleotide with a fluorescent 5' reporter dye and a 3' quencher is added to the PCR mixture. Fluorescence in the reporter is suppressed by the proximity of the quencher. The oligonucleotide anneals to one strand of the PCR product and is digested by *Taq* polymerase during the primer extension phase of PCR. The technique is very specific because *Taq* polymerase digests the oligonucleotide only when there is precise base pairing between the oligonucleotide and the target. *Taq* polymerase digestion liberates the fluorescent 5' reporter dye from the 3' quencher so that the amount of fluorescence increases with rounds of PCR (Fig. 18.9B). Quantification of the starting amount of template by either method requires that the researcher develop a set of standard curves with known amounts of starting nucleic acids. These curves fit a logistic growth model because the amount of fluorescence exceeds the machine's upper limit of detection (Fig. 18.9C). The X-intercept of each standard curve is estimated from regression of the linear portion of each curve (Fig. 18.10A), and these intercepts are then regressed on the $\log_{10}$ of the starting copy number to develop a single standard linear model for estimating the starting amount of nucleic acids in samples from individual arthropods (Fig. 18.10B).

Rapid Techniques for Detection of Pathogen-Specific Sequences

When many vectors are to be screened for pathogens, procedures that involve additional labor and costs associated with restriction enzyme and agarose gel analyses may be prohibitive to use. A series of novel techniques has been developed that signal the presence of a pathogen genome with minimal prepara-

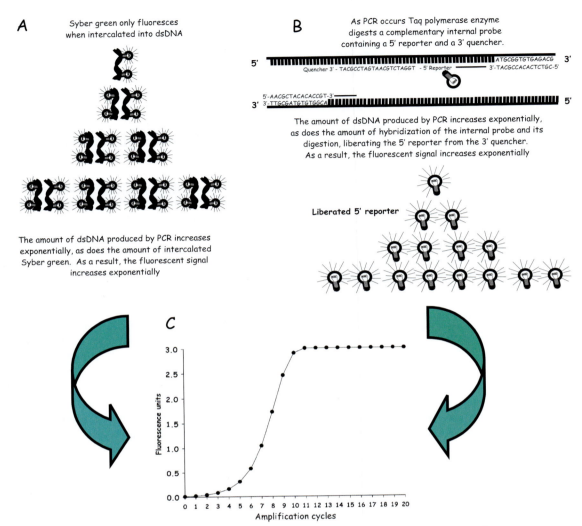

A Syber green only fluoresces when intercalated into dsDNA

The amount of dsDNA produced by PCR increases exponentially, as does the amount of intercalated Syber green. As a result, the fluorescent signal increases exponentially

B As PCR occurs Taq polymerase enzyme digests a complementary internal probe containing a 5' reporter and a 3' quencher.

The amount of dsDNA produced by PCR increases exponentially, as does the amount of hybridization of the internal probe and its digestion, liberating the 5' reporter from the 3' quencher. As a result, the fluorescent signal increases exponentially

Liberated 5' reporter

C

FIGURE 18.9 Quantitive PCR with either (A) Syber green or (B) Taqman procedures.

tion. The reagents required for these procedures are expensive, but their costs are declining rapidly as companies release new products. We have already discussed the Taqman procedure (Fig. 18.9B). Its application towards pathogen detection is illustrated in Figure 18.11A. A species- or strain-specific oligonucleotide, precisely complementary to the target sequence and with a fluorescent 5' reporter dye and a 3' quencher, is added to the PCR mixture. The oligonucleotide anneals to one strand of the PCR product and is digested by *Taq* polymerase during primer extension. However, if all or part of the oligonucleotide does not match the template, the oligonucleotide is not digested by *Taq* polymerase during primer extension. Fluorescence signals the presence of a pathogen species. No fluorescence is produced when there is imprecise base pairing between the oligonucleotide and the target.

Molecular beacons (Fig. 18.11B) are oligonucleotides that contain the complementary region to a pathogen species- or strain-specific DNA sequence. 5' and 3' to this region of the beacon are sequences that maintain the oligonucleotide in a hairpin configuration, with the fluorescent 5' reporter dye adjacent to the 3' quencher. If the beacon oligonucleotide encounters and anneals to pathogen DNA present in an arthropod, the fluorescent 5' reporter dye is pulled away from the 3' quencher. Fluorescence signals the presence of a pathogen species.

Dye-labeled oligonucleotide ligation (DOL) fluorescence resonance energy transfer (FRET) is illustrated in Figure 18.11C. Two oligonucleotides, one with a 5' fluorescent reporter, the second with a different 3' fluorescent reporter, are mixed with amplified template DNA and a ligase. As with Taqman, the oligonu-

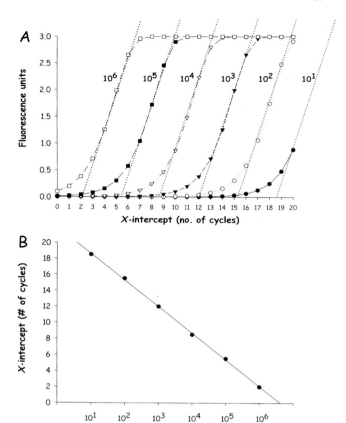

FIGURE 18.10 Derivation of a standard curve for quantitative PCR.

"probe" oligonucleotide is designed to hybridize to the pathogen species-specific template sequence and immediately adjacent to the invader oligonucleotide. The probe oligonucleotide has a 5' flap that does not anneal to the template. When there is precise base-pairing between the probe and the template DNA, this ternary complex is a substrate for Flap endonuclease, and site-specific cleavage removes the 5' flap from the probe. The cleaved probe then anneals to a third, "signal" oligonucleotide, which has a 3' region complementary to the 5' flap of the probe, with the exception of the final 3' nucleotide, which again forms a 3' overhang. The remainder of the signal oligonucleotide is designed to form a hairpin loop and has a 5' reporter dye adjacent to an internal or 3' quencher. Flap endonucleases cleave the 5' region of the signal, liberating the 5' reporter dye from the 3' quencher and allowing fluorescence. This mixture is heated, to denature all oligonucleotides, and then cooled to allow (1) the ternary invader–template–probe complex to reform and thus generate more free flaps, and (2) the new and existing 5' flaps to anneal to more of the signal oligonucleotide, thus liberating more reporter. Thus, there is an amplification of the signal in each cycle. Fluorescence signals the presence of a pathogen species. No fluorescence is produced when there is imprecise base-pairing between the probe oligonucleotide and the target.

The ligase chain reaction (LCR) is another technique for pathogen detection that does not require PCR for initial amplification of pathogen nucleic acids (Fig. 18.12). Four oligonucleotides and a thermophilic ligase isolated from *Thermus thermophilus aquaticus* are mixed directly with DNA isolated from an arthropod. Two of the oligonucleotides are precisely complementary to sequences immediately 5' and 3' of the *upper* strand of the pathogen species-specific template sequence. The other two oligonucleotides are precisely complementary to sequences immediately 5' and 3' of the *bottom* strand of the pathogen sequence. The solution is heated to denature the template DNA and the oligonucleotides. The solution is then cooled to allow annealing of the target to the oligonucleotides. *Taq* ligase covalently links the oligonucleotide pairs to one another if the nucleotides at the junction are precisely base-paired to the target. This, in turn, generates a new copy of both strands of the pathogen target sequence. As with PCR there is a geometric, twofold amplification of the pathogen target sequence during each heating/cooling cycle. The presence of ligated oligonucleotides signals the presence of a pathogen species. No ligation occurs when there is imprecise base-pairing between the oligonucleotides and the target.

cleotides are designed to precisely base-pair to the pathogen target sequence. The DNA ligase will covalently link the two oligonucleotides, but only if the nucleotides at the junction are exactly complementary to the target. The combination of the 5' and 3' fluorophores causes the reaction to produce a unique fluorescent color that signals the presence of pathogen DNA.

The invader assay is the most recent tool for pathogen detection (Fig. 18.11D) and has the advantage of not requiring PCR for initial amplification of pathogen nucleic acids. Three oligonucleotides and a "Flap endonuclease" are mixed directly with DNA isolated from an arthropod. Flap endonuclease cleaves branched DNA structures containing single-stranded 5' flaps. The first "invader" oligonucleotide is designed to hybridize to the sequence immediately 3' to the pathogen-specific sequence and to have a single 3' overhang that does not anneal to the template. The

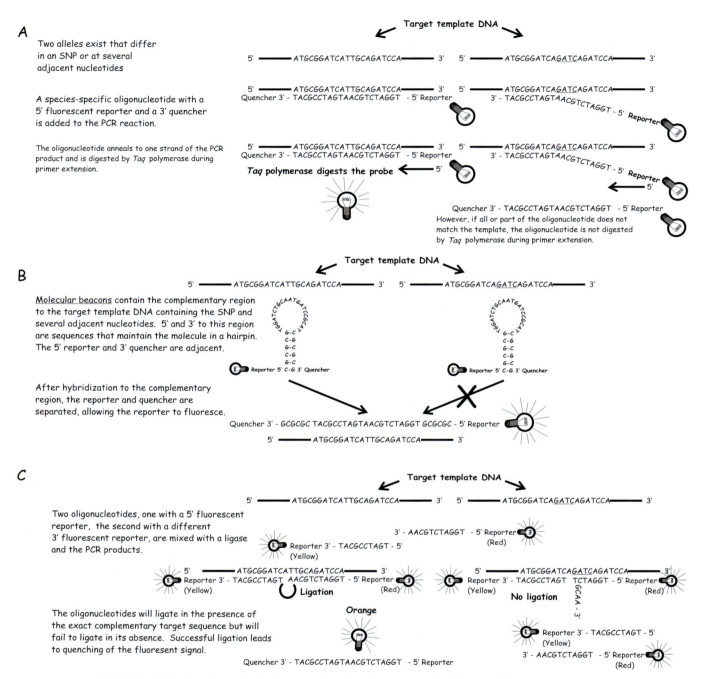

FIGURE 18.11 Rapid methods for detection of pathogen-specific DNA or RNA sequences using (A) Taqman, (B) molecular beacons, (C) DOL-FRET, and (D) invader assay.

D)

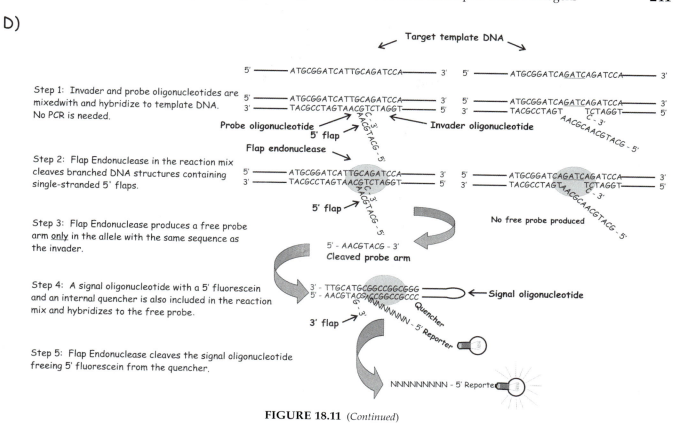

FIGURE 18.11 (*Continued*)

PATHOGEN ADAPTATION TO DISEASE VECTORS

In Chapter 1, evidence was presented that hematophagy, the habit of blood feeding, has evolved at least 21 times in disparate arthropod taxa. Hematophagy was probably quickly exploited by vertebrate pathogen species, both as a means for increased mobility and, more importantly, as a means to find and occupy novel vertebrate hosts. Great benefits accrued to pathogen species or populations that possessed the correct combination of morphological, physiological, and biochemical characters necessary to survive in hematophagous hosts. These pathogens abruptly acquired many adaptations beneficial to their own survival and dispersal. These included physiological and behavioral mechanisms evolved for host location and precise morphological and neurosensory mechanisms for proximal location of blood within the vertebrate vascular system. In addition, vector-borne pathogens became transmitted via saliva that contained immunomodulatory components that increased the chances of their transmission and survival in a new vertebrate host.

However, gaining access to an arthropod vector was only half of the adaptive process. The pathogen was next faced with the problem of invading the arthropod gut, disseminating to various organs, and eventually ending up in an organ of transmission without compromising host survival. Most pathogens targeted the arthropod salivary gland for oral transmission (e.g., *Plasmodium* sporozoites, arboviruses). Alternatively, some pathogens escaped through arthropod waste or secreted products onto the surface of a vertebrate host (e.g., *Trypanosoma cruzi*, relapsing fever *Borrelia* spp.). Many arboviruses and rickettsia exploited the female reproductive system to become transmitted either transovarially or via transovum. Similarly, some arboviruses exploited male reproduction and became venereally transmitted.

Simultaneously, pathogen adaptation in a hematophagous host required either being undetected or minimizing the impact on host survivorship and fecundity or, if the first two strategies weren't possible, evading destruction by the arthropods' innate immune system. Thus, a useful way of understanding the evolution of any pathogen–vector system is through a consideration of how pathogens progressively adapted to

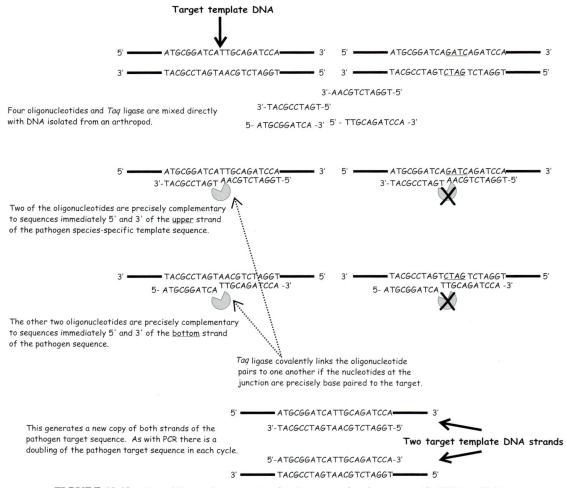

FIGURE 18.12   Use of ligase chain reaction for detection of pathogen-specific DNA or RNA.

and exploited specific morphological, physiological, and biochemical aspects of a hematophagous host without simultaneously compromising the fitness of that host.

We purposely avoid the terms *coevolution* or *coadaptation* in discussing this adaptation because these terms imply that a pathogen–arthropod association is either mutually beneficial or antagonistic and has evolved via stepwise, compensatory pathways. The concept of *host specificity* is also implicit in the conceptual framework of coevolution or coadaptation, suggesting that a pathogen has narrowly adapted to a specific host species. We believe that the evolution of host specificity would have been maladaptive, restricting a pathogen to one or a narrow set of vector species to maintain itself in natural enzootic cycles. Brooks and McLennan (1993), in an extensive review of literature on host–pathogen evolution, provide abundant empir-

ical evidence to dismiss the concept of host specificity. They conclude that the perception of host specificity usually arises in associations with limited historical opportunities for a pathogen to transfer to alternate hosts. Klompen et al. (1996) argued that host specificity in ectoparasites is often an artifact of oversampling of common hosts and inadequate sampling of rare or uncommon hosts. We, therefore, prefer the term *cospeciation* (Brooks and McLennan 1993), which implies that pathogen phylogenies parallel host phylogenies simply because the pathogens have not had the opportunity for host transfer. An example of cospeciation between microorganism symbiont species and *Cryptocercus* roaches is shown in Figure 18.13. The phylogeny of the endosymbiont follows almost entirely the phylogeny of its arthropod host. This implies that evolution and speciation of the endosymbiont follows the evolution and speciation of its arthropod host. Note

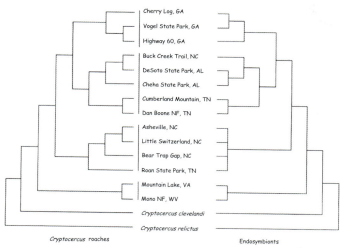

FIGURE 18.13 *Cryptocercus* species are subsocial, xylophagous cockroaches that live in temperate forests. Like other cockroaches, *Cryptocercus* harbor endosymbiotic bacteria in their fat bodies. There is complete concordance between the host and endosymbiont phylogenetic trees. Divergence estimates based on endosymbiont DNA sequences suggested that the palaearctic and nearctic *Cryptocercus* diverged 70–115 million years (Myr) ago and the eastern and western U.S. species diverged 53–88 Myr ago. Redrawn from Clark et al. (2001).

that in this case there may be limited opportunity for the endosymbiont to transfer among hosts. In a sense, each *Cryptocercus* host is an island, and different endosymbiont species are isolated on different islands. The extent to which this example applies to arthropod-borne pathogens depends on the extent to which pathogens cospeciate with their vector hosts. Pathogen species capable of moving from one arthropod species "island" to another are unlikely to show evolutionary patterns of cospeciation, while pathogens that are capable of propagation and transmission by only one or a few vector species are more likely to demonstrate cospeciation.

Pathogen adaptation to life in an hematophagous arthropod host seems to fall into two broad categories. Arboviruses, most *Rickettsia*, some bacterial, *Leishmania*, and *Trypanosoma* spp. evolved to invade, replicate within, and become released from their invertebrate host without being detected or by having a such a *small impact* on host survival that the vector did not react to their presence. Alternatively, the vertebrate pathogen may *negatively impact* fitness and reproduction (Berry et al. 1987, 1988) in its arthropod vector. In some protozoan (e.g., *Plasmodium* and nematodes), the arthropod frequently attempts to kill or reject the parasite. In these cases the parasite has had to evolve adaptations to avoid detection or destruction. In extreme, probably

recently evolved, vector–parasite associations, interactions are not benign for either the vector or the parasite. Consider the interactions between *Rickettsia rickettsii*, the etiological agent of louse-borne typhus, and its primary vector, *Pediculus humanus*, the human-body louse. The midgut epithelial cells of the louse are destroyed by the rickettsial infection, and the louse dies. Rickettsia are transmitted only through the feces of the louse or via ingestion of all or part of the louse body. Inhalation of dust or dried body parts is also a common route of infection.

From the pathogen's perspective, a competent vector has what the pathogen needs to replicate, survive, and be transmitted and doesn't mount a defense against the pathogen. Alternatively, an incompetent vector doesn't have what the pathogen needs and, even if it does, mounts such an effective defense that the pathogen must fight to use the vector. Plasticity in vector competence is consistent with a general genetic model in which multiple structural or biochemical factors in the arthropod vector must be present for successful completion of a pathogen's life cycle. The absence of any one of these factors renders a vector species incompetent. Alternatively, if a pathogen is deleterious to vector survival or reproduction, the vector may well have evolved an active resistance for rejection or destruction of a pathogen. Some vector species may be fixed for the presence of resistance factors, while others may lack one or more of these factors.

Arbovirus Evolution and Adaptation

As discussed in Chapter 30, an arbovirus must first attach and penetrate the mosquito's midgut epithelial cells and then replicate to a high titer within those cells. For some arboviruses, (e.g., bunyaviruses and orbiviruses), there is now evidence that proteolytic processing of virion surface proteins is necessary for efficient vector–midgut cell interaction (Xu et al. 1997). Recent research in Black's laboratory indicates that the midgut trypsins in *Ae. aegypti* are necessary for infection of midgut epithelial cells (Molina et al., unpublished). Interestingly, this proteolytic cleavage is not a prerequisite for successful infection of cells in secondary target organs, only for the midgut epithelial cells that are exposed to the proteolytic milieu of the midgut lumen. Laminan has been proposed to be a mosquito cell receptor for alphaviruses (Ludwig et al. 1996). Once inside the cell, host factors are required for uncoating and then translation of nonstructural viral proteins (e.g., RNA-dependent polymerases). Host factors are probably involved in proper transcription and translation of viral structural proteins. In the final stages, host factors are undoubtedly involved in proper packaging and assembly of virus. Next, the arbovirus must pass through the basal lamina surrounding the midgut and infect and replicate in surrounding tissues. Finally, the arbovirus must infect and possibly replicate within the salivary gland before it can be shed into the lumen of the glands for final transmission in the subsequent bite.

Manfred Eigen was a chemist and mathematician and the 1967 Nobel laureate in chemistry who also developed a model for the evolution of haploid genomes. The corollary model for diploid, sexually reproducing organisms was developed by Sewall Wright (Chapter 32) and is known as the shifting balance model. Because haploid genomes generally have low rates of recombination or reassortment, Eigen's model depended heavily upon high mutation rates for the generation of novel genetic variants. These variants are called *quasi-species*.

The quasi-species model is useful in understanding the evolution of a variety of haploid pathogens, particularly arboviruses. There are four basic premises of the quasi-species model. First, genetic variation is generated at all loci by a high mutation rate. Second, gene-by-gene interactions (pleiotropic effects) are widespread throughout the pathogen genome. Third, there are distinct relationships between multilocus genotypes and fitness. And fourth, the unit of selection is an individual haplotype.

As an example, consider an ~12-kb RNA viral genome. For the sake of simplicity, assume that there are four nonsynonymous substitutions that impact the fitness of the virus. Each corresponds to a second codon position. The substitutions are located at nucleotide positions 121, 467, 7364, and 9783. There are, therefore, $4 \times 4 \times 4 \times 4 = 256$ possible viral haplotypes at each of these four positions. Following a round of replication, the relative abundance of these 256 viral genomes can be represented and most easily visualized topographically (Fig. 18.14A). Relative abundances are determined by the relative mutation rates of each position, differential rates of transitions and transversions, and the fact that some genotypes cannot be replicated.

The units of selection are the 256 haplotypes. For each of these we need to estimate the relative fitness. Assume that we have derived the following fitness matrix:

	121	467	7364	9783
A	0.25	1.00	0.80	1.00
C	1.00	0.00	0.00	0.00
G	0.25	1.00	1.00	0.10
T	1.00	0.00	0.00	0.00

Each of the four columns of this matrix corresponds to the critical nucleotide position in the genome. The four rows correspond to the possible nucleotide identity at each of the four genome positions. The numbers indicate the fitness of the genome when that nucleotide exists at that position. The fitness of a particular four-nucleotide genotype is determined by summing fitness values across all positions. Relative fitness is calculated for each position by subtracting the minimum observed fitness from each total and then dividing this by the maximum observed fitness. Thus, for example,

$$W_{AAAA} = \frac{(0.25 + 1.00 + 0.80 + 1.00) - 0.25}{4} = 0.70$$

Next, we perform this calculation for all 256 possible four-locus genotypes and plot the relative fitness isoclines of the 256 genotypes (Fig. 18.14B). Notice that there are four genotypes of maximum fitness, $W_{CAGA} = W_{CGGA} = W_{TAGA} = W_{TGGA} = 0.9375$, and four additional genotypes with a lower fitness of $W_{CAAA} = W_{CGAA} = W_{TAAA} = W_{TGAA} = 0.8875$. Following error-prone replication, there follows a period of mass selection (Fig. 18.14C). During this period, only those genotypes with optimal fitnesses survive in the competition for cell surface receptors and access to the host cell's transcription and translation machinery. Notice that the final abundances of each of the eight most fit genotypes (Fig. 18.14C) do not necessarily correspond to their initial relative abundances following replication (Fig. 18.14A). Instead, the final abundances correlate with the relative fitness topography (Fig. 18.14B). Pathogens with fitness haplotypes of lower values don't survive or don't compete successfully. The eight surviving haplotypes are referred to as *quasi-species*.

The mutation selection cycle is repeated during each round of viral replication. The adaptive landscape changes constantly for a pathogen species through time due to changes in the host (individual and population immunity) and density-dependent changes in the pathogen population. In addition, new mutations arise frequently, creating new opportunities for selection. A quasi-species sitting at an adaptive peak in one generation cannot remain at that peak in future generations. In the case of arboviruses that alternate between vertebrate and invertebrate hosts, the adaptive landscape can change abruptly (Fig. 18.15). In this situation, the error-prone polymerase can be adaptive, providing an array of mutations for an alternative (possibly previously unexplored) topography.

Genetic recombination and reassortment are phenomena that also promote rapid arbovirus evolution. Recombination can occur when there is *coinfection* of different arboviral genomes that would allow crossing

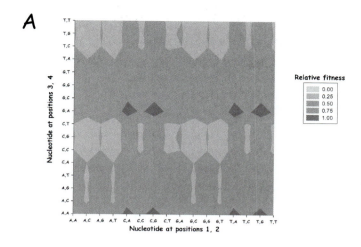

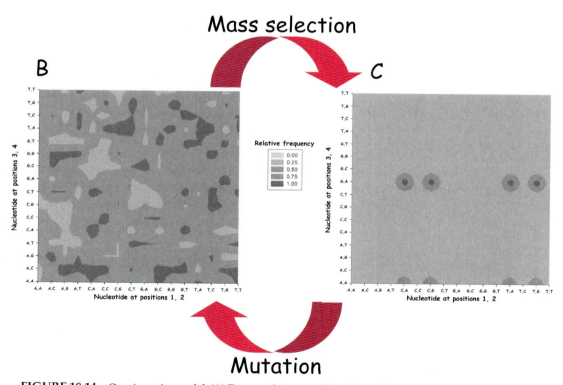

FIGURE 18.14   Quasi-species model: (A) Topographic map of the relative fitnesses of the 256 possible four-nucleotide genotypes; (B) topographic map of the frequencies of the 256 genotypes generated during one round of error-prone replication; (C) topographic map of the frequencies of the 256 genotypes after one round of selection. Relative survival of specific genotypes is dictated by the relative fitness patterns in part A.

over between different arboviral genomes. This phenomenon has been demonstrated with a variety of arboviruses (Holmes et al. 1999; Pletnev et al. 2002; Uzcategui et al. 2001). Reassortment occurs in multipartite arboviral genomes (e.g., bunya- and orbiviruses) when part of a viral genome established in one host recombines with another viral genome established in another host in the same zoonotic cycle. Reassortment has been demonstrated for a variety of arboviruses and may be especially important in zoonotic diseases (Beaty and Bishop 1988). We demonstrate in the last section of this chapter that molecular epidemiology can be used to identify transformation, reassortment, and recombination events.

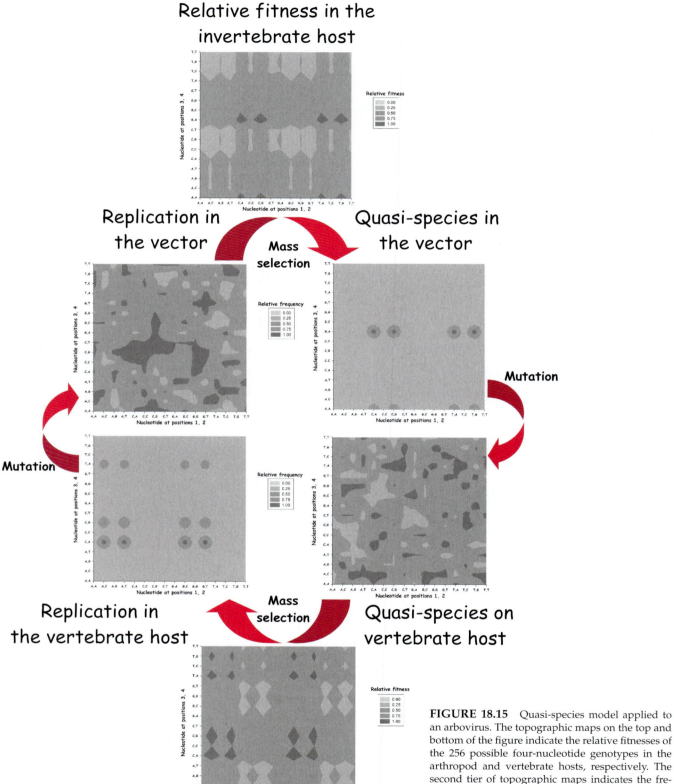

FIGURE 18.15 Quasi-species model applied to an arbovirus. The topographic maps on the top and bottom of the figure indicate the relative fitnesses of the 256 possible four-nucleotide genotypes in the arthropod and vertebrate hosts, respectively. The second tier of topographic maps indicates the frequencies of the 256 genotypes generated during one round of error-prone replication and selection in the arthropod vector. The third tier of maps indicates the genotype frequencies generated during one round of error-prone replication and selection in the vertebrate host.

Bacterial and Rickettsial Evolution and Adaptation

In fleas infected with the *Yersinia pestis* bacillus, the bacilli accumulate in the flea's proventriculus, a toothed, chitinized valve between the fore- and midguts. The massive accumulation of bacilli forms a plug that disrupts the feeding process of the flea vector. When an infected flea attempts to feed on a new vertebrate host, this blockage causes the flea to regurgitate the bacteria into the next vertebrate host. Because the flea cannot feed, it becomes permanently hungry and contacts many more hosts than it would normally otherwise, and increased disease transmission results.

Transmission by flea bite appears to be a relatively recent adaptation that distinguishes *Y. pestis* from closely related enteric bacteria. Hinnebusch (1996) showed that blockage was dependent on the hemin storage (*hms*) locus in *Y. pestis*. The *hms* mutants established long-term infection of the flea's midgut but failed to colonize the proventriculus. Thus, the *hms* locus markedly alters the course of *Y. pestis* infection in its insect vector, leading to a change in blood-feeding behavior and to efficient transmission of plague. Hinnebusch et al. (2002) showed that a plasmid-encoded phospholipase D (PLD), previously characterized as *Yersinia* murine toxin (Ymt), was required for survival of *Y. pestis* in the midgut of the rat flea *Xenopsylla cheopis*. Intracellular PLD activity appeared to protect *Y. pestis* from a cytotoxic digestion product of blood plasma in the flea gut. By enabling colonization of the flea midgut, acquisition of this PLD may have precipitated the transition of *Y. pestis* to obligate arthropod-borne transmission.

These patterns may, however, be specific for the flea species. Engelthaler et al. (2000) used a quantitative PCR assay to quantify *Y. pestis* loads in fleas and bacteremia levels in mice used as sources of infectious blood meals. *Xenopsylla cheopis* achieved higher infection rates, developed greater bacterial loads, and became infectious more rapidly than *Oropsylla montana*. Their results suggest that at the time of flea feeding, host blood must contain >10^6 bacteria/mL to result in detectable infections in these fleas and >10^7 bacteria/mL to cause infection levels sufficient for both species to eventually become capable of transmitting *Y. pestis* to uninfected mice. *Y. pestis* colonies developed primarily in the midguts of *O. montana*, whereas infections in *X. cheopis* often developed simultaneously in the proventriculus and the midgut.

Relapsing fever *Borrelia* spirochetes have exploited the water elimination anatomy (the coxal glands) and physiology of soft ticks as a means for transmission. Spirochetes are found in the excreted coxal fluid and are infectious either through the wound at the tick feeding site or through other skin lacerations or mucous membranes. Schwan and Piesman (2000) showed that *B. burgdorferi* undergoes changes in expression of important outer surface proteins in the midgut of its vector, *Ixodes scapularis*. The authors speculated that these changes may be important in the development of virulence of *B. burgdorferi* for the vertebrate host.

Transformation and recombination are phenomena that promote rapid prokaryotic pathogen evolution. *Transformation* is the insertion of new genetic information into a bacterial genome. This appears to be the way that genes encoded on plasmids (e.g., antibiotic resistance) are moved among bacterial species. It also appears to be a common mechanism by which pathogens with "new" antigens enter a host population. Recombination occurs when there is crossing over between new genes introduced on plasmids with other genes in the pathogen genome.

Plasmodium Evolution and Adaptation

After an anopheline mosquito takes a gametocytemic blood meal from a vertebrate host, the gametocytes must exflagellate and then fuse to form an ookinete in the lumen of the mosquito gut. The ookinete must then successfully penetrate the mosquito's midgut epithelial cells before forming an oocyst between the midgut cells and the basal lamina. Vinetz et al. (2000) showed that the *Plasmodium* ookinete produces chitinolytic activity that allows the pathogen to penetrate the chitin-containing peritrophic matrix. Allosamidin, a chitinase inhibitor, placed in the blood meal prevented the pathogen from invading the midgut epithelium. The authors cloned and sequenced the gene and determined that the protein has catalytic and substrate-binding sites characteristic of family 18 glycohydrolases.

A number of factors have been identified that block *Plasmodium* invasion and development. Invertebrates do not have the capability of producing antibodies, but they are capable of both cellular and humoral reactions to combat infections (Chapter 27). More recently, it has been shown that mosquitoes also express several elements of vertebrate-specific immune responses (Barillas-Mury et al. 1999). It has been shown that ookinetes penetrating midgut follicular epithelial cells induce various *Anopheles* species to produce defensin, a gram-negative bacteria–binding protein, NO synthetase, and initiate other enzymatic pathways that may ultimately lead to apoptosis (Han et al. 2000). In general, this response is so potent that only a small percentage of mature ookinete reach the basal lamina to form oocysts. The oocyst must next undergo sporogony and burst through the basal lamina. Strains of *An. gambiae* have been artificially selected to encapsulate, melanize, and thereby destroy oocysts (Collins et al. 1986).

The sporozoites must next infect the salivary gland and make their way into the lumen of the glands for final transmission in the subsequent bite. Rossignol et al. (1984) showed that *P. gallinaceum* sporozoites destroy segments of the salivary glands of *Ae. aegypti* mosquitoes and sought to determine whether salivary function was impaired. If so, then prolonged intradermal probing could occur because of the role of saliva in locating blood vessels. Uninfected *Ae. aegypti* mosquitoes probed for a shorter period than did either sporozoite-infected or saliva-deprived mosquitoes. Salivary apyrase activity (Chapter 1) was reduced by 60% following maturation of sporozoites. Interestingly, apyrase activity is confined to those regions of the salivary glands invaded by sporozoites. Sporozoite-infected and uninfected mosquitoes produced equal volumes of saliva. The authors concluded that sporozoite infection impairs the vector's ability to locate blood vessels by affecting the quality of salivary product, thereby increasing potentially infective host contacts. In a follow-up study, Rossignol et al. (1986) tested whether salivary gland pathology also leads to an increased biting rate. With an olfactometer they compared, for 5 days, relative daily biting rates between sporozoite-infected and uninfected *Ae. aegypti*. Infected mosquitoes exhibited a significant increase in olfactometer response that was also reflected in a decreased egg output. The authors concluded that if duration of contact with a host is limited, then infected mosquitoes may make more attempts at probing before being successful, and thus enhance transmission.

Filaria Evolution and Adaptation

Microfilariae, each in a saclike sheath, are ingested from the peripheral blood of the vertebrate host during the act of feeding. Diurnal patterns in the presence of the microfilariae in the vertebrate host may be more prevalent during defined periods of a day. In *Wuchereria bancrofti*, for instance, microfilariae are present in significant numbers in the human peripheral circulatory system between 10:00 pm and 2:00 am, which overlaps with the time of maximal biting activity of the vector *Cx. pipiens*. In endemic areas of the South Pacific, certain *Wuchereria* strains exhibit a diurnal presence in the peripheral blood and are referred to as *subperiodic*. Daytime-feeding mosquitoes are the major vectors in these areas. Thus, periodicity may have evolved in the filarid species as a response to the biting activity of vectors. Host-mediated cues, including reduced arterial oxygen tension and lowered temperature, may also be involved.

McGreevy et al. (1978) showed that the microfilariae of *W. bancrofti* and *Brugia pahangi* are killed by the shredding action of the cibarial and pharyngeal armatures and other papillae and spines in the foregut of mosquitoes. *Anopheles* species (the primary vectors in nature) have well-developed cibarial armatures and killed 36–96% of the ingested microfilariae. *Culex pipiens* has a poorly developed cibarial armature and killed only 6% of the microfilariae. *Ae. aegypti* and *Ae. togoi* lack cibarial armatures but have the remaining foregut structures and killed only 2–22% of the microfilariae. Having passed through the cibarial and pharyngeal armatures, the microfilariae shed their sheaths and migrate through the walls of the midgut. Migration through the midgut wall can occur in minutes, though live and possibly unsuccessful microfilariae may be found in the mosquito gut up to 4 days after ingestion. Practically all microfilariae passing through the gut migrate into the thorax within 12 hours. Beerntsen et al. (1995) demonstrated a barrier to midgut infection for *B. malayi* in experiments designed to assess numbers of microfilariae ingested and midgut penetration by microfilariae in susceptible and refractory strains of *Ae. aegypti*. Refractory mosquitoes ingested significantly fewer microfilariae than susceptible mosquitoes, and significantly fewer numbers of microfilariae penetrated through refractory midguts as compared to susceptible midguts. In 16.7% of the refractory midguts, no microfilariae were able to penetrate the midgut; in three refractory mosquitoes, over 250 microfilariae were ingested, but none penetrated the midgut. These results indicate that permissiveness of the midgut for penetration by microfilariae can determine not only pathogen intensity, but also prevalence of infection.

Following midgut penetration, development of microfilariae takes place within the large thoracic flight muscles, where the larvae become slightly shorter and much thicker. In *Dirofilaria immitis* the microfilariae undergo development in the Malpighian tubes rather than in the thoracic muscles. As with *Plasmodium*, refractory mosquitoes sense and treat the pathogen as a foreign body and can encapsulate and melanize microfilariae larvae in the hemolymph or in the malphigian tubules (Chen and Laurence 1985). A number of internal changes and two molts occur during development in vector tissues, ultimately resulting in infective third-stage larvae. The infective larvae migrate with little difficulty to the proboscis; when the mosquito feeds on a vertebrate, these larvae emerge through the wall of the mosquito's labellum.

Trypanosome Evolution and Adaptation

Once trypomastigotes of *Trypanosome brucei* are ingested they undergo developmental changes into the procyclic form and then the mesocyclic form in the tsetse fly gut. The mesocyclic form then penetrates the midgut wall and migrates through the anterior end of the gut into the salivary glands, where it undergoes further differentiation to the epimastigote, premetacyclic, and metacyclic trypomastigotes before oral transmission by the fly. Little is understood about vector competence in the tsetse host. However, many experiments have documented that developmental changes in the fly are essential for transmission. Only the metacyclic trypomastigotes are infectious, and this appears to be associated with establishment of the basic antigen surface on metacyclic trypomastigotes in the tsetse fly salivary glands.

Small numbers of trypomastigotes of *Trypanosoma cruzi* circulate in the peripheral circulatory system of humans. Once taken up in the triatomine blood meal, the organism replicates as a epimastigote in the mid- and hind gut of the bug. One to two weeks after infection, metacyclic trypomastigotes appear in the hindgut. *Trypanosoma cruzi* exploits the rapid flow of urine that washes out the triatome rectum during blood feeding. In this way literally thousands of metacyclic trypomastigotes are delivered to the vertebrate host's skin. *Trypanosoma* inhabit the proteolytic gut of their host and have dense surface glycolipidic coats that cannot be digested by their vector's enzymes. Adaptations of these pathogens to survive in a proteolytic environment probably preadapted them for survival in the vertebrate macrophage phagolysosome.

E. S. Garcia et al. (1994) reported that *R. prolixus*, with salivary glands infected with *T. rangeli*, pierced host skin more often and drew less blood and at a lower rate than controls when feeding on a rabbit. None of these differences was observed when feeding was performed through a membrane feeder. Salivary gland homogenates from infected insects, at 30 days after feeding/infection, had a significantly lower amount of total protein/salivary gland pair and less anticoagulant activity. Also, infected salivary glands exhibited significantly reduced apyrase activities and reactive nitrogen groups. The authors concluded that salivary infection of *T. rangeli* impairs the ability of the vector to locate blood vessels by affecting salivary antihemostatic properties, thus enhancing the possibility

of intradermal inoculation of pathogens into the mammalian host.

Leishmania Evolution and Adaptation

Little is understood about vector competence for *Leishmania* in sand flies. The requirement for a sugar meal by the infected sand fly was demonstrated more than 50 years ago as being essential for *Leishmania* to become infective to vertebrates, but the underlying reason remains obscure. Like Trypanosomes, *Leishmania* also undergo developmental changes within their sand fly vector. The fly ingests amastigotes in its blood meal. These are often in a macrophage but are also free in the blood. In the gut they transform into stumpy promastigotes, and as they do they migrate to the anterior end of the gut. As with the trypanosomes, *Leishmania* inhabit the proteolytic gut of their host and have dense, indigestible surface glycolipidic coats that probably protect them from the vector's enzymes. The promastigotes transform further into necto- or haptomonads forms of the paramastigote that then gives rise to the metacyclic promastigote that is infective to the next vertebrate host.

Schlein et al. (1991) showed lysis of the chitin layer in the anterior region of the peritrophic membrane in histological sections of infected flies. This lysis permitted the forward migration of a concentrated mass of pathogens. At a later stage the pathogens concentrated in the proventriculus that then lost its cuticular lining. They showed that chitinase and N-acetylglucosaminidase were secreted by cultured *L. major* promastigotes. Activity of both enzymes was also observed in *L. donovani*, *L. infantum*, *L. braziliensis*, *Leptomonas seymouri*, *Crithidia fasciculata*, and *Trypanosoma lewisi*. Subsequently, the chitinase also destroyed the valves associated with the feeding pump, presumably reversing the normal flow of pumping and injecting the pathogens into the vertebrate host. The infected sand flies have difficulty feeding, which, as with *Y. pestis* and *Plasmodium*, similarly increased transmission of the pathogens.

Pimenta et al. (1997) determined that the peritrophic matrix might inadvertently protect *Leishmania* from the hydrolytic activities of the sand fly midgut. They added exogenous chitinase to the blood meal, which completely blocked peritrophic matrix formation. Surprisingly, the absence of the peritrophic matrix was associated with the loss of midgut infections. The chitinase was not directly toxic to the pathogen, nor were midgut infections lost due to premature expulsion of the blood meal. Most pathogens were killed in chitinase-treated flies within the first 4h after feeding.

Early pathogen mortality was reversed by soybean trypsin inhibitor. Allosamadin, the chitinase inhibitor discussed earlier, led to a thickening of the peritrophic matrix and also prevented the early pathogen mortality seen in infected flies. Susceptibility to gut proteases was extremely high in transitional-stage pathogens, while amastigotes and fully transformed promastigotes were relatively resistant. The authors suggest that the peritrophic matrix creates a barrier to the rapid diffusion of digestive enzymes and limits the exposure of pathogens to these enzymes during the time when they are especially vulnerable to proteolytic damage.

Charlab and Ribeiro (1993) showed that salivary gland homogenates of female *L. longipalpis* inhibit the *in vitro* multiplication of promastigotes of *Leishmania mexicana amazonensis*. The effect seems to be cytostatic, since promastigote viability 24 h after exposure ranged from 55% to 100%. The cells cultivated in the presence of saliva were characterized by a very slender shape, with cell bodies that were almost two times as long as controls. The results suggest that vector saliva could influence the development of *Leishmania* pathogens within the vector by inhibiting their growth and triggering them to a differentiation pathway. Charlab et al. (1995) showed that *L. mexicana amazonensis* sensitivity to saliva is correlated with the phase of promastigote *in vitro* growth and can be decreased by the addition of hemin to the culture medium.

General Patterns in Pathogen Adaptation to Hematophagous Arthropods

There is abundant evidence that arthropod-borne pathogens have adapted in numerous independent ways to survive and, in many cases, to replicate in and escape from their arthropod host. The preceding examples display a continuum of arthropod/pathogen interactions from the very minimal interactions seen among arboviruses and their hosts, to the defensive reactions elicited by *Plasmodium* and filaria, to the lethal interactions seen between some *Rickettsia* and their lice vectors. As an explanation of this continuum, we propose that the severity of arthropod/pathogen interactions may reflect the number of generations over which these interactions have existed. This would suggest that arboviruses, most *Rickettsia*, and many trypanosome associations have existed for possibly millions and millions of generations, whereas *Plasmodium* and filaria associations with anopheline vectors may have been more recent, and *Rickettsia* and lice associations may be extremely recent.

The number of generations would be greatly accelerated in arboviruses, rickettsia, and many bacteria that undergo many replication cycles within one arthropod body. In the case of RNA viruses, the process would be accelerated even further by the error-prone RNA–dependent DNA polymerase. Many parasitology textbooks discuss the variety of trypanosomes kinetoplastids that are found in the guts of other, non-blood-feeding arthropods. Arthropod–trypanosome adaptation may have been well advanced by the time Cimicids and Triatomes became hematophagous. Conversely, *Plasmodium falciparum* may have existed only as a species and have been associated with humans, and possibly *An. gambiae*, within the last 10,000 years (Volkman et al. 2001; Tishkoff et al. 2001).

These patterns suggest two general models for pathogen adaptation to hematophagous arthropods. In the first model, we propose that hematophagous arthropods constantly come into contact with a wide diversity of potentially pathogenic organisms. In most circumstances the arthropod doesn't have what the pathogen needs to either replicate, survive, and/or be transmitted. Alternatively, the pathogen might elicit such a strong reaction from the arthropod that it is destroyed. In either case, this terminates the interaction. Rarely, possibly accidentally, interactions give rise to a productive, nonhostile initial interaction between the pathogen and arthropod, and the processes of pathogen adaptation to that arthropod host are allowed to proceed. In the second model, hematophagous arthropod lineages evolve from non-blood-feeding ancestral lineages. We know that many protozoa, rickettsia, viruses, trypanosomes, and bacteria (e.g., the mycoplasmas) are frequently found in non-hematophagous arthropods. When these lineages subsequently gave rise to hematophagous arthropods, the pathogen was already adapted to the arthropod host. In these cases, pathogens became transmitted to the vertebrate host and may, in rare cases, have caused pathogenic effects in the vertebrate host. This may be the case with many new and emerging pathogens.

Using Molecular Epidemiology to Understand the Evolution of Pathogens

The evolution of the influenza virus (family Orthomyxoviridae) that caused the 1918 "Spanish influenza" pandemic, in which more than 20 million people died, provides an excellent recent example of the power of molecular epidemiology to understand the evolution of pathogens. Fragments of RNA from the 1918 strain were obtained from lung samples of flu victims preserved in pathology museums or frozen in the Alaskan permafrost. These were subject to RT-PCR. These RNA pieces yielded the complete sequences of genes encoding three crucial flu virus proteins: hemagglutinin, neuraminidase, and nonstructural protein. There were no obvious features in these sequences that suggested why the 1918 virus was so virulent. Phylogenetic reanalysis of hemagglutinin gene sequences from humans, birds, and pigs by Gibbs et al. (2001) suggests that the 1918 virus hemagglutinin gene was a recombinant and that the recombination event occurred at about the same time as the Spanish flu pandemic. The course of events suggested by Gibbs et al. (2001) is shown in Figure 18.16. They proposed that the influenza virus switched from its original bird hosts to mammals and that there was initially independent evolution of human and swine influenza genomes. Early in the 1900s there were one or more recombination events between the globular domain of the hemagglutinin gene in the swine influenza genome and the stalk domain of the human hemagglutinin gene that gener-

ated the virulent 1918 influenza genome. A population genomics approach (Chapter 32) was used to define the actual location of the recombination (Fig. 18.17). Sometime within the last 50 years there was an additional recombination between the 1918 influenza genome with the original swine lineage, and it appears that the 1957 (Asian) and 1968 (Hong Kong) strains arose in a very similar way. Following the 1918 epidemic, the reassortant virus and the original swine lineages went extinct.

Molecular epidemiology and population genomics can also be used to identify mechanisms of pathogenesis. The avian influenza virus (subtype H5N1) that infected 18 Hong Kong residents in 1997, killing 6, was not a reassortant. Nucleotide sequencing showed that all of its genes were of avian virus origin. This highly lethal chicken virus had spread from chickens to people but had not adapted to spread from person to

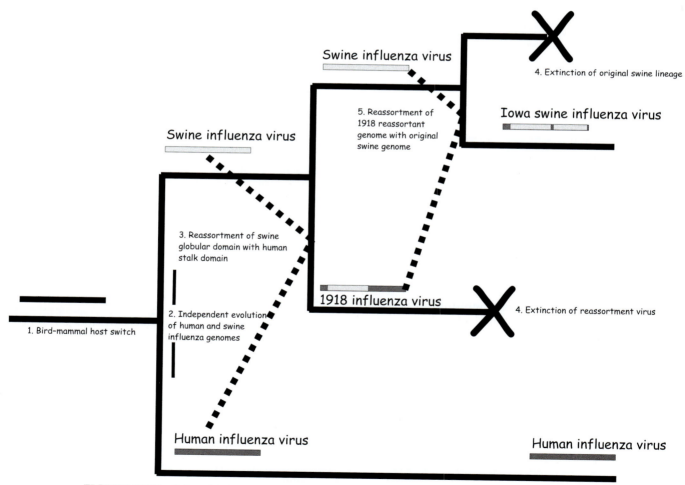

FIGURE 18.16 Inferred history of the influenza virus (family Orthomyxoviridae) that caused the 1918 "Spanish influenza" pandemic. Redrawn from Gibbs et al. (2001).

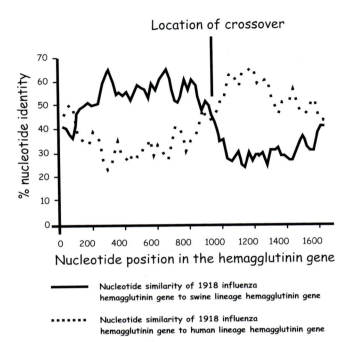

FIGURE 18.17 A graphical approach toward mapping the location of a crossover event in the hemagglutinin gene in the swine influenza genome and the stalk domain of the human hemagglutinin gene that generated the virulent 1918 influenza genome. Redrawn from Gibbs et al. (2001).

person, and killing all the chickens in Hong Kong stopped the epidemic. In 2001, another H5N1 virus appeared in the live chicken markets in Hong Kong. This virus has not infected anyone, and it differs from the 1997 H5N1 virus in its internal genes. Nevertheless, all of the chickens in Hong Kong were again slaughtered as a precaution. Using RT-PCR and infectious clones, Hatta et al. (2001) reconstructed some of the H5N1 viruses that killed the six people in Hong Kong in 1997 in an attempt to find out why this avian virus was so virulent for humans. They were able to divide the H5N1 viruses into two groups with high or low pathogenicity in mice. As with Gibbs et al. (2001), they demonstrated that hemagglutinin is partially responsible for the difference in virulence between the two groups of viruses. Strains with an isoleucine or serine at hemagglutinin position 227 had reduced virulence. In addition, the gene encoding one of the internal polymerase proteins was also partially responsible for the difference in virulence between the two groups of viruses.

SUMMARY

Arthropod-borne pathogens are not static entities, as shown by the recent emergence (or in many cases reemergence) of known species and the appearance of altogether-new species. This chapter discussed epidemiological, principles and tools that are critical to the development of molecular epidemiological methods for arthropod-borne diseases. Molecular genetic techniques available for the rapid detection and identification of these pathogens were reviewed in relation to their importance to understanding molecular epidemiology. Detection and diagnosis of these pathogens require the understanding of specific epidemiological principles in order to assess their impact on both host and environment. We used several examples to demonstrate the applications of molecular and epidemiological techniques.

Readings

Abderrazak, S. B., Oury, B., Lal, A. A., Bosseno, M. F., Force-Barge, P., Dujardin, J. P., Fandeur, T., Molez, J. F., Kjellberg, F., Ayala, F. J., and Tibayrenc, M. 1999. *Plasmodium falciparum*: Population genetic analysis by multilocus enzyme electrophoresis and other molecular markers. *Exp. Parasitol.* 92: 232–238.

Aviles, H., Belli, A., Armijos, R., Monroy, F. P., and Harris, E. 1999. PCR detection and identification of *Leishmania* parasites in clinical specimens in Ecuador: A comparison with classical diagnostic methods. *J. Parasitol.* 85: 181–187.

Bachellier, S., Saurin, W., Perrin, D., Hofnung, M., and Gilson, E. 1994. Structural and functional diversity among bacterial interspersed mosaic elements (BIMEs). *Mol. Microbiol.* 12: 61–70.

Barillas-Mury, C., Han, Y. S., Seeley, D., and Kafatos, F. C. 1999. *Anopheles gambiae* Ag-STAT, a new insect member of the STAT family, is activated in response to bacterial infection. *Embo. J.* 18: 959–967.

Barlough, J. E., Madigan, J. E., DeRock, E., and Bigornia, L. 1996. Nested polymerase chain reaction for detection of *Ehrlichia equi* genomic DNA in horses and ticks (*Ixodes pacificus*). *Vet. Parasitol.* 63: 319–329.

Bashiruddin, J. B., Camma, C., and Rebelo, E. 1999. Molecular detection of *Babesia equi* and *Babesia caballi* in horse blood by PCR amplification of part of the 16S rRNA gene. *Vet. Parasitol.* 84: 75–83.

Beaty, B. J., and Bishop, D. H. 1988. Bunyavirus–vector interactions. *Virus Res.* 10: 289–301.

Beerntsen, B. T., Severson, D. W., Klinkhammer, J. A., Kassner, V. A., and Christensen, B. M. 1995. *Aedes aegypti*: A quantitative trait locus (QTL) influencing filarial worm intensity is linked to QTL for susceptibility to other mosquito-borne pathogens. *Exp. Parasitol.* 81: 355–362.

Berry, W. J., Rowley, W. A., Clarke, J. L., 3rd, Swack, N. S., and Hausler, W. J., Jr. 1987. Spontaneous flight activity of *Aedes trivittatus* (Diptera: Culicidae) infected with trivittatus virus (Bunyaviridae: California serogroup). *J. Med. Entomol.* 24: 286–289.

Berry, W. J., Rowley, W. A., and Christensen, B. M. 1988. Spontaneous flight activity of *Aedes trivittatus* infected with *Dirofilaria immitis*. *J. Parasitol.* 74: 970–974.

Billinis, C., Koumbati, M., Spyrou, V., Nomikou, K., Mangana, O., Panagiotidis, C. A., and Papadopoulos, O. 2001. Bluetongue virus diagnosis of clinical cases by a duplex reverse transcription-PCR: A comparison with conventional methods. *J. Virol. Methods* 98: 77–89.

Boerlin, P., Peter, O., Bretz, A. G., Postic, D., Baranton, G., and Piffaretti, J. C. 1992. Population genetic analysis of *Borrelia burgdorferi* isolates by multilocus enzyme electrophoresis. *Infect. Immun.* 60: 1677–1683.

Bremer, C. W., and Viljoen, G. J. 1998. Detection of African horsesickness virus and discrimination between two equine orbivirus serogroups by reverse transcription polymerase chain reaction. *Onderstepoort J. Vet. Res.* 65: 1–8.

Breniere, S. F., Telleria, J., Bosseno, M. F., Buitrago, R., Bastrenta, B., Cuny, G., Banuls, A. L., Brewster, S., and Barker, D. C. 1999. Polymerase chain reaction–based identification of New World *Leishmania* species complexes by specific kDNA probes. *Acta Trop.* 73: 283–293.

Brooks, D. R., and McLennan, D. A. 1993. *Parascript: Parasites and the Language of Evolution.* Smithsonian Institution Press, Washington, DC.

Burt, F. J., Leman, P. A., Smith, J. F., and Swanepoel, R. 1998. The use of a reverse transcription-polymerase chain reaction for the detection of viral nucleic acid in the diagnosis of Crimean-Congo haemorrhagic fever. *J. Virol. Methods* 70: 129–137.

Campbell, W. P., and Huang, C. 1996. Detection of California serogroup bunyaviruses in tissue culture and mosquito pools by PCR. *J. Virol. Methods* 57: 175–179.

Charlab, R., and Ribeiro, J. M. 1993. Cytostatic effect of *Lutzomyia longipalpis* salivary gland homogenates on *Leishmania* parasites. *Am. J. Trop. Med. Hyg.* 48: 831–838.

Charlab, R., Tesh, R. B., Rowton, E. D., and Ribeiro, J. M. 1995. *Leishmania amazonensis*: Sensitivity of different promastigote morphotypes to salivary gland homogenates of the sand fly *Lutzomyia longipalpis*. *Exp. Parasitol.* 80: 167–175.

Chen, C. C., and Laurence, B. R. 1985. The encapsulation of the sheaths of microfilariae of *Brugia pahangi* in the haemocoel of mosquitoes. *J. Parasitol.* 71: 834–836.

Chiurillo, M. A., Sachdeva, M., Dole, V. S., Yepes, Y., Miliani, E., Vazquez, L., Rojas, A., Crisante, G., Guevara, P., Anez, N., Madhubala, R., and Ramirez, J. L. 2001. Detection of *Leishmania* causing visceral leishmaniasis in the Old and New Worlds by a polymerase chain reaction assay based on telomeric sequences. *Am. J. Trop. Med. Hyg.* 65: 573–582.

Collins, F. H., Sakai, R. K., Vernick, K. D., Paskewitz, S., Seeley, D. C., Miller, L. H., Collins, W. E., Campbell, C. C., and Gwadz, R. W. 1986. Genetic selection of a *Plasmodium*-refractory strain of the malaria vector *Anopheles gambiae*. *Science* 234: 607–610.

Desquesnes, M., McLaughlin, G., Zoungrana, A., and Davila, A. M. 2001. Detection and identification of *Trypanosoma* of African livestock through a single PCR based on internal transcribed spacer 1 of rDNA. *Int. J. Parasitol.* 31: 610–614.

Deubel, V., Huerre, M., Cathomas, G., Drouet, M. T., Wuscher, N., Le Guenno, B., and Widmer, A. F. 1997. Molecular detection and characterization of yellow fever virus in blood and liver specimens of a nonvaccinated fatal human case. *J. Med. Virol.* 53: 212–217.

Dolan, S. A., Dommaraju, C. B., and DeGuzman, G. B. 1998. Detection of *Francisella tularensis* in clinical specimens by use of polymerase chain reaction. *Clin. Infect. Dis.* 26: 764–765.

Dolina, M., and Peduzzi, R. 1993. Population genetics of human, animal, and environmental *Yersinia* strains. *Appl. Environ. Microbiol.* 59: 442–450.

Engelthaler, D. M., Gage, K. L., Montenieri, J. A., Chu, M., and Carter, L. G. 1999. PCR detection of *Yersinia pestis* in fleas: Comparison with mouse inoculation. *J. Clin. Microbiol.* 37: 1980–1984.

Fahrimal, Y., Goff, W. L., and Jasmer, D. P. 1992. Detection of *Babesia bovis* carrier cattle by using polymerase chain reaction amplification of parasite DNA. *J. Clin. Microbiol.* 30: 1374–1379.

Favia, G., Cancrini, G., Carfi, A., Grazioli, D., Lillini, E., and Iori, A. 2001. Molecular identification of *Borrelia valaisiana* and HGE-like *Ehrlichia* in *Ixodes ricinus* ticks sampled in northeastern Italy: First report in Veneto region. *Parassitologia* 43: 143–146.

Figueroa, J. V., Chieves, L. P., Johnson, G. S., and Buening, G. M. 1992. Detection of *Babesia bigemina*–infected carriers by polymerase chain reaction amplification. *J. Clin. Microbiol.* 30: 2576–2582.

Fischer, P., Liu, X., Lizotte-Waniewski, M., Kamal, I. H., Ramzy, R. M., and Williams, S. A. 1999. Development of a quantitative, competitive polymerase chain reaction—enzyme-linked immunosorbent assay for the detection of *Wuchereria bancrofti* DNA. *Parasitol. Res.* 85: 176–183.

Gage, K. L., Gilmore, R. D., Karstens, R. H., and Schwan, T. G. 1992. Detection of *Rickettsia rickettsii* in saliva, hemolymph and triturated tissues of infected *Dermacentor andersoni* ticks by polymerase chain reaction. *Mol. Cell Probes* 6: 333–341.

Galen, R. S., and Gambino, S. R. 1975. *Beyond Normality: The Predictive Value and Efficiency of Medical Diagnoses*. Wiley, New York.

Gallego, M., Pratlong, F., Riera, C., Munoz, C., Ribera, E., Fisa, R., Rioux, J. A., Dedet, J. P., and Portus, M. 2002. Isoenzymatic identification of *Leishmania* isolates from repeated clinical human leishmaniasis episodes in Catalonia (Spain). *Trans. R. Soc. Trop. Med. Hyg.* 96: 45–47.

Garcia, E. S., Mello, C. B., Azambuja, P., and Ribeiro, J. M. 1994. *Rhodnius prolixus*: Salivary antihemostatic components decrease with *Trypanosoma rangeli* infection. *Exp. Parasitol.* 78: 287–293.

Garcia, S., Crance, J. M., Billecocq, A., Peinnequin, A., Jouan, A., Bouloy, M., and Garin, D. 2001. Quantitative real-time PCR detection of Rift Valley fever virus and its application to evaluation of antiviral compounds. *J. Clin. Microbiol.* 39: 4456–4461.

Georges, K., Loria, G. R., Riili, S., Greco, A., Caracappa, S., Jongejan, F., and Sparagano, O. 2001. Detection of haemoparasites in cattle by reverse line blot hybridization, with a note on the distribution of ticks in Sicily. *Vet. Parasitol.* 99: 273–286.

Gibbs, M. J., Armstrong, J. S., and Gibbs, A. J. 2001. Recombination in the hemagglutinin gene of the 1918 "Spanish flu." *Science* 293: 1842–1845.

Hamano, S., Horio, M., Miura, S., Higo, H., Iihoshi, N., Noda, K., Tada, I., and Takeuchi, T. 2001. Detection of kinetoplast DNA of *Trypanosoma cruzi* from dried feces of triatomine bugs by PCR. *Parasitol. Int.* 50: 135–138.

Han, Y. S., Thompson, J., Kafatos, F. C., and Barillas-Mury, C. 2000. Molecular interactions between *Anopheles stephensi* midgut cells and *Plasmodium berghei*: The time bomb theory of ookinete invasion of mosquitoes. *Embo. J.* 19: 6030–6040.

Hatta, M., Gao, P., Halfmann, P., and Kawaoka, Y. 2001. Molecular basis for high virulence of Hong Kong H5N1 influenza A viruses. *Science* 293: 1840–1842.

Hinnebusch, B. J., Rudolph, A. E., Cherepanov, P., Dixon, J. E., Schwan, T. G., and Forsberg, A. 2002. Role of *Yersinia* murine toxin in survival of *Yersinia pestis* in the midgut of the flea vector. *Science* 296: 733–735.

Holmes, E. C., Worobey, M., and Rambaut, A. 1999. Phylogenetic evidence for recombination in dengue virus. *Mol. Biol. Evol.* 16: 405–409.

Hoti, S. L., Vasuki, V., Lizotte, M. W., Patra, K. P., Ravi, G., Vanamail, P., Manonmani, A., Sabesan, S., Krishnamoorthy, K., and Williams, S. A. 2001. Detection of *Brugia malayi* in laboratory and wild-caught *Mansonioides* mosquitoes (Diptera: Culicidae) using Hha I PCR assay. *Bull. Entomol. Res.* 91: 87–92.

Huang, C., Slater, B., Campbell, W., Howard, J., and White, D. 2001. Detection of arboviral RNA directly from mosquito homogenates by reverse-transcription-polymerase chain reaction. *J. Virol. Methods* 94: 121–128.

Kahl, O., Gern, L., Gray, J. S., Guy, E. C., Jongejan, F., Kirstein, F., Kurtenbach, K., Rijpkema, S. G., and Stanek, G. 1998. Detection of *Borrelia burgdorferi* sensu lato in ticks: Immunofluorescence assay versus polymerase chain reaction. *Zentralbl. Bakteriol.* 287: 205–210.

Kelly, D. J., Dasch, G. A., Chan, T. C., and Ho, T. M. 1994. Detection and characterization of *Rickettsia tsutsugamushi* (Rickettsiales: Rickettsiaceae) in infected *Leptotrombidium* (*Leptotrombidium*) *fletcheri* chiggers (Acari: Trombiculidae) with the polymerase chain reaction. *J. Med. Entomol.* 31: 691–699.

Kirvar, E., Ilhan, T., Katzer, F., Hooshmand-Rad, P., Zweygarth, E., Gerstenberg, C., Phipps, P., and Brown, C. G. 2000. Detection of *Theileria annulata* in cattle and vector ticks by PCR using the Tams1 gene sequences. *Parasitology* 120: 245–254.

Klompen, J. S., Black, W. C., Keirans, J. E., and Oliver, J. H., Jr. 1996. Evolution of ticks. *Annu. Rev. Entomol.* 41: 141–161.

Kramer, L. D., Chiles, R. E., Do, T. D., and Fallah, H. M. 2001. Detection of St. Louis encephalitis and western equine encephalomyelitis RNA in mosquitoes tested without maintenance of a cold chain. *J. Am. Mosq. Control Assoc.* 17: 213–215.

La Scola, B., Fournier, P. E., Brouqui, P., and Raoult, D. 2001. Detection and culture of *Bartonella quintana*, *Serratia marcescens*, and *Acinetobacter* spp. from decontaminated human body lice. *J. Clin. Microbiol.* 39: 1707–1709.

Le Blancq, S. M., Cibulskis, R. E., and Peters, W. 1986. *Leishmania* in the Old World: 5. Numerical analysis of isoenzyme data. *Trans. R. Soc. Trop. Med. Hyg.* 80: 517–524.

Lew, A. E., Bock, R. E., Minchin, C. M., and Masaka, S. 2002. A msp1alpha polymerase chain reaction assay for specific detection and differentiation of *Anaplasma marginale* isolates. *Vet. Microbiol.* 86: 325–335.

Linssen, B., Kinney, R. M., Aguilar, P., Russell, K. L., Watts, D. M., Kaaden, O. R., and Pfeffer, M. 2000. Development of reverse transcription-PCR assays specific for detection of equine encephalitis viruses. *J. Clin. Microbiol.* 38: 1527–1535.

Ludwig, G. V., Kondig, J. P., and Smith, J. F. 1996. A putative receptor for Venezuelan equine encephalitis virus from mosquito cells. *J. Virol.* 70: 5592–5599.

Machado, E. M., Alvarenga, N. J., Romanha, A. J., and Grisard, E. C. 2000. A simplified method for sample collection and DNA isolation for polymerase chain reaction detection of *Trypanosoma rangeli* and *Trypanosoma cruzi* in triatomine vectors. *Mem. Inst. Oswaldo Cruz* 95: 863–866.

Mahan, S. M., Peter, T. F., Simbi, B. H., and Burridge, M. J. 1998. PCR detection of *Cowdria ruminantium* infection in ticks and animals from heartwater-endemic regions of Zimbabwe. *Ann. NY Acad. Sci.* 849: 85–87.

Martin-Sanchez, J., Lopez-Lopez, M. C., Acedo-Sanchez, C., Castro-Fajardo, J. J., Pineda, J. A., and Morillas-Marquez, F. 2001. Diagnosis of infections with *Leishmania infantum* using PCR-ELISA. *Parasitology* 122: 607–615.

Masake, R. A., Njuguna, J. T., Brown, C. C., and Majiwa, P. A. 2002. The application of PCR-ELISA to the detection of *Trypanosoma brucei* and *T. vivax* infections in livestock. *Vet. Parasitol.* 105: 179–189.

McGreevy, P. B., Bryan, J. H., Oothuman, P., and Kolstrup, N. 1978. The lethal effects of the cibarial and pharyngeal armatures of mosquitoes on microfilariae. *Trans. R. Soc. Trop. Med. Hyg.* 72: 361–368.

Meiyu, F., Huosheng, C., Cuihua, C., Xiaodong, T., Lianhua, J., Yifei, P., Weijun, C., and Huiyu, G. 1997. Detection of flaviviruses by reverse transcriptase-polymerase chain reaction with the universal primer set. *Microbiol. Immunol.* 41: 209–213.

Moody, A. 2002. Rapid diagnostic tests for malaria parasites. *Clin. Microbiol. Rev.* 15: 66–78.

Moreli, M. L., Aquino, V. H., and Figueiredo, L. T. 2001. Identification of Simbu, California and Bunyamwera serogroup bunyaviruses by nested RT-PCR. *Trans. R. Soc. Trop. Med. Hyg.* 95: 108–113.

Nichol, S. T., Spiropoulou, C. F., Morzunov, S., Rollin, P. E., Ksiazek, T. G., Feldmann, H., Sanchez, A., Childs, J., Zaki, S., and Peters, C. J. 1993. Genetic identification of a hantavirus associated with an outbreak of acute respiratory illness. *Science* 262: 914–917.

Norris, D. E., Johnson, B. J., Piesman, J., Maupin, G. O., Clark, J. L., and Black, W. C. T. 1997. Culturing selects for specific genotypes of *Borrelia burgdorferi* in an enzootic cycle in Colorado. *J. Clin. Microbiol.* 35: 2359–2364.

Noyes, H., Pratlong, F., Chance, M., Ellis, J., Lanotte, G., and Dedet, J. P. 2002. A previously unclassified trypanosomatid responsible for human cutaneous lesions in Martinique (French West Indies) is the most divergent member of the genus *Leishmania ss*. *Parasitology* 124: 17–24.

Okabayashi, T., Hagiya, J., Tsuji, M., Ishihara, C., Satoh, H., and Morita, C. 2002. Detection of *Babesia microti*–like parasite in filter paper–absorbed blood of wild rodents. *J. Vet. Med. Sci.* 64: 145–147.

Paranjpe, S., and Banerjee, K. 1998. Detection of Japanese encephalitis virus by reverse transcription/polymerase chain reaction. *Acta Virol.* 42: 5–11.

Parkes, R., Lo, T., Wong, Q., Isaac-Renton, J. L., and Byrne, S. K. 2001. Comparison of a nested polymerase chain reaction-restriction fragment length polymorphism method, the PATH antigen detection method, and microscopy for the detection and identification of malaria parasites. *Can. J. Microbiol.* 47: 903–907.

Pfeffer, M., Proebster, B., Kinney, R. M., and Kaaden, O. R. 1997. Genus-specific detection of alphaviruses by a seminested reverse transcription-polymerase chain reaction. *Am. J. Trop. Med. Hyg.* 57: 709–718.

Pfeffer, M., Linssen, B., Parke, M. D., and Kinney, R. M. 2002. Specific detection of chikungunya virus using a RT-PCR/nested PCR combination. *J. Vet. Med. B Infect. Dis. Vet. Public Health* 49: 49–54.

Picken, R. N. 1992. Polymerase chain reaction primers and probes derived from flagellin gene sequences for specific detection of the agents of Lyme disease and North American relapsing fever. *J. Clin. Microbiol.* 30: 99–114.

Pimenta, P. F., Modi, G. B., Pereira, S. T., Shahabuddin, M., and Sacks, D. L. 1997. A novel role for the peritrophic matrix in protecting *Leishmania* from the hydrolytic activities of the sand fly midgut. *Parasitology* 115: 359–369.

Pletnev, A. G., Putnak, R., Speicher, J., Wagar, E. J., and Vaughn, D. W. 2002. West Nile virus/dengue type 4 virus chimeras that are reduced in neurovirulence and peripheral virulence without loss of immunogenicity or protective efficacy. *Proc. Natl. Acad. Sci. USA* 99: 3036–3041.

Rauter, C., Oehme, R., Diterich, I., Engele, M., and Hartung, T. 2002. Distribution of clinically relevant *Borrelia* genospecies in ticks assessed by a novel, single-run, real-time PCR. *J. Clin. Microbiol.* 40: 36–43.

Rossignol, P. A., Ribeiro, J. M., and Spielman, A. 1984. Increased intradermal probing time in sporozoite-infected mosquitoes. *Am. J. Trop. Med. Hyg.* 33: 17–20.

Rossignol, P. A., Ribeiro, J. M., and Spielman, A. 1986. Increased biting rate and reduced fertility in sporozoite-infected mosquitoes. *Am. J. Trop. Med. Hyg.* 35: 277–279.

Roux, V., and Raoult, D. 1999. Body lice as tools for diagnosis and surveillance of reemerging diseases. *J. Clin. Microbiol.* 37: 596–599.

Rubio, J. M., Benito, A., Berzosa, P. J., Roche, J., Puente, S., Subirats, M., Lopez-Velez, R., Garcia, L., and Alvar, J. 1999a. Usefulness of seminested multiplex PCR in surveillance of imported malaria in Spain. *J. Clin. Microbiol.* 37: 3260–3266.

Rubio, J. M., Benito, A., Roche, J., Berzosa, P. J., Garcia, M. L., Mico, M., Edu, M., and Alvar, J. 1999b. Seminested, multiplex polymerase chain reaction for detection of human malaria parasites and evidence of *Plasmodium vivax* infection in Equatorial Guinea. *Am. J. Trop. Med. Hyg.* 60: 183–187.

Salotra, P., Sreenivas, G., Pogue, G. P., Lee, N., Nakhasi, H. L., Ramesh, V., and Negi, N. S. 2001. Development of a species-specific PCR assay for detection of *Leishmania donovani* in clinical samples from patients with kala-azar and post-kala-azar dermal leishmaniasis. *J. Clin. Microbiol.* 39: 849–854.

Schlein, Y., Jacobson, R. L., and Shlomai, J. 1991. Chitinase secreted by *Leishmania* functions in the sandfly vector. *Proc. R. Soc. Lond. B Biol. Sci.* 245: 121–126.

Schwan, T. G., and Piesman, J. 2000. Temporal changes in outer surface proteins A and C of the lyme disease–associated spirochete, *Borrelia burgdorferi*, during the chain of infection in ticks and mice. *J. Clin. Microbiol.* 38: 382–388.

Shayan, P., Biermann, R., Schein, E., Gerdes, J., and Ahmed, J. S. 1998. Detection and differentiation of *Theileria annulata* and *Theileria parva* using macroschizont-derived DNA probes. *Ann. NY Acad. Sci.* 849: 88–95.

Shi, P. Y., Kauffman, E. B., Ren, P., Felton, A., Tai, J. H., Dupuis, A. P., 2nd, Jones, S. A., Ngo, K. A., Nicholas, D. C., Maffei, J., Ebel, G. D., Bernard, K. A., and Kramer, L. D. 2001. High-throughput detection of West Nile virus RNA. *J. Clin. Microbiol.* 39: 1264–1271.

Tanaka, M., Onoe, S., Matsuba, T., Katayama, S., Yamanaka, M., Yonemichi, H., Hiramatsu, K., Baek, B. K., Sugimoto, C., and Onuma, M. 1993. Detection of *Theileria sergenti* infection in cattle by polymerase chain reaction amplification of parasite-specific DNA. *J. Clin. Microbiol.* 31: 2565–2569.

Tishkoff, S. A., Varkonyi, R., Cahinhinan, N., Abbes, S., Argyropoulos, G., Destro-Bisol, G., Drousiotou, A., Dangerfield, B., Lefranc, G., Loiselet, J., Piro, A., Stoneking, M., Tagarelli, A., Tagarelli, G., Touma, E. H., Williams, S. M., and Clark, A. G. 2001. Haplotype diversity and linkage disequilibrium at human G6PD: Recent origin of alleles that confer malarial resistance. *Science* 293: 455–462.

Toure, F. S., Mavoungou, E., Kassambara, L., Williams, T., Wahl, G., Millet, P., and Egwang, T. G. 1998. Human occult loiasis: Field evaluation of a nested polymerase chain reaction assay for the detection of occult infection. *Trop. Med. Int. Health* 3: 505–511.

Uzcategui, N. Y., Camacho, D., Comach, G., Cuello de Uzcategui, R., Holmes, E. C., and Gould, E. A. 2001. Molecular epidemiology of dengue type 2 virus in Venezuela: Evidence for in situ virus evolution and recombination. *J. Gen. Virol.* 82: 2945–2953.

Vinetz, J. M., Valenzuela, J. G., Specht, C. A., Aravind, L., Langer, R. C., Ribeiro, J. M., and Kaslow, D. C. 2000. Chitinases of the avian malaria parasite *Plasmodium gallinaceum*, a class of enzymes necessary for parasite invasion of the mosquito midgut. *J. Biol. Chem.* 275: 10331–10341.

Volkman, S. K., Barry, A. E., Lyons, E. J., Nielsen, K. M., Thomas, S. M., Choi, M., Thakore, S. S., Day, K. P., Wirth, D. F., and Hartl, D. L. 2001. Recent origin of *Plasmodium falciparum* from a single progenitor. *Science* 293: 482–484.

Warrilow, D., Northill, J. A., Pyke, A., and Smith, G. A. 2002. Single rapid TaqMan fluorogenic–probe based PCR assay that detects all four dengue serotypes. *J. Med. Virol.* 66: 524–528.

Wasieloski, L. P., Jr., Rayms-Keller, A., Curtis, L. A., Blair, C. D., and Beaty, B. J. 1994. Reverse transcription-PCR detection of La Crosse virus in mosquitoes and comparison with enzyme immunoassay and virus isolation. *J. Clin. Microbiol.* 32: 2076–2080.

Watts, K. J., Courteny, C. H., and Reddy, G. R. 1999. Development of a PCR- and probe-based test for the sensitive and specific detection of the dog heartworm, *Dirofilaria immitis*, in its mosquito intermediate host. *Mol. Cell. Probes* 13: 425–430.

Webb, L., Carl, M., Malloy, D. C., Dasch, G. A., and Azad, A. F. 1990. Detection of murine typhus infection in fleas by using the polymerase chain reaction. *J. Clin. Microbiol.* 28: 530–534.

Weigle, K. A., Labrada, L. A., Lozano, C., Santrich, C., and Barker, D. C. 2002. PCR-based diagnosis of acute and chronic cutaneous leishmaniasis caused by *Leishmania* (*Viannia*). *J. Clin. Microbiol.* 40: 601–606.

Xu, G., Wilson, W., Mecham, J., Murphy, K., Zhou, E. M., and Tabachnick, W. 1997. VP7: An attachment protein of bluetongue virus for cellular receptors in *Culicoides variipennis*. *J. Gen. Virol.* 78: 1617–1623.

Zhang, S., Li, B. W., and Weil, G. J. 2000. Paper chromatography hybridization: A rapid method for detection of *Onchocerca volvulus* DNA amplified by PCR. *Am. J. Trop. Med. Hyg.* 63: 85–89.

19

Surveillance of Vector-Borne Diseases

CHESTER G. MOORE AND KENNETH L. GAGE

INTRODUCTION

Surveillance is "an organized system of collecting data" about the phenomenon under study (Bowen and Francy 1980). For a vector-borne disease, there are at least four different components of a complete surveillance program: (1) detection of disease in humans or domestic animals, (2) surveillance of vectors, (3) surveillance of pathogen activity in wild vertebrate hosts, and (4) study of weather patterns related to pathogen transmission (Bowen and Francy 1980). In this chapter, we deal only with the second of these components. For this chapter, then, a surveillance system is any procedure or group of procedures that collects the estimates of vector population density we need to predict, prevent, or control vector-borne disease.

Surveillance programs for vector-borne disease are carried out to anticipate and to prevent or control disease in humans or domestic animals. The specific type of surveillance system and the methods to be used are determined by the objective of the overall program. Thus, it is crucial to know what question(s) we seek to answer before designing the surveillance program. For example, if the objective is to *prevent* disease in humans, a surveillance program that only records the occurrence of human cases will be of little or no use. Instead, we need to measure the *predictors* of human cases. Depending on the type of information desired, different collection methods and equipment may be required. We must know which methods and equipment to use for a given purpose.

Programs may be either *intensive* or *extensive*. Intensive studies collect detailed data on one or very few species in a very restricted geographic area. Extensive studies collect data on many species over a large geographic area (Southwood 1978). While research projects frequently are intensive, vector-borne disease prevention and control programs generally involve extensive surveillance programs. In this chapter we concentrate on those surveillance methods that are used in prevention and control programs. However, we also point out methods that are useful in a research setting when they help to answer specific questions about the ecology of vector-borne disease.

DESIGN OF SURVEILLANCE PROGRAMS

Absolute vs. Relative Density

Absolute population estimates of density per unit area (e.g., number per hectare) are essential for the proper construction of life budgets and similar studies (Southwood 1978). However, relative estimates such as catch per unit of effort (e.g., females per trap night, larvae per dip) are adequate for most surveillance systems. In a few cases, animal products (e.g., pupal skins, puparia of muscoid flies) can be sampled to provide a rough population index. If simultaneous data are collected on absolute and relative density, correction factors can be calculated that permit the computation of absolute densities (Southwood 1978).

Sampling Bias and Assumptions

Certain assumptions are implicit in all sampling and trapping procedures. The impact of different assumptions frequently becomes evident only when two or more sampling methods are applied simultaneously. We must recognize hidden assumptions when designing a surveillance program. For example, are all physiologic stages and ages equally collected by this method? Does this trap have uniform sampling efficiency in all environments? Does the knockdown method really collect all the mosquitoes in a room?

More complex assumptions are reflected in decisions to sample only particular organisms. For example, we assume houseflies are not involved in transmitting any arboviral diseases of humans or domestic animals. Therefore, these insects are rarely processed for arbovirus isolation. However, when houseflies were studied during a vesicular stomatitis outbreak in Colorado, numerous VSV (Indiana) isolates were obtained (Francy et al. 1988). The actual role of houseflies in the dynamics of this disease remains speculative, however.

Sampling systems are usually biased in one way or another. If these biases are known and understood, they can be put to good use (intentional bias). However, when the biases in a system go unrecognized (unintentional bias), they can severely affect the interpretation of the data collected by that system. For example, entrance and exit traps can be used in malaria studies to sample only that portion of the Anopheline population that is entering or leaving houses. Gravid traps can be used to sample only the gravid *Culex pipiens* population. This increases the likelihood of arbovirus isolation, since these females (at least in anautogenous species) have previously taken at least one blood meal.

On the other hand, if the objective is to estimate absolute density but all traps have been placed along fencerows or other flight corridors, inflated estimates will be generated. Traps placed along flight corridors will collect more specimens than traps located in the middle of an open field or in similar habitats, thus artificially increasing the catch. Frequently, a vector surveillance program requires the placement and retrieval of many traps, put out and retrieved 1–5 days each week. It is tempting to place the traps in easily accessible sites for rapid servicing, without consideration of the habits or behavior of the vector populations we wish to sample. This will produce inaccurate data.

The placement of mosquito traps near larval habitats may result in collections of larger proportions of young individuals. This is undesirable if the specimens are to be processed for arbovirus isolation, because young individuals are unlikely to be infected.

The number of traps and sampling sites in a surveillance system depends on the biology and behavior of the vector being monitored and on the size of the area under surveillance. Cost is often important in surveillance. For a given cost, the most accurate estimates are obtained by taking many small samples rather than by taking fewer large ones (Southwood 1978). The cost of using different sampling systems may also vary, and this will be a critical issue for long-term surveillance programs.

Handling and Processing of Specimens

For some applications, specimens only need to be handled so they are identifiable. For other applications (e.g., blood meal identification, ovarian dissection, pathogen isolation), specimens must be handled in special ways. It is very important that those responsible for fieldwork clearly understand how specimens should be handled.

TICKS AND MITES

Tick Collection (Ixodidae)

Ticks are actually large mites (Order Parasitiformes). Many of the techniques used for collecting these ectoparasites also can be used to collect smaller mite taxa. Nonetheless, ticks are usually discussed separately from other mite groups, and we will continue the practice in this chapter. The most commonly used methods for collecting hard ticks (Ixodidae) are dragging, examination of captured hosts, and using attractants to lure ticks to a trap or collecting area (Falco and Fish 1992). Some ixodid ticks also can be collected from nest materials, as described later for argasid ticks. No single method is likely to apply for sampling all species or life cycle stages. Often a combination of techniques must be employed to gather the desired information.

Dragging (or *flagging*) involves pulling a cloth drag across vegetation where unfed ticks are questing for hosts. The questing ticks will cling to the passing drag and can be collected with forceps or by hand-picking at the end of each pass. Dragging is useful for collecting all three questing stages (larvae, nymphs, and adults) of many ixodid species, including *Amblyomma americanum* and members of the *Ixodes ricinus* complex (Milne 1943; Semtner and Hair 1973; Falco and Fish 1988). Other species can be sampled by dragging only during certain life cycle stages. For example, immature *Dermacentor andersoni* or *D. variabilis* are rarely encountered on drags, but questing adults of these species can be collected in large numbers by dragging (Philip 1937; Sonenshine et al. 1966). Conversely, larvae of the one-host tick *Boophilus microplus* can be collected by dragging (Zimmerman and Garris 1985), but unfed nymphs and adults remain on their ungulate hosts between feedings and must be collected by other means. Off-host stages of other hard ticks, especially those species of *Ixodes* inhabiting nests, roosts, caves, or cliff habitats, are unlikely to be encountered on drags because the unfed ticks remain sequestered in these habitats until their hosts return.

Drags usually consist of a piece of cloth (about 1 meter square) attached on one end to a pole. A rope is then attached to both ends of the pole and the drag is pulled across vegetation where ticks are questing (Fig. 19.1). Weights are sometimes sewn into the side of the drag opposite from the pole. This ensures that the drag stays in close contact with the vegetation (Falco and Fish 1988, 1992). Dragging is typically done for a standard amount of time or distance to give a relative density estimate of the number of ticks in the area. Another variant of the dragging technique uses a "flag" consisting of a cloth attached to one end of a long pole. This flag is then swept across vegetation where ticks are questing. Flags are most useful when ticks are questing on or near dense brush where it is difficult to pull a typical drag across this kind of vegetation.

Cylinder-shaped drags and hinge-type flag devices also have been described. These samplers were found slightly less efficient than traditional tick drags for collecting larvae of *Boophilus microplus* (Zimmerman and Garris 1985). Although cylinder and hinge flag samplers were less efficient than traditional drags in pasture environments, their heavier weight and greater sturdiness might make them superior for use in areas with dense brush (Zimmerman and Garris 1985).

FIGURE 19.1   Tick drag, a standard technique in tick surveillance. Photo by L. Carter, CDC.

Ixodid ticks also can be collected by using host animals or carbon dioxide as attractants (J. G. Wilson et al. 1972; Koch and McNew 1981, 1982; Falco and Fish 1992). Appropriate host animals are placed in cages near areas where ticks are likely to seek hosts. The animals are collected after a given period and examined for ticks. Traps baited with live animals also can be surrounded with sticky tape that captures the ticks as they crawl toward the host (Koch and McNew 1981).

Many kinds of CO_2 traps have been described for collecting various ixodid ticks, including species of *Amblyomma* and *Ixodes* (Koch and McNew 1981; Kinzer et al. 1990; Falco and Fish 1992). Koch and McNew (1981) compared different trap designs. They found that a simple collecting device consisting of a 365- to 370-g cube of dry ice placed on a white cotton cloth (0.7 by 0.9 m) was more effective than rabbit-baited traps, foam bucket CO_2 traps baited with dry ice, and dry chemical traps. Kinzer et al. (1990) compared CO_2-baited traps with flagging and found that significantly more *A. americanum* larvae, nymphs, and adults could be collected with CO_2-baited traps than by flagging. These authors also found that traps baited with CO_2 detected adult *A. americanum* activity several weeks earlier than flagging. Falco and Fish (1992) found that drag sampling was inferior to a CO_2 sampling device for collecting *I. dammini* larvae. However, the reverse was true for nymphal collections.

Sampling with CO_2 traps should be done for a standard period, after which ticks can be collected from the cloth and counted. One disadvantage of CO_2 traps is that they operate over a limited range (Koch and McNew 1981; Falco and Fish 1992) compared to the area sampled by dragging or trapping and examining host animals. It is also difficult to determine what the effective ranges of these collecting devices are and how the sampling range is influenced by meteorological factors.

Ixodid ticks also can be collected from host animals that have been trapped (Sonenshine et al. 1966; Main et al. 1982; M. L. Wilson and Spielman 1985; Falco and Fish 1992; Gage et al. 1992). The same trapping techniques described later for collecting flea hosts can be used to collect tick hosts. In some instances, trapping may be the only practical means of collecting enough ticks of a certain stage. For example, immature *D. variabilis* are rarely collected by dragging but often can be found in large numbers on small mammal hosts (Sonenshine et al. 1966; Gage et al. 1992).

Other techniques can be used to collect bird, lizard, or large- to medium-sized mammal hosts. Avian hosts of ticks can be captured for examination by using either mist nets or ground traps (Manweiler et al.

1990). Lizard hosts, such as western fence lizards harboring *Ixodes pacificus* larvae and nymphs, can be captured by noosing (Manweiler et al. 1990). Pets or livestock can be examined for ticks without anesthesia. However, it may be necessary to restrain large livestock, such as cattle or horses, to make collecting easier and safer. Tick hosts also can be collected by shooting the host animal and then examining it for ticks.

Dead hosts can be examined directly for ticks. Live wild animals should be humanely killed or anesthetized in metofane, halothane, or other suitable anesthetic, as described later for fleas, and then examined for ticks. Ticks can be removed from anesthetized or dead host animals with forceps. Collection procedures should be standardized as much as possible. This minimizes sampling variations related to differences between the techniques of individual collectors, the species of host or tick being sampled, the method of trapping employed, or other factors.

Another method for collecting ticks from captured live hosts is to hold hosts in the laboratory until attached ticks have completed engorgement and detach from the host (Mather and Spielman 1986; Gage et al. 1990). Infested hosts are usually housed in wire mesh cages that are placed within a cloth bag or over a pan of water. When the ticks have completed engorgement, they will fall from the host, where they can be collected from the bag or pan of water.

Ticks also can be collected from dead hosts by first digesting the host's skin in trypsin and then dissolving the remaining skin and hair in potassium hydroxide (KOH). This method destroys soft tissues but leaves the exoskeletons of ticks and other arthropod ectoparasites intact. Exoskeletons are removed from the digest solution by filtration and then collected for identification and counting (Henry and McKeever 1971). This method presumably collects all ectoparasites present on the host and is, therefore, more efficient than hand-picking with forceps. Few workers, however, attempt such a labor-intensive collection procedure. Besides being very time-consuming, this method, obviously, also precludes sampling the ticks or other ectoparasites for pathogenic microorganisms.

Tick Collection (Argasidae)

Sampling soft ticks (Argasidae) presents challenges that are somewhat different from those encountered while sampling hard tick populations. Some stages of certain argasid species remain on their hosts for long periods. Therefore, they can be collected using the techniques described earlier for examining hosts for ixodid ticks. Larvae and nymphs of the spinose ear tick (*Otobius megnini*) can be collected from the ear canals

of their hosts, where they spend several weeks feeding and developing before leaving the host after the final nymphal feeding (Cooley and Kohls 1944; Oliver 1989). Larvae of most *Argas* and some *Ornithodoros* species also attach to their avian hosts for several days and can be collected directly from these animals. Argasids, however, usually feed for only short periods during each stage of the life cycle. Some, such as *Otobius* adults, *Ornithodoros moubata* larvae, and probably all *Antricola* adults, do not feed at all.

Most argasids, therefore, must be collected off their hosts using methods other than direct examination of host animals. Many such methods have been described, and a few that are widely applicable are described later. One such method is the CO_2 trap described by V. I. Miles (1968) for collecting *Ornithodoros parkeri* from ground squirrel burrows (Fig. 19.2). This CO_2 trap is fashioned from a cardboard mailing tube that has a hose connected to one end that leads to a CO_2 source. The other end is fitted with a screen funnel that is open on both ends. Ticks attracted to the trap will climb toward the small end of the funnel and fall into the trap, where they remain until collected. Burrows are sampled by inserting the mailing tube trap and hose into a burrow for a few hours or overnight. This allows sufficient time for ticks to be attracted to and enter the trap.

Another means of sampling argasid ticks in animal burrows has been described by Pierce (1974). *Ornithodoros moubata* was collected from warthog dens by wrapping a cloth around a spade and rubbing this "lure" on the roof of den entrances. The ticks clung to the cloth on the spade and could be collected by hand picking. Pierce also collected these ticks by sorting soil

FIGURE 19.2 Tick trap of Miles (1968). See text for details. Photo by L. Carter, CDC.

samples from the floors of warthog dens through wire mesh.

Argasid ticks often can be collected from nest material, bat guano on cave floors, undersides of tree bark, underneath loose soil or rock layers on cliff faces, or other comparable habitats (Hopla and Loye 1983; Cooley and Kohls 1944). Small amounts of these samples may be examined for ticks by sorting the materials in a pan and using forceps or an aspirator to remove any ticks encountered. Larger amounts of nest material or other samples can be placed in a Berlese funnel or modified Tullgren apparatus (Baker and Wharton 1952; Krantz 1978). These use light and heat from a light bulb located at the top of the funnel-shaped devices. Ticks and other arthropods are driven from the sample material toward the bottom of the funnel, where they fall into a jar containing 70% ethanol or other collecting solution. Samples that contain small-grained material also can be sorted through soil sieves to separate ticks from debris (Hopla and Loye 1983).

Mite Collection

Mites are extremely diverse in their habits and the habitats they occupy. This has necessitated the development of a wide variety of sampling techniques designed to sample specific species or life cycle stages. There are, however, a few general sampling techniques, including those described earlier for ticks, that can be adapted for most sampling situations encountered by medical entomologists.

Like ticks, mites can be collected directly from captured, shot, or restrained host animals. Some mites, such as chiggers (Trombiculidae), attach firmly to the surface of the host's body and feed for periods of many hours to a few days. As was described for ticks, these mite species can be removed from the host with fine forceps. Alternatively, the hosts can be held over water until the mites complete engorgement, detach, and drop from the host into water, where they can be collected. It is also possible to collect chiggers or other fur- or feather-dwelling mites by digesting the host's skin as was described earlier for ticks (Henry and McKeever 1971).

Other mites, such as laelapids, dermanyssids, or macronyssids infesting mammals, do not attach as firmly to their hosts as do chiggers. These can be collected by holding the anesthetized or dead hosts over a white enamel pan and vigorously brushing these animals with a pocket comb, rat-tail file, toothbrush, or similar device. This will dislodge mites from the host's body so that they fall into the pan. They can be collected with forceps, a small brush, a wetted appli-

cator stick, or a Singer aspirator (Krantz 1978). Dead mammalian or avian hosts also can be shaken vigorously in a detergent solution, which will dislodge any loosely attached mites and cause them to float free. The mites can then be recovered by pouring the detergent through a Buchner funnel that contains filter paper for trapping the mites (Henry and McKeever 1971; Krantz 1978).

Nesting materials, soil debris, or other habitats of different off-host stages of various mite species also can be examined by collecting the material and placing it in a Berlese funnel or modified Tullgren apparatus as described earlier for ticks (Baker and Wharton 1952; Krantz 1978). This technique is especially useful for many nest-dwelling species and for the nonparasitic nymphal and adult stages of chiggers, including those *Leptotrombidium* species whose larvae are vectors of scrub typhus. Unfed chigger larvae also can be collected on black plates placed on the forest floor where these mites are questing for hosts (Hubert and Baker 1963; Upham et al. 1971).

Some mites, including rhinonyssids associated with birds and some halarachnids parasitizing mammals, actually live as internal parasites of their hosts. These must be collected via postmortem examinations of various tissues. Other mites also remain on their hosts for most of their life cycle. Some of these mites, including species in the families Sarcoptidae, Psoroptidae, Demodicidae, as well as others, infest the skin of mammals and cause mange or related skin disorders. Other species, many of which are members of the superfamily Analgoidea, cause skin conditions, feather loss, or irritability in birds (Krantz 1978). Infestations of these mites are often suspected on the basis of hair loss, feather loss, or skin lesions in host animals, but definitive diagnosis requires identifying the mites themselves.

For those species infesting the skin, identification typically involves microscopic examination of skin scrapings (Meleney 1985). Scabies mites, for example, can be collected by dipping a knife or similar tool in light mineral or machine oil and then using this tool to scrape skin from the suspected infestation. These skin scrapings, still suspended in the oil, are then placed on a microscope slide, overlaid with a cover slip, and examined with a compound microscope under low power. Scrapings also can be digested in potassium hydroxide or sodium hydroxide for a few hours to free the mites from the skin. The mites are then collected from the digested material by using a sugar flotation method (Meleney 1985). Other species, such as psoroptic mites, dwell on the surface of the host's skin and can be tentatively identified with a hand lens. Definitive identification, however, requires collecting the mites by scraping and then examining the washed and stained scrapings with a microscope (Meleney 1985). Quill mites of the family Syringophilidae are internal parasites of the quill portion of feathers. They can be removed from within the quill by dissection (Krantz 1978).

Some species of astigmatid mites cause dermatitis and allergic conditions in humans (Spieksma and Spieksma-Boezeman 1967). Probably the most important of these is the house-dust mite (*Dermatophagoides pteronyssus*). Although this mite is not parasitic, it is an important agent of allergic rhinitis and asthma in humans. House-dust mites can be sampled by examining dust vacuumed from surfaces using a specially modified vacuum apparatus (Arlian et al. 1983). The density of mites collected can be expressed as the number of mites per gram of dust examined. Konishi and Uehara (1990) also have described a double-antibody sandwich enzyme-linked immunosorbent assay (ELISA) that can identify antigens of *D. pteronyssus* and *D. farinae* in house-dust samples. The assay is very sensitive and can detect as little as 0.17µg of antigen, approximately equivalent to the amount of antigen found on 0.5 mites.

FIGURE 19.3 Standard 1-pint larval dipper used in mosquito surveys. CDC file photo.

FIGURE 19.4 CDC ovitrap. A hardboard paddle or paper strip is placed inside the container with about 1 to 1-1/2 inches of water. CDC file photo.

MOSQUITOES (DIPTERA: NEMATOCERA: CULICIDAE)

Eggs

Egg sampling methods fall into two general categories: samples from natural habitats, and artificial samplers. Artificial samplers, such as ovitraps, are placed at sampling stations to attract and sample the ovipositing female mosquito population. Artificial samplers can be used to increase the number of specimens collected or to achieve greater uniformity in a sampling system.

Methods for sampling natural habitats vary with the species of interest and the type of larval habitat. The standard 1-pint larval dipper (Fig. 19.3) is useful in assessing the oviposition activity of *Culex* and other raft-producing groups. Dipping is less effective for species that lay their eggs singly (*Anopheles*) on the water surface, and it is ineffective for species that oviposit on moist substrates (*Aedes*). For smaller habitats, a soup ladle can be used. Variation in results among technicians can seriously undermine the accuracy and usefulness of data collected from dipping. Several egg separators have been devised to separate *Aedes* eggs from soil and plant matter (Service 1993).

Oviposition traps sample the gravid population. This can be an advantage for many epidemiologic studies if the species is known to feed only on humans (e.g., *Aedes aegypti*) or the proportion of blood meals from humans is known. Traps can be separated on the basis of whether or not they retain the ovipositing females or allow them to escape. Female-retaining traps are discussed later in the section on sampling adults. Non-female-retaining oviposition samplers include the CDC ovitrap (Fig. 19.4) and bamboo pot (for *Aedes aegypti* and other container inhabiting *Aedes*), and oviposition pans (Fig. 19.5) (Horsfall et al. 1973; Reiter 1986) for *Culex* species. Ovitraps for *Culex* usually are larger and usually have an attractant or infusion. Trap placement is likely to be very important in determining response by ovipositing females. Nearby competing habitats can reduce the number of eggs collected.

FIGURE 19.5 Oviposition pan, consisting of a large plastic pan filled with an appropriate attractant, such as a hay infusion. Photo by C. G. Moore, CDC.

Larvae and Pupae

The standard device for collecting mosquito larvae from open water is the 1-pint dipper (Fig. 19.3). Many other devices have been used, and several larval concentrators have been developed (Service 1993). Turkey basters, soup ladles, tea strainers, and similar devices are useful for sampling small larval habitats, such as tree holes and scrap tires. Specialized techniques are needed for larvae of groups such as *Coquillettidia*, which attach to the roots of aquatic plants. Laird (1988) gives detailed descriptions of larval habitats and collecting methods.

Adults

A full discussion of the various traps and methods available is beyond the scope of these guidelines. Readers wishing more detailed information should consult Service (1993) or Moore et al. (1993).

Adults of many mosquito species are inactive during the day, resting quietly in dark, cool, humid places. An index of the population density can be obtained by carefully counting or collecting the adults found in a resting station. Sampling resting adults usually provides a more representative sample of the population: teneral, postteneral unfed, blooded, and gravid females as well as males. Population age structure also is more representative. However, different species and different gonotrophic stages may prefer different types of resting sites.

Sampling resting populations is usually time consuming, especially when looking for natural resting sites. Artificial resting stations may be constructed when suitable natural resting stations are not available. Aspirators can be used to collect resting adult mosquitoes. In addition, specimens can be collected with a sweep net, or they can be killed or immobilized by several materials (for example, pyrethroids, chloroform, triethylamine). "Knock-down" collections are often used to collect mosquitoes resting inside dwellings (e.g., *Anopheles*, *Aedes aegypti*). White sheets or other materials are spread on the floor, and a pyrethroid spray is used to kill or knock down the mosquitoes resting on the walls.

Nonattractant traps, such as the malaise trap, give a more representative sample of the population than attractant traps, but they only sample the airborne population. A representative sample is not always desirable. For virus studies, it is better to bias collections toward collection of physiologically old females. Representative samples are highly desirable for general ecological studies and essential for estimating absolute population density. Examples of nonattractant traps include the malaise trap, the ramp trap, truck traps, sticky traps, and suction traps. For details on these traps, consult Service (1993).

Animal-baited and CO_2-baited traps disproportionately attract host-seeking females. Usually, this is the population segment of greatest interest in vector-borne disease surveillance. The bait species or chemical attractant is important in trap performance. Often there is significant interhost variability in attractiveness, which may affect trap performance. Other considerations are the duration of collection (especially human landing/biting collections), time of day (especially important for species with a narrow host-seeking window). A final consideration is the need to decide whether to let mosquitoes feed or not (e.g., will specimens be used for blood meal identification?). CO_2-baited traps rely on the sublimation of dry ice (occasionally bottled CO_2) to provide the attractant, imitating CO_2 release by the host in animal-baited traps. Another material, 1-octen-3-ol, has been used either alone or with CO_2 as an attractant in bait traps (Kline et al. 1991).

Landing/biting collections, usually involving humans, horses, or other domesticated vertebrates, are used to sample selected portions of the mosquito population, particularly in studies to incriminate specific vectors or in other research applications (Service 1993). When using human bait, consideration must be given to the potential health risks involved. During epidemics, these activities should be restricted to individuals who are naturally immune or who are receiving appropriate prophylaxis. Landing/biting collections are often used when the entomological inoculation rate (EIR) is being estimated. The EIR is the product of the human-biting rate and the proportion of female mosquitoes that are infective (able to transmit the pathogen), and it is expressed as the average number of infective bites per person per unit time. The EIR was a good predictor of *Plasmodium falciparum* malaria prevalence in Africa (Beier et al. 1999).

Many animal-baited traps have been designed (Service 1993). These are used mostly for special studies rather than for routine surveillance. One important application for these traps is to determine the probable vector(s) of a particular virus or other agent to a given host [for example, eastern equine encephalitis (EEE) or western equine encephalitis (WEE) in horses]. Drop nets and tent traps (e.g., Mitchell et al. 1985) normally are left open or are suspended above the bait (human or animal). After a set period, the openings are closed or the net lowered and the trapped mosquitoes are collected (Service 1993). The Magoon trap is similar in principle to the tent trap, but it is sturdier in design, which provides some restraint for larger bait animals. Mosquitoes enter the trap but cannot escape, and they can be collected periodically. Several variations have been proposed.

Entrance/exit traps have a long history of use in malaria research. A variation with application to mosquito-borne encephalitis studies is the sentinel chicken shed (Rainey et al. 1962). The trap consists of a portable chicken shed and one or more removable mosquito traps. Mosquitoes attempting to enter the shed to feed are collected in the traps and can be removed the following morning.

The lard can trap (Fig. 19.6) is an economical, portable mosquito trap, made from a 12-inch (or larger) lard can (Bellamy and Reeves 1952). It is very effective in capturing *Cx. tarsalis* and *Cx. nigripalpus*.

The trap has inwardly directed screen-wire funnels on each end. It utilizes about 3 pounds of dry ice (wrapped in newspaper or insulated mailer) placed inside the can. The lard can trap also can be baited with a live chicken or other animal. An inner, double-screened enclosure can be used to prevent feeding by the trapped mosquitoes.

The Reiter gravid mosquito trap (Reiter 1983, 1987; Fig. 19.7) samples female *Culex* mosquitoes as they come to oviposit. It, therefore, is selective for females that have already taken at least one blood meal (Reiter et al. 1986). Since they have fed at least once, these individuals are more likely to be infected. This reduces the work involved in processing mosquito pools for virus isolation or dissecting females for oocysts, sporozoites, or filariae. Because of the introduced bias in age and physiologic state, infection rates will, on average, be higher than those obtained from light-trap catches.

Many mosquito species are attracted to light, making it possible to sample adult populations between dusk and dawn. Light traps probably work by disrupting the normal behavior of flying mosquitoes. Mosquito species respond differently to these traps. Some species are not attracted to light at all and may even be repelled (for example, *Cx. p. quinquefasciatus*). Light traps only sample the flying population. The catch is influenced by many factors, including light source, wavelength, and intensity. Competing light sources (including moonlight), fan size and speed, and presence or absence of screens also affect trap performance.

Trap placement (height, location in relation to trees and other cover, proximity to breeding sites, etc.) can have a marked effect on the species and numbers of mosquitoes collected. Some trial-and-error placement

FIGURE 19.6   Lard can trap of Bellamy and Reeves (1952). Photo by C. G. Moore, CDC.

FIGURE 19.7   Reiter gravid mosquito trap. Photo by R. S. Nasci, CDC.

FIGURE 19.8 New Jersey (or Mulhern) light trap. CDC file photo.

FIGURE 19.9 CDC light trap of Sudia and Chamberlain (1962). Photo by C. G. Moore, CDC.

is frequently involved in locating good trap placement sites. The light trap is usually suspended from a tree or post, so the light is approximately 5–6 feet above the ground. Place light traps 30 feet or more from buildings, in the tree line or near trees and shrubs. Do not place light traps near other lights, in areas subject to strong winds, or near industrial plants that emit smoke or fumes. Operate traps on a regular schedule from one to seven nights per week, from just before dark until just after daylight. Dry ice, as an added attractant, increases collections of many mosquito species, including *Culex tarsalis* and *Cx. nigripalpus*. A small block (3–4 pounds) of dry ice, placed in a padded shipping envelope or wrapped tightly in newspaper, is suspended beside the light trap. Because differences have been noted in the reactions of different species of mosquitoes, light-trap collections should be used with other population-sampling methods.

Light traps for mosquitoes fall into two general categories: large, New Jersey–type light traps (Mulhern 1942; Fig. 19.8) and small, portable light traps such as the CDC light trap (Sudia and Chamberlain 1962; Fig. 19.9). The New Jersey–type trap depends upon a 110-volt source of electric power, while the CDC-type trap is battery operated. Several modifications of the CDC light trap are also commercially available. A new trap system, using "counterflow" geometry, has recently been marketed. These traps use a catalytic combustion unit to convert propane gas to CO_2, water vapor, and heat, which serve as attractants. A major attraction of these traps is the ready availability of propane gas in many areas. Like most other traps, not all species are readily attracted to the propane-powered traps (Kline 2002).

OTHER BITING NEMATOCERA

Sand Flies (Psychodidae)

Larvae of sand flies are found in decaying organic matter in animal burrows, treeholes, and other protected sites. Sampling may involve digging up the burrow or extracting all material from the bottoms of treeholes.

A common method of collecting adult sand flies is the Chaniotis trap (Chaniotis 1978). This is a small, battery-operated light trap with the bulb near the bottom of the collecting chamber (Fig. 19.10). Traps are placed near animal burrows, tree buttress roots, tree holes, or other locations suspected of harboring adult flies. Sticky plates consist of flat sheets of Plexiglas or other material coated with a thin layer of castor oil or other similar medium. Sand flies are caught in the oil and can be retrieved the following day. Large-diameter, disposable plastic Petri dishes have also been used successfully as sticky traps (Al-Suhaibani 1990). Sticky traps are unlikely to produce specimens suitable for virus isolation because of interference of the oil with the isolation system (e.g., cell culture). Landing-biting collections can be made for human-biting species.

Ceratopogonid Midges (Culicoides spp.)

Larvae of ceratopogonids are found in habitats with high moisture, and some habitats may be entirely under water. Mud or other likely habitats can be sampled with a hand trowel, shovel, or other suitable device. Larvae can be collected via various salt flotation methods.

FIGURE 19.10 Chaniotis trap for sand flies. Photo by C. G. Moore, CDC.

BRACHYCEROUS FLIES OF MEDICAL AND VETERINARY IMPORTANCE

Horseflies and Deerflies (Tabanidae)

Tabanid flies deposit their eggs on structures in or near water (emergent vegetation, sticks, rocks). These structures can be sampled to estimate oviposition. Larval tabanids can be collected from wet soil at the edges of marshes, roadside ditches, rice field borrow pits, and similar locations. Larvae of some species are found in drier habitats, usually in shady locations.

Adults can be collected in a variety of traps, such as the malaise trap, the Manitoba fly trap, and the canopy trap. Adding a chemical attractant such as carbon dioxide, ammonia, or octenol can increase trap efficiency.

Horn Flies and Stable Flies (Muscidae)

Horn flies (*Haematobia irritans*) and stable flies (*Stomoxys calcitrans*) are hematophagous pests of veterinary importance. The stable fly may be an occasional vector of equine infectious anemia virus (Foil and Issel 1991). Stable flies are important pests of humans along Atlantic and Gulf Coast beaches of the United States. Adult stable flies and horn flies can be sampled via sweep nets or via aspirating flies coming to animal baits. Several traps have been used, including the Williams (or Alsynite) trap, a variety of the sticky trap, for stable flies (Williams 1973).

Houseflies

Housefly larvae are found in manure, garbage, and various types of decomposing organic matter. Larvae can be quantitatively sampled in these habitats by a variety of methods.

The "Scudder grill" (Scudder 1947) (Fig. 19.11) is a widely used method to estimate adult housefly densities. It consists of a wooden frame, generally painted white, with 16–24 1.9-mm (0.75″) wooden slats 61 cm (24″) in length. The slats are arranged on a frame with approximately 2 cm (0.8″) between slats. The grill is lowered over an area where flies have congregated. As the disturbed flies begin to return to the area, the number of flies landing in a 30-second period is recorded. Several similar counts are made at each site, and an average is computed.

Estimates of fly abundance also can be obtained by counting the numbers of flies on available surfaces (e.g., walls, feeding troughs, stables). Flies can be collected in baited traps (Fig. 19.12), on sticky paper, or

Adult ceratopogonids are collected by means of several types of light traps (Holbrook 1985; Weiser-Schimpf et al. 1990), including several of those described earlier for mosquitoes.

Blackflies (Simuliidae)

Larval blackflies can be collected from natural habitats in streams (rocks, vegetation, or other submerged structures) or by placing strips of plastic tape in the stream (Davies and Crosskey 1991).

Adults of human-feeding species are generally collected at human bait, because there are few if any useful traps for these groups. An aluminum plaque trap for ovipositing females has been found useful in certain situations (Davies and Crosskey 1991). Light traps are not particularly attractive, and the few blackflies in the collections have to be removed from all the other insects.

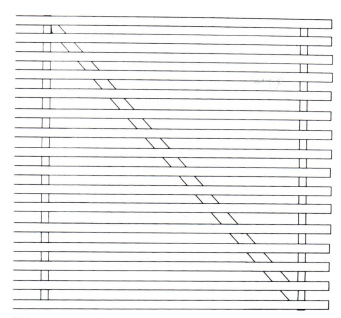

FIGURE 19.11   Scudder fly grid. After Pratt and Moore (1993).

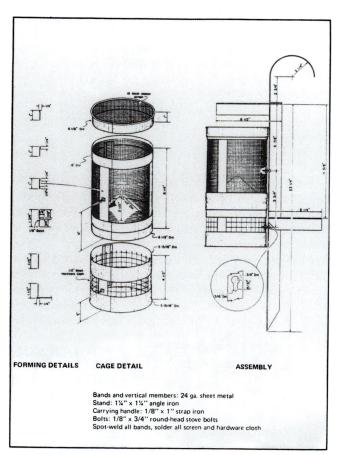

FORMING DETAILS   CAGE DETAIL   ASSEMBLY

Bands and vertical members: 24 ga. sheet metal
Stand: 1¼" x 1¼" angle iron
Carrying handle: 1/8" x 1" strap iron
Bolts: 1/8" x 3/4" round-head stove bolts
Spot-weld all bands, solder all screen and hardware cloth

FIGURE 19.12   Bait trap for houseflies. From Pratt and Moore (1993).

by sweep nets. The same technique(s) must be used throughout a study or control program so that data can be compared (World Health Organization 1991). Sweep nets can be used to collect adult flies, but standardization may be a problem.

Tsetse

Tsetse (*Glossina* spp.) adults can be collected by using a wide variety of traps. Since both males and females seek a blood meal, both can be collected at baits (Colvin and Gibson 1992). Electrocuting nets have been used to sample tsetse as well as to protect train passengers. Tsetse depend on both visual and olfactory cues in locating hosts, and these behaviors have been extensively exploited to develop tsetse control programs. Tsetse rely on shape, orientation, brightness, contrast, movement, and color to distinguish hosts. They can thus distinguish host animals against complex visual backgrounds. Carbon dioxide is the most effective attractant, but butanone, 1-octen-3-ol, and some phenol derivatives in cattle urine are also attractive (Colvin and Gibson 1992). The latter compounds can be economically used in trapping programs.

FLEAS (ORDER SIPHONAPTERA)

Removing Fleas from Captured Animals

Probably the most common method for collecting mammal fleas is to trap or shoot host animals and then to examine them for fleas (Holland 1949; Stark and Kinney 1969; Haas et al. 1973; Bahmanyar and Cavanaugh 1976; Campos et al. 1985). Nonlethal traps (live traps), such as the various commercially available box or cage varieties (Fig. 19.13A), are best for collecting fleas from small rodents because fleas tend to leave dead hosts as the carcasses cool.

Hosts captured in live traps should be anesthetized with metofane, halothane, or other suitable anesthetic agents before further processing (Clark and Olfert 1986). The anesthetized animal can then be placed in a white enamel pan (a depth of 20 cm or more is recommended). The animal is brushed vigorously from the tail end forward with a toothbrush, pocket comb, or other similar instrument (Fig. 9.14). The dislodged fleas fall from the host to the bottom of the pan, where they can be removed with forceps or a wetted applicator stick and placed in labeled vials for further processing. Any bedding material placed in traps to provide warmth for the host also should be examined for fleas.

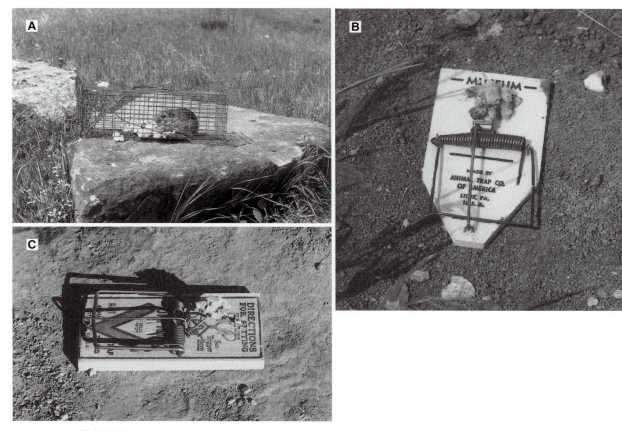

FIGURE 19.13   Traps for sampling mammals: (A) Wire trap for obtaining live animals; (B) and (C) snap traps. Photos by L. Carter, CDC.

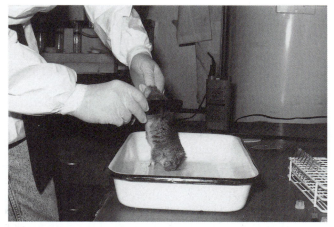

FIGURE 19.14   Technique for removing ectoparasites from dead or anesthetized animals. Photo by L. Carter, CDC.

Snap traps (Fig. 9.13B and C) also can be used to collect hosts for flea collection, but these traps should be checked every couple of hours or so. This ensures that the hosts are collected before the fleas begin to leave the cooling carcasses. Larger rodents, such as *Rattus* spp., may only be injured, rather than killed, in snap traps, and these injured animals can drag traps a considerable distance away from the original trap site. This can be prevented by tethering each trap to the ground using a piece of strong but flexible wire. Hosts killed in snap traps or by other means can be examined directly for fleas. However, it is preferable that carcasses be treated with insecticides or anesthetic agents before examination. This prevents live fleas from escaping onto the investigator or into the laboratory.

Other small mammal species often must be captured using different trapping techniques. For example, the preferred means for trapping shrews is to construct pit traps. These traps consist of a smooth-sided can or jar, buried so the top is level with the ground. Pit traps are placed in presumed runways, and animals fall into them while making their nightly rounds. Shrews do not survive well in captivity, and animals are often dead when collectors arrive. The fleas, however, should remain in the pit trap or on the host (Holland 1949).

Collecting fleas from birds also requires special techniques. Ground-dwelling birds can be captured in traps. Other techniques, such as mist netting, may be needed to live-capture adults of other bird species (Sonenshine and Stout 1970; Manweiler et al. 1990). If nests are accessible, nestling birds can be removed from their nests and examined for fleas (Hopla and Loye 1983). Fleas also can be collected from birds as well as from large and medium-sized mammals by shooting the hosts and then inspecting the carcasses for fleas (Holland 1949).

Collection of Fleas from Burrows

Fleas can be collected from burrows by using a swabbing device. It consists of a flexible steel cable or hard rubber hose with a piece of white flannel cloth attached to the end (Vakhrousheva et al. 1989; Beard et al. 1991) (Fig. 19.15). The swab is forced down the burrow entrance, where fleas will mistake it for their normal hosts and cling to the cloth. The cloth is then removed from the burrow and inspected for fleas. If desired, the cloth can be placed in a plastic bag and held for later examination. Freezing, anesthetization, or insecticides will kill fleas within the bags. V. I. Miles (1968) also has reported capturing fleas in burrows

FIGURE 19.15 Use of the burrow swab to collect fleas. Photo by L. Carter, CDC.

using the CO_2 trap described earlier for collecting argasid ticks (Fig. 19.2).

Collection of Fleas from Nests and Nesting Material

Many fleas spend more time in the nest of their host than on the host itself. Mammalian and avian nest material can be examined for fleas by sorting the contents in an enamel pan similar to that described earlier for brushing fleas from hosts. Although some fleas are poor jumpers, it is a good idea to treat the material with insecticides before sorting, in order to prevent them from escaping the pan. Nest material also can be loaded into a Berlese funnel or modified Tullgren apparatus, as described later for collecting argasid ticks and mites (Baker and Wharton 1952; Krantz 1978). Once the fleas reach the bottom of the funnel, they fall into a jar containing a saline solution or alcohol. Eggs, larvae, and pupae also can be collected from nest material, debris, or other material by sorting the samples through a series of sieves (Hopla and Loye 1983).

Hopla and Loye (1983) also used black cardboard cards coated with honey to collect fleas (*Ceratophyllus celsus*) gathered at the entrances of cliff swallow nests (Fig. 19.16). The fleas apparently confuse the shadow created by the black squares with the approach of their swallow hosts and jump onto the cards, where they become trapped in the sticky honey. The honey-coated squares with the captured fleas are then covered with onionskin paper and transported to the laboratory. The fleas can be removed from the cards by soaking the samplers in a dish filled with water. This will dissolve the honey and allow the fleas to float to the surface, where they can be removed easily with forceps.

Flea Indices

The most basic data obtained from flea and rodent surveys are the number of fleas of different species found on various species of hosts. These raw data can be used to calculate various indices, including the specific flea index (number of fleas per host of given flea and host species). Burrow, nest, or house indexes can be similarly calculated. The percentage of hosts infested with a particular species of flea also is important. This information can be used with other rodent and vector surveillance data to estimate human risks in areas with murine typhus or plague. For example, specific flea indices of greater than 1 for *X. cheopis* infestations on rats from plague-endemic areas are thought to represent dangerously high levels of risks

FIGURE 19.16 Black cardboard traps for collecting fleas. The fleas become trapped in the honey that coats the cards. Photo by C. E. Hopla.

for the occurrence of both rat plague epizootics and human plague epidemics (Pollitzer 1954).

Many factors affect the reliability of flea indices (Pollitzer 1954). These include host age, host species, trapping techniques, areas selected for sampling, and the tendency for a few hosts to be heavily infested while many hosts have few or no fleas (high variance to mean ratios for sample data). To obtain reliable indices for comparison between different survey sites, all trapping and ectoparasite collection procedures must be standardized. Schwan (1984) has described a sequential sampling method that can be used to determine how many host animals must be sampled before a reliable flea index is obtained for a given host/flea relationship. He reported that examination of as few as 20 Nile grass rats (*Arvicanthis niloticus*) was sufficient to yield reliable specific flea indexes for infesta-

tions of either *Dinopsyllus lypusus* or *Xenopsylla cheopis bantorum* on these animals.

MISCELLANEOUS GROUPS OF MEDICAL AND VETERINARY IMPORTANCE

Triatomine Bugs (Order Hemiptera)

Members of the reduviid subfamily Triatominae, especially various species of *Panstrongylus*, *Rhodnius*, and *Triatoma*, transmit *Trypanosoma cruzi*, the etiologic agent of Chagas' disease, or American trypanosomiasis. Many surveillance techniques have been described for sampling these bugs in sylvatic habitats, human dwellings, or peridomiciliary structures. One common method for collecting triatomine bugs in houses or out-buildings is to use pyrethroid sprays as "excitorepellents" to flush bugs from crevices in walls, thatch roofs, or other hiding places (Piesman et al. 1985; Rabinovich et al. 1990; Garcia-Zapata and Marsden 1992, 1993). As bugs are flushed from their hiding places they can be manually collected using forceps. Results of these surveys are usually expressed as the number of bugs captured per person-hour or as the number of bugs collected per house or other structure. When the latter measure is used, the searches should be conducted until no more bugs can be found in each house surveyed.

The presence of triatomine bugs in houses or other man-made structures also can be determined indirectly by identifying the distinctive fecal deposits left by these insects (Schofield et al. 1986). Typically, this is done by spreading a sheet of paper above a bed or other areas where bugs are likely to defecate after taking their nighttime blood meals (Garcia-Zapata and Marsden 1992, 1993). Triatomine infestations also can be detected directly or indirectly by using Gomez–Nunez boxes. These sampling devices consist of a piece of paper (approximately 8″ × 12″) that is folded accordion fashion and inserted into a cardboard frame (Garcia-Zapata and Marsden 1992). Bugs are attracted to the boxes as potential resting places or refuges and defecate on the paper sampler during their visit. Although the bugs themselves are occasionally removed from these samplers, infestations usually are identified by the presence of fecal deposits on the paper inserts. A similar device for identifying triatomine infestations in chicken coops or other peridomiciliary structures has been described by Garcia-Zapata and Marsden (1993). This simple but effective sampler consists of a section of bamboo open on one or both ends that contains strips of paper folded accordion fashion. Bugs enter these devices and defecate on the paper, as just described for the Gomez–Nunez box.

Triatomines also can be collected from sylvatic habitats, where they seek refuge between blood meals. Likely habitats include palm trees, hollow logs, bird or mammal nests, mammal burrows, and caves (D'Allesandro et al. 1984; Ekkens 1984). Any bugs encountered usually can be manually removed from these refuges with forceps, but considerable effort is required to excavate burrows or collect from other, difficult sites. M. A. Miles et al. (1981) identified probable triatomine habitats by tracking captured mammals to their burrows, nests in hollow logs, or other refuges. Although bugs seldom existed in large numbers in any given nest, roost, or burrow, this technique was the most effective means of locating triatomines in forest ecotopes. Nest material and debris also can be placed in a pan for examination. This latter method is especially useful for collecting small nymphs (Ekkens 1984).

The winged adult triatomine bugs can be collected by using light traps or by shining a light source onto a light-colored collecting area, such as a sheet, piece of canvas, or white board. The bugs can be manually gathered using forceps (Whitlaw and Chaniotis 1978; Ekkens 1981).

Lice (Anoplura)

There are two species of lice of medical importance: the head louse (*Pediculus capitis*) and the pubic, or "crab," louse (*Pthirus pubis*). The eggs of these insects are attached by a cement to the body hairs of the human host. Demonstrating the eggs (or "nits") is the most practical way to check for infestations. In elementary schools, children in grades K–2 provide the best estimators of louse activity (Juranek 1985). Infestation rates decline with age. Suspected nits should be examined microscopically to exclude dandruff, hair spray, and other similar artifacts (Epstein and Orkin 1985). Eggs of *P. capitis* are normally found at the back of the head or around the ears. The eggs of *P. pubis* are found in the pubic hairs or, less commonly, in the eyebrows, beard, or moustache (Epstein and Orkin 1985).

SUMMARY AND CONCLUSIONS

Surveillance for the insect vectors of diseases, although complex in its own right, is only one part of an overall disease surveillance program. The particular trapping or sampling tools and schedules to be used depend on the disease and vector(s) of interest and the particular question(s) to be answered. Proper planning before embarking on a surveillance program will ensure that meaningful data are collected.

As our understanding of the vector–pathogen–vertebrate host relationship in any particular disease changes, there will likely be a need for new procedures to collect, for example, individuals in only a certain physiologic or behavioral state. Such specialized sampling methods are common for mosquitoes. They are less common or absent for some other groups. Just as new laboratory diagnostic methods allow us to answer new questions (or finally answer questions that have been unanswered for years), the availability of new collection methods may let us answer previously unanswerable questions. Thus, the equipment available for surveillance defines, at least in part, what information can be collected to help us predict vector or disease activity.

Readings

Al-Suhaibani, S. 1990. *Field and Laboratory Studies of Sand Flies in Larimer County, Colorado*. Ph.D. thesis, Colorado State University.

Arlian, L. G., Woodford, P. J., Bernstein, I. L., and Gallagher, J. S. 1983. Seasonal population structure of house dust mites, *Dermatophagoides* spp. (Acari: Pyroglyphidae). *J. Med. Entomol.* 20: 99–102.

Bahmanyar, M., and Cavanaugh, D. C., 1976. *Plague Manual*. World Health Organization. Geneva, Switzerland.

Baker, E. W., and Wharton, G. W. 1952. *An Introduction to Acarology*. Macmillan, New York.

Beard, M. L., Rose, S. T., Barnes, A. M., and Montenieri, J. A. 1992. Control of *Oropsylla hirsuta*, a plague vector, by treatment of prairie dog burrows with 0.5% permethrin dust. *J. Med. Entomol.* 29: 25–29.

Beier, J. C., Killeen, G. F., and Githure, J. I. 1999. Short report: Entomologic inoculation rates and *Plasmodium falciparum* malaria prevalence in Africa. *Am. J. Trop. Med. Hyg.* 61: 109–113.

Bellamy, R., and Reeves, W. C. 1952. A portable mosquito trap. *Mosq. News* 12(4): 256–258.

Bowen, G. S., and Francy, D. B. 1980. Surveillance. In *St. Louis Encephalitis* (T. P. Monath, ed.). American Public Health Association, Washington, DC, Chap. 10.

Bram, R. A. (ed.) 1978. *Surveillance and Collection of Arthropods of Veterinary Importance*. U.S. Department of Agricultural Handbook No. 5.

Campos, E. G., Maupin, G. O., Barnes, A. M., and Eads, R. B. 1985. Seasonal occurrence of fleas (Siphonaptera) on rodents in a foothills habitat in Larimer County, Colorado, USA. *J. Med. Entomol.* 22: 266–270.

Chaniotis, B. N. 1978. Phlebotomine sand flies (Psychodidae). In *Surveillance and Collection of Arthropods of Veterinary Importance* (R. A. Bram, ed.). U.S. Department of Agricultural Handbook No. 5, pp. 19–30.

Clark, J. D., and Olfert, E. D. 1986. Rodents (Rodentia). Part 4. Special medicine: Mammals. In *Zoo and Wild Animal Medicine* (M. E. Fowler, ed.). Saunders, Philadelphia, pp. 727–748.

Colvin, J., and Gibson, G. 1992. Host-seeking behavior and management of tsetse. *Annu. Rev. Ent.* 37: 21–40.

Cooley, R. A., and Kohls, G. M. 1944. The Argasidae of North America, Central America and Cuba. *Amer. Midl. Nat.* Monograph No. 1. University Press, Notre Dame, IN.

D'Allessandro, Barreto, P., Saravia, N., and Barreto, M. 1984. Epidemiology of *Trypanosoma cruzi* in the oriental plains of Colombia. *Am. J. Trop. Med. Hyg.* 33: 1084–1095.

Davies, J. B., and Crosskey, R. W. 1991. *Simulium — Vectors of onchocerciasis*. Vector control Series, Training and Information Guide, Advanced Level (WHO/VBC/91.992). WHO, Division of Tropical Diseases, Geneva, Switzerland.

Ekkens, D. B. 1981. Nocturnal flights of *Triatoma* (Hemiptera: Reduviidae) in Sabino Canyon, Arizona. I. Light collections. *J. Med. Entomol.* 18: 211–227.

Ekkens, D. 1984. Nocturnal flights of *Triatoma* (Hemiptera: Reduviidae) in Sabino Canyon, Arizona. II. *Neotoma* lodge studies. *J. Med. Entomol.* 21: 140–144.

Epstein, E., Sr., and Orkin, M. 1985. Pediculosis: Clinical aspects. In *Cutaneous Infestations and Insect Bites* (M. Orkin and H. I. Maibach, eds.). Marcel Dekker, New York, Chap. 21.

Falco, R. C., and Fish, D. 1988. Prevalence of *Ixodes dammini* near the homes of Lyme disease patients in Westchester County, New York. *Am. J. Epidemiol.* 127: 826–830.

Falco, R. C., and Fish, D. 1992. A comparison of methods for sampling the deer tick, *Ixodes dammini*, in a Lyme disease endemic area. *Exp. Appl. Acarol.* 14: 165–173.

Fay, R. W., and Eliason, D. A. 1966. A preferred oviposition site as a surveillance method for *Aedes aegypti*. *Mosq. News* 26: 531–535.

Gage, K. L., Burgdorfer, W., and Hopla, C. E. 1990. Hispid cotton rats (*Sigmodon hispidus*) as a source for infecting immature *Dermacentor variabilis* (Acari: Ixodidae) with *Rickettsia rickettsii*. *J. Med. Entomol.* 27: 615–619.

Gage, K. L., Hopla, C. E., and Schwan, T. G. 1992. Cotton rats and other small mammals as host for immature *Dermacentor variabilis* (Acari: Ixodidae) in central Oklahoma. *J. Med. Entomol.* 29: 832–842.

Garcia-Zapata, M. T. A., and Marsden, P. D. 1992. Control of the transmission of Chagas' disease in Mambai, Goias, Brazil (1980–1988). *Am. J. Trop. Med. Hyg.* 46(4): 440–443.

Garcia-Zapata, M. T. A., and Marsden, P. D. 1993. Chagas' disease: Control and surveillance through use of insecticides and community participation in Mambai, Goias, Brazil. *Bull. Pan Amer. Health Org.* 27: 265–279.

Haas, G. E., Martin, R. P., Swickard, M., and Miller, B. E. 1973. Siphonaptera–mammal relationships in north central New Mexico. *J. Med. Entomol.* 10: 281–289.

Henry, L. G., and McKeever, S. 1971. A modification of the washing technique for quantitative evaluation of the ectoparasite load of small mammals. *J. Med. Entomol.* 8: 504–505.

Holbrook, F. R. 1985. A comparison of three traps for adult *Culicoides variipennis* (Ceratopogonidae). *J. Am. Mosq. Control Assoc.* 1(3): 379–381.

Holland, G. P. 1949. *The Siphonaptera of Canada*. Publication 817, Technical Bulletin 70. Dominion of Canada, Department of Agriculture. Ottawa, Canada.

Hopla, C. E., and Loye, J. E. 1983. The ectoparasites and microorganisms associated with cliff swallows in west central Oklahoma. I. Ticks and fleas. *Bull. Soc. Vector Ecol.* 8: 111–121.

Horsfall, W. R., Fowler, H. W., Moretti, L. J., and Larsen, J. R. 1973. *Bionomics and Embryology of the Inland Floodwater Mosquito*, Aedes vexans. University of Illinois Press, Urbana, IL.

Hubert, A. A., and Baker, H. J. 1963. Studies on the habitat and population of *Leptotrombidium* (*Leptotrombidium*) *akamushi* and *Leptotrombidium* (*Leptotrombidium*) *deliensis* in Malaya. *Am. J. Hyg.* 78: 131–142.

Juranek, D. 1985. *Pediculus capitis* in school children: Epidemiologic trends, risk factors, and recommendations for control. In *Cutaneous Infestations and Insect Bites* (M. Orkin and H. I. Maibach, eds.). Marcel Dekker, New York, Chap. 23.

Kinzer, D. R., Presley, S. M., and Hair, J. A. 1990. Comparative efficiency of flagging and carbon dioxide–baited sticky traps for collecting the lone star tick, *Amblyomma americanum* (Acarina: Ixodidae). *J. Med. Entomol.* 27: 750–755.

Kline, D. L. 2002. Evaluation of various models of propane-powered mosquito traps. *J. Vector Ecol.* 27: 1–7.

Kline, D. L., Wood, J. R., and Cornell, J. A. 1991. Interactive effects of 1-octen-3-ol and carbon dioxide on mosquito (Diptera: Culicidae) surveillance and control. *J. Med. Entomol.* 28(2): 254–258.

Koch, H. G., and McNew, R. W. 1981. Comparative catches of field populations of lone star ticks by CO_2-emitting dry-ice-, dry-chemicals-, and animal-baited devices. *Ann. Entomol. Soc. Amer.* 74: 498–500.

Koch, H. G., and McNew, R. W. 1982. Sampling of lone star ticks (Acari: Ixodidae): Dry ice quantity and capture success. *Ann. Entomol. Soc. Amer.* 75: 579–582.

Konishi, E., and Uehara, K. 1990. Enzyme-linked immunosorbent assay for quantifying antigens of *Dermatophagoides farinae* and D. *pteronyssus* (Acari: Pteroglyphidae) in house dust samples. *J. Med. Entomol.* 27: 993–998.

Krantz, G. W. 1978. *A Manual of Acarology*, 2nd ed. Oregon State University Book Stores, Corvallis, OR.

Laird, M. 1988. *The Natural History of Larval Mosquito Habitats*. Academic Press, New York.

Main, A. J., Carey, A. B., Carey, M. G., and Goodwin, R. H. 1982. Immature *Ixodes dammini* (Acari: Ixodidae) on small animals in Connecticut, USA. *J. Med. Entomol.* 19: 655–664.

Manweiler, S. A., Lane, R. S., Block, W. M., and Morrison, M. L. 1990. Survey of birds and lizards for ixodid ticks (Acari) and spirochetal infection in northern California. *J. Med. Entomol.* 27: 1011–1015.

Mather, T. N., and Spielman, A. 1986. Diurnal detachment of immature deer ticks (*Ixodes dammini*) from nocturnal hosts. *Am. J. Trop. Med. Hyg.* 35: 182–186.

Meleney, W. P. 1985. Mange mites and other parasitic mites. In *Parasites, Pests, and Predators* (S. M. Gaafar, W. E. Howard, and R. E. Marsh, eds.). Elsevier Science, Amsterdam, pp. 317–346.

Miles, M. A., deSouza, A. A., and Povoa. M. 1981. Chagas' disease in the Amazon Basin. III. Ecotopes of ten triatomine bug species (Hemiptera: Reduviidae) from the vicinity of Belem, Para State, Brazil. *J. Med. Entomol.* 18: 266–278.

Miles, V. I. 1968. A carbon dioxide bait trap for collecting ticks and fleas from animal burrows. *J. Med. Entomol.* 5: 491–495.

Milne, A. 1943. The comparison of sheep-tick populations (*Ixodes ricinus* L.). *Ann. Appl. Biol.* 30: 240–250.

Mitchell, C. J., Darsie, R. F., Monath, T. P., Sabattini, M. S., and Daffner, J. 1985. The use of an animal-baited net trap for collecting mosquitoes during western equine encephalitis investigations in Argentina. *J. Am. Mosq. Control Assoc.* 1(1): 43–47.

Moore, C. G., McLean, R. G., Mitchell, C. J., Nasci, R. S., Tsai, T. F., Calisher, C. H., Marfin, A. A., Moore, P. S., and Gubler, D. J. 1993. *Guidelines for Arbovirus Surveillance in the United States*. U.S. Dept. of Health and Human Services, Centers for Disease Control, Fort Collins, CO.

Mulhern, T. D. 1942. New Jersey light trap for mosquito surveys. *N.J. Agric. Exp. Sta., Circ. 421* [Reprinted in *J. Am. Mosq. Control Assoc.* 1(4): 411–418; 1985].

Oliver, J. H. 1989. Biology and systematics of ticks (Acari: Ixodidae). *Annu. Rev. Ecol. Syst.* 20: 397–430.

Philip, C. B. 1937. Six years' intensive observation on the seasonal prevalence of a tick population in western Montana. *Publ. Health Rep.* 52: 16–22.

Pierce, M. A. 1974. Distribution and ecology of *Ornithodoros moubata porcinus* Walton (Acarina) in animal burrows in East Africa. *Bull. Entomol. Res.* 64: 605–619.

Piesman, J., Sherlock, I. A., Mota, E., Todd, C. W., Hoff, R., and Weller, T. 1985. Association between household triatomine density and incidence of *Trypanosoma cruzi* infection during a nine-year study in Castro Alves, Bahia, Brazil. *Am. J. Trop. Med. Hyg.* 34: 866–869.

Pollitzer, R. 1954. *Plague*. World Health Organization, Geneva.

Rabinovich, J. E., Wisnivesky-Colli, C., Solarz, N. D., and Gurtler, R. E. 1990. Probability of transmission of Chagas' disease by *Triatoma infestans* (Hemiptera: Reduviidae) in an endemic area of Santiago del Estero, Argentina. *Bull. World Health Org.* 68: 737–746

Rainey, M. B., Warren, G. V., Hess, A. D., and Blackmore, J. S. 1962. A sentinel chicken shed and mosquito trap for use in encephalitis field studies. *Mosq. News* 22(4): 337–342.

Reiter, P. 1983. A portable, battery-powered trap for collecting gravid *Culex* mosquitoes. *Mosq. News* 43: 496–498.

Reiter, P. 1986. A standardized procedure for the quantitative surveillance of certain *Culex* mosquitoes by egg raft collection. *J. Am. Mosq. Control Assoc.* 2(2): 219–221.

Reiter, P. 1987. A revised version of the CDC gravid mosquito trap. *J. Am. Mosq. Control Assoc.* 3(2): 325–327.

Reiter, P., Jakob, W. L., Francy, D. B., and Mullenix, J. B. 1986. Evaluation of the CDC gravid trap for the surveillance of St. Louis encephalitis vectors in Memphis, Tennessee. *J. Am. Mosq. Control Assoc.* 2(2): 209–211.

Schofield, C. J., Williams, N. G., Kirk, M. L., Garcia-Zapata, M. T. A., and Marsden, P. D. 1986. A key for identifying faecal smears to detect domestic infestations of triatomine bugs. *Rev. Soc. Bras. Med. Trop.* 19: 5–8.

Schwan, T. G. 1984. Sequential sampling to determine the minimum number of host examinations required to provide a reliable flea (Siphonaptera) index. *J. Med. Entomol.* 21: 670–674.

Scudder, H. I. 1947. A new technique for sampling the density of housefly populations. *Pub. Hlth. Rep.* 62: 681–686.

Semtner, P. J., and Hair, J. A. 1973. The ecology and behavior of the lone star tick (Acarina: Ixodidae). V. Abundance and seasonal distribution in different habitat types. *J. Med. Entomol.* 10: 618–628.

Service, M. W. 1993. *Mosquito Ecology. Field Sampling Methods*, 2d. ed. Elsevier Applied, New York.

Sonenshine, D. E., and Stout, I. J. 1970. A contribution to the ecology of ticks infesting wild birds and rabbits in the Virginia–North Carolina Piedmont (Acarina: Ixodidae). *J. Med. Entomol.* 41: 296–301.

Sonenshine, D. E., Atwood, E. L., and Lamb, Jr., J. T. 1966. The ecology of ticks transmitting Rocky Mountain spotted fever in a study area in Virginia. *Ann. Entomol. Soc. Amer.* 59: 1234–1262.

Southwood, T. R. E. 1978. *Ecological Methods with Particular Reference to the Study of Insect Populations*, 2d. ed. Chapman and Hall, New York.

Spieksma, F. Th. M., and Spieksma-Boezeman, M. I. A. 1967. The mite fauna of house dust with particular reference to the house dust mite *Dermatophagoides pteronyssus* (Trouessart 1897) (Psoroptidae: Sarcoptiformes). *Acarologia* 9: 226–241.

Stark, H. E., and Kinney, A. R. 1969. Abundance of rodents and fleas as related to plague in Lava Beds National Monument, California. *J. Med. Entomol.* 6: 287–294.

Sudia, W. D., and Chamberlain, R. W. 1962. Battery-operated light trap, an improved model. *Mosq. News* 22: 126–129.

Upham, R. W., Jr., Hubert, A. A., Phang, O. W., bin Mat, Y., and Rapmund, G. 1971. Distribution of *Leptotrombidium (Leptotrombidium) arenicola* (Acarina: Trombiculidae) on the ground in west Malaysia. *J. Med. Entomol.* 8: 401–406.

Vakhrousheva, Z. P., Gorchakov, A. D., Kolupaeva, N. A., and Chernykh, E. G. 1989. The use of flannel flags for collecting fleas at the entrance of burrows of steppe rodents. *Meditsinskaya Parazitologiya i Parazitarnye Bolenzi* 1: 54–57.

Weiser-Schimpf, L., Foil, L. D., and Holbrook, F. R. 1990. Comparison of New Jersey light traps for collection of adult *Culicoides variipennis* (Diptera: Ceratopogonidae). *J. Am. Mosq. Control Assoc.* 6(3): 537–538.

Whitlaw, J. T., and Chaniotis, B. N. 1978. Palm trees and Chagas' disease in Panama. *Am. J. Trop. Med. Hyg.* 27: 873–881.

Williams, D. F. 1973. Sticky traps for sampling populations of *Stomoxys calcitrans*. *J. Econ. Ent.* 66(6): 1279–1280.

Wilson, J. G., Kinzer, D. R., Sauer, J. R., and Hair, J. A. 1972. Chemo-attraction in the lone star tick (Acarina: Ixodidae): I. Response to carbon dioxide administered via traps. *J. Med. Entomol.* 9: 245–252.

Wilson, M. L., and Spielman, A. 1985. Seasonal activity of immature *Ixodes dammini* (Acari: Ixodidae). *J. Med. Entomol.* 22: 408–414.

World Health Organization. 1991. *The Housefly*. Vector Control Series, Training and Information Guide, Intermediate Level (WHO/VBC/90.987). WHO, Division of Tropical Diseases, Geneva, Switzerland.

Zimmerman, R. H., and Garris, G. I. 1985. Sampling efficiency of three dragging techniques for the collection of nonparasitic *Boophilus microplus* (Acari: Ixodidae) larvae in Puerto Rico. *J. Econ. Entomol.* 78: 627–631.

III

PHYSIOLOGY OF INSECTS AND ACARINES

HENRY H. HAGEDORN

20

Vector Olfaction and Behavior

MARC J. KLOWDEN AND LAURENCE J. ZWIEBEL

INSECTS BEHAVING BADLY

The number of insects inhabiting our planet dwarfs that of all other animal life. Although only about a million species have been identified of the 5–30 million that are estimated to exist, relatively few of these are known to be vectors of pathogens that cause disease that affect humans or domestic animals. Indeed, of the approximately 3,500 species of mosquitoes worldwide, only a few are known to vector human parasites. The evolutionary path to becoming a vector is a complicated one, requiring a suite of adaptations that are arrived at through a coevolution of both the parasite and the insect host. To paraphrase Tolstoy's famous line in *Anna Karenina*, all successful vectors are alike, but all unsuccessful vectors are unsuccessful for different reasons. Common to successful vectors are the anatomical and physiological adaptations that are necessary to acquire and transmit parasites, as well as the specialized behavior patterns that allow them to find the hosts on which they feed. This discussion of vector behavior and olfaction will principally consider mosquitoes, because most is known about these important vectors, but the principles certainly apply to other groups of vectors as well.

Finding a specific host in an environment that abounds with distracting stimuli is not a trivial task. At the very least, a mosquito cannot be a vector unless it first locates an infected host, then acquires the parasite and remains alive while it becomes infective, and subsequently encounters an uninfected host into which the parasite enters. The behaviors that lead up to the acquisition of the parasites as well as those that persist throughout its lifetime are absolutely critical to an insect's vector status.

A good example of the consequences of these critical behavior patterns is the vector status of two species of mosquitoes within the *Anopheles gambiae* complex that differ in the degree to which they feed on humans. *Anopheles gambiae sensu stricto* prefers humans, while *An. quadriannulatus* prefers to feed on cattle. The behavioral preference of *An. gambiae* for humans is one of the reasons this insect is such an important vector of malaria and the closely related *An. quadriannulatus* is not.

Behavior is the interface between an organism's genome and the environment that allows an animal to respond to changing environmental stimuli. The overall movements we observe in animals essentially consist of the actions of individual muscles whose contractions are precisely modulated by the central nervous system (CNS). When expressed together or in a specific sequence, these contractions embody the displacements in space that are interpreted as the expression of particular behaviors. These specific behaviors are programmed into the CNS as so-called fixed action patterns, and when released at appropriate times they provide a selective advantage to those individuals expressing them. Thus, they are subject to natural selection enforcing their retention and shaping their development, and the genes that code for them are maintained in the gene pool when the individual survives and reproduces.

In relatively simple animals such as insects, fixed action patterns are expressed by the CNS as stereotyped behaviors. How a particular stereotyped behavior in insects might be expressed at a particular time is illustrated in Figure 20.1. When environmental information reaches sensory receptors, the receptors

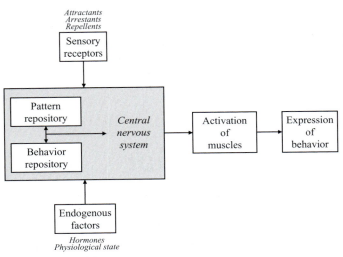

FIGURE 20.1 A model for the expression of behavior patterns in insects. The central nervous system receives information from sensory receptors that monitor the environment and from those that monitor physiological state. The code generated by these receptors is evaluated and compared to a template residing in memory. If the pattern matches, an associated behavior consisting of a preprogrammed sequence of muscle contractions is selected.

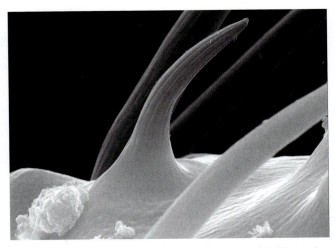

FIGURE 20.2 Scanning electron micrograph (11,679×) of a grooved peg sensillum from the antenna of an adult *An. gambiae* female mosquito.

translate it into a message pattern that is sent to the CNS. If the pattern that is received matches the template stored in a pattern bank, an associated fixed action pattern is selected from a behavior bank and the insect executes the specific behavior. Endogenous factors or physiological state can modify the behavioral expression and allow it to occur when biologically appropriate. Sensory receptors are thus the locks on the expression of the fixed action patterns that key environmental stimuli can unlock and cause the expression of behavior.

THE PERIPHERAL OLFACTORY APPARATUS

Olfaction plays a critical role in host-seeking and host-selection behaviors in flying vectors such as mosquitoes and related Diptera. In these insects, external chemical stimuli are detected by various types of structures. In particular, the segmented antennae and, to a lesser degree, the maxillary palps bear numerous hollow, hairlike cuticular sensilla that are responsible for detection of stimuli that elicit the behavioral patterns after processing by the CNS.

A variety of antennal sensilla have been described in several mosquitoes (Fig. 20.2), but they are best understood physiologically in *Ae. aegypti*. Studies of *Ae. aegypti* and *An. stephensi* suggest that more than 85% of the sensilla on the antennae function as olfactory structures. There is also evidence that various

other sensillar subtypes function as thermoreceptors, hygroreceptors, and mechanoreceptors. Several characteristically large bristles on a number of flying insect antennae have been shown to act as mechanosensory structures that are essential in the detection of wind currents and function in controlling the upwind movement of flight. In addition to sensillar diversity, there is good evidence that many insects form characteristic zonal arrays of particular sensillar types. In this manner the structure of the antennae may be best suited to survey the spatial distribution of odorants in their immediate environment by differentially stimulating the olfactory receptor cells along a proximal-to-distal line. An ability to distinguish local or microenvironmental odorant gradients might be critical in establishing fine level control of odorant source location.

The presence of numerous olfactory sensilla on the maxillary palps of several mosquito species suggests that they may also be secondary olfactory organs. In both sexes, four types of sensilla are observed on the palps as well as bristle-like sensilla chaetica and thin-walled peglike processes (sensilla basiconica), both of which contain neuronal material and may therefore be classified as sensory structures. In addition, a single mechanosensory campaniform sensillum has been described in some anophelines near the distal margin of the third segment of the palps whose dendrite contains a tubular body.

Variations in sensillar morphology contrast with the uniformity of the subcuticular sensillar structure (Fig. 20.3). The external cuticle of olfactory sensilla can form either a single- or double-walled structure that contains several pores through which the odorant molecules enter the aqueous lymph that fills the hollow internal cavity. Lying beneath the cuticle are often one or more primary olfactory receptor cells that are characteristically bipolar neurons with dendrites that point into the sensillum shaft. In this manner the olfactory receptor cell maximizes the surface of dendrite membrane that may be exposed to odorants.

In addition to the olfactory receptor cell, there are usually four auxiliary or sheath cells consisting of specialized epidermal cells that surround the entire receptor neuron apart from its axonal projection. Such auxiliary cells, often termed *trichogen* and *tormogen* cells, are thought to be responsible for ensuring the structural and electrophysiological integrity of the receptor cell as well as synthesis of the various components of the aqueous sensillar lymph. The lymph fills the lumen and provides the medium through which odorants interact with the signal transduction machinery on the dendritic membrane. The molecular constituents of the sensillar lymph carry out a wide variety of biochemical activities that are essential to ensure the ability of vectors to adequately sense and mount robust responses to their chemical environments.

MOLECULAR EVENTS IN OLFACTION

Vertebrates and invertebrates are remarkably similar with regard to the molecular mechanisms

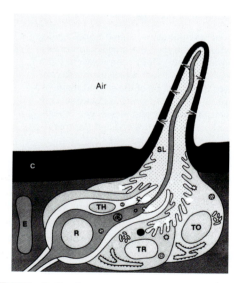

FIGURE 20.3 Stylized overview of the cellular organization of a typical insect sensillum. One or several bipolar ORNs [R] send an axon to higher structures in the CNS (dashed arrow) and a dendrite (solid arrow) to the peripheral region of the sensillum lumen (SL). Three auxiliary cells (Th: thecogen; To: tormogen; and Tr: trichogen) surround the ORNs and border the lumen. The cuticle (C) is shown as dense black, and undifferentiated epidermal cells (E) are shown as white shapes.

through which they are able to sense chemicals and other environmental cues. In all systems studied thus far, chemosensory signal transduction is mediated by a distinctive family of receptors that transverse the plasma membrane seven times and are coupled to heterotrimeric GTP-binding proteins (G-proteins) that lie on the cytoplasmic face of the membrane (Fig. 20.4). In the case of insect olfactory systems, odorant receptors are located on the dendrites of olfactory receptor cells that extend into the hollow space formed by distinctive sensory hairs that populate the antennae and maxillary palps. Once a receptor has bound its target, it undergoes a conformational change that facilitates the binding and activation of its cognate G-protein trimer. The activated G-protein complex then releases one of its subunits (Gα-GTP), which in turn rapidly induces downstream effector enzymes that synthesize one of a group of small molecules collectively known as *second messengers*, to reflect their role in relaying the signals that arise from extracellular chemicals.

Second messengers include molecules such as cyclic AMP (cAMP) and inositol 1,4,5-triphosphate (IP$_3$), which play essential roles in transducing chemical signals from the environment into the electrical signals that are the language of nervous systems. This is largely carried out through the action of cAMP or IP3-gated-cation channels that lie within the plasma membrane of the olfactory receptor cell and specifically open in response to the increase in the local cytoplasmic concentration of second messengers. In this manner, the rapid influx of Ca^{2+} or Na^{2+} ions into the dendrite cytoplasm of olfactory receptor cells results in sufficient depolarization of the membrane potential to induce an action potential that generates the electrical message that is sent along the axons to the higher-order centers of the CNS.

Several proteins involved in fundamental aspects of olfactory signal transduction have been molecularly cloned and characterized in both vertebrate and invertebrate model systems. These include components of the biochemical machinery necessary for signal transduction, such as olfactory-specific G-protein subunits, adenylate cyclase, cyclic nucleotide, and IP$_3$-gated channels, as well as other inositol carrier proteins. In addition, odor receptors have been identified in *Drosophila melanogaster* as well as in *An. gambiae*. Comparison of the sequenced genome of both *D. melanogaster* and *An. gambiae* has shown that a similar number of putative olfactory receptor genes exists in both insects, which may be reflective of a wider range of disease vectors. Interestingly, while there have been numerous studies in several systems that have identified many candidate olfactory receptors, the vast majority of these genes must be included in the broad category of "orphan" receptors, signifying the lack of

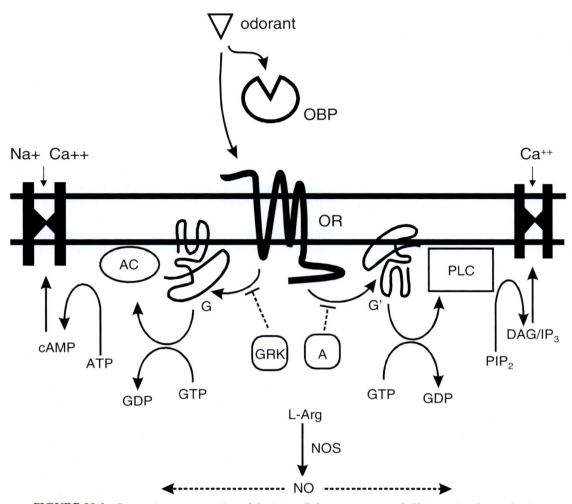

FIGURE 20.4 Composite representation of the intracellular components of olfactory signal transduction pathways. Here, a 7-transmembrane odorant receptor protein (OR) lying within the ORN dendrite interacts directly with odorants or alternatively in the context of odorant-binding protein (OBP) complexes. In both cases, subsequent interactions with heterotrimeric G-protein complexes (G/G') activate downstream effector enzymes adenyl cyclase (AC) and phospholipase C (PLC). This leads to the synthesis of the second messenger's cyclic AMP (cAMP), diacylglycerol (DAG), and inositol 1,4,5 triphosphate (IP3), which regulate several cation (Na$^+$, Ca^{2+}) channels that carry the transduction current. Another potential second messenger. Signaling is terminated (dashed lines) by decoupling of OR/G-protein complexes by arrestins (A) and G-protein-coupled receptor kinases (GRKs).

evidence showing a direct binding or other functional relationship between a putative receptor and any particular odorant.

Furthermore, a broad class of water-soluble proteins may have a vital role in the olfactory process. These odorant-binding proteins are expressed at high levels and are believed to act as odorant carriers to facilitate the solubilization of hydrophobic odorants and improve their delivery to receptors.

Termination of the sensory response is another important aspect of olfactory signal transduction. This involves the termination of the second-messenger pathways and the desensitization of receptor cells.

Both processes involve the action of specialized receptor kinases as well as arrestins that decouple the active G-protein complexes as well as actually reduce the numbers of receptors available on the surface of the receptor cell. In this manner the neurons become less sensitive to local odorant concentrations, thereby requiring higher levels of stimulation to maintain response levels. In addition to these events, the rapid degradation of free odorants in the sensillar lymph is carried out by several classes of degradative enzymes that are synthesized and secreted at high levels by olfactory receptor cells and the trichogen and tormogen cells. These enzymes are collectively known as *bio-*

transformation enzymes, to reflect their role in altering the chemical conformation of odorants. They provide, in a single reaction, a mechanism that effectively reduces the local concentration of odorants that could be bound by olfactory receptor proteins and facilitates their direct removal. Specifically, these include several classes of esterases, cytochrome P450, and glutathione S-transferase activities. In this manner, odorants that are presented to the olfactory receptor proteins on the dendritic outer membrane reflect as accurate a "real-time" representation of the airborne local environment as possible.

The importance of the CNS in interpreting, integrating, and responding to olfactory stimulation cannot be overstated. However, it must be stressed that apart from a growing appreciation of the organization of these processing centers, there is a profound lack of real understanding of the mechanisms that ultimately lead to these behaviors. A considerable degree of processing occurs between the G-protein-mediated events of signal transduction that take place in receptor cell dendrites and the coordinate neuronal activity that subsequently follows in the antennal lobe and, in turn, in the higher-order structures, such as the mushroom bodies. In this manner, sensory information is coupled to both behavioral responses and the processes of learning and memory, although we are only just beginning to acquire an understanding of the molecular-, cellular-, and system-level events that comprise these pathways. These questions remain at the forefront of basic neurobiology, and the relative simplicity of insect brains makes them uniquely suited to address these questions. Moreover, our understanding of the role of integrative processing in mediating the olfactory-driven behaviors, such as host preference, that are central to vectorial capacity may point to novel control strategies.

HOST-SEEKING BEHAVIOR

To ingest a meal of blood successfully, female mosquitoes must undergo a series of behaviors that will bring them in contact with a host. Little is known about the exact behaviors involved during this process and how specific stimuli may activate them. For that reason, the term *host-seeking behavior* has come to refer broadly to all the behaviors involved in the identification of a host and the movement of the mosquito that enable it to feed. Host-seeking behavior terminates once feeding is initiated and blood ingestion begins. Blood ingestion and the continuation of feeding are governed by midgut stretch receptors and additional stimuli that come from the blood itself as well as close-range stimuli from the host.

Because the blood within a host is not detectable from a distance, mosquitoes have evolved the ability to associate the metabolic byproducts of living things with the presence of blood and use these to cue in on a host. Many of the kairomones, which are chemical by-products produced by the host that are used in host location, have been identified, and their roles in long- and short-range attraction understood. CO_2 is the most universal attractant, with changes in its concentration probably being most responsible for the change in mosquito behavior. Increases in the firing rate of CO_2 receptors occur when concentrations of the gas increase by as little as 0.01%, and behavioral changes in mosquitoes occur with increases of 0.03%. Hosts produce other volatile components, such as lactic acid, acetone, and octenol, but these generally act as synergists in the presence of CO_2. There are species differences in sensitivity to CO_2 that may be related to host choice. For example, within the *An. gambiae* complex, the sibling species, *An. arabiensis*, *An. quadri-annulatus*, and *An. gambiae s.s.*, show marked differences in their preferred hosts and their responses to CO_2.

Volatile products are released from a host as discontinuous filamentous packets that move downwind and are further broken up by air currents, much like the dispersion pattern of smoke from a chimney. The challenge presented to the host-seeking mosquito is how to move progressively between the odor filaments while approaching the host without losing the trail. Upon entering an odor plume, mosquitoes continue to fly in a straight line. But if they subsequently leave the plume, they engage in rapid turns, a strategy that presumably increases the probability that they will reenter the plume or a neighboring one. The movements appear to be governed by an anemotaxis (movement upwind) that is guided by endogenous behavioral programs and modulated by host stimuli. The process begins relatively close to the host; mosquito host-seeking does not appear to be initiated until females are approximately 20–35 m downwind.

Visual cues, such as color and shape, are also important in the final approach to a host. Tsetse flies are attracted by both odors and visual cues. The contrast of a host against its background is important among the visual stimuli, enabling the tsetse to distinguish distant targets. Indeed, the pattern of stripes in zebras have been proposed to have evolved as a visual defense against tsetse feeding, obliterating the zebra's sharp visual edge and interfering with the visual pattern the fly receives.

Once host-seeking is successful and feeding commences, it is driven largely by phagostimulants in the blood. Adenine nucleotides are phagostimulants that continue the fixed action patterns associated with blood ingestion until stretch receptors triggered by abdominal distention finally terminate feeding. There is evidence for the heritability of these fixed action patterns, for it is possible to select for strains more efficient in their feeding.

There are hazards associated with a life devoted to obtaining blood. One is nutritional: Vertebrate blood lacks certain nutritive precursors, forcing vectors that feed exclusively on blood to evolve relationships with symbiotic microorganisms that supply the missing components (Chapter 23). A second hazard is behavioral. Ectoparasites that live continuously on a vertebrate host are seldom affected by the host's behavior, but host-seeking insects must contend with the defensive behavior of their hosts. A tiny animal approaching a host many times its size is taking a considerable risk. Host defensive behavior can account for a large degree of mosquito mortality and undoubtedly represents a strong selective pressure for the evolution of physiological mechanisms in the mosquito that reduce unnecessary host interactions. In many mosquito species that feed during the day or at crepuscular periods when host activity might discourage their attempts to feed, mechanisms have evolved to prevent the females from taking excessive risks when they are not necessary for reproduction. So, although the activation of host-seeking behavior is essential if the female is to ingest the protein that blood provides for the development of her eggs, there are times during her life when the behavior is inhibited endogenously. These include a failure to engage in host-seeking behavior immediately after adult eclosion, during the adult reproductive diapause, outside a circadian window of activity, and following a blood meal that initiates and sustains egg maturation. The frequency of blood ingestion is important because it determines the potential frequency of parasite transmission. Mosquitoes that feed more often have a greater chance of distributing the parasites they contain.

Teneral (newly emerged) females generally do not feed on blood for the first 1–2 days after eclosion. In *Culex* species, juvenile hormone (JH) is released after eclosion and stimulates the development of host-seeking behavior, although this mechanism does not appear to operate in *Ae. aegypti*. The failure of *Ae. aegypti* females to feed soon after eclosion may reflect non-JH-dependent developmental events, such as the maturity of sensory receptors, low nutritional reserves, and the initial softness of the cuticle that must first undergo hardening. Some species enter diapause as adults, and a reduction in JH titers is responsible for the entrance into the adult reproductive diapause, during which time they fail to feed. In *Culex pipiens*, the absence of host-seeking behavior during the adult diapause coincides with an absence of lactic acid sensitivity in antennal receptors and may therefore represent a behavioral inhibition based on the reduced sensitivity of peripheral receptors.

Once the female mosquito develops the competence to express host-seeking behavior, it appears to be the default behavior displayed whenever she is exposed to host stimuli during her species-specific circadian window of activity. When initially discouraged by defensive host behavior, the mosquito may show a variable degree of persistence in attempting to feed again. This persistence is related to the nutritional state of the mosquito. When discouraged, sugar-fed females are more likely to continue approaching a host than those that have been starved, and larger adults resulting from inadequate larval diets are more persistent than smaller adults.

In *Ae. aegypti*, there are at least two physiological mechanisms that regulate host-seeking behavior after blood is ingested. Immediately after a large blood meal, the mosquitoes withdraw their mouthparts and fly away, and any response to host stimuli is subsequently reduced. This initial refractory behavior is caused by the abdominal distention produced by the blood meal. Small blood meals below a distention threshold of 2–3 μL in *Ae. aegypti*, which normally ingest meals of about 5 μL, do not affect subsequent host-seeking behavior, but larger meals inhibit host-seeking. When distention is reduced by absorption and excretion of the meal, host-seeking behavior resumes. The abdominal distention after feeding does not affect the sensitivity of peripheral lactic acid receptors used to identify a host, but it may directly affect the activation of host-seeking behavior through the central nervous system.

Aedes aegypti that have ingested large blood meals continue to show an inhibition of host-seeking until after oviposition, even though the abdominal distention is reduced as the blood is being utilized. This second, more long-term inhibition of host-seeking results from another mechanism, which is hormonally based and unrelated to the size of the blood meal that was ingested as long as it was sufficient to initiate oogenesis. This inhibition is caused by the neuropeptide *Aedes aegypti* head peptide I (Aea-HP-I) that is released during egg maturation by way of several intermediary steps and generally lasts until oviposition occurs. Both hemolymph transfused from gravid donors as well as injected purified peptide are able to inhibit the host-seeking behavior of nongravid females. Aea-HP-I may decrease the sensitivity of sensory receptors and inhibit host-seeking behavior by reducing the perception of host stimuli that would otherwise activate the mosquito. It is not produced in anopheline mosquitoes and has no effect on their behavior when injected. After oviposition, host seeking behavior returns for the next gonotrophic cycle.

Although these two physiological mechanisms can potentially limit the host-seeking behavior of *Ae. aegypti* to a single time in each gonotrophic cycle, there are abundant examples of mosquitoes feeding multiply during a single gonotrophic cycle. Such multiple feeding and increased vector–host contact can significantly increase the potential for the acquisition and transmission of parasites. Understanding the factors that contribute to multiple feeding can help one better understand the sometimes explosive outbreaks of mosquito-borne diseases.

One factor that contributes to multiple feeding is the size of the blood meal that a mosquito is able to ingest. Mosquitoes that take small meals as a result of defensive host behavior are likely to continue to engage in host-seeking until their distention thresholds have been exceeded or until eggs begin to develop and the second inhibition is triggered. Host behavior can discourage mosquito feeding attempts, causing a reduction in the size of the blood meal and a shift in available hosts so that less defensive ones are subsequently fed upon.

Many parasites have evolved the means to manipulate the physiology and behavior of their hosts to effect their more efficient dissemination. Parasitic infection in mosquitoes can often affect the rate of blood ingestion. In *Ae. aegypti, Plasmodium gallinaceum* reduces the apyrase activity of the saliva (Chapter 28), making it difficult to locate blood, and the mosquito consequently probes longer in the host. Similarly, *Ae. triseriatus* infected with La Crosse virus probe more frequently, and *Ae. aegypti* infected with dengue virus require longer to feed and may be less sensitive to defensive behavior of the host. These increases in the duration of probing provide increased opportunities for parasite transmission in saliva. Some species of filarial worms develop in the flight muscles of the mosquito host. They can affect the spontaneous flight activity of the mosquito, and a large infection can significantly reduce its flight capacity. Heavily infected mosquitoes are consequently less likely to serve as vectors of filariae than those with smaller infections.

Parasites can be even more aggressive in their manipulation of host-seeking behavior. The ciliated protozoan *Lambornella clarki* infects the ovaries of tree hole mosquitoes, and the parasite is able to alter the behavior of the mosquito for its own benefit. While infected, host-seeking behavior is inhibited and oviposition behavior is activated, but the mosquito deposits protozoa rather than eggs. By altering the behavior of its host, the protozoan directs its own dissemination. It is not known whether the parasite produces substances that mimic the hormones the mosquito itself

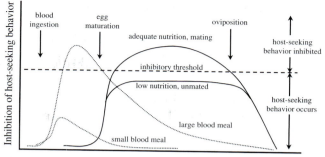

FIGURE 20.5 Behavioral inputs and thresholds and their net result on host-seeking behavior. Both blood ingestion (dotted lines) and egg development (solid lines) inhibit host-seeking behaviors when their inhibitory effects exceed a threshold (dashed line).

normally uses to alter behavior or whether it manipulates the mosquito to produce its own substances at the wrong time.

Whether the female is mated can have a significant effect on multiple feeding behavior that occurs while eggs are developing. Although mated females are inhibited from seeking a host during egg development, unmated females are not. The increase in the host-seeking behavior of unmated females occurs in the absence of the male accessory gland substances that are normally transferred during copulation. Male mosquitoes of several species are attracted to mammals, where they aggregate and intercept females arriving to feed, and the reduced inhibition of host-seeking behavior in the unmated, gravid females may provide a mechanism for their return to the host for mating so the eggs they carry can be laid. The behavior is further modulated by the decay of the effect of the male accessory gland secretions after transfer to the female.

The physiological mechanisms that regulate host-seeking behavior and, in theory, limit blood ingestion to once during each gonotrophic cycle may therefore be modulated and lead to multiple feeding. As summarized in Figure 20.5, an inhibitory threshold in the mosquito can be postulated above which endogenous stimuli inhibit host-seeking behavior. After a large blood meal, abdominal distention provides inputs that exceed this threshold and inhibit host-seeking behavior until the distention is gradually reduced, along with the blood volume. Small blood meals fail to reach this threshold at all and the mosquito continues to feed. Egg development provides other inputs that exceed the inhibitory threshold when the mosquito is adequately nourished and mated, but not when the female has an inadequate nutritional history or has failed to mate. It is this physiological modulation that undoubtedly is responsible for the examples of multiple feeding behavior often seen in field populations.

OTHER MOSQUITO BEHAVIORS

It is difficult to enumerate all the behaviors that mosquitoes express because, like Heisenberg's uncertainty principle, which at the quantum scale recognizes that measurements alone influence the object being measured, our observations of mosquito behaviors influence the very behaviors being displayed. For example, an examination of all the behaviors that comprise what we call host-seeking is complicated by the bias introduced by the observer's presence and the various stimuli that accompany it. As a result, most mosquito behaviors in the field are not studied directly but are instead inferred by their endpoints. Thus, host-seeking behaviors themselves are never measured but are generally inferred to have occurred when females are found near a host or within an attractant trap. With these considerations, the remainder of this chapter discusses the physiological regulation of several mosquito behaviors identified by their end result rather than the actual complex process by which they are expressed.

Regulation of Immature Behaviors

Knowledge of larval behavior is essential to predicting where the insects are most likely to be found for purposes of census and management. It is important to understand when, where, and how immatures feed when attempting to control them with microbial agents. Their susceptibility to predation by natural enemies is related to the behavioral thresholds that release defensive behaviors when larval mosquito predators are present. Mosquito larvae have little choice about the ephemeral aquatic habitats in which they develop; their distribution is determined largely by the choice of oviposition sites by gravid females, and the characteristics of these sites can significantly affect larval development. For example, when *Anopheles* larvae are reared in deeper pools of water, their eclosion rates decline significantly, engendered by the increased demand on their metabolic reserves made by diving. Although the repertoire of behaviors for immatures has been catalogued and described much more completely than for adults, the mechanisms involved in the control of these behaviors have not been as thoroughly identified. Of the larval behaviors that have been recognized, the control of only two, feeding and alarm behavior, have been examined.

Larval feeding behavior has been classified into four modes, including gathering, filtering, scraping, and shredding, each of which is elicited in response to certain food types. As in other animals, feeding behavior of larval mosquitoes is stimulated by the properties of the food, including the various components identified as phagostimulants. The rate of food ingestion by larvae is generally enhanced by yeast extracts, nucleotides, and general viscosity of the water. Water temperature, larval instar, and prior starvation also influence feeding rates. Larvae may also aggregate in response to gradients of temperature and nutrients within the water.

The alarm reaction is a defensive response to vibration and changes in light intensity that allows larvae and pupae to swim away from threatening situations. Diving is less likely to occur when pupae rest in a concave meniscus near a vertical surface or when in contact with submerged vertical surfaces. The net result of this response to stimuli is to maintain the pupae among vegetation that may be less accessible to predators and at the same time conserve pupal energy by reducing the need to display the alarm response. Differences in the response thresholds of larvae are heritable and can be modified by their nutritional state. Larvae that are maintained on a suboptimal diet are less likely to display the alarm response when stimulated by light.

Regulation of Other Adult Behaviors

Swarming Behavior

Mating in many mosquito species takes place while in swarms. Swarming is not required for all mosquitoes; some species mate soon after they emerge and have no need to swarm. Males first aggregate over conspicuous physical markers at specific times of the day, generally at dusk and dawn, displaying stereotyped flight behavior. In many species, swarms occur during the crepuscular periods at both sunrise and sunset. The timing of swarm formation is a function of the interaction between the circadian rhythmicity of flight activity and the response to light intensity, occurring through a window of illumination thresholds that initiate and terminate swarm formation. Behavior of a male within a swarm is determined more by the characteristics of the marker on which they swarm than an interaction with other members of the swarm. Females enter the swarm, presumably also responding to visual stimuli, and shortly afterward leave *in copula*. The short-range attraction of males to females is largely based on the species-specific wing beat frequency of the female, and the swarm provides the opportunity for the sexes to come together in order for this short-range attraction to be effective.

Johnson's organ in the pedicel of the male antenna receives the sound vibrations from the wing beat to identify conspecific females. For these vibrations to be perceived, a substantial portion of the hairs on the antennae must be extended. In *Ae. aegypti*, this extension occurs within 24 h after eclosion, and the hairs remain extended, allowing the male to mate whenever he perceives the proper wing beat. In contrast, many *Anopheles*, *Culex*, and *Mansonia* will extend their antennal hairs only at the times of the day when swarming actually occurs. The extension is under nervous control and occurs as a result of the swelling of an annulus at the base of each whorl as it become hydrated with increasing pH from the activity of surrounding support cells.

There are many devices on the market that purport to repel female mosquitoes that are attempting to feed on humans. These devices frequently advertise that the female is unusually sensitive to the wing beat of the male when she is carrying eggs and is repelled by that sound. Actually, of course, it is the male that responds to wing beat frequency and the female that is oblivious to it. Laboratory and field tests have shown these devices do not work. It may be possible, however, to take advantage of the male's response to wing beat frequency to reduce the field populations of males of some species. In a study that used broadcasted sound in the field, sufficient numbers of *Culex tritaeniorhynchus* males were actually removed from a natural population to accomplish a reduction in the insemination rate of females.

Mating Behavior

Once the sexes come into contact, they engage in a complex sequence of behaviors leading to the act of copulation, during which sperm are transferred. This behavior is particularly significant if genes are to be introduced into mosquito populations by genetically modified males. The females of most mosquito species are incapable of being inseminated for the first few days after they eclose, although they may still engage in copulatory behavior. This age-dependent insemination may be a mechanism that minimizes inbreeding by allowing dispersal from the breeding site before insemination occurs. It is the female that controls the refractory period, allowing neither the physical coupling nor the cues that are necessary for male ejaculation. A posteclosion production of juvenile hormone (JH) appears to cause the development of mating competence in the female, acting on the terminal abdominal ganglion, which innervates the reproductive system.

The physical behaviors involved in mating include the male grasping the female and introducing his aedeagus into the female tract, depositing sperm and male accessory gland secretions into the bursa copulatrix. Sperm subsequently move into the spermathecae, where they reside until the female releases them when fertilization occurs.

Once the female mosquito does mate, she may be refractory to subsequent inseminations for life. By failing to allow the claspers of subsequent potential mates to grasp her cerci, she prevents any more sperm from entering. Products of the male accessory glands (MAGs) that are transferred in the semen of the initial mate eliminate the tendency of the female to respond to males. The active principle in MAG substances was at first called *matrone*, but the use of the term conveys the incorrect notion that a single component is responsible for a range of diverse processes in many different species. Until all the diverse MAG components are identified, it is probably better to simply refer to them as *MAG substances*. The MAG substances may be species specific; and in several species of anophelines, MAG substances have no effect at all on mating behavior.

In those mosquitoes that do mate only once and are refractory to subsequent mating for the rest of their lives, some conditions may fail to invoke this refractoriness. Such times include the several hours immediately after an initial mating and before the MAG substances have had time to have an effect. If copulation is interrupted or if the females mate with a previously depleted male so that the transfer of MAG substances is not complete, the females may continue to mate.

There is better evidence for the occurrence of multiple mating in anophelines, but it is relatively infrequent. In field studies of anopheline mating behavior, relatively low rates of polyandry were identified. The male accessory glands of anophelines produce a mating plug that blocks the oviduct and prevents reinsemination of the female for 1–2 days. A small proportion of field-collected mosquitoes retain double plugs as evidence that multiple mating, if not necessarily insemination, has occurred. Another mechanism may prevent insemination once the plug is dissolved, because, as previously mentioned, MAG substances themselves fail to induce mating refractoriness in several anophelines.

Male Terminalia Rotation

Male mosquitoes eclose with genitalia that are improperly positioned for mating. Before they are able to successfully copulate, their terminalia must rotate through 180°, either clockwise or counterclockwise. This behavior occurs in male *Ae. aegypti* within 1–2 days after they eclose, allowing the genitalia of the male to fit properly with those of the female. Two pairs of opposing muscles in the rotating segments are responsible for movement, with one set apparently contracting at random to produce the random direction of rotation. The initiation of contraction does not appear to be controlled by the nervous system, but it may be a result of normal muscle maturation. Decapitation of the newly emerged male or the transection of the ventral nerve cord prior to rotation does not prevent normal rotation from occurring, but it is blocked by any damage to the arthrodial membrane between the segments that rotate.

Circadian Rhythmicity

Many of the behaviors that mosquitoes display are governed by a diel periodicity. Rhythmic adult activity makes it more likely that the female will be exposed to host stimuli for feeding, and to males for mating, at certain times of the day. Although some species show few or no rhythms associated with development, in others there are rhythms of hatching, larval-pupal ecdysis, adult eclosion, host-seeking, oviposition, and general activity. These rhythms are generally entrained by photoperiod and temperature, and appear to be governed by multiple endogenous circadian pacemakers. Females that have mated often show an altered periodicity that eliminates the activity peak normally associated with the time of day that mating occurs.

Oviposition Behavior

Larval and adult mosquitoes occupy profoundly different ecological niches. For her offspring to survive, the female must alter her usual terrestrial preferences and return to an aquatic habitat when ready to lay eggs. For this to occur, another behavioral program must be called into play that will allow the female to shift her responsiveness from terrestrial hosts to aquatic stimuli before the eggs can be laid. Preoviposition behavior, the long-range attraction to oviposition site stimuli, is mediated by visual and antennal receptors and, in *Ae. aegypti*, is modulated by a hemolymph-borne factor produced during egg development.

When female mosquitoes that had no eggs were transfused with hemolymph from a gravid female, they engaged in preoviposition behavior as if they were gravid. The presence of oviposition site stimuli and this preoviposition factor are not sufficient to trigger the behavior, however. A signal that mating has occurred that comes from the male accessory gland substances that are transferred during copulation is also required. In the absence of mating, gravid females fail to display preoviposition behavior, which prevents them from expending energy on an activity that would

be worthwhile only if eggs could be fertilized. Once initiated by egg development, preoviposition behavior continues as long as eggs remain within the ovaries and is terminated by nervous signals associated with egg deposition.

After preoviposition behavior brings the gravid female near the oviposition site, other, closer-range stimuli prevail to induce oviposition behavior, which involves the actual deposition of eggs and consists of another complex repertoire of behaviors. These stimuli are assessed by tarsal chemoreceptors that can gauge substrate texture, water salinity, and temperature. When integrated by the brain, motor programs controlled by the terminal abdominal ganglion are stimulated and expressed by the muscles of the oviduct, resulting in the deposition of eggs. Eggs may be laid singly, as in most *Aedes*, *Anopheles*, *Toxorhynchites*, and *Psorophora*, or in rafts, as in *Culex*, *Culiseta*, and *Mansonia*. The eggs that are laid in rafts are secured by interdigitation of the exochorionic pegs on their surface; there is no evidence in mosquitoes for the adhesive substances that other insects produce in their female accessory glands.

Oviposition behavior is often either blocked or considerably reduced until the female mates, with male accessory gland substances again involved in the release of the behavior. If the gravid female is decapitated, eggs may be oviposited indiscriminately, suggesting that the brain normally exerts an inhibitory influence on the muscles associated with oviposition.

FUTURE PROSPECTS

We are beginning to understand the complex physiological underpinnings of behavior in mosquitoes, especially for those behaviors that contribute to the transmission of parasites. Virtually nothing is known about the hierarchy of decision making when the mosquito is presented with several stimuli that can potentially initiate several different behaviors. Although we can predict, based on its physiological state, when a mosquito might engage in host-seeking behavior, it is not known how other stimuli, such as those from an oviposition site or from a brief interaction with a male, can "distract" the female from her original behavioral program. For example, prior carbohydrate feeding can influence whether *Cu. nigripalpus* mosquitoes choose to feed on blood or nectar when given a choice of both. Certainly, in the real world, mosquitoes are bombarded with environmental stimuli from many different sources. In these cases, the CNS must be able to weigh the priorities of the stimuli and express the most adaptive behavior, a process honed during millions of years of evolution to rely on a physiological state for the proper outcome. Increasing our basic understanding of the mechanisms by which such behaviors are generated may eventually result in methods that would effectively impair or otherwise disrupt these events. In turn, such advances may provide novel approaches to reduce the vectorial capacity of many vectors of disease agents.

Readings

Allan, S. A., Day, J. F., and Edman, J. D. 1987. Visual ecology of biting flies. *Annu. Rev. Entomol.* 32: 297–316.

Bentley, M. D., and Day, J. F. 1989. Chemical ecology and behavioral aspects of mosquito oviposition. *Annu. Rev. Entomol.* 34: 401–421.

Bowen, M. F. 1991. The sensory physiology of host-seeking behavior in mosquitoes. *Annu. Rev. Entomol.* 36: 139–158.

Brown, M. R., Klowden, M. J., Crim, J. W., Young, L., Shrouder, L. A., and Lea, A. O. 1994. Endogenous regulation of mosquito host-seeking behavior by a neuropeptide. *J. Insect Physiol.* 40: 399–406.

Charlwood, J. D., and Jones, M. D. R. 1980. Mating in the mosquito *Anopheles gambiae* s.l. II. Swarming behaviour. *Physiol. Entomol.* 5: 315–320.

Costantini, C., Sagnon, N. F., della Torre, A., Diallo, M., Brady, J., Gibson, G., and Coluzzi, M. 1998. Odor-mediated host preferences of West African mosquitoes, with particular reference to malaria vectors. *Am. J. Trop. Med. Hyg.* 58: 56–63.

Dekker, T., and Takken, W. 1998. Differential responses of mosquito sibling species *Anopheles arabiensis* and *An. quadrimaculatus* to carbon dioxide, a man or a calf. *Med. Vet. Entomol.* 12: 136–140.

Downes, J. A. 1969. The swarming and mating flight of Diptera. *Annu. Rev. Entomol.* 14: 271–298.

Edman, J. D., and Kale II., H. W. 1971. Host behavior: Its influence on the feeding success of mosquitoes. *Ann. Entomol. Soc. Amer.* 64: 513–516.

Egerter, D. E., and Anderson, J. R. 1989. Blood-feeding drive inhibition of *Aedes sierrensis* (Diptera: Culicidae) induced by the parasite *Lambornella clarki* (Ciliophora: Tetrahymenidae). *J. Med. Entomol.* 26: 46–54.

Ernst, K. D., Boeckh, J., and Boeckh, V. 1977. A neuroanatomical study on the organization of the central antennal pathways in insects. *Cell Tissue Res.* 176: 285–306.

Gillett, J. D. 1979. Out for blood: Flight orientation up-wind in the absence of visual cues. *Mosq. News* 39: 221–229.

Hansson, B. S., and Anton, S. 2000. Function and morphology of the antennal lobe: New developments. *Annu. Rev. Entomol.* 45: 203–231.

Hildebrand, J. G., and Shepherd, G. M. 1997. Mechanisms of olfactory discrimination: Converging evidence for common principles across phyla. *Annu. Rev. Neurosci.* 20: 595–631.

Kellogg, F. E. 1970. Water vapour and carbon dioxide receptors in *Aedes aegypti. J. Insect Physiol.* 16: 99–108.

Klowden, M. J. 1995. Blood, sex and the mosquito: The mechanisms that control mosquito blood-feeding behavior. *BioScience* 45: 326–331.

Klowden, M. J. 2001. Sexual receptivity in *Anopheles gambiae* mosquitoes: Absence of control by male accessory gland substances. *J. Insect Physiol.* 47: 661–666.

Klowden, M. J., and Briegel, H. 1994. Mosquito gonotrophic cycle and multiple feeding potential: Contrasts between *Anopheles* and *Aedes* (Diptera: Culicidae). *J. Med. Entomol.* 31: 618–622.

Klowden, M. J., and Lea, A. O. 1978. Blood meal size as a factor affecting continued host-seeking by *Aedes aegypti* (L.). *Am. J. Trop. Med. Hyg.* 27: 827–831.

Klowden, M. J., and Lea, A. O. 1979. Humoral inhibition of host-seeking in *Aedes aegypti* during oocyte maturation. *J. Insect Physiol.* 25: 231–235.

Koella, J. C., Sorensen, F. L., and Anderson, R. A. 1998. The malaria parasite, *Plasmodium falciparum*, increases the frequency of multiple feeding of its mosquito vector, *Anopheles gambiae. Proc. R. Soc. Lond. B Biol. Sci.* 265: 763–768.

Krupnick, J. G., and Benovic, J. L. 1998. The role of receptor kinases and arrestins in G protein–coupled receptor regulation. *Annu. Rev. Pharmacol. Toxicol.* 38: 289–319.

Laurent, G., Wehr, M., and Davidowitz, H. 1996. Temporal representations of odors in an olfactory network. *J. Neurosci.* 16: 3837–3847.

Malnic, B., Hirono, J., Sato, T., and Buck, L. B. 1999. Combinatorial receptor codes for odors. *Cell* 96: 713–723.

McIver, S. B. 1982. Sensilla of mosquitoes (Diptera: Culicidae). *J. Med. Entomol.* 19: 489–535.

Merritt, R. W., Dadd, R. H., and Walker, E. D. 1992. Feeding behavior, natural food, and nutritional relationships of larval mosquitoes. *Annu. Rev. Entomol.* 37: 349–376.

Moore, J. 1993. Parasites and the behavior of biting flies. *J. Parasitol.* 79: 1–16.

Murlis, J., Elkinton, J. S., and Carde, R. T. 1992. Odor plumes and how insects use them. *Annu. Rev. Entomol.* 37: 505–532.

Nielsen, E. T., and Haeger, J. S. 1960. Swarming and mating in mosquitoes. *Misc. Publ. Entomol. Soc. Am.* 1: 72–95.

Roth, L. M. 1948. A study of mosquito behavior. An experimental laboratory study of the sexual behavior of *Aedes aegypti* (Linnaeus). *Am. Midl. Nat.* 40: 265–352.

Sih, A. 1986. Antipredator responses and the perception of danger by mosquito larvae. *Ecology* 67: 434–441.

Spielman, A. 1964. The mechanics of copulation in *Aedes aegypti. Biol. Bull.* 127: 324–344.

Sutcliffe, J. F. 1994. Sensory bases of attractancy: Morphology of mosquito olfactory sensilla—a review. *J. Am. Mosq. Cont. Assoc.* 10: 309–315.

Takken, W., and Knols, B. G. 1999. Odor-mediated behavior of Afrotropical malaria mosquitoes. *Annu. Rev. Entomol.* 44: 131–157.

Vickers, N. J., Christensen, T. A., Baker, T. C., and Hildebrand, J. G. 2001. Odour-plume dynamics influence the brain's olfactory code. *Nature* 410: 466–470.

Vosshall, L. B., Amrein, H., Morozov, P. S., Rzhetsky, A., and Axel, R. 1999. A spatial map of olfactory receptor expression in the *Drosophila* antenna. *Cell* 96: 725–736.

Walker, E. D., and Edman, J. D. 1985. The influence of host defensive behavior on mosquito (Diptera: Culicidae) biting persistence. *J. Med. Entomol.* 22: 370–372.

Zacharuk, R. Y. 1980. Ultrastructure and function of insect chemosensilla. *Annu. Rev. Entomol.* 25: 27–47.

21

The Adult Midgut

Structure and Function

JAMES E. PENNINGTON AND MICHAEL A. WELLS

INTRODUCTION

The gut plays a pivotal role in all physiological processes. It is the site of nutrient digestion and absorption and also where parasites or viruses enter the vector for subsequent transfer to a vertebrate host. In hematophagous insects, the midgut is the site of blood meal digestion and absorption. Two basic patterns of blood digestion are found: (a) *batch* (mosquitoes, sand flies, and fleas), in which digestion proceeds simultaneously over the entire surface of the blood bolus; and (b) *continuous* (higher Diptera — tsetse fly — and Hemiptera — kissing bugs), in which the blood meal is held in the anterior midgut, where no digestion occurs, and portions are passed to the middle and posterior midgut, where digestion and absorption occur. The blood meal is normally separated from the midgut epithelium by an extracellular layer known as the *peritrophic membrane* or *peritrophic matrix* (Chapter 22).

Vertebrate blood is a nutritionally unique meal. Protein, the principal nutrient, accounts for almost 20% of the wet weight of blood. Compared to protein, carbohydrate and lipid contribute almost insignificantly to the mass of blood. Therefore, blood-sucking insects must use amino acids derived from blood meal proteins to synthesize the lipid and carbohydrate stores involved in egg production. In addition, amino acids derived from blood meal proteins are an important source of energy to fuel the demanding processes of egg production.

The size of the blood meal depends on several factors: the species of insect, temperature, age, mating status, previous feeding history, source of the blood, and the stage of the gonotrophic cycle. In general, temporary ectoparasites take large blood meals that can be two to ten times body weight. Such large blood meals impair mobility, but the mass of the blood meal is rapidly reduced by very efficient diuresis (water excretion) since 80% of the mass of the blood meal is water. The large volume of the blood meal could also place great stress on the gut, due to extensive distension, but the nature of the cellular junctions in the gut and the elasticity of the gut musculature and abdominal wall are adapted to withstand this stress.

The time required for digestion of a blood meal varies greatly and depends on the species of insect, the temperature, the blood source, the blood meal size, and several other factors. The time can vary from as short as 4 hours (*Pediculus humans*), 48 hours (*Glossina m. morsitans*), to hours (mosquitoes), to more than 300 hours (*Triatoma infestans*).

An important event in blood digestion is lysis of erythrocytes, because the red cells contain the majority of the protein in the blood meal. Some insects achieve lysis by mechanical means; for example, fleas push the blood bolus back and forth over projecting spines in the proventriculus, and mosquitoes possess a cibarial armature that ruptures red blood cells. Chemical hemolysis may be induced by material secreted in the saliva (*Cimex lectularis*), but more commonly such hemolysins are produced in the midgut.

Rhodnius proxlius produces a small peptide in the anterior midgut, the site of blood meal storage, which causes lysis. In the tsetse fly the hemolysin is produced in the digestive portion of the midgut.

STRUCTURE

One of the few areas of vector biology that has received significant attention across the range of blood-feeding insects is the structure of the gut. We summarize the general features of the structure of the gut of some species and give a brief overview of the function of the various gut subdivisions, when known. The insect gut contains a foregut, a midgut, and a hindgut. Figure 21.1 shows schematic illustrations of the organization of the digestive systems in four major types of blood-feeding insects. Drawing blood from the host into the gut involves one or more pumps— the cibarial and pharyngeal pumps. The esophagus connects the proboscis, which is inserted into the host, and the cardia, which is the opening to the midgut. In

Diptera the esophagus also opens into the crop, a blind sac (diverticulum) that stores sugars in mosquitoes.

The midgut is often divided into an anterior and posterior region. Salivary gland secretions are injected into the blood of the host, where they inhibit blood coagulation and prevent vasoconstriction—effects that aid the insect in withdrawing blood from the host (Chapter 28). The Malpighian tubules are the insect kidney. They filter the hemolymph and secrete the filtrate into the hindgut. The hindgut plays an important role in ion reabsorption (Chapter 26).

The midgut is the site of action for blood meal digestion and absorption and is the focus of our discussion. The midgut is a simple epithelium composed of a single layer of cells lying on a basolateral membrane that separates it from the hemolymph. There are three or four different types of cells present. The majority of the cells are digestive epithelial cells (Fig. 21.2). These cells have a typical microvillar structure on the lumenal surface. The digestive cells are responsible for the synthesis and secretion of digestive enzymes and usually have prominent secretory vesicles. In addition

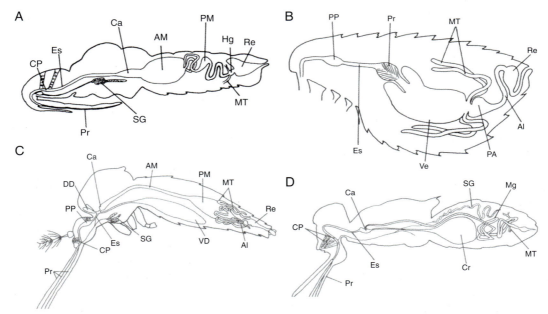

FIGURE 21.1 (A) Triatomid (kissing) bug alimentary canal. Based on Harwood and James, 1979, *Entomology in Human and Animal Health*, 7th ed. New York, Macmillan. AM = anterior midgut, Ca = cardia, CP = cibarial pump, Es = esophagus, Hg = hindgut, Pr = proboscis, PM = posterior midgut, Re = rectum, SG = salivary gland. (B) Flea alimentary canal. Based on Harwood and James, 1979, AI = anterior intestine, PP = pharyngeal pump, Es = esophagus, MT = Malpighian tubules, PA = pyloric ampulla, Pr = proventriculus, Re = rectum, Ve = ventriculus. (C) Mosquito alimentary canal. From R. E. Snodgrass, 1959, *The Anatomical Life of the Mosquito*. Smithsonian Misc. Coll. Vol. 139, No. 8, Washington, DC. AI = anterior intestine, AM = anterior midgut, Ca = cardia, CP = cibarial pump, DD = dorsal diverticulum, Es = esophagus, MT = Malpighian tubules, PP = pharyngeal pump, Pr = proboscis, PM = posterior midgut, Re = rectum, SG = salivary gland, VD = ventral diverticulum (crop). (D). Tsetse fly alimentary canal. Modified from Glasgow 1963. Ca = cardia, Cr = crop, CP = cibarial pump, Es = esophagus, Pr = proboscis, SG = salivary gland, Mg = midgut, MT = Malpighian tubules.

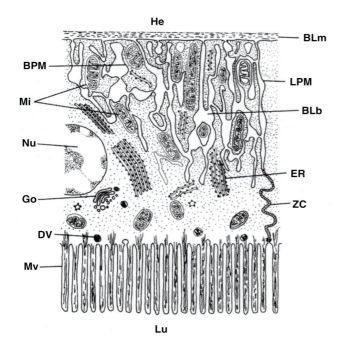

He
BLm
BPM
LPM
Mi
BLb
Nu
ER
Go
ZC
DV
Mv

Lu

FIGURE 21.2 Ultrastructure of a generalized insect midgut epithelial cell. BLm = basal lamina, He = hemocoel, LPM = lateral plasma membrane, BLb = basal labyrinth, BPM = basal plasma membrane, DV = dense vesicle, ER = endoplasmic reticulum, Go = Golgi, Lu = lumen, Mi = mitochondria, Mv = microvillus, Nu = nucleus, ZC = zonula continua. Adapted from Berridge, 1970. A structural analysis of intestinal absorption. In *Insect Ultrastructures* (A. C. Neville, ed.). Symposia of the Royal Society of London, No. 5.

to these cells, there are also absorptive cells, although they are often difficult to distinguish ultrastructurally from digestive cells; in some species one cell may serve both functions.

Distributed throughout the midgut in small numbers are endocrine and regenerative cells. Immunocytochemical studies have shown that the endocrine cells produce several peptides, but to date the function of these midgut peptides has not been elucidated. It is certainly tempting to speculate that they are involved in coordinating the digestion of the blood meal. This represents an exciting area of current research. The regenerative cells are undifferentiated cells, which differentiate to replace midgut cells that are sloughed off.

HEMIPTERA

The hemipteran midgut is divided into two parts, an anterior midgut (the stomach) and a posterior midgut (the intestine). The anterior midgut is not involved in protein digestion and absorption, but it has several important roles. It is the site of storage of the meal, which is slowly dispensed into the posterior midgut. It is also the site of water and ion absorption. Water is excreted from the hemolymph via the Malpighian tubules, but the ions are stored intracellularly in the anterior midgut as spherites. Eventually, the spherites dissolve, releasing the ions into the hemolymph. The anterior midgut is the site of hemolysis, and carbohydrate digestion also occurs in the anterior midgut. In adult insects lipid is digested and absorbed in the posterior midgut, but lipid storage droplets accumulate in the anterior midgut.

The intestine is the site of protease synthesis and secretion and the site of protein digestion and amino acid absorption; however, most of these activities reside in the anterior portion of the intestine. As discussed in the section on digestive enzymes, Hemiptera are unusual in that they have an acidic pH in the midgut and use cathepsins, which are normally lysosomal enzymes, as part of their protein digestion mechanism. The posterior portion of the intestine appears to be the major site for nutrient absorption. Midgut endocrine cells are also concentrated in the posterior intestine.

DIPTERA

Culicidae

The midgut of mosquitoes is simple, consisting of only an anterior midgut, which is not involved in blood digestion, and the posterior midgut, where all the digestion and absorption take place. The cells in the posterior midgut show all the characteristics of digestive epithelium, and there is no clear distinction between digestive and absorptive cells. (Figure 21.3 shows a detailed drawing of the organization of the *Aedes aegypti* digestive system.) Regenerative and endocrine cells are present, but apparently cell division does not occur in adults. There are more than 500 endocrine cells in the midgut of *Ae. aegypti*, making it the largest endocrine organ in the mosquito, but the function(s) of the endocrines remain to be established.

In *Ae. aegypti* the midgut is not fully differentiated at eclosion, and it requires about 3 days before the midgut has matured to the point where the female takes a blood meal. In the interim, she feeds on nectar or sugar water. In *Anopheles*, a similar maturation may occur, but females consume a blood meal shortly after eclosion.

Psychodidae

The midgut of the sand fly is essentially the same as that of the mosquito and follows the same developmental maturation following eclosion.

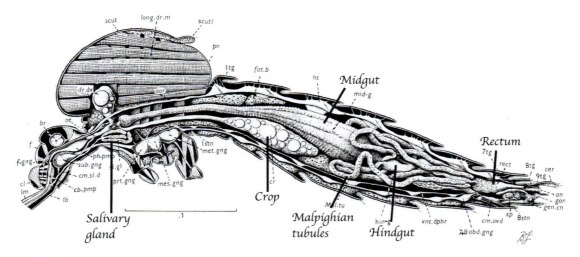

FIGURE 21.3 Cross section of an adult female *Ae. aegypti* showing the organization of the digestive system. Taken from Jobling (1987). Reproduced with permission from the Welcome Trust.

Simuliidae

The blackfly midgut has not been studied, except for peritrophic membrane formation.

Muscidae

Only the stable fly, *Stomoxys calcitrans*, has been examined. The midgut is divided into three regions— the anterior, central, and posterior midgut. The blood meal is stored in the anterior midgut, and it is also the site of water and ion resorption. The central midgut is the site of digestive enzyme synthesis and secretion. The posterior midgut is the site absorption of digestion products. Most lipids seem to be absorbed here, because the cells transiently accumulate numerous lipid vesicles during blood meal digestion.

Glossinidae

In most aspects, the midgut of the tsetse fly is similar to that of *Muscidae*. The major difference is the presence of mycetomes between the anterior and central midgut, whose cells contain endosymbiotic bacteria.

Siphonaptera

The midgut of fleas, many of which feed for long periods of time compared to the temporary ectoparasites discussed earlier, have some specializations of the midgut. Unlike the temporary ectoparasites, there is no regional specialization of the midgut, and it undergoes more extensive distension during blood feeding.

It is interesting to note the hypertrophy of the midgut during blood digestion, because it can grow up to 20 times its unfed length.

DIGESTIVE ENZYMES

The salivary glands of blood-sucking insects produce practically no digestive enzymes; rather, the digestive enzymes are produced by the midgut. The midguts of blood-sucking insects have developed specialized systems for the efficient digestion of protein-rich blood meals. Some specialization occurs in the form of unique systems for the regulation of expression of digestive enzyme (Chapter 36). Other optimizations occur in the digestive enzymes themselves that are unique compared to their well-characterized vertebrate counterparts. As would be expected from the nature of blood-sucking insects and their protein-rich meal, there is considerable research on digestive proteases and protein digestion.

Numerous digestive proteases have been characterized in several species of blood-sucking insects. Typically, these proteases are classified according to several factors: substrate cleavage site, active site, and substrate specificity. Proteases that release amino acids from the termini of proteins are classified as exoproteases; endoproteases generally cleave proteins at the carboxyl side of specific amino acid residues. The active sites of proteases are classified according to the catalytic agent present. Metalloproteases contain a metal ion in the active site; serine, cysteine, and aspartic proteases are named according to the amino acid residue present in the active site of the enzyme. Table 21.1 lists the general properties of common proteases.

In most blood-sucking insects the pH of the midgut is alkaline, and complete digestion of proteins requires the action of both endoproteases (trypsin and chymotrypsin) and exoproteases (carboxypeptidase and aminopeptidase) (Fig. 21.4).

Proteases act in a cooperative manner in which endoproteases hydrolyze protein, creating additional termini and smaller peptides for exopepdidases to digest. Superficially, the enzymes in this digestive system seem to correspond to those used in vertebrate digestion. In fact, as these enzymes have been isolated they have been named according to their well-characterized vertebrate counterparts. Several insect proteases, however, have been shown to have much broader substrate specificities than their presumed vertebrate counterparts. For example, chymotrypsin from *Ae. aegypti* has been shown to have activity on substrates normally processed by elastases. Insects may have developed enzymes with broader substrate specificities to increase the efficiency of digestion.

Hemiptera, on the other hand, have a slightly acidic midgut pH and do not have the "serine protease" digestive system typically found in most insects.

Hemiptera instead use digestive enzymes classified as *cathepsins*. Cathepsins are normally lysosomal enzymes involved in intracellular protein degradation. Cathepsins also occur in insect eggs and degrade egg storage proteins.

In some blood-sucking insects, such as *Stomoxys calcitrans*, digestive enzymes are stored as *zymogens* (inactive precursors) in secretory vesicles and are secreted into the midgut lumen in response to the meal. After secretion, zymogens are activated by a proteolytic hydrolysis of a short peptide from the N-terminus of the enzyme. In some anopheline mosquitoes there is detectable trypsin activity in the midgut prior to the blood meal, but in *Ae. aegypti* there is neither an active trypsin nor a zymogen in the midgut prior to the blood meal. Regardless of whether the midgut contains a stored zymogen or an active enzyme or neither, in all blood-sucking insects there is an increase in digestive enzyme activity in the midgut following a blood meal. The regulation of the synthesis of digestive enzymes in the midgut is described in the Chapter 36.

Digestive proteases also play a role in the activation of disease agents in the midguts of blood-sucking insects. In the case of the malaria parasite, the activity of a chitinase secreted by *Plasmodium* is necessary to facilitate the invasion of the parasite into the midgut. This chitinase enzyme is activated by the action of an insect protease, possibly mosquito trypsin, in a manner similar to the activation of proteolytic zymogens.

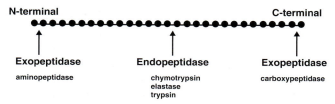

FIGURE 21.4 Mode of action of various proteases on a protein substrate.

CARBOHYDRATE DIGESTION

Carbohydrate is an important metabolite as a fuel for flight in many vector insects. Although the princi-

TABLE 21.1 Classification of digestive proteases found in the midguts of blood-sucking insects. Classification is based on the vertebrate system, but several insect proteases have been shown to have broader specificity

Protease Type	Enzyme Name	Active Site Type	Substrate Specificity/Preference
Exopeptidases:			
	Aminopeptidase	Metallo	N-terminal amino acids
	Carboxypeptidase A	Metallo	C-terminal amino acids
	Carboxypeptidase B	Metallo	C-terminal amino acids
Endopeptidases:			
	Trypsin	Serine	Carboxyl side of Lys and Arg
	Chymotrypsin	Serine	Carboxyl side of Phe
	Elastase	Serine	Carboxyl side of Leu, Ala
	Cathepsin L	Cysteine	Carboxyl side of Arg
	Cathepsin D	Aspartic	Phe-Phe bonds

pal nutrient obtained from vertebrate blood is protein, there is a small amount of carbohydrate present as monosaccharides, as well as oligosaccharides found on glycoproteins and glycolipids. In addition, several blood-sucking insects can supplement blood feeding with plant nectar or other plant fluids that provide additional carbohydrate. In dipterans, sugar solutions obtained from plant sources are typically stored in an organ known as the *crop,* but sugars are delivered to the gut for digestion and absorption. Even though there has been little investigation into the digestion of carbohydrates in blood-sucking insects, digestive glycosidases' of various specificities have been observed in insects such as the sand fly, *Lutzomyia longipalpis.*

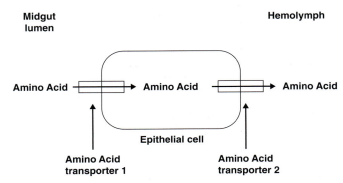

FIGURE 21.5   Two amino acid transporters (1 and 2) in the membranes of epithelial cells are needed to move amino acids from the lumen of the midgut to the hemolymph.

LIPID DIGESTION

The principal sources of lipid in a blood meal are derived from erythrocyte cell membranes and from lipid carried by plasma lipoproteins. In fact, there is so little lipid available in blood that mosquitoes use amino acids derived from the blood meal to produce lipids for storage in eggs. Even though various midgut lipase activities have been described in numerous insects, lipid digestion has been studied little in blood-sucking insects. Triacylglycerol lipases, which hydrolyze ester bonds on lipoprotein-bound triacylglycerol, releasing fatty acids, have been described in the midgut lumen of *Ae. aegypti* and *S. calcitrans.*

An important facet of lipid digestion that has received little attention is the role of phospholipases in the lysis of erythrocytes and subsequent release of cytoplasmic protein. The release of the detergent-like fatty acids from phospholipid by phospholipases could be an important means of disrupting cell membranes.

AMINO ACID ABSORPTION

The blood meal supplies the insect with proteins, whose amino acids can be used to support egg production. Once the proteins have been degraded to their constituent amino acids in the lumen of the gut, they must be absorbed by the midgut epithelial cells and transferred to the hemolymph, from which they can be taken up by the fat body and other tissues. This transport requires two types of transmembrane amino acid transporters: (1) in the apical membrane facing the midgut lumen and (2) in the basal membrane facing the hemolymph (Fig. 21.5).

Only two insect amino acid transporters, both from the midgut of larval *Manduca sexta,* have been charac

terized. Both require K^+ and operate at a high pH, which is characteristic of the midgut of Lepidoptera. These transporters seem to be unique; no vertebrate analogues have been described, and they are not expected to be widely distributed among insects. As described in more detail in Chapter 36, the genome of *Drosophila melanogaster* contains homologues of vertebrate amino acid transporters, so it is reasonable to suppose such transporters will also be found in mosquitoes and other vectors. Characterizing such transporters is an important area of future research.

CARBOHYDRATE ABSORPTION

Insects do not have an active transport system for glucose uptake in the midgut. Rather, they use a unique mechanism to create a concentration gradient in glucose between the midgut lumen and the hemolymph that ensures efficient passive transport (Fig. 21.6).

LIPID ABSORPTION AND TRANSPORT

Lipid absorption represents an interesting problem, because lipids are insoluble in water and must pass from the aqueous lumen, through the hydrophobic cell membrane of the midgut, and into the aqueous cytoplasm. Nothing is known about the mechanism of lipid absorption in insects, and, in fact, this remains a controversial subject in vertebrate systems. All insects require cholesterol or other sterol in the diet that can be converted to cholesterol, because insects cannot make cholesterol. As noted earlier, several species accumulate fat droplets in midgut cells during blood

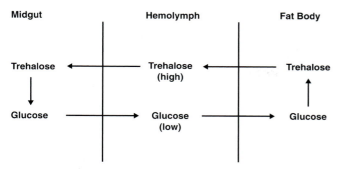

Midgut	Hemolymph	Fat Body

FIGURE 21.6 The uptake of glucose from the lumen of the midgut is driven by a concentration gradient created by maintaining a low concentration of glucose in the hemolymph. This is accomplished by converting glucose to trehalose, a disaccharide in which two glucose molecules are joined through their reducing carbons. This means that trehalose is the major sugar in the hemolymph and glucose is present at low concentration. When trehalose leaks into the midgut, it is converted to glucose by the enzyme trehalase in the midgut lumen, and the released glucose moves back into the hemolymph.

meal digestion, but the origin of this lipid is unclear. It might represent lipid absorbed from the digestion of blood meal lipids, or it might represent lipid made from amino acids in the midgut. This represents an interesting area for future research.

Insects transport lipids in the hemolymph as part of a lipoprotein complex called *lipophorin*. In most insects, fatty acids derived from dietary lipids are transported in lipophorin as diacylglycerol. This is true of Hemiptera. The blood-feeding insects from the higher Diptera probably transport diacylglycerol. On the other hand, mosquitoes and blackflies use triacylglycerol to transport fatty acids—this is the same mechanism used by vertebrates. Thus, although the structure of lipophorin is the same in these lower Diptera as it is in other insects, they have evolved the capacity to carry triacylglycerol. How this occurs is unknown, and it represents an important area of research in which at least some blood-feeding insects differ from other insects.

Readings

Chapman, R. F. 1998. *The Insects: Structure and Function.* Cambridge University Press, Cambridge, UK, Chaps. 3 and 4.

Turunen, S. 1985. Absorption. In *Comprehensive Insect Physiology, Biochemistry and Pharmacology,* Vol. 4 (G. A. Kerkut and L. I. Gilbert, eds.). Pergamon Press, New York, pp. 241–278.

Applebaum, S. W. 1985. Biochemistry of digestion. In G. A. Kerkut and L. I. Gilbert (eds.) *Comprehensive insect physiology, biochemistry and pharmacology:* Volume 4, Pergamon Press, New York, pp. 279–312.

Jobling, B. 1987. *Anatomical Drawings of Biting Flies,* British Museum (Natural History), London.

Clements, A. N. 1992. *The Biology of Mosquitoes,* Vol. 1: *Development, Nutrition and Reproduction.* Chapman and Hall, London.

Lehane, M. J. 1991. *Biology of Blood-Sucking Insects.* Harper Collins Academic, London.

Lehane, M. J., and Billingsley, P. F. (eds.) 1996. *Biology of the Insect Midgut.* Chapman and Hall, London.

Dow, J. A. T. 1986. Insect midgut function. *Adv. Insect Physiol.* 19: 188–328.

Terra, W. R. 1990. Evolution of digestive systems of insects. *Annu. Rev. Entomol.* 35: 181–200.

Billingsley, P. F. 1990. The midgut ultrastructure of hematophagous insects. *Annu. Rev. Entomol.* 35: 219–248.

Canavoso, L. E., Jouni, Z. E., Karnas, K. J, Pennington, J. E., and Wells, M. A. 2001. Fat metabolism in insects. *Annu. Rev. Nutr.* 21: 23–46.

22

The Peritrophic Matrix of Hematophagous Insects

MARTIN DEVENPORT AND MARCELO JACOBS-LORENA

INTRODUCTION

The peritrophic matrix (PM) is a semipermeable extracellular layer that lines the digestive tract of most insects, separating ingested food from the absorptive/secretory intestinal epithelium. The midgut is the only part of the alimentary tract not protected by a chitinous cuticle, and it is where food digestion and absorption take place. By lining the midgut, the PM is thought to protect the insect from pathogens, abrasion, toxic compounds, and in certain cases also to facilitate digestion. The PM is found in most insects, as well as many other arthropod classes and some other phyla, but has been studied most extensively in insects of medical importance as well as pests of crops and livestock. This chapter focuses on insects of medical importance, although other insects are discussed when relevant. Consequently, statements made here might not apply to PMs of every insect.

The PM is most commonly referred to as either the *peritrophic membrane* or the *peritrophic matrix* and, to a lesser extent, *peritrophic envelope*. It was first described in a caterpillar in 1762 by Lyonet. *Peritrophic* comes from the Greek words *peri*, for "around", and *trophic*, for "food". Since the PM invariably surrounds the food bolus, this is an appropriate name. The term *peritrophic membrane* was first proposed in 1890 by Balbiani, who referred to this structure as a "membranous sac that directly encloses the food bolus inside the gut." Nowadays the word *membrane* signifies a lipid bilayer, which the PM is not, and so we prefer to use the term *peritrophic matrix*. The word *matrix* correctly describes a non-membranous structure, although views on this issue vary.

TYPE 1 AND TYPE 2 PERITROPHIC MATRIX

There are two types of PM, type 1 (PM1) and type 2 (PM2), which are defined by their site of synthesis (Fig. 22.1). Each insect species and each developmental stage (larva versus adult) produces either PM1 or PM2 but not both. Most hematophagous insects, such as mosquitoes, blackflies, and sand flies, secrete a PM2 during larval life and a PM1 during adult life. The two types of PM have a number of distinct properties, which include morphology, composition, assembly, and, in some respects, function, which are discussed later. Both types of PM comprise chitin, proteins, and proteoglycans. Virtually nothing is known about the relationship at the molecular level between the two types of PM of the same organism. Both types of PM may appear as homogeneous or laminated structures when ultrathin sections are viewed via electron microscopy.

PM1

Formation of the PM1 in adult hematophagous insects starts within minutes of blood meal ingestion (Table 22.1). The PM1 is synthesized by the majority of the posterior midgut epithelial cells and forms a sac-like structure containing the ingested meal

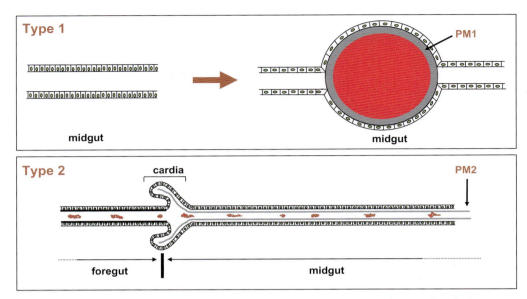

FIGURE 22.1 Schematic diagrams showing the sites of PM1 and PM2 synthesis within the insect gut. The PM1 surrounds a blood meal, and the PM2 surrounds food particles.

TABLE 22.1   Time of PM1 Formation in Some Medically Important Blood-Feeding Insects

Insect Species	First Detected	Mature	Reference
Aedes aegypti	5 h	12–17 h	Stohler (1957)
		5–8 h	Freyvogel and Stäubli (1965)
	4–8 h	12 h	Perrone and Spielman (1988)
Anopheles gambiae		13 h	Freyvogel and Stäubli (1965)
Anopheles stephensi		32 h	Freyvogel and Stäubli (1965)
	12 h	48 h	Berner *et al.* (1983)
Simulium ornatum	2–10 min	12–24 h	Reid and Lehane (1984)
Simulium equinum	2–10 min	12–24 h	Reid and Lehane (1984)
Simulium lineatum	2–10 min	12–24 h	Reid and Lehane (1984)
Simulium vitattum	20 min	6 h	Ramos *et al.* (1994)
Phlebotomus longipes	<24 h	48 h	Gemetchu (1974)
Phlebotomus perniciosus	30 min	36 h	Walters *et al.* (1993)
Lutzomyia spimicrassa	1–3 h	12–36 h	Walters *et al.* (1995)

The times of peritrophic matrix type 1 formation in mosquitoes (*Aedes, Anopheles*), black flies (*Simulium*) and sand flies (*Phelebotomus, Lutzomyia*) are listed. It should be emphasized that "first detection" and "maturation" are based on highly subjective criteria and that in most cases an exhaustive time course of PM formation has not been attempted. Moreover, in any given experiment the parameters measured vary largely from individual to individual. Thus, the listed values serve only as guidelines.

Refernces: Stohler, H. 1957. *Acta. Trop.* 14: 302–352; Freyvogel, T. A., and Staubli, W. 1965. *Trop.* 22, 118–147; Perrone, J. B., and Spielman, A. 1988. *Cell Tissue Res.* 252: 473–478; Berner, R., Rudin, W., and Hecker, H. 1983. *J. Ultrastr. Res.* 83: 195–204; Reid, G. D., and Lehane, M. J. 1984. *Ann. Trop. Med. Parasit.* 7: 527–539; Ramos, A., Mahowald, A., and Jacobs-Lorena, M. 1994. *J. Exp. Zool.* 268: 269–281; Gemetchu, T. 1974. *Ann. Trop. Med. Parasit.* 68: 111–124; Walters, L. L., Irons, K. P., Guzman, H., and Tesh, R. B. 1993. *J. Med. Entomol.* 30: 179–198; Walters, L. L., Irons, K. P., Guzman, H., and Tesh, R. B. 1995. *J. Med. Entomol.* 32: 711–725.

(Fig. 22.1). The thickness of PM1 is typically in the range of 1–20 μm. PM1 is the most common type of PM in adult blood-sucking insects, such as mosquitoes, black-flies and sand flies. Notable exceptions are the tsetse fly, *Glossina* sp., and the stable fly *Stomoxys calcitrans*, which secrete PM2. In hematophagous insects that secrete a PM1, no PM is present in the gut prior to the blood meal. Accordingly, no PM can be detected in insects kept in the laboratory on a sucrose diet. In these insects PM1 is produced in direct response to an ingested blood meal. In certain other, nonhematophagous insects, PM1 may be produced constitutively.

BEFORE A BLOOD MEAL

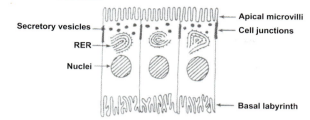

AFTER A BLOOD MEAL

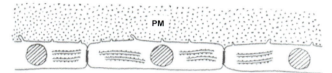

FIGURE 22.2 Schematic diagram of ultrastructural changes in PM1-secreting cells after blood feeding. Apical microvilli and basal labyrinth largely disappear to accommodate the flattening of the epithelial cells that accompanies the dramatic distention of the midgut. As the PM1 forms, secretory vesicles disappear and the RER whorls unfold. Not all of these changes occur in all insects.

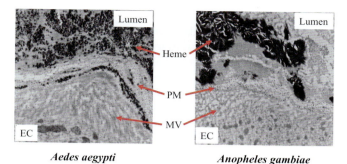

Aedes aegypti   *Anopheles gambiae*

FIGURE 22.3 Electron micrographs showing the PM1 from *Aedes aegypti* and *Anopheles gambiae*. The gut lumen is at the top and the epithelial cell (EC) is at the bottom. The PM appears as a layered electron-lucent structure, and in both cases it is labeled with immunogold particles that react specifically with the PM proteins AEIMUC1 (*Ae. aegypti*) and Ag-Aper1 (*An. gambiae*) (see text). Some labeling over the microvilli also occurs, which represents newly secreted protein. Black electron-dense heme aggregates are indicated. In *Ae. aegypti* the aggregates are small and are also seen within the PM, whereas in *An. gambiae* the heme aggregates are very large and are observed only on the lumenal side of the PM. Magnification is ×7000. Pictures kindly provided by Dr. Hisashi Fujioka; Case Western Reserve University.

Ingestion of a blood meal causes marked morphological changes of PM1-secreting midgut cells (Fig. 22.2). The columnar cells become stretched out and flattened. On the lumenal side, microvilli decrease in size and number to allow distention of the epithelium. On the basal side, unwinding of the extensive labyrinth network allows expansion of this surface. Cells are securely connected to each other at their lateral surfaces by cell junctions, such as septate junctions, gap junctions, and hemidesmosomes. Several lines of evidence strongly suggest that physical distention of the midgut epithelium by the blood meal is the signal for initiating PM secretion. For instance, midgut distention by injection of air through the anus can trigger PM1 formation. As digestion proceeds, the cells gradually return to their original shape and the microvilli and the basal labyrinth reorganize.

Mosquito midgut epithelial cells have prominent Golgi apparatus and abundant rough endoplasmic reticulum (RER), in accordance with their role as protein secretory cells. However, cell morphology differs between anopheline and culicine mosquitoes, suggesting different modes of PM1 formation. In *Aedes aegypti*, RER cisternae are often assembled into characteristic whorls resembling fingerprints. After a blood meal, the whorls unfold, correlating with activation of protein synthesis. Thus mRNAs encoding PM proteins (rather than the proteins themselves) are thought to be stored in sugar fed guts and their translation induced by blood feeding. Consistent with this view, mRNA encoding the *Ae. aegypti* PM protein AEIMUC (see below) is present before and after blood feeding, while

the corresponding protein can only be detected after blood feeding (Fig. 22.3). A gene encoding a second *Ae. Aegypti* PM protein, Aa-Aper50, is expressed and translated only after blood feeding.

By contrast, midgut epithelial cells of anophelines contain a large number of apical, secretory granules before the blood meal. Upon blood feeding, the apical granules disappear, their contents presumably having been released into the lumen. Thus, at least some of the PM proteins of anophelines are thought to be stored prior to blood feeding. In agreement with this model, two cloned *Anopheles gambiae* genes, *Ag-Aper1* and *Ag-Aper14* (see later), are both expressed and translated prior to blood feeding and colocalize to secretory vesicles lying beneath the epithelial cell apical region. After blood feeding, both *Ag-Aper1* and *Ag-Aper14* localize to the PM and are depleted from the epithelial cells (Fig. 22.3). Approximately 60 hours after a blood meal, when most of the blood has been digested, *Ag-Aper1* is again found in epithelial cell vesicles. However, Berner found that no PM is formed in the presence of an mRNA-polymerase inhibitor, suggesting that not all PM components are stored before the blood meal.

In sand flies the epithelial cells exhibit similar ultrastructural changes upon blood feeding. Conspicuous whorls or large linear formations of RER are found in starved sand flies. After blood feeding, dome-shaped apical surfaces of the epithelial cells disappear, presumably by stretching, and RER whorls unwind to become parallel to the flat apical surface. The cytoplasm is rich in ribosomes. Sand fly PMs exhibit clear species-specific differences in secretion, morphology, and rate of formation/degradation.

In blackflies (Simulids) secretion of PM components occurs rapidly, and they can be detected between the individual microvilli as early as two minutes after blood feeding. The epithelial cells are rich in RER and contain a small number of electron-dense apical vesicles. A fully differentiated PM is detectable as early as six hours after blood feeding (Table 22.1). *Simulium vittatum* has two major PM proteins (see PM components), neither of which is present in sugar fed flies. These proteins are rapidly induced by blood feeding, and their pattern of accumulation correlates with the thickening of the PM.

In summary, the secretory activity after blood ingestion is brought about mainly by secretion of stored PM proteins in *Anopheles* and mainly by stimulating PM protein synthesis in *Aedes*. Consequently, it might be expected that the assembly of the PM is quicker in anophelines, although this does not appear to be the case (Table 22.1). In other hematophagous insects, the relative contribution of each mechanism—release of stored proteins and activation of protein synthesis—appears to vary.

PM2

The PM2 is an open-ended sleevelike structure that lines the entire midgut and hindgut and is excreted with the feces through the anus. It is produced from the cardia, a specialized organ located at the junction of the cuticle-lined foregut and midgut (Fig. 22.1) and is usually 0.1–2 μm thick. In contrast with the PM1, the PM2 is constitutively secreted independent of the feeding status of the insect. Adults of many higher Diptera, the majority of which are non-blood-feeding, secrete a PM2.

The cardia is continuous with the gut epithelium and is formed by the double folding of this epithelium upon itself, creating a protected pocket from which the PM2 is secreted (Fig. 22.1). The structure of the cardia is variable and is characteristic of each insect species. In some insects different cell types can be identified microscopically within the cardia. Also, different layers seem to originate from different regions of the cardia, suggesting that each cell type is specialized in the secretion of defined PM2 components. This differentiation of cells is especially pronounced in the species more advanced in evolution and has been observed in the tsetse fly. In contrast, the appearance of the cardia cells of *Ae. aegypti* larvae does not differ, although the PM is layered. The morphology of the PM2-secreting cells in the adult fruit fly, *Drosophila*

melanogaster, was studied in detail by King. Here, the cardia can be divided into six zones according to their morphology (Fig. 22.4). The *D. melanogaster* PM2 has four distinct layers, and each layer is secreted by distinct formation zones.

Peritrophin-15, a PM2 protein of *Lucilia cuprina* (see later), is synthesized by cells throughout the cardia epithelium, but there appears to be a gradient of peritrophin-15 concentration within the cells from the anterior region to the posterior region of the cardia. The secretory cells have abundant RER, Golgi and vesicles, consistent with a secretory function associated with constitutive PM production. These cells package peritrophin-15 into membrane-bound vesicles, which move to the base of microvilli and apparently fuse with the cell membrane, where the contents

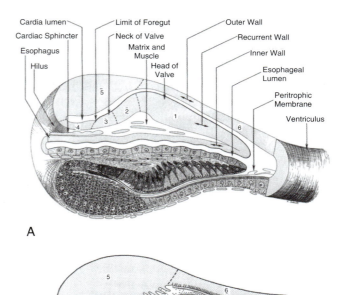

FIGURE 22.4 Schematic diagram of the cardia and its PM2 formation zone in *Drosophila melanogaster*. (A) Basic morphology of the cardia as seen in sagittal section. Anterior is to the left. The epithelium has been numbered 1–6 according to the morphology of the cells in that zone. (B) Stylized diagram of peritrophic matrix formation, showing the association of the four layers of the peritrophic matrix with specific regions of cardia epithelium (not drawn to scale). L1, the innermost layer of the peritrophic matrix, forms adjacent to the foregut cuticle (cu); L2 forms over zone 4; L3, over zone 5; and L4, over zone 6. Reproduced from King, Cellular organization and peritrophic membrane formation in the cardia (proventriculus) of *Drosophila melanogaster, Journal of Morphology*, by permission of Wiley-Liss, a division of John Wiley and Sons, Inc., copyright 1988.

of the vesicles are released into the intermicrovillar spaces by exocytosis. The protein then associates with the PM lying across the tips of the epithelial cell microvilli. At the position in the cardia where peritrophin-15 is secreted there is already a defined but immature PM.

DISSECTION OF THE PERITROPHIC MATRIX

When first secreted, the PM1 is soft and fragile, and cannot be physically isolated by dissection. Over time, the PM1 matures and its thickness gradually increases. This process can be followed by examining sections of fixed guts by light or electron microscopy. The final thickness of the PM1 varies from one individual to another and even spatially around the same PM1. Physical isolation of the PM by dissection is possible only when the PM is "mature." This occurs within a narrow range of time after the blood meal, which is determined empirically and is different for each insect species (Table 22.1). Toward the end of blood digestion the PM1 starts to degrade before it is eventually excreted with the remnants of the blood meal.

To induce PM1 formation, insects are usually fed with blood from an anesthetized animal. Alternatively, blood (or an artificial meal) can be fed by use of a water-jacketed membrane feeder covered with "Parafilm," sausage skin or an animal skin. Membrane feeding is useful to test the effect of a specific substance mixed into the blood meal, such as an antibody, a chitinase or an inhibitor (see later). To avoid contamination of PMs with components foreign to the mosquito (e.g., blood proteins), PM formation can be induced by feeding a protein-free meal containing a suspension of latex beads. The latex beads take the place of the red blood cells in providing bulk and maintaining the distention of the midgut. The addition of 0.1% low-melting agarose to the artificial latex meal can also help to maintain volume. The buffer for the protein-free meal should contain 150 mM NaCl and ATP as a phagostimulant (1–10 mM) with a pH of 7.0.

To dissect the PM1, the midgut is removed from cold-immobilized insects, and the epithelium is peeled away using a fine watchmaker's forceps to release the round mass of blood contained by the thick PM1 (Fig. 22.5). PMs are usually dissected in a saline solution, such as PBS. However, a number of PM proteins may be soluble in this solution. To minimize such losses and to increase PM rigidity, the dissection can be done in 50–95% ethanol in PBS. Dissected PMs from anopheline mosquitoes disintegrate quickly in water and should be kept in salt solution with a high concentration of divalent cations such as $CaCl_2$.

Since the PM2 is secreted continuously, no special treatment of the insect is necessary for dissection of the

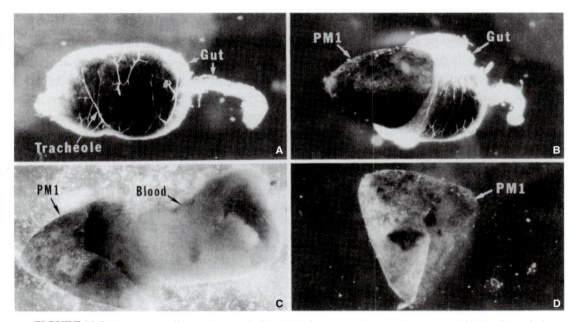

FIGURE 22.5 Dissection of the midgut and of the PM1 from an *Aedes aegypti* mosquito. (A) The distended midgut surrounds the PM1. (B) The partially "peeled" gut epithelium exposes the underlying PM1. The PM1 contains the blood meal. (C) Breaking of the PM1 releases the inner blood mass. (D) The same PM1 fragment as in (C) after removal of the blood.

PM2. However, to avoid contamination with food particles, one may feed mosquito larvae with agar to clear out intestinal contents. After the gut is removed from the abdomen, the protruding PM2 can be pulled out of the gut lumen (Fig. 22.6).

PERITROPHIC MATRIX COMPONENTS

Proteins, including glycoproteins and proteoglycans, are the major PM components. In addition, chitin, while less abundant, is probably an important structural component (Fig. 22.7). Difficulties in studying the PM components include contamination by food remnants and limited amount of material that can be obtained for study. The contamination problem is especially true for PM1, since its formation is usually induced by a blood meal rich in proteins and other components from the vertebrate host.

Chitin

Chitin, a linear polymer of N-acetylglucosamine (GlcNAc), is thought to be an important structural component of the PM. Chitin may provide a scaffold onto which proteins and other components attach, thus providing strength and a framework for assembly. Chitin polymers are released into the midgut lumen, where they are thought to be self-organized into fibers and interlocked by protein molecules (Fig. 22.7). Estimates of PM chitin content range from 3% to 40%, although the tests used to measure it may not be accurate.

There is strong evidence that chitin plays a major role in maintaining PM1 structure. For instance, the addition of exogenous chitinases to the PM1 of *Ae. aegypti*, *An. stephensi*, and *An. gambiae* dramatically affects their structural integrity. Also, Polyoxidin D, which inhibits insect chitin synthase activity, completely inhibits PM1 production in *Ae. aegypti*. Glutamine synthetase and chitin synthetase, two genes

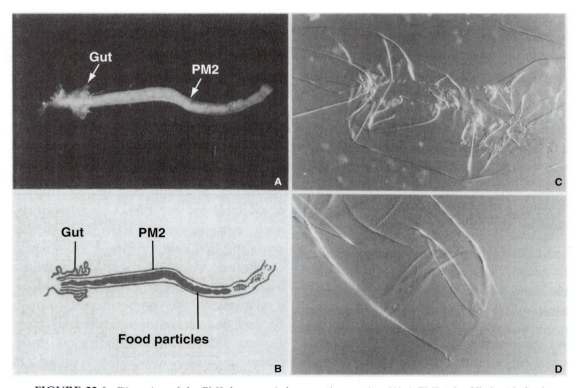

FIGURE 22.6 Dissection of the PM2 from an *Aedes aegypti* mosquito. (A) A PM2 tube filled with food particles, as seen under the dissecting microscope. The retracted gut wall covers the left end of the tube. (B) Schematic representation of the image in panel (A). (C) and (D) are micrographs of a PM2 after removal of the food contents, observed by differential interference contrast (Nomarski) optics. (C) A wrinkled portion of the PM2. (D) An open end of the PM2.

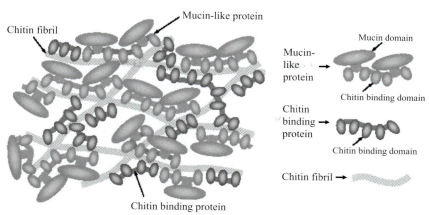

FIGURE 22.7 Schematic model of the PM molecular structure. Chitin fibers form a scaffold onto which various proteins, glycoproteins, and proteoglycans attach. The structure of the PM is maintained primarily by PM proteins through their multiple chitin-binding domains. High-affinity binding between the proteins and chitin fibrils minimizes the exposure of the proteins to digestive proteases and protects the chitin from degradation by chitinases. The mucin domains found in some of the PM proteins are highly resistant to protease degradation and are likely to shield other PM components from digestive enzymes in the midgut. Reproduced from Wang and Granados. 2001. *Arch. Insect Biochem. Physiol.* 47: 110–118. Reprinted by permission of Wiley-Liss Inc., a subsidiary of John Wiley & Sons, Inc.

involved in chitin synthesis in the gut of *Ae. aegypti*, have recently been cloned. Transcription of both genes is induced in response to blood feeding and no PM1 could be detected when the activity of glutamine synthetase was inhibited. Chitin synthetase mRNA is localized to the lumenal side of the midgut epithelial cell.

The role of chitin in PM2 structure is less clear. Edwards and Jacobs-Lorena (2000) showed that incubation of dissected larval PM2s of *Ae. aegypti* or *An. gambiae* with chitinase did not change PM structure when observed by means of light microscopy, although it did have a limited but significant effect on its permeability (see later). Also, Polyoxin D, which inhibits adult PM formation, does not affect the appearance or permeability of the larval PM significantly. Similarly, Calcofluor white (which binds to chitin) and Polyoxin D have no effect on *L. cuprina* larval PM2, although Calcofluor does have a pronounced effect on PM structure of the PM1 of *Trichoplusia ni* larvae. This may reflect differences in accessibility of the different reagents to the site of PM synthesis or the amount of chitin present in the PM. Since PM2s are synthesized in the protected pocket of the cardia, accessibility of the reagents to their molecular targets (proteins and proteoglycans) may sterically hinder access of chitinase or Calcofluor to chitin. In 2000, Tellam and Eisemann thoroughly re-examined the chitin content of *L. cuprina* larval PM2. They concluded that, although no single test was completely conclusive, chitin is at most 5.3% of the PM2 of this

species. Despite having no discernible effect on structure, Calcofluor and Polyoxin D cause larval mortality in *L. cuprina* but not in mosquitoes. Thus, although present at low levels, chitin may still be critical for function. In addition to the chitin backbone, protein–protein interactions are likely to play a major role in PM structure.

PM Proteins

Tellam placed PM2 proteins into four classes, according to their solubility in a series of buffers of increasing ionic strength and denaturing ability. The first group of proteins is easily removed using physiological buffers or high-ionic-strength buffers and may represent contaminating proteins from the gut lumen and digestive enzymes. Proteins of the second class are removed by relatively gentle detergents. The third group is extracted with strong denaturing agents (e.g., urea, SDS) under nonreducing conditions and are termed integral/intrinsic membrane proteins or *peritrophins*. The fourth group of proteins are nonextractable by treatment with detergents or denaturing agents, and they are poorly understood, although they can make up a substantial portion of the total PM proteins.

Two-dimensional polyacrylamide gel electrophoresis (2D-PAGE) indicated that PM1s of *Ae. aegypti* and *An. gambiae* have about 20–40 major proteins, about 15 of which appear to have the same mobility. A similar analysis suggested that the PM1 of blackflies has a relatively simple composition, with only 2 major proteins. The PM2 from adult *Glossina morsitans morsitans* (tsetse fly) appears to have approximately 40 proteins.

Proteins with Chitin-Binding Domains

A common feature of PM proteins is the occurrence of cysteine-containing domains similar to the chitin-binding domains of chitinases from several animals and microorganisms, including that of the *An. gambiae* gut-specific chitinase (Fig. 22.8). Typically, these domains are 60–70 amino acids in length and contain six cysteines that form three internal disulfide bonds, plus several conserved aromatic/hydrophobic amino acids. There is evidence that disulfide bonds in PM proteins contribute to protein stability and resistance to proteolysis in the gut. In several instances there is direct experimental evidence that recombinant PM proteins with chitin-binding domains do bind chitin.

A PM1 protein from the mosquito *An. gambiae*, Ag-Aper1, has been identified. This protein comprises two tandem chitin-binding domains separated by a short linker (Fig. 22.8). Multiple chitin-binding domains have been observed in a number of PM proteins from other insects, and such proteins may have a role in establishing PM structure by cross-linking chitin fibrils and creating a three-dimensional molecular meshwork (Fig. 22.7). Another possibility is that the chitin-binding domains bind to the GlcNAc-containing oligosaccharide moities of other PM proteins, thus providing an alternative means for establishing a three-dimensional network. The protein/chitin network is likely to be important to maintain the PM's inherent physical properties, such as strength, elasticity, and porosity.

A second PM1 protein, Ag-Aper14, has recently been identified from the *An. gambiae* adult PM1. Ag-Aper14 has a structure similar to that of peritrophin-15 of *L. cuprina* and its homologues in *Chrysomyia bezziana* and *D. melanogaster*, which represent a class of small PM2 proteins that contain a single chitin-binding domain (Fig. 22.8). A proposed function for these proteins is capping the ends of individual chitin polymer chains, possibly protecting the chitin fibrils from degradation by exochitinases or regulating the length of the chitin polymer.

The amino acid sequences of PM proteins are generally poorly conserved, even between homologues from relatively closely related insects. However, the six-cysteine architecture of the chitin-binding domains and their associated aromatic/hydrophobic amino acid residues are better conserved. It is proposed that the three disulphide bonds of the chitin-binding domain may constrain the polypeptide to present the aromatic amino acids on the protein surface for interactions with the sugar residues within the chitin fibril. This structure would allow considerable freedom at other amino acid positions and hence explain the low sequence identity overall.

Mucin-Like PM Proteins

A number of the cloned PM protein genes have both chitin-binding and mucin-like domains. As in the mammalian intestinal mucins, mucin-like domains are characterized by numerous serine and threonine residues, which are thought to be sites of extensive O-linked glycosylation. Mammalian intestinal mucins have been well studied, and they have a number of primary functions, including protection from abrasion, hydrolytic enzymes, heavy metals, and pathogens, while allowing the passage of digestion products for absorption by the intestinal epithelium. It is likely that the mucin-like domains of insect PM proteins share a number of these functions.

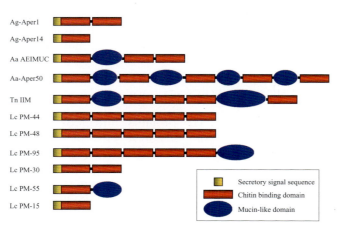

Ag-Aper1
Ag-Aper14
Aa AEIMUC
Aa-Aper50
Tn IIM
Lc PM-44
Lc PM-48
Lc PM-95
Lc PM-30
Lc PM-55
Lc PM-15

Secretory signal sequence
Chitin binding domain
Mucin-like domain

FIGURE 22.8 Structure of peritrophic matrix proteins from different insects. The names of the proteins are preceded by the insect species, abbreviated as follows: Aa, *Aedes aegypti*; Ag, *Anopheles gambiae*; Tn, *Trichoplusia ni*; Lc, *Lucilia cuprina*. Note that the sequence of the chitin-binding domains is poorly conserved among different proteins.

References: Ag-Aper1 (Shen, Z. and Jacobs-Lorena, M. 1998. *J. Biol. Chem.* 273: 17665–17670); Aa AEIMUC (Rayms-Keller, A., *et al.* 2000. *Insect Mol. Biol.* 9: 419–426); Tn IIM (Wang, P., and Granados, R. R. 1997. *J. Biol. Chem.* 272: 16663–16669); Lc PM-44 (Elvin *et al.*, 1996. *J. Biol. Chem.* 271: 8925–8935); Lc PM-48 (Schorderet, S., *et al.* 1998. *Insect Biochem. Mol. Biol.* 28: 99–111); Lc PM-95 (Casu, R., *et al.* 1997. *Proc. Natl. Acad. Sci. USA* 94: 8939–8944); Lc PM-15 (Wijffels, G., *et al.* 2001. *Biol. Chem.* 276: 15527–15536); Lc PM-30 and Lc PM-55 (unpublished; see Tellam, R. L., *et al.* 1999. *Mol. Biol.* 29: 87–101).

The first chitin and mucin domain–containing PM protein to be identified was insect intestinal mucin (IIM) from the larval PM1 of the lepidopteran *T. ni*. IIM contains several chitin-binding domains as well as two mucin-like domains (Fig. 22.8). Such mucin domains are resistant to protease degradation and are likely to shield other PM components from digestive enzymes. Recently a metal-inducible PM2 protein, AEIMUC1, was identified from *Ae. aegypti* larvae. This protein contains three chitin-binding domains and one mucin-like domain (Fig. 22.8). The same protein is also present in adult PM1 (Fig. 22.3). This is the first time that a PM protein has been identified in both the larval PM2 and adult PM1. Another *Ae. aegypti* adult PM1 protein, Aa-Aper50, containing multiple putative chitin-binding and mucin-like domains has also been identified.

Other PM Proteins

A cDNA termed *Ag-Lper1*, which encodes a secreted glutamine-rich PM2 protein from *An. gambiae* larvae, was recently identified (unpublished data). An antibody against the recombinant *Ag-Lper1* protein recognizes the PM2 and cytoplasmic vesicles in a subset of the cardia cells. The amide group of the glutamine side chain has extensive hydrogen-bonding capacity and may function in the interaction with other PM components. It might also be cross-linked by transglutaminases. Glutamine-rich proteins may constitute a new family of PM structural proteins, possibly belonging to the group of insoluble PM proteins (see earlier).

ORIGINS OF THE PM

It is speculated that the PM evolved from a mucosal lining of the insect intestine, in which case the midgut epithelial cells of ancestral insects must have been lined with a mucous layer similar to that found in vertebrates. Vertebrate gastrointestinal mucus is a gel-like substance composed of mucin aggregates. At some point in time the peritrophins evolved from the mucins by acquiring their chitin-binding domains. At the same time, or possibly earlier, secretion of chitin from the midgut epithelial cells must also have evolved in order to allow the characteristic chitin-protein network to develop. Early PMs would have had more of the properties of mucus, after which some of the peritrophins presumably lost their mucin domains, allowing the chitin-protein network of more recent PMs to develop. If this theory is correct, then secretion of the PM by the whole midgut epithelium would be the ancestral condition, and secretion by specialized regions (e.g. cardia) would have evolved more recently.

PERMEABILITY

Because the PM completely separates the food bolus from the secretory/absorptive epithelium, it must be sufficiently permeable to allow digestive enzymes to cross and reach the food bolus and the hydrolytic products of digestion to diffuse in the opposite direction to be absorbed by the epithelial cells. Here again, it is useful to discuss the PM1 and PM2 separately.

For technical reasons, much more is known about the permeability of PM2 than of PM1. For instance, the most direct way to determine the permeability of the PM is to add a marker of a particular molecular size into the food and determine its ability to traverse the PM. This kind of experiment is relatively straightforward with the constitutively expressed PM2, such as in mosquito larvae. However, a similar experimental approach would not be possible for the adult mosquito, since at the time of food ingestion the PM1 does not exist. Moreover, there are technical constraints that need to be considered. For instance, many marker molecules, including dyes, colloidal gold, and proteins, tend to aggregate under certain conditions, meaning that PM porosity may not correspond to the nominal particle size of the test substance. The shape of the molecule used on this kind of experiment and other physical properties (charge, hydrophobicity, etc.) might substantially affect permeability. Degradation of the test substance (e.g., a protein) in the harsh environment of the intestine is another factor to be considered. Dextrans appear to be a choice marker, but even then the possibility that the marker is adsorbed to other proteins (e.g., to blood albumin) should be taken into account. Another important factor is that pore sizes found within the PM at any one time may be variable, and also the permeability of the PM may be dynamic. Thus, one should be careful when extrapolating the results with test substances to "real-life" situations.

Several approaches have been used to study PM permeability in nonhematophagous insects. The permeability of the PM2 to different substances varies somewhat from organism to organism and from larva to adult. The PM2 has always been found to be impermeable to ferritin (450 kDa) but usually permeable to molecules around 30–40 kDa (e.g., horseradish peroxidase, 40 kDa). In adult tsetse flies, myoglobin (17 kDa) and horseradish peroxidase (40 kD) penetrate the PM2, while hemoglobin (68 kDa) may or may not penetrate.

Edwards and Jacobs-Lorena (2000) used a noninvasive *in vivo* assay for estimating larval PM2 permeability for the mosquitoes *Ae. aegypti* and *An. gambiae*. Permeability was determined by feeding the larvae with FITC-labeled dextrans (size range 4.4–2,000 kDa) and measuring their appearance in the gastric caeca (tracer dextrans must traverse the PM to reach the caeca). In both cases larval PM was found to be permeable to dextrans of up to 148 kDa. Additionally, the same permeability assay was used to measure disruption of the PM. Disruption was defined as the ability of labeled dextrans of 2,000 kDa, a size approaching that of virus particles, to traverse the PM. Dithiothreitol (a reducing agent) and, to lesser extent, chitinase were effective in disrupting the PM. However, Polyoxin D (an inhibitor of chitin synthesis), pronase (a nonspecific protease), and Calcofluor (a chitin-binding compound) did not alter the permeability significantly. Dithiothreitol is presumed to disrupt protein folding and protein–protein interactions by breaking disulfide bonds. Interestingly, none of these treatments affected the appearance of the PM.

An alternate method for measuring the porosity of the PM is to study the spatial distribution of digestive enzymes of known molecular weight within the endoperitrophic and ectoperitrophic spaces. Smaller enzymes may pass freely through the PM into the endoperitrophic space (the space enclosed by the PM), whereas larger enzymes may not penetrate the PM and are retained in the ectoperitrophic space (the space between the PM and the midgut epithelium) (see digestion, next section). Based on these kinds of data it is possible to obtain estimates of pore sizes within the PM.

In adult mosquitoes, blood meal digestion proceeds faster in the absence of a PM (see next section). The observation that the mosquito midgut secretes a chitinase in response to a blood meal can be interpreted in light of these results. It is possible that the mosquito chitinase plays a role in controlling PM porosity and thickness. Final PM thickness during digestion would result from a balance between PM synthesis by the midgut epithelium and degradation by the secreted chitinase.

DIGESTION

Extensive studies of the compartmentalization of digestive enzymes and the fluid fluxes within the larval midgut of *Rynchosciara americana* (Diptera), which possesses a PM2, have led Terra (1990) to propose a model for the role of the PM in digestion and enzyme recycling. According to this model, initial digestion of large molecules occurs inside the PM by the action of small enzymes, such as trypsin and amylase, which can freely pass across the PM. Intermediate digestion then occurs in the ectoperitrophic space by enzymes such as aminopeptidase that are too large to cross the PM. Final digestion occurs on the surface of the midgut cells by membrane-bound enzymes, the terminal products of which are absorbed by the epithelial cells. As food progresses posteriorly within the endoperitrophic space, the smaller products resulting from digestion and the free enzymes traverse the PM. The enzymes and partly digested molecules are then displaced anteriorly, propelled by fluid fluxes in the ectoperitrophic space (Fig. 22.9). The unbound enzymes are now free to diffuse back across the PM into the endoperitrophic space, thus preventing the loss of enzymes by excretion. Based on this model it follows that treatments that disrupt the formation of

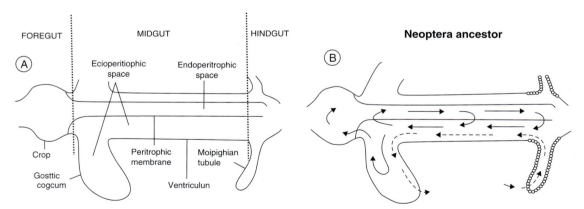

FIGURE 22.9 Generalized diagram of the insect gut (A) and representation (B) of fluid fluxes (dashed arrows). Reproduced with permission from Terra (1990). *Annual Review of Entomology*, vol. 35, by Annual Reviews Inc.

the PM should also have adverse effects on digestion and presumably on insect survival. There are extensive experimental data to support this function of the PM in nonhematophagous insects, and they may also be applicable to mosquito larvae.

Limited information on the spatial distribution of some of the digestive enzymes is available for adult mosquitoes. The best-studied digestive enzymes in mosquitoes are the trypsins. Trypsin 1, the major trypsin, is approximately 30 kDa in size, and it has been demonstrated by immunofluorescence that it can pass through the adult PM1 of *An. gambiae*. In addition to trypsin, aminopeptidase is found on both sides of the PM in *Ae. Aegypti*.

The rate of digestion in adult mosquitoes is faster in the absence of the PM. This was assayed by measuring total protein content in the mosquito guts at different times after feeding with blood plus chitinase (compared with blood alone) or with blood containing anti-PM antibody (compared with pre-immune serum). Therefore, treatments that prevent PM formation (chitinase) or alter its structure (antibodies) increase the rate of digestion. It seems counterintuitive that the mosquito produces a structure (the PM) that delays digestion and thus might reduce fitness. Therefore, the adult mosquito PM may have an evolutionary function other than digestion.

BINDING OF TOXIC COMPOUNDS

Hematophagous insects use hemoglobin as their major protein source. Its digestion generates a large amount of heme, which is highly toxic to cells because it generates free radicals that can disrupt membrane lipids or intracellular molecules. Therefore, inactivation or removal of these toxic compounds from the gut lumen is an important part of digestion. Pascoa *et al.* (2002) have shown that the PM1 of the mosquito *Ae. aegypti* binds heme. The amount of bound heme increases in parallel with the progression of blood digestion and accounts for the majority of heme present within the blood meal. Furthermore, a PM1 formed in plasma-fed mosquitoes can bind hemin (a derivative of heme) *in vitro* and the level of binding is saturable, suggesting that the PM has specific binding sites for hemin. Thus, by binding heme, it appears that the *Ae. aegypti* PM protects the midgut epithelium from its toxic effects. When heme is bound to the PM it has a light brown color. A similar color has also been described for the PM1 of sand flies. Preliminary data suggest that the *Ae. aegypti* peritrophin AEIMUC1 may bind to heme. Expression of AEIMUC1

mRNA in adult guts is induced by a blood meal but also by exposure to heavy metals. Apart from chitin binding, this is the first report of a specific function for a PM protein.

When viewed by means of electron microscopy, the midguts of *Ae. aegypti* contain a large number of small electron-dense granules derived from heme that are seen throughout the PM and within the endoperitrophic space adjacent to the PM (Fig. 22.3). The appearance of heme aggregates in the midgut of anophelines is much different, suggesting a different structure. Here, the aggregates are larger and less granular and lie adjacent to the PM1 in the endoperitrophic space (Fig. 22.3). It is possible that some components of the PM catalyze heme aggregation, which would explain why the aggregates are seen next to the PM. In the sand fly, *Lutzomyia spinicrassa*, the majority of the heme is also retained within the endoperitrophic space.

THE PM AS A BARRIER TO PATHOGEN INVASION

Pathogens transmitted by hematophagous insects invariably enter the insect with the blood meal. The pathogen must then make contact with the midgut epithelium, as this is the only region of the insect digestive tract without a cuticlular covering. In the majority of the cases, the pathogen traverses the epithelium and continues to develop in the insect body cavity. Since the PM completely surrounds the blood meal and separates the ingested parasite from the epithelium, the PM constitutes a potential barrier for pathogen invasion.

Circumstantial evidence suggests that the PM plays a role in protecting the midgut from pathogens. Obviously, this is not a barrier in insects that are effective vectors. When considering whether or not the PM is a barrier for pathogen invasion, it is critical to consider the time course of PM formation and to relate it to pathogen development. Three possible scenarios can be envisioned: (1) The pathogen attaches and/or traverses the gut as soon as it is ingested and before the PM has formed; (2) the pathogen first develops in the gut lumen and then traverses a mature PM to reach the epithelium; and (3) the pathogen moves toward the gut epithelium after the PM breaks down.

In the first case (pathogen attaches/traverses gut soon after ingestion) the PM probably does not constitute a significant barrier. Viruses and most microfilariae belong to this category. For instance, *Onchocerca volvulus*, the causative agent of river blindness, is transmitted by the blackfly, whose PM1 matures relatively fast (Table 22.1). Lewis states that microfilariae remain alive in the gut of blackflies for many hours (up to 24) after an infectious blood meal. However, Laurence demonstrated that most of the ingested microfilariae that traverse the gut do so within 30 min of ingestion and thereafter move only slowly until 2–4 h after the blood meal, by which time the PM is quite thick and further movement is blocked. In this experiment approximately 25% of the ingested microfilariae remained trapped within the PM and died. Microfilariae of *Brugia malayi* also penetrate the mosquito gut soon after ingestion with the blood meal and before PM1 formation.

The malaria parasite (*Plasmodium*) belongs to the second category. After ingestion, the gametocytes go through a sexual cycle in the midgut lumen and develop into motile ookinetes that invade the midgut epithelium, a process that takes 16–24 h, depending on the *Plasmodium* species. Major invasion of the epithelial cells occurs about 24 h after the blood meal, at a time when the PM1 is fully mature (Table 22.1). Thus, in this case the PM1 does constitute a barrier, which the parasite must traverse (Fig. 22.10). Indeed, Billingsley and Rudin found that the ability of *Ae. aegypti* to transmit *Plasmodium gallinaceum* was reduced when the thickness of the PM increased, indicating that the PM does act as a partial barrier to *Plasmodium* development. However, infectivity in *An. stephensi* was unaffected by the absence of the PM, suggesting that here the PM does not represent a significant barrier to invasion.

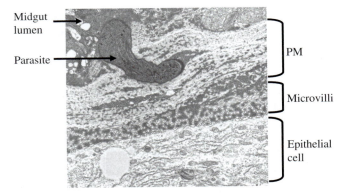

FIGURE 22.10 Electron micrograph of a *Plasmodium gallinaceum* ookinete traversing the PM1 of *Aedes aegypti*. A fibrous PM lines the microvilli of the midgut epithelium. To cross the PM, the anterior end of the parasite seems to insert itself between the PM layers. Reproduced with permission from Torii *et al.* (1992) *J. Protozool.*, 39:449–454. Picture kindly provided by Dr. Hisashi Fujioka, Case Western Reserve University.

How does *plasmodium* traverse the PM? Early studies indicated that the malaria parasite *P. gallinaceum* secretes a chitinase to penetrate the PM of *Ae. aegypti*. Allosamidin (a chitinase inhibitor) completely blocks development of the ookinete to the oocyst stage in the mosquito midgut, providing further support for the role of chitinase in penetration of the PM. *P. gallinaceum* chitinase activity increases by treatment with midgut proteases, and an anti-trypsin antibody blocks parasite development, with no deleterious effect on the mosquito. This suggests that the enzyme is secreted as an inactive pro-enzyme and reveals sophisticated mechanisms of adaptation of the parasite to the PM and digestive enzymes secreted by its host. However, recent analysis of *P. gallinaceum* chitinase activity indicates that ookinetes express more than one chitinase. One of these chitinases, PgCHT1, has been cloned, and, although it is secreted as a pro-enzyme, the ookinete itself appears to be capable of processing this into the fully active form. Moreover, it was discovered that a *P. falciparum* chitinase, PfCHT1, may not be synthesized as a pro-enzyme. Yet when either of these genes is disrupted, they significantly impair the ability of their respective parasite to infect the mosquito. It still remains to be shown if the mosquito midgut chitinase or other proteases secreted by the parasite also play a role in penetration of the PM by *Plasmodium*.

Chitinase activity has also been demonstrated in several trypanosomatids. This suggests that *Trypanosoma brucei*, the causative agent for African sleeping sickness, may use the same mechanism as *Plasmodium* to escape the preformed PM2 of the tsetse fly. Interestingly, the protozoon *Babesia microti* apparently uses the contents of a specialized organelle to traverse the PM1 of the tick *Ixodes dammini*.

The development of *Leishmania* in sand flies has served as an example of the third category, namely, movement of the parasite to the gut epithelium after PM breakdown. The initial steps of the *Leishmania* life cycle in the sand fly include the transformation of amastigotes into promastigotes, which divide rapidly within the blood meal. PM breakdown after several days was believed to be the trigger for the next phase of parasite development, attachment to the gut epithelium and anterior migration. However, Schlein *et al.* (1992) have shown that *Leishmania* may actually rely on chitinase to penetrate the PM, in a similar way as the malaria parasite does. Subsequently, Schlein *et al.* presented evidence that damage of the chitin-containing cuticular lining of the cardiac valve might enhance transmission of *Leishmania* by causing regurgitation of the parasites and deposition in the host tissue. Furthermore, the mature PM in some sand flies has an opening at the posterior end, which may be advantageous for *Leishmania* establishment. Thus, this mode of transmission may not occur. Interestingly, the PM may also play a role by protecting the parasite from the action of digestive enzymes, as suggested by the decreased survival of *Leishmania* in the absence of a PM.

The mature PM is impermeable to virus-sized particles and thus can protect from viral infections. Calcofluor increases the infection of a Lepidopteran to baculvirus-induced infection. In order to infect the larval midgut of the lepidopteran *T. ni*, *T. ni* granulosis virus (TnGV) uses a proteolytic enzyme (enhancin) that degrades the PM mucin protein IIM. Furthermore, disruption of the PM greatly enhances the effectiveness of bacterial toxins or viral infections.

In summary, there is strong evidence that for at least some pathogens (e.g., malaria, *Leishmania*, viruses) the PM is a partial barrier that they have evolved ways to traverse. These observations can be interpreted in two different ways. One is that the insect evolved the PM to protect itself from the pathogens. Another is that the PM evolved to perform other functions and evolutionarily predates vector–parasite interactions. The latter hypothesis implies that the parasites evolved means to traverse the PM, as opposed to insects evolving a PM to protect themselves from parasites. The available evidence does not allow the distinction between the two hypotheses.

CONCLUSION

Much progress has been made since the writing of this chapter for the first edition of this book. At that time, there was a good understanding of the morphological aspects of PM formation but hardly any information on the molecular constitution of this structure. Since then a number of PM proteins have been characterized by 2D-PAGE and quite a few genes encoding PM proteins have been cloned and sequenced, thereby providing valuable structural information about PM proteins and leading to the identification of characteristic structural motifs. These include chitin-binding domains that may function in the three-dimensional assembly of the PM, glutamine-rich proteins that by action of cross-linking enzymes may provide a tough backbone, and mucin-like domains that may function in protection against foreign organisms and digestive enzymes. Furthermore, additional information has been obtained concerning possible roles played by the PM in the physiology of insects. The PM2 appears to play an important role in compartmentalization of digestion. Moreover, evidence has surfaced that the PM1 of hematophagous insects has a protective role against the harmful effects of heme-containing compounds. While the *Plasmodium* parasite protects itself through heme polymerization, the hematophagous insects may "use" the PM1 to sequester and aggregate heme.

Many important questions remain to be answered. The requirement that the PM be permeable to enzyme-sized molecules implies an ordered structure and suggests that it is not just an amorphous barrier that surrounds the food bolus. How is this structure formed? A first step in addressing this question is to define the players: Discover what the constituent proteins are and how they are modified. The sequencing of the *An. gambiae* genome should facilitate discovery of genes encoding PM proteins. A second step is to understand how the PM proteins interact with each other to produce the three-dimensional structure. Screening for protein–protein interactions using techniques such as the powerful yeast two-hybrid system may provide useful information.

An additional subject that may have far-reaching consequences concerns the mechanisms of pathogen–PM interactions. For instance, while it is clear that *Plasmodium* chitinase is necessary for it to traverse the PM, virtually nothing is known about how secretion of the enzyme is triggered. Is signaling between parasite and PM required? The understanding of this type of issue could lead to the development of new approaches for interfering with parasite development in the mosquito and thus contain the spread of malaria. Research on the structure and function of the PM promises to provide answers to many questions of importance to the fields of entomology and medicine.

Readings

The entire June 2001 issue of the *Archives of Insect Biochemistry and Physiology* (Vol. 47, No. 2) is dedicated to the "Biological, Biochemical and Molecular Properties of the Insect Peritrophic Membrane." It contains many good and current reviews focusing on different areas relating to the periotrophic matrix.

Eisemann, C. H., and Binnington, K. C. 1994. The peritrophic membrane: Its formation, structure, chemical composition and permeability in relation to vaccination against ectoparasitic arthropods. *Int. J. Parasitol.* 24: 15–26.

Lehane, M. J. 1997. Peritrophic matrix structure and function. *Annu. Rev. Entomol.* 42: 525–550.

Peters, W. 1992. Peritrophic membranes. In *Zoophysiology*, Vol. 130 (D. Bradshaw, W. Burggren, H. C. Heller, S. Ishii, H. Langer, G. Neuweiler, and D. J. Randall, eds.). Springer-Verlag, Berlin.

Richards, A. G., and Richards, P. A. 1971. Origin and composition of the peritrophic membrane of the mosquito, *Aedes aegypti. J. Insect Physiol.* 17: 2253–2275.

Tellam, R. L. 1996. The peritrophic matrix. In *The Insect Midgut* (M. J. Lehane and P. F. Billingsley, eds.). Chapman and Hall, London.

Tellam, R. L., Wijffels, G., and Willadsen, P. 1999. Peritrophic matrix proteins. *Insect Biochem. Mol. Biol.* 29: 87–101.

Further Readings

Berner, R., Rudin, W., and Hecker, H. 1983. Peritrophic membranes and protease activity in the midgut of the malaria mosquito, *Anopheles stephensi (Liston)* (Insecta: Diptera), under normal and experimental conditions. *J. Ultrastr. Res.* 83: 195–204.

Billingsley, P. F. 1990. The midgut ultrastructure of hematophagous insects. *Annu. Rev. Entomol.* 35: 219–248.

Billingsley, P. F., and Rudin, W. 1992. The role of the mosquito peritrophic membrane in blood meal digestion and infectivity of *Plasmodium* species. *J. Parasitol.* 78: 430–440.

Edwards, M. J., and Jacobs-Lorena, M. 2000. Permeability and disruption of the peritrophic matrix and caecal membrane from *Aedes aegypti* and *Anopheles gambiae* mosquito larvae. *J. Insect. Physiol.* 46: 1313–1320.

Langer, R. C., and Vinetz, J. M. 2001. Plasmodium ookinete-secreted chitinase and parasite penetration of the mosquito peritrophic matrix. *Trends Parasitol.* 117(6): 269–272.

Laurence, B. R. 1966. Intake and migration of the microfilariae of *Onchocerca volvulus (Leuckart)* in *Similium damnosum* Theobald. *J. Helminthol.* 40: 337–342.

Pascoa, V., Oliveira, P. L., Dansa-Petretski, M., Silva, J. R., Alvarenga, P. H, Jacobs Lorena, M., and Lemos, F. J. A. 2002. *Aedes aegypti* peritrophic matrix and its interaction with heme during blood digestion. Insect Biochem. *Mol. Biol.* 32: 517–523.

Lewis, D. J. 1953. *Simulium damnosum* and its relation to onchocerciasis in Anglo-Egyptian Sudan. *Bull. Entomol. Res.* 43: 597–644.

Schlein, Y., Jacobson, R. L., and Messor, G. 1992. Leishmania infections damage the feeding mechanism of the sandfly vector and implement parasite transmission by bite. *Proc. Natl. Acad. Sci. USA* 89: 9944–9948.

Tellam, R. L., and Eisemann, C. 2000. Chitin is only a minor component of the peritrophic matrix from larvae of *Lucilia cuprina*. *Insect Biochem. Mol. Biol.* 30: 1189–1201.

Terra, W. R. 1990. Evolution of digestive systems of insects. *Annu. Rev. Entomol.* 35: 181–200.

Vuocolo, T., Eisemann, C. H., Pearson, R. D., Willadsen, P., and Tellam, R. L. 2001. Identification and molecular characterisation of a peritrophin gene, peritrophin-48, from the myiasis fly *Chrysomyia bezziana. Insect Biochem. Mol. Biol.* 31: 919–932.

23

Vector Nutrition and Energy Metabolism

GUOLI ZHOU, PATRICIA Y. SCARAFFIA, AND MICHAEL A. WELLS

INTRODUCTION

Three groups of blood-feeding insects can be distinguished based on their blood-feeding habits: (1) only the larvae feed on blood (rare, e.g., maggots); (2) they feed only on blood throughout the life cycle (tsetse flies, triatomines, lice); (3) only adults feed on blood (mosquitoes, blackflies, sand flies, fleas). There is also a difference in which sex feeds on blood: (1) both sexes are obligatory blood feeders (fleas); (2) both sexes are optional blood feeders (stable flies); (3) only the female is an optional blood feeder (mosquitoes). Insects that are obligatory blood feeders throughout the life cycle invariably harbor symbionts, suggesting that the blood meal may not be nutritionally adequate.

Nearly all of the work on nutrition and energy metabolism in blood-feeding insects has focused on mosquitoes, which is also our focus. "The reproductive capacity of female mosquitoes is affected by their nutrition in both larval and adult stages. Their reproductive potential is established by the end of the larval stage; their subsequent exploitation of that potential depends upon their nutrition during the adult stage" (Clements 1992). Understanding the biochemical mechanisms that validate these statements is essential for nearly all studies in mosquito biology. For example, if female mosquitoes must accumulate minimal energy reserves (lipid and glycogen) before they can mature a batch of eggs, their behavior will be directed toward obtaining those reserves, whether from a sugar source or from blood. Depending on the preferred energy source, such behavior impacts disease transmission,

because mosquitoes cannot transmit disease if they feed only on nectar. In addition, the size of the energy reserves directly determines longevity; mosquitoes with minimal energy reserves may not live long enough to find a host or to meet the time required for a pathogen to develop. The size of the energy reserves also affects the disposition of the blood meal. If the energy reserves are minimal, the blood meal components may be deaminated and used to increase the energy reserves rather than egg production, meaning that more than one blood meal may be required to mature the first batch of eggs. Such behavior can increase disease transmission because more hosts are bitten.

Studies carried out primarily in the laboratory of Hans Briegel in Switzerland have defined quantitatively the relationship between dietary intake and the accumulation of energy reserves. We summarize these studies first.

LARVAL NUTRITION

Under laboratory conditions, larvae accumulate reserves in proportion to their body size, and the larger female larvae accumulate more reserves than the smaller males. The fourth and final instar seems to be a critical period for nutrient accumulation, with more than 70% of the total protein and 80% of the total lipid deposited during this period. In large part, the amount of reserves present at metamorphosis is dependent on the availability of nutrients during the fourth instar.

Aedes aegypti is notable for the large amount of lipid it accumulates as compared to other culicine and anopheline mosquitoes. *Ae. aegypti* also displays a significant degree of nutritional plasticity during the fourth instar, because it is able to respond to starvation and continue feeding and prolonging development. It is also notable that mosquito larvae can pupate with significantly different sizes and nutritional reserves.

Mosquito larvae are aquatic, and the nutritional quality of the environment varies widely. Mosquito larvae fall into two ecological groups. (1) *Anopheles* spp., for example, inhabit groundwater (ponds, marshes, etc.) and exhibit rapid development and little ability to adjust their development time to accommodate variations in availability of nutrients. This behavior may be a response to the pressure exerted by predators that can consume the larvae if they persist in the larval stage for prolonged periods. (2) Many *Aedes* and *Culex* inhabit small containers, such as tree holes and man-made containers. In these environments, which are relatively predator-free and nutrient-poor, the larvae develop slowly and can adjust the length of the larval period to the availability of nutrients. *Ae. aegypti* is particularly adept at adjusting its larval development to the availability of nutrients. It can develop as fast as any species in a rich nutrient environment, but it can develop in nutrient-poor environments better than most other species.

ADULT NUTRITION

Almost all studies have been done on either *Ae. aegypti* or *Anopheles* spp. The results for *Ae. aegypti* showed: (1) the extent of nutrient storage was dependent on relative body mass—larger females had greater energy reserves; (2) when fed to repletion, large females ingested more blood and produced more eggs than smaller females; (3) in small females as little as 2% of the blood meal protein amino acids were converted to egg proteins, most are deaminated, and even in large females at most 20% of the blood protein amino acids were converted to egg proteins. Some of the α-keto acids derived from deamination of the blood meal protein amino acids may be used as the carbon source for egg lipid synthesis, and another portion may be used for energy production, but no quantitative measurements have been made. In addition, some of the α-keto acids may be used to synthesize energy reserves (lipid or glycogen + trehalose) of the female, but, again, there are no quantitative measurements.

When large females took small blood meals, they needed to use some of their energy reserves to support egg production, but with large meals there was apparently no use of the female's energy reserves for egg production. Under optimal conditions, with large females taking large blood meals, no more than 50% of the blood meal protein amino acids were converted to egg components (proteins plus lipids). Under suboptimal conditions, with small females taking small blood meals, no more than 30% of the blood meal protein amino acids were converted to egg components (proteins plus lipids).

For *Anopheles* spp., the extent of nutrient storage was also dependent on relative body mass: larger females had greater energy reserves. Anophelines showed somewhat similar results to those described for *Ae. aegypti*, except that they were less efficient than *Ae. aegypti* in converting blood protein amino acids into egg components. In this regard, they behaved as if they were small *Ae. aegypti*. Thus, at most 20% of meal protein amino acids were converted into egg components, but about 33% of the protein amino acids were converted to "maternal" energy stores, more than occurred in *Ae. aegypti*. The fate of the rest of the blood meal protein amino acids is unknown. Some anophelines have been reported to take multiple meals in a single gonadotrophic cycle under laboratory conditions, but a quantitative analysis of the fate of the different meals has not been reported.

From these studies we can see that the fate of the amino acids derived from meal protein is complex. Some of the amino acids are used to synthesize egg proteins; however, the majority of the amino acids are deaminated and the resulting α-keto acids are used for the synthesis of egg lipid, the synthesis of maternal energy reserves, or energy production (Fig. 23.1).

An important factor in regulating the fate of amino acids in the blood meal is sugar feeding. The roles of

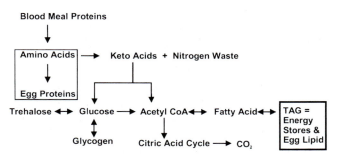

FIGURE 23.1 Possible fates of blood meal protein amino acids. In this diagram TAG stands for triacylglycerol, which can be found in fat body, midgut, or eggs. Glycogen can be found in the fat body, muscle, or midgut, and trehalose is present in the hemolymph. Utilization of acetyl-CoA in the citric acid cycle can occur in any tissue; however, nothing is known about the activity of the pathways depicted here in blood-fed mosquitoes.

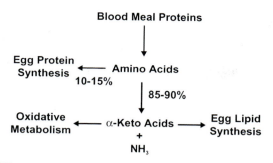

FIGURE 23.2 The metabolic consequence of amino acid deamination is the production of α-keto acids and ammonia. The α-keto acids can be used for lipid synthesis or for energy production; however, a mosquito needs to deal with all the ammonia produced.

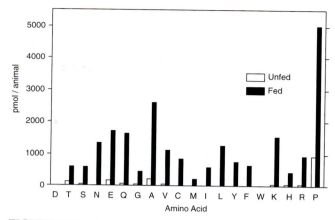

FIGURE 23.3 Changes in hemolymph-free amino acid composition 24 h following a blood meal in *Ae. aegypti*. The single amino acid code is used. While the concentration of many amino acids increases following feeding, note the very large increase in proline concentration.

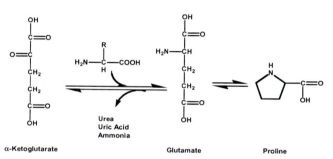

FIGURE 23.4 Proline can serve as a temporary storage reservoir for ammonia released from blood meal amino acids.

sugar feeding are (1) to facilitate previtellogenic follicle development; (2) to reduce the size of the minimal blood meal that will support egg production; and (3) to increase the number of eggs produced. However, sugar feeding does not increase the efficiency of utilization of protein amino acids in the blood meal for the production of egg components beyond 50% (see earlier).

The conclusion from these studies is that the fate of the protein amino acids in the blood meal is intimately tied to the energy status of the female at the time she takes the meal. If she has energy reserves above a critical minimum value, she will develop eggs; if not, the meal amino acids will be utilized to increase her energy reserves. The energy reserves depend on larval nutrition and on sugar feeding by the adult.

METABOLIC PROBLEMS ASSOCIATED WITH BLOOD MEAL METABOLISM

Based on what we know about amino acid metabolism in vertebrates, it appears that mosquitoes might have a difficult time dealing with a large increase in the concentration of amino acids following blood meal protein digestion, especially if >80% is converted to α-keto acids. Such massive deamination produces a large amount of ammonia, which the mosquito excretes as a combination of ammonia, uric acid, and urea (Fig. 23.2).

Conversion of amino acids to α-keto acids might occur in either the midgut or the fat body or both. A possible clue as to what is happening is provided by the analysis of free amino acids in the hemolymph following a blood meal in *Ae. aegypti* (Fig. 23.3).

The significant increase in proline concentration at 24 h is followed by a steady return to its pre-blood-

meal level by 96 h. What might be the importance of this massive production of proline? Proline is produced from glutamate, which in turn is the major product of transamination reaction involving α-ketoglutarate (Fig. 23.4).

The only fate of proline is its conversion into protein; otherwise it is an innocuous compound, although it might serve as a reservoir for ammonia (Fig. 23.4). This process could store the large amount of ammonia in a metabolicly "safe" form until it can be converted to urea or uric acid for secretion. As ammonia is recovered from proline, the α-ketoglutarate can then be used for energy production or for the synthesis of more proline. It should be noted that the synthesis of large amounts of uric acid and urea requires the input of substantial energy, perhaps as much as 15–20% of the caloric content of the blood meal proteins! In comparison, the interconversion of α-ketoglutarate and proline proceeds without

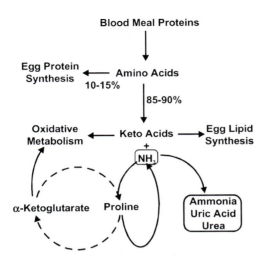

FIGURE 23.5 A possible proline cycle in mosquitoes allows temporary storage of ammonia derived from amino acid deamination in a nontoxic form.

significant energy cost. The tissue(s) in which the α-keto acids are converted to glucose and lipid are unknown. These results suggest that a proline cycle operates in a mosquito that allows massive deamination of amino acids for energy production while storing the toxic ammonia as proline (Fig. 23.5).

ENERGY AND REPRODUCTION

Ovarian follicles must reach a certain stage of development before a blood meal can support egg production. In small females with low energy reserves, additional energy is required to complete maturation of the follicles before a blood meal can support egg development. This energy can be derived from a sugar meal or from a blood meal, but sugar meals are more effective than blood meals in building energy reserves. Often, females take a sugar meal before the first blood meal in order to accumulate sufficient energy reserves to support egg development following a blood meal (Chapter 25). In the field, mosquitoes typically emerge with little energy reserves for flight and survival, and sugar feeding must occur during the first few days, with blood feeding occurring later. Without a sugar meal, males survive 2–3 days and females survive 3–5 days. Field-collected mosquitoes, large or small, usually have energy reserves far below those seen in laboratory insects and often are close to the minimal reserves seen just before death in starved laboratory mosquitoes. Thus, the notion that sugar availability is not limiting to fitness is unsupported by current data,

but much more work is needed in this area. An interesting case is given by *Ae. provocans*, whose preferred nectar sources are hedgerow plants. Recent agronomic practices that eliminated these plants may account for the scarcity of this species in southwestern Ontario, Canada.

The mixed-diet foraging lifestyle, sugar vs. blood meal, presents an unusual dilemma for the mosquito, which takes few, but large, protein meals. An appropriate partitioning of feeding on sugar or blood has critical impact on reproduction. Sugar is used to accumulate the necessary reserves to support egg development, which then requires a blood meal.

FLIGHT METABOLISM

Insect flight muscle is the most metabolically active tissue known: oxygen consumption increases 50- to 100-fold over resting muscle; in vertebrates the increase is 7- to 14-fold. Flight can be powered by sugar, lipid, or proline metabolism. At maximum output, a locust flight muscle hydrolyzes ATP at a rate of 18 μmol/sec, which means that the entire energy-rich phosphate pool (ATP and phosphoarginine) of the muscle can sustain flight for 1 second; hence, there must be prodigious ATP synthesis during flight, which can last several hours.

In Diptera and Hymenoptera, carbohydrate is the main substrate used to fuel flight. Unfed mosquitoes, 1–2 days old, have sufficient reserves to fly for several hours on a flight mill at a rate of about 1 km/h. Under these conditions, the rate of ATP hydrolysis for a mosquito, adjusted to the same body weight as the locust, is about 12 μmol/sec. The glycogen stores of muscle can sustain flight for a few minutes, but prolonged flight utilizes the glycogen stores of the fat body and the sugar is transported to muscle as trehalose. The mosquito must oxidize about 2.3 μg of glycogen/sec in order to produce the required ATP. During flight the mosquito first consumes sugar in the crop, if present, and then glycogen. After mosquitoes are flown to exhaustion, during which time they deplete their glycogen reserves, they can fly immediately after a sugar meal, but cannot fly for several hours after a blood meal, presumably reflecting the time required to convert blood meal protein amino acids into glucose and trehalose.

Proline is the main substrate that fuels flight in some insects by a pathway that involves partial oxidation of proline in flight muscle (Fig. 23.6). Originally described in the tsetse fly, *Glossina morsitans*, and the Colorado potato beetle, *Leptinotarsa decemlineata*, the

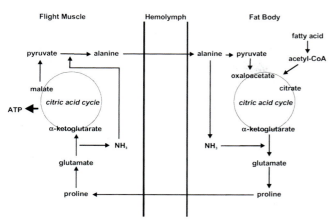

FIGURE 23.6 Use of proline to fuel flight metabolism. Proline is produced in the fat body from glutamate, which is derived from the citric acid cycle intermediate α-ketoglutarate. The proline is transported to flight muscle, where the carbon is converted back to α-ketoglutarate. The α-ketoglutarate then goes through part of the citric acid cycle to form malate. Decarboxylation of malate yields pyruvate, which is converted to alanine in the flight muscle. Then alanine is transported back to the fat body, where it is converted to pyruvate. In the fat body pyruvate is converted to oxaloacetate, which condenses with acetyl-CoA derived from fatty acid oxidation to form citrate. The citrate is converted to α-ketoglutarate to complete the cycle. The partial oxidation of proline in flight muscle yields 14 ATP.

oxidation of proline has been found in additional insects. In some insects both glucose and proline can be used to fuel flight. For example, in *Ae. aegypti* we have observed that the free proline concentration in the thorax drops 40–60% following 30–60 min of flight, suggesting that, in addition to glucose, proline may serve as a fuel for flight.

AUTOGENY

Autogeny refers to egg production in the absence of protein intake by the adult female, whereas *anautogeny* refers to egg production that is dependent on protein intake (blood meal) (Chapters 9 and 25). Autogeny has been reported in several species, and it is not uncommon to find both autogenous and anautogenous forms within the same species, e.g., *Culex pipens*, which live in different habitats. Autogeny shifts the responsibility for gathering nutrients for egg production from the adult stage to the larval stage. It is likely that storage proteins in the larva and perhaps proline play critical roles in providing the amino acids required for egg production. In some, but not all, autogenous mosqui-

toes, sugar feeding by the adult is required before ovarian development can begin.

SUMMARY

The manner in which the female mosquito digests a blood meal clearly has important implications in all aspects of mosquito biology. The importance of larval nutrition in determining the level of energy reserves at eclosion and its implications for the use of the initial meal taken by a female, whether sugar or blood, have important consequences in behavior and ecology. If a mosquito must take several meals, particularly blood, before she can mature a batch of eggs, then the chance of her reproducing decreases, while at the same time the chance for disease transmission increases. Because so much of the amino acids derived from blood meal protein amino acids are deaminated and used for purposes other than egg protein synthesis, it is important to know what those purposes are. It is also important to determine how the energy reserves of the female are used during a gonadotrophic cycle.

Readings

Beenakkers, A. M. Th., van der Horst, D. J., and van Marrewijk, W. J. A. 1985. Biochemical processes directed to flight muscle metabolism. In *Comprehensive Insect Physiology, Biochemistry and Pharmacology*, Vol. 10 (G. A. Kerkut and L. I. Gilbert, eds.). Pergamon Press, New York, pp. 451–486.

Briegel, H. 1990. Metabolic relationships between female body size, reserves, and fecundity of *Aedes aegypti. J. Insect Physiol.* 36: 165–172,

Briegel, H. 1990. Fecundity, metabolism, and body size in *Anopheles* (Diptera: Culicidae), vectors of malaria. *J. Med. Entomol.* 27: 839–850,

Clements, A. N. 1992. *The Biology of Mosquitoes*, Vol. 1: *Development, Nutrition and Reproduction.* Chapman and Hall, London.

Dean, R. L., Locke, M., and Collins, J. V. 1985. Structure of the fat body. In *Comprehensive Insect Physiology, Biochemistry and Pharmacology*, Vol. 3 (G. A. Kerkut and L. I. Gilbert, eds.). Pergamon Press, New York, pp. 155–210.

Foster, W. A. 1995. Mosquito sugar feeding and reproductive energetics. *Annu. Rev. Entomol.* 40: 443–474.

Keeley, L. L. 1985. Physiology and biochemistry of the fat body. In *Comprehensive Insect Physiology, Biochemistry and Pharmacology*, Vol. 3 (G. A. Kerkut and L. I. Gilbert, eds.). Pergamon Press, New York, pp. 211–248.

Lehane, M. J. 1991. *Biology of Blood-Sucking Insects.* Harper Collins Academic, London.

Mullins, D. E. 1985. Chemistry and physiology of the hemolymph. In *Comprehensive Insect Physiology, Biochemistry and Pharmacology*, Vol. 3 (G. A. Kerkut and L. I. Gilbert, eds.). Pergamon Press, New York, pp. 355–400.

Timmermann, S. E., and Briegel, H. 1999. Larval growth and biosynthesis of reserves in mosquitoes. *J. Insect Physiol.* 45: 461–470.

24

Mosquito Endocrinology

HENRY H. HAGEDORN

INTRODUCTION

Despite the small size of the adult female mosquito, its endocrinology is one of the better-studied among insects, and the studies are focused mostly on how hormones control reproduction. There is a long history of investigation of mosquito reproduction, starting with the demonstration, in 1945, by T. S. Dentinova that decapitation of a mosquito shortly after a blood meal prevented egg development. Her experiment started a long search for the factors involved that culminated 53 years later in the sequence of a peptide that is released from neurosecretory cells in the brain that starts egg development. In fact, however, the development of a batch of eggs involves many factors working together to control a complex process. Small it may be, but the mosquito is not simple.

BACKGROUND

The so-called "classic scheme" that was developed on the basis of experiments on large holometabolous insects such as *Hyalophora cecropia* and *Manduca sexta* showed the influence of three hormonal factors on the molting of insects, 20-hydroxyecdysone, juvenile hormone (JH), and an ill-defined factor called the *brain hormone* (Fig. 24.1). Each of these factors was thought to have a specific role to play: The brain hormone stimulated ecdysone production, ecdysone stimulated production of a new cuticle, and JH determined whether the new cuticle was larval, pupal, or adult. A better understanding of the endocrinology of these factors was obtained when their chemistry became known.

Ecdysone was isolated in the 1950s, JH in the 1960s, while the peptide brain hormones that regulate the production of ecdysone and juvenile hormone have only recently been sequenced and studied in detail. As more became known, it became obvious that the endocrinology of insects is considerably more complex than the classic scheme suggested.

Ecdysone

The "molting hormones" consist of a small group of ecdysteroids characterized by a steroid nucleus having numerous hydroxyl groups that render the ecdysteroids water soluble, an unusual characteristic for a steroid hormone (e.g., estrogen). The structure of these molecules is shown in Figure 24.2. The sources of ecdysone are the prothoracic gland in immature insects, the ovary in the adult, and the testis in some insects. Other tissues, particularly the epidermis, have been implicated in the synthesis of ecdysone in some insects, including the larval mosquito. The active molecule in most insects is 20-hydroxyecdysone. The hormone is secreted as the relatively inactive ecdysone that is enzymatically activated to 20-hydroxyecdysone in a number of tissues, but especially in the fat body. In a few insects (some Hemiptera and Hymenoptera) the active hormone is makisterone A, which has an extra carbon atom on the side chain.

20-Hydroxyecdysone controls a number of aspects of insect life, including molting, wandering behavior, the development of the nervous system, reproduction, and the production of sex pheromones, to name a few disparate examples. The mode of action of 20-hydroxyecdysone involves the activation of gene

317

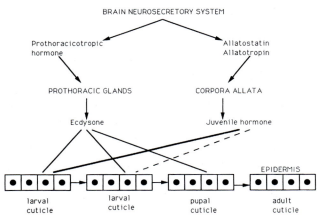

FIGURE 24.1 The "classic scheme" for endocrine control of molting and metamorphosis of holometabolous insects suggested that whether the epidermis produced larval, pupal, or adult cuticle depended on the hormones to which it was exposed. 20-Hydroxyecdysone (ecdysone) stimulated the production of a new cuticle in each case. But if the JH titer was high, a larval cuticle was produced; if it was low, a pupal cuticle was produced; if it was absent, an adult cuticle was produced. Also evident is the role of peptide hormones produced by neurosecretory cells in the brain.

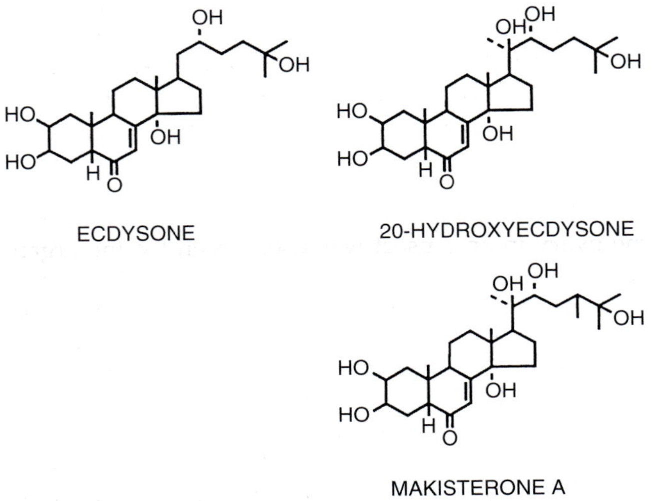

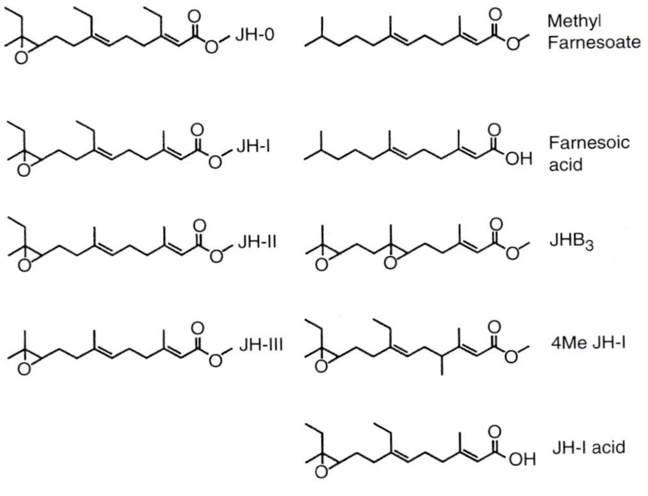

FIGURE 24.3 Nine forms of JH have been found in insects. JH-I was the first to be identified. JH-O, JH-I, JH-II, and JH-III are all present during various stages of lepidopteran development. JH-III is the form found in most insects, including the mosquito. JHB3 is the bis-epoxide of JH that is found in some cyclorraphan Diptera.

ECDYSONE

20-HYDROXYECDYSONE

MAKISTERONE A

FIGURE 24.2 Structural formulae for the ecdysteroids. Ecdysone is the product of the prothoracic glands and is enzymatically converted to 20-hydroxyecdysone, which is the active form of the hormone. Makisterone A is the active hormone in some Hemiptera and Hymenoptera.

transcription, as discussed in detail for the mosquito in Chapter 25.

Juvenile Hormone (JH)

In contrast to the relatively few members of the ecdysteroid family of hormones, the JHs found in insects are much more varied. The JH is the only

hormone known that is a terpenoid. Shown in Figure 24.3 are the nine forms of the JH molecule that are candidates for hormonal status: JH I, II, and III, JH 0, 4-methyl JH, JHB3, MF, FA, and the acid of JH I. Of these, JH III is the major form found in most insects, including mosquitoes. JHs 0, I, and II are found only in Lepidoptera. JHB3, the bis-epoxide of JH, is present and active in several cyclorrhaphous dipterans but has not been identified in other dipteran groups. The corpora allata are the source of the JH in insects, but there is one report that the accessory glands of *Aedes aegypti* can synthesize JH. The mode of action of JH is not well understood and may be variable.

Peptides

Many peptides that are hormones are products of specialized nerve cells called *neurosecretory cells*. That these peptide hormones are produced by cells in the nervous system is significant because the release of at least some of these peptides is photoperiodically regulated, while others are under different kinds of nervous control. It has become clear that in many cases neurosecretory peptides may also be produced by normal nerve cells and function as neurotransmitters or neuromodulators within the nervous system. The mode of action of most peptide hormones often involves the stimulation of secondary messenger systems, such as cAMP, in target cells.

It is clear that peptides have two general roles: (1) They can regulate the activity of glands producing other hormones (i.e., ecdysone and juvenile hormone), and (2) they can also directly affect physiological processes (i.e., diuresis, lipid and carbohydrate levels, heart rate, etc.). It has also become increasingly evident that a given peptide can have multiple roles. For example, a large group of related peptides has been isolated that are called *allatostatins*. They were originally described as inhibitors of JH synthesis by the corpora allata, and in some insects (cockroaches, crickets) they do have this function; however, in other insects they cause muscle contraction in the gut and ovary and may have no effect on the corpora allata.

The brain hormone mentioned earlier is a peptide produced by neurosecretory cells in the brain. It activates the production of ecdysone by the prothoracic glands of larval insects. It is called the *prothoracicotropic hormone*, or PTTH. It is a dimer composed of two identical chains of 109 amino acids each. This peptide is not homologous to any other peptide hormone yet sequenced. Hybridization of PTTH DNA to tissue sections shows reaction with only a single pair of dorsolateral neurosecretory cells in the brain that are identical to those identified by bioassay and antibodies against PTTH.

The pace of discovery of neurosecretory peptides having hormonal effects suggests that we have exposed only the tip of the iceberg. Many new peptides have been identified by their biological activity; some have been isolated and sequenced. The genes coding for peptides are being isolated and studied for clues as to their relationships and evolution.

EGG DEVELOPMENT IN THE MOSQUITO

Previtellogenesis

When the female *Ae. aegypti* ecloses, the follicles of the ovaries hold 100–120 ovarioles, each containing a germarium and a follicle with eight germ cells surrounded by a layer of follicle cells (Fig. 24.4). Over the next 2–3 days, one of the germ cells differentiates to become the oocyte and the rest become nurse cells, or trophocytes. During this period, the follicles double in size and the follicle cells multiply to keep up with the growing follicle. The follicles then enter what is termed the *resting stage*, where they remain until a blood meal is taken.

Vitellogenesis

Once these tissues have reached the resting stage, the female is ready to take a blood meal that induces the vitellogenic phase of egg development. The oocyte grows rapidly as it develops the yolk that will be needed by the developing embryo. The nurse cells provide some of the components of the yolk, including ribosomes and other cellular machinery needed for early development of the embryo. Messenger RNAs may also be contributed by the nurse cells, as occurs in *D. melanogaster* and many other eukaryotes. Most of the yolk is not made by the oocyte or nurse cells but is taken up from the hemolymph by receptor-mediated endocytosis. The vitellogenic phase ends with the production of the chorion by the follicle cells. The eggs may be retained until the female finds a suitable oviposition site.

The Endocrinology of Egg Development

There is considerable evidence that the events that occur during previtellogenesis and vitellogenesis are under hormonal control. Although the first experiment to suggest this was the simple decapitation experiment of Detinova in 1945, the implications were, at the time, profound because the concept of compounds that were released from an organ that had effects at a distance was rather new. Major advances in the field were made when Arden Lea (1967) developed methods for experimental removal, and reimplantation, of several of these organs in mosquitoes, such as the neurosecretory cells of the brain and the corpora allata, thereby expanding considerably our understanding of the endocrinology of egg development. The discovery that the ovary in any insect was itself an endocrine organ was novel at the time, but quite consistent with a similar role for the ovary in other animals. As analytical techniques advanced, it became possible to determine the chemistry of the putative hormones and to measure changes in amounts of these molecules during development.

Using these techniques it was shown that the factor from the brain was a neurosecretory peptide originally known as the egg development neurosecretory hormone but now known as the ovarian ecdysteroidogenic hormone (OEH). It stimulates the follicle cells of the ovary to produce ecdysone, which, after conversion to 20-hydroxyecdysone by 20-monooxygenase, regulates many of the events during vitellogenic development of the eggs. JH from the corpora allata was found to be essential for previtellogenic egg development.

Our current understanding of the changing levels of these hormones in *Ae. aegypti* is summarized in Figure 24.5. JH levels rise after eclosion, fall after a blood meal, and then rise again. After a blood meal OEH is released and 20-hydroxyecdysone levels rise but then

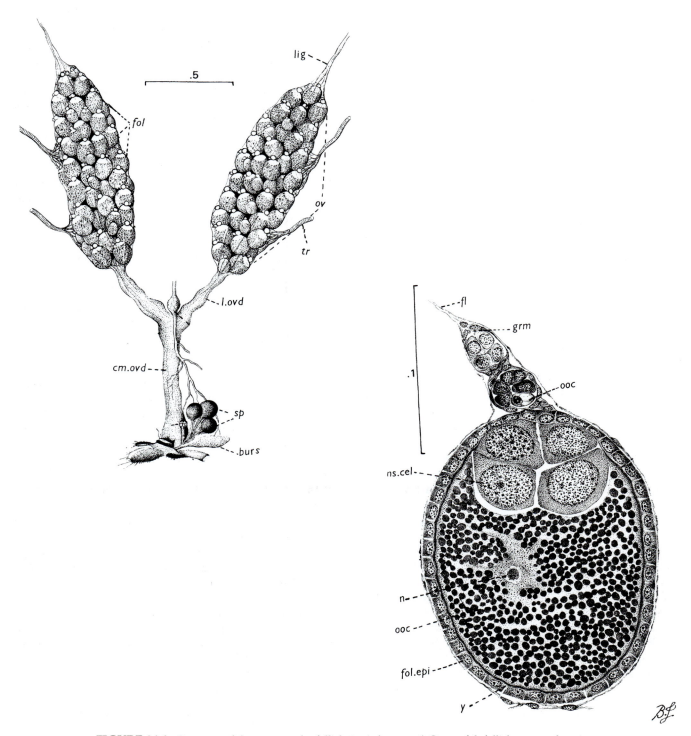

FIGURE 24.4 Structure of the ovary and a follicle in *Aedes aegypti*. Ovary: fol, follicle; cm. ovd, common oviduct; lig, ligament; ov, ovary; tr, trachea; l. ovd, lateral oviduct; sp, spermatheca; burs, bursa copulatrix: fl, filament; grm, germarium; ooc, oocyte; ns.cel; nurse cell; n, nucleus; fol.epi, follicular epithelium; y, yolk. Modified from Jobling, B. *Anatomical Drawings of Biting Flies.* British Museum (Natural History), London. With permission.

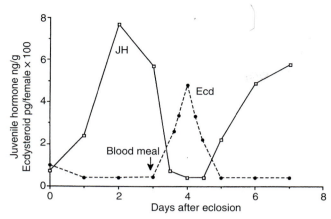

FIGURE 24.5 Changes ·in juvenile hormone and 20-hydroxyecdysone levels in the female *Aedes aegypti* from eclosion of the adult through a blood meal and the production of the first batch of eggs. Not shown is the release of the ovarian ecdysteroidogenic hormone. Data from Shapiro, A. B., et al. 1986. *J. Insect Physiol.* 32: 867–877, and Greenplate, J. T., et al. 1985. *J. Insect Physiol.* 31: 323–329.

fall just before the second rise in JH. It is apparent that 20-hydroxyecdysone is present during a period when JH levels are low.

REGULATION OF THE REGULATORS

Hormones regulate complex physiological processes. In order to regulate a process, a hormone must not only be synthesized, it must also be degraded, and the level of the response must be adjusted within some range. Adjusting the amounts of a hormone present can regulate a process rather precisely. To understand the physiological role of a hormone one must understand its chemistry, its source in the body, its target tissues, when it is present in the hemolymph, and its titer changes.

Juvenile Hormone

During previtellogenesis, JH has been shown to regulate a number of important events. It is necessary for the growth of the oocyte to the resting stage. It affects the fat body as the amount of DNA doubles in each cell and a modest amount of ribosomal RNA is made. In the fat body and gut JH induces the transcription of genes for ribosomal proteins and trypsin, respectively, that are later expressed after a blood meal. JH also promotes competence in both the ability of the ovary to respond to OEH and the ability of the fat body to respond to 20-hydroxyecdysone. In *Ae. aegypti* JH

induces the female to mate and in *Culex pipiens* it stimulates the female to search for a blood meal. Thus, the target tissues of JH after eclosion must include the ovarian follicles, the fat body, the gut, and the brain.

The levels of JH are high after eclosion, fall rapidly after a blood meal, and then rise again (Fig. 24.5). This pattern must reflect differential activities of synthesis and degradation. JH is synthesized in the corpora allata, which are attached to the aorta in the prothorax of the mosquito (Fig. 24.6). Readio and Meola (1999) incubated corpora allata from *Cx. pipiens* with a radioactive precursor and showed that synthesis of JH was high after adult eclosion but fell after a blood meal (Fig. 24.7). What controls the level of synthetic activity of the corpora allata? Allatostatins and allatotropins have been isolated from *Ae. aegypti*, but their function in the mosquito is unknown. Clearly, the synthetic activity of the corpora allata is regulated, but how remains an open question.

As a lipid, JH is not soluble in water, and it is bound in the hemolymph to specific binding proteins that transport it to the target tissues and protect it from degradation. However, the binding proteins do not protect the hormone from all enzymes. As shown in Figure 24.5, JH is rapidly degraded after a blood meal. There is evidence that even small amounts of JH are detrimental after a blood meal; therefore its removal is important for normal egg development. Two different enzymes appear at this time, JH-epoxidase and JH-esterase, both of which specifically degrade JH (Fig. 24.8), even that bound to binding proteins. As a result of the activity of these two enzymes, only trace levels of JH are seen by 24 h after a blood meal. Thus, the changing levels of JH are due to changes in both synthesis and degradation.

The lipid solubility of JH allows it to pass through cuticle. Topical application has been exploited as a way of studying its effects in the laboratory and has also been utilized for mosquito control. Simply adding the JH mimetic methoprene to water containing larvae kills them because the hormone mimetic disrupts normal eclosion of the adult. This technique is widely used to control certain kinds of mosquitoes and other insects with aquatic larvae.

Ensuring Previtellogenesis

It appears that there are three different mechanisms ensuring that previtellogenic events occur. The diet of the adult mosquito consists of nectar and blood. Newly eclosed adults usually take a nectar meal, which is required for normal previtellogenic egg development. Feinsod and Spielman (1980) have suggested that the nectar meal stimulates JH production

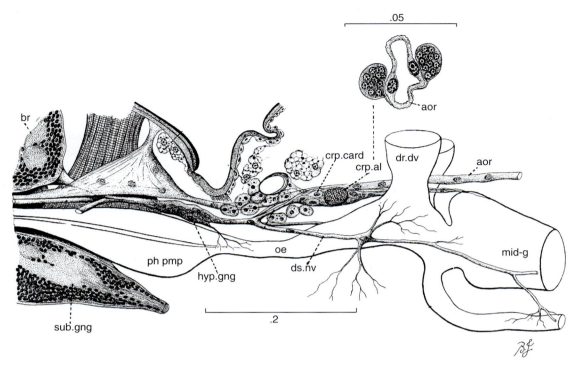

FIGURE 24.6 Morphology of the endocrine organs of the corpora allata and corpora cardiaca in the neck and prothorax of the mosquito, *Aedes aegypti*: br, brain; sub.gng, suboesophageal ganglion; ph pmp, pharyngeal pump; hyp.gng, hypocerebral ganglion; oe, oesophagus; gs.nv, gastric nerve; mid-g, midgut; crp.card, corpus cardiacum; crp.al, corpus allatum; dr.dv, dorsal diverticulum; aor, aorta. The upper diagram shows a cross section of the corpora allata. Modified from Jobling, B. *Anatomical Drawings of Biting* Flies. British Museum (Natural History), London. With permission.

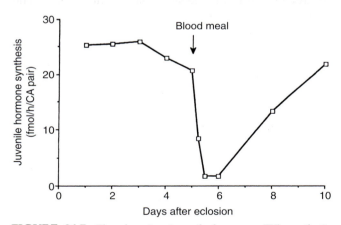

FIGURE 24.7 The changing juvenile hormone (JH) synthetic activity of the corpora allata (CA) was measured by incubating CA of female *Culex pipiens* in a medium containing [³H]-methionine to label newly synthesized JH. Synthesis of JH was found to be high in newly eclosed female, to decline after a blood meal, and then to increase again by the third day after a blood meal. Data from Readio J., et al. 1988. *J. Insect Physiol.* 34: 131–135.

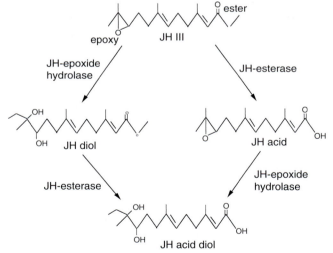

FIGURE 24.8 Degradation pathways of JH. Two enzymes, JH-epoxide hydrolase and JH-esterase degrade the epoxy and esterase groups, respectively, and inactivate the hormone. Although both pathways to the acid diol are possible, it is likely that the JH-epoxide acts first, because levels of this enzyme rise earlier than those of the JH-esterase after a blood meal in *Culex pipiens*.

by the corpora allata. There is some evidence that the accessory glands of males also produce JH (Borovsky et al. 1994). If so, a female that is unable to obtain a nectar meal might obtain the requisite JH by mating. In the wild, mosquito larvae are often starved. Previtellogenic follicle development in adults from such larvae is delayed, and the first blood meal taken by these animals may stimulate previtellogenic oocyte development rather than deposition of yolk; presumably the normal endocrine response to the blood meal is bypassed. Thus, there are several ways that the previtellogenic growth of the oocyte is ensured. Such redundancy is not unusual in physiological systems.

Ovarian Ecdysteroidogenic Hormone (OEH)

Ovarian ecdysteroidogenic hormone (OEH) was first identified in *Ae. aegypti* by Lea (1967) as a neurosecretory hormone produced by the medial neurosecretory cells of the brain that is stored in neurohemal cells that lie near the corpora cardiaca (Fig. 24.6). The hormone was found to stimulate the ovary to produce ecdysone via an effect on intracellular cAMP levels in the ovary. Mark Brown and colleagues (1994) isolated a peptide from an extract of 6 million heads of *Ae. aegypti*, and from its sequence isolated a cDNA coding for it. Expression of the gene in bacteria produced a peptide that stimulated egg development when 0.15 ng was injected into resting-stage females, and also stimulated ecdysone synthesis when incubated with ovaries. An antiserum against the peptide identified cells in the medial neurosecretory cells of the brain. Later studies showed that some axons from these cells pass through the corpora cardiaca and end on the anterior midgut. Immunostaining endocrine cells were also found in abdominal ganglia and in the posterior midgut, indicating multiple sources of OEH. Axons from the cells in the abdominal ganglia had release sites in perivisceral organs. Interestingly, after a blood meal, immunostaining material was not released from the neurohemal cells near the corpora cardiaca nor from axons on the anterior midgut but was released from the perivisceral organs.

What causes the release of OEH after a blood meal is not clear. It may be a two-stage process because there is evidence that a factor from the resting-stage ovary itself is necessary for the release of OEH. If so, this factor would signal that the process of previtellogenesis is complete. When a blood meal is taken it is possible that release from the perivisceral organs occurs as a result of changes in the midgut after a blood meal, such as stretching of the gut, or a chemical signal from the meal itself. OEH is not similar in sequence or structure to PTTH, although both stimulate ecdysone production by different tissues.

20-Hydroxyecdysone

The prothoracic glands are the source of 20-hydroxyecdysone in larval insects. These glands usually degenerate in the adult, and the follicle cells of the ovary are the source of the hormone in the adult. In many insects large amounts of ecdysteroids are stored in the developing oocyte as inactive conjugates. In these insects the embryo molts as it grows, and the stored ecdysteroids are activated and regulate these molts. In mosquitoes and some other Diptera, the ecdysone produced by the follicle cells enters the hemolymph and is important in the vitellogenic stage of egg development.

The rise in 20-hydroxyecdysone levels after the blood meal stimulates the vitellogenic phase of egg development, during which the oocyte develops the yolk needed by the developing embryo. Under the regulation of this hormone, the fat body produces large amounts of the major yolk protein, vitellogenin, which travels through the hemolymph to the oocyte, where it is taken up by receptor-mediated endocytosis (Chapter 25). The fat body also produces a serine carboxypeptidase and a cathepsin that are also taken up by the oocyte, where they are important in degradation of the vitellogenin during embryogenesis. A huge number of ribosomes and other cellular machinery are produced to support the synthesis of vitellogenin. The follicle cells of the oocyte produce the vitelline envelope proteins that form one of the layers of the chorion. The follicle cells also produce an enzyme, dopa decarboxylase, that is needed to harden and darken the chorion after oviposition. After taking a blood meal, host seeking is inhibited. There is some evidence that 20-hydroxyecdysone is involved in this inhibition (Bowen and Loess-Perez 1989), perhaps via an effect on the neuropeptide hormone Aea-HP-I (Chapter 20). Thus, the target tissues of 20-hydroxyecdysone are the fat body, the follicle cells and oocyte, and the brain.

In contrast to JH, ecdysone is a member of a large group of steroid hormones that are widely distributed in the animal kingdom. Ecdysteroids are common in many invertebrates, except the deuterosomes (echinoderms and chordates). The steroid hormones of vertebrates include estrogens and corticosteroids. The active form of ecdysone in most arthropods is 20-hydroxyecdysone (Fig. 24.2). Because it has a number of hydroxyl groups, the molecule is easily transported in the hemolymph. There is little evidence for proteins that specifically bind, and protect, these hormones in

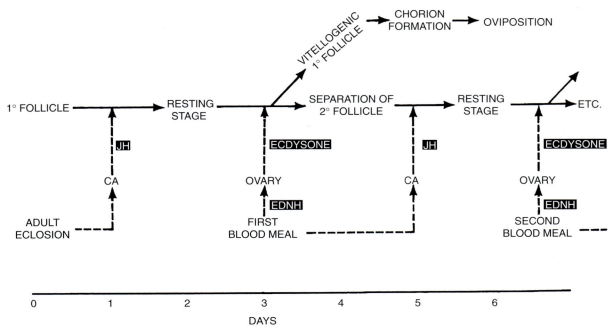

FIGURE 24.9 Mosquitoes can take several blood meals, each of which produces a batch of eggs. The endocrine basis for cyclic egg development is shown. Each blood meal results in the release of hormones that ensure not only the vitellogenic growth of one follicle but also the previtellogenic growth of the next.

insects. There are a number of enzymes that inactivate ecdysteroids, but they have not been studied in mosquitoes.

THE SECOND BLOOD MEAL

Egg production is cyclic in mosquitoes, meaning that after the first batch of eggs is laid, a second blood meal begins the process of egg development again (Fig. 24.9). The morphological basis for Figure 24.9 is the clear separation of the previtellogenic and vitellogenic stages of follicle growth by the need for a blood meal. The physiological basis for this relationship is the fact that after ecdysis, JH regulates previtellogenic growth of the follicle and that after the blood meal 20-hydroxyecdysone regulates vitellogenic growth. At a deeper level is the fact that within a day or two after a blood meal, the tissues involved lose their competence to respond to 20-hydroxyecdysone (and probably OEH) and must be exposed to JH to regain competency. Thus, each blood meal results in the release of hormones that ensure not only the vitellogenic growth of one follicle but also the previtellogenic growth of the next. The rise in 20-hydroxyecdysone not only stimulates the primary follicle to become vitellogenic, it also stimulates the

appearance of the secondary follicle. Similarly, the second rise in JH resets the system for the next blood meal by stimulating the events of previtellogenesis in the secondary follicle. This issue of tissue competence to respond to hormones is important because it indicates that the state of the target tissue is just as important as the presence of the hormone. "Competence" in molecular terms probably means the presence of appropriate hormone receptors and may also mean that the structure of the DNA in appropriate genes changes to allow an interaction with hormone receptor complexes.

Logically, one might expect the fat body to retain the large number of ribosomes and endoplasmic reticulum made in response to the first blood meal for reuse during the second cycle of egg development. Instead, almost all of these organelles are destroyed and recreated anew after the second blood meal. In this way the mosquito conserves scarce energy reserves obtained from the blood meal to ensure survival to the next blood meal.

There is considerable evidence that some mosquitoes may require several blood meals to complete full development of a single batch of oocytes. Such behavior has important implications for vector biologists because it increases the chances for transmission of pathogens. The effects of multiple meals on the endocrinology of egg development have not been investigated.

OTHER MOSQUITOES

Studies at the Shanghai Institute of Entomology by Zhu et al. (1980) on *Cx. pipiens pallens* and by Lu and Hagedorn (1986) on *An. albimanus* have obtained similar results to those described earlier for *Ae. aegypti*, suggesting that the endocrinology of egg development is similar in a wide variety of mosquitoes. Studies on the autogenous mosquito *Ae. atropalpus*, which develops a batch of eggs without a blood meal, provide an interesting contrast to the anautogenous *Ae. aegypti*. M. S. Fuchs and colleagues have shown that all of these major hormonal events also take place in the autogenous mosquito, *Ae. atropalpus*, including the involvement of JH during previtellogenic growth, OEH stimulation of ecdysone production by the ovary, and 20-hydroxyecdysone stimulation of vitellogenin synthesis. One major difference is that eclosion, rather than the blood meal, becomes the stimulus for the release of OEH. Another interesting difference is that *Ae. atropalpus* is much more sensitive to 20-hydroxyecdysone than is *Ae. aegypti*, which may indicate either some differences in hormone metabolism or more basic differences in the physiology of egg development in autogenous mosquitoes.

THE DIAPAUSING MOSQUITO

A number of adult mosquitoes diapause during the winter, including most *Anopheles*, *Culex*, and *Culiseta* in temperate regions. Diapause is induced by short day length and/or cool temperatures. It appears to be terminated by increasing day length. Diapausing mosquitoes show a reduced interest in host seeking, a cessation of oocyte development at an early previtellogenic stage, and an increased amount of lipid in the fat body. The endocrine basis of diapause is not well understood. From our earlier discussion about the role of JH in the early stages of egg development and feeding behavior, it is not surprising that topical application of JH to diapausing mosquitoes can induce host seeking and oocyte development to the resting stage. Readio et al. (1999) showed that the activity of the corpora allata was low in diapausing *Cx. pipiens*. But JH is most effective when it is applied to mosquitoes caught in December, when diapause is naturally terminating. This is a good example of the importance of the changing responsiveness of the target tissue to hormones. It also suggests that diapause is more than just

a lack of JH. Egg development involves several other hormones, as we have seen, and it is not known how the release of OEH or ecdysone is related to the termination of diapause. In contrast, in *D. melanogaster*, diapause can also be ended by topical application of JH, but this hormone is not involved in the normal process of diapause termination. Instead, 20-hydroxyecdysone titers rise as diapause is terminated.

Most diapausing mosquitoes do not feed, because they do not show host-seeking behavior. However, a small percentage of diapausing *An. freeborni* take blood meals, but the oocytes do not enter vitellogenesis. Instead, the reserves from the meal are used to increase lipid levels, thereby presumably improving survival. This has been called *gonotrophic dissociation*. It seems that in gonotrophic dissociation all of the normal endocrine regulatory processes of egg development must somehow be inactivated.

ENDOCRINE SYSTEMS

In its basic features, the regulation of egg development in the mosquito is similar to that seen in the frog or chicken. In the frog, seasonal changes in day length signal the beginning of the reproductive season. In response, the peptide gonadotropins, follicle-stimulating hormone (FSH) and luteinizing hormone (LH), are released from the pituitary and stimulate the production of the steroid hormone estrogen by the follicle cells of the ovary. In response to estrogen, the liver produces vitellogenin, which is taken up by the growing oocyte. The competence of the liver to respond to estrogen is controlled by thyroxin, to which the liver is exposed during metamorphosis. The remarkable similarities this has led to the regulation of egg development in the mosquito provide some insights into the endocrine control of physiological systems.

The first similarity is the role of the brain as a master control via release of peptide hormones, demonstrating the importance of integrating environmental changes with physiology. In the case of the frog, the brain is responding to the changing photoperiod; in the case of the mosquito, it is responding to eclosion and the blood meal. Second, in both systems the master controls are peptide hormones released from the brain. Peptide hormones typically regulate cellular response by activating an enzymatic pathway, in this case steroid production by the follicle cells in both the frog and the mosquito. Third, steroid hormones are used in both systems to stimulate a massive synthesis of protein by the liver or fat body. Steroids typically regulate the expression of genes, in this case the vitellogenin genes in the fat body or liver. And finally, the issue of tissue competence is evident in both systems. In both, competence is regulated by a separate hormone, thyroxin or JH. Why not simply use one hormone instead of such a hierarchical system involving at least three? It is likely that these complex systems allow feedback and fine regulation of the response and may also promote amplification of the response.

PUTTING ENDOCRINES IN THEIR PLACE

It may seem from this discussion that these various hormones are in charge of molting and reproduction. In fact, one of the first scientists to study the role of endocrines in metamorphosis discovered that this was not the case. Stefan Kopeć in 1922 found that the brain produced a factor that was necessary for metamorphosis of the salt marsh caterpillar, *Lymantria dispar*. But he also found that the target tissues had to be capable of responding. Careful studies of metamorphosis have shown that the target tissues become capable of responding a few hours before the hormone is released into the hemolymph. This is believed to be due to the programmed appearance of receptors in the target cells and has been described as "competence" in the earlier discussion. The key points are that each cell of the body has its own physiological rhythms and that the integration of the whole is not a function of the hormone by itself. Equally important is the state of target cells that have mechanisms capable of regulating the appearance of biologically important molecules that is integrated with the production and release of hormones.

PARASITES OF THE MOSQUITO

Parasites of insects, including mosquitoes, have been shown to use, or to manipulate, the hormones of the host to their own advantage. Two cases are known where parasites of mosquitoes use the ecdysone that appears after a blood meal to regulate the production of reproductive spores. One of these is a microsporidian, *Amblyospora* sp., which has a life cycle that alternates between copepods and mosquito larvae. In the male adult mosquito, the parasite sporulates, killing the host. In females, the parasite does not sporulate until after the blood meal, when the rise in 20-hydroxyecdysone levels stimulate sporulation. These spores enter the developing oocytes, resulting in vertical transmission to the next generation of mosquitoes (Lord and Hall 1983).

The second case involves a fungus, *Coelomomyces stegomyiae*, which also alternates between mosquito larvae and copepods. Mosquitoes that survive the infection and become adults have fungal hyphae throughout their body cavities. Spores do not develop until after a blood meal, and again it is 20-hydroxyecdysone that stimulates sporulation. In this case, however, the spores fill the developing follicles and the female deposits spores instead of eggs. Thus, the fungus uses the female mosquito as a mechanism for dissemination (Lucarotti 1992).

CONCLUSIONS

It should be clear that the endocrinology of mosquitoes is an area of considerable importance to vector biologists because hormones impact reproduction, feeding, behavior, nutrition, and parasite development. An exciting challenge for the future will be to build a firm genetic foundation for our understanding of physiological systems. With genetic transformation and genomic sequences in hand this is now feasible.

Readings

Brown, M. R., and Lea, A. O. 1989. Neuroendocrine and midgut endocrine systems in the adult mosquito. *Adv. Dis. Vector Res.* 6: 29–58.

Clements, A. N. 1992. *The Biology of Mosquitoes.* Vol. 1, Chap. 6. Chapman and Hall, London.

Gilbert, L. I., Granger, N. A., and Roe, R. M. 2000. The juvenile hormones: Historical facts and speculations on future research directions. *Insect Biochem. Molec. Biol.* 30: 17–644.

Hagedorn, H. H. 1994. The endocrinology of the adult female mosquito. *Adv. Dis. Vector Res.* 10: 109–148.

Mitchell, C. J. 1988. Occurrence, biology and physiology of diapause in overwintering mosquitoes. In *The Arboviruses: Epidemiology and Ecology*, Vol. 1 (T. P. Monath, ed.). CRC Press Boca Raton, Florida, pp. 191–217.

Further Readings

Borovsky, D., Carlson D. A., Hancock, R. G., Rembold, H., and Van Handel, E. 1994. De novo biosynthesis of juvenile hormone III and I by the accessory glands of the male mosquito. *Insect Biochem. Molec. Biol.* 24: 437–444.

Bowen, M. F., and Loess-Perez, S. 1989. A reexamination of the role of ecdysteroids in the development of host-seeking inhibition in blood-fed *Aedes aegypti* mosquitoes. In *Host-Regulated Developmental Mechanisms in Vector Arthropods* (D. Borovsky and A. Spielman, eds.). University of Florida Press, Vero Beach, pp. 286–291.

Brown, M. R., and Cao, C. 2001. Distribution of ovary ecdysteroidogenic hormone I in the nervous system and gut of mosquitoes. *J. Insect Sci.* 1: 3. Available online at www.insectscience.org/1.3.

Brown, M. R., Graf, R., Swiderek, K. M., Fendley, D., Stracker, H. T., Champagne D. E., and Lea, A. O. 1998. Identification of a steroidogenic neurohormone in female mosquitoes. *J. Biol. Chem.* 273: 3967–3971.

Feinsod, F. M., and Spielman, A. 1980. Nutrient-mediated juvenile hormone secretion in mosquitoes. *J. Insect Physiol.* 26: 113–117.

Hagedorn, H. H., O'Connor, J. D., Fuchs, M. S., Sage, B., Schlaeger, D. A., and Bohm, M. K. 1975. The ovary as a source of alpha ecdysone in an adult mosquito. *Proc. Natl. Acad. Sci. USA* 72: 3255–3259.

Hagedorn, H. H., Shapiro, J. P., and Hanoaka, K. 1979. Ovarian ecdysone secretion is controlled by a brain hormone in an adult mosquito. *Nature* 282: 92–94.

Hoffmann, J. A., and Lagueux, M. 1985. Endocrine aspects of embryonic development in insects. In *Comprehensive Insect Physiology, Biochemistry and Pharmacology*, Vol. 1 (G. A. Kerkut and L. I. Gilbert, eds.). Pergamon Press, New York, pp. 435–460.

Hotchkin, P. G., and Fallon, A. M. 1987. Ribosome metabolism during the vitellogenic cycle of the mosquito *Aedes aegypti*. *Biochim. Biophys. Acta* 924: 352–359.

Jenkins, S. P., Brown, M. R., and Lea, A. O. 1992. Inactive prothoracic glands in larvae and pupae of *Aedes aegypti*: Ecdysteroid release by tissues in the thorax and abdomen. *Insect Biochem. Molec. Biol.* 22: 553–559.

Lassiter, M. T., Apperson, C. S., Crawford, C. L., and Roe, R. M. 1994. Juvenile hormone metabolism during adult development of *Culex quinquefasciatus*. *J. Med. Entomol.* 31: 586–593.

Lea, A. O. 1967. The medial neurosecretory cells and egg maturation in mosquitoes. *J. Insect Physiol.* 13: 419–429.

Lord, J. C., and Hall, D. W. 1983. Sporulation of *Amblyospora* (Microspora) in female *Culex salinarius*: Induction by 20-hydroxyecdysone. *Parasitology* 87: 377–383.

Lu, Y. H., and Hagedorn, H. H. 1986. Egg development in the mosquito *Anopheles albimanus*. *Int. J. Invert. Reprod. Develop.* 9: 79–94.

Lucarotti, C. J. 1992. Invasion of *Aedes aegypti* ovaries by *Coelomomyces stegomyiae*. *J. Invert. Pathol.* 690: 176–184.

Masler, E. P., Fuchs, M. S., Sage, B., and O'Connor, J. D. 1980. Endocrine regulation of ovarian development in the autogenous mosquito, *Aedes atropalpus*. *Gen. Comp. Endocrinol.* 41: 250–259.

Niu, L. L., and Fallon, A. M. 2000. Differential regulation of ribosomal protein gene expression in *Aedes aegypti* mosquitoes before and after the blood meal. *Insect Mol. Biol.* 9(6): 613–623.

Noriega, F. G., Shah, D. K., and Wells, M. A. 1997. Juvenile hormone controls early trypsin gene transcription in the midgut of *Aedes aegypti*. *Insect Molec. Biol.* 6: 63–66.

Readio, J., Chen, M-H., and Meola, R. 1999. Juvenile hormone biosynthesis in diapausing and nondiapausing *Culex pipiens*. *J. Med. Entomol.* 36: 355–360.

Veenstra, J. A., and Costes, L. 1999. Isolation and identification of a peptide and its cDNA from the mosquito *Aedes aegypti* related to *Manduca sexta* allatotropin. *Peptides* 20(10): 1145–1151.

Zhu, X. X., Chen, Z.-F., and Cao, M.-X. 1980. Endogenous molting hormone level and vitellogenin synthesis during adult stage of the mosquito *Culex pipiens*. *Cont. Shanghai Inst. Entomol.* 1: 63–68.

25

Vitellogenesis of Disease Vectors, From Physiology to Genes

ALEXANDER S. RAIKHEL

INTRODUCTION

Vitellogenesis is a central event of egg maturation. Yolk protein precursors (YPPs) are produced in large amounts by female tissues, mostly by the fat body, accumulated by developing oocytes, and eventually used as a source of nutrients for developing embryos. During vitellogenesis, female oviparous animals devote most of their resources to synthesizing YPPs. Vector arthropods obtain these resources from digestion of protein-rich food, blood. Hematophagous arthropods require a blood meal to activate numerous genes essential for digestion of the blood, synthesis of yolk protein precursors, and, ultimately, production of eggs. As a consequence of feeding on blood, they may acquire pathogens and serve as vectors for numerous human diseases. Elucidation of the physiological and molecular mechanisms underlying the blood meal regulation of events and genes essential for vitellogenesis and egg maturation is of paramount importance in our efforts to develop novel approaches to vector and pathogen control.

The strict requirement of a blood or protein meal for the initiation of vitellogenesis results in a highly regulated cyclicity of egg production as each cycle is tightly coupled with food intake. Using protein-rich food enables vectors to achieve a high productivity of eggs. A unique set of adaptations at the morphological, physiological, and genetic levels is required in order to produce enormous amounts of YPPs and to load them into rapidly developing oocytes. These events are best understood in the yellow fever mosquito, *Ae. aegypti*, and are reviewed in this chapter. Unique features of vitellogenesis in other major groups of arthropod vectors are also discussed.

AEDES AEGYPTI AS A MODEL

The reproductive physiology of the female yellow fever mosquito, *Aedes aegypti*, has been the focus of intensive research for several decades. A number of factors have contributed to its becoming the model insect vector of choice. (1) It is an important vector of several human tropical diseases worldwide. (2) *A. aegypti* is a floodwater desert mosquito, so its eggs are adapted to withstand desiccation for long periods. The ability to store eggs and synchronize hatching makes it much easier to rear and manipulate in the laboratory than other disease vectors, so it has served as a model for medically important hematophagous insects. With the advent of novel techniques such as transgenesis, these features make *Ae. aegypti* an invaluable model insect vector. (3) The blood meal triggers events leading to synchronous oogenesis, greatly facilitating experimental dissection of the complex physiological events associated with insect reproduction. This synchronicity also involves the tissue-, sex-, and stage-specific activation of a number of genes, making it an excellent model for studying the molecular mechanisms of gene regulation in insect vitellogenesis.

THE VITELLOGENIC CYCLE

Aedes aegypti mosquitoes use the resources from digested vertebrate blood to synthesize yolk protein precursors (YPPs) for provisioning the developing oocytes, a process referred to as *vitellogenesis*. It requires the precise coordination of a number of complex physiological activities of the fat body and the ovary (Fig. 25.1).

When the adult emerges from the pupal stage, the paired corpora allata (CA) begin to secrete juvenile hormone (JH), and its titers peak at 2 days after eclosion (Fig. 25.1). A rise in JH titer in turn activates events that prepare the mosquito for vitellogenesis: The fat body is remodeled into an efficient protein factory and becomes responsive to 20-hydroxyecdysone (20E); the ovary becomes equipped with the endocytic machinery for massive and specific protein uptake and

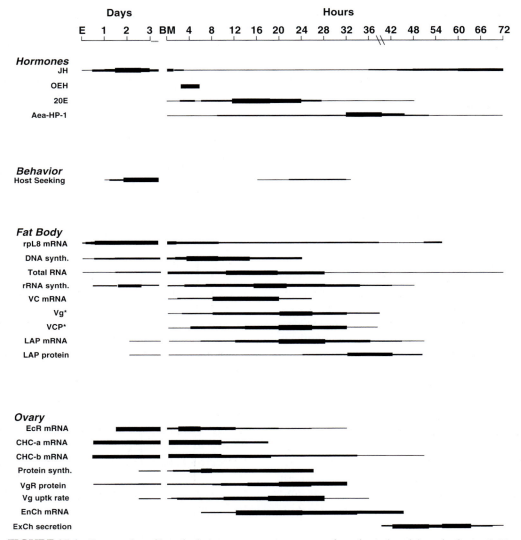

FIGURE 25.1 Temporal profiles of relative concentrations, rates of synthesis (synth.), and other activities with respect to relative hormone titers in *Aedes aegypti* after eclosion (E) and after a replete blood meal (BM). JH, juvenile hormone-III; OEH, ovarian ecdysteroidogenic hormone I; 20E, 20-hydroxyecdysone; Aea-HP-I, *A. aegypti* head peptide-I; rpL8, ribosomal protein L8; VC, vitellogenin convertase; Vg*, vitellogenin protein and mRNA (profiles are similar); VCP*, vitellogenic carboxypeptidase protein and mRNA (profiles are similar); LAP, lysosomal aspartic protease; CHC, clathrin heavy chain, isoform a or b; VgR, vitellogenin receptor; Vg uptk rate, rate of vitellogenin uptake by oocyte; EnCh, endochorion gene *15-a*; ExCh, exochorion. —, State-of-arrest of indefinite duration. \\, Scale of time intervals changes from 4 h to 6 h per tick mark. From Sappington and Raikhel (1999), with permission.

becomes responsive to ovarian ecdysiotropic hormone (OEH) to produce ecdysone; and the female becomes receptive to mating and responsive to host stimuli. JH production is then halted by a CA inhibitory factor released from arrested-stage ovaries and, together with degradation of existing JH by a specific esterase, results in a gradual decline in titers throughout the period of arrest. When the mosquito takes a blood meal, JH titers rapidly decline to a background level due to a dramatic increase in JH esterase activity.

If the blood meal is large enough (about 5 μL), stretch receptors in the posterior midgut inhibit host-seeking behavior. Smaller blood meals may not cause enough distension to trigger this inhibition, and host seeking continues until a sufficient volume of blood is ingested to activate the stretch receptors. A full blood meal results in continuation of vitellogenesis, which enters its trophic phase. It is characterized by rapid synthesis of YPPs in the fat body and rapid growth of the oocyte. It begins when OEH is released from the head and activates ovarian ecdysone synthesis, about 4–6 h after the blood meal (Fig. 25.1). OEH (referred to in the earlier literature as egg development neurosecretory hormone, or EDNH) is produced in the medial neurosecretory cells of the brain during the previtellogenic period; it then is transported to the corpora cardiaca (CC), where it is stored. The release of OEH from the CC requires a combination of several signals: neural stimuli from a distended midgut, increased levels of amino acids in the hemolymph as the blood meal is digested, and an unidentified hormone secreted by the ovaries (the OEH-releasing factor). After OEH release, ecdysteroid titers increase sharply to a peak at 18–24 h after the blood meal, followed by a rapid decline between 24 and 30 h after the blood meal. 20E regulates the transcription of various genes in both the fat body and ovary.

The vitellogenic phase ends between 30 and 36 h after the blood meal. During this time, termination of both YPP synthesis in the fat body and uptake by the oocyte occurs in synchrony. The fat body's protein synthetic machinery is degraded, and the cells are remodeled for storage of lipid and glycogen in preparation for the next gonotrophic cycle. The oocyte ceases to internalize YPPs when it loses contact with the hemolymph due to the closing of interfollicular channels and the fusion of the endochorionic plaques. The outer layer of the eggshell is secreted and the egg is prepared for oviposition (Fig. 25.1).

Vitellogenic Events in the Fat Body

The fat body is the major metabolic organ of insects, analogous to the vertebrate liver. It is a diffuse organ, occurring as sheets or lobes of tissue lining the body cavity, primarily in the abdomen. Its hormonally regulated functions are diverse, changing to meet the specific requirements of different phases of the life cycle. In the mosquito, during the previtellogenic phase, JH signals the fat body to prepare for vitellogenesis. Fat body trophocytes (the major fat body cell type) respond to JH by increasing ploidy level from 2n to 4n, while another 20% become octoploid by the third day after eclosion. In addition, ribosomes begin to accumulate. Total RNA increases by about 50% over the first 3 days after eclosion, with the highest rate of transcription occurring within the first 2 h. Not only is rRNA produced, but mRNA increases twofold over the first 3 days as well. These are likely mRNAs of housekeeping genes. For example, mRNA that encodes the ribosomal protein L8 is neither immediately translated nor degraded; rather, it accumulates and is translated only after a blood meal, during a second cycle of ribosome accumulation.

A low rate of YPP gene transcription and synthesis begins in the fat body as early as 30 minutes after the blood meal. By 3 h after the blood meal, secretory granules carrying vitellogenin and other YPPs appear in the trophocytes. The trophocyte nucleolus becomes multilobed, and ribosome proliferation resumes (Fig. 25.1). Sometime between 4 and 8 h after the blood meal, ribosomal RNA production begins to accelerate, peaking at a three- to fourfold-higher rate 12–18 h later and then decreasing to very low rates by 48 h. By 12 h, rough endoplasmic reticulum occupies most of the cytoplasm. At 3 h, Golgi complexes begin to increase in number, peaking about 18–24 h. By this time they contain cisternae two to three times wider than those found during the initiation phase of vitellogenesis.

Several species of YPPs are synthesized and secreted by the vitellogenic *Ae. aegypti* fat body. By far the most abundant is vitellogenin, a large protein that serves as the major source of amino acids for the future embryo. It is homologous to the vitellogenins of many other oviparous animals, including most insects, nematodes, and vertebrates. The other YPPs are the 53-kDa vitellogenic carboxypeptidase (VCP) and a 44-kDa vitellogenic cathepsin B (VCB), which is a thiol protease precursor. Both VCP and VCB are proenzymes internalized by the oocyte and activated at the onset of embryogenesis to digest vitellin (the crystallized storage form of vitellogenin). All three YPPs follow the same kinetics, with expression peaking at 24–28 h PBM. The level of the major lipid carrier protein, lipophorin, in the hemolymph increases dramatically as well. Lipophorin is also accumulated by *Ae. aegypti* oocytes.

The biosynthesis and posttranslational processing of *Aedes* vitellogenin has been studied in detail. The 6.5-kb vitellogenin mRNA transcript is translated in the RER as a 224-kDa pre-pro-vitellogenin. The pre-pro-vitellogenin is cotranslationally glycosylated and posttranslationally phosphorylated to produce a 250-kDa pro-vitellogenin intermediate (Fig. 25.2). The

pro-vitellogenin is rapidly cleaved in the rough endoplasmic reticulum into two fragments of 190 kDa and 62 kDa by vitellogenin convertase, a fat body–specific 140-kDa member of the subtilisin-like proprotein convertase family of enzymes. The pro-vitellogenin is cleaved at the dibasic amino acid motif, RYRR, which is preceded by a predicted β-turn (Fig. 25.2B). The

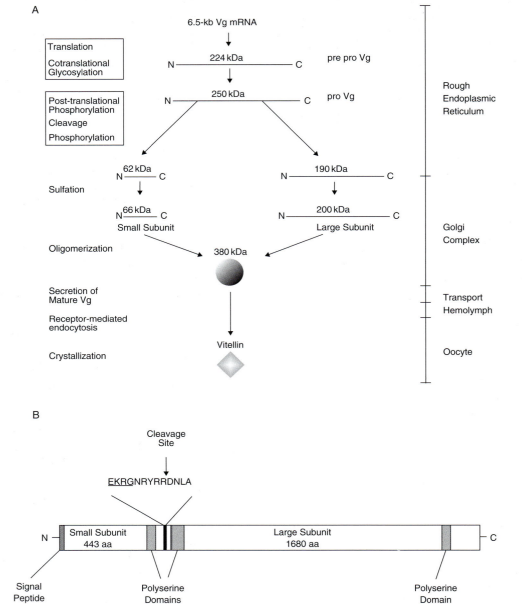

FIGURE 25.2   Co- and posttranslational processing (A) of *Aedes aegypti* pre-pro-vitellogenin (B), from synthesis in the fat body to crystallization as vitellin in the oocyte. The cleavage site is indicated by an arrow in (B) and the amino acid sequence (bold) in the context of a β-turn (underlined), which is recognized by the cleavage enzyme, vitellogenin convertase, is presented at the base of the arrow. From Sappington and Raikhel (1999) with permission.

combination of a RX^R/K^R motif near a β-turn is required for recognition and cleavage by the subtilisin-like enzyme from the convertase. Both vitellogenin peptides are modified further in the Golgi complex, where they undergo additional phosphorylation and tyrosine sulfation, increasing in size to that of the mature large (200 kDa) and small (66 kDa) subunits (apoproteins) of vitellogenin. These subunits oligomerize to form the mature 380-kDa vitellogenin, which is packaged into secretory granules (Fig. 25.2A). VCP and VCB follow the same pathway through the rough endoplasmic reticulum and Golgi complex as vitellogenin, being packaged in and released from the same secretory granules (Fig. 25.3).

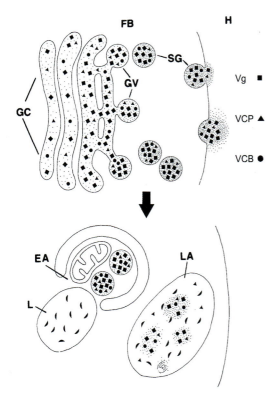

FIGURE 25.3 Secretory pathway of yolk protein precursors in the vitellogenic fat body of *Aedes aegypti*. (A) Fat body (FB) trophocytes synthesize three yolk protein precursors: vitellogenin (Vg), vitellogenic carboxypeptidase (VCP), and vitellogenic pro-cathepsin B (VCB). All three proteins are processed simultaneously through Golgi complexes (GC), packed together into secretory granules, which are secreted into the hemolymph (H), and then delivered to the oocyte (GV, Golgi condensing vacuoles; SG, secretory granules). (B) At the termination of vitellogenesis, the cessation of biosynthetic activity of trophocytes is characterized by formation of early (EA) and late (LA) autophagosomes that digest secretory granules and other organelles (L, lysosome). From Snigirevskaya et al. (1997a) with permission.

Secretory granules empty their contents into the hemolymph almost immediately after leaving the Golgi complex. The amount of vitellogenin transcript and protein in the fat body peaks between 24 and 30 h post–blood meal (PBM) and then declines rapidly, with vitellogenin synthesis halting altogether by 36 h. The mechanism for terminating vitellogenin gene transcription is not yet clear. The interruption of vitellogenin production in the fat body does not depend simply on turning off YPP gene transcription, but involves the active degradation of synthetic and secretory organelles in the trophocytes by the lysosomal system (Fig. 25.3). At the end of vitellogenesis, beginning between 30 and 36 hours after the blood meal, the biosynthetic machinery in the fat body is broken down by lysosomal enzymes. The number of lysosomes in the trophocyte cytoplasm increases many times from 24 to 36 h, and the activities of several lysosomal enzymes, including lysosomal aspartic protease, increase dramatically between 24 and 27 h, remaining high through at least 48 h. Mosquito lysosomal aspartic protease has been studied in detail at the molecular level. Enhanced fat body–specific transcription of the lysosomal aspartic protease gene begins about 6 h after the blood meal, with mRNA levels peaking at 24–30 h and declining rapidly after 36 h. 20E negatively regulates lysosomal aspartic protease mRNA translation, so the protein does not begin to accumulate until 20E titers start to fall after 24 h (Fig. 25.1).

Molecular Endocrinology of Vitellogenesis

The hemolymph titers of ecdysteroids in female mosquitoes are correlated with the rate of YPP synthesis in the fat body (Fig. 25.1). The ecdysteroid titers are only slightly elevated at 4 h after a blood meal; however, they rise sharply at 6–8 h and reach their maximum level at 18–24 h. Numerous studies have clearly established that the ecdysteroid control of vitellogenesis is a central event in the blood meal–activated regulatory cascade leading to successful egg maturation. The understanding of the molecular basis of ecdysteroid control of egg maturation in mosquitoes is important for future development of novel approaches to mosquito control.

Analysis of ecdysone effects on polytene chromosome puffing patterns in the late larval and prepupal salivary gland of *Drosophila* have suggested that the initial activation of a small number of early ecdysone-inducible genes leads to subsequent inducement of a large number of late target genes. Elucidation of this genetic hierarchy at the molecular level has led to the identification of the ecdysone receptor as a heterodimer of two nuclear receptors, the ecdysone

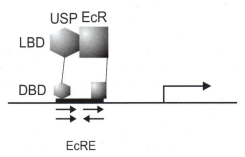

FIGURE 25.4 Two nuclear receptors, ecdysteroid receptor (EcR) and ultraspiracle (USP), form the functional insect ecdysteroid receptor. Each consists of several domains characteristic of the nuclear receptor superfamily of transcription factors. Their ligand-binding domains (LBD) are involved in dimerization, while DNA-binding domains (DBD) bind to the sequence-specific nucleotide sequences called *ecdysone response element* (EcRE). EcRE may contain either direct or inverted repeats (indicated by arrows) of the consensus sequence AGGTTA with a single or multiple spacer nucleotide. Only EcR binds 20E via its LBD.

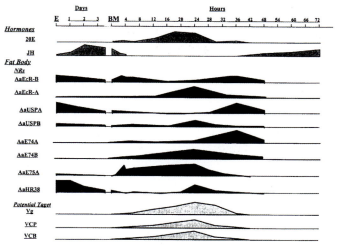

FIGURE 25.5 Transcript profiles of major genes of the ecdysteroid regulatory hierarchy during the first cycle of egg maturation in the mosquito *Aedes aegypti*. The previtellogenic period begins at eclosion (E) of the adult female. During the first 3 days after eclosion, both the fat body and the ovary become competent for subsequent vitellogenesis. The female then enters a state of arrest; yolk protein precursors, Vg (vitellogenin) and vitellogenic carboxypeptidase (VCP) are not synthesized during the previtellogenic period. Only when the female mosquito ingests blood (BM) is vitellogenesis initiated. *Hormones*: hormonal titers of juvenile hormone (JH) and ecdysteroids (20E) in *A. aegypti* females. *Fat Body*: relative levels of RNAs for nuclear receptors (*NRs*) determined by RT-PCR. AaEcR-A, ecdysone receptor isoform A; AaEcR-B, ecdysone receptor isoform B; AaUSPA, ultraspiracle isoform A; AaUSPB, ultraspiracle isoform B; AaE74B, E74 isoform B; AaE74A, E74 isoform A; AaE75A, E75 isoform A; AaHR38, mosquito homologue of *Drosophila* HR38. Modified from Raikhel et al. (1999) with permission.

receptor (EcR) and ultraspiracle (USP), the insect homologue of the vertebrate retinoid X receptor (Fig. 25.4). Furthermore, these studies have shown that the action of the ecdysone-receptor complex is indeed mediated by transcription factors such as *BR-C*, *E74*, and *E75* that are the early genes involved in regulation of late gene expression (e.g., vitellogenin). The ecdysone-mediated regulatory network is further refined by the presence of genes that are involved in setting up the timing and stage specificity of gene activation by this genetic hierarchy.

Consistent with the proposed role of 20E in activating mosquito vitellogenesis, experiments using an in vitro fat body culture have shown that physiological doses of 20E (10^{-7}–10^{-6} M) activate two *YPP* genes, vitellogenin and *VCP*. Transcripts of two different isoforms of EcR, EcR-A and EcR-B, are present in pre- and vitellogenic ovaries and fat bodies. Transcripts encoding two different USP isoforms are differentially expressed in these tissues as well (Fig. 25.5). The mosquito EcR-USP heterodimer has been shown to bind to various EcR response elements to modulate ecdysteroid regulation of target genes. These elements are present in vitellogenin and *VCP* genes.

Using the protein synthesis inhibitor, cycloheximide, has demonstrated that the activation of *YPP* genes in the mosquito fat body requires protein synthesis. Several genes of the ecdysteroid regulatory hierarchy are conserved between vitellogenesis in mosquitoes and metamorphosis in *Drosophila*. The *Ae. aegypti* homologue of the *Drosophila E75* gene, AaE75, a representative of the early gene level in the ecdysteroid response hierarchy (Fig. 25.6), is expressed in

the ovary and fat body following a blood meal. Similar to *Drosophila*, there are three AaE75 isoforms in *Ae. aegypti* that are inducible by 20E. Interestingly, in the mosquito fat body, AaE75 transcripts show two peaks, with a small peak coinciding with the first peak of 20E. The correlation between midvitellogenic expression of the AaE75 and vitellogenin genes suggests that the *YPP* genes are direct targets of AaE75 (Fig. 25.5). Indeed, analysis of the vitellogenin gene regulatory regions has identified an E75 binding site within a region required for high-level vitellogenin expression and a close correlation between endogenous site-specific binding activity and AaE75 expression. These findings suggest that AaE75 mediates ovary and fat body ecdysteroid responses and that ecdysteroid-triggered regulatory hierarchies, such as those implicated in the initiation of metamorphosis, are reiteratively utilized in the control of the reproductive ecdysteroid response.

Two isoforms of the *Drosophila* transcription factor E74 homologue, which share a common C-terminal

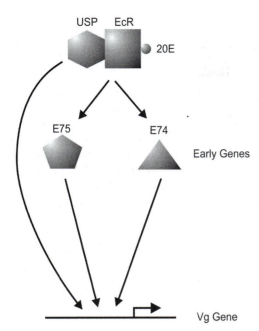

FIGURE 25.6 Direct–indirect regulation of the vitellogenin precursor genes by 20-hydroxyecdysone in the fat body of *Aedes aegypti*. After binding 20E, the EcR-USP heterodimer activates early genes, *E74* and *E75*, as well as directly acts on the vitellogenin gene allowing its expression. In turn, the products of *E74* and *E75* genes act as powerful activators of vitellogenin gene transcription. From Raikhel et al. (2002) with permission.

DNA–binding domain yet have unique N-terminal sequences, are present in the mosquito *Ae. aegypti*. They exhibit a high level of identity to DmE74 isoforms A and B and show structural features typical for members of the Ets transcriptional factor superfamily. Furthermore, both AaE74 isoforms bind to a *Drosophila* E74 binding site with the consensus motif C/AGGAA. The AaE74B transcript is induced by a blood meal–activated hormonal cascade in fat bodies and peaks at 24 h after the blood meal, the peak of vitellogenesis (Fig. 25.5). In contrast, AaE74A is activated at the termination of vitellogenesis, exhibiting a peak at 36 h in the fat body and 48 h in the ovary. These findings suggest that AaE74A and AaE74B isoforms play different roles in the regulation of vitellogenesis in mosquitoes, as an activator and a repressor of YPP gene expression. The vitellogenin gene regulatory portion contains E74 binding sites within the region required for the high level of vitellogenin expression.

Analysis of the 5'-upstream regulatory region of the mosquito vitellogenin gene has unexpectedly revealed the presence of putative binding sites for EcR-USP along with those to the early gene, E74 and E75, suggesting a complex regulation of this gene through a combination of direct and indirect hierarchies. The vitellogenin gene contains the functionally active EcRE that binds the heterodimer EcR-USP. A direct repeat with a 1-bp spacer (DR-1) with the sequence AGGC-CAaTGGTCG is the major part of the EcRE in the vitellogenin gene. Thus, the *Aedes aegypti* vitellogenin gene is the target of a direct–indirect regulation by 20E (Fig. 25.6).

In *Ae. aegypti*, a preparatory, previtellogenic, developmental period is required for the mosquito fat body to attain competence for 20E responsiveness as well as for the adult female to attain competence for blood feeding. *βFTZ-F1*, the orphan nuclear factor implicated as a competence factor for stage-specific response to ecdysteroid during *Drosophila* metamorphosis, likely serves a similar function during mosquito vitellogenesis. A homologue of *Drosophila βFTZ-F1* is expressed highly in the mosquito fat body during pre- and postvitellogenic periods when ecdysteroid titers are low. *AaFTZ-F1* expression is inhibited by 20E in vitro and superactivated by its withdrawal. After the in vitro superactivation of *AaFTZ-F1* expression in the previtellogenic mosquito fat body, a secondary 20E challenge results in superinduction of the early *AaE75* gene and the late target VCP gene. Moreover, a highly conserved homologue of *Drosophila HR3* (*AHR3*) is expressed in the vitellogenic tissues of the female mosquito. The expression of *AHR3* correlates with the titer of 20E, peaking at late pupa and adult vitellogenic female 24 h after the blood meal, preceding *AaFTZ-F1* expression peaks. The orphan nuclear receptor *DHR3*, an early–late gene with induction characteristics including properties of both early and late genes, is one of the key genes in the ecdysone-mediated genetic regulatory network. Recent studies have revealed that during insect metamorphosis *DHR3* has a dual role in repressing the early genes while activating *βFTZ-F1*. Thus, the regulation and function of *AaFTZ-F1* during mosquito reproduction closely resembles that shown at the onset of *Drosophila* metamorphosis, and *FTZ-F1* is therefore part of a conserved and broadly utilized molecular mechanism controlling the stage specificity of ecdysteroid response.

Structure and Function of Regulatory Regions of Yolk Protein Genes

In *Ae. aegypti*, vitellogenin is encoded by a family of five genes. The vitellogenin A1 gene has been cloned and analyzed. In this gene, a 6369-bp vitellogenin-coding region is interrupted by two short introns. *Drosophila* and *Aedes* transformation, as well as DNA-binding assays, have been used to identify *cis*-regulatory sites in the vitellogenin gene regulatory region, responsible for stage- and fat body–specific activation of this gene via a blood-meal-triggered cascade. These analyses revealed three regulatory regions in the 2.1-kb upstream portion of the vitellogenin gene that are sufficient to bring about the characteristic pattern of vitellogenin gene expression (Fig. 25.7). The proximal region, adjacent to the basal transcription start site, contains binding sites to several transcription factors: EcR/USP, GATA (GATA transcription factor), C/EBP (CAAT-binding protein) and HNF-3/FKH (hepatocyte nuclear factor 3/forkhead transcription factor). This region is required for the correct tissue- and stage-specific expression. It appears that a combinatorial action of these transcription factors is essential to bring a fat body–specific expression. EcR/USP acts as a timer allowing the gene to be turned on. However, the level of expression driven by this regulatory gene region is low. The median region carries the sites for early gene factors E74 and E75. It is responsible for a stage-specific hormonal enhancement of the vitellogenin expression. The addition of this region increases expression of the gene in the hormonally controlled manner.

Finally, the distal region of the 2.1-kb upstream portion of the vitellogenin gene is characterized by multiple response elements for a transcription factor called GATA. In transgenic experiments utilizing both *Aedes* and *Drosophila*, this GATA-rich region has been found to be required for the extremely high expression levels characteristic of the vitellogenin gene (Fig. 25.7).

In the mosquito, VCP is the second most abundant YPP. The VCP gene is expressed in synchrony with the vitellogenin gene: Transcripts of both appear within 1 h post–blood meal, reach maximal levels at 24 h, and rapidly decline to background levels by 36 h (Figs. 25.1

and 25.5). 20-Hydroxyecdysone activates both of these genes in fat body organ culture and in an in vitro cell transfection assay. Furthermore, the same hormone concentration, 10^{-6} M, is required for maximal activation of both these genes. In both genes most of the putative regulatory elements are located in the 2-kb upstream region of the transcription initiation site. Comparative analyses have revealed conservation of putative binding sites for transcriptional factors responsible for tissue- and stage-specific expression in putative regulatory regions of both genes, including EcRE, E74, GATA, C/EBP, and HNF-3/FKH.

Another level in regulation of the *YPP* gene expression in the mosquito fat body is provided by the control of transcription factors themselves. Levels of transcripts and active factors critical for transcription of *YPP* genes such as EcR, USP, E74, E75, and GATA are greatly enhanced during the vitellogenic cycle, themselves being targets of the blood meal–mediated regulatory cascade. This additional control provides yet another mechanism for amplification of *YPP* gene expression levels.

Molecular Mechanisms of the Previtellogenic Arrest

Anautogeny is one of most fundamental phenomena underlying the vectorial capacity of mosquitoes. It involves numerous adaptations from host-seeking behavior to a complete repression of egg maturation prior to a blood meal. The latter is poorly understood at the biochemical and molecular biological levels. Molecular studies of mosquito vitellogenesis have made it possible to elucidate two key mechanisms involved in the *YPP* gene repression in the fat body of the previtellogenic female mosquito.

An important adaptation for anautogeny is the establishment of previtellogenic developmental arrest (the state of arrest) preventing the activation of *YPP* genes in previtellogenic competent females prior to blood feeding. When an isolated fat body dissected from the previtellogenic female mosquito is incubated in vitro in the presence of physiological doses of 20E, the *YPP* genes are activated. In contrast, only the injection of 20E at supraphysiological dosages could stimulate some expression of these genes in vivo. Therefore, an inhibition of the 20E-signaling pathway is the essential part of the state of arrest, which may be maintained in vivo by undetermined factors. The disappearance of these factors or the appearance of additional, unidentified factors secreted in response to a blood meal may play a crucial role in the release of vitellogenic tissues from the state of arrest.

In *Ae. aegypti*, both AaEcR and AaUSP proteins are abundant in nuclei of the previtellogenic female fat

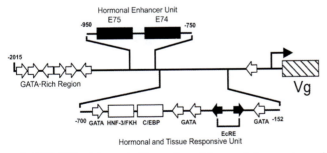

FIGURE 25.7 Schematic illustration of the regulatory regions of the *Aedes aegypti* vitellogenin gene. Numbers refer to nucleotide positions relative to the transcription start site. C/EBP, response element of C/EBP transcription factor; EcRE, ecdysteroid response element; E74 and E75, reponse elements for respective early gene product of the ecdysone hierarchy; GATA, reponse element for GATA transcription factor; HNF-3/FKH, response element for HNF3/forkhead factor; Vg, coding region of the vitellogenin gene. From Kokoza et al. (2001) with permission.

body at the state of arrest. In contrast, the EcR-USP heterodimer capable of binding to the specific EcREs is barely detectable in these nuclei. Studies have shown that the ecdysteroid receptor is a primary target of the 20E signaling modulation in mosquito target tissues at the state of arrest. One possible mechanism through which the formation of ecdysteroid receptor can be regulated is a competitive binding of other factors to either EcR or USP. Indeed, data show that at this stage, AaUSP exists as a heterodimer with the orphan nuclear receptor AHR38. AHR38, the mosquito homologue of *Drosophila* DHR38 and vertebrate NGFI-B/Nurr1 orphan receptors, is a repressor that disrupts the specific DNA binding of the ecdysteroid receptor and interacts strongly with AaUSP. However, in the presence of 10^{-6} M 20E, EcR can efficiently displace AHR38 and form an active heterodimer with USP. The latter event actually occurs after a blood meal (Fig. 25.8).

A second key element of the repression system is represented by a unique mosquito GATA transcription factor (AaGATAr) that recognizes GATA-binding motifs in the upstream region of the *YPP* genes, vitellogenin and *VCP*. Analyses of tissue and temporal distribution of the AaGATAr transcript have demonstrated that it is expressed only in the fat body of adult female mosquitoes. Furthermore, the active form of AaGATAr exists in nuclei of the fat body at the state of arrest as well as during postvitellogenic stages when *YPP* genes are shut off. AaGATAr acts as a repressor of *YPP* genes by occupying GATA-binding sites in regulatory regions of *YPP* genes. AaGATAr not only inhibits basal *YPP* gene activity, but it can also overcome ecdysone-dependent activation. Impairing ecdysone-dependent activation by AaGATAr during previtellogenesis is essential because at that period, detectable amounts of ecdysteroids as well as the components of the ecdysone receptor, AaEcR and AaUSP, are present. Thus, a tissue-specific repressor of GATA transcription factors is involved in the repression of *YPP* genes in the fat body of the *Ae. aegypti* female at the state of arrest. It has been observed that within 0.5 h after a blood meal, the GATA-binding sites of the vitellogenin gene are relieved from the AaGATAr binding, and another GATA factor, presumably an activator, binds these sites (Fig. 25.9). This event coincides

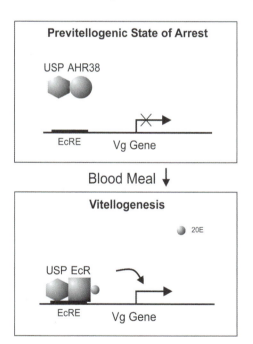

FIGURE 25.8 Schematic representation showing that the ecdysteroid receptor is a primary target of the 20E signaling modulation in target tissues at the state of arrest in *Aedes aegypti*. At the previtellogenic state of arrest, AaUSP exists as a heterodimer with the repressor AHR38 that prevents formation of the functional ecdysteroid receptor. After a blood meal, EcR displaces AHR38 and forms a functional heterodimer with USP capable of activating ecdysteroid-regulated genes. From Raikhel et al. (2002) with permission.

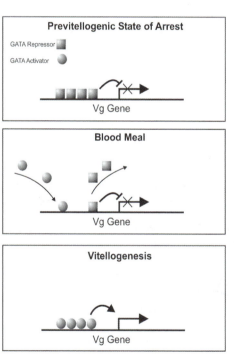

FIGURE 25.9 Schematic representation of the events of repression and derepression of the vitellogenin gene during the state of arrest and blood meal–dependent activation via GATA transcription factors in *Aedes aegypti*. At the state of arrest, GATA-binding motifs in the upstream region of the vitellogenin gene are occupied by a unique GATA repressor transcription factor that prevents the gene from being activated. Unknown factors released as a result of blood feeding result in replacement of GATA repressor by GATA activator and derepression of the gene. From Raikhel et al. (2002) with permission.

with the start of the vitellogenin gene transcription. Future studies should elucidate further this intriguing event in derepression and activation of *YPP* genes.

Ovarian Development and Yolk Protein Accumulation

In the mosquito, each of the two ovaries contains about 75 ovarioles of the meroistic polytrophic type. At eclosion, each ovariole is composed of an apical germarium and a single primary follicle. The follicle consists of one oocyte and seven nurse cells, which are interconnected by cytoplasmic bridges. These germ cells are surrounded by a monolayer of somatic follicular epithelial cells, which are also interconnected by cytoplasmic bridges.

Only one follicle in each ovariole matures per cycle. Secondary follicles separate from the germarium during maturation of primary follicles, but their development is arrested at an early stage and it does not resume until after oviposition of the fully developed eggs and a second blood meal.

During vitellogenesis, the oocyte accumulates massive amounts of YPPs and stores them in yolk bodies. Internalization is via receptor-mediated endocytosis: Membrane-bound receptors specifically bind YPPs that accumulate in clathrin-coated pits that pinch off into the cytoplasm as clathrin-coated vesicles. At adult eclosion, however, the oocyte cortex (the region of the cytoplasm just inside the plasma membrane) is undifferentiated and incapable of YPP internalization. Under the influence of rising JH titers, clathrin-coated pits and vesicles, microvilli, and endosomes appear at the oocyte surface, making it competent to internalize YPPs by 2–3 days after eclosion. At the molecular level, the increase in transcription of genes encoding receptors for yolk proteins and clathrin have been demonstrated, and mRNAs encoding receptors and clathrin are produced in the nurse cells and then transferred to the oocyte cytoplasm, where they accumulate. These mRNAs are translated into proteins only in the oocyte. Receptor protein increases threefold between the first and third day after eclosion, and it amasses at the oocyte surface. The follicle doubles in size, from about 50 to 100 μm in length, and the ovary enters a state of arrest until the female takes blood.

At the onset of vitellogenesis, the follicle cells pull away from the surface of the oocyte, creating a perioocytic space, remaining in contact with the oocyte only via the latter's microvilli. Narrow channels about 20 nm wide, along with a few larger intercellular spaces, appear between follicle cells, allowing the hemolymph to penetrate to the perioocytic space and to come in contact with the oocyte surface.

Vitellogenin, VCP, VCB, and lipophorin pass through the basal lamina surrounding the follicle and then through the narrow interfollicular channels, concentrating in the perioocytic space. For the first three of these a common pathway has been demonstrated. They bind to the oocyte surface in the subdomains located between and at the base of microvilli, an area characterized by a dense extracellular glycocalyx and numerous clathrin-coated pits in which they accumulate. Clathrin triskelions attach to the cytoplasmic tail of endocytotic receptors via an adaptor protein and then attach to one another to form a polyhedral lattice that deforms the plasma membrane into a depression. This pit grows as more receptor/clathrin complexes are recruited to it through the mediation of the self-assembling clathrin triskelions. Eventually it is sealed off, creating an intracellular clathrin-coated vesicle that is also internalized (Fig. 25.10). This process, referred to as *receptor-mediated endocytosis*, was first discovered in *Ae. aegypti* oocytes by Roth and Porter in 1964 and is now recognized as the universal mechanism employed by eukaryotic cells for internalizing specific macromolecules.

The receptor for vitellogenin (VgR) in *Ae. aegypti* has been extensively characterized biochemically and molecularly. The *VgR* gene is transcribed in the nurse cells, and the 7.3-kb mRNA transcript is translated in the oocyte. *VgR* is a large 205-kDa protein that, like other endocytotic receptors, is anchored in the plasma membrane by a single-pass hydrophobic transmembrane helix. The cytoplasmic tail is relatively short and contains a leucine-isoleucine internalization signal that presumably binds the clathrin adaptor protein. Over 90% of the molecule is extracellular, and its domain structure clearly identifies it as a member of the low-density lipoprotein receptor (LDLR) family, consisting of discreet clusters of repetitive elements arranged in a characteristic manner (Fig. 25.11). However, the mosquito VgR is about twice as large as vertebrate VgRs and LDLRs, harboring two clusters of cysteine-rich Class A repeats instead of one. Class A domains are implicated in the binding of numerous ligands in several other LDLR family members. Little is known about the specific interaction between vitellogenin and VgR at the molecular level except that it involves both vitellogenin subunits and that phosphorylation of vitellogenin is required. Binding affinity of vitellogenin is high, with an estimated dissociation constant (K_d) of 15 nM in enriched solutions of solubilized VgR.

Lipophorin functions as a yolk protein precursor in the mosquito *Ae. aegypti* and it is also internalized via receptor-mediated endocytosis. The cDNA encoding the oocyte receptor for lipophorin (LpRov) has been

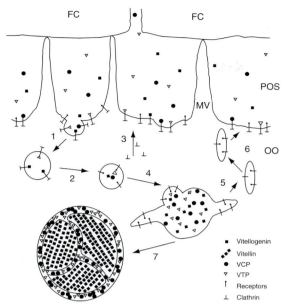

FIGURE 25.10 Receptor-mediated endocytosis and postendocytotic routing of the *Aedes aegypti* yolk protein precursors, vitellogenin (vitellogenin), vitellogenic carboxypeptidase (VCP), and vitellogenic pro-cathepsin B (VCB). 1, Receptor/ligand complexes accumulate in clathrin-coated pits that pinch off to form coated vesicles. Putative receptor(s) for VCP and VCB have not yet been isolated or characterized. 2, Clathrin is released from the receptors and the vesicle becomes an early endosome. Acidification of the endosome causes release of ligands from their receptors. 3, The released clathrin triskelions are presumably recycled to the cell surface. 4, The endosome fuses with other endosomes to create a transitional yolk body. Within this structure, ligands are segregated from receptors that accumulate in peripheral tubular arms. 5, The peripheral arms containing the receptors pinch off as tubular compartments and recycle to the surface (6). 7, Delipidated vitellogenin molecules begin to crystallize as vitellin (Vn), and at some point the transitional yolk body no longer fuses with endosomes. In the mature yolk body, crystalline Vn occupies the bulk of the organelle's interior, whereas VCP and VCB are confined to the peripheral spaces between Vn crystalloids and the organelle membrane. Based on Snigirevskaya et al. (1997a,b). From Sappington and Raikhel (1999) with permission.

cloned. It encodes a protein that is nearly twice as small as VgR, with a predicted molecular mass of 128.9 kDa. The deduced amino acid sequence of the cDNA shows that it encodes a protein homologue of the LDL receptor superfamily and that it harbors eight cysteine-rich ligand-binding repeats at the N-terminus, similar to the vertebrate very low-density lipoprotein receptor (VLDLR) and the chicken vitellogenin receptor. It is only distantly related to the mosquito VgR (18.3%) (Fig. 25.11). Experiments have shown that the mosquito LpRov specifically binds lipophorin. The receptors for VCP and VCB have not been identified, but multiple ligands are characteristic of VLDLR-type

receptors, and it is possible that LpRov serves as a receptor for VCP and VCB.

Soon after internalization, the coated vesicle loses its clathrin mantle, becoming an early endosome (Fig. 25.10). The receptors remain anchored in the vesicle membrane as the YPPs are released, presumably through acidification of the endosome lumen. In the absence of vitellogenin, endosomes containing nonspecific proteins that were internalized by the oocyte are routed directly to lysosomes for degradation, but the presence of vitellogenin somehow signals the endosomes to fuse with one another, creating transitional yolk bodies. Released YPPs concentrate in the lumen of the transitional yolk bodies, while membranous tubular compartments bud off to recycle the VgR, and probably LpR, back to the oocyte surface. At some point, transitional yolk bodies cease fusing with endosomes and begin the transformation to a mature yolk body. Vitellogenin condenses in the core of the yolk body as its load of lipids is removed, and it eventually crystallizes as vitellin. Within the mature yolk body, the vitellin occurs as two to four separate crystalloid bodies, each surrounded by and interspersed with a noncrystalline matrix. VCP and VCB, which are intermixed with vitellogenin in endosomes and transitional yolk bodies, are segregated from the vitellin as it crystallizes, collecting in the mature yolk bodies as a cap. By 48 h after a blood meal, VCP and VCB are evenly distributed in the matrix surrounding the crystalline vitellin (Fig. 25.10).

The length of the oocyte increases eightfold, and the volume several hundredfold, from 6 to 36 h after a blood meal as YPPs are internalized at increasing rates for 30 h. In addition to the necessary synthesis of large quantities of the endocytotic machinery required to sustain high rates of YPP internalization, the oocyte must stockpile a sizable supply of ribosomes and many species of mRNAs for use by the future embryo. These activities place a tremendous demand on the synthetic capacity of the follicle. In most insects with meroistic ovaries, the nurse cells manufacture the necessary rRNA, which is transferred to the oocyte via cytoplasmic bridges.

The rate of vitellogenin uptake by oocytes declines precipitously from 30 to 36 h after a blood meal, because access of hemolymph, and therefore of YPPs, to the surface of the oocyte is restricted by two concurrent events. Permeability of the follicular cell layer is eliminated by sealing junctions that form between follicle cells beginning at about 32 h. In addition, the endochorionic plaques fuse into an impermeable envelope around the oocyte. During fusion of the endochorion, the follicle elongates, and the nurse cells degenerate between 36 and 48 h.

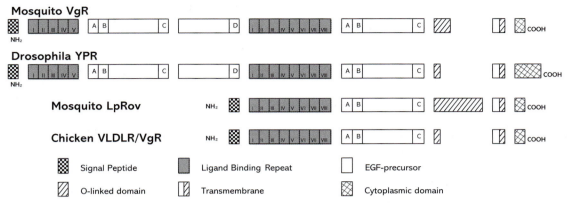

FIGURE 25.11 Schematic modular domain alignment of *Aedes aegypti* ovarian vitellogenin (VgR) and lipophorin receptors (LpRov) with other LDLR superfamily members. Schematic alignment of mosquito VgR with *Drosophila* yolk protein receptor (YPR) and of mosquito LpRov with chicken VLDLR/VgR. The cysteine-rich repeats in the ligand-binding domains and EGF-precursor domains are assigned the Roman numerals I–VIII and the letters A–D, respectively. Gaps beween domains are added to highlight the alignment of respective domains. From Cheon et al. (2001) with permission.

Secretory activity of the follicle cells decreases greatly between 30 and 36 h as synthesis of endochorion proteins is completed. Rapid secretion of exochorion proteins by the follicle cells begins at 38–42 h and is essentially completed by 60 h. Unlike the endochorion proteins, most exochorion proteins are glycosylated, and they range in size from 25 to 50 kDa. The exochorion comprises ~80% of the chorion by weight and has a highly sculptured appearance arising from uneven deposition.

As JH titers rise between 48 and 72 h, the secondary follicles (which form at ~20 h in response to high 20E titer) increase in length and develop to the previtellogenic state of arrest. A female that cannot oviposit immediately after egg maturation (72 h) may take a second blood meal, which is digested normally. However, the secondary follicles are prevented from breaking the state of arrest, presumably by an oostatic hormone, until the eggs have been oviposited.

CYCLORRHAPHAN DIPTERA

Yolk Proteins

Female-specific proteins that serve as yolk proteins in the Cyclorrhapha fundamentally differ from vitellogenins of most other insects. They are relatively small proteins in the size range of 40–51 kDa. Up to five yolk polypeptides have been described in *Calliphora erythrocephala*, *Sarcophaga bullata*, *Musca domestica*, and *Phormia regina*. Antigenic cross-reactions between

species indicate homologous relationships among these yolk proteins. Comparisons of deduced amino acid sequences have shown that the three yolk polypeptides of *Drosophila melanogaster* are related to each other and to other cyclorrhaphan yolk proteins. They exhibit no similarity to vitellogenins, resembling instead a family of vertebrate lipases, particularly in a noncatalytic lipid-binding domain. Yolk protein genes of the cyclorrhaphan dipteran, *Drosophila melanogaster*, are among the most studied.

The primary translation products are close to the size of each mature polypeptide of 40–51 kDa. Posttranslational processing includes deletion of signal polypeptides, glycosylation and phosphorylation, and tyrosine sulfation. The cyclorrhaphan yolk proteins have not been reported to be lipoproteins. In *C. erythrocephala*, yolk polypeptides associate as subunits of a larger native protein of 210 kDa. In *M. domestica*, 48-, 45-, and 40-kDa subunits are presumably organized into a 283-kDa secreted hemolymph native molecule that is transformed into a 281-kDa oocyte form.

The iron transport protein transferrin, a blood protein occurring in a wide variety of animals, is used by dipterans for the import of iron into vitellogenic follicles. In the flesh fly, *Sarcophaga peregrina*, transferrin is selectively deposited in yolk. A 65-kDa transferrin was identified by 45% sequence similarity to the transferrin of the moth *Manduca sexta* and by its content of 1 mole of Fe^{2+} per mole of isolated protein. In *D. melanogaster*, transferrin plays a role in defense against bacteria. It is likely that transferrin deposited in the *S. peregrina* yolk has a similar function in the embryo.

Sites of Yolk Protein Synthesis

Unlike the case with mosquitoes, the yolk proteins of the adult females of cyclorrhaphan Diptera are synthesized in both the fat body and the ovaries. For *D. melanogaster* and *M. domestica*, it has been shown that the production of yolk proteins by the ovarian follicle cells is initiated simultaneously with the onset of yolk protein uptake. There are no structurally obvious differences in the yolk proteins synthesized in each tissue. It is possible that the dual synthesis of yolk proteins is an adaptation to a rapid vitellogenic phase of oogenesis. However, a similarly rapid vitellogenesis is accomplished in other insects, such as mosquitoes, utilizing just a single site of yolk protein production. Interestingly, the stable fly (*Stomoxys calcitrans*) and tsetse fly (*Glossina austeni*) produce yolk proteins only in their follicle cells. It appears that the ability to make yolk proteins in the female fat body has been lost in these insects.

Accumulation of Yolk

The cyclorrhaphan Diptera yolk proteins made in the fat body enter the follicle through the interfollicular spaces, as in other insects, whereas those from the follicle cells are unidirectionally secreted toward the oocyte into the periooocytic space. As in mosquitoes, yolk proteins are then accumulated in developing oocytes via receptor-mediated endocytosis. The genetic approach, which so far has been unavailable for insects of medical importance, has been used to study the yolk protein receptor–mediated endocytosis in *D. melanogaster*. The yolk protein receptor is the product of the *yolkless* (*yl*) gene that encodes a member of the low-density lipoprotein receptor family. Mutations in *yl* fail to accumulate yolk in the oocyte. Strikingly, despite the difference between the yolk proteins of higher Diptera and the vitellogenins of lower insects, the yolk protein and vitellogenin receptors are very closely related between *Drosophila* and mosquitoes (Fig. 25.11). They share 40% identity and 57% similarity and are more related to each other than to any other members of the low-density lipoprotein receptor family. Searches of the completed *Drosophila* genome revealed no obvious potential vitellogenin genes. Therefore, it is of interest to understand how similar receptors recognize such different ligands. Use of an *yl-lacZ* fusion confirmed that the *yl* RNA is transcribed in the nurse cells and transported to the oocyte. The Yl protein begins to accumulate at the surface of the oocyte in the cortex just as yolk uptake begins at stage 8–9 of oogenesis. The *yl* mutants fail in this cortical accumulation of yolk protein receptor, which remains instead distributed throughout the cytoplasm.

Regulation of Vitellogenesis in Higher Diptera

In higher Diptera of medical and veterinary importance, eggs are produced synchronously in batches, usually in response to a blood or protein meal. The development of one batch of eggs is completed before another batch begins to develop. In these flies, for example, *C. erythrocephala* and *M. domestica*, the newly eclosed female produces mature oocytes only after meat feeding. At this point, a single follicle matures synchronously in each ovariole and a batch of eggs is produced.

It is common to all the higher Diptera that ecdysteroids stimulate yolk protein production in the fat body of males and females, but the sensitivity of this response, the nature of JH involvement, and the importance of the ovary in the process vary among species. There is an integral link between diet and the effects of JH and ecdysone on the progress of oogenesis and yolk protein production in *Musca*, *Calliphora*, *Sarcophaga*, and *Phormia*. *M. domestica* females maintained on sugar do not undergo vitellogenesis but arrest their follicles at an early stage of oogenesis. An intake of protein-containing food leads to a peak of ecdysteroid production. The ecdysteroid peak corresponds with oocytes passing through vitellogenesis, and the concentrations of circulating yolk also change, with a slight delay, in relation to the ecdysteroid levels. Thus, it seems likely that the protein feeding controls the production of vitellogenin in the fat body via ecdysteroid levels and that at the same time vitellogenesis begins and yolk uptake is triggered in the ovaries. As in mosquitoes, the oocytes must be released from a developmental arrest to proceed with development.

If the corpus allatum-cardiacum complex is removed soon after eclosion, the female is unable to respond to a protein diet. Removal of the complex or the ovary prevents the increase in ecdysteroid that normally follows a protein feed. This could be overcome by treatment of the flies with both hormones, suggesting that in *Musca*, diet operates via JH and ecdysone levels. The situation is rather similar in *Phormia regina*, where removal of the CA soon after eclosion arrests ovarian development. There is a window early in the cycle of response to a protein meal where the CA, and hence JH, is needed. The CA is needed for females to generate normal ecdysteroid levels in the hemolymph. It is likely that JH prepares the previtellogenic fat body to synthesize yolk proteins and the ovary to produce ecdysteroid that activates the *yp* genes. This is essentially similar to the events in *Ae. aegypti*.

The link between protein feeding and the cycle of egg production has been investigated in detail in *P. regina*, in which a midgut hormone has been isolated. The hormone is released from the midgut shortly after the onset of a protein meal, and its target is the neurosecretory cells of the brain. These then stimulate the whole cascade of events needed to produce a batch of eggs.

VITELLOGENESIS IN HEMATOPHAGOUS HEMIPTERA

Endocrine Control of Vitellogenesis

The Chagas' disease vector, *Rhodnius prolixus*, commonly called a *kissing bug*, is one of the best-studied insects with respect to the hormonal control of its reproduction. The vitellogenic role of juvenile hormone was one of the historical discoveries made by the founder of insect physiology, V. B. Wigglesworth, on this insect, which was his favorite experimental model. Classical experiments of removal of the CA, the source of JH, followed by reimplantation of the gland and monitoring of oocyte growth, eventually led to the understanding of the role that JH plays in vitellogenesis and oogenesis of *Rhodnius* as well as in other insect species. In fact, in most insects JH, not 20E, is the hormone regulating these reproductive events. The fundamental discovery of Wigglesworth concerning the key role of the CA in vitellogenesis of *R. prolixus* has been confirmed in other hematophagous hemipterans, such as *Triatoma infestans* and *Triatoma protracta*. Presently, it is well established that in these insects, JH released by the corpora allata after blood feeding activates vitellogenin gene expression and production of vitellogenin in the fat body. Despite decades of research, however, the molecular nature of JH action remains elusive. The effect of JH on yolk protein uptake in *Rhodnius* is one of the best studied. The juvenile hormone acts on the follicular epithelium of the ovarian follicles, causing the formation of wide spaces between epithelial cells. This JH-controlled event, called *patency*, is important for allowing vitellogenin and other yolk proteins to reach the oocyte.

Vitellogenin and Vitellin

In the kissing bug, the major yolk protein, vitellogenin, and its oocyte form, vitellin, are large phosphorylated lipoglycoproteins. They are composed of two types of apoproteins (subunits) of 180 and 50 kDa. Thus, the size and apoprotein composition of vitellogenin is similar to those of most of other insects. Likewise, the fat body is the major source of vitellogenin in *R. prolixus*. Recently, it has been shown that epithelial cells of vitellogenic follicles also synthesize vitellogenin. Unlike higher Diptera, in which the yolk protein synthesis is initiated simultaneously in both the fat body and the ovary, only late vitellogenic follicles are involved in the vitellogenin production of *Rhodnius*. The composition of ovarian vitellogenin is similar to that of the fat body.

Interestingly, three heterogeneous populations of vitellin have been identified in *Rhodnius* oocyte extracts and named VT_1, VT_2, and VT_3. They differ dramatically in their phosphorylation levels, lipid content, and sugar composition. VT_1 is the least phosphorylated and VT_3 is the most heavily phosphorylated. All three vitellins contain the same neutral lipids, but they are heterogeneous with respect to phospholipid composition. High-mannose oligossacharides are present in VT_2 and VT_3, while VT_1 also contains glucose resembling vitellogenin. The increase in the level of phosphorylation and the decrease in the content of glucose in VT_2 and VT_3 are attributed to postendocytic processing. Vitellogenin of ovarian origin is taken up by oocytes after the spaces between the follicle cells are already closed and uptake of vitellogenin from the hemolymph is interrupted. It is likely stored as VT_1, maintaining glucose in its oligossacharide structure.

Heme Metabolism and Egg Development

Blood-feeding insects, such as *Rhodnius*, ingest enormous amounts of blood in a single meal, usually comprising several times the animal's weight. A special problem generated by having vertebrate blood as the sole food source is the large amount of free-hemin that is produced upon digestion of hemoglobin. Wigglesworth was the first to notice the protein-associated reddish pigment present in *Rhodnius* hemolymph and eggs. He proposed that it was produced by partial digestion of hemoglobin from the vertebrate host. Hemin is a powerful generator of free-radical reduction products of dioxygen that are capable of causing biological injury through peroxidation of lipids, proteins, and DNA. Recently, a unique heme-binding protein has been isolated from *Rhodnius* hemolymph that acts as an antioxidant. The *Rhodnius* heme-binding protein (RHBP) is a 15-kDa polypeptide with a highly helical structure that binds heme noncovalently. RHBP is synthesized by the fat body; it is not a degradation product of hemoglobin. RHBP isolated from hemolymph is not saturated with heme and readily binds additional heme molecules. RHBP has also been isolated from ovaries. It is similar to its hemolymph counterpart. However, the oocyte protein is fully saturated and is not capable of binding additional heme. Although RHBP is not a sex-specific protein and is found in the hemolymph of males, which are also obligatory blood feeders, this protein is the second most abundant protein in *Rhodnius* eggs. It corresponds to about 1% of total protein, on a mass basis, and to approximately half of the amount of vitellin, on a molar basis. The reduction of heme-RHBP in the hemolymph, by experimentally changing the diet, decreased the number of eggs laid. By increasing the concentration of heme-RHBP in the hemolymph, the number of eggs produced increased in a dose-dependent manner, indicating that RHBP is essential for normal production of eggs.

In the oocyte, RHBP is found inside the mature yolk bodies, which suggests that RHBP is internalized from the hemolymph via receptor-mediated endocytosis. Indeed, the uptake of RHBP by *Rhodnius* ovaries has been found to be selective and specific. The main yolk protein, vitellogenin, does not compete with RHBP for the uptake. It appears that vitellogenin and RHBP bind to independent binding sites of the oocyte membranes and are then directed later to the same yolk compartment. The putative receptors for vitellogenin and RHBP have not been characterized. The capacity of the ovary to take up RHBP from the hemolymph varies during the days following the blood meal. It increases up to day 2, remains stable until day 5, and then decreases up to the end of oogenesis. The egg RHBP serves an important role in the development of the *Rhodnius* embryo as a source of heme, or iron.

VITELLOGENESIS IN TICKS

Gonotrophic Cycles

Reproduction in ticks varies drastically between soft (Argasidae) and hard (Ixodidae) ticks. Soft ticks require a few hours to feed to repletion. They mate off of the host before or after feeding, and one mating is sufficient for several gonotrophic cycles. Vitellogenesis is induced by feeding, but mating is required for oviposition. For example, unmated blood-fed *Ornithodoros moubata* females show levels of vitellogenin in the hemolymph similar to that of mated females after feeding, but they do not usually oviposit eggs unless mating occurs. The soft tick female lays several hundred eggs approximately 10–12 days after engorgement, surviving to reproduce again many times.

Hard ticks are divided into two groups, metastriate and prostriate ticks. Metastriate ixodid ticks are among the most highly specialized hematophagous arthropods. They feed on a host from several days to months, depending on stage and species. Initiation of vitellogenesis in ixodid ticks requires both mating and feeding. Mating occurs on the host, and the female starts feeding before the male becomes attracted to her. Mating triggers the female's rapid feeding stage, during which the female undergoes dramatic morphophysiological changes, allowing her to consume an enormous quantity of blood. After the completion of feeding, the female drops off the host, lays thousands of eggs, and dies after one gonotrophic cycle. Metastriate tick males require a blood meal for sperm production. The prostriate ticks of the genus *Ixodes* appear to be phylogenetic intermediates between argasids and the other ixodids. Prostriates have retained many characteristics similar to argasids that are related to reproduction, such as having several gonotrophic cycles and the ability to mate without blood feeding.

Sites of Vitellogenin Synthesis

Although early ultrastructural studies of tick oocytes implicated them in the synthesis of their own yolk proteins, further electron microscopical observations of intense endocytosis during vitellogenesis strongly suggest the extraovarian origin of yolk protein.

Unlike insects, the fat body of ticks consists of highly dispersed strands of cells attached to branches of the tracheal system and internal organs. It is most abundant around the tracheal trunks, near the spiracles. After a blood meal, the fat body cells increase in volume extensively but do not increase in cell number. The fine structure of the fat body cells in adult female ticks reveals that the RER develops and protein granules appear in the fat body cells shortly after a blood meal. The fat body has been established as the site of vitellogenin synthesis in the soft ticks *O. moubata* and *O. parkeri* and in the prostriate and metastriate ticks. However, the possibility that vitellogenin is synthesized by the midgut and ovary of hard ticks cannot be ruled out completely, because most researchers use immunodetection methods. More detailed analysis using in situ hybridization and other molecular techniques to determine the presence of vitellogenin mRNA in these tissues are needed to clarify this question.

Regulation of Vitellogenesis

Our understanding of tick endocrinology is much less advanced than that of the insects. This is largely due to the tick's condensation of the central nervous system to a single synganglion, which presents obstacles to experimentation. In ticks, as in hematophagous insects, vitellogenesis is induced primarily by blood feeding. Early researchers assumed that juvenile hormone had a role in regulation of reproduction in ticks similar to that of insects. Recent studies, however, have not been able to find direct evidence for the presence of farnesol, methyl farnesoate (JH precursors), or JH biosynthesis in adult hard and soft ticks. These data raise the possibility that there is a fundamental difference in the endocrine regulation of development and vitellogenesis in ticks relative to insects.

The involvement of the neurosecretory system in regulation of vitellogenesis is clearer. Experiments utilizing transplantation and ligation of the synganglion and injections of synganglion extracts have established the key role of the synganglion in oogenesis and vitellogenesis in several species of ticks. Chinzei and Taylor (1994) proposed a model for the regulation of vitellogenesis in *Ornithodoros* in which blood feeding activates the synganglion to produce the vitellogenesis-inducing factor, a peptide secreted into the hemolymph. This factor then stimulates an organ in the posterior half of the tick to secrete the fat body–stimulating factor, possibly an ecdysteroid that, in turn, activates the fat body to synthesize vitellogenin. The vitellogenin is carried to the ovary by the hemolymph and processed and incorporated into the oocytes as vitellin.

Recent studies give support to a role for ecdysteroids in the stimulation of vitellogenin synthesis in ticks. Ecdysteroid titers in the hemolymph of blood-fed females of several tick species correlate with vitellogenin peaks and egg-laying cycles. The fat body reportedly contains ecdysone 20-mono-oxygenase, which is capable of converting ecdysone into 20E. Moreover, 20E induces synthesis and secretion of vitellogenin in cultured fat bodies of *Dermacentor variabilis*. Interestingly, in ticks, in addition to or instead of the ovary, the epidermis could be the source of ecdysteroids. Both ecdysone receptor (EcR) and its heterodimeric partner retinoic X receptor (RXR) homologue have been cloned from the hard tick *Amblyomma americanum*. Although they are phylogenically distantly related to insect EcRs and USPs, the tick EcR-RXR heterodimer acts as the functional ecdysteroid receptor.

Properties of Vitellogenin

Vitellogenins and vitellins of several soft and hard ticks have been characterized biochemically. Tick vitellogenin is a conjugated, phosphorylated protein containing a heme moiety, lipids, and carbohydrate. Although only a partial sequence of tick vitellogenin is known, it is sufficient to establish its homology with other vitellogenins, including those of nematodes and insects. Vitellogenin biosynthesis and posttranslational modifications have been elucidated in the soft tick *O. moubata*. After pro-vitellogenin is translated from the mRNA, it undergoes processing by the addition of heme, carbohydrate, lipid, phosphorylation, and enzymatic cleavage. The processed polypeptides form vitellogenin-1, which aggregates as a dimer to form vitellogenin-2. Both the monomer (vitellogenin-1) and the dimer (vitellogenin-2) are released into the hemolymph and carried to the ovary, where they are incorporated into the oocytes after undergoing additional processing to form vitellin. The molecular weight of monomer vitellogenin-1 was estimated by gel filtration to be 300 kDa, whereas the molecular weights of vitellogenin-2 and Vn were estimated to be 600 kDa. Vitellogenin-2 is most likely an immediate precursor of Vn. Vitellins with a molecular weight of 480 kDa and of 480 and 370 kDa with multiple subunits have been characterized from eggs of ixodid ticks.

The most distinguishable feature of tick vitellogenins is that they are heme-proteins. Wigglesworth described the presence of a brown pigment derived, as he suggested, from host hemoglobin, in the eggs of *Ornithodoros* and *Ixodes* ticks. Indeed, ticks are not capable of synthesizing their own heme and they acquire it by degrading hemoglobin from the ingested blood. Heme must then be conjugated to the synthesized vitellogenin. After internalization by the oocyte, vitellin retains heme in a conjugated form. In vitro measurement of the binding of heme to vitellin showed that each vitellin molecule binds as many as 31 heme molecules. The association of heme with vitellogenin and vitellin strongly inhibits heme-induced lipid peroxidation, suggesting that binding of heme is an important antioxidant mechanism that protects the tick organism from oxidative damage. This mechanism also allows the safe storage of heme obtained from a blood meal inside eggs for future use. Thus, vitellin has multiple roles in ticks. It provides the amino acids needed for synthesis of new proteins by the embryo, and it is also a heme reservoir and an antioxidant protective molecule.

In the tick *Boophilus microplus*, an aspartic proteinase, yolk cathepsin, plays a role in vitellin degradation during embryogenesis. However, a novel aspartic proteinase has been recently characterized from the eggs of this tick. It is capable of binding heme with a 1:1 stoichiometry and is likely involved in regulating the rate of tick vitellin degradation. The rate of vitellin degradation must match that of heme utilization to avoid the formation of a potentially dangerous pool of free heme.

CONCLUSIONS AND FUTURE DIRECTIONS

Arthropod vectors transmit many diseases that are among the most threatening in modern times. Malaria is particularly devastating, taking a heavy toll on the human population by infecting over 300 million and killing over 2 million people each year. Diseases caused by vector-borne viruses, most importantly dengue fever, are reaching disastrous levels in some regions of the world as well. There is an urgent need to explore every possible avenue for developing novel control strategies against many vector-borne diseases.

Anautogeny is the most fundamental phenomenon underlying the strict control of vitellogenesis by intake of blood or protein meal. As a consequence, anautogeny is the sole reason that hematophagous arthropods are vectors of numerous pathogens. Yet it is most poorly understood at the biochemical and molecular biological levels. Two molecular mechanisms that repress the expression of yolk protein genes before a blood meal have been elucidated in the mosquito *Ae. aegypti*. More studies should be done to further elucidate molecular and genetic details of these mechanisms in this mosquito. It is also important to understand how universal these mechanisms are among other groups of arthropod vectors, particularly those distantly related to mosquitoes.

In hematophagous arthropods, strikingly different biochemical mechanisms have evolved dealing with the consequences of blood feeding and using products of hemoglobin digestion for embryogenesis. Continuing biochemical and molecular studies of these mechanisms that are unique to hematophagous arthropods are essential.

The elucidation of the molecular mechanisms underlying tissue- and stage-specific yolk protein gene expression is of great importance to the current efforts directed toward utilizing molecular genetics to develop novel strategies in vector control. Significant progress has been achieved in the biochemical and molecular characterization of key genes expressed in the fat body of the vector model, the mosquito, *Ae. aegypti*, during vitellogenesis. This progress provides the framework for further studies of molecular pathways leading to expression of fat body-specific genes as well as of the molecular basis for anautogeny in other vectors.

Currently, considerable effort is being directed toward using molecular genetics to develop control strategies targeting the arthropod vector itself and its ability to transmit pathogens. The powerful promoters of blood meal–activated fat body–specific genes are among the best candidates for construction of chimeric genes incorporating immune factors or other physiologically important molecules. These chimeric genes would be highly expressed in the fat body, and their products secreted to the hemolymph in response to a blood meal, an ideal milieu for targeting a pathogen. A transgenic mosquito with blood meal–activated systemic immunity has been engineered by using the vitellogenin promoter and coding region of defensin gene (*Def*). In this transgenic mosquito, the *Def* is activated by a blood meal, and the fat body produces and secretes large amounts of biologically active defensin. This work has provided proof of the concept that antipathogen effector molecules can be stably transformed into vectors and expressed in a precisely timed fashion engineered to maximally effect a pathogen. This success has clearly demonstrated the importance of fundamental research of the key physiological events triggered by a blood meal.

Acknowledgments

The author is grateful to Mr. G. Attardo for his help in making diagrams and figures, to Drs. V. Kokoza and J. S. Zhu and Mr. G Attardo for critical reading of the manuscript, and to Ms. Megan Ackroyd for editing the manuscript.

Readings

Bownes, M., and Pathirana, S. 2002. The yolk proteins of higher diptera. In *Progress in Vitellogenesis* (A. S. Raikhel and T. W. Sappington, eds.). Vol. XII of *Reproductive Biology of Invertebrates* (K. G. Adiyodi and R. G. Adiyodi, eds.). Wiley, New York.

Brown, M. R., Graf, R., and Swiderek, K. 1995. Structure and function of mosquito gonadotropins. In *Molecular Mechanisms of Insect Metamorphosis and Diapause*. Industrial Publishing and Consulting, Tokyo, pp. 229–238.

Chinzei, Y., and Taylor, D. 1994. Hormonal regulation of vitellogenin biosynthesis in ticks. In *Advances in Disease Vector Research*, Vol. 10 (K. F. Harris, ed.). Springer-Verlag, New York, pp. 1–22.

Raikhel, A. S., and Dhadialla, T. S. 1992. Accumulation of yolk proteins in insect oocytes. *Annu. Rev. Entomol.* 37: 217–251.

Sappington, T., Oishi, K., and Raikhel, A. S. 2002. Structural characteristics of insect vitellogenins. In *Progress in Vitellogenesis* (A. S. Raikhel and T. W. Sappington, eds.). Vol. XII, part A in *Reproductive Biology of Invertebrates* (K. G. Adiyodi and R. G. Adiyodi, eds.). Science Publishers, Inc. pp. 69–102.

Taylor, D., and Chinzei, Y. 2002. Vitellogenesis in ticks. In *Progress in Vitellogenesis* (A. S. Raikhel and T. W. Sappington, eds.). Vol. XII of *Reproductive Biology of Invertebrates* (K. G. Adiyodi and R. G. Adiyodi, eds.). Science Publishers, Inc. pp. 175–200.

Telfer, W. 2002. Insect yolk proteins: A progress report. In *Progress in Vitellogenesis* (A. S. Raikhel and T. W. Sappington, eds.). Vol. XII of *Reproductive Biology of Invertebrates* (K. G. Adiyodi and R. G. Adiyodi, eds.). Science Publishers, Inc. pp. 29–68.

Thummel, C. S. 1996. Flies on steroids—*Drosophila* metamorphosis and the mechanisms of steroid hormone action. *Trends Genetics* 12: 306–310.

Further Readings

Braz, G. R., Abreu, L., Masuda, H., and Oliveira, P. L. 2001. Heme biosynthesis and oogenesis in the blood-sucking bug, *Rhodnius prolixus. Insect Biochem. Molec. Biol.* 31: 359–364.

Buszczak, M., Freeman, M. R., Carlson, J. R., Bender, M., Cooley, L., and Segraves, W. A. 1999. Ecdysone response genes govern egg chamber development during mid-oogenesis in *Drosophila. Development* 126: 4581–4589.

Cheon, H. M., Seo, S. J., Sun, J., Sappington, T., and Raikhel, A. S. 2001. Molecular characterization of the VLDL receptor homolog mediating binding of lipophorin in oocytes of the mosquito, *Aedes aegypti. Insect Biochem. Molec. Biol.* 31: 753–760.

Cho, W. L., Tsao, S. M., Hays, A. R., Walter, R., Chen, J. S., Snigirevskaya, E. S., and Raikhel, A. S. 1999. Mosquito cathepsin B-like protease involved in embryonic degradation of vitellin is produced as a latent extraovarian precursor. *J. Biol. Chem.* 274: 13311–13321.

Guo, X., Xu, Q., Harmon, M. A., Jin, X., Laudet, V., Mangelsdorf, D. J., and Palmer, M. J. 1998. Isolation of two functional retinoid X receptor subtypes from the Ixodid tick, Amblyomma americanum (L.) *Mol. Cell. Endocrinol.* 30: 139(1–2),45–60.

Kokoza, V. A., Ahmed, A., Cho, W.-L., Jasinskiene, N., James, A., and Raikhel, A. S. 2000. Engineering blood meal–activated systemic immunity in the yellow fever mosquito, *Aedes aegypti. Proc. Natl. Acad. USA* 97: 1624–1629.

Kokoza, V. A., Martin, D., Mienaltowski, M., Ahmed, A., Morton, C., and Raikhel, A. S. 2001. Transcriptional regulation of the mosquito *Aedes aegypti* vitellogenin gene by a blood-meal-triggered cascade. *Gene* 274: 47–65.

Logullo, C, Vaz, I. D., Sorgine, M. H., Paiva-Silva, G. O., Faria, F. S., Zingali, R. B., De Lima, M. F., Abreu, L., Oliveira, E. F., Alves, E. W., Masuda, H., Gonzales, J. C., Masuda, A., and Oliveira,

P. L. 1998. Isolation of an aspartic proteinase precursor from the egg of a hard tick, *Boophilus microplus. Parasitology* 116: 525–532.

Martin, D., Piulachs, M. D., and Raikhel, A. S. 2001. A novel GATA factor transcriptionally represses yolk protein precursor genes in the mosquito *Aedes aegypti* via interaction with CtBP corepressor. *Molec. Cell. Biol.* 21: 164–174.

Martin, D., Wang, S.-F., and Raikhel, A. S. 2001. The vitellogenin gene of the mosquito *Aedes aegypti* is a direct target of the ecdysteroid receptor. *Molec. Cell. Endocrinol.* 173: 75–86.

Neese, P. A., Sonenshine, D. E., Kallapur, V. L., Apperson, C. S., and Roe, R. M. 2000. Absence of insect juvenile hormones in the American dog tick, *Dermacentor variabilis* (Say) (Acari: Ixodidae), and in *Ornithodoros parkeri* Cooley (Acari: Argasidae). *J. Insect Physiol.* 46: 477–490.

Raikhel, A. S., Kokoza, V. A., Martin, D., Ahmed, A., Zhu, J. S., Wang, S. F., Li, C., Sun J., Sun, G., Dittmer, N. T., Cho, K. H., and Attardo, G. 2002. Molecular biology of mosquito reproduction: From basic studies to genetic engineering of antipathogen immunity. *Insect Biochem. Molec. Biol.* 32: 1275–1286.

Sappington, T. W., and Raikhel, A. S. 1998. Molecular characteristics of vitellogenins and insect vitellogenin receptors. *Insect Biochem. Mol. Biol.* 28: 277–300.

Schonbaum, C. P., Perrino, J. J., and Mahowald, A. P. 2000. Regulation of the vitellogenin receptor during *Drosophila melanogaster* oogenesis. *Mol. Biol. Cell* 11: 511–521.

Sorgine, M. H., Logullo, C., Zingali, R. B., Paiva-Silva, G. O., Juliano, L., and Oliveira, P. L. 2000. A heme-binding aspartic proteinase from the eggs of the hard tick *Boophilus microplus. J. Biol Chem.* 275: 28659–28665.

Sun, G., Zhu, J., Li, C., Tu, Z., and Raikhel, A. S. 2002. Two isoforms of the early E74 gene, an Ets transcription factor homologue, are implicated in the ecdysteroid hierarchy governing vitellogenesis of the mosquito, *Aedes aegypti. Molec. Cell. Endocrinol.* 190: 147–157.

Zhu, J. S., Miura, K., Chen, L., and Raikhel, A. S. 2000. AHR38, a homolog of NGFI-B, inhibits formation of the functional ecdysteroid receptor in the mosquito, *Aedes aegypti. EMBO J.* 19: 253–262.

26

Osmotic and Ionic Regulation
by Mosquitoes

KLAUS W. BEYENBACH AND HENRY H. HAGEDORN

INTRODUCTION

Many studies have been done on osmotic and ionic regulation in both larval and adult mosquitoes, because they face very different problems in maintaining their osmotic and ionic balance. Larval mosquitoes can live in water of diverse composition and that imposes unique demands on them physiologically. Adult mosquitoes, like other hematophagous insects, face quite different problems when they take a blood meal that floods their hemolymph with large quantities of salt and water that must be eliminated.

SALT AND WATER TRANSPORT
IN LARVAL MOSQUITOES

Why should osmotic and ionic regulation by larval mosquitoes be of interest to the vector biologist? The distribution of mosquitoes is largely based on the availability of suitable sites for larval growth. Because larval mosquitoes are aquatic, the need for water is essential. However, water sources vary considerably in composition, from the nearly distilled water resulting from rain, to salt marshes and alkaline lakes. Although 95% of mosquito species live in freshwater, the distribution of some important vector and pest mosquitoes depends on their ability to cope with extreme situations, which has been the subject of many investigations. Mosquito larvae have developed physiological mechanisms that allow them to invade niches that are unusual in terms of salt composition. *Aedes campestris*

larvae have been collected from a lake in Canada where sodium was 0.48M, bicarbonate was 0.38M, and pH was 10.2! Some species can live equally well in freshwater or seawater (e.g., *Aedes campestris* and *Aedes dorsalis*) or in acid and alkaline lakes (e.g., *Aedes taeniorhynchus*). Rock pool mosquitoes, such as *Opifex fuscus* from New Zealand, are regularly exposed either to high saline concentrations as water evaporates from the pool or to almost freshwater as rain dilutes the pool. Brackish-water species live in areas, such as salt marshes, where freshwater mixes with seawater.

Insects must be able to regulate the composition of the fluid bathing their internal organs since cells are usually tolerant of only a narrow range of ionic and osmotic conditions. How do mosquito larvae regulate the osmolarity of their hemolymph? The ability of mosquito larvae to survive under varying conditions can be measured by examining hemolymph osmolarity in waters with different salt compositions. In larval mosquitoes that are able to live in diverse conditions, the composition of the hemolymph changes only slightly as the concentration of the external medium increases. An extreme example of this ability is *Opifex fuscus*, just mentioned, which lives in salty rock pools that vary widely in salt composition. Figure 26.1 shows the change in hemolymph osmolality as the concentration of the seawater in which the larvae were growing was increased from freshwater to 200% seawater (1.12 molar NaCl). The dashed line shows what the osmolaraity would have been if hemolymph osmolality matched the concentration of the water. The graph demonstrates that in dilute waters (from 0–30%

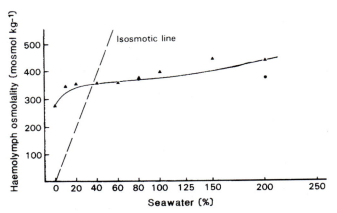

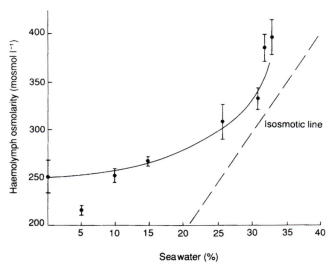

FIGURE 26.1 Mean osmolality of *Opifex fuscus* hemolymph taken from larvae reared in different concentrations of seawater. The isosmotic line indicates what the hemolymph osmolality would be if it were allowed to rise with the change in external seawater. Modified from Nicolson, S. W. 1972. *J. Entomol.* (A) 47: 101–108.

FIGURE 26.2 Osmolarity of larval *Aedes aegypti* hemolymph reared in different concentrations of seawater. Isosmotic line as in Figure 26.1. From Edwards, H. A. 1982. *J. Exp. Biol.* 101: 143–151.

seawater), the osmolarity of the hemolymph was maintained above that of the water. As the concentration of the seawater was increased from 40% to 200%, the osmolality of the hemolymph was kept nearly constant and much lower than that of the ambient water. Over the entire range, the osmolality of the hemolymph was maintained between 300 and 400 mosmol. The obvious question raised by this display of physiological athletics is how they do it.

The problem for larval mosquitoes is that while feeding they drink the solution in which they live, although some regulate the amount taken in. Salts and water also pass through the cuticle, which acts as a semipermeable membrane, so the water tends to flow toward the region with the highest salt concentration. In freshwater, both feeding and the flow of water into the animal across the cuticle would result in excess water that the larva has to eliminate. In water with a high saline concentration, the opposite is true, and the internal salt concentration rises. Basically, the mosquito meets these challenges either by eliminating excess salts via the Malpighian tubules and rectum or by uptake of needed salts via the rectal papillae.

Freshwater Mosquitoes

Aedes aegypti larvae are restricted to freshwater habitats. The larvae osmoregulate in dilute media, but osmoregulation breaks down if the osmolarity of the water nears that of the hemolymph. Figure 26.2 shows data for larvae of *Aedes aegypti* reared in dilute seawater. The curve in this graph is very different from the one in Figure 26.1. As the solute content increased from

0% to 20% seawater, the osmolality of the hemolymph was maintained near 250 mosmol. But as the concentration of the water rose above 250 mosmol (25% seawater), regulation failed and the hemolymph and water osmolalites rose together. The larvae died when the solute concentration reached 50% seawater.

In a freshwater species such as *Aedes aegypti*, a larva drinks about 300 nL of water each hour. Water movement into the animal across the cuticle is greatest via the anal papillae and can represent 33% of the body weight of the animal per day, or about 40 nL/h, assuming a body weight of 3 mg. Thus, larvae in freshwater are stressed by water overload and scarcity of salts. They compensate by taking up salts from the external water via the anal papillae that surround the anus and extend from the posterior end of the body. Ninety percent of the active exchange of sodium and chloride occurs across the anal papillae, while uptake of potassium is passive. Loss of sodium from the hemolymph to the water is reduced because the sodium binds to larger molecules in the hemolymph, such as proteins; this is also known to occur in dehydrating terrestrial insects. Elimination of excess water is achieved in the Malpighian tubules and rectum.

Whatever the route of entry, excess water ends up in the hemolymph. From there, active transport of potassium by the Malpighian tubule into the tubule lumen drives the passive flow of water and other solutes into the lumen by osmosis. As we shall see, the situation in the adult is quite different in the kinds of ions secreted. The osmotic concentration of the primary urine secreted by larval Malpighian tubules is close to that of the hemolymph, but the concentration of potassium is higher and that of sodium is correspondingly lower than that found in the hemolymph. In the rectum, needed salts are actively pumped back into the hemolymph. The recovery of potassium is especially important because it drives the activity of the tubules. The result is that the animal produces a dilute urine, thereby eliminating the excess water load. *Aedes aegypti* fails to regulate when the salt concentration of the water rises above the normal hemolymph osmolarity (Fig. 26.2), because it cannot produce a urine that is more concentrated than the hemolymph. It apparently lacks the ability to reverse the direction of transport and move ions into the rectum from the hemolymph.

Saline-Water Mosquitoes

When the saline-water species *Aedes detritus* is placed in distilled water, it produces a urine that is more dilute than the hemolymph (Fig. 26.3); in seawater its urine is much more concentrated than the hemolymph. Thus, mosquitoes that can live in both dilute or saline waters, such as *Aedes detritus*, can turn on or off various mechanisms for salt uptake or elimination. The anal papillae provide one such mechanism. In freshwater, larvae of *Aedes aegypti* face a salt deficit that is met by uptake of salt via the anal papillae. In saline-water species the uptake of ions via the papillae is regulated. For example, in *Aedes campestris*, ion uptake by papillae is negligible in seawater but high in freshwater.

In contrast to larvae that live in freshwater, species adapted to live in saline water actually face a deficit of water because they are constantly losing water via the urine and by diffusion across the cuticle. Salt water species recover water in part by drinking. Some species, such as *Aedes campestris* and *Culex tarsalis*, increase their rate of drinking when transferred from freshwater to seawater. Although drinking the saline solution supplies the water needed, it also adds to the salt load. Most of the excess salt is eliminated via the urine. How is this done?

The Malpighian tubules of saline-water larvae produce a primary urine that is similar in osmotic pressure to the hemolymph but high in potassium, just as in larvae living in freshwater. However, mosquitoes that live in waters with high concentrations of sulfate, such as *Aedes campestris*, actively transport sulfate into the lumen of the Malpighian tubule. As shown in Figure 26.4, the level of sulfate transport is related directly to the concentration of sulfate in the water in which the mosquito lives; sulfate transport can be developed within a few hours after exposure. This could result from an increase in the activity of existing sulfate transport proteins or from the synthesis of new transport proteins. In laboratory experiments using *Aedes taeniorhynchus*, it has been possible to select for

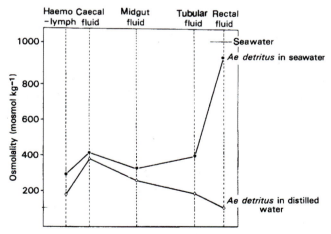

FIGURE 26.3 Osmolality of hemolymph and gut fluids from *Aedes detritus* reared in seawater or distilled water. Also indicated is the osmolality of seawater. Modified from Ramsay, J. A. 1950. *J. Exp. Biol.* 27: 145–157.

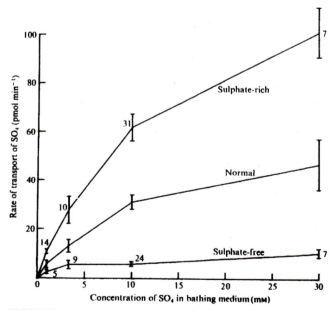

FIGURE 26.4 Change in sulfate uptake by Malpighian tubules of larval *Aedes taeniorhynchus* reared in seawater plus 89 mM sulfate, normal seawater, and sulfate-free seawater. At the fourth larval instar the Malpighian tubules were removed and incubated in a medium containing the indicated concentration of SO_4. From Maddrell, S. H. P., and Phillips, J. E. 1978. *J. Exp. Biol.* 72: 181–202.

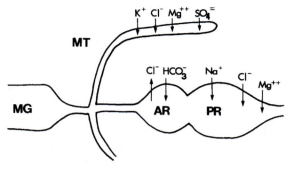

FIGURE 26.5 Diagram of the excretory system of *Aedes taeniorhynchus*. Demonstrated pathways of ion transport and their probable locations are shown. AR = anterior rectum, MG = midgut, MT = Malpighian tubule, PR = posterior rectum. From Bradley, T. J. 1984. *Lecture Notes on Coastal and Estuarine Studies*, Vol. 9, *Osmoregulation in Estuarine and Marine Animals*. Springer-Verlag, Berlin, pp. 35–50.

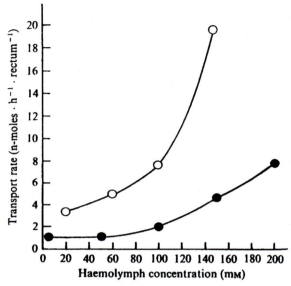

FIGURE 26.6 Relationship between the concentration of chloride (open circles) or sodium (closed circles) in artificial hemolymph bathing ligated recta of *Aedes taeniorhynchus* and the rate of transport of that ion. From Bradley, T. J., and Phillips, J. E. 1977. *J. Exp. Biol.* 66: 97–110.

populations of mosquitoes with an increased ability to deal with highly saline conditions. It is apparent that the capacity of the Malpighian tubules to transport ions can evolve in response to environmental stress.

Morphologically, the rectum in some saline-water species is divided into an anterior and a posterior region. As the urine enters the anterior rectum of *Aedes taeniorhynchus*, ions in short supply, such as potassium, are recovered. As shown in Figure 26.5, the posterior region of the rectum secretes all of the major ions commonly found in saline waters (Na$^+$, Mg^{++}, and Cl$^-$) into the lumen, corresponding to the concentration of the ions in the water. Figure 26.6 shows that the ability of the rectum to secrete Na$^+$ and Cl$^-$ is inducible and that the degree of activation is a function of the amount of the ion in the water. Thus, the presence of a distinct posterior rectum is unique to the saline-water species, and the activity of the posterior part of the rectum permits the excretion of a concentrated urine that allows the animal to maintain a stable hemolymph osmolarity in the face of large changes in salinity of the water.

Brackish-Water Mosquitoes

Brackish-water species, such as *Culiseta inornata* and *Culex tarsalis*, are found in salt marshes where freshwater and saltwater mix. *Culiseta inornata* cannot live in saline concentrations above 80% seawater. This species seems to be physiologically adapted for the medium range of salt concentrations. These mosquitoes are capable of regulating their hemolymph osmotic composition when living in solutions more dilute than 40% seawater, but they can also survive the increase in hemolymph osmotic pressure that occurs between 40% and 80% seawater (Fig. 26.7).

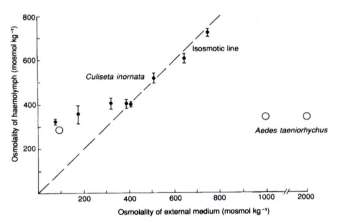

FIGURE 26.7 Osmolality of hemolymph of *Culiseta inornata* reared in different salt concentrations in water. 1000 mosmol/kg is equivalent to 100% seawater. Isosmotic line as in Figure 26.1. Also shown are three points (○) for *Aedes taeniorhynchus* reared under similar conditions. From Garrett, M., and Bradley, T. J. 1984. *J. Exp. Biol.* 113: 133–141.

The recta of *Culiseta inornata* and *Culex tarsalis* are simple, in that they do not have the divided regions that are apparently characteristic of saline-water *Aedes* spp. The urine in the rectum is neither concentrated nor diluted relative to the hemolymph, suggesting that these species are unable to alter the composition of the urine as it passes through the rectum, except for the recovery of needed ions such as potassium. However, the levels of proline in both the intracellular and extra-cellular (hemolymph) compartments of *Culex tarsalis* increase dramatically (up to 50 times) as the animal is exposed to higher concentrations of seawater. Tre-halose, a dimer of glucose, also doubles in concentra-tion in the hemolymph. Interestingly, proline levels rise in response to increases in NaCl concentration of the water, while trehalose levels rise in response to increases in the osmolarity of the water. Thus, it appears that this species uses a mechanism that is dif-ferent from those we have seen in freshwater and salt-water species for dealing with salt load. By increasing amino acid and sugar concentrations, the osmotic pressure of the hemolymph increases to a level that approximates that of the water, and this opposes the loss of water from the hemolymph. The NaCl levels of the hemolymph are modulated within 24h, possibly by NaCl excretion via the anal papillae. These mos-quitoes have, therefore, apparently developed a differ-ent mechanism for avoiding changes due to increased salinity of water that does not require specialized ion transport by the Malpighian tubules or rectum.

The relative abilities of fresh-, saline-, and brackish-water mosquitoes to regulate hemolymph osmolarity can be seen by comparing Figures 26.1, 26.2, and 26.7. Another measure of this ability can be obtained by comparing the maximum salt tolerance for *Aedes aegypti* (500 mosm), *Culiseta inornata* (800 mosm), and *Aedes taeniorhynchus* (3,000 mosm).

Ultrastructure Recapitulates Physiology

How do the Malpighian tubules and the rectum transport water and ions? There has been a consider-able amount of work done on how the structure of these tissues correlates with their function.

Malpighian tubules produce the primary urine by actively pumping potassium across the apical cell membrane into the tubule lumen from the cytoplasm of the cell. K^+-channels exist on the basal membrane (hemolymph side). Water follows by osmosis as K^+ is transported from hemolymph to tubule lumen. There is some evidence that channels specific for the move-ment of water, called *aquaporins*, are present in insect Malpighian tubules. Chloride follows the charge gradient caused by the pumping of potassium into the lumen, either via channels in the basal and apical membrane or perhaps by a paracellular route between the Malpighian tubule cells. Between these cells are smooth septate junctions that form a beltlike structure around the lumen side of the cells. These septate junc-tions form a barrier to some molecules. For example, it has been demonstrated in the kissing bug, *Rhodnius prolixus*, that they restrict the entry of large molecules,

such as hemolymph proteins, and charged organic molecules, such as some amino acids. However, as dis-cussed later for the adult mosquito tubule, septate junctions may be under hormonal control that dra-matically changes their permeability. The evidence for movement of molecules into the lumen via the para-cellular route is well established for nymphs of *Rhodnius prolixus* and for adult mosquitoes (see later) but has not been established for tubules of larval mosquitoes.

Each tubule contains two cell types, the *principal* cell and the *stellate* cell (Fig. 26.8). All of the physiological functions of the tubule have been attributed to the principal cell; and the role of the stellate cells in larvae is less well understood. The basal cell membrane is exposed to the hemolymph and is highly folded. The many folds in the basal membrane increase its surface, thereby increasing the efficiency of the cell. On the lumenal side of the cell, the membrane has many microvilli (Fig. 26.8) that often contain mitochondria. This membrane contains the potassium pumps, and the mitochondria provide the ATP to energize the V-type proton pump located in the apical membrane close by.

It is especially interesting to compare the ultra-structure of the rectum of freshwater and saltwater mosquito larvae. As already mentioned, the rectum of the saltwater species, *Aedes campestris*, is divided into

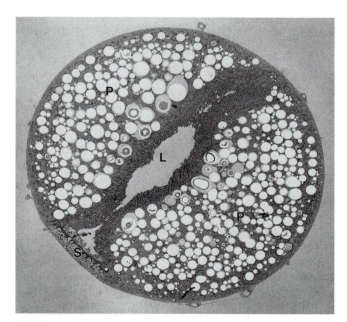

FIGURE 26.8 Structure of Malpighian tubules of larval *Aedes tae-niorhynchus*. The tubule is composed of two cell types, the principal cell (P) and the stellate cell (S). The brush border around the lumen (L) is formed by microvilli of the principal cell. Figure courtesy of T. J. Bradley.

two sections, the anterior and posterior rectum. The anterior region is similar in structure and function to the rectum of the freshwater mosquito, *Aedes aegypti*. Figure 26.9 shows the anterior rectum of *Aedes campestris*. Both the apical and basal membranes are infolded to form microvillar-like structures. The apical membrane infolding extends to about 20% of the total cell depth. The basal infoldings form a canalicular system that extends throughout the cell. Mitochondria are randomly distributed throughout the cell. The contrast with the posterior rectum of *Aedes campestris* is striking (Figs. 26.10A, B). Here, the apical membrane is much more highly infolded (Fig. 26.10A upper), extending to 60% of the total cell depth. The lamellae are thicker and more tightly packed. About 90% of the mitochondria are found in the apical region associated with the lamellae. The basal membrane is infolded to form canals as in the anterior rectum. The most remarkable aspect of this region is that the cell membranes are thrown up into folds or ridges, with the opposing cells forming a complex channel with tree-like branches that penetrate deeply into the cytoplasm (Fig. 26.10 lower).

The structure of the cells of the posterior rectum suggest that the most metabolically active region is the apical portion of the cell, with its highly infolded lamellae and numerous mitochondria. This hypothesis has been confirmed by studies of the movement of ions across the rectal epithelium, as discussed in the next section. Although the function of the complicated, tree-like structure of the basal membrane is not known, it may provide the surface area needed to allow the ions in the hemolymph to penetrate deeply into the cell, where they would flow rapidly into the cytoplasm to replace the ions being pumped into the lumen.

Several conclusions can be drawn from this brief discussion of how mosquito larvae survive in waters of diverse osmotic and ionic composition. First, the physiological mechanisms used vary considerably. For example, the use of amino acids and sugar to adjust osmotic pressure of the hemolymph in a brackish-water species, *Culex tarsalis*, and the development of the unique posterior rectum in some saline-water species. Second, inducible pumps are found in both the rectum and Malpighian tubules in several different saltwater species that respond to changes in ion concentration in the water. Third, in laboratory experiments using *Aedes taeniorhynchus*, it has been possible to select for populations of mosquitoes with increased ability to deal with high saline conditions. Thus, selective pressure in the field is likely to lead to the development of populations of mosquitoes with differing abilities to deal with local conditions.

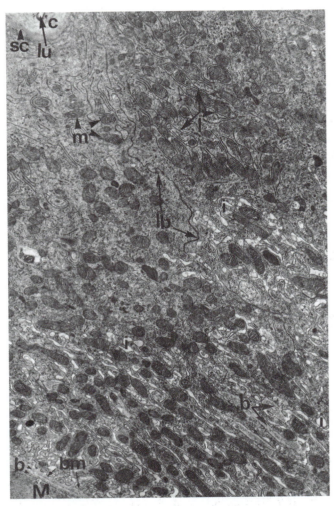

FIGURE 26.9 Ultrastructure of anterior rectum of *Aedes campestris* maintained in hyperosmotic medium. The lateral border is relatively straight and unspecialized, while the apical and basal membranes are infolded. Mitochondria are evenly distributed. b, basement membrane; bm, basal plasma membrane; c, cuticle; i, basal infolding; l, apical lamellae; lb, lateral border; lu, lumen; M, muscle; m, mitochondria; sc, subcuticular layer. Magnification, ×12,900. From Meredith, J., and Phillip, J. E. 1973. *Z. Zellforschung Mikr. Anat.* 138: 1–22.

SALT AND WATER TRANSPORT IN MALPIGHIAN TUBULES OF THE ADULT YELLOW FEVER MOSQUITO

As in the larva, Malpighian tubules in the adult mosquito mediate the first step in the regulation of the volume and composition of the extracellular fluid compartment, the hemolymph. The regulation begins in the distal, blind end of the tubule, where a secretion of the hemolymph is presented to the tubule lumen for

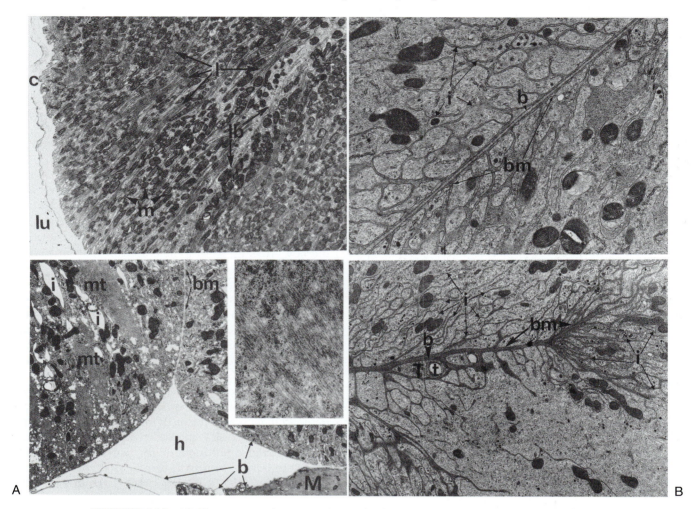

FIGURE 26.10 (A) Ultrastructure of posterior rectum of *Aedes campestris*. Above is shown the apical border of the posterior rectal epithelium at ×5,000. Below is shown the basal border of the posterior rectal epithelium at ×6,000. The inset shows a magnified view of the microtubules at ×24,000. (B) Ultrastructure of the posterior rectum of *Aedes campestris* showing views of the basal borders between two posterior rectal cells. Above ×11,000. Below ×7,000. h, hemocoel; mt, microtubular bundle; T, tracheal cell; t, trachea. Other abbreviations as in Figure 26.9. Figures courtesy of J. Meredith.

modification downstream (Fig. 26.11). The secretion (primary urine) consists largely of NaCl, KCl, and water that are moved from the hemolymph to the tubule lumen by a process known as *tubular secretion* (Table 26.1). Since the osmotic pressure of secreted fluid is close to that of the hemolymph, 354 mosm/kg, tubular secretion appears to be similar to glomerular filtration in the vertebrate kidney, where an ultrafiltrate of the plasma is produced. However, ultrafiltration and tubular secretion are different processes. In glomeruli of vertebrates, ultrafiltration is driven by the hydrostatic pressure in general and the blood pressure

in particular, and what separates filterable from nonfilterable solutes is the geometry and electrical charge of filtration pores in the glomerular basement membrane. Thus, filtration resembles a sieving process where the behavior of filtered solute and water is passive. In contrast, tubular secretion is driven by active transport mechanisms in epithelial cells that remove solutes from the hemolymph and transfer them into the tubule lumen. For example, epithelial cells lining the distal segment of the Malpighian tubule in *Aedes aegypti* use the energy from ATP to transport Na^+ and K^+ from the hemolymph to the tubule lumen.

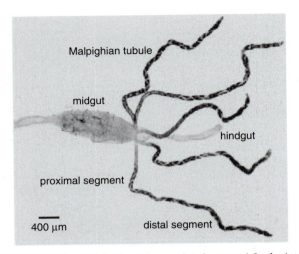

FIGURE 26.11 Malpighian tubules of *Aedes aegypti*. In the intact mosquito the tubules are folded alongside the mid- and hindgut. The blind (distal) end of the tubule begins at the rectum and extends forward to the midgut, where it folds upon itself to return to the junction of the mid- and hindgut, where fluid secreted by the tubule enters the intestine.

TABLE 26.1   Composition of excretory fluids in the yellow fever mosquito *Aedes aegypti*

Source	Osmolality (mOsm/kg H$_2$O)	[Na$^+$] (mM)	[K$^+$] (mM)	[Cl$^-$] (mM)
Hemolymph	354	96	6.5	135
Diuretic urine excreted from rectum *in vivo*, after onset of the blood meal	309	175	4.2	132
Fluid secreted *in vitro* by distal Malpighian tubules				
—under control condition	nm	94	91	161
—stimulated by mosquito natriuretic peptide (MNP)	nm	178	17	185
—stimulated by cAMP*	nm	148	31	181

* cAMP, cyclic adenosine monophosphate, the second messenger of MNP which is released into the hemolymph shortly after the onset of the blood meal (from Petzel, D. H. et al. 1987. Am. J. Physiol. 253: R701–R711; Wheelock G. H. et al. 1988. Arch. Insect Biochem. Physiol. 7:75–89).

Cl$^-$ and water follow passively by electrodiffusion and osmosis, respectively (Beyenbach 1995). The net effect of transepithelial secretion of NaCl, KCl, and water is the production of a luminal fluid volume (secretion) into which toxins, metabolic wastes, and other unwanted solutes can subsequently be "dumped" for excretion.

In the desiccating environment of the terrestrial habitat, all animals are concerned with the conservation of body water. Evaporative water loss must be minimized across the body wall, as must be renal water loss in the excretion of toxins and metabolic wastes. The strategy adopted by terrestrial insects is the reabsorption of solute and water from the tubule lumen without reabsorbing metabolic wastes and toxins. What is not reabsorbed from the secretion as it flows through the proximal Malpighian tubule, hindgut, and rectum is destined for excretion. As KCl, NaCl, and water are reabsorbed, toxins and wastes remaining behind may precipitate out of solution, thereby "freeing" additional water for reabsorption in the quest to conserve body water.

There are occasions when the excretory system of the terrestrial insect must deal with onslaughts of salt and water, as in hematophagous insects such as the yellow fever mosquito, *Aedes aegypti*. Feeding on blood is part of the reproductive cycle in the female, tapping a convenient source rich in nutrients, minerals, and energy for egg production (Clements 1992). But in consuming a blood meal up to four times her own body weight, the female mosquito takes on (1) a large payload that compromises flying and maneuverability, and (2) quantities of salt and water more than those needed for egg production. To deal with both problems, the female begins to urinate even before finishing her meal (Fig. 26.12). So brisk is the rate of urine flow that the NaCl concentration in the urine excreted from her rectum is similar to the NaCl concentration in fluid secreted upstream in the distal Malpighian tubule (Table 26.1). Three causes account for the similarities between excreted urine and fluid secreted upstream in the distal Malpighian tubule: (1) the sharp increase in the rate of NaCl and water secretion in the distal Malpighian tubule, (2) high flow rates down the tubule that reduce the contact time for reabsorbing NaCl and water along its way to the exterior, and (3) probably also the inhibition of NaCl and water absorption in proximal segments of the tubule, the hindgut, and rectum.

The initial diuresis of the blood meal rids the mosquito of excess Na$^+$, Cl$^-$, water, and therefore weight (Fig. 26.12). Later, as the rate of NaCl excretion falls, the rate of KCl excretion increases, reflecting the digestion of red blood cells. The mechanism and regulation of secretion in distal segments of Malpighian tubules of the *Aedes aegypti* have been studied in some detail. These are reviewed following a brief consideration of the experimental methods for investigating Malpighian tubules in vitro.

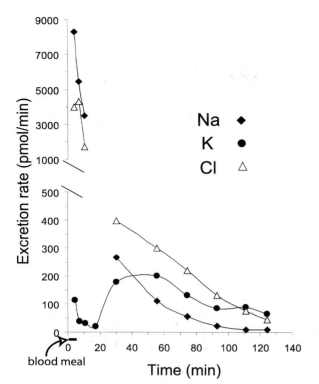

FIGURE 26.12 Blood meal diuresis in *Aedes aegypti*. The mosquito was allowed to feed on blood at time 0 min. Urine droplets exiting from the rectum were collected for measurements of volume and composition. The product of the urine excretion rate and ion concentration yields excretion rates in units of 10^{-12} mol/min (pmol/min). Within 2 minutes of the onset of the blood meal the first urine expelled from the rectum is rich in Na^+ and Cl^-. As the rate of Na^+ excretion falls, the rate of K^+ excretion rises, reflecting the K^+ load of digested red blood cells. Redrawn from Williams, J. C., et al. 1983. *J. Comp. Physiol.* 153: 257–265.

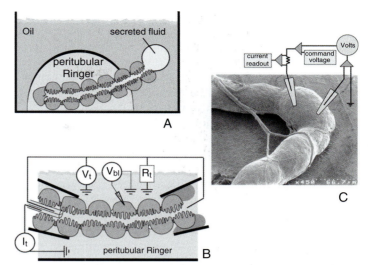

FIGURE 26.13 Methods for studying transport across Malpighian tubules. The Ramsay method (A) yields fluid secretion rates and the ionic composition of secreted fluid. The composition of the bathing medium (peritubular Ringer) is known. In vitro microperfusion of isolated tubules (B) yields measurements of transepithelial and transmembrane voltage and resistance for the study of electrogenic and electroconductive transport pathways. V, voltage; R, resistance, I, current; bl, basolateral membrane of principal cell; t, transepithelial. Two-electrode voltage clamp method (C) of a principal cell identifies ionic currents across basolateral and apical membranes of principal cells. A single principal cell is impaled with current and voltage microelectrodes. Note tracheal tubes invading the tubule for respiratory support. From Beyenbach, K. W., and Dantzler, W. H. 1990. *Methods Enzymol.* 191: 167–226, and Masia, R., et al. 2000. *Am. J. Physiol.* 279: F747–F754.

The Study of Malpighian Tubules in Vitro

The Ramsay assay of fluid secretion is the most popular method for investigating fluid secretion in Malpighian tubules in vitro (Ramsay 1953). As shown in Figure 26.13, a 3-mm-long distal (terminal) segment of the tubule is deposited in a droplet of Ringer solution under oil, and the open end of the tubule is pulled into the oil so that fluid secreted by the epithelial cells and flowing out of the open end of the tubule can accumulate as a droplet separate from the Ringer in which the tubule is bathed. The growth of the secreted droplet can be followed with measurements of its diameter to yield the secreted volume and the secretion rate with time. Secreted fluid can be collected and analyzed for its composition. After a control period, potential stimulators and inhibitors can be added to

the peritubular Ringer droplet to evaluate diuretic and antidiuretic effects. The Ramsay method provides important information about transport mechanisms and its regulation. Moreover, the Ramsay assay is widely used as bioassay for screening large numbers of fractions in the search for new diuretic or antidiuretic hormones.

Since the major solutes secreted by the distal Malpighian tubule of *Aedes aegypti* are Na^+, K^+, and Cl^-, their transepithelial movement is affected by electrical as well as chemical potentials. The Ramsay assay yields transepithelial chemical potentials from the analysis of ion concentrations in the peritubular Ringer and secreted fluid. However, the Ramsay method is not well suited for measurements of transepithelial electrical potentials. Transepithelial voltages (V_t) are best measured in isolated, perfused Malpighian tubules under conditions when the fluid perfusing the tubule lumen is the same as the bathing medium substituting for the hemolymph (Fig. 26.13B). A transepithelial voltage measured under these conditions, i.e., in the absence of transepithelial chemical potentials, reflects an active transport function of the tubule. It is therefore considered to be an "active transport voltage." Active transepithelial transport voltages in the Malpighian tubules of *Aedes aegypti* are on average 53 mV and lumen-positive, reflecting the active transport of cations, in particular the transepithelial secretion of Na^+ and K^+ from the peritubular medium (or hemolymph) into the tubule lumen. Cable analysis in isolated perfused tubules yields the transepithelial resistance (R_t). Since the electrical response of the tubule is observed immediately upon the application of potential transport stimulators and inhibitors, measures of V_t and R_t have been useful to screening large numbers of extract fractions in the search for new diuretic and antidiuretic hormones and neuropeptides (Petzel et al. 1985).

The in vitro microperfusion method of Malpighian tubules can be supplemented with microelectrode measurements of the voltage (V_{bl}) and the fractional resistance of the basolateral membrane of principal cells to yield additional information about events taking place at the basolateral cell membrane facing the peritubular bath or hemolymph and at the apical membrane facing the tubule lumen (Fig. 26.13B).

Whereas the method of in vitro microperfusion yields an extracellular perspective of transepithelial transport, two-electrode voltage clamp methods of principal cells allow an intracellular vantage point for observing transport across basolateral and apical membranes (Fig. 26.13C). In brief, an isolated Malpighian tubule is placed on the bottom of the perfusion bath. A principal cell is impaled with a voltage and a current electrode. The cell can then be clamped at any voltage by injecting current into the cell. The current needed to hold the cell at a certain clamp voltage is carried by ions entering or leaving the cell. Specific ionic currents can then be identified in ion-replacement experiments and further studied in the absence and presence of potential diuretic and antidiuretic agents (Masia et al. 2000).

The Distal Malpighian Tubule of the Adult Female Mosquito

Malpighian tubules of the adult male mosquito are only one-half the size of their counterparts in the female, and they secrete Na^+, K^+, Cl^-, and water in vitro at rates only one-sixth those of female tubules. The structural dimorphism and the amplification of secretory functions in the female apparently reflect the excretory pressures to eliminate the large volumes of salt and water of the blood meal. Principal cells are the dominant cell type in Malpighian tubules (Fig. 26.14). In tubules of the female *Aedes aegypti* they are large fusiform cells with a central nucleus positioned below the basolateral membrane (Fig. 26.14A). Principal cells are further characterized by a large number of intracellular spherites that in *Drosophila* are composites of Ca^{++}, Mg^{++}, K^+, proteoglycans, and glycosaminoglycans (Fig. 26.14B). Also unique to principal cells is a tall brush border, the microvilli of which are home to mitochondria (Fig. 26.14C). As is shown later, mitochondria positioned in microvilli provide the energy to fuel the active transport steps for moving K^+ and Na^+ across the

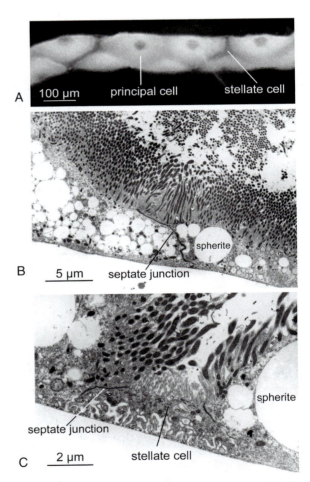

FIGURE 26.14 Distal Malpighian tubule of the adult female *Aedes aegypti*. (A) The tubule consists of principal and stellate cells in a ratio of approximately 6:1. (B) Principal cells are characterized by a large number of intracellular spherites and a tall brush border with microvilli containing mitochondria. (C) Stellate cells are small, thin, and transparent, allowing a view of the tubule lumen. A septate junction between two principal cells is shown in (B) and between principal and stellate cells in (C). For details about intracellular spherites, see Wessing, A., et al. 1992. *J. Insect Physiol.* 38: 543–554.

apical cell membrane from the cytoplasm into the tubule lumen.

Fewer than 20% of tubule cells are stellate cells. The basolateral membrane of stellate cells shows many infoldings (Fig. 26.14C). The apical membrane facing the tubule lumen gives rise to short, thin microvilli that are devoid of mitochondria. In Malpighian tubules of *Aedes aegypti*, stellate cells are too small for measurements with intracellular microelectrodes. For this reason, their function must be deduced from extracellular approaches. Transepithelial current peaks in the vicinity of stellate cells in Malpighian tubules of *Drosophila* are thought to provide a pathway for transepithelial Cl⁻ secretion (O'Donnell and Spring 2000). In *Aedes* Malpighian tubules, a high density of Cl⁻ channels has been found in the apical membrane. The Cl⁻ channels could be part of a transcellular pathway for Cl⁻ secretion under control conditions. However, Cl⁻ takes primarily a paracellular route through septate junctions when the tubule is stimulated by the diuretic peptide leucokinin (see later).

The Basic Secretory Transport System in the Distal Malpighian Tubule

Under control conditions, i.e., in the absence of diuretic agents, the isolated distal Malpighian tubule of the adult female *Aedes aegypti* secretes fluid spontaneously at a rate of 0.4 nL/min (Fig. 26.15A). A transepithelial electrical resistance of 11.4 kΩcm tubule length reflects a moderately tight epithelium poised to secrete electrolytes at modest transport rates under control, nondiuretic conditions (Beyenbach 1995).

The concentrations of NaCl and KCl in secreted fluid are approximately equimolar, between 90 and 95 mM (Fig. 26.15A, Table 26.1). Measurements of transepithelial electrochemical potentials show that the secretion of Na⁺ and K⁺ into the tubule lumen is active, against their respective electrochemical potentials. The energy for moving Na⁺ and K⁺ "uphill" into the tubule lumen is provided by ATP, which is the reason why the active transport pathway for both Na⁺

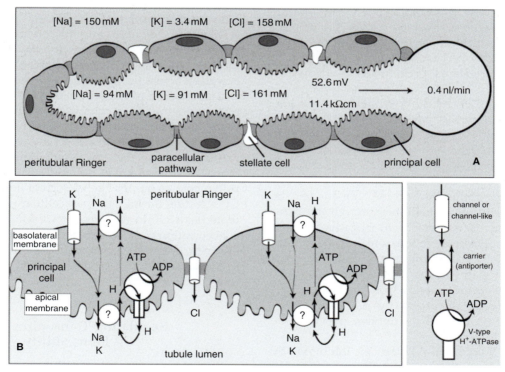

FIGURE 26.15 Epithelial transport mechanisms of the distal Malpighian tubules of *Aedes aegypti*. (A) Under control conditions, the isolated tubule secretes a fluid consisting largely of NaCl, KCl, and water. (B) K⁺ enters principal cells from the hemolymph through K⁺-channels, and Na⁺ enters presumably in exchange for H⁺. Proton electrochemical potentials across the apical membrane generated by the vacuolar-type H⁺-ATPase drive the extrusion of Na⁺ and K⁺ into the tubule lumen via H⁺/cation⁺ exchange transport. From Beyenbach, K. W., 2001. *News Physiol. Sci.* 16: 145–151.

and K$^+$ passes through principal cells (Fig. 26.15B). Transepithelial secretion of Cl$^-$ is passive, down the transepithelial Cl$^-$ electrochemical potential, which does not require ATP. Cl$^-$ may pass into the tubule lumen through stellate cells or through the paracellular, septate junctional pathway (Fig. 26.15B).

The reader probably expects the ATP-driven Na/K-ATPase to play a dominant role in transepithelial ion transport in Malpighian tubules, as it does in most eucaryotic cells. Although Malpighian tubules of *Aedes aegypti* display evidence for the presence of the Na/K-ATPase, this Na/K-pump plays a minor role in transepithelial transport. Instead, the vacuolar-type H$^+$-ATPase plays the central role powering the transepithelial secretion of Na$^+$, K$^+$, Cl$^-$, and probably other solutes (Beyenbach 2001). The V-type H$^+$-ATPase is an electrogenic proton pump located in the apical membrane of principal cells (Fig. 26.15B). Protons secreted by the V-type H$^+$-ATPase into the microenvironment of the luminal brush border return to the cytoplasm in exchange for Na$^+$ and K$^+$ via H$^+$/K$^+$ and H$^+$/Na$^+$ antiporters (exchange transporters) that have yet to be isolated (Fig. 26.15B). Whether a single antiporter (H$^+$/cation$^+$) can accept both Na$^+$ and K$^+$ or whether two separate antiporters are in operation remains to be determined. Likewise, it is unknown whether antiport is electroneutral with a stoichiometry of 1:1 or electrogenic with a stoichiometry of 2:1, or other ratios. Together, V-type H$^+$-ATPase and H$^+$/cation$^+$ form a continuous bioenergetic H$^+$ cycle fueled by the hydrolysis of ATP.

The extrusion of H$^+$ by the V-type H$^+$-ATPase into the tubule lumen carries positive current that must return to the cytoplasmic face of the proton pump. The current passes first through the paracellular, septate junction and then through K$^+$ channels in the basolateral membrane of principal cells (Fig. 26.15B). Positive current passing from the tubule lumen to the hemolymph (peritubular medium) through the paracellular pathway is equivalent to negative current moving in the opposite direction, which is the mechanism for secreting Cl$^-$ into the tubule lumen through the paracellular pathway of septate junctions (Fig. 26.15B). Positive current passing from the hemolymph into principal cells is carried by K$^+$ through K$^+$-channels in the basolateral membrane, which is the mechanism for K$^+$ entry.

When the V-type H$^+$-ATPase is inhibited by bafilomycin or by withholding ATP, transepithelial secretion of NaCl, KCl, and water comes to a halt. In addition, all membrane and transepithelial voltages collapse to values near zero, pointing to the central role of the apical membrane proton in powering transepithelial transport (Beyenbach et al. 2000).

Getting Rid of the NaCl Load of the Blood Meal

Within a minute or two of feeding on blood, the female *Aedes aegypti* commences a potent diuresis that rids the insect of the unwanted NaCl and water of the meal. As shown in Table 26.1 and Figure 26.12, the first urine to be excreted from the rectum is rich in NaCl. Gorging on blood is sensed by stretch receptors near the stomach or in the abdominal wall that are thought to cause the release of diuretic hormones into the hemolymph. The diuretic hormone causing the initial NaCl diuresis in *Aedes aegypti* is the mosquito natriuretic peptide (MNP). MNP probably belongs to the CRF-like diuretic peptides in insects, namely, that family of insect diuretic peptides resembling vertebrate sauvagine, urotensin, and corticotropin-releasing factor (Coast 1998).

In isolated Malpighian tubules, MNP brings about the selective stimulation of transepithelial NaCl secretion, as shown in Figure 26.16A. Cyclic AMP is likely to be the second messenger of MNP, because MNP increases intracellular cAMP levels in the tubule and cAMP mimics the effects of MNP on transepithelial NaCl and fluid secretion. Eletrophysiological studies show that MNP and cAMP increases the Na$^+$ conductance of the basolateral membrane in principal cells (Fig 26.16B). Na$^+$ entry into the cell (now also carrying positive current returning to the proton pump) increases the competitive status of intracellular Na$^+$ for extrusion across the apical membrane. Thus, the selective stimulation of transepithelial Na$^+$ is brought about by the opening of Na$^+$ channels in the basolateral membrane and, consequently, increased rates of Na$^+$ entry into principal cells followed by increased rates of Na$^+$ extrusion across the apical membrane (Fig. 26.16B). Since the stimulation of transcellular active Na$^+$ transport increases intraepithelial current flow (Fig. 26.15B), Cl$^-$ secretion through the paracellular pathway is stimulated as well. In the presence of MNP or cAMP, the concentration of K$^+$ in secreted fluid drops by a factor of seven. The decrease is merely a dilution effect due to the sevenfold increase in NaCl and water excretion (Figs. 26.15 and 26.16).

Regulating Transepithelial Cl$^-$ Permeability

As already discussed, one way to increase urine flow (diuresis) is to increase the rate of transepithelial Na$^+$ secretion via stimulation with the MNP transport. Another way to induce diuresis is to increase the transepithelial permeability to Cl$^-$, which is the mechanism of action of the diuretic peptide leucokinin. As

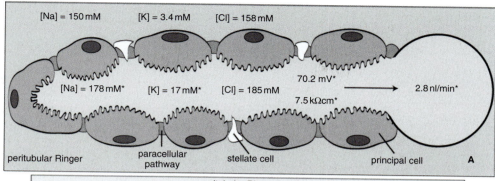

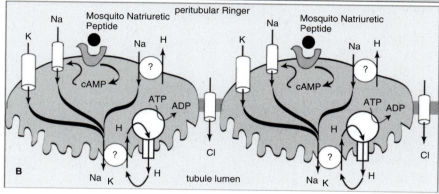

FIGURE 26.16 Mechanism of action of mosquito natriuretic peptide (MNP) in isolated distal Malpighian tubules of *Aedes aegypti*. (A) MNP increases the rate of fluid secretion sevenfold over control rates (see Fig. 26.15) by selectively stimulating transepithelial Na^+ and not K^+ secretion. MNP hyperpolarizes the transepithelial voltage and decreases the transepithelial resistance, consistent with the stimulation of an electroconductive transport pathway. (B) MNP increases intracellular cAMP concentrations, and cAMP goes on to activate Na^+ channels in the basolateral membrane of principal cells. * = significantly different from control, Figure 26.15. From Petzel, D. H., et al. 1987. *Am. J. Physiol.* 253: R701–R711, and Sawyer, D. B., and Beyenbach, K. W. 1985. *Am. J. Physiol.* 248: R339–R345.

shown in Figure 26.17A, leucokinin significantly stimulates the rate of transepithelial fluid secretion, with only small changes in the concentrations of K^+, Na^+, and Cl^- in secreted fluid. The nonselective stimulation of both NaCl and KCl, secretion would be expected from the increase in the septate junctional permeability to Cl^-, the anion common to both K^+ and Na^+. Furthermore, leucokinin causes the transepithelial voltage to drop to values close to zero, reflecting an electrical short-circuit that is caused by the increase in the paracellular Cl^- conductance (Fig. 26.17). The increase in the paracellular Cl^- permeability is corroborated by the substantial drop in the transepithelial resistance in the presence of leucokinin (Pannabecker et al. 1993).

The leucokinins mediate diuresis in several insect species, always mobilizing Ca^{++} as second messenger. In Malpighian tubules of *Aedes aegypti*, Ca^{++} residing in intracellular stores is insufficient to mediate the effects of leucokinin. Extracellular Ca^{++} and the activation of

Ca^{++} channels in the basolateral membrane are needed to mediate the full effects of leucokinin (Yu and Beyenbach 2002). The rise in cytoplasmic Ca^{++} concentrations and/or the depletion of intracellular Ca^{++} stores may then activate Ca^{++} channels in the basolateral membrane of principal cells. How cytosolic Ca^{++} goes on to increase septate junctional Cl^- permeability remains to be determined. Both epithelial and endothelial tight junctions are known to display permselectivity like septate junctions, and tight junction proteins such as the paracellins have been found in association with regulatory proteins and the cytoskeleton.

The leucokinins are a family of C-terminal amides that were first discovered by Holman and coworkers to stimulate the frequency of smooth muscle contractions in the cockroach *Leucopheae* (Holman et al. 1987). They were subsequently found also to possess diuretic potency in Malpighian tubules of *Aedes aegypti*. Since then, diuretic effects of leucokinins have been

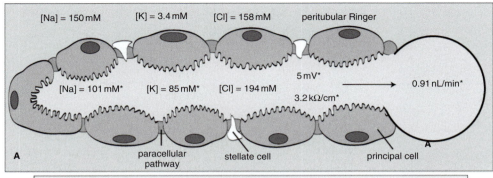

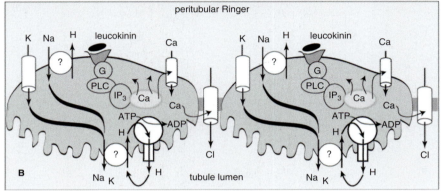

FIGURE 26.17 Mechanism of action of leucokinin in isolated distal Malpighian tubules of *Aedes aegypti*. (A) Leucokinin increases the rate of fluid secretion twofold over control rates without significant changes in the concentrations of Na⁺, K⁺, and Cl⁻ in secreted fluid (see Fig. 26.15). At the same time, the transepithelial voltage and resistance plummet to low values. The effects on electrolyte secretion and electrophysiology point to the increase in paracellular Cl⁻ permeability (conductance). (B) The effects of leucokinin are mediated in part by G-protein (G), phospholipase C (PLC), inositol triphophate (IP3), and Ca²⁺. * = significantly different from control, Fig. 26.15. From Pannabecker, T. L., et al. 1993. *J. Membr. Biol.* 132: 63, and Yu, M. J., and Beyenbach, K. W. 2002. *Am. J. Physiol.* 283: F508.

observed in the house cricket, locust, tobacco hornworm, fruit fly, and housefly. As many as 30 leucokinins have now been isolated and sequenced in four orders and nine species of insects, including three leucokinin-like peptides of the yellow fever mosquito (Veenstra et al. 1997).

Spontaneous Regulation of K⁺ Secretion

A third way to increase rates of transepithelial fluid secretion is to increase the rate of transepithelial K⁺ secretion. So far, a hormone or extracellular peptide that stimulates K⁺ secretion across Malpighian tubules has not been found. Such a kaliuretic hormone seems not needed in view of the high K⁺ permeability of the basolateral membrane in principal cells of the *Aedes* Malpighian tubule. More than half of the electrical conductance of this membrane, 64%, is due to the presence of K⁺ channels that are open most of the time, even under control conditions. A high K⁺ conductance of the basolateral membrane in *Aedes* Malpighian tubules

allows intracellular K⁺ to be near electrochemical equilibrium with hemolymph K⁺, as in the Malpighian tubules of the ant, *Formica*. Thus, an increase in the hemolymph K⁺ concentration would be expected to immediately raise intracellular K⁺, thereby stimulating K⁺ extrusion across the apical membrane into the tubule lumen. A decrease in hemolymph K⁺ concentration would be expected to have the opposite effect. Accordingly, the regulation of transepithelial K⁺ secretion is spontaneous; it is intrinsic to the tubule on account of the high K⁺ conductance of the basolateral membrane. Positive correlations between the rate of fluid secretion and hemolymph [K⁺] observed in Malpighian tubules of several insect species are consistent with an intrinsic mechanism for regulating transepithelial K⁺ secretion.

Diuretic Effects of Serotonin

Serotonin, also known as 5-hydroxytryptamine (5-HT), a peptide, is the first diuretic hormone identified in insects. In the Malpighian tubules of the blood-feeding insect *Rhodnius prolixus*, 5-HT binds to the 5-HT-2 receptor, which activates a Gs-protein and a cAMP-dependent protein kinase, leading to the inhibition of the Na/K ATPase. The inhibition of the Na/K ATPase in basolateral membranes of principal cells increases the competitiveness of cytoplasmic Na^+ for secretion into the tubule lumen. Indeed, inhibition of the Na/K ATPase with ouabain increases rates of transepithelial Na^+ secretion in *Rhodnius prolixus* Malpighian tubules.

5-HT is also a diuretic in the larval mosquito, where it stimulates fluid secretion via increased concentrations of intracellular cAMP in the Malpighian tubules. Although 5-HT has been observed to stimulate fluid secretion in adult mosquito Malpighian tubules, its function as a diuretic agent in the adult mosquito is questionable. 5-HT transcript and peptide have been identified in tracheolar cells and the hindgut but not in the Malpighian tubules of adult *Ae. aegypti* (Pietrantonio et al. 2001).

Other Diuretic and Antidiuretic Agents

Studies in insects other than mosquitoes have identified two additional peptides that affect fluid secretion in Malpighian tubules. The first is the cardio-accelerator peptide CAP-2b, which can stimulate or inhibit fluid secretion in Malpighian tubules in different species. In the Malpighian tubules of *Drosophila*, CAP-2b stimulates fluid secretion via cGMP, the stimulation of NO synthase, the production of NO, and the elevation of intracellular Ca^{++} concentration. In contrast, CAP-2b inhibits fluid secretion in Malpighian tubules of *Rhodnius*, also via a cGMP-mediated signaling pathway.

The second diuretic peptide is calcitonin-like and named Drome-DH31. It has been isolated from *Drosophila melanogaster*. Drome-DH31 stimulates transepithelial fluid secretion by stimulating the apical membrane V-type H^+-ATPase via cAMP (Coast et al. 2001).

The Movement of Water Across Malpighian Tubules

As in other epithelia, transcellular and paracellular pathways have been observed for the passage of water across Malpighian tubules. Evidence for paracellular water flow is obtained by comparing transepithelial solute and water fluxes. If the transepithelial secretion of mannitol, sucrose, or dextran increases in proportion to the rate of transepithelial water flow, an extracellular, paracellular pathway must be assumed, because cell membranes are impermeable to these

three solutes. Using this experimental approach in studies of the Malpighian tubules of *R. prolixus*, it was estimated that septate junctions have a physiological pore size of 11 Å, large enough for water with a molecular diameter of 2 Å. Evidence was also found for a transcellular water transport through water channels (Echevarria et al. 2001). Serotonin (5-HT) increases transepithelial fluid secretion across isolated Malpighian tubules of *Rhodnius prolixus* from values near zero at rest to more than $50\,nL/cm^2$ via a pathway that can be inhibited by $HgCl_2$ and organomercurials. Mercury-related compounds are known to block aquaporin water channels, and a search of water channels in *Rhodnius* Malpighian tubules has yielded a peptide of 286 amino acids, named RP-mip. The peptide is thought to form six membrane-spanning domains typical of aquaporin water channels. Fitting RP-mip to a dendrogram of water channels indicated an ancient peptide. In situ hybridization analysis of mRNA revealed RP-mip along the entire length of the Malpighian tubule. The search of water channels in Malpighian tubules of *Aedes aegypti* has so far yielded positive results in tracheal tubes associated with Malpighian tubules, but not in Malpighian tubules per se.

Studies of Malpighian tubules with the methods of molecular biology, cell physiology, and biophysics have revealed the mechanisms of diuresis and antidiuresis in unprecedented detail, but they have also uncovered complexities. At least four levels of complexity can be observed: (1) dose-dependent effects of diuretic neuropeptides on transcellular and paracellular transport pathways suggestive of receptor subtypes and different signaling pathways, (2) cell-type-specific signaling in one set of genetically defined principal cells and not in another, (3) synergistic effects of two or more diuretic agents, and (4) coordination and integration of transport activities in Malpighian tubules and the gut by a single diuretic or antidiuretic agent. Further study is needed to unravel these complexities, and genomic and proteomic approaches are needed to define the genetic controls for regulating transport across Malpighian tubules.

Acknowledgments

Thanks are due to Dr. T. J. Bradley for providing Figure 26.8 and to Dr. J. Meredith for Figures 26.9 and 26.10.

Readings

Beyenbach, K. W. 2001. Energizing epithelial transport with the vacuolar H^+-ATPase. *News Physiol. Sci.* 16: 145–151.

Bradley, T. J. 1987. Physiology of osmoregulation in mosquitoes. *Annu. Rev. Entomol.* 32: 439–462.

Clements, A. N. 1992 *The Biology of Mosquitoes.* Volume 1, *Development, Nutrition and Reproduction*; Volume 2, *Sensory Reception and Behavior.* Chapman and Hall, London.

Coast, G. M. 1998. Insect diuretic peptides: Structure, evolution and actions. *Am. Zool.* 38: 442–449.

Nicolson, S. W. 1993. The ionic basis of fluid secretion in the insect Malpighian tubule: Advances in the last ten years. *J. Insect Physiol.* 39: 451–458.

O'Donnell, M. J., and Spring, J. H. 2000. Modes of control of insect Malpighian tubules: Synergism, antagonism, cooperation and autonomous regulation. *J. Insect Physiol.* 46: 107–117.

Further Readings

Beyenbach, K. W. 1995. Mechanism and regulation of electrolyte transport in Malpighian tubules. *J. Insect Physiol.* 41: 197–207.

Beyenbach, K. W., Pannabecker, T. L., and Nagel, W. 2000. Central role of the apical membrane H^+-ATPase in electrogenesis and epithelial transport in Malpighian tubules. *J. Exp. Biol.* 203: 1459–1468.

Clark, T. W., and Bradley, T. J. 1996. Stimulation of Malpighian tubules from larval *Aedes aegypti* by secretagogues. *J. Insect Physiol.* 42: 593–602.

Coast, G. M., Webster, S. G., Schegg, K. M., Tobe, S. S., and Schooley, D. A. 2001. The *Drosophila melanogaster* homologue of an insect calcitonin-like diuretic peptide stimulates V-ATPase activity in fruit fly Malpighian tubules. *J. Exp. Biol.* 204: 1795–1804.

Echevarria, M., Ramirez-Lorca, R., Hernandez, C. S., Gutierez, A., Mendez-Ferrer, S., Gonzalez, E., Toledo-Aral, J. J., Ilundain, A. A., and Whittembury, G. 2001. Identification of a new water channel (Rp-MIP) in Malpighian tubules of the insect *Rhodnius prolixus*. *Eur. J. Physiol.* 442: 27–34.

Garrett, M. A., and Bradley, T. J. 1987. Extracellular accumulation of proline, serine and trehalose in the haemolymph of osmoconforming brackish-water mosquitoes. *J. Exp. Biol.* 129: 231–238.

Holman, G. M., Cook, B. J., and Nachman, R. J. 1987. Isolation, primary structure and synthesis of leucokinin-VII and -VIII: The final members of the new family of cephalomyotropic peptides isolated from head extracts of *Leucophaea maderae*. *Comp. Biochem. Physiol.* 88: 31–34.

Masia, R., Aneshansley, D., Nagel, W., Nachman, R. J., and Beyenbach, K. W. 2000. Voltage-clamping single cells in intact Malpighian tubules of mosquitoes. *Am. J. Physiol.* 279: F747–F754.

Pannabecker, T. L., Hayes, T. K., and Beyenbach, K. W. 1993. Regulation of epithelial shunt conductance by the peptide leucokinin. *J. Membr. Biol.* 132: 63–76.

Patrick, M. L., and Bradley, T. J. 2000. Regulation of compatible solute accumulation in larvae of the mosquito *Culex tarsalis*: Osmolarity versus salinity. *J. Exp. Biol.* 203: 831–839.

Petzel, D. H., Hagedorn, H. H., and Beyenbach, K. W. 1985. Preliminary isolation of mosquito natriuretic factors. *Am. J. Physiol.* 249: R379–R386.

Pietrantonio, P. V., Jagge, C., and McDowell, C. 2001. Cloning and expression analysis of a 5HT7-like serotonin receptor cDNA from mosquito *Aedes aegypti* female excretory and respiratory systems. *Insect Molec. Biol.* 10: 357–369.

Veenstra, J. A., Pattillo, J. M., and Petzel, D. H. 1997. A single cDNA encodes all three *Aedes* leucokinins, which stimulate both fluid secretion by Malpighian tubules and hindgut contractions. *J. Biol. Chem.* 272: 10402–10407.

Yu, M. J., and Beyenbach, K. W. 2002. Leucokinin activates Ca^{2+}-mediated signal pathway in principal cells of *Aedes aegypti* Malpighian tubules. *Am. J. Physiol.* 283: F499–F508.

Immune Responses of Vectors

CAROLINA BARILLAS-MURY, SUSAN PASKEWITZ, AND MICHAEL R. KANOST

INTRODUCTION

Insects mount efficient antimicrobial responses to pathogens through activation of the *innate* immune system. These innate responses are fast and transient, and they differ from vertebrate *adaptive* immune responses in that they do not result in genomic rearrangement of epitope-specific-recognition receptor molecules or clonal expansion of specific cell populations, nor do they induce long-lasting pathogen-specific "memory" cells. Recent studies have revealed that the innate immune system is ancient and that the signal transduction pathways mediating responses to pathogens have been conserved throughout evolution, all the way from insects to vertebrates, including humans. There is also increasing evidence that the initial interaction of a microorganism with the vertebrate innate immune system plays a critical role in guiding the adaptive immune responses in the direction appropriate for that particular pathogen.

Most of our current understanding of the insect defense responses derives from the combined use of biochemical techniques to purify and characterize the activities of individual protein components, mostly from large insects, such as lepidopteran moths, and also from *Drosophila melanogaster*, the use of the powerful genetic tools from *D. melanogaster*, and the cloning of many immune-related genes from multiple insect species. In this chapter we present an overview of our current understanding of the overall organization and regulation of the insect immune system (Fig. 27.1) and use it as a frame of reference to discuss the defense responses of insect disease vectors and how these responses may impact vector competence

and thus disease transmission. We will begin by describing the mechanisms that allow insects to recognize pathogens and initiate defense reactions and discussing the importance of the regulated activation of protease cascades to modulate these responses. We will then examine some of the main effector mechanisms, such as phagocytosis and synthesis of antimicrobial peptides. Special emphasis will be given to melanotic encapsulation responses, for this mechanism is involved in defense responses to parasites and filarial worms. We will finish by examining the defense responses of midgut epithelial cells to ookinete invasion and discussing future directions in the field, pointing out some of the fundamental questions that still remain unanswered.

PATHOGEN RECOGNITION AND ACTIVATION OF IMMUNE RESPONSES

Deciding When to Activate the Immune System

For a defense response to be triggered by an infected organism, the pathogen has to be detected and self-damage avoided by distinguishing it from self-tissues. The pattern-recognition theory proposed by Charles Janeway states that immune recognition is mediated by the products encoded by the genomes of two organisms (that of the host and that of the pathogen) acted upon by conflicting selective pressures. The development of a recognition system confers a selective advantage to the host, while avoiding recognition confers a

TABLE 27.1   Pattern-Recognition Proteins in Insects

Protein Family	Mass (kDa)	Domain Structure	Polysaccharides Recognized
β-1,3-Glucan-recognition protein/gram-negative-bacteria–binding protein	53	N-terminal glucan-binding domain, C-terminal glucanase-like domain	β-1,3-Glucan, lipoteichoic acid, lipopolysaccharide
C-type lectin	15–36	1–2 C-type lectin domains	Lipopolysaccharide
Hemolin	8	4 Ig domains	Lipopolysaccharide, lipoteichoic acid
Peptidoglycan-recognition protein	19	Bacteriophage lysozyme-like domain	Peptidoglycan

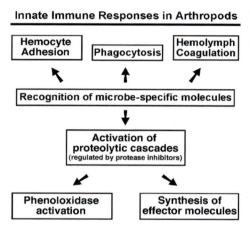

FIGURE 27.1 Diagrammatic representation of the overall organization of the innate immune system in arthropods.

selective advantage to the pathogen. As a result of these interactions, the host innate immune system has evolved pattern-recognition receptors that interact with pathogen-associated molecular patterns, which are essential for pathogen survival. Examples of pathogen-associated molecular patterns are lipopolysaccharides and teichoic acids from gram-negative and gram-positive bacteria, respectively; viral double-stranded RNAs; and mannans present in yeast cell walls. This theory predicts that pattern-recognition receptors and the signal transduction pathways they activate are ancient and that they arose early and have been conserved throughout evolution.

Pathogen-Recognition Molecules Characterized in Insects

Several types of insect plasma proteins have been shown to function as pattern-recognition molecules (Table 27.1). They bind to one or more polysaccharide

or lipopolysaccharide components on the surface of microbial cells. These interactions stimulate immune responses, such as phagocytosis, nodule formation, and phenol oxidase activation. However, the molecular mechanisms that link the "recognition" step to these protective responses have not yet been worked out in detail. Many of the insect pattern-recognition proteins that have been characterized are able to bind to more than one type of microbial molecular pattern, which makes them well suited for surveillance and to detect infection by a broad spectrum of potential pathogens. These proteins are synthesized primarily by the fat body but may also be produced by hemocytes or other tissues.

Hemolin is a pattern-recognition protein that has so far been identified only in lepidopteran insects. It is a 47-kDa protein composed of four immunoglobulin domains. The immunoglobulin domains in hemolin are most similar to those found in cell adhesion molecules of insects and vertebrates. There is no apparent ortholog of hemolin in the *D. melanogaster* genome, but the first four immunoglobulin domains in neuroglian, a membrane-bound cell adhesion protein identified in *D. melanogaster*, align well with the hemolin amino acid sequence. It seems likely that hemolin and neuroglian have evolved from a common ancestral gene. Hemolin is present at a low level in healthy larvae, and its concentration is greatly increased after challenge with bacteria. Hemolin binds to the surface of bacteria through interactions with lipopolysaccharide on gram-negative bacteria and with lipoteichoic acid on gram-positive bacteria and also binds to hemocytes. The ability of hemolin to bind both bacteria and hemocytes suggests that it may play a role facilitating the attachment of bacteria to the hemocyte surface, as a first step in phagocytosis or nodule formation.

A family of proteins that bind to peptidoglycan in bacterial cell walls was first identified in hemolymph of the silkworm, *Bombyx mori,* and later in other insects, including the mosquito *Anopheles gambiae,* and it has now been found in mammals. These peptidoglycan-recognition proteins are similar in sequence to lysozymes encoded by bacteriophage. Lysozyme is an enzyme that hydrolyzes glycosidic bonds in peptidoglycans. The insect and mammalian peptidoglycan-recognition proteins bind peptidoglycan but lack lysozyme enzymatic activity. In *B. mori*, peptidoglycan-recognition proteins are required for activation of the phenol oxidase cascade by grampositive bacteria. Like hemolin, the synthesis of peptidoglycan-recognition proteins is stimulated by bacterial challenge. *Bombyx mori* contains at least two peptidoglycan genes coding for recognition proteins, and in *D. melanogaster* 12 genes related to peptidoglycan-recognition proteins are expressed. Some homologues of the *D. melanogaster* peptidoglycanrecognition proteins appear to be intracellular or membrane-spanning proteins rather than plasma proteins. It seems likely that members of the peptidoglycan-recognition proteins family may bind other ligands besides peptidoglycan, and they may have several roles in immunity in addition to phenol oxidase activation.

Another family of pattern-recognition molecules bind to β-1,3-glucans in fungal cell walls and to lipopolysaccharides from gram-negative bacteria. They have been named β-1,3-glucan-recognition proteins or gram-negative-bacteria-binding proteins. Sequences of the β-1,3-glucan-recognition proteins contain two domains. The carboxyl-terminal domain is related to the sequence of a variety of glucanases, enzymes that hydrolyze glycosidic bonds in glucans, but the β-1,3-glucan-recognition proteins lack enzymatic activity. As in the case of peptidoglycanrecognition proteins, a protein that functioned to hydrolyze a specific polysaccharide has evolved in insects to bind the polysaccharides and act as a recognition molecule. The amino-terminal domain of approximately 15 kDa appears to be unique to invertebrates. This domain binds tightly to β-1,3 glucans in the β-1,3-glucan-recognition proteins isolated from *B. mori* and *Manduca sexta* (tobacco hornworm). At least two genes from the β-1,3-glucan-recognition protein family are present in the genomes of species in which efforts have been made to identify these proteins (*D. melanogaster*, *M. sexta*, and *An. gambiae*). The β-1,3-glucan-recognition proteins from *B. mori* and *M. sexta* have been shown to function as pattern-recognition molecules in activation of the prophenol oxidase cascade, and it is likely that other members of this family have additional functions in immunity as well.

Lectins are proteins that contain specific carbohydrate-binding sites. Several lectins are known to act as pattern-recognition molecules in mammalian and insect innate immune responses. Because lectins are often multimeric proteins, they have the ability to cross-link and agglutinate cells that have a specific carbohydrate ligand on their surface. Lectin activity has been detected in the hemolymph and midguts of many insect species, but only a few lectins have been characterized at the molecular level.

Lectins from insect plasma that have been biochemically characterized are from the C-type (calciumdependent) lectin family. Two calcium ions are bound at specific sites in each C-type carbohydrate recognition domain, and one of the calcium ions participates directly in forming the carbohydrate-binding site. In vertebrates, C-type lectins such as plasma mannose–binding protein are pattern-recognition molecules that function as opsonins (*opsonins* are proteins that coat pathogens, rendering them recognizable as foreign, thus facilitating clearance by phagocytes) and in activation of the complement system. The *D. melanogaster* genome contains numerous C-type lectins genes and C-type lectins, whose expression is induced by wounding or infection, that have been identified in *An. gambiae*. However, the functions of these proteins remain to be discovered. C-type lectins have been cloned and characterized biochemically from the flesh fly, *Sarcophaga peregrina*, the American cockroach (*Periplaneta americana*), and three lepidopterans, *B. mori*, *M. sexta*, and *Hyphantria cunea*. The cockroach and lepidopteran C-type lectins bind to the carbohydrate moieties of lipopolysaccharides from gram-negative bacteria, stimulating phagocytosis and nodule formation by hemocytes and activation of prophenol oxidase. Expression of the lectin genes is induced in response to infections, resulting in higher lectin concentrations in hemolymph.

PHAGOCYTOSIS

Phagocytosis, the process by which a cell engulfs a small particle, is the initial response of arthropod hemocytes to bacteria and other small invaders. Plasmatocytes and granular cells are the primary hemocyte types that take part in phagocytosis. Phagocytosis requires the action of actin filaments, connected by adaptor molecules to cell surface receptors, and can be thought of as a specialized form of receptor-mediated endocytosis. The process begins with attachment of the particle to the hemocyte surface, probably through an interaction with hemocyte membrane proteins. Scavenger receptors are one type of transmembrane protein that may fulfill the receptor function. They have been well characterized in mammalian macrophages, and they have been identified in *D. melanogaster* and in *An. gambiae*. Attachment may be enhanced by plasma pattern-recognition proteins that function as opsonins by binding to the microbial surface and to some hemocyte receptors. Experimental tests for opsonin activity demonstrate that phagocytosis occurs at a greater rate with particles treated with the putative opsonin than with control particles. Hemolin, β-1,3-glucan-recognition proteins and C-type lectins have been suggested to function as opsonins in insects. A protein with sequence similarity to complement factor C3 has recently been demonstrated to act as an opsonin in phagocytosis of gram-negative bacteria by an *An. gambiae* cell line.

Attachment of a foreign particle to the hemocyte surface stimulates a signal transduction pathway that results in formation of pseudopodia that engulf the particle. Endocytosis then occurs, and the ingested particle is trapped in an intracellular vesicle known as a *phagosome*. The phagosome fuses with lysosomes, which contain enzymes that are believed to be involved in the intracellular killing of the pathogen. However, little is known regarding the actual mechanisms used in killing phagocytosed organisms in arthropods.

ACTIVATION OF PROTEOLYTIC CASCADES

Proteases and Immunity

An interesting parallel between vertebrate and invertebrate immunity involves the reliance on cascades of serine proteases activated in response to infection or wounding. In vertebrates, the complement and coagulation cascades are the two best examples of sequential activation of multiple proteases and other substrates, ultimately resulting in localized and highly regulated responses. In invertebrates, phenol oxidase activation, hemolymph coagulation, and synthesis of at least some antimicrobial peptides appear to be regulated by similar cascades. Many serine proteases have been cloned from *An. gambiae*, and a number of these have characteristics that suggest involvement in immunity. One example is a protease that has been named Sp22D (Sp = serine protease and 22D refers to its chromosomal location). Sp22D is a putative recognition protein with multiple domains, including chitin binding, mucin-like, and scavenger receptor cysteine rich-like domains, as well as the serine protease catalytic domain. The domain organization of Sp22D suggests that this molecule could bind to pathogen molecular patterns. This interaction could activate the protease domain and initiate an immune response. Sp22D is expressed in multiple body parts and during

much of development, most intensely in hemocytes. The protein appears to be posttranslationally modified. Its integral form is secreted in the hemolymph, whereas a smaller form, potentially generated by proteolytic processing, remains intracellular. Bacterial challenge results in low-level RNA induction, but the protein does not bind to bacteria, nor is its processing affected by bacterial infection. High affinity binding of Sp22D to chitin has been observed, and it has been suggested that it may respond to exposure of naked chitin during tissue remodeling or damage.

Several proteases involved in immunity have an N-terminal "clip" domain (disulfide knotted domain) followed by a catalytic domain at the C-terminus. The clip domains are identified by a characteristic pattern of cysteines. Their function is not clear, but they may be involved in protein–protein interactions for example, between the serine protease and its substrate or its activator. Alternatively, they may act as catalytic domain repressors or as antimicrobial peptides. This protease family includes several enzymes involved in dorsal ventral patterning in *D. melanogaster* (e.g., easter, snake) that function in a sequential activation cascade leading to proteolytic activation of pro-spätzle. Spätzle belongs to the cysteine-knot family of growth factors and cytokines and activates the Toll pathway both during embryonic development and in response to infection. Genetic analysis has revealed that easter and snake are not involved in immune activation of spätzle, but other members of the clip domain serine proteases are likely to serve this role.

It is also of interest to note that the coagulation cascade in the horseshoe crab involves clip domain serine proteases. This cascade can be triggered by both fungal and bacterial surface molecules. All of the immune-related clip domain proteases are constitutively expressed in the hemolymph, where they can rapidly activate defense responses. Taken together, these observations suggest that clip domain proteases may be widely used in arthropod responses to pathogens and parasites.

A number of other *An. gambiae* serine proteases have been identified that exhibit changes in mRNA levels following pathogen injection, parasite infection, or wounding (e.g., Sp14D1 and 14D2, Sp14A, ISP13, ISPL5, and several recently identified expressed sequence tags).

Activation of Phenol Oxidases

Phenol oxidases play a central role in catalyzing the reactions leading to melanin formation. Melanin is a brown-to-black, electron-dense, chemically-resistant pigment that results from the enzyme-mediated polymerization of polyphenols. Melanin is an important constituent of encapsulation reactions in mosquitoes and is also involved in cuticle and egg chorion formation.

Most insect tyrosinase-type phenol oxidase (PO) enzymes share an important characteristic: They occur in an inactivated zymogen form. Figure 27.2 illustrates the proteolytic cascade leading to PO activation and melanin formation. Clearly, an enzyme that generates highly reactive and potentially toxic quinones, such as indole–5,6-quinone, must be closely regulated. In vitro, phenol oxidase can be activated by many things, including isopropanol, chloroform, fatty acids, detergents, urea, and acid–base shock, but when in vivo serine proteases are thought to function in this capacity. Protease activation of prophenol oxidase is cation (Ca^{++}) dependent and results in the loss of a small peptide from prophenol oxidase. Once activated, phenol oxidase enzymes tend to be "sticky," which could limit their dissemination throughout the insect. In some insects, phenol oxidase can also be regulated by specific peptide inhibitors, only a few of which have been identified. Finally, a gene named DOX A2 has been cloned from both *A. gambiae* and *D. melanogaster*. This gene was at one time thought to be a diphenol oxidase structural gene but is now believed to be part of a regulatory subunit of the 26S catabolic proteasome. The fact that this gene was originally identified by its effect on melanin formation by isozymes in

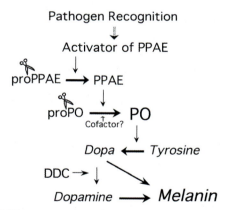

FIGURE 27.2 Simplified diagram of phenoloxidase (PO) activation and melanin formation during melanotic encapsulation. The steps between pathogen recognition and activation of pro-phenoloxidase activating enzyme (PPAE) have not been defined (dashed arrow). Proteolytic processing of proPPAE (scissors) generates active PPAE, which in turn cleaves pro-phenoloxidase (proPO) (scissors) to active PO. PO catalyzes the conversion of tyrosine to dopa, which is converted to dopamine by dopa decarboxylase (DDC). Other substrates and enzymes may be involved but have not been clearly implicated in melanization of parasites by vectors.

SDS PAGE gels suggests that the proteasome may be involved in phenol oxidase regulation.

Prophenol Oxidase–Activating Enzymes

Prophenol oxidase–activating enzymes have been purified from several insects, including *D. melanogaster*, *M. sexta*, *B. mori*, and a beetle, *Holotrichia diomphalia*. In each case, the activator proved to be a member of the serine protease family. In the tobacco hornworm, the silkworm, and the beetle, the activating proteases were characterized by having one or more N-terminal "clip" domains followed by a trypsin-like catalytic domain at the C-terminus. As with phenol oxidase, the prophenol oxidase–activating enzymes are thought to occur in the insect in an inactive zymogen form. Thus, they too must be activated, most likely by another protease. This sequence of reactions is referred to as the *phenol oxidase cascade*. None of the prophenol oxidase–activating enzymes has been definitively identified in mosquitoes or other vectors, although several clip domain serine proteases that exhibit changes in transcription during immune responses have been cloned from *An. gambiae*.

Cofactors Required for Prophenol Oxidase Activation

During the purification of prophenol oxidase–activating enzymes from both *H. diomphalia* and *M. sexta*, additional protein cofactors were discovered that were necessary for phenol oxidase activation. Although the prophenol oxidase–activating enzymes could act on synthetic substrates, they alone did not activate prophenol oxidase in reconstitution experiments in vitro. The purified *H. diomphalia* cofactor is structurally similar to a *D. melanogaster* clip domain serine protease called *masquerade* but does not have certain amino acids that are usually critical for proteolytic activity. The masquerade-like protein and the phenol oxidase–activating enzyme are sufficient to activate prophenol oxidase in an in vitro system. Cofactors were not necessary for activation of prophenol oxidase by the *B. mori* or *D. melanogaster* prophenol oxidase–activating enzymes. Whether cofactors are necessary for mosquito prophenol oxidase activation is not yet known.

Initial Triggers of the Prophenol Oxidase Cascade in Mosquitoes

We can be reasonably confident that prophenol oxidase, a prophenol oxidase–activating enzyme, and an activator of prophenol oxidase–activating enzyme are involved in parasite/pathogen encapsulation and melanization in mosquitoes. What is still missing is an understanding of the molecular patterns in eucaryotic parasites that initiate this response, the mosquito pattern-recognition receptors that interact with them, and the mechanism by which those activated recognition molecules trigger the remainder of the cascade.

Serpins and Regulation of Proteolytic Cascades

The immune responses activated by proteases are regulated in vivo to limit them to localized areas and to protect the host from harmful effects of over-stimulation of the immune system. In mammals, the proteolytic cascades that activate blood clotting and the complement system are regulated by plasma proteins called *serpins*, which are specific serine protease inhibitors. Arthropod plasma also contains serpins, 45- to 50-kDa proteins that are similar in sequence to mammalian serpins such as antithrombin and plasminogen activator inhibitor. Serpins contain an exposed reactive site loop that interacts with the active site of a target serine protease. As the protease attempts to cleave the serpin reactive site loop, it is trapped in a stable serpin–protease complex. Therefore, once a serine protease is activated, it typically has a short life before it is inhibited by a specific serpin, which limits the magnitude of the protease cascade response. The sequence of amino acids in the reactive site loop determines a serpin's inhibitory specificity. Most serpins in mammalian blood and arthropod hemolymph have probably evolved to regulate the activity of a specific endogenous serine protease (such as inhibition of thrombin by the serpin antithrombin). However, there can be redundancy in serpin function; more than one serpin may inhibit an individual protease.

In arthropods, serpins have been most thoroughly investigated in the horseshoe crab, *Tachypleus tridentiatus*, and the lepidopteran insect *M. sexta*. Three serpins have been identified in hemocyte granules of the horseshoe crab. Upon degranulation, the serine protease zymogens and the serpins are released from the hemocytes, and the proteases are activated by limited proteolysis. Each serpin inactivates one type of protease, to turn off the cascade and limit the coagulation response to a localized area near a wound or infection. In *M. sexta*, serpins are present in plasma and function in a similar manner to regulate the phenol oxidase cascade. There are at least five serpin genes in *M. sexta*. In one of these genes, alternate exon splicing results in production of serpins with 12 different reactive site loop sequences and corresponding diverse inhibitory activities. There are more than 20 serpin genes in the *D. melanogaster* genome, but a physio-logical function is apparent for only one of them. A mutation in one *D. melanogaster* serpin gene in the *nec* locus results in constitutive activation of the Toll pathway and expression of the drosomycin gene. This occurs as a result of the lack of regulation of a protease in the pathway that leads to proteolytic activation of spätzle.

Little is known at this time about the functions of serpins in vector species, but it can be predicted that, as in *M. sexta*, *D. melanogaster*, and mammals, each vector species may have a family of serpin genes to regulate a variety of serine protease activities. Eight serpin cDNAs from *An. gambiae* have been identified so far, but their physiological functions have not been described. In addition to serpins, insect hemolymph contains lower-molecular-weight protease inhibitors from several other families. These also are likely to exist in vector species and to have specific functions in regulating proteases activated during immune responses.

TRANSCRIPTIONAL ACTIVATION OF EFFECTOR MOLECULES

Antimicrobial Peptides

Injection of bacteria or fungi into the insect hemocoel results in the induction of several antimicrobial peptides, and the composition of the resulting antimicrobial "cocktail" secreted into the hemolymph varies depending on the type of pathogen injected. Many proteins and peptides active against gram-negative and gram-positive bacteria and fungi have been characterized from a variety of insect species. Table 27.2 illustrates the properties of the different classes of antimicrobial peptides characterized so far in *D. melanogaster*. Defensins are active against gram-positive bacteria and have also been characterized in several mosquito species, including *An. gambiae* and *Ae. aegypti*. Cecropins are active against both gram-positive and gram-negative bacteria and have been identified in *Ae. aegypti* and *Ae. albopictus*. The insect fat body, an organ with the combined functions of vertebrate adipocytes and hepatocytes, is the main site of transcriptional activation and synthesis of these effector molecules. However, they can also be synthesized by epithelial cells lining structures, such as the trachea, midgut, and cuticle in insects and the lungs, epidermis, and gastrointestinal tract in vertebrates, that come in contact with the environment and that could be a port of entry for pathogens. For example, defensin expression is induced in the midgut of *An. gambiae* in response to malaria infection.

TABLE 27.2   Antimicrobial Peptides from *Drosophila melanogaster*

Antimicrobial Peptide	Active Against	Structure
Drosomycin	Fungi	Amphipathic helix linked to antiparallel sheet
Defensin	Gram-positive bacteria and some fungi	
Cecropin	Gram-positive and gram-negative bacteria and some fungi	Two helices
Diptericin	Gram-negative bacteria	Linear polypeptides with overrepresentation of proline and/or glycine residues
Drosocin	Gram-negative bacteria	
Attacin	Gram-negative bacteria	
Metchnikowin	Gram-positive bacteria and some fungi	

Signal Transduction Pathways

As the regulatory regions of the genes coding for these peptides were characterized, it became apparent that regulatory elements, similar to those found in genes activated by the innate immune system in vertebrates, were present and essential for their transcriptional activation in response to pathogens. Genetic analysis in *D. melanogaster* has defined two main signal transduction cascades that regulate transcription of antimicrobial peptides, the Toll and the immunodeficiency pathways. Flies in which both pathways have been disrupted are unable to activate any known antimicrobial peptides when challenged, indicating that both pathways are essential. However, other cascades could also participate in modulating these responses.

The genes participating in the Toll pathway were first characterized for their role in establishing the dorsoventral polarity in *D. melanogaster* embryos and were later found also to participate in the transcriptional regulation of the antifungal peptide Drosomycin. The Toll cascade has remarkable similarities in overall organization to the vertebrate pathway mediating activation of the nuclear factor-B (NF-κB) in response to interleukin-1 (Fig. 27.3). In both flies and vertebrates a cytokine, spätzle or interleukin-1, binds to its receptor, the Toll and interleukin-1 receptor, respectively, which have a region of sequence homology in their intracytoplasmic domain. Activation of the Toll and interleukin-1 receptor in turn leads to degradation of a repressor protein, cactus and inhibitor of κB (I-κB), respectively, and activation of the transcription factors Drosophila immune factor (Dif) and NF-κB, respectively, both belonging to the Rel family of receptors.

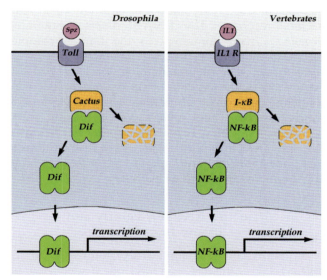

FIGURE 27.3 Comparison of the *Drosophila* Toll and vertebrate interleukin-1 signal transduction pathways. Interaction of the cytokine [spätzle (Spz) or interleukin1 (IL-1)] with its receptor leads to degradation of a repressor [cactus or I-B], which is followed by nuclear translocation of a Rel family transcription factor [Dif or NF–κB] and induction of mRNA expression of effector genes.

Rel Family of Transcription Factors

Three members of the Rel family have been characterized in *D. melanogaster*, Dorsal, Relish, and Dif, and one from *An. gambiae*, Gambif1 (with highest homology to Dorsal). Disruption of the Relish and Dif genes has been shown to impair the induction of antimicrobial peptides. However, although both Dorsal and Gambif1 translocate to the nucleus of fat body cells in response to infection, *dorsal* mutants induce all known antimicrobial genes normally. The Imd pathway has been defined genetically and involves the *D. melanogaster* homologues of vertebrate I-κB kinase and I-κB kinase , which mediate activation of Relish. But the receptor, the ligand that activates it, and other genes participating in this cascade remain to be defined.

STAT Family of Transcription Factors

Gene disruption experiments in mice have shown that several members of the Signal Transducers and Activators of Transcription (STATs) family of transcription factors have nonredundant functions in activating gene expression in response to cytokines. A member of this receptor family from *D. melanogaster*, D-STAT, participates in embryonic development. *Anopheles gambiae* STAT, Ag-STAT, was the first insect member of this family shown to be involved in antibacterial responses. The cDNA of thioester-containing protein 1, with homology to the human complement C3-chain, was recently characterized in *D. melanogaster*. Thioester-containing protein 1 mRNA expression is up-regulated in response to bacterial challenge, and constitutive high levels of expression were observed in flies with a gain-of-function mutation of the JAK kinase hopscotch (*hoptum-1*), the kinase that activates Drosophila-STAT. These results indicate that thioester-containing protein 1 expression is probably regulated by the STAT pathway.

The Toll Receptor Family and Innate Immunity

A family of receptors with homology to *D. melanogaster* Toll has recently been identified in humans (10 have been found so far) and named Toll-like receptors (TLR1 through TLR10). Gene disruption experiments in mice have demonstrated that they play a critical role in the initial responses of the vertebrate innate immune system to specific pathogen molecules. Disruption of TLR4, TLR2, and TLR5 leads to an inability to mount responses to bacterial lipopolysaccharides from gram-negative bacteria, to soluble peptidoglycans and bacterial lipoproteins of gram-positive bacteria, and to bacterial flagellin, respectively. These receptors are thought to be activated as they interact with microbial components containing pathogen-associated molecular patterns, either directly or bound to pathogen-recognition molecules. For example, lipopolysaccharides bind to lipopolysaccharide-binding protein, which interacts with the receptor CD14 to form a complex that activated TLR4. The recognition system is complex, for recent evidence indicates that the responses vary in different cell types, that accessory proteins can modulate the responses, and that TLRs have been found to be capable of signaling as heterodimers. This is an active area of study, because elucidating the initial decisions of the innate immune system, which define the direction the acquired immune responses will take, is at the core of understanding human defense responses to pathogens and has opened the potential for the beginning of a whole new area in human immune diagnosis and therapeutics.

In *D. melanogaster*, eight members of the Toll family have been characterized (Toll, 18-wheeler, and Toll-3 through Toll-8). Toll and Toll-5 mediate Drosomycin expression when activated forms are transfected into cell lines. Several Toll receptors have been cloned from other insect species, including *Drosophila pseudobscura*, *Drosophila similans*, *Glossina palpalis palpalis* (tsetse), *An. gambiae*, *Ae. aegypti*, and the grasshopper *Schistocerca americana*. Phylogenetic analysis revealed separate clustering of Toll proteins from insects and vertebrates, suggesting that these two groups share a common ancestor but have evolved independently.

MELANIZATION RESPONSES

Melanotic Encapsulation

Encapsulation is a common immune reaction in insects against parasites or pathogens that can also be induced by injection of experimental materials into the hemocoel. This type of reaction can either be cellular or acellular in nature. Cellular encapsulation can occur in two forms: (1) as nodule formation against small particulate material (e.g., bacteria), and (2) as capsule formation, which often occurs if the foreign object exceeds the size of phagocytic hemocytes. Nodules appear as compacted masses of bacteria enmeshed with hemocytes, the whole of which usually becomes infiltrated with the pigment melanin. Many nodules are then walled off under a final layer of hemocytes. The second type of cellular capsule occurs in its simplest form when a large foreign object becomes covered by certain types of hemocytes, which eventually form flattened layers completely enclosing the object. This type of encapsulation is common in Lepidoptera. In other insects, cellular encapsulation takes different forms. Sometimes, hemocytes may lyse or degranulate at the surface of the pathogen, resulting in formation of an inner core of melanin and necrotic cells, which then becomes surrounded by multiple layers of unmelanized cells. In other systems, large masses of melanized hemocytes form an irregular layer over the pathogen's surface.

The other major type of encapsulation, acellular encapsulation, is uncommon in insects but is of particular interest to vector biologists because it occurs primarily among certain nematoceran Diptera (e.g., sand flies, mosquitoes). In these insects a low number of circulating hemocytes is common, and layers or clusters of these cells usually are not seen around targets. Instead, capsules take the form of a sheath of melanin deposited on the surface of foreign objects. The melanin layer often appears to originate from the hemolymph but without obvious hemocyte contact with the target. This type of acellular defense is often called *humoral encapsulation* or *humoral melanization*. The term *melanotic encapsulation* has been used to refer both to melanized cellular capsules and to humoral encapsulation and is therefore confusing and not recommended for use.

Encapsulation Response to Protozoan Parasites

Most parasitic protozoans that are transmitted by insects remain within the lumen of the digestive tract, where encapsulation is unlikely to occur. An important exception is the malaria parasite, which develops at several sites within the mosquito and must pass from the midgut through the hemolymph to gain access to the salivary glands. Interestingly, one of the first demonstrations of melanization of a parasite by an insect was published by Sir Ronald Ross, who won a Nobel Prize for discovery of the role of the mosquito in transmission of malaria. During his investigations, Ross described blackened sporozoites visible within the oocyst capsule on the midgut of some mosquitoes. These structures later became known as Ross's black spores. Subsequently, other investigators described the occurrence of melanized ookinetes, oocysts, and even sporozoites within the salivary glands (Fig. 27.4). In recent times, melanization of malaria parasites has been observed only rarely during field studies of natural parasite/mosquito interactions.

The phenomenon received more attention after strains of the African malaria vector, *Anopheles gambiae*, were genetically selected to enhance this characteristic. In these "refractory," or "resistant," strains, ookinetes of many species of *Plasmodium* are melanized and killed. Ookinetes are encapsulated between the basal lamina of the midgut and the midgut epithelial cells and are not directly in the hemocoel; thus, the involvement of the hemocytes in this response is not clear. Quantitative trait loci mapping indicates that the response is influenced by two or three genes, but these have not yet been identified.

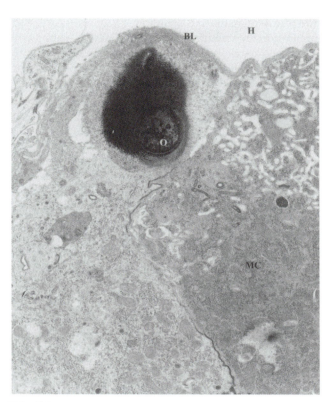

FIGURE 27.4 Melanized ookinete (*Plasmodium cynomolgi*) on the midgut of a refractory strain of *Anopheles gambiae*. H = hemocoel, MC = midgut cell, O = ookinete, BL = basal lamina.

Responses to Microfilarial Worms

Mosquitoes, tabanids, and blackflies all transmit important nematode parasites of humans. Neither tabanids nor blackflies are known to encapsulate these worms, but there are several species of mosquitoes that can respond in this way. In some instances mosquitoes apparently use this type of immune response to destroy parasites acquired in nature. This observation is important because it may suggest that one of the reasons some species of mosquitoes that feed on humans do not transmit particular pathogens relates to their ability to mount effective immune responses against the pathogen, although there are undoubtedly multiple mechanisms underlying vector competence.

There are numerous reports of the melanization of worms in mosquitoes following ingestion or injection into the hemocoel and during in vitro incubation in hemolymph. The filarial nematodes *Brugia malayi*, *Brugia pahangi*, and *Dirofilaria immitis* have been used in experiments with mosquitoes in the genera *Aedes*, *Armigeres*, and *Anopheles*.

A combination of humoral melanization and cellular encapsulation has been reported for certain of these mosquito–microfilariae systems. First, the worms are enclosed in an acellular layer of melanin, perhaps with the active participation of hemocytes. Subsequently, hemocyte adherence occurs, resulting in a single-layered cellular capsule over the melanized core. In other mosquito/worm combinations, ultrastructural observations indicate that hemocytes first contact the worm surface and then lyse or degranulate. Contact and degranulation are followed by formation of the melanized capsule, and cell debris is visible in these capsules. In some cases, hemocytes also are involved in termination of the reaction. These cells may form a layer over the melanized parasite or may be involved in production of a double membrane structure that encloses parasites, melanized capsules, and cellular debris. Interestingly, this membrane has properties that suggest it may be similar to mosquito basal lamina. It should also be noted that tissues other than the hemolymph, such as the thoracic muscle cells and Malpighian tubule cells, have been reported to function in melanizing intracellular parasitic nematodes.

How Are Parasites Killed or Restrained by Encapsulation?

Encapsulation may kill parasites in two ways. Some intermediates formed during melanin production are highly reactive and may be toxic to the parasites. Alternatively, the formation of the capsule may result in inability to obtain nutrients or exchange wastes.

Inhibition of Antibacterial Responses in Ticks

Several human bacterial pathogens, notably *Borrelia recurrentis* in body lice and *Borrelia burgdorferi* in *Ixodes ricinus*-group ticks, are known to disseminate into the hemocoel of the vector, where they would be accessible for nodule formation, without eliciting a defense response. To date, however, little is known concerning the immune system of these vectors or their relatives. Hemocytes of some tick species are able to encapsulate experimental materials and to phagocytose bacteria. Tick hemolymph often turns dark upon collection, suggesting the presence of melanizing enzymes. However, bioassays of *Ixodes* and *Dermacentor* hemolymph have not detected phenol oxidase, one of the most important enzymes in insect melanization. Additional studies will be needed before we can determine how vector-borne bacterial pathogens are able to evade the immune responses of their arthropod hosts.

Biochemistry of Melanin Formation

Much of what is known about melanin formation is derived from studies of cuticular melanization in model insects. But, increasingly, these studies are being applied to defense responses of insect vectors as well. The enzymes involved in melanine formation will be discussed, with special emphasis on those mediating melanin formation in response to pathogens, which are shown as a simplified diagram in Figure 27.2.

Phenol Oxidases

The phenol oxidase enzymes are thought to be central to melanin formation. Phenol oxidases are copper-binding enzymes that occur in plants, fungi, and animals. In insects, two types of phenol oxidase have been recorded: the laccases and the tyrosinases. Laccases are involved in cuticular sclerotization and can act on some of the same substrates (*p*- and *o*-diphenols but not monophenols) as the tyrosinase-type phenol oxidases (*o*-diphenols and monophenols). Laccase-type phenol oxidases are not known to act during immune responses. Tyrosinase-type phenol oxidases are further subdivided into two main groups. Phenol oxidases that catalyze the hydroxylation of tyrosine, a monophenol, to dopa are called *monophenol* oxidases. The *diphenol* oxidases act on *o*-diphenols, including dopa and its relatives, to produce reactive quinones. Diphenol and monophenol oxidases activities may occur within the same enzyme. The reactive quinones formed by these enzymes probably undergo additional structural rearrangements, some of which may be enzyme mediated, before final polymerization into melanin. During encapsulation, polyphenol derivatives probably also cross-link with proteins from the hemolymph or hemocytes, but nothing is known about these interactions.

Mosquito Phenol Oxidases

Phenol oxidase enzymes are involved in many processes in insects, so it is not surprising that multiple phenol oxidase genes have been discovered in mosquitoes. In *An. gambiae*, for example, six distinct genes have been cloned and characterized as to expression sites and gene features. The genes show distinct expression profiles in the mosquito, spanning from embryo to adult in an overlapping manner. These phenol oxidases are not induced in a cell line in response to bacterial challenge. Interestingly, 20-hydroxyecdysone may be involved in regulation of some of these genes. Three phenol oxidase genes are known from *Armigeres subalbatus*, and two have been reported from *Ae. aegypti*. In *An. subalbatus*, studies using the Sindbis virus to knockout gene expression have pinpointed one of the phenol oxidase genes as important in melanization of filarial worms. The specific enzyme(s) involved in melanization of *Plasmodium* in *An. gambiae* has not yet been identified. Multiple, closely-related enzymes with overlapping pattern of expression make this a challenging system for investigation.

Other Enzymes That May Be Involved in Melanization

In addition to the members of the phenol oxidase cascade just described, other enzymes may be involved in melanin production during encapsulation. For example, dopa decarboxylase converts dopa into dopamine. Dopamine is a neurotransmitter but also may be acted on by phenol oxidase to produce reactive quinones. Dopa decarboxylase activity is elevated during filarial worm infections in mosquitoes. Dopa decarboxylase mutants of *D. melanogaster* show a reduced ability to melanize parasitoids under high-temperature regimes that render dopa decarboxylase nonfunctional.

Dopachrome-conversion enzyme mediates a structural rearrangement of dopachrome (formed nonenzymatically from dopa), resulting in the formation of an indole. The indole then is oxidized to an indole quinone, which can polymerize to form melanin. A similar function is attributed to dopachrome tautomerase in vertebrates. Although dopachrome-conversion enzyme occurs in adult mosquitoes and other insects, its involvement in immunity has not been documented.

Other enzymes that may be involved in melanization include phenylalanine hydroxylase, which converts phenylalanine to tyrosine, and tyrosine hydroxylase, which converts tyrosine to dopa.

IMMUNE RESPONSES OF MOSQUITO MIDGUT CELLS TO MALARIA INFECTION

Mosquitoes acquire malaria through blood feeding, and thus the midgut is the first organ to come in contact with the invading parasites. Infection of *An. gambiae* with *P. berghei* (murine malaria) results in transcriptional activation of several genes in the midgut, such as the gram-negative binding protein, defensin, and galactose-specific lectin and nitric oxide synthase, which are also induced in response to bacterial challenge and have significant sequence homology to molecules participating in vertebrate immunity. Nitric oxide synthase induction in response to malaria infection was first reported in *An. stephensi*, and it was also shown that dietary provision of the synthase substrate,

L-arginine, reduced *Plasmodium* infection, while administration of a nitric oxide synthase inhibitor significantly increased the number of parasites that developed in the midgut following infection. Recently, a member of the transforming growth factor cytokine family, which is expressed in the midgut, has also been characterized in this species.

These defense responses are observed in mosquito strains that are susceptible to *Plasmodium* infection and thus are efficient disease vectors. Why are these responses unable to kill the parasites? The responses might not have the intensity required to achieve sterile immunity. There may be trade-offs between immunity and fecundity. Fitness costs of an encapsulation response could include toxicity, disruption of biological pathways that function in more than one way, and diversion of resources away from other functions. As an example, melanin formation requires tyrosine as a precursor. Tyrosine is also important for egg chorion formation. As a possible result, delayed oviposition has been observed in mosquitoes actively encapsulating parasites. Alternatively, the natural responses may have the potential to kill *Plasmodium*, but parasites may have evolved to be at the right place at the right time. To examine this second possibility, the temporal-spatial relationships between parasite invasion and activation of specific effector molecules have to be considered.

Given that all mosquitoes seem to have the potential ability to respond to infections with an encapsulation response, it is curious that such defenses are not universally applied against malaria parasites and worms. Why do vectors remain competent? There may be several reasons for this. First, the parasite may have little effect on the mosquito's survival or fecundity. For many years, researchers believed that infections of malaria parasites did not have substantial effects on the mosquito, based primarily on laboratory investigations. More recently, some investigations have reported reduced fecundity, life span and flight ability in infected vectors in the field. Depending on the severity of these effects, there could be selection pressure for resistance to infection. However, even if there are effects on these traits, there may be little evolutionary pressure to increase resistance, since infection rates in the field may be very low.

THE TIME BOMB MODEL OF OOKINETE INVASION OF MIDGUT EPITHELIAL CELLS

Studies on the cell biology of *P. berghei* ookinete invasion of midgut cells and the epithelial responses of *An.*

stephensi midgut cells to this process revealed that ookinetes invade columnar epithelial cells with microvilli (Fig. 27.5). The invaded cells are damaged and protrude toward the midgut lumen immediately following invasion. Ookinetes are capable of extensive lateral movements, crossing several cells before reaching the basal lamina, and they can also move laterally in the space between the basal lamina and the epithelial cells. Sometimes oocysts form several cell diameters away from the invasion site, underneath healthy uninvaded cells. Some characteristic changes are observed in invaded cells, such as increased expression of nitric oxide synthase, a substantial loss of microvilli, and genomic DNA fragmentation. Ookinetes release the glycosylphosphatidylinositol-anchored surface protein Pbs21 and secrete a subtilisin-like serine protease as they migrate across the cytoplasm of the cell (Fig. 27.6). These observations indicate that parasites inflict severe irreversible damage, leading to subsequent cell death. The parasite-induced damage is repaired by an actin string-purse-mediated restitution mechanism. As a final step, the invaded cells degenerate and "bud off" into the midgut lumen, and the integrity of the epithelium is preserved. Assuming that the natural mechanisms that lead to death of the invaded cells are also lethal for the invading parasite, the proposed model predicts that, if the cell could respond faster (or more strongly?) or if the motility of the parasite could be reduced, it might be possible to alter the outcome of invasion and thus prevent disease transmission.

FUTURE DIRECTIONS

In recent years there has been tremendous progress in our understanding of the overall organization and regulation of the insect immune system. It is now clear that insects have evolved a pathogen-recognition system that enables them to activate sophisticated networks of signaling pathways that coordinate the expression of multiple effector genes. However, many questions still remain unanswered. In the area of pathogen recognition, we have some knowledge of how lepidopteran, dipteran, and orthopteran insects detect bacterial and fungal infection, and we have characterized several types of pathogen-recognition molecules that carry out this function. However, most of the information available comes from work on large lepidopteran species and *D. melanogaster*, and there are few such data for vector species. Furthermore, we lack an understanding of how insects recognize protozoan and metazoan parasites and viruses. Perhaps pattern-recognition proteins similar to those that detect

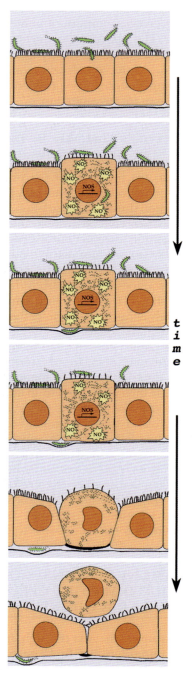

FIGURE 27.5 Diagrammatic representation of the time bomb model of ookinete invasion. Parasite invasion results in increased expression of nitric oxide synthase, loss of surface microvilli, cell protrusion to the luminal side, and genome fragmentation. The cell damage is irreversible, leading to nuclear rupture and cell death. The midgut epithelia is repaired by "budding off" the damaged cells to the midgut lumen by a restitution mechanism consisting of an actin-mediated movement of the neighboring cells that does not involve cell division.

FIGURE 27.6 Immunofluorescence staining of a *Plasmodium berghei* ookinete invading an *An. stephensi* midgut epithelial cell. The surface protein Pbs21 is stained in light blue, a subtilisin-like serine protease (PbSub2) in red, and actin in green (labeled phalloidin). The parasite-released PbSub2 in the cytoplasm of the cell has emerged on the basal side, immediately underneath a muscle fiber.

bacteria and fungi also are able to interact with molecular patterns present in these organisms; alternatively, it is possible that molecules involved in their recognition represent a new class or receptors yet to be discovered. Because many vector-borne diseases are caused by RNA viruses that have double-stranded (ds) RNA intermediate forms in their replication cycle, the fact that experimental introduction of dsRNA into the cytoplasm of insect cells results in gene silencing suggests that insects have the ability to mount antiviral responses. These responses could be important in determining an arthropod's capacity to transmit a given virus. This will undoubtedly be an active area of research in the near future.

The serine protease pathways that are activated in immune responses are likely to be quite complex. For example, three out of the 16 known serpin activities in *M. sexta* plasma are known to inhibit phenol oxidase activation. The remaining serpins may inhibit proteases from pathways that have not yet been discovered. The serine proteases represent one of the largest gene families in the *D. melanogaster* genome, and many of them may be present in hemolymph and have immune functions. A challenge in the coming years will be to work out the details of the protease cascades through a combination of genetic and biochemical methods and to gain an understanding of how they are regulated.

The genome sequencing of *An. gambiae* has already been completed, and comparative analysis with the complete *D. melanogaster* genome sequence will be very informative. We will gain new insight by comparing the overall organization of gene clusters and determining the extent to which orthologs are shared between these two species. One of the major challenges for the near future is to develop experimental strategies to bridge the gap between gene sequence information and the determination of gene function. A combination of genome-wide analysis of gene expression, using DNA microarray technology, biochemical characterization of specific components participating in pathogen recognition, studies of signal transduction pathways, and the identification of antiparasitic activities will all broaden our understanding of the immune system of insect vectors and how these responses affect vector competence.

The discovery of the human Toll-like receptors (TLRs) illustrates how information obtained from a model insect has opened a new area of study in vertebrate immunity. Considering the ancient nature of the innate immune system and the extent of the evolutionary conservation of several components between insects and vertebrates, understanding the interactions of pathogens with the vectors that transmit them could also bring new insight to the responses of vertebrate epithelial cells and the vertebrate innate immune system to these agents.

Readings

Bayne, C. J. 1990. Phagocytosis and nonself recognition in invertebrates. *Bioscience* 40: 723–731.

Brey, P. R., and Hultmark, D. (eds.) 1998. *Molecular Mechanisms of Immune Responses in Insects*. Chapman and Hall, London.

Dimopoulos, G., Mueller, M., Levashina, E. and Kafatos, F. C. 2001. Innate immune defense against malaria infection in the mosquito. *Curr. Opin. Immunol.* 13: 79–88

Gillespie, J. P., Kanost, M. R., and Trenczek, T. 1997. Biological mediators of insect immunity. *Annu. Rev. Entomol.* 42: 611–643.

Imler, J. L., and Hoffmann, J. A. 2000. Signaling mechanisms in the antimicrobial host defense of *Drosophila*. *Curr. Opin. Microbiol.* 3(1): 16–22.

Imler, J. L. and Hoffmann, J. A. 2001. Toll receptors in innate immunity. *Trends Cell. Biol.* 11(7): 304–11.

Further Readings

Barillas-Mury, C., Wizel, B. and Han, Y.-S. 2000. Mosquito immune responses and malaria transmission: Lessons from insect model systems and implications for vertebrate innate immunity and vaccine development. *Insect Biochem. Molec. Biol.* 30: 429–442.

Bulet, P., Hetru, C., Dimarcq, J. L., and Hoffmann, D. 1999. Antimicrobial peptides in insects: Structure and function. *Dev. Comp. Immunol.* 23: 329–344.

Chen, C., and Billingsley, P. F. 1999. Detection and characterization of a mannan-binding lectin from the mosquito, *Anopheles stephensi* (Liston). *Eur. J. Biochem.* 263: 360–366.

Dimopoulos, G., Casavant, T. L., Chang, S. R., Scheetz, T., Roberts, C., Donohue, M., Schultz, J., Benes, V., Bork, P., Ansorge, W., Soares, M. B., and Kafatos, F. C. 2000. *Anopheles gambiae* pilot gene discovery project: Identification of mosquito innate immunity genes from expressed sequence tags generated from immune-competent cell lines. *Proc. Natl. Acad. Sci. USA* 97: 6619–6624.

Gorman, M. J., and Paskewitz, S. M. 2001. Serine proteases as mediators of mosquito immune responses. *Insect Biochem. Molec. Biol.* 31: 257–262.

Han, Y. S., Thompson, J., Kafatos, F. C., and Barillas-Mury, C. 2000. Molecular interactions between *Anopheles stephensi* midgut cells and *Plasmodium berghei*: The time bomb theory of ookinete invasion of mosquitoes. *EMBO J.* 19(22): 6030–6040.

Hoffmann, J. A., Kafatos, F. C., Janeway, C. A., and Ezekowitz, R. A. 1999. Phylogenetic perspectives in innate immunity. *Science* 284(5418): 1313–1318.

Ihle, J. N. 1996. STATs: Signal transducers and activators of transcription. *Cell* 84: 331–334.

Kanost, M. R. 1999. Serine protease inhibitors in arthropod immunity. *Dev. Comp. Immunol.* 23: 291–301.

Levashina, E. A., Langley, E., Green, C., Gubb D., Ashburner, M., Hoffmann, J. A., and Reichhart, J. M. 1999. Constitutive activation of Toll-mediated antifungal defense in serpin-deficient *Drosophila*. *Science* 285: 1917–1919.

Levashina, E. A., Moita, L. F., Blandin, S., Vriend, G., Lagueux, M., and Kafatos, F. C. 2001. Conserved role of a complement-like protein in phagocytosis revealed by dsRNA knockout in cultured cells of the mosquito *Anopheles gambiae*. *Cell* 104: 709–718.

Luo, C., and Zheng, L. 2000. Independent evolution of Toll and related genes in insects and mammals. *Immunogenetics* 51(2): 92–98.

Medzhitov, R., and Janeway, C. A., Jr. 1997. Innate immunity: The virtues of a nonclonal system of recognition. *Cell* 91(3): 295–298.

Paskewitz, S. M., and Gorman, M. J. 1999. Mosquito immunity and malaria parasites. *Am. Entomologist* 45: 80–94.

Werner, T., Liu, G., Kang, D., Ekengren, S., Steiner, H., and Hultmark, D. 2000. A family of peptidoglycan recognition proteins in the fruit fly *Drosophila melanogaster*. *Proc. Natl. Acad. Sci. USA* 97: 13772–13777.

Wilson, R., Chen, C., and Ratcliffe, N. A. 1999. Innate immunity in insects: The role of multiple, endogenous serum lectins in the recognition of foreign invaders in the cockroach, *Blaberus discoidalis*. *J. Immunol.* 162: 1590–1596

Yu, X.-Q., Zhu, Y.-F., Ma, C., Fabrick, J. A., and Kanost, M. R. 2001. Pattern recognition proteins in *Manduca sexta* plasma. *Insect Biochem. Molec. Biol.* in press.

28

Blood-Feeding Arthropod Salivary Glands and Saliva

JESUS G. VALENZUELA

INTRODUCTION

Salivary glands produce saliva, a fluid composed of a variety of molecules that are injected into the mammalian host when an insect attempts to blood feed. These salivary compounds locally modify vertebrate host physiology. This physiologic change favors the blood feeder by increasing the chances of a successful blood meal acquisition. This change also affects establishment of the parasite or pathogen transmitted by the blood feeder. The structure and function of blood-feeding arthropod salivary glands has been extensively reviewed, so this chapter focuses on the secretory fluid produced by salivary glands of disease vectors, the importance of saliva in vector–host interaction, the role of saliva in transmission of vector-borne pathogens, and the use of salivary proteins as an alternative method to control vector-borne diseases.

Salivary glands are composed of a single-cell epithelial layer. There are two basic structural types: the *tubular* and the *alveolar* (the reservoir type can also be defined as a third category). In the tubular type, a layer of cells surrounds a central duct; these cells store the secretory products in their apical vesicles. Saliva is then secreted into the duct and delivered to the mouthparts with the help of a salivary pump. Alveolar glands resemble grapes, and the cells are organized around branching ducts. The mosquito salivary gland is an example of the tubular type, while salivary glands from ticks are alveolar. Salivary glands, in general, have distinct types of cells that have different functions. In female mosquitoes, cells from the anterior lobe are associated with secretion of enzymes relevant to sugar feeding, while cells from the posterior lobe are associated with secretion of molecules relevant to blood feeding.

ROLE OF SALIVA IN VECTOR–HOST INTERACTIONS

The saliva of blood-feeding arthropods was believed for many years to be inert, utilized only to lubricate the insect mouthparts. This concept has changed with the demonstration that the saliva of blood feeders is composed of potent pharmacologically active products that alter the host hemostatic, inflammatory, and immune systems. Blood-feeding arthropods attempting to feed on a vertebrate host cause damage to the skin; a primary response to the loss of blood and tissue injury is activation of the hemostatic system, whose primary role is to avoid blood loss when injury occurs. It has three main functions: vasoconstriction, platelet aggregation, and the blood-coagulation cascade. This system is highly redundant. When platelets are active, they release thromboxane A2, a potent vasoconstrictor that decreases the blood flow; in addition, platelets change the composition of their membranes (to negatively charged groups), to allow the formation of protein complexes for the blood-coagulation cascade. The endpoint of the blood-coagulation cascade is the formation of thrombin. Active thrombin cleaves fibrinogen into fibrin, to form the blood clot. Thrombin also activates platelets by cleaving a receptor in the surface of platelets.

Blood feeders need to counteract not only the efficiency of the hemostatic system but also its redundancy. Saliva contains compounds capable of counteracting the host hemostatic response as well as its redundancy. The strategies used by various blood feeders to accomplish these tasks are listed in Table 28.1. The main activities can be subdivided into three

groups: vasodilators, inhibitors of platelet aggregation, and inhibitors of the blood-coagulation cascade.

Vasodilators

Vasoconstriction reduces blood flow in the site of the injury and consequently reduces the chances for blood feeders to obtain a blood meal. The problem of vasoconstriction has been solved by blood-feeding

TABLE 28.1 Salivary Antihemostatic Activities from Blood-Feeding Arthropods

Source[a]	Vasodilator	Anticlotting	Antiplatelet	Reference
Aedes aegypti	Sialokinin I and II	Factor Xa inhibitor	Apyrase	Champagne et al. (1995) Stark and James (1998)
Anopheles albimanus	Peroxidase	Anophelin (antithrombin)	Apyrase Anophelin	Ribeiro and Valenzuela (1999) Valenzuela et al. (1999) Francischetti et al. (1999)
Boophilus microplus	PGE2	Antithrombin		Horn et al. (2000)
Cimex lectularius	Nitrophorin	Tenase inhibitor	Novel apyrase	Valenzuela et al. (1998a) Valenzuela et al. (1998b)
Ctenocephalides felis			Apyrase	Cheeseman (1998)
Culicoides varipennis	22-kDa protein	Factor Xa inhibitor	Apyrase	Perez de Leon et al. (1998)
Glossina m. morsitans		Novel thrombin inhibitor	ADP hydrolytic activity, small peptide	Capello et al. (1996)
Ixodes scapularis	6-keto-PGF1___ (prostacyclin)	Ixolaris, novel anticoagulant	Apyrase	Ribeiro et al. (1998) Francischetti et al. (2002)
Lutzomyia longipalpis	Maxadilan	Novel anticoagulant	Apyrase	Ribeiro et al. (1989) Charlab et al. (1999)
Ornithodorus moubata		Tick anticoagulant peptide (TAP)	Apyrase	Ribeiro et al. (1991)
Phlebotomus papatasi	AMP and adenosine		Novel apyrase	Ribeiro et al. (1999) Valenzuela et al. (2001)
Rhodnius prolixus	Nitrophorins	Nitrophorin 2, Factor X complex formation inhibitor	Apyrase and RPPAI	Champagne et al. (1995) Francischetti et al. (2000)
Simulium vittatum	SV erythema protein	Anti-Factor Xa antithrombin, inhibitor of Factor V	Apyrase	Cupp et al. (1998) Abebe et al. (1996)
Triatoma pallidipennis		Triabin (antithrombin)	Pallidipin	Noeske-Jungblut et al. (1995)
Triatoma infestans	Endothelium-dependent vasodilator	Anticoagulant affecting Factors V and VIII	Mn^{2+}- and Co^{2+}-dependent apyrase	Ribeiro et al. (1998)

[a] Blood-feeding arthropod saliva or salivary gland homogenate (SGH).

References: Champagne, D. E. 1995. Proc. Natl. Acad. Sci. USA 31: 694–698; Stark, K. R., James, A. A. 1998. J. Biol. Chem. 14: 20802–20809; Ribeiro, J. M. C., Valenzuela, J. G. 1999. J. Exp. Biol. 202: 809–816; Valenzuela, J. G., et al. 1999. Biochemistry 38: 11209–11215; Francischetti, I. M. B., et al. 1999. Biochemistry 38: 16678–16685; Horn, F., et al. 2000. Arch. Biochem. Biophys. 384: 68–73; Valenzuela, J. G., et al. 1998a. J. Biol. Chem. 273: 30583–30590; Valenzuela, J. G., Ribeiro, J. M. C. 1998b. J. Exp. Biol. 201: 2659–2664; Cheeseman, M. T. et al. 2001. Insect Biochem. Mol. Biol. 31: 157–164; Perez de Leon, A. A., et al. 1998. Exp. Parasitol. 88: 121–130; Cappello, M., et al. 1996. Am. J. Trop. Med. Hyg. 54: 475–480; Ribeiro, J. M. C., et al. 1988. J. Parasitol. 74: 1068–1069; Francischetti, I. M. B., et al. 2002. Blood in press; Charlab, R., et al. 1999. Proc. Natl. Acad. Sci. USA 96: 15155–15160; Ribeiro, J. M. C., et al. 1989. Science 243: 212–214; Ribeiro, J. M. C., et al. 1991. Comp. Biochem. Physiol. 100: 109–112; Valenzuela, J. G. 2001. J. Exp. Biol. 204: 229–237; Ribeiro, J. M. C., et al. 1999. J. Exp. Biol. 202: 1551–1559; Champagne, D. E., et al. 1995. J. Biol. Chem. 270: 8691–8695; Francischetti, I. M. B., et al. 2000. J. Biol. Chem. 275: 12639–12650; Cupp, M. S., et al. 1998. J. Exp. Biol. 201: 1553–1561; Abebe, M., et al. 1996. J. Med. Entomol. 33: 173–176; Noeske-Jungblut, C. 1995. J. Biol. Chem. 270: 28629–28634; Ribeiro, J. M. C., et al. 1998. J. Med. Entomol. 35: 599–610.

arthropods with an incredible variety of strategies. Several insects use vasodilators to relax blood vessels. Within the Diptera, the mosquito *Aedes aegypti* injects small peptides called *sialokinins* into its host. These molecules act directly in the endothelium, producing nitric oxide, which acts on smooth muscle cells, causing vasorelaxation. The *Anopheles albimanus* salivary vasodilator is a 67-kDa peroxidase; it belongs to the myeloperoxidase gene family and is able to degrade potent vasoconstrictors such as norepinephrine and thromboxane A2. In nematoceran Diptera, the saliva of the New World sand fly, *Lutzomyia longipalpis*, contains the most potent vasodilator known today. This 6-kDa peptide, named *maxadilan*, acts on the pituitary adenylate cyclase–activating peptide (PACAP) type I receptors, activating an adenylate cyclase and causing vasodilation. In the Old World sand fly, *Phlebotomus papatasi*, the strategy is completely different; the saliva contains large amounts of free adenosine and adenosine monophosphate (AMP), which are potent vasodilators as well as inhibitors of platelet aggregation. Other examples of vasodilators are the salivary prostacyclin from the tick *Ixodes scapularis* and the vasodilator from the black fly, *Simmulium vittatum*, a novel 15-kDa salivary protein that acts by opening ATP-dependent K$^+$ channels. Hemipteran have quite different and interesting vasodilators. The kissing bug, *Rhodnius prolixus*, and the bedbug, *Cimex lectularius*, have colored proteins named *nitrophorins*. Nitrophorins are hemoproteins that bind the gas nitric oxide (NO) in the salivary glands, which is released when the saliva is injected into the vertebrate host. NO is a potent vasorelaxant, acting directly in smooth muscle cells by activating a guanylate cyclase. The nitrophorin from *R. prolixus* belongs to the family of lipocalins, while the one from *C. lectularius* belongs to the inositol phosphatase family of proteins. It is interesting to note that these two independent families of proteins have acquired the function to bind heme and NO.

Platelet Aggregation Inhibitors

Platelet aggregation is one of the first lines of defense against blood loss. Activated platelets have several effects. They release vasoconstrictor components (thromboxane A2 and 5-HT) and additional agonists of platelet aggregation (ADP), form the platelet plug, and serve as a template for the blood-coagulation cascade. Therefore, it is in the best interest of blood feeders to counteract this branch of the hemostatic system. There are several physiologic agonists for platelet aggregation. The three most important are adenosine diphosphate (ADP), thrombin, and colla-

gen. Concentrations of ADP outside of cells are several orders of magnitude lower than in the cytoplasm. After tissue injury, large amounts of ADP are released; platelets are activated and release even more ADP, causing more platelets to be activated.

All blood feeders studied to date, with the exception of the tick *Amblyoma americanum*, contain a salivary enzyme named *apyrase*. This enzyme hydrolyzes ADP into AMP and orthophosphate, thereby inhibiting platelet aggregation. The apyrase from the mosquito *Ae. aegypti* belongs to the family of 5′-nucleotidases and is magnesium or calcium dependent, with a molecular weight of approximately 68 kDa. The bedbug, *C. lectularius*, has a different apyrase; this calcium-dependent 36-kDa enzyme does not have similarities to mosquito apyrase or to other known proteins in searched databases. Of interest, the apyrases from the Old World sand fly, *P. papatasi*, as well as from the New World sand fly, *L. longipalpis*, belong to the novel family of *Cimex* apyrases.

Besides apyrases, blood feeders have found other solutions to block platelet aggregation. For example, the hemipterans *R. prolixus* and *C. lectularius* use nitrophorins to release NO, which is a potent inhibitor of platelet aggregation, while the tick *Ixodes scapularis* uses prostacyclin to block platelet aggregation. Collagen is also a potent agonist of platelet aggregation. The saliva of *R. prolixus* and *Triatoma pallidipennis* inhibits collagen-induced platelet aggregation. The platelet aggregation inhibitor from *R. prolixus* is a lipocalin that inhibits platelets by binding and sequestering the small amount of ADP that potentiates collagen-induced aggregation. This is a novel strategy discovered in blood feeders, where a salivary protein is acting as an ADP "mop" to block platelet aggregation. Thrombin, another important agonist for platelet aggregation, acts by cleaving the thrombin receptor on platelets. This cleavage results in the activation and aggregation of platelets. Inhibitors of thrombin-induced platelet aggregation have been isolated from the saliva of *An. albimanus* (anophelin), in the hemipteran *Triatoma pallidipennis* (triabin), and in the tsetse fly, *Glossina morsitans*. Other examples of anticoagulants are listed in Table 28.1.

Blood-Coagulation Cascade Inhibitors

The blood-coagulation cascade involves the sequential activation of proenzymes, resulting in thrombin activation, which cleaves fibrinogen into fibrin. Polymerization of fibrin results in blood clot formation. Inhibitors of different steps of the blood-coagulation cascade have been characterized from the saliva of different blood feeders. *Aedes aegypti* has a direct inhibitor of the coagulation factor Xa. This 54-kDa salivary protein is similar to the serpin superfamily of serine protease inhibitors. *Anopheles albimanus* anticoagulant Anophelin, is a specific and tight binding inhibitor of thrombin that binds to the catalytic site of thrombin as well as to the regulatory exosite. Messages similar to this novel 6-kDa peptide are also found in the salivary glands of *An. gambiae*, *An. stephensi*, and *An. arabiensis*. The salivary anticoagulant from the tsetse fly, *Glossina morsitans*, is also a novel 3.5-kDa antithrombin peptide.

ROLE OF SALIVA IN THE INFLAMMATORY RESPONSE

Inflammation is the response to localized injury involving such cells as neutrophils, monocytes, and lymphocytes as well as various mediators, such as chemokines, plasma enzymes, lipid inflammatory mediators, and cytokines. Because the inflammatory response represents a threat for blood-feeding arthropods when they attempt to probe and feed, modulation of this response is beneficial. Inhibitors of the inflammatory response have been described in some blood-feeding arthropods, particularly ticks, which usually stay on the host for a long period of time and have to avoid these types of host responses. For example, a metallo-dipeptidyl carboxypeptidase, which cleaves bradykinin, a molecule involved in pain induction, was characterized in the saliva of the tick *I. scapularis*.

Another important part of the inflammatory response is the *complement cascade*. The complement cascade has two pathways: classical and alternate. From the complement cascade many inflammatory mediators are formed, such as C5a, and C3a; therefore, modulation or inhibition of the complement cascade may be advantageous for blood feeders in avoiding inflammatory reactions. The tick *I. scapularis* has a salivary protein that inhibits the alternative pathway of complement. This 18-kDa protein, Isac (*I. scapularis* anticomplement), is a novel protein that inhibits formation of C3 convertase, thereby acting as a regulator of the complement cascade. An anaphylotoxin-inactivating activity was also reported from the saliva of *I. scapularis*. Anaphylotoxins C3a and C5a are inflammatory mediators derived from the complement cascade. Anaphylotoxins are chemotactic to neutrophils and can cause histamine release from mast cells and basophils.

The main function of neutrophils is engulfing and killing invading microorganisms by secreting inflammatory mediators such as leukotrienes and oxygen radicals. They migrate to the inflammatory site by the action of chemokines. Tick saliva was reported to inhibit key proinflammatory activities of neutrophils,

such as aggregation of these cells when activated by anaphylotoxins, release of enzymes, production of oxygen radicals, and the engulfing of bacteria.

Cell recruitment is an important step during the inflammatory response; most of this recruitment is by chemokines derived mostly from macrophages. The chemokine interleukin-8 (IL-8) is a potent chemoattractant for neutrophils. IL-8 triggers a G-protein-mediated activating signal that results in neutrophil adhesion and subsequent transendothelial migration to the injury site. Anti-IL-8 activity was recently reported from saliva of various ticks. Histamine, another very potent inflammatory mediator, is a vasoactive factor that binds to H1 and H2 receptors, causing edema and erythema by dilating and increasing the permeability of small blood vessels. Histamine is mostly released from mast cells and basophils via an IgE-dependent mechanism. Antihistaminic activities have been reported in kissing bugs and ticks. In the saliva of the kissing bug, *R. prolixus*, salivary nitrophorin binds histamine with high affinity. This hemoprotein carries NO, but NO is displaced in the presence of histamine, and histamine binds to the Fe^{3+} heme group. *Rhodnius prolixus* salivary nitrophorin can therefore scavenge histamine produced at the site of injury. The tick *Rhipicephalus appendiculatus* also has a salivary histamine-binding protein of novel sequence. It is interesting to note that the structure of this high-affinity histamine-binding protein is a lipocalin, which is the structural backbone for *R. prolixus* nitrophorins. Tick histamine-binding proteins were shown to outcompete histamine receptors for the ligand and to decrease the effects of histamine.

ARTHROPOD SALIVA AND THE VERTEBRATE HOST IMMUNE SYSTEM

Modification of the host immune response by salivary components of blood-feeding arthropods has been extensively studied in ticks and sand flies. Ticks remain on the host for a long period of time to complete their blood meal, compared with most blood feeders (days rather than minutes). This long interaction with the host generates a strong response to the tick bite and, more specifically, to the proteins the tick injects into the host. It is well documented that immunity to tick bites or to tick salivary secretion results in rejection of the tick. Tick rejection, a combination of humoral and cellular responses to salivary proteins, usually occurs in nonnatural hosts and is rarely seen in natural host–tick associations. In natural associations, it seems that ticks have coevolved with the host to overcome its immune response. Tick bites or tick salivary gland homogenates can suppress the host immune system, and this suppression may allow the tick to stay longer and complete the blood meal. Table 28.2 summarizes the salivary activities and isolated molecules affecting host immunity to saliva of ticks and other blood-feeding arthropods.

ROLE OF SALIVA IN PARASITE TRANSMISSION

Many parasites or pathogens are delivered from the salivary glands to the skin of the host via salivary fluid. Even when parasites or pathogens are not in the salivary gland, they come into contact with saliva in the insect mouthparts or at the site where the parasite is deposited on the skin. Saliva of blood feeders changes the physiology of the host, and parasites and pathogens take advantage of this change (Table 28.3). *Leishmania major* parasites in the presence of salivary gland extract were found to be more infective than parasites injected alone. Similar results were obtained with the saliva of *L. longipalpis* and *Leishmania brasiliensis*, a parasite usually not very infective in animal models. A similar effect of saliva is also observed in other pathogen–parasite interactions. For example, the bite of the mosquito *Ae. triseriatus* potentiates vesicular stomatitis New Jersey virus infection in mice, while *Ae. triseriatus*, *Ae. aegypti*, and *Culex pipiens* feeding enhances infection of Cache Valley virus in mice. Tick saliva also affects pathogen transmission (Table 28.3). Saliva of the tick *Dermacentor reticulatus* promotes vesicular stomatitis virus growth in vitro, while saliva of the tick *R. appendiculatus* enhances Thogoto virus transmission. Other examples are listed in Table 28.3. Another effect in parasite transmission is seen in *P. papatasi* saliva, which increases uptake of *Leishmania* parasites by macrophages, the immune system cells in which the parasite develops. When a sand fly is probing and feeding, it salivates into the skin of the host, but the saliva is also ingested with the imbedded blood. Saliva may thus have some effect on the parasite in the midgut. Saliva may have a cytostatic effect on Leishmania parasites and may stop parasite multiplication and promote parasite development in the gut.

VACCINE STRATEGIES: USING SALIVARY PROTEINS TO CONTROL PARASITE TRANSMISSION

Because saliva can modify the host hemostatic, inflammatory, and immune responses, it was proposed that saliva of blood feeders is a good target for vaccine strategies to control vector-borne diseases. It has been reported that preexposure to blood-feeder bites or injection of saliva into the vertebrate host generates a strong immune response to salivary proteins that affects the establishment of the parasite or pathogens delivered by the vector (Table 28.4).

For vaccination using vector salivary proteins, there are at least two strategies that can be followed: (1) Isolate the salivary immunosuppressor, which affects establishment of the parasite and prepares antibodies against this component to neutralize its activity; and (2) target salivary proteins that generate strong immune responses, which can kill the pathogen indirectly. Combining these two strategies, which are not mutually exclusive, may give optimal results. Each of the strategies is described here in terms of two approaches using different sand flies. *Lutzomyia longipalpis* saliva has an exacerbating effect in *L. major* infection. Maxadilan, the salivary vasodilator from this sand fly, was demonstrated to be a very potent immunosuppressor. It has been suggested that neutralization of this molecule may lead to blockage of the exacerbating effect of the saliva of this sand fly on parasite infection. Maxadilan exacerbates *L. major* infection to the same degree as whole saliva and animals vaccinated with maxadilan were protected against *L. major* infection. A strong humoral and cellular response (Th1 type) was developed against maxadilan in mice injected with this salivary protein. It is possible that antibodies to maxadilan neutralized the immunosuppressor effect of this molecule, thereby blocking establishment of the parasite on the vertebrate host, and that the cellular response to maxadilan might also be responsible for this protective effect, as suggested by Morris et al. (2001).

An exacerbating effect of *P. papatasi* saliva on *L. major* infection in naïve animals has also been seen. This effect was abrogated in mice previously exposed to sand fly salivary gland homogenate. This was the first demonstration of the "proof of principle" that saliva from sand flies can be used as vaccine to control Leishmania infection. It is important to note that Old World sand flies do not have maxadilan in their saliva but instead have salivary adenosine as an immunomodulator of the vertebrate host immune response. Animals previously exposed to uninfected *P. papatasi* sand fly bites were protected against subsequent challenge of bites by Leishmania-infected sand flies. This suggested that protection was due both to a strong delayed-type hypersensitivity (DTH) response induced by sand fly bite (salivary proteins) and to high levels of interferon at the bite site (see Kamhawi 2000). Multiple exposure to bites by *P. papatasi* in either mouse or human generates strong antibodies to salivary proteins as well as a strong DTH response. This

TABLE 28.2   Blood-Feeding Arthropod Salivary Immunomodulatory Activities

Source[a]	Effect on Host Immune System	Reference
Phlebotomus papatasi saliva and adenosine	Down-regulation of nitric oxide synthase gene in activated macrophages	Katz et al. (2000)
Phlebotomus papatasi	Down-regulates Th1 and upregulates Th2 in naïve mice	Mbow et al. (1998)
Lutzomyia longipalpis-maxadilan	Inhibits TNF-α and induces IL-6 in mouse macrophages	Soares et al. (1998)
Phlebotomus duboscqi	Chemotactic to peritoneal monocytes	Anjili et al. (1995)
Aedes aegypti	SGH inhibition of TNF-α release from mast cells	Bissonnette et al. (1993)
Aedes aegypti	SGH modulation of mouse cellular immune response	Cross et al. (1994)
Simulium vittatum	SGH modulation of mouse immune response	Cross et al. (1993)
Aedes aegypti and *Culex pipiens* feeding	Modulation of mouse Th1 and Th2 cytokine response	Zeidner et al. (1999)
Rhodnius prolixus SGH	Reduction of mouse splenic lymphocytes proliferative responses (spontaneous and mitogen induced)	Kalvachova et al. (1999)
Dermacentor reticulatus	Suppression of INF-γ activity	Hajnicka et al. (2001)
Rhipicephalus appendiculatus	Decreased mRNA production for IFN-γ, TNF-α, IL-1, IL-13, IL-5, IL-6, IL-7, IL-8	Fuchsberger et al. (1995)
Dermacentor reticulatus	Decrease natural killer cell activity	Kubes et al. (1994)
Ixodes ricinus–infested mice Cultured lymphocytes from *Ixodes ricinus*-infested mice.	Reduction of T-cell responsiveness Increase in IL-4 expression, low levels of INF-γ, and high levels of IL-10	Ganapamo et al. (1996)
Dermacentor andersoni	Suppression of T-cell response by a 36-kDa salivary protein	Bergman et al. (2000)
D. andersoni-infested mice	IL-1, IL-2, INF-γ, and TNF suppression on macrophages. Reduction of lymphocyte adhesion molecules (LFA-1) and (VLA-4)	Macaluso and Wikel (2001)
Ixodes scapularis saliva	Inhibition of T-cell proliferation and suppression of IL-2; L-2 binding protein	Gillespie et al. (2001)
Rhipicephalus sanguineus	Inhibition of mouse T-cell proliferation and macrophage activity	Ferreira and Silva (1998)
Rhipicephalus appendiculatus	Immunoglobulin binding protein	Wang and Nuttall (1995, 1999)
Rhipicephalus appendiculatus	Suppression of transcription and secretion of IL-1, TNF-α, and IL-10 on mouse macrophage cells	Gwakisa et al. (2001)
Ixodes ricinius SGH	Polarization toward Th2 response and suppression of T-cell response	Kovar et al. (2001)
Ixodes ricinus	Modulates macrophage and T-cell response	Leboulle et al. (2002)
Amblyomma americanum salivary glands	Presence of macrophage migration inhibitory factor	Jaworski et al. (2001)
Dermacentor reticulatus *Amblyoma variegatum* *Rhipicephalus appendiculatus* *Haemaphysalis inermis* *Ixodes ricunus*	Anti-IL-8 activity	Hajnicka et al. (2001)

[a] Blood-feeding arthropod saliva or salivary gland homogenate (SGH).

References: Katz, O., et al. 2000. *Am. J. Trop. Med. Hyg.* 62: 145–150; Mbow, M. L. 1998. *J. Immunol.* 161: 5571–5577; Soares, M. B., et al. 1998. *J. Immunol.* 160: 1811–1816; Anjili, C. O., et al. 1995. *Acta Trop.* 60: 97–100; Bissonnette, E. Y., et al. 1993. *Parasite Immunol.* 15: 27–33; Cross, M. L., et al. 1993. *J. Med. Entomol.* 30: 928–935; Cross, M. L., et al. 1994. *Am. J. Trop. Med. Hyg.* 51: 690–696; Zeidner, N. S., et al. 1999. *Parasite Immunol.* 21: 35–44; Kalvachova, P., et al. 1999. *J. Med. Entomol.* 36: 341–344; Hajnicka, V., et al. 2001. *Parasite Immunol.* 23: 483–489; Fuchsberger, N., et al. 1995. *Exp. Appl. Acarol.* 19: 671–676; Kubes, M., et al. 1994. *Immunology* 82: 113–116; Ganapamo, F., et al. 1996. *Immunology* 87: 259–263; Bergman, D. K., et al. 2000. *J. Parasitol.* 86: 516–525; Macaluso, K. R., Wikel, S. K. 2001. *Ann. Trop. Med. Parasitol.* 95: 413–427; Gillespie, R. D., et al. 2001. *J. Immunol.* 166: 4319–4326; Ferreira, B. R., Silva, S. J. 1998. *Vet. Immunol. Immunopathol.* 64: 279–293; Wang, H., Nuttall, P. A. 1995. *Parasite Immunol.* 17: 517–524; Wang, H., Nuttall, P. A. 1999. *Cell. Mol. Life Sci.* 56: 286–295; Kovar, L., et al. 2001. *J. Parasitol.* 87: 1342–1348; Gwakisa, P., et al. 2001. *Vet. Parasitol.* 99: 53–61; Leboulle, G., et al. 2002. *J. Biol. Chem.* 277(12): 10083–10089; Jaworski, D. C., et al. 2001. *Insect Mol. Biol.* 10: 323–331; Hajnicka, V., et al. 2001. *Parasite Immunol.* 23: 483–489.

TABLE 28.3   Effect of Saliva of Blood-Feeding Arthropods on Pathogen Transmission

Blood-Feeding Arthropod	Effect of Saliva on Pathogen Transmission
Lutzomyia longipalpis salivary gland homogenate	Enhancement of *Leishmania major* infection (Titus and Ribeiro 1988)
Lutzomyia longipalpis saliva	Enhancement of *L. brasiliensis* infection (Lima and Titus 1996)
Phlebotomus papatasi salivary gland homogenate	Enhancement of *L. major* infection (Theodos et al. 1991; Belkaid et al. 1998)
Aedes triseriatus bite	Potentiates vesicular stomatitis New Jersey virus infection in mice (Limesand et al. 2000)
Aedes triseriatus, Aedes aegypti Culex pipiens	Enhancement of Cache Valley virus infection in mice (Edwards et al. 1998)
Phlebotomus papatasi saliva	Macrophage uptake of Leishmania parasites (Zer et al. 2001)
Lutzomyia longipalpis saliva	Cytostatic effect on Leishmania parasites (Charlab and Ribeiro 1993)
Dermacentor reticulatus salivary	Promotes virus growth in vitro gland extract (Hajnicka et al. 1998) and nucleocapsid production (Kocakova et al. 1999)
Rhipicephalus appendiculatus	Enhancement of Thogoto virus transmission (Jones et al. 1989) and increases *Theilera parva* infection in lymphopcytes (Shaw et al. 1993)
Dermacentor reticulatus Ixodes ricinus Rhipicephalus appendiculatus	Enhancement of tick-borne encephalitis virus transmission (Labuda et al. 1993)

References: Titus, R. G., Ribeiro, J. M. C. 1988. *Science* 239: 1306–1308; Lima, H. C., Titus, R. G. 1996. *Infect. Immun.* 64: 5442–5445; Theodos, C. M., et al. 1991. *Infect. Immun.* 59: 1592–1598; Belkaid, Y., et al. 1998. *J. Exp. Med.* 188: 1941–1953; Limesand, K. H., et al. 2000. *Parasite Immunol.* 22: 461–467; Edwards, J. F., et al. 1998. *J. Med. Entomol.* 35: 261–265; Zer, R., et al. 2001. *Int. J. Parasitol.* 31: 810–814; Charlab, R., Ribeiro, J. M. C. 1993. *Am. J. Trop. Med. Hyg.* 48: 831–838; Hajnicka, V., et al. 1998. *Parasitology* 116: 533–538; Kocakova, P., et al. 1999. *Acta Virol.* 43: 251–254; Jones, L. D., et al. 1989. *J. Gen. Virol.* 70: 1895–1898; Shaw MK et al. 1993. *Infect. Immun.* 61: 1486–495; Labuda, M., et al. 1993. *Med. Vet. Entomol.* 7: 193–196.

TABLE 28.4   Effect of Preexposure to Bite, Saliva, or Isolated Salivary Protein on Pathogen Transmission

Blood-Feeding Arthropod	Effect of Preexposure
Rabbits preexposed to *Dermacentor andersoni*	Protection from *Franciscella tularensis* infection (Bell et al. 1979)
Mice preexposed to *Ixodes scapularis*	Protection from *Borrelia burgdorferi* infection (Wikel et al. 1997)
Guinea pigs preexposed to *Ixodes scapularis*	Protection from *Borrelia burgdorferi* infection (Nazario et al. 1998)
Guinea pigs preexposed to *Rhipicephalus appendiculatus*	Protection from *Thogoto* virus infection (Jones and Nuttall 1990)
Mice preexposed to salivary gland of *Phlebotomus papatasi*	Protection from *L. major* infection (Belkaid 1998)
Mice preexposed to *P. papatasi* bite	Protection from *L. major* infection (Kamhawi et al. 2000)
P. papatasi Sp15 protein and DNA vaccination	Protection from *L. major* infection (Valenzuela et al. 2001)
Maxadilan (*Lutzomyia longipalpis*) vaccination	Protection from *L. major* infection (Morris et al. 2001)

References: Belkaid, Y., et al. 1998. *J. Exp. Med.* 188: 1941–1953; Bell, J. F., et al. 1979. *Am. J. Trop. Med. Hyg.* 28: 876–880; Jones, L. D., et al. 1990. *J. Gen. Virol.* 71: 1039–1043; Kamhawi, S., et al. 2000. *Science* 290: 1351–1354; Morris, R. V., et al. 2001. *J. Immunol.* 167: 5226–5230; Nazario, S., et al. 1998. *Am. J. Trop. Med. Hyg.* 58: 780–785; Valenzuela, J. G., et al. 2001. *J. Exp. Med.* 194: 331–342; Wikel, S. K., et al. 1997. *Infect. Immun.* 65: 335–338.

strong cellular response, known as *Harara*, is primarily a CD4-dependent immune response, with accumulation of proinflammatory cells at the site of the bite 24–48 hours after antigen (salivary protein) presentation to T-cells. This strong cellular response helps the sand fly to probe and feed faster, because blood flow is greater in these sites than in normal skin sites. It has been proposed that this strong cellular response may work against the parasites the sand flies deliver because of the inflammatory cells brought to the site, particularly macrophages. Active macrophages in a DTH site release cytokines, reactive NO, and oxygen radicals, making this site inhospitable for *Leishmania*. It is also well documented that immunity to *Leishmania* parasites is cellular and thus not antibody mediated.

We recently isolated the *P. papatasi* salivary protein responsible for protection to *L. major* infection, a

15-kDa salivary protein of novel sequence named SP15. Animals vaccinated with either SP15 protein or DNA plasmid containing the SP15 gene were protected against *L. major* infection when animals were challenged with parasite plus saliva. Because SP15-vaccinated animals generated a strong antibody response to SP15 protein as well as a strong DTH reaction, we wanted to determine whether the protection was antibody mediated or due only to DTH. For our investigation, we used B-cell-deficient mice that cannot generate antibodies but can nonetheless produce a cellular response. B-cell-deficient mice that received the SP15 DNA plasmid vaccine generated a strong cellular response to salivary proteins but did not produce antibodies; this group of mice was protected from a *L. major* infection. These data suggested that cellular responses to salivary proteins were sufficient for protection against *Leishmania* infection, probably by killing the parasite indirectly, and that antibodies to salivary proteins were not necessary for this protection. Because preexposure to sand fly saliva or bites or to the DNA plasmid vaccination protected mice from *L. major* infection, it may be possible that human immunity to sand fly salivary proteins may also protect against leishmaniasis (see Valenzuela et al. 2001).

FROM CLASSICAL TO CURRENT: MOLECULAR BIOLOGY TECHNIQUES TO STUDY VECTOR SALIVARY PROTEINS AND MESSAGES

Because saliva is so important in blood feeding, host immunity, and pathogen transmission, characterization of the salivary components responsible for these activities is integral both to understanding their mechanism of action and to developing new strategies to block pathogen transmission. Isolation of salivary components has been a challenge for many years because the amount of salivary protein in many blood feeders (particularly mosquitoes, sand flies, fleas, and midges) is very small. As a result, large numbers of salivary glands are needed to produce the amount of material necessary to perform classical biochemical purification. Many salivary proteins with antihemostatic, inflammatory, and immunomodulatory activities have been isolated following classical biochemical and molecular biology approaches. Yet compared to the variety of activities present and reported from the saliva of different blood-feeding arthropods, there is little about these molecules in the literature. The traditional protocol is purification of a salivary activity from large amounts of either saliva or salivary gland

homogenate supernatant (hundreds to thousands of salivary gland pairs). Recent advances in molecular techniques allow simultaneous study of multiple proteins and genes. The vector biology field is experiencing this same change of focus in studying the messages (genes) and proteins present in salivary glands of blood-feeding arthropods.

Our laboratory is using an approach that consists of massive sequencing of full-length salivary gland cDNA libraries in conjunction with a proteomic approach to sequencing the most abundant proteins secreted from the salivary glands of blood-feeding arthropods. With this high-throughput approach, a number of sequences that matched proteins with predicted functions have been discovered and also many novel DNA sequences. The complexity of secreted salivary proteins is very low (only small numbers of secreted proteins are detected in Coomassie blue–stained gels, for example), which makes the proteomic analysis very accessible. Because salivary glands are composed of few cells, the messages expressed in the salivary glands code for secretory proteins. As a result, the most abundant messages sequenced represent secretory proteins. Proteomic analysis of saliva from vectors shows the variety of the proteins present in saliva of blood-feeding arthropods as well as the types of molecules common to different families of blood feeders.

One proteomic approach used to study salivary proteins from disease vectors involves obtaining sequence information from proteins separated by one-dimensional gel electrophoresis. The aminoterminal sequence obtained is compared to the predicted translated products from the salivary gland cDNA database (created using the massive sequencing approach) using an algorithm similar to the BLOCK program. The N-terminus sequence usually matches a cDNA whose translated product is present in the salivary gland library. By computer analysis, information for multiple sequences resulting from a single one-dimensional PAGE band can be resolved. A protein band in a one-dimensional gel usually has at least one sequence or signal (mixed signal) that cannot be usually analyzed. When these sequences are analyzed using the BLOCK program and translated sequences from the library, the program can separate or resolve the different sequences present in a protein band.

FUNCTIONAL GENOMICS OF VECTOR SALIVARY GLAND MESSAGES

A functional genomic approach is the search for a function for a particular gene(s) or cDNA with predicted or unknown function. Now that we are in the postgenomic era, the information generated by massive sequencing projects will no doubt represent new challenges, in terms of predicting functions of novel sequences. In terms of functional approaches for salivary messages, we concentrated first on sequences with similarities to proteins with known function that may be related to blood feeding or may have an effect on host response. Our approach was to demonstrate that the predicted activity (by sequence analysis) existed in the saliva of the blood feeder and/or to express the recombinant protein to test for the predicted activity. One advantage of the functional approach is that most of the cDNA used to produce the salivary gland libraries represent full-length messages. These messages are thus immediately available for subcloning in various recombinant expression systems. Examples of activities and sequences discovered using this comprehensive approach include *P. papatasi* salivary apyrase; the novel family of salivary anticoagulants from *L. longipalpis*; adenosine deaminase from *Culex* and *Aedes* mosquitoes; the novel tissue factor pathway inhibitor from saliva of *I. scapularis*; salivary nucleosidase from the *Ae. aegypti* mosquito; the D7 family of salivary proteins of *An. arabiensis, An. stephensi, Ae. aegypti, C. quinquefasciatus, P. papatasi*, and *L. longipalpis; An. stephensi* and *C. quinquefasciatus* apyrases; the antigen-5 family of proteins of *Anopheles, Aedes, Culex*, and *Phlebotomus*; the salivary angiopoietin family of insect proteins in *Ae. aegypti*; the salivary kininase from *I. scapularis*; novel lipocalins from the kissing bug, *R. prolixus*; and the first salivary protein that blocks pathogen transmission from the saliva of *P. papatasi*.

CONCLUSION

The salivary gland is a vital organ from the point of view of the pathogen, because it is where it stays before delivery to the vertebrate host. This organ is also important because it produces saliva. This salivary secretion, which is important for mouthpart lubrication, solubilization of food, and other important aspects of insect physiology, also contains potent pharmacologically active components that modulate the host hemostasis, inflammation, and immune systems, which help disease vectors to probe and feed better. Saliva also directly affects establishment of the pathogen delivered by the vector. Although the importance of saliva in vector-borne diseases has been clearly demonstrated, the practical aspects of using saliva as an alternative approach to vector-borne disease control are only now being addressed. Challenges remain in isolating and characterizing the salivary molecules that

can be used to control pathogen transmission and to gain a better understanding of how these salivary molecules affect the host immune response and, consequently, establishment of the pathogen. New strategies to study salivary components are emerging, and high-throughput approaches to isolate and sequence messages (genes) and proteins are under way. These new approaches are giving us a welcome flood of information that will impact the way salivary proteins from vectors of pathogens are studied and used to control vector-borne diseases. For now, we need to focus on the biology of the isolated sequences and on more robust methods to express recombinant proteins in the active form. This will enable us to test both predicted and novel functions. Cloning novel salivary cDNA into different expression vectors in a high-throughput manner is now an accessible protocol. It will be challenging to develop high-throughput expression systems to test many of the novel predicted activities and to find more robust and natural models of transmission of disease vectors to test salivary proteins as vaccines to control pathogen transmission.

Readings

Kamhawi, S. 2000. The biological and immunomodulatory properties of sand fly saliva and its role in the establishment of *Leishmania* infections. *Microbes Infect.* 2: 1765–1773.

Ribeiro, J. M. C. 1995b. Insect saliva: Function, biochemistry and physiology. In: *Regulatory Mechanisms of Insect Feeding* (R. F. Chapman, and G. de Boer, eds.). Chapman and Hall, London, pp. 74–97.

Sauer, J. R., McSwain, J. L., Bowman, A. S., and Essenberg, R. C. 1995. Tick salivary gland physiology. *Annu. Rev. Entomol.* 40: 245–267.

Wikel, S. K. 1996. Host immunity to ticks. *Annu. Rev. Entomol.* 41: 1–22.

Further Readings

Belkaid, Y., Kamhawi, S., Modi, G., Valenzuela, J., Noben-Trauth, N., Rowton, E., Ribeiro, J., and Sacks, D. L. 1998. Development of a natural model of cutaneous leishmaniasis: Powerful effects of vector saliva and saliva preexposure on the long-term outcome of *Leishmania major* infection in the mouse ear dermis. *J. Exp. Med.* 188: 1941–1953.

Belkaid, Y., Valenzuela, J. G., Kamhawi, S., Rowton, E., Sacks, D. L., and Ribeiro, J. M. C. 2000. Delayed-type hypersensitivity to *Phlebotomus papatasi* sand fly bite: An adaptive response induced by the fly? *Proc. Natl. Acad. Sci. USA* 97: 6704–679.

Charlab, R., Valenzuela, J. G., Rowton, E. D., and Ribeiro, J. M. C. 1999. Toward an understanding of the biochemical and pharmacological complexity of the saliva of a hematophagous sand fly *Lutzomyia longipalpis. Proc. Natl. Acad. Sci. USA* 96: 15155–15160.

Edwards, J. F., Higgs, S., and Beaty, B. J. 1998. Mosquito feeding-induced enhancement of Cache Valley Virus (Bunyaviridae) infection in mice. *J. Med. Entomol.* 35: 261–265.

Francischetti, I. M. B., Ribeiro, J. M. C., Champagne, D., and Andersen, J. 2000. Purification, cloning, expression, and mechanism of action of a novel platelet aggregation inhibitor from the salivary gland of the blood-sucking bug, *Rhodnius prolixus*. *J. Biol. Chem.* 275: 12639–12650.

Henikoff, S., Henikoff, J. G., and Pietrokovski, S. 1999. Blocks+: A nonredundant database of protein alignment blocks derived from multiple compilations. *Bioinformatics* 15: 471–479.

James, A. A. 1994. Molecular and biochemical analysis of the salivary glands of vector mosquitoes. *Bull. Institute Pasteur* 92: 133–150.

Jones, L. D., Hodgson, E., and Nuttall, P. A. 1989. Enhancement of virus transmission by tick salivary glands. *J. Gen. Virol.* 70: 1895–1898.

Kamhawi, S., Belkaid, Y., Modi, G., Rowton, E., and Sacks, D. 2000. Protection against cutaneous leishmaniasis resulting from bites of uninfected sand flies. *Science* 290: 1351–1354.

Limesand, K. H., Higgs, S., Pearson, L. D., and Beaty, B. J. 2000. Potentiation of vesicular stomatitis New Jersey virus infection in mice by mosquito saliva. *Parasite Immunol.* 22: 461–467.

Morris, R. V., Shoemaker, C. B., David, J. R., Lanzaroand, G. C., and Titus, R. G. 2001. Sandfly maxadilan exacerbates infection with

Leishmania major and vaccinating against it protects against L. major infection. *J. Immunol.* 167: 5226–5230.

Nazario, S., Das, S., de Silva, A. M., Deponte, K., Marcantonio, N., Anderson, J. F., Fish, D., Fikrig, E., and Kantor, F. S. 1998. Prevention of *Borrelia burgdorferi* transmission in guinea pigs by tick immunity. *Am. J. Trop. Med. Hyg.* 58: 780–785.

Ribeiro, J. M. C. 1995a. Blood-feeding arthropods: Live syringes or invertebrate pharmacologists? Infect. Agents Dis. 4: 143–152.

Titus, R. G., and Ribeiro, J. M. C. 1988. Salivary gland lysates from the sand fly *Lutzomyia longipalpis* enhance Leishmania infectivity. *Science* 239: 1306–1308.

Valenzuela, J. G., Belkaid, Y., Garfield, M. K., Mendez, S., Kamhawi, S., Rowton, E. D., Sacks, D. L., and Ribeiro, J. M. C. 2001. Toward a defined anti-Leishmania vaccine targeting vector antigens: Characterization of a protective salivary protein. *J. Exp. Med.* 194: 331–342.

VECTOR GENETICS

WILLIAM C. BLACK IV

29

Systematic Relationships Among Disease Vectors

WILLIAM C. BLACK IV AND STEPHEN C. BARKER

INTRODUCTION

Systematics is the study of evolutionary relationships among species. *Phylogenetics* is a procedure for estimating these relationships based upon data from extant and fossil taxa. Phylogeneticists use these data to construct an *evolutionary tree*. The tree's trunk contains the most ancient hypothetical ancestors. First-order branches represent more recently derived ancestral species, and the terminal branches and leaves represent extant taxa. The goal of systematics is to derive trees that most accurately represent the true evolutionary relationships among extant taxa. Given an accurate phylogeny, vector biologists can then overlay information on anatomy, development, cytogenetics, behavior, host preference, ecology, life history, biogeography, or the range of pathogens transmitted. Such a process can provide important insights into the evolution of these characters as well as the vector species themselves. When did a particular morphological character arise? How many times during the course of evolution did adaptation to a particular habitat occur? Does the ability to vector a particular pathogen occur within a single evolutionary lineage, or has the ability arisen many times independently over the course of evolution?

We begin this chapter with the premise that phylogenetic relationships can be most accurately and objectively assessed with molecular genetic data sets, and we restrict our discussion to nucleic acid sequences. We will discuss how these data are obtained and how sequences are most objectively aligned, and then we will provide a basic description of the three most commonly used approaches that can be taken toward estimating phylogenetic relationships. We will finish with examples of how molecular systematics has been used to understand the evolution of disease vectors.

Obtaining DNA Sequence Data

Systematics is an old science. In contrast, molecular systematics is a recent science dating back only to the early 1960s, when systematists, working with biochemists, used differences in melting temperatures (T_m) of hybrid DNA molecules between two species as a general measure of genetic differences. Early molecular systematists experimentally measured the T_m *between pair of species* and subtracted from that *the average T_m for DNA in each species* to derive $\odot T_m$, a measure of overall DNA divergence. $\odot T_m$ was observed to be small for closely related species but to increase among dissimilar species. This method allowed for estimation of differences among a wide variety of taxa. However, the method was impossible to use on small taxa that were represented by a few individuals and from taxa from which DNA extraction was inefficient. In addition, it was difficult with early technology to assess $\odot T_m$ among closely related species.

DNA hybridization remained the only means for assessing DNA similarity until the early 1980s, when recombinant DNA technology allowed investigators to build genomic and cDNA libraries. These could be probed with a gene obtained from one species to

389

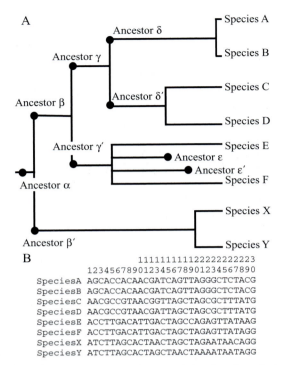

A

Ancestor δ — Species A / Species B

Ancestor γ

Ancestor β

Ancestor δ′ — Species C / Species D

Ancestor γ′ — Species E
Ancestor ε
Ancestor ε′ — Species F

Ancestor α

Species X
Species Y

Ancestor β′

B
```
                    11111111112222222223
          12345678901234567890234567890
SpeciesA  AGCACCACAACGATCAGTTAGGGCTCTACG
SpeciesB  AGCACCACAACGATCAGTTAGGGCTCTACG
SpeciesC  AACGCCGTAACGGTTAGCTAGCGCTTTATG
SpeciesD  AACGCCGTAACGATTAGCTAGCGCTTTATG
SpeciesE  ACCTTGACATTGACTAGCCAGAGTTATAAG
SpeciesF  ACCTTGACATTGACTAGCTAGAGTTATAGG
SpeciesX  ATCTTAGCACTAACTAGCTAGAATAACAGG
SpeciesY  ATCTTAGCACTAGCTAACTAAAATAATAGG
```

FIGURE 29.1 (A) A representation of the "true" evolutionary relationships among eight extant and nine ancestral species. (B) A taxon-×-character data set containing the DNA sequences of a particular, homologous gene in all eight extant species.

extract a similar gene from a new species. DNA sequencing technology became widely available at the same time and allowed the assemblage of DNA sequence data sets (e.g., Fig. 29.1B). However, at that time this technology was slow and extremely expensive.

It wasn't until Kerry Mullis invented the polymerase chain reaction in the mid-1980s that the molecular systematics of diverse organisms became feasible. Commercially available thermostable polymerases and programmable thermal cyclers became available in the early 1990s, and for the first time systematists could amplify and compare genes from virtually any organism, including museum specimens and even fossils. Significantly for vector biology, Chris Simon and others published a large list of oligonucleotide primers with which to amplify various regions of the mitochondrial genome in arthropods. At the same time, various publications catalogued oligonucleotide sequences with which to amplify all or portions of the nuclear ribosomal RNA cistron. Currently, oligonucleotide sequences for a large diversity of single-copy nuclear genes are available (Hillis et al. 1996).

During the past decade, systematists have acquired access to a vast and growing number of genes with

which to infer phylogenetic relationships. At the end of this chapter we will describe a number of studies of molecular phylogenetic relationships among vectors. These studies used a variety of gene sequences to infer phylogenetic relationships. One lesson that molecular systematists have learned over the past decade is that no one sequence is adequate for examining phylogenetic relationships among all species of a particular taxonomic group. Genes with high substitution rates may most accurately estimate relationships among recently evolved taxa but will not be useful among ancient taxa. Conversely, genes with low substitution rates accurately infer phylogenetic relationships among anciently derived taxa but will be uninformative among recent taxa.

Assembling Molecular Systematic Datasets

The first step in assembling a molecular systematics data set is to collect specimens of all of the species the investigators propose to analyze. The specimens can be from museum collections, from field collections preserved in alcohol, or from live material. The investigators should always consult with traditional, morphologically based, systematists or sources in determining which taxa should be examined to assess the phylogeny in a particular group. This includes a prudent choice and representative sampling of in-group and out-group taxa, as explained later.

The investigators next select a pair of oligonucleotide primers from the large numbers currently available. The investigators should be able to infer from the literature or by consulting colleagues which genes are likely to be useful in assessing genetic differences at the proposed taxonomic level. For example, mitochondrial genes might be an excellent choice if the investigator is examining relationships among closely related species or among populations of a species. Conversely, if interested in higher-level systematic questions, they might consider the use of various ribosomal RNA genes (Chapter 33) or other conserved nuclear genes.

The investigators next attempt to amplify the gene from the organisms they have collected. If successful, the investigators then sequence the gene either using commercially available gene-sequencing facilities or by setting up a sequencing facility in their own laboratory. One issue is whether the investigators should clone the amplified sequences prior to sequencing or should sequence the PCR products directly. We strongly advocate the latter approach. The sequencing ladder, or *chromatograph*, usually displays the predominant nucleotide at any particular position, whereas cloned sequences frequently contain rare or uncommon nucleotides in particular positions.

The next task is to assemble this information into a data set. As you'll learn later, accurate alignment of nucleotides is critical in accurately assessing phylogenetic relationships. There are a number of software packages (e.g., CLUSTALW) that can be used for automated alignment of sequences. However, the authors advocate the use of these packages only for initial alignment. The final alignment should be done manually on the basis of secondary structure when working with nuclear or mitochondrial ribosomal RNA or based on amino acids and codons when working with protein-coding genes.

Terminology

As with any scientific discipline, systematics involves a standard set of jargon. We begin with a brief description of that terminology to facilitate later discussion. Figure 29.1A is a representation of the "true" evolutionary relationships among eight extant species. In practice, the actual evolutionary relationships are *never* known; they are estimated from characters measured in extant or fossil species. However, for the purposes of introducing terminology, we pretend that we have a time machine and have tracked the precise evolutionary relationships. The phylogeny in Figure 29.1A is interpreted in the following manner. At some point in the past, ancestral species α split and gave rise to two new ancestral taxa, β and β'. Through time, β and β' became reproductively isolated. Later, taxon β gave rise to two new ancestral taxa, γ and γ'. Eventually ancestral taxa γ' gave rise to the two extant species that we know today as E and F, which have existed without speciation for a long period of time, and two species (ancestors ϵ and ϵ') that went extinct. Thus, ancestors of ϵ' underwent a period of rapid speciation (*anagenesis*). About the same time that taxa γ' speciated, taxon γ gave rise to two ancestral species δ and δ'. Taxon δ' gave rise to the extant taxa C and D. Species δ failed to speciate (*stasigenesis*) until very recently, when it gave rise to the two sister species we now call taxon δ.

Figure 29.1B is a taxon-by-character data set containing the DNA sequences of one particular gene in all eight species. *Homology* of characters is one of the most fundamental and important assumptions of phylogenetics. In building a taxon-by-character matrix we implicitly assume that the eight genes in Figure 29.1B are homologous; they were all derived from the same common ancestral gene in ancestor α. While we cannot explicitly test this assumption, molecular systematists often believe this assumption is justified because the gene was extracted using the same DNA probe or was amplified with PCR using the same oligonucleotide primers. This assumption extends to the level of individual characters in the data set. There are 30 characters in the data set in Figure 29.1B. The alignment of these 30 characters among the eight taxa further assumes that the first nucleotide in all taxa was derived from the same first nucleotide in ancestor α'.

A *character state* is the condition of a homologous character in each taxon. Molecular character states are usually discrete. There can be multiple character states (e.g., A, C, G, T, gap, or 1 of 21 amino acid characters) or two character states (e.g., character present or absent). Character states can be ordered (e.g., A gave rise to G gave rise to C) or unordered (e.g., a nucleotide mutates with equal frequency to any other nucleotide). Modern phylogenetic studies use *out-group taxa* to *order* characters in the data set. Species considered to be moderately or closely related to the species under study (*in-group taxa*) are usually selected as outgroups. Again, the molecular systematist must work closely with traditional, morphologically based, systematists that are very familiar with the taxa under study to make prudent choices of in-group and out-group taxa. Choice of distantly related out-group taxa can lead to incorrect inferences about the ancestral state of characters.

Phylogenetic Inference

Now let's consider the evolutionary relationships among the eight species in Figure 29.1A with respect to the nucleotide characters that currently exist in each species (Figure 29.1B). Species A–F constitute the in-group taxa among which we are interested in inferring evolutionary relationships, while the out-group taxa we have chosen are species X and Y. *Plesiomorphic* refers to the ancestral state of a character. *Apomorphic* is the descendant state of a character. For example, the "A" at 6, 12, 23, and 26; a "C" at 10; a "G" at 29; and a "T" at nucleotide 2 are all plesiomorphic states of these characters. Correspondingly, a "C, G, or T" at 6, 12, 23, and 26; an "A or T" at 10; an "A, T, or C" at 29; and an "A, C, or G" at nucleotide 2 are all apomorphic states of these characters.

Characters of in-group taxa that are shared with ancestors are referred to as *symplesiomorphic*. In a phylogenetic analysis, we would intuitively place in-group taxa with more symplesiomorphic characters at the base of the evolutionary tree, while in-group taxa that had more apomorphic characters would be further from the base. Notice that species E and F have nine symplesiomorphic characters with species X and

Y (an "A" at positions 22 and 26; a "C" at positions 8, 14, and 18; and a "T" at positions 4, 5, 11, and 24). In contrast, species A and B are only symplesiomorphic for a "C" at nucleotide 8, while species C and D are symplesiomorphic for the "G" at nucleotide 7, a "T" at 15, and a "C" at 18. Therefore, intuitively, E and F are more ancestral ("basal") than C and D, which are more ancestral than A and B.

Characters shared among a group of species that evolved from some common ancestor are referred to as *synapomorphies*. The "A" at nucleotides 12, 23, and 25 and the "C" at 10 all represent synapomorphies for ancestors of β'. The "C" at 2, the "GA" at characters 6 and 7, and the "T" at 10 are synapomorphies for ancestors of γ', while the "CC" at characters 5 and 6, the "AC" at 10/11, and the "T" at 14 are synapomorphies for ancestors of γ. Finally, the "A" at 2, the "C" at 22, the "G" at 4, and the "T" at characters 8, 26, and 29 are synapomorphies for ancestors of δ', while the "A" at 4, the "C" at 26 and 29, the "G" at 2 and 22, and the "T" at 18 are synapomorphies for ancestors of δ.

Characters are generally classified into one of three categories of phylogenetically informativeness: *informative*, *uninformative*, or *misinformative*. All of the characters listed in the prior two paragraphs are phylogenetically informative because they allow taxa to be positioned with respect to one another. Characters that do not vary among taxa (e.g., characters at 1, 3, 9, 16, 20, and 28) and characters that evolved in a species after it arose (autapomorphies also called a *derived character* or a *specialization*; e.g., the "C" at 19 in species E or at 27 in species X) are phylogenetically uninformative.

Of greatest concern are characters that are shared among species that arose from *different* ancestors. These are referred to as *homoplasies*. These are shared but independently derived or convergent characters. They are phylogenetically misinformative because they actually cause us to misconstrue common ancestry. Examples of homoplasious characters are the "G" at position 7 shared among species C, D, X, and Y; the "C" at position 8 shared among species A, B, X, and Y; and the "G" at position 13 shared between species C and Y. If we were to assess phylogenetic relationships based upon only these three characters, then we would *incorrectly* infer that C and D shared an immediate ancestor with X and Y, or that A and B share an immediate ancestor with X and Y, or that C and Y share a recent common ancestor. In each case these characters would cause the phylogeneticist to make an incorrect inference. *An important aspect of modern phylogenetic analysis is that inferences are never based upon one or a few characters but rather are inferred from analysis of many characters.*

Cladistics

We have to this point assumed the impossible—that we have perfect knowledge of evolutionary histories among the eight species. Phylogenetic inference is the process of estimating evolutionary relationships based upon a multitude of characters in extant taxa. For many years phylogenetic reconstruction was based upon intuitive inferences made by experts in a particular taxonomic group. Often these inferences were correct, but as frequently they were based upon assessment of one or a few characters that the taxonomist believed to be especially important.

In 1950 a German entomologist, Willi Hennig, published *Phylogenetic Systematics* (Hennig 1996). This work revolutionized systematic practice as applied to traditional morphological and life history characters. This treatise formally developed all the terminology and concepts that we have just discussed. These ideas and the general approach to phylogenetic reconstruction are now referred to as *cladistics*.

Hennig suggested that objective, accurate phylogenies cannot be estimated by examining one or a few characters. Instead, phylogenies must be estimated by examining and codifying as many characters as possible. In this way taxa with abundant symplesiomorphic characters are placed along basal branches, while terminal branches are associated with taxa that have accumulated apomorphic characters. Relationships among in-group taxa are inferred based upon numbers of synapomorphic characters.

However, cladistics still lacked a means to assemble this information into an objective phylogeny. In fact, many alternative trees could be assembled from a given a set of characters. A critical next step in the history of cladistics was the development, in the 1960s, of the parsimony criterion by Jim Farris and others. Specifically, *maximum parsimony* (MP) assumes that from the vast array of possible trees, the true evolutionary phylogeny (Figure 29.1A) is represented by the trees that require the fewest number of steps to construct.

The algorithm for identifying the shortest (most parsimonious) tree or trees is illustrated in Figure 29.2. The maximum number of potential trees among *T* taxa is:

$$\text{No. of trees} = \Pi\,(2i - 5) \qquad i = 3, T \qquad (1)$$

Thus, for eight taxa there are 10,395 potential trees. The goal of an MP analysis is to identify the trees that require the fewest numbers of *state changes*, or *steps*. To determine the number of steps associated with each tree, we count the number of state changes for each character. For example, notice that for characters 1 and

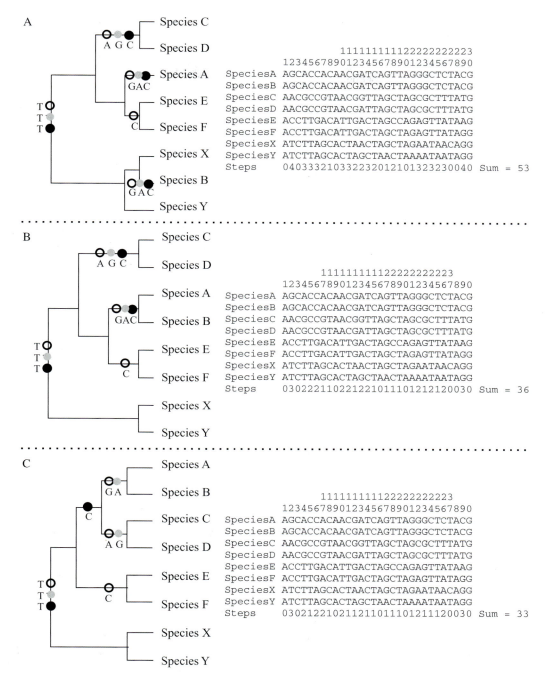

FIGURE 29.2   Algorithm for identifying maximum parsimony (MP) trees among the eight extant taxa in Figure 29.1A using the taxon-by-character data set in Figure 29.1B.

3, there are no changes in the data matrix, so we record a zero under these columns of the data matrix for each tree in Figure 29.2. For character 2, there are four state changes (open circles) in tree A and three state changes in trees B and C. These numbers are recorded under column 2. For character 4, there are three state changes

(gray circle) in tree A, two state changes in trees B and C; for character 5, there are three state changes (black circle) in tree A, two state changes in tree B, and one state change in tree C. This process continues across all 30 characters, and then we sum the total number of state changes for each tree. Tree A requires 53 changes,

B requires 36 changes, and C requires 33 changes. Thus, tree C (also the true evolutionary tree) is the most parsimonious tree.

This was a fairly laborious procedure just to determine the number of steps in three alternate trees. There are 10,392 additional trees that must also be examined, and therefore it comes as no surprise that most cladistic analysis is usually performed on a computer. A number of software packages have been written (e.g., PAUP, Phylip, Hennig) that generate and estimate the length of alternate trees and report the trees of minimal length. In practice, a complete analysis of all possible trees is not feasible. Molecular systematic studies typically involve 50–100 taxa. According to Eq. (1) this would require searches among 3×10^{74} to 2×10^{182} trees. This is computationally prohibitive for even the fastest computers. Swofford et al. (1996) provide an excellent discussion of alternative tree-searching algorithms.

An issue associated with cladistic analysis is that frequently there are many *equally parsimonious trees*. MP analysis of the data set in Figure 29.1B actually yields three equally parsimonious trees (Fig. 29.3). Notice that all three trees have equivalent branching patterns (topologies) for six of the taxa but vary in the branching order between species E and F. Tree 1 indicates that they arose from a single common ancestor (are *monophyletic*). Tree 2 provides no evidence for common ancestry; the relationship between E and F is unresolved. This is referred to as a *polytomy*. Tree 3 indicates that F arose from β′ and that E arose from a subsequent ancestor. The topology of tree 3 suggests incorrectly that E and F arose from a single common ancestor (are *polyphyletic*). The cladist is faced with a problem: Which of the three equally parsimonious trees is correct?

A number of different approaches have been developed. One solution is to derive a *consensus tree*, a tree that represents a common topology among all trees. There are various types of consensus trees. A *strict* consensus tree contains only those branches shared among all trees. All other branches would be resolved as polytomies. Tree 2 in Figure 29.3 is, in fact, the strict consensus tree for all three equally parsimonious trees. Alternatively, a *majority* consensus tree can be resolved that indicates the branches that appear in the majority of trees. Often, the frequency with which branches appear among the equally parsimonious trees appears on or near each branch. Since all three trees have the same topology for six of the eight taxa, branches involving these six species would have a frequency of 100%; however, no two trees agree on the branching order between species E and F, so there would be no majority consensus. This polytomy is still unresolved.

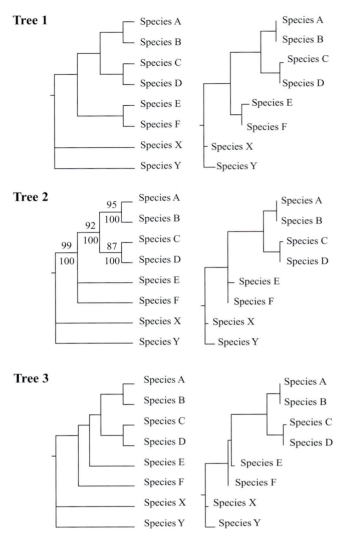

FIGURE 29.3 Three equally parsimonious trees among the eight extant taxa in Figure 29.1A. Bootstrap scores appear above branches and consensus scores appear below branches in tree 2.

Bootstrapping (Fig. 29.4) is a popular alternative approach for resolving these conflicts. Bootstrapping randomly samples characters from the original data set to create a new data set. MP trees are determined for the permuted data set, and the computer records the number of times each branch is resolved. This procedure is repeated 100–1,000 more times. The final product is a single bootstrap tree that indicates the branches that appear in the majority of bootstrap replications and the frequency with which these appear among the replications. Bootstrap trees indicate the consistency with which a data set supports a particular branch. Examine the bootstrap data set generated in line 3 of Figure 29.4. In this data set, 11 of the

Bootstrap Analysis

1. Original data set with *n* nucleotides.

```
                    11111111112222222222 3
           1234567890123456789012345 67890
SpeciesA AGCACCACAACGATCAGTTAGGGCTCTACG
SpeciesB AGCACCACAACGATCAGTTAGGGCTCTACG
SpeciesC AACGCCGTAACGGTTAGCTAGCGCTTTATG
SpeciesD AACGCCGTAACGATTAGCTAGCGCTTTATG
SpeciesE ACCTTGACATTGACTAGCCAGAGTTATAAG
SpeciesF ACCTTGACATTGACTAGCTAGAGTTATAGG
SpeciesX ATCTTAGCACTAACTAGCTAGAATAACAGG
SpeciesY ATCTTAGCACTAGCTAACTAAAATAATAGG
```

2. Pick n random numbers from 1,...,n.

9,24,9,12,23,19,7,7,9,23,5,26,1,12,2,25,2,18,7,18,1,28,11,16,18,4,8,9,10,22

3. Build a new data set with the *n* randomly chosen characters

```
           2 121     2 2 1 2 1 1 2111   12
           94923977935612252878181 6848902
SpeciesA ACAGGTAAAGCCAGGTGTATAACATACAAG
SpeciesB ACAGGTAAAGCCAGGTGTATAACATACAAG
SpeciesC ACAGGTGGAGCTAGATACGCAACACGTAAC
SpeciesD ACAGGTGGAGCTAGATACGCAACACGTAAC
SpeciesE ATAGGCAAAGTAAGCTCCACAATACTCATA
SpeciesF ATAGGTAAAGTAAGCTCCACAATACTCATA
SpeciesX ATAAATGGAATAAATATCGCAATACTCACA
SpeciesY ATAAATGGAATAAATATCGCAATACTCAC A
```

Nucleotides 3, 6, 13-15, 17, 20, 21, 27, 29, and 30 are never sampled. Nucleotides 4, 5, 8, 10, 11, 16, 19, 22, 24-26; and 28 are sampled once, while 1, 2, 12, and 23 are sampled twice. Nucleotides 7 and 18 are sampled thrice and 9 was sampled four times.

4. Repeat the bootstrap process many (100–1,000) more times.

5. Perform phylogenetic analysis on each data set.
   Parsimony — generate the minimal-length trees/bootstrap.
   Maximum likelihood — generate 1 minimal-length tree/bootstrap.
   Distance — generate 1 distance matrix/bootstrap;
             collapse each matrix to build 1 tree/bootstrap.

6. Derive a consensus tree among all replications.
   In how many replications is a particular branch supported?
   To what extent does a derived tree depend upon a few characters?
   Does the entire data set support particular trees or branches?

FIGURE 29.4 Bootstrapping algorithm.

characters in the original data set are not represented, whereas four are sampled twice, two are sampled thrice, and character 9 is represented four times. If a branch depends critically upon one or a few characters, then the majority of bootstrap data sets are not likely to contain that character and the bootstrap support for that branch will be low. Conversely, if a large number of characters support a particular branch, then every bootstrapped data set is likely to contain one or several characters that support a particular branch. Tree 2 in Figure 29.3 is the bootstrap tree. The percentage of 1,000 bootstrap replications that supported a branch is indicated above that branch. Branches with less than 50% support are displayed as polytomies. Note that, once again, the polytomy between E and F is unresolved.

Branch Lengths

The trees that we have inferred to this point estimate only the overall evolutionary relationships or topology among these species. Another important, albeit contentious, aspect of phylogenetic reconstruction involves the relative branch lengths among the eight taxa. The trees on the left and right sides of

Figure 29.3 have the same topology. However, branch lengths among the trees on the right side are proportional to the relative number of character state changes along each branch. The trees indicate small branch lengths between A and B, C and D, E and F, and the clade containing E and F. In contrast, the clade containing A–D and the clade containing all in-group taxa both have relatively long branches. Based upon these trees we might conclude that species E and F arose first but that the gene hasn't changed a great deal since they arose. We would further conclude that ancestor γ didn't arise until much later and then gave rise to ancestor δ and more recently to ancestor δ'. The gene hasn't changed a great deal since A and B or C and D arose. Unfortunately, this interpretation doesn't accurately represent the true evolutionary relationship. Species E and F arose first, but the particular gene we have chosen to examine hasn't changed since the time of divergence (a process referred to as *stasis*). Stasis might occur if the gene is subject to intense purifying selection (Chapter 32) and the substitution rate is low. Species C and D arose from δ' before A and B arose from δ. Furthermore, only A and B are recently derived.

Distance is a product of rate and time, an important issue to consider when estimating branch lengths. Therefore, we don't know if a short branch indicates that two species recently arose or if they became isolated long ago but that the gene we have chosen to examine has been subject to intense purifying selection and thus has had a slow substitution rate. Conversely, a long branch might indicate that two species became isolated long ago, but it could also arise between recent species in which the substitution rate of the gene we have chosen to examine is high (*anastasis*). The gene might be subject to directional or diversifying selection or subject to minimal purifying selection. There are few options in dealing with this rate-versus-time issue. Some phylogeneticists have proposed the model of a molecular clock, in which they assume that mutations occur at regular, fixed intervals (like seconds ticking away on a clock). Furthermore, if the time since divergence among taxa can be estimated by dating of fossils, then the rate of the clock can be estimated and the genetic distance can be used to estimate the actual time since divergence.

Under the simplifying assumption of a molecular clock, branch lengths would strictly reflect the time since divergence. However, a major problem with this assumption is that the speed of the molecular clock not only varies greatly among genes but also differs greatly among taxa. Data collected on thousands of genes demonstrate that some have a very slow substitution rate (e.g., ribosomal proteins) while other genes have a substitution rate that is greater by one to two orders of magnitude (e.g., some cytokines, introns, pseudogenes). The reasons for this are diverse, but differing intensities of purifying selection among genes undoubtedly account for much of the variation. Another problem is that substitution rates for one particular, homologous, gene can vary greatly among taxa. For example, the substitution rate in the mitochondrial genome is ~10-fold higher in invertebrates than in vertebrates. The DNA hybridization experiments discussed earlier demonstrated that substitution rates are higher among rodents than among primates. A large proportion of this variation in substitution rates among taxa probably reflects generation time (e.g., rodent generation time is 10- to 20-fold shorter that primate generation time). Whatever the cause, a generalized molecular clock model seems implausible without a great deal of calibration. This also suggests why examination of several different genes is prudent. Genes with high substitution rates may accurately estimate relationships among recently evolved taxa but will contain too many homoplasious characters to accurately estimate relationships among ancient taxa. Conversely, genes with low substitution rates are less likely to have homoplasious characters among ancient taxa but will be uninformative among recent taxa.

In a cladistic approach, branch lengths are generally proportional to the numbers of substitutions between DNA molecules from a pair of sister taxa. All character state changes among nucleotides are counted as a single step. This is referred to as *Fitch parsimony*. Figure 29.5A illustrates the problem with this treatment. The four nucleotides constituting DNA consist of two chemical groups called *purines* (adenine and guanine) and *pyrimidines* (cytosine and thymine). A substitution of a purine to another purine or of a pyrimidine to another pyrimidine (both are defined as *transitions*) involves only minor chemical changes. In contrast, a substitution of a purine to a pyrimidine (or *vice versa*) involves repairing a DNA strand (a *transversion*). We would expect, and abundant data show, that transitions are much more frequent than transversions. We might therefore, decide to differentially weight the steps among nucleotide substitutions. For example, a transversion might be counted as two steps instead of one. Referred to as *weighted parsimony*, this involves the construction of a step matrix. For nucleotides, a step matrix would indicate the number of steps we would apply to changes from one nucleotide to another. Figure 29.5b indicates that there are 12 possible character state changes among nucleotides. We might choose to apply different weights to all of these steps, or we might prefer to assign equal weight to both

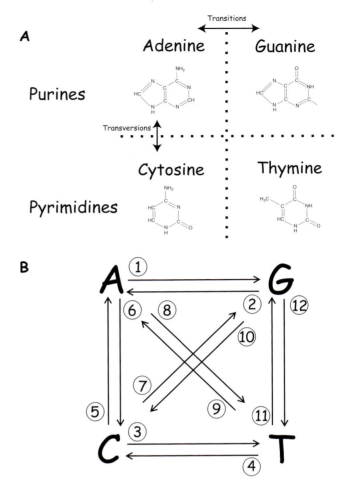

FIGURE 29.5 (A) Classification of mutations. (B) Twelve possible character state changes among nucleotides.

changes between two particular nucleotides (e.g., same number of steps from A to G as from G to A). We can envision a large number of possibilities, and thus one of the issues associated with weighted parsimony is determining the most objective and accurate step matrix.

Phenetics

Ten years after Hennig's publication of *Phylogenetic Systematics* (but at least 6 years before it was translated into English in 1996), concern with the intuitive nature of phylogenetics led to the rise of an alternative, *phenetic* approach to systematics. This approach was codified in *Principles of Numerical Taxonomy* (Sokal and Sneath 1963). Pheneticists basically proposed that taxa should be grouped and classified according to overall similarity, as measured by defined rules. Like the cladists, they recognized that the most accurate and objective classifications would arise from analyses based on the largest numbers of characters. They initially stated that every character should be assigned equal weight. Classification was then based upon quantitative measures of the overall (phenetic) similarity or distance between the taxa. The overall patterns of character correlations could then be used to infer phylogenetic pathways.

The phenetics approach is illustrated in Figures 29.6–29.8. The five most common measures of genetic distances among nucleotides and their associated assumptions are shown in Figure 29.6. The numbers i in the $P(i)$ term correspond to the 12 possible substitutions listed in Figure 29.5B. Note that each of the five distance measures treats these probabilities differently.

The Jukes–Cantor distance assumes that all 12 substitutions are equally likely. This is analogous to Fitch parsimony, where all substitutions are given equal weight. The Kimura two-parameter distance subdivides the 12 substitutions into transitions and transversions. The Tamura model is similar to the Kimura two-parameter in this regard but further considers the G/C and A/T content of the gene under consideration. The Tajima–Nei model takes a different approach and considers the probability that any nucleotide will change to one specific nucleotide. What is the probability that a G will be substituted by an A? The Tamura–Nei model is an extension of the Tamura model but subdivides transitions into purines and pyrimidines and considers the frequency of each nucleotide separately.

The Jukes–Cantor distance is calculated simply by counting the numbers of substitutions between the genes in a pair of taxa. For example, there are no substitutions between A and B:

$$D_{JC} = -(3/4) \ln [1 - (4/3)(0/30)] = 0.000$$

there is one substitution between C and D, so

$$D_{JC} = -(3/4) \ln [1 - (4/3)(1/30)] = 0.034$$

and there are two substitutions between E and F, so

$$D_{JC} = -(3/4) \ln [1 - (4/3)(2/30)] = 0.070$$

There are always $n(n-1)/2$ pairwise distances among n taxa. Therefore, in our example there would be $8(7)/2 = 28$ pairwise distances. As these are computed they are placed in a distance matrix. The Jukes–Cantor distance matrix is shown on the bottom of Figure 29.6. Values along the principal diagonal are zeros (the distance between a taxon and itself is zero). A distance matrix is symmetrical; that is, corresponding distances (A–B vs. B–A) above and below the principal diagonal are equal.

The other distance measures in Figure 29.6 are more complicated to calculate. The Kimura two-parameter distance requires counting numbers of transitions and transversions between taxa. The Tamura model counts transitions and transversions and, in addition, requires determination of the G–C content. The Tajima–Nei model requires determination of the frequency of each nucleotide but, like Jukes–Cantor distance, only further requires counting numbers of substitutions. The Tamura–Nei distance is the most complicated, requiring the estimation of eight different components for each pair of taxa. Computation of distance matrices is usually completed using the same software packages listed earlier for MP analysis.

The next task is to turn the information in the distance matrix into a phylogeny. This is accomplished

Jukes–Cantor Model

$P(1) = P(2) = P(3) = \ldots = P(12) = \lambda = $ rate of substitution

$D_{JC} = -(3/4) \ln (1 - (4/3)p)$, where $p = n_d/n = $ no. substitutions/no. nucleotides

Kimura Two-Parameter Model

$P(1) = P(2) = P(3) = P(4) = \alpha = $ transition rate

$P(5) = P(6) = P(7) = \ldots = P(12) = \beta = $ transversion rate

$D_{K2P} = -(1/2)\ln (1 - 2P - Q) - (1/4)\ln (1 - 2Q)$,

where $P = n_s/n$, where $n_s = $ no. transitions, and $Q = n_v/n$, where $n_v = $ no. transversions

Tamura Model

$P(1) = P(4) = \theta\alpha = $ G/C content α

$P(2) = P(3) = (1 - \theta)\alpha$

$P(6) = P(7) = P(10) = P(11) = \theta\beta$

$P(5) = P(8) = P(9) = P(12) = (1 - \theta)\beta$

$D_T = -2\theta(1 - \theta) \ln (1 - P/(2\theta(1 - \theta)) - Q) - \frac{1}{2}(1 - 2\theta(1 - \theta))\ln (1 - 2Q)$

Tajima–Nei Model

$P(1) = P(7) = P(11) = \Delta\delta = P(N => G)$, where N = any other nucleotide

$P(2) = P(5) = P(9) = \Delta\alpha = P(N => A)$

$P(3) = P(8) = P(12) = \Delta\beta = P(N => T)$

$P(4) = P(6) = P(10) = \Delta\gamma = P(N => C)$

$D_{TJN} = -b \ln (1 - p/b)$,

where $b = \frac{1}{2}(1 - \sum_{i=1-4} f_i + p^2/c)$ and $c = \frac{1}{2} \sum_{i=1-3} \sum_{j=1-4} (f_{ij}/f_j f_j)$

$f_i, f_j = $ respective frequencies of the ith and jth nucleotides

$f_{ij} = $ frequency of nucleotide pairs i and j at a homologous site in an alignment of two species

Tamura–Nei Model

$f_A, f_C, f_G, f_T = $ respective frequencies of A, C, G, and T

$f_R, f_Y = $ respective frequencies of purines and pyrimidines

$\alpha_1, \alpha_2 = $ respective probability of purine and pyrimidine transitions

$P_1, P_2 = $ respective frequencies of purine and pyrimidine transitions

$P(1) = \alpha_1 f_G$	$P(5) = P(9) = \beta f_A$
$P(2) = \alpha_1 f_A$	$P(6) = P(10) = \beta f_C$
$P(3) = \alpha_2 f_T$	$P(7) = P(11) = \beta f_G$
$P(4) = \alpha_2 f_C$	$P(8) = P(12) = \beta f_T$

$D_{TN} = -(2f_A f_G/f_R) \ln [1 - (f_R/2f_A f_G)P_1 - (1/2f_R)Q] - (2f_C f_T/f_Y) \ln (1 - (f_Y/2f_C f_T) P_2 - (1/2f_Y) Q) - 2[f_R f_Y - (f_A f_G f_Y/f_R) - (f_C f_T f_R)/f_Y]\ln [1 - (1/2f_R f_Y)Q]$

	Species A	Species B	Species C	Species D	Species E	Species F	Species X	Species Y
Species A	0.000	0.000	0.441	0.383	0.730	0.647	1.207	1.648
Species B	0.000	0.000	0.441	0.383	0.730	0.647	1.207	1.648
Species C	0.441	0.441	0.000	0.034	0.824	0.730	1.057	1.057
Species D	0.383	0.383	0.034	0.000	0.730	0.647	0.931	1.207
Species E	0.730	0.730	0.824	0.730	0.000	0.070	0.441	0.572
Species F	0.647	0.647	0.730	0.647	0.070	0.000	0.330	0.441
Species X	1.207	1.207	1.057	0.931	0.441	0.330	0.000	0.147
Species Y	1.648	1.648	1.057	1.207	0.572	0.441	0.147	0.000

FIGURE 29.6 Methods for calculating the five most common measures of genetic distances among nucleotides. The Jukes–Cantor distance matrix among the eight extant taxa in Figure 29.1A using the taxon-by-character data set in Figure 29.1B is shown on the bottom.

with tree-building algorithms. The simplest of these is called the unweighted pair group method with arithmetic mean (UPGMA) (Fig. 29.7). The numbers in this matrix are the $\odot T_m$ values among five primate species. The first step in transforming this distance matrix into a tree is to locate the minimum distance. This is the distance between humans and chimps. These are located in a cluster by placing chimps and humans along symmetrical branches the lengths of which are equal to half the distance. The second step is to collapse the matrix by a row and column. The distance between the new (Human–Chimp) cluster and each of the unclassified taxa is calculated as the average of the distance between these taxa and humans and chimps. For example, the distance between the (Human–Chimp) cluster and Gorillas = $(1.51 + 1.57)/2 = 1.54$. That value is placed in the new matrix.

Once the next matrix is calculated, we repeat all of the same steps of the algorithm. The smallest distance is between (Human–Chimp) and Gorillas. We attach gorillas to the human–chimp group with a branch equal to half the distance $(1.54/2 = 0.77)$, the branch that connects this to (Human–Chimp) is the difference in branch lengths $(0.77 - 0.73 = 0.04)$. We compute the next matrix by averaging distances between the (Human–Chimp–Gorilla) cluster and all others. This is completed until one element remains in the matrix. For n taxa there are $n - 1$ iterations of the algorithm. The product is the phylogeny shown at the bottom of Figure 29.7.

Consider some of the assumptions we have made in using UPGMA. The distances among species are displayed in the tree as averages. This implicitly assumes that substitution rates between the two species are equal. This assumption is also implicit in the next step, in which we average the difference between a cluster and the remainder of species in the matrix. The distances among species and clusters are displayed as differences in branch lengths. This step implicitly assumes that distances between the taxa are additive and independent. Neither of these assumptions is probably valid. Thus, just as there are several different genetic distance measures, there are various tree-building algorithms.

Saitou and Nei (1987) developed a more realistic and much more frequently used algorithm called neighbor-joining (Fig. 29.8). We begin with the same matrix; however, the decision as to which taxa to join is based upon a rate-corrected distance derived using the average distance between a pair of taxa and all other taxa. The pair of taxa with the smallest rate-corrected distance are joined to a common node 1. However, note that the branch lengths are based upon average distances of the joined taxa with all other

species; unlike UPGMA, they are not independent of the other distances.

This neighbor-joining algorithm is repeated again. But in contrast to UPGMA, distances are corrected based upon estimated branch lengths, not averages. Also note that new distances are based upon distances to nodes rather than taxa. The product of $n - 1$ iterations is the phylogeny shown at the bottom of Figure 29.8. The most obvious difference between the two phylogenies is that the orangutans have shorter distances to gorillas, chimps, and humans than they have to rhesus monkeys. The greater distance between orangutans and gorillas, chimps, and humans in UPGMA resulted from averaging the rhesus monkeys with only orangutans rather than taking into account that rhesus monkeys are actually more different from *all* of the other primates in this study.

Distance neighbor-joining trees constructed with the simplest (Jukes–Cantor) and the most complex (Tamura–Nei) distance measures are shown in Figure 29.9A. Bootstrap values are shown over each branch. Bootstrapping in phenetic analysis is completely analogous to MP bootstrap analysis (Fig. 29.4). A distance matrix is constructed from each bootstrapped data set, a phylogeny is constructed from each matrix and the frequency with which individual branches appear is recorded.

The major difference between the two trees lies in the problematic association between species E-F. The species are polyphyletic using Jukes–Cantor distances but are monophyletic with strong bootstrap support using the Tamura–Nei distances. The Jukes–Cantor tree is most similar to the MP trees that we derived earlier (Fig. 29.3). This is predictable because equal weight is applied to all nucleotide substitutions in computing the Jukes–Cantor distance and in performing Fitch Parsimony (each nucleotide change counted as one step in Fig. 29.2 and 29.3).

The Tamura-Nei distances between species E-F are relatively small because the two species differ by only two transitions. The Tamura–Nei neighbor-joining tree is similar to the true evolutionary tree in Figure 29.1 in several aspects. Species C and D arise earlier than do species A and B. Species E and F are monophyletic, and distances between them are greater than the distance between A and B or between C and D.

Despite many obvious similarities with cladistics, phenetics has been attacked by cladists over several issues. Cladists argue that organismal relationships cannot be deduced from overall similarity, only from similarity among synapomorphic characters. Furthermore, cladistic methods attempt to distinguish among sources of similarity and thereby account for specified character state distributions in terms of phylogenetic

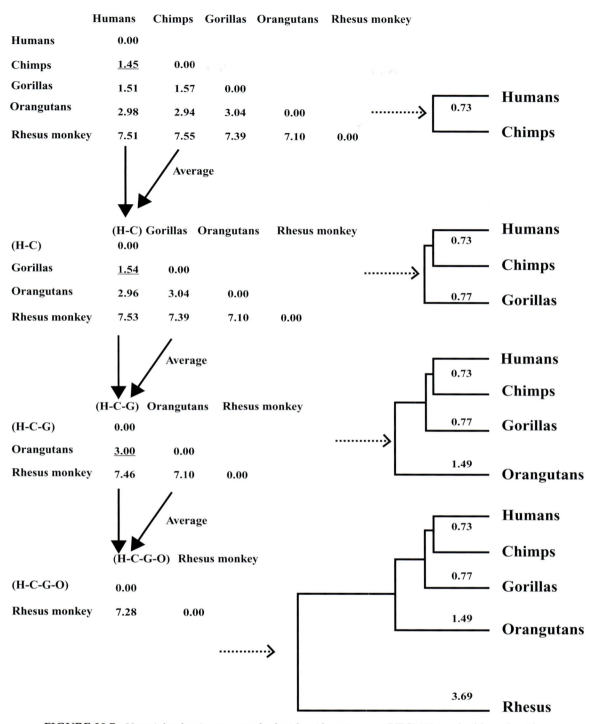

FIGURE 29.7 Unweighted pair group method with arithmetic mean (UPGMA) tree-building algorithm.

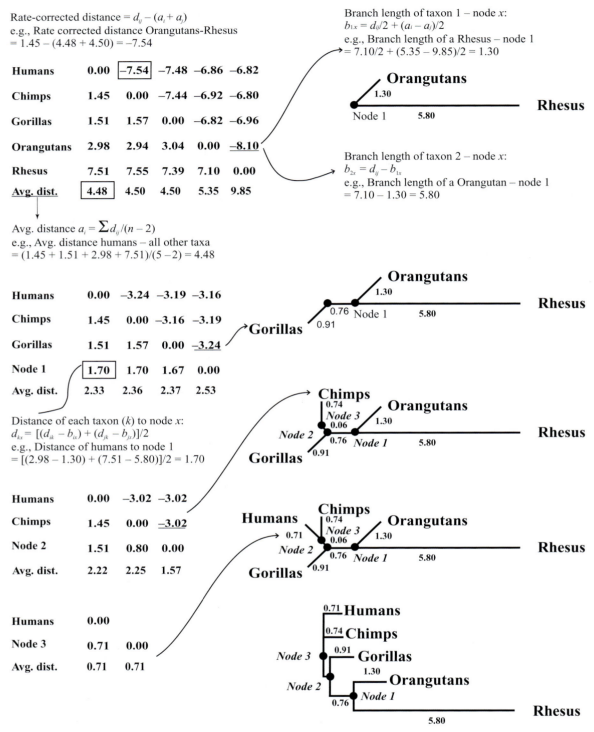

Rate-corrected distance $= d_{ij} - (a_i + a_j)$
e.g., Rate corrected distance Orangutans-Rhesus
$= 1.45 - (4.48 + 4.50) = -7.54$

Humans	0.00	−7.54	−7.48	−6.86	−6.82
Chimps	1.45	0.00	−7.44	−6.92	−6.80
Gorillas	1.51	1.57	0.00	−6.82	−6.96
Orangutans	2.98	2.94	3.04	0.00	−8.10
Rhesus	7.51	7.55	7.39	7.10	0.00
Avg. dist.	4.48	4.50	4.50	5.35	9.85

Avg. distance $a_i = \sum d_{ij}/(n-2)$
e.g., Avg. distance humans – all other taxa
$= (1.45 + 1.51 + 2.98 + 7.51)/(5-2) = 4.48$

Branch length of taxon 1 – node x:
$b_{1x} = d_{ij}/2 + (a_i - a_j)/2$
e.g., Branch length of a Rhesus – node 1
$= 7.10/2 + (5.35 - 9.85)/2 = 1.30$

Orangutans
1.30
Node 1   5.80
Rhesus

Branch length of taxon 2 – node x:
$b_{2x} = d_{ij} - b_{1x}$
e.g., Branch length of a Orangutan – node 1
$= 7.10 - 1.30 = 5.80$

Humans	0.00	−3.24	−3.19	−3.16
Chimps	1.45	0.00	−3.16	−3.19
Gorillas	1.51	1.57	0.00	−3.24
Node 1	1.70	1.70	1.67	0.00
Avg. dist.	2.33	2.36	2.37	2.53

Distance of each taxon (k) to node x:
$d_{kx} = [(d_{ik} - b_{ix}) + (d_{jk} - b_{jx})]/2$
e.g., Distance of humans to node 1
$= [(2.98 - 1.30) + (7.51 - 5.80)]/2 = 1.70$

Orangutans
1.30
0.76 Node 1   5.80
Rhesus
Gorillas
0.91

Humans	0.00	−3.02	−3.02
Chimps	1.45	0.00	−3.02
Node 2	1.51	0.80	0.00
Avg. dist.	2.22	2.25	1.57

Chimps
0.74
Node 3   **Orangutans**
0.06   1.30
Node 2
0.76   *Node 1*   5.80   **Rhesus**
0.91
Gorillas

Humans	0.00
Node 3	0.71	0.00
Avg. dist.	0.71	0.71

Chimps
0.74
Humans   *Node 3*   **Orangutans**
0.71   0.06   1.30
Node 2
0.76   *Node 1*   5.80   **Rhesus**
0.91
Gorillas

0.71 **Humans**
0.74 **Chimps**
Node 3   0.91 **Gorillas**
1.30
Node 2   **Orangutans**
0.76   *Node 1*
5.80   **Rhesus**

FIGURE 29.8   Neighbor-joining (NJ) tree-building algorithm. *Source*: Saitou and Nei (1987).

A

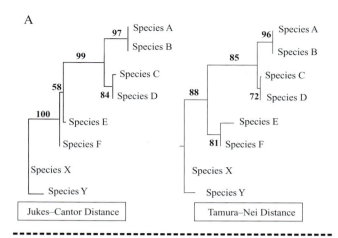

B

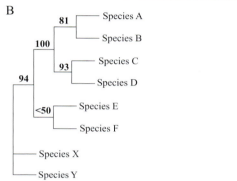

FIGURE 29.9 (A) Distance neighbor-joining trees constructed with the simplest (Jukes–Cantor) and the most complex (Tamura–Nei) distance measures among the eight extant taxa in Figure 29.1A using the taxon-by-character data set in Figure 29.1B. (B) ML tree among the same taxa. Bootstrap values are shown over each branch.

history. The use of out-groups ascribed polarity to characters, allowing estimation of the correct branch order from ancient to more recent taxa. As alluded to earlier, cladists have focused almost exclusively on assessing the topological component of evolutionary trees (*cladogenesis*) rather than branch lengths (*anagenesis*). The ultimate goal of cladistics has been to develop organismal classifications based on correctly inferred cladogenetic histories.

Pheneticists have countered these attacks by presenting many of the same issues we have raised in this section. How reliably can synapomorphies be identified? How are character conflicts resolved when putative clades identified by different presumptive synapomorphies disagree? Isn't choosing out-group taxa an inherently subjective process? Cladistics can be applied only to discrete character sets and not to quantitative measures of differences among taxa. While the debate between the two schools has at times become

acrimonious, the arguments and discussion have also been very productive, causing proponents of both schools to develop new methods to overcome deficiencies. In practice, as in our example, cladistic and phenetic methods when applied to large molecular genetic data sets usually arrive at very similar inferred phylogenies.

Maximum Likelihood

The ability of any phylogeny reconstruction method to estimate the true evolutionary tree is referred to as its *consistency*. Felsenstein (1978) published a critical paper in which he showed that under some (very reasonable) circumstances, parsimony methods can be "positively misleading" because as the numbers of nucleotides in a data set increases, MP is more and more likely to encounter homoplasious characters. This is because of the limited number of character states (A, C, G, T) for each DNA character. Thus, as we examine more and more DNA sequences we are more and more likely to infer an incorrect phylogeny. This generates a paradoxical situation in which fewer characters are preferable to many because only by sampling few enough characters is there any chance of sampling informative rather than misinformative characters and thereby inferring the true evolutionary tree. This paradox has come to be known as the *Felsenstein zone of uncertainty*. A very readable explanation and an example of this paradox appear in Swofford et al. (1996).

Felsenstein (1981) published a series of papers describing an alternative, maximum likelihood (ML) method for phylogenetic inference. This method is similar to parsimony, in that it evaluates all possible trees. But ML differentiates among trees based upon the probability that the tree under consideration would produce the observed data. The tree that best explains the observed patterns of variation in the data set (rather than the tree of shortest length) is the tree of maximum likelihood. The procedure for ML inference is illustrated in Figure 29.10. The tree in this figure is one of the 10,395 possible trees. Just as in MP analysis, each character in the tree is assessed. However, an important and unique aspect of ML analysis is that *all* characters are treated as informative. Thus, even constant or autapomorphic characters carry information. This property is evident in the Hasegawa, Kashino, and Yano (HKY) substitution probability matrix shown in Figure 29.11. The HKY matrix is one of the most utilized matrices in ML analysis. It assigns eight distinct probabilities for the 16 possible substitutions. Notice that values in the principal diagonal are not zero and that the matrix is not symmetrical. Notice also

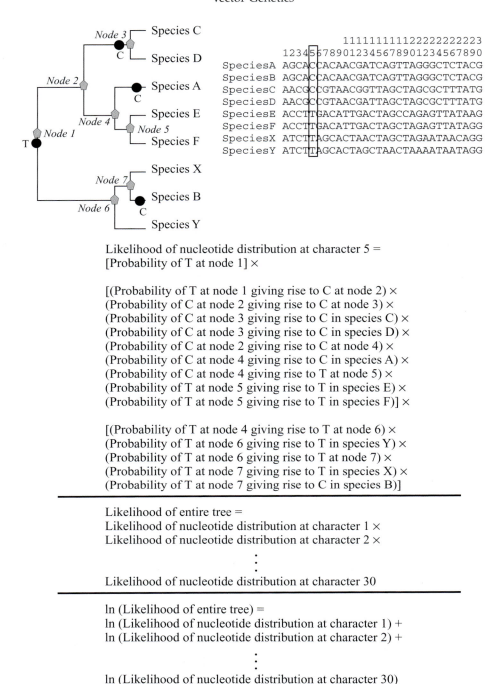

Likelihood of nucleotide distribution at character 5 =
[Probability of T at node 1] ×

[(Probability of T at node 1 giving rise to C at node 2) ×
(Probability of C at node 2 giving rise to C at node 3) ×
(Probability of C at node 3 giving rise to C in species C) ×
(Probability of C at node 3 giving rise to C in species D) ×
(Probability of C at node 2 giving rise to C at node 4) ×
(Probability of C at node 4 giving rise to C in species A) ×
(Probability of C at node 4 giving rise to T at node 5) ×
(Probability of T at node 5 giving rise to T in species E) ×
(Probability of T at node 5 giving rise to T in species F)] ×

[(Probability of T at node 4 giving rise to T at node 6) ×
(Probability of T at node 6 giving rise to T in species Y) ×
(Probability of T at node 6 giving rise to T at node 7) ×
(Probability of T at node 7 giving rise to T in species X) ×
(Probability of T at node 7 giving rise to C in species B)]

Likelihood of entire tree =
Likelihood of nucleotide distribution at character 1 ×
Likelihood of nucleotide distribution at character 2 ×
.
.
.
Likelihood of nucleotide distribution at character 30

ln (Likelihood of entire tree) =
ln (Likelihood of nucleotide distribution at character 1) +
ln (Likelihood of nucleotide distribution at character 2) +
.
.
.
ln (Likelihood of nucleotide distribution at character 30)

FIGURE 29.10   Maximum likelihood (ML) method for phylogenetic inference.

that the probabilities incorporate many of the same variables we considered in Figure 29.6.

Figure 29.10 illustrates the ML procedure for only one of the 30 characters (character 5). The likelihood is estimated beginning at node 1, the node associated with ancestor φ. A "T" is the plesiomorphic state of character 5. We first determine the probability of a T at node 1 [Prob(T ⇔ T), Fig. 29.11]. Next, we proceed along the branches to node 2. We calculate the probability that a T at node 1 will give rise to a C at Node 2 [Prob(T ⇒ C)]. Proceeding along to node 3, we calculate the probability that a C at node 2 gives rise to a C at node 3 [Prob(C ⇔ C)]. Continuing along this path, we estimate the probability that a C at node 3

Substitution probability matrix (Hasegawa, Kishino, and Yano, 1985)

μt = average number of substitutions
Π_x = frequency of nucleotide X
R = purine frequency
Y = pyrimidine frequency
K = transition rate

$A \leftrightarrow A$	$C \rightarrow A$	$G \rightarrow A$	$T \rightarrow A$
$\Pi_A + \Pi_A[Y/R]e^{-\mu t} + (\Pi_G/R)e^{-\mu t[1+R(K-1)]}$	$\Pi_C(1 - e^{-\mu t})$	$\Pi_G + \Pi_G[Y/R]e^{-\mu t} - (\Pi_G/R)e^{-\mu t[1+R(K-1)]}$	$\Pi_T(1 - e^{-\mu t})$
$A \rightarrow C$	$C \leftrightarrow C$	$G \rightarrow C$	$T \rightarrow C$
$\Pi_A(1 - e^{-\mu t})$	$\Pi_C + \Pi_C[R/Y]e^{-\mu t} + (\Pi_T/Y)e^{-\mu t[1+Y(K-1)]}$	$\Pi_G(1 - e^{-\mu t})$	$\Pi_T + \Pi_T[R/Y]e^{-\mu t} - (\Pi_T/Y)e^{-\mu t[1+Y(K-1)]}$
$A \rightarrow G$	$C \rightarrow G$	$G \leftrightarrow G$	$T \rightarrow G$
$\Pi_A + \Pi_A[Y/R]e^{-\mu t} - (\Pi_A/R)e^{-\mu t[1+R(K-1)]}$	$\Pi_C(1 - e^{-\mu t})$	$\Pi_G + \Pi_G[Y/R]e^{-\mu t} + (\Pi_A/Y)e^{-\mu t[1+R(K-1)]}$	$\Pi_T(1 - e^{-\mu t})$
$A \rightarrow T$	$C \rightarrow T$	$G \rightarrow T$	$T \leftrightarrow T$
$\Pi_A(1 - e^{-\mu t})$	$\Pi_C + \Pi_C[R/Y]e^{-\mu t} - (\Pi_C/Y)e^{-\mu t[1+Y(K-1)]}$	$\Pi_G(1 - e^{-\mu t})$	$\Pi_T + \Pi_T[R/Y]e^{-\mu t} + (\Pi_C/Y)e^{-\mu t[1+Y(K-1)]}$

FIGURE 29.11 Hasegawa, Kashino, and Yano (HKY) substitution probability matrix is one of the most utilized matrices in ML analysis. *Source*: Hasegawa et al. (1985).

gives rise to a C at species C [Prob(C ⇔ C)]. Going back to node 3, we also estimate the probability that a C at node 3 gives rise to a C at species D [Prob(C ⇔ C)]. From here we back down to node 2 and proceed to node 4.

Probabilities are calculated for each branch of the path until all paths have been covered. The final step in assessing the likelihood of character 5 is to multiply all of the various probabilities together. We then proceed on to character 6 and calculate the probability of character changes along the same tree. This procedure continues until we have covered all 30 characters. The likelihood of the entire tree is estimated by multiplying together the likelihoods of each of the 30 characters. By taking the natural log of the likelihood of the entire tree we derive a log-likelihood score for the entire tree. We would then proceed to analyze and calculate a log-likelihood score for the next tree. This could be repeated until all 10,395 possible trees had been examined, but it usually involves using the alternative tree-searching algorithms discussed earlier.

The ML procedure is obviously calculation intensive, so much so that the computer programs listed earlier are essential for performing the ML procedure. The ML tree of our example data set is shown in Figure 29.9B. Bootstrap values are shown over each branch.

Bootstrap analysis in ML analysis is completely analogous to bootstrap analysis of MP trees (Fig. 29.4). An ML tree is constructed for each bootstrapped data set, and the frequency with which individual branches appear is recorded.

The topology of the ML tree is extremely similar to the Tamura–Nei neighbor-joining tree. A comparison of the parameters used in the HKY matrix in Figure 29.11 with those used in computing the Tamura–Nei distance in Figure 29.6 indicates why this outcome is likely. However, one major difference between the two trees is that there is low bootstrap support for the branch containing species E–F in the ML tree. Notice also that the ML tree indicates that the origins of the branches leading to species E and to species F arise at the same time as ancestors γ and γ′ and that the branches leading to species C and D arise before the branches leading to species A and B. The only major inaccuracy in the ML tree is the relatively long branch lines between species A and B.

Summary

We have now introduced the terminology of molecular systematics and have described three alternative methods for inferring phylogenetic relationships among species based upon molecular genetics data sets. Which of these three methods should be used in routine molecular systematic analyses? The authors strongly advocate the use of all three methods. The rare instances in which differences in topology and resolution among the methods occur may yield insights into the quality of the nucleic acid datasets.

CURRENT STATUS OF THE MOLECULAR SYSTEMATICS OF DISEASE VECTORS

No systematics studies of disease vectors had been published by the time the first edition of this text went to press in 1995. At this writing, a large number of molecular systematic studies of a variety of disease vector taxa have been completed or are under way. Here, we review those advances, present a working hypothesis (consensus phylogeny) for each group, highlight gaps in our knowledge, and give examples of how phylogenies can provide insight into the evolution of certain anatomical, developmental, cytogenetic, behavioral, host preference, ecological, life history, and biogeographical characters. It is highly desirable that taxonomic schemes reflect accurately the phylogeny of the organisms they contain. We will indicate how knowledge of evolutionary relationships among species is affecting their taxonomic classification (Chapter 33).

Ticks (Order Arachnida, Suborder Ixodida)

Dr. Harry Hoogstraal was one of the first taxonomists to publish hypotheses about the phylogenetic relationships of ticks (Hoogstraal and Aeschlimann 1982). However, this phylogeny was not inferred from a formal analysis of genotypic or phenotypic features. Rather, Hoogstraal and Aeschlimann's phylogenetic hypothesis was based on their intuitive ideas about the evolution of vertebrate hosts of ticks, host specificity, morphology, life history, reproduction, and karyotypes. Hoogstraal's phylogeny, which forms the basis of Western classification of the ~850 species of ticks, was not tested until the mid-1990s, when Black and Piesman (1994) sequenced ~460 bp from the 3' end of the mitochondrial 16s rRNA gene (rDNA) in 36 hard- and soft-tick species and a mesostigmatid mite, *Dermanyssus gallinae*, was used as an out-group. This study contained many problems associated with an inappropriate use of a rapidly evolving gene to infer ancient relationships, inadequate taxon sampling, and use of a single, distantly related outgroup.

When this study was initiated in 1991, oligonucleotide primers that worked well for PCR in insects often did not work for ticks. Ticks are arachnids belonging to the arthropod subphylum Chelicerata, while insects belong to the other subphylum, Mandibulata Chapter 2. These subphyla have not shared a common ancestor for ~450 million years. Our laboratories independently discovered that the metastriate ticks, in fact, have a different arrangement of genes in the mitochondrial genome and that this prevented many of the initial PCR attempts. However, since the mid-1990s, there have been many analyses of the phylogenetic relationships of ticks, particularly hard ticks (Ixodidae), and a consensus about much of hard-tick evolution has emerged (Figs. 29.12 and 29.13).

Tick Origins and the Relationship Between Soft and Hard Ticks

All studies performed to date support the monophyly of the family Ixodida relative to the other parasitiform mite suborders and of its constituent families Ixodidae (hard ticks—node 1, Fig. 29.12), Argasidae (soft ticks, node 3), and Nutalliellidae (node 2). The family Holothyrida appears to be most closely related to the Ixodida (Dobson and Barker 1999; Klompen et al. 2000). The Holothyrida occur only in areas

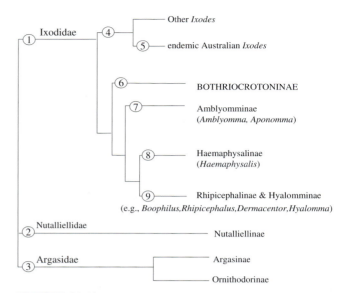

FIGURE 29.12 Working hypothesis for the phylogenetic relationships of ticks (Suborder Ixodida). This hypothesis was proposed on the basis of analyses of DNA sequences and phenotypes. *Sources*: Black and Piesman (1994), Crampton et al. (1996), Black et al. (1997), Klompen et al. (2000).

assumed to have been part of the Gondwana supercontinent, such as Australia, New Zealand, South America, and various islands in the Indian Ocean from Madagascar to New Guinea. Recent studies on holothyrid and opilioacarid feeding behavior suggest that representatives of these suborders are scavengers, feeding on crushed insects but not on live prey. Ticks may therefore be derived from scavengers, not nest predators, a scenario often proposed for the evolution of parasitism in mites.

Another general observation is that branch lengths have generally been much longer among argasid taxa than among the ixodids. This suggests that argasid taxa and the ancestral lineages leading to these taxa have existed for a longer period of time than the ixodid taxa. Several molecular studies are consistent with the idea that hard ticks probably arose in the early Cenozoic ~60 mya. Therefore, the argasids probably arose even earlier in the Cretaceous, and this is supported by recent fossil evidence.

The phylogenetic placement of Nutalliellidae is still unresolved. This is due largely to the fact that *Nutalliella namaqua*, the only species in the Nutalliellidae, has not been collected for many years. Attempts to amplify DNA from museum specimens by both of our laboratories resulted only in the amplification of DNA from fungi that infect the specimens either before or after their death.

Prostriata (Ixodes *spp.*)

Morphology, behavior (mating off-host), and mitochondrial markers (2 tRNA genes tRNA-Glu and tRNA-Ser[AGN]) indicate that prostriate ticks (node 4) are monophyletic. However, two distinct clades occur within the Prostriata: those taxa found in Australia (node 6) and those found elsewhere (node 5). This suggests that the genus *Ixodes* may be paraphyletic.

At a lower taxonomic level, investigations of the subgenera and species complexes in the genus *Ixodes* (node 5) have confirmed morphological studies and extended our knowledge of these groups. Examination of the ITS-2 region of nuclear rDNA has confirmed that two members of the subgenus *Ixodes* belong in separate subgenera. The latter analysis placed the South American species *Ixodes loricatus* and *I. luciae*, referred to informally as the "luciae group," in a clade appearing basal to the subgenera (in ascending order) *Pholeoixodes* and *Trichotoixodes*, as well as remaining members of the subgenus *Ixodes*. Other studies indicate that species in the *Ixodes ricinus* complex in the subgenus *Ixodes* need to be added to and removed from the 15-member complex in order for it to be monophyletic and that auricular shape, as a variable ridge, may have evolved multiple times in the subgenus *Ixodes* (Fukunaga et al. 2000).

Metastriate Ticks

Morphology, behavior (mating on host), and the unusual order of mitochondrial genes strongly support monophyly of metastriate ticks (node 7).

Amblyomminae

Studies involving the position of the subfamily Amblyomminae within the Metastriata strongly suggest that the group is polyphyletic, with branches that have Laurasian (North American) and Gondwanan origins. Indeed, a major lineage of ticks was discovered from phylogenetic analyses of nucleotide sequences (Dobson and Barker 1999). The subfamily Bothriocrotinae (node 8), has been created for the five known species (genus *Bothriocroton*) of this lineage. These species were formerly in the genus *Aponomma*. Other members of *Aponomma* arise on a common clade within the remaining *Amblyomma* species.

Haemaphysalis

Several molecular studies suggest that *Haemaphysalis* (node 10) is polyphyletic, with the primitive species *Ha. inermis* arising on a different branch than other members of the genus. In support of the molecular-based findings, the morphology of *Haemaphysalis inermis* is somewhat dissimilar to that of other *Haemaphysalis* spp.

Rhipicephalinae, Rhipicephalus, *and* Hyalomma

All molecular studies that have sampled members of the subfamily Hyalomminae consistently place it within the Rhipicephalinae (Fig. 29.13), in contrast to a basal relationship for Hyalomminae, as previously depicted (Hoogstraal and Aeschlimann 1982). Thus, the subfamily Hyalomminae should be reduced, possibly to a genus, within the subfamily Rhipicephalinae (node 11).

The Rhipicephalinae is the most genus-rich subfamily in the Ixodida, having eight genera. The genus *Boophilus* (five spp.) and the genera *Margaropus* (three spp.) and *Anomalohimalaya* (three spp.) arise within the *Rhipicephalus* lineage (~75 spp.) (Fig. 29.13). Thus, the genus *Rhipicephalus* is paraphyletic. The name *Rhipicephalus* Koch, 1844, has priority over the name *Boophilus* Curtice 1891, so the *Rhipicephalus* spp. cannot be placed into the genus *Boophilus*, which instead suggests that the genus *Boophilus*, and possibly *Margaropus* and *Anomalohimalaya*, should be reduced to subgenera or species groups of *Rhipicephalus*. By reducing *Boophilus* to a subgenus of *Rhipicephalus*, the names *Boophilus annulatus* and *Boophilus microplus* could still

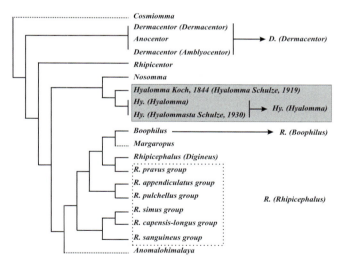

FIGURE 29.13 Working hypothesis of the phylogeny of the Rhipicephalinae and Hyalomminae, from Murrell et al. (2001). The taxa in plain text at the ends of broken lines have not been studied with molecular markers; their phylogenetic positions were inferred from the morphological analyses of Klompen et al. (1997). Shading indicates the subfamily Hyalomminae *sensu strictu* that Murrell et al. (2000) proposed should be sunk into the Rhipicephalinae. Arrows indicate proposed taxonomic changes for the taxa that are apparently paraphyletic (i.e., incomplete or unnatural taxa). Taxa in brackets are subgenera.

be used without contradicting the rules and practice of the International Code of Zoological Nomenclature. Alternatively, the seven or so species-groups of *Rhipicephalus* could be elevated to genera. Both studies that have examined phylogenetic relationships species among *Boophilus* suggest that *Rhipicephalus* is paraphyletic, with *Boophilus* arising on more than one branch within the *Rhipicephalus* (Murrell et al. 1999; Beati and Keirans 2001). The genus *Dermacentor* includes approximately 34 species. Molecular studies suggest that the monotypic genus *Anocentor* arises within *Dermacentor* (Fig. 29.13).

Phylogenetic Insight into the Biology and Biogeography of Ticks

Accurate phylogenetic trees allow the systematist to interpret character evolution in and biogeographical origins of tick lineages. What follows are three examples of the ways phylogenies can be used to interpret the evolution of ticks and their historical biogeography (Murrell et al. 2001). Prior to accurate phylogenies, one could only speculate about the origin and direction of phenotypic evolution.

Evolution of Ornamentation in Rhipicephalinae

Members of Rhipicephalinae tend to be a uniform reddish brown in coloration. However, a number of *Dermacentor* species have quite distinctive pigmentation and structural coloration. *Dermacentor rhinocerinus* in particular has striking metallic gold structural coloration (Fig. 29.14A). This apparently arose early during *Dermacentor* evolution and was lost only in *Anocentor nitens*. Ornamentation apparently evolved once in the basal lineage, leading to *Nosomma monstrosum*. Ornamentation arose again in the basal lineage, leading to two African *Rhipicephalus* species, *R. pulchellus* and *R. maculatus*. Both have light tan lines and patterns on a dark reddish brown background, giving them the appearance of wood inlay.

Evolution of Life Cycles in Ixodidae

In a three-host life cycle, each of the three mobile stages (larvae, nymphs, and adults) leaves the host to molt. In some ticks, molting from larvae to nymphs takes place without leaving the host: the two-host ticks [*Rhipicephalus* (*Digineus*) spp. and some *Hyalomma* spp]. In the most extreme case, all of the mobile stages

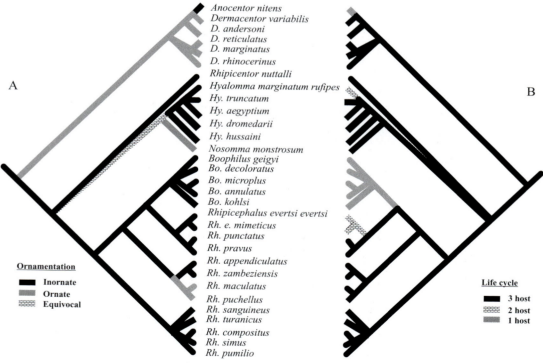

FIGURE 29.14   A strict consensus of the eight shortest MP trees from a total-evidence analysis of nucleotide sequences from four genes (12S, COI, ITS2, 18s rRNA) and morphology (Murrell et al. 1999). The number of hosts required to complete the life cycle has been mapped onto this tree.

remain on one host (all *Boophilus* spp., *A. nitens* and *D. albipictus*).

The molecular phylogeny of rhipicephaline ticks reveals that a life cycle with a reduced number of hosts (one or two hosts) has evolved many times: *A. nitens*, *Hyalomma m. rufipes*, *R evertsi*, and in the *Boophilus* spp. (Figure 29.14B). It is unclear whether the evolution of two-host life cycles in the two subspecies of *Rhipicephalus evertsi* and the one-host life cycles of the *Boophilus* spp. were part of the same evolutionary event. Indeed, despite being phylogenetically close to *Boophilus* spp., *R. evertsi* is apparently more closely related to species from the *R. pravus* group, which contains three-host ticks (Figure 29.13). In one analysis, *R. evertsi* was the sister group to the *Boophilus* spp. This suggests that the one-host life cycle may be a modification of a two-host life cycle rather than having evolved directly from a three-host life cycle.

Biogeography of Ticks

When the biogeographic regions inhabited by rhipicephaline ticks are mapped onto their phylogeny,

it becomes obvious that the ancestors of the Rhipicephalinae were Afrotropical (Fig. 29.15). Indeed, the ancestors of the three major clades of ticks probably all lived in Africa. Anna Murrell, a recent Ph.D. student in Dr. Barker's laboratory has hypothesized that the ancestor of the *Dermacentor–Anocentor* lineage evolved in the Afrotropical region (Murrell et al. 2001). Dispersal into Eurasia was probably in the Eocene (50 mya), after which Africa became isolated. Then, most dispersal and cladogenesis occurred in the Palearctic and Nearctic regions, where most *Dermacentor* species are found today. Only two species from the *Dermacentor–Anocentor* lineage are found in Africa today, suggesting that there has been little speciation in this lineage in Africa or that species from this lineage have gone extinct. During the Oligocene (~35 mya), dispersal between Eurasia and the Nearctic was possible; much later (2.5 mya), dispersal between the Nearctic and Neotropical regions could have occurred via the Isthmus of Panama. Murrell's hypotheses of an African origin of the genus *Dermacentor* contrasts with previous ideas about its origin, which have centered on arguments over whether the genus evolved in the

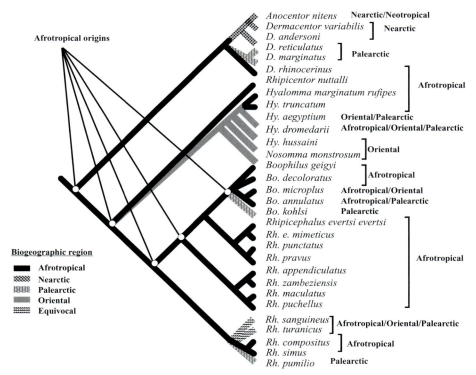

FIGURE 29.15 Biogeography of ticks of the subfamily Rhipicephalinae s.1. (Murrell et al. 1999). Information on the biogeography of ticks was from Camicas et al. (1998). Introductions to biogeographical regions for *B. microplus*, *B. annulatus* (cattle tick), and *R. sanguineus* (dog ticks) were excluded.

Nearctic region and moved into the Palearctic and Oriental regions or whether the *Dermacentor* spp. evolved in the Palearctic and then dispersed to Nearctic and Oriental regions (Balashov 1994; Crosbie et al. 1998).

Murrell suggests that the *Nosomma–Hyalomma* lineage evolved from an ancestor that lived in the Oriental region, perhaps in the early Miocene (~19 mya). Movement from Asia to Eurasia then became possible, and there was probably dispersal back into Africa after Africa and Eurasia were joined by a land bridge 14 mya.

The *Boophilus–Rhipicephalus* lineage probably evolved and radiated in Africa, i.e., while Africa was mostly isolated from the Palearctic and Oriental regions before the formation of the land bridge between Africa and Eurasia (14 mya). Dispersal and radiation into Eurasia and Asia probably occurred after the land bridge formed between Africa and Eurasia in the Miocene. The phylogeny indicates that the genus *Boophilus* also evolved in Africa. The *Rhipicentor* lineage (two spp.) appears to have evolved in and then remained in Africa, although it is also possible that species from this lineage evolved in, or dispersed to, other regions but then went extinct. *Anomalohimalaya* spp. live in the Palearctic and Oriental regions and *Margaropus* and *Cosmiomma spp.* in Afrotropical regions, but until the phylogenetic position of these genera are resolved their historical biogeography cannot be inferred.

Order Diptera, Suborder Nematocera

Phylogenetic relationships among families in the primitive Diptera suborder Nematocera have been estimated with suites of morphological characters (Oosterbroek and Courtney 1995) and nucleotide sequences from the 18S and 5.8S nuclear ribosomal DNA (rDNA) (Miller et al. 1997) and 28S rDNA (Pawlowski et al. 1996). The working hypothesis appears in Figure 29.16. All concepts support the hypothesis that hematophagy evolved four times independently in primitive Diptera. The morphological and 18S data sets are congruent in identification of Chaoboridae (phantom midges) as a sister group to Culicidae and in placement of Corethrellidae as a basal clade to Chaoboridae–Culicidae. Phantom midges inhabit freshwater habitats and are often found with mosquitoes. Phantom midge larvae are predaceous on aquatic arthropods. Little is known about adult biology; however, none are blood feeders.

The 28S data set supported monophyly of these three families but consistently indicated Chaoboridae–Corethrellidae as sister taxa. Phylogenies

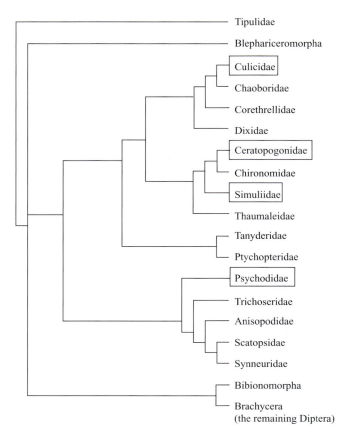

FIGURE 29.16 Working hypothesis of the phylogenetic relationships of the major Nematocerous Dipteran families based upon molecular genetic data sets. Hematophagous families appear in boxes.

of higher-order relationships among these three families and Chironomidae, Ceratopogonidae, Dixidae, Psychodidae, and Simulidae are incongruent in all three studies. Each study cites several independent lines of support for the higher-order relationships derived from their respective phylogenies, but all studies also indicate that these relationships were supported by few characters or lack strong bootstrap support. The phylogenetics of the important vector families Ceratopogonidae, Psychodidae, and Simulidae have not been examined in sufficient detail to warrant further discussion.

Mosquitoes (Family Culicidae)

The family Culicidae, which includes all mosquitoes, is divided into three subfamilies, *Anophelinae* (node 1, Fig. 29.17), *Toxorhynchitinae* (node 2), and *Culicinae* (node 3) (Knight and Stone 1977; Knight 1978). *Anophelinae* includes three genera, the Neotropical *Chagasia* (four species), the Australasian *Bironella* (nine species),

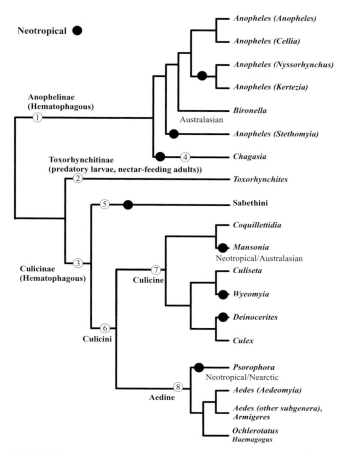

Neotropical ●

Anophelinae
(Hematophagous)
①

Anopheles (Anopheles)

Anopheles (Cellia)

Anopheles (Nyssorhynchus)

Anopheles (Kertezia)

Bironella

Australasian

Anopheles (Stethomyia)

Toxorhynchitinae
(predatory larvae, nectar-feeding adults))
②

④ Chagasia

Toxorhynchites

⑤ Sabethini

Culicinae
(Hematophagous)
③

Culicine
⑦

Coquillettidia

Mansonia
Neotropical/Australasian

Culiseta

Wyeomyia

Deinocerites

Culex

Culicini
⑥

Aedine
⑧

Psorophora
Neotropical/Nearctic

Aedes (Aedeomyia)

Aedes (other subgenera),
Armigeres

Ochlerotatus
Haemagogus

FIGURE 29.17 Working hypothesis of the phylogenetic relationships among families, subfamilies, and genera in the family Culicidae. Black dots indicate that a taxon is primarily Neotropical in distribution.

and the nearly cosmopolitan *Anopheles*, with some 386 species grouped in six subgenera. *Toxorhynchitinae* includes a single genus, *Toxorhynchites*, with 65 species. *Culicinae* is by far the largest subfamily, subdivided into 10 tribes, 30 genera, 109 subgenera, and including about 2,610 described species. With changes in taxonomic status and description of additional taxa, the total number of genera, subgenera, and species in Culicidae currently stands at approximately 37, 128, and 3,209, respectively. The current taxonomy of Culicidae is well documented at Walter Reed Biosystematics Unit Web site (http://wrbu.si.edu/www/culicidae/cataloggeneraentry.html).

Based on the fossil record and biogeographic inference, it has been suggested that mosquitoes evolved by the Jurassic, ~210 million years ago. Continental breakup at that time led to fragmentation and geographical isolation of populations. New Zealand has been in its present position of isolation for approxi-

mately the last 50 million years. With the exception of three species, *Aedes notoscriptus*, *Aedes australicus*, and *Culex quinquefasciatus*, the present-day mosquito fauna of New Zealand is relict and endemic. This provides circumstantial evidence that the genus *Aedes* existed prior to the island's separation from Australia and that it was probably widely dispersed during the Cretaceous, which began 145 mya (Belkin 1968). Fossils of the family Culicidae (*Culex*, *Aedes*) and its sister family Chaoboridae are well known from the Eocene (Tertiary) and Oligocene, which began 60 and 55 mya, respectively. Munstermann and Conn (1997) reviewed the impact of molecular biology and cladistic analysis on systematics of selected taxa of Culicidae, with particular emphasis on *Aedes* and *Anopheles* species.

Relationships Among Mosquito Subfamilies

The relationship of Culicidae subfamilies has been examined with nucleotide data sets using rDNA genes (Pawlowski et al. 1996; Miller et al. 1997) and a single-copy nuclear gene *white* (Besansky and Fahey 1997). All three studies were congruent in placement of Anophelinae as the basal clade in Culicidae (Fig. 29.17, node 1). Furthermore, the 18S and *white* genes were consistent in placing *Toxorhynchitinae* (node 2) as basal to the *Culicinae* (node 3).

Anophelinae

Krzywinski et al. (2001) examined two mitochondrial genes and a region of the 28S nuclear ribosomal gene to reconstruct a phylogeny of Anophelinae. The in-group consisted of all three genera of Anophelinae and five of six subgenera of *Anopheles*. Out-group taxa consisted of six genera of Culicinae. Anophelinae (node 1) was determined to be monophyletic. The four species of *Chagasia* (node 4) are Neotropical in distribution and basal to the subfamily. The genera *Anopheles* and *Bironella* and subgeneric clades *Nyssorhynchus* and *Kerteszia* within the *Anopheles* were also shown to be monophyletic. The authors suggested that lack of resolution of *Bironella* and *Anopheles* clades, or basal relationships among subgeneric clades within *Anopheles*, indicates rapid diversification. The fact that the four basal clades are Neotropical or Australasian suggests that Anophelinae probably was Gondwanan in origin.

Toxorhynchitinae

All analyses are consistent in placing Toxorhynchitinae (node 2) as basal to the Culicinae (node 3). However, these studies did not address the key question as to whether Toxorhynchitinae arose within

Anophelinae or arose as a separate lineage from a common ancestor with Anophelinae. This issue may become resolved in the future through examination of additional gene sequences and intensive sampling of primitive and derived members of both Toxorhynchitinae and Anophelinae. However, it is also quite possible that ancestral taxa are extinct in either or both subfamilies and that the issue will never be adequately resolved. Members of the subgenera *Ankylorhynchus* and *Lynchiella* are Neotropical, while members of the subgenus *Afrorhynchus* are all Afrotropical, and the subgenus *Toxorhynchites* is cosmopolitan. A molecular systematic investigation of *Toxorhynchites* has not been completed but could assist in the interpretation of the biogeographical origin of Toxorhynchitinae and therefore the evolution of the Culicinae.

Culicinae

Relationships among tribes, genera, and species in Culicinae (node 3) have also been evaluated using cladistic analysis. Within Culicinae, the Neotropical tribe Sabethini (node 5) is basal to Culicini and Aedini. All data sets support a monophyletic relationship between Culicini and Aedini. Many subgeneric relationships within Sabethini, Culicini, and Aedini may be paraphyletic and warrant taxonomic revision. Judd (1996) examined 59 morphological characters in 37 taxa within the tribe Sabethini. Cladistic analysis, using *Eretmapodites quinquevittatus* and *Haemagogus spegazzinii* as out-groups, supported Sabethini as a monophyletic group but strongly suggested paraphyletic relationships among species in at least three genera (*Runchomyia*, *Tripteroides*, and *Wyeomyia*). Besansky and Fahey (1997) performed a thorough taxon sampling of variation in the *white* gene among taxa in tribes Culicini, Sabethini, and Aedini in the Culicinae. Their analysis also supported placement of Sabethini as basal to Culicini and Aedini (node 6).

The tribe Culicini probably originated in the Neotropics (Node 6). The Culicine genera (node 7) *Wyeomyia* and *Deinocerites* and the *Coquillettidia* subgenus *Rhynchotaenia* are all Neotropical. Miller et al. (1996) examined sequence divergence in the entire internal transcribed spacer (ITS) among 14 species in four subgenera of the genus *Culex*. Species in the subgenera *Culex*, *Lutzia*, and *Neoculex* were monophyletic. There was low bootstrap support for monophyly of species in the subgenus *Culex*, but only single species were examined in the subgenera *Lutzia* and *Neoculex*. Some relationships among species and species complexes were also examined.

The Aedine taxa (node 8) also probably originated in the Neotropics. *Aedes* has recently been subdivided

into two genera, *Aedes* and *Ochlerotatus*, based upon well-defined synapomorphies (Reinert 2000). Molecular phylogenies support *Psorophora* (all Neotropical or Nearctic) as basal to *Aedes* and *Ochlerotatus*. Wesson et al. (1992) sequenced the 5.8S-28S half of the internal transcribed spacer of the rDNA cistron (ITS2) to examine phylogenetic relationships among seven species in three genera (*Aedes*, *Haemagogus*, and *Psorophora*) of Aedini. The resolved phylogeny provided evidence for biogeographical relationships among Aedini species; one clade contained Old World *Aedes* species (*Ae. aegypti*, *Ae. simpsoni*, *Ae. vexans*, and *Ae. albopictus*), a second clade contained the New World taxa (*Ochlerotatus triseriatus*, *Haemagogus mesodentatus*, and *Psorophora verox*). Like Wesson et al. (1992), the *white* gene placed Old and New World *Aedini* in separate clades with high bootstrap support.

Fleas (Order Siphonaptera)

Fleas apparently descended from a common ancestor and thus are a monophyletic lineage. However, the nearest relatives of the fleas are the snow fleas (the family Boreidae of the Mecoptera, the scorpion flies) (Fig. 29.18). Thus, the Order Mecoptera is a paraphyletic taxon, containing some of the descendents of the most recent common ancestor of the fleas and scorpion flies. There are two main taxonomic alternatives. The fleas could be treated as a suborder of Mecoptera. Alternatively, the other two families, the Nannochoristidae and the Boreidae, could each be elevated to Order. This option has the advantage of leaving the Order Siphonaptera intact.

Otherwise, our knowledge of the phylogeny and evolution of fleas is patently rudimentary. There are two groups working on the phylogeny and

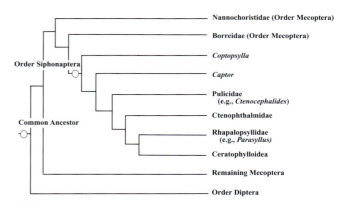

FIGURE 29.18 Working hypothesis of the phylogenetic relationships of the flea (Order Siphonaptera) and scorpion flies (Order Mecoptera) based on the analyses of nucleotide sequences from four genes. *Source*: Whiting (2002a,b).

evolution of this group: Michael Whiting's groups at Brigham Young University, Utah (Whiting 2002a,b), and a group at the Zoological Institute, St. Petersburg, Russia.

Lice (Order Phthiraptera)

Traditionally, lice have been taxonomically placed into two orders: the chewing lice (Order Mallophaga) that feed either on the blood or the skin feathers of birds and mammals, and the sucking lice (Order Anoplura) that feed on the blood of mammals. However, there is now a strong consensus that the Mallophaga, which contains three suborders, Amblycera, Ischnocera, and Rhyncopthirina, is paraphyletic and that the Anoplura arose from an Rhyncopthirina-like ancestor (Fig. 29.19). Thus, treatment of Mallophaga and Anoplura as separate orders is unjustified, and both are treated as lineages in a single order, Phthiraptera.

The putative sister group of the lice, the Psocoptera (book lice, bark lice, and psocids) have mandibles and feeding habits that are similar to those of chewing lice.

So it seems that the ancestor of the lice was a Psocoptera-like insect. Perhaps these insects lived in the same environments as mammals and birds (e.g., leaf litter of the forest floor) and thus became associated with these mammals or birds by chance. Associations with these mammals or birds may have aided dispersal and so parasitism may have evolved. Dry skin is in some ways similar to leaf litter and other vegetable matter, so perhaps the mouthparts of the Psocoptera-like insects were preadapted for feeding on the skin and feathers of mammals and birds.

Figure 29.19 suggests that the Rhyncopthirina (only three species in one genus that infest African and Asian elephants, warthogs, and the red river hog), is a highly derived lineage of lice. Figure 29.19 also suggests that lice have switched from birds to mammals on several occasions. It is not clear whether mammals or birds were the archetypal hosts of lice.

Conclusions

Fossils resembling arthropods first appeared during the late Proterozoic, about 600–540 million years ago.

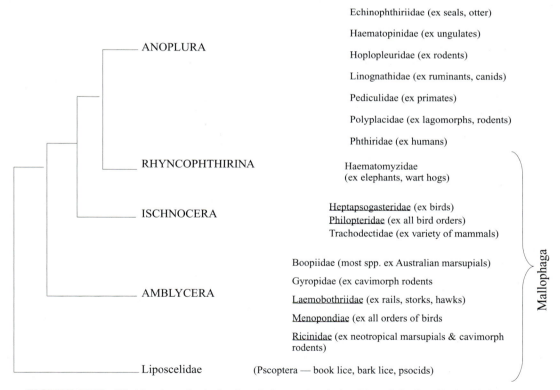

FIGURE 29.19 Working hypothesis for the phylogenetic relationships of the lice (Order Phthiraptera). Note that the Mallophaga (chewing lice) is paraphyletic (i.e., incomplete and unnatural), since it does not contain all of the descendents of the most recent common ancestor of the chewing lice (ancestor A). *Sources*: Clay (1968), Lyal (1985), Barker (1994).

Today, the phylum Arthropoda contains ~80% of all extant, metazoan animal species and arthropods have come to occupy virtually every marine, freshwater, terrestrial, and aerial habitat on earth. Arthropods are also essential components of most of the major food chains. In many ecosystems arthropods consume and recycle detritus and the associated bacteria, algae, and fungi. Many species consume either the living or dead tissues of terrestrial and aquatic plants and animals. Many are voracious predators or parasites. Given this huge taxonomic and ecological diversity, it is not surprising that a small proportion of arthropods have evolved the ability to utilize a rich and abundant source of nutrients, the blood of vertebrates.

Black (2002) and Chapter 2 trace the adaptations associated with vertebrate host location: proximal localization of shallow capillary beds, blood acquisition through either telmophagy or solenophagy, the role of saliva in disrupting hemostasis, pain, and inflammation at the feeding site, and immunomodulation. Also traced were the adaptations for processing the blood meal through both digestion and excretion and the use of the digested blood in maturing eggs for the subsequent generations. All of these adaptations for hematophagy in arthropods have arisen on at least 17 independent occasions. In every case, the prehematophagous ancestral lineages faced a common set of problems. Arthropod mouthparts have been modified in a wide variety of ways to derive a common set of phlebotomist tools, initially to permit telmophagy and eventually to enable solenophagy. Biochemical adaptations have occurred in the saliva of hematophagous arthropods to overcome the common problems of hemostasis, vasoconstriction, pain s ensation, and inflammation. As with mouthparts, virtually every arthropod examined to date has enlisted a different biochemical solution for each of these problems.

The overriding message in considering all of these adaptations in light of the phylogenies presented in this chapter is that no general, consistent morphological, physiological, or biochemical adaptations for hematophagy have been detected among all of the hematophagous arthropod lineages. However, as a general theme, arthropods, when faced with a common set of problems associated with gaining access to vertebrate blood, have taken up many independent but ultimately convergent paths.

References

Balashov, Y. S. 1994. Importance of continental drift in the distribution and evolution of ixodid ticks. *Ent. Rev.* 73: 42–50.

Barker, S. C. 1994. Phylogeny and classification, origins, and evolution of host associations of lice. *Int. J. Parasitol.* 24: 1285–1291.

Beati, L., and Keirans, J. E. 2001. Analysis of the systematic relationships among ticks of the genera *Rhipicephalus* and *Boophilus* (Acari: Ixodidae) based on mitochondrial 12S ribosomal DNA gene sequences and morphological characters. *J. Parasitol.* 87: 32–48.

Belkin, J. N. 1968. Mosquito studies (Diptera: Culicidae). VII. The Culicidae of New Zealand. *Contributions Am. Entomol. Ins.* 3: 1–182.

Besansky, N. J., and Fahey, G. T. 1997. Utility of the white gene in estimating phylogenetic relationships among mosquitoes (Diptera: Culicidae). *Mol. Biol. Evol.* 14: 442–454.

Black, W. C. 2002. Evolution of arthropod disease vectors. In *Evolution of Infectious Diseases* (C. L. Greenblatt, ed.). Oxford University Press, London.

Black, W. C., and Piesman, J. 1994. Phylogeny of hard- and soft-tick taxa (Acari: Ixodida) based on mitochondrial 16S rDNA sequences. *Proc. Natl. Acad. Sci U.S.A.* 91: 10034–10038.

Black, W. C., Klompen, J. S., and Keirans, J. E. 1997. Phylogenetic relationships among tick subfamilies (Ixodida: Ixodidae: Argasidae) based on the 18S nuclear rDNA gene. *Mol. Phylogenet. Evol.* 7: 129–144.

Camicas, J. L., Hervy, J. P., Adam, F., and Morel, P. C. 1998. *The Ticks of the World. (Acarida, Ixodida) Nomenclature, Described Stages, Hosts, Distribution.* Orstom, Paris.

Clay, T. 1968. Contributions towards a revision of *Myrsidea* Waterston. III (Menoponidae: Mallophaga). *Bull. Brit. Mus. (Natural History) Entomol.* 21: 203–243.

Crampton, A., McKay, I., and Barker, S. C. 1996. Phylogeny of ticks (Ixodida) inferred from nuclear ribosomal DNA. *Int. J. Parasitol.* 26: 511–517.

Crosbie, P. R., Boyce, W. M., and Rodwell, T. C. 1998. DNA sequence variation in *Dermacentor hunteri* and estimated phylogenies of *Dermacentor* spp. (Acari: Ixodidae) in the New World. *J. Med. Entomol.* 35: 277–288.

Dobson, S. J., and Barker, S. C. 1999. Phylogeny of the hard ticks (Ixodidae) inferred from 18S rRNA indicates that the genus *Aponomma* is paraphyletic. *Mol. Phylogenet. Evol.* 11: 288–295.

Felsenstein, J. 1978. Cases in which parsimony and compatability methods will be positively misleading. *Systematic Zool.* 27: 401–410.

Felsenstein, J. 1981. Evolutionary trees from DNA sequences: a maximum-likelihood approach. *J. Mol. Evol.* 17: 368–376.

Fukunaga, M., Yabuki, M., Hamase, A., Oliver, J. H., Jr., and Nakao, M. 2000. Molecular phylogenetic analysis of ixodid ticks based on the ribosomal DNA spacer, internal transcribed spacer 2, sequences. *J. Parasitol.* 86: 38–43.

Hasegawa, M., Kishino, H., and Yano, T. 1985. Dating of the human–ape splitting by a molecular clock of mitochondrial DNA. *J. Mol. Evol.* 22: 160–174.

Hennig, W. 1996. *Phylogenetic Systematics.* University of Illinois Press, Champaign-Urbana, IL.

Hillis, D. M., Mable, B. K., and Moritz, C. 1996. Applications of molecular systematics: the state of the field and a look to the future. In *Molecular Systematics* (D. M. Hillis, C. Moritz, and B. K. Mable, eds.). Sinauer Associates, Sunderland, MA, pp. 515–543.

Hoogstraal, H., and Aeschlimann, A. 1982. Tick host specificity. *Bull. Soc. Entomol. Suisse* 55: 5–32.

Judd, D. D. 1996. Review of the systematics and phylogenetic relationships of the Sabethini (Diptera: Culicidae). *Systematic Entomol.* 21: 129–150.

Klompen, J. S. H., Oliver J. H., Keirans, J. E., and Homsher, P. J. 1997. A re-evaluation of relationship in the Metastriata (Acari: Parasitiformes: Ixodidae). *Systematic Parasitology* 38: 1–24.

Klompen, J. S. H., Black, W. C., Keirans, J. E., and Norris, D. E. 2000. Systematics and biogeography of hard ticks: a total evidence approach. *Cladistics* 16: 79–102.

Knight, K. L. 1978. *Supplement to a Catalog of the Mosquitoes of the World.* Entomological Society of America, College Park, MD.

Knight, K. L., and Stone, A. 1977. *A Catalog of the Mosquitoes of the World (Diptera: Culicidae),* 2nd ed. Entomological Society of America, College Park, MD.

Krzywinski, J., Wilkerson, R. C., and Besansky, N. J. 2001. Evolution of mitochondrial and ribosomal gene sequences in Anophelinae (Diptera: Culicidae): implications for phylogeny reconstruction. *Mol. Phylogenet. Evol.* 18: 479–487.

Lyal, C. H. C. 1985. Phylogeny and classification of the Psocodea, with particular reference to the lice (Psocodea, Phthiraptera). *Systematic Entomol.* 10: 145–165.

Miller, B. R., Crabtree, M. B., and Savage, H. M. 1996. Phylogeny of fourteen *Culex* mosquito species, including the *Culex pipiens* complex, inferred from the internal transcribed spacers of ribosomal DNA. *Insect Mol. Biol.* 5: 93–107.

Miller, B. R., Crabtree, M. B., and Savage, H. M. 1997. Phylogenetic relationships of the Culicomorpha inferred from 18S and 5.8S ribosomal DNA sequences. (Diptera: Nematocera). *Insect Mol. Biol.* 6: 105–114.

Munstermann, L. E., and Conn, J. E. 1997. Systematics of mosquito disease vectors (Diptera, Culicidae): impact of molecular biology and cladistic analysis. *Annu. Rev. Entomol.* 42: 351–369.

Murrell, A., Campbell, N. J., and Barker, S. C. 1999. Mitochondrial 12S rDNA indicates that the Rhipicephalinae (Acari: Ixodida) is paraphyletic. *Mol. Phylogenet. Evol.* 12: 83–86.

Murrell, A., Campbell, N. J., and Barker, S. C. 2000. Phylogenetic analyses of the rhipicephaline ticks indicate that the genus *Rhipicephalus* is paraphyletic. *Mol. Phylogenet. Evol.* 16: 1–7.

Murrell, A., Campbell, N. J., and Barker, S. C. 2001. A total-evidence phylogeny of ticks provides insights into the evolution of life cycles and biogeography. *Mol. Phylogenet. Evol.* 21: 244–258.

Oosterbroek, P., and Courtney, G. 1995. Phylogeny of the nematocerous families of Diptera (Insecta). *Zool. J. Linn. Soc.* 115: 267–231.

Pawlowski, J., Szadziewski, R., Kmieciak, D., Fahrni, J., and Bitta, G. 1996. Phylogeny of the infraorder Culicomorpha (Diptera: Nematocera) based on 28S RNA gene sequences. *Systematic Entomol.* 21: 167–178.

Reinert, J. F. 2000. New classification for the composite genus *Aedes* (Diptera: Culicidae: Aedini), elevation of subgenus *Ochlerotatus* to generic rank, reclassification of the other subgenera, and notes on certain subgenera and species. *J. Am. Mosq. Control Assoc.* 16: 175–188.

Saitou, N., and Nei, M. 1987. The neighbor-joining method: a new method for reconstructing phylogenetic trees. *Mol. Biol. Evol.* 4: 406–425.

Sokal, R. R., and Sneath, P. H. A. 1963. *Principles of Numerical Taxonomy.* W. H. Freeman, San Francisco.

Swofford, D. L., Olsen, G. J., Waddell, P. J., and Hill, D. M. 1996. Phylogenetic inference. In *Molecular Systematics* (D. M. Hillis, C. Moritz, and B. K. Mable, eds.). Sinauer Associates, Sunderland, MA, pp. 407–514.

Wesson, D. M., Porter, C. H., and Collins, F. H. 1992. Sequence and secondary structure comparisons of ITS rDNA in mosquitoes (Diptera: Culicidae). *Mol. Phylogenet. Evol.* 1: 253–269.

Whiting, M. 2002a. Mecoptera is paraphyletic: multiple genes and phylogeny of Mecoptera and Siphonaptera. *Zoologica Scripta* 31: 31–40.

Whiting, M. F. 2002b. Phylogeny of the holometabolous insect orders: molecular evidence. *Zoologica Scripta.* 31: 3–15.

30

Genetics of Vector Competence

WILLIAM C. BLACK IV AND DAVID W. SEVERSON

INTRODUCTION

Vector *competence* refers to the intrinsic permissiveness of an arthropod vector for infection, replication, and transmission of a vertebrate pathogen. In this chapter we assume that genetic factors in the vector condition, at least in part, the ability of a pathogen to be transmitted. This assumption has been validated in the few vector–pathogen systems in which it has been tested. In some cases it may be manifest as differences in competence of two closely related vector species for a particular pathogen. For example, *Anopheles stephensi* is highly susceptible to *Plasmodium berghei,* whereas *An. gambiae* is nearly incompetent. Similarly, extracted salivary glands from a competent vector, *An. dirus,* are infected by *P. knowlesi* when they are transplanted into an incompetent vector, *An. freeborni,* but are not infected when transplanted from *An. freeborni* into *An. dirus* (Rosenberg 1985). The human head louse *Pediculus humanis capitis* does not become infected with *Rickettsia prowazekii,* the etiological agent of louse-borne typhus, whereas its sister subspecies *P. h. corporis* becomes readily infected and in fact dies. In other cases, different populations of a single vector species may differ in their ability to transmit a particular pathogen. This is the case among different populations and subspecies of *Aedes aegypti.* When populations of West African *Ae. aegypti formosus* are orally infected with Yellow Fever virus in the laboratory, only ~30% become capable of oral transmission. In contrast, no fewer than ~60% of *Ae. aegypti aegypti* collected from most locations throughout the world transmit flaviviruses when orally infected (Tabachnick et al. 1985).

Plasticity in vector competence is consistent with a general genetic model in which multiple structural or biochemical factors must be present for successful completion of a pathogen's life cycle. The absence of any one of these factors renders a vector species incompetent. Alternatively, if a pathogen is deleterious to vector survival or reproduction, the vector may well have evolved an active mechanism for rejection or destruction of a pathogen. Some vector species may be fixed for the presence of resistance factors, while others may lack one or more of these factors. Variation in vector competence among populations of a single species could arise due to genetic factors' becoming fixed in some populations and at variable frequencies or altogether absent in others.

In this chapter we discuss how to test for and quantify the extent to which vector competence is under genetic control. We explain how to map and thereby determine the magnitude and number of genetic factors that control vector competence. Finally, we discuss different approaches to identify the mechanisms and genes that control vector competence. We focus on three pathogen–vector systems: the competence of culicine mosquitoes for arboviruses and for filarial worms and the competence of anopheline and culicine mosquitoes for *Plasmodium* species. However, the concepts, issues, and technology developed in this chapter can be applied broadly to any of the many vector–pathogen systems.

MECHANISMS OF
REFRACTORINESS

Arboviruses

Figure 30.1A is a generalized diagram indicating six potential barriers that an arbovirus faces in being transmitted by an arthropod. Assuming that a mosquito has taken a viremic blood meal from a vertebrate host, the virus must first attach and penetrate the mosquito's midgut epithelial cells and then replicate to a high titer within those cells. Factors that block either of these two events constitute a *midgut infection barrier* (MIB). Next, the virus must pass through the basal lamina surrounding the midgut and infect and replicate in surrounding tissues. Anything that prevents these events leading to a disseminated infection acts as a *midgut escape barrier* (MEB). Finally, the arbovirus must infect and possibly replicate within the salivary gland before it can be shed into the lumen of the glands for final transmission in the subsequent bite. Factors that suppress these phases serve as a *transmission barrier* (TB).

We can further subdivide any of these barriers at a finer, biochemical level (Fig. 30.2). A vector protein or substance produced by a protein is required for each step of viral penetration and replication in midgut tissues. It may be that every vector species has most of these factors but that only competent species have *all*. Alternatively, it may be that susceptible populations have all of the required factors, while those with intermediate susceptibility are polymorphic for a required factor and refractory populations lack the factor altogether. For example, an MIB for a positive-sense RNA virus may exist in the lumen of the midgut and involve reduced ability to cleave proteins on the viral coat before attachment or penetration can occur. In addition, there may be receptors on the surface of the midgut cell that are necessary for attachment of the virus. Once inside the cell, host factors may be necessary for uncoating and then translation of nonstructural viral proteins (e.g., RNA-dependent polymerases). Host factors are probably involved in proper transcription and translation of viral structural proteins. In the final stages, host factors are undoubtedly involved in proper packaging and assembly of virus.

Plasmodium

Analogous barriers exist for arboviral and *Plasmodium* transmission (Fig. 30.1B). A mosquito takes a gametocytemic blood meal from a vertebrate host; the male gametocyte must exflagellate and then fuse with the female gametocyte to form a motile ookinete. The ookinete must then successfully escape from the blood meal through the peritrophic matrix and penetrate the mosquito's midgut epithelial cells before lodging between the midgut cells and the basal lamina, where it completes development to the oocyst stage. Factors that block any phase of ookinete development and cell penetration constitute an MIB. The oocyst must next undergo sporogony, and the sporozoites must pass through the basal lamina. Anything that prevents these events acts as an MEB. Finally, the sporozoites must infect the salivary glands and make their way into the lumen of the glands for transmission to a new vertebrate host during subsequent blood feeding. Prevention of salivary gland infection would constitute a TB.

A number of factors have been identified that block *Plasmodium* invasion and development. It has been shown that ookinetes penetrating midgut follicular epithelial cells induce various *Anopheles* species to produce defensin, a gram-negative-bacteria–binding protein, and nitric oxide synthetase and initiate other enzymatic pathways that may ultimately lead to apoptosis (Han et al. 2000). In general, this MIB is so potent that only a small percentage of mature ookinetes reach the basal lamina to form oocysts. In both *Anopheles gambiae* and *Aedes aegypti*, a large percentage of *Plasmodium gallinceum* ookinetes are lysed within the midgut epithelial cells. An MEB has been observed in strains of *An. gambiae* that have been artificially selected to encapsulate, melanize, and thereby destroy oocysts (Collins et al. 1986). Susceptible vector species may have failed to develop one or all of these resistance factors, whereas refractory species may have potent or constitutively expressed resistance factors or, as with viruses, lack factors that are essential for *Plasmodium* survival.

Filarial Worms

Barriers to the parasitic nematodes responsible for human lymphatic filariasis, *Wuchereria bancrofti*, *Brugia malayi*, and *B. timori*, have limited analogy to those confronting arboviruses and *Plasmodium* (Fig. 31.1C). As with other pathogens, microfilariae circulating in the peripheral blood are taken up by the mosquito host during blood feeding. However, because microfilariae are relatively large (200–300 µm), they use a "brute force" approach to move relatively quickly from the blood meal and through the midgut epithelium into the hemocoele. Once in the hemocoele, the microfilariae migrate to the thorax, where they complete development to infective L3-stage larvae intracellularly within the indirect flight muscles. This process requires about 2 weeks. Thereafter, the L3-stage larvae, measuring up to 1.5 mm in length, move from the

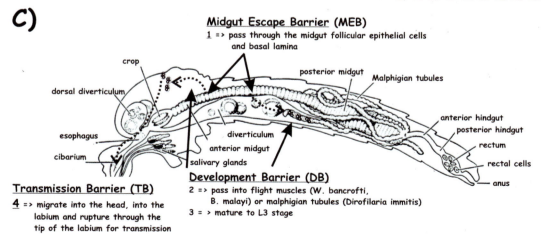

A)

Midgut Infection Barrier (MIB)
1 => establish an infection in the midgut epithelium
2 => replicate in the midgut epithelium cells

crop
dorsal diverticulum
posterior midgut
Malphigian tubules
esophagus
cibarium
diverticulum
anterior midgut
salivary glands
salivary ducts
anterior hindgut
posterior hindgut
rectum
rectal cells
anus

Midgut Escape Barrier (MEB)
3 => pass through the basal lamina
4 => replicate in other organs and tissues

Transmission Barriers (TB)
5 => infect salivary glands (SIB)
6 => escape into the lumen of the salivary gland (SEB)

B)

Midgut Infection Barrier (MIB)
1a. Microgamont
2. Exflagellation
1b. Macrogamont
3. Fertilization
4. Ookinete penetration

crop
dorsal diverticulum
posterior midgut
esophagus
cibarium
diverticulum
anterior midgut
salivary glands
anterior hindgut
posterior hindgut
rectum
rectal cells
anus

Transmission Barrier (TB)
8. Sporozoite invasion
9. Sporozoite release

Midgut Escape Barrier (MEB)
5. Oocyst formation
6. Sporogony
7. Release of sporozoites

C)

Midgut Escape Barrier (MEB)
1 => pass through the midgut follicular epithelial cells and basal lamina

crop
dorsal diverticulum
posterior midgut
Malphigian tubules
esophagus
cibarium
diverticulum
anterior midgut
salivary glands
anterior hindgut
posterior hindgut
rectum
rectal cells
anus

Transmission Barrier (TB)
4 => migrate into the head, into the labium and rupture through the tip of the labium for transmission

Development Barrier (DB)
2 => pass into flight muscles (W. bancrofti, B. malayi) or malphigian tubules (Dirofilaria immitis)
3 => mature to L3 stage

FIGURE 30.1 Barriers to transmission of (A) arboviruses, (B) *Plasmodium* spp., and (C) filaria in a mosquito.

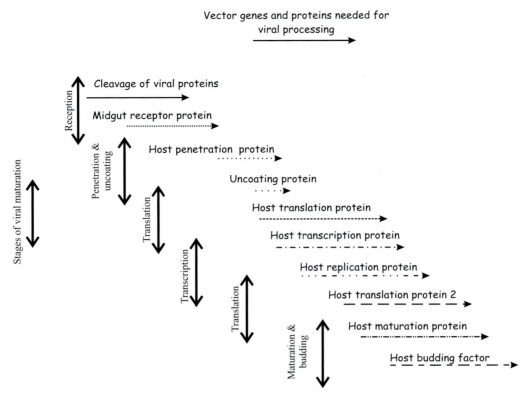

FIGURE 30.2 Hypothetical biochemical and genetic barriers to transmission of arboviruses in an arthropod vector.

thoracic musculature to the head. When the infected mosquito bloodfeeds, the larval enter the labium and escape by rupturing through the tip of the labium (Chapter 1) and are deposited on the skin surface. They then gain entrance to the vertebrate host through the puncture made by the mosquito. In contrast to arboviruses and *Plasmodium*, filarial worms are nonreplicative within the mosquito host, and the salivary glands do not play a direct role in their transmission.

Limitations on filarial worm invasion and development include: (1) factors that limit escape from the blood meal and block penetration through the midgut epithelium, which collectively can be considered as an MEB, and (2) recognition within the hemocoele triggering a humoral immune response that may result in melanotic encapsulation. Apparently those microfilariae that successfully evade the humoral response and invade the thoracic musculature will complete development to the L3 stage. Escape from the blood meal can be influenced by the rate of blood clot formation and formation of the peritrophic matrix that forms around the blood meal. Indeed, it has been shown with strains of *Ae. aegypti* that genetic factors play a signif-

icant role in the ability of *B. malayi* microfilariae to escape the blood meal and successfully penetrate across the midgut epithelium (Beerntsen et al. 1995). Genetic control over humoral resistance factors has also been documented in both *Culex pipiens* and *Ae. aegypti*. Laboratory strains have been developed that are highly susceptible to or are completely refractory to filarial worm infection.

PHYSIOLOGICAL GENETICS

Genome refers to the entire complement of DNA contained in a haploid cell of a species or individual. A genetic *locus* is a specific location in that genome. This could be the specific location of a gene or of a nucleotide in a gene. A large array of nucleotide variation has been shown to exist in genes when they are sampled from natural populations. Each novel nucleotide sequence is considered a novel *allele*. Variation among alleles consists primarily of *synonymous substitutions* that do not change the primary amino acid sequence. The majority of *nonsynonymous substitutions* involve changes among amino acids of similar size, function, or charge. This pattern of variation is consistent with the *neutral theory of molecular evolution* (Kimura 1991; Chapter 32). This model assumes that most mutations that occur are deleterious to the fitness of an individual and are removed through *purifying selection*. Rarely, nonsynonymous substitutions yield proteins that differ in function without being deleterious. In addition, mutations will occasionally arise that destroy the function of the gene. The latter might involve the introduction of a premature stop codon, a frameshift mutation, deletion of a critical part of a gene, or mutations in upstream promoters or enhancers. Such mutations would be tolerated in most vector species because they are diploid. One functional copy of a gene may be sufficient for vector survival (such genes are referred to as *haplosufficient*).

In classical genetics, genes that fail to be transcribed or translated or to encode functional proteins segregate as *recessive* alleles relative to *dominant* alleles that encode a functional protein. Only individuals that inherit two copies of nonfunctional alleles will display the absence of gene function. Individuals with one (*heterozygous*) or two (*homozygous*) functional copies will display a normal phenotype. Dominant genetic effects might be seen in situations where a receptor is involved. Presence of the receptor, whether encoded by one or two functional copies of the gene, is sufficient to enable pathogen attachment or invasion. Only vectors without any functional receptors (homozygotes recessives) are refractory. This pattern was seen with transovarial transmission of LaCrosse encephalitis by *Aedes triseriatus* (Graham et al. 1999). Alleles that encode functional proteins that differ, albeit slightly, in function are said to be *codominant* or *additive*. With alleles of additive effect, homozygotes for one allele may have a different phenotype than homozygotes for the other allele. Furthermore, heterozygotes may display a different, possibly intermediate, phenotype. In *Ae. aegypti*, alleles at loci on chromosomes 2 and 3 are additive in their effects on a dengue MIB (Bosio et al. 2000).

A point that is often overlooked in genetic studies of vector competence is that *many* genes are probably involved in the complete development (e.g., Fig. 30.2) or destruction of a pathogen. All vector competence studies to date have involved *spontaneous mutations* or *natural polymorphisms* in genes involved in pathogen transmission. In every case investigators collected mosquitoes from the field, found variation in vector competence, and were able to increase or decrease vector competence alleles by artificial selection. The neutral model is probably a useful predictor of the types of allelic variation that exist at genes involved in infection, replication, and transmission of a

pathogen. Synonymous and the majority of nonsynonymous substitutions have no effect on vector competence. Rarely, nonsynonymous substitutions yield proteins that either enhance or reduce the ability of the protein to affect pathogen development. Mutations that destroy gene function may reduce vector competence if the encoded protein was essential for pathogen survival (e.g., a viral receptor) or might enhance vector competence if the encoded protein was involved in destroying or reducing pathogen survival.

However, it is highly likely that the majority of genes involved in pathogen replication and transmission (Fig. 30.2) contain spontaneous mutations that don't influence phenotype. Reliance on spontaneous mutations in these genes won't, therefore, identify them as important in the pathogen's life cycle. Another extremely important approach to studying the genetics of vector competence is to generate *induced mutations* using ionizing radiation, chemical mutagens, or transposable elements to cause defects in other genes involved in pathogen development. This approach has been used to obtain a generalized reduction in vector competence for filaria (Rodriguez et al. 1992). The problem with this approach is that treatment of a vector with mutagens creates many genetic lesions, some of which may reduce vector competence but via mechanisms that reduce overall fitness and therefore would not survive in nature. A second problem is that a phenotypic reduction in vector competence doesn't allow the investigator to identify the genes in which lesions have occurred. *Targeted mutagenesis* of a gene presumptively involved in vector competence would be preferable. This could be accomplished with Sindbis viral vectors (Chapter 39) or via transposable elements (Chapter 40).

QUANTITATIVE GENETICS

Determination and Quantification of Genetic Effects

Quantitative genetics estimates the genetic and environmental effects associated with variation in quantitative rather than discrete traits (Falconer and MacKay 1996). These traits include size, weight, morphometric characters, growth rate, longevity, fecundity, bristle number, and many components of vector competence. In treating vector competence as a quantitative trait we assume that it is under the control of several genetic loci and subject to environmental effects that are not controlled by the investigator. We have already suggested the involvement of multiple genetic loci in all three vector pathogen systems (Fig. 30.1 and 30.2). Further, the researcher may not be able to control many factors that affect vector competence (e.g., pathogen titer, gonadotrophic stage, longevity, body size, larval nutrition, local temperature or humidity effects in an insectary), and these would be manifest as environmental effects.

Measuring Quantitative Genetic Traits

An important aspect of the quantitative genetics of vector competence concerns the ways in which we measure susceptibility. Competence for arboviruses might be measured in terms of the number of infectious particles we find in the midgut, head, or saliva. Susceptibility to *Plasmodium* might be measured in terms of the number of oocysts on the gut wall or as the number of encapsulated and melanized oocysts or by the number of sporozoites in the saliva. Vector competence for filarial worms might be measured in terms of number of microfilariae ingested and number penetrating the midgut, number that reach the appropriate developmental tissue, or number that develop to infective L3-stage larvae. Any types of finite counts are referred to as *meristic characters*.

Almost always, viral particle numbers are estimated by dilution assays and/or antigen capture assays (e.g., ELISA), where the quantity of virus is correlated with the intensity of staining or fluorescence in enzyme activity assays. Antigen capture assays might also be used to estimate sporozoite infection rates, because frequently too many sporozoites are present to make a direct count feasible. In these cases, the measures of pathogen density would be continuously distributed along a linear or logarithmic colorimetric scale. These are referred to as *continuous characters*.

Alternatively, we may only be interested in determining whether the midgut, head, or saliva becomes infected. In this case, we would only determine whether a tissue is infected (scored "+" or "1") or uninfected ("−" or "0"). This approach is frequently used in vector competence. Referred to as a *threshold model*, this assumes the presence of an unobservable continuous variable (often referred to as "liability") that underlies the binary expression of the phenotype. Almost any quantitative character can be treated as a threshold character even though the liability is distributed as an unobservable continuous quantitative trait. A probit or logistic model is used when the liability is assumed to be normally distributed. In applying this model to vector competence we assume that below a certain liability threshold a vector remains uninfected but that above that threshold the vector can become infected.

Quantitative Genetic Experimental Designs

Two general experimental designs are used to quantify genetic and environmental effects. These are (1) measured responses to artificial selection and (2) patterns of inheritance in defined pedigrees. In all three vector–pathogen systems, the first evidence for a genetic effect came from artificial selection experiments. Miller and Mitchell (1991) showed that strains of *Ae. aegypti* could be artificially selected that were either highly susceptible or highly refractory to flavivirus transmission. Collins et al. (1986) selected strains of *An. gambiae* that were highly susceptible to developing oocysts and were similarly able to select strains that encapsulated and melanized the majority of oocysts. Strains of both *An. gambiae* and *Ae. aegypti* that are refractory to *P. gallinaceum* via an ookinete lysis mechanism have also been selected (Collins et al. 1986; Thathy et al. 1994; Vernick et al. 1995). Therefore, *Plasmodium* refractoriness is likely obtained through multiple genetic mechanisms. Macdonald (1962) was able to select an *Ae. aegypti* strain susceptible to the filarial worm, *Brugia malayi*, and Beerntsen et al. (1995) demonstrated that this strain could be directionally selected for very high susceptibility or complete refractoriness. The ability to select permissive and refractory strains provides *prima facie* evidence that vector competence is under genetic control. A response to selection suggests that alleles exist at one or more of the genes involved in vector competence and that these alleles vary in activity. By breeding only those vectors that are refractory or susceptible, we isolate and increase refractory or susceptible alleles in individual genetic lines or strains.

Artificial selection experiments by themselves tell us little about the type or magnitude of genetic effects, the numbers of genetic loci involved, or the phenotypic effects of alleles at those loci. These parameters are instead estimated using quantitative genetic analyses of patterns of phenotypic inheritance in defined pedigrees. Most of these methods assume a general model:

$$V_p = V_g + V_e \qquad (1)$$

where V_p is the observed variance in phenotype observed among a group of individuals, V_g is the portion of V_p that can be accounted for by genetic effects, and V_e is the remaining portion of V_p. V_e is generally considered to encompass all of the factors not under control by the investigator. There are many experimental designs that can be used to estimate V_p, V_g, and V_e in a phenotypic character. One approach is illustrated in Figures 30.3 and 30.4. The crossing design

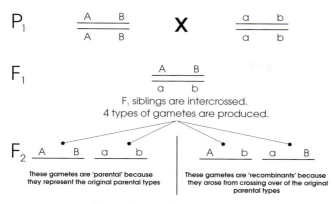

P_1 A B   X   a b
    A B   a b

F_1 A B
    a b

F_1 siblings are intercrossed.
4 types of gametes are produced.

F_2 A   B   a   b   A   b   a   B

These gametes are 'parental' because they represent the original parental types

These gametes are 'recombinants' because they arose from crossing over of the original parental types

FIGURE 30.3 Schematic of an F_1 intercross and hypothetical segregation of alleles at two loci in the intercross.

shown in Figure 30.3 is referred to as an F_1 *intercross*. Two parents that differ in the phenotype of interest are crossed to produce multiple F_1 offspring. The F_1 brothers and sisters are intercrossed to produce a set of F_2 offspring. Figure 30.4 indicates the way that two loci, both of which control the phenotype of interest, would segregate in an F_1 intercross. We assume that all P_1 individuals are genetically homogeneous and therefore V_g accounts for none of the variation in V_p in either of the P_1 strains. Similarly, $V_g \simeq 0$ among F_1 offspring because they are assumed to be uniformly heterozygous for the alleles inherited from the P_1 parents. However, there are nine different genotypes segregating among the F_2 offspring. Thus, potentially a large portion of V_p among F_2 individuals could be attributable to V_g. This design allows us to estimate V_e from both the P_1 strains and the F_1 generation. We can then subtract V_e from V_p to obtain an estimate of V_g using Eq. (1).

As an example of how this might be applied in the study of quantitative genetics of vector competence, assume that the *A* allele (Fig. 30.4) contributes 15 units of some measure of susceptibility, while the *a* allele contributes only 1. In contrast, the *B* allele contributes 30 units of susceptibility, while *b* allele contributes only 5. Further assume that the P_1 mother is uniformly homozygous for *A* and *B* alleles, while the P_1 father is uniformly homozygous for *a* and *b* alleles. Using the numbers listed under the additive model in Figure 30.4, $V_e = 10.95$ in both the P_1 strains. The mean phenotype of the P_1 mother is 90, while the mean phenotype of the P_1 father is 12. Next, we measure the phenotype distribution of the F_1 offspring and find that the mean phenotype of the F_1 offspring is 51, while the variance is again 10.95. Brothers and sisters are mated to one another to generate the F_2 offspring. We measure the phenotype among 20 offspring, and, for

the purposes of illustration, the first 16 offspring have each of the 16 genotypes shown in Figure 30.4, while the remaining 4 are all genotype *AaBb*. Notice that the mean phenotype among the F_2 offspring has not changed; however, the variance has increased dramatically to 368.25. Thus, from Eq. (1), $V_g = 368.25 - 10.95 = 357.31$.

These results are often summarized as the *broadsense heritability* (h^2), the proportion of V_p that is accounted for by V_g. In this example, $h^2 = 357.31/368.25 = 0.97$. Additional information is obtained from an F_1 intercross. Notice that the mean phenotype of the F_1 and F_2 generations is exactly the average of the two P_1 strains. In addition, if we plot the mean phenotype against the genotype at each locus, we observe a straight line (Fig. 30.5A). Furthermore, plotting the mean phenotype against the combined effects of both loci (Fig. 30.5B), we also observe a straight line. This indicates that the alleles at the two loci are additive in their contribution to the overall phenotype and that the two loci are also additive in their combined effect on the overall phenotype. If we assume complete additivity, equal effects of all genes, and no linkage, it can be shown that the number of genes (n) that control the phenotype of interest is approximated by the equation (Falconer and MacKay 1996)

$$n = \frac{D^2}{8V_g^2} \tag{2}$$

where D is the difference in phenotype between the two P_1 parents. In our example,

$$n = \frac{(90-12)^2}{8(357.31)} = 2.13$$

an estimate very close to the number of loci in our model.

To model a case of dominance we could instead assume that having one or two *A* alleles contributes 30 units of susceptibility and having one or two *B* alleles contributes 60, but that the *a* and *b* alleles contribute none. The mean phenotype of the P_1 mother is again 90, but the mean phenotype of the P_1 father is 0. $V_e = 10.95$ in both P_1 strains. V_p of the $F_1 = 10.95$; however, the mean phenotype of the F_1 offspring is now 90. The mean phenotype among the F_2 offspring is 69, while $V_p = 835.16$. Thus, $V_g = 835.16 - 10.95 = 824.21$, $h^2 = 0.99$, and

$$n = \frac{(90-0)^2}{8(824.21)} = 1.23$$

In contrast with the additivity model, the mean phenotype of the F_1 and F_2 generations has shifted toward

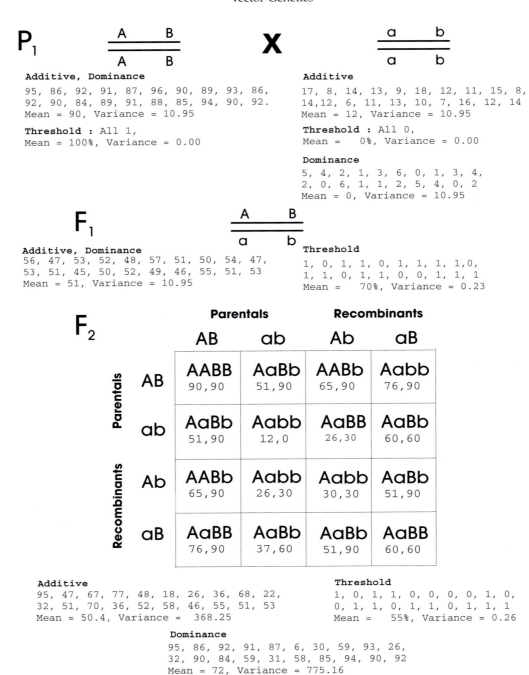

FIGURE 30.4 Hypothetical distribution of phenotypes conditioned by alleles at *A* and *B* loci in an F_1 intercross. Phenotype distributions are listed for cases described in the text in which alleles are additive or dominant or when the phenotype is treated as a threshold character.

the phenotype of the P_1 parents carrying the dominant alleles. In addition, if we plot the mean phenotype against the genotype at each locus, we fail to see a straight line at individual loci (Fig. 30.5C) or for the combined effect of both loci (Fig. 30.5D). These patterns indicate that alleles at these loci are dominant in

their effects on phenotype. The effective number of alleles is no longer accurate because we have violated the assumption of additivity in Eq. (2).

We have to this point assumed that susceptibility can be measured on a continuous quantitative scale. We could instead treat susceptibility as a threshold

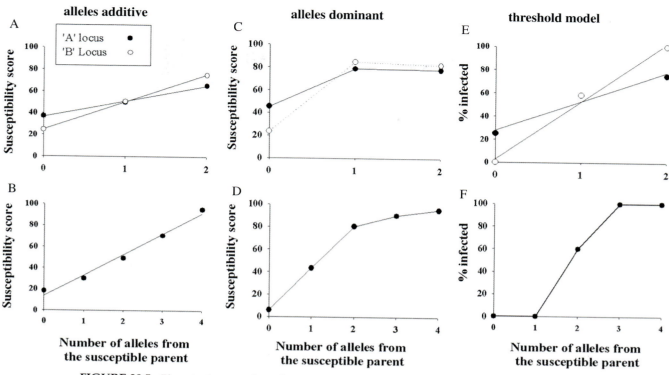

FIGURE 30.5 Phenotypic means from data listed in Figure 30.4 plotted against genotypes at the *A* and *B* loci (A, C, E) or when alleles at both loci are considered together (B, D, F). Means are plotted when alleles are additive (A, B) or dominant (C, D) or when the phenotype is treated as a threshold character (E, F).

character by assuming that the susceptibility threshold is approximately 50. Vectors with a lower susceptibility score do not transmit, while all vectors with a susceptibility score greater than 50 are capable of transmission. Phenotypes are scored as "1" if the vector is susceptible and "0" if refractory. Under these conditions, all of the P_1 maternal strain can transmit whereas none of the paternal strain can transmit. $V_p = 0$ in both parental strains. In the F_1 generation, 14 individuals (70%) would be susceptible and $V_p = 0.233$. In the F_2 generation, 11 individuals (55%) are susceptible and $V_p = 0.261$. Thus, $V_g = 0.261 - 0.233 = 0.027$, $h^2 = 0.10$. The broad sense heritability is much smaller than previous estimates with additivity and dominance because V_g accounts for a small part of V_p when lumping individuals with susceptible scores between 0 and 49.9 into a single "0" category and individuals with susceptible scores between 50 and 100 into the "1" category. When plotting the mean phenotype against the genotype at each locus (Fig. 30.5E) or for both loci (Fig. 30.5F), we fail to see a straight line. These patterns show that we detect only a small proportion of the genetic variance when measuring the outcome, rather than the continuous underlying susceptibility score.

V_g can be further subdivided into three components: V_a, V_d, and V_i. V_a and V_d, are the portion of V_g attributable to additive and dominant genetic effects, respectively. V_i is the variance due to *epistasis* or *interlocus interactions* and can be a significant portion of V_g if the trait of interest is affected by multiple loci. V_e can be further subdivided into $V_{e(g)}$ and $V_{e(l)}$, the portion of V_e attributable to *general* and *local environmental effects*, respectively. For example, assume we are analyzing the genetic and environmental components of vector competence among three strains of a container-breeding mosquito. Assume that vector competence is strongly correlated with adult size. We set up four replicate larval-breeding containers for each strain, with equal amounts of eggs, water, and food in each container. We place all of the containers from strain 1 on a top shelf in the insectary, strain 2 containers on a lower shelf, and strain 3 containers on the bottom shelf. In doing this we assume that all containers experience equal temperature, light, and humidity. However, in reality, upper shelves experience higher temperatures and receive more light than lower shelves and containers on the left side of a shelf are exposed to a lower humidity than those on the right side. $V_{e(g)}$ would be the portion of V_p that would be attributable to differences in temperature, light, and humidity among all containers. In contrast, $V_{e(l)}$ would be the portion of V_e attributable to local environmental factors that cause V_p among individuals within one container to be less than V_p among all containers. This might occur if the eggs we placed in each "replicate" container came from different parents. V_p in this case would arise because eggs in one container are more genetically homogeneous than the eggs among containers. Similarly, if larvae in one container experience greater sources of mortality than larvae in another container, then this might cause larval density to shift among containers. In this scenario, some containers would be expected to produce larger adults than others. These examples of V_e are for illustrative purposes only. Inflation or overestimation of V_e arose from bad assumptions by the investigator or from poor experimental design. In reality, we would randomize container placement using a completely randomized, block, or Latin squares design. We would also mix the sources of eggs or first-instar larvae to avoid placing siblings in a common container.

In total, V_p can then be partitioned into five components (Falconer and MacKay 1996):

$$V_p = V_g + V_e = V_a + V_d + V_i + V_{e(g)} + V_{e(l)} \qquad (3)$$

These components are most frequently measured using an experimental design referred to as a *half-sibling analysis*. This analysis involves mating a father to several mothers. This process is repeated with several different fathers. The phenotype of all F_1 individuals is measured in each full-sibling family. One of the advantages of this approach is that parents can be taken directly from a field population rather than from strains selected for opposite phenotypes. Variance components are estimated using the *analysis of variance (ANOVA)* design shown in Table 30.1.

TABLE 30.1 Determination of Sire, Dam, and Error Variance Components in Half- and Full-Sib Families

Source	d.f.	Mean Square	Mean Square Components
Among fathers	$s-1$	MS_{Sire}	$\sigma^2_{Error} + k\sigma^2_{Dam}\,dk\sigma^2_{Sire}$
Among mothers mated to one father	$s(d-1)$	MS_{Dam}	$\sigma^2_{Error} + k\sigma^2_{Dam}$
Among progeny of a single mother	$sd(k-1)$	MS_{Error}	σ^2_{Error}

s = number of fathers, d = number of mothers/father, k = number of offspring/mother.

The variation in phenotype among full siblings estimates σ^2_{Error}. The variation in mean phenotype among the offspring of mothers mated to one father estimates $\sigma^2_{Error} + k\sigma^2_{Dam}$. Thus, σ^2_{Dam} can be estimated by subtracting σ^2_{Error} and dividing by number of offspring measured in each full-sibling family. The variation in mean phenotype among the offspring of one father estimates

$$\sigma^2_{Error} + k\sigma^2_{Dam}\,dk\sigma^2_{Sire}$$

Thus, σ^2_{Sire} is estimated by subtracting $\sigma^2_{Error} + k\sigma^2_{Dam}$ and dividing by the product of number of mothers mated to a father and the offspring measured in each full-sibling family. It can be shown that

$$V_a = 4\sigma^2_{Sire}$$

$$V_d = \frac{4}{3}\sigma^2_{Error}\,\tilde{}\,\sigma^2_{Dam}$$

$$V_{e(g)} = \sigma^2_{Dam} - \sigma^2_{Sire}$$

$$V_{e(l)} = \sigma^2_{Erro} - \sigma\sigma^2_{Sire}$$

These variance components lead to a different estimate of heritability, referred to as *narrow-sense heritability* (also designated h^2). This measures the proportion of V_p that is accounted for by V_a. This is considered a much more useful statistic than broad-sense heritability because V_a indicates that amount of V_p that is due to genetic factors that can be selected. Remember that dominant alleles mask or hide the presence of recessive alleles in heterozygous individuals. When we select an individual with a dominant phenotype, we don't know if we are selecting a homozygous or heterozygous dominant genotype.

Bosio et al. (2002) used a half-sib design to measure the amount ($TCID_{50}$) of dengue 2 virus in the midguts (MIB) and heads (MEB) of *Aedes aegypti aegypti* from Puerto Rico. In the midgut, $V_p = 6.70$, $V_a = 2.74$, $V_d = 0.00$, $v_{e(l)} = 3.96$, and $V_{e(g)} = 0.00$. The narrow- and the broad-sense $h^2 = 0.41$. In the mosquito head, $V_p = 4.23$, $V_a = 0.58$, $V_d = 2.28$, $V_{e(l)} = 0.80$, and $V_{e(g)} = 0.57$. The narrow-sense $h^2 = 0.14$, and the broad-sense $h^2 = 0.68$. From this information we can conclude that V_p in MIB is due to genetic factors and that alleles and loci are additive in their effects on MIB. In contrast, note the large difference between the narrow- and broad-sense h^2 for MEB. This is due to the large V_d component. We again conclude that V_p in MEB is due to genetic factors but that alleles and/or loci are dominant in their effects on MEB.

It is important to understand that heritability estimates reflect genetic variation underlying phenotypic distributions in the family or population under analysis. It can be shown (Falconer and MacKay 1996) that the broad-sense heritability

$$h^2 = \frac{2pq[a + (q-p)d]^2}{\sigma^2} \quad (4)$$

where p is the frequency of an allele x affecting a phenotype and q are the cumulative frequencies of alternative alleles. σ^2 is V_p, and a and d are, respectively, the additive and dominance affects of allele x. Thus, both broad- and narrow-sense h^2 are strongly influenced by p and q. If $p = 1$ or $p = 0$ in populations, then it is likely that $h^2 \cong 0$. Thus, one cannot generalize estimates of broad- and narrow-sense h^2 across populations or families. Heritability estimates must be determined independently for each population being studied.

GENETIC MAPS

Having (1) determined that our vector competence phenotype is under genetic control, (2) quantified the portion of V_p due to V_g versus V_e, and (3) identified how alleles at vector competence loci are inherited, we are now in a position to map the locus or loci that control vector competence. This means that we wish to locate loci that control vector competence relative to loci of a known location in the vector genome. The locations of these genes are known either through a *linkage map* or through a *physical map*. A linkage map indicates the frequency with which markers along a chromosome recombine during meiosis. A locus has a position in a linkage map relative to the frequency with which it recombines with other loci during meiosis. In contrast, a physical map represents, at varying levels of resolution, the nucleotide sequence of an entire genome. A locus has a precise location in a physical map, defined by its position in the linear arrangement of nucleotides along a chromosome.

At one time, linkage maps existed for only a few vectors. In contrast, during the past decade, linkage maps have become available for a large variety of vector species. This is due entirely to a revolution in molecular genetics that dramatically increased the numbers and types of genetic markers that could be analyzed in vector genomes. This led to the evolution of *intensive*, or *saturated*, linkage maps.

Physical maps for vectors have historically been based on low-resolution cytogenetic approaches. The most useful physical mapping done to date has been accomplished with anopheline mosquitoes, wherein banding patterns displayed in polytene chromosomes have been used to identify chromosome

rearrangements and also to determine the chromosome location of cloned DNA fragments by *in situ* hybridization. With culicine mosquitoes, polytene chromosomes have not proven useful for physical mapping. Physical mapping in these mosquitoes has depended on differential staining of metaphase chromosomes. The resolution obtained with these procedures is at best several kilobases and more often several megabases. Because of contemporary technical advances, high-resolution physical mapping has advanced to the level that the nucleotide sequence of even large genome organisms can rapidly and accurately be determined.

INTENSIVE LINKAGE MAPPING

The units of a linkage map are centimorgans (cM), a unit of recombination frequency between alleles at a pair of loci within a single round of meiosis. Linkage maps have most frequently been generated using F_1 intercross designs (Fig. 30.3) or recurrent backcrosses. Suppose that we examine 100 gametes and that there are 75 gametes with parental genotypes and 25 gametes with recombinant genotypes (Fig. 30.3). The linkage distance between loci A and $B = 25/100 = 25\,\mathrm{cM}$. Alternatively, if there were equal numbers of gametes with parental and recombinant genotypes, the linkage distance between loci A and B would be $50/100 = 50\,\mathrm{cM}$. Loci A and B are then said to be *unlinked*. An estimate of $50\,\mathrm{cM}$ is the upper limit of recombination distance between two loci because there cannot be more gametes with recombinant genotypes than gametes with parental genotypes. However, linkage maps are frequently much greater than $50\,\mathrm{cM}$ due primarily to the physical size of individual chromosomes (e.g., Fig. 30.6). Loci physically located near opposite ends of a chromosome can be shown to be "linked" based on pairwise centimorgan estimates among intervening loci. This is because *map distances are roughly additive along a chromosome*.

Two- and three-point test crosses were for decades the only means to determine gene order along a chromosome. Test crosses were used to determine the relative order of genes that controlled phenotypes such as eye and body color, bristle number, and other morphological characters. However, *during the last 20 years a new paradigm has emerged in genetic linkage mapping, where hundreds or thousands of markers could be mapped simultaneously in a single cross*. This revolution was driven in large part by the development of genetic markers that are abundant in eukaryotic genomes and the discovery that the majority of these markers are variable among individual organisms. *Thus, starting from any pair of P_1 parents, a geneticist could reliably expect*

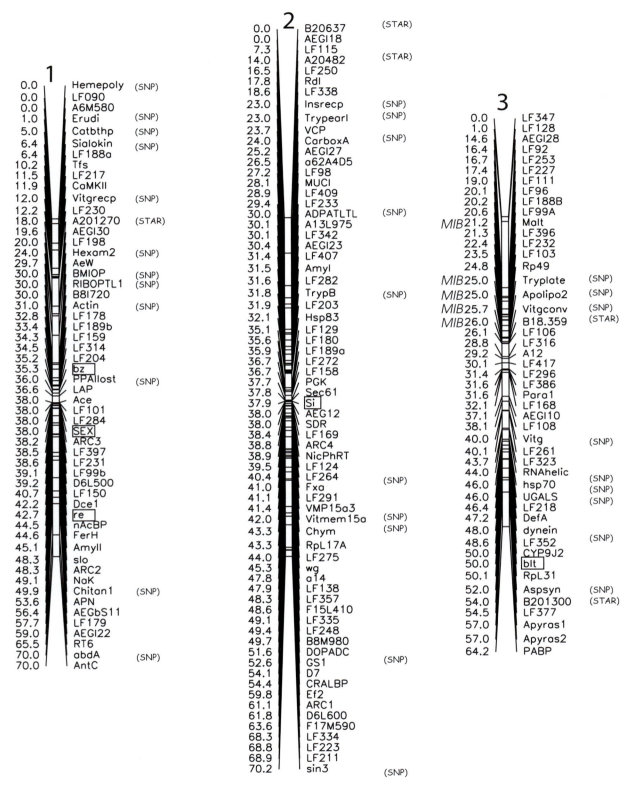

FIGURE 30.6 Current linkage map of *Aedes aegypti*. There are 55, 69, and 48 markers, respectively, on chromosomes 1, 2, and 3. "SNP" appears by loci mapped using SNP variation. STAR loci are indicated. Morphological markers appear in boxes. The remaining markers were mapped using RFLP variation (Severson et al. 2002). Markers linked to a QTL illustrated in Table 30.2 and Figure 30.21 are labeled "MIB."

alleles to be segregating at most loci distributed throughout the genome.

Genetic Marker Technology

Restriction Fragment Length Polymorphisms (RFLPs)

Beginning in the 1980s, recombinant DNA technology and Southern analysis provided a way of detecting a much greater diversity of genetic polymorphisms in genes than was possible with morphological or biochemical markers. The first DNA-based, mosquito genetic map was constructed for *Ae. aegypti* (Severson et al. 1993). This map was developed using molecular techniques that identify specific regions of the genome using *restriction fragment length polymorphisms* (RFLPs) identified using Southern blotting and radiolabeled probe hybridization. Briefly, genomic DNA from individuals from segregating populations is cleaved with DNA sequence-specific restriction endonucleases, size fractionated with agarose gel electrophoresis, and transferred to nylon membranes. The membranes are hybridized with labeled, low-copy-number genomic DNA or cDNA clones. RFLPs specific to each individual are identified by autoradiography (Fig. 30.7). At present, the *Aedes aegypti* linkage map (Fig. 30.6) consists of 145 RFLP loci that cover 205 cM and also includes six morphological marker loci (Severson et al. 2002). A limitation to these markers is that the Southern blotting process allows analysis of a maximum of 25–30 markers per individual mosquito because of the limits on DNA quantities. However, these numbers are more than adequate to perform a standard 10-cM genome scan for quantitative trait analysis (see Quantitative Trait Locus Mapping later). In addition, because most of these markers are cDNA clones, they are extremely useful for comparative genome analysis among culicine mosquito species (Chapter 31).

The Polymerase Chain Reaction

The polymerase chain reaction (PCR) was invented by Kerri Mullis in 1985 as a tool to quickly amplify a gene sequence represented once or only a few times in a large and complex mixture of genes. Starting with a sample of the total DNA from an organism, within 2–3 hours a particular region of the genome can be amplified over 1–100 million times.

In the 18 years since its invention, PCR has changed the way that almost all molecular genetic determinations are done. Prior to PCR, cloning a gene from an organism required construction of a genomic library, screening the library with a homologous probe, isolating clones that cross-hybridized with the probe,

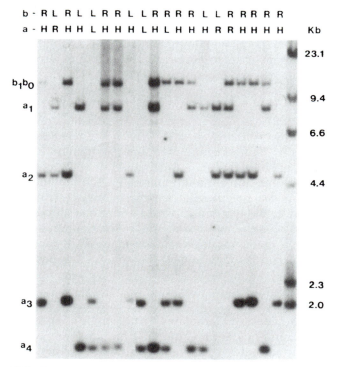

FIGURE 30.7 Autoradiograph illustrating RFLP alleles segregating at loci LF99a (20.6 cM — chromosome 3 — Fig. 30.6) and LF99b (39.1 cM — chromosome 1 — Fig. 30.6) in an F_1 intercross between the Liverpool and red-eye strains (Severson et al. 1993). Markers at LF99a are labeled a_1–a_4, while alleles at LF99b are labeled b_1b_0. Genotypes at the LF99a locus are labeled at the top of the gel as "R" for red-eye P_1 genotype, "L" for Liverpool P_1 genotype, or "H" for heterozygote (note that there are two heterozygote patterns). Genotypes at the LF99b locus are labeled at the top of the gel as "R" for red-eye P_1 genotype or "L" for Liverpool P_1 genotype. Note that this is the same marker that is physically mapped in Figure 30.25.

purifying these clones, and then analyzing them. This procedure required months. In contrast, PCR directly amplifies a sequence in a few hours using oligonucleotide primers that anneal to conserved regions that flank a target gene sequence. The amplified gene is visualized on an agarose gel and can then be cloned or analyzed with restriction enzyme digestion or even sequenced directly. Genes that were previously analyzed using Southern blots and hybridization with gene-specific probes are now analyzed directly following PCR amplification. Modifications of the original PCR procedures are now found in all fields of genetics, including genome mapping, population genetics, gene expression, molecular taxonomy, and systematics as well as genetic screening, forensics, and disease diagnosis. The power of PCR in mapping is that the technique can amplify a DNA sequence from a minute amount of template DNA. The amplified DNA can then be analyzed for nucleotide variation using methods we discuss later. This permits the vector geneticists to perform hundreds of determinations on a single organism.

PCR reactions are generally done in small (500-μL) microcentrifuge tubes. From 25 to 100 μL of a PCR reaction mixture is placed in the tube. The reaction mixture generally contains 10 mM Tris-HCl (pH 8.5–9.0), 50.0 mM KCl, 1.5 mM MgCl$_2$, 1% Triton X-100, 0.01% gelatin (w/v), 200 μM dNTPs, 1 μM of each primer, and 1 unit of a thermostable DNA polymerase such as *Taq* polymerase. Template DNA (1–10 ng) is placed in the reaction mixture, overlaid with oil and placed on a thermal cycler. The thermal cycler is generally a metal block that holds many PCR reaction tubes. The temperature of the block can be programmed to follow a precise cycle of heating and cooling.

The first step of the PCR process involves melting the double-strand template DNA to single strands at a temperature of 95–98°C for 0.5–1 minute (Fig. 30.8A). For the second step, the temperature of the reaction is lowered to a point at which a pair of flanking oligonucleotide primers can anneal to the template DNA (Fig. 30.8B). The flanking primers are designed so as to anneal with high specificity to conserved sequences that flank a gene. If the annealing temperature is too high, the primers will not anneal; if the temperature is too low, the primers will anneal nonspecifically to various regions of the genome and produce several false products. The annealing step is allowed to proceed for 0.5–1 minute. The third and final step involves raising the temperature to 72°C (Fig. 30.8C). This is the optimal temperature for the thermostable DNA polymerase and permits rapid extension of the primer to build a DNA strand that is complementary to the template strand to which the primer annealed. The extension step is allowed to proceed for 2.5–3 minutes. At the end of the extension step, the number of double-stranded DNA molecules is doubled. The thermal cycler then returns to 95–98°C to melt the newly synthesized strands, and the temperature is lowered to allow primer annealing and than raised to promote extension. This cycle is repeated 20–30 times. Because the number of DNA molecules is doubled at each cycle there will be 2^{20}–2^{30} (10^6–10^9) molecules (Fig. 30.8D).

Random Amplified Polymorphic DNA (RAPD) PCR and Arbitrarily Primed (AP) PCR

Williams et al. (1990) described RAPD-PCR, a process that uses a 10-oligonucleotide primer of random sequence but with a minimum guanine–cytosine content of 60%. Welsh and McClelland (1990) described an analogous technique, AP-PCR, that employs a longer primer but with no known similarity to sequences in the template DNA (e.g., the M13 universal sequencing primer). With both techniques, a relatively low annealing temperature (37–40°C) is used. Low temperatures allow the primer to anneal to arbitrary regions of the genome, which may not be fully complementary to the primer (Fig. 30.9, Step 1). A low annealing temperature is maintained throughout all cycles in RAPD-PCR but is used in only the initial cycles with AP-PCR. In order for amplification to occur, the primer must anneal on complementary strands of the template DNA and the 3′ ends of the annealed primers must face each other (Fig. 30.9, Step 2). Furthermore, the primer annealing sites must be separated at a distance of no greater than 3,000 bp because this is the maximum size that can be amplified with routine PCR. In practice, amplified fragments are rarely greater than 2,000 bp. The requirement that the 3′ ends of the annealed primers face one another suggests that annealing sites are exact or similar inverted repeats. The observation that single substitutions, especially in the 3′ end of the primer, can change amplified banding patterns implies that annealing in RAPD-PCR must be precise. The products of RAPD-PCR from an individual's genome are typically a series of fragments that vary in intensity and in size from 200 to 2,000 bp. Polymorphisms occur as the presence or absence of a specific fragment among individuals (Fig. 30.9—agarose gel). Absence of a fragment presumably occurs because amplification cannot proceed on DNA strands from either of the homologous chromosomes in an individual. This can occur through point mutations at one or both primer annealing sites on a DNA strand, inversions surrounding a site, or insertions that separate the annealing sites at a greater distance than can be amplified (bottom of Fig. 30.9). Arbitrarily primed PCR is not quantitative, and consequently it is unknown whether individuals whose DNA yields a specific fragment are heterozygous (have one copy) or homozygous (two copies) for an amplifiable allele. Alleles at arbitrarily primed loci therefore segregate as dominant markers, although SSCP analysis (see later) can often reveal codominant polymorphisms in dominant "band present" alleles (Antolin et al. 1996).

A problem that we have encountered when using RAPD markers in mapping is that, presumably because of high substitution rates, bands of equivalent mobility cannot be amplified in all families and loci mapped in earlier studies are not always polymorphic in later studies. We have modified RAPD markers to overcome this problem by developing

A Melt native template DNA to single strands (98°C)

B Primers anneal to flanking regions (50 - 60°C)

5'-AACGCTACACACCGT-3'
TTGCGATGTGTGGCA

ATGCGGTGTGAGACG
3'-TACGCCACACTCTGC-5'

C Primers extend to build the complementary strand (70°C)

5'-AACGCTACACACCGT-3'——→
3'-TTGCGATGTGTGGCA

ATGCGGTGTGAGACG
←——— 3'-TACGCCACACTCTGC-5'

AACGCTACACACCGT
TTGCGATGTGTGGCA

ATGCGGTGTGAGACG
TACGCCACACTCTGC

AACGCTACACACCGT
TTGCGATGTGTGGCA

ATGCGGTGTGAGACG
TACGCCACACTCTGC

D 20 - 30 cycles

E The numbers of complete gene copies doubles every cycle

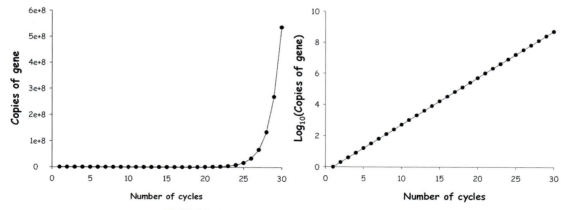

FIGURE 30.8 Steps of the polymerase chain reaction (PCR).

STEP 1. PRIMER ANNEALING

CHROMOSOMES (2N = 6)

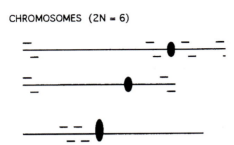

STEP 2: PRIMER EXTENSION - TAQ POLYMERASE

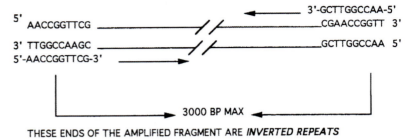

THESE ENDS OF THE AMPLIFIED FRAGMENT ARE *INVERTED REPEATS*

INVERTED REPEAT DNA SEQUENCES ARE FOUND IN:
A) HETEROCHROMATIC REGIONS
    TELOMERES
    CENTROMERES
B) THE ENDS OF TRANSPOSABLE ELEMENTS

AMPLIFIED REGIONS ARE NOT RANDOMLY LOCATED

MUTATIONS CAUSING LOSS OF RAPD BAND PHENOTYPE

1. POINT MUTATION(S)

PRIMER CANNOT ANNEAL

AACCGGTTCG _____ // ____ 3'-GCTTGGCCAA-5'
                                  *TT*AACCGGTT 3'
TTGGCCAAGC _____ // ____ *AA*TTGGCCAA 5'

MUTATION AT ONE OF THE PRIMER ANNEALLING SITES

NO AMPLIFICATION ON ONE STRAND = segregates as a recessive allele

2. INVERSION OF EITHER PRIMER ANNEALING SITE

PRIMERS NO LONGER FACE ONE ANOTHER
NO AMPLIFICATION ON ONE STRAND = segregates as a recessive allele

3. INSERTION OF LARGE (2-3 KB) FRAGMENT

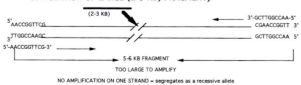

5-6 KB FRAGMENT
TOO LARGE TO AMPLIFY

NO AMPLIFICATION ON ONE STRAND = segregates as a recessive allele

FIGURE 30.9 Steps of the random amplified polymorphic DNA PCR. Examples of RAPD-PCR products fractionated on an agarose gel are shown. The three classes of mutations leading to gain or loss of a band are illustrated at the bottom of the diagram.

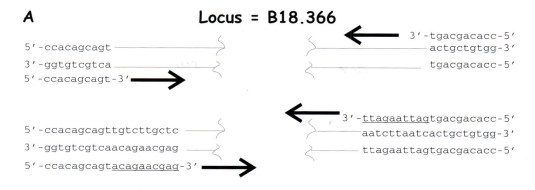

A   Locus = B18.366

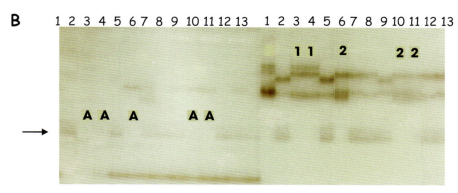

FIGURE 30.10 (A) Conversion of an RAPD band to a sequence-tagged RAPD (STAR) locus. The RAPD band indicated with an arrow in (B) was scraped from the gel, reamplified, cloned, and sequenced; and primers were built that had the original 10 nucleotides of the RAPD and the proximal ten 3' nucleotides obtained from the sequence. RAPD products from 13 *Aedes triseriatus* mosquitoes from an F_1 intercross are labeled A (absence) in the left half of the gel. STAR products from the same 13 mosquitoes are labeled 1 or 2 on the right half of the gel. Genotypes at the RAPD locus could only be scored as to presence or absence of a band, whereas four distinct genotypes are present at the STAR locus.

sequence-tagged RAPDs (STARs) (Fig. 30.10). STAR markers are generated by flooding a polymorphic RAPD band that had been resolved on an SSCP gel with 20 µl TE. After 30 sec, the gel and TE are scraped from the glass plate into a sterile microcentrifuge tube containing 80 µl TE. This is briefly vortexed and centrifuged at 17,000 × *g* for 10 sec, and 2 µl is used as template DNA in a RAPD-PCR reaction with the original 10-oligonucleotide primer. The amplified product is analyzed with agarose gel electrophoresis to confirm that it is approximately the same size as the scraped fragment. If so, the PCR product is purified, cloned, and sequenced. Forward and reverse STAR primers are designed to contain the original 10-oligonucleotide primer and the next 10 nucleotides of the sequence (Fig. 30.10A). We have found that STAR loci can consistently be amplified among all families and that the alleles frequently segregate as codominant rather than

dominant markers (Fig. 30.10B). We have mapped many STAR loci in *Aedes aegypti* (Fig. 30.6).

Amplified Fragment Length Polymorphisms (AFLPs)

Vos et al. (1995) developed AFLP-PCR as an alternative method to amplify arbitrary regions of genomes. It is a technique based on the selective amplification of a subset of genomic restriction fragments using PCR (Fig. 30.11). DNA is digested with restriction endonucleases, and double-stranded DNA adapters are ligated to the ends of the DNA fragments to generate template DNA for PCR amplification. The sequence of the ligated adapters and the adjacent restriction site serve as primer binding sites for subsequent amplification of the restriction fragments by PCR. Selective nucleotides extending into the restriction fragments are added to the 3' ends of the PCR

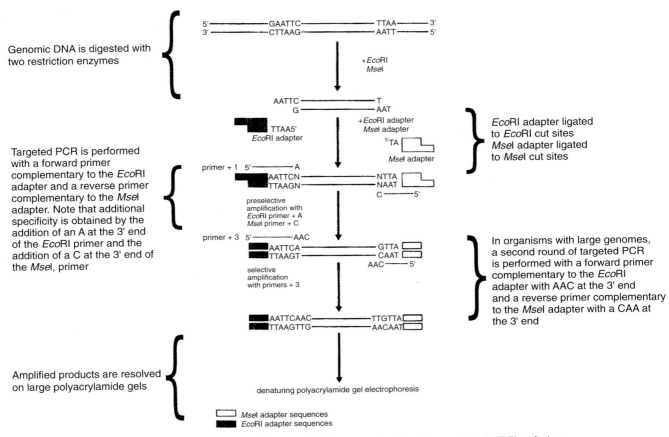

Genomic DNA is digested with two restriction enzymes

Targeted PCR is performed with a forward primer complementary to the EcoRI adapter and a reverse primer complementary to the MseI adapter. Note that additional specificity is obtained by the addition of an A at the 3' end of the EcoRI primer and the addition of a C at the 3' end of the MseI, primer

Amplified products are resolved on large polyacrylamide gels

EcoRI adapter ligated to EcoRI cut sites
MseI adapter ligated to MseI cut sites

In organisms with large genomes, a second round of targeted PCR is performed with a forward primer complementary to the EcoRI adapter with AAC at the 3' end and a reverse primer complementary to the MseI adapter with a CAA at the 3' end

denaturing polyacrylamide gel electrophoresis

MseI adapter sequences
EcoRI adapter sequences

FIGURE 30.11 Illustration of the amplified fragment length polymorphism (AFLP) technique.

primers such that only a subset of the restriction fragments is recognized. Only restriction fragments in which the nucleotides flanking the restriction site match the selective nucleotides will be amplified. The subset of amplified fragments is then analyzed via denaturing polyacrylamide gel electrophoresis.

As with RAPD- and AP-PCR, polymorphisms identified using AFLP are typically inherited as dominant markers in a Mendelian fashion. However, they are reported to be more reproducible than RAPD or AP polymorphisms. The technique was developed primarily as an alternative to RAPD markers for intensive genome mapping in plants. Plant genomes can have large amounts of repetitive DNA, and geneticists observed that too many bands were amplified using RAPD- or AP-PCR. AFLP is highly specific and reproducible for the DNA sequences that it amplifies. It has had limited use in vector linkage mapping, probably because AFLPs are more expensive than RAPD- or AP-PCR, codominant marker systems are readily available and vector genomes tend not to be overabundant in repetitive DNA.

Microsatellite Markers

A large proportion of the genomes of eukaryotes and some prokaryotes consists of repetitive DNA (Chapter 31). When total genomic DNA is isolated from an organism and differentially sedimented in an ultracentrifuge on a cesium chloride (CsCl) gradient, a variety of lighter and heavier bands appear as "satellites" around the main band of genomic DNA. When satellite bands were extracted from a CsCl gradient and characterized by cloning and sequence analysis, it was discovered that satellite DNA consists of a complex mixture of repetitive elements. Some of these repetitive elements are very large (>10 kb), many are of a moderate size (100–800 bp), and a distinct fraction consisted of small repeats 20–30 bp in length; still others consisted of only di-, tri-, or tetranucleotide repeats. It was subsequently determined that these short repeats are arranged in tandem, sequential arrays that are highly dispersed throughout a genome. Collectively, these di-, tri-, or tetranucleotide repeat elements are referred to as *microsatellites*. Alleles at microsatellite loci tend to be highly variable *in length* among individuals in a population and are generally classified according to the size of the repeat unit, the number of repeat units per array, and the genomic location of the tandem arrays.

Length polymorphisms at microsatellite loci appear to arise through exceptionally high rates of a mutational process called *slipped-strand mispairing* (SSM) or *slippage replication* (Fig. 30.12). Alleles at microsatellite loci tend to segregate as codominant markers because the investigator can detect the presence of alleles of different lengths from each parent (Fig. 30.12—gel). This feature and the high rate of length polymorphisms make microsatellites ideal markers for linkage mapping. An intensive linkage map of *An. gambiae* has been constructed from microsatellite loci (Zheng 1999; Wang et al. 1999). Similar maps of culicine mosquitoes have not been constructed due to an overall paucity of microsatellites in these genomes (Fagerberg et al. 2001). However, the TAG loci shown in Figure 30.12 are examples of one type of microsatellite locus found in *Ae. aegypti*.

Single-Nucleotide Polymorphisms (SNPs)

Because SNPs are highly abundant and can and are being adapted for high-throughput analysis, it seems clear that they will become the genetic markers of choice (Landegren et al. 1998; Brookes 1999). A number of techniques are available that vary in their sensitivity, technical difficulty, and cost.

In the late 1980s, single-strand conformation polymorphism (SSCP) analysis provided a means to detect 95–99% of all nucleotide substitutions in a gene (Fig. 30.13A) (Orita et al. 1989). Following PCR, the double-stranded product is melted to single strands at 95°C for 5 min and then placed immediately into an ice bath (0–4°C) so that stable single-strand duplexes are formed from intrastrand base pairing. The conformation of the single-strand duplexes is extremely sensitive to the primary DNA sequence. Orita et al. (1989) reported that up to 95% of all point mutations in a gene affect its single-strand conformation. Variation in the conformation of the intrastrand duplexes is visualized in a nondenaturing gel retardation assay. Following PCR and the heating and rapid cooling process, the chilled products are loaded onto a 5% polyacrylamide 1X TBE gel containing 5% glycerol and fractionated at room temperature for 16 hours, using low amperage to ensure no increases in temperature that disrupt the intrastrand pairing conformation. The gel is then removed, and the products are visualized using silver staining. As an example, the segregation of SSCP genotypes at the *blood meal–induced ovarian protein (BMIOP)* locus in an F_1 intercross of *Ae. aegypti* is shown in Figure 30.13B.

As an illustration of the utility of SSCP, Fulton et al. (2001) designed primers to amplify ~500-bp fragments from 94 *Ae. aegypti* cDNAs in GenBank. These primer pairs amplified 94 loci, 57 (61%) of which segregated in a single F_1 intercross family among 83 F_2 progeny. This produced a dense linkage map of one marker

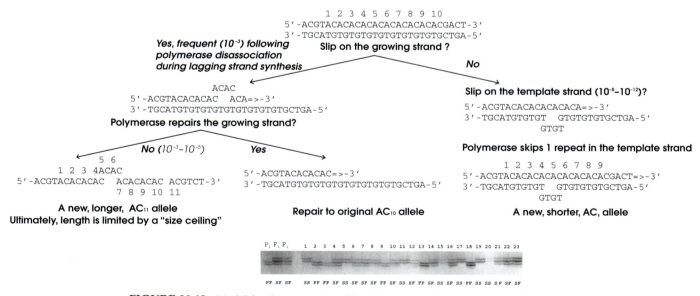

FIGURE 30.12 Model for the generation of length polymorphisms at microsatellite loci.

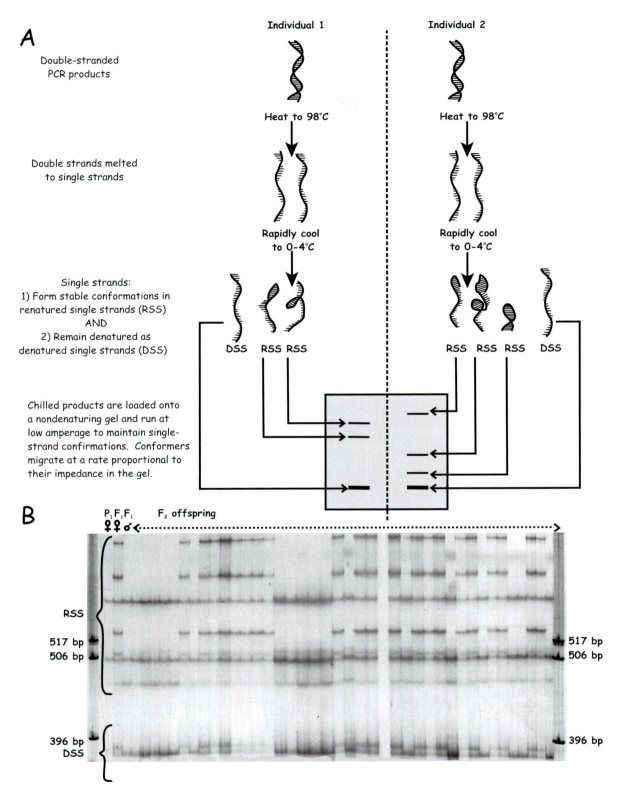

FIGURE 30.13 (A) Illustration of the single-strand conformation polymorphism (SSCP) technique. (B) Segregation of SSCP genotypes at the *BMIOP* (*blood meal–induced ovarian protein*) locus in an F$_1$ intercross family.

every 2 cM distributed over a total length of 134 cM. These results suggest that SNPs are abundant within mosquito genomes and that SSCP-based markers provide an opportunity to examine large numbers of loci within individuals. A drawback, however, is that this technique is labor intensive and requires the use of sequencing-size slab gels and the manual scoring of genotypes.

Another SNP genotyping method called either melting-curve SNP (McSNP—Fig. 30.14) or dynamic allele-specific hybridization (DASH) has been adapted for use with *Ae. aegypti*. Briefly, the technique combines the classic approach for discriminating alleles, restriction fragment digestion, with DNA fragment detection by melting-curve analysis (Akey et al. 2001). A gel-free system (Hybaid DASH system) performs melting-temperature calculations for PCR products directly in 96-well microplates with ~15-minute total processing time (Fig. 30.14). Thereafter, samples are automatically scored relative to their genotype at the target SNP. McSNP assays have been developed for 34 *Ae. aegypti* sequences.

Of interest, an SNP associated within an existing restriction site is often found, so the McSNP assay can be performed directly. However, even in instances where the SNP is not within a restriction site, a usable

site can be created by employing one of the PCR primers to introduce a novel base that converts one of the SNP allele regions into a restriction site (Fig. 30.15). McSNP works best if the difference in melting temperature between the undigested and digested alleles is 5°C or higher. Other methods of SNP detection are discussed in Chapter 18.

Statistical Aspects of Intensive Linkage Mapping

Modern statistical algorithms have been developed that analyze recombination frequencies simultaneously among many loci. They are designed to perform multipoint linkage comparisons using all possible two-point frequencies to identify the most likely linear orders for each linkage group. Provided that a sufficient number of marker loci with completely random genomic distributions are examined, the number of linkage groups will be equal to the haploid chromosome number. These algorithms require accurate information on the segregation of alleles in an F_1 intercross, a *recombinant inbred line*, or a *backcross to a recurrent parent*. That is, we must have the correct genotype information for each of the P_1 and F_1 parents. However, given the difficulties in mating behaviors associated with some species, the investigator may only know the genotype of one P_1 parent (usually the female); the genotype of the other P_1 parent is inferred from the F_1 parents or from the observed genotype frequencies in the F_2 progeny. Note also that with DNA-based markers, up to four distinct alleles can be segregating at each locus in a given cross. That is, the P_1 parents can both be heterozygous for unique alleles (for example, see Fig. 30.7).

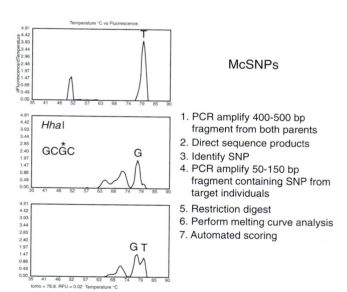

1. PCR amplify 400-500 bp fragment from both parents
2. Direct sequence products
3. Identify SNP
4. PCR amplify 50-150 bp fragment containing SNP from target individuals
5. Restriction digest
6. Perform melting curve analysis
7. Automated scoring

FIGURE 30.14 Melting-curve detection of single-nucleotide polymorphisms (McSNP). DNA fragment detection by melting-curve analysis (Akey et al. 2001) is also known as dynamic allele specific hybridization (DASH). A gel-free system performs melting-temperature calculations for PCR products directly in 96-well microplates. Thereafter, samples are automatically scored relative to their genotype at the target SNP. The melting-temperature profiles of a T/T homozygote, a G/G homozygote, and a G/T heterozygote are shown from top to bottom.

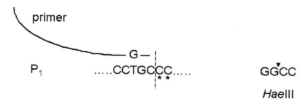

FIGURE 30.15 An SNP associated within an existing restriction site is often found, so the McSNP assay can be performed directly. However, even in instances where the SNP is not within a restriction site, a usable site can be created by employing one of the PCR primers to introduce a novel base that converts one of the SNP allele regions into a restriction site.

Loci for which the P_1 parents do not share any common alleles are considered to be *completely informative*, because we can unambiguously assign the correct lineage in their backcross or intercross progeny. An additional caveat of DNA-based markers is that for some loci the P_1 parents can share a common allele. Such loci are considered to be *partially informative*, because we can only unambiguously assign genotypes as parental or recombinant for alleles that are not shared (Fig. 30.16). Note that recombination frequencies can still be calculated using partially informative marker loci, but these estimates have large standard errors associated with them. This limits the accuracy with which correct linear orders can be determined for such markers. Multipoint linkage comparisons that include partially informative marker data should be viewed with caution and used only as a mechanism for preliminary linkage map development.

A recurrent backcross entails crossing F_1 daughters to their father. The reciprocal backcross, between a son and mother, is impossible in most insects since females retain sperm in their spermathecae. Recurrent backcrosses are difficult for many vectors because it requires that the father be kept alive and viable until his daughters are reproductive (20–30 days in the case of mosquitoes).

An F_1 intercross has already been described in Figure 30.3. However, analysis of segregation in such a cross is not limited to F_2 offspring. Additional meioses allow for additional recombination among closely linked markers, which in turn provide for fine-scale and high-resolution estimates of gene order among markers. When an F_1 intercross family is continued beyond the F_2 generation by intercrossing the F_2 offspring, that family is known as an *advanced intercross line* (AIL) (Darvasi and Soller 1995). *Aedes aegypti* AILs have been initiated from single-pair mating between *P. gallinaceum*–susceptible and *P. gallinaceum*–refractory parents, from *B. malayi*–susceptible and *B. malayi*–refractory mosquitoes, and from MEB+ and MEB– strains. All F_1 progeny from these crosses were then intermated enmasse and allowed to oviposit. For each generation thereafter, a large number of individuals ($>10^3$) were reared and allowed to mass mate and oviposit. This ensures a large effective population number contributing to the AIL. This strategy was continued through the F_8 generation.

Figure 30.17 demonstrates the effectiveness of the AIL strategy in increasing the observed recombination frequencies. Recombination frequencies between individuals at the F_2 and F_8 generations were compared

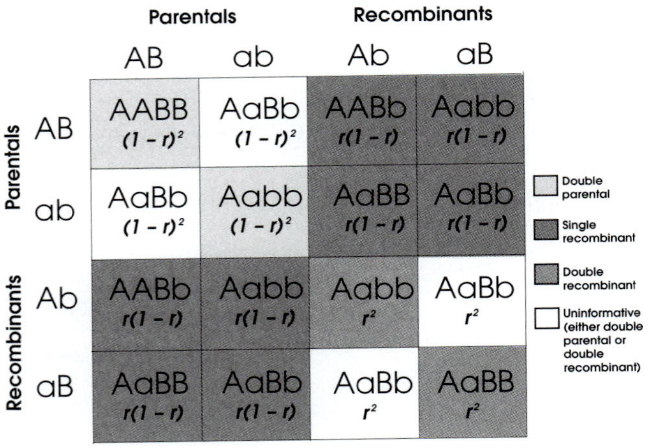

FIGURE 30.16   Example of a partially informative cross in an F_1 intercross family. Inheritance of alleles at the A and B loci are as shown in Figures 30.3 and 30.4. Notice that an F_2 individual with genotype *AaBb* could arise from either two parental games or two recombinant gametes. Thus, *AaBb* individuals must be excluded from the estimation of recombination rates.

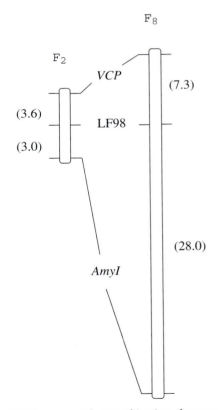

FIGURE 30.17   Increased recombination frequencies among markers between loci *VCP–LF98–Amy1* (27.2–31.5 cM, Chromosome 2) arising from six rounds of meiosis between F_2 and F_8 generations in an advanced intercross line (AIL). Notice that the centimorgan distance increase from 6.6 cM estimated in the F_2 generation to 35.3 cM estimated in the F_8 generation to.

across a genome region known to contain a QTL determining *P. gallinaceum* susceptibility (Severson et al. 1995). Notice the considerable map expansion that occurred with the F_8 population, although the actual level of expansion varies nonrandomly. This phenomenon is consistent across the entire genome. This apparently is observed with other organisms, yet, most importantly, AIL provides a several-fold increase in statistical resolution for interpreting complex traits (Darvasi and Soller 1995).

Genetic linkage mapping entails an additional conceptual issue for the interpretation of reported map distances. That is, more than one crossover event can occur within individual chromosomes during meiosis. Note that in instances where two crossovers occur between two loci, the resultant haploid gametes will retain the parental dilocus genotype (the *haplotypes* of the respective parents). These *double recombination* events will not be evident; therefore, the true recombination frequency among the resultant progeny will be underestimated. Furthermore, double recombinations occur nonrandomly because the occurrence of

one crossover physically restricts the opportunity for another to occur nearby. This is termed *interference*. In general, the greater the physical distance between two marker loci, the greater the likelihood that multiple crossovers will occur between them.

Recombination frequency estimates inherently assume complete interference. Most computer programs for linkage mapping also include mapping functions that allow us to adjust the recombination frequency estimates for double recombination. The *Haldane mapping function* assumes that no interference occurs and produces the largest centimorgan-map-distance estimates. The *Kosambi mapping function* assumes that an intermediate level of interference occurs and will, therefore, produce somewhat smaller centimorgan-map-distance estimates. Note that both mapping function estimates will be larger than the associated recombination frequencies.

As genotypes are determined at each genetic marker, they are entered into a locus × F_2 or BC offspring matrix (Fig. 30.18). The genotypes of the F_1 parents and the F_2 or BC progeny are entered for each

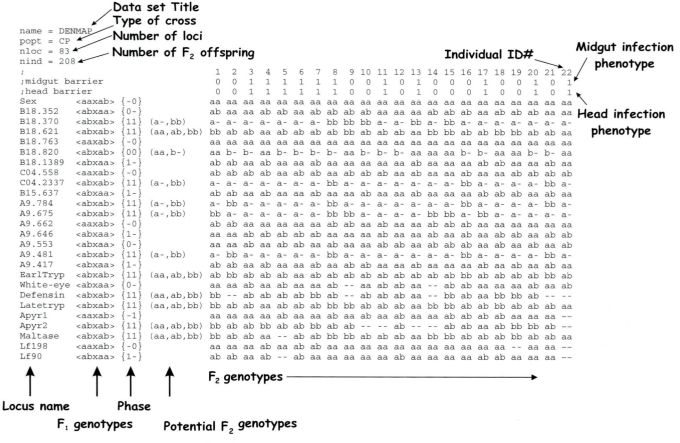

FIGURE 30.18 Structure of a JoinMap 2.0 data matrix.

of the loci using a five-letter code. In general, an "A" or "aa" represents the P_1 maternal genotype and a "B" or "bb" represents the P_1 paternal genotype. A "C," "ab" or "b-" represents the P_1 paternal genotype when a dominant allele is present at a locus (e.g., RAPD or AFLP markers) or when genotypes at a codominant locus are only partially informative. Conversely, a "D," "ab", or "a-" represents the P_1 maternal genotype with dominant alleles or partially informative genotypes. An "H" or "ab" represents either of the F_1 parental genotypes at a fully informative locus. The data set in Figure 30.18 was generated in an F_1 intercross that produced 208 F_2 offspring for mapping. This data set has 83 loci. Figure 30.18 shows genotypes for only the first 22 F_2 individuals and 26 loci.

With 83 loci, there are 3,403 estimates of pairwise recombination distances, and manual calculation of linkage distances becomes too laborious and inaccurate. Commonly used software packages such as MAP-MAKER (Lander et al. 1987) and JoinMap (Stam and Ooijen 1995) have been developed that utilize the format in Figure 30.18 to (1) determine, for each pair of loci, which dilocus genotypes are informative, (2) estimate pairwise distances from informative genotypes, and (3) estimate the log odds ratio (LOD) scores for each of these estimates. The LOD score is the probability that the linkage estimate could have arisen by random chance. The equation is

$$\text{LOD} = -1 \times \text{Log10 (probability that an event occurred by random chance)} \quad (5)$$

For example, if we flip a coin 10 times, the probability that it will land heads all 10 times is 0.000977 and

$$\text{LOD} = -1 \times \text{Log10}(0.000977) = 3.01$$

A partial output of linkage distances and associated LOD scores are shown in Figure 30.19. Notice that closely linked markers generally have high LOD scores and as recombination rates approach 50%, the LOD scores decline. However, this association is also significantly affected by the size of the family and the actual number of F_2 offspring with informative dilocus genotypes. *The most limiting component for statistical confidence in any mapping estimate is the sample size,* which obviously influences the number of informative meioses available for recombination calculations. Notice, for example, that *defensin* and *late trypsin* dilocus genotype ratios estimate ~9.7% recombination and that this estimate arises from completely informative dilocus genotypes. The LOD score is 28.7. However, with *defensin* and *TAG66.268*, we observed no recombination, but the estimate arises from only partially informative dilocus genotypes. The LOD score is accordingly lower (23.0).

Both the MAPMAKER and JoinMap computer programs use the two-point linkage estimates calculated across all loci to assign them to linkage groups. Both programs have default values for assignments, or the investigator can set their own preferences. However, the strategies for determining linear orders within linkage groups vary between the two. The MAP-MAKER program uses an iterative process to computationally examine possible linear orders for a given group of marker loci and calculates a likelihood estimate at each step, until the likelihood converges to a maximum (Lander and Green 1987). This method is guaranteed to find the "best" linear order for any given data set. However, the iterative process, even though optimized with this program, can be extremely time-consuming, particularly as the number of loci examined increases. The program does allow one to initially select a subset of loci for a given linkage group for the initial likelihood analysis and identify their "best" linear order. Thereafter, the remaining loci are sequentially added (manually by the investigator) to compute their "best" position within the previously calculated framework order. This can reduce the computing time considerably and still identify the best maximum-likelihood linear order.

The JoinMap program requires no trial-and-error interaction by the user and is designed for speed in preparing a linkage map (see http://weeds.ngh.harvard.edu/goodman/doc/joinmap_manual.html). The program starts by selecting the marker pair that contains the most linkage information. Criteria for selection of this pair includes the magnitude of the associated LOD score, which increases with both decreasing recombination values and increasing sample sizes and is influenced by marker phase, and also that the pair is linked to at least two other markers in the data set. In a stepwise manner, the program then chooses the next marker to be added based on its total linkage information relative to the markers already included. After each stepwise marker addition, the program alternates (1) local "reshuffling" of markers and recalculation of the maximum-likelihood estimates with (2) global "reshuffling" of markers as a series of moving intervals and recalculation of the maximum-likelihood estimates. This process is continued until all markers are added and the final maximum-likelihood estimate for the linear orders is accepted. Note that since all possible marker order permutations are not examined, the computing time is significantly reduced using the JoinMap program. However, one caution is that it is not guaranteed to produce the "best" linear order for a data set. An important utility of this program is that it can be used to integrate independent data sets from

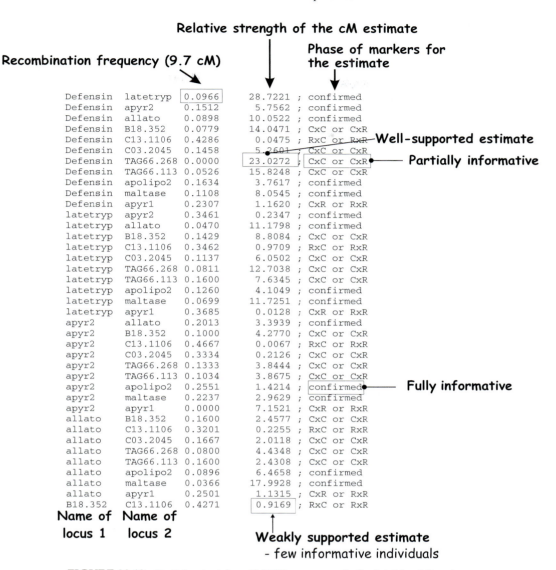

Recombination frequency (9.7 cM)

Relative strength of the cM estimate

Phase of markers for the estimate

Defensin	latetryp	0.0966	28.7221	;	confirmed
Defensin	apyr2	0.1512	5.7562	;	confirmed
Defensin	allato	0.0898	10.0522	;	confirmed
Defensin	B18.352	0.0779	14.0471	;	CxC or CxR
Defensin	C13.1106	0.4286	0.0475	;	RxC or RxR
Defensin	C03.2045	0.1458	5.2601	;	CxC or CxR
Defensin	TAG66.268	0.0000	23.0272	;	CxC or CxR
Defensin	TAG66.113	0.0526	15.8248	;	CxC or CxR
Defensin	apolipo2	0.1634	3.7617	;	confirmed
Defensin	maltase	0.1108	8.0545	;	confirmed
Defensin	apyr1	0.2307	1.1620	;	CxR or RxR
latetryp	apyr2	0.3461	0.2347	;	confirmed
latetryp	allato	0.0470	11.1798	;	confirmed
latetryp	B18.352	0.1429	8.8084	;	CxC or CxR
latetryp	C13.1106	0.3462	0.9709	;	RxC or RxR
latetryp	C03.2045	0.1137	6.0502	;	CxC or CxR
latetryp	TAG66.268	0.0811	12.7038	;	CxC or CxR
latetryp	TAG66.113	0.1600	7.6345	;	CxC or CxR
latetryp	apolipo2	0.1260	4.1049	;	confirmed
latetryp	maltase	0.0699	11.7251	;	confirmed
latetryp	apyr1	0.3685	0.0128	;	CxR or RxR
apyr2	allato	0.2013	3.3939	;	confirmed
apyr2	B18.352	0.1000	4.2770	;	CxC or CxR
apyr2	C13.1106	0.4667	0.0067	;	RxC or RxR
apyr2	C03.2045	0.3334	0.2126	;	CxC or CxR
apyr2	TAG66.268	0.1333	3.8444	;	CxC or CxR
apyr2	TAG66.113	0.1034	3.8675	;	CxC or CxR
apyr2	apolipo2	0.2551	1.4214	;	confirmed
apyr2	maltase	0.2237	2.9629	;	confirmed
apyr2	apyr1	0.0000	7.1521	;	CxR or RxR
allato	B18.352	0.1600	2.4577	;	CxC or CxR
allato	C13.1106	0.3201	0.2255	;	RxC or RxR
allato	C03.2045	0.1667	2.0118	;	CxC or CxR
allato	TAG66.268	0.0800	4.4348	;	CxC or CxR
allato	TAG66.113	0.1600	2.4308	;	CxC or CxR
allato	apolipo2	0.0896	6.4658	;	confirmed
allato	maltase	0.0366	17.9928	;	confirmed
allato	apyr1	0.2501	1.1315	;	CxR or RxR
B18.352	C13.1106	0.4271	0.9169	;	RxC or RxR

Well-supported estimate

Partially informative

Fully informative

Name of locus 1   Name of locus 2

Weakly supported estimate - few informative individuals

FIGURE 30.19 Partial output from JMREC, a program in the JoinMap 2.0 package.

various mapping populations (Stam and Ooijen 1995). It uses markers common to the individual populations as "bridges" to integrate markers that are unique to specific populations.

QUANTITATIVE TRAIT-LOCUS MAPPING

Having completed a linkage map, we are finally in a position to identify genome regions that contain loci that condition vector competence. With the emergence of intensive linkage mapping came the ability to simultaneously scan entire genomes to test for and identify regions that are statistically associated with pheno-

typic traits such as vector competence. This process came to be known as *quantitative trait-locus (QTL) mapping*. Intensive linkage mapping and QTL analyses are so closely allied that researchers usually accomplish both simultaneously by incorporating P_1 parents that differ in vector competence phenotypes (e.g., one parent from a susceptible strain is crossed with another parent from a refractory strain). The F_1 offspring are intercrossed and the phenotypes of individual F_2 offspring are stored in the same matrix alongside the marker genotype data used in intensive linkage mapping. For example, notice that in the 6th and 7th lines of the matrix in Figure 30.18, the MIB and MEB phenotypes of each mosquito are recorded. The 6th line indicates whether an F_2 female had an infected (scored as "1") or uninfected midgut ("0"). The 7th line indicates whether an F_2 female had an infected or uninfected head.

There are various computer programs that utilize this database format to methodically test each locus for a statistical association between the genotypes at marker loci and the phenotypes in the F_2 offspring. Associations between genotypes at each locus and phenotypes can be assessed by a contingency χ^2 analysis if analyzing threshold characters or by ANOVA procedures when analyzing meristic or continuous characters. Contingency χ^2 analyses for two of the loci in Figure 30.6 are shown in Table 30.2.

The null hypothesis in contingency χ^2 or ANOVA tests is that midgut infection or escape barrier rates are equal in each genotype class. Thus, marginal probabilities for columns are the frequencies of each genotype at a locus, and marginal probabilities for rows are the overall midgut infection or escape barrier rates. We reject the null hypothesis at the *Late Trypsin* locus (Table 30.2). Notice that 86% of F_2 females with an "aa" genotype (both alleles inherited from the P_1 mother) have an infected midgut, 41% of "ab" genotype individuals (one allele inherited from the P_1 mother) have an infected midgut, but only 13% of "aa" F_2 individuals (no alleles inherited from the P_1 mother) have an infected midgut. In this particular cross, the P_1 mother was highly susceptible while the P_1 father was refractory. Not only are these values significantly different, but the midgut infection rates are also in the anticipated direction. F_2 individuals homozygous for alleles from the susceptible parent are 45% more susceptible than heterozygous individuals, which are in turn 28% more susceptible than F_2 individuals homozygous for alleles from the refractory parent. These values are plotted in Figure 30.20A. Compare this with Figures 30.5A,B. These analyses illustrate several important factors associated with the inheritance of dengue susceptibility. We can first of all infer that a locus that affects midgut infection rates is located in proximity to the *Late Trypsin* locus. Second, we can infer that alleles at the susceptibility locus are additive in their effect on phenotype.

Contrast the association between phenotypes and genotypes at the *Late Trypsin* locus with the association with genotypes at the *Apyrase 2* locus, some 32 cM away (Fig. 30.20A). In this case, we *accept* the null hypothesis at the *Apyrase 2* locus. Infected midguts occur in 44% of F_2 females with an "aa" genotype and 33% of "ab" females, and 31% of "aa" F_2 individuals have an infected midgut. The contingency χ^2 analysis tells us that midgut infection rates are homogeneously distributed among genotypes at the *Apyrase 2* locus (Table 30.2). We can infer that a locus that affects

midgut infection rates is not located in proximity to the *Apyrase 2* locus.

A contingency χ^2 analysis or ANOVA procedure can be repeated exhaustively among all loci in a study. However, a major problem with this approach is that it treats each locus as an independent factor. In reality, we know that alleles at loci do not segregate independently. All marker loci will be statistically associated with phenotype if they are even in proximity to the actual QTL. Furthermore, the most likely location of the QTL cannot be inferred using this approach. This

TABLE 30.2   Contingency χ^2 Analysis of Dengue Midgut Infection Phenotypes and Genotypes at the *Late Trypsin* and *Apyrase 2* Loci

| Phenotype | Marker Genotype (*Late Trypsin*) | | | |
	aa	ab	bb	Total
Midgut uninfected ("0")				
Observed	1	23	13	37 (0.61)
Expected	4.2	23.7	9.1	
Cell χ^2	2.5	0.0	1.7	
(e.g., for the first cell: expected $= 0.61 \times 0.11 \times 61 = 4.2$				
Cell $\chi^2 = (1 - 4.2)^2/4.2 = 2.5$				
Midgut infected ("1")				
Observed	6	16	2	24 (0.39)
Expected	2.8	15.3	5.9	
Cell χ^2	3.8	0.0	2.6	
Total	7	39	15	61
	0.11	0.64	0.25	
MIR	0.86 (6/7)	0.41	0.13	

Contingency $\chi^2 = 2.5 + 0.0 + 1.7 + 3.8 + 0.0 + 2.6 = 10.6$
Degrees of freedom $= (\text{rows} - 1)(\text{columns} - 1) = (2 - 1)(3 - 1) = 2$
Missing genotype or phenotype data for 23 individuals
$P < 0.001$

| Phenotype | Marker Genotype (*Apyrase 2*) | | | |
	aa	ab	bb	Total
Midgut uninfected ("0")				
Observed	9	30	9	48 (0.65)
Expected	10.4	29.2	8.4	
Cell χ^2	0.2	0.0	0.0	
Midgut infected ("1")				
Observed	7	15	4	26 (0.35)
Expected	5.6	15.8	4.6	
Cell χ^2	3.8	0.0	2.6	
Total	16	45	13	74
	0.22	0.61	0.18	
MIR	0.44	0.33	0.31	

Contingency $\chi^2 = 0.2 + 0.0 + 0.0 + 0.3 + 0.0 + 0.1 = 0.7$
Missing genotype or phenotype data for 10 individuals
$P \geq 0.05$

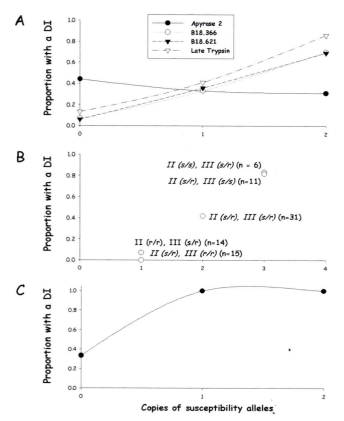

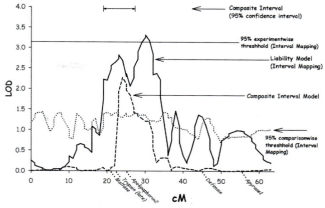

FIGURE 30.21 Plot of LOD values associated with MIB along chromosome III in *Aedes aegypti*. Names of a few markers are listed to orient QTL positions relative to Figure 30.6. LOD values estimated by interval mapping (IM) using a heterogeneous residual variance model for binary traits (Xu et al. 1998) appear as a solid line. LOD values estimated by composite interval mapping (Zeng 1994) appear as a dashed line. Comparisonwise 95% thresholds appear as a dotted line, while 95% experimentwise thresholds are represented by the straight line along the top of each graph. Intervals at which the CIM LOD estimated in *Zmapqtl* (QTL Cartographer 1.13) (Basten et al. 1997) that exceeded the top 950 CIM LOD scores in 1,000 permutations are bracketed.

FIGURE 30.20 Phenotypic means plotted against genotypes at (A) markers on chromosome 3 for MIB in *Aedes aegypti*, (B) combined markers on chromosomes 2 and 3 for MIB in *Aedes aegypti*, and (C) markers on chromosome 1 linked to a QTL controlling an MEB in *Aedes aegypti*.

has prompted the development of much more stringent and exact methods for testing for the presence of a QTL and, most importantly, for estimating the linkage position of the QTL.

Interval mapping (Lander and Botstein 1989) is a method that assesses pairs of adjacent markers to test for statistical associations of an interval with the phenotype. The null hypothesis (H_0) is that there is no association between the phenotype of an individual and the genotype at a particular locus. The alternative hypothesis (H_a) is that there *is* an association between the phenotype of an individual and that genotype. QTL mapping uses LOD scores, as discussed earlier, to test the likelihood that a particular association arose by random chance. In this case, LOD = $\log_{10}$(probability of H_a/probability of H_0). For example, if the probability of H_a equals the probability of H_0, then LOD = $\log_{10}(1) = 0$. But if the probability of H_a exceeds the probability of H_0, then LOD $\gg 0$. Alternatively, if the prob-

ability of H_a equals 1000 times the probability of H_0 then LOD = 3.0. In interval mapping, the LOD score for each interval is plotted on the *y*-axis and the linkage map is plotted on the *x*-axis (Fig. 30.21). Interval mapping assesses LOD values all along a chromosome and then estimates the most likely location of the QTL. For example, in Figure 30.21 we would infer the existence of two QTL controlling a dengue MIB, one located at ~20 cM and a second at ~30 cM (notice that this corresponds to the locations of significant χ^2 analysis in Fig. 30.6).

A legitimate question concerns the magnitude of LOD scores that should be considered valid. Currently, permutation tests are used to assess the consistency with which a data set supports LOD scores for each interval. Churchill and Doerge (1994) proposed a method whereby, after the LOD scores have been calculated for the original data set, phenotypes of F_2 individuals are randomly permuted on the original F_2 individual genotypes. LOD values are then estimated and stored for the permuted data set. This permutation procedure is repeated 1,000 more times. The software then reports for each interval the 900th, 950th, and 990th largest LOD values. It also reports the 900th, 950th, and 990th largest experimentwise thresholds. The 95% comparisonwise or interval-wise and experimentwise thresholds for the threshold model are shown in Figure 30.21. The comparisonwise threshold tends to overestimate the number of QTLs. Note that the liability model LOD 95% crosses the comparisonwise threshold at five locations (~20, ~30, ~38, ~45, and 55 cM). In contrast, the experimentwise threshold tends to be overly conservative and, therefore, underestimates the number of QTLs. Note that the liability model LOD 95% barely crosses the experimentwise threshold at ~30 cM.

Depending on their magnitude of effect and linkage relationships, LODs at different map intervals may covary with one another, and this can upwardly bias LOD estimates at individual intervals. *Composite interval mapping* (CIM) (Zeng 1994) adjusts LOD scores at individual intervals using a variable number of markers (n_p) to control for effects of other intervals in the map and a variable window size (w_s) to adjust for the effects of intervals in proximity to the interval under analysis. Notice in Figure 30.21 that while interval mapping (IM) estimates up to 5 QTLs, CIM infers a single QTL at ~25 cM. Permutation tests can also be applied to CIM. The 95% confidence interval for the CIM analysis is shown as a bracketed arrow at the top of Figure 30.21. Taken together, IM and CIM suggest the existence of a QTL controlling a dengue MIB at ~25 cM. *As in development of a primary linkage map, the most critical factor in determining the accuracy of QTL placement is the sample size.*

QTL MAPPING IN VECTORS

QTL mapping techniques have been successfully used to examine several complex phenotypes in *An. gambiae* and *Ae. aegypti*. Although QTL mapping can be applied to virtually any phenotype, most studies to date with mosquitoes have addressed aspects of competence to transmit particular pathogens, including dengue virus, *Plasmodium*, and *B. malayi*.

With *An. gambiae*, QTL studies have largely focused on the *Plasmodium* melanotic encapsulation resistance phenotype. Three QTLs with additive genetic effects have been identified that collectively account for ~70% of the variance in melanization of *P. cynomolgi* B ookinetes (Zheng et al. 1997). The major QTL (*Pen1*) is located on the right arm of chromosome 2, while the minor QTLs are located on the left arm of chromosome 2 (*Pen2*) and the left arm of chromosome 3 (*Pen3*). The *Pen1* interval has also been shown to play the major role in the melanization response of *An. gambiae* to intrathoracically injected CM-Sephadex beads (Gorman and Paskewitz 1997).

Insecticide resistance, with its associated limitations for controlling vector populations, has become a significant global problem. Although insecticide resistance is often determined by single-gene effects, some resistance mechanisms reflect complex inheritance patterns. For example, DDT resistance in some *An. gambiae* populations is due to metabolism by increased glutathione S-transferase activity. Two QTLs that account for >50% of the variance in susceptibility to DDT have been identified (Ranson et al. 2000). One of these (*rtd1*) is located on the right arm of chromosome 3 and is recessive with respect to susceptibility. The second (*rtd2*) is located on the left arm of chromosome 2 and has an additive genetic effect.

With *Ae. aegypti*, QTL studies have been conducted for vector competence to dengue virus, *B. malayi*, and *P. gallinaceum*. With *B. malayi*, two QTLs determining susceptibility have been identified (Severson et al. 1994). The major QTL (*fsb1*) is located on chromosome 1 and accounts for 22–43% of the variance in susceptibility. It is recessive with respect to susceptibility. A second QTL (*fsb2*) is located on chromosome 2 and accounts for <10% of the variance in susceptibility. It has an additive effect with respect to susceptibility. With some genetic backgrounds, these QTLs also indicate epistatic interactions, wherein an individual mosquito must be homozygous for recessive alleles at *fsb1* and also carry at least one allele for susceptibility at *fsb2* in order to support development of the parasite. An additional QTL (*idb*) that influences *B. malayi* midgut penetration and subsequent intensity has also been identified (Beerntsen et al. 1995). With *P. gallinaceum*, two QTLs influencing development to the oocyst stage have been identified (Severson et al. 1995). The major QTL (*pgs1*) is located on chromosome 2 and accounts for 49–65% of the variance. A second, minor QTL (*pgs2*) is located on chromosome 3 and accounts for 10–14% of the variance. Both QTLs exhibit a partial dominance effect with respect to refractoriness and appear to act independently. Of interest, the *pgs1* and *idb* QTLs both reside within the same genome interval as *fsb2*. This suggests that either several independent genes that determine vector competence are clustered in the same genome region on chromosome 2 or that single genes from this region could influence susceptibility to a wide range of pathogens.

QTLs determining both MIB and MEB to dengue virus infection have been identified (Bosio et al. 2000). Two QTLs located on chromosomes 2 and 3 account for ~30% of the variance in MIB, and both appear to have additive and independent genetic effects relative to refractoriness to dengue infection (Figs. 30.20A,B). Alleles at a locus on chromosome 1 of *Aedes aegypti* that encode an MEB are recessive (Fig. 30.20C) relative to susceptible alleles at that locus.

PHYSICAL MAPPING

Contemporary physical mapping is all about order. That is, our goal is to determine the linear physical relationships among DNA sequences for the individual chromosomes within a genome. Physical maps are important for (1) studies of genome organization and evolution (Chapter 31); (2) localization and isolation of genes of interest by *positional or map-based cloning*; and (3) for their role as a scaffold for the assembly of sequence data obtained in systematic genome-sequencing efforts. All physical maps involve the ordering of clones, usually by some form of probe hybridization assay or via landmark sequences that define PCR primers for known genome sequences. The PCR-based landmarks are commonly termed *sequence-tagged sites (STSs)*. The STS designation can be applied to any single copy sequence, including random genomic sequences obtained from any cloned source, microsatellite markers, and cDNA sequences. Because the cDNAs represent expressed genes, they are often termed *expressed sequence tags, or ESTs*. Our specific level of interest for physical mapping can vary from the macro level covering many megabases of DNA to the micro level of the individual nucleotide positions.

Physical Mapping in Anophelines

Polytene chromosomes are routinely resolved in anopheline mosquitoes and provide an excellent framework for physical mapping (Fig. 30.22; Sharakhov et al. 2001). A standard procedure is the direct *in situ* hybridization of cloned sequences to the polytene chromosome. For *An. gambiae*, these have included cloned RAPD fragments, microsatellites, cDNAs, cosmids, and BAC clones (della Torre et al. 1996; Zheng et al. 1996). An example with a RAPD sequence is shown in Figure 30.23 (Dimopoulos et al. 1996). In addition, a collection of 54 microdissected divisions of *An. gambiae* polytene chromosomes was prepared and provides an extremely useful tool for physical mapping (Zheng et al. 1991). For this procedure, microdissected chromosome sections are digested with a frequent cutting restriction enzyme and ligated with an adapter sequence so that the individual section pools can be PCR amplified. The

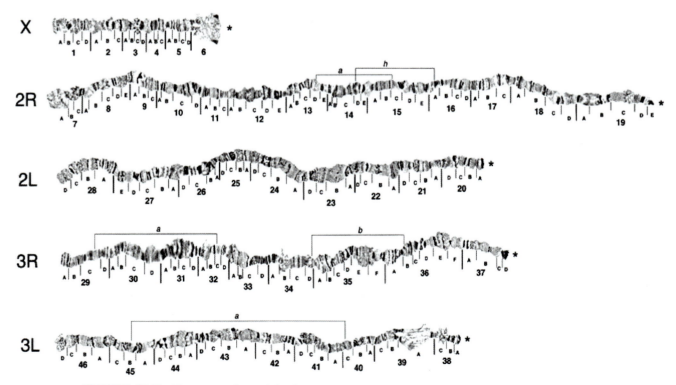

FIGURE 30.22 Photomap of *Anopheles funestus* polytene chromosomes. Brackets represent inversions found in the specimens from coastal Kenya. Stars indicate centromeric regions. From Sharakhov et al. (2001).

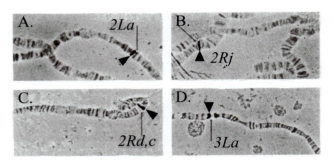

FIGURE 30.23 *In situ* hybridization of RAPD-derived STSs that map near *Anopheles gambiae* paracentric inversion breakpoints. *In situ* hybridization to nurse cell polytene chromosomes for R7 (A), Rll (B), R12 (C), and R17 (D). Note the placement of these in Figure 30.24. The arrowheads point to the hybridization signal, while the thin arrows point to the location of the breakpoints of the respective inversions. From Dimopoulos et al. (1996).

physical position of unknown clones can then be determined either by hybridization to dot blots of the pools immobilized on filters or by PCR assaying the individual pools for STSs. Physical mapping of RAPD and microsatellite sequences that had also been mapped genetically provided for the integration of the physical and genetics maps of *An. gambiae* (Dimopoulos et al. 1996; Zheng et al. 1996). Note, as shown in Figure 30.24 (Dimopoulos et al. 1996), that the genetic and physical distances do not reflect linear correlations. A small genetic distance can represent a large physical distance, and *vice versa*.

Physical Mapping in Culicines

Physical mapping in culicine mosquitoes has required a different approach, because their polytene

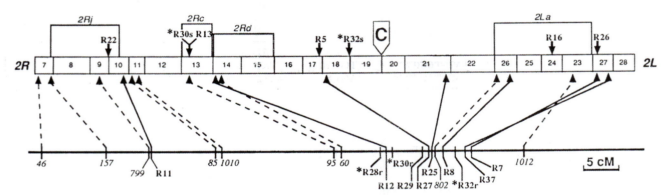

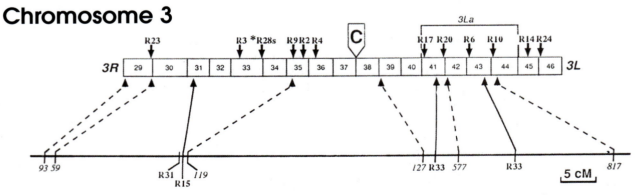

FIGURE 30.24 Integration of the genetic linkage maps and the physical map of *Anopheles gambiae* autosomes. The maps incorporate RAPDs (continuous lines and codes beginning with R) and selected microsatellites (dashed lines and simple numbers; Zheng et al. 1996). The centimorgan scale of the linkage maps is shown by a bracket. In the cytogenetic maps, chromosomal divisions are shown as numbered boxes. The centromere (C) is flagged. From Dimopoulos et al. (1996).

chromosomes are poorly resolved and therefore largely irreproducible. This likely relates to the larger percentage of repetitive DNA inherent to culicine genomes that restricts the ability to cleanly separate individual chromosomes. The physical map for *Ae. aegypti* was instead developed using *fluorescent in situ hybridization (FISH)* combined with digital-imaging microscopy to position random cosmid clones on metaphase chromosomes (Fig. 30.25) (Brown et al. 1995). This map presently includes ~350 cosmid and BAC clones. While standard FISH mapping has a resolution of ~1 Mb, high-resolution mapping can be achieved by hybridization to interphase nuclei or even to extended chromatin fibers. The use of contrasting fluorescent tags also allows one to directly compare linear positions of multiple clones simultaneously. The physical and genetic maps for *Ae. aegypti* have been integrated (Fig. 30.26; Brown et al. 2001). This process included FISH mapping cosmids that were determined via STS screening assays to contain the gene for some of the cDNAs that had been mapped as RFLP markers and also included the direct hybridization of some cDNA clones using an ultrasensitive FISH technique. An example of FISH mapping with a cDNA initially mapped genetically to single loci on chromosomes 1 and 3 is shown in Figure 30.25. The disparity in relative distances between the physical and genetic

maps is particularly evident with *Ae. aegypti* (Fig. 30.26). The centromere regions of each chromosome contain *heterochromatin*, or highly condensed DNA with few active genes. The distal chromosome arms contain *euchromatin* DNA that readily decondenses and includes the active genes.

Positional Cloning

Our abilities to develop genetic and physical maps and to identify QTL regions containing genes that influence complex phenotypic traits sets the stage for efforts to isolate the genes associated with these QTLs. The general strategy for positional or map-based cloning of genes for QTLs is fairly straightforward, while its actual implementation remains a challenge. *The greatest challenge to success in map-based positional cloning is the ability to examine a population with a sample size with suitable power to detect the QTL with a high degree of confidence. This effort is completely dependent on identifying individuals carrying informative meioses. For these reasons, development of AILs is a critical part of map-based positional cloning.*

The first step is to map the target genes in a large segregating population using DNA-based markers. Most primary genome scans for QTL identification involve a set of marker loci spaced at ~10 to 20-cM intervals across the entire genome. At this level one can expect to detect QTLs that account for at least 10% of the phenotypic variance with a population size of 150–200 individuals. However, as we attempt to increase our power of resolution for positional cloning, the necessary sample size increases significantly. For example, the number of individuals required to have a 95% chance of recovering at least one recombinant between markers that are c map units apart is ca. $3/c$. At the 0.1- and 0.01-cM map distances we would therefore require at least 3,000 and 30,000 individuals, respectively. Note that this number can be reduced by using an AIL mapping population, but it will still likely require at least 1,000 individuals.

Once a QTL is located to a specific interval, the markers flanking that interval are checked for their physical map positions. This provides us an indication of the physical distance within which we have confirmed the location for the target gene. Most primary genome scans only delimit a QTL to a physical location containing >10 Mb of genomic DNA. Our interest then is to identify additional genetic markers within the QTL interval because this will provide us a mechanism to genetically reduce the physical size of our target gene search. One approach to identifying new markers within a QTL is called *bulked segregant analysis* (BSA), wherein DNA from selected individuals

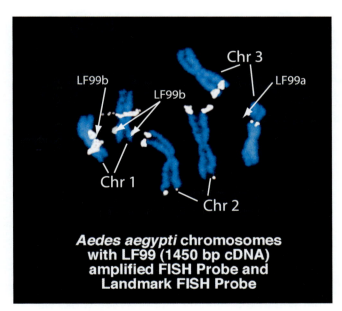

FIGURE 30.25 Mapping of a cDNA clone using ultrasensitive FISH techniques. A photomicrograph of FISH with a labeled cDNA clone to *Aedes aegypti* metaphase chromosomes is depicted. There are two LF99 loci (Fig. 30.6). In the original photo the LF99 cDNA signal is colored green and the landmark signals are red. From Brown et al. (2001).

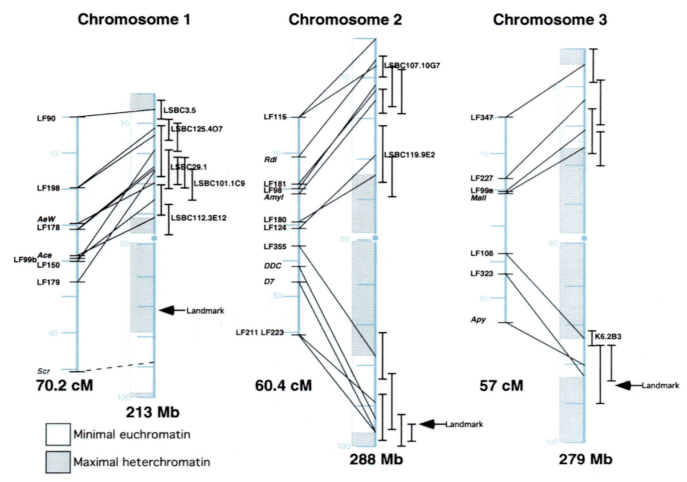

FIGURE 30.26 Integration of the genetic linkage maps and the physical map of *Aedes aegypti*. Heavy lines link the genetic loci to their appropriate locations on the physical map. When a cosmid clone was found using STS primers for a specific cDNA locus, its position is indicated at the centerpoint of the FISH % FLpter standard deviation error bars on the physical map. FLpter refers to the fractional length of the chromosome from the telomere of the p arm. FLpter is commonly used to measure the location of a probe. The range of the measurement is from 0 to 1, with the p arm telomere at 0.00 and the q arm telomere at position 1.00. Euchromatic and heterochromatic regions are indicated by open and shaded blocks, respectively. From Brown et al. (2001).

from a segregating population is pooled based on their genotypes at the QTL flanking markers. One pool consists of DNA from individuals homozygous for one of the parental genotypes, and a second pool consists of those of the other parental genotype. DNA markers that show polymorphism between the two pools are likely to map near or within the QTL. The BSA approach has been used successfully to identify new marker loci within QTL for *Ae. aegypti* susceptibility to both *B. malayi* and *P. gallinaceum* (Severson et al. 1999).

The next step is to use the markers flanking the minimum QTL interval as starting points for *chromosome walking* across the interval. For a chromosome walk, we use an STS-based genomic library screening strategy for developing a series of overlapping contigs

of large-fragment clones. Briefly, we start by using sequence data for the genetic markers to develop an STS assay that allows us to rapidly identify a cosmid or BAC clone that contains the marker sequence. We next sequence both ends of the positive clones and use the data to perform additional STS assays. The process is repeated until we have "walked" across the entire interval. By using a combination of chromosome walking and development of new genetic markers from within the genomic clones as we walk, we can also continually narrow the size of the QTL interval.

Finally, once an acceptable minimum physical contig is identified, our attention is directed toward identifying a candidate gene located within it. One method is to sequence the entire contig and identify all genes located within it. Each of these must then be evaluated as to its likelihood as a candidate gene and then tested to prove its identity via a suitable phenotypic complementation assay. Another method is to survey the existing gene databases for likely candidates based on their putative protein functions and then to determine if they map to the QTL interval or to directly identify expressed genes within the interval. We recently used the direct approach to identify a candidate mucin-like gene that maps within the *pgs1* QTL determining *P. gallinaceum* susceptibility in *Ae. aegypti* (Morlais and Severson 2001). Although no candidate genes for a vector have been confirmed as the actual target within a QTL, there are several ongoing positional cloning efforts directed toward QTLs associated with vector/pathogen interactions, including, *Ae. aegypti*/dengue, *Ae. aegypti*/*B. malayi*, *Ae. aegypti*/*P. gallinaceum*, and *An. gambiae*/*P. cynomolgi*. Association mapping and population genomics (Chapter 32) are additional methods to confirm that candidate genes condition vector competence.

Genomic Databases

A new paradigm in our ability to understand and dissect virtually any aspect of vector genetics is imminent. The recent complete sequence determinations for the *Drosophila melanogaster* and *Anopheles gambiae* genome sets a new standard for contemporary genomics that will now be applied to other vector species. Indeed, the *An. aegypti* genome sequencing effort is currently under way, and other species, including *Culex pipiens* are likely to follow. Updated information on these and other projects is readily available at the Mosquito Genomics web server (http://klab.agsci.colostate.edu) and the AnoDB (http://konops.imbb.forth.gr/AnoDB).

Readings

Akey, J. M., Sosnoski, D., Parra, E., Dios, S., Hiester, K., Su, B., Bonilla, C., Jin, L., and Shriver, M. D. 2001. Melting curve analysis of SNPs (McSNP): A gel-free and inexpensive approach for SNP genotyping. *Biotechniques* 30: 358–362, 364, 366–367.

Antolin, M. F., Bosio, C. F., Cotton, J., Sweeney, W., Strand, M. R., and Black, W. C. t. 1996. Intensive linkage mapping in a wasp (*Bracon hebetor*) and a mosquito (*Aedes aegypti*) with single-strand conformation polymorphism analysis of random amplified polymorphic DNA markers. *Genetics* 143: 1727–1738.

Basten, C. J., Weir, B. S., and Zeng, Z.-B. 1997. *QTL Cartographer: A Reference Manual and Tutorial for QTL Mapping*. Department of Statistics, North Carolina State University, Raleigh, NC.

Beerntsen, B. T., Severson, D. W., Klinkhammer, J. A., Kassner, V. A., and Christensen, B. M. 1995. *Aedes aegypti*: A quantitative trait locus (QTL) influencing filarial worm intensity is linked to QTL for susceptibility to other mosquito-borne pathogens. *Exp. Parasitol.* 81: 355–362.

Bosio, C. F., Fulton, R. E., Salasek, M. L., Beaty, B. J., and Black, W. C. t. 2000. Quantitative trait loci that control vector competence for dengue-2 virus in the mosquito *Aedes aegypti*. *Genetics* 156: 687–698.

Brookes, A. J. 1999. The essence of SNPs. *Gene* 234: 177–186.

Brown, S. E., Menninger, J., Difillipantonio, M., Beaty, B. J., Ward, D. C., and Knudson, D. L. 1995. Toward a physical map of *Aedes aegypti*. *Insect Mol. Biol.* 4: 161–167.

Brown, S. E., Severson, D. W., Smith, L. A., and Knudson, D. L. 2001. Integration of the *Aedes aegypti* mosquito genetic linkage and physical maps. *Genetics* 157: 1299–1305.

Churchill, G. A., and Doerge, R. W. 1994. Empirical threshold values for quantitative trait mapping. *Genetics* 138: 963–971.

Collins, F. H., Sakai, R. K., Vernick, K. D., Paskewitz, S., Seeley, D. C., Miller, L. H., Collins, W. E., Campbell, C. C., and Gwadz, R. W. 1986. Genetic selection of a *Plasmodium*-refractory strain of the malaria vector *Anopheles gambiae*. *Science* 234: 607–610.

Darvasi, A., and Soller, M. 1995. Advanced intercross lines, an experimental population for fine genetic mapping. *Genetics* 141: 1199–1207.

della Torre, A., Favia, G., Mariotti, G., Coluzzi, M., and Mathiopoulos, K. D. 1996. Physical map of the malaria vector *Anopheles gambiae*. *Genetics* 143: 1307–1311.

Dimopoulos, G., Zheng, L., Kumar, V., della Torre, A., Kafatos, F. C., and Louis, C. 1996. Integrated genetic map of *Anopheles gambiae*: Use of RAPD polymorphisms for genetic, cytogenetic and STS landmarks. *Genetics* 143: 953–960.

Fagerberg, A. J., Fulton, R. E., and Black, W. C. 2001. Microsatellite loci are not abundant in all arthropod genomes: Analyses in the hard tick, *Ixodes scapularis*, and the yellow fever mosquito, *Aedes aegypti*. *Insect Mol. Biol.* 10: 225–236.

Falconer, D. S., and MacKay, T. F. C. 1996. *Introduction to Quantitative Genetics*, 4th ed. ••, New York.

Fulton, R. E., Salasek, M. L., DuTeau, N. M., and Black, W. C. 2001. SSCP analysis of cDNA markers provides a dense linkage map of the *Aedes aegypti* genome. *Genetics* 158: 715–726.

Gorman, M. J., and Paskewitz, S. M. 1997. A genetic study of a melanization response to Sephadex beads in *Plasmodium*-refractory and -susceptible strains of *Anopheles gambiae*. *Am. J. Trop. Med. Hyg.* 56: 446–451.

Graham, D. H., Holmes, J. L., Higgs, S., Beaty, B. J., and Black, W. C. 1999. Selection of refractory and permissive strains of *Aedes triseriatus* (Diptera: Culicidae) for transovarial transmission of La Crosse virus. *J. Med. Entomol.* 36: 671–678.

Han, Y. S., Thompson, J., Kafatos, F. C., and Barillas-Mury, C. 2000. Molecular interactions between *Anopheles stephensi* midgut cells and *Plasmodium berghei*: The time bomb theory of ookinete invasion of mosquitoes. *Embo. J.* 19: 6030–6040.

Kimura, M. 1991. The neutral theory of molecular evolution: A review of recent evidence. *Jpn. J. Genet.* 66: 367–386.

Landegren, U., Nilsson, M., and Kwok, P. Y. 1998. Reading bits of genetic information: Methods for single-nucleotide polymorphism analysis. *Genome Res.* 8: 769–776.

Lander, E. S., and Botstein, D. 1989. Mapping Mendelian factors underlying quantitative traits using RFLP linkage maps. *Genetics* 121: 185–199.

Lander, E. S., and Green, P. 1987. Construction of multilocus genetic linkage maps in humans. *Proc. Natl. Acad. Sci. USA* 84: 2363–2367.

Lander, E. S., Green, P., Abrahamson, J., Barlow, A., Daly, M. J., Lincoln, S. E., and Newburg, L. 1987. MAPMAKER: An interactive computer package for constructing primary genetic linkage maps of experimental and natural populations. *Genomics* 1: 174–181.

Macdonald, W. W. 1962. The genetic basis of susceptibility of infection with semiperiodic *Brugia malayi* in *Aedes aegypti*. *Ann. Trop. Med. Parasitol.* 56: 373–382.

Miller, B. R., and Mitchell, C. J. 1991. Genetic selection of a flavivirus-refractory strain of the yellow fever mosquito *Aedes aegypti*. *Am. J. Trop. Med. Hyg.* 45: 399–407.

Morlais, I., and Severson, D. W. 2001. Identification of a polymorphic mucin-like gene expressed in the midgut of the mosquito, *Aedes aegypti*, using an integrated bulked segregant and differential display analysis. *Genetics* 158: 1125–1136.

Orita, M., Iwahana, H., Kanazawa, H., Hayashi, K., and Sekiya, T. 1989. Detection of polymorphisms of human DNA by gel electrophoresis as single-strand conformation polymorphisms. *Proc. Natl. Acad. Sci. USA* 86: 2766–2770.

Ranson, H., Jensen, B., Wang, X., Prapanthadara, L., Hemingway, J., and Collins, F. H. 2000. Genetic mapping of two loci affecting DDT resistance in the malaria vector *Anopheles gambiae*. *Insect Mol. Biol.* 9: 499–507.

Rodriguez, P. H., Lazaro, C., and Castillon, R. 1992. Reduction in the susceptibility of *Aedes aegypti* to *Brugia malayi* infection after treatment with ethyl methanesulfonate. *J. Am. Mosq. Control Assoc.* 8: 416–418.

Rosenberg, R. 1985. Inability of *Plasmodium knowlesi* sporozoites to invade *Anopheles freeborni* salivary glands. *Am. J. Trop. Med. Hyg.* 34: 687–691.

Severson, D. W., Mori, A., Zhang, Y., and Christensen, B. M. 1993. Linkage map for *Aedes aegypti* using restriction fragment length polymorphisms. *J. Hered.* 84: 241–247.

Severson, D. W., Mori, A., Zhang, Y., and Christensen, B. M. 1994. Chromosomal mapping of two loci affecting filarial worm susceptibility in *Aedes aegypti*. *Insect Mol. Biol.* 3: 67–72.

Severson, D. W., Thathy, V., Mori, A., Zhang, Y., and Christensen, B. M. 1995. Restriction fragment length polymorphism mapping of quantitative trait loci for malaria parasite susceptibility in the mosquito *Aedes aegypti*. *Genetics* 139: 1711–1717.

Severson, D. W., Zaitlin, D., and Kassner, V. A. 1999. Targeted identification of markers linked to malaria and filarioid nematode parasite resistance genes in the mosquito *Aedes aegypti*. *Genet. Res.* 73: 217–224.

Severson, D. W., Meece, J. K., Lovin, D. D., Saha, G., and Morlais, I. 2002. Linkage map organization of expressed sequence tags and sequence tagged sites in the mosquito, *Aedes aegypti*. *Insect Mol. Biol.* (in press).

Sharakhov, I. V., Sharakhova, M. V., Mbogo, C. M., Koekemoer, L. L., and Yan, G. 2001. Linear and spatial organization of polytene chromosomes of the African malaria mosquito *Anopheles funestus*. *Genetics* 159: 211–218.

Stam, P., and Ooijen, J. W. V. 1995. JoinMap (tm) version 2.0: Software for the calculation of genetic linkage maps. CPRO-DLO, Wageningen, The Netherlands.

Tabachnick, W. J., Wallis, G. P., Aitken, T. H., Miller, B. R., Amato, G. D., Lorenz, L., Powell, J. R., and Beaty, B. J. 1985. Oral infection of *Aedes aegypti* with yellow fever virus: Geographic variation and genetic considerations. *Am. J. Trop. Med. Hyg.* 34: 1219–1224.

Thathy, V., Severson, D. W., and Christensen, B. M. 1994. Reinterpretation of the genetics of susceptibility of *Aedes aegypti* to *Plasmodium gallinaceum*. *J. Parasitol.* 80: 705–712.

Vernick, K. D., Fujioka, H., Seeley, D. C., Tandler, B., Aikawa, M., and Miller, L. H. 1995. Plasmodium gallinaceum: A refractory mechanism of ookinete killing in the mosquito, *Anopheles gambiae*. *Exp. Parasitol.* 80: 583–595.

Vos, P., Hogers, R., Bleeker, M., Reijans, M., van de Lee, T., Hornes, M., Frijters, A., Pot, J., Peleman, J., Kuiper, M., et al. 1995. AFLP: A new technique for DNA fingerprinting. *Nucleic Acids Res.* 23: 4407–4414.

Wang, R., Kafatos, F. C., and Zheng, L. 1999. Microsatellite markers and genotyping procedures for *Anopheles gambiae*. *Parasitol. Today* 15: 33–37.

Welsh, J., and McClelland, M. 1990. Fingerprinting genomes using PCR with arbitrary primers. *Nucleic Acids Res.* 18: 7213–7218.

Williams, J. G., Kubelik, A. R., Livak, K. J., Rafalski, J. A., and Tingey, S. V. 1990. DNA polymorphisms amplified by arbitrary primers are useful as genetic markers. *Nucleic Acids Res.* 18: 6531–6535.

Xu, S., Yonash, N., Vallejo, R. L., and Cheng, H. H. 1998. Mapping quantitative trait loci for binary traits using a heterogeneous residual variance model: An application to Marek's disease susceptibility in chickens. *Genetica* 104: 171–178.

Zeng, Z. B. 1994. Precision mapping of quantitative trait loci. *Genetics* 136: 1457–1468.

Zheng, L. 1999. Genetic basis of encapsulation response in *Anopheles gambiae*. *Parassitologia* 41: 181–184.

Zheng, L. B., Saunders, R. D., Fortini, D., della Torre, A., Coluzzi, M., Glover, D. M., and Kafatos, F. C. 1991. Low-resolution genome map of the malaria mosquito *Anopheles gambiae*. *Proc. Natl. Acad. Sci. USA* 88: 11187–11191.

Zheng, L., Benedict, M. Q., Cornel, A. J., Collins, F. H., and Kafatos, F. C. 1996. An integrated genetic map of the African human malaria vector mosquito, *Anopheles gambiae*. *Genetics* 143: 941–952.

Zheng, L., Cornel, A. J., Wang, R., Erfle, H., Voss, H., Ansorge, W., Kafatos, F. C., and Collins, F. H. 1997. Quantitative trait loci for refractoriness of *Anopheles gambiae* to *Plasmodium cynomolgi* B. *Science* 276: 425–428.

31

Genome Evolution in Mosquitoes

DAVID W. SEVERSON AND WILLIAM C. BLACK IV

INTRODUCTION

The mosquito family, Culicidae, is a wonderful example of evolutionary success, with mosquitoes inhabiting virtually every conceivable habitat on the planet. Individual species exist that are well adapted for both larval and adult survival, even in extreme environments. Across a broad environmental spectrum, we can easily find and observe mosquito species adapted to breeding in snow-melt pools, stagnant ponds, small artificial and natural containers, and even high-salinity pools. We see incredible species-specific variation in many behavioral, morphological, and physiological characteristics. Nonetheless, all extant mosquito species derive from a common evolutionary lineage (Rohdendorf 1974).

From this ancestral blueprint and following about 250 million years of separate evolution, there currently exist more than 3,500 mosquito species divided among three subfamilies, Anophelinae, Culicinae, and Toxorhynchitinae. Over the last 40 years extensive cytogenetic investigations have been made of many mosquito genera and species (Kitzmiller 1976; G. B. White 1980; Rai and Black 1999). However, one could easily argue that these species contain many common genes and that the speciation process is likely driven by divergence among a relatively small subset of genes (Wu 2001). That is, we classify them as mosquitoes because they are much more similar to each other than they are different.

With only a single exception, all mosquito species share a genome complement of $2n = 6$ chromosomes. However, this does not suggest that the genomes of mosquitoes have remained static. Within the family,

both heteromorphic and homomorphic sex chromosomes have evolved, and there is extensive evidence for translocations and inversions. Furthermore, there is eightfold variation in genome size among species. Investigations of the present-day structure and organization of these chromosomes provide us with insights concerning mosquito systematics and perhaps a basis for understanding what makes them biologically distinct. The genomics era offers immense potential because the complete genome sequence will soon be determined for at least a small number of individual mosquito species. For example, the *Anopheles gambiae* genome-sequencing project is well advanced, and a similar project for *Aedes aegypti* is anticipated. In this chapter, we present an overview of our current knowledge of mosquito chromosome organization and evolution, with an emphasis on comparative genomics. *Comparative genomics* refers to the study of whole genome structure and organization, both within and among species. Contemporary technical advances have made it possible to extend the classical comparative approach in evolutionary biology to the discrete units that define genome structure and function.

CHROMOSOME MORPHOLOGY AND EVOLUTION

Phylogenetic relationships among members of the family Culicidae are discussed in detail in Chapter 29. Briefly, all studies to date consistently support Chaoboridae–Corethrellidae as sister taxa to Culicidae. All analyses support *Anophelinae* as the basal clade in Culicidae and are consistent in placing

449

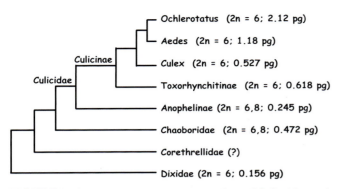

FIGURE 31.1 Genome sizes in some members of Culicoidea and the proposed phylogeny (Chapter 29).

Toxorhynchitinae as basal to the *Culicinae*. Within *Culicinae*, the tribe *Sabethini* is basal to *Culicini* and *Aedini*. All data sets support a monophyletic relationship between *Culicini* and *Aedini*. Many subgeneric relationships within *Sabethini*, *Culicini*, and *Aedini* may be paraphyletic and warrant taxonomic revision. If we overlay cytogenetic information onto this phylogeny, several patterns emerge (Fig. 31.1).

Basic Chromosome Number

Chromosomal karyotypes have been established for 19 of the ~37 genera, 35 of the 128 subgenera, and >200 of approximately 3,500 species (G. B. White 1980; Rai et al. 1982; Rai and Black 1999). One of the most remarkable findings of this karyotypic survey is that despite the ancient origin of the group and despite extensive repatterning of the genome involving translocations and inversions (Matthews and Munstermann 1994), the basic chromosome number ($2n = 6$) has remained unchanged. The only exception, *Chagasia bathana* ($2n = 8$) of the subfamily Anophelinae, possesses three autosome pairs and a heteromorphic pair of sex chromosomes (Kreutzer 1978). Interestingly, this is the most primitive member of *Anophelinae* and therefore of Culicidae (Figs. 29.16 and 29.17). All other anophelines possess two pairs of generally metacentric chromosomes of unequal size and one pair of heteromorphic sex chromosomes that often show extensive polymorphism in overall length and of the quantity and quality of heterochromatin differentiation among various species (G. B. White 1980). The position of the centromeres in the heteromorphic X and Y chromosomes in Anophelinae varies from subtelocentric or acrocentric to submetacentric and metacentric (Baimai et al. 1993a,b, 1995).

In contrast, species of the subfamilies Toxorhynchitinae and Culicinae all possess three pairs of homomorphic metacentric and/or slightly submetacentric chromosomes: a pair of small chromosomes, a pair of large chromosomes, and a pair of intermediate-sized chromosomes (Rai 1963; McDonald and Rai 1970; Rai et al. 1982; Rao and Rai 1987a). In Culicine mosquitoes, no sex chromosome dimorphism is evident. In species in which linkage group–chromosome correlations have been made, the shortest chromosome usually contains the sex-determining locus (McDonald and Rai 1970; Baker et al. 1971). Differences clearly exist in overall lengths and arm ratios of individual chromosomes, both within and between species, but can be easily overlooked if careful measurements of each arm of a chromosome are not made (Rai 1980; Rai et al. 1982). Total chromosomal length varies almost fivefold, from 8.2 µ in *Anopheles quadrimaculatus* to 39.3 µ in *Aedes alcasidi*. Within the genus *Aedes*, there is a threefold variation in chromosome length.

Conservation of chromosome number in Culicidae does not imply that linear orders have been rigidly maintained within individual chromosomes. Matthews and Munstermann (1994) indicated that while groups of allozyme loci remained conserved as linked and collinear blocks in a variety of Culicine taxa, the overall genome organization reflects numerous translocations and inversions. The extensive variation in chromosome number in most Diptera taxa studied does not predict the extreme conservation found in Culicidae. For example, chromosome number ranges from $n = 3$ to 7 in the genus *Drosophila*, and from $n = 3$ to 8 in the genus *Glossina*. In the family Muscidae, species with 5, 7, and even 9–10 pairs are known, although the chromosome number in most species examined is $2n = 12$. Nevertheless, certain other Dipteran families, such as Simulidae and Sarcophagidae, also show extensive conservation of chromosome number, although some exceptions do occur (M. J. D. White 1973). No logical explanation exists for the extraordinary conservation of the haploid chromosome number in Culicidae. The chromosomal karyotype data from *Culicidae* in general support M. J. D. White's (1973) suggestion that there may be some kind of barrier that maintains chromosome number in the Diptera. Nevertheless, we know nothing about the actual nature of such a barrier.

Centromere and Telomere Structure

At present, we have little information concerning the nature of the chromosome regions specialized for chromosome movement during cell division and chromosome end capping. The centromere serves as the central component for spindle fiber attachment during mitosis and meiosis. The molecular organization of centromeres in mosquitoes has not been determined. The ends of each chromosome typically consist of short, repetitive sequences called *telomeres* that are essential for chromosome stability. Telomeres provide an end cap that prevents chromosomes from fusing to each other. In most organisms studied the telomere consists of six to eight bp repeats that are added to the ends by the enzyme telomerase. In contrast, telomere organization in *An. gambiae* apparently comprises a complex that consists of a tandem array of 800 bp repeats that is independent of any telomerase activity (Roth et al. 1997). This strategy for telomere organization may be the norm among dipterans, for it has been observed in both *Drosophila* and *Chironomus* (Biessmann et al. 1990; Lopez et al. 1996).

Evolution of Sex-Determining Systems in Culicidae

Current dogma suggests that sex evolved independently in many organisms as an allelic difference at a gene located on a homologous pair of autosomes. Sexual dimorphism is evidenced among individuals heterozygous or homozygous at the sex-determining locus. While several theories have been postulated for the evolution of heteromorphic X and Y chromosomes, it seems likely that an early step involves the restriction in recombination between the sex-determining region and genes controlling male and female sex function (Charlesworth 1991, 1996). Once recombination between a proto-X and proto-Y is restricted, the Y chromosome is subject to degeneration through the progressive fixation of mutations. The absence of genetic exchange and corresponding loss of genetic activity then facilitates the stable insertion and accumulation of transposable elements and other repetitive sequences.

As indicated earlier, the Culicidae reflect an interesting dichotomy in sex chromosome evolution. That is, evolution of a heteromorphic Y-chromosome may have occurred only once or possibly may have been reversed in the evolution of sex chromosomes in Culicidae. The primitive Nematocera families Tipulidae and Dixidae possess homomorphic sex chromosomes. However, the sister families Chaoboridae–Corethrellidae contain genera with homomorphic (*Eucorethra, Corethrella, Chaoborus*) and heteromorphic (*Mochlonyx*) sex chromosomes (Rao and Rai 1987a). If homomorphy is ancestral in Culicidae, then it was retained in the lineages leading to Toxorhynchitinae and Culicinae, while heteromorphy probably evolved early in the evolution of Anophelinae and was retained in all taxa. Alternatively, if as proposed by Rao and Rai (1987a), Culicidae arose from a *Mochlonyx*-like ancestor, then Anophelinae retained heteromorphic sex chromosomes, while homomorphic sex chromosomes evolved through euchromatinization or loss of the Y in Toxorhynchitinae and Culicinae.

The lack of sex chromosome dimorphism among the culicines implies that sex is determined as a gene at a single autosomal locus, perhaps not necessarily associated with a particular chromosome across species. Females are homozygous recessive at this locus (*mm*) while males are heterozygous (*Mm*) for a dominant allele (Gilchrist and Haldane 1947; McClleland 1967). Indeed, chromosomes containing the male-determining *M* allele or the recessive female-determining *m* allele behave as any of the other autosomes, in that both are well populated with active genes and recombination occurs freely between them. In several *Aedes* species, the *m*-allele-containing chromosome is not only one of the shortest chromosome pairs (representing linkage group I), but also can be identified by the presence of Giemsa C-banding intercalary heterochromatin (Rai and Black 1999). The intercalary heterochromatin is not located on the shortest chromosome pair in *Armigeres subalbatus* and *Toxorhynchites splendens*, but instead is located on one of the larger chromosome pairs. It likely remains indicative of the *m*-allele-containing chromosome, however, because the sex-determination locus maps genetically with linkage group III markers in *Ar. subalbatus* (Ferdig et al. 1998). The sex-determination locus in *Culex tritaeniorhynchus* appears to be fluid in chromosome location, for it is located on chromosome 1 in some populations and on chromosome 3 in others (Baker et al. 1971; Mori et al. 2001). Also, in *C. tritaeniorhynchus* an unknown mechanism completely restricts recombination in females across the entire genome.

Despite the differences in sex chromosome structure within the Culicidae, it appears that all mosquito species still follow Haldane's rule for species hybrids (Presgraves and Orr 1998). Haldane's rule states that hybrids between closely related species will preferentially reflect inviability or infertility in the heterogametic sex. Interspecific comparisons across a wide range of *Anopheles* and *Aedes* species clearly support Haldane's rule, for in nearly every instance either the F_1 males are affected or both sexes are affected (Table 31.1). The F_1 females of such matings are rarely affected.

Genome Size

Mosquito genomes reflect considerable diversity in complexity that is determined by the relative proportions and distribution of single-copy, middle-repetitive, and highly-repetitive sequences. The

TABLE 31.1 Summary of Results of Hybridization Experiments Among Sister Species in Anophelinae and Culicinae

		Females Affected	Males Affected	Both Sexes Affected
Aedes	Sterility	0	11	10
	Inviability	1	1	11
Anopheles	Sterility	0	56	20
	Inviability	3	21	40

Source: Presgraves and Orr (1998). The authors classified outcomes as producing either hybrid sterility or inviability. The outcome of a reciprical cross was scored as B if both sexes were affected, M if males were affected, F if females were affected, and N if neither sex was affected.

haploid genome size, commonly termed the *C*-value, is measured in picograms (1×10^{-12} g), where 1 picogram (pg) equals 0.98×10^6 kilobases (kb) of double-stranded DNA. Estimates of mosquito genome sizes have been obtained using quantitative cytophotometry of Feulgen-stained primary spermatocytes, and in a few instances through analyses of renaturation kinetics of nuclear DNA (Palmer and Black 1997). Genome size varies about eightfold across mosquito species, likely due to species-specific complexities in repetitive sequences, yet also possibly reflecting considerable variation within an individual species.

Genome sizes have been estimated for 44 species belonging to 13 genera of mosquitoes and related Diptera (Table 31.2). Genome size is generally small in Anophelinae (0.24–0.29 pg). The single species examined in subfamily Toxorhynchitinae, *Toxorhynchites splendens*, has an intermediate-size genome of 0.62 pg, as do *Sabethes cyaneus* and *Wyeomyia smithii* (Sabethini). The haploid genomes of *Culex* species examined ranged from 0.54 to 1.02 pg and those of *Culiseta* species (Culicini) from 0.92 to 1.25 pg. *Armigeres subalbatus* and *Haemagogus equinus* (Aedini) contained 1.24 pg and 1.12 pg, respectively. At the generic level, the cosmopolitan genus *Aedes* showed more than threefold variation in nuclear DNA amounts, with the Polynesian spp. *Aedes cooki* and *Ae. pseudoscutellaris* possessing the lowest genome size of 0.59 pg, while *Ochlerotatus zoosophus* has the highest genome size of 1.9 pg.

In the context of phylogenetic relationships, the genome estimates suggest a trend toward an increase in genome size during the evolution of the Culicidae (Fig. 31.1). Black and Rai (1988) demonstrated that all classes of repetitive DNA sequences increased linearly in amount with total genome size. Furthermore, linear

regression analysis of 28 species belonging to 11 genera of the superfamily Culicoidea showed a highly significant positive correlation ($r = 0.87$; $p < 0.0001$) between total chromosomal length and haploid genome size (Rao and Rai 1987b). Nevertheless, the eightfold variation in haploid genome size was accompanied by only a 4.5-fold variation in the total chromosomal length, indicating that DNA amounts have increased almost twice as much as the increase in chromosomal size.

Organization at the Sequence Level

Individual eukaryote chromosomes each contain a single DNA molecule that is intimately associated with histone and nonhistone proteins. The DNA molecule consists of single-copy, middle-repetitive, and highly repetitive sequences, the specific composition of which can vary greatly both within and among chromosomes. The tightly compacted chromosome structure that we are all familiar with is, however, only evident during mitosis and meiosis. At other times in the cell cycle, the individual chromosomes exist in a relatively dispersed or less compacted state. The DNA and protein complex is referred to as *chromatin*, and a simple light microscope examination of metaphase chromosomes stained with various dyes, such as Feulgen and Giemsa, reveals that some chromosome regions stain much darker than other regions. The dark-staining regions represent very highly condensed chromatin, commonly termed *heterochromatin*, while the lighter-staining regions are termed *euchromatin*. Most of the active genes are located within the euchromatin, while heterochromatin is likely to be inactive and to remain highly condensed. Heterochromatin can be further subdivided into constitutive and facultative. *Constitutive* heterochromatin is always highly condensed and inactive, while *facultative* heterochromatin can vary between highly condensed and less condensed and active forms, depending on life stage and tissue types. One general feature of heterochromatin is that it is largely composed of repetitive DNA sequences. All mosquito species contain blocks of heterochromatin around the centromere regions of all chromosomes, although there are differences among species (Black and Rai 1999). In Anophelinae, the Y-chromosome seems to be completely heterochromatic, and considerable portions (one-half to three-fourths of the total length) of the X-chromosomes also seem to be heterochromatic. As indicated earlier, the female-determining sex allele seems to be associated with a heterochromatin band in some of the Culicinae.

Studies using DNA reassociation kinetics have provided information on genome organization in

TABLE 31.2 Haploid Genome Size (picogram DNA) in 38 Species Belonging to Nine Genera of Mosquitoes and Related Taxa

Family	Subfamily	Tribe	Genus	Subgenus	Species	Haploid Genome Size (pg) ± SE
Dixidae			*Dixa*		*obscura*	0.16[a]
Chaoboridae	Corethrellinae		*Corethrella*		*brakeleyi*	0.47 ± 0.02[b]
	Chaoborinae		*Mochlonyx*		*velutinus*	0.55 ± 0.02[b]
	Chaoborinae		*Chaoborus*	Chaoborusa	*americanus*	0.40 ± 0.02[b]
Culicidae	Anophelinae		*Anopheles*	Anopheles	*labranchiae*	0.23[a]
			An.		*atroparvus*	0.24[a]
			An.		*quadrimaculatus*	0.25 ± 0.01[b]
			An.		*freeborni*	0.29[a]
			An.	Cellia	*stephensi*	0.24[a]
			An.	Cellia	*gambiae*	0.27[e]
	Toxorhynchitinae		*Toxorhynchites*		*splendens*	0.62 ± 0.02[b]
	Culicinae	Sabethini	*Sabethes*		*cyaneus*	0.79 ± 0.02[b]
			Wyeomyia		*smithii*	0.86 ± 0.01[b]
		Culicini	*Culex*	Culex	*pipiens*	1.02 ± 0.19[a]
			Cx.	Culex	*pipiens*	0.54 ± 0.01[b]
			Cx.		*quinquefasciatus*	0.54 ± 0.01[b]
			Cx.		*restuans*	1.02 ± 0.04[b]
		Culisetini	*Culiseta*	Culicella	*litorea*	0.92[a]
			Cu.	Culicella	*morsitans*	1.21 ± 0.04[b]
			Cu.	Climacura	*melanura*	1.25 ± 0.01[b]
		Aedini	*Haemagogus*	Haemagogus	*equinus*	1.12 ± 0.02[b]
			Armigeres	Armigeres	*subalbanus*	1.12 ± 0.03[b]
			Ochlerotatus		*canadensis*	0.90 ± 0.02[c]
			Oc.		*communis*	1.01 ± 0.05[c]
			Oc.		*caspius*	0.99[a]
			Oc.		*stimulans*	1.44 ± 0.04[c]
			Oc.		*excrucianus*	1.50 ± 0.03[c]
			Oc.		*triseriatus*	1.52 ± 0.06[c]
			Oc.		*zoosophus*	1.90 ± 0.06[c]
			Ae.	Aedes	*cinereus*	1.21 ± 0.03[c]
			Ae.	Howardina	*bahamensis*	1.38 ± 0.03[c]
			Aedes	Stegomyia	*pseudoscutellaris*	0.59 ± 0.01[c]
			Ae.		*cooki*	0.59 ± 0.03[c]
			Ae.		*polynesiensis*	0.73 ± 0.02[c]
			Ae.		*aegypti*	0.81 ± 0.03[c]
			Ae.		*malayensis*	0.94 ± 0.03[c]
			Ae.		*hebrideus*	0.97 ± 0.03[c]
			Ae.		*seatoi*	0.97 ± 0.02[c]
			Ae.		*alcasidi*	0.97 ± 0.02[c]
			Ae.		*unilineatus*	1.06 ± 0.04[c]
			Ae.		*metallicus*	1.09 ± 0.03[c]
			Ae.		*heischii*	1.12 ± 0.04[c]
			Ae.		*katherinensis*	1.28 ± 0.02[c]
			Ae.		*pseudoalbopictus*	1.29 ± 0.03[c]
			Ae.		*flavopictus*	1.33 ± 0.02[c]
			Ae.		*albopictus*	
			(Geographic Populations)			
				Koh Samui, Thailand		0.62 ± 0.02[d]
				Korea		0.69 ± 0.03[d]
				Tananareve, Madagascar		0.78 ± 0.03[c]
				Sri Lanka		0.92 ± 0.05[d]
				Pontianak, Indonesia		1.07 ± 0.04[c]
				Ndo Ndo Creek, Solomon Islands		1.12 ± 0.06[d]
				Tananareve, Madagascar		1.15 ± 0.03[d]
				Hong Kong		1.26 ± 0.03[c]
				Mauritius		1.32 ± 0.04[c]
				Saigon, Vietnam		1.36 ± 0.04[d]
				Taipei, Taiwan		1.48 ± 0.05[d]

(continues)

TABLE 31.2 (*continued*)

Family	Subfamily	Tribe	Genus	Subgenus	Species	Haploid Genome Size (pg) ± SE
			(Geographic Populations)			
			Malaysia			
				Gertak Sanguul		0.64 ± 0.02[d]
				Malaysia		0.81 ± 0.03[d]
				Perak Road		0.83 ± 0.03[d]
				Sabah		0.85 ± 0.02[d]
			Singapore			
				Kent Ridge		0.75 ± 0.02[d]
				Amoy		1.29 ± 0.06[d]
			India			
				Calcutta		0.86 ± 0.03[c]
				Kolar		0.94 ± 0.03[c]
				Hardwar		0.96 ± 0.02[d]
				Delhi		1.02 ± 0.01[c]
				Pune		1.07 ± 0.62[c]
				Shalimar Bagh		1.42 ± 0.05[d]
			Hawaii			
				Makiki		0.75 ± 0.03[d]
				Oahu		1.24 ± 0.03[c]
				Manoa		1.47 ± 0.06[d]
			Japan			
				Nagasaki		0.76 ± 0.03[d]
				Saga		0.80 ± 0.02[d]
				Kabeshima		0.82 ± 0.03[d]
				Ebina		0.85 ± 0.03[d]
				Seburi		1.11 ± 0.04[d]
				Zama		1.16 ± 0.05[d]
				Tokyo		1.29 ± 0.03[c]
			Brazil			
				Cariacica		0.98 ± 0.04[d]
				Santa Tereza		1.18 ± 0.02[d]
			United States			
				Chambers County, TX		1.03 ± 0.03[d]
				Chicago, IL		1.11 ± 0.09[d]
				Jacksonville, FL		1.13 ± 0.10[d]
				Memphis, IN		1.23 ± 0.13[d]
				Houston 203, TX		1.33 ± 0.08[d]
				Indianapolis, IN		1.34 ± 0.09[d]
				Milford, DE		1.46 ± 0.05[d]
				New Orleans, LA		1.48 ± 0.26[d]
				Brazoria County, TX		1.50 ± 0.05[d]
				Evansville, IN		1.59 ± 0.11[d]
				Savannah, GA		1.65 ± 0.07[d]
				Houston 61, TX		1.66 ± 0.08[d]

[a] Jost and Mameli (1972).
[b] Rao and Rai (1990).
[c] Rao and Rai (1987).
[d] Kumar and Rai (1990).
[e] Besansky and Powell (1990).

mosquitoes. By genome organization, we are referring to the amounts, complexity, and dispersion of repetitive elements in a genome. The principle of DNA reassociation kinetics is relatively simple. The DNA is heated, to denature it into the two complementary single-stranded molecules, and then cooled and allowed to reform double-stranded molecules. Hydroxyapatite chromatography separates single- and double-stranded DNA molecules so that the reassociation process can be monitored. The amount of reassociation is reported as a function of the initial DNA concentration (C_0) and the time required to reassociate,

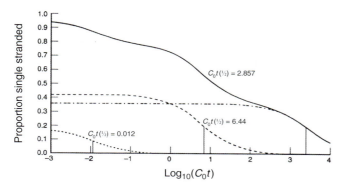

FIGURE 31.2 Reassociation of a eukaryotic genome plotted as a function of the $\log_{10}C_0t$ values. The solid line represents the composite curve. Dashed and dotted lines represent reassociation curves of the individual unique (-•-), moderately repetitive (—), and highly repetitive (····) components. From Palmer and Black (1997).

commonly abbreviated as C_0t (Fig. 31.2). The key to its utility is that multiple- or high-copy-number sequences will reassociate much faster than low- or single-copy sequences, and the resulting C_0t curve can be used to estimate the amount and complexity (variety of difference sequence types) of the various fractions.

The results of these analyses indicate that in the Anophelinae the majority (60–80%) of the genome is composed of single-copy sequences, while middle-repetitive and highly repetitive sequences are generally more prevalent in the Culicinae (Black and Rai 1988; Warren and Crampton 1991; Besansky and Powell 1992).

Two basic forms of genome organization have been described that generally apply to all eukaryotes. The first type, termed *short-period interspersion*, describes a pattern wherein single-copy sequences, 1,000–2,000 bp in length, alternate regularly with short (200–600 bp) and moderately long (1,000–4,000 bp) repetitive sequences. This characterizes genome organization in the majority of animal species and seems to be the norm for the culicine species (Black and Rai 1988), although the genus *Culex* appears to have an intermediate interspersion pattern (Cockburn and Mitchell 1989). The second type of genome organization, termed *long-period interspersion*, describes a pattern of long (≥5,600 bp) repeats alternating with very long (≥13,000 bp), uninterrupted stretches of unique sequences. Based on observations in *Anopheles quadrimaculatus* (Black and Rai 1988) and *A. gambiae* (Besansky and Powell 1992), genome organization in the Anophelinae follows the long-period interspersion pattern. This pattern is also characteristic of the fruitfly, *Drosophila melanogaster*.

The differences in genome organization between the Anophelinae and Culicinae have a significant influence on the utility of polytene chromosomes for physical mapping. Polytene chromosomes are the product of repeated chromosomal replication in the absence of cytokinesis, and they are evident in several tissues, including, most prominently, the salivary glands of larvae and the ovarian nurse cells of adult females. The individual chromatids remain aligned, revealing an easily visible and scorable banding pattern. Polytene chromosome banding patterns have been best characterized in *Drosophila* and provide important tools for the in situ hybridization and mapping of nucleic acid probes. Among the eight mosquito genera in which polytene chromosome morphology has been studied, only *Anopheles* spp. possess a chromocenter. All other genera (*Aedes*, *Culex*, *Mansonia*, *Toxorhynchites*, *Orthopodomyia*, *Wyeomyia*, *Sabethes*) lack a distinct chromocenter (G. B. White 1980; Munstermann et al. 1985). In the Anophelinae, high-quality preparations of polytene chromosomes are readily obtained and have proven invaluable for species identification and physical mapping. However, polytene chromosomes have not proven useful for mapping in the Culicinae because they do not spread easily. This likely reflects the repetitive nature of their genomes, which promotes ectopic pairing. Therefore, while cytogenetic studies of the Anophelinae have taken advantage of polytene chromosomes, the Culicinae have relied on examinations of metaphase chromosomes (Chapter 30).

Genes: Single-Copy and Multigene Families

GenBank (release 129) holds 229,199 Culicidae sequences (Table 31.3). The overwhelming majority of these sequences are from the *An. gambiae* genome project, either in the form of shotgun sequence accessions or bacterial artificial chromosome (BAC) end sequence data in the GSS database or as partial random cDNA sequence reads in the EST database. If we exclude the *An. gambiae* sequences, the remainder

TABLE 31.3 Culicidae Sequences Available from GenBank (Release 129)

Culicidae total	229,199
Culicidae minus *Anopheles gambiae* shotgun sequences	159,474
Culicidae and expressed sequence tags (EST database)	95,622
Culicidae and genome survey sequences (GSS database)	60,372
Culicidae minus *Anopheles gambiae*	3,826
Culicidae and *Aedes aegypti*	2,110

TABLE 31.4 Distributions of Multigene Families in *Saccharomyces cerevisiae, Drosophila melanogaster,* and *Caenorhabdites elegans*

Family Size	Yeast	Drosophila	C. elegans
1	4,768	10,786	12,858
2	415	404	665
3	56	113	188
4	23	46	93
5	9	21	71
6–10	19	52	104
11–20	8	26	57
21–50	0	11	33
51–80	0	0	5
>80	0	1	3
Number of gene families	530	674	1,219
Number of unique gene types*	5,298	11,460	14,077

* One gene family is counted as one unique gene type.
Source: Gu et al. (2002).

of the Culicidae is represented by only 3,826 sequences, of which 2,110 are from *Ae. aegypti*. Further, a number of the deposited sequences are multiple submissions of the same gene. Some targeted efforts have been directed toward genes expressed in particular tissues, such as antenna, salivary glands, midgut, Malpighian tubules, and fat body. In addition, it is clear that some genes are represented as members of large gene families, such as the serine proteases and cytochrome P450s.

The initial sequence assembly and gene annotation efforts for *An. gambiae* estimate a total of about 15,212 genes, although the actual number is probably somewhat lower. The NCBI and EBI data analysis was to be completed and available to the public in 2002. While we await the completed sequence assembly and gene annotation, we can get some insight on what we might expect for mosquitoes from other completed genome projects. A recent estimate of the total number of annotated genes for *Drosophila melanogaster* is 14,332, although the estimate is still tentative because our ability to identify open reading frames and their corresponding gene associations remains an inexact science (Harrison et al. 2002). Based on observations in yeast, *Drosophila melanogaster*, and *Caenorhabditis elegans*, the majority of genes are represented as single copies in the genome, with most multigene families consisting of only two or three members (Table 31.4). Gene families consisting of 50 or more members are rare.

Repetitive Sequences

Repetitive sequences can be broadly characterized into the moderate repeat elements that may occur in upwards of several hundreds of copies per genome and the highly repeated elements represented by several thousands of copies per genome. Repetitive elements described in mosquitoes would include the multigene family groups, microsatellites, minisatellites, ribosomal DNA (rDNA), and transposable elements.

As indicated in Chapter 30, the microsatellite repeat class has been used very effectively as a highly polymorphic, PCR-based genetic marker system in the Anophelinae, but it proved to be of limited utility for the Culicinae, largely due to their close association with other repeat classes in the latter group of mosquitoes. Briefly, these generally include di-, tri-, and tetranucleotide repeats that reflect locus-specific length polymorphisms in repeat number. In addition, another genetic marker system, RAPD-PCR (Chapter 30), takes advantage of 10-base-pair (bp) random repeat sequences that often occur within a few hundred base pairs of each other in an inverted configuration; similar short repeats are also known to exist in tandem in mosquito genomes and could be considered as examples of minisatellite elements.

The ribosomal RNA genes are arranged in multiple tandemly repeated units that vary from about 40 to 1,000 copies among mosquito species (Kumar and Rai 1993). The basic organizational unit of rDNA is an external transcribed spacer (ETS), the 18S RNA gene, an internal transcribed spacer (ITS 1), the 5.8S RNA gene, an internal transcribed spacer (ITS 2), the 28S RNA gene, and an intergenic nontranscribed spacer (IGS) (Fig. 33.5). In most mosquitoes, the rDNA is localized on a single chromosome, with the exception of *Ochlerotatus triseriatus*, where it is located on chromosomes 1 and 3 (Kumar and Rai 1993). For most culicines, the rDNA is localized on chromosome 1, with some exceptions: It is localized to chromosome 2 in *Aedes mediovittatus* and *Haemagogus equinus* and to chromosome 3 in *Armigeres subalbatus* and *Tripteroides bambusa*. With most anophelines, the rDNA is localized to the X-chromosome, but is located on both the X and Y chromosomes in *Anopheles quadriannulatus, Anopheles melas, Anopheles merus*, and *Anopheles quadrimaculatus* (Collins et al. 1989; Kumar and Rai 1990).

An interesting feature of rDNA genes and other tandemly arrayed repetitive sequences is that the multiple copies do not evolve independently, but instead evolve in concert and tend to become homogenized within species over time (Hillis and Dixon 1991; Elder and Turner 1995; Liao 1999) (Fig. 33.3). Although concerted evolution appears to be a universal phenomenon, the mechanisms that promote it remain the subject of debate. Two general hypotheses for a mechanism are most commonly suggested that both result from DNA replication and repair processes. First, unequal crossovers between daughter chromosomes could rapidly increase or decrease the frequencies of individual repeat types, since it can result in the duplication or elimination of many repeats in a single event. The second mechanism involves gene conversion, in which DNA repair machinery converts one of the alleles at a locus to the other allele in heterozygous individuals. Of significance, although the individual repeat sequences reflect considerable homogeneity, the DNA sequences that flank them are often not affected by concerted evolution and retain high levels of polymorphism. As discussed in Chapter 33, the ITS and IGS regions of rDNA reflect considerable inter- and intraspecific polymorphisms and provide important tools for species identification.

Satellite DNA is an example of highly repetitive DNA and takes its name from the observation that it is often evident as a distinct band from the remainder of the nuclear DNA in CsCl gradients (Chapter 30). It typically consists of very long tracts of relatively simple tandem repeats. Although little is known about satellite DNA in mosquitoes, it is likely that these sequences are present in the centromere and telomere regions of mosquito chromosomes and that they may play a role in chromosome maintenance and replication (Csink and Henikoff 1998).

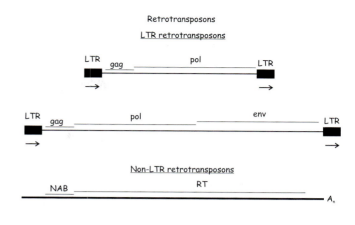

FIGURE 31.3 Transposable elements. LTR retrotransposons have long terminal repeats (LTRs) at their termini and coding sequences similar to the *gag* and *pol* genes of a retrovirus. Some, like the *gypsy* element of *Drosophila*, also have an *env*-like gene. Non-LTR retrotransposons have no terminal repeats and usually have two coding sequences. The first, NAB, codes for nucleic acid–binding protein, and the second RT, a protein with reverse transcriptase activity. Transposons have terminal inverted repeats and encode transposase. Redrawn from Wright and Finnegan (2001).

Transposable Elements

Transposable elements (TEs) are moderately repetitive DNA sequences that have the ability to move about the genome and to duplicate themselves in the process (Kidwell and Lisch 2001). The elements may be autonomous, wherein they carry genes necessary for their own movement or transposition, or they may be nonautonomous and depend on gene products from related autonomous elements for their movement. There are two general classes of TEs: The class I elements are the retrotransposons that transpose by reverse transcription of an RNA intermediate; the class II elements are the transposons that transpose directly from DNA to DNA (Fig. 31.3). The retrotransposons are divided into two groups, those that have long terminal repeats (LTRs) and those without the repeats (non-LTRs), which includes the long interspersed nuclear elements (LINEs) and the short interspersed nuclear elements (SINEs).

Transposable elements have been shown to represent a significant portion of the middle repetitive DNA of most eukaryote genomes, and indeed this is the case for mosquitoes. A large number of repetitive elements have been identified in both anopheline and culicine genomes (Table 31.5). For example, the SINE element

Feilai is highly abundant in the *Ae. aegypti* genome, with about 59,000 copies that represent about 2% of the entire genome (Tu 1999). Also of interest, various TEs are frequently found clustered together. *Pony* transposon elements represent about 1.1% of the *Ae. aegypti* genome and are often found near *Feilai* and other transposable elements (Tu 2000). Most transposable elements are found in A+T-rich noncoding regions of the genome. In general, TEs are very abundant in genome areas with little or no recombination; therefore, they may represent a significant fraction of heterochromatin DNA. Most of these TEs are inactive, which is a reflection of the basic nature of these sequences. That is, because the effects of active TEs in a genome are usually detrimental, they are subject to high mutation rates and are effectively silenced.

Transposable elements continue to play a significant role in genetic variation and overall genome evolution (Kidwell and Lisch 2001). Because of their mobile and replicative activity, the insertion or excision of TEs within exons or regulatory sequences of genes can have profound effects on gene activity (McClintock 1984). Insertion of TEs into genes often results in null mutations but can on occasion result in novel

TABLE 31.5 Repetitive Elements of Anophelines and Culicines

Repetitive Elements	Drosophila Homolog	Anophelines	Culicines
Retro elements			
LINEs or nLTR			
		LINE-like	
		Q	*Q*-like
		RT1	
		RT2	
		T1	
	Bilbo		*Bilbo*-like
	DMCR1A	*DMCR1A*-like	
	G		*G*-like
	Jockey	*JUAN*-like	*JUAN*
	LINEJ1	*LINEJ1*-like	*LINEJ1*-like
	R2B	*R2B*-like	*R2B*-like
	TARTB1	*TARTB1*-like	
			Lian
	I	*MosquI*-like	*MosquI*
		I-like	
SINEs		*Feilai*-like	*Feilai*
LTR elements			
	Gypsy	*Accord*-like	
		Bloodi-like	
		DM412-like	
		HMSBeagle-like	
		Mdg1-like	*Mdg1*-like
		Mdg3-like	
		Quasimodo-like	
		Tabor-like	
		Tirant-like	
		Ty3-like	
		Ulyss-like	
		Zam-like	
	Copia	*DM1731*-like	
		Moose	
		Ty1-like	
		ZebeedeeI-like	*ZebeedeeI*
		ZebeedeeII-like	*ZebeedeeII*
			MosqCopia
	Pao	*Batumi*-like	
		Bel-like	*Bel*-like
		Diver-like	
		Ninja-like	
		Pao-like	
		Roo-like	*Roo*-like
DNA elements	*Het-A*	*Odysseus*	
	mariner	*mariner*	*CM-gag*
	Tc1	*Pegasus*	
		FB4-like	
		Ikirara	
		Paris-like	
		Quetzal	
		Uhu-like	
MITEs			*Dufu*
			Kaizoku
			Pony
			Wujin
			Wukong
			Wuneng
			Youzi
Satellites	*HSATII*	*HSATII*-like	

LINE, long interspersed element; LTR, long terminal repeat name; MITE, miniature inverted-repeat transposable element; SINE, short interspersed element. *Source*: Condensed from Severson et al. (2001).

phenotypes (Kidwell and Lisch 1997). Although the mechanism is unclear, there is evidence suggesting that TEs can be activated when an organism is subjected to stress, where *stress* is defined as environmental changes that drastically reduce fitness (Capy et al. 2000). One potential consequence of this activation is that it may create novel genotypes that have fitness benefits under stressful conditions.

Comparative Genomics

Detailed genome comparisons among mosquitoes and other insects have been limited by the general lack of suitable cytogenetic or genetic tools with broad applicability across species (Heckel 1993). Comparative examinations of morphological mutant and isozyme loci among the sheep blowfly, *Lucilia cuprina*, the medfly, *Ceratitis capitata*, the housefly, *Musca domestica*, and *Drosophila melanogaster* suggested that linkage groups have generally remained conserved over the evolutionary history of these higher Diptera (Foster et al. 1981). Linkage group correlations among the higher Diptera and the lower Diptera, which includes the Culicidae, were relatively weak (Weller and Foster 1993). A synthesis by Matthews and Munstermann (1994) of available isozyme linkage data representing mosquito species across all three subfamilies provided considerable insight into chromosome evolution in mosquitoes. It became evident that groups of genes do remain physically linked within the mosquito genera. Linkage group conservation, even across diverse genera, was sufficient to permit identification of six syntenic arrangements of enzyme loci, which is suggestive of ancestral linkage groups. Further, this analysis suggested that these six syntenic groups probably represent whole chromosome arms and that karyotype evolution in mosquitoes may have largely involved gross structural changes, including Robertsonian translocations and paracentric inversions.

Development of DNA-based genetic markers provides us with a suitable basis for comparative genome mapping. Homologous genes can be identified directly by DNA-DNA cross-hybridization and by nucleotide sequence analysis. Comparative linkage maps, based on homologous genes as markers, can be used effectively to predict the location of a particular gene in one species based on knowledge of its position in another species (Nadeau 1989). Comparative maps have important implications for both exploring gene functions and understanding chromosome evolution in mosquito species.

Within the Anophelinae, a general pattern for conservation of linkage groups, and particularly chromosome arms, was implicated in early investigations by apparent similarities observed in the banding patterns of polytene chromosomes among various species (Coluzzi et al. 1970; Green and Hunt 1980; Kitzmiller et al. 1967). Although the X chromosome remained conserved across *Anopheles* species, chromosomes 2 and 3 showed evidence for whole-arm translocations that generated polymorphisms in arm associations (Green and Hunt 1980). Subsequently, correlation between chromosome arms and genetic linkage groups was confirmed for species within the subgenus *Cellia* (Hunt 1987). Comparative in situ DNA hybridizations that identify 23 common loci in *An. gambiae* and *An. albimanus* clearly indicate a general pattern of whole-arm conservation across species, with considerable rearrangement evident when the individual-arm homologues are compared (Fig. 31.4). Because *An. gambiae* and *An. albimanus* represent members of evolutionarily diverse subgenera, *Cellia* and *Nyssorhynchus*, it seems likely that this pattern holds across most if not all anophelines and that more closely related species may reflect higher frequencies of linkage order conservation within homologous arms.

An extensive genetic linkage map, based largely on random cDNA clones or cDNA clones of known genes, has been developed for *Aedes aegypti* (Severson et al. 2002; Fulton et al. 2001). These sequences provide single-locus physical landmarks that are often highly conserved across mosquito species and, therefore, are extremely useful for comparative mapping. Indeed, they have provided us the basis for the development of comparative linkage maps for several culicine species. The first comparative RFLP marker linkage map was developed for *Aedes albopictus* (Severson et al. 1995). This map included 18 RFLP markers that provided broad coverage of the *Ae. aegypti* genome. Although the observed recombination frequencies between markers varied across the two species, the linkage group associations and linear orders of the markers remained statistically identical. The level of conservation observed between *Ae. aegypti* and *Ae. albopictus* is probably not surprising, for they are both members of the subgenus *Stegomyia*. However, the comparative map developed for a distant genus, *Armigeres subalbatus*, also showed considerable conservation with *Ae. aegypti* (Ferdig et al. 1998). For 25 of the 26 RFLP markers examined, the linkage groups and linear orders were nearly identical to those of *Ae. aegypti*. The two species do reflect an inversion polymorphism across the centromere region of chromosome 2, and the sex determination locus maps to different chromosomes.

We generally would predict that conserved linkage associations should decrease as the phylogenetic

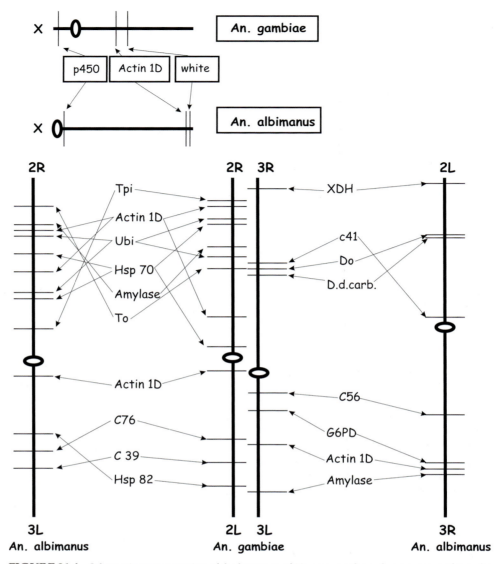

FIGURE 31.4 Schematic representation of the locations of 23 genes on three chromosomes of *Anopheles gambiae* and *An. albimanus* to show that rearrangements have occurred. Redrawn from Cornel and Collins (2000).

distance increases between species. Indeed, comparative linkage maps for *Culex pipiens* and *Ochlerotatus triseriatus* do reflect extensive chromosomal rearrangements when compared to *Ae. aegypti*, *Ae. albopictus*, or *Ar. subalbatus* (Mori et al. 1999; Anderson et al. 2001). As predicted by isozyme marker data (Matthews and Munstermann 1994), for both *Cx. pipiens* and *Oc. triseriatus*, chromosomes 2 and 3 largely reflect whole-arm translocations when compared with *Ae. aegypti* chromosomes 2 and 3, with several apparent within-arm inversion polymorphisms. The *Ae. aegypti/Cx. pipiens* comparative maps are shown in Figure 31.5. As one

would expect, near complete linkage conservation was observed between *Cx. pipiens* and *Culex quinquefasciatus*, another species within the subgenus *Culex* (Mori et al. 2001).

Linear orders for marker loci on chromosome 1 seem to be remarkably conserved across a diverse group of species, including *Ae. aegypti*, *Ae. albopictus*, *Ar. subalbatus*, *Cx. pipiens*, and *Cx quinquefasciatus*, although the observed recombination frequencies vary considerably. In contrast, chromosome 1 in *Oc. triseriatus* reflects considerable rearrangement when compared to these species (Anderson et al. 2001). Also of

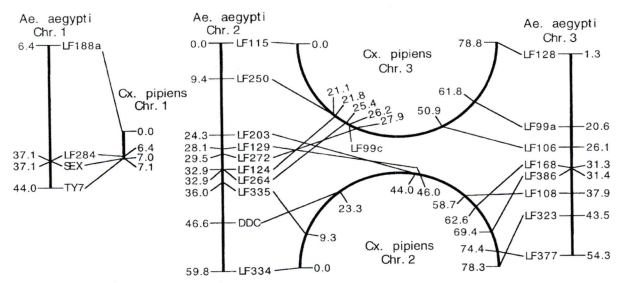

FIGURE 31.5 Comparative genetic maps of *Culex pipiens* and *Aedes aegypti* based upon common RFLP loci. Map units are in Kosambi distances (cM). From Mori et al. (1999).

interest, one marker (LF90) on the p-terminus of chromosome 1 and another (LF377) on the q-terminus of chromosome 3 in *Ae. aegypti* (Severson et al. 2002) remain unlinked in our studies of *Oc. triseriatus*. The chromosome locations for these loci therefore remain undefined in *Oc. triseriatus*. The only culicine species examined to date that clearly shows interchromosomal rearrangements that involve chromosome 1 is *Aedes togoi* (Tadano 1984; Matthews and Munstermann 1994). Our preliminary RFLP marker comparisons for *Ae. togoi* and *Ae. aegypti* have confirmed this phenomenon.

Finally, the availability of the complete sequence for the *Drosophila melanogaster* genome (Adams et al. 2000) provides us an immediate and powerful tool for comparisons with other insects, particularly other dipterans, including mosquitoes. Comparative investigations to date clearly indicate, however, that the *Drosophila* genome information will be of only limited utility for predicting gene locations and orders in mosquitoes. That is, while blocks of orthologous genes have remained physically linked following the divergence of the *Drosophila* and mosquito lineages, the linear orders of these genes have been extensively reshuffled. This paradigm holds for both anophelines (Bolshakov et al. 2002) and culicines. The complete annotation of the *An. gambiae* genome represents an exciting milestone in mosquito genomics and will undoubtedly spur similar projects for additional species, most likely *Ae. aegypti* and *Cx. pipiens*. It seems inevitable that the rapid growth in contemporary genomics will ultimately unveil the detailed evolutionary histories of a number of mosquito species and allow for unprecedented direct comparisons across diverse taxa.

Readings

Adams, M. D., Celniker, S. E., Holt, S. E., et al. 2000. The genome sequence of *Drosophila melanogaster*. *Science* 287: 2185–2195.

Anderson, J. R., Grimstad, P. R., and Severson, D. W. 2001. Chromosomal evolution among six mosquito species (Diptera: Culicidae) based on shared restriction fragment length polymorphisms. *Mol. Phylogenet. Evol.* 20: 316–321.

Baker, R. H., Sakai, R. K., and Mian, A. 1971. Linkage group–chromosome correlation in *Culex tritaeniorhynchus*. *Science* 171: 585–587.

Baimai, V., Rattanarithikul, R., and Kajchalao, U. 1993a. Metaphase karyotypes of *Anopheles* of Thailand and Southeast Asia. I. The Hyrcanus group. *J. Am. Mosq. Control Assoc.* 9: 59–67.

Baimai, V., Rattanarithukul, R., and Kajchalao, U. 1993b. Metaphase karyotypes of *Anopheles* of Thailand and Southeast Asia. II. The Maculatus group, Neocellia series, subgenus *Cellia*. *Mosq. Syst.* 25: 116–123.

Baimai, V., Rattanarithikul, R., and Kijchalao, U. 1995. Metaphase karyotypes of *Anopheles* of Thailand and Southeast Asia: IV. The Barbirostris and Umbrosus species groups, subgenus *Anopheles* (Diptera: Culicidae). *J. Am. Mosq. Control Assoc.* 11: 323–328.

Besansky, N. J., and Powell, J. R. 1992. Reassociation kinetics of *Anopheles gambiae* (Diptera: Culicidae) DNA. *J. Med. Entomol.* 29: 125–128.

Biessmann, H., Carter, S. B., and Mann, J. M. 1990. Chromosome ends in *Drosophila* without telomeric DNA sequences. *Proc. Natl. Acad. Sci. USA* 87: 1758–1761.

Black, W. C. IV, and Rai, K. S. 1988. Genome evolution in mosquitoes: Intraspecific and interspecific variation in repetitive DNA amounts and organization. *Genet. Res.* 51: 185–196.

Bolshakov, V. N., Topalis, P., Blass, C., Kokoza, E., della Torre, A., Kafatos, F. C., and Louis, C. 2002. A comparative genomic analysis of two distinct Diptera, the fruitfly, *Drosophila melanogaster*, and the malaria mosquito, *Anopheles gambiae*. *Genome Res.* 12: 57–66.

Capy, P., Gasperi, G., Biemont, C., and Bazin, C. 2000. Stress and transposable elements: Coevolution or useful parasites? *Heredity* 85: 101–106.

Charlesworth, B. 1991. The evolution of sex chromosomes. *Science* 251: 1030–1033.

Charlesworth, B. 1996. The evolution of chromosomal sex determination and dosage compensation. *Current Biol.* 6: 149–162.

Cockburn, A. F., and Mitchell, S. E. 1989. Repetitive DNA interspersion patterns in Diptera. *Arch. Insect Biochem. Physiol.* 10: 105–113.

Collins, F. H., Paskewitz, S. M., and Finnerty, V. 1989. Ribosomal RNA genes of the *Anopheles gambiae* species complex. *Adv. Disease Vector Res.* 6: 1–28.

Coluzzi, M. M., Cancrini, G., and DiDeco, M. 1970. The polytene chromosomes of *Anopheles superpictus* and relationships with *Anopheles stephensi*. *Parassitologia* 12: 101–112.

Cornel, A. J., and Collins, F. H. 2000. Maintenance of chromosome arm integrity between two *Anopheles* mosquito subgenera. *J. Heredity* 91: 364–370.

Csink, A. K., and Henikoff, S. 1998. Something from nothing: The evolution and utility of satellite repeats. *Trends Genet.* 14: 200–204.

Elder, J. F., Jr., and Turner, B. J. 1995. Concerted evolution of repetitive DNA sequences in eukaryotes. *Quar. Rev. Biol.* 70: 297–320.

Ferdig, M. T., Taft, A. S., Severson, D. W., and Christensen, B. M. 1998. Development of a comparative genetic linkage map for *Armigeres subalbatus* using *Aedes aegypti* RFLP markers. *Genome Res.* 8: 41–47.

Foster, G. G., Whitten, M. J., Konovalov, C., Arnold, J. T. A., and Maffi, G. 1981. Autosomal genetic maps of the Australian sheep blowfly, *Lucilia cuprina dorsalis* R.-D. (Diptera: Calliphoridae), and possible correlations with the linkage maps of *Musca domestica* L. and *Drosophila melanogaster* (Mg.). *Genet. Res. Camb.* 37: 55–69.

Fulton, R. E., Salasek, M. L., DuTeau, N. M., and Black IV, W. C. 2001. SSCP analysis of cDNA markers provides a dense linkage map of the *Aedes aegypti* genome. *Genetics* 158: 715–726.

Gilchrist, B. M., and Haldane, J. B. S. 1947. Sex linkage and sex determination in a mosquito, *Culex molestus*. *Hereditas (Lund.)* 33: 175–190.

Green, C. A., and Hunt, R. H. 1980. Interpretation of variation in ovarian polytene chromosomes of *Anopheles funestus* Giles, *A. parensis* Gillies, and *A. aruni*? *Genetica* 51: 187–195.

Gu, Z., Cavalcanti, A., Chen, F.-C., Bouman, P., and Li, W.-H. 2002. Extent of gene duplication in the genomes of *Drosophila*, nematode, and yeast. *Mol. Biol. Evol.* 19: 256–262.

Harrison, P. M., Kumar, A., Lang, N., Snyder, M., and Gerstein, M. 2002. A question of size: The eukaryotic proteome and the problems defining it. *Nuc. Acids Res.* 30: 1083–1090.

Heckel, D. G. 1993. Comparative linkage mapping in insects. *Annu. Rev. Entomol.* 38: 381–408.

Hillis, D. M., and Dixon, M. T. 1991. Ribosomal DNA: Molecular evolution and phylogenetic inference. *Quar. Rev. Biol.* 66: 411–453.

Hunt, R. H. 1987. Location of genes on chromosome arms in the *Anopheles gambiae* group of species and their correlation to linkage data for other anopheline mosquitoes. *Med. Vet. Entomol.* 1: 81–88.

Kidwell, M. G., and Lisch, D. R. 1997. Transposable elements as sources of variation in animals and plants. *Proc. Natl. Acad. Sci. USA* 94: 7704–7711.

Kidwell, M. G., and Lisch, D. R. 2001. Perspective: Transposable elements, parasitic DNA, and genome evolution. *Evolution* 55: 1–24.

Kitzmiller, J. B. 1976. Genetics, cytogenetics, and evolution of mosquitoes. *Adv. Genetics* 18: 315–433.

Kitzmiller, J. B., Frizzi, G., and Baker, R. H. 1967. Evolution and speciation within the *maculipennis* complex of the genus *Anopheles*. In *Genetics of Insect Vectors of Disease* (J. W. Wright and R. Pal, eds.). Elsevier, Amsterdam, pp. 151–210.

Kreutzer, R. D. 1978. A mosquito with eight chromosomes: *Chagasio bathana* Dyar. *Mosq. News* 38: 554–558.

Kumar, A., and Rai, K. S. 1990. Chromosomal localization and copy number of 18S + 28S ribosomal RNA genes in evolutionarily diverse mosquitoes (Diptera, Culicidae). *Hereditas* 113: 277–289.

Kumar, A., and Rai, K. S. 1993. Molecular organization and evolution of mosquito genomes. *Comp. Biochem. Physiol.* 106B: 495–504.

Liao, D. 1999. Concerted evolution: Molecular mechanism and biological implications. *Am. J. Hum. Genet.* 64: 24–30.

Lopez, C. C., Nielsen, L., and Edstrom, J.-E. 1996. Terminal long tandem repeats in chromosomes from *Chironomus pallidivittatus*. *Mol. Cell. Biol.* 16: 3285–3290.

Matthews, T. C., and Munstermann, L. E. 1994. Chromosomal repatterning and linkage group conservation in mosquito karyotype evolution. *Evolution* 48: 146–154.

McClelland, G. A. H. 1967. Speciation and evolution in *Aedes*. In *Genetics of Insect Vectors of Disease* (J. W. Wright and R. Pal, eds.). Elsevier, New York, pp. 277–311.

McClintock, B. 1984. The significance of responses of the genome to challenge. *Science* 226: 792–801.

McDonald, P. T., and Rai, K. S. 1970. Correlation of linkage groups with chromosomes in the mosquito, *Aedes aegypti*. *Genetics* 66: 475–485.

Mori, A., Severson, D. W., and Christensen, B. M. 1999. Comparative linkage maps for the mosquitoes (*Culex pipiens* and *Aedes aegypti*) based on common RFLP loci. *J. Heredity* 90: 160–164.

Mori, A., Tomita, T., Hidoh, O., Kono, Y., and Severson, D. W. 2001. Comparative linkage map development and identification of an autosomal locus for insensitive acetylcholinesterase-mediated insecticide resistance in *Culex tritaeniorhynchus*. *Insect Mol. Biol.* 10: 197–203.

Munstermann, L. E., Marchi, A., Sabatini, A., and Coluzzi, M. 1985. Polytene chromosomes of *Orthopodomyia pulcripalpis* (Diptera, Culicidae). *Parassitologia* 27: 267–277.

Nadeau, J. H. 1989. Maps of linkage and synteny homologies between mouse and man. *Trends Genet.* 5: 82–86.

Palmer, M. J., and Black IV, W. C. 1997. The importance of DNA reassociation kinetics in insect molecular biology. In *The Molecular Biology of Insect Disease Vectors: A Methods Manual* (J. Crampton, C. B. Beard, and C. Louis, eds.). Chapman and Hall, New York, pp. 172–194.

Presgraves, D. C., and Orr, H. A. 1998. Haldane's rule in taxa lacking a hemizygous X. *Science* 282: 952–954.

Rai, K. S. 1963. A comparative study of mosquito karyotypes. *Ann. Entomol. Soc. Amer.* 56: 160–170.

Rai, K. S. 1980. Evolutionary cytogenetics of Aedine mosquitoes. *Genetica* 52/53: 281–290.

Rai, K. S., and Black IV, W. C. 1999. Mosquito genomes: Structure, organization, and evolution. *Adv. Genet.* 41: 1–33.

Rai, K. S., Pashley, D. P., and Munstermann, L. E. 1982. Genetics of speciation in aedine mosquitoes. In *Recent Advances in Genetics of Insect Disease Vectors* (W. M. W. Steiner, W. J. Tabachnick, K. S. Rai, and S. K. Narang, eds.). Stipes, Champaign, IL, pp. 84–129.

Rao, P. N., and Rai, K. S. 1987a. Comparative karyotypes and chromosomal evolution in some genera of *Nematocerous* (Diptera: Nematocera) families. *Ann. Entomol. Soc. Amer.* 80: 321–332.

Rao, P. N., and Rai, K. S. 1987b. Inter- and intraspecific variation in nuclear DNA content in *Aedes* mosquitoes. *Heredity* 59: 253–258.

Rohdendorf, B. 1974. *The Historical Development of Diptera.* University of Alberta Press, Edmonton, Alberta.

Roth, C. W., Kobeski, F., Walter, M. F., and Biessmann, H. 1997. Chromosome end elongation by recombination in the mosquito *Anopheles gambiae. Mol. Cell. Biol.* 17: 5176–5183.

Severson, D. W., Mori, A., Kassner, V. A., and Christensen, B. M. 1995. Comparative linkage maps for the mosquitoes, *Aedes albopictus* and *Ae. aegypti,* based on common RFLP loci. *Insect Mol. Biol.* 4: 41–45.

Severson, D. W., Brown, S. E., and Knudson, D. L. 2001. Genetic and physical mapping in mosquitoes: Molecular approaches. *Annu. Rev. Entomol.* 46: 183–219.

Severson, D. W., Meece, J. K., Lovin, D. D., Saha, G., and Morlais, I. 2002. Linkage map organization of expressed sequence tags and sequence tagged sites in the mosquito *Aedes aegypti.* Insect Mol. Biol. 11: 371–378.

Tadano, T. 1984. A genetic linkage map of the mosquito *Aedes togoi. Jpn. J. Genet.* 59: 165–176.

Tu, Z. 1999. Genomic and evolutionary analysis of *Feilai,* a diverse family of highly reiterated SINEs in the yellow fever mosquito, *Aedes aegypti. Mol. Biol. Evol.* 16: 760–772.

Tu, Z. 2000. Molecular and evolutionary analysis of two divergent subfamilies of a novel miniature inverted repeat transposable element in the yellow fever mosquito, *Aedes aegypti. Mol. Biol. Evol.* 17: 1313–1325.

Warren, A. M., and Crampton, J. M. 1991. The *Aedes aegypti* genome: Complexity and organization. *Genet. Res. Camb.* 58: 225–232.

Weller, G. L., and Foster, G. G. 1993. Genetic maps of the sheep blowfly *Lucilia cuprina*: Linkage-group correlations with other dipteran genera. *Genome* 36: 495–506.

White, G. B. 1980. Academic and applied aspects of mosquito cytogenetics. In *Insect Cytogenetics* (R. L. Blackman, G. M. Hewitt, and M. Ashburner, eds.). Blackwell Scientific, London, pp. 245–274.

White, M. J. D. 1973. *Animal Cytology and Evolution*, 3rd ed. Cambridge University Press, Cambridge, U.K.

Wright, S., and Finnegan, D. 2001. Genome evolution: Sex and the transposable element. *Current Biol.* 11: R296–R299.

Wu, C.-I. 2001. The genic view of the process of speciation. *J. Evol. Biol.* 14: 851–865.

32

Population Genetics of Disease Vectors

WILLIAM C. BLACK IV AND WALTER J. TABACHNICK

INTRODUCTION

Population genetics is the study of genetic variation within and among natural populations of a species and the forces that act to differentiate or homogenize extant populations in nature (Tabachnick and Black 1994). It includes studies of the flow of genetic information among individuals in one population and among individuals from different populations. Coupled with the growing ability to characterize a species genome, population genetics provides a powerful framework to differentiate genomewide and locus-specific effects and thereby test for associations between phenotypic variation and gene function in natural populations. This latter area has come to be known as *population genomics* (Black et al. 2001) and *association mapping* (Mackay, 2001).

Population genetics has its foundations in the works of Sir Ronald A. Fisher, Sewell Wright, and J. B. S. Haldane. Each contributed to the development of the theoretical basis for the discipline in the 1930s. In the late 1930s and 1940s, the field was integrated with evolutionary and systematic biology through the works of Theodosius Dobzhansky and, later, Ernst Mayr, George G. Simpson, and G. Ledyard Stebbins. In the 1960s, Mootoo Kimura showed that genetic polymorphisms that are neutral in terms of natural selection not only can become established and are in fact abundant. This area came to be known as *neutralism* and established the foundation of modern biochemical and molecular population genetics.

As early as the 1950s, J. B. Kitzmiller and G. B. Craig, Jr., recognized the importance of population genetics

in understanding arthropod disease vectors. Vector population genetics studies initially relied on Craig's analysis in the early 1960s of morphologic mutants in populations of *Aedes aegypti*. In spite of these early attempts, the difficulty of population genetic analyses using morphological variants prevented much progress. A major breakthrough occurred with isozyme electrophoresis and its first application to the population genetics of *Drosophila pseudoobscura* populations in 1966 by R. C. Lewontin and J. L. Hubby. Soon after, there were similar studies by L. Bullini and colleagues, at the University of Rome, on species of *Aedes*. Craig and Leonard E. Munstermann, at the University of Notre Dame, began population genetic studies of African *Ae. aegypti*, and W. J. Tabachnick and Jeffery R. Powell, at Yale University, also began analyzing populations of *Ae. aegypti* from throughout the world. M. L. Cheng, at the University of California—Los Angeles, and S. Miles, at the London School of Tropical Medicine and Hygiene, conducted studies of the *Culex pipiens* complex. Research principally with *Ae. aegypti* has proceeded to the present, and other species of *Aedes* have also accumulated a large body of information. Population genetic studies of several species of *Anopheles* have been conducted by M. Coluzzi, J. Seawright, G. Lanzaro, N. Besansky, T. Lehman, and J. Conn. Since then, there have been population genetic investigations in a diversity of other arthropod vectors, including many additional species of mosquitoes (de Sousa et al. 1999; Kambhampati et al. 1990; Onyabe and Conn 2001; West and Black 1998), *Glossina* spp. (Krafsur et al. 2001), sand flies (Esseghir et al. 1997; Munstermann et al. 1998; Soto et al. 2001), Ceratopogonids (biting midges) (Tabachnick 1990), and ticks (Norris et al. 1996; Delaye et al. 1997; Kain et al. 1999).

In Chapter 29 the use of molecular phylogenetics to estimate the origins of species was described. As an example, two sister species, *Aedes aegypti* (species A) and *Ae. simpsoni* (species B) shared a common, extinct ancestor (*Ae.* species γ) (Fig. 32.1A). Sometime in the past, populations of *Ae.* species δ became isolated from species δ', possibly because of geographic barriers or through specialization to different ecological niches or through phenological or seasonal adaptation. Through time these populations diverged genetically

from their ancestors, eventually developing either prezygotic and/or postzygotic isolation mechanisms; from that point on, the genomes of the two populations began to evolve independently. The population, destined to become *Ae. aegypti*, grew and members dispersed to new, suitable habitats. Eventually, through a variety of mechanisms, *Ae. aegypti* became widespread throughout the tropical and subtropical world (Fig. 32.1B).

Systematics employs homologous sets of genetic characters among individual members of different species to resolve phylogenetic relationships. In population genetics, homologous sets of genetic characters

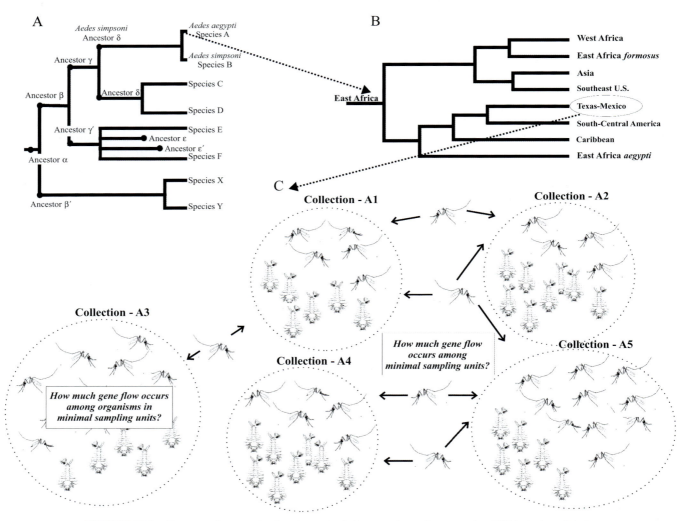

FIGURE 32.1 Conceptual connection between population genetics and systematics. (A) Evolution of *Aedes aegypti* as a species; (B) evolution of *Ae. aegypti* worldwide populations; (C) local patterns of gene flow among local *Ae. aegypti* collections.

within and among individual populations are the subject of analysis. In addition, the genetic relationships among populations (Fig. 32.1B) or pedigrees among individual members of a population are estimated. The genetic relationships of *Ae. aegypti* populations in Figure 32.1B were compiled in a series of now-classical papers by Tabachnick, G. P. Wallis, Powell, and Munstermann (reviewed in Tabachnick 1991). This phylogeny was estimated using the phenetic techniques discussed in Chapter 29 by applying measures of genetic distance appropriate for population genetics.

Population genetics seeks to study mechanisms that maintain genetic homogeneity among extant populations and the forces that cause populations to genetically diverge. For the purposes of this chapter we define a *genetic population* as a group of individuals among which there are no barriers to mating or to *gene flow*. Ecologists or epidemiologists often refer to a population as a collection of individuals that are spatially clustered around finite resources needed for survival, reproduction, or development. We define each of these spatial clusters as a *collection*, as opposed to a population. Collections may consist of genetically heterogeneous groups of individuals. These groups might correspond to families, harems, or socially structured assemblages among which there is limited gene flow.

When characterizing an individual collection of seemingly like organisms, a population geneticist is interested in the degree to which the organisms in the collection mate at random with respect to one another and if individuals from different collections mate with one another or are isolated to a degree (Fig. 32.1C). If individuals within a collection are more genetically homogeneous than individuals from other collections, then a collection is actually a genetic population and the population geneticist will likely try to define the nature of barriers to gene flow among collections. Does distance or do geological features act as barriers to gene flow? A variety of studies have focused on these aspects of *Ae. aegypti* population genetics (Apostol et al. 1996; Conn et al. 1999; Gorrochotegui-Escalante et al. 2000).

MEASURING ALLELE AND GENOTYPE FREQUENCIES IN NATURAL POPULATIONS

Genetic populations are characterized according to the frequencies of alleles, haplotypes, or genotypes among its individual members. To accomplish this the geographic landscape over which genetic populations are spatially distributed needs to be determined. Put another way, the geographic scale over which individuals are *panmictic* (mate at random) has to be estimated. *Nested spatial sampling* is a technique often employed to determine the geographic scale of a population.

Figure 32.2 illustrates a nested spatial design for a study of gene flow among *Ae. aegypti* collections in east central Mexico. Here, the goal is to establish the geographic range over which *Ae. aegypti* populations are panmictic based upon 15 collections in coastal southeastern Mexico. This can be addressed by answering several specific questions. Do the northern and southern mountain ranges (indicated with heavy lines) act as barriers to gene flow? Does the point of land extending to the coast act as a barrier to gene flow? In this study the landscape could be subdivided into three regions: A, to the north of the mountains; B, inland sites bounded by mountains to the north and south; and C, to the south of the mountains. Five collection sites were analyzed in each of the three regions. With this type of design the geneticist can determine if mosquitoes within a collection mate at random or to what degree collections are genetically distinct. Do all collections in a region represent a single genetic population? Nested spatial sampling also allows tests for distance as a factor in separating populations. In other words, are populations in geographic proximity more genetically similar than populations that are further apart? Over what geographic distance are populations panmictic? Questions at a higher geographic scale can also be addressed: Are all collections from regions A, B, and C genetically distinct from one another? Are all collections in regions A and C separate from B? Are all collections from regions A and C similar to but distinct from B? A seasonal, or temporal, component can also be considered by sampling at various time points, or seasons.

Having decided on the nested spatial design to address the goals of the project, a sampling protocol of the arthropod vector must be considered. What life stage should be sampled? Adults will have had a greater chance to disperse, and sampling immatures may lead to oversampling of brothers and sisters in individual families. Generally, we advocate sampling adults where possible, but often this is impractical (or impossible). In the case of *Aedes aegypti*, Apostol et al. (1994) showed that on average an *Ae. aegypti* female lays fewer than 10 eggs per oviposition site, suggesting that most oviposition containers hold multiple sibships. Similar studies have not been performed with other arthropod vectors, but the possibility of oversampling siblings in spatial clusters of immature stages should always be considered. Such an event is often manifest as a deficiency of observed heterozygotes (the phenomenon known as *Wahlund's effect* is

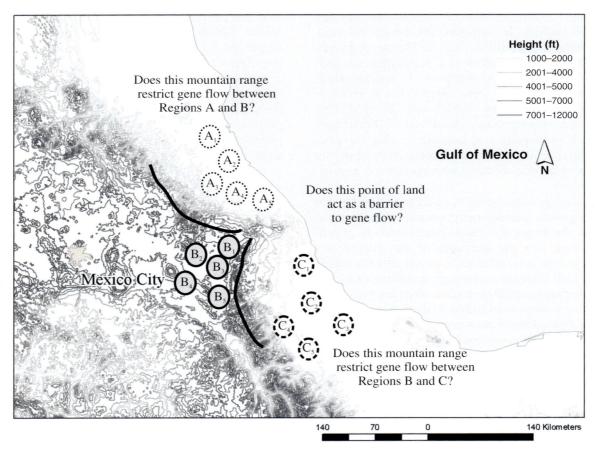

FIGURE 32.2 Nested spatial sampling design for a study of gene flow among *Ae. aegypti* collections in east central Mexico.

discussed later in this chapter). At a minimum, immatures should be sampled from several dispersed traps or over several different areas of a collection site.

Having resolved the life stage to be sampled, the investigator is faced with this question: How many individual organisms should be sampled per collection? Although one might assume that very large collections of hundreds to thousands of organisms will yield a highly accurate estimate of allele, haplotype, or genotype frequencies, this would require impractical amounts of effort, money, and time to process and analyze. More importantly, the actual increase in precision is not worth the effort. Small collections (5–10 organisms) are inexpensive but will yield inaccurate and perhaps misleading estimates. So what is an optimal sample size? This issue can be addressed with standard statistical sampling theory.

Figure 32.3 illustrates how allele and genotype frequencies and their standard errors are estimated at a genetic locus among 50 diploid individuals in a col-

lection. The frequency of the slow allele = $p_{(slow)}$ = 0.29 and has a standard deviation (s) of 0.045 (Fig. 32.3). With a sample size of 10 organisms,

$$s^2 = [0.029(1 - 0.29)]/(2 \times 10)$$

and s = 0.102. Multiplying this by 1.96 we estimate the size of the 95% confidence interval to be 0.199. In other words, by chance alone 95% of 10 organism collections with $p_{(slow)}$ = 0.29 will be estimated as having a frequency between 0.091 and 0.489. Sample sizes less than ~50 diploid organisms or ~100 haploid organisms have a large 95% confidence interval due to the chance effects of sampling.

Conversely, sample sizes greater than 50 diploid or 100 haploid genomes don't yield appreciably smaller 95% confidence intervals. Figure 32.4 indicates the relationship between sample size and the accuracy of the allele or genotype frequency estimates relative to the actual frequency in the population. For example, doubling the sample size from 50 to 100 diploid indi-

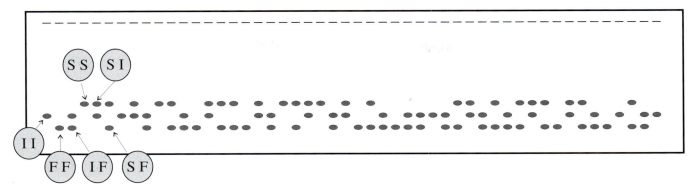

1) There are 3 alleles, slow (S), intermediate (I), fast (F). There are n $(n + 1)/2$ potential genotypes with n alleles. $n = 3$ homozygotes (SS, II, FF) and $n(n - 1)/2 = 3$ heteroxygotes (SI, SF, IF) = 6 genotypes total (all depicted in the first 6 lanes, reading left to right).

2) Count the number of all six genotypes: 4 SS, 5 II, 7 FF, 9 SI, 12 SF, and 13 IF

3) Count the number of times each allele appears in each genotype and multiply this by the number of times that that genotype appears. For example, the S allele occurs twice in an SS homozygote and once in an SF or SI heterozygote.

	#	S	I	F
SS	4	8	0	0
II	5	0	10	0
FF	7	0	0	14
SI	9	9	9	0
SF	12	0	12	12
IF	13	13	0	13
Total	50	29	32	39

Alleles = 29 (S)+32 (I)+39 (F) = 100. Should be 2N, where N is the number of diploid individuals sampled.

4) Allele frequency $p_{(slow)} = 29/100 = 0.29$ $p_{(intermediate)} = 32/100 = 0.32$ $p_{(fast)} = 39/100 = 0.39$

5) Allele frequencies are binomially distributed variables.

The sampling variance of an allele frequency p.

$$s^2_p = p(1 - p)/2N$$
$$s^2_{p(slow)} = 0.29 \times (1 - 0.29)/(2 \times 50) = 0.00206$$

The standard deviation of an allele frequency p.

$$s_p = \sqrt{p(1 - p)/2N}$$
$$s_{p(slow)} = 0.04538$$

FIGURE 32.3 Estimation of allele and genotype frequencies from a gel. The alleles, represented by the band phenotypes, segregate as three codominant markers with slow, intermediate, or fast mobility during electrophoresis.

viduals decreases the size of the 95% confidence intervals around the true gene frequency by only 5%. Increasing sample size from 100 to 500 diploid individuals decreases the size of the 95% confidence intervals by only another 7%, so the great increase in sample size is not worth the cost. We generally advocate sample sizes of ~50 organisms when sampling a diploid (e.g., nuclear) genetic marker and 60–100 organisms when sampling haploid (e.g., mitochondrial, or Y-linked) genetic markers.

GENE FLOW IN COLLECTIONS

Having made collections in a nested spatial design and processed genetic markers in the collected organisms, one might initially determine the minimal geographic scale at which individuals mate at random. An individual collection is an obvious scale at which to begin to address this question. Did organisms in a collection arise by a process of random mating?

A population model of random mating is used to develop null hypotheses with which to address this question. This model, also known as the Hardy–Weinberg model, is illustrated in Figure 32.5. This is a discrete-generation model that considers the survival and transfer of genes from one generation to the next. Simply put, the model states "in a large randomly-mating population, in the absence of selection, migration, and mutation, gene and genotypic frequencies will remain constant from generation to generation." Hence, implicit in this model are five broad assumptions. (1) The model assumes that members of an infinitely large diploid, sexually reproducing parent

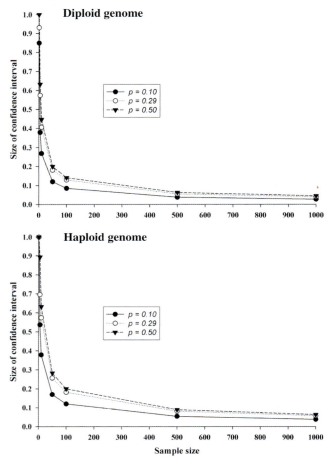

FIGURE 32.4 Relationship between sample size and the accuracy of the allele frequency estimates in diploid and haploid genomes.

population produce, through meiosis, an infinitely large gamete population. (2) Parents are assumed to mate randomly, irrespective of genotypes. (3) The haplotypes of gametes and the genotypes of zygotes, larvae, and adults don't respectively influence their ability to fuse, survive, mate, or reproduce. (4) No new genetic variation arises in the population through immigration. (5) No new genetic variation arises in the population through mutation (assumption 5).

This model can be illustrated by evaluating two generations of a randomly mating population (Fig. 32.6). With three alleles there are six genotypes and 36 mating types. All mating types between like genotypes (e.g., SI × SI) are unique, but the remainders are represented twice (e.g., the array of gametes produced by SI × FF is the same as the array produced by FF × SI). Thus, there are 36 mating types – 6 unique mating types = 30/2 = 15 duplicate mating types or 21 distinct mating types. Under assumption 2, the frequency with which each of these mating types occurs is the product of the frequencies of each genotype in the population.

Each mating produces offspring following Mendelian inheritance since there is no selection, mutation, or migration. Hence, in Figure 32.6 where the frequency of the slow, intermediate, and fast alleles are 0.29, 0.32, and 0.39, respectively, and the original genotype frequencies are 0.08, 0.10, 0.14, 0.18, 0.24, and 0.26 for SS, II, FF, SI, IF, and SF, respectively, then, for example, SS × SS matings represent $0.08 \times 0.08 = 0.0064$ of all matings, and all offspring are SS with frequency 0.0064 in the next generation. Similarly, SF × SF matings represent $0.24 \times 0.24 = 0.0576$ of all matings, but this ~5.8% is distributed in the next generation as one-fourth SS, one-half SF, and one-fourth FF due to Mendelian segregation. Figure 32.6 shows that the gene and genotypic frequencies are constant under all five assumptions. Thus, the Hardy–Weinberg model predicts that in a diploid, sexually reproducing species, allele and genotype frequencies will remain constant from one generation to the next when all five assumptions are true.

Needless to say, the calculations illustrated in Figure 32.6 are laborious. The Hardy–Weinberg model can be more easily handled by exploiting assumption 1, i.e., treating the population as an infinitely large gene pool. Table 32.1 illustrates how the calculations in Figure 32.6 can be simplified and how observed genotype frequencies can be tested for conformity to the Hardy–Weinberg model with a χ^2 goodness-of-fit test. Another implicit outcome of the Hardy–Weinberg model is that if genotype frequencies are not in Hardy–Weinberg proportions in generation 0, they will be in Hardy–Weinberg proportions in generation 1 (if all other assumptions continue to be valid). This outcome will not necessarily be true for sex-linked loci. Nevertheless, genotypes at sex-linked loci rapidly approach Hardy–Weinberg proportions in a few generations.

The Hardy–Weinberg model provides a very useful null hypothesis with which to test the validity of the five assumptions. The problem is that if any *one* assumption is false, then genotypes will not be in Hardy–Weinberg proportions. It is often difficult to sort out which of the assumptions are invalid in a particular population.

RANDOM ASSORTMENT AND INDEPENDENT SEGREGATION OF LOCI IN POPULATIONS: LINKAGE DISEQUILIBRIUM

The Hardy–Weinberg model makes predictions about allele and genotype frequencies at a single

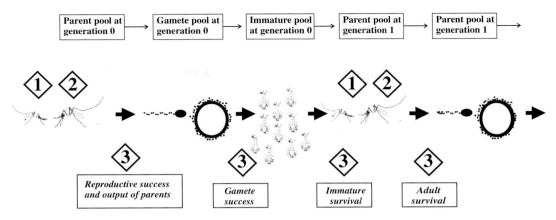

| Parent pool at generation 0 | → | Gamete pool at generation 0 | → | Immature pool at generation 0 | → | Parent pool at generation 1 | → | Parent pool at generation 1 | → |

① ② ① ②

③ ③ ③ ③

| Reproductive success and output of parents | | Gamete success | | Immature survival | | Adult survival |

Assumptions

① **Population sizes are infinitely large.**

② **Adults mate irrespective of genotypes.**

③ **The haplotypes of gametes and the genotypes of zygotes, larvae, and adults don't respectively influence their ability to fuse, survive, mate, or reproduce.**

④ **No migration among populations.**

⑤ **The locus under examination doesn't have a high mutation rate.**

FIGURE 32.5 Discrete generations and the five assumptions implicit in the Hardy–Weinberg model.

genetic locus. However, a genome is made up of many loci, some of which assort together because they are closely linked on the same chromosome (Chapter 30). How are alleles at different loci expected to segregate in a population? Consider the segregation of three codominant alleles at locus A with two alleles at locus B (Fig. 32.7). Hardy–Weinberg assumptions apply at both loci. P_{xy} is the observed frequency of gametes containing A_x and B_y alleles, and their expected frequency is the frequency of allele x at locus A multiplied by the frequency of allele y at locus B. For example, if P_{IF} is the observed frequency of $A_{intermediate}$ B_{fast} gametes then:

$$D_{IF} = P_{IF} - p_{(A\ intermediate)}\ p_{(B\ fast)} \qquad (1)$$

where D_{IF} is defined as the *linkage disequilibrium coefficient*. If the probability of sampling an $A_{intermediate}$ B_{fast} gamete equals the product of their independent frequencies ($p_{(A\ intermediate)}\ p_{(B\ fast)}$), then $D_{IF} = 0$ and $A_{intermediate}$ and B_{fast} alleles are said to be in *linkage equilibrium*. Otherwise, $A_{intermediate}$ and B_{fast} alleles are said to be in *linkage disequilibrium* if they occur together more often ($D_{IF} > 0$) or less often ($D_{IF} < 0$) than predicted by their independent frequencies.

In Figure 32.8 we apply this framework to the same discrete-generation model developed in Figure 32.5. For example, assume that the males are initially all A_{slow} B_{slow} double homozygotes and that females are all

A_{fast} B_{fast} double homozygotes. The frequency of A_{slow} B_{slow} sperm = 1.00 even though the frequencies of A_{slow} and B_{slow} alleles in the overall population are both 0.50. Thus, $D_{SS} = 1.00 - (0.5 \times 0.5) = 0.75$. The same is true of D_{FF} in eggs. Sperm fertilize eggs to produce A_{slow} B_{slow}/A_{fast} B_{fast} zygotes that mature to adults. The frequency that alleles at the two loci occur together in the next generation of gametes will be a function of the linkage relationship between the loci. There are four potential gamete types: A_{fast} B_{fast}, A_{slow} B_{slow}, A_{fast} B_{slow}, and A_{slow} B_{fast}. The latter two gamete types will arise only through recombination during meiosis in the parents. The frequency of these gametes is actually the centimorgan distance between the loci. This distance is derived by measuring recombination frequencies between alleles at the two loci, as demonstrated in Chapter 30. A_{fast} B_{fast} and A_{slow} B_{slow} gametes are maintained when there is no recombination during meiosis in the parents.

This scenario in Figure 32.8 can be translated into a mathematical model. The frequency of $A_{intermediate}$ B_{fast} gametes in generation 1 is:

$$P_{IF(1)} = (1 - r)\ P_{IF(0)} + rp_{(A\ intermediate)}\ p_{(B\ fast)} \qquad (2)$$

where r is the frequency of recombination (cM distance/100). Equation (2) states that the frequency of $A_{intermediate}$ B_{fast} gametes in generation 1 ($P_{IF(1)}$) equals the

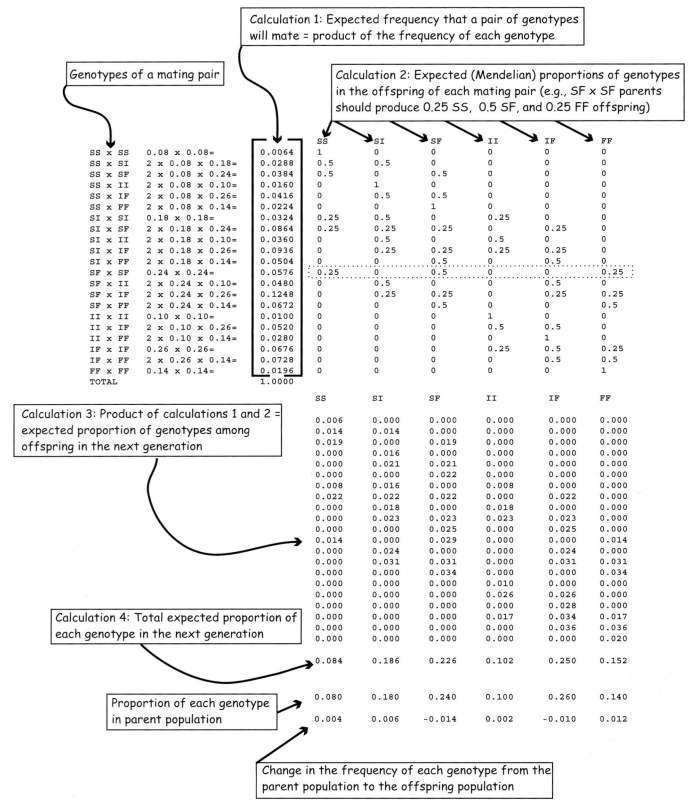

FIGURE 32.6 Calculation of the expected proportion of mating frequencies among parental genotypes and resulting genotype frequencies among offspring in the Hardy–Weinberg model.

**TABLE 32.1   Hardy–Weinberg Model, Assuming Infinitely
Large Population Size**

If parents release their gametes into a random pool:

Probability of sampling a slow allele	=	$ps = s$
Probability of sampling an intermediate allele	=	$pi = i$
Probability of sampling a fast allele	=	$pf = f$

The probability of gametes combining at random = products of their frequencies:

SS homozygote	=	s^2
SI heterozygote	=	$2si$
SF heterozygote	=	$2sf$
II homozygote	=	i^2
IF heterozygote	=	$2if$
FF homozygote	=	f^2

Using the data in Figure 32.3:

Probability of sampling a slow allele	$= s =$	0.29
Probability of sampling an intermediate allele	$= i =$	0.32
Probability of sampling a fast allele	$= f =$	0.39
		1.00

Predicted genotype frequencies with equal survival and random mating of genotypes:

	Expected Rate		Expected Rate × 50
SS homozygote =	$0.29^2 =$	0.0841	4.21
SI heterozygote =	$2 \times 0.29 \times 0.32 =$	0.1856	9.28
SF heterozygote =	$2 \times 0.29 \times 0.39 =$	0.2262	11.31
II homozygote =	$0.32^2 =$	0.1024	5.12
IF heterozygote =	$2 \times 0.32 \times 0.39 =$	0.2496	12.48
FF homozygote =	$0.39^2 =$	0.1521	7.61
Total		1.0000	50.00

The predicted genotype frequencies with random survival and random mating of genotypes can be compared with the observed values using a χ^2 goodness-of-fit test ([(observed − expected)^2/expected] summed over all genotypes):

		Expected Number	Observed Number	χ^2 Calculation		χ^2
SS homozygote	=	4.21	4	$(4 - 4.21)^2/4.21$	=	0.010
SI heterozygote	=	9.28	9	$(9 - 9.28)^2/9.28$	=	0.008
SF heterozygote	=	11.31	12	$(12 - 11.31)^2/11.31$	=	0.042
II homozygote	=	5.12	5	$(5 - 5.12)^2/5.12$	=	0.003
IF heterozygote	=	12.48	13	$(13 - 12.48)^2/12.48$	=	0.022
FF homozygote	=	7.61	7	$(7 - 7.61)^2/7.61$	=	0.055
Total	1.0000	50.00	50	calculated χ^2	=	0.141

The number of degrees of freedom (d.f.) in a χ^2 goodness-of-fit test = no. of genotypes − no. of alleles = 6 − 3 = 3 d.f. Critical χ^2 with 3 d.f. and an α value of 0.05 = 7.81. Therefore, we conclude that the observed genotype frequencies $\cong$ expected genotype frequencies under the five assumptions in Figure 32.5.

Observed genotype frequencies are tested for conformity to the Hardy–Weinberg model with a χ^2 goodness-of-fit test.

2 loci: <u>**Locus A alleles**</u>   <u>**frequency**</u>   <u>**Locus B alleles**</u>   <u>**frequency**</u>

A_{slow} $p_{(A\ slow)}$ B_{slow} $p_{(B\ slow)}$

$A_{intermediate}$ $p_{(A\ intermediate)}$ B_{fast} $p_{(B\ fast)}$

A_{fast} $p_{(A\ fast)}$

	A_{slow}	$A_{intermediate}$	A_{fast}
B_{slow}	*Obs.* $= P_{slow,slow}$ *Exp.* $= p_{(A\ slow)}\,p_{(B\ slow)}$	*Obs.* $= P_{intermediate,slow}$ *Exp.* $= p_{(A\ intermediate)}\,p_{(B\ slow)}$	*Obs.* $= P_{fast,slow}$ *Exp.* $= p_{(A\ fast)}\,p_{(B\ slow)}$
B_{fast}	*Obs.* $= P_{slow,fast}$ *Exp.* $= p_{(A\ slow)}\,p_{(B\ fast)}$	*Obs.* $= P_{intermediate,fast}$ *Exp.* $= p$	*Obs.* $= P_{fast,fast}$ *Exp.* $= p_{(A\ fast)}\,p_{(B\ fast)}$

Interpretation: **The probability of sampling an $A_{intermediate}B_{fast}$ gamete should equal the product of their independent frequencies ($p_{(A\ intermediate)}p_{(B\ fast)}$). If this equality exists, then alleles at the A and B loci are said to be in linkage equilibrium.**

FIGURE 32.7   Linkage equilibrium with two loci. Obs. = observed number; Exp. = expected number assuming independence.

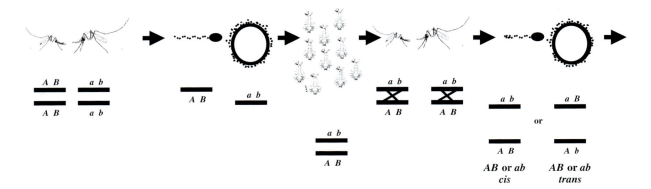

In one generation, the frequency with which one pair of alleles at separate loci occur together in gametes will be a function of:

1) The frequency of the alleles in the population.
2) The linkage relationship of the loci.
3) The current frequency of the gametes.

FIGURE 32.8   Linkage relationships in the Hardy–Weinberg model from Figure 32.5. The model begins with all males being homozygotes for A and B alleles and females being homozygotes for a and b alleles at both loci.

frequency of $A_{intermediate}$ B_{fast} gametes in generation 0 ($P_{FF(0)}$) that did not recombine $(1 - r)$ added to the frequency that $A_{intermediate}$ and B_{fast} occur independently in generation 0 ($p_{(A\ intermediate)}$ $p_{(B\ fast)}$) through recombination (r). Substituting Eq. (1) into Eq. (2) we get

$$D_{IF(1)} = (1 - r)D_{IF(0)} \qquad (3)$$

which states that the disequilibrium between alleles at the A and B loci will decrease by $(1 - r)$ each genera-

tion. Equation (3) can be generalized over many generations to

$$D_{IF(t)} = (1 - r)^{t}D_{IF(0)} \qquad (4)$$

After t generations the initial disequilibrium between alleles will have declined by $(1 - r)^{t}$. If we plot D_{IF} across generations, we see that disequilibrium declines within just a few generations (Fig. 32.9A).

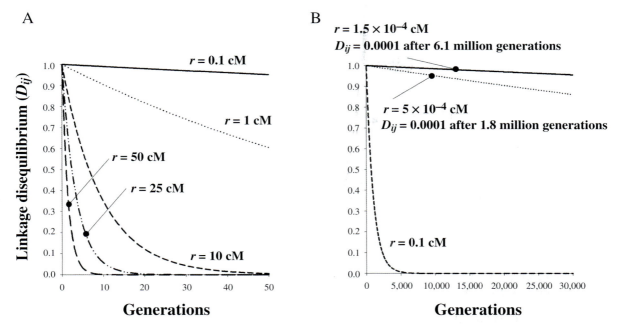

FIGURE 32.9 The rate with which linkage disequilibrium declines over generations as a function of the recombination rate r. The decline is rapid (A) for large recombination rates but very slow (B) when considering segregating sites within an ~500-bp gene, which, for example, in *Aedes aegypti* are expected to have rates of $1.5 - 5 \times 10^{-4}$ cM/bp.

However, the rate of decline depends largely on r. Disequilibrium declines slowly among tightly linked loci (small values of r) (Fig. 32.9B).

The resolution of base pairs/cM varies from 1.0 to 3.4 Mbp/cM along different arms of the *Ae. aegypti* genome (Brown et al. 2001), which equals $0.294 - 1.000 \times 10^{-6}$ cM/bp. The expected frequency of recombination among segregating sites within a gene that, for example, consists of 500 bp is $1.5 - 5 \times 10^{-4}$ cM/bp. Solving for t in Eq. (4) when $D_{ij} = 0.0001$, we estimate that from $1.8 - 6.1$ million generations (or assuming ~12 generations/year, 150,000–510,000 years) would be required to bring segregating sites within a gene into linkage equilibrium. The large range of years reflects different recombination rates found in different parts of the *Ae. aegypti* genome (Chapter 30). Despite this seemingly long time frame, you will soon see that most segregating sites within the *early trypsin* gene of *Ae. aegypti* are in linkage equilibrium.

Thus, this model predicts that ultimately no linkage disequilibrium should be detected even among tightly linked loci within long-established populations. Many studies have validated this prediction, even when examining single-nucleotide polymorphisms (SNPs) within a single genetic locus. However, linkage disequilibrium may be detected in haploid genomes (e.g., mtDNA, cpDNA), on haploid sex chromosomes (e.g., the Y chromosome), in haploid organisms in which recombination is rare or unknown (e.g., bacteria, viruses), or in recently established populations.

Assumption 1: Population Sizes Are Infinitely Large

The Hardy–Weinberg model assumes large (potentially infinite) sample sizes. In reality, populations are often small and may be founded by one or a few individuals. It should be intuitively obvious that this can drastically change allele frequencies and therefore genotypic frequencies between generations in the same manner as described earlier when discussing small sample sizes. The behavior of allele frequencies in small, finite populations can be accurately modeled using either a binomial or multinomial probability function. For example, assume that only five adults mate every generation or that many adults mate every generation but only five zygotes survive. Thus, there are $2N = 10$ genomes sampled every generation. The probability that a particular allele, say, A_{slow}, survives in one generation is its frequency $p_{(A\ slow)}$. The number of copies of A_{slow} that survive in the 10 genomes sampled will follow a binomial distribution with:

$$\text{Probability } (A_{slow} \text{ is sampled } x \text{ times in } 10 \text{ selections}) = {}_xC_{10}s^x(1-s)^{n-x} \tag{5}$$

where ${}_xC_n = n!/x!(n-x)!$

Assume, for example, at generation 0 the "A" collections in Figures 32.1 and 32.2 consist of one large, panmictic gene pool and $p_{(A\ slow)} = 0.50$. At generation 0, we drop large domes over each of the collection sites (A_1–A_5) so that the outcome of picking individuals in one collection has no influence on the outcome of sampling individuals from the other four collections. We then follow each A_x collection over 100 generations and know that only five individuals gave rise to the new collection every generation. Thus, in every generation there are 10 alleles sampled. The outcome of this scenario can be modeled with the binomial probability density function (Table 32.2). Notice that by adding probabilities in Table 32.2 there is a 37.5% chance that $p_{(A\ slow)}$ will decrease, a 37.5% chance that it will increase, and a 25% chance that $p_{(A\ slow)}$ will remain the same. Thus, there is a 75% chance that $p_{(A\ slow)}$ will change from one generation to the next in any of the "A" collections, simply due to the small sampling of alleles each generation. The outcome of repeating this experiment for 100 generations is shown in Figure 32.10A. In just over 10 generations, A_{slow} has gone to extinction in two of the A collections and has become fixed in the other three collections. We see the same result if we repeat this experiment but double the number of individuals surviving through one generation to 10 (Fig. 32.10B). However, notice that it takes a greater amount of time for A_{slow} to go to fixation or extinction. When increasing the number of surviving individuals to 25 (Fig. 32.10C), after 100 generations A_{slow} is still segregating in two collections, and with 50 surviving individuals per generation A_{slow} is segregating in all five collections (Fig. 32.10D).

This result was generalized by Fisher and Wright, who showed that the general consequences of repeated finite sampling is the loss of alleles and het-erozygotes at a locus and lower genetic variation across loci. Once an allele is lost, there can be no further sampling effect because there is no more variation. The conditions for the models just presented are somewhat artificial because we selected gametes one at a time. Random sampling will almost always be at the level of diploid parents or zygotes rather than gametes to continue the next generation. The same principles apply with diploid genotypes. The only difference is that we would use a multinomial rather than a binomial distribution because there will be more than two classes of outcomes (a minimum of three genotypes with two alleles).

The influence of small population size or founding effects on vector population genetics variation has been observed in many species. The heterozygosities of North American populations of the *Culicoides variipennis* complex (biting midge vector of bluetongue virus) vary greatly from region to region. Heterozygosity is significantly higher in populations of *Culicoides sonorensis* from California than from those sampled in Colorado or in comparison from *Culicoides variipennis* from New England. This is largely attributable to the large year-round populations of *Culicoides* in California as compared to the small isolated populations in temperate clines that essentially disappear during cold winters (Tabachnick 1996).

The outcome of these models is often referred to as *random genetic drift*. The critical point about genetic drift is that in the absence of any adaptation, allele frequencies can change dramatically among collections due simply to small numbers of founding members. Alleles can differ in frequency among collections for entirely random (nonmechanistic, nonselectionist, nonadaptive) reasons. The drift model therefore acts as a null model when testing hypotheses of adaptation. How do we know that the differences in phenotype,

TABLE 32.2   Expected Frequency Distribution of the Number of Alleles Surviving from One Generation to the Next When the Frequency of the Allele to 0.50 and Only N = 5 Individuals Survived from One Generation to the Next

Prob(0 alleles surviving) =	$_0C_{10}(0.50^0 \times 0.50^{10}) =$	0.0010, where $_0C_{10}$	$= 6!/0!(6-0)! = 1$
Prob(1 allele surviving) =	$_1C_{10}(0.50^1 \times 0.50^9) =$	0.0098, where $_1C_{10}$	$= 6!/1!(6-1)! = 10$
Prob(2 alleles surviving) =	$_2C_{10}(0.50^2 \times 0.50^8) =$	0.0439, where $_2C_{10}$	$= 6!/2!(6-2)! = 45$
Prob(3 alleles surviving) =	$_3C_{10}(0.50^3 \times 0.50^7) =$	0.1172, where $_3C_{10}$	$= 6!/3!(6-3)! = 120$
Prob(4 alleles surviving) =	$_4C_{10}(0.50^4 \times 0.50^6) =$	0.2051, where $_4C_{10}$	$= 6!/4!(6-4)! = 210$
Prob(5 alleles surviving) =	$_5C_{10}(0.50^5 \times 0.50^5) =$	0.2461, where $_5C_{10}$	$= 6!/5!(6-5)! = 252$
Prob(6 alleles surviving) =	$_6C_{10}(0.50^6 \times 0.50^4) =$	0.2051, where $_6C_{10}$	$= 6!/6!(6-6)! = 210$
Prob(7 alleles surviving) =	$_7C_{10}(0.50^7 \times 0.50^3) =$	0.1172, where $_7C_{10}$	$= 6!/6!(6-6)! = 120$
Prob(8 alleles surviving) =	$_8C_{10}(0.50^8 \times 0.50^2) =$	0.0439, where $_8C_{10}$	$= 6!/6!(6-6)! = 45$
Prob(9 alleles surviving) =	$_9C_{10}(0.50^9 \times 0.50^1) =$	0.0098, where $_9C_{10}$	$= 6!/6!(6-6)! = 10$
Prob(10 alleles surviving) =	$_{10}C_{10}(0.50^{10} \times 0.50^0) =$	0.0010, where $_{10}C_{10}$	$= 6!/6!(6-6)! = 1$

The expected frequencies follow a binomial probability density function.

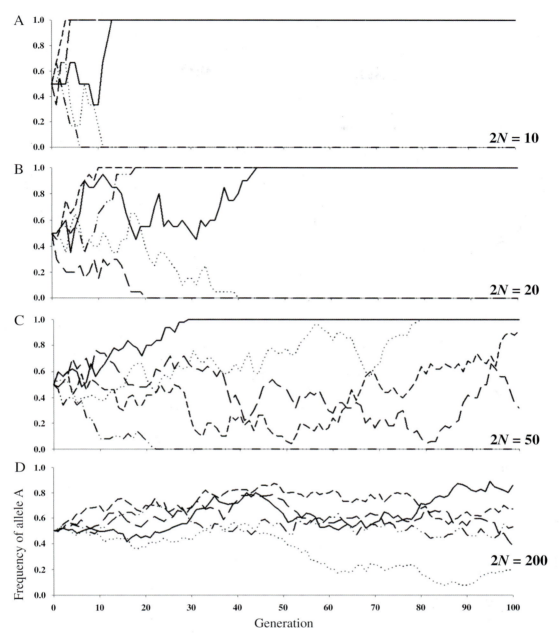

FIGURE 32.10 Genetic drift as a function of N_e.

genotype, or haplotype that we discover among field populations are attributable to adaptation? Small founding population size alone will generate large variances in allele frequencies.

Assumption 2: Adult Mate Irrespective of Genotypes

The three types of nonrandom mating in natural populations are assortative mating, inbreeding, and obligate outcrossing. *Assortative mating* refers to the choice of mates based on a specific phenotype controlled by a specific genotype. *Positive assortative mating* occurs when individuals consistently choose mates that are genetically *similar* to themselves usually based upon recognizable phenotypes (e.g., height, weight, color). *Negative assortative mating* occurs when individuals consistently choose mates that are genetically *different* from themselves. Mating with members of the opposite sex is the perfect example of negative assortative mating. Males can successfully mate only with females.

Inbreeding refers strictly to mating among relatives. Since relatives share genes, it is analogous in its effects to positive assortative mating. Inbreeding may not involve mate choice but can instead be the result of limited ability to disperse or limited local resources. In these situations, the effects are similar to those arising from small sample size. A major evolutionary impact of inbreeding is to expose lethal and deleterious recessive alleles to selection resulting from the loss of alleles and increases in homozygotes. If a species or a population has an abundance of lethal or deleterious recessive alleles at a particular locus, then the probability increases that these will become homozygous in inbred individuals and thereby are exposed to negative selection. *Obligate outcrossing* refers to mating (usually obligatory) between nonrelatives.

Assortative mating and inbreeding do not change allele frequencies across generations but have strong impacts on genotypic frequencies in collections. Basically, positive assortative mating and inbreeding lead to an excess of homozygotes (or a deficiency of heterozygotes) relative to Hardy–Weinberg expectations. The major difference between positive assortative mating and inbreeding is that positive assortative mating acts quickly and only *on those genes affecting mate selection*. In contrast, inbreeding acts slowly; however, it acts *on the whole genome*. Negative assortative mating leads to an excess of heterozygotes (or a deficiency of homozygotes). Similarly, the major difference between negative assortative mating and obligate outcrossing is that negative assortative mating acts quickly and on only those genes affecting mate selection. Obligate outcrossing acts slowly and on the whole genome.

In studying nonrandom mating in populations, it is useful to track relationships or pedigrees among alleles in gametes that successfully join to form a zygote. This relationship is conveniently described by the inbreeding coefficient F that Wright originally defined as

$$F = 1 - (H_{obs}/H_{exp}) \qquad (6)$$

where H_{obs} is the observed number of heterozygotes and H_{exp} is the expected number of heterozygotes, assuming Hardy–Weinberg. Note that if $H_{obs} = H_{exp}$, then genotypes are in Hardy–Weinberg proportions and $F = 0$. However, if $H_{obs} < H_{exp}$ (fewer heterozygotes than expected, more homozygotes than expected), then $F > 0$. If $H_{obs} > H_{exp}$ (more heterozygotes than expected, fewer homozygotes than expected), then $F < 0$.

Wright developed F as a *genetic correlation* between alleles. Two alleles are said to be *identical in state*

(abbreviated as *iis* in the literature) if they possess identical nucleotide sequences. Two alleles are said to be correlated if they arose in a common ancestor. They are then *identical by descent (ibd)*. Alleles that are *ibd* are also called *autozygous*. If alleles are *not ibd*, they are *allozygous*.

Under random mating, $F = 0$, indicating that there is no genetic correlation between uniting gametes. Under positive assortative mating and inbreeding, $F > 0$ because alleles in uniting in gametes are *ibd* more often than expected by random chance. Under negative assortative mating or obligate outcrossing, alleles in uniting gametes are allozygous more often than expected by random chance and $F < 0$. Consider a population in which A and a alleles occur with respective frequencies of p and q. If there is nonrandom mating, then gametes with alleles *ibd* have a greater or lesser chance of uniting to form a zygote. This situation is specified by the following equation:

$$\text{Prob}(A_1A_2) = p^2(1 - F) + pF \qquad (7)$$

The probability of sampling an *AA* homozygote is the frequency of the *A* allele squared (the frequency expected under Hardy–Weinberg) multiplied by the probability that A_1 and A_2 are not *ibd* $(1 - F)$, in addition to the probability of sampling two alleles that are *iis* multiplied by the probability that both are *ibd* (F). The same logic applies to aa homozygotes such that

$$\text{Prob}(a_1a_2) = q^2(1 - F) + qF \qquad (8)$$

However, with *Aa* heterozygotes:

$$\text{Prob}(Aa) = 2pq(1 - F) + 0 \qquad (9)$$

because *A* and *a* are not *iis* and therefore can never be *ibd*.

As an example, assume that $p = 0.7$ and $q = 0.3$ in a collection in which $F = 0.25$. Applying Eqs. (7)–(9), the frequency of *AA* homozygotes will be $(0.7)^2 \times (1 - 0.25) + (0.7 \times 0.25) = 0.5425$, *Aa* heterozygote frequency is $2 \times 0.7 \times 0.3(1 - 0.25) = 0.3150$, and *aa* homozygote frequency $= (0.3)^2(1 - 0.25) + (0.3 \times 0.25) = 0.1425$. If 100 organisms were sampled in this population, there would be $0.3150 \times 100 = 32$ *Aa* heterozygotes observed instead of the $2 \times 0.7 \times 0.3 \times 100 = 42$ expected. Thus, $F = 1 - (32/42) \cong 0.25$.

RANDOM GENETIC DRIFT AND INBREEDING

The similarity between genetic drift and inbreeding is easy to understand at an intuitive level. Recall that inbreeding is likely among individuals in nature when there is a limited ability to disperse or limited local resources. Therefore, any two alleles in a collection may be *ibd* due simply to limited population size.

Let's return to the discrete-time model in Figure 32.5. Assume that there are N parents at generation 0. At any locus, we assume that each allele at any one locus is either not *iis* or *ibd* such that there are $2N$ unique gametes produced by N parents at generation 0. This also assumes that each parent produces equal numbers of gametes or that the number of gametes that each parent produces follows a random (Poisson) distribution. If these assumptions are valid, then the frequency of each unique allele in the zygote population will be $1/2N$. Accordingly, F in consecutive generations can be specified by the following equation:

$$F_{t=1} = 1/2N_{t=0}(1) + [1 - (1/2N_{t=0})]F_{t=0} \qquad (10)$$

The correlation between loci in uniting gametes in generation 1 ($F_{t=1}$) is the probability that the two gametes arose from the *same* parent in generation 0 ($1/2N_{t=0}$) and the probability that, if the two gametes arose from *different* parents in generation 0 ($1 - 1/2N_{t=0}$), they were *ibd* in the population at generation 0 ($F_{t=0}$). Technically, the event of one gamete fertilizing another gamete from the same parent can happen in one generation only in species that have the ability to self. The exact solution for a sexually reproducing species is

$$F_{t=1} = 1/(2N+1)_{t=0}(1) + \{1 - [1/(2N+1)_{t=0}]\}F_{t=0} \quad (11)$$

However, this correction turns out to be of minor importance for $N \geq 20$, so we will work with the more general form. Equation (10) can be expanded to

$$1 - F_t = (1 - 1/2N)^t(1 - F_0) \qquad (12)$$

After t generations, the initial lack of correlation among uniting gametes will decline by $[1 - (1/2N)]^t$. When N is small $[1 - (1/2N)] \ll 1$, the decline is rapid. When N is large, $[1 - (1/2N)] \cong 1$, the decline is slow (Fig. 32.11).

It is more common to see Eq. (12) presented in terms of H_{obs} and H_{exp}. Recall that $F = 1 - (H_{obs}/H_{exp})$ or $1 - F = (H_{obs}/H_{exp})$, so $(H_{obs}/H_{exp}) = (1 - 1/2N)^t$ or $H_{obs} = (1 - 1/2N)^t H_{exp}$. In other words, heterozygosity decreases as t increases and the rate of decline and observed heterozygosity increase as N decreases. Through time, we can apply the equation

$$H_t = (1 - 1/2N)^t H_0 \qquad (13)$$

which can be approximated with the equation

$$H_t = H_0 e^{-t/2N} \qquad (14)$$

Equation (14) is useful for predicting the outcome of inbreeding. For example, we might be interested in determining how much time is required to reduce

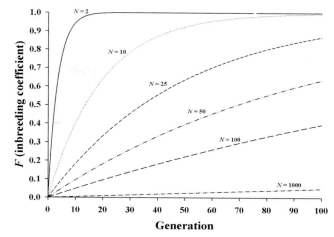

FIGURE 32.11 Inbreeding as a function of N_e from Eq. (12).

heterozygosity of a population by half. This can be reworded mathematically as: What is t when $H_t = \frac{1}{2}H_0$? We can use Eq. (14) to show that $t = 1.39N$. It takes $1.39N$ generations to reduce heterozygosity by half. If $N = 10$, then $t \cong 14$ generations. Considered over many loci, this result says that the average number of heterozygous loci per individual is halved in 14 generations. With full sib mating, $N \cong 2$ and $t = 2.77$, or for every three generations of sib mating, heterozygosity is halved.

It should be evident that N is critical in determining the rates of inbreeding. N is often designated N_e and is defined as the *effective population size*. The actual "census" size of a population always overestimates N_e because not all individuals reproduce. Common reasons include (1) unequal numbers of males and females, (2) individuals who are too young or too old to reproduce, and (3) unequal family sizes. For these reasons, N_e can change dramatically over time. N_e is estimated in three distinct ways. Most commonly, the *inbreeding* N_e is estimated by the change in the H_{obs}/H_{exp} ratio through time. N_e can also be measured from the change in the variance of allele frequencies over time or space and is referred to as the *variance effective size* when estimated in this way. The *eigenvalue* N_e is the rate of loss in heterozygosity in specific age classes over time. Under most conditions, all three methods have estimated approximately the same value in field studies.

In deriving Eq. (14) we made the tacit (and unreasonable) assumption that N_e remains constant over time. If population size is not constant, then it stands to reason that N_e will also not be constant. In deriving Eq. (12) from Eq. (11) we used the recurrence equation:

$$1 - F_t = (1 - 1/2N_{t-1})(1 - F_{t-1}) \qquad (15)$$

Or, by analogy,

$$1 - F_{t+1} = (1 - 1/2N_t)(1 - F_t)$$

and

$$1 - F_{t+2} = (1 - 1/2N_{t+1})(1 - F_{t+1})$$

By substitution:

$$1 - F_{t+2} = (1 - 1/2N_{t+1})(1 - 1/2N_t) \\ (1 - 1/2N_{t-1})(1 - F_{t-1}) \qquad (16)$$

The terms containing N are a harmonic mean rather than an arithmetic mean of N_e. The harmonic mean of N_t is thus used to estimate the inbreeding N_e and is calculated as

$$1/N_e = 1/t \sum_{i=0...t}(1/N_{ei}) \qquad (17)$$

The harmonic mean best represents the true impact of inbreeding. For example, if a mosquito population is founded by a single inseminated female that then produces 120 offspring that then produce 10,000 offspring, then $1/N_e = 1/3(1/2 + 1/120 + 1/10,000) = 0.170$, or $N_e \cong 6$ mosquitoes, while the arithmetic average is 3,374 mosquitoes. Thus, the harmonic mean more accurately reflects the amount of genetic diversity maintained in a population.

Wahlund's Effect

Focus again on the "A" collections in Figures 32.1 and 32.2. Recall that at generation 0, we isolated collection sites A_1–A_5 and only N_e individuals gave rise to the next generation. We then followed each A_x collection over 100 generations. In every generation there were $2N_e$ alleles sampled. The outcome of running that model for many generations was that A_{slow} went to either fixation or extinction in all five collections (Fig. 32.10). Note that since extinction is as likely as fixation, the overall average allele frequency will remain $p_{(A slow)} = 0.5$. If we sample the entire set of A collections after 100 generations, we would find few or no heterozygotes, even though the Hardy–Weinberg model would predict:

$$2p_{(A slow)}(1 - p_{(A slow)}) = 2 \times 0.5 \times (1 - 0.5) = 0.5$$

or 50% heterozygotes. F calculated among all five collections $= 1 \times [1 - (0/0.5)] = 1$, indicating that all of the alleles are *ibd*. Such a strong genetic correlation could occur through inbreeding enforced by reproductive isolation or because of random genetic drift within each collection. The reduction in observed heterozygotes due to gene frequency differences among populations is referred to as *Wahlund's effect*. Wahlund's effect occurs in any situation in which partially or wholly reproductively isolated units differing in gene frequency are sampled as a single unit. When this

occurs, Wahlund's effect is manifest as a deficiency of observed heterozygotes (or an excess of homozygotes).

When examining the A population in Figures 32.1 and 32.2, there may be no indication of genetic differences or separate populations among the A_x collections. Only a large F is observed. However, because nested spatial sampling was performed, one can subdivide F for the entire "A" population into two components. First, one can examine the correlation among uniting gametes from individuals (I) in each A_x collection or subpopulation (S). We will label this F_{IS}. Second, the geneticist can also examine the correlation among uniting gametes in each collection relative to uniting gametes in the total "A" population (T). We will label this F_{ST} and will label the inbreeding coefficient among all individuals in the total population as F_{IT}. These three inbreeding coefficients are formally called *Wright's F-statistics* for subdivided populations.

F_{IS} was defined by Wright as "the average over all subpopulations of the correlation between uniting gametes (I) relative to those of their own subpopulation (S)." This is Eq. (6) for each collection. F_{ST} was defined as "the correlation between random gametes within a subpopulation (S) relative to gametes in the entire population (T)." It is a measure of nonrandom mating among collections but is also commonly used as a metric of genetic distance. F_{IT} was formally defined as "the correlation between gametes that unite to produce individuals (I), relative to gametes of the entire population (T)." It is an overall measure of nonrandom mating among and within collections. F_{IS} in collection A_x is calculated as

$$F_{IS(A_x)} = 1 - H_{obs(A_x)}/[2p_{(A slow, x)}(1 - p_{(A slow, x)})] \quad (18)$$

Over all five collections:

$$F_{IS} = 1 - \left[\left(\sum_{x=1...5} H_{obs(A_x)} \right) / H_{exp(S)} \right]/5 \qquad (19)$$

Where

$$H_{exp(S)} = \left[2 \sum_{x=1...5} p_{(A slow, x)}(1 - p_{(A slow, x)}) \right]$$

Table 32.3 shows an example of the calculation of F_{ST} over four alleles in the "A" collections in Figures 32.1 and 32.2. Over all collections, F_{ST} is calculated as a ratio of heterozygotes:

$$F_{ST} = 1 - H_{exp(S)}/H_{exp(T)} \qquad (20)$$

where $H_{exp(T)} = 2\bar{p}(1 - \bar{p})$ and $\bar{p}$ is the weighted average allele frequency in the total population. Equation (20) can be manipulated to show that

$$F_{ST} = \sigma_p^2/\bar{p}(1 - \bar{p}) \qquad (21)$$

F_{IT} can also be calculated as an inbreeding coefficient with the following modifications over all collections:

TABLE 32.3 Example of F_{ST} Calculation from Population Allele Frequency Data

	Frequencies of Most Common Allele At:			
	Locus A	Locus B	Locus C	Locus D
Collection A_1	0.300	0.200	0.500	1.000
Collection A_2	0.250	0.500	0.100	0.500
Collection A_3	0.500	0.100	0.600	0.000
Collection A_4	0.700	0.300	0.200	0.000
Collection A_5	0.450	0.250	0.350	0.400
Weighted average allele frequency ($\bar{p}$)	0.440	0.270	0.350	0.380
σ^2_p	0.025	0.018	0.034	0.138
$H_{exp(T)}$ $2\bar{p}(1-\bar{p})$	0.493	0.394	0.455	0.471
$2p(1-p)$ Collection A_1	0.420	0.320	0.500	0.000
$2p(1-p)$ Collection A_2	0.375	0.500	0.180	0.500
$2p(1-p)$ Collection A_3	0.500	0.180	0.480	0.000
$2p(1-p)$ Collection A_4	0.420	0.420	0.320	0.000
$2p(1-p)$ Collection A_5	0.420	0.420	0.320	0.000
$H_{exp(S)}$ $2\Sigma_{x=1...5}p(1-p)/5$	0.429	0.355	0.370	0.125
Equation (20): $F_{ST}=1-H_{exp(S)}/H_{exp(T)}$ Average = 0.290	0.129	0.110	0.187	0.733
Equation (21): $F_{ST}=\sigma^2_p/\bar{p}(1-\bar{p})$	0.129	0.110	0.187	0.733

$$F_{IT} = 1 - \left(\sum\nolimits_{x=1...5} H_{obs(A_x)}\right)/H_{exp(T)} \quad \text{or}$$
$$1 - F_{IT} = \left(\sum\nolimits_{x=1...5} H_{obs(A_x)}\right)/H_{exp(T)} \tag{22}$$

Recall that

$$F_{IS} = 1 - \left[\left(\sum\nolimits_{x=1...5} H_{obs(A_x)}\right)/H_{exp(S)}\right]/5 \quad \text{or}$$
$$1 - F_{IS} = \left[\left(\sum\nolimits_{x=1...5} H_{obs(A_x)}\right)/H_{exp(S)}\right]/5$$

and that

$$F_{ST} = 1 - H_{exp(S)}/H_{exp(T)} \quad \text{or}$$
$$1 - F_{ST} = H_{exp(S)}/H_{exp(T)}$$

Thus,

$$(1 - F_{IS})(1 - F_{ST}) = \left[\left(\sum\nolimits_{x=1...5} H_{obs(A_x)}\right)/H_{exp(S)}\right]$$
$$\left[H_{exp(S)}/H_{exp(T)}\right] = \left(\sum\nolimits_{x=1...5} H_{obs(A_x)}\right)/H_{exp(T)} = 1 - F_{IT}$$

Thus,

$$(1 - F_{IT}) = (1 - F_{IS})(1 - F_{ST})$$

or, multiplying and rearranging terms,

$$F_{IT} = F_{IS} + F_{ST} - F_{IS}F_{ST} \tag{23}$$

The correlation among uniting gametes in a subdivided population is the sum of the correlation among uniting gametes in collections and a correlation among the uniting gametes among all collections minus their covariance. Weir and Cockerham (1984) improved methods for estimating Wright's F-statistics that account for small and unequal sample sizes. F-statistics calculated via Weir and Cockerham (1984) are f, θ, and F, analogous, respectively, to F_{IS}, F_{ST}, and F_{IT}.

Wahlund's effect can estimated by substituting F_{ST} into Eq. (7):

$$\text{Prob}(A_1A_2) = \bar{p}^2(1 - F_{ST}) + \bar{p}F_{ST} \tag{24}$$

The probability of sampling an AA homozygote in a population subdivided into reproductively isolated collections is the average frequency of A squared (the expected frequency under Hardy–Weinberg) multiplied by the probability that A_1 and A_2 are allozygous, in addition to the probability of sampling alleles *iis* twice multiplied by the probability that both sampled alleles are *ibd* due to population subdivision. Equation (24) reduces to

$$\text{Prob}(A_1A_2) = \bar{p}^2 + \sigma^2_p \tag{25}$$

With continuous gene flow (i.e., random mating) among collections, $\sigma^2_p = 0$ and $\text{Prob}(A_1A_2) = \bar{p}^2$ (Hardy–Weinberg expectations). However, if there is reduced gene flow among collections, then $\sigma^2_p > 0$ and there will be an excess of homozygotes (because a variance is always greater than 0). Wahlund's effect is proportional to the variance of allele frequencies among collections.

As discussed in Figure 32.2, subdivided populations are usually not limited to one level of nesting or to one geographic scale. F_{ST} can be further subdivided to study levels of differentiation at different geographic scales. For example, in Figure 32.2, F_{ST} (all collections) can be subdivided into F_{ST} (collections within regions) + F_{ST} (regions within the total). This permits us to ask: What is the amount of gene flow among collections within each region A, B, or C as compared with the amount of gene flow among these three regions? If the mountain ranges or point of land do not act as barriers to gene flow, then we would expect F_{ST} (collections within regions) $\geq F_{ST}$ (regions within total). Alternatively, F_{ST} (all collections) can be subdivided into F_{ST} (collections within regions) + F_{ST} (regions A/C versus B) to ask: What is the amount of gene flow among collections within regions A and C versus B as compared with the amount of gene flow between A/C compared with B regions? Specifically, if the mountain ranges isolated collections in region B from those in regions A and C, then we would expect F_{ST} (regions A/C versus B) > F_{ST} (collection within regions). Black et al. (1988b) used a nested spatial design to examine gene flow among *Ae. albopictus* populations in Kuala Lumpur and Kuala Trengganu in Peninsular Malaysia and Sarawak and Sabah in Borneo. Collections were nested within these four areas, and areas were nested within either Malaysia or Borneo. Fifty-two percent of the total variance among all collections was attributable to differences between Malaysia and Borneo. In all analyses, the amount of variance among collections within the four areas was consistently four to five times as large as the amount of variance among areas in either Malaysia or Borneo. Almost all of the variance within either Malaysia or Borneo regions was attributable to local differentiation, suggesting that genetic drift is an important component of the natural breeding structure of this species. This indicates that the large amounts of local differentiation found in U.S. populations were not a consequence of recent colonization (Black et al. 1988a). De Merida et al. (1999) partitioned the variance in mitochondrial haplotype frequencies among northern and southern Guatemalan *An. albimanus* collections made at the same sites during wet and dry seasons. Approximately 99% of the variation in haplotype frequencies arose among individuals in a collection, and the remaining variation arose among sites within a season. Negative variation was estimated among seasons, indicating that there was more variation among sites within seasons than among seasons. There was no variation in haplotype frequencies between wet and dry season collections in either northern or southern Guatemala.

Wahlund's effect also influences linkage disequilibrium (Table 32.4). Ohta (1982) developed a method for partitioning linkage disequilibrium observed in an overall population (D_{IT}^2) into linkage disequilibrium arising through genetic drift (D'_{IS}^2) (Wahlund's effect) and linkage disequilibrium arising through consistent epistasis between alleles (D'_{ST}^2). D_{ST}^2 estimates the average maximum disequilibrium that can arise due to genetic drift, and D_{IS}^2 is the average disequilibrium due to epistasis. D_{IS}^2 is D_{ij} squared [Eq. (1)] averaged over all collections. Her methods can assess whether disequilibrium observed in collections consistently supports epistasis or instead arises through genetic drift. When disequilibrium arises by random drift,

$$D_{IS}^2 < D_{ST}^2 \quad \text{and} \quad D'_{IS}^2 > D'_{ST}^2 \tag{26}$$

If disequilibrium is consistent for specific allele pairs in collections, then

$$D_{IS}^2 > D_{ST}^2 \quad \text{and} \quad D'_{IS}^2 < D'_{ST}^2 \tag{27}$$

TABLE 32.4 Wahlund's Effect Influences Linkage Disequilibrium

Locus	Allele	Collection 1	Collection 2	Total
A	i	0.1	0.7	$\frac{1}{2}(0.1 + 0.7) = 0.40$
B	j	0.8	0.1	$\frac{1}{2}(0.8 + 0.1) = 0.45$

- Individuals in collections 1 and 2 are reproductively isolated.
- Assume that we do not perceive these collections and instead sample them as one large collection.
- Further assume that we sample equal numbers of individuals from each of the two collections.
- Alleles A_i and B_j are in linkage equilibrium in the two collections, so if we sample 100 individuals from each collection, the number of A_iB_j individuals will be $0.1 \times 0.8 \times 100 = 8$ individuals in collection 1 and $0.7 \times 0.1 \times 100 = 7$ individuals in collection 2 and 15 A_iB_j individuals total.
- However, we would expect $0.40 \times 0.45 \times 200 = 36 A_iB_j$ individuals.
- Thus, Wahlund's effect creates a deficiency of A_iB_j individuals.

Unequal systematic disequilibrium arises if selection for specific allele pairs occurs only in a few collections.

Assumption 3: No Migration Among Populations

We know that small, finite population size leads to inbreeding within isolated collections. Furthermore, any reproductive isolation among small collections causes genetic drift that eventually leads to alleles' becoming fixed or extinct in collections. We are now in a position to examine the role of migration as a force that homogenizes allele frequencies among collections. To do so, we remove the large domes over each of the collection sites (A_1–A_5) so that allele frequencies among collections are no longer necessarily independent. We allow organisms (and thus alleles) to migrate freely among collections A_1–A_5.

Having violated assumption 4, we now have a slightly different model than that originally presented in Figure 32.5. This is called Wright's "island model," in which the frequency of an allele in a collection i is p_i and the average frequency of an allele among collections is $\bar{p}$. We assume there is no differential migration among genotypes, so among the migrants the frequency of the allele is also $\bar{p}$. The migration rate is defined as m, the probability that a randomly selected gene in any collection arose from a migrant. This implies that m is equally distributed among all pairs of collections.

Consider an allele at generation t. The allele is either "native" (arose within that collection) or "foreign" (arose through immigration into that collection). The probability of selecting a native allele in generation 0 is $p_{i(t=0)}$, while the probability that it didn't arise through immigration generation 0 is $1 - m$. Therefore, the probability of selecting a native allele is $p_{i(t=0)}(1 - m)$. Similarly, the probability of selecting a foreign allele is $\bar{p}$ and the probability that the allele arose through immigration is m. Thus, the probability of selecting an immigrant allele is $\bar{p}m$. In consecutive generations the frequency of an allele will be

$$p_{i(t=1)} = p_{i(t=0)}(1 - m) + \bar{p}m \qquad (28)$$

The probability of selecting a particular allele in generation 1 is the sum of the probability of selecting a native allele in generation 0 and the probability that that allele will arise by immigration in generation 0. Equation (28) expands to

$$p_{i(t)} = \bar{p}_i + p_{i(0)}(1 - m)^t \qquad (29)$$

Because $m > 0$, $1 - m < 0$, and $\lim\limits_{t \to inf.} p_{i(0)}(1 - m)^t = 0$,

equation 29 implies that over time, $p_{i(t)}$ converges to $\bar{p}_i$. Because $(\mu + \gamma) > 0$, $1 - (\mu + \gamma) < 0$ and $\lim\limits_{t \to inf.} p_{i(0)}[1-(\mu+\gamma)]^t = 0$, through time, $p_{i(t)}$ converges to $\gamma/(\mu + \gamma)$. In other words, collections that vary in their initial frequency p_i will all eventually converge to $\bar{p}$ through migration (Fig. 32.12A).

The island model assumes that N in any collection is sufficiently large to avoid genetic drift. We can lift that assumption using Eq. (11). Recall that alleles arising from outside a collection are always assumed to be allozygous with regard to alleles from a collection. Thus, the probability that two alleles that unite to form a gamete arose from native parents is $(1 - m)^2$. Multiplying this probability by Eq. (11) we obtain

$$F_{t=1} = \{1/2N_{t=0} + [1 - (1/2N_{t=0})]F_{t=0}\}(1 - m)^2 \quad (30)$$

This reduces to

$$F_{ST} = (1 - 2m)/(4N_e m + 1 - 2m)$$

or, for small values of m,

$$F_{ST} \cong 1/(4N_e m + 1) \qquad (31)$$

This equation provides a means to estimate the effective migration rate ($N_e m$), defined as the number of reproducing migrants exchanged per generation among collections in the island model from F_{ST}. Rearranging Eq. (31), we get

$$N_e m \cong (1 - F_{ST})/4F_{ST} \qquad (32)$$

Figure 32.13 is a graph of the relationship between F_{ST} and $N_e m$. The important implication of this equation is that in principle the number of reproducing migrants exchanged per generation among collections in the island model need only be four to six individuals to effectively homogenize collections.

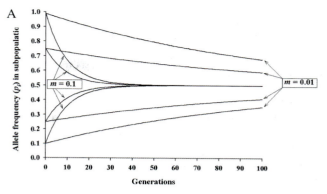

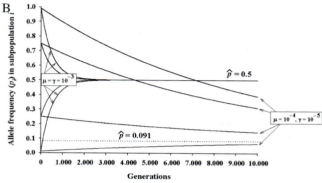

FIGURE 32.12 (A) The role of migration in homogenizing allele frequencies among collections in Wright's island model [Eq. (28)]. (B) The role of mutation in homogenizing allele frequencies among collections in Wright's island model [Eq. (33)].

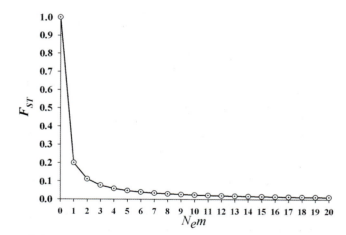

FIGURE 32.13 Relationship between F_{ST} and $N_e m$ [Eq. (32)].

Table 32.5 lists $N_e M$ values that have been estimated from F_{ST} for a variety of disease vectors. Notice that a range of values are reported for various species. This is because the authors used nested spatial sampling and reported values for different geographic scales.

TABLE 32.5   Estimates of F_{ST} and N_eM in Various Vector Species

Species	Locations of Study	F_{ST}	N_eM	Isolation by Distance?	Ref.
Anopheles gambiae	Mali, West Africa	0.006	>40	Y	Taylor et al. (2001)
	Kenya and Senegal	0.008	>3	Y	Lehmann et al. (2000)
Anopheles albimanus	Central America	0.020–0.700	<1–12	Y	De Merida et al. (1999)
Aedes aegypti	Puerto Rico	0.020–0.025	10–12	N	Apostol et al. (1996)
	Northeastern Mexico	0.066	3.5	Y	Gorrochotegui et al. (2000)
Aedes albopictus	Southeastern United States	0.278–0.361	<1	N	Kambhampati et al. (1990)
	Malaysia	0.316–0.605	<1	N	Black et al. (1988)
Aedes albifasciatus	Central Argentina	0.027–0.098	2–9	Y	de Sousa et al. (1999)
Ochlerotatus cataphylla	North central Colorado	0.059	4	N	West and Black (1998)
Ochlerotatus pullatus	North central Colorado	0.058–0.099	2–4	N	West and Black (1998)
Ochlerotatus hexodontus	North central Colorado	0.033–0.061	4–7	N	West and Black (1998)
Glossina pallidipes	Zimbabwe, Mozambique, Kenya, and Ethiopia	0.485	<1	Y	Krafsur and Wohlford (1999)
Glossina morsitans centralis	Botswana, Namibia, Zambia, and Tanzania	0.150–0.225	<1–1.5	Y	Krafsur et al. (2001)
Ixodes pacificus	Geographic range in western United States	0.160	1.3	Y	Kain et al. (1997)

Only studies that explicitly tested for isolation by distance are included.

Assumption 4: Loci Under Examination Don't Mutate

Next, we consider the possibility that new alleles enter, or leave, a collection through forward and reverse mutation, respectively. μ is the probability of a foreword mutation ($\alpha_i \rightarrow \alpha$) and γ is the probability of a reverse mutation ($\alpha_I \leftarrow \alpha$). Many of the assumptions used when examining the effects of mutation are analogous to those used in the migration model, except that new alleles arise in or leave collections through mutation rather than migration. The recurrence equation for mutation is easily understood at an intuitive level:

$$p_{t=1} = p_{t=0}(1 - \mu) + (1 - p_{t=0})\gamma \quad (33)$$

The frequency of an allele in generation 1 is a function of its frequency in generation 0 multiplied by the probability that it did not mutate $(1 - \mu)$, in addition to the frequency of alternative alleles in the population in generation 0 multiplied by the probability γ that they mutated to the allele under consideration. This can be extended over t generations to

$$p_{i(t)} = p_{i(0)}[1 - (\mu + \gamma)]^t + \gamma/(\mu + \gamma) \quad (34)$$

Because $(\mu + \gamma) > 0$, $1 - (\mu + \gamma) < 1$ and $\lim_{t \to inf} p_{i(0)}[1 - (\mu + \gamma)]^t = 0$, through time, $p_{i(t)}$ converges to $\gamma/(\mu + \gamma)$. Regardless of the initial allele frequency

in a collection, mutation will cause alleles to converge to the frequency $\gamma/(\mu + \gamma)$ at equilibrium (Fig. 32.12B).

We can again lift the assumption that N_e in any collection is sufficiently large to avoid genetic drift by using Eq. (11) and applying assumptions about *ibd* and *iis* with regard to alleles from a collection. Thus, the probability that two alleles that unite to form a gamete are *iis* and *ibd* = $(1 - \mu)^2$. Multiplying this probability by Eq. (11) we obtain

$$F_{t=1} = \{1/2N_{t=0} + [1 - (1/2N_{t=0})]F_{t=0}\}(1 - \mu)^2 \quad (35)$$

The correlation between loci in uniting gametes in generation 1 is the probability that the two gametes arose from the same parent $(1/2N)$ in generation 0 and the probability that, if the two gametes arose from different parents $(1 - 1/2N)$, they were *ibd* in the population at generation 0 ($F_{t=0}$). This correlation is multiplied by the probability that neither allele mutated in the previous generation. This reduces to

$$F_{ST} = (1 - 2\mu)/(4N_e\mu + 1 - 2\mu)$$

or, for small values of μ,

$$F_{ST} \cong 1/(4N_e\mu + 1) \quad (36)$$

Comparing Eqs. (31) and (36), we note that both N_em and $N_e\mu \cong (1 - F_{ST})/4F_{ST}$. How can we estimate parameters N_em and $N_e\mu$ from the same statistic, F_{ST}? The answer is that we can't! We cannot know if a new allele arrives in a collection through mutation or migration, so F_{ST} is actually estimated assuming that neither allele immigrated or mutated in the previous generation, or

$$F_{t=1} = \{1/2N_{t=0} + [1 - (1/2N_{t=0})]F_{t=0}\}(1 - m)^2(1 - \mu)^2 \quad (37)$$

which reduces to

$$F_{ST} \cong 1/[4N_e(\mu + m) + 1] \quad (38)$$

Equation (38) has important implications. Estimates of N_em from Eq. (32) always implicitly assume negligible mutation. With phenotype or isozyme markers this wasn't a problem because population geneticists could assume $\mu \sim 10^{-6}$. However, recall that the RAPD, AFLP, and microsatellite markers, discussed in Chapter 30, have been estimated to have $\mu \sim 10^{-3}$. In these cases, N_em and $N_e\mu$ could easily be of the same magnitude and thus overestimate the actual N_em. Another issue concerns use of microsatellite loci in which μ and $\gamma \sim 10^{-3}$ (Fig. 29.12). Thus, $p_{i(t)} = \gamma/(\mu + \gamma) \cong 0.5$. This implies that regardless of migration, microsatellite alleles in different populations should converge to equal frequencies. As a consequence, $F_{ST} \cong 0$ and $N_em \cong \infty$. This is a situation referred to as *size homoplasy*. Lehmann et al. (1996) discuss these issues with regard to microsatellite alleles in *Anopheles gambiae*.

Use of Eq. (38) to estimate N_em assumes that the conditions of the island model are valid. In particular, the assumption that m is equally distributed among all pairs of collections can be highly problematic, depending on the geographic locations of collections. For example, this assumption would probably be invalid in our original study design (Fig. 32.2). While the assumption might be valid *within* each of the three regions A, B, and C, the very nature of our questions suggests that it is not valid among regions. Finite geographical barriers such as mountain ranges, rivers, lakes, oceans, or deserts form obvious barriers to gene flow. m within a region is unlikely to be the same as m between regions.

Often geographic distance alone constitutes a barrier to gene flow. Collections that are in proximity to one another will exchange more migrants than collections that are farther apart. In this case, the assumption of uniform migration is violated: m will be negatively correlated with geographic distance, and collections are said to be genetically *isolated by distance*. This can be tested using a standard linear regression model, where Slatkin's linear F_{ST} [$F_{ST}/(1 - F_{ST})$] among all possible pairwise combinations of collections is regressed upon pairwise linear or natural logarithm geographic distances among collections (Rousset 1997). The Mantel test is used as a conservative esti-

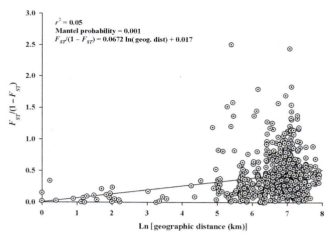

FIGURE 32.14 Testing for isolation by distance among 38 collections of *Aedes aegypti* from throughout Mexico.

mate of the significance of the regression coefficient because pairwise distance estimates are nonindependent by their very nature (Fig. 32.14). When a significant relationship is detected, an island model is no longer appropriate, and N_em cannot be evaluated among all collections because it varies according to the proximity of collections. Conn et al. (1999) used this approach to demonstrate isolation by distance among collections of *An. darlingi* in Venezuala, Bolivia, and Brazil.

Assumption 5: Selection and Adaptation

We have to this point assumed that the genotype or haplotype present in a particular life stage does not influence the survival or reproduction of that stage. Models of *viability selection* predict allele and genotype frequencies when this assumption is false. Models of *sexual selection* predict these frequencies when mates choose or exclude mates based upon their genotype. Models of *fecundity selection* predict outcomes when fecundity varies among parental genotypes. *Gametic selection* and *meiotic drive* models predict allele and genotype frequencies when the haplotypes of gametes produced and/or the zygotes that arise are not in Mendelian proportions.

Viability Selection

Assume that there are two alleles at a locus, A and a, with respective frequencies of p and q. In the model in Figure 32.5 there will be three genotypes AA, Aa, and aa. If these three genotypes differ in their ability to survive, we refer to the differential survivorship among these three genotypes as their *relative fitnesses* (designated w_{xx}). These three genotypes survive in ratios of $w_{AA}:w_{Aa}:w_{aa}$. Therefore, assuming that genotypes are in Hardy–Weinberg equilibrium, the ratio of genotypes in a population is $p^2w_{AA}:2pqw_{Aa}:q^2w_{aa}$, and the ratio of *A* and *a* alleles is, respectively:

$$[p^2w_{AA} + pqw_{Aa}]:[pqw_{Aa} + q^2w_{aa}] \quad \text{or}$$
$$p(pw_{AA} + qw_{Aa}):q(pw_{Aa} + qw_{aa}) \tag{38}$$

Terms in parentheses are referred to as the *marginal fitness* of A:

$$w_A = pw_{AA} + qw_{Aa} \tag{39}$$

To estimate or predict values of *p*, these ratios have to be divided by the average fitness ($\bar{w}$), where $\bar{w} = p^2w_{AA} + 2pqw_{Aa} + q^2w_{aa}$. Within this framework, we can develop recurrence equations to predict allele frequencies:

$$p_{t+1} = p_t(p_tw_{AA} + q_tw_{Aa})/\bar{w} = p_tw_A/\bar{w} \tag{40}$$

$$q_{t+1} = q_t(p_tw_{Aa} + q_tw_{aa})/\bar{w} = q_tw_a/\bar{w} \tag{41}$$

These recurrence equations are useful in predicting allele frequencies under a variety of different condi-tions. Consider a situation in which a new allele *A* has just entered a population and has a higher relative fitness than the existing allele *a*. *A* slowly goes to fixation, and the population gradually approaches the optimal fitness of 1.0 when the population is fixed for *A* (Fig. 32.15A). The larger the differential among $w_{AA}:w_{Aa}:w_{aa}$, the faster *A* goes to fixation and the faster the average fitness of the population is optimized (Fig. 32.15B). This result can be generalized for many rela-tive fitness conditions. The greater the fitness differen-tial among the three genotypes, the more rapidly weaker genotypes are eliminated.

The reciprocal arrangement yields a similar result. For example, consider a situation in which a new allele *a* has just entered a population and has a higher rela-tive fitness than the existing allele *A*. *A* slowly goes to extinction and the population gradually approaches the optimal fitness 1.0 when the population is fixed for *a* (Fig. 32.15C). This result can also be generalized for many relative fitness conditions. The greater the fitness differential among the three genotypes, the faster the least fit allele goes to extinction.

Consider a different situation, in which the new allele *A* is dominant to the existing allele *a*. *A* slowly

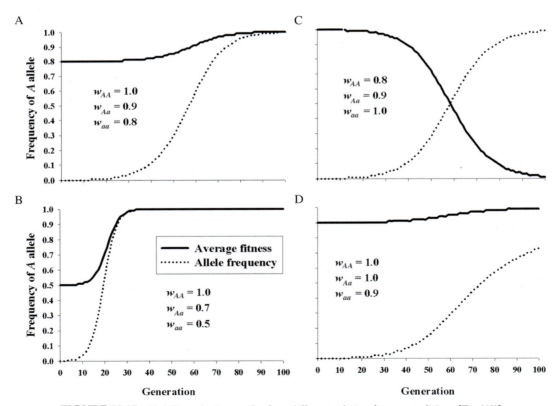

FIGURE 32.15 Viability selection under four different relative fitness conditions [Eq. (40)].

goes to fixation and the population only gradually approaches the optimal fitness. This occurs because the less fit allele is hidden from natural selection in heterozygotes. When the new allele A is dominant to the existing allele a, A only gradually approaches fixation, even when there are very large fitness differentials (Fig. 32.15D). If, instead, the new, more fit allele a is recessive to the existing allele A, A slowly goes to

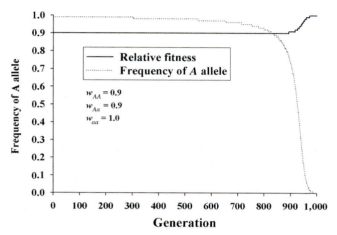

FIGURE 32.16 Viability selection on a less fit recessive allele [Eq. (40)].

extinction and the population only gradually approaches the optimal fitness (Fig. 32.16). This occurs because only (aa × aa), (Aa × aa), or (Aa × Aa) crosses produce aa individuals, and a selection differential only exists between aa homozygotes and the other two genotypes. However, aa × aa crosses are initially very rare $(0.001 \times 0.001)^2 = 10^{-12}$, as are Aa × aa crosses $(2 \times 0.999 \times 0.001)(0.001 \times 0.001) = 2 \times 10^{-9}$ and Aa × Aa crosses $(2 \times 0.999 \times 0.001)^2 = 4 \times 10^{-6}$.

Equations (40) and (41) also predict the outcome of a situation in which the heterozygote has a higher fitness than either of the two homozygotes (defined as *overdominance*). A new allele a enters a population and has a higher relative fitness than the existing allele A. A slowly declines in frequency as the population gradually approaches the optimal fitness. However, A never goes to extinction because Aa heterozygotes have a fitness advantage over aa homozygotes. Instead, the frequency of A reaches a stable equilibrium frequency $\hat{p}$ (Fig. 32.17A). If A has just entered a population and has a lower relative fitness than the existing allele a, then A will slowly increase in frequency as the population gradually approaches the optimal fitness. A has a lower relative fitness than a; however, Aa heterozygotes have a fitness advantage over aa homozygotes. Eventually, the frequency of A reaches $\hat{p}$

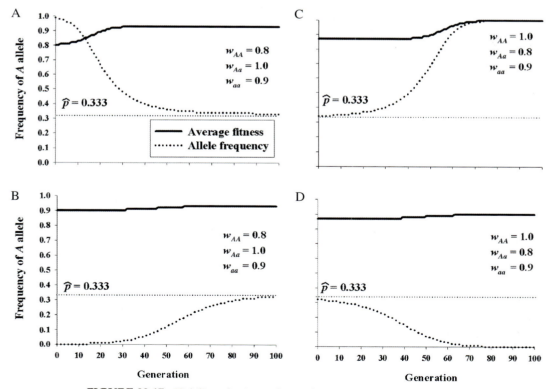

FIGURE 32.17 Viability selection with overdominance and underdominance.

(Fig. 32.17B). This result can be generalized for many conditions of overdominance and initial values of p_0. We can estimate $\hat{p}$ by manipulating Eq. (40) to derive

$$\Delta p = (p_t q_t / \bar{w}[p_t(w_{AA} - w_{Aa}) + q_t(w_{Aa} - w_{aa})] \quad (42)$$

At equilibrium $\Delta p = 0$ and Eq. (41) reduces to

$$\hat{p} = w_{aa} - w_{Aa} / (w_{AA} - 2w_{Aa} + w_{aa})$$

With dominance,

$$\hat{p} = (w_{aa} - w_{Aa}) / (-w_{Aa} + w_{aa}) = 1$$

With overdominance, $\hat{p}$ is determined by the relative fitnesses of the three genotypes ($w_{AA} < w_{Aa} > w_{aa}$),

$$\hat{p} = (0.9 - 1.0) / (0.8 - 2(1) + 0.9) = 0.3333$$

Equations (40) and (41) also predict the outcome of a situation in which the heterozygote has a *lower* fitness than either of the two homozygotes (called *underdominance*, or *heterozygote inferiority*). If a new allele A has been introduced into a population and has a higher relative fitness than the existing allele a, then A slowly increases in frequency as the population gradually approaches the optimal fitness of 1.0 (Fig. 32.17C). However, the outcome of underdominance is dependent upon initial conditions. If, instead, the new allele A is introduced at a frequency less than $\hat{p}$, then A slowly decreases in frequency as the population gradually approaches the local optimal fitness of 0.9 (Fig. 32.17D). With heterozygote inferiority, ($w_{AA} > w_{Aa} < w_{aa}$),

$$\hat{p} = (0.9 - 1.0) / (0.8 - 2(1) + 0.9) = 0.3333$$
$$= \text{saddle point, or instability point}$$

There are relatively few examples of heterozygote inferiority in nature, unless a species lives in a coarse-grained environment where there is strong selection for alternative homozygous genotypes. However, models of underdominance are useful for assessing strategies involving chromosome translocations in genetic control. A translocation occurs when a chromosome arm is broken (usually by ionizing radiation) and then fuses with a nonhomologous arm. Strains bearing translocations can be introduced into natural pest populations to create a genetic load due to the continued production of less fit genotypes. When translocation heterozygotes mate, they produce 16 types of zygotes. Only 1/16 of zygotes are homozygous wild type, 1/16 are translocation homozygotes, and 4/16 of zygotes are translocation heterozygotes. However, the remaining 10/16 of zygotes are aneuploid and die during embryogenesis. A mating between two translocation-homozygous individuals produces only translocation-homozygous offspring. A cross between a wild-type parent and a translocation-homozygous parent produces only translocation-heterozygous offspring. A cross between a translocation-heterozygous parent and a translocation-homozygous parent produces 50% translocation-heterozygous and 50% translocation-homozygous offspring. Therefore, translocation heterozygotes have a lower fitness than wild-type organisms or organisms that are translocation homozygotes.

In what has come to be known as the *fundamental theorem of natural selection*, Fisher showed that with regard to viability selection:

$$\Delta w = 2\sigma^2_w / \bar{w} \quad (43)$$

This shows that fitness changes will always be zero or greater because $\sigma^2_w / \bar{w} \geq 0$. Therefore, fitness will increase or remain the same over time. Natural selection will not allow a population to decrease in average fitness. Furthermore, the rate of increase in fitness is proportional to the genetic variance in fitness. If individuals in a population display a wide array of fitnesses and fitness is under genetic control, then the rate with which natural selection can remove the less fit individuals from a population will be large. In contrast, if individuals in a population display a narrow range of variation in fitness, then the rate will be small with which natural selection removes less fit individuals. A common way to present Fischer's fundamental theorem of natural selection is by plotting the mean fitness as a function of allele frequencies. The plot is called an *adaptive topography* (Fig. 32.18).

So far we've only considered loci at which there are two alleles. This is a highly artificial situation because we know there can be a virtually continuous array of alleles present at any locus in a natural population. However, the mathematics of viability selection become complex with multiple alleles. Consider the well-known relationship between allele frequency data on the human β-globin gene, malaria susceptibility, and sickle cell anemia in West Africa. There are >100 alleles of the β-globin gene (most of them are very rare). The three most common alleles in West Africa are $Hb\beta^A$, $Hb\beta^C$, and $Hb\beta^S$. $Hb\beta^S / Hb\beta^S$ homozygotes have sickle-cell disease, an affliction that leads to severe anemia (low red blood cell count), with very painful, potentially lethal episodes. The molecular basis for the disease is the abnormal tertiary structure of the globin protein, and red blood cells from individuals with the disease assume sickle shapes under conditions of low oxygen tension, when their hemoglobin forms long crystals. The blood merozoite stage of *Plasmodium falciparum* is destroyed in $Hb\beta^S / Hb\beta^S$ individuals, rendering these individuals effectively resistant to malaria. Indirect evidence for the relationship between the elevated frequency of sickle-cell disease and malaria is that their geographic distributions are very similar. $Hb\beta^A$ homozygotes are susceptible to *Plasmod-*

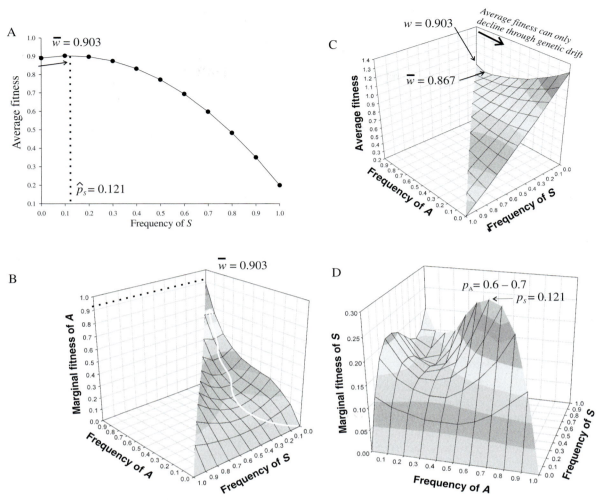

FIGURE 32.18 Adaptive topography for average fitness as a function of the frequencies of $Hb\beta^A$, $Hb\beta^C$, and $Hb\beta^S$ alleles [Eq. (44)]. (A) Adaptive topography in the presence of malaria when the $Hb\beta^S$ allele first appears; (B) adaptive topography as a function of $w_{Hb\beta A}$ in the presence of malaria when the $Hb\beta^S$ allele first appears; (C) adaptive topography in the presence of malaria when the $Hb\beta^C$ allele first appears; (D) adaptive topography as a function of $w_{Hb\beta S}$ in the presence of malaria.

ium falciparum malaria, which kills ~20% of infected children under the age of 5 in sub-Saharan Africa. $Hb\beta^A/Hb\beta^S$ heterozygotes are said to have the "sickling trait." They are resistant to malaria but do not have severe anemia. An $Hb\beta^C$ allele entered (or evolved in) Africa more recently. $Hb\beta^A$ and $Hb\beta^C$ homozygotes and heterozygotes are very similar in phenotype. The viabilities of the six genotypes have been calculated:

$$\bar{w} = p_A^2 w_{AA} + p_S^2 w_{SS} + p_C^2 w_{CC} + 2p_A p_A w_{AS}$$
$$+ 2p_A p_C w_{AC} + 2p_S p_C w_{SC} = 0.89 p_A^2 + 0.20 p_S^2 \quad (44)$$
$$+ 1.31 p_C^2 + (1)2p_A p_S + (0.89)2p_A p_C + (0.70)2p_S p_C$$

By entering all possible permutations of p_A, p_S, p_C into Eq. (44) we can obtain fitness topographies (Fig.

32.18). Note that in fitness or adaptive topographies, organisms with high fitness are located at "peaks" while least fit individuals are positioned within "valleys."

The original human population was fixed for allele A but, in the presence of malaria, had a marginal fitness of 0.89 (Fig. 32.18A). When $Hb\beta^S$ appeared, it increased because the fitness of $Hb\beta^A/Hb\beta^S$ was 1. $Hb\beta^S$ homozygotes were very rare and didn't affect the marginal fitness of $Hb\beta^S$. The $\hat{p}_s$ was

$$(w_{AA} - w_{AS})/(w_{AA} - 2w_{AS} + w_{SS}) =$$
$$(0.89 - 1.0)/(0.89 - 2 + 0.2) = 0.121$$

The $\hat{p}_A$ was $1 - \hat{p}_S = 0.879$; $\bar{w}$ at equilibrium was 0.903 (Fig. 32.18B). When a mutation gave rise to $Hb\beta^C$

(Fig. 32.18C), the possibility of its increasing was dependent upon its relative marginal fitness. The

$$w_C = p_A w_{AC} + p_S w_{SC} + p_C w_{CC}$$
$$= (0.879 \times 0.89) + (0.121 \times 0.70) = 0.867$$

when C was rare, but $\bar{w}$ at equilibrium had a greater value of 0.903. Even though there were higher fitness peaks, the population would have had to decline in fitness to reach these higher peaks. Fisher's fundamental theorem shows that this is impossible by natural selection alone. Therefore, C could not have established itself by selection alone. However, also notice that it would only require a minimal amount of genetic drift for a small, isolated population to cross the fitness threshold and end up under a new peak. From there, natural selection alone would move the population up to the new peak. Notice also that p_A can range from 0.6 to 0.7 when $\hat{p}_S = 0.121$ (Fig. 32.18D). This would suggest that other alleles, including $Hb\beta^C$, could vary between 0.18 and 0.28. Thus, a combination of drift and selection could explain the common frequency of other $Hb\beta$ alleles.

Frequency-Dependent Viability Selection

Although we've defined only a single fitness for each genotype, a genotype may have different fitnesses according to environments, seasons, population densities, or the frequencies of other alleles. The latter is called *frequency-dependent selection*; the frequency of an allele determines its relative fitness. We have already discussed this situation with regard to assortative mating, in which the fitness of a particular genotype may depend upon its frequency. Under positive assortative mating, a genotype will be most fit when it is most common because there is a greater chance of genetically similar mates encountering one another. Under negative assortative mating, a genotype will be most fit when it is rare because there is a greater chance of genetically *dissimilar* mates encountering one another. For example, consider a locus with two alleles A and a at frequencies of p and q, respectively. Under negative assortative mating, their relative fitnesses might be: $w_{AA} = 1 - p^2$, $w_{Aa} = 1 - (2pq)$, and $w_{aa} = 1 - q^2$. With these conditions it can be shown that the overall fitness is optimized when $p = 0.5$ (Fig. 32.19). Frequency-dependent selection maintains stable polymorphisms at the locus.

Sexual Selection

Sexual selection is a widely observed phenomenon that violates Hardy–Weinberg assumptions 2 (requiring adults to mate irrespective of genotypes) and 5 (all genotypes to have equal reproductive success). The

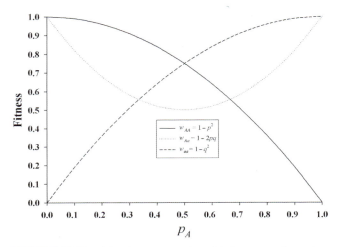

FIGURE 32.19 Adaptive topography with frequency-dependent selection.

earliest natural history literature documented that in many species certain individuals that were either larger or more colorful or possessed enlarged structures for combat were more successful in attracting or obtaining mates. This phenomenon was termed *sexual selection*.

Darwin first proposed sexual selection as a means through which natural selection occurs because traits that were advantageous for mating would obviously be subject to positive selection. However, Darwin recognized an immediate paradox, because many of the characters subject to sexual selection do not typically increase and may often decrease survival in the individuals who have them. Having a big, brightly colored tail is likely to attract predators and/or make it harder to escape. The benefits of sexual selection must offset the costs to viability. The reason that low-viability traits evolve is that they significantly increase mating success and therefore increase reproduction. A male with a small, dull-colored tail might survive much better, but if he doesn't attract females his genes will not be passed to the next generation, and a dull tail would ultimately be maladaptive.

The next question that arose was: Why do traits that increase mating success but decrease survival evolve only in males? In most species males can increase their reproduction by attracting more mates, but a female only needs to attract one male to become fertile, and once she carries the male's sperm she doesn't increase her reproduction by attracting more males until after she produces more offspring. In many vector species females store sperm from their first copulation and use this to fertilize eggs throughout the remainder of their life. If a male can attract many females, he can potentially be the father to many offspring. Males in many vector species emerge before females and are developmentally ready for copulation as females emerge. Males compete for females but are typically short lived in these species.

Female choice is another important aspect of sexual selection. Why should females evolve a preference for more ornate or combative males? If sexual selection occurs through female choice of males, then there must be simultaneous evolution of male traits and female choice for these traits. Female choice has been demonstrated to occur in a large number of species of birds, reptiles, fish, and insects. Thornhill and Alcock (2001) provide a complete treatment of sexual selection in insects. Aside from McLain et al. (1985), there has been very little work on sexual selection in arthropod vectors.

Fecundity Selection

An implicit part of assumption 5 is that all genotypes produce either equal numbers of zygotes or that production of zygotes per genotype follows a Poisson distribution. Fecundity selection operates on the products of a mating pair rather than the fitness of the individual parents. We assume that individuals mate at random but that there is a differential number of zygotes produced per genotype pair. This is often represented by fecundity matrices:

		Male Genotype		
		AA	Aa	aa
Female	AA	ϕ_1	ϕ_2	ϕ_3
Genotype	Aa	ϕ_4	ϕ_5	ϕ_6
	aa	ϕ_7	ϕ_8	ϕ_9

where ϕ_x represents the number of offspring produced by a specific pair of parental genotypes. If all ϕ_x are equal, then genotypes in the zygotes will follow Hardy–Weinberg proportions. Otherwise, the frequencies of genotypes arising from specific matings have to be multiplied by ϕ_x to predict outcomes from one generation to the next.

Theoretical population geneticists have attempted to use the same framework for fecundity selection that we developed for viability selection. But no derivations analogous to the fundamental theorem of natural selection have been derived. The mathematics becomes rapidly complex, so computer algorithms have had to be developed to model the outcomes of fecundity selection. Most of these indicate that the adaptive topography of fecundity selection contains many fitness peaks and valleys associated with the various values of ϕ_x. There are many conditions in which fecundity decreases rather than increases as a product of the fecundity selection. We know little about fecundity selection in arthropod vectors.

Meiotic Drive

We have assumed that the haplotypes inherited by a gamete follow the rules of Mendelian genetics and that gametes fuse independent of their haplotypes to form diploid zygotes. However, there are a number of situations in which the fitness of a gamete is governed by the alleles that it carries. The most common outcome of gametic selection is *segregation ratio distortion* (SD), in which the observed ratio of genotypes differs from the predicted Mendelian ratio. For example, a cross between an *Aa* heterozygote and an *aa* homozygote should produce progeny genotype ratios of $1Aa:3aa$. If the observed ratio varies significantly from $1:3$, then SD has occurred. If there is an excess of *Aa* genotypes, then this would suggest a mechanism known as *meiotic drive* for the *A* allele. Applying all other assumptions in Figure 5, a model of meiotic drive predicts

$$p_{t+1} = p_t^2 + 2kp_tq_t \tag{45}$$

Notice that if $k = \frac{1}{2}$, then Hardy–Weinberg proportions are obtained. But if $k > \frac{1}{2}$, *A* goes to fixation; and if $k < \frac{1}{2}$, *a* goes to fixation. The outcomes of this model are simple: The driven allele goes to fixation. So, why do we encounter cases of meiotic drive in nature?

The best-known examples of meiotic drive are the *t* alleles in mice and the SD system in *Drosophila melanogaster*. Meiotic drive has also been documented in *Ae. aegypti* (Wood and Ouda 1987). All of these systems are genetically complex. Part of the complexity stems from the existence of autosomal and X-chromosomal modifiers that suppress the segregation effects of SD.

Neutralism and Molecular Evolution

When allozymes were first beginning to be characterized in natural populations in the 1960s, geneticists were baffled by the large amount of revealed genetic variation. In particular, some assumed that all or most genetic variation must be adaptive; therefore, the increased variation revealed by allozymes suggested that the environment occupied by a species must contain a bewildering array of selection regimes to maintain such a high numbers of adaptive polymorphisms. How can all of this genetic variation be maintained? Most of the viability selection models we have reviewed predict that suboptimal genotypes will be selected against. In order for all of this variation to be adaptive, there would have to be a great deal of genetic death in each generation and in each environment to eliminate less fit genotypes. In the 1980s, recombinant DNA technology revealed still higher levels of genetic variation, and the most recent techniques, discussed in Chapter 30, have revealed yet another level of diversity.

The obvious alternative explanation for this variation is that most molecular variation has no impact on fitness. In 1968, Mootoo Kimura introduced an important theoretical paper predicting that polymorphisms that are neutral on their effects on fitness will not only become established but should in fact be abundant (Kimura 1968). This was the seed that initiated the concept of *neutral evolution*, or *neutralism*. Stated simply, most polymorphisms, whether at the level of amino acids or nucleic acids, have no or minimal effects on the phenotype and thus ultimately have no measurable impact on fitness. Kimura suggested that the vast majority of amino acid and nucleic acid substitutions arose through random mutation and that their ultimate fate was subject largely to the effects of random genetic drift. Neutralism was initially controversial because it seemed antithetical to Darwin's theory of natural selection. Discussion over the implications of the various views has come to be known as the *selectionist–neutralist debate*.

Kimura pointed out that natural selection only acts upon phenotypic variation that impacts fitness. There are a number of reasons to expect neutral polymorphisms at the levels of both amino acids and nucleic acids. The genetic code is degenerate. Assuming no codon usage bias, most mutations at the third position of codons generally have no effect on amino acid substitutions. Furthermore, many amino acids are functionally similar (Jackson et al. 1996). Genomics has repeatedly taught us that prokaryotic and eukaryotic genomes have enormous functional redundancy (e.g., humans have seven different isozymes for lactate dehydrogenase). These genomes have several enzymes that perform the same function; therefore it is likely that eliminating or reducing the activity of one redundant gene would have little effect on the fitness of an organism.

The neutral theory of molecular evolution provides a context for testing hypotheses of adaptation. Genetic variation observed in a particular gene is compared to what should arise purely by mutation. The alternative hypothesis is that genome-wide or locus-specific variation within and between populations is adaptive, arising from selection that reduces or maintains genetic variation.

There are four models of selection at a molecular level. *Purifying selection*, an important component of neutralism, *rapidly eliminates* any nucleic acid substitutions that reduce the fitness of an organism. The combined influences of neutral evolution and purifying selection are frequently cited to explain the overall variation observed in a gene. Purifying selection homogenizes populations by removing unfit alleles that might arise in a population by migration or mutation. *Positive, negative, or directional selection* occur when an allele, new to a species through mutation or new to a population through migration, confers a relative fitness advantage to an organism and thereby increases in frequency. Positive selection can explain clines of variation among geographic or seasonal populations or differences in allele frequencies among nearby populations that differ in some important environmental variable. Lenormand et al. (1998) is an excellent examination of positive clinal selection with regard to insecticide resistance in *Culex pipiens*.

Balancing selection maintains genetic variation and can occur in number of distinct ways. Overdominance is the most common explanation, but a more plausible mechanism may be through frequency-dependent selection, as discussed earlier. Balancing selection also occurs when mutations arise that confer a fitness advantage in an organism but only in certain environments or circumstances. For example, a novel allele for insecticide resistance arises and is subject to positive selection in one environment but has a lower fitness in the absence of that insecticide and is subject to negative selection in an insecticide-free environment. Curtis et al. (1978) provides an example of this phenomenon in *Anopheles stephensi*.

Diversifying selection may be the force that causes some genes to be highly variable and to encode hypervariable proteins. It has been suggested that diversifying selection acts upon vector salivary gland genes. If a population of vertebrate hosts develops immunity to one or more components of vector saliva that are essential in acquiring a blood meal, then current and future generations of the vector will be unable to obtain blood from that host population. There would be selection for molecular mechanisms that would allow, or possibly actively generate, mutations in salivary gland genes. The maxadilan gene in sand flies is highly variable, so much so that it is rare to find two flies with identical maxadilan genotypes (Lanzaro et al. 1999). As with frequency-dependent selection, arthropod vectors with new alleles and genotypes at a locus might have a higher fitness than individuals with more common genotypes and alleles. Genes in a pathogen that are susceptible to host immunity may be subject to *diversifying selection* (Tan and Riley 1996), as are genes that confer resistance to a parasite or pathogen (e.g., the major histocompatibility complex genes of vertebrates (Hughes 2000)). In practice, without understanding the mode of action of all or parts of an enzyme, it is difficult to differentiate the forces of diversifying selection and neutral evolution.

Neutral theory provides several testable predictions about the fate of neutral substitutions in a gene. The probability that a new, neutral allele eventually becomes fixed is its initial frequency, calculated as $1/2N_e$. New alleles can become easily fixed in small-N_e populations but have a limited chance of becoming established in large-N_e populations. The rate with which mutual mutations are fixed $= \mu'$. μ' is also the rate with which new alleles arise per gene per generation, and it is independent of population size because while the average number of new, neutral mutations per generation is $2N_e\mu'$, the rate with which neutral mutations are fixed is $1/2N_e \times 2N_e\mu' = \mu'$. The average time between consecutive substitutions is $1/\mu'$ (if $\mu' =$ fixation per generation, $1/\mu' =$ generation per fixation). Kimura showed that the average time to fixation of a neutral allele is $4N_e$ generations and that average time to extinction of a neutral allele is $(2N_e/N)\ln(2N)$ generations, where $N =$ total population size. In other words, a long time is required for a neutral allele to be fixed. A short time is required for a new neutral allele to go to extinction. Neutralism also predicts that drift will reduce heterozygosity at a rate of $1/2N_e$ per generation [note that this is also predicted by Eq. (12)]. Furthermore, if a mutation yields an allele that is not *iis* with existing alleles, then upon reaching a mutation–drift equilibrium (the number of new alleles arising through mutation equals the number of new alleles lost to genetic draft), then $H_{exp} = 1/[4N_e\mu' + 1]$ and the average expected heterozygosity is $4N_e\mu'/[4N_e\mu' + 1]$.

Neutral evolution also predicts the outcome for "nearly neutral" substitutions. If a selection coefficient s_{xx} is estimated as $1 - w_{xx}$, then neutral mutations don't necessarily have $s = 0$. They can be *nearly neutral* (s very small). Kimura showed that μ' is proportional to $4N_es$. He showed that if $4N_es > 10$, selection will offset drift and that if $4N_es < 0.1$, drift will offset selection. However, if $0.1 < 4N_es < 10$, a balance will occur between selection and drift.

For example, if $s = 0.025$ (strong selection), N_e would have to be <1 for $4N_es < 0.1$. Thus, N_e would have to be very small to allow drift to offset strong selection. Alternatively, N_e would only have to be 100 or more for $4N_es > 10$. N_e wouldn't have to be very large to allow strong selection to offset drift. N_e would have to

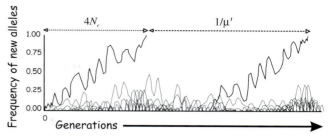

FIGURE 32.20 Kimura's concept of the fate of new, neutral alleles over the course of time.

be maintained between 1 and 100 for $0.1 < 4N_es < 10$. N_e would have to stay relatively small to maintain a balance between strong selection and drift. Alternatively, if $s = 2.5 \times 10^{-4}$ (weak selection), N_e would have to be <100 for $4N_es < 0.1$. N_e could remain relatively large and the effects of genetic drift need only be slight to offset the effects of weak selection. N_e would have to be $> 10,000$ for $4N_es > 10$. N_e would have to be very large to allow weak selection to offset drift. N_e would have to be between 100 and 10,000 for $0.1 < 4N_es < 10$. N_e would have to remain large to maintain a balance between selection and drift.

Kimura's concept of neutralism over the time course of evolution is illustrated in Figure 32.20. Mutation is constantly generating new alleles. Most (99.9%) of these mutations are eliminated immediately by purifying selection. However, neutral mutations may result in novel alleles. The majority of these new alleles will have a short time to extinction, but notice that at any given point in time in Figure 32.20 a variety of alleles exist at varying frequencies. At a predictable period of time a new neutral mutation will appear that for reasons largely predicted by N_e become established and eventually fixed in the population. There is an extended time, proportional to $4N_e$, required for these new neutral alleles to go to fixation. The frequency with which new alleles arise that will eventually go to fixation is proportional to $1/\mu'$ generation per fixation. Kimura's intuition about variation in nucleotides and amino acids has been largely proven correct by the available empirical data.

POPULATION GENOMICS

During the last decade of the 20th century, the discovery that thousands of genes, of known function and position in a genome, can be analyzed simultaneously in single organisms provided new insights into population genetics. Four technologies were critical in developing population genomic approaches in arthropod disease vectors. First, PCR (Chapter 30) allowed for the amplification of many loci from small amounts of genomic DNA typically isolated from a vector (e.g., mosquitoes, blackflies, sand flies, biting midges, chiggers, and larval and nymphal ticks). Second, a variety of highly polymorphic markers, discussed in Chapter 30, were found to be abundant in arthropod genomes. Third, techniques for detecting SNPs in genes allowed genetic variation to be examined in even the most conserved genes. Finally, the statistical algorithms discussed in Chapter 30 allowed recombination to be simultaneously analyzed among many genetic markers segregating in one or a few genetic families, thereby allowing vector geneticists to rapidly and inexpensively construct linkage maps of the relative positions of markers in a vector genome (Chapters 30 and 31).

Vector population geneticists can now choose from among many genetic markers with which to analyze variation throughout an entire genome. In the case of *An. gambiae* (and soon for *Ae. aegypti*), we have the sequence of the entire genome! Thus, it is no longer necessary to use one or a few loci to draw inferences regarding F_{IS}, F_{ST}, F_{IT}, D_{ij}, D_{IS}^2, D_{ST}^2, D'_{IS}^2, D'_{ST}^2, N_e, $N_e(m + \mu)$, or w_{xx} among and within vector populations. Population genetic analyses can now be conducted by sampling whole vector genomes and estimating population genetic statistics at two levels: across variable loci dispersed throughout the genome and among SNPs that segregate within individual genes (commonly called *segregating sites*).

Most importantly, the ability to sample genomes means that the actual sampling distributions of population genetic statistics can now be derived from many individual loci or among many segregating sites within a gene. Genome sampling enables population geneticists, for the first time, to distinguish effects that act upon the whole genome (genetic drift, migration, and inbreeding) from those that act upon individual loci or nucleotides (selection, mutation, recombination, and assortative mating). The term *population genomics* describes the process of simultaneous sampling of numerous variable loci within a genome and the inference of locus-specific effects from the sample distributions (Black et al. 2001).

Probably the most novel and exciting aspect of population genomics is that it allows, for the first time, the detection and identification of specific nucleotides that control the expression of phenotypes or quantitative traits in whole organisms from the field. This means that the contribution of individual nucleotide variation in determining an organism's phenotype can be assessed in natural populations. This process has come to be known as *quantitative trait nucleotide (QTN) mapping* and has been rapidly embraced by genetic epidemiologists as a powerful tool for identifying heritable genetic predisposition to disease in humans (Risch 2000). *Drosophila* researchers have used QTN mapping to identify SNPs in ectodermally expressed genes that control bristle number in *Drosophila* (Mackay 2001). Domestic animal and plant breeders use QTN mapping to identify SNPs associated with increased yield or other desirable characters as selectable markers for more rapid crop and animal improvement (Mackay 2001).

For basic research, QTN mapping provides a novel and necessary interface between population genetics and molecular biology. *A priori* hypotheses concerning the adaptive significance of a segregating site requires knowledge of the basic biochemistry and physiology of the gene product. Population geneticists can study variation in individual genes and identify genes and regions of a gene that are apparently subject to selection. However, without an understanding of the molecular structure, function, and interaction of amino acids and a clear picture of how a gene functions in the metabolism or development of an organism, the adaptive significance of a segregating site will remain obscure. Conversely, while the molecular biologist may understand in detail the structure, function, and physiology of a gene product, only insights gained through population genomics can indicate whether variation found in a gene is subject to natural selection, which parts of a gene or protein are subject to selection, and the mode of selection acting on that gene. Ultimately, this interface will yield a means for predicting mechanisms of molecular adaptation.

QTN Mapping in *Aedes aegypti*

The *Ae. aegypti* genome contains from 10^8 to 10^9 bp (Brown et al. 2001), and from the many population genetic and mapping studies we know that RFLPs and SNPs are abundant. Thus, there are potentially millions of segregating sites in the *Ae. aegypti* genome. QTN mapping fundamentally entails performing many statistical tests of associations between a phenotype (e.g., presence or absence of an MIB in a mosquito) and genotypes at segregating sites within a genome. If one were to sample all of these polymorphisms and apply, independently to each, a statistical test of association (e.g., χ^2) at $\alpha = 0.05$, we would expect a false-positive association between a phenotype and a genotype once every 20 segregating sites. If, conservatively, 1% of the genome consists of segregating sites, then we would expect from 50,000 to 500,000 [1% of $(10^8–10^9\,bp) \times \alpha = 0.05$] segregating sites to be significantly associated with a phenotype by random chance alone. Random sampling on a genomewide basis, with, for example, RAPD, AFLP, or microsatellite markers (Chapter 30) would decrease the number of independent tests but would still not minimize the chances of finding false-positive associations.

QTN mapping, therefore, requires that an investigator take aggressive steps to minimize the likelihood of detecting false positives. However, statistical inference theory teaches that we cannot minimize the likelihood of false positives without simultaneously increasing the likelihood of false negatives (decreasing the *power* of a statistical test). For example, one could use a Bonferroni correction to reduce the likelihood of false positives by adjusting α to such a low level that only test statistics that exceed critical values once every 50,000–500,000 tests are considered significant. However, in so doing, we would also decrease the likelihood of detecting polymorphisms that are truly associated with a phenotype. Steps can be taken to minimize the likelihood of detecting false positives without compromising our ability to detect true genotype–phenotype associations. We illustrate these steps using a current study to assess the role of *early trypsin* (ET) in conditioning midgut susceptibility to dengue virus infection in *Ae. aegypti*.

Step 1: A Priori Hypothesis Testing with Candidate Genes

The foregoing discussion predicts that a random sampling of any genome will generate many false-positive associations. This suggests that it is prudent to establish *a priori* hypotheses regarding specific gene regions or genes that are most likely to condition the phenotype of interest. Genetic mapping of QTL (Chapter 30) may be used to identify a specific genome region that is likely to contain candidate genes. Furthermore, physiological or genetic studies may suggest specific genes or gene products within a QTL that are likely to condition the phenotype of interest.

Little is known of the early events of flavivirus infection in insect midgut epithelial cells. For other arboviruses (e.g., bunyaviruses and orbiviruses), there is a prerequisite for proteolytic processing of virion surface proteins for efficient vector-midgut cell interaction. When *Ae. aegypti* females are fed a dengue infectious blood meal that also contains soybean trypsin inhibitor, there is a significant reduction in the rate of midgut infection, suggesting that midgut trypsins may be involved in proteolytic processing of dengue virus-surface proteins.

In *Aedes* spp. a small amount of ET is produced immediately following ingestion of free amino acids, but the most abundant trypsins are not produced if the meal does not contain free amino acids. ET is apparently part of a unique signal transduction system (Noriega et al. 2001) in which a large pool of transcribed ET message resides in the midgut of newly eclosed adults that is induced by free amino acids. Its function may be to "taste" the incoming meal to determine if there is sufficient protein to support a gonadotrophic cycle. If so, the signal transduction pathway activates the transcription of other genes to commence blood meal digestion. When ET expression is knocked out using Sindbis virus constructs, dengue transmission is completely disrupted. Furthermore, linkage mapping of the locations of specific midgut trypsin genes in the *Ae. aegypti* genome showed that alleles of ET cosegregate with the midgut susceptibility phenotype (Bosio et al. 2000), suggesting the possibility that these alleles may vary in their abilities to process dengue virus-surface proteins. ET is therefore a candidate gene for flavivirus midgut susceptibility, and the specific *a priori* hypothesis to be tested is that different alleles of ET vary in their efficiency for cleaving dengue-virus surface proteins and, thus, indirectly condition midgut susceptibility to DEN viruses in *Ae. aegypti*.

Step 2: Large Sample Sizes Increase Statistical Power

Assume that because of environmental effects, dominance, and epistatic interactions, a substitution at a QTN in ET causes on average a 1% difference in overall phenotype. For example, 70% of vectors with a G/G nucleotide at the ET QTN develop a midgut infection when orally infected with DEN2 virus, while 69% of vectors with a G/T develop a midgut infection, and 68% of vectors with a T/T develop a midgut infection. A question arises as to how many organisms would have to be sampled to detect a significant difference between organisms with these three genotypes.

Our null hypothesis (H_0) is that midgut infection rate (MIR)(G/G) = MIR(G/T) = MIR(T/T). Statistical *power* is the probability that a test rejects H_0 when it is false. Figure 32.21 shows a series of *power curves* for heterogeneity χ^2 tests of H_0. Sample size (N) is plotted along the x-axis and power is plotted on the y-axis. Curve B is the power curve for heterogeneity χ^2 tests of the alternative hypotheses H_a: MIR(G/G) 1% > MIR(G/T) 1% > MIR(T/T). This curve shows that there is >50% probability of accepting H_0 when N is less than 5,000 organisms. In fact, N would need to be 22,330 organisms to have a 95% probability of

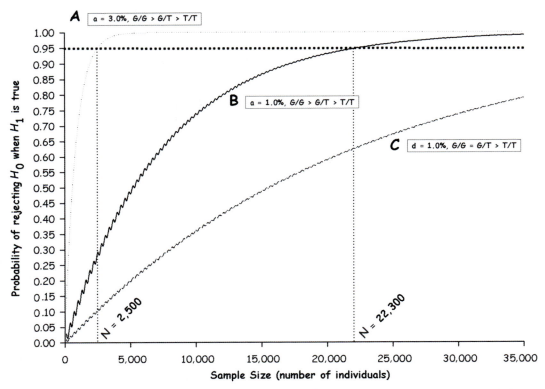

FIGURE 32.21 A series of *power curves* for heterogeneity χ^2 tests of H_0. Sample size (N) is plotted along the x-axis and power is plotted on the y-axis. Curve (A) is the power curve when H_a: MIR(G/G) 3% > MIR(G/T) 3% > MIR(T/T). Curve (B) is the power curve for heterogeneity χ^2 tests of the alternative hypotheses H_a: MIR(G/G) 1% > MIR(G/T) 1% > MIR(T/T). Curve (C) is the power curve when there is dominance among the QTN substitutions or assuming an allele with G at the QTN is dominant to an allele with T at the position, H_a: MIR(G/G) > MIR(G/T) 1% = MIR(T/T).

rejecting H_0 with a Type I error rate of $\alpha = 0.05$ (we reject H_0 in one of 20 tests when H_0 is true). Another way of saying this: We would have to sample 22,330 organisms to be able to detect MIR(G/G) 1% > MIR(G/T) 1% > MIR(T/T). Curve A is the power curve when H_a: MIR(G/G) 3% > MIR(G/T) 3% > MIR(T/T). In this case, we would have to sample 2,500 organisms to be able to detect this difference. Thus, the larger the additive genetic effect, the fewer the number of organisms needed to detect the QTN.

The general relationship between the magnitude of the additive genetic effect and sample size is illustrated in Figure 32.22A. Notice that the larger the additive genetic effect (x-axis), the smaller the sample size necessary to detect the QTN. Figure 32.22A also shows power curves for $\alpha = 0.01$ and $\alpha = 0.001$. Sample sizes needed to detect significant associations increase as α decreases. One would need to increase sample sizes to reduce the likelihood of rejecting H_0 when it is true. Another way of interpreting these curves is that if our sample size is 1,000 individuals, we should only expect to be able to detect QTNs at which the additive genetic effect is 5% or greater.

Curve C in Figure 32.21 is the power curve when there is dominance among the QTN substitutions or assuming an allele with G at the QTN is dominant to an allele with T at the position, H_a: MIR(G/G) > MIR(G/T) 1% = MIR(T/T). In this case we would have to sample 67,200 organisms to be able to detect this difference. This is simply because there is less variation in phenotype among the three genotypes. Figure 32.22B shows that the larger the dominance effect (x-axis), the smaller the sample size necessary to detect the QTN. However, notice in comparing Figures 32.22A and B that much larger sample sizes are required to detect QTNs with dominance effects. If our sample size is again 1,000 individuals, we should only expect to be able to detect QTNs at which the dominance effect is 7.5% or greater.

Step 3: Independent Segregation Among Polymorphic Sites

Linkage disequilibrium among segregating sites in a candidate gene is another potential source of false positives in testing phenotype–genotype associations.

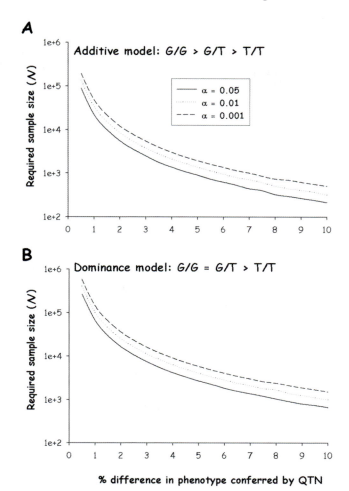

FIGURE 32.22 General relationship between the magnitude of the (A) additive or (B) dominance genetic effects and sample size for $\alpha = 0.05$, $\alpha = 0.01$, and $\alpha = 0.001$.

If one or several segregating sites are in disequilibrium with the QTN, then one would incorrectly infer that these sites control the phenotype of interest. Disequilibrium would therefore elevate the rate of false positives. Thus, accurate detection of locus-specific effects depends critically upon there being linkage equilibrium among genes and nucleotide substitutions within genes.

Linkage equilibrium is achieved through extensive and continuous recombination and/or through selective forces acting independently on different sites in a gene. If separate mutations are correlated either because they do not recombine or because they are both involved in creating essential secondary structure in a gene product or, for example, because their gene products interact within the same metabolic pathway, then we cannot differentiate forces acting upon individual segregating sites within a gene.

Linkage disequilibrium was tested for by amplifying the entire ET gene and a 64-bp intron in the 5' end

of the gene (828 bp total) in two overlapping halves in 1,566 individual *Ae. aegypti* collected from throughout Mexico. The presence of a midgut infection or escape barriers (MIB and MEB in Chapter 30) was determined in each mosquito. Single-strand confirmation polymorphism was used to distinguish unique genotypes (Chapter 30), and each unique genotype was sequenced to determine the segregating sites associated with each SSCP genotype. A total of 56 segregating sites were detected among 66 unique SSCP genotypes.

To test for independence among the 56 segregating sites Ohta's linkage disequilibrium analysis [Eqs. (26) and (27)] was performed on all pairs of segregating sites (Fig. 32.23). Most (95%) of the variation in D_{IT}^2 was due to genetic drift or nonsystematic epistasis. A pocket of disequilibrium was observed among segregating sites 132–374. It may be that these segregating sites are of recent origin or that specific combinations of nucleotides at these sites are maintained by selection. However, the overall absence of disequilibrium in the remainder of the gene suggests that extensive recombination disrupts disequilibrium among segregating sites within the ET gene in populations of *Ae. aegypti*.

There are several circumstances that might produce disequilibrium among segregating sites. First, recombination rates are not uniformly distributed across chromosomes and, in general, are much lower among genes located within or near centromeres and telomeres than among genes in other chromosome regions (Begun and Aquadro 1992; Fig. 30.26). Second, *selective sweeps* occur when a gene with a large *positive* impact on fitness is closely linked to one or more genes that are neutral or nearly neutral in their effects on fitness. If a locus is subject to rapid positive selection, then proximal neutral genes will also be carried along ("swept" or "hitchhike") to fixation if selection is rapid enough that recombination cannot independently assort the proximal genes (Hudson 1994). For example, we would expect that genes linked to an insecticide-resistance allele would have lower genetic variability due to strong selection for resistance than genes that assort independent of the resistance gene. Third, *background selection* is a corollary phenomenon in which a gene with a negative impact on fitness is closely linked to one or many neutral genes. As the locus under selection is rapidly swept to extinction, so are the linked neutral genes (Hudson 1994). For example, we would expect the genes linked to an insecticide-susceptibility allele to have lower genetic variability following extinction of the original susceptibility alleles (Roush and McKenzie 1987).

An assumption of independent evolution may also be invalid among genes located within chromosomal

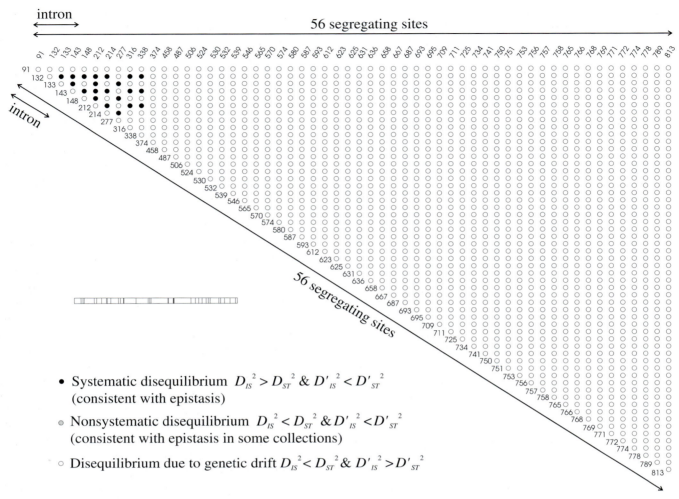

- Systematic disequilibrium $D_{IS}^2 > D_{ST}^2 \ \& \ D'^2_{IS} < D'^2_{ST}$ (consistent with epistasis)

- Nonsystematic disequilibrium $D_{IS}^2 < D_{ST}^2 \ \& \ D'^2_{IS} < D'^2_{ST}$ (consistent with epistasis in some collections)

- Disequilibrium due to genetic drift $D_{IS}^2 < D_{ST}^2 \ \& \ D'^2_{IS} > D'^2_{ST}$

FIGURE 32.23   Analysis of linkage disequilibrium in the *Early Trypsin* gene of *Aedes aegypti*.

inversions, because inversions eliminate or severely suppress recombination. We have already discussed in Chapters 29–31 the fact that many anopheline vector species have floating paracentric inversions. A possible advantage of inversions may lie in their ability to maintain favorable (epistatic) groups of alleles at loci contained in the inversion. A number of recent studies have investigated the distribution of nucleotide diversity within and between chromosomal arrangements at loci closely linked to the inversion breakpoints in *Drosophila* (Andolfatto and Kreitman 2000). These studies arrived at different conclusions regarding the level of exchange between arrangements, suggesting that the rate may vary by genome location, the age of the inversion, and the genes located in proximity to the breakpoints. However, all studies suggest that genetic exchange between genes contained in inversions is suppressed at inversion breakpoints relative to other

genomic regions. In general, these studies found an absence of shared polymorphism between arrangements and a low diversity of alleles at the junction region and little evidence for genetic exchange between karyotypic classes.

Step 4: Reducing False Positives by Sampling Panmictic Units

The extent to which locus-specific QTN effects can be accurately detected depends greatly on the assumption that migration, drift, and inbreeding homogeneously influence variation throughout a genome. Genomewide processes should affect all loci in a similar way on average, but it should be remembered that the "on average" is just that, in the statistical sense. Although genomewide processes affect all loci similarly, every population statistic is subject to sampling error. Even loci subject to identical evolutionary processes will vary in their statistical properties because of sampling error and also due to the stochastic nature of vector populations affected by environmental and demographic variability.

A priori we would expect that (1) genetic variation should be uniformly distributed among populations with high gene flow, (2) genetic drift will cause populations to differentiate and will reduce heterozygosity when they are maintained or established by low N_e, and (3) inbreeding will cause individuals to become homozygous at the majority of loci throughout their genomes. Although evolutionary processes can in theory be unambiguously classified as either genomewide or locus-specific, in practice the two are not always easy to disentangle. We have discussed neutralism and the fact that some alleles may act as if they are not under selection when they arise or have occurred in small populations. Also consider that when a new mutation arises in a population it is in complete disequilibrium with all other segregating sites on the same chromosome. The extent to which disequilibrium is disrupted is a function of meiosis, generation length, and the physical distance between markers. Any factors that reduce recombination will confound locus-specific and genome-wide effects. Again, locus-specific effects (selective sweeps and background selection) acting on the new allele and associated loci are confounded by genome-wide processes.

We have shown that homozygotes become common under inbreeding and drift. Under these conditions, selection against recessive deleterious mutations also becomes common, as does fixation of slightly deleterious alleles. In these cases, drift and inbreeding may generate locus-specific effects. Numerous studies have documented the large genetic loads that insect genomes carry, including recessive alleles that are lethal or deleterious when homozygous. This is certainly true of *Ae. aegypti*, in which Craig, Rai, and colleagues were able to isolate a large number of spontaneous mutants from inbred field strains (e.g., Bhalla and Craig 1967; Petersen et al. 1976). Suppose in a small population that lethal or deleterious alleles at several loci reach a high frequency by genetic drift. A period of rapid purifying selection would occur in the population as individuals homozygous for these recessive alleles were selected against. Proximal genomic regions would be subject to strong background selection, and heterozygosity in that genomic region would also be greatly reduced. Similarly, if an allele is uncommon (<5%) in a natural population, then homozygotes for that allele will be very rare (<0.25%). Suppose again that in a small population the uncom-

mon alleles at one or several loci do reach high frequency by genetic drift. Homozygotes that may have an unusually high fitness in the small population, and the uncommon allele will become fixed through positive selection, causing associated alleles at linked loci to increase in frequency by hitchhiking. The implication for population genomics is that many loci and proximal genomic regions with high loads of deleterious and lethal alleles may be much more susceptible to reduced heterozygosity during inbreeding or population bottlenecks. In practice this could be difficult to distinguish from a general reduction in heterozygosity due to increased drift.

Another important caveat to an assumption of genome-wide effects concerns *epistatic interactions* among alleles at different loci. Rapid adaptive evolution may result from beneficial combinations of epistatically acting alleles that drift to high frequency in small, subdivided populations. This occurs because normally rare or uncommon alleles may suddenly reach high frequencies. This in turn gives rise to new combinations of multilocus genotypes in individual organisms that encode a novel array of phenotypes. Thus, while we would expect genetic variability to decrease when a population is founded by few individuals, genetic variance due to epistasis may actually increase genotypic and phenotypic variance following a severe population bottleneck (Wade and Goodnight 1991). The implication is that novel epistatic interactions may cause correlated (nonindependent) shifts among multiple parts of a genome, and proximal genomic regions could be subject to selective sweeps and background selection. While this has yet to be documented at the level of individual nucleotides, epistasis would be manifest as significant disequilibrium among substitutions at different loci.

Because of all of these caveats, it is not safe to assume uniform genome-wide processes for QTN mapping in *Ae. aegypti*. For example, an association could be detected between MIR and an ET SNP that has no effect on MIR because both happen to be at a high frequency in regions of Mexico where *Ae. aegypti* have a high MIR for other reasons. One approach to take in testing for this possibility is to analyze variation in nuclear or mitochondrial genes that are unlikely to affect MIR and then determine if, in fact, there is a statistical association between these "neutral" genes and the MIR phenotype. If genome-wide effects are uniform among all mosquitoes sampled, then inbreeding, migration, and drift should influence all parts of the genome equally, and this will be observed in associations between variation in "neutral" markers and MIR. Alternatively, if genomewide effects are heterogeneous among all mosquitoes sampled, then inbreeding, migration, and drift will influence parts of the genome differently and may generate a *spurious* association between variation in the "neutral" marker and MIR.

In *Ae. aegypti* in Mexico, variation in the mitochondrial genome (mtDNA) was analyzed with respect to MIR. A more thorough test of genomewide effects might be to analyze a series of RAPD bands (e.g., Apostol et al. 1996). In using mtDNA we assume that variation in the mitochondrial genome represents genome-wide effects and does not influence MIR. In other words, if populations differ in MIR and also differ in the frequencies of different mtDNA haplotypes, then one might erroneously conclude there is an association between mtDNA variation and MIR. More importantly, similar false associations are expected when correlating variation in ET with MIR. A series of computer programs called PGENOME (Lozano-Fuentes and Black, unpublished) display estimates of F_{ST} and θ (Weir and Cockerham 1984) for individual SNPs among mosquitoes with or without an MI. Because

$$\chi^2_{[(\#collections - 1)(\#genotypes - 1)d.f.]} = 2NF_{ST}$$

we can perform a parametric test of H_0: MIR(N_1/N_1) = MIR(N_1/N_2) = MIR(N_2/N_2), where N_1 and N_2 are alternative nucleotides at a segregating site. PGENOME also uses a permutation test (Doerge and Churchill 1996) to determine the consistency with which a data set supports each of the estimated θs. PGENOME randomly assigns the ET sequence of one mosquito and its MIR phenotype to another mosquito. This is repeated for each mosquito until all phenotypes–genotypes in the original datasets are randomly permuted. PGENOME then recalculates θ between mosquitoes with or without an MIR and stores the permuted θ. PGENOME permutes the data set and stores the permuted θ 10,000 more times and then ranks the 10,000 permuted θs for each segregating site.

Figure 32.24 is output from PGENOME. θs between mosquitoes with and without MIR at each of the 24 segregating sites are plotted as filled circles along the length of the mitochondrial NADH dehydrogenase subunit 4 gene. The 9,500th permuted θs are plotted as open circles at each of the segregating sites to indicate how frequently the original θ exceeds the 95% permuted θ. This indicates how frequently we could expect to recover the original θ by chance.

Figure 32.24A indicates that the θ from the original data set exceeds the 95% permuted θ in four of the 24 segregating sites in the mtDNA. If variation in the mtDNA indeed has no influence on MIR, we would incorrectly infer the existence of four QTNs in the mtDNA. However, we know that DEN susceptibility

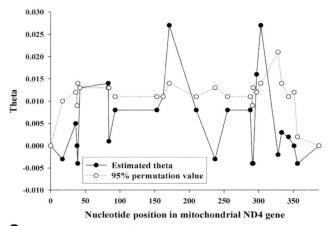

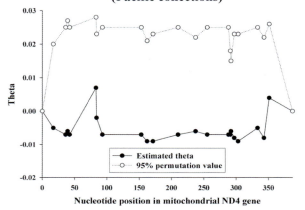

FIGURE 32.24 Associations between mtDNA genotypes and dengue midgut susceptibility phenotypes in *Aedes aegypti*.

varies widely among geographic regions of Mexico, as do the frequencies of mtDNA markers (Gorrochotegui-Escalante et al. 2000). This means that we could also expect to find a spurious genotype–phenotype association with ET SNPs that also don't influence MIR. Earlier studies of mtDNA and RAPD markers (Gorrochotegui-Escalante et al. 2000) showed that *Ae. aegypti* Mexico populations consist of four genetic groups comprising populations in the northeast, Yucatan, Quintana Roo, and along the Pacific coast. The analysis shown in Figure 32.25A was therefore repeated in each of these four regions. In every region, the original θ never exceeded the 95% permuted θ (e.g., Fig. 32.24B for the Pacific). We therefore proceeded to test for ET QTN in each of these four geographic regions.

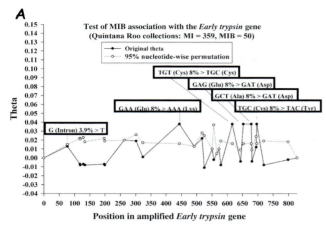

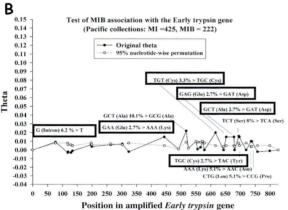

FIGURE 32.25 Associations between *Early Trypsin* genotypes and dengue midgut susceptibility phenotypes in *Aedes aegypti* in the Yucatan (A) and (B) Pacific collections in Mexico.

Step 5: Replication Across Panmictic Units

Subdividing analyses across panmictic units has another advantage. Despite the four steps we have taken to this point, spurious associations are still likely. However, if the QTN that we have detected is valid (it really does cause a barrier to midgut infection), then it should be detected in each of the panmictic units. Another way of considering this is that if there are 20 segregating sites in a gene, then with $\alpha = 0.05$ we would expect 1 QTN (0.05×20) to be detected in each panmictic unit by chance. However, the likelihood of the same segregating site being detected as a QTN in two panmictic units is $(1/20)^2 = 0.0025$, or $(1/20)^3 = 0.000125$ in three panmictic units. Thus, replication of QTN analyses across multiple panmictic units is a useful method for confirming a QTN.

Replication can also be used to qualitatively assess the validity and repeatability of the inferred phenotype–genotype association. If two nucleotides are segregating, then there are three possible phenotype–genotype associations: $MIR(N_1/N_1)$, $MIR(N_1/N_2)$, and $MIR(N_2/N_2)$. For three segregating nucleotides there are six possible phenotype–genotype associations; and if all four nucleotides are segregating, then there are 10 potential genotypes. If a QTN is valid, then the way that the specific genotypes both qualitatively and quantitatively control a phenotype should be repeatable across panmictic sampling units. For example, if $MIR(N_1/N_1)$ 5% > $MIR(N_1/N_2)$ 5% > $MIR(N_2/N_2)$ in one panmictic sampling unit, then this same relationship should be approximated across all panmictic sampling units.

Quantitative Trait Nucleotides (QTNs) Mapping in Population Genomics

We tested for QTNs among 56 segregating sites in ET between mosquitoes with or without a MIR in each of the four geographic regions (sample sizes of 409, 647, 306, and 204, respectively). The original θ exceeded the 95% permuted θ in 6, 10, 1, and 0 sites of the 56 sites, respectively, in each population. Using the values in Figure 32.22, with these sample sizes we expect to detect QTNs only when the additive genetic effects are equal to or greater than 10%, 8.5%, 7.5%, and 6.0%, respectively. If there are dominant genetic effects, then these sample sizes would only detect effects that are equal to or greater than 15%, 13%, 12%, and 10%, respectively.

No QTN was detected in common among all four panmictic sampling units; however, the magnitude of additive or dominance effects would have had to be 10–15% or greater for a QTN to be detected with the small sample size in the Northeast. One common QTN was detected in the Yucatan and Quintana Roo and along the Pacific coast. Five additional common QTNs were detected between the Quintana Roo and Pacific coast collections (Fig. 32.25). While these additional QTNs may be valid, we will focus on only the one QTN that was common in the Northeast, Yucatan, and Pacific collections.

Figure 32.26 plots the MIR phenotype of mosquitoes in the Yucatan and Quintana Roo collections against the genotype at this QTN. In Quintana Roo, MIR(G/G) 7% > MIR(G/T) 1% > MIR(T/T), while among Pacific collections MIR(G/G) 8% > MIR(G/T) 2% > MIR(T/T). This suggests that there is both additivity and dominance among alleles at this QTN. Genotype G/G is slightly dominant to G/T and T/T. Power calculations with H_a: MIR(G/G) 8% > MIR(G/T) 2% > MIR(T/T) indicate that sample sizes would have had to be at least 400 to identify this QTN. Therefore, it is

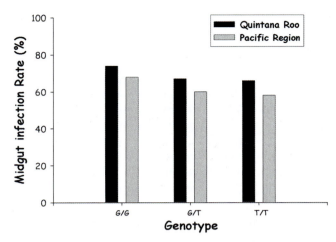

FIGURE 32.26 MIR phenotype of mosquitoes in (A) Quintana Roo and (B) Pacific collections plotted against the genotype at this QTN. In Quintana Roo, MIR(G/G) 7% > MIR(G/T) 1% > MIR(T/T), while among Pacific collections MIR(G/G) 8% > MIR(G/T) 2% > MIR(T/T).

understandable that this QTN wasn't detected in the Northeast and was barely detected in the Yucatan.

The QTN at position 91 is located at the 5′ end of the ET intron and is five nucleotides from the GT intron splice site. The sequence of the 5′ end of the intron to the QTN is 5′-GTAAG<u>G</u>-3′. A "T" substitution at position 6 generates an alternative GT splice site, and we have evidence from laboratory cDNA experiments that this occurs in mosquitoes with G/T or T/T genotypes. It will be necessary to perform experiments to determine if these genotypes differ in trypsin activity. Alternatively, the QTN at position 91 may be in disequilibrium with downstream or upstream regulators that may influence ET expression. Notice in Figure 32.23 that site 91 is adjacent to a pocket of disequilibrium among other segregating sites in the 5′ end of the ET gene. In any case, detection of a QTN associated with disruption or alternative splicing in a gene is not unusual. The majority of QTNs that, for example, control bristle number in *Drosophila* (Mackay 2001) or are associated with heritable genetic predisposition to disease in humans (Risch 2000) also tend to involve insertions, deletions, premature stops, or missense mutations.

The population genomics approach offers an opportunity to assess the dynamics of population-level factors and the influences these factors have on specific genetic polymorphisms. The analyses we have described provide an alternative means to investigate genotypic effects on phenotypic variation. The ET example provides supporting evidence for the impor-

tance of ET variation and effects on MIB that can now be extended to other tests to more directly test this relationship using other genetic techniques.

CONCLUSION

We have introduced the most fundamental aspects of classical population genetics and in most cases have illustrated how these fundamentals have been applied in studies of vector populations. Very little or no research on arthropod vectors has been completed in subject areas where we provided no examples. These are fruitful areas for future investigation. Vector population genetics is still in its infancy. The challenges and opportunities are great. In the beginning of the efforts to utilize molecular markers in population genetic analyses, vector biologists were interested in measuring genetic variation and characterizing those features of vector biology that influence how genetic variation is structured within and between populations. The great majority of this work used neutral markers whose variation was influenced largely by stochastic and demographic factors related to population size, drift, and migration. We end this chapter with the next challenge. As candidate genes are identified and their roles in impacting biologically important traits, such as those involved in vectorial capacity, have been assessed, population geneticists will be able to identify features of a species biology and habitat that control the variation in these genes.

Understanding the complex dynamics of selective regimes interconnected with drift, breeding structure, migration, and mutation, and the impact on vector-borne disease epidemiology is the ultimate objective. We will need to address the impact of chromosomal structure and gene regulation of phenotypic variation, with attention to environmental factors. It is through such studies that we can understand vector-borne disease epidemiology, risks for vector infection, and pathogen transmission to vertebrates, and can develop new strategies that utilize biological and environmental approaches to shape vector genetic variation to reduce human and domestic animal disease.

Readings

Andolfatto, P., and Kreitman, M. 2000. Molecular variation at the In(2L)t proximal breakpoint site in natural populations of *Drosophila melanogaster* and *D. simulans*. *Genetics* 154: 1681–1691.

Apostol, B. L., Black, W. C., Reiter, P., and Miller, B. R. 1994. Use of randomly amplified polymorphic DNA amplified by polymerase chain reaction markers to estimate the number of *Aedes aegypti* families at oviposition sites in San Juan, Puerto Rico. *Am. J. Trop. Med. Hyg.* 51: 89–97.

Apostol, B. L., Black, W. C., Reiter, P., and Miller, B. R. 1996. Population genetics with RAPD-PCR markers: The breeding structure of *Aedes aegypti* in Puerto Rico. *Heredity* 76: 325–334.

Begun, D. J., and Aquadro, C. F. 1992. Levels of naturally occurring DNA polymorphism correlate with recombination rates in *D. melanogaster. Nature* 356: 519–520.

Bhalla, S. C., and Craig, G. B., Jr. 1967. Bronze, a female-sterile mutant of *Aedes aegypti. J. Med. Entomol.* 4: 467–476.

Black, W. C., Ferrari, J. A., Rai, K. S., and Sprenger, D. 1988a. Breeding structure of a colonising species: *Aedes albopictus* (Skuse) in the United States. *Heredity* 60: 173–181.

Black, W. C., Hawley, W. A., Rai, K. S., and Craig, G. B., Jr. 1988b. Breeding structure of a colonizing species: *Aedes albopictus* (Skuse) in peninsular Malaysia and Borneo. *Heredity* 61: 439–446.

Black, W. C., Baer, C. F., Antolin, M. F., and DuTeau, N. M. 2001. Population genomics: Genome-wide sampling of insect populations. *Annu. Rev. Entomol.* 46: 441–469.

Bosio, C. F., Fulton, R. E., Salasek, M. L., Beaty, B. J., and Black, W. C. 2000. Quantitative trait loci that control vector competence for dengue-2 virus in the mosquito *Aedes aegypti. Genetics* 156: 687–698.

Brown, S. E., Severson, D. W., Smith, L. A., and Knudson, D. L. 2001. Integration of the *Aedes aegypti* mosquito genetic linkage and physical maps. *Genetics* 157: 1299–1305.

Conn, J. E., Rosa-Freitas, M. G., Luz, S. L., and Momen, H. 1999. Molecular population genetics of the primary neotropical malaria vector *Anopheles darlingi* using mtDNA. *J. Am. Mosq. Control Assoc.* 15: 468–474.

Curtis, C. F., Cook, L. M., and Wood, R. J. 1978. Selection for and against insecticide resistance and possible methods of inhibiting the evolution of resistance in mosquitoes. *Ecol. Entomol.* 3: 273–287.

Delaye, C., Beati, L., Aeschlimann, A., Renaud, F., and de Meeus, T. 1997. Population genetic structure of *Ixodes ricinus* in Switzerland from allozymic data: No evidence of divergence between nearby sites. *Int. J. Parasitol.* 27: 769–773.

De Merida, A. M., Palmieri, M., Yurrita, M., Molina, A., Molina, E., and Black, W. C. 1999. Mitochondrial DNA variation among *Anopheles albimanus* populations. *Am. J. Trop. Med. Hyg.* 61: 230–239.

de Sousa, G. B., Panzetta de Dutari, G. P., and Gardenal, C. N. 1999. Genetic structure of *Aedes albifasciatus* (Diptera: Culicidae) populations in central Argentina determined by random amplified polymorphic DNA-polymerase chain reaction markers. *J. Med. Entomol.* 36: 400–404.

Doerge, R. W., and Churchill, G. A. 1996. Permutation tests for multiple loci affecting a quantitative character. *Genetics* 142: 285–294.

Esseghir, S., Ready, P. D., Killick-Kendrick, R., and Ben-Ismail, R. 1997. Mitochondrial haplotypes and phylogeography of Phlebotomus vectors of *Leishmania major. Insect Mol. Biol.* 6: 211–225.

Gorrochotegui-Escalante, N., Munoz, M. L., Fernandez-Salas, I., Beaty, B. J., and Black, W. C. 2000. Genetic isolation by distance among *Aedes aegypti* populations along the northeastern coast of Mexico. *Am. J. Trop. Med. Hyg.* 62: 200–209.

Hudson, R. R. 1994. How can the low levels of DNA sequence variation in regions of the *Drosophila* genome with low recombination rates be explained? *Proc. Natl. Acad. Sci. U.S.A.* 91: 6815–6818.

Hughes, A. L. 2000. Evolution of introns and exons of class II major histocompatibility complex genes of vertebrates. *Immunogenetics* 51: 473–486.

Jackson, B. M., Drysdale, C. M., Natarajan, K., and Hinnebusch, A. G. 1996. Identification of seven hydrophobic clusters in GCN4 making redundant contributions to transcriptional activation. *Mol. Cell Biol.* 16: 5557–5571.

Kain, D. E., Sperling, F. A., and Lane, R. S. 1997. Population genetic structure of *Ixodes pacificus* (Acari: Ixodidae) using allozymes. *J. Med. Entomol.* 34: 441–450.

Kain, D. E., Sperling, F. A., Daly, H. V., and Lane, R. S. 1999. Mitochondrial DNA sequence variation in *Ixodes pacificus* (Acari: Ixodidae). *Heredity* 83: 378–386.

Kambhampati, S., Black, W. C., IV, Rai, K. S., and Sprenger, D. 1990. Temporal variation in genetic structure of a colonising species: *Aedes albopictus* in the United States. *Heredity* 64: 281–287.

Kimura, M. 1968. Genetic variability maintained in a finite population due to mutational production of neutral and nearly neutral isoalleles. *Genet. Res.* 11: 247–269.

Krafsur, E. S., and Wohlford, D. L. 1999. Breeding structure of *Glossina pallidipes* populations evaluated by mitochondrial variation. *J. Hered.* 90: 635–642.

Krafsur, E. S., Endsley, M. A., Wohlford, D. L., Griffiths, N. T., and Allsopp, R. 2001. Genetic differentiation of *Glossina morsitans centralis* populations. *Insect Mol. Biol.* 10: 387–395.

Lanzaro, G. C., Lopes, A. H., Ribeiro, J. M., Shoemaker, C. B., Warburg, A., Soares, M., and Titus, R. G. 1999. Variation in the salivary peptide, maxadilan, from species in the *Lutzomyia longipalpis* complex. *Insect Mol. Biol.* 8: 267–275.

Lehmann, T., Hawley, W. A., and Collins, F. H. 1996. An evaluation of evolutionary constraints on microsatellite loci using null alleles. *Genetics* 144: 1155–1163.

Lehmann, T., Blackston, C. R., Besansky, N. J., Escalante, A. A., Collins, F. H., and Hawley, W. A. 2000. The Rift Valley complex as a barrier to gene flow for *Anopheles gambiae* in Kenya: The mtDNA perspective. *J. Hered.* 91: 165–168.

Lenormand, T., Guillemaud, T., Bourguet, D., and Raymond, M. 1998. Evaluating gene flow using selected markers: A case study. *Genetics* 149: 1383–1392.

Mackay, T. F. 2001. Quantitative trait loci in *Drosophila. Nat. Rev. Genet.* 2: 11–20.

McLain, D. K., Rai, K. S., and Rao, P. N. 1985. Ethological divergence in allopatry and asymmetrical isolation in the South Pacific *Aedes scutellaris* subgroup. *Evolution* 39: 998–1008.

Munstermann, L. E., Morrison, A. C., Ferro, C., Pardo, R., and Torres, M. 1998. Genetic structure of local populations of *Lutzomyia longipalpis* (Diptera: Psychodidae) in central Colombia. *J. Med. Entomol.* 35: 82–89.

Noriega, F. G., Edgar, K. A., Goodman, W. G., Shah, D. K., and Wells, M. A. 2001. Neuroendocrine factors affecting the steady-state levels of early trypsin mRNA in *Aedes aegypti. J. Insect Physiol.* 47: 515–522.

Norris, D. E., Klompen, J. S., Keirans, J. E., and Black, W. C. 1996. Population genetics of *Ixodes scapularis* (Acari: Ixodidae) based on mitochondrial 16S and 12S genes. *J. Med. Entomol.* 33: 78–89.

Ohta, T. 1982. Linkage disequilibrium due to random genetic drift in finite subdivided populations. *Proc. Natl. Acad. Sci. U.S.A.* 79: 1940–1944.

Onyabe, D. Y., and Conn, J. E. 2001. The distribution of two major malaria vectors, *Anopheles gambiae* and *Anopheles arabiensis*, in Nigeria. *Mem. Inst. Oswaldo Cruz* 96: 1081–1084.

Petersen, J. L., Larsen, J. R., and Craig, G. B., Jr. 1976. Palp-antenna, a homeotic mutant in *Aedes aegypti. J. Hered.* 67: 71–78.

Risch, N. J. 2000. Searching for genetic determinants in the new millennium. *Nature* 405: 847–856.

Roush, R. T., and McKenzie, J. A. 1987. Ecological genetics of insecticide and acaricide resistance. *Annu. Rev. Entomol.* 32: 361–380.

Rousset, F. 1997. Genetic differentiation and estimation of gene flow from F-statistics under isolation by distance. *Genetics* 145: 1219–1228.

Soto, S. I., Lehmann, T., Rowton, E. D., Velez, B. I., and Porter, C. H. 2001. Speciation and population structure in the morphospecies *Lutzomyia longipalpis* (Lutz & Neiva) as derived from the mitochondrial ND4 gene. *Mol. Phylogenet. Evol.* 18: 84–93.

Tabachnick, W. J. 1990. Genetic variation in laboratory and field populations of the vector of bluetongue virus, *Culicoides variipennis* (Diptera: Ceratopogonidae). *J. Med. Entomol.* 27: 24–30.

Tabachnick, W. J. 1991. The yellow fever mosquito: Evolutionary genetics and arthropod-borne disease. *Am. Entomol.* 37: 14–24.

Tabachnick, W. J. 1996. *Culicoides variipennis* and bluetongue-virus epidemiology in the United States. *Annu. Rev. Entomol.* 41: 23–43.

Tabachnick, W. J., and Black, W. C. 1994. Making a case for molecular population genetic studies of arthropod vectors. *Parasitol. Today* 11: 27–30.

Tan, Y., and Riley, M. A. 1996. Rapid invasion by colicinogenic *Escherichia coli* with novel immunity functions. *Microbiology* 142: 2175–2180.

Taylor, C., Toure, Y. T., Carnahan, J., Norris, D. E., Dolo, G., Traore, S. F., Edillo, F. E., and Lanzaro, G. C. 2001. Gene flow among populations of the malaria vector, *Anopheles gambiae*, in Mali, West Africa. *Genetics* 157: 743–750.

Thornhill, R., and Alcock, J. 2001. *The Evolution of Insect Mating Systems.* Lightning Source, New York.

Wade, M. J., and Goodnight, C. J. 1991. Wright's shifting balance theory: An experimental study. *Science* 253: 1015–1018.

West, D. F., and Black, W. C. 1998. Breeding structure of three snow pool *Aedes* mosquito species in northern Colorado. *Heredity* 81: 371–380.

Weir, B. S., and Cockerham, C. C. 1984. Estimating *F*-statistics for the analysis of population structure. *Evolution* 38: 1358–1370.

Wood, R. J., and Ouda, N. A. 1987. The genetic basis of resistance and sensitivity to the meiotic drive gene D in the mosquito *Aedes aegypti* L. *Genetica* 72: 69–79.

Molecular Taxonomy and Systematics of Arthropod Vectors

WILLIAM C. BLACK IV AND LEONARD E. MUNSTERMANN

INTRODUCTION

Taxonomy is the theory and practice of classifying organisms (Mayr and Ashlock 1991). It is a science whose origin dates back over two centuries and thus draws from extensive accumulated information on morphology, biogeography, and habitat distributions of a large proportion of extant species. Historically, taxonomy has implicitly adopted the *morphological species concept*. Species were identified on the basis of one or several consistently distinguishing morphological characters. This approach permitted the construction of hundreds of taxonomic keys for species identification. Those same keys, or updated versions of them, remain the foundation of most species identification.

However, as knowledge of species became more detailed, more examples appeared where morphologically identical organisms differed greatly in behavior, physiology, and genetics. This suggested that these identical organisms might belong to different species. Furthermore, some animal and many plant species were discovered to display a continuous range of morphological variation in taxonomic characters throughout their distribution. When sampled along a transect, organisms in proximity to one another were morphologically similar, but organisms at greater and greater distances from one another along the transect became morphologically distinct. This lead to a realization that morphologically different organisms can indeed belong to the same species. For these reasons, in the last century the morphological species concept came to be gradually replaced by the more stringent *biological species concept*, in which species were considered valid only when they exist as reproductively isolated gene pools (Mayr and Ashlock 1991). With the biological species concept, morphologically identical, *sympatric* (overlapping in geographic range) organisms are classified as separate species when they can be shown to be reproductively isolated. Conversely, morphologically distinct organisms can be a single species, especially when *allopatric* (distributed in different geographic ranges) or when the morphological features constitute intraspecies genetic polymorphisms.

Immediate criticism arose concerning the biological species concept. Implicit in the concept is that a methodology exists for identifying reproductively isolated gene pools. The phylogenetic species concept was developed to address this criticism as a possible criterion for delineating species. With this criterion, valid species have a monophyletic ancestry (Chapter 29). When organisms identified morphologically as a single species are polyphyletic (several genetic lineages), then this becomes evidence for the presence of several different species. Conversely, when organisms identified morphologically as one or more species are monophyletic, then these morphologically distinct organisms may require reevaluation and perhaps designation as a single species.

The phylogenetic species concept was proposed decades prior to the technology discussed in Chapter 29. At the time, the concept was criticized as circular logic; the same morphological criteria used to define species were then used to assess phylogenetic relationships. However, with the advent of biochemical and nucleotide analysis methods, the phylogenetic species concept became testable, because these data sets were assessed independently of morphological data sets. Avise and Ball (1990) added an important addendum to the phylogenetic species concept called the *concordant species concept*. They argued that phylogenies derived from single genes reflect only the phylogeny of that gene rather than the species phylogeny. This effect is mitigated, however, when several genes are examined and the single-gene effects become minimal. In the following, the concordant species model becomes particularly relevant when examining relationships among closely related, recently derived species, such as races within malaria mosquito, *Anopheles gambiae*.

Correct species identification is critical to medical entomologists during suppression of disease outbreaks and in associated epidemiological and ecological studies. However, the problems elicited by the various species concepts apply to a large number of disease vector species. Many closely related vector species are difficult or impossible to identify with

morphological characters. In these cases, the taxonomic community has turned to the molecular phylogenetic and concordant species concepts for providing the necessary identifications. Molecular taxonomy is the classification and identification of organisms based on protein or nucleic acid characters rather than morphological characters.

WHO NEEDS MOLECULAR TAXONOMY?

The number of hematophagous arthropod species is large (Table 33.1). For most of the species in the taxa listed, morphological characters are clear and the species designation is conclusive. Furthermore, the vector species composition in a geographic region is often well known, and correct identifications are obtained without the need of highly trained taxonomists. Once a correct identification has been made, control programs can interrupt disease transmission by accessing critical information on adult and larval

TABLE 33.1   Numbers of Species in the Major Arthropod Groups that are Reported to Feed on Vertebrates (from Lane and Crosskey, 1995). Tick and Mite Species were Estimated with Assistance from Dr. Hans Klompen

Order or Suborder	Group	Number of species described worldwide
Class Insecta		
Anoplura	Sucking lice (all families)	490
Mallophaga	Chewing lice (all families)	3,000
Hemiptera	Bed bugs (Cimicidae)	108
	Kissing bugs (Reduviidae, Triatominae)	120
Diptera	Biting midges (Ceratopogonidae)	1,300
	Mosquitoes (Culicidae)	3,600
	Sand flies (Psychodidae, Phlebotominae)	700
	Black flies (Simuliidae)	1,460
	Horse and Deer flies (Tabanidae)	3,500
	Filth flies (Muscidae, Glossinae)	23
	Filth flies (Muscidae, Stomoxyinae)	49
	Louse flies (Hippoboscidae), Bat flies (Nycteribiidae, Streblidae)	600
Siphonaptera	Fleas (all families)	2,200
Lepidoptera	Moths (Noctuidae)	3
All orders		17,150
Subclass Acari		
Ixodida		
	Ixodidae	660
	Argasidae	140
Mesostigmata		4,500
Prostigmata		10,500
Astigmata		10,000
All Orders		25,800

habitats, host preference, ecology, vector competence, and histories of earlier epidemics and, ultimately, apply the control tactics that are appropriate for that species. Furthermore, correct species identification is required for monitoring of vector or reservoir populations for the prevalence of infected individuals. Prevention and control are often focused directly on the vector species of concern (*species sanitation*). Certainly, basic research on the ecology and geographic and seasonal distributions of vector species also requires accurate identification.

However, for certain species or species groups in each of the taxa listed in Table 33.1, taxonomic problems are prevalent. Every family contains genera and subgenera whose member species are difficult or impossible to separate based on morphology. The taxonomic problems in many of these families, detailed in Lane and Crosskey (1995), fall into three categories:

1. Closely related but reproductively isolated species that are morphologically identical
2. Species that can be easily distinguished but only during a single life stage
3. Minute arthropods that require careful preparation before identification is possible

The taxonomy of arthropod vectors in these three categories can be a daunting task, requiring extensive time, training, and experience. This can severely limit the scope of epidemiological or ecological studies, to say nothing of the inability to react quickly in control or abatement programs. In these cases, molecular approaches to taxonomy may be required to improve the speed and accuracy of vector species identification.

We begin by giving several examples of medically important taxa in all three categories and how an inability to identify these species has impacted epidemiological and ecological studies. We then discuss different genetic and biochemical techniques and describe the taxonomic procedures to follow in testing their validity. With each technique we discuss how molecular taxonomy can be and has been used for identification.

Species Complexes

When two populations of a species become reproductively isolated either through geographic separation or as a consequence of the evolution of prezygotic (e.g., behavioral) or postzygotic (e.g., sterility or lethality in F_1 offspring) isolation mechanisms, they gradually accumulate genetic differences. Given sufficient time, these genetic differences may eventually become manifest as subtle morphological differences. Over an extended time, these may become manifest as distinct morphological differences. However, character differences arising from recent or incipient (ongoing) reproductive isolation are frequently difficult to use in routine identification.

Closely related but reproductively isolated species that appear morphologically identical are often referred to as *cryptic* or *sibling species*, and a group of cryptic species is referred to as a *species complex*. The abbreviation *s.l.* follows the name of a species complex, and *s.s* follows the names of the individual member species of that complex. These abbreviations stand for *sensu lato* (Latin meaning "in the broad sense") and *sensu stricto* (Latin for "in the strict sense"). The name of the original, albeit polyphyletic (Chapter 29), species followed by *sensu lato* becomes the name of the species complex. The name of the original species followed by *sensu stricto* applies to only one of the species in the group and is given either to the first organisms collected in the species group or to the most broadly distributed species. In polytypic species *sensu* is often followed by an author designation indicating that a specimen was identified based on the characters described by that author and implies that not all descriptions of that species agree.

Cryptic species were not considered a practical problem until the discovery that closely related species can differ greatly in their vector competence, preferred host, larval habitat, and adult feeding behavior. The most striking historical example involves the *Anopheles maculipennis s.l.* complex in Europe. Previous to the 1920s, this complex was described as a single species with a broad Palearctic range. At that time, malariologists knew that the European distribution of malaria was largely coastal and did not coincide with the widespread and often dense inland distribution of *An. maculipennis*. This phenomenon came to be known as *anophelism without malaria*. The first description of a distinct taxon within *An. maculipennis s.l.* appeared in 1924, when Van Thiel described *An. atroparvus* in Holland. It was a short-winged, nonhibernating form of *An. maculipennis*, and he determined that this species was responsible for indoor malaria transmission during winter. In 1926, Falleroni found two colors of eggs within *An. maculipennis s.l.* collected in Italy. The race associated with the dark gray eggs was described as *An. labranchiae*, and the race with the silvery white eggs was *An. messeae*.

The 1920s saw the development of the precipitin test that permitted the identification of a blood meal with respect to its animal source. This method was used in the 1930s to demonstrate that *An. maculipennis s.l.* was zoophilic (had a feeding preference for animals other than humans) in malaria-free regions, but was anthropophilic (feeding preference for humans) in malarious regions. When feeding preference was analyzed in Falleroni's species, only *An. labranchiae* was anthropophilic and, hence, the only species responsible for malaria transmission. Further investigation divided the dark gray egg taxon into *An. labranchiae, An. atroparvus,* and *An. sacharovi.* The silvery egg taxon was subdivided into *An. maculipennis s.s, An. messae, An. melanoon,* and *An. subalpinus.* These seven taxa were found to be partially or fully reproductively incompatible, indicating that they were suitably delineated biological species. Later, distinctive differences among them were identified in mating patterns and in inversion patterns of the *polytene chromosomes* of the larval salivary gland (explained below). After the taxonomy of this group was clarified, a study of species distributions became possible. Generally, the seven species were found to occupy disjunct geographic regions, although several did not overlap in distribution. Of the seven, only two, *An. labranchiae* and *An. sacharovi,* are important malaria vectors throughout their ranges. *Anopheles sacharovi* is the only species that can be identified as an adult; the others must be identified on the basis of egg morphology or polytene chromosomes. Note that when we speak of *An. maculipennis s.l.* we refer broadly to all seven members of the species complex, but when we refer to *An. maculipennis* s.s. we are referring strictly to the *An. maculipennis* biological species component.

Perhaps the most important vector species complex is *An. gambiae s.l.,* the primary malaria and filariasis vector in Africa. Inland, the larvae are found in small sunlit, freshwater pools throughout most of the species range. However, along the coasts these larvae can be found in brackish intertidal swamps. Larvae occurring in coastal habitats are morphologically indistinguishable from larvae collected at inland sites, but adults and eggs obtained by rearing these larvae are distinctive. During laboratory studies on the inheritance of dieldrin resistance, Davidson (1956) demonstrated F_1 hybrid male sterility within and among geographical populations of *Anopheles gambiae s.l.* Subsequent studies of the Mendelian inheritance of F_1 sterility within and among *An. gambiae s.l.* populations demonstrated the existence of six reproductively isolated cryptic species (Coluzzi et al. 1979). These six species were named *An. gambiae s.s., An. arabiensis, An. quadriannulatus, An. bwambae, An. merus,* and *An. melus.* Years of intensive field collections have found *An. gambiae s.s.* and *An. arabiensis* to be widespread throughout Africa. They are often sympatric and are the two most important vectors of malaria and bancroftian filariasis. *Anopheles gambiae s.s.* is more anthropophilic and endophilic (preferring to feed indoors) than *An. arabiensis. Anopheles quadriannulatus* is zoophilic and not a

vector of either malaria or filariasis. *Anopheles bwambae* is known only from mineral water springs at the base of the Ruwenzori Mountains, Bwamba County, Uganda.

These six species were identifiable only by polytene chromosome inversion patterns (Fig. 33.1). Inversion patterns are identified by cytogenetic analysis of polytene chromosomes in the ovarian nurse cells of half-gravid females or in the salivary glands of fourth-instar larvae. Polytene chromosomes are the product of a process termed *endomitosis* (chromosome replication without nuclear or cytoplasmic fission) and commonly occur in certain insect tissues. When endomitosis occurs in insect orders (e.g., Diptera, Collembola) that have a somatic pairing of homologous chromosomes, polytenization occurs. The chromosomes appear as thick, ropy structures with distinct "bands and puffs" that correspond to different

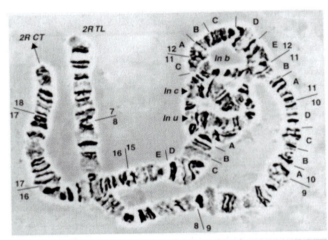

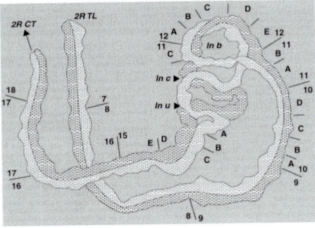

FIGURE 33.1 Polytene chromosome preparation from the ovaries of a half gravid *Anopheles gambiae s.s.* The analysis indicates the presence of a triple inversion heterozygote. From Toure et al. (1998).

densities and conformations of euchromatin and heterochromatin. The linear arrangements of these landmarks have been used as characters to define cryptic species in a number of anopheline species groups (Table 33.2). In insects bearing an inversion on only one homologous chromosome, i.e., heterozygous for the inverted segment, loop structures are visible when complementary regions align. These loop structures are clearly visible with compound microscopy and provide key features for cytogenetic analyses (Fig. 33.1).

During meiosis, recombination is reduced or eliminated among the loci contained within the inversion. As a result, gametes arising from an inversion heterozygous mosquito are orthologous for genes arising from its parents. Recombination will occur only within parents that are homozygous for an inversion. Analyses of polytene chromosomes in *An. gambiae s.s.* from West Africa revealed a complex of paracentric (outside the centromere) inversions, with the majority occurring on the right arm of the second chromosome (2R) (Fig. 33.1). The spatial distribution of these inversions shows a strong association with ecological and climatic zones, and their distributions are not random, even on a microgeographic level (Coluzzi et al. 1979; Bryan et al. 1982).

Where several of these inversions occur in sympatry, their relative frequencies change seasonally, most likely in response to annual fluctuations in climate (Toure et al. 1998). These observations are consistent with a hypothesis that multilocus genotypes contained within and maintained by inversions are adaptive, especially with respect to survival under varying degrees of aridity (Coluzzi 1982). Extensive field studies of *An. gambiae s.s.* in areas of sympatry indicate that 2R inversion genotypes are seldom in Hardy–Weinberg equilibrium due to a deficiency or absence of inversion heterozygotes. This result suggests barriers to gene flow among mosquito subpopulations with different 2R inversion genotypes. On this basis, *An. gambiae s.s.* was subdivided into five chromosomal inversion forms: Bamako, Bissau, Forest, Mopti, and Savannah (Coluzzi 1984). Savannah is considered the typical 2R form of *An. gambiae s.s.*, with the broadest geographical distribution, occurring throughout sub-Saharan Africa. Bamako is found in Mali and Northern Guinea, associated with the upper Niger River and its tributaries. Forest is a forest-breeding

TABLE 33.2   Examples of Well-known Disease Vector Species Complexes (Morphologically Identical) and Species Groups (Morphologically Similar but Can Be Distinguished at One or More Life Stages)

Order	Family	Tribe	Genera + Species groups/species complexes
	Class Insecta		
Anoplura — Sucking lice (all families)			*Pediculus humanus capitus* P. h. humanus
Diptera			
	Biting midges (Ceratopogonidae)		*Culicoides variipennis*
	Mosquitoes (Culicidae)	Subfamily *Anophelinae*	
			Anopheles albitarsis An. culicifacies An. darlingi An. dirus An. gambiae An. hyrcanus An. leucosphyrus An. maculatus An. maculipennis An. marshalli An. nuneztovari An. pseudopunctipennis An. punctulatus An. quadrimaculatus
	Subfamily *Culicinae*	Tribe *Sabethini*	*Trichoprosopon sp.*

(continues)

TABLE 33.2 *(continued)*

Order	Family	Tribe	Genera + Species groups/species complexes		
		Tribe *Culicini*			
			Culicine genera		
				Culex pipiens	
				Cx. sitiens	
				Cx. vishnui	
			Aedine genera		
				Aedes aegypti	
				Ae. simpsoni	
				Ae. scutellaris	
				Ochlerotatus communis	
				Oc. excrucians	
				Oc. mariae	
				Oc. punctor	
	Sand flies (Psychodidae, Phlebotominae)		Genus *Lutzomyia*		
				(Dampfomyia) delpozoi	
				(D.) dreisbachi	
				(Helcocertomyia) lanei	
				(Lutzomyia) arugaoi	
				(L.) baityi	
				(L.) longipalpis	
				(L.) migonei	
				(Nyssomyia) oswoldoi	
				(N.) pilosa	
				(Psathyromyia) shannoni	
				(Psychodopygus) rupicola	
				(P.) saulensis	
				(Trichopygomyia) verrucarum	
	Black flies (Simuliidae)				
				Simulium damnosum	
				S. exiguum	
				S. metallicum	
				S. neavei	
				S. ochraceum	
				S. venustum	
	Tsetse flies (Muscidae, Glossinae)				
				Glossina fusca	
				G. morsitans	
				G. palpalis	
	Subclass Acari				
Ixodida					
	Ixodidae (hard ticks)				
				Ixodes ricinus/persulcatus	
				Ix. holocyclus	
				Ix. crenulatus	
				Ixodes pavlovskyi	
				Haemaphysalis erinacei	
				Dermacentor marginatus	
				Rhipicephalus turanicus	
				Boophilus microplus	
				Hyalomma asiaticum	
				Hy. marginatum	
	Argasidae (soft ticks)				
				Ornithodoros moubata	
				Argas reflexus	

form and is thought to represent the ancestral inversion form. Mopti is found in Mali, Guinea, Ivory Coast, and Burkina Faso; and because of its association with flooded plains and irrigated fields, it maintains a continuous life cycle even throughout the dry season. Bissau is endemic to The Gambia. Population studies have revealed a low frequency of hybrids between some forms (e.g., Mopti and Savannah, Bamako and Savannah) and complete reproductive isolation between others (e.g., Bamako and Mopti).

Recent research has explored the reproductive status of the five chromosomal forms of *An. gambiae s.s.* using molecular techniques to examine the variation among 12 genes located throughout the *An. gambiae s.s.* genome. This is a clear application of the concordant species concept, and the results illustrate why this concept is important. These studies suggested the presence of partial barriers to gene flow between some chromosomal forms in the Ivory Coast and other West African countries to the north and west but that some introgression was indicated between Savannah and Mopti types in Benin and countries to the east. Collectively, these studies indicate the need for a broader geographical sampling of *An. gambiae s.s.*, increased research on mechanisms of prezygotic reproductive isolation, and field-based studies of survival and fecundity in hybrids to test for postzygotic reproductive isolation. However, current evidence has led to proposals to elevate the Savannah and Mopti forms to species-level designations.

Table 33.2 lists most of the currently recognized disease vector species complexes (morphologically identical) or species groups (morphologically similar). All of the examples share two important features. First, members of species complexes differ dramatically in their vector potential. Second, in almost every case, identification of species within the complex requires access to specific developmental stages, detection of subtle morphological differences, or specialized techniques.

Morphologically Similar Species

Closely related, morphologically similar species usually can be separated at one or more life stages if adequate taxonomic keys are available. However, a problem frequently encountered is that congeneric species cannot be distinguished at all life stages. For example, some members of *An. maculipennis s.l.* can be identified only as eggs. Many boreal species of the aedine genus *Ochlerotatus* are easily distinguished by genitalic structures of adult males or as fourth-instar larvae. Several New World sand flies of the genus *Lutzomyia* are distinguished only by subtle differences

in male genitalia. Keys for early instars and pupae are nonexistent. Trapping methods often preclude the capture of males. Furthermore, examination of genitalic structures requires slide mounts and a compound microscope. For many culicine mosquitoes, identification of adult females is hampered by loss of diagnostic scales or setae that occurs as a consequence of aging or the collection process.

Similarly, many species of ticks and hematophagous mites are separable only as nymphs or adults. Larval keys are rare. Some species cannot be distinguished in both sexes. For example, *Boophilus annulatus* and *B. microplus*, the primary vectors of babesiosis in cattle, can be separated only by the presence of a caudal appendage in males. Morphological characters in the females are variable; immature specimens cannot be separated at all.

Epidemiological or ecological studies become difficult in species that can be identified only during one stage of their development. Sometimes in these cases, the specimens must be matured in the laboratory until they reach a stage that can be identified. Alternatively, specimens must be mated and the offspring reared to procure an accurate species diagnosis. In the context of emergency situations, these practices are too slow to be practical.

Small Species

Many arthropod vectors are small and require preservation, clearing, and mounting for microscopic analysis of characters. This is true for most medically important mites, chewing and sucking lice, biting midges, sand flies, blackflies, and fleas. The preparation process usually destroys the microorganisms associated with these specimens, and therefore information on the pathogens they may carry is lost in the preparative process. If biological control organisms are used in association with these specimens, they too will be eliminated.

THE MOLECULAR APPROACH TO TAXONOMY

Generally, molecular approaches are appropriate only when taxonomic ambiguities arise, due to factors just outlined. Under these circumstances, molecular taxonomy becomes an essential tool for the following reasons: (1) Members of species complexes can be identified, (2) morphologically similar species can be identified at any life stage, (3) current technology allows visualization of biochemical and protein markers in small arthropods, and, in all three cases, (4) associated microorganisms can be identified, whether pathogens (Chapter 18) or biological control agents (Chapter 43). Nonetheless, *the practice of developing molecules as taxonomic tools is not quick, nor is the effort trivial. Molecular taxonomy requires the same rigor and careful attention to intraspecific variation that has been practiced for centuries with morphologically based taxonomy.* The fundamental difference between molecular and morphological characters is that the former consist of an array of identifiable single gene or base-pair differences, whereas the latter are the product of a variety of genetic components, including developmental genes, structural genes, and epistatic interactions among multiple loci. Environmental effects and their interaction genotypes may also affect morphological variation, but these are less likely to be evidenced at the molecular level.

METHODS IN MOLECULAR TAXONOMY

A variety of methods are available to reveal nucleotide- or protein-level differences among individual insects. This chapter focuses on relatively inexpensive techniques (no more than $5/determination) that can be completed in 24–48 h. These methods can supplant an array of techniques that have been applied to resolve taxonomic ambiguities, including scanning electron microscopy, banding patterns of polytene chromosomes (cytotaxonomy), cuticular hydrocarbons, immunology, and nucleotide sequencing. Lane and Crosskey (1995) provide an excellent review of these methods and their applications.

Nucleic Acid–Based Taxonomy

A eukaryotic genome consists of a complex mixture of unique and repetitive gene sequences (Chapter 31). Some of these sequences code for proteins or ribosomal and transfer RNAs involved in transcription and translation. Noncoding regions may contain regulatory or structural elements or may be merely "parasitic" or "junk" DNA. The goal of molecular taxonomy is to locate genomic regions that are most likely to vary among closely related species. In general, these will be genes or regions that mutate or evolve quickly.

Noncoding regions are often targeted for use in molecular taxonomy because they are assumed to be less affected by adaptive selection. Noncoding repetitive regions are especially useful as species-diagnostic probes because they are easier to detect in hybridization assays due to high copy number and because they can diverge quickly among closely related species.

DNA sequences selected as identification targets may vary among species by a presence–absence phenotype or by specific nucleic acid sequence (dependent on the resolving power of the technique chosen).

Specimen Preparation

Time spent in specimen preparation for DNA-based analyses varies enormously, depending on the detection techniques to be employed. The most rapid technique is the *squash blot*, in which many insects (or parts of many insects) are simultaneously crushed directly onto nitrocellulose or nylon filters to be probed with species-specific markers (Gale and Crampton 1987, 1988). A more sensitive approach has used the polymerization chain reaction, PCR (Chapter 30), where even small body parts of arthropods (e.g., legs, heads, or abdomens) provide sufficient DNA to make species determinations. Scott et al. (1993) demonstrated successful PCR amplification with a single mosquito leg. Alternatively, investigators often want to store specimen DNA for use in additional analyses. A variety of procedures are available for isolating DNA from individual insects (Black and DuTeau 1997).

DNA Hybridization

In hybridization assays, a labeled, single-stranded, species-specific piece of DNA (the probe) is mixed with single-stranded DNA (the target) that has been immobilized on a nylon or nitrocellulose filter. The target DNA originates from field-collected individuals of unknown species identity. The mixture is incubated for a short time with a combination of high-temperature and salt-concentration conditions sufficient to ensure that the probe and target strands can hybridize. The temperature and salt conditions control the stringency, or accuracy, of a hybridization reaction. Stringency of the hybridization must be sufficiently great that the probe does not hybridize to arbitrary regions of the target genomic DNA and thereby lead to false-positive identifications. However, the stringency must also be sufficiently low to allow hybridization between the probe and any complementary target sequences, thereby preventing false negatives. Generally, stringency conditions are adjusted so that the probe will hybridize to regions that are ~90% complementary in sequence. Following hybridization, the filter containing the immobilized target DNA is washed in a series of low-, medium-, and high-stringency washes and dried. The washes remove probe that has not bound to a complementary piece of target DNA. Hybridization events are detected by exposure of the filter to X-ray film for radioactively-labeled probes or through enzyme detection for probes labeled with specific ligands. Results are recorded as the presence or absence of a hybridization signal at each location of the target DNA on the filter. The filter is then stripped of the label at a high temperature or by soaking in an alkaline solution; this disrupts base pairing between the probe and the target DNA. If necessary, the filter DNA can then be hybridized again with another probe. Table 33.3 lists the steps involved in hybridization assays.

Requirements for a Species-Specific Probe

Strategies for isolating species-specific probes are reviewed in Post and Crampton (1988) and Post et al. (1992). The tactic most commonly used is illustrated in Figure 33.2. Individual genomic libraries are constructed for each of the species in a complex. These libraries are then lifted to a set of replicate filters and probed with labeled whole genomic DNA from each complex member. This tactic accomplishes two things. First, regions of whole genomic DNA probe are labeled according to their relative abundances. Highly repetitive DNA is abundantly labeled, middle-repetitive DNA is moderately labeled, and unique-sequence DNA is not detected. Therefore, only clones with repetitive DNA give a strong hybridization response with whole genomic DNA. Second, a repetitive-sequence clone that contains species-specific sequences or copy number will hybridize differentially on replicate filters (where each filter is probed with DNA from each species of the complex).

A good candidate for a species-specific probe is a clone that provides a strong hybridization signal (indicating high copy number) in one species but gives little or no signal when probed with genomic DNA from other species. Clones meeting this standard are isolated, purified, labeled, and tested as probes on filters containing DNA from each species in the complex. The ideal, species-specific probe will hybridize strongly with only one member of a species complex. Probes to be avoided are those that show a moderate amount of hybridization with nontarget species or that hybridize weakly or inconsistently with the target species. Subtractive hybridization techniques have been described for isolating species-specific sequences (Clapp et al. 1993).

The described cloning procedures are biased toward repetitive DNA. This bias is useful for two reasons. First, repetitive-DNA probes ensure a strong hybridization signal when screening field-collected material. This is important in identifying small arthropods. More importantly, repetitive elements can undergo a process described as *concerted evolution* by means of mechanisms collectively known as *molecular drive* (Fig. 33.3). During the speciation process, populations become reproductively isolated and begin accumulating unique sequences. Mutations within repetitive regions of the genome have been shown to evolve in a "concerted" fashion rather than independently, with the consequence that a mutation found in one repetitive element is distributed throughout all elements. Dover and Tautz (1986) discussed the many

TABLE 33.3   General Conditions for the Performance of Hybridization Assays in Molecular Taxonomy

1. Target DNA is placed onto a nylon or nitrocellulose filter either through squashing part or the whole of an insect directly onto a filter sitting on filter paper saturated with 10% SDS (sodium dodecyl sulphate) or by placing extracted DNA directly onto a fresh hybridization filter.

2. Target DNA is denatured by setting the hybridization filter on filter paper saturated with 0.1 M NaOH/1.5 M NaCl for 5 min. The hybridization filter is blotted dry and neutralized by placing it on filter paper saturated with 0.5 M Tris-HCl pH 7.0/3.0 M NaCl for 5 min. The hybridization filter is blotted dry and baked for 2 hrs at 80°C under vacuum to adhere the DNA to the filter.

3. Baked filters are sealed in a plastic bag with 10 ml of pre-hybridization solution (50% deionized formamide, 0.2% polyvinyl-pyrrolidone (M.W. 40,000), 0.2% bovine serum albumin, 0.2% ficoll (M.W. 400,000), 0.05 M Tris-HCl pH 7.5, 1.0% SDS, 1.0 M NaCl, 0.1% sodium pyrophosphate, 1% sonicated calf thymus DNA, and 10% dextran sulfate (M.W. 500,000) and 100 μg/ml denatured and sonicated calf thymus DNA). The filter is pre-hybridized at 42°C for 6 hrs or overnight.

4. The probe DNA is labeled with CTP or ATP (cytosine or adenosine triphosphate) α-labeled with ^{32}P or biotin using nick translation or random primer labeling. The labeled probe is mixed into 1 ml of 100 μg/ml denatured and sonicated calf thymus DNA and the solution is boiled for 10 min to denature the probe. This is then placed directly into 3 ml of the pre-hybridization buffer described in step 3. The bag is opened and the probe solution is placed in the bag with the pre-hybridization solution and filter. The bag is resealed and hybridization proceeds for 24 hrs at 42°C.

5. Following hybridization, the filter is washed twice for 5 min in 1× wash buffer (0.3 M NaCl, 0.06 M Tris-HCl (pH 8.0), 2 mM EDTA) at room temperature, twice in 1× buffer with 1% SDS for 30 min at 60°C, and twice in 0.1× wash buffer for 30 min at room temperature. The filter is allowed to dry at room temperature. For radioactively labeled probes, filters are loaded into X-ray cassettes with film and exposed overnight. For probes labeled with ligands, enzyme detection is used following manufacturer's instructions.

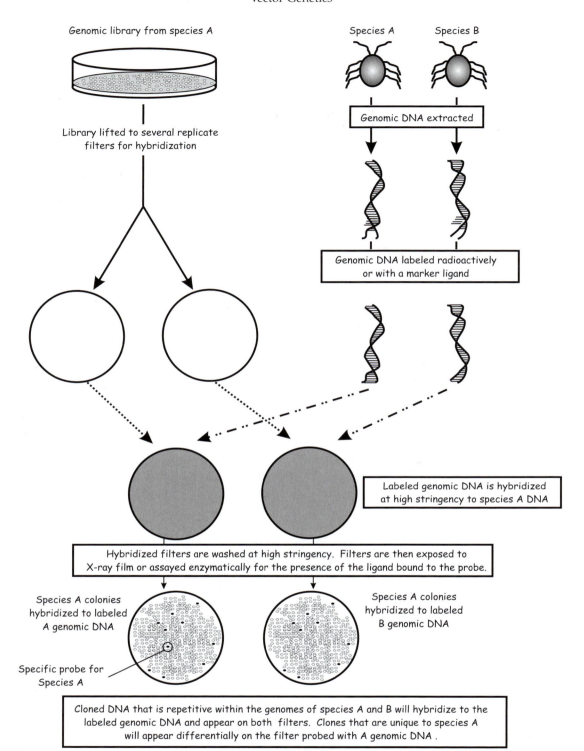

FIGURE 33.2   Protocol for identification of species-specific probes in genomic libraries. Adopted from Post and Crampton (1988).

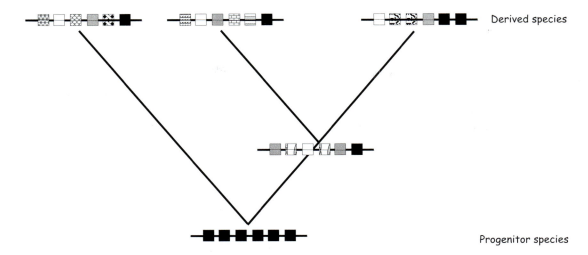

Independent evolution among members of a multigene family: All family members evolve independent of one another in species derived from the progenetor species.

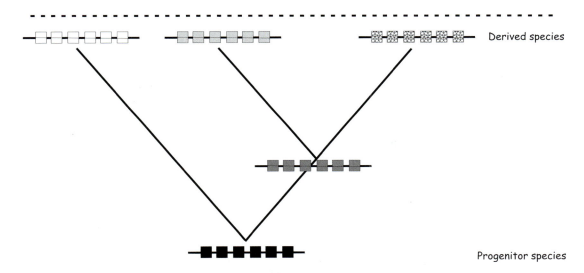

Concerted evolution among members of a multigene family: All family members evolve in concert with one another as new species arise from the progenetor pecies.

FIGURE 33.3 Concerted evolution is commonly observed among members of a multigene family. New variants that arise following speciation are not confined to one member of the multigene family but spread through the family as a result of several molecular processes, chiefly unequal crossing over and gene conversion.

molecular mechanisms (chiefly unequal crossing over, gene conversion, and slippage replication) that may explain this commonly observed phenomenon. For taxonomic purposes, the importance of repetitive elements is their ability to diverge rapidly in sequence and abundance following the reproductive isolation of populations.

The foregoing techniques have been applied successfully for the development of species-specific probes in several vector complexes. Post et al. (1992) described probes for identification of species in the *An. gambiae* complex. Audtho et al. (1995) developed probes for identification of *An. dirus* sibling species; Post and Flook (1992) identified members of the *Simulium damnosum* complex. Hill et al. (1992) reported a refinement in which species-specific oligonucleotides were identified by sequencing species-specific clones and the oligonucleotides subsequently used as probes.

They also described nonradioactive, synthetic DNA probes that provided species identifications in less than 48 hours.

Advantages and Disadvantages of DNA Hybridization Techniques

There are several advantages of DNA hybridization techniques: (1) They involve minimal preparation of the field-collected specimen; (2) they have proved sensitive in all the previously mentioned studies; (3) the time needed to visualize results is short and becoming shorter as methods improve. There are also disadvantages: (1) Probe isolation is expensive and can be difficult and time-consuming; (2) construction of genomic libraries is costly. However, the development of both is a one-time investment and innovations such as subtractive hybridization procedures may obviate the need for libraries and ease probe isolation.

The major operational disadvantage of DNA hybridization techniques arises when there are several species that must be identified in a complex. For each species, the filter must be stripped, prehybridized, and reprobed to identify all the species arrayed on a filter. Placing insects between two filters prior to crushing them can make duplicate filters; this allows simultaneous probing of two membranes. Nonradioactive approaches that use species-specific oligonucleotides can reduce the number of separate hybridizations required. Each probe can be labeled with a different ligand (e.g., biotin, digoxigenin), which binds to reporter molecules (e.g., streptavidin-labeled alkaline phosphatase or horseradish peroxidase) that yield different colors in their respective reactions. Alternatively, with fluorescence detection methods, oligonucleotides can be labeled with different fluorescent molecules that appear as different colors under varying wavelengths of light. Either approach can reduce the processing to a single hybridization step. Development of nonradioactive techniques will also allow application of these techniques in situations where radioactive label is not available or not permitted.

Polymerase Chain Reaction (PCR)

Polymerase chain reaction can amplify a DNA sequence from a minute amount of template DNA. The amplified DNA can then be analyzed by various methods to determine species identity. Therefore, PCR-based techniques are extremely useful for small arthropods. Furthermore, PCR permits the investigator to retain most of the specimen, which can then be examined for pathogens, parasites, biological control agents, gut contents, age, or other epidemiologically informative characters.

Unlike probe-hybridization approaches, amplified regions need not be repetitive. This greatly increases the number of genes or regions that can be exploited for identification. In designing PCR systems for taxonomy, the goal is to identify regions of the genome that have species-specific sequences or are species-specific in size (Fig. 33.4). First, complete sequence information in the chosen region must be available from all species to be differentiated. This may require cloning the sequence from a library or the use of PCR primers derived from other taxa that anneal to regions that flank the target sequence but are conserved across distant taxa. The cloned or amplified regions are then sequenced. From the complete sequence information, primer "A" is designed so that it anneals to a conserved sequence in the target region, and the second primer, "B," is chosen to anneal only to a region unique to a species. Ultimately, PCR contains multiple reverse primers, each complementary to a single member of the complex. A further requirement for detection is that each amplified product be unique in size. To accomplish this, the reverse primer is built in a variable region that is unique to one species. The reverse primer is then built to a unique region for each additional species. If deletions and insertions have developed between primer annealing sites in the evolution of two species then the second species-specific primer can occur in the same region as the first species, but size polymorphisms can still be visualized.

During PCR, primer "A" will anneal with template DNA from all species; however, the species-specific reverse primer, if designed correctly, will anneal only to the flanking region in the species with the complementary sequence. Geometric amplification occurs only if both primers can anneal to DNA of the target species. The PCR reaction mix contains all primers, but only those complementary to the template from the unknown field specimens will result in amplification. Furthermore, they will amplify a fragment that has a unique size for each species. This size difference can be visualized on ethidium bromide–stained agarose gels following electrophoresis of the amplified product (Fig. 33.4).

Collins and Paskewitz (1996) reviewed applications of this technique in differentiation of members of the *An. gambiae* complex. Cornel et al. (1996) described the development of this technology for members of the *An. quadrimaculatus* complex occurring in North America, as did Walton et al. (1999) for the *An. dirus* complex in southeast Asia. Toma et al. (2000) described this technology for members of the *Culex vishnui*

1) Amplify a genomic region containing species specific sequences in each member of the species complex.

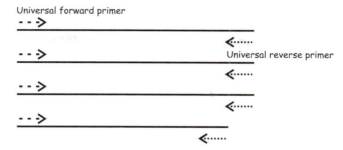

2) Obta n complete nucleotide sequences in all species. Locate sequences that are species specific and are located in regions that will yield products of a detectably different size.

3) For each species build a primer that is complementary to the species specific region.

4) Perfcrm PCR with a mixture of the 5 primers: the universal forward primer and the 4 reverse, species ipecific primers.

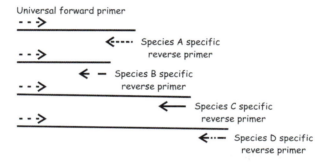

5) Separate products of the PCR reaction on agarose gels. Stain gel with ethidium bromide and vizualize bands under ultraviolet light. Each species produces a band that is unique in size.

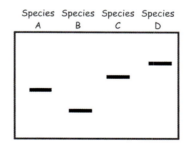

FIGURE 33.4 Steps in developing species-specific PCR primers as a tool for discriminating closely related species.

subgroup in Asia. For the *Cx. pipiens* complex and the associated species, *Cx restuans* and *Cx. salinarius*, Crabtree et al. (1995) developed species-specific PCR primers; however, insufficient variation was detected to differentiate between *Cx. pipiens s.s.* and *Cx. quinquefasciatus*. Later, they used genomic subtractive hybridization to identify a region of nucleic acid heterology between the genomes of the two latter species (Crabtree et al. 1997).

Clearly, repetitive DNA sequences are often unique to given species. However, although designing species-specific primers in these regions may be possible, repetitive regions frequently vary more in number than in kind among closely related taxa. Consequently, a greater effort in sequencing and searching is necessary to identify species-specific sequences in repetitive regions. Furthermore, the PCR is so sensitive that if a repetitive region occurred even in a few copies in a nontarget species, the primers would no longer be specific. Since the standard PCR is not quantitative, the numbers of copies in the initial template DNA is not reflected in the amount of product at the end of the reaction.

In the search for species-specific sequences, multigene families are another alternative worth explo-

ration. Members of multigene families are frequently distributed in an array of tandem repeats. Conserved coding regions of family members are separated by noncoding, spacer regions. The ribosomal RNA genes that code for the RNA components of ribosomes are an example of a multigene family with this structure (Fig. 33.5). Histone genes, globin genes, and actin and myosin genes are other examples. Each of these multigene families has conserved coding regions that flank variable, noncoding regions. With multigene families, a conserved primer (e.g., primer "A" in Fig. 33.4) is

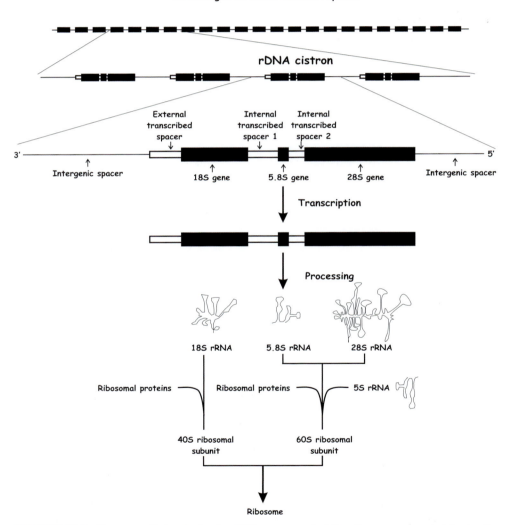

FIGURE 33.5 The genes that code for the RNA component of the ribosome are coded for by the rDNA cistron. Individual rDNA cistrons form members of the rDNA multigene family. These members are distributed as tandem repeats in one or more locations in the genome of an organism. The rDNA cistron alternates between conserved rRNA coding genes, variable internal transcribed spacers, and a highly variable intergenic spacer.

constructed in a conserved area of the coding region, while species-specific primers (reverse primers in Fig. 33.4) can be designed in the more variable non-coding regions.

The rDNA cistron has been used extensively in this manner (Collins and Paskewitz 1996). The rDNA genes occur as 150–1,000 tandem repeats in arthropods. Each repeat contains 18S, 5.8S, and 28S rRNA genes alternating with a large intergenic spacer (IGS) region. Located between the 18S and 5.8S genes is the first internal transcribed spacer (ITS1), and between the 5.8S and the 28S genes is the ITS2. Each member of the rDNA multigene family is connected by the IGS that is 3–5 kb (or greater) in length. Members of this multigene family seem to evolve through concerted evolution and molecular drive (Fig. 33.3). The IGS frequently contains tandem direct or nested repeats that vary in size. While the rRNA coding regions are central to the construction and functioning of the ribosome and conserved within species, the ITS and the IGS regions frequently vary intraspecifically in sequence and length. Following the logic described earlier, the scheme developed in the PCR diagnosis for members of the *An. gambiae* complex uses a universal primer in the 28S gene, and the species-specific reverse primers are located in the IGS region (Collins and Paskewitz 1996). An alternate universal primer has been designed that is located in the 18S gene, with species-specific reverse primers located in the ITS1 or ITS2.

A variant of this method developed in the past decade eliminates the species-specific reverse primers. Primers are developed in conserved regions of the 18S, 5.8S, or 28S genes that encompass and amplify regions of the ITS1 and ITS2. The amplified regions are then digested with restriction enzymes that recognize 4-bp motifs. The resulting restriction fragment length polymorphisms (RFLPs) are often diagnostic for species. In discriminating species of the *Anopheles punctulatus* complex, Beebe and Saul (1995) were the first to apply this technology to vector taxonomy. West et al. (1997) used the RFLP technique to develop a taxonomic key for 13 boreal species in the aedine genus *Ochlerotatus*. Poucher et al. (1999) developed an identification key for the 17 most common *Ixodes* tick species in the United States (Fig. 33.6).

Advantages and Disadvantages of Targeted PCRs

DNA extracted from specimens stored in alcohol or dried usually provides adequate template for PCRs. Primers, or at least primer sequences, can be sent to other laboratories and reaction conditions can be standardized to promote reproducibility across laboratories. The precautions to take when setting up a PCR

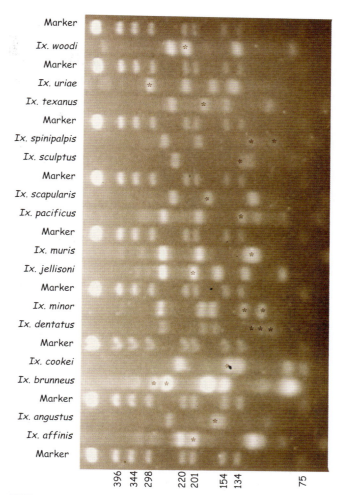

FIGURE 33.6 Application of PCR RFLP for construction of a molecular taxonomic key for identification of 17 *Ixodes* species of the United States. PCR primers amplified a ~900-bp fragment that extended from the 3′ half of the 5.8S gene (5′-YTG CGA RAC TTG GTG TGA AT-3′) to the 5′ end of the 28S gene (5′-TAT GCT AA RTT YAG SGG GT-3′) (Y = Cm, R = A/G; S = C/G). PCR product was double digested with *MspI* and *CfoI* restriction enzymes. Digested fragments were size-fractionated with electrophoresis on a 2.5% metaphor agarose. From Poucher et al. (1999).

laboratory and the techniques used to set up PCRs are easily taught. Concerns about contamination by DNA from distant taxa are minimal because primers are species specific.

The disadvantages of PCR techniques are that the thermal cycler and electrophoresis equipment are fairly expensive. Development of primers requires advance knowledge of complete sequence information in target taxa and closely related species. However, as more sequence information accumulates from diverse taxonomic groups, this is steadily becoming less of an issue. Minute amounts of template DNA can become contaminants by being passed among tubes, and this is an ever-present danger, particularly when several scientists or projects are under way in the same laboratory. A negative control must be associated with each trial, consisting of a tube containing all components of a reaction and handled simultaneously and in an identical manner to all treatment tubes.

Black and Munstermann (1996) described in detail the use of banding patterns resulting from RAPD-PCR, AP-PCR (arbitrarily primed), or SSCP (Chapter 30) analysis of the products of targeted PCR products. Since then, increased application of these methods has revealed that the banding patterns can often be very complex for analysis and difficult to assess for homology among taxa. In contrast, targeted PCRs or PCR-RFLPs produce amplified products where sizes are easily compared on agarose or PAGE gels.

Proteins as Taxonomic Markers

Isozyme electrophoresis was first used as a tool for recognizing species differences in the middle 1960s. The principle of isozyme electrophoresis consists of separating proteins on the basis of electrical charge. In an electric field, proteins of highest charge move fastest toward the pole of opposite charge. In a homogeneous gel matrix, the proteins of identical charge move together at the same rate toward the cathode or anode. Because of their aqueous solubility and intrinsic levels of amino acid variability, enzymes have proven the most serviceable of proteins in taxonomy and population genetics. For a successful separation of the enzyme and visualization on a gel, the following conditions must be met:

1. Enzymes must retain their activity after electrophoresis. Therefore, before the final preparation for electrophoresis, the enzymes must be preserved in the living organism or frozen at temperatures lower than −40°C. Enzymes are generally separated on nondenaturing (native) gels that are chilled during the electrophoretic process; this avoids chemical or heat denaturation, respectively.
2. The buffer, gel pore size, and electrical field must be optimized and standardized to permit the orderly migration and separation of proteins in the gel.
3. Following electrophoresis, the gel contains many soluble insect proteins separated by size and charge. The activity of a particular enzyme is visualized by immersing the gel in a histochemical solution that contains the substrate for that enzyme linked through several chemical reactions to a dye. The location of the enzymatic reaction on the gel is demarcated by precipitation of the dye. Applications are possible only for enzymes for which a his-

tochemical stain has been developed. At present, approximately 350 histochemical stains are available for elucidating specific enzymes (Manchenko 1994).

Protein Electrophoresis

The amino acids that contribute most to an enzyme's charge are those with unbalanced numbers of carboxyl (—COOH) and amino (—NH$_2$) side groups. At a pH greater than the isoelectric point of a protein, the hydrogen ions disassociate from the carboxyl groups, leaving the protein with a net negative charge. In an electric field, these proteins migrate to the cathode. In a mixture of enzymes with a difference of a single carboxyl group (addition or subtraction of an aspartic acid or glutamic acid), the enzymes are readily separated on a gel when subjected to an electric field (Fig. 33.7).

The operational application of electrophoresis involves two steps. The first consists of the mechanics of the process; the second is the interpretation and translation of the banding patterns into numerical data. The mechanics of electrophoresis can be delineated into five stages: gel preparation, specimen preparation, enzyme separation, enzyme visualization with histochemical staining, and gel archiving. Interpretation requires measurement of band movement, interpreting band morphologies into genotypes, estimating allele frequencies, and comparing species by genetic distance values and dendrograms. Procedures for this last step were discussed in Chapter 32.

Gel Preparation

Three types of gels form matrices sufficient for separation of enzymes in electrophoresis: starch, polyacrylamide, and cellulose acetate. The most popular and widely applied has been the starch gel; the electrophoretic apparatus is inexpensive, thick slabs are readily sliced for multiple enzyme assays, and the system is very easy to use. The disadvantage is that starch gels do not store well and, for some enzymes, provide resolution of lower quality. Starch gel electrophoresis procedures are well summarized in Hillis et al. (1996).

Small quantities of enzyme can be visualized on cellulose acetate gels; therefore, this technique is excellent for very small organisms. Although the gels are easy to preserve, they are generally quite small, and resolution of gel phenotypes can be difficult. Kazmer (1991) described a system for analysis of small arthropods using cellulose acetate gel electrophoresis (CAGE) with isoelectric focusing.

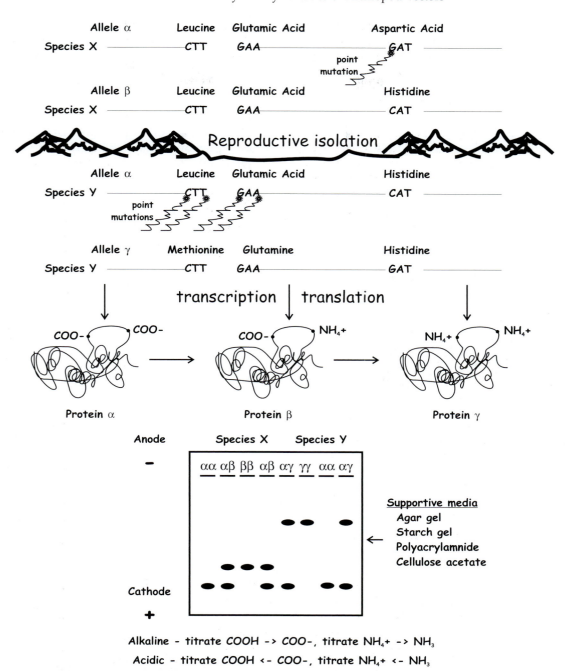

FIGURE 33.7 Genetic polymorphisms revealed by protein electrophoresis on a nondenaturing gel. Point mutations in the gene that cause nonsynonymous substitutions produce proteins with altered amino acid sequences. When these substituted amino acids carry different charges, the mobility of the protein in an electrical field will be altered. Different types of substitutions will be detected under different pH conditions during electrophoresis.

Polyacrylamide gel electrophoresis (PAGE) was among the first systems developed for separating proteins and now is the most common. However, during the 1970s and mid-1980s they became less commonly used because the gels cannot be sliced for multiple enzyme assays on the same sample, the acrylamide has toxic properties, and the electrophoretic apparatus is relatively expensive. More recently, however, commercial PAGE systems are available that allow electrophoresis to be performed on multiple gels simultaneously, and thin gels permit very high resolution and require small sample quantities. Specific details on the pouring, loading, and running of native PAGE gels are provided in Table 33.4; methods for preparation of native PAGE gels for population genetic or taxonomic work are not available in general references.

Histochemical Staining

Histochemical stains for 350 enzymes appear in Manchenko (1994). Although Singh and Rhomberg (1987) reported on genetic variability in 117 enzyme loci in *Drosophila melanogaster*, only 30 or fewer are routinely examined in a typical electrophoretic study. The most useful enzymes have been esterases, glycolytic pathway enzymes, and a handful of kinases, proteases, and transaminases.

Several strategies are used to locate the position of each enzyme in the gel matrix. However, the general scheme is as follows: (1) The gel is placed in a shallow pan or tray with the appropriate histochemical solution. This usually consists of the enzyme substrate and cofactors, such as magnesium, required for enzyme activity. (2) Many enzymes, such as the dehydrogenases, remove a proton that is passed to a cofactor, usually NAD+ or NADP+. (3) The cofactor is reduced *in situ* to NADH or NADPH. (4) To produce a visible band at the site of the enzyme activity, a tetrazolium dye is reduced, in turn, by NADH or NADPH. A colored precipitate is immediately formed at the site of enzyme activity. After a few minutes to a half hour, the gels are fixed in a methanol–acetic acid fixative until ready to hydrate and dry. As an example, an activity stain for phosphoglucomutase, a glycolysis enzyme, is illustrated in Figure 33.8. Gels stained with a variety of different stains are illustrated for the housefly *Musca domestica* and *Aedes albopictus* in Figure 33.9.

Data Quality Control and Storage

Complete records of migration patterns are essential for interpreting results or reestablishing electrophoretic conditions. Every newly initiated study

---

TABLE 33.4   Instructions for Pouring, Loading, and Running Native PAGE Gels

---

Tris-citrate (TC) buffer

To make 1 liter of 40× TC Buffer, dissolve 94.1 g (0.78 M) Tris and 45 g citric acid (monohydrate) into 700 ml distilled water. After the salts are dissolved, titrate the solution to pH 7.1 with 1 M citric acid and bring the entire solution to volume. Dilute this buffer 1:39 with distilled water to make the electrode buffer.

Tris-borate-EDTA (TBE) buffer

To make 1 liter of 10× TBE Buffer, dissolve 98.3 g (0.81 M) Tris, 12.5 g (0.20 M) boric acid, and 5.58 g (15 mM) EDTA (disodium) in distilled water sufficient to bring the volume to 1 liter. The pH will be 8.8–9.0; adjustment of pH will be unnecessary. Dilute this buffer 1:9 with distilled water to make the electrode buffer.

Acrylamide stock

To make 1 liter of acrylamide stock solution, dissolve 380 g acrylamide and 20 g N,N',-bis-acrylamide in 900 ml distilled water. After the salts are dissolved completely (30 to 40 minutes), filter the solution through 3–4 layers of Whatman #1. Bring to 1 liter volume.

To pour 100 ml of polyacrylamide gel solution:

(1) Pour the 12.5 ml of the acrylamide stock solution into a beaker with either 5 ml of the TC concentrate and 82.5 ml distilled water or 10 ml of the TBE concentrate and 77.5 ml distilled water.

(2) Mix briefly and then de-gas the gel solution by applying vacuum for 5 min until bubbles no longer form.

(3) Add 500 µl photo-flo, 150 µl TEMED N,N,N'N'-tetramethyl-ethylenediamine and stir briefly.

(4) Add 500 µl of 10% (w/v) ammonium persulfate and stir briefly. The gels must be poured immediately and teflon slot formers ("combs") placed in position at the top of the gel before polymerization. These slots act as wells into which insect homogenates will be loaded. The gel must set for approximately 1 hr before the combs are removed.

(5) A "pre-run" at 200 V is performed after polymerization to remove catalysts and to stabilize the buffer through the gel.

Specimen preparation

From 0.5–2 mg of tissue are sufficient to resolve enzymes in 12 to 16 or more gels. Single specimens (in the case of sand flies, *Culicoides*, or mosquitoes) are placed in a 25 µl loading buffer 20% sucrose, Triton X-100 (0.5%), Tris-citrate pH 7.0 electrode buffer, and bromophenol blue tracking dye (trace amount). The specimen is homogenized with a pestle fitted to the tube, and the tubes centrifuged in a microfuge (3 min, 12,000× g). The supernatant is removed from the tube with a 25 µl gas chromatography syringe and 1–2 µl placed in one well of each gel. The high density provided by the sucrose solution causes the homogenate to sink to the bottom of the well. Specimens are kept at 0–4°C throughout this preparation to insure enzyme stability.

Electrophoresis

After the specimens are loaded, electrode buffer cooled to 0°C is circulated around the gel plates to maintain an even and cold environment. Constant voltage is applied to suit the desired length of the separation time. Usually around 300 V are sufficient with either buffer system to complete enzyme separation in 2 to 3 hr. The gels are then removed to trays for histochemical staining.

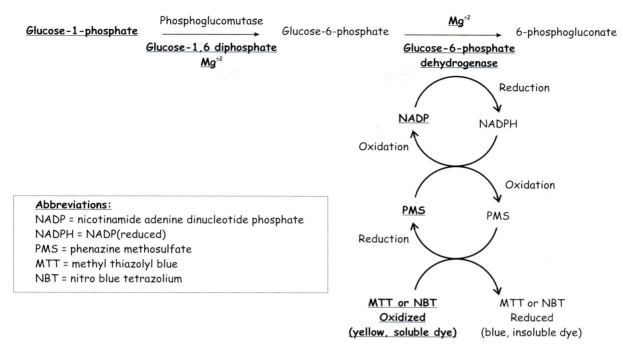

FIGURE 33.8 Steps in the activity stain for phosphoglucomutase. Underlined items in bold are components that are added to the stain mixture. All other items are reaction products. This series of reactions is initiated at the locations on the gel where phosphoglucomutase occurs. The location of the reaction is visualized by the presence of reduced, blue insoluble MTT or NBT dye.

necessarily requires comparisons from many gels. The conditions for each electrophoretic run must be kept constant throughout a study. To control for inevitable day-to-day variation, homogenates from standard laboratory strains are loaded on all gels. The mobility of enzymes of each sample is measured with respect to this reference standard.

The migration of bands can be recorded on data forms for later comparison. However, differences among bands on the same and different gels can be subtle, and a permanent record of the gel is often desirable. This is best accomplished by photographing the gel; reserving gels between dialysis membranes is an inexpensive alternative (Table 33.5). Digital storage by scanning either the fresh or dried gel is a preferred mode. Scanned images have the further capacity of electronic enhancement of low enzyme activity (very light bands) not visible to the human eye on the original gels (Mukhopadhyay et al. 1997, 2001).

Biochemical Taxonomy (for Earlier References see Munstermann 1988)

Biochemically based dichotomous keys for identifying mosquito species are rare. Most studies have compared only a few species in a closely related group, where, in most cases, formal keys are unnecessary. Unfortunately, the deficiency in electrophoretic keys has impeded the recognition and application of electrophoresis as a general identification technique. Moreover, as larger numbers of species are compared, more enzyme assays per individual specimen may become necessary. Fortunately, improved technology has mitigated the latter problem, in terms of both increased sensitivity and cost.

The very first biochemical key for mosquitoes was that of Miles (1979) for five of the species of the *An. gambiae* complex. Miles' keys have proven to be of exceptional value under local conditions and have had more extensive application for *An. gambiae* identifications in Tanzania and the Grand Comoros Island. Using single or several electrophoretic enzyme phenotypes for distinguishing mosquito sibling species has a fairly long history. Among the first enzymes to be used was alkaline phosphatase for separating the sibling species pair *An. labranchiae*–*An. atroparvus*. The sibling species complex in the North American *An. quadrimaculatus s.l.* required a key based on enzymatic characters (Narang et al. 1989). Subsequently, multiple enzyme assays were applied to a species set that included six European species of the *An. maculipennis* complex, three of the *Ae. mariae* complex, and several

A   Phosphoglucomutase (<u>Musca domestica</u>)

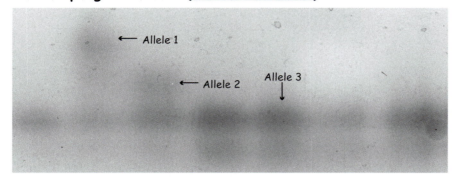

B   Amylase -fast locus (<u>Musca domestica</u>)

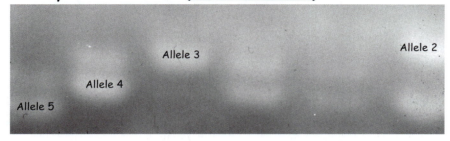

C   Esterase (<u>Aedes albopictus</u>)

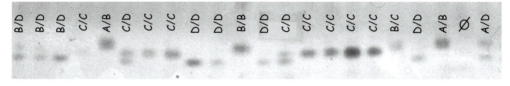

D   Isocitrate dehydrogenasee (<u>Aedes albopictus</u>)

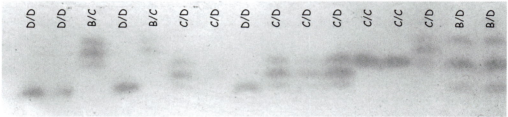

FIGURE 33.9   Photographs of isozyme gels. (A) Phosphoglucomutase allozymes in the housefly (*Musca domestica*). (B) Allozymes at the fast Amylase + locus in *Musca domestica* (allele designations follow Black and Krafsur 1985). (C) Esterase allozymes and genotypes in the *Aedes albopictus*. (D) Isocitrate dehydrogenase allozymes and genotypes in the *Aedes albopictus*. Allele designations follow Black et al. (1988).

other species groups (Bullini and Coluzzi 1982). North American researchers described diagnostic isoenzymes for members of the *Ae. triseriatus* species group, *Ae. atropalpus* species group, *Ae. scutellaris* species group, and local *Culex* species groups. More recently, discovery of new species and redefinition of species relationships based on enzyme variation has been undertaken in several phlebotomine sand fly taxa, including *Phlebotomus papatasi* (Ghosh et al. 1999), *Lutzomyia longipalpis* (Lampo et al. 2000), and *L. shannoni* (Mukhopadhyay et al. 2001). Dujardin et al. (1999) used biochemical markers to differentiate members of the medically important tribe Rhodniini (Hemiptera: Reduviidae). They revealed three main species groups

TABLE 33.5   Storage of PAGE Gels Using Dialysis Membranes

Equipment:

Two 3–6 mm thick Plexiglass frames. Recommended outer dimension is 5 cm wider and taller than the gel. A square is cut from the center of each piece of Plexiglass such that the frame will fit evenly around the entire gel with a minimum of 6–12 mm to spare along all edges of the gel.

Two dialysis membranes cut to the outside dimensions of the frame.

Six black spring steel clamps with chrome handles.

Procedure:

(1) Rinse fixative (1 : 5 : 5 glacial acetic acid, methanol, water) from gels and submerse in deionized water for at least 20 min.
(2) Cut two sheets of dialysis membrane to size of frame, 2 sheets per gel.
(3) Rinse a sheet of dialysis membrane in deionized water. Place over lower frame, smooth to prevent the lodging of bubbles between frame and membrane.
(4) Place trimmed gel gently on wet membrane with no bubbles between gel and membrane.
(5) Make labels on laser printer, 16 pt type font. Label lists at minimum the gel code number and enzyme system. Print directly or photocopy printer output onto transparency sheets.
(6) Place trimmed transparent label at upper left of gel: gel number and enzyme.
(7) Rinse second dialysis sheet, place over gel to edges of frame; remove all bubbles.
(8) Clamp the frames together with the 6 spring steel clamps. Two at each the base and the top and one on each side.
(9) Allow to dry under an incandescent lamp.
(10) Remove the dried gel and membrane. Trim and place in a notebook.

within the genus *Rhodnius*. Monteiro et al. (2002) examined allozyme relationships among 10 species of Rhodniini and demonstrated paraphyly (Chapter 29) of *Rhodnius*, including *Psammolestes*. In all of the studies just cited, formal electrophoretic keys were rarely presented, with the exception of those for nine species of the *An. annulipes* group and four *Aedes* species of the *Communis* species group (Brust and Munstermann 1992).

Advantages and Disadvantages of Protein Electrophoresis

The foregoing studies demonstrate the utility of protein electrophoretic techniques for taxonomic problems involving closely related species. When larger numbers of more distantly related species are studied, electrophoretic methods can have definite advantages over morphological methods of identification under some circumstances. However, correct identifications depend on the extent of prior information available on population structure of each species being identified.

A disadvantage of protein electrophoresis relative to analysis of nucleic acids is that the expression of proteins can differ in amount or in the patterns of expression across life stages. Less enzyme is expressed in earlier life stages (correlated with insect volume), and, occasionally, different forms of the enzyme (isozymes) are expressed. This may require dichotomous keys specific for each stage of development. Perhaps the major disadvantage of using proteins in taxonomic work is that they are very easily degraded and enzyme activity is rapidly lost. Protein instability necessitates either maintaining collected specimens alive and in good condition awaiting return to the laboratory or storage of specimens at low temperatures (−40°C) in the field. In contrast, dried specimens and material preserved in alcohol are frequently used for nucleic acid analyses.

INTRASPECIFIC VARIATION IN MOLECULAR TAXONOMY

Regardless of the chosen technique, good taxonomic practice dictates that the amount of intraspecific information in a character be estimated, described, and incorporated into the taxonomic description of a species. A taxonomic character that appears to be useful for separation of taxa in one geographic region may overlap between taxa in another region. Therefore, when developing molecular markers for sampling populations from diverse geographic and temporal settings, the variation in pattern or sequence among these collections must first be examined. The primary question must be posed: Is the amount of variation within the species less than the variation among the closely related taxa? If so, then the markers or patterns identified are unique to a species. If the answer is no, then too much overlap with closely related groups is present to rely on that marker or pattern. Molecular biologists must recognize that taxonomic classifications are rarely infallible. When working with closely related taxonomic groups, the taxa under examination may not necessarily be separate species. Quite possibly, they have been incorrectly or unnecessarily described as different species in the taxonomic literature. Developing species-diagnostic markers impli-citly involves testing the species status of the taxa compared.

Another consideration when examining closely related taxa is that a taxonomic gradient or cline will be discovered. Species with broader distributions often occur along a latitudinal, longitudinal, climatic, or altitudinal gradient. As discussed earlier, taxonomists sampling populations at ends of geographic gradients can erroneously split a single species into two, even though gene flow occurs freely among populations distributed through the intermediate regions. This pattern can be revealed only through extensive geographic sampling.

In the development of species-specific molecular markers, care must be taken to consider the function and application of the proposed markers, to sample taxa in a geographically dispersed design, and to examine the taxonomic literature. One danger is to develop molecular markers from laboratory colonies or single field collections and then to assume that they are representative of the entire species. These markers, especially if they originate from laboratory colonies, have originated from a limited sample of the actual genetic variability in the field. The sequences already discussed as useful in separating closely related taxa are workable because they mutate quickly. Sequences that mutate quickly also vary intraspecifically. Molecular taxonomists must estimate intraspecific variation when working with closely related taxa. This dictates taking a population genetics approach rather than a strict taxonomic one.

As an example, Wesson et al. (1993) examined gene flow among populations of *Ixodes dammini* and *I. scapularis* using sequence data from the rDNA ITS1 and ITS2 regions. By comparing variability of their markers within and among taxa, they found complete overlap in sequences among these taxa. This supports reclassification of *I. dammini* as a junior synonym of *I. scapularis* (Oliver et al. 1993). However, if they had simply accepted the existing taxonomic status of these and gathered sequence information from single laboratory colonies, they might have developed primers that had no relevance to gene flow in the field. Good molecular markers will also require time, patience, and extensive testing in the field. Scott et al. (1993) reported that their

markers have been tested in a large number of laboratories throughout the range in which members of the *An. gambiae* complex occurs.

CONCLUSION

Good molecular taxonomy, already being practiced in a number of groups of medically important arthropods, has improved the accuracy of identifications in a timely fashion during outbreaks of disease. Coupled with the diagnostic techniques being advanced for detection and identification of pathogens and parasites (Chapter 18), vector molecular taxonomy will permit vector capacity to be analyzed in local populations and even among individuals within populations. This level of resolution is unprecedented in the history of vector-borne disease.

Readings

Audtho, M., Tassanakajon, A., Boonsaeng, V., Tpiankijagum, S., and Panyim, S. 1995. Simple nonradioactive DNA hybridization method for identification of sibling species of *Anopheles dirus* (Diptera: Culicidae) complex. *J. Med. Entomol.* 32: 107–111.

Avise, J. C., and Ball, R. M. J. 1990. Principles of genealogical concordance in species concepts and biological taxonomy. *Oxford Surv. Evol. Biol.* 7: 45–67.

Beebe, N. W., and Saul, A. 1995. Discrimination of all members of the *Anopheles punctulatus* complex by polymerase chain reaction—restriction fragment length polymorphism analysis. *Am. J. Trop. Med. Hyg.* 53: 478–481.

Black, W. C., and DuTeau, N. M. 1997. RAPD-PCR and SSCP Analysis for insect population genetic studies. In *The Molecular Biology of Insect Disease Vectors: A Methods Manual* (J. Crampton, C. B. Beard, and C. Louis, eds.). Chapman and Hall., New York, pp. 361–373.

Black, W. C., and Krafsur, E. S. 1985. Electrophoretic analysis of genetic variability in the house fly (*Musca domestica* L.). *Biochem. Genet.* 23: 193–203.

Black, W. C., and Munstermann, L. M. 1996. Molecular taxonomy and systematics of arthropod vectors. In *Biology of Disease Vectors* (W. C. Marquardt and B. J. Beaty, eds.). University Press of Colorado, Boulder, pp. 438–470.

Black, W. C., Hawley, W. A., Rai, K. S., and Craig, G. B., Jr. 1988. Breeding structure of a colonizing species: *Aedes albopictus* (Skuse) in peninsular Malaysia and Borneo. *Heredity* 61: 439–446.

Brust, R. A., and Munstermann, L. E. 1992. Morphological and genetic characterization of the *Aedes (Ochlerotatus) communis* complex (Diptera: Culicidae) in North America. *Ann. Entomol. Soc. Am.* 85: 1–10.

Bryan, J. H., M.A., D. D., V., P., and M., C. 1982. Inversion polymorphism and incipient speciation in *Anopheles gambiae s. str.* in The Gambia, West Africa. *Genetica* 59: 167–176.

Bullini, L., and Coluzzi, M. 1982. Evolutionary and taxonomic inferences of electrophoretic studies in mosquitoes. In *Recent Developments in the Genetics of Insect Disease Vectors, A Symposium Proceedings* (W. W. M. Steiner, W. J. Tabachnick, K. S. Rai, and S. Narang, eds.). Stipes, Champaign, IL, pp. 465–482.

Clapp, J. P., McKee, R. A., Allen-Williams, L., Hopley, J. G., and Slater, R. J. 1993. Genomic subtractive hybridization to isolate species-specific DNA sequences in insects. *Insect Mol. Biol.* 1: 133–138.

Collins, F. H., and Paskewitz, S. M. 1996. A review of the use of ribosomal DNA (rDNA) to differentiate among cryptic *Anopheles* species. *Insect Mol. Biol.* 5: 1–9.

Coluzzi, M. 1982. Spatial distribution of chromosomal inversions and speciation in Anopheline mosquitoes. In *Mechanisms of Speciation: Proceedings from the International Meeting on Mechanisms of Speciation* (C. Barigozzi, ed.). Alan R. Liss, New York.

Coluzzi, M. 1984. Heterogeneities of the malaria vectorial system in tropical Africa and their significance in malaria epidemiology and control. *Bull. WHO* 62: 107–113.

Coluzzi, M., Sabatini, A., Petrarca, V., and Di Deco, M. A. 1979. Chromosomal differentiation and adaptation to human environments in the *Anopheles gambiae* complex. *Trans. R. Soc. Trop. Med. Hyg.* 73: 483–497.

Cornel, A. J., Porter, C. H., and Collins, F. H. 1996. Polymerase chain reaction species diagnostic assay for *Anopheles quadrimaculatus* cryptic species (Diptera: Culicidae) based on ribosomal DNA ITS2 sequences. *J. Med. Entomol.* 33: 109–116.

Crabtree, M. B., Savage, H. M., and Miller, B. R. 1995. Development of a species-diagnostic polymerase chain reaction assay for the identification of *Culex* vectors of St. Louis encephalitis virus based on interspecies sequence variation in ribosomal DNA spacers. *Am. J. Trop. Med. Hyg.* 53: 105–109.

Crabtree, M. B., Savage, H. M., and Miller, B. R. 1997. Development of a polymerase chain reaction assay for differentiation between *Culex pipiens pipiens* and *Cx. p. quinquefasciatus* (Diptera: Culicidae) in North America based on genomic differences identified by subtractive hybridization. *J. Med. Entomol.* 34: 532–537.

Davidson, G. 1956. Insecticide resistance in *Anopheles gambiae* Giles: A case of simple Mendelian inheritance. *Nature* 178: 861–863.

Dover, G. A., and Tautz, D. 1986. Conservation and divergence in multigene families: Alternatives to selection and drift. *Philos. Trans. R. Soc. Lond. B Biol. Sci.* 312: 275–289.

Dujardin, J. P., Chavez, T., Moreno, J. M., Machane, M., Noireau, F., and Schofield, C. J. 1999. Comparison of isoenzyme electrophoresis and morphometric analysis for phylogenetic reconstruction of the Rhodniini (Hemiptera: Reduviidae: Triatominae). *J. Med. Entomol.* 36: 653–659.

Gale, K. R., and Crampton, J. M. 1987. DNA probes for species identification of mosquitoes in the *Anopheles gambiae* complex. *Med. Vet. Entomol.* 1: 127–136.

Gale, K. R., and Crampton, J. M. 1988. Use of a male-specific DNA probe to distinguish female mosquitoes of the *Anopheles gambiae* species complex. *Med. Vet. Entomol.* 2: 77–79.

Ghosh, K. N., Mukhopadhyay, J. M., Guzman, H., Tesh, R. B., and Munstermann, L. E. 1999. Interspecific hybridization and genetic variability of *Phlebotomus* sand flies. *Med. Vet. Entomol.* 13: 78–88.

Hill, S. M., Urwin, R., and Crampton, J. M. 1992. A simplified, nonradioactive DNA probe protocol for the field identification of insect vector specimens. *Trans. R. Soc. Trop. Med. Hyg.* 86: 213–215.

Hillis, D. M., Mable, B. K., and Moritz, C. 1996. Applications of molecular systematics: The state of the field and a look to the future. In *Molecular Systematics* (D. M. Hillis, C. Moritz, and B. K. Mable, eds.). Sinauer Associates, Sunderland, MA, pp. 515–543.

Kazmer, D. J. 1991. Isoelectric focusing procedures for the analysis of allozyme variation in minute arthropods. *Ann. Entomol. Soc. Am.* 84: 332–339.

Lampo, M., Feliciangeli, M. D., Marquez, L. M., Bastidas, C., and Lau, P. 2000. A possible role of bats as a blood source for the *Leishmania* vector *Lutzomyia longipalpis* (Diptera: Psychodidae). *Am. J. Trop. Med. Hyg.* 62: 718–719.

Lane, R. P., and Crosskey, R. W. 1995. *Medical Insects.* Chapman and Hall, London.

Manchenko, G. P. 1994. *A Handbook of Detection of Enzymes on Electrophoretic Gels.* CRC Press, Boca Raton, FL.

Mayr, E., and Ashlock, P. D. 1991. *Principles of Systematic Zoology*, 2nd ed. McGraw-Hill, New York.

Miles, S. J. 1979. A biochemical key to adult members of the *Anopheles gambiae* group of species (Diptera: Culicidae). *J. Med. Entomol.* 15: 297–299.

Monteiro, F. A., Lazoski, C., Noireau, F., and Sole-Cava, A. M. 2002. Allozyme relationships among 10 species of Rhodniini, showing paraphyly of *Rhodnius* including *Psammolestes. Med. Veterin. Entomol.* 16: 83–90.

Mukhopadhyay, J., Rangel, E. F., Ghosh, K., and Munstermann, L. E. 1997. Patterns of genetic variability in colonized strains of *Lutzomyia longipalpis* (Diptera: Psychodidae) and its consequences. *Am. J. Trop. Med. Hyg.* 57: 216–221.

Mukhopadhyay, J., Ghosh, K., Ferro, C., and Munstermann, L. E. 2001. Distribution of phlebotomine sand fly genotypes (*Lutzomyia shannoni*, Diptera: Psychodidae) across a highly heterogeneous landscape. *J. Med. Entomol.* 38: 260–267.

Munstermann, L. E. 1988. Biochemical systematics of nine nearctic *Aedes* mosquitoes (subgenus *Ochlerotatus*, *annulipes* group B). In *Biosystematics of Hematophagous Insects* (M. W. Service, ed.). Clarendon, Oxford, England, pp. 133–147.

Narang, S. K., Kaiser, P. E., and Seawright, J. A. 1989. Dichotomous electrophoretic taxonomic key for identification of sibling species A, B, and C of the *Anopheles quadrimaculatus* complex (Diptera: Culicidae). *J. Med. Entomol.* 26: 94–99.

Oliver, J. H., Jr., Owsley, M. R., Hutcheson, H. J., James, A. M., Chen, C., Irby, W. S., Dotson, E. M., and McLain, D. K. 1993. Conspecificity of the ticks *Ixodes scapularis* and *I. dammini* (Acari: Ixodidae). *J. Med. Entomol.* 30: 54–63.

Post, R. J., and Crampton, J. M. 1988. The taxonomic use of variation in repetitive DNA sequences in the *Simulium damnosum* complex. In *Biosystematics of Hematophagous Insects* (M. W. Service, ed.). Clarendon, Oxford, England, pp. 133–147.

Post, R. J., and Flook, P. 1992. DNA probes for the identification of members of the *Simulium damnosum* complex (Diptera: Simuliidae). *Med. Vet. Entomol.* 6: 379–384.

Post, R. J., Flook, P. K., and Wilson, P. K. 1992. DNA analysis in relation to insect taxonomy, evolution and identification. In *Insect Molecular Science. 16th Symposium of the Royal Entomological Society of London* (J. M. Crampton and P. Eggleston, eds.). Academic Press, London, pp. 21–34.

Poucher, K. L., Hutcheson, H. J., Keirans, J. E., Durden, L. A., and Black, W. C. 1999. Molecular genetic key for the identification of 17 *Ixodes* species of the United States (Acari: Ixodidae): A methods model. *J. Parasitol.* 85: 623–629.

Scott, J. A., Brogdon, W. G., and Collins, F. H. 1993. Identification of single specimens of the *Anopheles gambiae* complex by the polymerase chain reaction. *Am. J. Trop. Med. Hyg.* 49: 520–529.

Singh, R. S., and Rhomberg, L. R. 1987. A comprehensive study of genic variation in natural populations of *Drosophila melanogaster.* I. Estimates of gene flow from rare alleles. *Genetics* 115: 313–322.

Toma, T., Miyagi, I., Crabtree, M. B., and Miller, B. R. 2000. Identification of *Culex vishnui* subgroup (Diptera: Culicidae) mosquitoes from the Ryukyu Archipelago, Japan: Development of a species-diagnostic polymerase chain reaction assay based on sequence variation in ribosomal DNA spacers. *J. Med. Entomol.* 37: 554–558.

Toure, Y. T., Petrarca, V., Traore, S. F., Coulibaly, A., Maiga, H. M., Sankare, O., Sow, M., Di Deco, M. A., and Coluzzi, M. 1998. The distribution and inversion polymorphism of chromosomally recognized taxa of the *Anopheles gambiae* complex in Mali, West Africa. *Parassitologia* 40: 477–511.

Walton, C., Handley, J. M., Kuvangkadilok, C., Collins, F. H., Harbach, R. E., Baimai, V., and Butlin, R. K. 1999. Identification of five species of the *Anopheles dirus* complex from Thailand, using allele-specific polymerase chain reaction. *Med. Vet. Entomol.* 13: 24–32.

Wesson, D. M., McLain, D. K., Oliver, J. H., Piesman, J., and Collins, F. H. 1993. Investigation of the validity of species status of *Ixodes dammini* (Acari:Ixodidae) using rDNA. *Proc. Natl. Acad. Sci. U.S.A.* 90: 10221–10225.

West, D. F., Payette, T., Mundy, T., and Black, W. C. 1997. Regional molecular genetic key of thirteen snow pool *Aedes* species (Diptera: Culicidae) in northern Colorado. *J. Med. Entomol.* 34: 404–410.

MOLECULAR BIOLOGY OF INSECTS AND ACARINES

ANTHONY A. JAMES

34

A Brief Introduction to Molecular Biology, Genomics, and the Transmission of Vector-Borne Pathogens

ANTHONY A. JAMES

INTRODUCTION

Molecular biology involves, in its broadest definition, the study of the fundamental properties of the macromolecules of living systems. It encompasses research to elucidate their structure and function as well as the processes that regulate their synthesis, location, activation (if necessary), and turnover. While great advances have been made in many areas of molecular biology, clearly the largest impact has come from the study of DNA and the genes that it encodes. As such, most biologists recognize molecular genetics and the study of gene expression as the core areas of research in molecular biology. Now coupled with genomics, the study of whole genomes, molecular biology as a science has led to an unparalleled acceleration in the rate at which we acquire information about living organisms. Many recent review articles and commentaries have raised expectations about what we hope to learn from molecular biology, genomics, and what has been called postgenomics, and it is valid for vector biologists to want to know what this means for them and how this information is going to translate into something that can be used to control vector-borne diseases. It is not possible in this short chapter to address all of the expectations or provide all of the answers. However, this chapter acquaints the nonspecialist with the thrust of the research approaches, the definition of some terms, and a number of the interesting challenges that lay ahead. The discussion begins with a description of how a genetics perspective has influenced molecular biological analyses and how these analyses have changed in response to a vast amount of new knowledge about vector genomes.

PHENOTYPES TO GENES: WHAT YOU SEE IS WHAT YOU GET

Genetics provides a good foundation on which to develop molecular biological analyses. This results mostly from the hypothesis that stipulates that if it is possible to recover mutations that affect a specific molecular biological process, then the process must have a legitimate significance to the organism. This approach creates an interesting challenge because molecular biologists must validate what they see (a phenotype) by identifying the responsible genes. Thus, these analyses start with the premise that phenotypes in organisms are made manifest as a result of the expression of genes. Phenotypes are important to vector biologists because these embody the physical, physiological, and behavioral characteristics that have considerable significance for the ability of an arthropod to serve as a vector. Differences in phenotypes among species are interpreted to result from differences in the numbers, types, organization, and expression patterns of multiple genes. Different phenotypes within a species likely result from allelic differences in one or more genes. Geneticists exploit these phenotypic differences as markers for a particular genetic makeup and then use the differences to characterize specific genes and how they produce the phenotype.

In insects, the fruit fly, *Drosophila melanogaster*, provides a tremendous source of information about the structure and function of genes, and much of the application of modern molecular biology to vector insects has relied on this organism as a model. Indeed, some of the first successes using molecular biological tools to change the phenotypes of vectors such as mosquitoes, depended to a great extent on the function of fruit fly genes and control DNA sequences in their distant relatives. The fruit fly affords a wealth of visible, biochemical, and developmental marker genes that are the envy of every vector biologist. Fortunately, the tools of molecular biology, specifically comparative gene studies, allow the rapid identification of homologous and similar genes from vector species. The lesson here is that a knowledge of *Drosophila* can provide a good perspective on what is potentially possible to achieve in vector species. It is important to keep in mind that not all arthropods are the same and that while it may be possible to find a set of homologous genes in the fruit fly and a vector, experiments that go beyond merely showing DNA sequence similarity are needed to properly assign a function to the vector genes.

The genetic and molecular analysis of some phenotypes is straightforward. These simple phenotypes result from the actions of single or a small number of genes often catalyzing metabolic events and exhibiting minimal effects of modifying genes. For example, mutations in the genes that encode enzymes responsible for synthesizing pigments deposited in the eyes result in mosquitoes with white eyes instead of the normal deep purple color (Fig. 34.1). Mutations in a single gene can be complemented by providing a wild-type copy of the gene, either by crossing it in during mating or by transformation techniques (Chapter 40), and we interpret this complementation to be proof that we have defined the specific character of the gene. Thus, we anticipate that for any phenotype we will be able to find a gene or a set of genes that is responsible for producing the phenotype. However, our ability to discover and characterize those genes depends entirely on how well our phenotype is defined. The example of eye color, that is, the alternate states of "color–no color," seems fairly obvious, but some phenotypes that clearly have a genetic basis, such as a

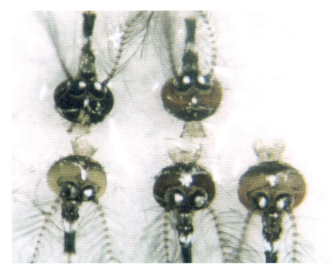

FIGURE 34.1 Simple phenotypes: heads of wild-type, mutant, and transformed *Aedes aegypti*. Mutations in the *kynurenine hydroxylase-white* gene cause the normally wild-type purple eye (upper left) to appear white (lower left). Mosquitoes transformed with a wild-type copy of the *cinnabar* gene of *Drosophila melanogaster* show complementation and eye color (upper right and lower middle and right).

preference for feeding on humans, may turn out to be difficult to define genetically. This phenotype may not turn out to be a simple dichotomy of "bites people–does not bite people" that is linked to one or a small number of genes; therefore, assigning specific phenotypic characteristics to individual genes may not be possible. Thus, we must consider what it means to call something a phenotype and recognize that some characters or behaviors, although possible to describe in words, may not have an easy definition at the genetic and molecular levels of analyses. Such complex phenotypes result from the actions of multiple genes and can manifest differences due to modifying genes and the general genetic background of a particular population.

Difficulties may result from the fact that some phenotypes result from the activity of multiple genes and therefore are polygenic traits. Polygenic traits may require cumbersome genetic methods and make it hard to describe them molecularly. For example, resistance or susceptibility to a specific pathogen is a phenotype of significance for vector biologists. However, when examined closely, we see that there is a potential for multiple parameters to be associated with a resistance phenotype, including the prevalence of infected individuals in the resistant population (*penetrance*), the mean intensity of infection in individual members of the population (*expressivity*), and the mode of action of the resistance mechanism. Phenotypes often are defined arbitrarily in these circumstances, and therefore it is difficult to assign a particular set of characteristics of the phenotype to a specific gene or small number of genes. However, tools are being developed to identify the genes associated with complex phenotypes.

Assuming that a phenotype can be defined sufficiently so that it can be evaluated by genetics, it is then possible to study the genes and gene products responsible for that phenotype. The approaches and tools of biochemistry can provide an entry into subsequent genetic and molecular biology studies. The identification and purification of active proteins and the use of protein sequence determination methods permit the deduction of the primary sequence of the DNA in which they are encoded. Genes responsible for eye color, salivary gland proteins, digestive enzymes, and components of eggs all have been discovered and are described in other chapters in this section. In general, these genes have been recovered from fragments of genomic DNA preserved in various bacteriophage or plasmid libraries, often preceded by or in parallel with the isolation of cDNAs that represent the expressed portions of the genes. These cDNAs can be used to recover and analyze the DNA adjacent to the expressed portions of the genes and define *cis*-acting and other control elements. This type of "molecular morphology" has been practiced successfully for the past 15 years or so in vector biology, with particular successes being achieved in mosquitoes. We have learned that mosquito genes, and the genes of ticks, sand flies, tsetse, and others, are similar to genes discovered in the fruit fly and other organisms. They have coding regions (*exons*) and intervening noncoding regions (*introns*) that are spliced out of the product RNA. They have 5'-end and 3'-end transcribed and untranslated regions and often have 5'-end enhancer-like sequences that confer sex-, tissue-, and stage-specific expression. Why, then, undertake molecular biology studies with a vector species when it would be so much easier to work on fruit flies? Of course, the answer is that vectors do things that fruit flies do not. Most specifically, they feed on blood and as a consequence transmit diseases. Therefore, there are a number of biological questions that can be addressed only by studying a vector species.

Hematophagy as a Phenotype

The process of locating a host and taking and digesting a blood meal for either nutritional or reproductive requirements sets vectors apart from many other arthropods. Clearly, this feature of these organisms involves genetically determined behavior, adaptations to counteract host hemostasis and immunity, and well-regulated digestive and reproductive gene expression (Table 34.1). The most easily described phenotypes associated with hematophagy include the expression of salivary gland proteins that inhibit platelet aggregation, prevent coagulation, and promote vasodilation (Chapter 28). The subsequent digestive phenotype includes the synthesis of a peritrophic matrix and in some organisms the expression of proteases in small amounts that precede the abundant expression of enzymes capable of metabolizing the blood meal (Chapters 21 and 22). This digestion is followed most often by the transport of amino acids to the fat body, where they are used to synthesize yolk and other proteins (Chapters 23, 25, and 37). Problems with metabolizing the large amount of iron present in the blood meal are mitigated by the genes whose products can maintain this metal in a nontoxic form (Chapter 23). All of these processes and their constituent genes are the subject of ongoing molecular biological research.

The more difficult phenotypes associated with hematophagy involve host location, preference, and other aspects of behavior. While it is clear that many mosquitoes feed on one species of vertebrate host more frequently than on others and even appear to prefer one individual within a species over others, the

TABLE 34.1 Examples of Simple and Complex Phenotypes Associated with Hematophagy in Mosquitoes

Phenotype	Simple or Complex*
Salivary gland gene expression:	
Vasoconstriction	Simple
Platelet antiaggregation	Simple
Anticoagulation	Simple
Midgut gene expression:	
Digestive enzymes	Simple
Transport proteins	Simple
Storage proteins	Simple
Regulatory proteins	Complex
Fat-body gene expression:	
Yolk products	Simple
Regulatory genes	Complex
Host seeking	
Orientation	Complex?
Odorant-binding proteins	Simple?
Odorant receptors	Complex?
Host preference	Complex?

* Refers to the number of genes that may be involved in a specific phenotype. Simple phenotypes have one or a few genes associated with a specific aspect of the phenotype; complex phenotypes likely involve many genes and associated modifier genes.

genetic basis for this behavior remains undefined. Efforts are being made to characterize genes that are involved in chemoreception (Chapter 20), but elucidating how the products of these genes result ultimately in behaviors will be a challenge. Other complex phenotypes include mating behavior, oviposition, physiological factors associated with overwintering or surviving dry seasons, and the ability to produce progeny antogenously. It is likely that multiple genes are involved in these phenotypes, and this requires the more complex analyses of quantitative trait loci (QTL) and genomics.

Vector Competence as a Phenotype

A striking observation about arthropod vectors of disease is the specificity of their interactions with pathogens. For example, mosquitoes do not transmit the parasites that cause leishmaniasis, trypanosomiasis, or onchocerciasis. This is observed despite the fact that mosquitoes undoubtedly feed on humans or animals that are infected with these pathogens. Furthermore, culicine mosquitoes are much better vectors of flaviviruses than are anophelines, whereas anophelines alone are responsible for transmitting human malaria. The genetic ability of a vector to support the development and propagation of a specific pathogen has been designated *vector competence*. Variations in vector competence within a vector species have provided the impetus for the meiotic mapping of putative genes responsible for this phenotype. Surprisingly, in some cases only a few genes appear to be involved in the ability of mosquitoes to transmit viral, malarial, or filarial pathogens (Chapters 27 and 30). A comprehensive review by Beerntsen et al. (2000) analyzes many of these data. Molecular biology and genetics offer powerful tools to determine the nature of genes involved in vector competence. Techniques such as QTL mapping and other high-definition mapping strategies (Chapter 31) allow investigators to define more precisely chromosomal loci that encode genes involved in vector competency, but the final identification of a specific gene will most likely depend on genomics and complementation by transformation.

GENES TO PHENOTYPES: WHAT DO YOU SEE WHEN YOU GET IT?

The recent sequencing of the human, *Anopheles gambiae*, and *Plasmodium falciparum* genomes has been accompanied by much enthusiasm in researchers for what lies ahead. Echoing the comments in the introduction of this chapter, it is a fair question to ask what lies ahead. In some vague way, many people feel that the complete sequence of an organism will lead somehow to a full understanding of that organism. If we define an organism as a complete manifestation of its genome in a resulting phenotype, it is fair to say that what lies ahead is a long way off. The intricacies of many seemingly simple genetic processes still elude us, and the ability to understand them in a more holistic or systems biological approach requires the development of sophisticated computational and analytical tools. However, research in genomics has spawned a surge of development of these tools in a number of related disciplines, and they are being incorporated into molecular biological analyses (Table 34.2).

Genomics consists of a palette of techniques that leads to the determination of the primary nucleotide sequence of the total genome of an organism. These techniques are now automated and considerably reduce the time it takes to acquire the sequence. Because of repetitive sequences, heterochromatin, and other features of the DNA, no genome sequence of a metazoan is likely to be "complete." There is likely to be some information absent; however, much can be done with what has been determined. What happens after the sequence has been obtained, postgenomics, is where the most exciting science takes place. This also presents the greatest challenges. Knowing the primary sequence of a gene is only the beginning of determining how it may be involved in a relevant phenotype. A good example of this is the analysis of mosquito

TABLE 34.2   New Disciplines in Molecular Biology and Genetics

Genomics	Study of whole genomes based on knowledge of the primary sequences
Bioinformatics	Tools and procedures to make coherent the primary sequence of a genome
Structural biology	Modeling of the actual structure of biological molecules to gain insights into their function
Proteomics	Study and knowledge of the complete expression profile of a genome
Transgenesis technology	Ability to extract genes from living organisms, manipulate those genes, and reinsert them into living organisms
Functional genomics	Study of phenotypes that result from complex interactions of individual genes

genes involved in olfaction. The sequencing of the genome of *An. gambiae* (and *D. melanogaster*) led to the identification of large families of both putative odorant-binding proteins and presumed olfactory receptors. While it has been shown that many of these genes are expressed in the antennae, an organ associated with olfaction, to date no ligand has been identified that binds to any of the mosquito proteins. There is clearly much work to be done.

Bioinformatics is a new field of science; its objective is to turn the raw sequencing data into useful information (Chapter 47). An essentially one-dimensional string of characters (the primary sequence of nucleotides) is meaningless until it is interpreted and analyzed. This analysis starts with the decoding of this information by identifying genes (coding and noncoding regions) and adds to it data from other research areas to build significance into the primary data. Bioinformatics requires biological, computational, and mathematical skills to practice it in a meaningful way. The utility of a genome project is enhanced significantly by bioinformatic procedures such as annotation (linking of the existing, available biological information about a specific gene or DNA sequence to the actual sequence entry), tools that allow effective searches (data mining), and display programs that enhance intuitive interactions of the investigator with the information database. These intuitive programs allow investigators to discover and analyze significant patterns in the data and lead to biologically relevant modeling of the living system. Ultimately, bioinformatics assists in taking the input sequence information and transforming it to an output that results in a meaningful interpretation of information.

One of the biggest challenges of functional genomics is to come up with ways to link the predictions arising from a bioinformatic analysis (*in silico*, as it has been called) to what is happening in the live organism (*in vivo*). The one-dimensional primary sequence has to integrate in some way into the three-dimensional world of expressed products (after all, this is what the organism does). Two branches of functional genomics, *structural genomics* and *proteomics*, forge some of these links. Proteins are the most conspicuous functional macromolecules. They are the principle molecules in a cell that actually "do" things. Their functional properties may involve a strong structural role as they form significant intra- and extracellular complexes. Alternately, but not exclusively, their main properties may be regulatory or catalytic, true "do-er" molecules. The specific properties of a given protein result from its DNA-encoded primary sequence of amino acids and how that sequence is organized in its secondary (folding on itself), tertiary

(folding in space), and quaternary (association with other proteins) structure. It is easy to appreciate that if it were possible to extrapolate from the primary sequence of DNA what the final functional form of the protein is; then the power of the genomics is enhanced greatly.

Structural biology provides a way to make the connections between primary encoded information and the complex three-dimensional world. Starting with purified proteins, structural biologists have accumulated a wealth of structural information using modern atomic imaging tools and advanced computational methods. Large databases of solved structures of proteins exist, and these have been used to model proteins whose structures are known only by primary sequence. While these predictive models have value, there is still a great need to solve directly the structure of proteins. As computational skills improve, small changes in amino acid sequence likely will reveal subtle differences in the conformation of proteins that could have profound effects on their function. Even if the differences in the functional properties are not great, they may be sufficient to help in designing novel and specific intervention strategies with drugs or genetics. Thus, there is a need for automated processes for determining structures, including expression and production of the target protein, protein crystallization, and collection and analysis of atomic data.

One of the complications of genomics is that individual genes may give rise to multiple expressed products as a result of alternate splicing of exons or posttranslational modification of the protein product. Thus, the *proteome*, the complete collection of translated products, will not match one-to-one with the genome. Indeed, the complexity of the proteome far exceeds the complexity of the genome. New techniques in protein biochemistry and sequence determination permit the direct-determination proteomes. It has been possible for a long time to resolve individual proteins derived from a complex mixture using two-dimensional (2-D) gel electrophoresis. With the deployment of highly sensitive and efficient protein-sequencing procedures and bioinformatics, 2-D gels have experienced a significant rebirth. One of the truly remarkable advances that has been developed utilizes mass spectrometry to measure the molecular weights of partial or full peptides. The specific amino acid composition of the peptide under investigation has a nearly unique molecular weight. Bioinformatic tools can be used to predict what the likely amino acids composition is and then search a database of full protein sequences for short stretches of amino acids that have that composition. Thus, short sequences of amino acids serve as a tag for identifying much larger proteins. This method, coupled with Edman degradation sequencing of peptides, allows the identification of the protein from which the peptides were derived as well as the gene. Indication of whether a gene gives rise to alternate expressed products becomes evident when proteins of different molecular weight are assigned to the same gene.

Automation of procedures such as the resolving of complex mixtures of proteins on 2-D gels, automatic isolation of individual species from the gels, followed by mass determination and data bank screening provide the real possibility to define the proteome. One of the remaining problems is the rapid determination of the presence and distribution of a specific protein *in vivo*. Antibodies are the gold standard for protein identification. They are specific and can be labeled in any number of ways to make them visible. However, the production of individual antibodies that would distinguish uniquely each component of the proteome will take a long time. Therefore, this is a research area where a novel technology could have a major impact.

We have seen how we might benefit from knowing the sequence of the genome as well as from identifying the expression products, but that gets us to the title of this section: We have it, but what do we see? Or more precisely, what does it mean? There are a number of ways to assign, in a meaningful way, a function to a gene or set of genes. A direct genetic approach would be to perform a knockout of the gene. In this approach various techniques, such as homologous recombination, RNA interference (RNAi), and mutations in regulatory genes, can result in the absence of a target gene product in an animal. Investigators predict (although it is more often just a hope) that they will see a phenotype in the knockout animal that is straightforward to interpret. For example, knockout a gene encoding an enzyme responsible for the pigment biosynthetic pathway, and the resulting animal has white eyes. One of the more confounding phenotypes is lethality. We all would be pleased if the gene we are studying had this phenotype because we would interpret this result as support for the conclusion that the gene must be important; however, investigators are now faced with the challenge of determining how and why a mutation in this gene results in death. It becomes important to know as much as possible about the expression properties of the gene in order to assign significance to the phenotype.

One of the properties of genes that should be determined is the developmental expression profile. In these studies, detection of mRNA is often used as a surrogate marker for gene expression. The techniques used include Northern blot analyses and gene amplification procedures. Such techniques can be used to determine sex-, stage-, and tissue-specific expression patterns. Once the expression profile has been elucidated, it is possible to compare that profile in a normal (wild-type) animal with what is seen in the mutant strain and determine when and where the divergence from normal expression occurred. This may help define the primary site and time at which the genetic lesion resulted in the lethal phenotype.

Analysis of the mRNA expression profile may not be sufficient to define the true expression profile of a given gene. mRNAs are translated into proteins, and the genetic lesion may be manifest in some posttranslational modification or processing of the proteins. Specific antibodies will allow immunoblot analyses to determine if changes have occurred in the electrophoretic properties of a protein. Immunolocalization studies of cells or whole tissues can reveal changes in protein behavior at the cellular level.

Even with all of the afore-described studies completed, it may still be difficult to conclude unequivocally why an alteration in a specific gene and gene product resulted in a particular phenotype. This may be because there are multiple and complementary genes that compensate partially or wholly for the loss of a single gene. Furthermore, regions of DNA flanking the gene may contain enhancer or other regulatory sequences that influence its expression, and these effects may not be evident in expression profile analyses. Studies of genes near heterochromatin provide evidence that the location of a gene or a chromosome may affect its expression properties. Thus, it could be possible to have a wealth of sequence and expression information about a gene and its products and still not be able to assign to it a meaningful phenotype.

Functional genomics provide a way to extract information about gene functions even if the specific phenotype associated with a gene is unclear. Recently, scientists have been able to determine that one type of resistance-to-malaria phenotype of *An. gambiae* is associated with elevated expression of genes involved in oxidative stress. This is a particularly interesting example because it shows that the overall physiological state of the insect (something that cannot be attributed to the action of a single or small number of genes) is potentially responsible for the phenotype. It is possible that a variation or mutation in one or a small number of regulatory genes is responsible for the phenotype, but the resistance phenotype still has to be defined as the function of multiple genes. We anticipate seeing more solutions to complex phenotypes using this type of approach.

THE LEGACY OF GENOTPES AND PHENOTYPES

Some of the most interesting phenotypes in vectors leave us wondering about the origin of the genes that are responsible for them. We want to know where these genes originated and how they have changed to serve a new role. Molecular biology has provided powerful tools for evaluating the evolutionary relationships among vector arthropods as well as providing insights into vector-specific phenotypes. Although still in its early stages, scientists have been able to use molecular diagnostic DNA markers and comparative genomics to sort out relationships among and within species. Differences in biting preferences (anthrophilicity) and vector competence are phenotypes that are evident among vector species (Chapter 30). However, exciting data are being accumulated that reveal the complexity of vector species and populations, all arising presumably from genetic differences that are amenable potentially to molecular and genetic analyses.

FINAL REMARKS

As stated in the introduction, I have aspired to acquaint the reader with some of the approaches being used in molecular biology study of vectors. While there is much to be discovered at a basic science level, the ultimate goal is to use these technologies to mitigate the incredible burden of vector-borne diseases. Genetic control strategies are being researched (Chapters 40 and 46), and it is likely that useful control methods will come out of this work.

Significant challenges remain. From a genomics perspective, one of the greatest challenges is going to be how to devise meaningful experiments that utilize the information from humans, mosquitoes, and pathogens. What types of experiments can be done that use the information from all of these organisms?

Another challenge will be to develop procedures in structural biology and modeling that will identify truly new targets for intervention. A new approach to finding drugs for treating malaria is an example of this. Are there yet-undiscovered biochemical pathways or hierarchies of gene expression that provide sites of attack against the vector? The integration of nongenomic data from ecological, behavioral, and other such disciplines with the genomic data is a challenge to those using bioinformatics. Intuitively, there should be much to discover should these data be integrated effectively. The outcome remains to be seen.

Many vectors and the diseases they transmit are endemic to developing countries, and it is expected that disease control specialists from these countries will benefit from the molecular and genomic advances. However, while the bionomic knowledge of the disease pathogens and their vectors is represented in disease-endemic countries, the technical knowledge and expertise of the molecular tools are often not there. Therefore, another challenge is to educate scientists with strong biological training in the tools of molecular biology and genomics. This coupled with global access to these tools should manifest a positive impact on disease control and prevention.

Readings

Beerntsen, B., James, A. A., and Christensen, B. 2000. Genetics of mosquito vector competence. *Microbiol. Molec. Biol. Rev.* 64: 115–137.

Hardy, J. L., Houk, E. J., Kramer, L. D., and Reeves, W. C. 1983. Intrinsic factors affecting vector competence of mosquitoes for arboviruses. *Annu. Rev. Entomol.* 28: 229–262.

Hoffman, et al. 2002. *Plasmodium*, human and *Anopheles* genomics and malaria. *Nature* 415: 702–709.

Kafatos, F. C. 2001. The future of genomics. *Molec. Aspects Med.* 22: 101–111.

Kafatos, F. C. 2002. A revolutionary landscape: The restructuring of biology and its convergence with medicine. *J. Molec. Biol.* 319: 861–867.

The October 4 (2002) issue of *Science* reporting the sequencing of the *Anopheles gambiae* genome. *Science* 298.

35

Cultured Cells as a Tool for Analysis of Gene Expression

ANN M. FALLON AND TIMOTHY J. KURTTI

INTRODUCTION

Low numbers of progeny and the relatively long life span of eukaryotes, compared to most bacteria, have limited genetic analysis of these complex organisms. In the past century, these limitations have been addressed in part by development of methods for in vitro propagation of somatic cells derived from particular tissues. Most of the approaches used today in insect cell culture have their origins in mammalian cell culture, with adaptations, such as the formulation of media based on the composition of hemolymph, specific to the system at hand. This chapter focuses on the culture of mosquito and tick cells. Basic information on the techniques used in animal cell culture can be found in Freshney (1994).

Early progress in mammalian cell culture was facilitated by naturally occurring, cancerous tumors, whose constituent cells are refractory to the factors that limit normal growth and cell division. For example, the well-known HeLa cell line was established from a cervical carcinoma. Over the years, it was gradually recognized that overgrowth of fibroblasts in mammalian cell culture obscured the presence of slower-growing cell types. Today, a wide range of vertebrate cell lines representing specific cell types is available. Coupled with the techniques and approaches known as *somatic cell genetics*, vertebrate cell lines have provided important insights into diverse molecular processes that cannot be studied easily in the intact organism.

In contrast to most vertebrates, insects typically have short life spans and produce many progeny. Experimental manipulation of insects, however, often is difficult due to their small size and the difficulty of obtaining gram quantities of dissected tissues. Insect cell culture provides a source of genetically homogeneous material that can be produced in sufficient quantity to offset the small size of many medically or agriculturally important species. For example, lepidopteran cells have been used extensively to elucidate molecular aspects of baculovirus replication, from the perspectives of both viral genes and host cell responses. Over the past decade, cell lines from *Drosophila melanogaster* and *Spodoptera frugiperda* have been exploited commercially for the development of expression systems for recombinant proteins that require posttranslational modification. These advances have supported increased interest in the metabolic processes of insect cells, relative to vertebrate cells.

CELL LINES DERIVED FROM MOSQUITOES AND TICKS

Arthropod cell lines are isolated from developmental stages that contain mitotically active cells. Cell lines from mosquitoes and ticks are typically established from embryos, while many lepidopteran lines have been established from pupal ovaries or imaginal discs. In contrast to the situation with vertebrate cell lines, which can often be identified as to tissue of origin, most insect cell lines currently in use for analysis of gene expression are of unknown tissue origin. In particular, we note that techniques for the isolation of mosquito and tick cell lines representing specific types of cells from the organism have not been developed. Likewise, the extent to which insect cells differentiate in vitro is largely unexplored.

Current interest in the potential manipulation of physiological processes in transgenic and para-transgenic arthropods as a means to prevent disease transmission to humans has stimulated the recent development of new cell lines, particularly from vector species. Aside from their advantages for expressing eukaryotic proteins, insect cell lines have been shown to display responses characteristic of the innate immune response after appropriate stimulation. This property has facilitated the purification and analysis of peptides and proteins produced by immune-activated cells, using protein chemistry as well as standard molecular techniques. Cell lines that produce immune factors often are loosely designated as hemocyte-like cell lines, but precise assignment of the tissue of origin of most insect cell lines awaits development of appropriate molecular markers.

In this chapter, we describe general approaches to the use of mosquito and tick cell cultures in vector biology, giving particular attention to procedures used commonly in our respective laboratories. Mosquito cell lines have been available since the late 1960s and have been used extensively in arbovirus research. The first tick cell lines were reported in 1975, and their applications are only more recently realized. We therefore have considerably more experience and published information with mosquito cells than with tick cells. In the following text, we compare and contrast properties of mosquito and tick cell lines, with the intention of providing the reader a feel for the diversity of culture conditions and the various practical uses of the cells. We encourage investigators who plan to incorporate cell culture into their research approach to spend some time in a laboratory where vertebrate or invertebrate cell culture is used routinely.

CELLS HAVE A LIFE

As you work with cells in culture, the single most important factor to keep in mind is that the cells are living entities. As such, they require certain conditions for optimal growth and viability. Because the condition of the cells, metabolically speaking, is often reflected in their appearance, the easiest way to monitor cultured cells is close observation under an inverted microscope. With practice, the investigator

will develop a knowledge of the cells and their behavior that will be invaluable in more refined applications, such as transfection and induction of the immune response. In short, the investigator will be able to assess visually whether cells are in good condition. This skill, acquired over a period of months, will amply reward the investigator, because cells that are not in an appropriate metabolic state may not cooperate when asked to perform experimentally. Parameters that can be assessed visually include presence of dividing cells, an increase in cell layer confluency with time, ability of cells to remain attached (in the case of an adherent line), and overall confluency.

A common problem is encountered when attempting to subculture a cell population that has become confluent and is no longer doubling at an appreciable rate. Growing cells secrete various metabolites, such as lactic acid, that cause a drop in the pH of the culture medium (as indicated by the pH indicator, phenol red). Conversely, an accumulation of dead or dying cells will make some media alkaline. With C7-10 mosquito cells, for example, failure to double during the first day after seeding suggests that the cells may be old and/or stressed. Sometimes a failure to grow can be remedied by simply refeeding the cells with fresh medium. Occasionally, a cell culture that fails to grow has become contaminated with bacteria, yeast, or fungi, which may be difficult to detect at the early stages. In this case, the experiment will most likely need to be aborted, for contaminants typically grow more quickly than arthropod cells. A particularly insidious problem with arthropod cell culture is the persistent infection with arboviruses, which can easily be overlooked because arthropod cells typically fail to show a cytopathic effect. If the lab has ready access to a vertebrate cell line, transfer of filtered medium to a susceptible vertebrate culture can assess the presence of arbovirus.

Since the early 1960s, when the first insect cell culture was established by T. D. C. Grace, a large number of cell lines have been established, and many of these lines are available from the laboratories in which they are in use. We emphasize that a means of storing cell stocks in liquid nitrogen is advisable. There are several reasons for doing so, including possible genetic change in the cells with continued maintenance in culture, the loss of cells due to contamination or toxic medium, cross-contamination of cultures with other lines, and lapses in grant funding and/or personnel. When stock cultures are kept in liquid nitrogen, they can be recovered after many years of storage. Looking toward the future, establishment of a central repository for vector arthropod cell lines may be of some benefit to the medical entomology community at large. In lieu of this, storing your cells in your own lab and in the lab of a colleague (using different liquid nitrogen sources) provides a simple but important means of recovering lines that may be inadvertently lost or contaminated.

Cell Lines and Culture Media

The mosquito cell lines commonly used in the Fallon laboratory are the C7-10 *Aedes albopictus* cell line (originally cultured by K. R. P. Singh), the Aag-2 *Aedes aegypti* cell line (originally cultured by J. Peleg), and the ASE-IV line from *Anopheles stephensi*, isolated by T. J. Kurtti. The C7-10 cell line is a clonal population that was isolated in the laboratory of Dr. Victor Stollar at the University of Medicine and Dentistry of New Jersey, while the Aag-2 line was designated as such after adaptation to Eagle's medium. Because the immediate ancestors of the C7-10 cells were the first mosquito cells to be used to obtain somatic cell mutants (Mento and Stollar 1978) and C7-10 cells were used to optimize transfection protocols (Fallon 1997), most of the work in the Fallon lab makes use of these cells. Relative to Aag-2 and ASE-IV cells, the C7-10 cells also have desirable physical properties, such as uniform growth in monolayers and a short doubling time of 18 h.

Mosquito cells can be cultured in a variety of media, including that which has been formulated for mammalian cells. We maintain mosquito cells in Eagle's medium supplemented with nonessential amino acids, glutamine, penicillin and streptomycin, and fetal bovine serum (FBS) at a final concentration of 5%. Although other sources of serum sometimes support mosquito cell growth, we have noted considerable lot-to-lot variation. Although we routinely treat FBS at 56°C to inactivate complement, we have not systematically examined whether this precaution is necessary. Sample bottles of serum can be screened for compatibility with cells before purchase and can be obtained sometimes at no cost from the distributor.

The Eagle's medium that we use with mosquito cells is buffered with bicarbonate, and our cells are grown in a 5% CO_2 atmosphere to maintain the pH of the medium at approximately 7. Use of Eagle's medium is not essential, but it does facilitate direct comparisons with vertebrate cells, which are typically grown in bicarbonate-buffered media. The pH of the culture medium can be monitored visually by noting the color of phenol red added as an indicator. Mosquito cells can withstand short periods at higher pH, which routinely occur when cells are manipulated in a laminar flow hood. To stabilize the pH during prolonged maintenance outside the incubator (for example, to measure transport across the cell membrane or to expose cells to an ultraviolet light source), we supplement the medium with HEPES at a final concentration of 10 mM. (We make a 1 M stock of the "free acid" form of HEPES, sterilize it by filtration, and add it directly to medium without adjusting the pH of the HEPES stock. Sodium or potassium salts of HEPES will be basic when dissolved and therefore are inappropriate for this application.)

Tick cells are cultured in Leibovitz's L15 medium or modifications of L15. This basal medium, like Eagle's medium, was originally formulated for the culture of vertebrate cells. In contrast to Eagle's medium, L15 was designed for the cultivation of mammalian cells in free gas exchange with normal atmosphere. The buffer system is based on the use of high amino acid levels rather than the CO_2–$NaHCO_3$ buffer system. Thus, this medium allows the culture of cells without the need for a CO_2 incubator. Basal L15 has been supplemented with FBS (10–20%) and tryptose phosphate broth (10%). The composition of tick hemolymph and nutritional requirements of insect and tick cells have been used to formulate a modification of L15, L15B (Munderloh and Kurtti 1989). This modified L15 allows the growth of tick lines in a medium supplemented with low levels of FBS (2–5%) and tryptose phosphate broth (0–5%). Tick cells can also be maintained (without growth) for several weeks in L15B without supplemental protein. A medium supplement that is useful for the culture of tick cells is a lipoprotein–cholesterol concentrate from bovine serum. A mixture of lipoprotein–cholesterol, phospholipids and fatty acids, this supplement is generally used to enhance the growth of animal cells in serum-free or serum-reduced media.

One final note about the composition of media for cell culture—the levels of exogenous precursors will affect the efficiency of radiolabeling. For example, to detect rare proteins, it may be desirable to carry out labeling procedures in a modified medium in which the corresponding nonlabeled precursor (amino acid) is present in reduced concentration or is even absent. Conversely, in some analyses, the cells may incorporate isotope too quickly to maintain linearity in the process under investigation. To rectify this situation, the medium must be supplemented with unlabeled precursor such that the ratio of unlabeled to labeled compound allows linear uptake of label over the desired time period. For example, in studies that involve incorporation of [³H]thymidine during the mosquito cell cycle, we typically add unlabeled thymidine at a concentration of 1 μM to Eagle's medium containing 5% fetal bovine serum. Labeled amino acids as well as nucleosides (but not bases and nucleotides) are readily taken up by cells in culture. With respect to

DNA and RNA precursors, it is useful to remember that cells prefer the precursor that contains sugar but lacks the acidic phosphates. The negatively charged nucleotides, which are used in many biochemical reactions, are not taken up by living cells but may be used, for example, with permeabilized nuclei.

Cells in Culture Undergo Changes

Life in culture medium places cells under continual selective pressure. Needless to say, only those cells that divide in the artificial environment of a culture medium survive to constitute a cell line. Many cells in adult insects do not divide but maintain their DNA as polytene chromosomes or polyploid nuclei. It is not clear whether such cells can become mitotic and generate cell lines when placed in culture. Although mosquito cell lines are relatively easy to establish from embryos and lepidopteran cell lines from embryos or imaginal discs, other desirable lines have been refractory to growth in culture. A recent description of procedures for establishment of insect cell lines is found in Lynn (2001).

Ideally, cells that are in use for experiments should be maintained in the exponential phase of growth by routine dilution into fresh medium. Lab jargon referring to this procedure includes the terms *splitting*, *subculturing*, *seeding*, or *passaging* the cells. Similarly, seeding cells into a tissue culture dish is often referred to as *plating*, and with adherent cells the percentage of the substrate that is covered with cells is a measure of confluency. When cells are subcultured, a proportion of the original population is transferred to a new flask containing fresh medium. Over a period of time, the flask will become crowded with progeny cells, requiring further subculture to maintain the cell line. Cells that are not subcultured in a timely fashion (monitor confluency of attached cells or turbidity in flasks with floating cells) will eventually die as nutrients from the medium are exhausted and waste products accumulate.

The essential process of subculturing provides a simple example of the selection that occurs during maintenance of a cell line in vitro. For example, by modifying the way cells are subcultured, one can "select" for cells that adhere more or less tightly to a plastic or glass substrate. If "floating" cells are routinely discarded before subculture, for example, over time you will select for the subpopulation of cells that adhere more firmly to the substrate. Treatment with drugs, hormones, or transfection agents can also affect adherence of cultured cells, and this should be monitored by direct observation, particularly if the subse-

quent processing procedures are designed for attached cells. Aside from the selection that is inevitable in tissue culture, cells can be mutagenized and specific mutant cells recovered using the procedures of somatic cell genetics described later.

Considerable patience may be required for establishment of a new cell line. In the early development of cell cultures from *D. melanogaster* embryos, it took approximately 3 years for the first continuous line to become established (Echalier and Ohanessian 1970). Early after a cell line is established, subculturing will be infrequent (monthly) and the dilution will be relatively low (i.e., 1:1 cells and fresh medium). As cells adapt to growth in culture, their population doubling time typically decreases, and they can be diluted more extensively for routine subculture. For example, we subculture the mosquito cells (which have doubling times of 18–36 h) using a 20-fold dilution. Tick cells grow more slowly than mosquito cells and have population doubling times of 3–5 days. Thus, we subculture them using 5- or 10-fold dilutions. Some lines can be diluted more, but several weeks of intermittent feeding (replacement of the culture medium with fresh medium) are needed before the flask will be crowded with progeny cells. It must be appreciated that the ability of the cells to continue growth in cultures plated with low numbers of cells is essential to the recovery of clones and studies in somatic cell genetics. Currently, clonal analyses of the type that have been carried out with the *Ae. albopictus* C7-10 mosquito cell line have not been extended to other mosquito cell lines or to any tick cell lines.

Cells that are plated at a relatively low density, such as 1×10^5 cells/mL, undergo a characteristic increase in cell number that can be represented by the curve shown in Figure 35.1. Immediately after plating, there is a slight lag in cell growth, roughly equivalent to a population doubling time. During this interval, the cell number may actually decrease slightly, relative to the number of cells plated. The lag phase is followed by a period of rapid (exponential) growth over the next several days. For most experiments, it is important to use cells that are in the exponential phase (between the paired arrows in Fig. 35.1). During exponential growth, the cells are growing and dividing in the presence of excess nutrients, and crowding is minimal. The physiology of such cells will support metabolic processes at optimal levels and will provide the most reproducible data when results are compared from several individual experiments. For example, cells that have stopped growing may not express a transfected gene at optimal levels; thus, it would not be valid to compare their expression levels with those of cells in

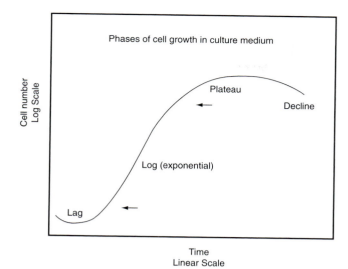

FIGURE 35.1 Typical growth curve for cultured cells. Note that the cell number increases exponentially and therefore is plotted on a log scale.

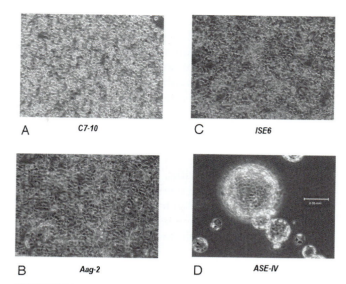

FIGURE 35.2 Morphology of various mosquito and tick cells. (A) C7-10 (*Aedes albopictus*) cells; (B) Aag-2 (*Aedes aegypti*) cells; (C) ISE6 (*Ixodes scapularis*) cells; (D) ASE-IV (*Anopheles stephensi*) cells. All panels were photographed under the same magnification. A scale bar is shown in panel D.

exponential growth. The population doubling time is calculated from the rate of growth during the exponential phase and can be used as a rough measure to estimate when cells plated at a given density will be ready for harvest. Finally, as the cells become crowded, they divide more slowly and enter the so-called plateau, or stationary, phase. The C7-10 cells do not stop growing abruptly; rather, they tend to form clumps, which detach from the monolayer and float in the medium. As nutrients are depleted, cells that are not subcultured die in 2–3 weeks.

Arthropod cell lines show considerable variation in their physical properties (Fig. 35.2). Typically, *Ae. albopictus* C7-10 cells grow as monolayers and start to form clumps only when they become crowded (Fig. 35.2A). These cells can be transferred from flasks directly into suspension culture, where they grow as cell clusters rather than as isolated individuals. The *Ae. aegypti* Aag-2 cells tend to be more adherent than C7-10 cells, and they grow in clusters, with a central group of cells surrounded by fewer cells and intervening spaces (Fig. 35.2B). Both Aag-2 and ISE6 cells from the tick, *Ixodes scapularis* (Fig. 35.2C), tend to be elongated relative to the more rounded C7-10 cells. The *An. stephensi* ASE-IV cells form large spheres that do not attach to the substrate (Fig. 35.2D). Because these cells adhere tightly to each other and do not adhere to culture dishes, we have not used them extensively for experimental investigation. Furthermore, the size of ASE-IV cell clusters varies substantially, and it can be

difficult to produce a series of plates, each containing an equivalent number of cells. Based on their ability to form hollow vesicles, the ASE-IV cells are considered to be epithelial cells. Although the cells cannot be dissociated and counted electronically, cell growth can be monitored by protein content. Alternatively, nuclei from lysed cells can be counted using a hemocytometer.

For many applications, such as following metabolic processes using radiolabeled precursors, cells that grow as monolayers on the surface of a plastic tissue culture dish are the most convenient. If the cells adhere tightly enough, protein synthesis, for example, can be monitored by adding a radioactive amino acid to the culture medium. After an appropriate incubation period, the medium can be removed by aspiration, and the cells (still adhering to the plate) can be washed gently with PBS. Finally, trichloroacetic acid can be added directly to the cells on the plate to precipitate protein, and radioactivity can be recovered by lysis in sodium hydroxide. If the cells do not adhere well to the plate, then procedures based on centrifugation will be required for washing the cells and recovering protein.

Tick cell lines offer a comparable in vitro system for the analysis of gene expression and somatic cell genetics of ticks. There are several lines from the vectors of tick-borne pathogens. These have been used to study tick-borne protozoa, bacteria, and viruses, but molecular genetic studies on the cells are rare (Kurtti et al. 1988). Most tick lines originate from primary cultures seeded with tissues from several hundred embryos. Primary cultures are incubated for several (6–12) months before cell proliferation is apparent. Cells with a variety of morphologies can proliferate, and these have been described, based on their appearance, as hemocyte-like, fibroblast-like, epithelial-like, or neuronal-like. Although we know that a variety of cell types can be grown in vitro, we still need to develop protocols for the selective culture of lines of specific cell types and the reagents that would identify tissue(s) of origin.

Tick and mosquito cell lines are often diploid, and karyology can be used to confirm the identity of a line with respect to species. Isozymes, such as lactate dehydrogenase, malate dehydrogenase, and malic enzyme, are also helpful to confirm cell line identity when these are compared with those from the arthropod species from which the cell line was isolated. Likewise, the small heat shock proteins have been used to confirm the species of origin of the C7-10 and Aag-2 cell lines as *Ae. albopictus* and *Ae. aegypti*, respectively. Cross-contamination of cultures can occur when several lines are maintained in the same laboratory. This can be minimized by keeping separate media for each line and by handling the slower-growing lines first when two or more lines are used simultaneously.

The Question of "Transformation"

Our understanding of cell growth in culture is based on extensive study of mammalian cell lines, which progress through several distinct stages during prolonged maintenance in culture. Although the boundaries at which these stages occur are somewhat arbitrary when applied to a cell population, they serve as a starting point for identifying differences between vertebrate and invertebrate cell lines that may reflect fundamental differences in response to as-yet-unknown growth factors.

The primary culture used to initiate a mammalian cell line derives directly from a tissue or tumor after treatment with enzymes or physical maceration. Typically, only a subset of the cells thus obtained will be capable of cell division. The normal cells that arise from a primary culture exhibit the properties of contact inhibition, serum dependence, a normal diploid karyotype, and a limited life span. Technically, these "normal" populations are designated as cell strains rather than cell lines. Cell strains eventually die out after a certain number of divisions (on the order of 20–50), which depends on the species and age of the

animal from which the cells were taken. First described by Leonard Hayflick (1965), the point at which the cell strain ceases to divide is sometimes called the *crisis period*.

Although many cells die during the crisis period, some cells in the culture may have undergone changes that allow them to divide indefinitely, and these will continue to produce progeny cells. At this point, the population is technically designated as a permanent or established cell line. Such immortalized cells are said to have undergone a process called *transformation*, which is characterized by a suite of phenotypic changes, including loss of contact inhibition. In the context of cell culture, transformation implies a loss of normal growth control similar to what occurs in cancerous tissues. Permanent cell lines can be produced directly from neoplastic tumors, or they can arise as variants of a cell strain during repeated subculture. In addition to the loss of contact inhibition, permanent cell lines typically are subtetraploid, have reduced serum requirements, and can be grown in suspension culture.

In the case of invertebrate cells, an equivalent progression from cell strain to permanent cell line has not been described for any species, nor is it clear whether invertebrate cells that divide indefinitely can be considered transformed. As an example, like transformed cells, the mosquito C7-10 cell line is capable of indefinite growth and can be grown in suspension culture; unlike transformed cells, its requirement for serum does not decrease, and the C7-10 cells retain a diploid karyotype. Insect cell cultures are routinely called cell lines, as opposed to cell strains.

Perhaps the closest invertebrate analogies to transformation have been observed in *D. melanogaster*, in which malignant, intermediate, and benign neoplasms have been described. We will not discuss these lines in detail, other than to note that some of these phenotypes seem to arise from specific mutations, as described in detail by Gateff (1978). Although most of these lines have received little attention, we note that the tumorous blood cell line, mbn-2, has been particularly valuable in studies on *Drosophila* innate immunity (Fallon and Sun 2001).

GENETICS IN A CULTURE DISH

The extensive studies that have been done with mammalian cells in culture have generated a suite of technological protocols collectively designated *somatic cell genetics*. These approaches are based on the observation that cells in culture can be mutagenized and variants recovered by adding selective agents to the medium. Ideally, it is possible to recover from a mutagenized population clones with altered phenotypes, each derived from a single cell. A classic example of this application is the selection of cells resistant to the thymidine analogue, 5-bromodeoxyuridine (BrdU), which acts both as a mutagen and as a selective agent. Cells can survive in the presence of BrdU only if they do not incorporate this analogue into DNA. TK(–) cells, which are deficient in the enzyme thymidine kinase (TK), for example, would be BrdU-resistant. As such, TK(–) mutants also were important in developing early transfection protocols, wherein the thymidine kinase gene from herpes simplex virus was reintroduced into TK(–) mammalian cells, and clones that acquired the transfected TK gene were selected with HAT medium, which contains hypoxanthine, aminopterin, and thymidine. Another classic mutation that has been extensively studied in mammalian cell lines is resistance to the guanine analogue, 8-azaguanine. Typically, this phenotype is due to mutation in the gene encoding hypoxanthine guanine phosphoribosyl transferase (HGPRT). Because this gene is X-linked, and therefore hemizygous in mammalian cells, recovery of this recessive mutant is relatively easy. We note that metabolic processes in arthropod cells are not necessarily similar to what has been described in mammalian cells. For example, the C7-10 mosquito cell line is naturally deficient in HGPRT and cannot incorporate radiolabeled hypoxanthine from the culture medium (Fallon 1996).

There are only a few examples of somatic cell mutants that have been derived directly from mutagenized populations of arthropod cells. In a series of seminal papers, Mento and Stollar described an approach for mutagenizing mosquito cells with ethylmethane sulfonate (EMS) and selecting lines resistant to BrdU (deficient in thymidine kinase activity), ouabain (which targets a sodium-potassium ATPase in the cell membrane), and α-amanitin (an inhibitor of RNA polymerase II). More recently, we have obtained lines resistant to the protein synthesis inhibitors cycloheximide and puromycin and to the metabolic inhibitors methotrexate and hydroxyurea. In the case of *D. melanogaster*, in which a wide variety of mutations are available, cell lines that express mutant phenotypes also can be obtained directly from mutant individuals.

So You Want to Get a Mutant Cell

In the following paragraphs, we will describe in generic terms a process by which mutant cells can be obtained from arthropod cells in culture; specific details are included in Mento and Stollar (1978), and

Generation of Somatic Cell Mutants

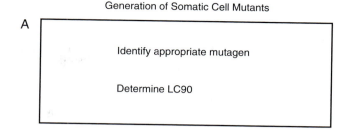

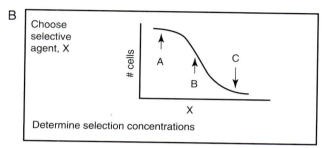

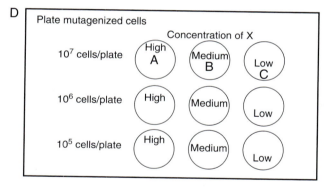

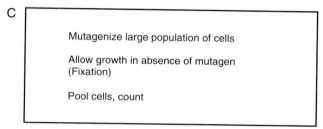

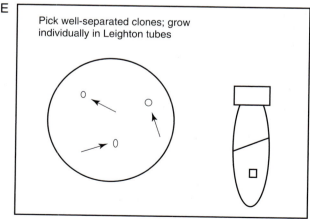

FIGURE 35.3 Schematic outline of the procedure for obtaining somatic cell mutants. Details can be found in the text.

the procedure is outlined in Figure 35.3. Preliminary considerations include a choice among mutagens and their mode(s) of action. Mosquito cells have been mutagenized with EMS, which causes point mutations. The cell population is treated with mutagen at a sufficient dose to kill roughly 90% of the cells (Fig. 35.3A). One also needs baseline information on the effect of the selective agent, compound "X", on the cells. To obtain this information, we seed cells at a standard concentration (2×10^5 cells in 2 mL of medium in 35-mm culture dishes) in the presence of various concentrations of X. Control plates contain solvent, without X. We monitor the growth of the cells until the control plates become confluent. At this time, cell numbers are determined for all of the plates, and growth as a function of X is plotted (Fig. 35.3B). Depending on the nature of X, a log scale may be needed to represent the concentrations tested. From this curve, an LC50 (lethal concentration at which 50% of the cells die) can be calculated. The LC50 should not vary by more than twofold between experiments. Cells that are not in exponential growth at the time of plating will produce higher levels of variability. For example, if "old" cells are plated, the lag time will be prolonged, and the toxicity of X will be overestimated. Because toxicity is a function of both cell density and drug concentration, when the cells are maintained in drug too long (after controls reach confluency), toxicity will be underestimated, because the treated cells will continue to grow, relative to control cells that have entered the stationary phase.

Plating of Mutant Cells and Selection

Once conditions for mutagenesis and sensitivity to the selective agent have been established, it is time to generate mutant cells and plate them out under selective conditions. After mutagenesis (Fig. 35.3C), cells should be allowed to grow as a population in the absence of mutagen for three to five doublings, to allow "fixation" of mutations. As mentioned earlier, growth of the desired mutant cell in the presence of X will be a function of cell density and drug concentration. In addition, the mutagenized population of cells will be stressed, and a considerable number of microscopically intact cells may not in fact be viable. To maximize the chance of obtaining the desired mutant cell, we prepare a total of nine 100-mm plates, using three cell densities (10^7, 10^6, and 10^5 cells/plate) at each of three concentrations of X (A, B, and C in Figs. 35.3B, D). The object of this exercise is to have at least one plate that yields well-separated clones, each of which presumably originated from a single cell. Particular care and attention to sterile technique need to be

observed when handling these cells, because appearance of clones takes 2–3 weeks. Once clones are visible to the eye, their location can be marked on the plate and the accumulation of cells observed under the microscope. When approximately 200 cells have accumulated, we remove the medium (taking care that the clones do not dry out; consider temporarily turning off the air blower in your laminar flow hood). The plate is tipped at an angle, and clones are recovered by touching them to a sterile 2-mm^2 filter paper, moistened in medium and handled with sterile forceps. To prevent residual cells from a disturbed clone from contaminating another clone, clones are picked from the bottom to the top of the dish.

The clones are transferred to individual Leighton tubes (or disposable, flat-sided culture tubes) containing 1.5 mL of culture medium and allowed to generate a population of cells. This step may take several weeks; the first appearance of growing cells is usually close to the filter paper. It is strongly recommended that one pick several clones, for not all of them will generate populations. Once a population is obtained, it is tested for its LC50 in the presence of X. Clones may be used directly, subcloned (to ensure an origin from a single cell), or even reselected for higher levels of resistance. To recover genes that become amplified under selective conditions, we have not found it necessary to start with mutagenized cells. Methotrexate-resistant mutant cells from which the dihydrofolate reductase gene has been cloned have been described in an earlier review (Fallon 1996).

Is Your Mutant Allele Dominant or Recessive?

A second important technique of somatic cell genetics is cell fusion, which in mammalian cells was used in early attempts to map genes to chromosomes. Although fusion sometimes occurs spontaneously, its frequency is low. Rates of cell fusion can be increased by treatment with UV-killed viral agents or by chemical agents that facilitate recovery of somatic cell hybrids at a reasonable frequency. A familiar application involving somatic cell fusion is the now-routine production of monoclonal antibodies, wherein a normal, activated antibody-producing B-cell is fused with a myeloma cell. The continuous cell line thus generated has the immortal-growth properties of the myeloma cell, and it produces the specific antibody characteristic of an individual B-cell.

Among the few examples of somatic cell fusion that have been described using arthropod cells is an analysis of cytopathic effect in *Ae. albopictus* mosquito cells by Stollar and coworkers. The basis for this work was the recovery of mosquito cell clones that differed in their response to infection with Sindbis virus (Tatem and Stollar 1986). Some clones showed cytopathology (CPE+), while others did not (CPE–). First, two mutant cell lines with different dominant phenotypes were derived. The CPE(+) cells were mutagenized and selected for resistance to ouabain; likewise, the CPE(–) line was selected for resistance to α-amanitin. After selection, the karyotype of these mutant cells was checked to ensure that they had retained the normal diploid karyotype with six chromosomes. Populations of these two cell clones were fused using polyethylene glycol, and cells doubly resistant to ouabain and α-amanitin were selected. As expected, the hybrid cells contained 12 chromosomes, while each parental cell clone had a diploid chromosome complement of 6. When tested with Sindbis virus, the hybrid cells were CPE(+), suggesting that this phenotype is due to the presence (rather than the absence) of a hypothetical gene product. Although the particular molecule responsible for CPE has not been identified, the recent advances in molecular biology and protein chemistry now make this a feasible undertaking.

ANALYSIS OF GENE EXPRESSION IN VECTOR-BORNE PATHOGENS

Mosquito- and tick-transmitted pathogens each live in two different worlds. They are able to adapt to their poikilothermic vectors and homeothermic vertebrate hosts. This capability involves differential gene expression and, in some cases, pronounced structural and functional changes in the pathogen as it shuttles between hosts having major physiological differences. The search for factors involved with pathogen transmission and infectivity has focused on identifying those genes whose expression is up- or down-regulated during the blood meal. Mosquito and tick cell lines can be used as model systems to study the regulation of these genes. In such studies, in vitro cultured pathogens are subjected to environmental (physicochemical) shifts that mimic those that occur in the vector during the blood meal. Vector-borne pathogens usually respond to simple environmental cues. These include changes in temperature, osmotic pressure, and pH. The effect of temperature on the expression of outer surface proteins by the Lyme disease spirochete, *Borrelia burgdorferi*, has been studied in tick cell culture systems. In some cases, basement membrane-like macromolecules can serve as cues for parasite development (e.g., the avian malaria parasite, *Plasmodium gallinaceum*). The effects of other physiological parameters that are known to change during the blood meal, e.g., pH, are largely untested. While cell lines offer the advantages of simplification and experimental manipulation, the events observed in a model system must be correlated with what takes place in vivo.

The range of microbes cultivated and analyzed in invertebrate cell cultures includes viral (e.g., arboviruses), prokaryotic (e.g., rickettsiae, ehrlichiae), and eukaryotic (e.g., malarial and babesial parasites) microorganisms. In addition, it is important to keep in mind that mosquitoes and ticks may harbor microbes (symbionts) that appear to reside solely within the vector. Several common considerations are involved in the in vitro isolation and use of cultured invertebrate cells for the analysis of gene expression by vector-borne pathogens. These include the preparation of the pathogen for inoculation into a cell culture, the selection of an appropriate cell culture system, and the confirmation of microbial identity and purity.

Preparation of Microbes for Cocultivation

The choice of inoculum varies with the pathogen, its availability, and its freedom from possible microbial contaminants. Infected tissues from the vector or the vertebrate host are most commonly used. Microbes transmitted vertically during the blood meal can be conveniently harvested from whole blood collected aseptically from an infected host. Methods are available for the selective purification of these microbes from blood. These commonly involve the concentration of the parasite or infected cells from whole blood by differential and/or gradient centrifugation. Malarial and babesial gametocytes and ookinetes can be prepared and separated from whole blood prior to their inoculation into cell cultures. Rickettsial parasites such as *Anaplasma marginale* in erythrocytes and *Anaplasma phagocytophilium* in granulocytes have also been prepared for cocultivation by this method. Transovarially transmitted microorganisms can be isolated from ovarian or developing embryonic tissues. In this case, infected tissues from surface-disinfected females or eggs are used. Commonly used disinfectants include aqueous solutions of household bleach, detergents (e.g., 0.05% benzalkonium chloride), and ethanol (70%). Embryonic cell lines chronically infected with rickettsiae (e.g., *Rickettsia peacockii*), which are transovarially transmitted, have been isolated from infected embryos that were used to initiate the primary cultures. The isolation of microbial parasites from gut tissues of mosquitoes is more problematic because of gut microorganisms and necessitates the use of mosquitoes reared axenically. The infection of gnotobiotic mosquitoes by membrane feeding has been used to obtain anopheline gut tissues infected with gut stages of the malarial parasite (ookinetes) for cocultivation with mosquito cells. In some cases the parasite can be conveniently transferred from an infected invertebrate or vertebrate cell line into an uninfected culture. For example, mammalian cells infected with anaplasma parasites (e.g., HL60 cells infected with *Anaplasma phagocytophilium*) can be used to inoculate tick cell cultures (e.g., ISE6 cell line of *Ixodes scapularis*), and vice versa. Such studies have been done to examine antigenic changes that occur in the organism during its transition from one host to the other.

Selection of the Cell Line or Culture System

If available, it would appear to be desirable to use a cell line from a vector known to be involved with the transmission cycle of the microbe. However, a "homologous" line may not always be extant or available, and lines from a related vector have given meaningful results. In addition, homologous vector cell lines may vary greatly in their ability to support microbial growth and development. Thus, the phenotype of the cell line (e.g., its ability to mount an innate immune response) may be more important than the species identity of the line. For example, lines from the tick *I. scapularis* vary greatly as substrates for the cocultivation of the Lyme disease spirochete, *Borrelia burgdorferi*. Some *I. scapularis* lines phagocytose and eliminate all of the spirochetes within 5–7 days. In contrast, other lines supported growth of mammalian infectious spirochetes for many transfers. If a microbe is to be continuously maintained in a cell culture system, attention to medium composition is needed. For example, the basal medium must be supplemented with N-acetyl glucosamine for spirochete growth. Rickettsial and bacterial microbes should be maintained on antibiotic-free medium. When microbial growth results in the destruction of the cell layer, it is important that the microbes be transferred to a culture of uninfected cells at an appropriate time, usually before the complete collapse of the cell layer.

Confirmation of Microbial Identity and Purity

When there is evidence that the pathogen has been successfully introduced into the culture system, you will need to confirm its identity and establish its purity. Vectors and cell cultures can be coinfected with more than one kind of microbe. Most notably, viruses can inadvertently be introduced with the inoculum, regardless of its origin, from the vector, mammalian host, or another cell culture. In addition, microbial contaminants can be introduced via medium supplements such as serum. Care must be taken to establish that the pattern of gene expression, and the biological features of an infected culture are attributable to the host cell and specific microbe under study and not to a contaminant.

The identity of the microbe in culture can be confirmed by several methods. Light or electron microscopic confirmation is imperative to establish a morphological correlation and to evaluate host cell–microbe relationships (e.g., the obligate intracellular nature of a parasite such as a rickettsia). Centrifugation of infected cells onto microscope slides followed by fixation and staining of the cells with appropriate dyes and/or antibodies works well. With prokaryotic microbes, PCRs using diagnostic primer sets can be used, such as those that amplify the ribosomal genes (or portions thereof). The amplification products can be sequenced either by direct PCR sequencing or after the amplicons have been cloned in *E. coli*. The latter method has the advantage that it can be used to indicate the purity of the microbial population in culture. Individual clones should give you the same DNA sequence.

Gene Expression in Transfected Cells

The process by which exogenous DNA is introduced into cultured cells is called *transfection*. Transfection produces "transformed" cells, but it is important to note that these cells have not undergone the neoplastic transformation discussed earlier in the context of establishment of a permanent mammalian cell line. A wide variety of transfection reagents have become available commercially, and the reagent currently used in the Fallon lab is CellFECTIN, from Invitrogen Life Technologies, Carlsbad, CA. Because methods for transfection of mosquito cells have been reviewed recently (Fallon 1997), here we will focus briefly on the more recent progress that has been made using antisense approaches.

Antisense gene expression provides a means of disrupting expression of a particular cloned gene. In recent studies, we have shown that transient expression of an antisense 20-hydroxyecdysone receptor (EcR) cDNA in C7-10 cells affects growth and survival. In addition, we were able to recover a second isoform of the EcR receptor from cells stably transfected with an antisense EcR. Although our use of antisense and related RNAi approaches with mosquito cells is still in its early stages, we anticipate that these techniques will provide a powerful tool for uncovering alternative pathways by which mosquito cells carry out specific metabolic functions.

Finally, cultured cells provide an important resource for analysis of candidate promoters, using standard methods of deletion analysis, electrophoretic mobility shift assays, and related tools. As we begin to explore genes that play fundamental roles in the basic biology of the cell, transfected cells will provide a tool for identification of redundant metabolic pathways that may enhance our understanding of vector arthropods themselves and host–pathogen interactions in these vectors.

PROSPECTS FOR THE FUTURE

Cell culture provides a largely unexploited tool for investigating both the basic metabolic processes of vector cells and the molecular interactions between arthropod hosts and the pathogens they transmit. In this review, we focused on what has been accomplished with mosquito and tick cell lines. Although tick cells typically grow more slowly and are more challenging to maintain relative to mosquito cells, we anticipate that approaches that have already been applied to mosquito cells will provide important comparative information on these diverse classes of vectors and the pathogens they transmit.

The approaches based on somatic cell genetics of mammalian cells, coupled with more recent transfection and cocultivation technologies, provide important tools for analysis of host–pathogen interactions. Information gained from these approaches is likely to provide new insights into how vector–pathogen interactions might be disrupted by novel chemical methods or in transgenic arthropods. As an example, we have shown that mosquito cell lines produce a variety of

immunity proteins, including defensins, cecropins, transferrin, and lysozyme (Fallon and Sun 2001; Chapter 27). We anticipate that the cell lines will continue to be useful as we explore the expression of genes encoding these proteins and the interactions among the proteins themselves in the context of the immune response.

Readings

Echalier, G., and Ohanessian, A. 1970. In vitro culture of *Drosophila melanogaster* embryonic cells. *In Vitro* 6: 162–172.

Fallon, A. M. 1996. Transgenic insect cells: Mosquito cell mutants and the dihydrofolate reductase gene. *Cytotechnology* 20: 23–31.

Fallon, A. M. 1997. Transfection of cultured mosquito cells. In: *The Molecular Biology of Insect Disease Vectors, A Methods Manual* (J. M. Crampton, C. B. Beard, and C. Lewis, eds.), Chapman and Hall, New York, pp. 430–443.

Fallon, A. M., and Sun, D. 2001. Exploration of mosquito immunity using cells in culture. *Insect Biochem. Molec. Biol.* 31: 263–278.

Freshney, R. I. 1994. *Culture of Animal Cells*. Wiley-Liss, New York.

Gateff, E. 1978. The genetics and epigenetics of neoplasms in *Drosophila*. *Biol. Rev.* 53: 123–168.

Hayflick, L. 1965. The limited in vitro lifetime of human diploid cell strains. *Exp. Cell Res.* 37: 614–636.

Kurtti, T. J., Munderloh, U. G., and Ahlstrand, G. G. 1988. Tick tissue and cell culture in vector research. *Adv. Dis. Vector Res.* 5: 87–109.

Lynn, D. E. 2001. Novel techniques to establish new insect cell lines. *In Vitro Cell. Dev. Biol. Animal* 37: 319–321.

Mento, S. J., and Stollar, V. 1978. Isolation and partial characterization of drug-resistant *Aedes albopictus* cells. *Somatic Cell Genet.* 4: 179–191.

Munderloh, U. G., and Kurtti, T. J. 1989. Formulation of medium for tick cell culture. *Exp. Appl. Acarol.* 7: 219–229.

Munderloh, U. G., Liu, Y., Chen, C., and Kurtti, T. J. 1994. Establishment, maintenance and description of cell lines from the tick *Ixodes scapularis J. Parasitol.* 80: 533–543.

Sang, J. H. 1981. *Drosophila* cells and cell lines. In: *Advances in Cell Culture 1* (K. Maramorosch, ed.), pp. 125–182.

Tatem, J., and Stollar, V. 1986. Dominance of the CPE(+) phenotype in hybrid *Aedes albopictus* cells infected with Sindbis virus. *Virus Res.* 5: 121–130.

36

Genomics and Gene Expression in Vectors

JUN ISOE, FERNANDO G. NORIEGA, AND MICHAEL A. WELLS

INTRODUCTION

Hematophagous insects use blood to obtain the nutrients required for survival and egg production. The process of blood feeding activates gene expression in the midgut, the site of blood meal protein digestion and absorption; in the fat body, where the egg protein, vitellogenin, is produced; and in the ovary, where eggs are produced. There is considerable interest in understanding the mechanisms whereby blood feeding induces gene expression and how the processes are controlled. These unique mechanisms, once understood, might offer new approaches for controlling vector insects and the transmission of disease.

In order to make the topic manageable within the space available, we focus our discussion on mosquitoes and use examples related closely to our research interests. Within this context, we focus on several ways to study gene expression. We describe approaches that depend on (1) the functional analysis of gene expression, (2) genomic analysis of gene expression, and (3) analysis of regulatory regions in mosquitoes.

FUNCTIONAL ANALYSIS OF GENE EXPRESSION

General Considerations

Gene expression during the life of the adult female mosquito is intimately related to two physiological processes: blood meal digestion and oogenesis. In anautogenous mosquitoes, such as *Aedes aegypti*, there are sets of genes that are expressed before and after each blood feeding. This gene expression is under the strict control of a hormonal cascade, the initiation or termination of which requires a blood meal. At emergence, the female *Ae. aegypti* mosquito is not competent to carry out either of these two physiological processes. She needs to complete the maturation of tissues, such as the midgut, fat body, and ovaries, before a blood meal can be taken and digested and eggs can be developed.

Before blood feeding, juvenile hormone (JH) plays a major role in the regulation of gene expression. Juvenile hormonal signals the completion of ecdysis to the adult stage and initiates reproductive processes. After a blood meal is taken, the JH levels drop dramatically and nutritional and different hormonal signals (such as ecdysone) become important as regulators of gene expression.

Regulation of the Expression of Protease Genes in Mosquitoes

Midgut protease genes are excellent models for studying gene expression. Among these genes, we find examples of stage, tissue, and sex specificity, constitutive and inducible expression, and transcriptional and translational regulation. Most of the protease genes are expressed abundantly and therefore are easy to study, an important feature when working with small organisms. *Aedes aegypti* and *An. gambiae* trypsin genes are among the most studied mosquito genes and will be used as examples to illustrate studies based on the functional analysis of gene expression.

In all the mosquito species studied to date, there is a biphasic expression of protease genes following blood feeding. Ingestion of the meal activates the first phase of protein synthesis, and small amounts of particular proteases (early proteases) are released into the midgut lumen. The enzymatic activity of these early proteases is essential for activating the transcription of the second group of proteases (late proteases) that are ultimately responsible for the digestion of blood meal proteins. Different strategies have evolved in different species of mosquito to implement this biphasic regulation. In some mosquitoes, the early proteases are stored as zymogens, while in others translational or transcriptional control are found. In most cases, regulation of late proteases is transcriptional.

Regulation of Expression of Trypsin Genes in *Ae. aegypti*

The midgut of female *Ae. aegypti* synthesizes two main trypsin forms following a blood meal (Fig. 36.1). *Early trypsin* is produced in nanogram amounts, appears in the midgut within 1h of the blood meal, and disappears by 6–8h after the blood meal. *Late trypsin* is produced in microgram amounts, begins to appear 8–10h after the blood meal, and accounts for most of the endoproteolytic activity present in the midgut during blood meal digestion. Early trypsin activity is an essential part of the signal transduction pathway that activates the transcription of the late trypsin gene. The addition of soybean trypsin inhibitor (STI) to a protein meal prevents transcriptional activation of the late trypsin gene, protein digestion, and egg development. This inhibitory effect can be overcome

by feeding a protein meal that has been digested partially ex vivo with bovine trypsin before adding STI. Somehow, the products released during predigestion mimic the activity of early trypsin in the mosquito midgut and are able to restore transcriptional activation of the late trypsin gene to control levels. The mechanism by which the activity of early trypsin in the lumen of the midgut is connected to regulation of late trypsin gene expression has proved difficult to elucidate. It is not a specific peptide or amino acid produced by the action of early trypsin, yet the activation of late trypsin transcription requires early trypsin enzymatic activity. Resolution of this enigma is an important goal in understanding the regulation of blood meal protein digestion.

Regulation of Early Trypsin Transcription and Translation

Transcription of the early trypsin gene is part of the normal postemergence maturation of the midgut in the adult female and is controlled by JH levels. The effect of JH on the transcription of early trypsin was studied using abdominal ligation, which involves the isolation of the midgut from the endocrine gland, the corpora allata (CA), that produces JH (Fig. 36.2). The separated abdomens then can be supplemented with hormone to study its effect on transcription. The levels of early trypsin mRNA in control, nonligated, insects and in ligated abdomens treated or not with JH were compared by Northern blot hybridization.

Although it has been shown that JH regulates the transcription of the early trypsin gene in a stage-dependent, dose-dependent, and "head-independent" manner, we do not yet know whether JH directly affects early trypsin gene expression or whether some other factor, produced in response to JH, mediates the effect.

In the unfed midgut there is neither active early trypsin nor early trypsin zymogen. Thus, although the early trypsin mRNA is present, it is not translated. After the first week of adult life, if females do not ingest a blood meal, there is a slow decrease in the steady-state level of early trypsin mRNA, but it remains readily detectable for up to a month after emergence. Feeding per se or filling of the midgut is not enough to stimulate early trypsin translation, because meals containing saline, latex beads, or sugar solutions that fill the midgut do not stimulate early trypsin protein synthesis. Several proteins of variable molecular weight and different amino acid sequences are able to induce early trypsin synthesis; therefore, we can exclude the possibility that the presence of a

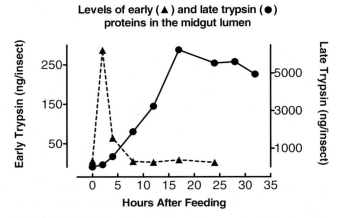

FIGURE 36.1 Mass of early (▲) and late trypsin (●) in the adult female *Aedes aegypti* following a blood meal.

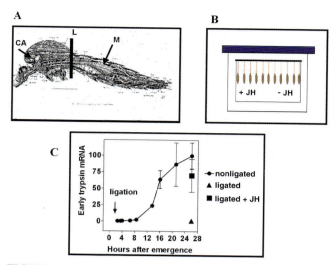

A

B

C

FIGURE 36.2 Effect of juvenile hormone (JH) on early trypsin transcription. (A) Midguts (M) were isolated from the corpora allata (CA) by abdominal ligation (L). (B) abdomens were incubated for 24 h in a wet chamber. Juvenile hormone (–JH) was added topically to the abdomens. Controls (–JH) were topically applied with acetone. (C) Nonligated insects (●) show the normal time course of early trypsin mRNA appearance in the midgut following emergence. Ligated animals (▲) failed to produce detectable amounts of early trypsin mRNA. Ligated insects treated with JH (■) produced nearly normal levels of early trypsin mRNA.

specific peptide derived from a blood protein is essential for induction. Feeding meals containing free amino acids or intrathoracic injection of amino acid mixtures induced early trypsin translation. These data suggest that the size of the free amino acid pool in the midgut somehow regulates early trypsin synthesis. Indeed, when ^{35}S-labeled amino acids are added to a protein meal, the labeled precursors are rapidly absorbed and incorporated into newly synthesized early trypsin. We currently favor the hypothesis that the level of amino acid charging of tRNAs regulates early trypsin translation. This hypothesis is attractive, considering that an increase in the size of the midgut amino acid pool is sufficient to activate synthesis of early trypsin.

Regulation of Late Trypsin Transcription and Translation

Several different proteins, e.g., albumin and gamma globulin, when fed individually or in mixtures to mosquitoes are able to induce late trypsin synthesis, but meals containing only amino acids, saline, or agarose are poor inducers of transcription. In addition, there are some proteins that fail to induce late trypsin transcription, e.g., collagen. Thus, transcription of the late

trypsin gene is dependent on both the quality and quantity of protein in the meal. The changes in the steady-state levels of late trypsin mRNA during the first 24 h postfeeding are also directly proportional to the concentration of protein in the meal.

As already indicated, the identities of the "activators" released by the action of early trypsin that promote late trypsin transcription are still unknown. In addition, the signaling system that transduces the lumenal signal into a cellular signal within the midgut cell is unknown, although preliminary data suggest that a cAMP-dependent process is involved. Also unknown are the regulatory elements within the late trypsin gene that control expression.

Regulation of Expression of Additional Digestive Proteases in *Aedes aegypti*

Less is known about the regulation of expression of other midgut proteases. A cDNA for a midgut chymotrypsin/elastase induced by blood feeding has been described. The chymotrypsin mRNA, absent in larvae, pupae, males, and newly emerged females, reaches detectable levels within 24 h postemergence and attains a maximum level 3–7 days after emergence. In this regard it resembles the early trypsin mRNA. Translation of the chymotrypsin mRNA is induced by feeding a protein meal, again like early trypsin; but in contrast to early trypsin there is a large increase in midgut chymotrypsin enzymatic activity 20–24 h after feeding.

A gut-specific *Ae. aegypti* carboxypeptidase A gene has been cloned and characterized. It appears to be regulated like late trypsin in that its mRNA accumulates to high levels only ~16–24 h after ingestion of a blood meal. An aminopeptidase with a similar late pattern of expression has also been found. It is tempting to speculate that there exists a group of genes whose transcription is regulated by a set of common regulatory elements. Additional trypsins, chymotrypsins, aminopeptidases, and carboxypeptidases have been identified in *Ae. aegypti* by cDNA or gene sequencing, but further research is required to define their patterns of expression and regulation.

Regulation of Expression of Digestive Proteases in *Anopheles gambiae*

The *An. gambiae* genome contains seven trypsin genes that are clustered together in the genome within 11 kilobases (kb) of DNA. The expression of two of the genes, *Antryp1* and *Antryp2*, is induced by blood feeding and is presumed to play a role in blood meal protein digestion. The role of the other genes, *Antryp3, -4, -5, -6, -7*, remains to be elucidated. *Antryp3, -4, -5, -6, -7* genes are expressed constitutively, because their mRNAs can be detected in unfed female mosquitoes. After blood feeding, the amount of these mRNAs drops to undetectable levels, and they do not reappear until the end of the gonadotrophic cycle. The blood meal–inducible trypsin, *Antryp1*, also is expressed constitutively at a low level in the gut of male and female pupae and adult female mosquitoes. Using *Antryp4*-specific antibodies, it was shown that this trypsin is synthesized and stored as a zymogen in the midgut epithelium of unfed females. Secretion and activation of this trypsin was shown to occur in the midgut lumen immediately after fluid ingestion and independent of the protein content of the meal. This might be the *An. gambiae* analogue of the *Ae. aegypti* early trypsin.

Regulation of Expression of Additional Digestive Proteases in *Anopheles gambiae* Females

The *An. gambiae* genome also contains two chymotrypsin genes, *Anchym1* and *Anchym2*, that are clustered in tandem within 6 kb of genomic DNA. Blood feeding induces the transcription of both *Anchym1* and *Anchym2*. An additional chymotrypsin-like serine protease gene has been described in *An. gambiae*. In this case expression is down-regulated after blood ingestion.

Expression of a carboxypeptidase gene has been described in the midgut of *An. gambiae*. The mRNA is found in the pupae and sugar-fed adult female mosquitoes, and its expression is induced approximately 10-fold within 3 h following a blood meal. By 24 h after a blood meal, mRNA abundance returns to a level close to that present before a blood meal.

Summary of Functional Analysis of Gene Expression

In this section we have described approaches based on studying either the expression of a specific mRNA or a specific protein or both. Applying this approach to trypsins in both *Ae. aegypti* and *An. gambiae* has established that blood feeding induces a complex array of regulatory processes, including both translational and transcriptional control. Further studies are required to identify the elements within the various genes that control expression, the transcription factors that carry out this control, and the signaling pathways within the cells that regulate the activity of the transcription factors.

GENOMIC ANALYSIS OF GENE EXPRESSION

Identifying New Genes

The first mosquito nuclear gene was cloned and sequenced in the 1980s. Since then, several new molecular biological tools have been developed that enable vector biologists to clone and sequence many genes of interest. Some of these genes may help us to understand important physiological and biochemical processes in vectors; others may encode proteins that are involved in the interaction between a vector and its pathogens, hopefully shedding new light on controlling disease.

We describe several molecular cloning strategies that are useful for discovering new genes. These might be vector genes encoding proteins with known functions but not previously described in the vector or genes with unknown proteins where the corresponding mRNA is expressed differentially in response to physiological, developmental, or environmental changes in the same organism. Polymerase chain reaction (PCR) approaches have been used extensively to clone fragments of genes using degenerate oligonucleotide primers that are designed based on the primary amino acid sequences of proteins derived either from the N-terminal sequence of a specific protein or from the conserved functional domains of homologous proteins. Polymerase chain reaction approaches also can be used to clone mosquito homologues of *Drosophila melanogaster* genes. Once these fragments have been cloned, cDNA and genomic DNA library screening is performed to clone the full-length transcription unit of the genes. The library screening may be done under high- or low-stringency conditions, resulting in the cloning of a corresponding target gene or other related genes, respectively. We will describe "high"- and "low"-stringency library screening later in the chapter. In addition, the recently developed DNA microarray high-throughput approach can be utilized to identify differentially expressed genes. A combination of all available techniques will undoubtedly enhance the successful cloning of the genes of interest.

PCR Approaches

The discovery of a thermostable DNA polymerase in the early 1990s profoundly changed molecular biological approaches for studying genes. Polymerase chain reaction approaches, including RT-PCR (reverse transcription coupled with PCR) and 5' and 3' RACE (rapid amplification of cDNA ends) have been used extensively to clone cDNA fragments of many genes. Both PCR approaches may utilize degenerate oligonucleotide primers, which are designed using either a functional domain of homologous proteins or the known N-terminal amino acid sequence of protein of interest. Thereafter, a library screening approach is performed to clone the complete transcription unit of genes. Several mosquito genes have been cloned and sequenced using these approaches.

Case Study: Exopeptidases

Exopeptidases cleave single amino acids from proteins or peptides either from the amino terminal (aminopeptidases) or carboxyl terminal (carboxypeptidases) ends. Aminopeptidase and carboxypeptidase enzymatic activities, which are responsible for the final stage of a protein meal digestion, have been described in the midgut lumen of several mosquito species, and several genes encoding exopeptidases have been cloned and sequenced from mosquitoes (Table 36.1).

The protein sequence of *Ae. aegypti* aminopeptidase N (*AaAP-I*) has recently been deposited in GenBank. The first digestive carboxypeptidase gene of a hematophagous insect was discovered and character-

ized in the blackfly, *Simulium vittatum*. Combined with *An. gambiae* (*AnCPA-I*) and *Ae. aegypti* (*AaCPA-I*) homologues, the molecular signatures present in the amino acid sequences suggest that all genes belong to the carboxypeptidase A family. Here, we briefly describe the procedures used in cloning of additional exopeptidases genes from *Ae. aegypti* using PCR approaches.

- Retrieve the amino acid sequences for carboxypeptidases from GenBank and perform a protein BLAST search (blastp) against the protein sequence databases (nr: all nonredundant GenBank CDS translations + PDB + SwissProt + PIR + PRF, http://www.ncbi.nlm.nih.gov/BLAST/).
- Perform multiple sequence alignments to locate conserved region of insect carboxypeptidases (e.g., PILEUP/PRETTY in GCG program). A part of the protein alignments is shown in Figure 36.3, with conserved domains underlined.
- Design degenerate primers based on the conserved primary amino acid sequences:

Sense strand: IHAREWI	5′ ATHCAYGCNMGNGARTGGAT 3′ 384-fold degeneracy
Antisense strand: DPNRNWN/D	5′ TYCCARTTNCKRTTNGGRTC 3′ 512-fold degeneracy

where, Y = T or C; H = A, C, or T; M = A or C; N = A, C, G, or T.

- Isolate total RNA from blood-fed midgut.
- Carry out first-strand cDNA synthesis using reverse transcriptase with oligo-dT-coupled adapter primer.
- Carry out PCR amplification of mosquito carboxypeptidase by either RT-PCR with two car-

TABLE 36.1 Dipteran Exopeptidase Sequences Deposited in GenBank (20001)

Accession No.		Organisms
AAB96576	(*AnCPA-I*)	*An. gambiae*
AAD47827	(*AaCPA-I*)	*Ae. aegypti*
P42788		*S. vittatum*
AAC05137		*Drosophila heteroneura*

```
AaCPA-I      nqy.dqVqll  egGhsfenRs  ikGVKvSykt  g.nPgifvEg  gIHAREWIsp
AnCPA-I      sehpkeVell  daGrshqnRt  mkGVKlSygp  g.rPgvflEg  gIHAREWIsp
Simulium     qehpehVepv  vgGksyegRe  irGVKvSykk  g.nPvvmvEs  nIHAREWIta
Drosophila   tkyphvVtlv  egGktyqgRs  ilGVKiSksq  sekPgiflEa  gIHAREWIns
Consensus    ------V---  --G-----R-  --GVK-S---  ---P----E-  -IHAREWI--

AaCPA-I      AtvaYilNeL  LTStdpkvrn  iAenYdWYmf  PsvNPDGYvY  ThkkdRlWRK
AnCPA-I}     AtvtYilNqL  LTSedakvra  lAekFdWYvf  PnaNPDGYaY  TfqvnRlWRK
Simulium     AtttYllNeL  LTSknstire  mAenYdWYif  PvtNPDGYvY  ThttdRmWRK
Drosophila   AaatYiiNqL  LTSnvdsikq  lAdnYnWYvi  PhaNPDGFvY  ThtndRmWRK
Consensus    A---Y--N-L  LTS-------  -A--Y-WY--  P--NPDGY-Y  T----R-WRK

AaCPA-I      TRtpy.sggC  fGaDPNRNWd  FHWaEqGtSn  rcnsdTYgGp  hAFSEVETkS
AnCPA-I      TRkay.gpfC  yGaDPNRNWd  FHWaEqGtSn  nacsdTYhGs  eAFSEVETrS
Simulium     TRspnpdslC  aGtDPNRNWn  FHWmEqGtSs  rpcteTYgGk  kAFSEVETrS
Drosophila   TRtpygs..C  fGaDPNRNWg  FHWnEvGaSn  sacadTYaGp  sAFSEIETlS
Consensus    TR------C   -G-DPNRNW-  FHW-E-G-S-  -----TY-G-  -AFSEVET-S
```

FIGURE 36.3 Multiple sequence alignments of conserved regions of insect carboxypeptidases.

boxypeptidase-specific degenerate primers or 3′ RACE, using one forward-degenerate primer and an adapter primer with cDNA as a template.

- Ligate the PCR products into a cloning vector (TA vector for Taq DNA polymerases and blunt-end vector for high-fidelity thermostable DNA polymerases).
- Verify that the cloned gene is the correct one via DNA sequencing.
- Screen cDNA and/or genomic DNA library to clone the transcription unit.

The results of sequencing using the 3′ RACE approach showed that we cloned *AaCPA-I* and a new carboxypeptidase gene (*AaCPA-II*) in *Ae. aegypti*. A full-length *AaCPA-II* cDNA was subsequently cloned by library screening and sequenced. The two carboxypeptidases showed 55.9% amino acid similarity. RT-PCR approaches failed to clone additional carboxypeptidases, probably due to the three nonsynonymous amino acid substitutions in the region of the reverse primer, therefore causing the unfavorable annealing of the reverse primer to DNA templates.

A similar approach was used to clone additional aminopeptidase genes from *Ae. aegypti*. Briefly, RT-PCR and 3′ RACE were used to clone a new gene, *AaAP-II*, which shares relatively low overall amino acid sequence similarity (45.0%) with *AaAP-I*. Genomic clones of both genes show that the location of the first intron in the coding region of both members is conserved. However, unlike *AaAP-I*, which has a small intron (56 bp), a transposable element has been inserted within the first intron in *AaAP-II*.

What We Can Do with *Drosophila melanogaster* Genome Sequences

Many genes are part of a gene family that arose most likely as a consequence of gene duplication events. If one is trying to measure the expression of a particular gene and is unaware of the presence of additional genes, then the results of expression analysis may be misleading because the probe used may detect the products of several genes. However, if one clones the additional genes in the family, a better experimental procedure, such as quantitative RT-PCR with gene-specific primers or Northern blot using less conserved regions such as 5′- or 3′-end untranslated regions, can be used to examine the pattern of expression of each gene. This also applies to protein expression for each gene. These approaches would lead to a better understanding of the role for each gene and gene product.

One way to find additional members of a gene family would be to take advantage of the recently completed genome sequence of *D. melanogaster*.

Genome projects for a wide variety of organisms, including human, yeast, fruit fly, nematodes, *Arabidopsis*, and a number of microorganisms, have been completed, generating a very large sequence database. A publicly available database greatly assists researchers working on other organisms, especially when the organism of interest is closely related to one of the model organisms in the database. The recent completion of sequencing of the *An. gambiae* genome is a significant milestone that makes it possible to do comparative work with a complete complement of vector genes. More distantly related organisms are usually less useful because orthologous or homologous protein sequences between the two species have accumulated a great number of nonsynonymous amino acid substitutions.

The entire genome of *D. melanogaster* has recently been sequenced, revealing approximately 13,600 putative genes. The database for the *Drosophila* genome is readily available from various Internet sites (e.g., FlyBase: http://flybase.bio.indiana.edu/, http://hedgehog.lbl.gov:8001/cgi-bin/annot/query), and protein and nucleotide sequences can be retrieved directly from the National Center for Biotechnology Information (http://www.ncbi.nlm.nih.gov/). Fortunately, mosquitoes and *D. melanogaster* are both dipteran insects; thus, homologous sequences between the two species retain relatively high similarity as compared to other, distantly related insects. In addition, previously cloned mosquito sequences and currently ongoing mosquito ESTs (expressed sequence tags) and GSSs (genomic survey sequences) projects also provide useful sequence information (Mosquito Genomics Web server: http://klab.agsci.colostate.edu/AnoDB; http://konops.anodb.gr/AnoDB/; and http://bioweb.pasteur.fr/BBMI/index.html). Therefore, thorough searches of the *Drosophila* genome database would allow us to design highly probable degenerate oligonucleotide primers for PCR amplification of target genes in mosquitoes. Here, we have taken advantage of the database to clone mosquito genes of interest to our lab and will describe two case studies.

Case Study 1: Angiotensin-Converting Enzymes

Angiotensin-converting enzymes (ACEs) are dipeptidyl carboxypeptidases. In addition to their role in peptide hydrolysis in the gut, in mammals ACE are known to possess a regulatory role for activating or inactivating bioactive peptides, such as prohormones. Mosquitoes are ideal model organisms because many physiological processes in various tissues are tightly regulated by JH, ecdysone, and peptide hormones (e.g., allatotropin). To regulate the activity of prohormones, it is possible that ACEs might be involved in the processing of peptide hormones. Although ACE activity has been previously shown in *An. gambiae*, the genes and gene products as well as the substrate for ACE in mosquitoes have not been determined. To take advantage of ACE genes that have been cloned and sequenced in *D. melanogaster* (Accession No. AAB02171), we initiated cloning of ACE genes in mosquitoes.

Case Study 2: Amino Acid Transporters

A dozen mammalian amino acid transporters (AATs) have been cloned recently and functionally characterized. Molecular cloning of insect AATs have not received much attention, except for a sodium/potassium AAT from *Manduca sexta* and one AAT that was found serendipitously in *D. melanogaster*. In mosquitoes, amino acids derived from digested blood meal proteins in the lumen of the midgut must pass two plasma membranes (apical and basolateral regions) via AAT transmembrane receptors in order to enter the hemolymph to be utilized by target tissues (e.g., fat body for vitellogenin synthesis). Amino acid transporters may act as important checkpoints for trafficking amino acids between different cell and tissues. Hormonal induction of amino acid transport activity has been well characterized in mammals, but it is still poorly understand at the molecular level in insects. To gain a better knowledge of digestive physiology in mosquitoes, we attempted to clone an AAT. Here is a brief procedure used in cloning these genes.

- Review literature of ACEs or AATs (PubMed: http://www.ncbi.nlm.nih.gov/Entrez/).
- Retrieve amino acid sequences of mammalian and insect ACEs or AATs from GenBank (protein sequence database).
- Conduct BLAST searches (blastp) against *Drosophila* genome sequence:
  Protein query — protein database
- Conduct BLAST searches (tblastn) against Arthropoda EST and GSS sequences:
  Protein query — translated database of the EST and GSS
- Retrieve amino acid sequences of putative insect ACEs and AATs from GenBank.
- Perform multiple sequence alignments (e.g., PILEUP / PRETTY in GCG program).

- Design degenerate oligonucleotide primers based on the conserved amino acid sequences.
- Perform PCR amplification of mosquito ACEs and AATs using genomic DNA or cDNA as a template.
- Clone into a cloning vector (TA vector for Taq DNA polymerases and blunt-end vector for high-fidelity thermostable DNA polymerases).
- Verify the identity of the cloned gene by DNA sequencing and BLAST searches.
- Screen cDNA and genomic DNA libraries to clone ACE and AAT transcription unit.

We have cloned some of the mosquito genes that are homologues of ACEs and AATs. However, we need to be cautious and keep in mind that a similarity between two sequences does not necessarily mean they are homologous to each other, until the function of the cloned gene products is experimentally proved. Nevertheless, a readily available *Drosophila* genome sequence and mosquito EST and GSS can greatly enhance the pace of cloning and, therefore, of the characterization of mosquito genes and gene products.

Low-Stringency Library Screening

Nucleic acid hybridization analyses, including library screening, genomic Southern blots, and Northern blots, have been used to isolate clones, to determine the copy number of the genes, and to examine the pattern of gene expression, respectively. Temperature and ionic strength are the major factors influencing hybridization of nucleic acids. The stringency of these conditions during the hybridization and washing steps and the region of genes used as a probe may be optimized to hybridize the probe either specifically to target gene sequence (high stringency) or to closely related genes as well (low stringency). For example, the expression pattern of three closely related vitelline envelope genes in *Ae. aegypti* has been examined under high-stringency conditions with high temperature, low Na^+ concentration, and 3'-end untranslated regions of the gene as probes, resulting in gene-specific hybridization. Alternatively, one may also perform hybridization at low-stringency conditions with a highly conserved and functional region of genes as a heterologous probe from different species, resulting in cross-hybridization.

Genome projects and knockout experiments of many model organisms, including *D. melanogaster*, *C. elegans*, and mouse, have shown that many genes have multiple copies as the result of gene-duplication events. Duplicated genes could evolve into new genes with novel functions as a consequence of relaxation of functional constraints. Alternatively, gene duplication could change one of the members of the gene family into a nonfunctional gene. Although functional domains of the primary amino acid sequences among members of a gene family may be conserved, the neutral theory of evolution suggests that the third codon position has a weak selective constraint because of synonymous codons coding for the same amino acid. Therefore, by considering synonymous and non-synonymous substitutions, additional members may be identified using a probe coding for functional domains with library screening at the low-stringency level.

Case Study 1: Trypsin-like Genes in Aedes Aegypti

In *Ae. aegypti* three trypsin-like genes have been cloned and sequenced, and the studies mentioned earlier suggest that two of them, *AaTRYP-I* (late) and *AaTRYP-III* (early), play a major role in blood meal digestion. A previously published *AaTRYP-II* (5G1) sequence is incomplete at the 5′ end of the coding region. To clone and sequence the complete open reading frame of the *AaTRYP-II*, low-stringency screening was conducted using genomic and cDNA phage libraries, in which the latter was constructed from blood-fed whole body. Briefly, 284 bp of the *AaTRYP-II* coding region was labeled by PCR using digoxygenin 11-dUTP and there were libraries screened. Surprisingly, *AaTRYP-II* cross-hybridized with five new trypsin-like genes and *AaTRYP-III* at low-stringency conditions. These new trypsin sequences, along with all the other digestive enzyme sequences described in this chapter, are shown in Table 36.2.

All of these newly discovered genes have a complete open reading frame with no internal stop codons or frameshifts, and they are also expressed in blood-fed whole body, indicating that these are not pseudogenes. The lowest overall nucleotide sequence identity between the probe region of *AaTRYP-II* and the target trypsin-like genes was 52.5% (*AaTRYP-VII*). However, an amino acid sequence motif, GGKDSCQGDSGGP, which is highly conserved among many serine proteases, is 100% identical between these two gene products, and the nucleotide identity encoding this motif is 89.7%, suggesting that this region might be strongly contributing to the hybridization between the probe and *AaTRYP-VII*. It would be of interest to determine whether the newly identified trypsin-like genes have any functional role in the digestion of a blood meal in *Ae. aegypti*. This library screening at low-stringency

TABLE 36.2 List of Digestive Enzymes Cloned in the *Aedes aegypti* Mosquito

	Previously Described	PCR Using Degenerate Primers	Library Screening at Low Stringency	EST
Chymotrypsin-like genes				
AaCHYM-I	AAB01218			
AaCHYM-II	AAF43707			
AaCHYM-III				X
Trypsin-like genes				
AaTRYP-I	AAA29356			
AaTRYP-II	P29787			
AaTRYP-III	P29786			
AaTRYP-IV			X	
AaTRYP-V			X	
AaTRYP-VI			X	
AaTRYP-VII			X	
AaTRYP-VIII			X	
AaTRYP-IX				X
Carboxypeptidases				
AaCPA-I	AAD47827			
AaCPA-II		X		
AaCPB-I				X
Aminopeptidases				
AaAP-I	AAK55416			
AaAP-II		X		

X indicates that the genes were cloned and sequenced from either PCR, library screening, or EST approaches.

conditions also implies that a region of the gene chosen as a probe might influence the outcome of studies (e.g., Northern blot analysis for gene expression). RT-PCR experiments with gene-specific primers may be performed in order to examine the detailed expression profile of all trypsin-like genes in *Ae. aegypti*.

Case Study 2: Vitellogenin Genes in Aedes Aegypti

Vitellogenins are the precursors of yolk proteins utilized in oviparous animals to provide nutrition for the developing embryo. Many organisms, including *Xenopus laevis*, *C. elegans*, *D. melanogaster*, and other insects, have multiple vitellogenin genes. The first vitellogenin gene (Vg-A1) was isolated by differential screening of the *Ae. aegypti* genomic DNA library. A subsequent genomic Southern analysis shows that the Vg-A1 probe cross-hybridized with other members in *Ae. aegypti*. Using the highly conserved coding region of Vg-A1 as a probe and low-stringency library screening, three additional members of the vitellogenin gene family, Vg-A2, Vg-B, and Vg-C, have been cloned and sequenced. Sequence analysis showed that Vg-A1 and Vg-A2 are allelic to each other; Vg-A1 and Vg-B are closely related and possibly arose by a recent gene duplication event; and Vg-C is distantly related to Vg-A1 and Vg-B and possibly arose by an earlier gene-duplication event.

Expressed Sequence Tags (EST) and Genomic Survey Sequence (GSS) Projects

Expressed sequence tags and GSSs are sequences derived from cDNA and genomic DNA libraries, respectively, from which individual plaques or colonies are picked randomly and sequenced. In general, both projects are done to identify new genes present in the organism of interest. One of the critical factors researchers face in a large-scale project aiming to identify new genes efficiently and cost-effectively is the genome size of the organism of interest. The genome size varies significantly among mosquitoes: *An. gambiae* has approximately 270 million bp, which is one of the smallest genome sizes among mosquitoes, making this species a good candidate for a whole-genome-sequencing project. As of March 2002, 4,445,304 genomic sequences containing 3,508,329,249 bp had been deposited to GenBank, revealing the first entire mosquito genome sequence. On the other hand, the *Ae. aegypti* genome (~820 million bp) is about four times larger than the *An. gambiae* genome. Assuming that the number of protein-coding genes in both species is approximately the same, the increased genome size of *Ae. aegypti* most likely involves expan-

sion of noncoding regions due to the high copy number of transposable elements, including *Feilai* and *Lian* — these elements are not present in *An. gambiae* genome. Therefore, ESTs may be the preferable approach to identify new genes in *Ae. aegypti*.

The level of transcription varies significantly among genes, for some genes are more highly expressed than others. Therefore, normalization procedures in which the more abundant cDNA sequences are depleted during library construction increase the probability of identifying new genes in EST projects. Genomic survey sequences may also assist the physical mapping of the genome of organisms.

Two large-scale mosquito EST projects have been carried out recently to facilitate gene discovery. cDNA clones were sequenced from an enriched Malpighian tubule and gut library in *Ae. aegypti* (1,356 sequences as of September 28, 2001) and from a normalized library of immune-competent cell lines in *An. gambiae* (6,037 sequences). These sequences are available from an EST database (http://www.ncbi.nlm.nih.gov/dbEST/). Our laboratory has undertaken a small-scale EST project from nonnormalized blood-fed midgut cDNA clones in *Ae. aegypti*, resulting in the identification of three new genes encoding digestive enzymes. A subsequent cDNA library screening resulted in the cloning of full-length clones for all genes — carboxypeptidase B (*AaCPB-I*), chymotrypsin-like (*AaCHYM-III*), and trypsin-like (*AaTRYP-IX*). The amino acid sequence of each new gene differed significantly from the previously cloned digestive enzymes in *Ae. aegypti*. It will be important to determine the role of each gene in the digestion of blood meal. Large-scale EST and GSS projects for mosquitoes and other arthropod vectors will undoubtedly identify many new genes, and a functional analysis of each gene and gene product will provide new insight for controlling vector-borne diseases. Clones from a normalized cDNA library that are used for an EST project may also be spotted for DNA microarray analysis for a large-scale expression study.

DNA Microarray

A new high-throughput technique called *DNA microarray* has been developed recently to identify genes that may be up-regulated or down-regulated in response to varying physiological conditions. Briefly, each clone from an ideally normalized cDNA library is amplified by PCR, and then the DNA clones are immobilized individually on a glass slide as probes. Then, first-strand cDNA is synthesized from two physiological conditions (e.g., total RNA extracted from midgut of unfed and blood-fed mosquitoes) using fluorescent nucleotides [e.g., unfed — Cy3 (yellow) — and fed — Cy5 (red)] and used as targets. The two targets are mixed and hybridized to the probes on the glass slide. If a particular gene is expressed equally in both states, both targets will hybridize and the spot will be green; if the gene is expressed only in the unfed state, then only the Cy3 target will hybridize and the spot will be yellow; finally, if the gene is expressed only in the fed state, then only the Cy5 target will hybridize and the spot will be red. Clones that are differentially expressed are sequenced and further analyzed. Several well-established DNA microarray protocols are available on the Internet (e.g., National Human Genome Research: http://www.nhgri.nih.gov/DIR/Microarray/protocols.html, *Drosophila*: http://quantgen.med.yale.edu/protocols.htm). A DNA microarray combined with an EST project will be a powerful approach to identify and study the pattern of gene expression in arthropod vectors.

ANALYSIS OF REGULATORY REGIONS IN MOSQUITOES

Regulatory DNA elements, including promoters, enhancers, and repressors, often reside in the untranscribed regions upstream (5') and/or downstream (3') of transcription units. These elements, such as ecdysone-responsive elements, serve as target sites for transcription factors, thereby modulating the level of the gene expression. Some highly expressed mosquito genes such as vitellogenins and trypsins are tissue, stage, and sex specific and are induced by blood feeding. Thus, a better understanding of regulatory DNA elements may provide information about the transcriptional regulation of important mosquito genes and might serve as useful promoters for transgenic constructs. The first step involved in the analysis of regulatory DNA sequences is to obtain the genomic sequence, including upstream and downstream sequences, for the genes of interest, either by genomic DNA library screening or by inverse PCR approaches. After determining the genomic DNA sequences flanking the coding region, other methods are used to narrow down the DNA regulatory elements potentially involved in the transcriptional regulation of genes. Here, we describe several approaches that assist in locating putative regulatory elements.

Comparative Sequence Analysis

Among distantly related insects, the amino acid sequence in the DNA-binding domain of homologous transcription factors are conserved, e.g., ecdysone

receptors, suggesting that the DNA elements they recognize also are conserved. Although *D. melanogaster* is not a hematophagous insect, the DNA sequence elements that play a regulatory role in gene expression have been identified for a wide variety of transcription factors. By taking advantage of the information already available from the *D. melanogaster* literature, putative regulatory DNA elements have been identified for several mosquito genes (e.g., vitellogenins, trypsins, vitellogenic carboxypeptidase, vitelline envelope). In addition, a useful database is available for predicting potential transcription factor–binding sites identified from other organisms (e.g., TRANSFAC; http://transfac.gbf.de/TRANSFAC/index.html).

Phylogenetic Footprinting Analysis

Phylogenetic footprinting analysis is an approach to search for common regulatory elements involved in the regulation of homologous gene transcription among evolutionarily closely related species. The analysis is based on the assumption that critical DNA elements have much higher selective constraints than other, noncoding sequences that allow frequent mutation and insertion/deletion events. Phylogenetically conserved putative regulatory elements possibly involved in the regulation of the vitellogenin gene expression have been identified by phylogenetic footprinting analysis. The first step is to understand phylogenetic relationships of the mosquitoes of interest, either from known mosquito systematics or from phylogenetic trees based on molecular markers. This is necessary in order to choose the closely and distantly related taxa to be included. For example, if *Ae. aegypti* were the species of interest, then *Ae. albopictus*, and *Ae. polynesiensis*, which all belong to the subgenus *Stegomyia*, are chosen as being closely related, while *Ae. atropalpus* and *Ae. triseriatus* (recently revised as *Ochlerotatus triseriatus*) would be chosen as distantly related species (see Table 36.3).

The second step involves the construction and screening of genomic DNA libraries in order to clone and sequence vitellogenin regulatory regions. This is followed by multiple alignments of the vitellogenin nucleotide sequences (PILEUP in GCG or ClustalW). The alignment can sometimes be improved manually. The third step is sequence analysis to locate blocks of conserved vitellogenin nucleotide sequences across the mosquito taxa examined (e.g., MEME program in the GCG). The phylogenetic footprinting analysis of three members of the vitellogenin gene family in *Aedes* mosquitoes revealed several invariant elements that are evolutionarily conserved (Fig. 36.4). Among these elements are the TATA-binding site and a GATA

TABLE 36.3   Mosquito Classification and the Nucleotide Length of Vitellogenin Gene 5′ Flanking Region Sequenced

	Species	5′ Flanking Region of Vg Sequenced		
Family Culicidae				
Subfamily Culicinae				
Tribe Aedeomyiini				
Genus *Aedes*				
Subgenus *Stegomyia*				
Aedes aegypti	Vg-A1 (2090)	Vg-B (5834)	Vg-C (1272)	
Aedes polynesiensis	Vg-A1 (321)	Vg-B (1150)	Vg-C (3575)	
Aedes albopictus	Vg-A1 (945)		Vg-C (99)	
Subgenus *Ochlerotatus*				
Aedes atropalpus		Vg-B (1053)	Vg-C (646)	
Subgenus *Protomacleaya*				
Aedes triseriatus			Vg-C (1988)	
Tribe Culicini				
Genus *Culex*				
Subgenus *Culex*				
Cules quinquefasciatus			Vg-C (2749)	
Subfamily Anophelinae				
Genus *Anopheles*				
Subgenus *Nyssorhynchus*				
Anopheles albimanus			Vg-C (1090)	
Subfamily Toxorhynchitinae				
Genus *Toxorhynchites*				
Subgenus *Toxorhynchites*				
Toxorhynchites amboinensis			Vg-C (2214)	

The number in parentheses indicates the nucleotide length of the 5′ flanking region counting from just upstream of the ATG translation initiation site.

transcription factor–binding site. There are several other conserved sequences evident in Figure 36.4 that do not correspond to the binding sites for known transcription factors. These may or may not represent functional regulatory sites, and this can be determined only by a functional assay, e.g., the gel mobility shift assays described later.

These putative regulatory elements identified either by phylogenetic footprinting or comparative sequence analysis need to be examined further by other experimental methods, such as a gel mobility shift and DNase I footprinting assays, that would reveal more information on their possible functional roles. Nevertheless, both comparative and phylogenetic footprinting analysis provide a starting point for molecular characterization for regulatory regions of genes of interest.

DNase I Footprinting

DNase I footprinting is a technique for determining a specific DNA sequence that might be the binding site for a transcription factor and is based on the fact that a protein bound to a segment of DNA interferes with the DNase I digestion of that region. In order to use this method, a protein extract or, ideally, a purified candidate DNA-binding transcription factor, is incubated with a radiolabeled DNA containing the putative regulatory element in question, allowing DNA–protein interaction to occur, and then partially digesting with DNase I such that, on average, every DNA molecule is cut in the unprotected portion of the sequence. A standard DNA sequencing gel is used to resolve the digestion products. A comparison of the DNase I digestion pattern in the presence and absence of the transcription factor allows the identification of a footprint (protected region).

Using a similar concept, potential regulatory regions containing a specific regulatory DNA element can be narrowed down by DNase I–hypersensitive site experiments prior to the DNase I footprinting assay. Chromatin structure in the nucleus usually restricts the access of transcription factors to regulatory DNA elements when inducible genes are in an inactive state; however, the chromatin structure must open up to allow transcription of active genes. In this open state, the DNA is sensitive to DNase I. Thus, comparing chromatin structure in the active and inactive states of

```
Ae. aegypti              .cGATTtGcT aGTGGTCATT GACTggcT.g atCGATAAGc gATtATGAAT
Ae. polynesiensis        .cGtTTcGcT aGTGGTCATT GACTAgcc.g atCGATAAGc aATcATGAAT
Ae. triseriatus          gaGATTaGaT cGTGGTCATT GACTAcgT.t gcCGATAAGc gATtATGAAT
Ae. atropalpus           ctGATTaGaT cGTGGTCATT GACTAgtT.c gaCGATAAGg aATtATGAAT
Cx. quinquefasciatus     ...ATcaGc. .GcatTgATa GACcAtcTgg atCGAT.... ..TcATtAtT
Consensus                --GATT-G-T -GTGGTCATT GACTA--T-- --CGATAAG- -AT-ATGAAT

Ae. aegypti              GAActaaatc tgATAaCcAT GGAct.agga CAAAgc.Gt. ..CcacTttg
Ae. polynesiensis        GAAtcgcgac tgATAaCcAT GGAct.agga CAAAgc.Gtc caCcaaTgcg
Ae. triseriatus          GAActacctc tgA.AtCgtT GGAt...gga CAAAattGat caCccaTgct
Ae. atropalpus           GAAacatatg c.ATAtCgAT cGAtggctcg aAAAat.Gct agCacaTgca
Cx. quinquefasciatus     GAAc...... ..ATAttgAT G.Att.cacg CAAAa..... ..........
Consensus                GAA------- --ATA-C-AT GGA------- CAAA---G-- --C---T---

Ae. aegypti              ggATGATCCa TATAAAAAGG AtcgaCTGGA ActCAAgtAG CACAGTTCaA
Ae. polynesiensis        gaATGATCCt TATAAAAAGG AatggCTGGA AggCAATCAG CACAGTTCtA
Ae. triseriatus          aaATGATCCt TATAAAAAGG ActtgCTGGA tttCAATCAG CACAGTTagA
Ae. atropalpus           aaATGATCCa TATAAAAAGG AtcagCTGGA AtcCAATCAG CACAGTTCgA
Cx. quinquefasciatus     ...TGtTCCa TATAAAgct tggtcCTGGA ActCAtTCAG CACAcTTCag
Consensus                --ATGATCC- TATAAAAAGG A----CTGGA A--CAATCAG CACAGTTC-A

Ae. aegypti              CTtTTCCTTT CattTCaTCt AA.AcCAA.. .aaaGA..cc AgTGAtccAG
Ae. polynesiensis        CTtTTCCaTT CaacTCgTCG AAtAcCAcaa actaGAgtcc tgcGAtccAG
Ae. triseriatus          CatcTCCTTT C....C.TCG AAtAgCcAga a...GAgg.. AtTGA...AG
Ae. atropalpus           CTcTTCCTTT CatcTCgTCG AAtttCAAgt aaagGAgtga AgTGAtctAG
Cx. quinquefasciatus     tTcTTCCTTT C...TCggtG AAcAttAAgc gaagGAttct AcTGtctcAa
Consensus                CT-TTCCTTT C---TC-TCG AA-A-CAA-- ----GA---- A-TGA---AG

Ae. aegypti              gG....AAtC ATG
Ae. polynesiensis        gG....AAtC ATG
Ae. triseriatus          tG....tAgC ATG
Ae. atropalpus           tG....AgcC ATG
Cx. quinquefasciatus     cGaactAAtC ATG
Consensus                -G----AA-C ATG
```

FIGURE 36.4 Nucleotide alignment of promoter region of vitellogenin gene (Vg-C) in mosquitoes. ATG translation initiation site and putative TATA box are underlined.

genes often identifies DNase I–hypersensitive sites, which are potential regulatory regions. Both DNase I footprinting and DNase I–hypersensitive site approaches have been widely used in studying gene expression in vertebrates and *D. melanogaster*, although these approaches have not yet been explored in mosquito genes. A subsequent gel mobility shift assay confirms the interaction between a specific DNA element and the transcription factor.

Gel Mobility Shift Assay

A gel mobility shift assay provides a simple and rapid method for identifying DNA–protein interactions, and it is ideal for confirming the interaction between a regulatory DNA element and its transcription factor. This assay is based on the observation that a complex of DNA and protein migrates through a gel more slowly than does free DNA. The assay is performed by incubating a radiolabeled DNA fragment containing the putative transcription factor–binding site with a purified protein of interest or a crude protein extract (such as nuclear or cellular extracts). The binding reactions are then analyzed on a nondenaturing polyacrylamide gel. The specificity of the DNA-binding protein for the putative binding site then is established by competition experiments using an excess of nonradiolabeled DNA fragments. The method has been used to examine the transcriptional regulation of several genes in mosquitoes. In *Ae. aegypti*, an ecdysone-responsive element, which is the binding site for the ecdysone receptor–ultraspiracle heterodimer, has been confirmed by the gel mobility shift assay. The GATA transcription factor has also been shown by this assay to interact with the appropriate element in the regulatory region of trypsin genes in *Anopheles* mosquitoes and vitellogenin genes in *Ae. aegypti*. Interestingly, the result of an independent study of vitellogenin regulatory regions by phylogenetic footprinting analysis also suggested the possible regulatory role of the GATA transcription factor in vitellogenin gene expression.

The approaches just described would assist in a better understanding of the interaction between transcription factors and their regulatory elements and therefore transcriptional regulation of mosquito genes. However, recent genomic analysis of mosquito genes revealed that several highly repetitive transposable elements often reside in the upstream regulatory region, complicating the interpretation of previously characterized regulatory regions of mosquito genes. Therefore, putative regulatory DNA elements found by either comparative or phylogenetic footprinting analysis must be experimentally tested by either DNase I footprinting or gel mobility shift assays.

Summary of Genomic Analysis of Gene Expression

In this section we described approaches that may be used to clone genes of interest. After detailed analysis of homologous sequences available from databases, PCR approaches would specifically amplify the genes of interest. Library screening at low stringency has allowed us to clone additional trypsin-like genes in *Ae. aegypti*. In addition, new, powerful techniques, such as EST projects and DNA microarray analysis, will undoubtedly reveal many new genes and the pattern of their expression. We also described approaches that one may use to identify regulatory elements that might be involved in the transcriptional regulation of genes.

Readings

Dimopoulos, G., Casavant, T. L., Chang, S., Scheetz, T., Roberts, C., Donohue, M., Schultz, J., Benes, V., Bork, P., Ansorge, W., Soares, M. B., and Kafatos, F. C. 2000. *Anopheles gambiae* pilot gene discovery project: Identification of mosquito innate immunity genes from expressed sequence tags generated from immune-competent cell lines. *Proc. Natl. Acad. Sci. U.S.A* 97: 6619–6624.

Harshman, L. G., and James, A. A. 1998. Differential gene expression I insects: Transcriptional control. *Annu. Rev. Entomol.* 43: 671–700.

Kafatos, F. C. 2001. The future of genomics. *Mol. Aspects Med.* 22: 101–111.

Nature Genetics (Vol. 21, 1999), has several articles on DNA microarrays on pp. 5–37.

Noriega, F. G., and Wells, M. A. 1999. A molecular view of trypsin synthesis in the midgut of *Aedes aegypti*. *J. Insect Physiol.* 45: 613–620.

37

Analysis of Stage- and Tissue-Specific Gene Expression in Insect Vectors

JINSONG ZHU, VLADIMIR KOKOZA, AND ALEXANDER S. RAIKHEL

INTRODUCTION

Most genes in metazoan organisms are under strict regulatory control, allowing their expression in a defined stage-, tissue-/cell-, and sex-specific manner. Synergistic action of various regulatory hierarchies is responsible for such defined expression of their target genes. In the process of gene activation, sequence-specific transcription factors interact with respective enhancer/promoter elements of hormonally controlled, tissue-specific genes. Studies in *Drosophila* and other model organisms have provided the foundation for our understanding of how these regulatory hierarchies come together to bring about expression of a gene in a precisely controlled manner. Much of this knowledge has been achieved through advanced studies of genetics and genetic transformation. More recently, studies in model organisms have greatly benefited from the sequencing of the organisms' genomes and subsequent development of functional genomic techniques.

The unique feature of many genes in arthropod vectors is that their expression is triggered by interaction with either a host or a pathogen. Some of these genes are characterized by a robust response to activation stimuli. Examples include genes encoding multiple salivary factors that are expressed after the onset of feeding on the host; midgut carboxypeptidases and trypsins that are expressed after intake of blood; and fat-body yolk protein precursors that are expressed after the onset of blood meal–activated egg development. Changes in expression of genes involved in

host-seeking behavior, chemoreception, or vector–pathogen interactions are subtler. In addition, most genes in both groups are expressed in a tissue- or even cell-specific manner. For example, in the mosquito *Aedes aegypti*, expression of genes encoding yolk protein precursors is limited to the female fat body. Molecular mechanisms governing the host- and pathogen-activated gene expression are of acute interest for vector molecular biologists. In vector arthropods, sequential activation of numerous genes essential for host-seeking behavior, feeding, digestion of the blood, synthesis of yolk protein precursors, and ultimately production of eggs is host dependent. Moreover, it is in the process of blood or protein feeding on a host that a vector acquires pathogens that it subsequently transmits to the next host. Thus, elucidation of the molecular mechanisms underlying the host-activated gene expression is of paramount importance in our efforts to develop novel strategies of vector and pathogen control.

For example, in order to achieve the most efficient disruption of mosquito-specific stages of the *Plasmodium* life cycle by genetically manipulating the mosquito host, the putative antipathogen effector molecule should be expressed during critical stages of vector–pathogen interaction. Moreover, tissue- and stage-specific expression of an effector molecule reduces a possible nonphysiological effect on the host and should minimize a negative selective pressure on a transgenic vector in future release programs.

The powerful promoters of blood meal–activated midgut- and fat-body-specific genes are best preferred candidates for construction of chimeric genes incorporating immune factors or other putative antipathogen effector molecules. These chimeric genes could be highly expressed in the midgut or the fat body in response to a blood meal, and their products could be secreted to the midgut lumen or the hemolymph, respectively. Therefore, understanding what combination of binding sites in the regulatory region of a gene is required for its host-specific activation in a precise stage- and tissue-specific manner represents an important and challenging task in studying transcriptional control in arthropod vectors.

Despite remarkable progress achieved during recent years in the molecular biology of vector insects, genetic transformation is still limited only to several species of mosquitoes. Furthermore, even for these species, transformation tools are lagging behind sophisticated approaches available for a model organism such as *Drosophila*. Moreover, genetics as an experimental tool is nonexistent for any arthropod vector.

The goal of this chapter, in addition to briefly introducing the major concepts of gene regulation, is to provide a description of some approaches that can be employed when undertaking a molecular analysis of the transcriptional regulatory mechanisms for a newly isolated gene from a vector insect. The examples we use are studies on the gene regulation during vitellogenesis in the mosquito *Ae. aegypti*, which is complicated by the unique nature of blood meal–dependent regulation of many genes and the lack of facile genetics.

REGULATION OF GENE EXPRESSION

Eukaryotic genomes comprise highly organized structural and functional units. As a consequence, regulation of a specific gene is a complex multistep process that requires a precise orchestration of interactions between chromatin-modification complexes, tissue- or cell-specific activators, and repressors, and the basal transcription machinery (Fig. 37.1A). In any given cell type of a multicellular eukaryotic organism, most of the genes are either permanently or temporarily shut down, and relatively few are expressed. The expression of these few genes either is constitutive or is governed via complex gene regulatory cascades.

Initiation of Transcription: Chromatin Remodeling

The structural organization of DNA in a eukaryotic cell plays a central role in the regulation of its genes.

The eukaryotic DNA is assembled into chromatin. The DNA is coiled around a histone octamer, forming a nucleosome. In the nucleosome most genes are repressed, except those for which transcription is brought about by specific positive regulatory mechanisms. Histones in the nucleosome occlude protein-binding sites at gene regulatory regions, thereby interfering with the interaction of the gene with numerous factors required for regulation of gene expression. Thus, the chromatin must be remodeled during the activation of a specific gene to allow interaction with regulatory and transcription factors. Initiation of transcription proceeds in two stages: (1) relief from the chromatin repression and (2) interaction of the gene promoter with polymerase II and accessory factors.

Relief from chromatin repression is triggered when activator proteins or other sequence-specific DNA-binding proteins recruit chromatin-modifying enzymes and direct them to precise "target" genes. The chromatin-modifying enzymes fall into two broad classes: ATP-dependent remodeling enzymes and histone acetyltransferases. Once these enzymes bind to a particular gene, they remodel the chromatin, opening up the binding sites of the DNA so that other activators and the general transcription machinery factors can bind to their appropriate sites on the promoter of the gene.

Promoter Structure and Initiation of Transcription

Transcription initiation of a eukaryotic gene involves the assembly of two types of complexes along the DNA. One, called the initiation complex, assembles at the core promoter region, which is the region of DNA surrounding the transcription start site (TSS). The core promoter binds general transcription factors (GTFs) and the RNA polymerase II, which contains multiple protein factors, including GTFs and the TATA-box-binding protein (TBP) and its associated factors (TAFs). The core promoter directs the polymerase II to initiate transcription at the correct start site. The GTFs are universal, and they function at most promoters of eukaryotic genes. However, in vivo, in the absence of regulatory proteins, the core promoter is generally inactive and fails to interact with GTFs.

Immediately upstream of the core promoter is the regulatory promoter. Clusters of transcriptional factor–binding sites, termed *regulatory modules*, are located in the regulatory promoter. If activation of the module results in higher expression for the gene, it is termed an *enhancer*; otherwise, it's called a *repressor*. Enhancer and repressor sites are also situated at

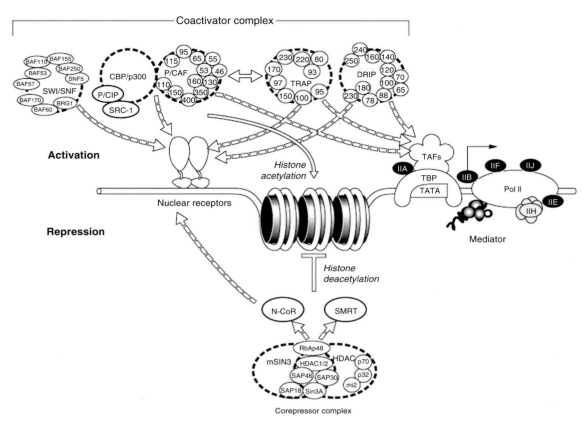

FIGURE 37.1A Regulation of nuclear receptor functions by multiple coactivator and corepressor complexes. Protein complexes implicated in hormonal regulation of gene transcription are shown. Unknown factors within the complexes are indicated by their apparent molecular sizes (in kilodaltons). Placement of the factors within each complex is arbitrary. Ligands of nuclear receptors induce recruitment of the coactivator complexes, leading to activation of transcription. The double-headed arrow suggests a sequential model in which the TRAP or DRIP complexes activate transcription initiation after the chromatin-remodeling step catalyzed by CBP/P/CAF or SWI/SNF complexes. The CBP/p300/P/CAF complexes, as well as the TRAP or the DRIP complexes, may provide a link between nuclear receptors and the core machinery (indicated by the dashed arrows). The mSin3/HDAC corepressor complexes, harboring deacetylase activities, are linked to nuclear receptors via N-CoR or SMRT in the absence of ligand. From Xu et al. (1999), with permission.

various distances upstream or downstream from the core promoter, sometimes as far as 100 kb or more (Fig. 37.1B).

Activator and Repressor Transcription Factors

Activation of a particular gene requires coordinated binding of multiple transcription factors to its promoter-enhancer region. Distinct signal transduction pathways regulate specific transcription factors associated with enhancer or repressor elements.

Early work on eukaryotic transcriptional control focused on positive regulatory DNA elements, known as *enhancers*, and on proteins, termed *activators*, that bind to them and stimulate transcription. Most

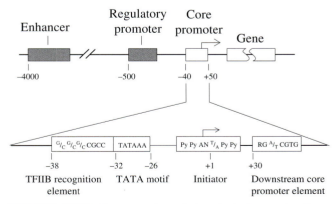

FIGURE 37.1B Sequence elements in a typical promoter. From Carey and Smale (2000), with permission.

activator proteins are modular, with distinct DNA-binding and activation domains. They are specific for a particular gene or gene family and, typically, couple transcription to the physiological needs of a cell. Activators are often latent and become functional in transcription during a physiological response. For example, nuclear hormone receptors are persistently associated with their respective enhancers of target genes and require only binding to a particular lipophilic signaling molecule to activate transcription.

Cell- and/or stage-specific activation of a gene requires binding of multiple transcription factors to a particular promoter-enhancer region. The factors bind cooperatively, as dictated by a unique composition and spatial arrangement of transcription factor–binding sites. Moreover, although many activators are expressed ubiquitously, others are restricted to certain cell types, regulating genes necessary for a particular cell's function of the cell. The assembly of these enhancer complexes, referred to as the *enhanceosomes*, is facilitated by protein–protein interactions between DNA-bound factors and protein-induced DNA bending. Two features of enhanceosome assembly are especially important for this type of transcriptional regulation, termed *combinatorial control*: Combinations of multiple transcription factors generate diverse patterns of regulation, and highly cooperative binding ensures the specificity of transcriptional control (Fig. 37.1C).

The interactions between the enhanceosome and the components of the general machinery are indirect and are linked by proteins called *coactivators*. Most eukaryotic transcription factors contain one or more transactivation domains involved in interactions with downstream coactivators and GTFs (Figs. 37.1A, 37.1C).

Transcription factors that inhibit transcription are called *repressors*. Some factors can play dual roles, being either activators or repressors, depending on cofactors they interact with in a particular cell type or at a particular time during the cell cycle. It is becoming increasingly clear that the transcription rate of a particular gene, either in a specific cell type or following exposure to a particular stimulus, is regulated by a relative balance of activator and repressor transcription factors.

This brief overview displays an enormous complexity of gene regulation in eukaryotic organisms. In the following section, we describe some methods that can be utilized for the study of gene expression in vector insects.

APPROACHES FOR STUDYING GENE EXPRESSION IN VECTOR ARTHROPODS

Analyzing Transcript Size and Profile

An important objective in the analysis of a newly cloned gene is elucidation of the mechanisms regulating its expression. A gene can be regulated by modulating (1) the transcription initiation rate, (2) the transcription elongation rate, (3) the pre-mRNA splicing pattern, (4) mRNA stability, (5) mRNA transport, or (6) the translation rate. Furthermore, these events can occur in a stage-, tissue-, or sex-specific manner. Therefore, a complete analysis of the gene activity in a given cell or tissue often is complicated and should include several techniques that could account for a number of the regulatory parameters just mentioned.

In most cases, studies are initiated by evaluating the state of mRNA for a gene of interest. Numerous techniques developed to analyze mRNA permit one to test for the presence of and to establish the sizes of transcripts, as well as to determine their levels within a given tissue or cell population. Most commonly used techniques include Northern blots and reverse-transcription polymerase chain reaction assay (RT-PCR). Each of these techniques can be used to determine quantitatively and qualitatively a specific RNA within a given tissue or cell population.

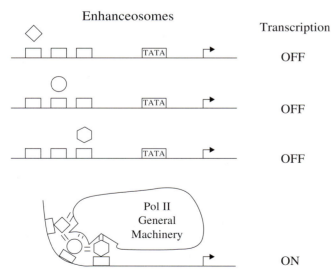

FIGURE 37.1C Enhancesome theory. Individual activators bind DNA weakly on their own. Together, they bind cooperatively, often with the help of architectural proteins that bend DNA. Once bound, they recruit the general machinery to generate synergistic transcription. From Carey and Smale (2000), with permission.

However, each technique has inherent technical advantages and limitations that could make it more or less amenable for a particular application. Ideally, one should utilize at least two techniques to study a particular gene transcript.

Northern Blot Analysis

Advantages of the Northern blot include the ease of this technique, which efficiently provides information not only about the presence of a transcript but also about its size. In this technique, sample RNA is separated on the basis of size via denaturing agarose gel electrophoresis, and then transferred to a solid support and immobilized. A radiolabeled or nonisotopically labeled RNA or cDNA nucleic acid (RNA or DNA) probe of a specific gene under study is then used to detect the message of interest via hybridization. The size of the transcript is determined based on the size of a known marker that is run on the same gel. These markers are typically stained with ethidium bromide or are radiolabeled RNA markers so that they can be visualized and provide an accurate sizing ladder in gels or on autoradiographs.

To normalize the input of different samples, blots may be hybridized with a probe for actin (or another housekeeping gene), the mRNA amount of which is relatively constant under most circumstances. Detection of rare transcripts by Northern blot analysis requires isolation of mRNA, which is usually achieved by the use of oligo-(dT)-selection of RNA. In this technique, isolation of mRNA is based on the presence of poly-A tails in eukaryotic mRNAs. Columns of immobilized oligo-dT are used to prepare enriched mRNAs.

Northern blot analysis is particularly valuable when utilized in conjunction with the tissue or organ isolation that is possible for most vector arthropods. For example, the Northern blot was used to discover where the gene for lipophorin (Lp) is expressed in *Aedes aegypti*. Lp, the major lipoprotein in insect hemolymph, plays an important role in vitellogenesis as a lipid carrier as well as a yolk protein precursor. It has been reported that, in *Ae. aegypti*, the Lp level increases upon ingestion of a blood meal, reaching its maximum levels by 40–48 h post–blood meal (PBM), when major events of egg yolk and lipid deposition have been completed. However, these determinations of Lp levels were made using the whole bodies of mosquitoes. When Lp mRNA levels were examined via Northern blot analysis in the dissected fat bodies and ovaries, it was shown that the Lp gene was expressed only in the fat body (Fig. 37.2). Furthermore, the fat-body levels of Lp mRNA reached their maximum at 18 h PMB (data not shown).

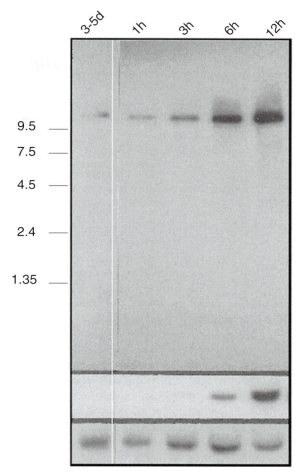

FIGURE 37.2 Lipophorin mRNA expression in the fat body of the female mosquito during the first vitellogenic cycle. Northern blot analysis of mRNA prepared from dissected fat bodies at the indicated times, 25 mosquito-equivalents loaded per lane. After hybridization with Lp cDNA, the membranes were striped and successively probed with actin cDNA and the yolk protein precursor vitellogenic carboxypeptidase (VCP) cDNA. An RNA ladder (in kilobases) is noted on the left; d, days post eclosion; h, hours post–blood meal. From Sun et al. (2000), with permission.

The Northern blot also allows detection of alternatively spliced transcripts or multiple transcripts generated from a single gene. For example, the Northern blot was used to discover different versions of the clathrin heavy chain (CHC) transcript in different mosquito cells. The CHC is one of the key molecules involved in yolk accumulation in mosquito oocytes. Mosquito CHC cDNA clones showed extensive congruence of DNA sequences in the protein-coding region, but they fell into two classes that differed in their 3'-untranslated region (UTR): One group exhibited a long 3'-UTR of 1.4 kb, whereas the other had a short 3'-UTR of 0.35 kb. Northern blot analyses revealed two types of CHC transcripts in mosquito tissues: an "ovarian" 6.5-kb mRNA found only in the ovaries of previtellogenic and vitellogenic females, and a "somatic" 7.5-kb mRNA present in the somatic tissues of females and whole-body preparations of males. Significantly, when the unique portion of the long 3'-UTR (1-kb fragment) was used as the probe in the blot, hybridization was observed to the somatic mRNA only, indicating that the major difference between the two mRNAs was the length of the 3'-UTR.

Reverse-Transcription Polymerase Chain Reaction (RT-PCR)

The RT-PCR technique is one of the most sensitive techniques for mRNA detection. In this assay, an RNA template is copied into a complementary DNA transcript (cDNA) using a retroviral reverse transcriptase. The cDNA sequence of interest is then amplified exponentially using the PCR. Detection of the PCR product is typically performed by using either radiolabeled probes or PCR primers.

Because of recent advances in PCR technology, it is now possible to detect nearly any RNA transcript, regardless of the amount of starting material or the relative abundance of the specific mRNA. The sensitivity of RT-PCR has implications for how it is used as a quantifying tool. Because amplification is exponential, small sample-to-sample concentration and loading differences are amplified as well; therefore, the PCR requires careful optimization when used for quantitative mRNA analysis. Pilot experiments must include selection of a quantitation method and determination of the exponential range of amplification for each mRNA under study. The RT-PCR technique can be used to quantify and compare transcript abundance across multiple samples. The use of an internal control, which has invariant expression within those samples, is essential for sample normalization. Results are expressed as ratios of the gene-specific signal to the internal control signal. This yields a corrected relative value for the gene-specific product in each sample. These values may be compared between samples for an estimate of the relative expression of target RNA in the samples.

For example, RT-PCR has been particularly useful in studying profiles of transcription factors involved in 20-hydroxyecdysone-mediated regulation of yolk protein gene expression in *Ae. aegypti*. Despite their key role in the regulation of gene expression, these genes are expressed at low levels that could be detected only using techniques such as RT-PCR. Sensitivity of several nuclear receptors to 20-hydroxyecdysone has been determined using the fat-body organ culture in combination with RT-PCR. In this case, RT-PCR has been instrumental in determining subtle differences in nuclear gene responses to the hormone (Fig. 37.3).

Correlation of Transcription and Translation

When working with gene expression, it is important to remember that transcription and translation events could be regulated differentially and thus that the transcript profile may not be a true reflection of the activity of a given gene product. A striking example of such regulation has been found in the mosquito fat body during the termination of yolk protein synthesis. In this tissue, the lysosomal system is involved in the postvitellogenic remodeling of the fat body during which the biosynthetic machinery of fat-body cells (ribosomes, rough endoplasmic reticulum, Golgi complexes, and yolk protein-containing secretory granules) are degraded. The activity of several lysosomal enzymes rises dramatically during remodeling, which takes place between 30 and 48h after the initiation of vitellogenesis by blood meal (PBM). Western immunoblot analysis of the lysosomal cathepsin-like aspartic protease (LAP) has shown that its protein profile exhibited a rise at 30h, with the peak at 36–38 h PBM. However, Northern analysis using the LAP cDNA has revealed that the kinetics of its transcript corresponds to that of the major yolk protein vitellogenin (Vg), peaking at 24h PBM (Fig. 37.4). Further studies have revealed that the LAP mRNA is under the translational inhibition of 20E, the falling titer of which allowed the LAP protein translation to proceed.

In Situ Hybridization

Northern blots and RT-PCR are both based on disruption of tissues or an entire organism for isolation and characterization of RNA; thus, they do not provide sufficient information concerning the gene under investigation if it is expressed in a cell population or even in a single cell within a tissue or an organ. This is particularly true for tissues and organs with highly heterogeneous cell populations, such as the nervous and reproductive systems. In many cases, analysis of tissue specificity of gene expression is also hampered by an extremely small size of many vector arthropods, from which isolation of individual organs and tissues is not practical. The in situ hybridization technique has been developed to localize mRNA of interest within a tissue, cell population, or organ of an organism.

In situ hybridization can be performed using cell culture, a dissected tissue, an organ, or an entire organism. Cellular structures are preserved during fixation, most commonly by use of paraformaldehyde.

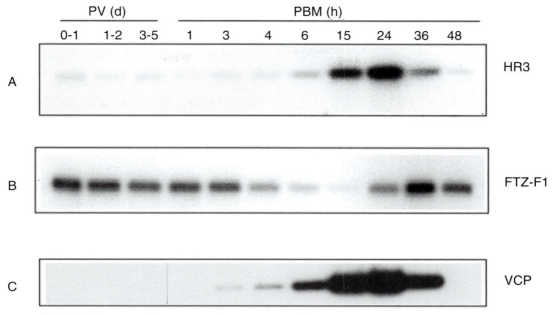

FIGURE 37.3 Temporal expression of AaFTZ-F1 in the fat body during the first vitellogenic cycle by RT-PCR/Southern analyses. (A) The early-late gene AHR3 expression profile. The 0.22-kb PCR product was generated with the AHR3-specific primers and hybridized with a ^{32}P-labeled AHR3-specific DNA probe. (B) AaFTZ-F1 expression profile. The 0.2-kb PCR fragment was generated with the AaFTZ-F1-specific primers Pr-3 and Pr-4 and hybridized with a ^{32}P AaFTZ-F1 specific probe. (C) Expression profile of the mosquito vitellogenic carboxypeptidase (VCP). VCP is shown as an independent control for RT-PCR/Southern analyses of an ecdysone-inducible late gene in the mosquito fat body. From Li et al. (2000), with permission.

Hybridization is performed utilizing an antisense mRNA probe to a transcript of interest on sections or *en toto* in the case of small organs or the cell culture. In currently used techniques, enzymatic colormetric reactions are used to visualize the hybridization results. The major drawback of this technique is its relative insensitivity; thus, nonabundant or low-abundant mRNAs are hard to detect.

Studies of transcript localizations in mosquito oocytes provide an example of effective use of in situ hybridization. In *Ae. aegypti*, a blood meal triggers massive production of yolk protein precursors by the fat body and their subsequent accumulation by developing oocytes via receptor-mediated endocytosis. The size of oocytes increases more than 300-fold within 36 h post–blood feeding, due mainly to yolk protein accumulation. Mosquito oocytes are extraordinarily specialized for receptor-mediated endocytosis, containing numerous coated vesicles. Three key genes involved in the receptor-mediated endocytosis of yolk protein precursors are the ovary-specific CHC, the Vg receptor (VgR), and the ovarian lipophorin receptor (AaLpRov). Cloning of these three genes has permitted the localization of the transcripts of these genes.

To understand how the transcripts were localized, one must first understand the basic anatomy of *Ae. aegypti* ovaries. Each paired mosquito ovary consists of about 75 ovarioles. In turn, each ovariole contains a single primary follicle and a germarium with underdeveloped secondary follicle that remains under arrest until full maturation of the primary follicle oocyte. The primary follicle is composed of a single oocyte and seven nurse cells surrounded by a monolayer of follicular cells.

The spatial distribution of these three transcripts in the ovary was determined by whole-mount in situ hybridization. For each transcript, the specificity of the hybridization was confirmed by using a control, sense cDNA probe, which for each reaction did not produce any hybridization signal in the mosquito ovaries at any stage of development (Fig. 37.5). The hybridization signal produced by the anti-sense-specific probes was observed in substantial amounts in oocytes of developing primary follicles, whereas nurse cells exhibited considerably lower levels of each transcript. No positive reaction was present in the somatic follicle cells of the primary follicles for any of the three genes. Figure 37.5 shows in situ hybridization of the

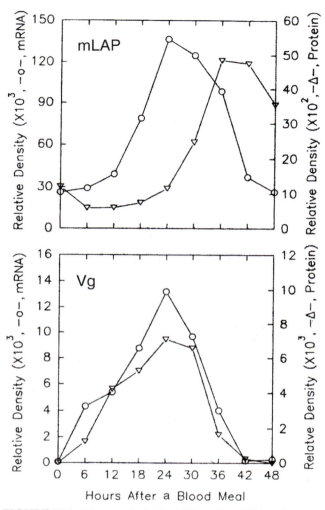

FIGURE 37.4 Kinetics of mLAP and Vg at the mRNA and protein levels in the mosquito fat bodies during vitellogenesis. Open circle, mRNA; open triangle, proteins. The upper panel is mLAP, and the lower panel is Vg. Numbers on the bottom indicate the amount of time after a blood meal when fat bodies were collected. From Cho and Raikhel (1992), with permission.

AaCHC transcript. Interestingly, the AaVgR, Aa clathrin heavy chain, and AaLpRov mRNA could also be detected in the germarium, where it is localized in germ-line cells as well as in secondary follicles separated from the germarium at mid-vitellogenesis. Thus, in situ hybridization has been critical in understanding that the expression of gene coding components of the oocyte receptor-mediated machinery occurs in the germ-line cells as a part of the developmental program of ovarian differentiation.

Determination of the Transcription Start Site

Identification of a gene's transcription start site (TSS) is essential for studying the mechanisms governing its transcription and for mapping its regulatory regions, particularly those situated in proximity to a TSS. The primer extension and S1 nuclease protection analyses are the most commonly used techniques for mapping the 5′ end of transcripts and determining the TSS.

In the primer extension assay, a specific end-labeled primer is hybridized to a complementary region of mRNA of the gene under study. This primer is then utilized by reverse transcriptase to synthesize a cDNA molecule based on the RNA as the template. The resulting cDNA is subsequently analyzed on a denaturing polyacrylamide gel. The length of the cDNA reflects the number of bases between the labeled nucleotide of the primer and the 5′ end of the RNA and, thus, the number of bases between the labeled nucleotide and the TSS. The primer extension assay has been used to determine the start sites of transcription in a number of *Ae. aegypti* genes: clathrin heavy chain (CHC), vitellogenic carboxypeptidase (VCP), lysosomal aspartic protease (LAP), and vitellogenin receptor (VgR).

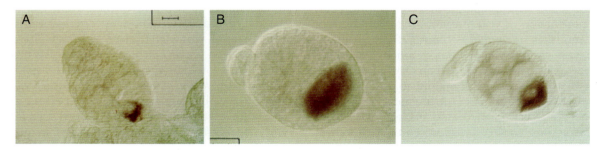

FIGURE 37.5 Spatial distribution of mosquito CHC mRNA in ovaries dissected from previtellogenic and vitellogenic females. Whole-mount preparations were hybridized in situ with a digoxigenin-labeled antisense single-stranded 3.9-kb *Ae. aegypti* CHC cDNA probe. (A) Ovary dissected from a 0- to 1-day-old previtellogenic female; (B) ovary dissected from a 3- to 4-day-old previtellogenic female; (C) ovary dissected from a vitellogenic female, 6 h PBM. From Kokoza and Raikhel (1997), with permission.

Because the primer extension reaction is often prone to terminate prematurely due to mRNA secondary structure, the S1 nuclease protection assay is often carried out in parallel, to confirm the position of a TSS. In the S1 nuclease protection assay, a fragment of cDNA (complementary to the RNA of interest) that extends beyond the 5′ end of the RNA is labeled at a restriction site within the RNA-complementary region. The labeled DNA is hybridized to RNA and then digested with the single-strand specific nuclease S1. The resultant fragment of protected cDNA is run on a denaturing gel to determine its size.

To perform either of the two assays mentioned, cDNA sequences that are close to the 5′ end of the transcripts are indispensable. When cDNA clones are first obtained, either by screening cDNA libraries or by RT-PCR, they are often truncated at one or both of the 5′ and 3′ ends. These truncated ends can be isolated with a procedure known as rapid amplification of cDNA ends (RACE), which amplifies nucleic acid sequences from a messenger RNA template between a defined internal site and unknown sequences at either the 3′ or the 5′ end of the mRNA.

3′ RACE takes advantage of the natural poly(A) tail in mRNA as a generic priming site for PCR amplification. In this technique, an mRNA is converted into cDNA using reverse transcriptase (RT) and an oligo-dT adapter primer. Specific cDNA is then directly amplified by PCR using a gene-specific primer (GSP) that anneals to a region of known exon sequences and an adapter primer that targets the poly(A) tail region. This permits the capture of unknown 3′-mRNA sequences that lie between the exon and the poly(A) tail.

5′ RACE, or "anchored" PCR, is a technique that facilitates the isolation and characterization of 5′ ends from low-level messages. First-strand cDNA synthesis is primed using a gene-specific antisense oligonucleotide (GSP1) that permits cDNA conversion of a specific mRNA and maximizes the potential for complete extension to the 5′ end of the message. Following cDNA synthesis, the first-strand product is purified from unincorporated dNTPs and GSP1. Terminal deoxynucleotidyl transferase (TdT) is used to add homopolymeric tails to the 3′ ends of the cDNA. Tailed cDNA is then amplified by PCR using a mixture of three primers: a nested gene-specific primer (GSP2), which anneals at the 3′ end to GSP1; and a combination of a complementary homopolymer-containing anchor primer and corresponding adapter primer, which permit amplification from the homopolymeric tail. This allows amplification of unknown sequences between the GSP2 and the 5′ end of the mRNA. Products generated by the 3′ and 5′ RACE procedures may be combined to generate a full-length cDNA for a gene of interest.

Molecular characterization of the mosquito clathrin heavy chain (CHC) gene presents a good example of isolating 5′ and 3′ ends of transcripts, followed by mapping the transcription initiation sites. When the mosquito CHC cDNA clones were first isolated from a cDNA library, all clones contained only the carboxyl half of the CHC protein-coding region and part of the 3′ untranslated region. In order to obtain the full-length CHC cDNA clone, 5′ RACE and 3′ RACE PCRs were carried out. Sequencing of the 5′ RACE products revealed two classes of cDNA (0.34- and 0.68-kb) encoding CHC isoforms, *Ae. aegypti* CHCa and *Ae. aegypti* CHCb, which differed in their 5′-UTRs and NH$_2$-terminal coding portions. Probes designed from both isoform-specific sequences were used to screen the genomic library, and two phage clones, CHC4B and CHC4D, were isolated. Restriction mapping and sequencing of restriction fragments hybridizing to probes from *Ae. aegypti* CHCa and *Ae. aegypti* CHCb demonstrated that the two clones overlap and that the two isoforms are coded by alternatively spliced exons (Fig. 37.6).

Primer extension and S1 nuclease protection experiments were conducted to determine the transcription start site of the ovarian mRNA-coded isoform *Ae. aegypti* CHCb. A 30-base oligonucleotide primer, complementary to the sequence of putative exon 1b, was hybridized to total previtellogenic ovarian RNA and extended using Superscript II reverse transcriptase. This resulted in a 552-bp extension product, placing the transcription start site 432 bp upstream of the available cDNA. In parallel, S1 nuclease protection experiments were performed. When a 522-nucleotide DNA probe, corresponding to a genomic fragment containing most of the putative exon 1b and extending upstream, was hybridized to total ovarian RNA, one strong S1-resistant product of 502 bp was detected. The size of exon 1b was calculated to be 692 bp, and the inferred start site position corresponds well with that predicted from the primer extension experiments (Fig. 37.7).

Functional Studies of Tissue- and Stage-Specific Gene Expression

When initiating analysis of transcriptional control of a particular gene, a critical goal is development of assays for measuring the activity of its relevant regulatory regions. With appropriate functional assays, it is possible to assess a region of the gene that would mimic the gene's endogenous expression pattern of the gene. Mutations can then be introduced into this

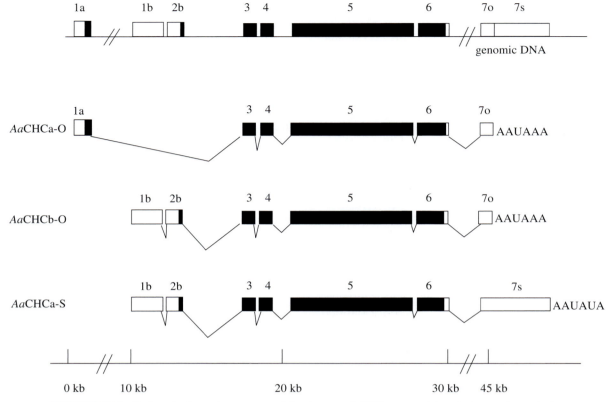

FIGURE 37.6 Deduced schematic structure of the *Ae. aegypti* CHC gene and its relation to alternatively spliced mRNAs. The *Ae. aegypti* CHC gene is depicted as boxes and lines representing exons and introns, respectively. Exons are numbered, and exon sequences containing untranslated mRNA sequences are represented as open boxes. Three putative transcripts generated from the *Ae. aegypti* CHC gene are shown. Two alternative polyadenylation signals AAUAAA and AAUAUA are indicated. From Kokoza and Raikhel (1997), with permission.

region to identify important regulatory elements and, ultimately, important transcription factors.

When studying tissue- and stage-specific gene expression, it is imperative to have easy access to cells or tissues of specific age and specialization. *In vitro* organ culture, a technique that relies on the use of an isolated tissue, is ideally suited for preliminary assessments of gene expression. For most vectors, the small size of their dissected organs or tissues permits cells good access to oxygen, which allows a considerable culturing time without apparent damage to the tissue. An *in vitro* organ culture is particularly useful for detecting transcripts when it is combined with such a sensitive technique for transcript detection as RT-PCR.

Numerous examples of *in vitro* organ culture are available in the literature. One of the most striking illustrations of the organ culture technique is how it was used in *Ae. aegypti* to analyze the effect of the steroid hormone 20E-hydroxyecdysone on the

expression of the genes for the ultraspiracle (USP) in the fat body. The USP is an obligatory dimerization partner for the ecdysteroid receptor (EcR). Two USP isoforms, USP-A and USP-B, occur in *Ae. aegypti*, and these two isoforms are expressed at different vitellogenic stages and apparently differentially controlled by this steroid hormone. In the fat body, USP-A mRNA is highly expressed during the pre- and late-vitellogenic stages, corresponding to a period of low ecdysteroid titer, while USP-B mRNA exhibits its highest levels during the vitellogenic period, correlating with a high ecdysteroid titer. 20-Hydroxyecdysone (20E) has opposite effects on USP isoform transcripts in *in vitro* fat-body culture. This steroid hormone up-regulates USP-B transcription, and its presence is required to sustain a high level of USP-B expression. In contrast, 20E inhibits activation of USP-A transcription (Figs. 37.8A and B). As a positive control, the effect of 20E on the VCP gene was monitored as well (Fig. 37.8C). Utilization of *in vitro* fat-body culture has proven to be an

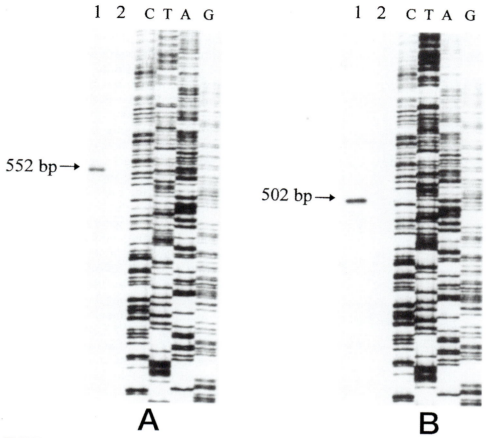

FIGURE 37.7 Mapping the transcription initiation site. (A) Primer extension was performed with a 30-mer antisense oligonucleotide complementary to position 91–121 of the *Ae. aegypti* CHCb cDNA. The primer was labeled, annealed to 10 µg of total RNA from previtellogenic ovaries (lane 1) or 10 µg of control tRNA (lane 2), and extended with Superscript II reverse transcriptase. The arrow indicates the position of the extended product. (B) S1 nuclease mapping. 10 µg of total RNA from previtellogenic ovaries (lane 1) or 10 µg of control tRNA (lane 2) were probed with a 5′-end-labeled 522-bp fragment of single-stranded DNA derived from the genomic sequence. The arrow shows the position of the protected fragment. The sequencing ladder (CTAG) prepared from M13mp18 DNA served as the size reference in both experiments. From Kokoza and Raikhel (1997), with permission.

effective technique for the analysis of several regulatory genes in this mosquito tissue.

Transient Transfection Assay

In general, most functional assays rely on so-called *reporter genes*, which are known genes whose transcription is regulated by the gene of interest and whose mRNA or protein level can be measured easily and accurately. If a regulatory region of interest is a promoter, it is placed immediately upstream of the reporter gene such that the promoter will drive reporter gene transcription. If the control region of interest is an enhancer or another control region that functions at a distance from the promoter, then a well-characterized promoter is usually placed upstream of the reporter gene, with the enhancer inserted upstream of this promoter or downstream of the reporter gene. The reporter gene and the control region of interest can be then studied via transient transfection assay or they can be stably integrated into the genome of an insect vector via germ line transformation (see later), and the precise activities of the control region can be determined.

Transient transfection assay has successfully been used to study the regulation of transcription. In this assay, plasmids containing the control region of interest are introduced into cells maintained in culture by a technique called *cell transfection*. This assay is considered to be transient because the plasmids are not integrated into the host cell genome. As such, mRNA or protein synthesis from the reporter gene must be

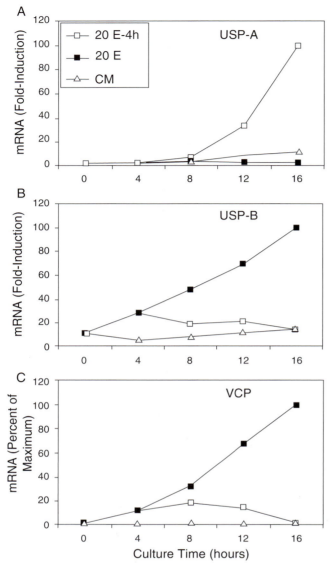

FIGURE 37.8 Effect of 20-hydroxyecdysone on the transcription of USP-A (A), USP-B (B), and VCP (C) in the previtellogenic fat bodies. Previtellogenic fat bodies dissected from female mosquitoes (3–5 days after eclosion) were incubated in culture media only (CM), with a 4-hour pulse treatment of 10^{-6} M 20E (20E-4h), with continuous 20E treatment (20E) for 16 hours. At every 4-hour interval, a group of 9 fat bodies were collected and subjected to RNA isolation, RT-PCR amplification, and Southern blot analysis. The intensity of bands from Southern blots was quantified by phosphorimaging. For USP-A and USP-B, the level of amplified transcript detected from previtellogenic fat bodies prior to any treatment was defined as 1 unit. For the VCP mRNA, the level of amplified transcript detected from previtellogenic fat body treated only with 20E for 16 hours was defined as 100 units. From Wang et al. (2000), with permission.

measured within a short time period after transfection, ranging from 1 to 3 days. The transient assay is rapid and simple to perform; however, there are several primary limitations of this assay. First, the plasmids are not in an appropriate chromatin configuration. This may cause a control region or element to function aberrantly. Second, because there are only a few differentiated cell lines available for insects, some essential tissue- or cell-specific transcription factors or cofactors may be absent in host cells in the transfection assay.

Analysis of the 5′ regulatory region of the mosquito vitellogenin gene provides an example of successful utilization of the transient transfection assay. Cloning and analysis of the 5′ upstream regulatory region of the mosquito Vg gene has revealed the presence of a putative binding site for the ecdysteroid receptor, suggesting direct involvement of this receptor for the steroid hormone, 20-hydroxyecdysone, in regulation of this gene. *Drosophila* S2 cells were used to investigate whether or not the putative binding represented a functional binding element for the ecdysteroid receptor (EcRE). The 2.1-kb Vg 5′ upstream regulatory region was placed in front of a promoterless luciferase gene, used as a reporter. Reporter gene responses were boosted to 2.5-fold induction when exogenous *Aedes* EcR expression vector was transfected into cells along with *Aedes* USP, indicating that the 2.1-kb region harbored elements responsive to 20E. Additional reporter plasmids were constructed, all containing different deletions in this 2.1-kb region. All these plasmids were tested for their 20E response. Based on their responses, the sequence between −348 bp and −102 bp of the Vg promoter was inferred to constitute the putative ecdysteroid-responsive region (ERR). To further confirm that the ERR was responsible for 20E responsiveness, a reporter plasmid, VgΔ$^{348–102}$-Luc, containing the entire 2.1-kb Vg regulatory region with a deletion between −348 bp and −102 bp, was tested in S2 cells, demonstrating that it lacked any responsiveness to 20E (Fig. 37.9). Thus, utilization of the transfection assay in the heterologous cell line has been an effective technique for identifying some properties of the mosquito gene.

Study of the Gene Expression Using Heterologous Transgenic Organisms

Using stable transformation in transgenic organisms offers unparalleled opportunities for the analysis of tissue- and stage-specific gene expression. Transgenesis has been developed recently for *Aedes* and *Anopheles* mosquitoes; previously this approach was amenable for routine experimentation in insects using only *Drosophila* P-element-mediated transformation.

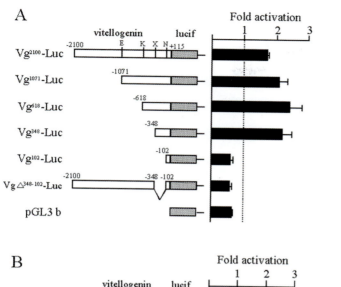

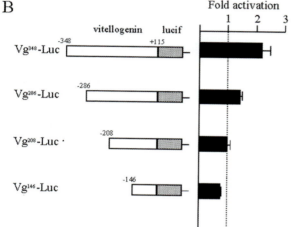

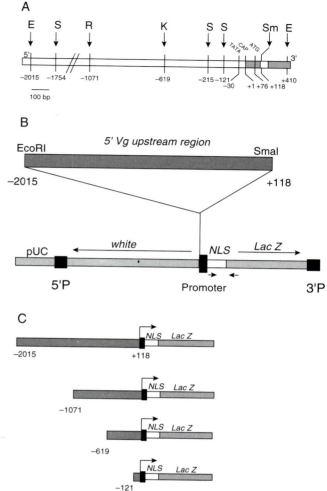

FIGURE 37.9 (A) Mapping of ecdysteroid-responsive region (ERR) in the Vg promoter. S2 cells were transfected with 25 ng pAc5-LacZ, 12.5 ng of each expression vector, pAc5-AaEcR and pAc5-AaUSP, together with 100 ng reporter plasmid Vg 2100-Luc (row 1), or reporter constructs with serial deletion of Vg promoter, Vg1071-Luc (row 2), Vg 618-Luc (row 3), Vg 348-Luc (row 4), Vg 102-Luc (row 5), VgD 348:102-Luc (row 6), or pGL3 (row 7). (B) Localization of the EcREs in the Vg promoter by progressive deletion analysis. The same as in (A), except different reporter constructs were used, Vg 348-Luc (row 1), Vg 286-Luc (row 2), Vg 208-Luc (row 3), and Vg 146-Luc (row 4). After transfection, cells were exposed to 1 mM 20E or ethanol as a control for 24 h. Results are expressed as fold-induction of luciferase activity from cell treated with 20E over that treated with ethanol. Transfection assays were performed in triplicates and repeated two to four times. From Martin et al. (2001), with permission.

FIGURE 37.10 (A) Schematic diagram of a 2.4-kb EcoRI-EcoRI fragment of genomic DNA containing the mosquito Vg gene 5′ upstream region with restriction enzyme map and positions of regulatory and coding sequences. (B) Structure of C4PLZ transformation vector with a 2.1-kb EcoRI-SmaI fragment of Vg 5′ upstream region used in this study to transform *Drosophila* yw67 mutant strain. (C) Extent of 5′ deletions of the upstream flanking sequences of the mosquito Vg gene that were inserted into C4PLZ vector and used for transformation of fruit flies. Numbers refer to the nucleotide position relative to the transcription start site. E, EcoRI; R, EcoRV; K, KpnI; S, Sau3A; Sm, SmaI; NLS, nuclear localization signal. The arrows below the diagram show the positions of primers used in RT-PCR to study expression of Vg-LacZ transgenes. From Kokoza et al. (2001), with permission.

Furthermore, transgenesis is not yet available for most vector arthropods. However, the striking conservation of transcriptional machinery, including tissue-specific transcription factors, permits utilization of *Drosophila* P-element-mediated transformation for the analysis of vector insect genes.

Drosophila P-element-mediated transformation has been utilized to identify the regulatory elements in the

Vg gene of *Ae. aegypti* that determine its tissue-specific expression. The 2.1-kb 5′ upstream putative regulatory region of the Vg gene was subcloned into the *D. melanogaster* P-element transformation vector, in which the Vg promoter was fused to a LacZ gene encoding for β-galactosidase as a reporter gene, which can be detected by X-gal staining (Figs. 37.10A, B). A series of deletions were made in the 5′ upstream region of the Vg gene in order to localize boundaries of the 5′ region required for tissue-, stage-, and sex-specific expression of the gene (Fig. 37.10C). The expression patterns of the LacZ gene driven by a 2.1-kb Vg upstream region were monitored by staining with X-gal in dissected flies and by quantitative RT-PCR. The 2.1-kb fragment was sufficient for a correct tissue- and stage-specific expression of the reporter in *Drosophila.* Flies with this transgene expressed β-galactosidase only in nuclei of the vitellogenic female fat body (Fig. 37.11). No expression was observed in nuclei of the larval fat body or other tissues (not shown).

Further analyses of transgenic flies using quantitative RT-PCR demonstrated that the level of LacZ gene expression in the fat body of transgenic flies depended on the length of the 5′ upstream Vg region. The 2.1-kb Vg-LacZ transgene exhibited a high level of expression (Fig. 37.12A). The same pattern of expression was observed for 1.2-kb Vg-LacZ transgene; however, the average expression level was much lower, reaching only 35–40% of that in the longer construct. Transgenic females containing the 0.7-kb construct of Vg-LacZ transgene demonstrated low fat body–specific expression, usually less than 10% of the 2.1-kb Vg-LacZ construct. In contrast, the shortest 0.2-kb Vg-LacZ transgene construct had low but detectable expression in all tested tissues. The three longest constructs were then tested for the timing of their expression in the transgenic flies. The expression of all three transgenes reached their respective maximal level in 1-day-old females and was maintained at the same high level in 7-day-old females (Fig. 37.12B).

These analyses using transgenic *Drosophila* demonstrate a remarkable conservation of the regulatory region of the Vg gene directing the expression of the reporter in a correct stage- and tissue-specific manner similar to the endogenous *Drosophila* yolk protein genes.

Using Homologous Transgenesis for the Analysis of Tissue- and Stage-Specific Gene Regulation

Despite technological advances of transgenesis in *Drosophila*, this organism cannot be used for the analysis of genes involved in host–vector interaction. For example, a unique feature of genes encoding yolk protein precursors in anautogenous mosquitoes is the requirement of a blood meal for their activation. Only after a blood meal is the major YPP gene, Vg, transcribed at a very high level in the female fat body. Recent progress in mosquito transformation has made it possible to test the regulatory mechanism

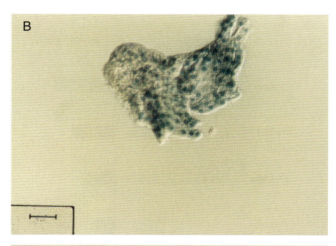

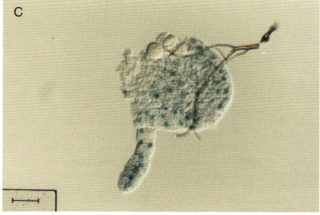

FIGURE 37.11 Histochemical detection of the β-galactosidase activity in 3- to 5-day-old female flies transformed by mosquito Vg-LacZ transgenes. (A) Transformed with 2.1-kb Vg-LacZ construct. Isolated fat body of transgenic flies carrying 2.1-kb Vg-LacZ construct (B) and 1.2-kb Vg-LacZ construct (C). From Kokoza et al. (2001), with permission.

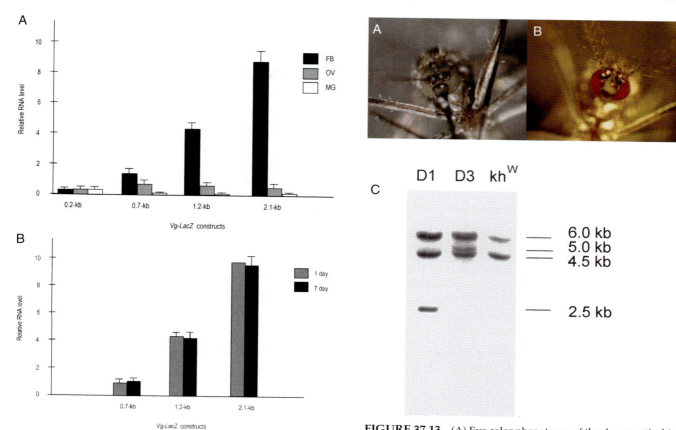

FIGURE 37.12 (A) Analysis of tissue specificity of Vg-LacZ expression in different transgenic lines. The bars show the results of RT-PCR experiments given as a relative expression level of β-galactosidase/rp49 RNA. RNA was isolated from fat body (FB), ovary (OV), and midgut (MG) of 1- to 3-day-old vitellogenic females. The quantification of the β-galactosidase/rp49 expression level was calculated using phosphorimager analysis of Southern blot RT-PCR hybridization signals. Data represent the mean of three independent experiments. (B) Relative transcription level of different Vg-LacZ transgenes during *Drosophila* adult life. The ratio obtained for the β-galactosidase/rp49 expression level of 1-day-old females of a 2.1-kb Vg-LacZ construct was arbitrarily assigned a value of 10, and the ratios obtained at other stages and for other constructs were normalized to it. From Kokoza et al. (2001), with permission.

FIGURE 37.13 (A) Eye color phenotypes of the *Ae. aegypti* white-eye khw host strain and (B) red-eye pH[cn][Vg-DefA] transgenic D3 strain. (C) Southern blot analysis of genomic DNA isolated from the two different strains D1 (2.1-kb Vg promoter) and D3 (1.1-kb Vg promoter). The DNA from the G10 generation was digested with the restriction enzymes XbaI-ApaI and hybridized with the DefA probe. A hybridization band of 2.5 kb corresponds to the transgene in the D1 strain and the one of 5.0 kb corresponds to the other in the D3 strain. From Kokoza et al. (2001), with permission.

responsible for this blood meal–dependent activation of the Vg gene.

The regulation of the *Aedes Vg* promoter was analyzed first by using a Hermes-mediated gene transformation in *Ae. aegypti*. Stable transgenic *Ae. aegypti* lines have been generated, containing a chimeric transgene with the 2.1-kb 5′ upstream region of the Vg gene and a coding region of the *Ae. aegypti* defensin A gene (*DefA*) as a reporter (Fig. 37.13). The *Hermes* transposable element as a vector and the *Drosophila* cinnabar gene as a marker were used to transform the white-eye *Aedes aegypti* host strain. Polymerase chain reaction

amplification of genomic DNA and Southern blot analyses, carried out through the ninth generation (G9), showed that the Vg-DefA transgene insertion was stable. The Vg-DefA transgene was strongly activated in the fat body by a blood meal. Similar to the endogenous Vg gene, the Vg-DefA transcript was upregulated only in fat bodies of blood-fed females. Likewise, the mRNA levels reached a maximum at 24 h post–blood meal, corresponding to the peak expression time of the endogenous Vg gene. Furthermore, it exhibited a developmental profile that mirrored that of the endogenous Vg gene, indicating that the 2.1-kb 5′ upstream region of the Vg gene contained information required for correct tissue- and stage-specific expression (Fig. 37.14). In summary, a 2.1-kb Vg upstream promoter fragment was sufficient for activation of the transgene by a blood meal–triggered

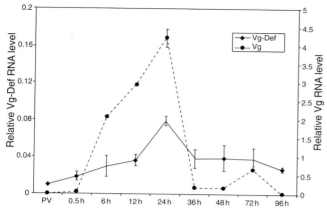

FIGURE 37.14 Developmental profiles of Vg-Defensin mRNA and Vg mRNA expression in the fat bodies of transgenic mosquitoes of D1 line at different time points after blood meal. The relative transcription level for both mRNAs was normalized to actin mRNA expression. Data represent the mean of three independent experiments. From Kokoza et al. (2001), with permission.

regulatory cascade in a tissue-specific manner. It was also adequate to drive a high level of the transgene expression.

Further progress utilization of homologous transformation of vectors depends largely on the development of a highly effective routine gene transformation for vector insects. The piggyBac transposable vector pBac[3xP3-EGFP afm] and the selectable marker green fluorescent protein, EGFP, under the control of the promoter 3xP3 expressed in the eyes has proven to be an efficient system for transformation of *Ae. aegypti*. Several stable transformant lines showed intense fluorescence in their eyes (Fig. 37.15A). Importantly, strong fluorescence could be detected in larval stages of wild-type Rockefeller strain of *Ae. aegypti*, permitting early screening of transformants (Fig. 37.15B). These results clearly indicate that the 3xP3-EGFP marker can be used for microinjection directly into wild-type mosquitoes.

CHARACTERIZATION OF BINDING SITES FOR DNA-BINDING PROTEINS

Electrophoretic Mobility Shift Assay (EMSA)

After delineating regulatory DNA sequence elements within a control region of interest, the identity of specific binding sites for transcription factors can be further characterized. The most common strategy begins with a database search to predict which pro-

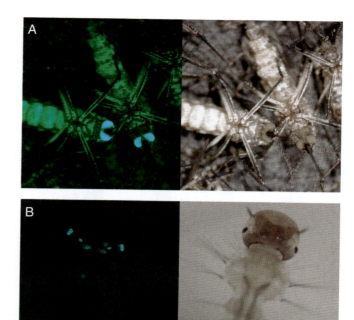

FIGURE 37.15 Expression of the 3xP3-EGFP selectable marker in the eyes of transgenic *Aedes aegypti*. (A) EGFP expression in adult eyes of transgenic *khw* mosquitoes (white-eye strain) transformed by pBac[3xP3-EGFP]Vg-DefA. (B) Expression and visibility of the EGFP selectable marker in the eyes of a black-eye hybrid third-instar larva resulted from the crossing of a *khw* transgenic female with a wild-type (Rockefeller/UGAL) untransformed male. The visibility of the EGFP expression in the mosquitoes with black-pigmented eyes was limited to the larval and pupal stages. The left panels are fluorescent images; the right panels are bright-light images. From Kokoza et al. (2001a), with permission.

teins or protein families might be capable of binding a particular DNA sequence. If a candidate binding site is identified, a protein–DNA interaction assay, such as an electrophoretic mobility shift assay (EMSA), can be performed with the recombinant or *in-vitro*-translated proteins to determine whether they are indeed capable of binding the putative binding site. If binding is observed, EMSA can be performed with extracts from appropriate tissues to determine whether the binding activities have the potential to be functionally relevant *in vivo*.

Many protein–DNA complexes are sufficiently stable that they remain bound during electrophoresis through a (nondenaturing) polyacrylamide gel. In EMSA, a selected restriction fragment or synthetic duplex oligonucleotide is labeled (to make a probe) and mixed with a protein (or crude mixture of proteins). If the DNA fragment binds to the protein, the complex migrates much slower in the gel than does the free probe. The presence of a slowly moving signal is indicative of a complex between the DNA probe and some protein(s). By incubating the probe and proteins in the presence of increasing amounts of competitor fragments or antibodies against the proteins, one can test for specificity and even gain some information about the identity of the binding protein.

Electrophoretic mobility gel shift assays have been used extensively in the analysis of transcription factors bound to regulatory regions of the mosquito Vg gene. An example of analytical EMSA is provided by the study of the binding properties of the ecdysone receptor (EcR) and the nuclear receptor ultraspiracle (USP) from *Ae. aegypti*. In this study, the DNA-binding properties of the EcR/USP heterodimer have been analyzed with respect to the effects of nucleotide sequence, orientation, and spacing between half-site consensus sequence AGGTCA in various natural and synthetic ecdysteroid response elements (EcRE). The mosquito ecdysteroid receptor has been shown to exhibit a broad binding specificity, forming complexes with inverted (IR) and direct (DR) repeats of sequence AGGTCA (the consensus half-site of nuclear receptor response elements) separated by spacers of variable length (Fig. 37.16). Competition experiments and direct estimation of binding affinity indicated that, given identical consensus half-sites and an optimal spacer, the EcR/USP heterodimer bound an IR with higher affinity than a DR. This study has predicted that the type of EcRE present in the regulatory region of an ecdysteroid-responsive gene can determine how the hormone 20-hydroxyecdysone governs its expression. Indeed, it has been established that the mosquito 20-ecdysteroid responsive and fat body–specific genes, Vg and VCP, contain a weak DR-1 EcRE in their regulatory regions.

Electrophoretic mobility gel shift assay can serve as an effective means for detecting the presence of functional transcription factors or complexes in tissues of a particular developmental stage. For example, using EMSA, the presence of specific binding to EcRE was detected in fat-body nuclear protein extracts (NEs) isolated from mosquitoes after a blood meal (Fig. 37.17). Given the protein complexity of nuclear extracts, controls are essential. In this case, the specificity of binding was confirmed by competition with the excess of unlabeled specific EcRE probe as well as by supershifts with antibodies against either *Aedes aegypti* EcR or *Drosophila* USP (Fig. 37.17). Nonspecific probes were used as additional controls. Thus, it was confirmed that the retarded band in EMSA of vitellogenic nuclei was due to binding of the EcR/USP complex.

Once the binding activity of a specific transcription factor's binding activity is detected in the fat-body nuclear extract, EMSA can be further utilized for the

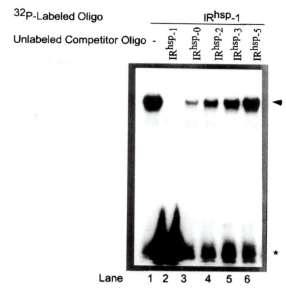

FIGURE 37.16 Effect of spacer length in the imperfect inverted repeats (IRhsps) on binding with AaEcR/AaUSP. EMSAs of *in vitro*–synthesized AaEcR/AaUSP incubated with 0.05 pmol of ^{32}P-labeled IRhsp-1 in the absence (lane 1) or presence of a 50-fold molar excess of unlabeled IRhsp-1 (lane 2) or its variants, with varying length of spacer nucleotides between the two half-sites (IR^{hsp}-0, IR^{hsp}-2, IR^{hsp}-3, IRhsp-5, lanes 3–6, see the following list for each sequence). The position of the AaEcR/AaUSP/^{32}P-IRhsp-1 complex is indicated by an arrowhead and that of the free probe by an asterisk. Oligonucleotides used in this experiment (only one strand of each probe is shown):

IR^{hsp}-0: agagacaagGGTTCATGCACTtgtccaa
IR^{hsp}-1: agagacaagGGTTCAaTGCACTtgtccaat
IR^{hsp}-2: agagacaagGGTTCAatTGCACTtgtccaa
IR^{hsp}-3: agagacaagGGTTCAaatTGCACTtgtccaa
IR^{hsp}-5: agagacaagGGTTCAaataaTGCACTtgtccaa

From Wang et al. (1998), with permission.

analysis of the factor's developmental profile in the tissue of interest. For example, when this approach was applied for the vitellogenic mosquito fat body using the native Vg EcRE, it was revealed that the intensity of the EcR/USP complex binding coincides with major vitellogenic events in this tissue (Fig. 37.18A).

The second complex detected in the vitellogenic fat-body nuclear extracts corresponded to a GATA-like factor. A complex could be detected using one of the GATA binding sites from the proximal Vg gene region in EMSA. The GATA-specific band was found in NE of vitellogenic fat bodies as early as 1 h after blood feeding, and the intensity of the band increased in parallel with the increasing level of the Vg gene expression, reaching the highest levels at the peak Vg gene expression and then decreasing and disappearing at

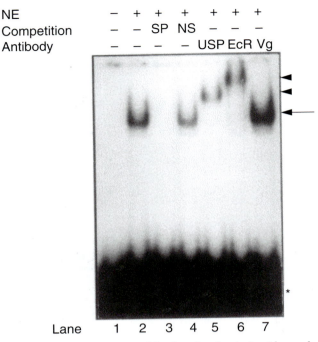

FIGURE 37.17 Detection of the functional ecdysteroid complex (EcR/USP) in the nuclear protein extracts (NEs) from mosquito fat body. Fat-body NEs were prepared from 500 female mosquitoes 24 h after a blood meal. NE was incubated with radiolabeled IR^{hsp}-1 probe (see Fig. 37.16). One hundred molar excess of unlabeled specific probe, IR^{hsp}-1 (SP) or a double-stranded oligonucleotide of unrelated sequence (NS) were used as controls (lanes 3 and 4, respectively). Supershift controls were performed using anti-USP (lane 5), anti-EcR (lane 6), and anti-Vg (lane 7) antibodies. The position of each specific supershift is indicated by an arrowhead. Note the lack of a supershift with nonspecific anti-Vg antibodies (arrow). The position of free probe is marked by an asterisk. From Miura et al. (1999), with permission.

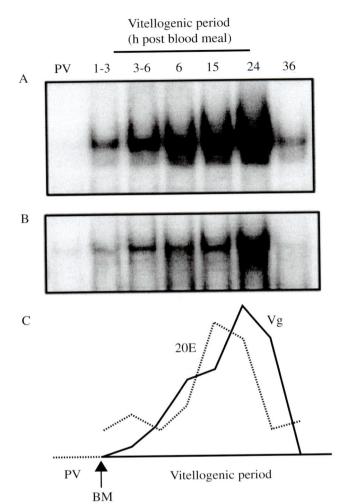

FIGURE 37.18 Developmental profile of the native AaEcR-AaUSP EcR (A) and AaGATA complexes (B) in *Ae. aegypti* fat-body nuclear extracts during vitellogenesis. Fat-body nuclear extracts were prepared throughout previtellogenesis (PV) and at different hours during the vitellogenic period and subjected to EMSA with ^{32}P-labeled VgEcRE and VgGATA binding sites, respectively. (C) Corresponding profiles of the 20E titer and the expression of the Vg gene in the fat body. BM, blood meal. From Kokoza et al. (2001), with permission.

the time of termination of vitellogenesis (Fig. 37.18B). This analysis has revealed that another level in regulation of Vg gene expression in the mosquito fat body is provided by the control of transcription factors themselves. It has been shown that levels of active factors critical for transcription of Vg gene greatly enhanced during the vitellogenic cycle and that transcription factors are themselves targets of the blood meal–mediated regulatory cascade.

Evaluating the Role of Protein–Protein Interactions in the Regulation of Gene Expression

As discussed earlier, transcriptional regulation of a particular gene requires coordinated binding of multiple transcription factors to its promoter-enhancer region. Furthermore, cofactors that do not directly bind DNA but rather interact with transcription factors play a vital role in this process. Therefore, understanding the interaction between transcription factors and their respective cofactors is essential in dissecting a regulatory hierarchy of a particular gene.

Several techniques have been used widely and efficiently to identify proteins that participate in this particular type of gene regulation. One technique, known as the *two-hybrid system*, is especially well designed for such investigations. It employs the yeast *Saccharomyces cerevisiae* as a physiologic test tube. This method relies on the modular properties of site-specific transcriptional activators. Two hybrid proteins are constructed: one, the "bait," which consists of the DNA-binding domain (DBD) of yeast Gal4 or bacterial LexA factors fused to one of the proteins of interest; the other, the "target," contains an activation domain of Gal4 or LexA fused with the second tested protein. The system exploits reporter genes under the control of the DNA-binding site specific to either Gal4 (UAS) or LexA factors. Binding of the bait alone to its response element does not activate the reporter genes, while interaction of the bait and the target brings about the activation. Therefore, protein–protein interaction can be detected and measured by the activity of reporter genes.

The latter application of the two-hybrid assay is similar to that of the GST-pulldown assay. In this assay, a bait protein is fused to glutathione S-transferase (GST) and an expressed fusion product is fixed to glutathione-agarose beads. The protein of interest is then tested for interaction, and results are visualized by immunoblotting with antibodies against the tested protein. The GST-pulldown assay is often used in conjunction with the two-hybrid approach to verify proteins that are isolated by it.

An example of using both techniques is provided by the study of the interaction of AaUSP, an obligatory partner of the ecdysteroid receptor, with a repressor AHR38, which belongs to the family of nuclear receptors. An important adaptation for anautogeny is the establishment of previtellogenic developmental arrest (the state of arrest), preventing the activation prior to blood feeding of yolk protein precursor genes in previtellogenic competent females. An inhibition of the ecdysteroid-signaling pathway is the essential part of the state of arrest, which may be maintained *in vivo* by undetermined factors. In *Ae. aegypti*, both EcR and USP proteins are abundant in nuclei of the previtellogenic female fat body at the state of arrest; however, the EcR-USP heterodimer capable of binding to the specific EcREs is barely detectable in these nuclei (Fig. 37.18A). A possible mechanism through which the formation of ecdysteroid-receptor activity can be regulated is a competitive binding of other factors to either EcR or USP. AHR38 inhibited ecdysone-dependent activation of a reporter gene mediated by the ecdysone-receptor complex in transient transfection assays. Analyses using two-hybrid tests and GST pulldown assays demonstrated that USP, a component of functional ecdysone-receptor complex, interacted directly with AHR38 (Fig. 37.19).

This hypothesis was supported by coimmunoprecipitation experiments, which were designed to detect protein–protein interactions *in vivo*. Fat-body nuclear

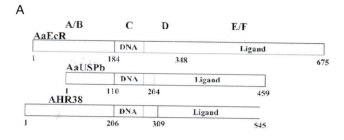
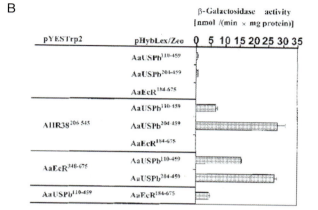
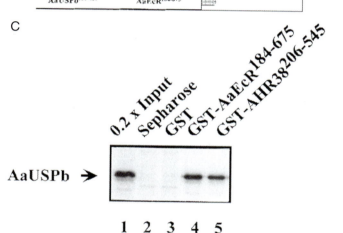

FIGURE 37.19 Yeast two-hybrid assay test of interaction of AHR38 with AaUSP. (A) Schematic representation of the domain structure of AaEcR, AaUSPb, and AHR38 used in generating fusion protein constructs for the yeast two-hybrid assays. (B) Yeast two-hybrid assays reveal interactions between AHR38 and AaUSP. Yeast cells were cotransformed with the indicated pHybLex/Zeo and pYESTrp2 constructs. β-Galactosidase activity was measured and normalized for protein concentration. The data represent means ± SD of three experiments. (C) GST assay showing that AHR38 binds directly to AaUSP. [35S]Methionine-labeled AaUSPb was incubated with glutathione-Sepharose beads (lane 2), beads bound with GST (lane 3), GST-AaEcR (lane 4), or GST-AHR38 (lane 5). The beads were washed in phosphate-buffered saline and collected by centrifugation. Bound proteins were eluted in SDS sample buffer, resolved by SDS-PAGE, and visualized by autoradiography. From Zhu et al. (2000), with permission.

proteins were extracted from previtellogenic and vitellogenic female mosquitoes. First, the USP-containing complexes were immunoprecipitated using the anti-USP antibodies and analyzed by sequentially immunoblotting with antibodies against EcR and AHR38. In the second test, antibodies against EcR and AHR38 were incubated with fat-body nuclear extracts independently. The precipitates were separated on a gel electrophoresis and detected by anti-USP antibody. Both of these tests showed that the USP interacted with AHR38, but not EcR, in the fat-body nuclei of previtellogenic female mosquitoes during the state of arrest. After blood meal, USP heterodimerizes with EcR, while active YPP synthesis occurred in the fat body (Fig. 37.20). These findings suggest that AHR38 is involved in modulating 20E signaling by direct interactions with USP in the fat body of *Ae. aegypti* during the vitellogenesis. This example clearly demonstrates the power of techniques for protein–protein interaction in solving some important issues relevant to vector biology.

In Vivo Detection of Protein Binding in Chromatin

Gene regulation is a complex process. Numerous regulator proteins are required for the accurate temporal and spatial regulation of each gene. Often these factors are assembled into multiprotein complexes, contributing to specific gene regulation events. Understanding how these factors are organized in the chromosome and how their functions are regulated *in vivo* is a challenging task. One of the most useful techniques for studying this level of gene regulation is a chromatin immunoprecitation (ChIP) assay. The ability to provide direct evidence that given regulatory proteins are associated "in time and space" with specific genomic regions is a key determinant for the merit of the ChIP assay.

The ChIP technique involves *in vivo* cross-linking of a particular transcription factor and its multiprotein cofactor complex to the DNA region at the time of their interaction. This is achieved by formaldehyde fixation of macromolecular chromosomal structures in living material, such as tissue culture cells, isolated tissues, or embryos. The cross-linked chromatin is purified by sonication in a high-detergent buffer and then used as a substrate for immunoprecipitation, with the antibodies directed against a protein of interest. To dissociate DNA from the immunoprecipitated complex, the cross-linking is fully reversed by extensive digestion with proteinase K and mild heat treatment. The specific genomic fragments corresponding to specific protein-binding sites are determined by PCR.

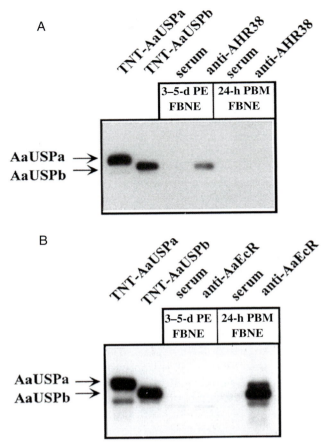

FIGURE 37.20 Coimmunoprecipitation analyses of interactions between AHR38 and AaUSP, AaEcR, and AaUSP in the fat body of *Ae. aegypti* at the previtellogenic and vitellogenic periods. Nuclear extracts were prepared from the fat bodies of 250 adult females for each time point. An aliquot equivalent to 100 mosquitoes was precleared with protein A-agarose and then incubated with rabbit polyclonal anti-AHR38 (A) or rabbit polyclonal anti-AaEcR antibodies (B) or the respective preimmune sera (not shown). The resulting immune complexes were then precipitated by the addition of protein A-agarose beads. After extensive washing, immune complexes were dissociated by boiling the beads in 1× SDS sample buffer. Protein samples were separated by SDS-PAGE, followed by immunoblot analyses with mouse anti-DmUSP antibodies. *In vitro* translated proteins (TNT-AaUSPa, -AaUSPb) were used as controls in Western blot analysis. From Zhu et al. (2000), with permission.

The ChIP assay has been used to analyze whether or not the ecdysteroid receptor binds 5′ regulatory region of the *Aedes* Vg gene in mosquito during vitellogenesis. Transient transfection assays and EMSA experiments suggested that the EcR-USP heterodimer directly bound to one of these EcREs (named VgEcRE1). Fat bodies dissected from vitellogenic female mosquitoes were treated with formaldehyde. Antibodies against EcR or USP were used to immunoprecipitate the sonicated chromatin fragments. The

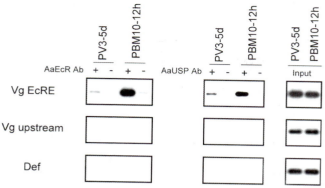

FIGURE 37.21 Transcriptional activation of Vg in response to a blood meal is associated with the binding of AaEcR-AaUSP to the Vg promoter. Cross-linked chromatin preparations from cells of fat body dissected before (PV3–5d) or after (PBM10–12h) blood meal were precipitated with antibodies against either AaEcR or AaUSP. Immunoprecipitated and input material was analyzed by PCR, with primers corresponding to the indicated promoters. VgEcRE, a 300-bp region containing ecdysteroid responsive element (VgEcRE1) of the Vg promoter; Vg upstream, a 300-bp region about 1 kb upstream from VgEcRE1; Def, a 300-bp region of the promoter of mosquito defensin gene, which is not regulated by 20E signaling.

EcR- or USP-bound DNA fragments were then analyzed by PCR, using primers flanking potential EcREs in the Vg gene. The same sequence corresponding to VgEcRE1 was revealed by PCR, indicating that this portion of the gene indeed interacted with both EcR and USP during vitellogenesis after blood meal activation of vitellogenesis (Fig. 37.21; J. Zhu and A. Raikhel, unpublished data). Thus, the ChIP assay confirms the *in vivo* existence of a direct interaction of the Vg gene with the ecdysteroid receptor suggested by investigations described in this chapter.

SUMMARY

In this chapter we have focused on issues that one faces when undertaking a transcriptional regulation analysis in an insect vector, and we have outlined recommended strategies for completing the analysis by illustrating with, in most circumstances, the study of the vitellogenin gene in *Ae. aegypti*. It was not our intention to cover all state-of-the-art techniques available. Besides, this step-by-step approach may not be applicable to all regulatory scenarios. However, we hope our introduction will enhance understanding of the complex gene regulation in insect vectors and help investigators who want to pursue the regulatory mechanisms for a new gene that has been identified.

Acknowledgments

The authors are grateful to Mr. G. Attardo for his help in making diagrams and figures and to Ms Ranae Wooley for editing the manuscript. We acknowledge support from the NIH.

Readings

Regulation of Gene Expression

Carey, M., and Smale, S. T. 2000. *Transcriptional Regulation in Eukaryotes.* Cold Spring Harbor Laboratory Press, Cold Spring Harbor, New York.

Courey, A. J. 2001. Cooperativity in transcriptional control. *Curr. Biol.* 11: R250–R252.

Kornberg, R. D., and Lorch, Y. 1999. Twenty-five years of the nucleosome, fundamental particle of the eukaryote chromosome. *Cell* 98(3): 285–294.

Malik, S., and Roeder, R. G. 2000. Transcriptional regulation through mediator-like coactivators in yeast and metazoan cells. *Trends Biochem. Sci.* 25: 277–283.

Naar, A. M., Lemon, B. D., and Tjian, R. 2001. Transcriptional coactivators complexes. *Annu. Rev. Biochem.* 70: 475–501.

Rachez, C., and Freedman, L. P. 2001. Mediator complexes and transcription. *Curr. Opin. Cell Biol.* 13(3): 274–280.

Roeder, R. G. 1996. The role of general initiation factors in transcription by RNA polymerase II. *Trends Biochem. Sci.* 21: 327–335.

Struhl, K. 1999. Fundamentally different logic of gene regulation in eukaryotes and prokaryotes. *Cell* 98(1): 1–4.

Wallrath, L. L., Lu, Q., Granok, H., and Elgin, S. C. 1994. Architectural variations of inducible eukaryotic promoters: Preset and remodeling chromatin structures. *Bioessays* 16(3): 165–170.

Wu, J., and Grunstein, M. 2000. 25 years after the nucleosome model: Chromatin modifications. *Trends Biochem. Sci.* 25(12): 619–623.

Xu, L., Glass, C. K., and Rosenfeld, M. G. 1999. Coactivator and corepressor complexes in nuclear receptor function. *Curr. Opin. Genet. Dev.* 9: 140–147.

Temporal and Tissue-Specific Gene Expression in Vector Arthropods

Cho, W. L., and Raikhel, A. S. 1992. Cloning of cDNA for mosquito lysosomal aspartic protease. Sequence analysis of an insect lysosomal enzyme similar to cathepsins D and E. *J. Biol. Chem.* 267(30): 21823–21829.

Dimopoulos, G., Casavant, T. L., Chang, S., Scheetz, T., Roberts, C., Donohue, M., Schultz, J., Benes, V., Bork, P., Ansorge, W., Soares, M. B., and Kafatos, F. C. 2000. *Anopheles gambiae* pilot gene discovery project: Identification of mosquito innate immunity genes from expressed sequence tags generated from immune-competent cell lines. *Proc. Natl. Acad. Sci. U.S.A.* 97(12): 6619–6624.

Edwards, M. J., Moskalyk, L. A., Donelly-Doman, M., Vlaskova, M., Noriega, F. G., Walker, V. K., and Jacobs-Lorena, M. 2000. Characterization of a carboxypeptidase A gene from the mosquito, *Aedes aegypti. Insect Mol. Biol.* 9(1): 33–38.

Li, C., Kapitskaya, M. Z., Zhu, J., Miura, K., Segraves, W., and Raikhel, A. S. 2000. Conserved molecular mechanism for the stage specificity of the mosquito vitellogenic response to ecdysone. *Dev. Biol.* 224(1): 96–110.

Kokoza, V. A., and Raikhel, A. S. 1997. Ovarian- and somatic-specific transcripts of the mosquito clathrin heavy-chain gene generated by alternative 5′-exon splicing and polyadenylation. *J. Biol. Chem.* 272(2): 1164–1170.

Kokoza, V., Ahmed, A., Cho, W.-L., Jasinskiene, N., James, A. A., and Raikhel, A. S. 2000. Engineering blood meal-activated systemic immunity in the yellow fever mosquito, *Aedes aegypti. Proc. Natl. Acad. Sci. USA* 97: 9144–9149.

Kokoza, V. A., Martin, D., Mienaltowski, M. J., Ahmed, A., Morton, C. M., and Raikhel, A. S. 2001a. Transcriptional regulation of the mosquito vitellogenin gene via a blood meal-triggered cascade. *Gene* 274(1–2): 47–65.

Kokoza, V. A., Ahmed, A., Wimmer, E. A., and Raikhel, A. S. 2001b. Efficient transformation of the yellow fever mosquito *Aedes aegypti* using the piggyBac transposable element vector pBac[3xP3-EGFP afm]. *Insect Biochem. Mol. Biol.* 31: 1137–1143.

Martin, D., Wang, S. F., and Raikhel, A. S. 2001. The vitellogenin gene of the mosquito *Aedes aegypti* is a direct target of ecdysteroid receptor. *Mol. Cell Endocrinol.* 173(1–2): 75–86.

Miura, K., Wang, S. F., and Raikhel, A. S. 1999. Two distinct subpopulations of ecdysone receptor complex in the female mosquito during vitellogenesis. *Mol. Cell. Endocrinol.* 156: 111–120.

Shen, Z., Dimopoulos, G., Kafatos, F. C., and Jacobs-Lorena, M. 1999. A cell surface mucin specifically expressed in the midgut of the malaria mosquito *Anopheles gambiae. Proc. Natl. Acad. Sci. U.S.A.* 96(10): 5610–5615.

Smartt, C. T., Kim, A. P., Grossman, G. L., and James, A. A. 1995. The Apyrase gene of the vector mosquito, *Aedes aegypti*, is expressed specifically in the adult female salivary glands. *Exp. Parasitol.* 81(3): 239–248.

Sun, J., Hiraoka, T., Dittmer, N. T., Cho, K. H., and Raikhel, A. S. 2000. Lipophorin as a yolk protein precursor in the mosquito, *Aedes aegypti. Insect Biochem. Mol. Biol.* 30: 1161–1171.

Wang, S.-F., Miura, K., Miksicek, R. J., Segraves, W. A., and Raikhel, A. S. 1998. DNA binding and transactivation characteristics of the mosquito ecdysone receptor–ultraspiracle complex. *J. Biol. Chem.* 273: 27531–27540.

Wang, S.-F., Li, C., Zhu, J., Miura, K., Miksicek, R. J., and Raikhel, A. S. 2000. Differential expression and regulation by 20-hydroxyecdysone of mosquito ultraspiracle isoforms. *Dev. Biol.* 218: 99–113.

Zhao, X., Smartt, C. T., Hillyer, J. F., and Christensen, B. M. 2000. A novel member of the RING-finger gene family associated with reproductive tissues of the mosquito, *Aedes aegypti. Insect Mol. Biol.* 9(3): 301–308.

Zhu, J., Miura, K., Chen, L., and Raikhel, A. S. 2000. AHR38, a homolog of NGFI-B, inhibits formation of the functional ecdysteroid receptor in the mosquito *Aedes aegypti. EMBO J.* 19(2): 253–262.

38

Gene Expression in Acarines

DEBORAH C. JAWORSKI

INTRODUCTION

Ticks comprise the superfamily Ixodoidea of the order Parasitiformes and subclass Acari (class Arachnida). Ticks differ significantly from insects in their anatomy and physiology (Chapters 3 and 4). For example, ticks generally lack a peritrophic membrane, and the midgut lumen contains no trypsins. The midgut lumen has a neutral pH, and little blood digestion occurs in this environment. The midgut cells engulf the blood meal, and most digestion occurs intracellularly. Furthermore, ticks have complex salivary glands that secrete cement components and substances that counteract host homeostasis and concentrate the blood meal. For hard ticks the salivary glands are the major osmoregulatory organs.

The Ixodoidea consists of three families: Argasidae, Nuttalliellidae, and Ixodidae. Of the more than 800 tick species identified, more than two-thirds of these ticks belong to the family Ixodidae. This chapter focuses on the soft ticks, Argasidae, and the hard ticks, Ixodidae. Nuttalliellidae exists as one species and only from female specimens. Major emphasis is placed on gene expression in Ixodidae. Although it is inadvisable to generalize about one genus of ticks and assume that it is true of other genera, a lack of research on gene expression makes this difficult to avoid.

TICK FEEDING

Hard-Tick Feeding

Ticks exist in two separate worlds, and it requires unique strategies to negotiate these worlds. The first environment often is referred to as *off-host*; there are no published studies on gene expression in the off-host tick. Ticks spend most of their lives off-host conserving energy by slowly digesting nutrients from the previous blood meal and drinking water from subsaturated air. Bursts of energy are spent by ticks climbing out of their microclimate in pursuit of a host in response to cues such as CO_2. The tick microclimate is found in the layers of vegetation closest to the ground, where the humidity is high. Ticks are sensitive to humidity, and each tick species has a specific *critical equilibrium humidity* (CEH). The CEH is the point at which the tick begins to gain water weight after a period of desiccation (this value must be above 95% relative humidity (RH) for the survival of most tick species). Once a tick attaches to a host and generates a small wound with its chelicerae, it must then rapidly convert to its on-host, or blood, environment. The tick essentially becomes part of the host by laying down cement to seal the tick-feeding site and forming a continuous tube from the host lesion through the mouthparts and into the midgut. Chemical communication begins when the tick responds to the events that lead to host homeostasis by secreting the contents of its salivary glands. Many of the salivary secretions have pharmacological actions. This chemical communication must continue for the adult hard tick for more than 6–10 days. After an initial phase of slow feeding, a rapid-feeding phase ensues when the tick takes in the bulk of the blood meal. A typical feeding period for a female ixodid tick is shown in Figure 38.1. Adult females of the lone star tick, *Amblyomma americanum*, have an expandable exocuticle, with deep folds that open to accommodate the blood meal. The blood meal is concentrated, and the female weight at dropoff (the time when the female leaves the host to lay eggs) is roughly one-tenth of that of the blood meal consumed;

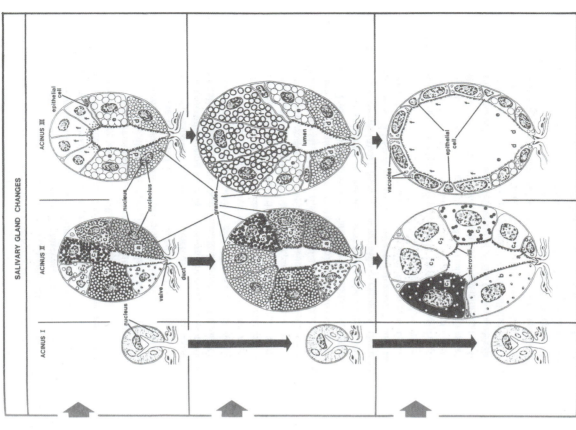

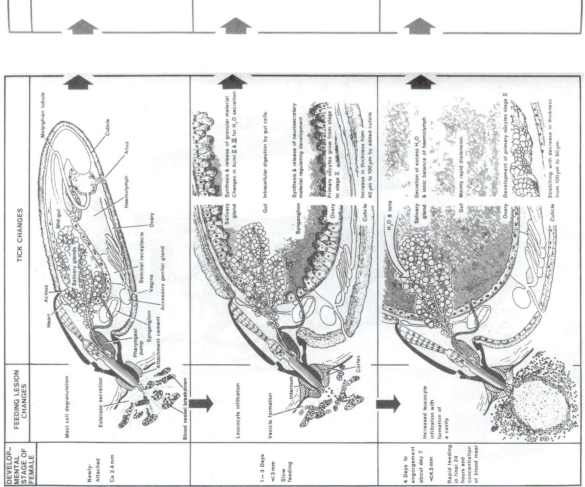

FIGURE 38.1 Selected Tick Feeding Events. *Left panel:* Flowchart of important events during the tick feeding period. *Right panel:* Salivary gland morphogenesis for female ticks. Salivary gland cell types are Type I (water balance acinus), Type II, and Type III (granular-secreting acini). Salivary gland diagrams are taken from Binnington et al. (1982). All tick genera differ slightly on these time points, and the values, in days, represent the lone star tick, *Amblyomma americanum.*

they ingest more than 5 mL of blood during feeding and weigh about 600 mg at repletion. Finally, two-thirds of the concentrated blood meal is converted to egg mass. While males do attach to the host and may initiate mate-guarding responses via their salivary secretions, their cuticle is not expandable and they cannot accommodate a large blood meal. Males spend most of their time on the host secreting saliva, imbibing blood, and moving about to inseminate attached females. Nymphal and larval hard ticks feed for 4–5 days on their hosts. A blood meal can sustain them for many months until the next host is found.

SOFT-TICK FEEDING

Soft ticks differ from hard ticks in their feeding strategy. Generally, soft ticks are found in the nests or beds of their hosts, and, unlike many of the hard ticks, they do not seek a different host in each life stage. Rather than remain attached to their host for days, soft ticks feed for about 1 hour and engorge to about 12 times their beginning weight. The cuticle of the soft tick has star-shaped folds that expand during engorgement, but they do not have the deep folds of the hard ticks. After feeding, the coxal glands (Chapter 4) function to concentrate the blood meal, and an egg mass is laid. Soft ticks alternate between periods of feeding and periods of egg laying. Soft ticks have been known to survive for more than 20 years without a blood meal.

MOLECULAR BIOLOGY OF TICKS

Tick physiology and biochemistry have been slow to enter the arena of molecular genetics. In the late 1980s, the first tick gene characterized and produced by recombinant DNA technology was the *BM86* (86-kiloDaltons [kDa], basement membrane protein) of the cattle tick, *Boophilus microplus*. BM86 has been the major component in a commercially successful tick vaccine used in Australia and Cuba. In 1990, *tick anti-coagulant protein* (TAP) was identified from the saliva of a soft tick as an inhibitor of factor Xa. It has been purified and cloned by many different methods and a great deal is known about its effects on mammalian hosts, but little is known about its expression in the tick. At about the same time, we began to exploit the phenomenon of host resistance to tick feeding and use molecular biolological techniques to identify immunoreactive clones from cDNA libraries made from mRNA isolated from tick salivary glands. In 1995, we reported our discovery of a secreted calreticulin;

the corresponding gene was the first ixodid tick salivary gland gene to be cloned and sequenced.

Published information on expressed tick genes is summarized in Table 38.1. Tissue and stage specificity, length of coding region, and what might be known of the function are summarized for each of these genes. The sequences of more than 90 tick genes (not including ribosomal genes and EST projects) had been deposited in Genbank at the time of this writing. In addition, there are published reports on tick proteins that have not been characterized at the cDNA level; these are not included in Table 38.1. For the most part, few expressed genes have been identified from each representative genus. *Boophilus microplus* has the distinction of having the most genes identified and published to date, 11, while *A. americanum* has 10 published genes. In general, little is known about gene expression in ticks, and even less is known about gene expression in other acarines.

Ticks have a system of differential gene expression where the activity in the cells goes from undetectable to highly expressed in only a few minutes. Reports of proteins induced by tick feeding are common, and it appears that blood feeding may induce many tick genes. A difficulty for this type of research is finding appropriate controls; for example, a gene that remains unchanged throughout the feeding process. Even the actin gene can be up-regulated during feeding. The on-host period is relatively long compared to other hematophagous arthropods; as a result, the feeding process has many layers of complexity, adding to the challenge of interpreting feeding studies. Defining or characterizing one factor does not always provide insight into the system. And for each factor discovered, there is likely to be an alternate factor that contributes to feeding by a similar mechanism. Additionally, as discussed later, each tick genus may have its own unique genes producing similar proteins. For example, each tick appears to have its own variation of a histamine-binding protein. Little primary sequence overlap has been observed in the expressed genes from tick cDNA libraries. We all seem to be finding unique sets of genes. This leads to questions such as: Are all the salivary gland proteins different from genus to genus? Given that *A. americanum* has a genome of 1.04×10^9 bp, there may be many unique genes.

MORPHOGENESIS AND GENE EXPRESSION IN THE TICK SALIVARY GLANDS

Morphological changes in the ixodid salivary glands during feeding have been studied via both light and electron microscopy. Female ixodid tick salivary

TABLE 38.1 Summary of Published Reports on Tick cDNAs

Gene/Protein	Tick Species	Size of cDNA/Protein*	Life Stage	Tissue*	Function
Ecdysteroid Receptor-RA 1	*Amblyomma americanum*	8239 bp 560 aa/64 kDa		SG	Receptor and ecdysteroid tested
Ecdysteroid Receptor-RA 2	*Amblyomma americanum*	1710 bp 570 aa/ kDa		SG	
Ecdysteriud Receptor-RA 3	*Amblyomma americanum*	6191 bp aa/48.4 kDa		SG	
Leucokinin Receptor	*Boophilus microplus*	1194 bp 397 aa/44 kDa	Eggs, larvae, nymphs, males and females		None published
Retinoid X Receptor-RXR 1	*Amblyomma americanum*	3190 bp 1200 ORF/44 kDa	Eggs, larvae, females — multiple transcripts in O1, O2	SG — Un to Repl	No activity for this receptor using retinoic acid
Retinoid X Receptor-RXR 2	*Amblyomma americanum*	1649 bp 1241 ORF/413 aa/ 45.5 kDa	Eggs, larvae, nymphs — multiple transcripts, F-O1, O2		
Anticoagulant protein (rTAP)	*Ornithodorus moubata*	60 aa			Antithrombosis, cleaves factor XA
Salivary gland antigen p29	*Haemaphysalis longicornis*	1027 bp 277 aa/29 kDa	Adult	Whole Ad, 3-day-fed females	None published, has some host protective value for tick challenge
Serine proteases HLSG-1, HSLsg-2	*Haemaphysalis longicornis*	HLSG-1/1200 bp 1040 bp ORF 346 aa/37.7 kDa	Adult, eggs	SG MG/Un, feeding	None published
BM 86	*Boophilus microplus*	1982 bp ORF 650 aa/86 kDa	Females	MG	Membrane protein — ticks feeding on vaccinated hosts lyse, turning a bright red
Cytochrome P450 (CYP41)	*Boophilus microplus*	518 aa/59 kDa	Larvae	Whole ticks	None published
Anticomplement (ISAC)	*Ixodes scapularis*	755 bp 163 aa/18.5 kDa	Females	SG, 3- to 4-day fed	Inhibits complement activity as measured by expression in COS cells
Macrophage migration inhibitory factor (MIF)	*Amblyomma americanum*	ORF 348 bp 116 aa/12.6 kDa	Females	MG, 3- to 4-day fed	Inhibits random macrophage migration
SALP16	*Ixodes scapularis*	575 bp 152 aa/16.4 kDa	Eggs, nymphs	SG, fed nymphs	None published
Calreticulin (CRT)	*Amblyomma americanum*	58 kDa	Female	3-day-fed salivary gland	None published
Immunoglobulin-binding proteins (IGBPs-MA, -MB, -MC)	*Rhipicephalus appendiculatus*	MA-29 kDa MB-25 kDa MC-21 kDa	Male	Partially fed	Possible mate guarding — ensure females can take adequate blood meal in host environment
Histamine-binding proteins (HBP-RaHBP1, RaHBP2)	*Rhipicephalus appendiculatus*	RaHBP-1 172 aa, 20 kDa RaHBP-2 171 aa, 20 kDa	Females	SG, partially fed	High affinity for histamine, as shown by recombinant proteins

(continues)

TABLE 38.1 (*continued*)

Gene/Protein	Tick Species	Size of cDNA/Protein*	Life Stage	Tissue*	Function
Histamine-binding protein (RaHBP-3)	*Rhipicephalus appendiculatus*		Males, larvae, nymphs		
Immunosuppressant protein	*Dermacentor andersoni*	772 bp, 36 kDa	Females	SG, partially fed	Host immunosuppression
Vacuolar ATPase, C subunit	*Amblyomma americanum*	1373 bp	Eggs, females	SG, 3-day fed	None published
Stearoyl CoA desaturase	*Amblyomma americanum*	1488 bp, 8000 bp	Eggs, larvae, nymphs, adults	SG, 3-day fed	None published
5′ nucleotidase	*Boophilus microplus*	135 kDa	Larvae, females	20- to 30-mg females	Enzymatic activity of cell lysates assayed
Octopamine-like G-protein	*Boophilus microplus*	1938 bp/419 aa	Larvae	Unfed, 10-days old	None published
Acetylcholinesterase gene	*Boophilus microplus*	Two possible mRNAs differ by six aa 561 bp/62 kDa	Larvae	Unfed, 10-days old	None published
Glutathione-S-transferase gene	*Boophilus microplus*	830 bp 223 aa/25.6 kDa	Larvae	Unfed	None published
Sodium channel gene	*Boophilus microplus*	Partial 2110 bp cDNA /Single point mutation changing Phe to Ile	Larvae	Unfed	None published
Peroxiredoxin	*Haemaphysalis longicornis*	939 bp, 222 aa, 26.6 kDa	Females	Whole	None published
cAMP-dependent protein kinase	*Amblyomma americanum*	Three isoforms/372, 422, and 462 aa			Appears to be down-regulated during tick feeding
Defensin	*Ornithodoros moubata*	4 kDa, 37 aa	Females	Fed, Hemolymph	Up-regulated in response to bacterial injection and blood feeding
Two cathepsin L–like cysteine proteinase genes	*Haemaphysalis longicornis*	Both 1200 bp			None published
Factor Xa	*Ornithodoros savignyi*	213 bp 60 aa/7 kDa		SG	Cell lysates inhibited fXa by 91%
Ki nuclear lupus autoantigen	*Rhipicephalus appendiculatus*	866 bp 245 aa	Females	SG, 2-day fed	None published
Ku p70 lupus auto antigen	*Rhipicephalus appendiculatus*	2135 bp 600 aa	Females	SG, 2-day fed	None published
Acetylcholinesterase gene	*Boophilus microplus*	ORF 1689 bp, 563 aa	Larvae	Unfed	None published

* *Abbreviations*: SG — salivary glands; MG — midgut tissue; O — ovaries; Ad — adult; Un — unfed; Repl — replete female; bp — base pair; kDa — kilodalton; ORF — open reading frame; aa — amino acid.

glands comprise three acinar cell types and a fourth type described from males. Secretions from Type IV acini of males are thought to facilitate spermatophore transfer during copulation; however, the male acini have been studied relatively little compared to females. Type I acini are involved mainly in water balance for the off-host tick, and their functional importance during feeding remains unclear, although some expansion of these glands is observed during feeding. The Type I acinus has been studied under conditions of desiccation and rehydration. It appears that when the tick is trying to rehydrate, the mitochondria in these cells are condensed. This suggests that these cells are active metabolically. Type II and Type III salivary gland acini of female ixodids contain granules and undergo extensive morphogenesis that is initiated upon attachment to the host and ends with their complete degeneration after detachment. Eight different cell types have been identified in the Type II and Type III acini. Type II acini contain *a*-, *b*-, and *c*-cells, while the Type III acini contain *d*-, *e*-, *f*-cells. Type II and Type III acini contain adlumenal and ablumenal interstitial cells as well. The *f*-cells of the Type III acini undergo transformation to facilitate water transport during the rapid engorgement phase of feeding. The tick concentrates the blood meal during this stage, so the transformation of *f*-cells to a water-transport epithelium is critical to the final 12–24h of feeding. During rapid engorgement, the salivary cells degenerate, and this degeneration is correlated with a rise in ecdysteroid titers in the ovaries of the female tick.

Elaborate morphogenesis of the salivary glands is seen during feeding, as evidenced by ultrastructural changes together with changes in the abundance of protein soon after attachment, indicating that differential gene expression is occurring. Differential gene activity is defined as a change in the number, type, and amounts of genes expressed in unfed- versus feeding-tick tissues. Soon after attachment, the gland protein content of female ticks increases and new polypeptides can be detected by SDS-polyacrylamide gel electrophoresis (PAGE). Using SDS-PAGE, tick gland proteins are separated by size as they migrate through the gel toward the positively charged electrode at the bottom of the gel. Synthesis of some polypeptides appears to be induced as feeding progresses, while others decrease. Furthermore, additional polypeptides are initiated by mating and are likely to be important for acquiring the critical quantity of blood that results in egg laying. Unmated females do not enter the rapid engorgement phase and do not lay eggs. Protein synthesis occurs soon after feeding is initiated and correlates with a rise in mRNA in feeding glands. The mRNA increases to more than five times the basal level by the second day of attachment, reaches a maximum in rapidly engorging females (after more than a week of feeding), and begins to decline 24h before repletion and detachment from the host. Further evidence of protein synthesis soon after attachment is indicated by surface area increases in the rough endoplasmic reticulum of the salivary gland granular acini. More detail concerning the underlying cellular mechanisms of fluid and protein secretion from the tick salivary glands can be found in Sauer et al. (1994).

The salivary secretions of ticks appear to be subject to strong evolutionary pressures exerted by the host responses to tick feeding. Due to such pressures, a wide range of proteins is secreted from the many granules found in the salivary glands. Histamine-binding proteins (HBPs) provide an illustration of how this may work. Histamines generally cause inflammation and irritation that increase the chance of host grooming, and this puts the tick at risk during feeding. Ticks secrete histamine-binding proteins during feeding in order to sequester the histamine that is released by the host cells. This presumably leads to less irritation and a lower probability that the host will remove the tick. Three different HBPs were characterized from the ixodid tick, *Rhipicephalus appendiculatus*. These proteins were found to be stage and sex specific, in that females produced two HBPs while males, larvae, and nymphs produced a different HBP. The two female HBPs were secreted at the same time, suggesting different possible roles for these proteins. The primary amino acid sequences of the female HBPs were more similar to each other than they were to the male HBP. Each of these proteins had high affinity for histamine, and molecular modeling revealed similar spatial structures. Recently, a serotonin- and histamine-binding protein from *Dermacentor reticulatus* was described that has 36–40% amino acid identity to the *R. appendiculatus* proteins. In addition, another HBP has been characterized from male *A. americanum*. These proteins all appear to have structural similarities despite their divergent amino acid sequences.

Host immune responses may affect the degree of specificity for proteins that are secreted into the feeding lesion. We know that certain ticks will feed on almost any type of host, while others feed on only one type of host. Ticks, such as *R. appendiculatus* and *B. microplus*, that feed on cattle are affected adversely by histamine. These cattle tick species do not feed well on rodents. In contrast, *D. reticulatus* feeds mainly on rodents, and it secretes proteins to manage the serotonin and histamine levels in the feeding lesion. From these examples, it is clear that ticks have developed salivary gland proteins with a great deal of specificity at the generic and specific levels.

MORPHOGENESIS AND GENE EXPRESSION IN TICK MIDGUT EPITHELIUM

The tick midgut undergoes major changes during feeding. Aside from light and electron microscope studies on the changes during feeding, little is known about the underlying physiology and biochemistry of digestion. The tick imbibes more than 100 times its unfed weight in blood. The lumen of the tick midgut has a pH near neutrality, and both hard and soft ticks digest the blood meal intracellularly. Blood meal digestion occurs via receptor-mediated endocytosis, pinocytosis, and phagocytosis. Proteases from the lumen have not yet been identified, and there is no evidence for digestion of blood components by the salivary secretions.

We have looked for specific protease activity in the midgut tissues using zymograms, gels impregnated with either gelatin or casein to which protein samples can be loaded and separated. After separation, the gels are renatured and incubated in buffers of varying pH. Proteases then are visualized with various stains that highlight the absence of the substrates. The results of our zymograms show that tick midgut proteases are active at a low pH, supporting the hypothesis that these proteases are active in lysosomes (Fig. 38.2). For example, we detected a midgut protease at pH 3.0 that was not active at pH 7.0.

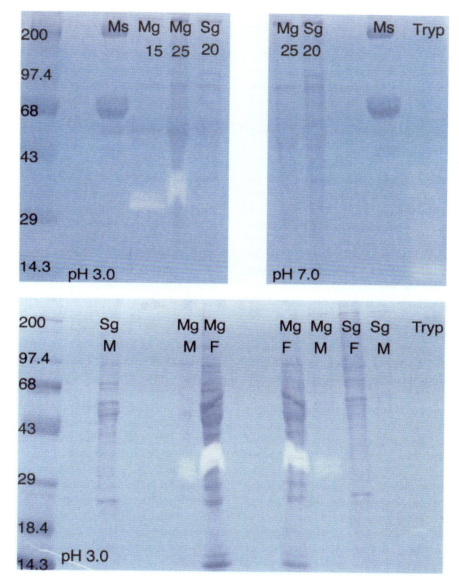

4–6% Novex zymogen gels (casein)

FIGURE 38.2 Specific proteases in tick midgut tissues. Zymograms (described in the chapter) indicate the specificity of proteases from tick midgut tissues. Light cleared areas represent proteases found in these tissues. (A) At pH 3.0, lanes 3–6, only midgut proteases were detected; at pH 7.0, lanes 7–11, only a trypsin control was reactive with the casein substrate. (B) At pH 3.0, lanes 5–6 and 9–10, show that the reaction is found in midguts of females and males. *Abbreviations*: Ms, *Manduca sexta* hemolymph (control); Mg, midgut epithelia; Sg, salivary glands; Tryp, trypsin control; M, male; F, female. Molecular weights are indicated on the figure.

Ticks require a digestive mechanism that is quite different from that of a blood-feeding insect (Chapter 21). Ticks use the salivary glands and coxal glands to concentrate by fluid secretion the nutrients found in the blood meal. Hemoglobin and other plasma proteins then are endocytosed by the midgut cells. Following a blood meal, midgut cells contain hemolysins, anticoagulant, proteases, and host complement. Albumin, transferrin, and immunoglobulins pass through the midgut intact, somehow avoiding digestion, and immunoglobulins are preferentially moved through the tick midgut cells.

Study of the tick midgut is likely to provide many insights for our understanding of pathogen transmission. The midgut is the first line of defense for the tick against the responses of the host to feeding. The blood meal contains a wide variety of host components that first must be engulfed by the cells of the midgut before a response can be generated to the host defenses. We assume that the midgut provides a form of innate immunity for the tick as well as serving the role of a metabolic storage compartment for fueling the processes that are stimulated by tick feeding. A pathogen must first penetrate the midgut to the hemocoel in order to reach the salivary glands. While little is known about innate immunity of acarines to ingested pathogens, there is some evidence indicating that ticks do have induced hemolymph proteins that can kill invading bacteria. And indeed the first truly successful tick vaccine for the treatment of Lyme disease emphasizes the importance of the tick midgut in transmission of disease agents. This vaccine is directed against an outer surface antigen of the *Borrelia burgdorferi* spirochete that is present in the tick. Spirochetes are not stimulated to move from the midgut to the salivary glands when ticks feed on a vaccinated host. Aside from the biochemical characterization of a few proteins and proteases, little is known about the midgut and the midgut genes that are induced by tick feeding.

The remainder of this chapter focuses on tick gene expression in more detail, using examples from the author's research. The main aim for our research has been to concentrate on the molecular events of the first several days of tick feeding, because it is during this time that most pathogens gain entry to the host via the saliva. From our early studies we identified several feeding-specific salivary gland polypeptides that are immunogenic. The genes encoding two proteins—calreticulin and a macrophage migration inhibitory factor (MIF)—and a cDNA that encodes a putative protein similar to a gamma-interferon-inducible lysosomal thiol reductase (GILT) are discussed.

CALRETICULIN

Calreticulin, a calcium sink, was identified using hyperimmune sera from rabbits that had been exposed to ticks and standard immunoscreening techniques. The sera were analyzed by immunoblots of salivary gland extracts that were obtained from unfed and partially fed *A. americanum* adults. RNA was prepared from tick salivary glands and used to make cDNA that was packaged into a bacteriophage. Plaques were grown on plates and were overlayed with filters that had been soaked previously in a solution that induces protein production. These filters were incubated with the hyperimmune serum to isolate immunoreactive clones.

We expressed the unknown cDNA (161A) in the Bluescript plasmid using the Lac Z promoter and made antibody to the purified fusion protein. Using this antibody we were able to demonstrate the appearance of this protein in the salivary glands of early-feeding females. Ticks feed well and suffer no adverse consequences from feeding on rabbits that have never served as hosts. In contrast, tick-immunized rabbits showed the signs of systemic Arthus reactions (antigen–antibody binding resulting in immune-mediated allergic responses) and locally at the bite sites. Indeed, these hosts were responding to secreted tick proteins. After complete sequencing of the clone, a strong similarity of the conceptual translation product to calreticulin was found during our database searches. We confirmed this finding by a positive reaction of our fusion protein with antibody made to the rabbit skeletal muscle calreticulin.

We since have determined that calreticulin is an immunodominant protein in the salivary glands of ticks. Calreticulin was isolated initially from the salivary glands of *A. americanum* and also was shown to be in the midgut and saliva. Subsequently, it was shown to be in *Dermacentor variabilis* (American dog tick) salivary glands and saliva. More recently, tick calreticulin has been used as an antigen for ELISA and immunoblot assays as an indication of tick bite exposure. Immunoblot assays show the presence of a calreticulin protein in the salivary glands and midgut of *Ixodes scapularis* and *I. pacificus* ticks (Fig. 38.3). Hyperimmune sera to midgut extracts identified a calreticulin clone from our midgut cDNA library. Primers based on the *A. americanum* nucleotide sequence amplified a calreticulin product from *I. scapularis* DNA.

What is the role of calreticulin in the tick? Based on its presence in tick saliva and its known function as a calcium sink, we initially suggested that calreticulin might be a cofactor for ADP-degrading enzymes such as apyrase. Elevated calcium in the lesion could acti-

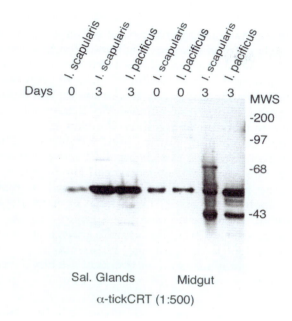

Sal. Glands   Midgut

α-tickCRT (1:500)

FIGURE 38.3 Calreticulin in *Ixodes scapularis* and *Ixodes pacificus* tissues. Samples prepared from the salivary glands and midgut epithelia of unfed (day 0) and partially fed females (day 3) were assayed for the presence of calreticulin. Protein from these samples were separated by gel electrophoresis and transferred to nitrocellulose filters. Filters were probed with a specific antisera to tick calreticulin at 1:500. Reactions were detected using a secondary antibody and chemiluminescent detection. Tick species is indicated above each lane. Molecular weights are indicated on the figure.

TABLE 38.2   Sequence Analysis of Selected Midgut cDNAs

Putative Gene	Identity/Positives (%)
Ubiquitin	96/100
Glucose-regulated protein	84/95 to *Rattus* sp.
26S Proteosome subunit	96/100
Macrophage migration inhibitory factor	48/66 to *Brugia malayi*
Cyclophilin A–binding protein	92/97 to *T. congolense*
Membrane protein	52/70 to *Drosophila melanogaster*
Hyaluronate lyase precursor	33/50
Gamma-interferon-inducible lysosomal thiol reductase (GILT)	32/48 to *Homo sapiens*
Serine proteinase	33/48
Intestinal mucin	—
Calreticulin	100/100 to *Amblyomma americanum*

vate ADP-degrading enzymes, leading to impaired platelet function. In other systems, calreticulin has been shown to have antithrombotic activity activated by the nitric oxide pathway. It remains to be shown that calreticulin has a role in blood feeding. Calreticulin is likely an example of a parasitic adaptation to evade the host response that was once simply a remnant of some other physiological process in the tick.

We suggested previously that the *f*-cells of the Type III salivary gland acini were likely to contain calreticulin. The *f*-cells rapidly enlarge and fill with secretory granules soon after tick attachment. They differentiate into typical glycoprotein-secreting cells with an extensive endoplasmic reticulum. These cells actively secrete for 2–4 days and then undergo autophagy to break down the protein synthesis mechanism. Finally, these cells are converted to a fluid transport epithelium. Antibody localization confirmed the early-differentiated *f*-cell as a source of calreticulin.

EXPRESSED MIDGUT PROTEINS

We began to characterize the genes that contribute to the digestion of blood meal. To do this we produced the first cDNA expression library from the explanted midgut tissue of 3-day-fed females. From our midgut cDNA library, we found many novel cDNAs (not shown) and many cDNAs that had sequence similarity to other genes in Genbank (Table 38.2).

The results of a typical mass excision procedure and several plasmid platings from these phagemids are displayed in Table 38.2. Approximately three-fifths of the cDNAs of the first 100 clones sequenced represented genes encoding ribosomal RNAs. In spite of the repetitive sequencing, the first midgut library was quite robust and provided a number of tick cDNAs. However, to minimize sequencing of the ribosomal genes, we produced a second midgut expression library (midgut explanted from partially fed females) that had been normalized with tick genomic DNA.

CALRETICULIN/MOBILFERRIN

We began our molecular studies of the tick midgut in order to learn more about blood meal digestion and pathogen transport in and out of this tissue. In the process, we found a midgut calreticulin, mobilferrin, that might be involved in blood meal processing. Mobilferrin is a cytosolic isoform of calreticulin that has been found to be important in mammals for managing iron in the diet. A tick has a large amount of iron to deal with in its blood meal, and recent studies show that calreticulin mRNA is increased with increased concentrations of iron in intestinal epithelial cells. Figure 38.4 illustrates a model of how mobilferrin functions in the rat intestinal tissue. Apomobilferrin complexes with the carboxy-terminus of alpha integrin. Iron is solubilized by mucin in the lumen and is transferred to the integrin–mobilferrin complex. This complex is internalized, and mobilferrin associates with β-2-microglobulin, flavin mono-oxygenase, and nicotinamide adenine dinucleotide phosphate to make a cytosolic complex called paraferritin. The flavin mono-oxgenase and nicotinamide adenine dinucleotide phosphate reduce the ferrous iron to the ferric state. In rats, paraferritin transfers the reduced iron to the mitochondria for incorporation into heme proteins and other end products.

The presence of calreticulin in ticks suggests an iron-metabolizing mechanism analogous to that in mammals. It is likely that tick calreticulin plays some role in the processing of heme proteins taken in with

the blood meal. The tick has a huge storage–excretion problem with its predominantly intracellular digestive system, and the adaptations to deal with its blood meal are likely to provide some insight into iron absorption and lysosomal digestion. Little is known about how the blood meal is processed in the tick cells; however, through our molecular analysis of the midgut tissue we have found several components that might contribute to blood meal digestion. We have identified a midgut calreticulin, mucin, and transferrin (Table 38.2). How does the tick digest the hemoglobin and other components of the blood meal? How is the iron overload in the tick midgut cells managed? Is there a hemoglobin receptor? And how is hemoglobin digested? There are no data that support answers to these questions. Through continued molecular analysis of midgut tissue proteins, we should be able to begin to learn more about the events of tick blood meal digestion.

MACROPHAGE MIGRATION INHIBITORY FACTOR

One of the most interesting cDNAs that we found encodes a putative protein with strong identity to mammalian macrophage migration inhibitory factor (MIF). Tick MIF is highly active by inhibiting random macrophage migration by more than 58%. We showed that MIF is present in the midgut and salivary glands of unfed and partially fed female lone star ticks. We have not been able to detect MIF in saliva; however, we have been able to demonstrate an effect in vivo of MIF peptide at the feeding site during feeding. Rabbits immunized with an MIF peptide exhibited less inflammation and edema at feeding sites in the first 6 days of tick feeding than a control rabbit. Unlike the previous work with calreticulin, the host effect was minimal for tick feedings on MIF-immunized animals, and the effects were seen locally at the feeding site. It was apparent that ticks on MIF-immunized rabbits fed more slowly and had less edema and erythema than control rabbits, which led us to hypothesize that secreted MIF was bound in the feeding site and inhibited from its normal function at the feeding site.

In another study, we measured the blood flow at the feeding site on day 6 using an optical Doppler tomography (ODT) device. Optical Doppler tomography, also termed Doppler optical coherence tomography (OCT), is a recently developed technique for imaging both tissue structure and flow velocity of moving particles in highly scattering media. We documented that in four of six treatment rabbits, blood flow to the feeding site was reduced in the peptide-immunized

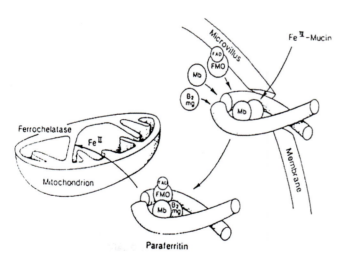

FIGURE 38.4 Postulated model for mucosal uptake of iron in intestinal absorptive cells. Apomobilferrin (Mb) "docks" on the carboxy-terminal region of alpha integrin. Iron solubulized by mucin is transferred to the intergrin–mobilferrin complex, and this is internalized. Flavin mono-oxygenase (FMO) becomes associated with the complex and reduces the iron in concert with nicotinamide adenine dinucleotide phosphate. Fe(II) is made available to the mitochondrion for incorporation into heme proteins and other end products. Other proteins appear to associate with the complex, including β-2-microglobulin, which was recently associated with iron overloading. From Umbreit et al. (1998).

TABLE 38.3   Measurements of Blood Flow in Tick Feeding Sites on MIF-Peptide—Immunized Hosts

Peptide Plus KLH			KLH Alone		
Flux control	Flux experiment	% Change	Flux control	Flux experiment	% Change
58.2	91.8	22.4	146.6	172.7	8.2
66.2	78.2	8.3	94.3	68.0	−16.2
73.3	55.0	−14.3	83.8	125.1	20.0
75.9	61.5	−10.0	128.7	172.4	15.0
83.6	57.6	−18.4	40.8	81.9	33.5
101.0	85.3	−8.0	73.6	39.2	−30.0

hosts (Table 38.3). Two of six control rabbits showed reduced blood flow to the feeding lesion. From these results, we concluded that those animals that were immunized with MIF had reduced inflammation.

The possibility that ticks could secrete cytokines had not been considered before MIF was discovered, but it was assumed that ticks secrete products that had an effect on host cytokines. For example, Sonenshine (1991) suggested that mammalian MIF may help keep macrophages at the tick lesion site. Given the long period of evolution of ticks and their successful coevolution with the host, the recruitment of some of their genes expressing cytokine-like molecules to a host evasion strategy seems feasible. MIF is a proinflammatory cytokine, principally countering the immunosuppressive effects of glucocorticoid. Thus, one possible role of tick MIF is to increase inflammation at the feeding site, although this is in contrast to the suggestion that the principal role of saliva is to reduce inflammation. An increase in blood flow that accompanies inflammation could benefit the tick, especially if other tick products, such as an anaphylatoxin inactivator, as proposed by Ribeiro (1987) inhibited other aspects of inflammation, such as pain. Secretion of prostaglandins by ticks is associated with increased blood flow into the feeding lesion, so the action to increase inflammation of another tick product is not inconceivable. Mammalian MIF is a known regulator of innate and acquired immunity, and it has a number of roles, including inducing inflammation in response to bacteria and viruses and activating macrophages and T-cells to release insulin from the pancreas. MIF stimulates macrophages to produce TNFα and nitric oxide when given in combination with IFN-γ. MIF is required for T-cell activation and antibody production by B-cells. In studies on the impact of MIF on rheumatoid arthritis, delayed-type hypersensitivity reactions were induced, and anti-MIF antibodies inhibited these reactions. Delayed-type hypersensitivity reaction is a

hallmark of tick rejection in a sensitized host. Determining the role of MIF in tick feeding should extend our understanding of tick feeding. While we suspect that MIF is a secreted saliva protein, we did isolate this cDNA from our midgut library, and it appears that MIF is more abundant in the midgut tissue.

GAMMA-INTERFERON-INDUCIBLE LYSOSOMAL THIOL REDUCTASE

We recovered an additional cDNA that encodes a product that has 32% amino acid similarity to human gamma-interferon-inducible lysosomal thiol reductase (GILT) (Table 38.2). GILT catalyzes the reduction of disulfide bonds at acidic pH to denature proteins, a critical condition for proteolysis in the lysosome. In mammalian systems, a precursor GILT protein is synthesized and transported to endosomes by mannose-6-phosphate receptors. It has been proposed that the primary function of GILT is to facilitate complete unfolding of proteins destined for lysosomal degradation. The finding of a cDNA encoding a GILT-like protein in ticks provides some evidence to support the hypothesis that digestion of the blood meal takes place in lysosomes under acid conditions.

SUMMARY

Looking at Table 38.2 we might conclude that we know quite a bit about gene expression in ticks; however, we have little information on the regulation of tick genes. There are no characterized promoters or regulatory elements or sequences that have been described for any of these genes. Major emphasis for tick molecular biology has been on novel salivary gland genes. Other genes described are those that could be useful for the development of acaricides (acetylcholinesterase, sodium channel, and p450 genes) or vaccine development (Bm86, p29). Few genes have been characterized that could impact vitellogenesis in the tick (ecdysteroid receptor/retinoic acid receptor), and one tick defensin gene has been described. Strong functional data are lacking, and little information is available concerning the impact of these genes on tick feeding.

Readings

Tick Genes

Baxter, G. D., and Barker, S. C. 1998. Acetylcholinesterase cDNA of the cattle tick, *Boophilus microplus*: Characterization and role in organophosphate resistance. *Insect Biochem. Molec. Biol.* 28: 581–589.

Baxter, G. D., and Barker, S. C. 1999. Isolation of a cDNA for an octopamine-like, G-protein coupled receptor from the cattle tick, *Boophilus microplus*. *Insect Biochem. Molec. Biol.* 29: 461–467.

Bergman, D. K., Palmer, M. J., et al. 2000. Isolation and molecular cloning of a secreted immunosuppressant protein from *Dermacentor andersoni* salivary gland. *J. Parasitol.* 86: 516.

Bior, A. D., Essenberg, R. C., and Sauer, J. R. 2002. Comparison of differentially expressed genes in the salivary glands of male ticks, *Amblyomma americanum* and *Dermacentor andersoni*. *Insect Biochem. Molec. Biol.* (In press).

Crampton, A. L., Baxter, G. D., et al. 1999. Identification and characterization of a cytochrome P450 and processed pseudogene from an arachnid: The cattle tick, *Boophilus microplus*. *Insect Biochem. Molec. Biol.* 29: 377–384.

Das, S., Marcantonio, N. et al. 2000. SALP16, a gene induced in *Ixodes scaplularis* salivary glands during tick feeding. *Am. J. Trop. Med. Hyg.* 6: 99–105.

Guo, X., Harmon, M. A., et al. 1997. Isolation of a functional ecdysteroid receptor homologue from the ixodid tick *Amblyomma americanum* (L.). *Insect Biochem. Molec. Biol.* 27: 945–962.

Guo, X., Xu, Q., et al. 1998. Isolation of two functional retinoid X receptor subtypes from the Ixodid tick, *Amblyomma americanum* (L.). *Molec. Cell. Endocrin.* 139: 45–60.

He, H., Chen, A. C., et al. 1999. Characterization and molecular cloning of a glutathione S-transferase gene from the tick, *Boophilus microplus* (Acari: Ixodidae). *Insect Biochem. Molec. Biol.* 29: 737–743.

He, H., Chen, A. C., et al. 1999. Identification of a point mutation in the para-type sodium channel gene from a pyrethroid-resistant cattle tick. *Biochem. Biophys. Res. Com.* 261: 558–561.

Hernandez, R., He, H., et al. 1999. Cloning and sequencing of a putative acetylcholinesterase cDNA from *Boophilus microplus* (Acari: Ixodidae). *J Med. Entomol.* 36: 764–770.

Holmes, S. P., He, H., et al. 2000. Cloning and transcriptional expression of a leucokinin-like peptide receptor from the Southern cattle tick, *Boophilus microplus*. *Insect Molec. Biol.* 9: 457–465.

Jaworski, D. C., Jasinskas, A., et al. 2001. Identification and characterization of a homologue of the proinflammatory cytokine macrophage migration inhibitory factor in the tick, *Amblyomma americanum*. *Insect Molec. Biol.* 10: 323–331.

Jaworski, D. C., Simmen, F. A., et al. 1995. A secreted calreticulin protein in ixodid tick saliva. *J. Insect Physiol.* 44: 369–375.

Joubert, A. M., Louw, A. I., et al. 1998. Cloning, nucleotide sequence and expression of the gene encoding factor Xa inhibitor from the salivary gland of the tick, *Ornithodoros savignyi*. *Exp. Appl. Acarol.* 22: 603–619.

Liyou, N., Hamilton, S., et al. 1999. Cloning and expression of exo-5–nucleotidase from the cattle tick, *Boophilus microplus*. *Insect Molec. Biol.* 8: 257–266.

Luo, C., McSwain, J. L., et al. 1997. Cloning and sequence of a gene for the homologue of the stearoyl CoA desaturase from salivary glands of the tick *Amblyomma americanum*. *Insect Molec. Biol.* 6: 267–271.

McSwain, J. L., Luo, C., et al. 1997. Cloning and sequence of a gene fro a homologue of the C subunit of the V-ATPase from the salivary gland of the tick *Amblyomma americanum*. *Insect Molec. Biol.* 6: 67–76.

Mulenga, A., Sugimoto, C., et al. 1999. Molecular cloning of two *Haemaphysalis longicornis* cathepsin L-like cysteine proteinase genes. *J. Vet. Med. Sci.* 61: 497–502.

Mulenga, A., Sugimoto, C., et al. 1999. Molecular characterization of a *Haemaphysalis longicornis* tick salivary gland-associated 29-kilodalton protein and its effect as a vaccine against tick infestation in rabbits. *Infect. Imm.* 67: 1652–1658.

Nakajima, Y., van der Goes van Naters-Yasui, A., et al. 2001. Two isoforms of a member of the arthropod defensin family from the soft tick, *Ornithodoros moubata* (Acari: Argasidae). *Insect Biochem. Molec. Biol.* 31: 747–751.

Neeper, M. P., Waxman, L., et al. 1990. Characterization of recombinant tick anticoagulant peptide. *J. Biol. Chem.* 265: 17746–17752.

Paesen, G. C. 2000. Tick histamine-binding proteins: lipocalins with a second binding cavity. *Biochim. Phys. Acta* 1482: 92–101.

Paesen, G. C., de A. Zanotto, P. M., et al. 1996. A tick homologue of the human DNA helicase II 70-kDa subunit. *Biochim. Biophys. Acta* 1305: 120–124.

Paesen, G. C., and Nuttall, P. A. 1996. A tick homologue of the human Ki nuclear autoantigen. *Biochim. Biophys. Acta* 1309: 9–13.

Palmer, M. J., McSwain, J. L., et al. 1999. Molecular cloning of a cAMP-dependent protein kinase catalytic subunit isoforms from the lone star tick, *Amblyomma americanum* (L.). *Insect Biochem. Molec. Biol.* 29: 43–51.

Rand, K. N., Moore, T., et al. 1989. Cloning and expression of a protective antigen from the cattle tick *Boophilus microplus*. *Proc. Natl. Acad. Sci. U.S.A.* 86: 9657–9661.

Sangamnatdej, S., Paesen, G. C., Slovak, M., and Nuttall, P. A. 2002. A high-affinity serotonin- and histamine-binding lipocalin from tick saliva. *Insect Molec. Biol.* 11: 79–86.

Tsuji, N., Kamio, T., et al. 2001. Molecular characterization of a peroxiredoxin from the hard tick *Haemaphysali longicornis*. *Insect Molec. Biol.* 10: 121–129.

Valenzuela, J. G., Charlab, R., et al. 2000. Purification, cloning and expression of a novel salivary anticomplement protein from the tick, *Ixodes scapularis*. *J. Biol. Chem.* 275: 18717–18723.

Wang, H. A. N., and P. A. 1999. Immunoglobulin-binding proteins in ticks: New target for vaccine development against a blood-feeding parasite. *Cell. Molec. Life Sci.* 56: 286–295.

Tick Feeding, Salivary Glands, and Midgut Epithelia

Bininington, K. C., Stone, B. F., and Kemp, D. H. 1982. Tick attachment and feeding: Role of the mouthparts, feeding apparatus, salivary gland secretions and the host response. In *Physiology of Ticks* (F. Obenchain and R. Galun, eds.). pp. 119–168.

Brossard, M., and Fivaz, V. 1982. *Ixodes ricinus* L: Mast cells, basophils and eosinophils in the sequence of cellular events in the skin of infested or reinfested rabbits. *Parasitology* 85: 583–592.

Coons, L. B., Rosell-Davis, R., and Tarnowski, B. I. 1986. Bloodmeal digestion in ticks. In *Morphology, Physiology, and Behavioral Biology of Ticks* (J. R. Sauer and J. A. Hair, eds.). pp. 248–279.

Dickinson, R. G., O'Hagen, J. E., et al. 1979. Studies on the significance of smooth muscle contracting substances of the cattle tick, *Boophilus microplus J. Austral. Entomol. Soc.* 18: 199–210.

Kemp, D. H., Hales, J. R., et al. 1983. Comparison of cutaneous hyperemia in cattle elicited by larvae of *Boophilus microplus* and by prostaglandins and other mediators. *Experentia* 39: 725–727.

Madden, R. D., Sauer, J. R., et al. 1996. Dietary modification of host blood lipids affect reproduction in the lone star tick, *Amblyomma americanum. J. Parasitol* 82: 203–209.

Ribeiro, J. M. C. 1987. Role of saliva in blood-feeding by arthropods. *Annu. Rev. Entomol.* 32: 463–478.

Raikhel, A. S. 1983. Intestine. In *Atlas of Ixodid Tick Ultrastructure* (A. S. Raikhel, translator, Yuri S. Balashov, ed.). Special Publication of the Entomological Society of America, Leningrad, pp. 59–107.

Sauer, J. R., McSwain, J. L., and Essenberg, R. C. 1994. Cell membrane receptors and regulation of cell function in ticks and blood-sucking insects. *Int. J. Parasitol.* 24: 33–52.

Sonenshine, D. 1991. *The Biology of Ticks.* vol 2. Oxford University Press, New York.

Wikel, S. K. 1996. Tick modulation of host cytokines. *Exp. Parasitol.* 84: 304–309.

Macrophage Migration-Inhibitory Factor (MIF)

Bucala, R. 1996. MIF rediscovered: Cytokine, pituitary hormone, and glucocorticoid-induced regulator of the immune response. *FASEB J.* 10: 1607–1613.

Bucala, R. 2000. A most interesting factor. *Nature* 408: 146–147.

Calandra, T., Bernhagen, J., et al. 1995. MIF as a glucocorticoid-induced modulator of cytokine production. *Nature* 377: 68–71.

Santos, L., Hall, P., et al. 2001. Role of macrophage migration inhibitory factor (MIF) in murine antigen-induced arthritis: Interaction with glucocorticoids. *Clin. Exp. Immunol.* 123: 309–314.

Mobilferrin

Umbreit, J. N., Conrad, M. E., Moore, E. G., and Latour, L. F. 1998. Iron absorption and cellular transport: The mobilferrin/paraferritin paradigm. *Seminars Hematol.* 35: 13–26.

Conrad, M. E., and Umbreit, J. 2000. Iron absorption and transport—an update. *Am. J. Hematol.* 64: 287–298.

Nunez, M. T., Osorio, A., Tapia, V., Vergara, A., and Mura, C. V. 2001. Iron-induced oxidative stress up-regulates calreticulin levels in intestinal epithelial (Caco-2) cells. *J. Cell Biol.* 82: 660–665.

Gamma-Interferon Inducible Lysosomal Thiol Reductase (GILT)

Arunachalam, B., Phan, U. T., Geuze, H. J. and Cresswell, P. 2000. Enzymatic reduction of disulfide bonds in lysosomes: Characterization of a gamma-interferon-inducible lysosomal thiol reductase (GILT). *Proc. Natl. Acad. Sci. U.S.A.* 97: 745–750.

Phan, U. T., Arunachalam, B. and Cresswell, P. 2000. Gamma-interferon-inducible lysosomal thiol reductase. *J. Biol. Chem.* 275: 25907–25914.

Honey, Duff, K. M., Beers, C., Brissette, W. H., Elliott, E. A. Peters, C., Maric, M., Cresswell, P., and Rudensky, A. 2001. Cathepsin S regulates the expression of cathepsin L and the turnover of gamma-interferon-inducible lysosomal thiol reductase in B lymphocytes. *J. Biol. Chem.* 276: 22573–22578.

Virus-Induced, Transient Expression of Genes in Mosquitoes

KENNETH E. OLSON, JONATHAN O. CARLSON, AND CAROL D. BLAIR

INTRODUCTION

The transposable elements *mariner, Minos, Hermes,* and *piggyBac* are promising tools for generating transgenic insects, especially insects such as mosquitoes that can be vectors of human and animal disease. However, the transgenesis of mosquitoes remains a laborious and time-consuming procedure for gene characterization, mutagenesis, and expression. The propagation and maintenance of multiple mosquito lines is difficult in even the most spacious, well-equipped insectaries. Additionally, the complex life cycle of some medically important mosquitoes makes routine transgenesis difficult, whereas other transient expression systems may more easily and rapidly answer important biological questions posed by researchers.

In this chapter, we discuss two virus-based systems that allow efficient, transient gene expression in mosquito vectors. Gene expression from these systems is termed *transient* because gene expression usually occurs within a single generation and the gene of interest is not integrated into the genome of the vector species and is not heritable. The first virus expression system is derived from a mosquito-borne RNA virus, Sindbis (SIN; genus: *Alphavirus,* family: *Togaviridae*), which allows expression of heterologous genes in *Ae. aegypti, Ochlerotatus triseriatus, Culex pipiens,* and *Anopheles gambiae.* The second virus expression system is derived from small, single-stranded DNA viruses of Culicine mosquitoes, termed *densoviruses* (*Parvoviridae*). This chapter begins with a short review of SIN

virus molecular biology and a discussion of current uses of SIN virus expression systems as important tools for studying basic virus–vector interactions, determining biological functions of mosquito genes, and identifying molecular strategies for interfering with the transmission of vector-borne diseases. A description of densovirus molecular biology and expression systems derived from the virus follows along with a discussion of their potential role as agents in vector control.

SINDBIS VIRUS MOLECULAR BIOLOGY

SIN viruses are enveloped RNA viruses approximately 70 nm in diameter that replicate exclusively in the cytoplasm of infected cells. The nucleocapsid of the virus is composed of a genomic RNA and 240 copies of a single virus-encoded capsid protein, arranged in an icosahedral lattice with $T = 4$ symmetry. The surrounding envelope contains 80 spikes composed of trimers of envelope glycoprotein (E1 and E2) heterodimers. The SIN virus genome is a positive-sense, single-stranded, nonsegmented RNA about 11.7 kilobases (kb) in length. This means that the genome can act as mRNA in the infected cell for immediate translation of the viral replication machinery. The 5′ end is capped with 7-methylguanosine, and the 3′ end contains a poly (A) tail. Genomic RNA sequences from a number of strains of SIN viruses [HRsp AR339, consensus AR339, SAAR 86 (South African), Girdwood

(South African), and MRE16 (Malaysia)] have been published and are available in the database.

The 5′ two-thirds of the genome is translated to form polyproteins (P123 or P1234) that contain the viral nonstructural proteins (nsP1–nsP4). These proteins form different viral replicase complexes that synthesize positive- or negative-sense genomic RNAs. A subgenomic (26S) mRNA, colinear with the 3′ one-third of the genome, also is translated into a polyprotein that contains the structural components of the virus. Capsid (C) protein subunits are autocatalytically cleaved from the polyprotein cotranslationally; the remaining polyprotein molecule is translocated into the endoplasmic reticulum (ER) of the host cell. Cleavage and modification of the envelope glycoproteins (E1 and PE2) is a multistep process that takes place during vesicular transport through the ER/Golgi complex. E1 association with PE2 through disulfide bridges is required for oligomer formation and the export of E1 from the endoplasmic reticulum. Glycoprotein spikes and nucleocapsids assemble at the plasma membrane to produce progeny virions. Two smaller, unpackaged polypeptides (E3 and 6K) are produced during SIN virus infections as cleavage products during glycoprotein processing. A noncoding region (NCR) at the 3′ end of genomic and subgenomic RNAs, contiguous with a poly(A) tail, contains characteristic repeated sequence elements and may play a role in host specificity, possibly through interactions with cellular proteins.

Full-length infectious cDNA clones of alphaviruses have been generated for SIN, Semliki Forest, Ross River, Venezuelan equine encephalitis, and O'nyong-nyong viruses. Each clone contains a cDNA representing genomic RNA. To rescue virus, genomic RNA is transcribed in vitro from the cDNA clone in the presence of a capping nucleotide analogue. The synthesized positive-sense genome is electroporated into a susceptible cell line to produce the virus. This technology has allowed researchers to manipulate RNA genomes and identify viral determinants of host range, virulence, replication, disassembly, assembly, and packaging.

Sindbis Virus Expression Systems:
An Overview

Infectious-clone technology also has led to the development of new expression systems for use in mosquitoes. These transducing systems express transiently heterologous RNAs and proteins in mosquito cell culture and in mosquito larvae and adults. The SIN virus expression systems benefit vector biology for a number of reasons. First, they allow researchers to observe the biological consequences of overexpressing gene products such as toxins, immune peptides, and single-chain antibodies in mosquitoes. Second, these expression systems are important tools for silencing gene expression in mosquitoes. SIN viruses have now been used to silence luciferase expression in a transgenic line of *Aedes aegypti*, to interfere with infection and transmission of dengue viruses in the same vector species, as well as to silence prophenoloxidase activity, thus preventing encapsulation of macroparasites in *Armigeres subalbatus*. Third, SIN virus expression systems can allow researchers to rapidly test the merits of a given gene-based genetic control strategy prior to use in transformed mosquitoes. SIN virus expression systems by no means lessen the importance of arthropod vector transformation; rather, they complement transformation studies by allowing researchers an opportunity to rationally decide what gene products or effector molecules merit further analysis in transgenic mosquitoes.

Double Subgenomic SIN (dsSIN) Virus
Expression System

The SIN-based transducing system most frequently used in mosquitoes is termed the double subgenomic SIN (dsSIN) expression system. Two dsSIN expression systems, pTE/3′2J and pMRE/3′2J, are currently in use. dsSIN plasmids contain an SP6 bacteriophage promoter that transcribes in vitro a genomic RNA consisting of SIN 5′ NCR, nsp1–nsp4 coding regions, an internal RNA subgenomic promoter sequence, the SIN structural genes, a second internal RNA subgenomic promoter sequence, and the 3′ NCR (Fig. 39.1). The origin of the envelope glycoproteins of dsSIN viruses, TE/3′2J and MRE/3′2J, are a mouse neurovirulent strain and a Malaysian strain of SIN, respectively. The first subgenomic RNA promoter or core promoter maps from position −19 to +5 relative to the subgenomic mRNA start. The second subgenomic RNA promoter contains the core sequence plus sequences −98 to +14 to produce enhanced activity. A polylinker sequence is inserted downstream of the duplicated promoter to facilitate the introduction of heterologous genes and is followed by the 3′ NCR of SIN virus (Fig. 39.1). Full-length genomic RNA is transfected into appropriate cells. Within the transfected cell, three dsSIN RNA species occur: the genomic RNA, the first subgenomic mRNA, which is translated to form the virus structural proteins, and a second subgenomic mRNA, which can be translated to form the heterologous protein. Virions containing the recombinant RNA genome are generated that are infectious for multiple replicative cycles.

Double Subgenomic (dsSIN) Viruses

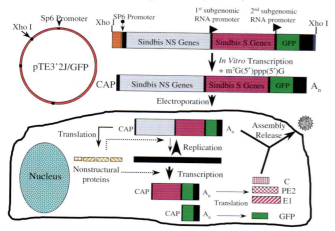

FIGURE 39.1 Outline of protocol for producing an infectious dsSIN virus designed to express GFP. NS = nonstructural, S = structural, GFP = green fluorescent protein, NCR = viral noncoding region, C = capsid, PE2 = viral envelope precursor glycoprotein E2, E1 = viral envelope glycoprotein E1, CAP = 7′ methylguanosine cap.

Expression of Proteins by dsSIN Viruses

When C6/36 (cultured *Aedes albopictus*) cells were infected with TE/3′2J/CAT virus at a multiplicity of infection (moi) >20, 100% of the cells expressed the reporter gene encoding chloramphenicol acetyl transferase (CAT; 8.3×10^5 polypeptides per cell) at 24 h postinfection. Approximately $4.0 \log_{10}$ 50% tissue culture infectious *doses* (TCID$_{50}$) of the TE/3′2J/CAT virus were inoculated into *O. triseriatus* mosquitoes. Titers of $>6.0 \log_{10} \text{TCID}_{50}$ per mosquito were detected within 4 days. Approximately 1.4×10^{10} polypeptides of CAT were detected within 2 days of infection. CAT activity peaked at day 6 and remained at peak levels to day 20. Immunofluorescence and CAT activity assays were used to localize CAT expression in infected mosquitoes and demonstrated that CAT was present in neural, midgut, and salivary gland tissues. The TE/3′2J dsSIN system has also been used to express green fluorescent protein (GFP) in *Ae. aegypti*, *Ae. triseriatus*, *Cx pipiens*, and *Anopheles gambiae* (Fig. 39.2).

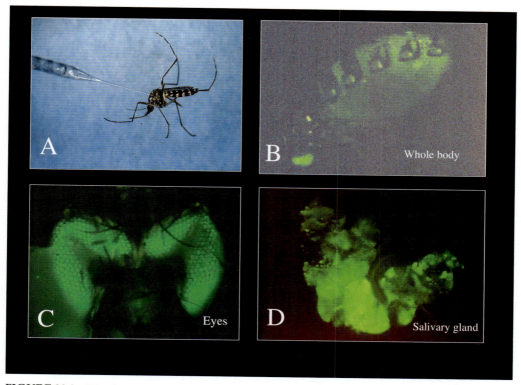

FIGURE 39.2 Distribution of GFP 4 days after intrathoracically inoculating *Aedes aegypti* with 10^3 plaque-forming units (pfu) of TE/3′2J/GFP virus. (A) Delivery of a dsSIN virus by intrathoracic injection of a mosquito. (B, C, D) Detection of GFP in a whole body, eyes, and salivary glands, respectively, of a female mosquito. (Image B approximately 40× magnification; images C and D 200× magnification)

The utility of dsSIN viruses as tools for gene expression in vector biology can be demonstrated by the following example. A TE/3'2J dsSIN virus was engineered to express a single-chain antibody that binds to the circumsporozoite protein of the avian malaria parasite, *Plasmodium gallinaceum*. The single-chain antibody (scFv) gene had been engineered to contain the Fab regions of heavy and light chains of a monoclonal antibody, N2H6D5, that blocked sporozoite invasion of *Aedes aegypti* salivary glands. Mosquitoes were intrathoracically injected with the scFv dsSIN virus and challenged several days later by allowing the mosquitoes to feed on parasitemic chickens. An examination of mosquitoes at 11 days postchallenge showed that the mean intensities of sporozoite infections of salivary glands in mosquitoes expressing N2scFv were reduced as much as 99.9% when compared to controls. This study clearly demonstrated that TE/3'2J dsSIN viruses can be used to rapidly test the potential of any protein-based, antiparasite strategy in mosquitoes and to assess whether a particular antiparasite effector gene warrants further development in transformed mosquitoes.

Aedes, *Culex*, and *Anopheles* mosquitoes are infected with TE/3'2J viruses by intrathoracic injections. A problem with this route of infection is that TE/3'2J viruses do not infect midgut epithelial cells, precluding gene expression in initial sites of virus/vector interactions. Moreover, TE/3'2J viruses poorly infect the midgut of *Aedes aegypti* by the oral route of infection. A second dsSIN virus system was engineered to overcome this problem by exchanging the structural genes of TE/3'2J virus with those of MRE16 SIN virus (Fig. 39.3A). MRE16 virus was isolated from *Culex* mosquitoes in Malaysia and has been shown to efficiently infect *Aedes aegypti* midguts by 4 days after ingestion of a blood meal containing >10^5 plaque-forming units (pfu)/mL of virus. Furthermore, MRE16 disseminates to secondary tissues in >90% of infected female mosquitoes. The new dsSIN system, termed MRE/3'2J, had a similar pattern of infection as MRE16 virus. Mosquitoes infected with MRE/3'2J/GFP virus showed enhanced expression of GFP in midgut tissues (Fig. 39.3B). Another advantage of MRE/3'2J viruses is that they can be used to obtain gene expression in larvae, pupae, and adult *Aedes aegypti* by feeding larvae C6/36 cells infected with the virus. This technique has been used to express GFP in late-instar larvae and pupae and to express antibacterial defensin proteins in adult mosquitoes.

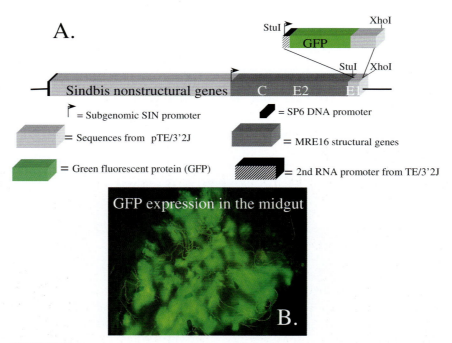

FIGURE 39.3 (A) Genome structure of virus cDNA found in MRE/3'2J/GFP plasmid. NS = nonstructural, S = structural, GFP = green fluorescent protein, NCR = viral noncoding region, C = capsid, E2 = viral envelope glycoprotein E2, E1 = viral envelope glycoprotein E1. (B) GFP expression in the midgut of *Aedes aegypti* 4 days after infection with an artificial blood meal containing 10^8 pfu/mL of virus.

The MRE/3'2J expression system demonstrates that it is possible to manipulate dsSIN expression systems to attain transient gene expression in a targeted mosquito tissue. Mining sequences from different SIN viruses and other alphaviruses should enhance the number of expression systems available for use with other vector species.

dsSIN Viruses as Tools to Test RNA-Mediated Interference Strategies

Initial studies using dsSIN virus expression systems to test viral interference strategies were based on observations by plant virologists that expression in plant cells of untranslatable mRNA or antisense RNA derived from a plant viral genome could inhibit infection by the homologous virus. Heterologous RNA plant virus expression systems were effective vehicles for delivery of interfering RNA.

Expression from the dsSIN vector TE/3'2J of RNA complementary to the La Crosse virus nucleocapsid mRNA was shown to interfere with LAC replication in C6/36 mosquito cells, and infection with TE/3'2J containing inserts of either positive-sense or antisense RNA from the dengue and yellow fever virus genomes inhibited replication of these flaviviruses in *Aedes aegypti*. In each case, the interfering RNA required >80% sequence identity or complementarity with the genome of the target virus. The most straightforward interpretation of these results was that the abundant, cytoplasmic transcripts from the second subgenomic promoter of dsSIN were able to directly block homologous virus replication by formation of RNA:RNA hybrids that prevented translation (antisense transcript) or transcription (untranslatable positive-sense transcript).

More recently, a phenomenon termed *posttranscriptional gene silencing* (PTGS) has been shown to account for plant virus–mediated interference with replication of heterologous viruses as well as with expression of endogenous genes and transgenes in plants. The dsRNA replicative intermediates of plant virus replication were implicated as the triggers of virus-induced PTGS. Plant geneticists have demonstrated the essential role of double-stranded (ds) RNA by showing that simultaneous expression of sense and antisense RNA in plant cells was more effective for gene silencing and virus resistance than expression of either single-stranded (ss) RNA alone.

The discovery that injection of dsRNA into *Caenorhabditis elegans* triggered sequence-specific PTGS (termed RNA interference [RNAi] in invertebrates) focused attention on dsRNA as an effector of gene silencing. It was rapidly shown that dsRNA can elicit PTGS/RNAi in a number of organisms, including *Drosophila melanogaster*.

Whether the mechanism of dsSIN virus–mediated viral interference involved direct blocking of viral gene expression or was an RNAi-like phenomenon has been explored by construction of dsSIN viruses that had inserted sequences derived from the genome of one or more of the four DEN virus serotypes. Intrathoracically injected *Aedes aegypti* mosquitoes developed sequence-specific resistance to challenge with homologous DEN viruses, and resistance was independent of the orientation of the effector RNA (Fig. 39.4). Expression of DEN RNA from any of several coding regions, with a minimum length of ~100 nucleotides (nt), prevented the accumulation of homologous genomic RNA, and C6/36 cells were highly resistant to DEN-2 virus when challenged at 2, 5, or 8 days after the initial dsSIN virus infections, even after the concentration of intracellular RNA containing a DEN-2 prM gene insert had sharply declined. Initiation of resistance occurred prior to or within the first 8 hours after challenging with DEN-2 virus. These observations are consistent with the mechanism of PTGS in plants and RNAi in *Drosophila* in which a cellular enzyme (designated Dicer in *Drosophila*) digests triggering dsRNA to 21- to 25-nt fragments that act as guide sequences to confer sequence specificity on a second nuclease complex that degrades cognate mRNA. Indeed, 21- to 25-nt RNA homologous to a dsSIN insert has been observed in midguts of *Ae. aegypti* and in C6/36 cells after recombinant dsSIN infection. It therefore seems probable that the efficacy of dsSIN expression systems as triggers for RNA-mediated viral interference is due to the formation of dsRNA replication intermediates in SIN virus–infected mosquito cells rather than to the production of abundant sense or antisense ssRNA alone.

dsSIN Viruses as Tools to Silence Mosquito Genes

Virus-induced gene silencing (VIGS) has been used not only to interfere with plant virus replication, but also to silence endogenous gene expression in plants. For example, plant geneticists have shown that infection of tobacco plants with potato virus X vectors that expressed either sense or antisense RNA derived from the *Nicotiana* phytoene desaturase gene posttranscriptionally suppressed expression of the endogenous gene. They also used VIGS to silence a GFP transgene in tobacco plants.

To test the hypothesis that dsSIN expression of antisense RNA could inhibit expression of a targeted mosquito gene, germ-line-transformed *Aedes aegypti*

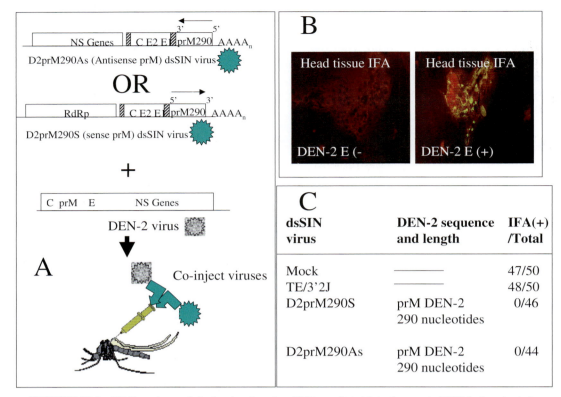

FIGURE 39.4 (A) Experimental design for detecting RNA-mediated interference to DEN-2 virus in *Aedes aegypti*. dsSIN viruses D2prM290As and D2prM290S contain a 290-nucleotide RNA derived from the premembrane (prM) coding region of the Jamaican 1409 strain of DEN-2 virus. The 290 nucleotide RNA was inserted in antisense or sense orientation with respect to the 2nd subgenomic RNA promoter in D2prM290As and D2prM290S, respectively. Mosquitoes were coinjected with 10^5 pfu of each dsSIN virus and 10^3 pfu of DEN-2 virus. (B) Mosquito heads were analyzed at 14 days postinfection for the presence of DEN-2 envelope (E) antigen by an indirect immunofluorescence assay (IFA; 100× magnification). (C) Results of IFA analysis of heads from mosquitoes injected with DEN-2 virus and either no dsSIN, TE/3'2J, D2prM290S, or D2prM290As virus.

that express firefly luciferase from the mosquito *Apyrase* promoter were intrathoracically inoculated with TE/3'2J/anti-luc that transcribes RNA complementary to the 5' end of luciferase mRNA. Mosquitoes infected with antisense-luc-expressing virus exhibited 90% reduction in luciferase expression compared with uninfected and control dsSIN-infected mosquitoes at 5 and 9 days postinoculation (Fig. 39.5). *Armigeres subalbatus* mosquitoes transduced with dsSIN virus that expressed a 600-nt antisense RNA targeted to the endogenous phenoloxidase gene showed significant reduction in hemolymph enzyme activity compared with controls. When these mosquitoes were challenged with *Dirofilaria immitis*, melanization of microfilariae was almost completely inhibited. Thus SIN virus expression vectors can be powerful tools in functional analysis of gene expression in mosquitoes.

MOLECULAR BIOLOGY OF MOSQUITO DENSOVIRUSES

The densonucleosis viruses, or densoviruses, are members of the subfamily *Densovirinae* of the family *Parvoviridae*. Most of the mosquito densoviruses described to date are in the *Brevidensovirus* genus. As with all parvoviruses, the densoviruses are nonenveloped icosahedral viruses with a diameter of 20–25 nm. The most extensively studied mosquito densovirus is the *Aedes* densonucleosis virus (AeDNV), which was isolated from a laboratory colony of *Aedes aegypti* mosquitoes in Kiev, Ukraine. Several other viruses have been isolated from persistently infected mosquito cell lines. Densoviruses are insect viruses, and they do not infect vertebrate organisms. It is thought that the viral life cycle in nature involves horizontal transmission among larvae in oviposition sites.

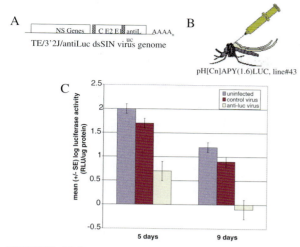

FIGURE 39.5 Virus-Induced Gene Silencing of a Transgene in *Aedes aegypti*. (A) Genomic RNA structure of TE/3'2J/antiLuc virus. This virus contains an RNA sequence derived from the 5' end of the luciferase gene and inserted in an antisense orientation with respect to the 2nd subgenomic RNA promoter. (B) Approximately 10^3 pfu/mL of virus was injected into the transgenic line pH[Cn]APY(1.6)luc43 that expresses luciferase in the salivary glands of adult females. (C) Mean $\log_{10}$ luciferase (LUC) activity, measured as relative luciferase units standardized to total protein concentration of the salivary glands, in APY-LUC43 mosquitoes 5 and 9 days after intrathoracic injection with either antiLuc virus or control virus. Uninoculated control mosquitoes were cold-anesthetized only.

Some infected females that emerge from infected pupae are capable of vertical transmission and thereby can spread the virus to new oviposition sites (Fig. 39.6).

The genome of AeDNV is a linear 4,000-base single-stranded DNA molecule with terminal inverted repeat sequences that are predicted to fold into T- or Y-shaped structures (Fig. 39.7A). These structures act as replication origins and packaging signals for the DNA. Replication and transcription take place in the nucleus of the infected cell using host DNA and RNA polymerases. An infectious plasmid clone of AeDNV has been constructed, and when it is transfected into C6/36 cells, an infection is initiated and production of virus results. As in most mammalian parvoviruses, all of the genes are coded on the same strand (Fig. 39.7A), and the negative-sense strand is the one that is predominantly packaged in particles. The virus protein (VP) gene codes for the structural proteins of the virus particle and occupies the 3' end of the positive-strand genome. The genes for the nonstructural (NS) proteins NS1 and NS2 are found in the 5' end of the positive strand. The *NS2* gene is contained within the *NS1* gene

but in a different translational reading frame. The NS1 protein is required for virus replication, while the function of the NS2 protein is unknown. The *NS1* and *NS2* genes are transcribed from the promoter (P_{NS}) located 7 map units from the 5' end. The VP gene is transcribed from the P_{VP} promoter located 60 map units from the 5' end.

Densovirus Transduction

The infectious clone of AeDNV (pUCA) has been used as the basis for construction of a densovirus transducing system. The only requirements for packaging are the terminal hairpin structures of the genome and that the length of the transducing genome be 4 kb or less. The central portion of the AeDNV genome can be replaced by a gene of interest, and the missing viral genes can be provided in *trans* by helper constructs. If the VP gene is the only one replaced (Fig. 39.7B, for example), the transducing genome will be capable of self-replication since it retains its *NS1* gene. If VP is provided in *trans*, the transducing genomes will be packaged. If both VP and NS1 are replaced, both proteins must be supplied in *trans* by a helper. The helper construct that has proven to be most useful is the full-length infectious clone, pUCA. The plasmids containing the transducing genome and pUCA are cotransfected into C6/36 cells, both genomes are replicated, and a mixture of wild-type virus and transducing particles is produced. It is also possible to provide the complementing gene products by using cell lines that have been stably transformed by constructs that constitutively produce VP and/or NS1 or by infecting transfected cells with a Sindbis virus vector that produces VP protein. In these cases it is possible to get pure populations of transducing particles, although the yields of particles have been disappointing and the systems need to be optimized.

Reporter genes encoding the *E. coli* β-galactosidase and GFP and have been cloned into the AeDNV genome and packaged into particles for delivery to mosquito cells or mosquitoes. One particularly useful transducing genome has the GFP gene fused to the *NS1* gene (Fig. 39.7B). This transducing genome produces an NS1-GFP fusion protein that has all of the functions of NS1 and can be visualized in living cells by fluorescence microscopy. When transducing particle preparations containing both wild-type and transducing genomes are added to water containing mosquito larvae, the larvae are efficiently infected, and the course of infection can be followed by observation of the larvae with a fluorescence microscope. The first signs of infection are usually in the anal papillae of the larvae (Fig. 39.8A). These are organs involved in the

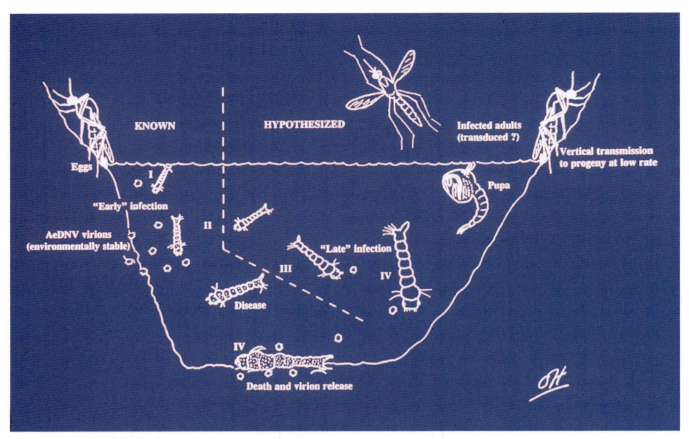

FIGURE 39.6   Life cycle of densoviruses in Culicine mosquitoes. Artwork produced by Dr. Steve Higgs.

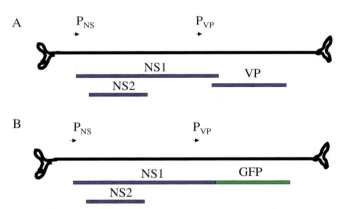

FIGURE 39.7   (A) Structure of the single-stranded DNA genome of a densovirus. P = viral DNA promoter, NS1 and NS2 = viral non-structural genes, VP = virus structural protein gene. (B) Structure of a transducing virus genome in which the VP gene has been replaced with a gene of interest, green fluorescent protein (GFP).

maintenance of the ionic strength of the hemolymph of the larva (Chapter 26). Sometimes the infected anal papillae are lost and the infection is cured, but more often the virus disseminates, resulting in a large number of tissues and cell types showing green fluorescence, including fat body, muscle, nerves, and hemocytes (Fig. 39.8B and C). If first- and second-instar larvae are infected, they usually die as larvae. Larvae infected at later stages often are able to pupate, and some emerge as infected adults. Multiple tissues are infected in the adult mosquitoes, including the ovaries in the female. The GFP gene has been detected by PCR in the larval offspring of infected females, thereby indicating the possibility of vertical transduction. Thus the AeDNV transducing system can be used to express genes of interest in multiple tissues and life stages in mosquitoes.

Potential Role for Densoviruses in Control of Vector-Borne Disease

Densoviruses have a number of properties that make them attractive for applications in the control of

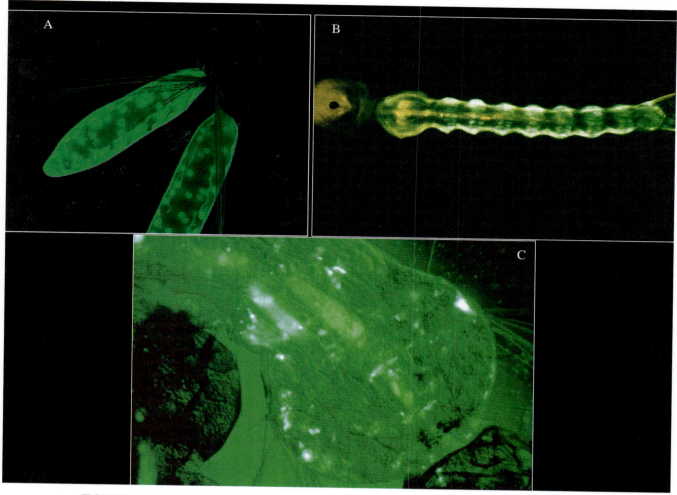

FIGURE 39.8 GFP expression in anal papillae of *Aedes aegypti* larva (A; 200× magnification), whole larva (B; 40×), and thorax of larva (C; 200×) following infection with densoviruses containing GFP transducing and wild-type (pUCA) genomes. Part A was provided by Drs. Martin Edwards and Marcelo Jacobs-Lorena.

vector-borne disease. (1) They have a narrow host range. AeDNV infects only mosquitoes of the genera *Aedes*, *Culex*, and *Culiseta*. Other insects, including chironomids, flies, bees, cockroaches, and lepidopterans, are not infected. Further testing in aquatic crustaceans, fishes, vertebrate cell lines, and rodents showed no evidence of infection or other deleterious effects. Thus, they are narrowly targeted to mosquitoes. (2) Since they are nonenveloped viruses, AeDNV are stable in the environment. They are resistant to extremes in temperature and pH and to organic solvents. (3) Their ability to function as transducing virus makes them attractive for expression of novel genes (e.g., RNAi constructs) to reduce vector potential in wild populations. (4) Densoviruses are pathogenic to mosquitoes,

and this makes them attractive candidates for biocontrol applications against mosquito populations.

The characteristics of viruses that are necessary for these different applications may differ considerably. Virus characteristics desirable for biocontrol applications are difficult to predict and could vary with the strategy to be employed. For simple mosquito control applications, a highly pathogenic and lethal virus with a narrow host range would seem to be the most desirable. This would allow rapid killing of target mosquitoes, and probably would not permit the virus to spread widely in the population beyond the application site. On the other hand, if introduction of a gene for resistance to a mosquito-borne pathogen into a mosquito population is the objective, a less pathogenic virus capable of efficient vertical transmission seems more desirable, since this should allow the virus to spread beyond the application site. Depending on the epidemiology of the particular mosquito-borne disease and the economics of implementation of a control program, a virus that kills most mosquitoes and renders the ones that remain resistant to the target pathogen might be the most effective. With these applications in mind, research to elucidate the molecular nature of cytopathogenicity and other virulence determinants and to isolate new mosquito densoviruses with different host range and phenotypic properties is under way.

Readings

Sindbis Virus Molecular Biology and Sindbis Virus Expression Systems: An Overview

Hahn, C. S., Hahn, Y. S., Braciale, T. J., and Rice, C. M. 1992. Infectious Sindbis virus transient expression vectors for studying antigen processing and presentation. *Proc. Natl. Acad. Sci. U.S.A.* 89: 2679–2683.

Olson, K. E. 2000. Sindbis virus expression systems in mosquitoes: Background, methods, and applications. In: *Insect Transgenesis: Methods and Applications* (A. M. Handler, and A. A. James, eds.), CRC Press, Boca Raton, FL, pp. 161–189.

Schlesinger, S., and Schlesinger, M. J. 2001. Togaviridae: The viruses and their replication. In: *Fields Virology*, 4th ed. (D. M. Knipe, and P. M. Howley, eds.), Lippencott, Williams, and Wilkins, Philadelphia, pp. 895–916.

Strauss, J. H., and Strauss, E. G. 1994. The alphaviruses: Gene expression, replication and evolution. *Microbiol. Rev.* 58: 491–562.

Expression of Proteins by dsSIN Viruses

Capurro, M. L., Coleman, J., Myles, K. M., Olson, K. E., Beerntsen, B. T., Rocha, E., Krettli, A. U., and James, A. A. 2000. Virusexpressed, recombinant single-chain antibody blocks sporozoite invasion of salivary glands in *Plasmodium gallinaceum*–infected *Aedes aegypti*. *Am. J. Trop. Med. Hyg.* 62: 427–433.

Cheng, L. L., Bartholomay, L. C., Olson, K. E., Lowenberger, C., Higgs, S., Beaty, B. J., and Christensen, B. M. 2001. Characterization of an endogenous gene expressed in *Aedes aegypti* using an orally infectious recombinant Sindbis virus. *J. Insect Sci.* 1: 10 (Online at www.insectscience.org/1.10/).

Higgs, S., Olson, K. E., Klimowski, L., Powers, A. M., Carlson, J. O., Possee, R. D., and Beaty, B. J. 1995. Mosquito sensitivity to a scorpion neurotoxin expressed using an infectious Sindbis virus vector. *Insect Mol. Biol.* 4: 97–103.

Higgs, S., Traul, D., Davis, B., Wilcox, C., and Beaty, B. 1996. Green fluorescent protein expressed in living mosquitoes without the requirement for transformation. *Biotechniques* 21: 660–664.

Olson, K. E., Higgs, S., Hahn, C. S., Rice, C. M., Carlson, J. O., and Beaty, B. J. 1994. Expression of chloramphenicol acetyltransferase in *Aedes albopictus* (C6/36) cells and *Aedes triseriatus* mosquitoes using double subgenomic recombinant Sindbis virus vectors. *Insect Biochem. Mol. Biol.* 24: 39–48.

Olson, K. E., Myles, K. M., Seabaugh, R. C., Higgs, S., Carlson, J. O., and Beaty, B. J. 2000. Development of a Sindbis virus expression system that efficiently expresses green fluorescent protein in

midguts of *Aedes aegypti* following per os infection. *Insect Mol. Biol.* 9: 57–65.

dsSIN Viruses as Tools to Test RNA-Mediated Interference Strategies

Adelman, Z. N., Blair, C. D., Carlson, J. O., Beaty, B. J., and Olson, K. E. 2001. Sindbis virus–induced silencing of dengue viruses in mosquitoes. *Insect Mol. Biol.* 10: 265–273.

Angell, S. M., and Baulcombe, D. C. 1997. Consistent gene silencing in transgenic plants expressing a replicating potato virus X RNA. *EMBO J.* 16: 3675–3684.

Bernstein, E., Caudy, A. A., Hammond, S. M., and Hannon, G. J. 2001. Role for a bidentate ribonuclease in the initiation step of RNA interference. *Nature* 409: 363–366.

Blair, C. D., Adelman, Z. N., and Olson, K. E. 2000. Molecular strategies for interrupting arthropod-borne virus transmission by mosquitoes. *Clin. Microbiol.* 13(4): 651–661.

Fire, A., Xu, S., Montgomery, M. K., Kostas, S. A., Driver, S. E., and Mello, C. C. 1998. Potent and specific genetic interference by double-stranded RNA in *Caenorhabditis elegans*. *Nature* 391: 806–811.

Gaines, P. J., Olson, K. E., Higgs, S., Powers, A. M., Beaty, B. J., and Blair, C. D. 1996. Pathogen-derived resistance to dengue-2 virus in mosquito cells by expression of the premembrane coding regions of the viral genome. *J. Virol.* 70: 2132–2137.

Hamilton, A. J., and Baulcombe, D. C. 1999. A species of small antisense RNA in posttranscriptional gene silencing in plants. *Science* 286: 950–952.

Higgs, S., Rayner, J. O., Olson, K. E., Davis, B. S., Carlson, J. O., Blair, C. D., and Beaty, B. J. 1998. Engineered resistance in *Aedes aegypti* to a West African and a South American strain of yellow fever virus. *Am. J. Trop. Med. Hyg.* 58: 663–670.

Kennerdell, J. R., and Carthew, R. W. 1998. Use of dsRNA-mediated genetic interference to demonstrate that frizzled and frizzled 2 act in the wingless pathway. *Cell* 95: 1017–1026.

Powers, A. M., Kamrud, K. I., Olson, K. E., Higgs, S., Carlson, J. O., and Beaty, B. J. 1996. Molecularly engineered resistance to California serogroup virus replication in mosquito cells and mosquitoes. *Proc. Natl. Acad. Sci. U.S.A.* 93: 4187–4191.

Ruiz, M. T., Voinnet, O., and Baulcombe, D. C. 1998. Initiation and maintenance of virus-induced gene silencing. *Plant Cell* 10: 937–946.

Waterhouse, P. M., Graham, M. W., and Wang, M. B. 1998. Virus resistance and gene silencing in plants can be induced by simultaneous expression of sense and antisense RNA. *Proc. Natl. Acad. Sci. U.S.A.* 95: 13959–13964.

Zamore, P. D., Tuschl, T., Sharp, P. A., and Bartel, D. P. 2000. RNAi: Double-stranded RNA directs the ATP-dependent cleavage of mRNA at 21 to 23 nucleotide intervals. *Cell* 101: 25–33.

dsSIN Viruses as Tools to Silence Mosquito Genes

Johnson, B. W., Olson, K. E., Allen-Miura, T., Rayms-Keller, A., Carlson, J. O., Coates, C. J., Jasinskiene, N., James, A. A., Beaty, B. J., and Higgs, S. 1999. Inhibition of luciferase expression in transgenic *Aedes aegypti* mosquitoes by Sindbis virus expression of antisense luciferase RNA. *Proc. Natl. Acad. Sci. U.S.A.* 96: 13399–13403.

Ratcliff, F. G., MacFarlane, S. A., and Baulcombe, D. C. 1999. Gene silencing without DNA. RNA-mediated cross-protection between viruses. *Plant Cell* 11: 1207–1216.

Shiao, S. H., Higgs, S., Adelman, Z., Christensen, B. M., Liu, S. H., and Chen, C. C. 2001. Effect of prophenoloxidase expression knockout on the melanization of microfilariae in the mosquito *Armigeres subalbatus*. *Insect Mol. Biol.* 10: 315–321.

Molecular Biology of Mosquito Densoviruses

Afanasiev, B. N., Galyov, E. E., Buchatsky, L. P., and Kozlov, Y. V. 1991. Nucleotide sequence and genomic organization of *Aedes* densonucleosis virus. *Virology* 185: 323–336.

Afanasiev, B. N., and Carlson, J. O. 2000. Densoviruses as gene transfer vehicles. In: *Contributions to Microbiology: Parvoviruses*. Karger AG. 4: 33–58.

Carlson, J., Olson, K. E., Higgs, S., and Beaty, B. J. 1995. Molecular genetic manipulation of mosquito vectors. *Annu. Rev. Entomol.* 40: 359–388.

Gorziglia, M., Botero, L., Gil, F., and Esparza, J. 1980. Preliminary characterization of virus-like particles in a mosquito (*Aedes pseudoscutellaris*) cell line (Mos. 61). *Intervirology* 13: 232–240.

Jousset, F.-X., Barreau, C., Boublik, Y., and Cornet, M. 1993. A parvolike virus persistently infecting a C6/36 clone of *Aedes albopictus* mosquito cell line and pathogenic for *Aedes aegypti* larvae. *Virus Res.* 29: 99–114.

Kimmick, M. W., Afanasiev, B. N., Beaty, B. J., and Carlson, J. O. 1998. Gene expression and regulation from the p7 promoter of the *Aedes* densonucleosis virus. *J. Virol.* 72: 4364–4370.

O'Neill, S. L., Kittayapong, P., Braig, H. R., Andreadis, T. G., Gonzalez, J. P., and Tesh, R. B. 1995. Insect densoviruses may be widespread in mosquito cell lines. *J. Gen. Virol.* 76: 2067–2074.

Ward, T. W., Kimmick, M. W., Afanasiev, B. N., and Carlson, J. O. 2001b. Characterization of the structural gene promoter of *Aedes aegypti* densovirus. *J. Virol.* 75: 1325–1331.

Densovirus Transduction

Afanasiev, B. N., Ward, T. W., Beaty, B. J., and Carlson, J. O. 1999. Transduction of *Aedes aegypti* mosquitoes with vectors derived from Aedes densovirus. *Virology* 257: 62–72.

Afanasiev, B., Kozlov, Y., Carlson, J., and Beaty, B. 1994. Densovirus of *Aedes aegypti* as an expression vector in mosquito cells. *Exp. Parasitol.* 79: 322–339.

Allen-Miura, T. M, Afanasiev, B. N., Olson, K. E., Beaty, B. J., and Carlson, J. O. 1999. Packaging of AeDNV-GFP transducing viruses by expression of viral proteins from a Sindbis virus expression system. *Virology* 257: 54–61.

Carlson, J., Afanasiev, B., and Suchman, E. 2000. Densoviruses as transducing vectors for insects. In: *Insect Transgenesis: Methods and Applications* (A. M. Handler, and A. A. James, eds.), CRC Press Boca Raton, FL, pp. 139–159.

Ward, T. W., Jenkins, M. S., Afanasiev, B. N., Edwards, M., Duda, B. A., Suchman, E., Jacobs-Lorena, M., Beaty, B. J., and Carlson, J. O. 2001a. *Aedes aegypti* transducing densovirus pathogenesis and expression in *Aedes aegypti* and *Anopheles gambiae* larvae. *Insect Mol. Biol.* 10: 397–406.

Stable Transformation of Vector Species

CRAIG J. COATES

INTRODUCTION

Since the advent of recombinant DNA technology in the 1970s, entomologists have yearned to have the ability to genetically transform and thus manipulate insect genomes. As described in this chapter, the ability to genetically transform an insect species allows for a vast array of manipulations that can be utilized for both basic and applied research purposes. Even as we enter the postgenomic era for some insect species, genetic transformation remains one of the most powerful enabling technologies for studies of insects at a molecular level. In this section, unlike the transient systems described in previous chapters, stable transformation occurs when a gene of interest is inserted into an insect chromosome such that it is transmitted through the germ line in a Mendelian manner. *Stable* in this context implies that the transgene is present in the nucleus of every cell in the organism and is maintained in subsequent generations.

While transient systems provide advantages in ease of use and speed, they also have significant limitations. Transient systems may be inappropriate for studies in which the gene of interest is not expressed in a desired tissue or developmental stage, due to viral tropism or limitations of the DNA delivery system. Viral expression systems are effective only in those species and tissues in which the virus can establish an efficient infection. Furthermore, the level of transient gene expression will vary among tissues and may differ even within experiments due to variations in viral infection, replication, and dissemination. Finally,

transient gene expression systems also may result in gene dosages and expression levels that are not physiologically relevant.

An ideal genetic transformation system provides an advantage over transient systems, in that a constant gene dosage is achieved, resulting in consistent gene expression at physiologically relevant levels. Furthermore, for studies of gene regulation, genetic transformation places the transgenes in the proper physical context of a chromosome. This allows all potential levels of transcriptional regulation to occur, including those involving changes in chromatin folding and structure. Finally, while transient systems have proven valuable for rapidly assessing the function and efficacy of a variety of recombinant gene constructs for antipathogen activity in disease vectors, transient gene expression systems will not be suitable for future applications in the generation of refractory strains, as discussed later.

Genetic transformation of insects was first demonstrated in the fruit fly, *Drosophila melanogaster*. Gerry Rubin and Alan Spradling (1982) used the *P* transposable element and a wild-type copy of the rosy gene (*ry+*) to generate stable transformants of a *ry-* mutant strain with restored eye color. Researchers attempting to transform other insect species have closely followed the *P* element paradigm. Continued developments in genetic transformation technologies in *D. melanogaster* have provided further models for researchers to follow in other insects. In should be noted that there still remains an extensive amount of knowledge to be gained from the drosophila system that can be applied to studies of disease vectors. It is important not to lose sight of this valuable model genetic organism because it will remain the predominant insect of choice for several years to come, due in no small part to the ease of rearing and to the vast genetic and DNA sequence resources.

Genetic transformation experiments in *D. melanogaster* have been utilized for gene-discovery purposes and for investigations of gene function, gene regulation, and interactions at the DNA, RNA, and protein levels. It is clear that the ability to perform similar experiments in disease vectors represents a critical step forward in our continuing understanding of the basic biology of these insects at a molecular level. Even as the power and sensitivity of molecular biology and recombinant DNA technologies continue to increase, genetic transformation remains the experimental procedure of choice for confirming the function of a gene or, more particularly, a specific allele of a gene. Furthermore, genetic transformation technologies allow the testing of the function of a gene, or gene regulatory sequences, from one organism, in a different species, or even crossing kingdom boundaries. There are several instances where researchers have taken a gene of interest from a disease vector and transformed this gene into *D. melanogaster*. While often productive, this approach has not always resulted in the production of meaningful data, indicating that the regulatory and functional characteristics of the gene were not conserved across species. While many biological processes are likely to be shared or at least to be similar among *Drosophila* and disease vectors, there will almost certainly be specific functions associated with blood feeding and pathogen transmission that can be adequately addressed only in a blood-feeding vector insect.

The use of genetic transformation technologies to perform gene-discovery experiments will be even more valuable in disease vectors than in *D. melanogaster,* since there are few available mutant strains and less genomic and expressed sequence tag (EST) information for these species. Insertional mutagenesis and enhancer-trapping techniques for disease vectors will provide additional opportunities to discover genes that are relevant to the unique physiology of these insects and to pathogen transmission. These so-called basic research applications of genetic transformation technologies will result in an increased understanding of the behavior, development, physiology, and disease transmission potential of a wide variety of disease vectors, which may in turn lead to novel control methods for both the vectors and the pathogens.

From an applied research perspective, genetic transformation technologies allow researchers to contemplate additional applications that have not been considered for *Drosophila* or other nonvector insects. Many of the proposed applications involve the modification of a competent disease vector genome such that it can no longer transmit a particular pathogen. These genetically modified strains are referred to as *transmission-blocking*, or *refractory*, strains. There are several pathogen targets currently being evaluated for the development of such strains. Similarly, there are a number of potential "refractory," or transmission-blocking, molecules that are being considered for activity against these pathogens, including endogenous and exogenous immune peptides, antisense and double-stranded RNA molecules, and single-chain antibodies.

DNA INTRODUCTION

All transgenic insect strains produced to this point have resulted from the microinjection of plasmid DNA molecules into preblastoderm embryos. The preblastoderm embryos of higher dipterans exist as a syncitium of dividing nuclei. At the onset of the blastoderm stage, cell membranes are formed, thereby preventing access of the introduced plasmid DNA to the nucleus. Most microinjection protocols attempt to optimize the delivery of the plasmid DNA such that it is at a maximal concentration near the germ-line, or pole cell, nuclei, which are typically located at the posterior end of the embryo. DNA delivery is attempted at the earliest possible stage of embryonic development in the hope that chromosomal integration events will thus occur early in the development of the germ line, such that the proportion of the total progeny that carries a transgene insertion is maximized.

In contrast to *Drosophila* spp., the embryos of many disease vectors cannot survive the removal of the chorion, a maternally derived structure analogous to an eggshell, which undergoes rapid melanization and hardening on exposure to air and protects the embryos from desiccation. The difficulties associated with microinjecting through the chorion results in the introduction of low amounts of DNA and poor survival rates following injection. One of the advances that assisted the development of a genetic transformation system for the human malaria vector, *Anopheles stephensi*, was the use of a chemical compound, pNpGB, that delays chorion hardening, thus increasing the period of time after egg collection during which satisfactory injections can be performed. It is apparent that the quality of injections is critical for these experiments. Needle design, consistency, and the mechanical process of DNA introduction in general have proven to be key factors in the success of genetic transformation experiments.

Other methodologies for DNA delivery, such as the use of solid needles coated with DNA, electroporation, and biolistics, have been attempted in both *Drosophila* and disease vectors. However, the amounts of DNA delivered into embryos using these methods are below those achieved using embryo microinjection with glass capillary needles. An alternative strategy being considered is the delivery of DNA into a life stage other than the embryo. It has been proposed that direct DNA delivery to the ovaries and DNA mixed with sperm for the purposes of artificial insemination may be viable alternatives. These proposals have some merit, in that this may result in the production of transgenic individuals one generation earlier than occurs through the use of current methodologies and would alleviate some of the problems associated with complicated mating systems.

DNA INTEGRATION USING TRANSPOSABLE ELEMENTS

Transposable element vectors have been used to integrate transgenes into the chromosomes of all transgenic insects produced to this point. While *P* has proven to be an excellent element on which to base a genetic transformation vector for *D. melanogaster*, it appears to have limited utility for other insects and has been used successfully only to transform closely related *Drosophila* species. However, there are a number of characteristics associated with the *P* element that highlight the potential use of transposable elements to spread refractory genes in disease vector populations. In the mid- to late 1970s, researchers crossing wild strains of *D. melanogaster* with established laboratory strains noted a phenomenon called *hybrid dysgenesis*, resulting in reduced fertility or sterility as well as somatic and germ-line mutations. It was eventually determined that these effects were being caused by the transpositional activity and random genomic insertions of the *P* transposable element. The *P* element is present in all wild populations and absent from laboratory populations established from collections made in the early 1900s. This situation is thought to be the result of a recent invasion, possibly by a horizontal transfer event or recent activation of an ancient element, and subsequent spreading throughout the wild population.

The ability of the *P* element to mobilize and insert in new genomic locations was recognized by Rubin and Spradling (1982) as a potential mechanism for inserting recombinant gene constructs into the *Drosophila* genome. The availability of existing laboratory strains, empty of *P* elements such that transposition into these genomes was not repressed, represented an advantage, as was the presence of mutations within these strains that could be used as the basis for marker gene rescue and the easy identification of transgenic individuals.

The successful genetic transformation of *D. melanogaster* provided the impetus for researchers to attempt germ-line transformation experiments in disease vector species. Researchers proceeded to use the *P* element for genetic transformation experiments in these other species, without significant success, except for an isolated transformation event in *An. gambiae*, as discussed later. In addition to DNA delivery and embryo manipulation problems, it was discovered that the *P* element did not have high levels of transpositional activity in the embryos of non-drosophilid insects. Needless to say, this revelation brought about an abrupt halt in experiments utilizing the *P* element for genetic transformation of disease vectors and a renewed search for alternative transposable elements.

Transposable elements are ubiquitous in nature and have been found in every genome that has been characterized systematically. While a number of different transposable elements have been utilized as genetic transformation vectors for insects, they are all members of the Class II family of transposable elements and share the common features of relatively small inverted terminal repeats (ITRs) and an open reading frame encoding a transposase protein. Class II elements transpose via a DNA-only intermediate, and they are thought to use a cut-and-paste (conservative) type of mechanism for movement, although alternative transpositional or integrative mechanisms have been proposed. In the cut-and-paste model of transposition, the transposase protein makes cuts at or very near the inverted terminal repeats, often in an apparent staggered fashion, and then makes a similar cut at the target site such that the transposable element can be pasted into its new genomic location (Fig. 40.1). Furthermore, these elements share the critical ability of the transposase proteins they encode being able to act in trans. Importantly for the purposes of genetic transformation, this allows the development of binary systems in which two plasmids are coinjected. The transposase protein is produced from one injected plasmid, the helper plasmid, where it can act on a separate donor plasmid that carries the transposable element with a marker gene and other transgenes. The transposase coding region on the donor plasmid is interrupted by the insertion of the transgenes, rendering this element incapable of catalyzing its own

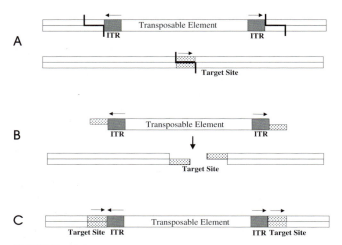

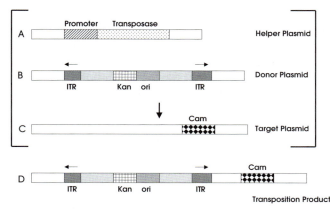

FIGURE 40.1 Cut-and-paste model for Class II transposable element transposition. (A) The transposable element is represented as a solid rectangle, with the inverted terminal repeats (ITRs) indicated as shaded areas. The two strands of the flanking DNA region and the target site (stippled area) are shown as thin rectangles; the staggered cut sites made by the transposase are represented as bold lines. (B) Intermediate stage with a staggered break in the target DNA able to form complementary base pairing with the overhangs of the transposable element–containing fragment. (C) Filling in and ligation result in the insertion of the transposable element into its new genomic location and the direct duplication of the target site on each side. Arrows indicate the relative orientations of the ITRs and target sites.

FIGURE 40.2 Plasmid-based transposition assay. Three plasmids are coinjected into preblastoderm embryos. The helper plasmid (A) contains the transposable-element open reading frame (transposase) under the control of a promoter and thus provides the transposase protein. The donor plasmid (B) contains the transposable element in which a kanamycin antibiotic resistance–gene (Kan) and an *E. coli* origin of replication (ori) have been inserted between the inverted terminal repeats. The target plasmid (C) contains a chloramphenicol antibiotic–resistance gene (Cam) and cannot replicate in *E. coli*. Successful transpositions of the transposable element from the donor plasmid to the target plasmid result in the production of transposition products (D) that confer kanamycin and chloramphenicol resistance and can replicate in *E. coli*, thus utilizing a triple selection method for quantifying transpositional activity.

movement. The helper plasmid typically consists of a completely or partially truncated transposable element, such that one or both of the ITRs are missing. Therefore, this element also is incapable of catalyzing its own movement. To maximize the amount of transposase protein produced, the transposase open reading frame often is fused with the promoter of an exogenous gene, for example, one encoding a heat-shock protein, which provides a high level of inducible transcriptional activation.

An increasing number of transposable elements from both vector and nonvector insect species have been isolated and characterized in some detail. Some transposable elements were discovered by chance as insertions within genes, often causing phenotypic changes. Other transposable elements were found as a result of homology-based searches using low-stringency DNA hybridization techniques. More recently, such searches have been performed using the polymerase chain reaction (PCR) and degenerate oligonucleotides designed to conserved regions of transposable elements from the same family. The typical progression of events following the discovery of a representative member of a transposable element family within a particular insect genome is the use of Southern blotting techniques or quantitative dot

blotting to determine the element copy number in the genome. PCR screening and DNA sequence determinations are used to characterize the extent of length and sequence variations, particularly within the transposase coding region. It is clear from several studies that, in general, the majority of transposable elements within a genome are inactive, presumably due to mutation and selective forces acting against insertional activity. While the specific mechanisms are not known, models of transposable-element invasions suggest a period of high transpositional activity followed by a period of stabilization and then stochastic loss.

The net result of several studies is the availability of a number of transposable elements, from several distinct families, that can be used for genetic transformation experiments. Given the amount of time and labor these experiments require, plasmid-based transposition or mobility assays (Fig. 40.2) simplify transposable-element choice and provide the researcher with some confidence in the activity of the transposable element to be used for a transformation experiment. Transposition assays provide an assessment of the relative activity levels of different transposable elements within the embryonic soma of a developing insect by determining the rate of transposition between two introduced plasmids. Clearly, if an element fails to function under these assay conditions it is unlikely that it would function effectively in a genetic transformation experiment. While a positive result in these assays is not a guarantee of success, there has been a good correlation between results from these assays and successful transformation experiments.

P

Not surprisingly, given the results in *D. melanogaster*, the *P* element was the first transposable element utilized in attempts to produce transgenic vector species. While genetic transformation experiments were performed in a number of species using this transposable element, only one isolated success was reported. Louis Miller and colleagues (1987) used a *P* element construct containing a selectable neomycin-drug-resistance marker gene to produce a transgenic line of *An. gambiae*. While molecular analyses confirmed the integration of the construct into a mosquito chromosome and that the neomycin-resistance phenotype was inherited in a stable manner, it was determined that the integration event was independent of the *P* element. In other words, while the presence of the *P* element DNA sequences and transposase may have assisted integration, this did not occur by the canonical cut-and-paste mechanism typical of *P* element transposition in *D. melanogaster*. This result, and the subsequent discovery of the lack of *P* element activity in nondrosophilid insects, led to an intense search for alternative transposable elements to use for genetic transformation experiments.

Minos

The *Minos* element, from the *Tc1-mariner* superfamily of transposable elements, was isolated from *D. hydeii* during a library screen for sequences flanking ribosomal RNA genes. Following the *P* element paradigm, *Minos* was used in a binary system to genetically transform the Mediterranean fruit fly, *Ceratitis capitata*. This represented the first genuine, completely transposable element–mediated, germ-line transformation success in a nondrosophilid insect species. In this successful experiment the *C. capitata white* gene was used to complement a white-eye mutant phenotype, partially restoring eye color in transgenic individuals. This experiment again demonstrates the value of pre-existing genetics, cloned genes, and laboratory strains with mutant phenotypes. In conjunction with the enhanced green fluorescent protein (EGFP) marker gene, *Minos* was subsequently used to genetically transform the human malaria vector, *An. stephensi*.

Hermes

The *hAT* family of transposable elements was defined when it was determined that significant regions of amino acid homology existed between transposable elements isolated from *D. melanogaster* (*Hobo*), maize (*Ac*), and snapdragon (*Tam3*). *Hermes*, a member of the *hAT* family isolated from the housefly, *Musca domestica*, was used to successfully transform the yellow fever mosquito, *Ae. aegypti*. Transformation was observed by the complementation of the white-eyed mutant phenotype of the *khw* mosquito strain with a wild-type copy of the *D. melanogaster cinnabar* (*cn+*) gene. Unlike Drosophilid species, which have both ommochrome and pteridine pigment pathways contributing to eye color, *Ae. aegypti* has only one pigment pathway contributing to eye color; thus, a single mutation in this pathway can lead to the mutant white-eye phenotype. This pathway is analogous to the ommochrome pathway in *Drosophila* spp., and the *khw* strain is so named because it lacks a functional *kynurenine hydroxylase* enzyme, which is the protein product of the *Drosophila cn+* gene. The *Hermes* transposable-element system, in conjunction with EGFP, was also used to produce transgenics of the southern house mosquito, *Culex quinquefaciatus*.

Mos1

The *Mos1* element from *D. mauritiana*, a member of the *Tc1-mariner* superfamily of transposable elements, was used to transform *Ae. aegypti* using the *cn+* gene from *D. melanogaster*, in a similar manner to the *Hermes* experiment described earlier. Additional genetic transformation experiments were performed in *Ae. aegypti* using a purified *Mos1* transposase protein in place of a helper plasmid, thereby demonstrating that this was a viable option for those species in which promoter utilization had not been well characterized. It was also thought that the use of a purified transposase protein may result in integrations earlier in the development of the germ line, leading to a greater proportion of transgenic progeny.

piggyBac

The *piggyBac* element, a member of the TTAA family of transposable elements, was first discovered as an insertion in a baculovirus that was being passaged through a *Trichoplusia ni* cell line. The insertion event was detected because it caused a reduction in polyhedra production in infected cells. *piggyBac* has since been more completely characterized and utilized as a genetic transformation vector for a number of insect species. *piggyBac* is now one of the more widely used genetic transformation vectors and, like *Minos*, *Hermes*, and *MosI*, appears to have a wide host range and thus limited requirements for host cell factors for transpositional activity. The *piggyBac* element has been used for the successful transformation of *An. albimanus*, *An. gambiae*, *An. Stephensi*, and *Ae. aegypti*.

Genetic transformation experiments involving disease vectors have typically resulted in 1–12% of the fertile Go individuals producing a small number of transgenic progeny. (In standard transformation nomenclature, the Go generation is that which has been injected with the DNA. These animals also are designated as "founders" because they are used to start potential transformed lines. Subsequent generations are designated G1, G2, G3, and so on.) The observed transformation rates are much lower than those detected when using *P* in *D. melanogaster*, which are typically over 20% and can be as high as 50%. The reasons for this are likely to vary in each situation and include ease of egg manipulation, survival following injection, and genome size and organization, all factors that are independent of the transposable element itself. Furthermore, due to the requirements for finer needles and the smaller size of the embryos, the volume of plasmid DNA introduced into embryos is generally lower in other species, which may be an indication of the primary cause of the low frequencies obtained. Plasmid-based transposition assay frequencies are not significantly different for individual transposable elements in *D. melanogaster* as compared to frequencies achieved in different species, yet these same transposable elements have been used for *D. melanogaster* transformation experiments and have produced transformants at similar rates to the *P* element. It is also possible that the exogenous promoters used to drive marker gene expression are poorly utilized in these species and that a majority of the insertion events recovered are the result of positive position effects. Positive position effects occur when a transposable-element construct inserts in a region of the genome that is under the influence of a local, endogenous enhancer, thus increasing the level of transcription from the transgene promoter. Furthermore, and perhaps most importantly, genome size and composition almost certainly play a significant role, with the *D. melanogaster* genome having a relatively simple structure with a reduced number of repetitive elements and heterochromatin compared to many disease vectors and other insects that have been transformed. From a simple consideration of viable integration sites versus genome size, many disease vector species are at a disadvantage. It seems likely that more integration events will occur in heterochromatic or transcriptionally inactive regions because other transposable elements may not share the apparent preference of *P* for inserting into the 5′-end regions of coding sequences, which are likely to be transcriptionally active.

TRANSFORMATION MARKERS

The availability of mutant strains and the corresponding wild-type gene was a clear advantage in the development of the *P* element system for generating transformants of *D. melanogaster*. Due to the intensity of genetic research performed on this model genetic organism, there are many mutant strains available with obvious phenotypic defects, and the biochemical processes and genes that control these pathways are well characterized. One of the major problems facing researchers working on other insects was the lack of suitable transformation markers and mutant strains to use as recipients. This led to an investigation of the use of insecticide-resistance genes as transformation markers, a problematic approach, given the existence of natural resistance alleles in these populations that resulted in false positive progeny during selection.

A large increase in the number of successful transformation experiments coincided with the discovery and use of the green fluorescent protein (GFP) as a marker for genetic transformation experiments. This is an important recent advance in the field of insect transformation because it removed the need for prior genetic and biochemical information, the existence of a mutant strain, or an understanding of the insect genome of interest. GFP appears to have universal potential; detection is nondestructive and can be performed at essentially any life stage, simplifying screening, rearing, and crossing procedures. A number of GFP variants with enhanced levels of fluorescence and modified absorption and emission spectra have been produced, further increasing the utility of these proteins for genetic transformation experiments. One of the first variants produced was the enhanced green fluorescent protein (EGFP), which has been used to transform several disease vector species, including *An. albimanus*, *Cx. quinquefaciatus*, *An. stephensi*, *Ae. aegypti*, and *An. gambiae*.

BASIC GENETIC TRANSFORMATION TECHNOLOGIES

While many of the initial germ-line transformation experiments in disease vectors were designed as tests of the proof of principle that transformation could be achieved in a particular species, researchers are now moving forward with basic research goals related to gene discovery and function as well as more applied goals related to reducing pathogen transmission. There are a number of methodologies that can be utilized in a disease vector species once a successful germ-line transformation system has been developed. These technologies have been utilized with great success in the *Drosophila* system and also are used in genetic transformation experiments in other model genetic organisms, such as the nematode, *Caenorhabditis elegans*, and the zebra fish, *Danio rerio*. Progress toward optimizing these techniques in disease vectors is ongoing, and substantial progress has been made in some instances. It is reasonable to assume that these technologies will transfer fairly easily from *Drosophila* to other insects, the main requirement being an increase and optimization of the efficiency and frequency of transposition, rather than substantial modifications of the methodology.

Insertional Mutagenesis

As the name implies, this technique uses the ability of transposable elements to insert into the genome and cause mutations. This is a natural result of endogenous transposable-element activity and in several cases it was this activity that led to the discovery and subsequent characterization of a transposable-element family. In essence, transposable-element vectors are introduced into the germ-line to produce transgenic lines. In some cases, an insertion event causes a dominant mutation, but in most cases any resulting mutation will be recessive and homozygous lines of each insertion event are produced and screened for changes in phenotype. A change in a particular phenotype implies that the insertion has interrupted or affected the expression of a gene that is involved in the production or maintenance of the "normal" phenotype or condition. The transposable-element insertion provides a molecular tag by which researchers can locate, identify, and characterize the gene involved.

One of the reasons for the success of this procedure in *D. melanogaster* is the preference of the *P* element to insert at or near the 5′ ends of genes. This often results in the inactivation of the gene or a modification in the expression profile of the gene. It is not clear whether the *Minos, Hermes, MosI,* or *piggyBac* elements have the same or different gene insertion preferences, for the relatively low number of insertion events recovered from genetic transformation experiments using these elements has not allowed such a determination to be made. The ability to perform successful insertional

mutagenesis screens in disease vectors would represent a significant advance, because there are relatively few characterized genes and mutant phenotypes available for further genetic studies in these species. Any such screen in disease vectors would likely focus on biological processes of major importance, such as blood-feeding, egg development, host-seeking behavior, pathogen transmission, and immune system function. One of the current difficulties in performing these experiments in disease vectors is an inability to easily produce large numbers of transgenic lines. As a result, efforts in several laboratories are directed at optimizing and improving the efficiencies of existing genetic transformation systems. Additional limiting factors are the cost and amount of insectary space required to rear and maintain large numbers of disease vector strains.

Gene Insertion/Functional Studies

Whether a gene has been isolated by traditional means, such as homology-based hybridization, or by insertional mutagenesis, there is often a need to confirm its function using methodologies that go beyond DNA sequence database comparisons. Ever increasingly accurate predications of gene function can be made based on a growing amount of DNA sequence data from a number of species, particularly those in which the whole genome sequence is known, yet there is no comparison to a functional study in which the gene is expressed in the genome of interest. These studies are often performed in conjunction with a previously created mutant strain, whether by transposable-element insertion or traditional mutagenesis techniques.

The function of genes isolated from disease vectors has been inferred in many cases by comparison to known sequences from the *D. melanogaster* genome and in some cases by genetic transformation of *D. melanogaster* with the isolated gene. While this has been successful in some cases for confirming gene function and is somewhat satisfactory, it will be a significant advantage to be able to confirm gene function in the same species or at least in another blood-feeding species, particularly for those genes involved directly in the blood-feeding process or pathogen transmission, for these are unlikely to be well conserved or even present in the *Drosophila* genome. Furthermore, even for those genes that have orthologous sequences in the *Drosophila* genome, these may not have the same function, because they may have been coopted or modified during evolution to allow for the development of blood-feeding capability.

Enhancer Trapping

Enhancer trapping is a technique that allows the identification of DNA sequences involved in the regulation of gene expression (Fig. 40.3). Enhancers are able to promote or regulate gene expression at a chromosome position distant from promoters and can be at the 5′ or 3′ end and even in the introns of genes. Some enhancers are able to influence the expression of several neighboring genes and often act in a tissue- or developmental stage–specific manner. The principle of enhancer trapping involves a transposable-element vector containing a reporter gene, often the bacterial gene, *lacZ*, in the case of *D. melanogaster*, under the control of a minimal promoter with weak activity such that in a typical transformed line there is low or undetectable levels of expression. However, if the transposable element inserts near an endogenous enhancer, the regulation of the reporter gene will come under the influence of that enhancer. Many disease vector species have high levels of endogenous bacteria and thus a high level of background *lacZ* activity, requiring the use of EGFP or another fluorescent reporter gene for this purpose. There is no requirement for enhancer-trap lines to be homozygous, for the reporter gene provides a dominant phenotype. As with insertional mutagenesis, the inserted transposable element provides a molecular tag to identify the chromosomal

region containing the putative enhancer and in many cases will also contain genes that have the same pattern of expression. More advanced constructs can be created that also allow these lines to be used for enhancer-driven ectopic expression of additional genes for further studies of gene function and expression.

This technique is of clear value for studies of disease organisms because it is likely that, at least at some level, genes expressed in the midgut and salivary glands and the induction of egg development following a blood meal will be controlled by enhancers. These tissues and processes are of particular interest because the midgut is the first point of contact between pathogens and the disease vector and, in most cases, must be successfully traversed or infected for parasite transmission to occur. In addition, in the case of malaria parasites, key developmental events occur within the lumen of the mosquito midgut. Furthermore, the salivary glands are the last point of residence for most pathogens prior to transmission to the human host, and these tissues must be invaded or infected by the pathogens for successful transmission to occur. Finally, egg development is critically important for those disease vectors that transmit pathogens vertically to their progeny and in general for the production of the next generation of disease vectors and population expansion. Both insertional mutagenesis and enhancer trapping technologies were facilitated in *D. melanogaster* by the creation of helper lines in which the transposase of the respective transposable element was produced at high levels from a chromosomal location. This helper strain can be crossed to other lines containing transposable-element insertions to remobilize the transposons to new genomic locations. It is likely that the creation of helper strains will be of great value in disease organisms, and efforts are continuing to produce these lines.

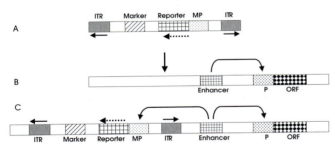

FIGURE 40.3 Enhancer trap system. (A) Enhancer trap vector. This construct consists of a transposable-element vector with inverted terminal repeats (ITRs), shown as shaded areas. The marker gene is used to identify transgenic individuals. The remainder of the construct consists of a reporter gene whose transcription is controlled by a minimal promoter (MP), such that the expression of the reporter gene will be low in the absence of a positive position effect. The dotted arrow indicates the direction of transcription. (B) Chromosomal target. The arrow indicates a possible insertion site for the enhancer trap vector. This schematic has a gene (ORF) under the control of a promoter (P), transcription being stimulated by the action of a 5′-end enhancer element, which is able to recruit additional transcription factors to the promoter. (C) Integration of the enhancer trap vector. Insertion of the vector near an endogenous enhancer element results in the stimulation of transcription of the reporter gene from the minimal promoter, mimicking the transcription profile of the endogenous gene.

Promoter–Reporter Gene Fusions

Promoter-reporter gene fusions are used to analyze the control of gene expression by cloned promoter fragments. These can be used to dissect promoter function at a broad or fine scale and can also be used to test enhancer function. Whether a promoter is identified by traditional gene cloning or by transgenic technologies using insertional mutagenesis or enhancer trapping, there is often a requirement to define more precisely the promoter activity, particularly if it is to be used as the basis for expressing a refractory gene in a disease vector.

A promoter–reporter gene fusion construct provides an opportunity to assess promoter function by observing the activity of the associated reporter gene. At a qualitative level this can be used to determine the tissue- or developmental stage–specific activity of a given promoter. The firefly luciferase gene is often used as the reporter in disease vectors because the protein it encodes, luciferin, has low background activity; the detection using a luminometer is highly quantifiable, sensitive, and linear in response over several orders of magnitude. Quantitative analyses can be confounded by the position effects that are experienced at different chromosomal insertion points. As discussed previously, transgene insertions occur at relatively random points in the genome and can become influenced by the surrounding chromosomal material, including enhancers, silencers, and the actual chromatin structure itself. These effects can modify the baseline activity of the inserted promoter such that it may not be at the normal endogenous level. More detailed analyses of promoters can be performed by deleting specific regions of the promoter and repeating the genetic transformation experiments such that the specific contribution to gene expression is determined for each region of the promoter. For this procedure to be effective, large numbers of transgenic lines need to be generated to remove, or at least ameliorate the position effects such that an average level of expression can be calculated for each promoter fragment. Alternatively, boundary elements that insulate regions of chromosomes from nearby position effects can be included in the transgene construct.

The ability to express transgenes in specific tissues or in response to particular environmental, behavioral, or developmental cues will be critical to our ability to produce strains of insects that are refractory to pathogens that they would normally transmit. In particular, there may be a requirement to limit the expression of transgene products to females at a certain time point during or after blood feeding or in a particular tissue that is critical to pathogen development. Enhancer-trap studies would provide regulatory elements that could be used to express transgene products in the desired manner. More traditional gene-discovery methods, such as homology-based cloning, differential cDNA library screening, antibody screening of expression libraries, and protein purification, have resulted in the identification and characterization of a number of tissue-, sex-, and stage-specific genes and regulatory sequences from disease vectors. For *Ae. aegypti* in particular, a number of tissue-specific promoters have been used to express the luciferase reporter gene or immune defense peptides, to assess promoter function with great success. Salivary gland, fat body, and midgut promoters have all been used in this manner in *Ae. aegypti*.

Gene Fusions

Gene fusions are somewhat similar to reporter gene fusions, except that in this case, in addition to the promoter, the coding region of the gene of interest is fused in frame with a reporter gene such that the two proteins are coexpressed and colocalized. Similar to promoter–reporter gene fusions, this procedure can be used to assess the tissue- or developmental stage–specific expression of a gene, with the added advantage of providing information about the cellular location of a target gene. The reporter gene can be visualized at the appropriate time point, and the endogenous protein will be colocalized with the reporter gene protein. There may be a requirement to test 5′- and 3′-end gene fusions, particularly for membrane-bound or cell-sorted proteins that contain signal peptides or GPI-anchor attachment sequences. Gene fusions can be used to observe the expression, movement, and degradation of a protein in real time if a nondestructive fusion protein such as EGFP is used. Gene fusion studies in disease vectors will be particularly useful to follow the expression and movement of proteins that interact with pathogens, particularly those that are involved in pathogen binding, cell invasion, and the immune defense system.

Applications of Transgenic Vectors

The primary application of a genetic transformation system is to provide the ability to manipulate the genome of the insect for basic research purposes, such as gene discovery, confirmation of gene function, and investigations of the control of gene regulation. The ability to perform these experiments will undoubtedly lead to new discoveries of the molecular mechanisms controlling a variety of processes of importance to vector insects and disease transmission, such as behavior, blood-feeding physiology, insect immunity, and pathogen development.

Beyond these so-called basic applications of the genetic transformation technology, it has been postulated that it will be possible to produce genetically engineered strains of disease vectors that are refractory to pathogen infection, development, or transmission and to use these strains to replace wild-type strains or to spread refractory genes into wild populations. The goal of this approach is to increase the number of refractory mosquitoes in the wild, leading to a decrease in pathogen transmission and a concomitant decrease in the incidence, and thus the cost, of disease. This premise is debated heavily in the case of malaria transmission, where there appears to be conflicting data from different geographic areas with respect to the rate of transmission and the corresponding incidence and severity of the disease. The key elements required are an efficient genetic transformation system, a refractory gene that produces a product active against the pathogen in question, suitable tissue-specific regulatory sequences to enable the expression of the refractory product in the correct location and at the correct developmental stage, and a mechanism to spread the refractory gene into wild mosquito populations.

As described earlier, an increasing number of disease vector species have been genetically transformed with a variety of transposable-element vectors. Furthermore, the expression of transgene products in a tissue-specific manner has also been demonstrated in some cases. Research beyond the scope of this chapter has demonstrated that suitable refractory mechanisms have been identified for both viral and parasitic pathogens of human health importance, and functional refractory genes are currently being incorporated into genetic transformation constructs. Large-scale testing of transgene expression and stability under mass rearing conditions has not been performed, and searches for suitable gene drive mechanisms are ongoing.

Given that *Ae. aegypti* was the first disease vector transformed, it is no surprise that an increased number of advanced transformation experiments have been completed in this species as compared to other disease vectors. It is also no small coincidence that *Ae. aegypti* is one of the easier disease vectors to be reared and manipulated in the laboratory, leading a larger number of laboratories conducting research on this species. Initial experiments in this species involved the use of the *Hermes* and *Mos1* elements to demonstrate that functional genetic transformation systems could be developed for this species. In both cases the *cinnabar* eye color gene from *D. melanogaster* was used to rescue a white-eye mutant strain. The phenotypic rescue was partial, and it was observed that the eye color varied between different lines, presumably as a result of position effects. The eye color gene provided a simple determination of heterozygotes and homozygotes based on the intensity of the eye color. Genetic crosses were performed demonstrating that the transgene was inherited in a Mendelian manner. Furthermore, in one transgenic line, the insertion was homozygous lethal; in several other lines, insertions were determined to be linked to the sex locus.

These initial preliminary results alone demonstrate the potential of genetic transformation systems. Insertions were occurring in transcriptionally distinct regions of the genomes based on the variations in eye color. For the line with the homozygous lethal phenotype, the transposable element can be used as a molecular tag to determine the identity of the essential gene that was disrupted. Several insertions were closely linked to the sex locus, again demonstrating the potential of the system for gene discovery and enhancer trapping.

The next stage of experiments in *Ae. aegypti* used fragments of the upstream regulatory regions from salivary gland genes in an effort to define or confirm those promoter regions that are required for salivary gland–specific gene expression. These experiments were successful, with luciferase expression being detected in both male and female salivary glands when a promoter from a gene involved in sugar feeding was used, the expression being limited to a specific region of the female salivary glands when a blood-feeding gene promoter was used. In both cases, the expression profile mirrored that of the endogenous gene. These experiments also revealed some aspects of the genetic transformation systems that need to be improved, mainly that each of the lines had different levels of transgene expression, as was observed for the eye color marker gene. Furthermore, in the case of one family, features of position-effect variegation were observed, in that individuals within the same family had different levels of transgene expression.

Further experiments in this organism involved the use of a promoter from the fat body–specific vitellogenin gene to more closely define the specific sequence elements involved in the expression of the endogenous gene. The vitellogenin promoter was also used to drive the expression of an immune peptide, Defensin A, isolated from *Ae. aegypti*, for the ultimate purpose of determining if the presence of the peptide in the mosquito hemolymph can prevent the migration of malaria parasites from the midgut to the salivary glands and thus block transmission. These experiments demonstrated that elements of the mosquito immune system can be engineered in a transgenic mosquito such that they became activated upon blood feeding rather than responding to an infection. Similar transgenic experiments have demonstrated the ability to transcribe the luciferase reporter gene in the midgut following a blood meal, using a carboxypeptidase promoter. Other experiments in progress for this species attempt to develop strains that constitutively express transposase proteins at high levels for the purposes of improving genetic transformation systems for insertional mutagenesis and enhancer trap protocols. Furthermore, experiments with active transposable elements that can catalyze their own movement, in cis, are ongoing to investigate the possibility of using transposable elements to spread refractory transgenes into wild populations.

Aedes aegypti is an important disease vector in its own right and will be a target for the development of refractory strains with transgenes directed against the yellow fever and dengue fever viruses. In conjunction with the avian malaria parasite, *Plasmodium gallinaceum*, this mosquito forms a model system for investigations of human malaria transmission. Furthermore, it is also generally expected that *Ae. aegypti* will continue to serve as a useful model for the development and optimization of genetic transformation technologies such that they can be rapidly applied to other disease vectors, such as the human malaria vector, *An. gambiae*.

SUMMARY

In a relatively short period of time, researchers working toward the genetic transformation of disease vectors have progressed from being unable to transform any species to being able to transform at least five species of mosquitoes that transmit pathogens. It is expected that this number will increase as more laboratories become involved in this research and attempt genetic transformation experiments with the newly available transposable elements and marker genes. It appears that one impediment to improving the efficiency of genetic transformation efforts in disease vectors is the mechanical introduction of DNA plasmids into embryos. If the amount of DNA introduced and the survival rate following injection could be improved, this would presumably increase the transformation efficiency. Other deficiencies include a lack of understanding of the different genome structures and characteristics as well as a need for a more specific identification of gene regulatory elements to enable efficient and physiologically relevant transgene expression.

The development of genetically engineered refractory strains will continue to occur at an increased pace, limited only by the researchers' imagination. Much work remains before effective field releases of these strains can be contemplated. Tests of the level and stability of transgene expression in large populations through multiple generations must be initiated. Effective gene-spreading mechanisms remain to be developed and tested. Experiments to test large-scale rearing capabilities, relative fitness, and compatibility with wild populations need to be performed to determine the feasibility of this approach for controlling disease transmission. Safety and risk assessment studies will also be required.

In summary, the number of genetic transformation experiments being performed in disease vectors is increasing, and exciting findings are being reported as a result of these experiments. It is expected that transposase-producing lines will be produced shortly and that lines designed for large-scale insertional mutagenesis and enhancer trap experiments will also be developed. The promise of an increased understanding of disease vectors at a molecular level will be fully delivered in the near future when these and related technologies are available routinely for these species.

Readings

Transposable Elements and Genetic Transformation

Atkinson, P. W., Pinkerton, A. C., and O'Brochta, D. A. 2001. Genetic transformation systems in insects. *Annu. Rev. Entomol.* 46: 317–346.

Cooley, L., Kelley, R., and Spradling, A. 1988. Insertional mutagenesis of the *Drosophila* genome with single *P* elements. *Science* 239: 1121–1128.

Franz, G., and Savakis, C. 1991. *Minos*, a new transposable element from *Drosophila hydei*, is a member of the Tc1-like family of transposons. *Nucleic Acids Res.* 19: 6646.

Gates, J., and Thummel, C. S. 2000. An enhancer trap screen for ecdysone-inducible genes required for *Drosophila* adult leg morphogenesis. *Genetics* 156: 1765–1776.

Handler, A. M. 2001. A current perspective on insect gene transformation. *Insect Biochem. Mol. Biol.* 31: 111–128.

Handler, A. M., and James, A. A. 2000. *Insect Transgenesis Methods and Applications.* CRC Press, Boca Raton, FL.

Horn, C., and Wimmer, E. A. 2000. A versatile vector set for animal transgenesis. *Dev. Genes Evol.* 210: 630–637.

Rubin, G. M., and Spradling, A. C. 1982. Genetic transformation of *Drosophila* with transposable element vectors. *Science* 218: 348–353.

Germ-Line Transformation of Disease Vectors

Allen, M. L., O'Brochta, D. A., Atkinson, P. W., and Levesque, C. S. 2001. Stable, germ-line transformation of *Culex quinquefasciatus* (Diptera: Culicidae). *J. Med. Entomol.* 38: 38701–38710.

Catteruccia, F., Nolan, T., Loukeris, T. G., Blass, C., Savakis, C., Kafatos, F. C., and Crisanti, A. 2000. Stable germ-line transformation of the malaria mosquito *Anopheles stephensi*. *Nature* 405: 959–962.

Coates, C. J., Jasinskiene, N., Miyashiro, L., and James, A. A. 1998. Mariner transposition and transformation of the yellow fever mosquito, *Aedes aegypti*. *Proc. Natl. Acad. Sci. U.S.A.* 95: 3748–3751.

Grossman, G. L., Rafferty, C. S., Clayton, J. R., Stevens, T. K., Mukabayire, O., and Benedict, M. Q. 2001. Germ-line transformation of the malaria vector, *Anopheles gambiae*, with the *piggyBac* transposable element. *Insect Mol. Biol.* 10: 597.

Jasinskiene, N., Coates, C. J., Benedict, M. Q., Cornel, A. J., Rafferty, C. S., James, A. A., and Collins, F. H. 1998. Stable transformation of the yellow fever mosquito, *Aedes aegypti*, with the *Hermes* element from the housefly. *Proc. Natl. Acad. Sci. U.S.A.* 95: 3743–3747.

Kokoza, V., Ahmed, A., Wimmer, E. A., and Raikhel, A. S. 2001. Efficient transformation of the yellow fever mosquito, *Aedes aegypti*, using the *piggyBac* transposable element vector pBac[3xP3-EGFP afm]. *Insect Biochem. Mol. Biol.* 31: 1137–1143.

Miller, L. H., Sakai, R. K., Romans, P., Gwadz, R. W., Kantoff, P., and Coon, H. G. 1987. Stable integration and expression of a bacterial gene in the mosquito *Anopheles gambiae*. *Science* 237: 779–781.

Nolan, T., Bower, T. M., Brown, A. E., Crisanti, A., and Catteruccia, F. 2002. *PiggyBac*-mediated germ-line transformation of the malaria mosquito *Anopheles stephensi* using the red fluorescent protein dsRED as a selectable marker. *J Biol Chem*. In press.

Perera, O. P., Harrell, R. A., and Handler, A. M. 2002. Germ-line transformation of the South American malaria vector, *Anopheles albimanus*, with a *piggyBac/EGFP* transposon vector is routine and highly efficient. *Insect Mol. Biol*. In press.

Applications of Transgenic Technologies for Disease Vectors

Carareto, C. M., Kim, W., Wojciechowski, M. F., O'Grady, P., Prokchorova, A. V., Silva, J. C., and Kidwell, M. G. 1997. Testing transposable elements as genetic drive mechanisms using *Drosophila* P-element constructs as a model system. *Genetica*. 101: 13–33.

Coates, C. J., Jasinskiene, N., Pott, G. B., and James, A. A. 1999. Promoter-directed expression of recombinant firefly luciferase in the salivary glands of *Hermes*-transformed *Aedes aegypti*. *Gene* 226: 317–325.

James, A. A. 2000. What's that buzz? Mosquitoes and fruit flies commingle. *Parasitol. Today* 16: 503–504.

James, A. A., Beerntsen, B. T., Capurro, M. D., Coates, C. J., Coleman, J., Jasinskiene, N., and Krettli, A. U. 1999. Controlling malaria transmission with genetically engineered, *Plasmodium*-resistant mosquitoes: Milestones in a model system. *Parassitologia* 41: 461–471.

Kokoza, V., Ahmed, A., Cho, W. L., Jasinskiene, N., James, A. A., and Raikhel, A. 2000. Engineering blood meal-activated systemic immunity in the yellow fever mosquito, *Aedes aegypti*. *Proc. Natl. Acad. Sci. U.S.A.* 97: 9144–9149.

Moreira, L. A., Edwards, M. J., Adhami, F., Jasinskiene, N., James, A. A., and Jacobs-Lorena, M. 2000. Robust gut-specific gene expression in transgenic *Aedes aegypti* mosquitoes. *Proc. Natl. Acad. Sci. U.S.A.* 97: 10895–10898.

Pinkerton, A. C., Michel, K., O'Brochta, D. A., and Atkinson, P. W. 2000. Green fluorescent protein as a genetic marker in transgenic *Aedes aegypti*. *Insect Mol. Biol.* 9: 1–10.

CONTROL OF INSECTS AND ACARINES

JANET HEMINGWAY

41

Chemical Control of Vectors and Mechanisms of Resistance

JANET HEMINGWAY AND HILARY RANSON

The economics of developing, safety testing, and marketing insecticides means that novel compounds are not developed specifically for the control of disease vectors. All insecticides are developed primarily for the agricultural markets, and formulations are subsequently developed for some of these agrochemicals for use in public health. This interreliance on the agricultural market effectively means that there are only limited numbers of insecticides that can be used for vector control, and vectors may already have been exposed to these insecticides through breeding and resting in agricultural areas before they are deployed in public health programs.

Extensive exposure of insect vectors to insecticides eventually selects for resistance to them. Hence, a good understanding of what chemical classes are available, their modes of action, and what resistance mechanisms are selected is essential if chemical control is to be used, either in isolation or as part of a wider integrated pest management program.

HISTORY

Early pesticides included natural botanicals, such as nicotine, rotenone, and pyrethrum, along with other chemicals, such as lime sulfur, arsenic, mercuric chloride, and soaps. The scientific development of insecticides began in 1867 with the formulation and use of the arsenical Paris green. In the 1920s the structures of many of the botanical insecticides, which had been

used since the early 1800s, were made known. But it was not until 1939 that Müller discovered the insecticidal properties of the first synthetic insecticide, DDT (dichlorodiphenyltrichloroethane). The potential of this new insecticide was demonstrated in 1943 when an epidemic of louse-transmitted typhus was controlled in Naples. Subsequently, DDT played a major role during World War II in controlling outbreaks of typhus, trench fever, and louse-borne relapsing fever by direct insecticidal dusting of both soldiers and civilians and their clothes. The major benefit of DDT, however, came in malaria control, where it was the central plank of the World Health Organization's global malaria eradication campaign. While agricultural use of DDT has now ceased in almost all areas, due to its environmental persistence and reduced efficacy against resistant insects, it still has a vital role in malaria control, where it remains part of our limited arsenal of cost-effective and safe insecticides for indoor residual spraying. In recognition of this role, it was specifically exempted from a total ban in the recent International POPs (persistent organic pollutants) treaty. Its residual properties allow it to be sprayed on the indoor surfaces of houses, where mosquitoes that come into contact with it die before they are able to transmit the malaria infection. After the discovery of DDT there was a rapid increase in the number of insecticides discovered; the list includes other organochlorine insecticides, some with similar modes of action to DDT, others, such as BHC (benzene hexachloride) with different actions. In 1945 the first phosphorothioate

627

(OP) insecticides were discovered, followed in 1953 by the carbamates and almost a decade later by the pyrethroids. These four insecticide classes still make up over 90% of the public health insecticide market. They have been joined by bacterial insecticides, such as *Bacillus thuringiensis israelensis* (Bti) and insect growth regulators (IGRs), although the use of such compounds is limited for many disease vectors due to their high cost and larval-specific actions.

In addition to insecticide development, formulations have now been developed that are more effective (such as wettable powders [WPs] and emulsifiable concentrates [ECs]). And in many countries there has been a move toward a more targeted use of insecticides, e.g., insecticide-impregnated bednets, curtains, or plastic sheeting to replace wholesale indoor residual spraying.

CLASSIFICATION OF PESTICIDES

Insecticides are classified according to their chemical structure. Figure 41.1 shows the basic chemical structures of the four main insecticide classes, with a few specific examples of each class.

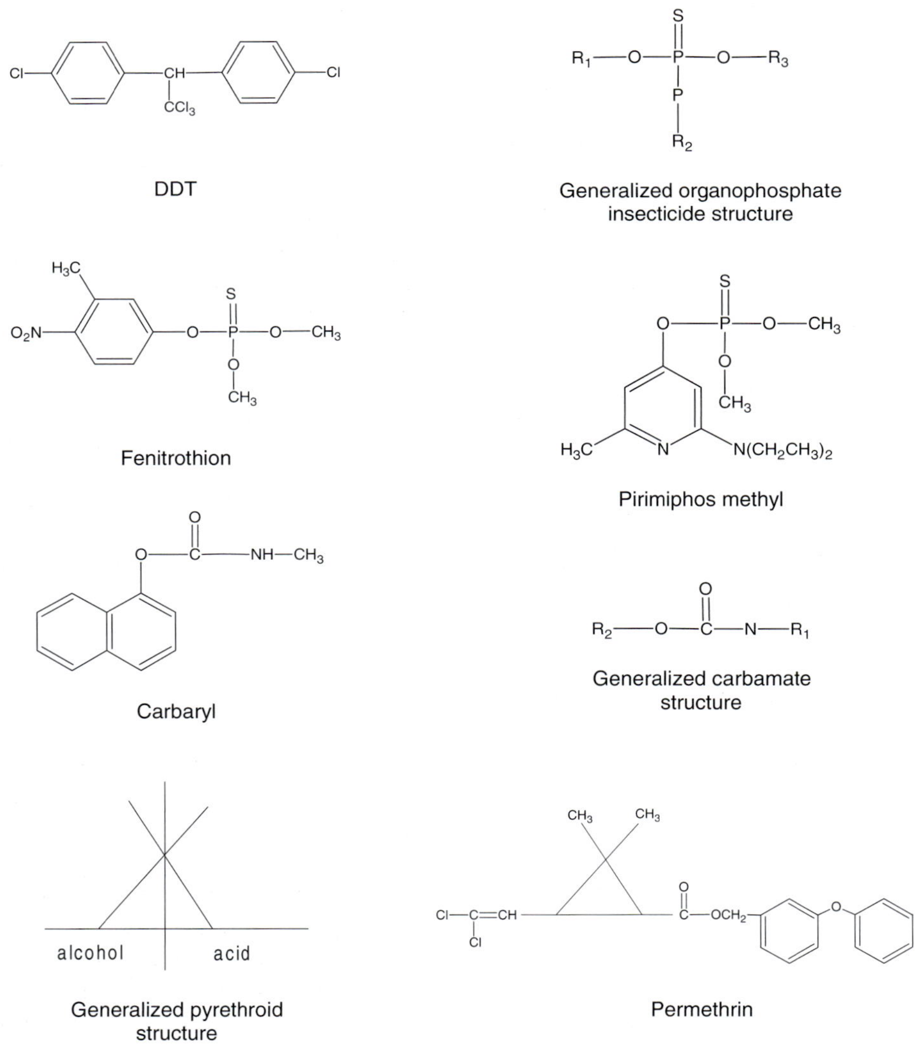

FIGURE 41.1 The basic chemical structure of the four main classes of insecticides. Organochlorine insecticides are represented by DDT. Organophosphate insecticides are represented by fenitrothion and pirimiphos methyl. Carbamate insecticides are represented by carbaryl. Pyrethroid insecticides are represented by permethrin.

Organochlorines

All chlorinated hydrocarbon insecticides are aryl, carbocyclic, or heterocyclic compounds with molecular weights ranging from 291 to 545. Most of these compounds are inhibitors of the normal functioning of the nervous system. DDT and its analogues act on the sodium channels of the nerve membrane, whereas benzene hexachloride (BHC) and cyclodienes such as dieldrin act on the GABA receptors. In spite of their similarity of chemical structures, insecticides in this group differ widely in their toxicity and stability. The chemical simplicity of the group means these insecticides are cheap and easy to manufacture, but their persistence in the environment, wildlife, and humans has drastically reduced their use since the 1970s.

Phosphorothioate Insecticides

These insecticides share a common general chemical structure, but they differ in their physical and pharmacological properties and, consequently, in their uses. These insecticides are usually less stable than the organochlorine insecticides. They are invariably administered as the insecticidally inactive phosphorothioate and through the action of monooxygenases (MFOs) in the presence of water are activated within the insect to the insecticidal organophosphate (Fig. 41.2). Phosphorothioate insecticides can vaporize quickly and often have associated with them a sulfurous "bad egg" smell. These insecticides are not stored in the body fat of animals, but are readily broken down and excreted through the kidneys.

This insecticide class acts by binding the enzyme acetylcholinesterase at the nerve junction. Once bound, this enzyme can no longer remove acetylcholine from the nerve–membrane junction, and the nerves continue to fire in an uncontrolled manner, eventually leading to paralysis and death of the insect. These insecticides, often referred to as *cholinesterase inhibitors*, act in a similar way on humans, where blood cholinesterase levels should be regularly checked in people coming in regular contact with the insecticide concentrate to ensure that inadvertent poisoning has not occurred. Examples of phosphorothioate insecticides used in vector control are temephos,

chlorpyrifos, malathion, fenitrothion, and pirimiphos methyl.

Carbamates

The carbamates have an identical mode of action to the organophosphates, but they are used in their insecticidally active form. Their mammalian toxicity can vary markedly. Again, all carbamates have a common general structure, which is limited by its requirement to act as a cholinesterase inhibitor.

Pyrethroid Insecticides

The insecticidally active components from pyrethrum flowers are known collectively as pyrethrins. Several synthetic pyrethrins were made in the late 1940s and 1950s. All of these compounds are unstable when exposed to UV light. Their lack of persistence makes them good, safe "knock-down" agents, and they are often still used in aerosols, frequently with the MFO synergist piperonyl butoxide, which increases their insecticidal activity and reduces their cost.

Pyrethroids were developed from the pyrethrins. The basic structure has acid and alcohol groups. This insecticide group has now been extensively developed through four chemical "generations," making it the most successful commercial insecticidal group. Some members of all generations are now stable in light and air, making them suitable as residual insecticides. They also function in the insect in extremely low quantities, making them relatively safe to handle at operational concentrations. Pyrethroids act in exactly the same way as DDT and its analogues. Examples of pyrethroids are permethrin, lambda-cyhalothrin, and deltamethrin.

Other Insecticides

Insect Growth Regulators (IGRs)

These compounds act on the highly species-specific insect hormonal systems that control molting and metamorphosis. They have the advantage of low mammalian toxicity, but they have several disadvantages, such as their species specificity, time required to kill, poor stability, and higher cost as compared to more conventional insecticides. Examples of IGRs are pyriproxyfen and buprofezin.

Juvenile hormones are used by insects to regulate growth and metamorphosis. Only a small amount of the hormone is required for the larva to metamorphose into a pupa. Adding synthetic juvenile hormone (e.g., methoprene) inhibits this molting process.

FIGURE 41.2 Monooxygenase-mediated reaction of a phosphorothioate to an active organophosphate insecticide.

Chitin Synthesis Inhibitors

The benzoylphenylureas interfere with the formation of chitin, a major constituent of the exoskeleton of insects. As vertebrates and most plants do not form chitin, these compounds are probably safe for humans, domestic animals, and plants. An example of this chemical group, which was developed in the early 1970s, is diflubenzuron.

Neonicotinoids and Chloronicotinyls

This insecticide group has yet to see extensive use for vector control, but the forerunner of this group, imidadoprid, is already the most widely used agricultural insecticide. Other members of this class are currently being released or are in the late stages of development, and it is likely that they will be marketed for public health use within the next decade. These insecticides target the nicotinic acetylcholine receptors in insect nerve junctions.

Bacteria

Although usually considered biological control agents, the toxins produced by bacteria such as *Bacillus thuringiensis israelensis* and *Bacillus sphaericus* can be considered as insecticides. These toxins disrupt the midgut lining of mosquito and blackfly larvae.

Mode of Action

The way that a particular insecticide affects its target is referred to as its *mode of action*. Various modes of action exist.

- Stomach poisons affect the insect when they are ingested and are absorbed through the digestive tract. These can be applied directly to the vector or as a systemic to the host.
- Contact insecticides enter the insect through the body wall or respiratory system.
- Systemics are applied to the host to combat parasites in or on the host, such as fleas and botflies.
- Fumigants are volatile and enter the insect's body through the respiratory system.
- Suffocants include some of the earlier "insecticidal" oils, which affect the respiratory system via the tracheae or respiratory siphon.
- Desiccants cause the vector to dry out by disruption of the waxy layer of the cuticle.
- Repellents make the insect avoid or leave a treated surface. These can alter the insect's feeding or oviposition behavior.
- Attractants are used as baits to lure the insects into a trap or treated area. Attractants include pheromones secreted by insects that influence the behavior of other members of the same species.
- Hormones inhibit the growth and development of the insect. They can interfere with the formation of the cuticle during larval development or inhibit the larval-to-pupal molt.

These modes of action are not always mutually exclusive. For example, several pyrethroids are both repellents and contact insecticides.

Formulations

The active ingredients in insecticides are very rarely used in their pure form. When formulated in low concentrations, they are much safer to handle and their insecticidal properties can be enhanced. Dry formulations include dusts, granules, wettable powders, and soluble powders. Liquid formulations include emulsifiable concentrates, solutions, flowables, and aerosols.

Adjuvants are added to the insecticide to change the physical and chemical characteristics of the formulation. Adjuvants can improve wetting characteristics, modify the rate of spray evaporation, improve the uniformity of the deposit, or reduce the phytotoxicity. Surface-active agents (surfactants) include wetting agents, emulsifiers, adhesives, and spreaders. Synergists can be added to enhance the activity of the insecticide, particularly where expensive insecticides are involved. For example, aerosols containing pyrethroids often contain the synergist piperonyl butoxide.

When choosing a formulation, certain factors should be considered: the type of pest to be controlled, application equipment required, whether the formulation will be used indoors or outdoors, coverage, and drift.

Dusts

Dusts or powders use an inert carrier, such as talc, clay, or diatomaceous earth. Dusts penetrate dense foliage better than sprays. They are also less likely to damage plants because they do not contain solvents, oils, or emulsifiers. Drift onto nontarget areas can be hazardous and also reduces the effectiveness of dusts. Dusts usually have a shorter residual effect than other formulations.

Granules

Granule formulations are made by applying a liquid formulation of an active ingredient to coarse particles of a porous material (e.g., clay, corn cobs, or walnut shells). Granules are much larger than dust particles and therefore are less prone to drift.

Wettable Powders

Wettable powders resemble dusts but are formulated to mix with water. They form a suspension rather than a true solution when mixed with water. Although wettable powders contain wetting agents or "spreader-stickers," they contain no solvents and therefore are less likely to cause plant injury than are emulsifiable sprays containing solvents. Constant agitation is necessary when using these formulations; otherwise, particles settle to the bottom of the spray tank. These were the early class of formulations used in residual house spraying for malaria control, but the formulations leave a residue on the house walls, and their abrasive properties on the spray nozzles make them less than ideal for this purpose.

Soluble Powders

Soluble powders are similar to wettable powders, except they form true solutions in water.

Baits

Baits are edible or attractive substances mixed with a toxicant that attracts pests and subsequently poison them.

Emulsifiable Concentrates

Emulsifiable concentrates are solutions of active ingredients and an emulsifying agent dissolved in an organic solvent that can then be mixed with water. After the water and solvent evaporate from the sprayed surface, the remaining insecticide adheres to the sprayed surface.

Solutions

Solutions with a high concentration of the active ingredient are diluted with oil or petroleum solvents. These are often used as ultralow-volume (ULV) sprays. A ULV concentrate is dispersed from specialized equipment without further dilution.

Flowables

Active ingredients that can be formulated only as a solid or a semisolid are ground and mixed with a liquid to form a suspension. Flowables can be added to water and need only moderate agitation to mix.

Aerosols

The active ingredient in a solvent solution is dispersed in very small droplets via a pressurized system. Aerosols are used in treating areas that can be difficult to reach with other applications.

Slow-Release Systems

Slow-release systems blend an active ingredient with a material from which it will evaporate or be released at a controlled rate. Microencapsulation is a type of slow-release formulation in which the active ingredient is enclosed in a material such as polyamide, neoprene, polyvinyl dicloride, or polyester. Once applied, the active ingredient can diffuse from the matrix. Other slow-release formulations include insecticide-impregnated mats and coils, where the active ingredient is released when it is heated.

VECTOR CONTROL AND RESISTANCE

Resistance is defined by the World Health Organization as "the development of an ability in a strain of some organism to tolerate doses of a toxicant that would prove lethal to a majority of individuals in a normal population of the same species." Resistance is a genetically inherited characteristic whose frequency increases in the vector population as a direct result of the selective effects of the insecticide.

To understand the processes by which resistance evolves, we must understand the mechanisms that produce and select for resistant individuals. Genetic and phenotypic variation affecting resistance arises in individuals as a result of mutation or gene duplication that modifies some normal physiological, morphological, or behavioral aspect of the phenotype. Such phenotypic changes typically enhance the process of insecticide detoxification, reduce the sensitivity of the nervous system to the insecticides, or increase the insects' ability to avoid contact with the toxicant. When insecticide is applied, individuals possessing such mutations have a considerable advantage over more susceptible individuals in the population. They have a higher probability of surviving insecticide treatment and, on average, will contribute more offspring than susceptible individuals to the next generation. As a result, the gene conferring resistance will increase in frequency in the population over time.

Insecticide resistance can be detected and investigated at many levels, from the molecular characterization of genes conferring resistance and their biochemical products, to the role gene products play in overcoming the toxic effects of insecticides, to studies of the ecological and evolutionary forces that affect the dynamics of genes conferring resistance in populations.

Detection of Resistance Within a Population

Insecticide Dose–Response Bioassays

Samples of insects are exposed to a range of doses of insecticide that produce a range of mortalities among the treated sample. Plotting the raw data of mortality against dose from such experiments should result in a sigmoid curve. To transform this into a straight-line response, log dosage is plotted against probit mortality (Fig. 41.3). This allows the investigator to establish the dose required to kill a given percentage of treated individuals. Strains are usually characterized by the dose that kills 50% (LD50) or 95% (LD95) of individuals. The slope of the response line is a measure of the population variability; the steeper the slope, the more homogeneous the strain. When a genetically heterogeneous field strain is subjected to a dose–response bioassay, the line should have a shallow slope. As selection with insecticide increases, the slope becomes steeper, because the insects become more homogeneous in their resistance levels.

The resistance levels of populations can be compared, using this methodology, by calculating the resistance ratio. For example, the resistance ratio (RR) at the LD50 dose is the LD50 dose of the resistant strain divided by the LD50 dose of the susceptible strain. Resistance ratios are usually calculated at either the LD50 value or the LD95 value.

Diagnostic Dose Bioassays

When working with field collections of insects, it is often impossible to collect sufficient specimens to produce a good log-dose probit-mortality line. To overcome this problem, for some species, e.g., anophe-

line mosquitoes, a predetermined insecticide dose has been established that is known to be lethal to susceptible individuals but that a high proportion of resistant individuals can tolerate.

A current list of recommended diagnostic doses for many insecticides for a number of arthropod disease vectors is available from the World Health Organization, as are the standard kits for resistance testing. Testing insects at these diagnostic dosages gives a crude measure of the amount of resistance in a population. However, it should be noted that due to the overlap in mortality lines of susceptible and resistant insects in many instances (see Fig. 41.3), this diagnostic dose often underestimates the real extent of the resistance.

Mechanisms of Resistance

Resistance is a genetically inherited characteristic that has an underlying physiological basis. Methods to measure these underlying mechanisms of resistance vary in their sophistication. Bioassays of a resistant strain with a range of insecticides from different classes can establish the resistance and cross-resistance profiles of the strain. Insecticide bioassays that involve preexposure to a synergist can be used to determine whether a specific resistance is blocked. For example, piperonyl butoxide is often used as a monooxygenase synergist and indicates that changes in this enzyme class underlie resistance. A number of simple biochemical assays are available to detect increased activity of the three enzyme systems (esterases, glutathione S-transferases, and monooxygenases) involved in insecticide metabolism. Other biochemical and PCR-based molecular methods are available to detect changes in sensitivity of the insecticide's target site within the insect.

Resistance arises as a result of a genetic change that alters the normal physiological, morphological, or behavioral attributes of a species. Resistance mechanisms can be divided into four broad categories: reduced penetration, metabolism, site insensitivity, and behavior.

Reduced Penetration

Many formulations of insecticides are designed to enter the insect through the cuticle. Cuticular changes that reduce the rate of penetration confer resistance to a number of insecticides. Reduced penetration alone usually confers only a low level of resistance. In combination with other resistance mechanisms, however, it can potentially result in a large nonadditive increase in resistance. By slowing the rate at which the insecti-

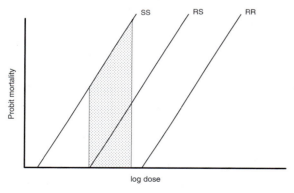

Shaded box indicates overlap of dose range between SS and RS mosquitoes

FIGURE 41.3 Probit mortality plot showing the overlap (shaded area) in the dose range between susceptible (SS) and heterozygous-resistant (RS) mosquitoes subjected to an insecticide. Note that homozygous-resistant mosquitoes (RR) fall outside the shaded area.

cide reaches its target site within the insect, other mechanisms can more effectively detoxify the insecticide en route.

Metabolic Resistance

A small group of enzymes or families of enzymes are involved in metabolic resistance. None of the enzymes involved is unique to resistant insects. Resistance is the result of a structural change in the enzyme molecule that increases its ability to detoxify or bind the insecticide and/or an increase in the amount of enzyme produced. While the enzyme families involved in resistance have been known for many years, their exact role in some instances has only recently been determined. The major enzyme systems involved in resistance and their roles are detailed next.

Monooxygenases: The cytochrome P450–dependent monooxygenases (MFOs) are a large group of oxidative enzymes with overlapping substrate specificities. Over 80 members of this large gene family are present in *Drosophila melanogaster*, and it is likely that similar numbers occur in other insects. These enzymes confer resistance primarily to pyrethroids and carbamates and to a lesser extent to organophosphates and organochlorines. They are also responsible for activating the phosphorothioate insecticides to the active organophosphate form. An indication of MFO involvement in resistance can be obtained by preexposing insects to the synergist piperonyl butoxide (PB) before insecticide exposure, although care needs to be taken in interpreting these data in isolation because there is recent evidence that PB also inhibits some esterases.

Esterases: These enzymes cleave carboxylester and phosphodiester bonds. They are extremely important in resistance to organophosphate insecticides and to a much lesser extent in pyrethroid resistance. One of the most thoroughly studied cases of resistance involving these enzymes is in the mosquito *Culex quinquefasciatus*. Resistance in this species is associated with elevated activity of one or two esterases. Esterase activity can easily be measured, using standard substrates, such as α- or β-naphthyl acetate, in individual insects. Elevation of this class of B-esterases inevitably produces some resistance, because by definition they are inhibited by 1×10^{-3} M paraoxon (the active analogue of parathion) and are similarly susceptible to inhibition by other oxon analogues of the phosphorothioate insecticides. These elevated esterases act effectively as an insecticide sink, rapidly binding and slowly metabolizing the insecticides. The genetic basis of the esterase elevation in *Culex* for most of these esterases is gene amplification. Often two esterases are

coamplified on the same piece of DNA (termed an *amplicon*) and are inherited as a single unit.

Resistance caused by esterases can be synergized by DEF (*S,S,S*-tributylphosphorotrithioate) in bioassays. A specific esterase-based resistance, which is limited to malathion and related insecticides that share a common carboxylester bond structure, can by synergized by TPP (triphenyl phosphate). This latter type of resistance is relatively common in *Anopheles* mosquitoes. The underlying genetic basis of this type of resistance is a mutation within the active site region of the enzyme rather than gene amplification. The mutation causes an increase in activity specifically with malathion. Recent data from *Lucillia cuprina* (a sheep blowfly) suggest that the normal function of this malathion carboxylesterase within the insect may be to regulate cascades of monooxygenase expression and that some of the mutations documented affect monooxygenase regulation, producing resistance via this route.

Glutathione S-Transferases: Glutathione S-transferases (GSTs) are primarily involved in DDT, organophosphate, and pyrethroid resistance. Their role in organophosphate resistance was determined several decades ago and is a classic conjugation reaction, with reduced glutathione being attached to the oxon analogue of the phosphorothioate insecticide to produce a more water-soluble product that is already tagged by the glutathione for export from the cell. The role of GSTs in DDT resistance was only elucidated in the 1990s. For many years it was known that the primary metabolites of DDT resulted from dehydrochlorination, and this was often correlated with an increase in GST activity, but it was not obvious how the two were connected. Clarke and Shaaman then demonstrated that the dehydrochlorination reaction was catalyzed by GSTs, which were using reduced glutathione as a cofactor rather than as a conjugate. Similarly, pyrethroid resistance has often been correlated with increased GST activity without any evidence of GS-conjugated metabolites. The role of GSTs in pyrethroid resistance was first detailed in 2001 by Vontas et al. They showed that the GSTs were effective at protecting the insect cells against lipid peroxide damage, which would otherwise result from pyrethroid-mediated free-radical activity.

Elevated levels of GSTs in *Anopheles* mosquitoes, where this system is best studied, are primarily due to changes in the regulation of several GST families.

Site Insensitivity

Resistance to several insecticide classes is conferred by alteration of the insecticide's target site. Such alterations have been observed in neuronal enzymes and receptors, which are the target sites of the majority of insecticides commonly used for vector control.

Insensitive acetylcholinesterase: Organophosphate and carbamate insecticides act by inhibiting the enzyme acetylcholinesterase (AChE). Many insect vectors have developed resistance to these compounds through structural modification of AChE. The enzyme in resistant individuals is less sensitive to inhibition than that from susceptible individuals. Simple biochemical assays, using the carbamates or oxon analogues of the phosphorothioates at fixed concentrations to measure enzyme inhibition, are available for this type of resistance mechanism. The structural modifications resulting in altered insecticide sensitivity can be conferred by a large number of point mutations, which can operate either singly or in combination.

Knockdown resistance: Knockdown resistance (*kdr*) received its name from the observation of insects following treatment with DDT or pyrethroids. Susceptible insects exposed to pyrethroids are rapidly paralyzed ("knocked down"). In certain strains of resistant insects this does not occur. This lack of knockdown is produced by mutations in the *para*-gated sodium channel gene, whose protein subunits make up the voltage-sensitive sodium channels on the nerve membranes. The most common resistance-associated mutation is situated on domain II of the protein and involves a leucine-to-phenylalanine, -serine, or -histidine substitution, although there are several other mutations in other domains that can produce the *kdr* phenotype. The *kdr* mechanism produces cross-resistance between DDT and pyrethroids and is the only resistance mechanism common to both insecticide groups. In *Anopheles gambiae* the leucine-to-phenylalanine mutation produces significant pyrethroid resistance, while the leucine-to-serine mutation at the same site produces stronger DDT resistance than pyrethroid resistance. *kdr* is the only resistance mechanism that is usually genetically recessive, with other metabolic and site insensitivity mechanisms producing a resistance phenotype that is intermediate between the homozygous-resistant and susceptible parents.

Cyclodiene resistance: Cyclodiene insecticides such as dieldrin target the neuronal gamma-aminobutyric acid (GABA) receptors. Resistance involving insensitivity of this target is produced by a single invariant amino acid substitution involving an alanine to a serine. This is a common, stable resistance mechanism, with resistant individuals persisting at high frequencies in many populations for many years after cyclodiene treatment has ceased. While cyclodienes are no longer exten-

sively used, this resistance mechanism may affect the fipronil class of insecticides, which share the same target site.

Behavior

Changes in behavior that result in reduced contact with the insecticide can enhance the probability of survival in an insecticide-treated environment. Such changes can involve a reduced tendency to enter sprayed houses or an increased tendency to move away from treated surfaces once contact is made. While a small number of behavioral resistances have been documented, this is a minor resistance mechanism in comparison to the other mechanisms already detailed.

POPULATION BIOLOGY OF RESISTANCE

The Evolution of Resistance

Because the number of classes of insecticides available for control of vectors is limited, it is vital that effective resistance management strategies are employed. The aim is either to develop control strategies designed to prevent or delay the onset of resistance in populations exposed to a pesticide for the first time or to develop management programs that cause existing resistance to decline. In order to do this, you first need to evaluate the factors that can influence the evolution of resistance, which are shown in Table 41.1. Resistance management strategies try to take advantage of the factors that are under human control to minimize the rate of evolution of resistance.

Genetic Factors

The number and frequency of resistance alleles in the population have a major impact on the evolution

TABLE 41.1 Factors Influencing the Rate of Evolution of Insecticide Resistance

Entomological	Generation time of insect
	Insect population size
	Life stage of insect exposed
	Heterogeneity of insect population
Environmental/chemical	Insecticide selection pressure
	Proportion of the population exposed
	Time span of exposure
	Insecticide dose rate
	Extent of prior exposure to insecticide
	Residual efficacy of insecticide
	Decay rate of insecticide

of resistance. Random genetic events generate mutant alleles, some of which confer resistance. The frequency of these "resistant" mutations in the absence of insecticide selection is dependent on the fitness costs of these mutations. Alleles with strong pleiotropic effects are generally selected against in the absence of insecticide pressure. When insecticide selection pressure is applied, the frequency of the resistant alleles increases. The rate at which this occurs is dependent on many variables, including nongenetic factors such as selection regime and population structure. However, the initial allele frequency and fitness costs of the "resistant" allele are important factors to consider, especially when insecticides are used at lower doses.

Knowledge of the dominance of the trait is important, for this can affect the outcome of resistance management strategies. For example, when resistance exhibits incomplete dominance, applying a dose of insecticide that kills all heterozygotes makes the resistance phenotype functionally recessive.

Biological, Ecological, and Operational Factors

The biological factors influencing the evolution of resistance include life history parameters specific to the vector species, and these are difficult to exploit for resistance management purposes. For example, resistance accumulates faster in species, such as mosquitoes, with short life cycles and abundant progeny than in tsetse flies, which produce few eggs and have a relatively long generation time. In some situations, resistance management can be assisted by the creation of refugia containing susceptible individuals. This is achieved by leaving some areas untreated with insecticides. And it can be beneficial if the susceptible individuals recolonize the treated areas and mate with the resistant individuals, thereby lowering the frequency of the "resistant" allele. However, in many disease control programs, the maintenance of a refugia population is unacceptable.

The operational factors include properties of the insecticide used and how it is applied. Ideally, this information should be readily available so that operational factors can be varied to optimize resistance management. However, in practice this is often not the case, and vital information, for example, on the past use of insecticides, is frequently lacking.

Resistance Management Strategies

Computer modeling has helped identify the key factors affecting the evolution of resistance. These models provide a simple means of predicting the efficacy of different strategies of pesticide use in resistance management schemes. However, their validity depends on the accuracy of the assumptions about the system being modeled. Often, vital statistics such as population size, migration rates, selection intensity, and the relative fitness of the "resistant" alleles are unknown, and this lack of information can undermine the predictions of the models.

Many models assume that resistance is controlled by a single locus with a resistant allele, R, and a susceptible allele, S, and that, as the frequency of the R allele increases, resistance increases. For polygenic traits, in which several genes contribute toward the resistance phenotype, the response to selection can be difficult to predict, for information about the linkage relationships of the resistance genes, the relative contribution of each gene to resistance, and any gene interactions that affect the level of resistance is often lacking. In such quantitative traits, rather than monitoring the change in frequency of individual resistance alleles at each locus, populations are often assessed in terms of the heritability of the trait. The heritability of a trait is the proportion of the variance in phenotype that is caused by genetic variation (as opposed to that caused by environmental variation).

Several models of resistance have been tested under laboratory conditions but rarely under field conditions. Recent advances in biochemical and molecular assays for detection of resistance alleles enable large numbers of individual insects to be assayed for multiple resistance mechanisms, and thus field trials of the computer models are now feasible. An ongoing trial in southern Mexico compares changes in frequencies of resistant alleles under mosaic and rotational resistant management strategies to single insecticide use.

Strategies Using Single Insecticides

Most models assume the initial frequency of the R allele is low and therefore the vast majority of the R alleles exist as heterozygotes. In this situation, choosing an insecticide application sufficiently high to kill all the heterozygotes has been advocated. The frequency of RR survivors is then assumed to be so low that they would be overwhelmed by and mate with SS immigrants. For this approach to succeed, all insects must receive the intended dose of insecticide so that no RS individuals survive; this is difficult to achieve under field conditions. Furthermore, the additional costs, both economic and environmental, associated with this strategy must be considered.

Strategies Using More Than One Insecticide

The measure of success of strategies employing two or more insecticides is whether resistance to all of the compounds when used together evolves more slowly than the combined time it would have taken for a population to evolve resistance to each one individually. Some models suggest that mixtures of insecticides with different modes of action, rotations in time, or spatial patterns of applications can be useful in managing resistance. The success of a mixture of insecticides relies on the fact that if the expected frequency of R alleles at two different genetic loci is low, the presence of individuals carrying an R allele at both loci will be extremely rare. The use of insecticides in rotational schemes in which they are applied in an alternating sequence is also based on the assumption that an individual usually does not carry R alleles at two different loci. If the frequency of individuals heterozygous or homozygous for the allele conferring resistance to one insecticide increases under selection with that insecticide, they will be killed when the switch is made to a second insecticide. In many models, when advocating the use of rotations it is assumed that the resistance genes have a selective disadvantage in the absence of insecticide. Limited laboratory experiments generally support this assumption, but the pleiotropic effects of resistance genotypes will not necessarily be the same for different populations.

Detection of Resistance Alleles

In order to plan effective resistance management assays and to monitor the success of existing control strategies, simple, reliable assays that can be performed on individual mosquitoes are needed. A number of biochemical and molecular assays for the common resistance mechanisms have been developed, as described earlier. These assays are simple to perform and can sometimes be carried out in the field without transporting insects to the laboratory. It should be noted, however, that many of the assays for detecting metabolic resistance rely on detecting increased enzymatic activity against model substrates in resistant individuals. All enzymes involved in detoxifying insecticides belong to large enzyme families, members of which have varying substrate specificities; without a detailed knowledge of the catalytic properties of the actual enzymes responsible for metabolizing the insecticides, the correlation between increased activity toward a model substrate and resistance is not proven.

Allele-specific PCR (polymerase chain reaction) assays have been developed to detect several resistance alleles, an example of which is the *kdr* assay that detects mutant sodium channel alleles. These assays are not readily adapted for field application, but they have the advantage of detecting heterozygotes that may be missed by phenotypic measurements that often fail to detect resistance in field populations until the resistance alleles are present at relatively high frequencies in the population. Using PCR-based assays, detection of heterozygotes provides advance warning of resistance-related compromises in control.

In many cases of metabolic resistance to insecticides, the exact molecular mechanism of resistance is unknown, and hence allele-specific assays to monitor resistance are not available. Recently, the rapid advances in genomics have paved the way for the identification of these resistance genes by positional cloning strategies (Chapter 36).

Stability of Resistance Alleles

A key factor undermining the success of resistance management strategies is the relative fitness of the resistant alleles in the absence of selection. If resistant genotypes are not selected against in the absence of insecticide selection, rotation of insecticides with different modes of action will do little to prevent the accumulation of R alleles. There have been relatively few long-term field studies of resistance gene frequencies, which are needed to assess the fitness costs of R alleles in natural populations, but a common prediction of models of resistance evolution is that the pleiotropic costs of resistance are reduced as selection pressure continues. This can occur either by replacement of the original R alleles with adaptive alleles with less deleterious effects on fitness or by the evolution of modifier genes at different loci from the resistance gene, which modify the fitness costs associated with resistance. Elegant studies of the evolution of resistance genes demonstrating both these predicted outcomes have been conducted for the sheep blowfly, *Lucilia cuprina*, and the mosquito *Culex pipiens*.

Status of Resistance in Arthropod Vectors

The status of resistance in arthropod vectors of disease was reviewed recently (Hemingway and Ranson 2000); however, information is incomplete because in many countries resistance surveys are absent or incomplete. Furthermore, the impact of resistance on the control of disease vectors and on disease transmission is often unknown. In some cases resistance has been implicated in control failures leading to increased incidence of disease. An example of this is in southern Africa, where the resistance of the malaria vector *Anopheles funestus* to pyrethroid insecticides has been accompanied by dramatic increases in malaria incidence in the region. In other cases, the presence of relatively high levels of resistance has apparently had little adverse effect on disease control.

For example, data from experimental huts in West Africa have shown that pyrethroid-resistant *An. gambiae* are less susceptible to the irritant effect of the insecticide and hence tend to require a higher dose of insecticide than susceptible insects. Thus, the percentage mortality in the resistant population was higher than predicted and the effect of resistance on insecticide-impregnated bed-net efficiency was lower than expected from laboratory experiments.

Recent evidence suggests that the physiological effects of insecticide resistance on the insect may have a more direct effect on disease transmission. A negative correlation has been observed between elevated esterase-based resistance in *Culex* mosquitoes and the insects' ability to act as a vector of *Wuchereria bancrofti* (McCarroll et al. 2000). It is not known how widely applicable this observation is or if similar interactions occur between resistance and disease transmission in other vector species, but, if substantiated, this could have major implications for resistance management and disease control. One hypothesis is that the observed effect was not a direct one of the esterases themselves, but was produced by the change in redox potential in the gut cells, where these esterases are massively overexpressed. This hypothesis fits well with recent genomic microarray-based data in *An. gambiae*, where the clusters of genes affecting the insects' ability to deal with oxidative stress have altered expression patterns in malaria-refractory and -susceptible insects.

Readings

Brooke, B. D., Kloke, G., Hunt, R. H., Temu, E. A., Koekemoer, L. L., Taylor, M. E., and Coetzee, M. 2001. Bioassay and biochemical analyses of the insecticide resistance in southern African *Anopheles funestus*. *Bull. Ent. Res.* 91: 265–273.

Chandre, F., Darriet, F., Duchon, S., Finot, L., Manguin, S., Carnevale, P., and Guillet, P. 2000. Modifications of pyrethroid effects associated with *kdr* mutation in *Anopheles gambiae*. *Med. Vet. Ent.* 14: 81–88.

Clarke, G. M. 1997. The genetic and molecular basis of developmental stability: The *Lucilia* story. *Trends Ecol. Evol.* 12: 89–91.

ffrench-Constant, R. H., Anthony, N., Aronstein, K., Rocheleau, T., and Stilwell, G. 2000. Cyclodiene insecticide resistance: From molecular to population genetics. *Annu. Rev. Entomol.* 45: 449–466.

Guillemaud, T., Lenormand, T., Bourguet, D., Chevillon, C., Pasteur, N., and Raymond, M. 1998. Evolution of resistance in *Culex pipiens*: Allele replacement and changing environment. *Evolution* 52: 443–453.

Hemingway, J., and Ranson, H. 2000. Insecticide resistance in insect vectors of human disease. *Annu. Rev. Entomol.* 45: 371–392.

Penilla, R. P., Rodrigues, A. D., Hemingway, J., Torres, J. L., Arredondo-Jimenez, J. I., and Rodriguez, M. H. 1998. Resistance management strategies in malaria vector mosquito control. Baseline data for a large-scale field trial against *Anopheles albimanus* in Mexico. *Med. Vet. Entomol.* 12: 217–233.

McCarroll, L., Paton M. G., Karunaratne, S. H. P. P., Jayasuryia, H. T. R., Kalpage, K. S. P., and Hemingway, J. 2000. Insecticide resistance status directly affects vectorial capacity in the mosquito *Culex quinquefasciatus*. *Nature* 407: 961–962.

Ranson, H., Jensen, B., Wang, X., Prapanthadara, L., Hemingway, J., and Collins, F. H. 2000. Genetic mapping of two loci affecting DDT resistance in the malaria vector *Anopheles gambiae*. *Insect Molec. Biol.* 9: 499–507.

Taylor, M., and Feyereisen, R. 1996. Molecular biology and evolution of resistance to toxicants. *Mol. Biol. Evol.* 13: 719–734.

Vontas, J. G., Small, G. J., and Hemingway, J. 2001. Glutathione S-transferases as antioxidant defence agents confer pyrethroid resistance in *Nilaparvata lugens*. *Biochem. J.* 357: 65–72.

42

Environmental Management for Vector Control

GRAHAM J. SMALL

*T*he environment plays a particularly important role in determining the distribution of vector-borne diseases. In addition to water and temperature, other factors such as humidity, vegetation density, patterns of crop cultivation, and housing may be critical to the survival of the different species of disease-carrying vectors. All those diseases are most serious in the poorest countries and among those living in the most difficult and impoverished conditions. They contribute to a vicious circle of poverty and the continued marginalization of people living in disease-ridden areas.

[Panel of Experts on Environmental Management for Vector Control (PEEM), WHO/FAO/UNEP/UNCHS]

INTRODUCTION

It has long been realized that one way in which vector-borne diseases can be controlled is to manage the environment of the vectors that transmit them, i.e., to destroy the habitat. Environmental management may include destruction of breeding sites by drainage, filling, impounding, or channeling streams and rivers into canals or by altering the vegetation and shade characteristics of the sites favored by the vectors. These types of methods are particularly applicable to mosquito control. Different species of mosquitoes, whether vectors of human or animal disease or simply a biting

nuisance, have distinct larval and pupal water quality requirements. For example, some favor temporary breeding sites, while others favor permanent water bodies; some are found in water with a high organic level, and others require low levels; and some require salinity, whereas salinity is fatal to others. Changing water quality in breeding sites can, therefore, have a dramatic effect on the distribution of mosquito species.

The basic definition of environmental management for vector control is "the planning, organization, carrying out, and monitoring of activities for the modification and/or manipulation of environmental factors or their interaction with humans with a view to preventing or minimizing vector propagation and reducing human–vector–pathogen contact" (WHO 1980). The extensive literature on environmental management for vector control has been well reviewed by other authors, notably Ault (1994) and C. F. Curtis (1990). In addition, reports by WHO (1982) and Mather and That (1984) contain both extensive bibliographies and practical information on implementing environmental control measures.

Even though environmental management for vector control was commonly used before and during World War II, it was largely ignored postwar, being replaced with synthetic insecticides to control vectors and drugs to treat the diseases. Environmental management has yet to be adopted on a wide scale, especially in Africa, where it could potentially have the greatest impact on vector-borne diseases. This is due not only to vector control programs being insecticide- or insecticide-impregnated bed-net-based, but also to the extreme poverty and lack of resources of the communities in

639

question. In these countries, environmental management has to be part of an integrated control program that has no net economic drain on individuals of the community and, ideally, should actually generate an income or some other resource that will benefit the community. It should also be noted that environmental management practices may adversely affect the flora and fauna of an area (Provost 1972, 1973), making environmental impact assessments during the planning stage essential.

All too often humankind has provided the breeding sites with the water quality that particular vector species require. This includes the digging of open-pit latrines (breeding *Culex quinquefasciatus* and blowflies), the provision of human-made habitats such as water-storage vessels and flower vases (breeding some *Aedes* spp.), and the spread of rice cultivation, with its associated irrigation (favored by a number of mosquito species). A number of environmental management programs have addressed these problems.

What current programs also highlight is that for an effective environmental management program to be put in place, it is essential first to have an intimate knowledge of the biology of the vector species to be controlled. In addition, an awareness-and-education program should be initiated to ensure public acceptance and assistance when implementing control measures.

In general, where methods of environmental management of vectors have been applied, these have been part of an integrated vector control program, including biological and/or chemical control components. These components and their potential constituent methods are summarized in Figure 42.1.

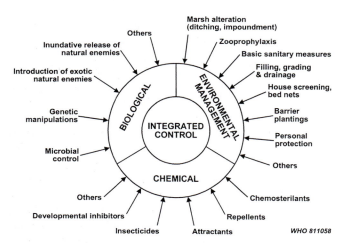

FIGURE 42.1 Diagram showing the components of integrated control (environmental management, chemical, biological) and their potential constituent methods. Reproduced by permission from World Health Organization (1982, p. 10).

This chapter focuses on environmental management programs that have been successful in controlling important arthropod vectors of human diseases. Because of its importance, much of the discussion of environmental management of arthropod vectors is on mosquito control.

HISTORICAL BACKGROUND

Even though the role of mosquitoes as vectors of disease was not proved until Ronald Ross' study during the 1890s of the relationship between *Anopheles* mosquitoes and malaria transmission, the association of certain diseases with mosquitoes has long been realized. Indeed, as far back as the 1st century BC, Virtuvius, a Roman engineer and architect, designed a drainage system to flush coastal marshes and recommended that houses not be built near these marshes because, as was the belief of the time, the mists coming from the marshes were harmful when mixed with the spirits of poisonous insects (mosquitoes) (Service 1978). Indeed, the word *malaria* is a corruption of the Italian *mala aria*, meaning "bad air," because the air of the marshes was believed to carry a noxious substance capable of causing disease.

In many parts of Europe, malaria was endemic until recent times. In England, for example, malaria, then termed *the ague*, was prevalent among populations living on or near river estuaries, where *Anopheles atroparvus*, the most competent of the native vectors, bred in the brackish water of the marshes. The decline of malaria in Europe can generally be attributed to the application of a classic method of environmental control, albeit unintentionally. The draining of marshes for land reclamation removed most of the breeding sites of the mosquito vectors that, along with a cooling of the climate during the latter part of the millennium, resulted in a reduction in the number of cases of malaria. The last indigenous case of malaria in England was reported in the 1950s (Reiter 2000).

The first person systematically to employ environmental management methods to control mosquito vectors was William Crawford Gorgas, a medical officer with the U.S. Army. In 1989, Gorgas was made chief sanitary officer for the city of Havana, a city subject to sporadic outbreaks of yellow fever vectored by *Aedes aegypti*. Having been convinced by the work of Carlos J. Finlay and the board headed by Walter Reed that *Aedes aegypti* was the vector of yellow fever in the city, Gorgas instituted a program in which all breeding sites of the mosquito were systematically eliminated. In this way, Havana was rid of *Aedes aegypti* and of yellow fever within just a few years.

Using the same methods Gorgas set up a successful vector control program during the building of the Panama Canal, resulting in the control of both malaria and yellow fever, which facilitated the successful completion of the canal in 1915 (Gibson 1989).

Integrated control of mosquito vectors was commonplace prior to World War II, with larviciding, using petroleum oil and Paris green (copper aceto-arsenite), complemented with modification of breeding places such as drainage, filling, and vegetation removal. The end of World War II saw the introduction of DDT and other synthetic insecticides, which had an immediate and powerful effect on mosquito populations. As a result of the success of these insecticides, together with the introduction of drugs for treatment and prophylaxis of some vector-borne diseases, environmental control methods were largely ignored. More recently, however, vector control seems to be have come full circle. Drug resistance is prevalent in the malaria parasites in many African countries, and insecticide resistance in the mosquito vectors is impacting vector control measures. In addition, public health and environmental issues over widespread use of insecticides has come to the fore. This has resulted in renewed interest in environmental control methods and of integrating these methods into vector control programs.

MOSQUITO CONTROL

Environmental Methods for Vector Control in Paddy Fields

Several important species of *Anopheles* that serve as vectors of malaria, and *Culex* spp. that vector Japanese encephalitis virus (JEV), are able to breed in paddy fields. Taking all mosquito vectors together, more than 135 pest and vector species of *Anopheles* and *Culex* have been associated with rice cultivation (Lacey and Lacey 1990). Considering the enormous areas of land under rice cultivation in Asia and the increasing areas under rice cultivation in Africa (especially West Africa), this presents a major problem for vector control.

Although chemical larviciding has been employed against mosquito vectors breeding in rice paddies, mainly against *Culex* vectors of JEV, it is generally accepted that, because of the huge scale of rice cultivation, chemical larviciding for vector control is uneconomic as a routine method. In addition, indiscriminate spraying of organophosphorus and carbamate insecticides for the control of rice pests has selected for insecticide-resistant populations of *Culex tritaeniorhynchus* in many countries. An outbreak of JE

in Korea during 1982, with 2,975 cases and 280 deaths, was directly attributed to insecticide resistance in *Cx. tritaeniorhynchus* populations in Korea (Rajagopalan et al. 1990). Resistance in rice pests is also widespread, with the result that routine application of insecticides to rice paddies is now less common. Insecticide spraying also kills many of the natural predators of mosquitoes, with the result that the larval population of vector species may actually be higher postspray (Service 1977; Reiter 1980).

An environmental control method employed successfully in several countries to control mosquitoes breeding in paddy fields is intermittent irrigation. This method, developed primarily to increase rice yields, was first suggested in 1922 and employed successfully in Indochina (Rajagopalan et al. 1990). The method is based on the principle of supplying only as much water as is needed for optimal plant growth. After rice seedlings are transplanted and have become established (typically 10–15 days), the fields are drained and then intermittently flooded with shallow water every 3–5 days. This water disappears over the next 24–48 hours through a combination of absorption, percolation, and evaporation (Luh 1984). The cycle of flooding and draining is repeated until the rice plants are mature (Fig. 42.2). The application of this method is, however, restricted to regions having sufficient rainfall for repeated flooding and having sandy loam soils that permit rapid draining of the paddy fields after each inundation. Intermittent irrigation has proved successful in controlling the malaria vector *An. labranchiae* in rice-growing regions of Portugal, the malaria vector *An. sinensis*, and the JE vector *Cx. tritaeniorhynchus* in the Henan province of China and can be used effectively in controlling some members of the *An. culicifacies* complex in India (Rajagopalan et al. 1990).

Environmental Methods for Vector Control in Salt Marshes

A number of important vector mosquito species breed in the brackish water of coastal marshes. These include: *An. melas* and *An. merus* in Africa; *An. sundaicus* in Asia; *An. labranchiae*, *An. atroparvus*, and *An. sacharovi* around the Mediterranean; and *An. albimanus*, *An. aquasalis*, and *An. grabhamii* in the Americas.

One approach to controlling mosquitoes breeding in salt marshes is to drain the marsh and, hence, remove the breeding sites. This method was used in the eastern United States. Construction of ditches for mosquito control, which lowered the water table in salt marshes, succeeded in controlling the mosquito problem but

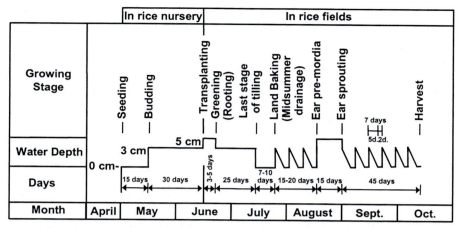

FIGURE 42.2 Schedule for intermittent irrigation of rice in use in the Jining area of China. Reproduced by permission from Curtis, C. F. (ed.) (1990, p. 134).

also resulted in the invasion of lower marshes by high marsh plants and substantially reduced the numbers of other invertebrates (Reimold 1977). The practice of ditching has now largely been abandoned for more environmentally friendly methods of control.

Excluding saltwater, which prevents marshes from becoming brackish, has also been employed successfully in the past. Banks and sluice gates are installed that exclude saltwater at high tides but allow freshwater to drain out during low tides. This method was effective in controlling *An. sundaicus* in Malaya and *An. sacharovi* in Italy (Mitchell 1996). However, it cannot be universally applied. Some mosquito species, such as *An. albimanus*, can breed in freshwater as well as in brackish water, and the existence of freshwater in marshes may provide the water quality required by local freshwater-breeding vectors.

Production of mosquitoes from salt marshes has also been controlled by preventing water from standing in the breeding sites long enough for mosquitoes to complete their development or by covering areas completely with water to prevent gravid female mosquitoes from finding oviposition sites. These methods are particularly applicable to the control of salt marsh *Aedes*, such as *Ae. taeniorhynchus*. Not only are they effective in controlling the local *Aedes* populations, but they can actually increase the populations of other ecologically important organisms living in the salt marshes (Provost 1977).

Loss of wetland habitat is now a subject for serious public debate, especially in the United States, which has lost more than 75% of its coastal wetlands through drainage and land reclamation projects (Zedler 1991).

All future salt marsh mosquito control programs based on environmental management will, therefore, have to be mindful not only of the effect that mosquitoes have on people living in and around the salt marsh but also of the impact that any proposed measures will have on salt marsh ecosystems.

Environmental Methods for Vector Control in Sanitation Systems

Installing sewerage is impractical in much of the developing world because of the expense of building the system and because it requires piped water for its operation. Therefore, where a sanitary system is built in these countries, it normally consists of pit latrines or cesspits. While pit latrines provide an effective solution for the disposal of fecal waste, they may also be the breeding sites for the vector of Bancroftian filariasis, *Cx. quinquefasciatus*, and for blowfly species, which present a health hazard through contamination of food with fecal pathogens.

Both oil and organophosphorus insecticides have been used to control these insects in pit latrines, with the organophosphorus insecticide chlorpyrifos being judged the most cost effective (Graham et al. 1972). However, repeat treatments of large numbers of latrines with insecticide represents a considerable and continuing financial drain. In addition, widespread and high-level organophosphorus resistance has developed in many populations of *Cx. quinquefasciatus*. *Bacillus sphaericus* has also been used against the larvae of *Cx. quinquefasciatus*. This bacterium is tolerant of pollution, persists for many weeks after application, and controls organophosphorus resistance of *Cx. quinquefasciatus*. However, *Bacillus sphaericus* has yet to find widespread use.

Several physical methods of environmental management in pit latrines have been applied successfully to *Cx. quinquefasciatus* and blowfly control. One of these is the pour-flush, or S-bend, latrine, in which water forms a seal, preventing the escape of odors from the pit latrine and preventing entry and exit of insects. Latrine floors incorporating the pour-flush system are now manufactured on a large scale in Asia (C. F. Curtis et al. 1990). A second system employs a tightly fitting concrete lid that prevents flies, mosquitoes, and other insects from either entering or exiting the pit except when it is in use (Winblad and Kilama 1985). Finally, efficient ventilated pit latrines have been developed in Zimbabwe (Fig. 42.3). In this type of latrine the squat hole or pedestal is enclosed in a brick-built structure with a tightly fitting concrete roof. A screened ventilation pipe emerges above the roof of the building and draws a current of air into the building, down through the squat hole of the pedestal into the latrine, and then out the ventilation pipe. The odors emerging from the pipe attract flies but the screen prevents them from entering, and the light attracts flies up the pipe, with the screen preventing them from emerging. The merits of these physical methods of control is that, despite the initial financial outlay, they will, if

built to specifications, last for many years without any further financial cost.

Layers of expanded polystyrene beads have been used successfully to control *Cx. quinquefasciatus* in wet pit latrines and cesspits. After placing unexpanded polystyrene beads into boiling water to expand them, they are poured into the latrine or cesspit. There, they spread to fit exactly the water surface and can prevent mosquito breeding for many years; the only limit to their effectiveness being flooding of the latrine, in which case the beads can float out (C. F. Curtis et al. 1990). This method of control has been tried in several countries, including Tanzania, Zimbabwe, and India, and was effective in suppressing *Cx. quinquefasciatus* populations. For example, treatment of 3,844 wet pit latrines and cesspits with polystyrene beads in Zanzibar town reduced the adult biting population in houses by 65% (Maxwell et al. 1999). However, it should be noted that the unexpanded beads have a limited shelf life and cannot be stockpiled for long periods in developing countries.

Environmental Methods for Vector Control in Water Impoundments

Water impoundment projects, such as the damming of rivers and streams, and the construction of irrigation systems, can dramatically alter the environment both in and around them. In particular, they can provide or enlarge the environment suitable for the breeding of major invertebrate disease vectors. Not only can these bodies of water increase the size and number of sites suitable for breeding, but, if poorly managed, they can also provide particular ecological niches, such as seepage areas, swamps, and backwaters, that attract other vector species. The growth of vegetation in and around the margins of water bodies after impoundment can promote the proliferation of mosquitoes.

In the Untied States, well-established methodologies have been put in place to control mosquitoes breeding in artificial lakes. These were first developed by the Tennessee Valley Authority following the construction of artificial lakes in the southeastern states of the country and the subsequent outbreaks of malaria caused by the vector, *An. quadrimaculatus*, breeding in them (USPHS and TVA 1947). In essence, these methodologies control seasonal water levels within the artificial lakes so as to limit the growth of aquatic and semiaquatic vegetation around the lake margins and hence deprive anopheline mosquitoes of habitat in which to breed. These methods, with some modifications to suit local conditions, have now been adopted in other temperate countries, although not always with

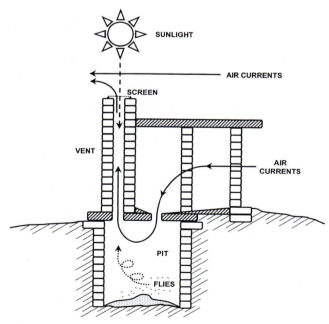

FIGURE 42.3 Diagram of a Blair ventilated improved pit (VIP) latrine showing the air currents and the movement of blowflies. Adapted from Curtis, C. F. (ed.) (1990, p. 178).

as much success. Application of the same methodologies in the tropics is more problematic, because they depend on the seasonality of the breeding of mosquitoes and the growth of plants, whereas in the tropical setting these may be continuous (WHO 1982).

Environmental Methods for the Control of *Aedes*

Aedes aegypti and *Ae. albopictus* are the principal vectors of dengue virus, which causes dengue fever (DF), and also the vectors of the much more serious dengue haemorrhagic fever (DHF), outbreaks of which have occurred in both southeast Asia (Halstead 1966) and the Caribbean (Tonn et al. 1982). *Aedes aegypti* is also the vector of urban yellow fever in Africa and the Americas.

Both *Ae. aegypti* and *Ae. albopictus* breed in mainly manmade habitats such as water-storage vessels, in flower vases, and in water collecting in discarded car tires, tins, and blocked gutters. A variety of methods have been developed for the control of *Aedes* spp., including chemical, biological, and environmental components, and, generally speaking, most vector control programs integrate several control methods from these components to suppress *Aedes* populations. Environmental measures include prevention of breeding in water-storage vessels by screening or regularly scrubbing, emptying, and careful disposing of discarded containers, cleaning gutters, and making holes in unneeded containers.

A good example of a program integrating different methods of control was established in Cuba during the 1980s. Following the DF and DHF epidemic of 1981, when there were an estimated 24,000 cases of severe haemorrhagic disease and 158 fatalities (Guzman et al. 1984), a two-phase program was established. During the first, "attack," phase, 15,000 specially mobilized Civil Defense staff dosed water containers with temephos sand granules and carried out residual spraying of fenthion and thermal fogging of malathion in and around houses. This phase successfully reduced the *Aedes* populations and the number of DF and DHF cases. In the "consolidation" phase, environmental methods were introduced, such as puncturing of old water containers, using artificial flowers in houses and cemeteries to avoid water in vases, and compulsory use of mosquito-proof plastic containers for water storage. Legal sanctions were also taken against households with *Aedes* breeding on their premises, which reinforced vigilance. In Singapore, much of the vector control is community based, with source reduction and health education being supported by law enforcement. Householders and businesses are charged with preventing *Aedes* breeding on their property, and vector control officers make regular inspections to ensure compliance (Chan 1985). Those found to have mosquitoes breeding are fined or, in some circumstances, jailed. Insecticidal spraying in Singapore is generally carried out only in the event of outbreaks of DF/DHF. Despite the success of this program in preventing significant transmission for 20 years, Singapore, along with other Asian countries, had a resurgence of dengue/DHF from 1990 to 1994 (Anonymous 1994).

OTHER ARTHROPOD VECTORS OF DISEASE AGENTS

Tsetse Flies

Tsetse flies are the vectors of African trypanosomiasis. Each year tsetse newly infect 20,000–25,000 people and there are about 55,000 deaths. Economic and social factors have made continued implementation of effective vector control strategies difficult. Successful vector control programs exist in fewer than 2% of the areas where the disease is endemic. Ground spraying of insecticides, aerial spraying, the use of cloth baits (cloth sprayed with insecticide), and the use of live baits (cattle sprayed with insecticide) have all shown some success.

Untreated traps have been used effectively for control of savannah species, with odors (e.g., *Glossina pallidipes*), and for control of riverine species, without odors (e.g., *G. palpalis* and *G. fuscipes fuscipes*). Their use is dependent on the existence of a cheap, effective trap/odor bait combination for the species involved. Such traps have proved to be effective in a community-based approach to control. The advantages of using untreated traps are that insecticide is eliminated and they are relatively cheap to construct (PAAT 2001).

Bush clearing and game destruction is one of the oldest methods of tsetse control, and bush clearing, either for tsetse control or as a result of human settlement activities, is still practiced. The environmental consequences of such methods, if practiced on a large scale, can be immense and include severe soil erosion and resource degradation in some areas (Dublin 1991). The planting of trees to stop soil erosion has resulted in tsetse reinfestation in some areas. Game destruction too removes a valuable resource, which can be utilized through game cropping or tourism, and may cause a switch in the tsetse to feeding mainly on livestock, with the additional effect that parasitemia in cattle increases, hence increasing infection rates of the flies (FAO 1977).

Although these methods are now rarely practiced specifically to control tsetse, increasing population pressure, especially in West Africa, has indirectly resulted in widespread bush clearing and game destruction for agricultural purposes. This has probably had more effect on tsetse populations than mankind's direct attempts to control tsetse.

Triatomine Bugs

Triatomine bugs (Chapters 5 and 51) are the vectors of Chagas disease, which exists only on the American continent and is caused by the flagellated protozoan *Trypanosoma cruzi*. The geographical distribution of Chagas disease extends from Mexico to the south of Argentina, and about 89 million people in Latin America are at risk of infection.

The basic methods for Chagas vector control are well established. Houses and peridomestic habitats are sprayed with wettable powder or flowable formulations of pyrethroids, such as deltamethrin, lambda cyhalothrin, and cyfluthrin (Chaper 41). These are approved for use in domestic situations and quickly produce a dramatic reduction in infestation rates. But treated houses remain vulnerable to reinfestation. In the long term, this can be addressed by gradually improving the quality of rural housing, rendering it less suitable for triatomine bugs. This has led to the inclusion within the Chagas disease vector control strategies of several countries, including Brazil, of improvement of dwellings, which are then less liable to harbor the triatomine bugs (Guhl and Vallejo 1999).

Sand Flies

The insect vectors of leishmaniasis, phlebotomine sand flies, are found throughout the tropical and subtropical regions of Africa, Asia, the Mediterranean, southern Europe (Old World), and South and Central America (New World). It is estimated that approximately 12 million people are currently infected and a further 367 million are at risk of acquiring leishmaniasis in 88 countries, 72 of which are developing countries, and 13 of those are among the least developed in the world (WHO 1991). The presence of leishmaniasis is strongly linked with poverty, economic development, and various environmental changes, such as deforestation, urbanization, migration of people into endemic areas, and the building of dams. The annual incidence rate is estimated to be 1–1.5 million cases of cutaneous leishmaniasis and 500,000 cases of visceral leishmaniasis, the two major clinical types of leishmaniasis (WHO 1991).

In endemic areas such as Bangladesh and India, sand fly control is often combined with malaria control, and in Brazil with malaria and Chagas disease control programs (WHO 1991; Al-Masum et al. 1995). This is a cost-effective method, given the expense of mass spraying with chemicals such as malathion, fenitrothion, and propoxur.

Environmental control measures against sand flies include the installation of fine mesh screens (less than 16-mesh) on doors. In endemic areas of zoonotic visceral leishmaniasis, elimination of wild animals and stray dogs is sometimes carried out, either by shooting or by using poisoned baits, to reduce the animal reservoirs of the disease and the number of potential hosts of the sand fly.

Ticks

Black-legged ticks (*Ixodes scapularis*) are responsible for transmitting Lyme disease bacteria to humans in the United States. On the Pacific coast, the bacteria are transmitted to humans by *Ixodes pacificus*. In the United States, Lyme disease is mostly localized to states in the northeastern, mid-Atlantic, and upper north-central regions and to several counties in northwestern California. In 1999, 16,273 cases of Lyme disease were reported to the Centers for Disease Control and Prevention (CDC 2001).

Strategies to reduce the abundance of ticks in endemic residential areas may include the removal of leaf litter, brush piles, and wood piles around houses and at the edges of yards and the clearance of trees and brush to admit more sunlight and to reduce the amount of suitable habitat for deer, rodents, and ticks. Tick populations have also been effectively suppressed through the application of pesticides to residential properties. Community-based interventions to reduce deer populations or to kill ticks on deer and rodents have not been extensively implemented, but they may be effective in reducing community-wide risk of Lyme disease. The effectiveness of deer-feeding stations equipped with pesticide applicators to kill ticks on deer and of other baited devices to kill ticks on rodents is currently under evaluation.

PERSPECTIVE

The successful completion of the Panama Canal in 1915 was made possible only when the link between mosquito vector breeding and human health was understood and environmental management measures were put in place to control mosquitoes. Malaria and yellow fever were brought under control not by the drugs and pesticides we use today, but by understanding the biology of the mosquito vectors and then removing their breeding sites. Since World War II, however, the increasing use of insecticides for vector control and of drugs and vaccines to treat vector-borne diseases has resulted in the abandonment of environmental management. This has led to the evolution and spread of drug resistance in some vector-borne parasites and to the evolution and spread of insecticide resistance in many mosquito vectors. At the same time, human populations, and the irrigated agriculture to support them, have expanded, providing an increased number of potential hosts for the disease agents and a vastly increased number of breeding sites for the mosquitoes that transmit them. The resurgence of malaria and other vector-borne disease has prompted many involved in the control of vector-borne diseases to reexamine environmental management practices.

Many of the environmental management methods for vector control in current use have been outlined in this chapter. These have for the most part been shown to be successful in small-scale and, more rarely, in large-scale and countrywide projects. The challenge now is to find ways to integrate these into vector control programs of countries across the tropics, especially in Africa. Vector control programs are often planned and instituted for an entire country and are almost entirely based on insecticide or insecticide-treated bednet control measures. In contrast, vector control solutions based on environmental management have to be tailored to local conditions and require the active involvement of local communities. In addition, with increasing attention being focused on the impact of man on the environment, any environmental management practices that adversely effect the environment, no matter what their benefits to human health, are likely to be unacceptable. It is, therefore, imperative that research be undertaken to develop new, sustainable, and environmentally benign methods of environmental management.

Readings

Al-Masum, M. A., et al. 1995. Visceral leishmaniasis in Bangladesh: The value of DAT as a diagnostic tool. *Trans. R. Soc. Trop. Med. Hyg.* 89: 185–186.

Anonymous. 1994. The dengue situation in Singapore. *Epidemiol. Bull.* 20: 31–33.

Ault, S. K. 1994. Environmental management: A re-emerging vector control strategy. *Am. J. Trop. Med. Hyg.* 50: 35–49.

CDC. 2001. Lyme Disease—United States, 1999. *CDC MMWR.* 50: 181–185.

Chan, K. L. 1985. *Singapore's Dengue Haemorrhagic Fever Control Programme: A Case Study on the Successful Control of* Aedes aegypti *and* Aedes albopictus *Using Mainly Environmental Measures as a Part of Integrated Vector Control.* Southeast Asian Medical Information Center, Tokyo, Publ. No. 45.

Curtis, C. F., ed. 1990. *Appropriate Technology in Vector Control.* CRC Press, Boca Raton, FL.

Curtis, C. F., et al. 1990. Insect proofing of sanitation systems. In: *Appropriate Technology in Vector Control* (C. F. Curtis, ed.). CRC Press, Boca Raton, FL., pp. 173–186.

Dublin, H. 1991. Dynamics of the Serengeti-Mara Woodlands: A historical perspective. *Forest Conservation History* 35: 169–178.

FAO. 1977. *The Environmental Impact of Tsetse Control Operations.* Food and Agriculture Organization of the United Nations, Rome.

Gibson, J. M. 1989. *Physician to the World: The Life of General William C. Gorgas.* University of Alabama Press, Tuscaloosa, AL.

Graham, J. E., et al. 1972. Studies on the control of *Culex pipiens fatigans. Mosq. News* 32: 399.

Guhl, F., and Vallejo, G. A. 1999. Interruption of Chagas disease transmission in the Andean countries: Colombia. *Mem. Inst. Oswaldo Cruz, Rio de Janeiro.* 94 (Suppl. I): 413–415.

Guzman, M. G., et al. 1984. Dengue haemorrhagic fever in Cuba. II. Clinical investigations. *Trans. R. Soc. Trop. Med. Hyg.* 78: 239–241.

Halstead, S. B. 1966. Mosquito-borne haemorrhagic fevers of South and Southeast Asia. *Bull. W.H.O.* 35: 3–15.

Lacey, L. A., and Lacey, C. M. 1990. The medical importance of riceland mosquitoes and their control using alternatives to chemical insecticides. *Am. Mosq. Control Assoc. Suppl.* 2: 1–93.

Luh, P-L. 1984. Effects of rice growing on the population of disease vectors. In: *Environmental Management for Vector Control in Rice Fields* (T. H. Mather and T. T. That, eds.). Food and Agriculture Organization of the United Nations, Rome, pp. 130–132.

Mather, T. H., and That, T. T. 1984. *Environmental Management for Vector Control in Rice Fields.* Food and Agriculture Organization of the United Nations, Rome.

Maxwell, C. A., et al. 1999. Can vector control play a useful supplementary role against bancroftian filariasis? *Bull. W.H.O.* 77: 138–143.

Mitchell, C. J. 1996. Environmental management for vector control. In: *The Biology of Disease Vectors* (Barry J. Beaty and William C. Marquardt, eds.). University Press of Colorado, Niwot, pp. 492–501.

PAAT. 2001. Choice and integration of control techniques. In *Training Manual for Tsetse Control Personnel, Volume 4: Use of Attractive Devices for Tsetse Survey and Control.* Electronic Training Resource of the Programme Against African Trypanosomiasis. http://www.fao.org/paat/html/1tm4_7.htm.

Provost, M. W. 1972. Environmental hazards in the control of disease vectors. *Environ. Entomol.* 1: 333–339.

Provost, M. W. 1973. Environmental quality and the control of biting flies. *Proceedings of a Symposium.* University of Alberta Defense Research, Edmonton, pp. 1–7.

Provost, M. W. 1977. Mosquito control in coastal ecosystem management. In: *A Technical Manual for the Conservation of Coastal Zone Resources* (J. R. Clark, ed.). Wiley, New York, pp. 666–671.

Rajagopalan, P. K. et al. 1990. Environmental and water management for mosquito control. In: *Appropriate Technology in Vector Control* (C. F. Curtis, ed.). CRC Press, Boca Raton, FL., pp. 121–138.

Reimold, R. J. 1977. Mangals and salt marshes of eastern United States. In: *Wet Coastal Ecosystems. Ecosystems of the World: Vol. 1* (V. J. Chapman, ed.). Elsevier Scientific, Amsterdam, pp. 157–166.

Reiter, P. 1980. The action of lecithin monolayers on mosquitoes. III. Studies in irrigated rice fields in Kenya. *Ann. Trop. Med. Parasitol.* 74: 41.

Reiter, P. 2000. From Shakespeare to Defoe: Malaria in England in the Little Ice Age. *Emerg. Infect. Dis.* 6: 1–11.

Service, M. W. 1977. Mortalities of the immature stages of species B of the *Anopheles gambiae* complex in Kenya: Comparison between rice fields and temporary pools. *J. Med. Entomol.* 13: 535–545.

Service, M. W. 1978. A short history of early medical entomology. *J. Med. Entomol.* 14: 603–626.

Tonn, R. J., et al. 1982. *Aedes aegypti*, yellow fever and dengue in the Americas. *Mosq. News* 42: 497–501.

U.S. Public Health Service and Tennessee Valley Authority. 1947. *Malaria Control on Impounded Water.* Government Printing Office, Washington, DC.

Winblad, U., and Kilama, W. 1985. *Sanitation Without Water.* Macmillan Education, London.

World Health Organization. 1980. *Environmental Management for Vector Control.* WHO Tech. Rpt. Series, No. 649, Geneva, Switzerland.

World Health Organization. 1982. *Manual on Environmental Management for Vector Control.* Offset Publication No. 66, Geneva, Switzerland.

World Health Organization. 1991. *Control of the Leishmaniases.* WHO Tech. Rpt. Series, No. 793, Geneva, Switzerland.

Zedler, J. B. 1991. The challenge of protecting endangered species habitat along the southern California coast. *Coastal Mgmt.* 19: 35–53.

Biological Control of Mosquitoes

JANET HEMINGWAY

INTRODUCTION

Biological control involves the reduction of a target pest population by a predator, pathogen, parasite, competitor, or toxin produced by a microorganism. Biological control usually has the advantage, over conventional broad-spectrum insecticides, of target host specificity with correspondingly little disruption of nontarget organisms in the environment. Furthermore, it is possible that some living biological control agents can provide long-term control after a single introduction.

Modern biological control efforts began in 1889 in California, when a predator was introduced to control a scale insect that was devastating the citrus industry. Since then, biological control agents have been used against a variety of agricultural pests in diverse locations. For cropping systems, biological control is now an important component of several integrated pest management schemes. However, biological control of arthropod vectors that transmit pathogens to vertebrates has lagged behind control of agricultural pests. Most efforts have concentrated on larval mosquitoes and blackflies.

Biological control became popular in the early 1900s when the mosquito fish, *Gambusia affinis*, was introduced in many countries to control mosquito larvae. Insecticides, toxic to both larvae and adults, largely replaced these early control methods in the 1940s and 1950s with the discovery and use of DDT. Interest in using biological control reemerged in the 1960s when concerns over large-scale insecticide use started to grow. The nematode *Romanomermis culicivorax* and the protozoan *Nosema algerae* were extensively studied

because they could infect and kill mosquito larvae. Large-scale field trial results varied. Poor control was attributed to the adverse effect of environmental factors on these biological agents. Nonetheless, *Romanomermis* was first marketed in 1976. High production costs, problems with storage and transport, and then the development of the pathogenic bacterium *Bacillus thuringiensis* subsp. *israelensis* eventually led to its commercial demise as a biological control agent.

Bacillus thuringiensis (*Bt*) was discovered in Japan in 1901 as a pest of silkworms. It was developed in 1915 in Germany for grain moth larval control. *B. thuringiensis* subsp. *israelensis* (*Bti*) was discovered in 1975. It is a bacterium belonging to the family *Bacillaceae*. Unlike the previous *thuringiensis* isolates, which had been successfully used to control several agricultural pests, *Bti* toxins killed mosquito and blackfly larvae. By 1981 *Bti* was available as a mosquito and blackfly control agent. Several firms now market *Bti*-based products. Strains of *Bacillus sphaericus* that are toxic to mosquitoes have also been developed commercially. These biological control agents are toxic due to the formation of crystal toxins. *Bacillus sphaericus*, for example, produces a binary toxin during its exponential growth phase. The two proteins within this toxin are both required to kill mosquito larvae.

These *Bacillus* spp. are effective when placed in a larval habitat, but early formulations were not persistent. Commercial products are currently dried powders or briquettes containing a mixture of dried spores and crystal toxins. *Bt* currently represents just under 1% of the total "agrochemical" market worldwide. Researchers continue to seek living agents that reproduce and maintain themselves in the mosquito larval environment and provide ongoing control of the target population. Formulations, such as slow-release briquettes and floating granules, have been improved to keep the *Bacillus* accessible to the insect larvae for longer periods. The fungus *Lagenidium giganteum* provides long-term control of mosquito larvae in unpolluted freshwater environments and has been approved by the U.S. Environmental Protection Agency (USEPA) for application as a mosquito control agent. Transgenic plants and other organisms that produce *Bt* toxins and reproduce in the environment have now been developed, although their major market is in the agricultural arena.

Biological control requires sound ecological information about the vector population and the biological control agent(s). This chapter covers fundamental principles of biological control, the ecology of selected biological control agents for mosquitoes, and considerations for the successful implementation of biological control strategies against vector populations.

FUNDAMENTAL PRINCIPLES OF BIOLOGICAL CONTROL

Population Ecology

Considerable research has focused on parameters affecting mosquito population size (Mogi 1993; Service 1985). Figure 43.1A depicts hypothetical vector population growth in the absence of exogenous mortality factors such as biological control agents and catastrophic mortality events. Population density increases

until it reaches K, the theoretical carrying capacity that the environment can support. Once K is surpassed, intraspecific competition for food, space, or other limiting resources causes population growth to decline. When the population drops below K, competitive pressure relaxes and the population increases.

Most populations do not fluctuate around their carrying capacity because other mortality factors are present. Mortality factors are often classified as density dependent (DD) or density independent (DI). With DD factors, mortality increases as the population becomes denser and decreases as the population becomes less dense. Density-dependent mortality factors tend to regulate a population around an average size and to resist changes to that average size. The intraspecific competition that maintains the population of Figure 43.1A around K is a DD factor. In Figure 43.1B, biological control agents function as DD mortality factors that maintain the population at a new equilibrium point below K. If a population is heavily regulated by DD factors, killing an extra percentage of larvae has no effect on the resulting adult population because of the compensating reduction in DD mortality. In fact, overcompensation, resulting in the emergence of more adults, is theoretically possible. This is one of the major reasons why larval biological control agents are not used for malaria control, because reduction in the density of adult insects transmitting the disease is the only real measure of success.

Density-independent mortality factors are characterized by mortality that is not proportionate to the population density. Such DI mortality reduces population size but does not promote population stability at a typical density. In Figure 43.1C, the hypothetical population growth curve (of Fig 43.1A) is shown with the addition of a single DI event that kills 80% of the population. The population can return to K in the absence of further DI mortality events. Density-independent factors can cause wide population fluctuations, depending upon their frequency and severity and the fecundity of the population subjected to them. Floods, temperature extremes, droughts, and insecticides frequently act as catastrophic DI mortality factors. Natural enemies also cause DI mortality. The percentage of mortality depends on the relative population sizes and on the presence of alternative hosts or prey. Density-independent mortality can be important for mosquitoes and can partially explain the large fluctuations in natural populations.

Most populations are subject to both DD and DI mortality factors. The relative strength of each mortality factor varies with the species, season, and location. Figure 43.1D depicts a hypothetical mosquito population in a rice field (Mogi 1993) in which both DD and

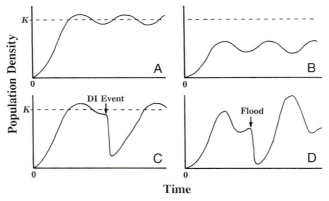

FIGURE 43.1 Population dynamics. (A) Population growth in the absence of external mortality factors. (B) Population growth with density-dependent regulation below the carrying capacity. (C) Population growth with a single density-dependent mortality factor. (D) A hypothetical mosquito population subjected to a combination of density-independent and density-dependent mortality factors.

DI mortality factors are operating. At time 0, mosquitoes begin colonizing the rice field. The population grows rapidly and then decreases as natural enemies colonize the field. Flooding is a major DI mortality event, flushing the fields of both mosquitoes and their predators. The mosquitoes then recolonize faster than the predators; hence, the mosquito population rebounds to higher than preflood levels. When the predators finally recolonize, they reduce mosquito numbers to a preflood state. Vector population resurgence can also be observed after applications of pesticides that kill both the biological control agents and the vector, once the insecticidal effects have been removed. In nature the balance in these relationships is often biased toward the pest species, which often have life history traits that enable them to rapidly recolonize and reproduce, while the biological control agents may require a minimum host density to support them.

Biological control agents can function as either DD or DI mortality agents. Pathogens and oligophagous predators (restricted to a few prey species) are more likely to provide DD mortality because their populations depend on the specific vector population. Unless these natural enemies can persist in the absence of the vector, they require a residual vector population to support them (at least if ongoing control is the goal). The minimum host density required by the pathogen or predator determines the size of the residual vector population. Natural enemies that provide DI mortality can maintain a vector population at low densities, depending upon relative population sizes and the availability of alternate prey or hosts. Microbial toxins function as DI mortality agents that can provide a large reduction in the vector population. With both DD and DI mortality, biological control will be successful only if the vectorial capacity of the adult vector population declines to an acceptable level.

Mosquito Ecology

The ecology of different mosquito species varies considerably (Chapter 9), but some generalizations can be made. Mosquito populations are often characterized by rapid increases and precipitous declines. The females are highly fecund, and most species have short generation times under tropical conditions. Thus, mosquito populations can quickly increase when the breeding season begins or rapidly rebound after a catastrophic event. Furthermore, the adults of many species disperse well and the females quickly recolonize habitats.

Larval habitats are diverse, ranging from ephemeral to permanent, from artificial to completely natural, from hoofprints to rice fields, although most species occupy only a well-defined subset of these breeding sites. Generally, mosquitoes avoid deep or rapidly flowing water. Some species prefer a sunny habitat, while others are found only in shaded locations. Furthermore, mosquitoes may shift preferred habitats as the seasons change and as new habitats become available. Larvae are not evenly distributed within favored breeding sites. They tend to aggregate within larger habitats and only colonize a subset of smaller sites such as containers.

Different mortality factors predominate in different habitats. In permanent ponds and rice fields, larvae often experience over 80% mortality, primarily from predators. In small-container habitats, such as tree holes, over 80% mortality can occur from overcrowding and competition and, more rarely, from predators or pathogens. Density-independent mortality factors, such as flooding, temperature extremes, and desiccation, can be important in any of these habitats.

How do these mosquito characteristics affect biological control efforts? The ideal mosquito biological control agent probably does not exist, as it should respond to any vector population increase by rapidly colonizing its habitat and quickly producing numerous progeny. It should be able to efficiently find all vectors, survive periods of mosquito absence, and function well in any habitat.

Mosquito natural enemies usually affect the larval stages. The diversity of larval habitats, feeding behavior, and physiology probably provides an insurmountable challenge to the development of a single control agent for all mosquito species. For example, *Culex quinquefasciatus* in polluted wastewater, *Aedes aegypti* in human potable water containers, and *Anopheles gambiae* in hoofprints are not accessible to any single agent. Within one genus, larvae can often exhibit marked species-specific differences in susceptibility to microbial control agents. No single agent or formulation provides adequate control.

TYPES OF BIOLOGICAL CONTROL

Natural biological control is vector reduction caused by naturally occurring biotic agents. Biological control agents are more likely to cause significant mortality in permanent habitats than in ephemeral habitats. Natural control provides greater than 80% larval reduction by DI and DD action in some habitats. Abiotic mortality factors (such as unfavorable temperatures) contribute even further to population reduction. Natural biological control alone does not usually reduce vector populations sufficiently to interrupt disease transmission. However, the explosion of a vector population that can occur after disruption of natural biological control (e.g., by use of a broad-spectrum pesticide) demonstrates the contribution that natural control can make to vector population suppression.

Applied biological control is planned human intervention to add biological control agents to a site (augmentation) or to protect the agents already present (conservation). Although applied biological control is less likely to harm nontarget organisms than conventional insecticides, it may alter, for example, the timing of a pathogen outbreak or introduce artificially high numbers of a native or exotic natural enemy into the environment.

Augmentation is the deliberate release of natural enemies into vector habitats to reduce the pest population. Two major strategies for augmentation are inoculation and inundation. With *inoculative* releases, small numbers of natural enemies are introduced that are expected to reproduce in the environment and provide long-term vector suppression over successive generations. The inoculative approach depends on a residual vector population or on an agent that persists in the absence of the host to support the biological control agent. Inoculation is a good strategy for container habitats, such as tree holes. Though the initial seeding of habitats is labor intensive, vector reduction results from a natural enemy that can persist during periods of host absence. Inoculations also work in large, accessible habitats. Spring release of mosquito-eating fish into rice fields when the fields are first flooded is a good example of inoculative biological control in vector management. Both indigenous and exotic biological control agents can be introduced in augmentative biological control.

For *inundative* releases, overwhelming numbers of organisms are released to produce an immediate decline in a vector population. Inundative releases are frequently performed with microbial pathogens and are analogous to the use of chemical insecticides. Indeed, inundative control often utilizes toxins produced by a microorganism in artificial culture. Subsequent reproduction of living pathogens can occur, but the reproductive rate is usually not sufficient to maintain control of the pest population. Inundative releases are most practical in large, accessible vector habitats and are often required to interrupt disease transmission.

Conservation manipulates the environment to optimize natural biological control by minimizing detrimental effects on natural enemies or by enhancing their efficacy. Introduced biological control agents must be conserved to be effective. To conserve natural enemies, habitats can be constructed to provide refugia during the winter or dry periods. Selective insecticides that are not toxic to the natural enemies can be used. Monitoring vector populations and applying chemical insecticides only when vector density warrants (rather than following a predetermined application schedule) can facilitate natural biological control.

Integrated pest management (IPM) is a model for most pest control. All available methods of pest management are examined, and an optimal combination of methods is designed to provide adequate control and minimize adverse environmental side effects. More specific than most other vector control methodologies, applied biological control is a prime candidate for inclusion in IPM programs. However, biological control alone is probably not sufficient for most vector control.

BIOLOGICAL CONTROL AGENTS

Pathogens

Invertebrate pathogens include viruses, bacteria, protists, fungi, and nematodes. Each group or species varies in its route(s) of infection, host specificity, infectivity and virulence, prevalence in nature, dispersal and persistence mechanisms, and sensitivity to biotic and abiotic parameters. Hundreds of different pathogens and potential pathogens have been reported that infect vector insects (Roberts et al. 1983).

Invertebrate pathogens affect their target host by three different, but sometimes interrelated, pathogenic strategies. The first strategy is *patent toxicity*. For example, the most successful vector control organism to date, *Bacillus* spp., produces highly specific toxins during sporulation. The living or dead spores with accompanying toxins are applied inundatively, essentially as chemical insecticides. The second strategy is *invasion*. The invading pathogens kill the host after entry into the body over a period of days or weeks. These pathogens can invade vital tissues, sap host resources, or produce a lethal lesion during exit from the host. Viruses, microsporans, fungi, and nematodes usually function in this manner. Microsporans and viruses invade the larval midgut following ingestion, whereas most fungi and nematodes penetrate the cuticle to enter the target insect. A third strategy involves *sublethal effects*. Insects surviving sublethal concentrations of toxic or invasive pathogens exhibit deformities, small size, or reduced fecundity as adults. Some viruses, protozoa, and fungi are transmitted to the next generation by sublethally infected female hosts, providing a mechanism for these pathogens to move to new habitats along with the adult insect. Transovarially transmitted pathogens of low pathogenicity can suppress a population over the course of many generations, but this is probably not a useful strategy against vectors.

Pathogens are subject to many environmental factors, such as salinity, pH, oxygen availability, ultraviolet (UV) radiation, adverse temperatures, and pollution. To obtain maximal immediate and long-term success, the specific sensitivities of a pathogen must be considered when planning an application. In cases of patent toxicity, reproduction may or may not occur in target larvae and the progeny may or may not exert further control on the target population. Establishment of these pathogens in the environment is not usually a realistic goal; thus, it is important to determine whether the toxin will remain active and whether the target insect will ingest sufficient toxin in the given habitat. For example, the toxic crystal *Bt* protein in commercial formulations is effective only when eaten by insects with an alkaline gut pH and the specific membrane structures required to bind the toxin. Not only must the insect have the correct physiology and be at a susceptible stage of development, but the bacterium must be eaten in sufficient quantity. Commercial *Bt* spores do not usually spread to other insects or cause disease outbreaks on their own. In contrast, many invasive pathogens often reproduce successfully, and their progeny can infect other larvae in the environment. When establishment and long-term vector population reduction are goals, the habitat and the pathogen's environmental constraints must be considered.

Natural recycling of pathogens in the environment frequently involves periods of relatively low pathogen prevalence interspersed with unpredictable epizootics. Favorable environmental conditions, the presence of infective pathogens, and a threshold density of the host population must coincide for an epizootic to occur. The initial introduction of high densities or infective-stage pathogens often induces an epizootic in host populations that are below the characteristic threshold density. The epizootic generally cannot sustain itself for a long period because the host population is rapidly decimated. Pathogens that are able to produce new infective stages even in the absence of a host or that require a low threshold of density are best suited for long-term vector suppression.

Viruses

Although insect viruses are quite useful in microbial control of lepidopteran insect pests, no known virus pathogens appear to have good potential as microbial control agents of vectors. Viruses affect only a small percentage of most vector populations. High doses of the viruses must be ingested to infect the host, with older larvae requiring even more virus than younger instars. Moreover, many can be transovarially transmitted, demonstrating that they are not effective in killing the current generation.

Natural mosquito populations sometimes harbor nuclear polyhedrosis viruses (NPVs), cytoplasmic polyhedrosis viruses (CPVs), and mosquito iridescent viruses (MIVs). Mosquito NPVs affect the larval mosquito midgut, and the natural prevalence of the virus can occasionally reach 70%. High doses are required to cause laboratory infections. Infected adults and pupae can be found, and tranovarial transmission is sometimes observed. CPVs also infect midgut cells. They stress, but do not directly kill, the affected larvae. MIVs affect *Aedes* larvae in the snowmelt pools of northern temperate forests. The virus replicates in the fat body, muscles, and hypodermis and usually gives the larvae an iridescent blue-green or violet appearance. Most infected larvae die before pupation, but a few live to adulthood and transmit the virus transovarially. It is difficult to infect mosquitoes in the laboratory.

Protista

The pathogenic protists are a diverse polyphyletic group with a wide variety of life cycles and pathogenic actions. There are perhaps more types of protists identified from mosquito larvae than any other type of pathogen. Members of the phylum Microspora are the most common and extensively studied mosquito pathogens. They produce a coiled polar filament in the spore (Fig. 43.2). When spores are ingested by an appropriate host, the polar filament is forcibly ejected. If the polar filament penetrates a gut cell, the sporoplast migrates through the filament to initiate an infection in the cell. Microspora infecting vector larvae produce single or polymorphic spore types.

Nosema and *Vavraia* spp. produce a single spore type, develop asexually, and easily infect larvae in the laboratory. Although these Microspora have a broad host range, they usually produce low mortality and do not persist in the environment. Therefore, they are not likely to be useful in vector control. *Nosema algerae* might attract more interest for use in anopheline control if an inexpensive in vitro production system could be devised to replace current in vivo methods and if formulation technology could improve the ability of spores to float and survive in aquatic habitats.

Microspora that produce polymorphic spores affect a wide range of mosquito species, infect transovarially

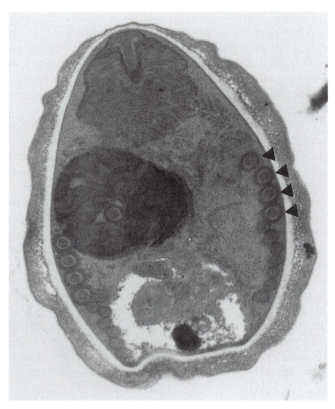

FIGURE 43.2 Ultrastructure of the spore of *Amblyospora* sp. The polar filament is seen as cross sections of a tubule that coils around the inner surface of the spore wall. Photo courtesy of T. Andreadis.

Fungi

Fungal pathogens generally infect a wide range of hosts; however, strain or pathotype differences in host range and virulence are well documented. To initiate an infection, viable spores must contact the host cuticle, recognize a host, attach, germinate, and penetrate the cuticle. Vegetative growth in the hemocoel leads to death of the host by starvation or invasion of vital organs. After host death, the next generation of infectious fungal forms is produced. The mosquito pathogens that are most promising kill their hosts within 1–3 days. Fungal activity is temperature dependent and can be limited by periodic desiccation of the habitat. Most fungi have a resistant spore stage or other method for surviving periods of host absence and unfavorable abiotic conditions. Three fungal pathogens or mosquito larvae have potential as microbial control agents: *Lagenidium giganteum*, *Coelomomyces* spp., and *Culicinomyces clavisporus*.

Lagenidium giganteum infects mosquito species and can recycle among hosts every 2–3 days (Figs. 43.5 and 43.6). Its asexual zoospore survives about a day, and the sexually produced oospore can survive up to 7 years under adverse environmental conditions (Kerwin and Washino 1988).

Inundation with zoospores can quickly reduce a high-density host population. Oospores can be particularly useful for control of low-density hosts because they germinate asynchronously and persist well in the environment. Organic pollution and high salinity rapidly inactivate zoospores, so this fungus is best used in unpolluted freshwater environments. Products based on mycelium and oospores of *Lagenidium* have been registered by the USEPA.

Coelomomyces spp. are obligately parasitic fungi that infect a broad range of mosquitoes and have frequently been described from natural epizootics. Attempts to infect mosquito larvae with this fungus in the laboratory were unsuccessful until the discovery of an obligate sexual generation in a copepod alternating with an asexual generation in the mosquito (Whistler et al. 1974). Each *Coelomomyces* species has an optimal copepod host. Severe infections can kill mosquito larvae; however, lightly infected mosquitoes can survive to adulthood, providing a mechanism for dispersal of the fungus. *Coelomomyces* spp. are attractive biological control agents because they are persistent, can kill over 90% of the vector population, and successfully establish in new locations. However, because of its complex life cycle, in vitro mass production of this fungus is impractical.

Culicinomyces clavisporus infects a very broad host range among mosquitoes and readily produces infectious conidia in artificial culture. It tolerates

or orally, consistently produce high mortality, and persist in the environment. A copepod serves as an intermediate host for *Amblyospora* spp., and other polymorphic species (some of which do not require intermediate hosts) can have complex life cycles. *Amblyspora connecticus* in *Aedes cantator* and the copepod *Acanthocyclops vernalis* (Figs. 43.3 and 43.4) produce regular epizootics in mosquito populations and can have a natural regulatory effect (Andreadis 1990). These protozoa could be introduced inoculatively as a component of an integrated control program.

The ciliates *Tetrahymena pyriformis* and *Lambornella clarki* provide natural control of mosquito larvae in tree holes and are potentially useful as applied biological control agents. The infective stages encyst on and penetrate the host. They then reproduce and kill the host larvae. Dead larvae can contain hundreds of mobile ciliates that forcibly exit through the cuticle. *Lambornella* has both parasitic and free-living forms. The free-living forms provide a means of persistence during periods of host absence and transform into parasitic forms in the presence of host larvae.

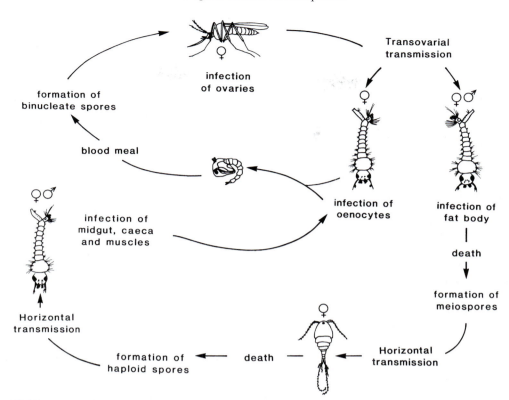

FIGURE 43.3 Life cycle of *Amblyospora* spp. The cycle starts at the top of the diagram with the adult female mosquito and moves clockwise. In response to taking a blood meal, *Amblyospora* spores already in the body of the female release the sporoplasm, which infects the oocytes of the ovary, thereby ensuring vertical transmission of the infection. Some infected female progeny support benign infections and transmit the infection transovarially to the next generation. In male larvae and other female larvae, the organism undergoes meiosis and forms thousands of haploid spores in the fat body, ultimately killing the host. When these haploid spores are ingested by a female copepod, an asexual cycle is initiated, leading to the production and release of haploid spores. Haploid spores eaten by a mosquito larva invade the tissues and enter a sexual phase in which the parasite returns to the diploid state. Female mosquitoes thus infected can then transmit the protistan transovarially to reinitiate the cycle. Adapted from T. Andreadis (1990) and the Society of Protozoologists.

FIGURE 43.4 Two mosquito larvae, one infected with *Amblyospora* sp. (left) and one healthy larva (right). Photo from the late W. R. Kellen.

freshwater and brackish water. *Culicinomyces* spores must be ingested by the larvae, where they adhere to the cuticle-lined foregut or hindgut, germinate, penetrate the cuticle, and proliferate in the larval body. Following death of the larva, the fungus sporulates on the surface of the dead host. Occasionally, a lightly infected larva survives to adulthood. Interest in developing this fungus is minimal because of its temperature-dependent activity, difficulties in storage (it can be dried and stored only temporarily), the high dosages required for field efficacy, and lack of adequate field persistence and recycling.

Nematodes

Although insect-parasitic nematodes are not microorganisms, their small size, persistence in the environment, and utility in biological control have resulted in their inclusion in most discussions of insect

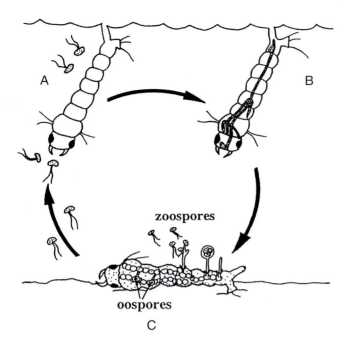

FIGURE 43.5 Life cycle of *Lagenidium giganteum*. (A) Motile bifla-gellated zoospores locate and encyst on the cuticle (usually on the head or siphon/anal gills) of a mosquito larva. (Zoospores are mag-nified compared to the size of the larva.) (B) Upon penetration, mycelia extend through the hemocoel to cause (C) the death of the larva, probably by starvation, within 34–72 h. For the next few days, the hyphae form sporangia, which form exit tubes to release asexual zoospores that continue the life cycle, or the fusion of two adjacent hyphal segments form sexual oospores. The oospores persist for months or years and then make new zoospores to reinitiate the cycle.

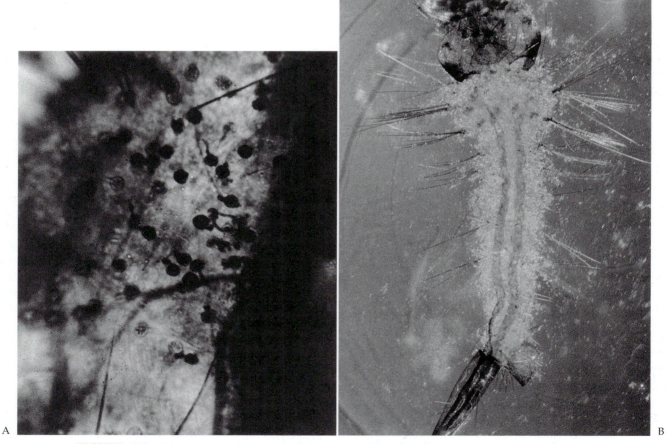

FIGURE 43.6 (A) Zoospores of *Lagenidium* encysted on the cuticle of a larva of *Culex tarsalis*. (B) *Culex tarsalis* larva filled with *Lagenidium* hypae. Light micrographs courtesy of J. Kerwin.

pathogens. Several species of Mermithid nematodes specifically parasitize mosquito or blackfly larvae, but only *Romanomermis culicivorax* has been developed for vector control. *Romanomermis* females lay eggs in the substrate of mosquito larval habitats. If the site desiccates and then rehydrates, the eggs develop into preparasites, which actively seek hosts for up to 2 days. The preparasite attaches to a host larva by means of a stylet and bores a hole through the cuticle. The nematode then burrows into the substrate, matures to the adult stage, mates, and lays eggs. The life cycle requires at least 1 month at favorable temperatures.

Romanomermis culicivorax (Fig. 43.7) has a relatively broad mosquito host range; however, it is active only in warm water and is inhibited by salinity. The number of nematodes per host affects the sex ratio, with more males being produced as the parasite density in the host increases. This and its relatively long life cycle make it most viable as an inundative biological control agent. A large-scale production system was devised for *Romanomermis* using *Culex pipiens* larvae as hosts and a product was marketed for a few years. Although this nematode is no longer commercially available in the United States, other parasitic nematodes of insects are produced commercially.

Predators

The most commonly studied predators of mosquitoes include many invertebrates, such as predaceous insects, hydra, and flatworms, and vertebrates, such as fish. Natural enemies with good searching abilities are often found at low densities because they keep their prey/host population at a low density. Predators are usually polyphagous (feed on many prey species), which is advantageous because they can survive the absence of the target species but may make them suboptimal for biological control. Predators not only seek and attack prey, but also can kill more prey than they consume. Most predators are more mobile than the pathogens and are usually less sensitive to water quality parameters.

Predators can respond to a surfeit of prey by reproducing or immigrating, which is called a *numerical response*. However, predators often reproduce more slowly than their prey; thus, a fecund prey population can potentially escape their influence. When the vector population overwhelms the predator population, deliberate introduction of predators can achieve the balance desirable for vector control.

The presence or absence of alternate prey affects the level of control achieved. Thus, prey species A, present at low densities, is sometimes ignored in favor of more abundant species B. As the density of species A increases, the predator can switch to preferentially feeding on A over B. This switching phenomenon is possibly caused by enhanced recognition of A or simply by increased contact.

The ideal predator for biological control of mosquitoes should: (1) consume large numbers of prey per predator, (2) quickly colonize new habitats, (3) prefer the vector but be able to thrive on alternate prey, (4) excel at finding prey, and (5) reproduce rapidly, producing large numbers of offspring. No predator fits this description exactly.

Invertebrate Predators

Invertebrate predators have been introduced less often than vertebrate predators (fish). Most invertebrate introductions have been experimental. Recognition and conservation of important invertebrate predators have often been considered the appropriate methods because of the high costs associated with their mass rearing. A few noninsect invertebrates have been studied for their potential as mosquito control agents. Hydra (Hydrozoa) ambush many invertebrate

FIGURE 43.7 Nematode *Romanomermis culicivorax* coiled within the thorax of a *Culex* sp. larva. Light micrograph courtesy of J. Petersen.

species in aquatic habitats. In temperate habitats they can reduce larval mosquito populations. Flatworms tolerate the presence of pesticides but are very sensitive to salinity and water temperature. Many species kill more prey than they consume. Predaceous copepods provide high mortality in container habitats and some species are easy to mass-produce in the laboratory.

The Coleoptera (beetles) and Trichoptera (caddis flies) are efficient predators of mosquito larvae; however, at naturally occurring densities they do not yield adequate control. Dragonflies reduce both larval and adult mosquito populations but are far from specific for mosquitoes.

Toxorhynchites has been released as a biological control agent of container-breeding mosquitoes. *Toxorhynchites* larvae prey on other mosquito larvae; the large adults feed only on nectar (Fig. 43.8). *Toxorhynchites* larvae kill more than they can consume. When prey are absent, the larvae can withstand long periods of starvation. Because the adult female seeks containers in which to oviposit, they naturally disperse to habitats that are difficult for humans to find and treat.

However, they prefer natural to artificial containers, so human intervention is needed to repeatedly treat artificial containers. Unfortunately, this mosquito has a lower fecundity and longer life cycle than most of its prey; hence, it is not an ideal biological control agent.

Copepods are microcrustaceans that are often found in the same habitats as mosquito larvae. Several species of copepods prey on mosquito larvae. *Mesocyclops aspericornis* controls *Aedes* and *Culex* spp. larvae in French Polynesia (Riviere et al. 1987; Lardeux et al. 1992), and Brazilian strains of *M. aspericornis* and *M. longisetus* are effective predators of *Ae. aegypti* larvae in laboratory tests (Kay et al. 1992).

Vertebrate Predators

Fish are the predators used most commonly for mosquito control. Small fish are more likely to eat larvae than large fish. In regions where human digenetic flukes infect fish as intermediate hosts, introduced fish must not support fluke development. Two species are frequently used in mosquito control: the mosquito fish, *Gambusia affinis* (see Fig. 9.17) and the

FIGURE 43.8 Predaceous *Toxorhynchites* larva with a partially ingested larva of a culicine mosquito. Photo courtesy of J. Lenly, Florida Medical Entomology Laboratory, Vero Beach.

guppy, *Poecilia reticulata*. *Poecilia* is similar to *Gambusia* in prey and habitat preferences; however, it survives better in water with low dissolved oxygen content. *Gambusia* are efficient predators in clear temperate waters, where they prey on mosquitoes and other invertebrates. They survive water temperatures as low as 13°C and can gradually adapt to water temperatures above 35°C and to saltwater. *Gambusia* are frequently maintained in fish farms over the winter months and released into mosquito habitats in the early spring, when mosquito breeding begins.

Gambusia are surface feeders attracted to moving prey and provide the best mosquito reduction in water with little vegetation. Despite the expense of rearing and transport, mosquito fish have reduced both the total cost and the number of insecticide applications required for mosquito control in Californian rice fields (Lichtenberg and Getz 1985).

Gambusia do not always reduce mosquito populations. They can adversely affect populations of other predators without causing much mosquito mortality. *Gambusia* has some other potential disadvantages. They eat zooplankton and allow algal blooms that are inappropriate for some habitats. Further, they can overwhelm or eat the progeny of native fish, endangered species such as the desert pupfish, or economically important fish. Because of these negative characteristics, indigenous fish should be used when possible.

Tilapia spp., *Aphyocypris chinensis*, and carp can be used for vector reduction and to provide food. Smaller fish are less likely to be eaten, although they are sometimes used by local people as bait. If the fish are eaten before the vector season is over, it negatively impacts biological control. Prolific fish such as *Tilapia* can better tolerate reductions in population. For example, in China, pisciculture practiced in rice fields helps reduce vector populations. The fish are harvested when the fields are drained. *Apocheilus blochi* tolerates polluted waters and can be used in pit latrines. In habitats that periodically dry out, the annual fish *Cynolebias* and *Nothobranchius* can be used. These species lay drought-resistant eggs.

EFFICACY AND SUSTAINABILITY OF BIOLOGICAL CONTROL IN DIVERSE HABITATS

Both abiotic and biotic factors can affect the efficacy and sustainability of biological control, and the suitability of a habitat must be verified at each new location.

Temperature can affect a pathogen's ability to infect and the speed with which it kills the host. Pathogens from temperate locations often do poorly in warm tropical waters, and pathogens from tropical regions can be inactivated in cold waters. Predators also have optimal temperature ranges. Within natural waters, pH is not a limiting factor for most multicellular natural enemies. Many freshwater species do not tolerate saltwater or brackish water. Organic matter or pollutants can also affect natural-enemy survival and performance. Decreased oxygen availability and the sedimentation of infective spores by particulate matter have been implicated in reduced biological control. The addition of toxic agents, such as chemical pesticides, can be detrimental to a natural-enemy population.

The habitat type, size permanence, and water depth all determine which biological control agents can successfully establish in a given location.

Biotic factors affecting control agent efficacy include the ecology of the agent itself, the natural history of the target species, and the presence of other, interacting species. Important characteristics of the biological control agent are its host specificity and adaptation, dispersal and host-finding abilities, infectivity or capture ability, persistence mechanisms, generation, time, and population age structure. Genetic engineering can be used to address some innate limitations of biological control agents. For the target host, the reproductive strategy, dominant type of population regulation, aggregation, diversity of breeding sites, and innate resistance to the biological control agent must be known and assessed against the characteristics of the control agent. Other species can function as competitors, as alternate hosts/prey available to serve as distraction or extra food supply, or as natural enemies of the biological control agent.

Development of resistance by the vector is a possibility that cannot be ignored. Laboratory strains have been selected to be more resistant to nematode attack and bacterial toxins. Resistance to bacterial toxins has been selected in field populations of Lepidoptera. Pathogens and predators that depend on a vector population face selective pressures to overcome escape mechanisms the vector may develop.

Commercial feasibility of biological control depends upon many factors. Biological agents capable of being produced, stored, and applied as easily as conventional insecticides but that yield long-term adequate suppression of vector populations would be ideal. The microbial toxins are commercially viable because they fill most of these requirements. Living agents, however, are frequently difficult to produce or store and are often ineffective in disease reduction.

Biosafety

Biological control agents should be evaluated in terms of their impact on nontarget organisms such as beneficial invertebrates. Most biological control agents of mosquitoes are relatively innocuous to other organisms.

Microbes must also be evaluated for their effect on vertebrates. Their toxic, carcinogenic, teratogenic, and allergenic properties must be determined on representative vertebrate species. To date, microbial pathogens of insects have proven to be relatively harmless to vertebrates.

Mass Production of Biological Control Agents

The *Bacillus* spp. and the fungi *Lagenidium* and *Culicinomyces* are facultative pathogens that can be grown in fermentors without loss of virulence. All viruses, most protists, and the nematode *Romanomermis* must be grown in living hosts. Invertebrate predators can sometimes be grown on an artificial medium or on easily reared prey.

Formulation, Storage, and Transport

Various formulations have been developed that stabilize pathogen viability and provide long-term storage. The physical form of the final product (e.g., powder, liquid, wettable powder, or granules) can be varied to suit the habitat. Granules effectively penetrate to the water's surface and thus are used to broadcast over vegetation. The size of particles may also influence whether they are ingested by the target larvae. Floating particles maintain a pathogen in the larval feeding zone longer than the infective spores alone. Many microbial formulations are similar to chemical formulations, to enhance acceptance with organizations that already use insecticides for vector control and to facilitate use in conventional application equipment.

Ideally, for long-term storage, biological control agents should be maintained at room temperature. *Bacillus* preparations and oospores of *Lagenidium* do not require special handling; however, many other pathogens must be stored at cool temperatures, properly aerated, and protected from contamination. Predators normally cannot be stored as long as some microbial pathogens.

Lightweight formulations are preferred for transport because they are less expensive to ship. Pathogens or predators that must be shipped and stored in an aqueous medium are less useful than those that can be desiccated. Fish, in particular, require an aqueous medium for transport.

SUMMARY AND FUTURE DIRECTIONS

Biological control has many perceived advantages. It is relatively natural and host specific, it can be self-perpetuating, and it does not adversely affect beneficial invertebrates. As attractive as biological control is in principle, in practice this technique has had limited success in vector control as compared to crop pest control. Several innovative genetic approaches are being applied to microbial organisms in an attempt to extend their usefulness as vector control agents. New technologies are aimed at combining the benefits of an inundation with an introduction. Where integrated vector management is introduced, combinations of control measures are essential to provide optimal control with minimal detrimental environmental impacts. Both applied biological control and conservation of natural biological control should be components of these programs.

Readings

Andreadis, T. G. 1990. Experimental transmission of a microsporidian pathogen from mosquitoes to an alternative copepod host. *Proc. Nat. Acad. Sci. U.S.A.* 82: 5574.

Kay, B. H., Cabral, C. P., Sleigh, A. C., Brown, M. D., Ribeiro, Z. M., and Vasconcelos, A. W. 1992. Laboratory evaluation of Brazilian *Mesocyclops* for mosquito control. *J. Med. Entomol.* 29: 599–602.

Kerwin, J. L., and Washino, R. K. 1988. Field evaluation of *Lagenidium giganteum*. Description of a natural epizootic involving a new isolate of the fungus. *J. Med. Entomol.* 25: 452–460.

Lardeux, F., Riviere, F., Sechan, Y., and Kay, B. H. 1992. Release of *Mesocyclops aspericornis* for control of larval *Aedes polynesiensis* in land crab burrows on an atoll of French Polynesia. *J. Med. Entomol.* 29: 571–576.

Litchtenberg, E. R., and Getz, W. 1985. Economics of rice-field mosquito control in California. *BioScience* 35: 292–297.

Mogi, M. 1993. Effect of intermittent irrigation on mosquitoes (Diptera: Culicidae) and larvivorous predators in rice fields. *J. Med. Entomol.* 30: 309–319.

Riviere, F., Kay, B. H., Klein, J. M., and Sechan, Y. 1987. *Mesocyclops aspericornis* and *Bacillus thuringiensis* var. *israelensis* for the biological control of *Aedes* and *Culex* vectors breeding in crab holes, tree holes and artificial containers. *J. Med. Entomol.* 24: 425–430.

Roberts, D. W., Daoust, R. A., and Wraight, S. P. (compilers). 1983. *Bibliography on Pathogens of Medically Important Arthropods: 1981.* World Health Organization, Geneva, Publication VBC/83.1.

Service, M. W. 1985. Some ecological considerations basic to biocontrol of Culicidae and other medically important insects. In *Integrated Control of Vectors*, Vol. 1. (M. Laird, and J. W. Miles, eds.). Academic Press, London, pp. 9–30.

Whisler, H. C., Zebold, S. L., and Shemanchuk, J. A. 1974. Alternative host for mosquito parasite *Coelomomyces*. *Nature* 251: 715–716.

44

Genetic Control of Vectors

ROGER J. WOOD

INTRODUCTION

In 1944–5, F. L. Vanderplank briefly eradicated sleeping sickness from a small region of Tanzania by exploiting the phenomenon of interspecific hybrid sterility. Into a population of the local vector *Glossina swynnertoni* he introduced wild-collected pupae of a closely related but nonvector species, *Glossina morsitans centralis*. The male hybrids were fully sterile, the female hybrids partially so. By his radical action Vanderplank took the first practical step toward defining the subject of this chapter. I review progress made in using genetic methods since that time and reflect on the lessons to be learned and their value in bringing about *population suppression* of an increasing number of pests. I also review the potentiality for achieving a more ingenious outcome, *population replacement*. The prospect of replacing disease vectors with harmless forms of the same species is an attractive idea, currently gaining support from rapid developments in recombinant DNA technology.

Reviews of potential categories of genetic control published in the 1970s included discussion of their operational feasibility against vectors of disease. Interest centered on the species specificity and nonpolluting properties of genetic methods, in contrast to insecticides. Apart from ecological considerations of ever-widening scope and complexity, the expanding horizon of resistance had created a fear that the chemical battle was being lost. If insecticides were to retain their value, some way had to be found to reduce dependence upon them by diversifying into nonchemical methods of control.

The choice of genetic method for a given species depends on the control strategy to be adopted, whether *focal*, i.e., adjusted to local needs, or in an *areawide* rolling program. Examples are given to illustrate the major principles, after which progress made so far in developing genetic techniques for controlling vectors of human disease is considered.

STERILE INSECT TECHNIQUE (SIT)

In terms of operational effectiveness on an extensive practical scale, only one genetic approach to control has so far enjoyed major success. This is the sterile-insect technique (SIT), or sterile-insect release method (SIRM), which is based on the release into the natural habitat of large numbers of the target pest, sterilized in a breeding factory. The released insects carry dominant lethal mutations in their reproductive cells, due to treatment with ionizing radiation or chemical mutagens. Any wild female mating with such a male lays sterile eggs.

Species vulnerable to SIT are those that can be bred in large numbers at economic densities under factory conditions and in which dominant lethal mutations can readily be induced (Fig. 44.1). The success of SIT depends on the sterile male's capacity to locate, attract, and mate with wild females, in direct competition with wild males. Their inevitable inequality in this respect, due to the impact both of artificial breeding and mutagenic treatment on their behavior, makes it imperative that the numbers released exceed the natural population density by a large margin, usually 10 times or more. Because of this, it is normal to precede SIT releases by closely targeted insecticide treatment (often formulated as a bait) to reduce the initial population

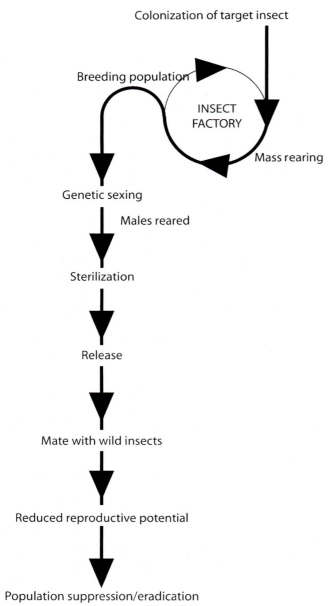

Colonization of target insect

Breeding population

INSECT FACTORY

Mass rearing

Genetic sexing

Males reared

Sterilization

Release

Mate with wild insects

Reduced reproductive potential

Population suppression/eradication

FIGURE 44.1   Flowchart of SIT.

conventional approaches. The released males search out the native females, and one of them may inseminate several. In some species a female is receptive to insemination only once; in other cases she may receive a mixture of fertile and sterile sperm. In either case the effect on the next generation depends on the proportion of sterile to fertile males and the competitiveness of the treated males.

Various factors govern the success of SIT. The decline in the population expected in the generation following release depends not only on the ratio of released males to wild males, but also on the residual fertility and relative competitiveness of the released males compared with wild males. The impact of immigration by fertilized females must also be considered, as well as the reproductive capacity of the species. Reproductive capacity is particularly difficult to estimate because it may increase as density is reduced. The impact of immigration also increases as density is reduced. Two examples of species highly vulnerable to SIT can be given to illustrate the potential of the technique.

Screwworm Fly

A dramatic demonstration of the value of SIT to deal with an emergency took place in Libya, North Africa, during the period 1988–91, following the discovery of unusual maggots in animal wounds. These had been laid by a species of fly formerly confined to the Americas, the New World screwworm fly, *Cochliomyia hominivorax*, long the object of control by SIT in the United States, Mexico, and territories to the south. It is supposed that their unexpected appearance in Libya, spotted by sharp-eyed staff at the Faculty of Veterinary Medicine, University of Tripoli, resulted from contaminated livestock imported from somewhere on the American continent. In Libya the camel became a host, as well as cattle, horses, sheep, goats, and donkeys. Medical authorities soon identified the first cases in humans. The fly already occupied tens of thousands of square kilometers. Under optimal conditions of temperature, a new generation was expected every 3 weeks. Uncontrolled movements of livestock were expected to spread the disease much more widely. The potential for economic and social disaster, especially in sub-Saharan Africa, galvanized the Libyan government and relevant U.N. Agencies, coordinated by the Food and Agricultural Organization in Rome, to jump into action without delay. A plan was drawn up with four parts: (1) surveillance to identify the limits of distribution, (2) treatment of wounds of infected animals and human beings, (3) spraying of livestock with recommended insecticide, and (4) release of large

density. Even so, the numbers released have to be high enough to create a discernible impact. To achieve this, factory-scale production must incorporate an appropriate degree of automation, with facilities for rapid mass sterilization, and continuous quality control. Prior to release, the sterilized insects are marked in some way so that they may later be identified in field traps designed to monitor the effect of releases on population density and structure.

Experience has shown that although SIT is expensive to set up, it can provide particularly good value against species that are difficult to control with numbers of sterile flies. Experience on the American continent predicted that the species would never be controlled on an areawide basis without SIT.

Then as now, the screwworm fly was bred extensively at a factory in Mexico, at Tuxtla Gutierrez, in the state of Chiapas. From this major production unit, with a capacity to produce up to 500 million flies per week, sterilized screwworm pupae were transported in cargo planes to Libya. Packed in boxes designed to open as they fell, they were loaded into light aircraft and released in a grid pattern across the infested territory. Releases began in December 1990, in numbers that rose to 40 million per week by May 1991, by which time new cases of screwworm infestation had declined to zero and the species had been given no chance to spread to neighboring countries. To make sure of the complete elimination of the pest from Africa, SIT releases were continued into 1992.

The lightning response to this crisis in Libya (1988–91), which averted the threat of an international agricultural tragedy of major proportion, was made possible by the widespread experience of SIT against screwworm on the American continent. To create a highly efficient breeding factory, technology had been developed and refined by trial and error since the first breeding unit was set up at Bithlo, Florida, in 1955–57. Much had been learned since those early pioneering days, enabling screwworm to be cleared totally from the United States and Mexico and almost completely throughout the rest of Central America to Panama. Success did not come easily, because of this species' capacity to find ways to circumvent the impact of SIT. But with experience, the release program was modified. Eradicating the screwworm fly from Mexico, 1972–86, cost US $330 million. This program represented US $1 billion in savings to the Mexican livestock industry. With Mexico free of screwworms, the U.S. livestock industry was at a significantly lower risk of reinfestation, which was estimated to have cost US $375 million annually in earlier years (Statement from Mexico-United States Commission for Screwworm Eradication, April 1988).

The screwworm story has established SIT as a most potent control method when applied intensively over a wide area. Without it, the screwworm cannot be eradicated. Equally, this technique used on its own will not eradicate the pest; SIT must be applied in appropriate combination with other methods of control, in an integrated program.

Tsetse Fly

Among other insects against which SIT has been applied persistently, in a series of trials in east, west,

and southern Africa starting in the 1970s, are species of *Glossina*, the tsetse flies. As vectors of animal trypanosomiasis, these flies are of great importance in restricting agricultural development. They are ideal subjects for SIT because of their low natural fecundity (they bear single live young) and poor potential for the population to recover. But SIT also depends on raising sufficient numbers of flies for release, all at the perfect stage at the same time, not easy with an insect of low fertility. Improvements in rearing technology have solved the problem, a major element of which has been the development of a membrane system for blood feeding. Early success with SIT was reported in a trial against *Glossina morsitans morsitans* on a commercial ranch in Tanzania. Later, four other *Glossina* species were eliminated from 3,000 and 1,500 km² areas of agropastoral land in two West African countries by integrating SIT with other techniques. Even so, there was limited interruption of disease transmission because other major vectors were left uncontrolled.

A more satisfactory outcome of tsetse control was achieved on the island of Unguja, Zanzibar, East Africa, home to just one vector of trypanosomiasis, *Glossina austini*. Situated within easy flying distance by light aircraft from the African mainland, the island was supplied with sterile flies from a breeding factory at Tanga in Tanzania. Aerial releases totalling 8.5 million individuals began in 1994, preceded by insecticide treatments. In 1997 the authorities were able to announce that the species had been eradicated from the island. Now, by building on the Zanzibar experience, an extensive campaign against *Glossina pallidipes* is being organized in Ethiopia, aimed at eradicating it from a 2.5 million–hectare area in the Southern Rift Valley. With international assistance, a new breeding factory is being commissioned. As in Zanzibar, the aim is to interrupt the transmission of animal trypanosomiasis.

To use SIT to control sleeping sickness in humans, also transmitted by tsetse flies, is unfortunately impractical with present techniques. Release of tsetse flies close to human habitation leads to an unacceptable increase in biting. This cannot be avoided by releasing only male flies, because they too take blood. It is a problem that has stimulated thoughts on the feasibility of "autosterilization," in which wild flies would sterilize themselves and no extra flies would need to be released. A trap would be developed to attract wild flies, sterilize them chemically, and then release them. So far "autosterilization" is just an idea, awaiting improvements in trap technology and the development of satisfactory safety measures to avoid hazards to human health. If it becomes practical to develop autosterilization, it promises to be particularly suitable for locally focused control measures.

STERILE-INSECT TECHNIQUE AGAINST VECTORS OF HUMAN DISEASE

Field Trials

When only the female of a vector species takes blood, the prospect of using SIT close to human habitation is theoretically more straightforward. Mosquitoes have long seemed a potential target for the technique. In 1970 a field trial against the mosquito *Culex pipiens quinquefasciatus* took place on the small island of Seahorse Key, off the west coast of Florida. The species was suppressed and finally eliminated within 12 weeks (six generations) by the daily release of 8,000–18,000 males sterilized by a pupal dip in an aqueous solution of thiotepa. The effect was only temporary because the island was not fully isolated from mosquito invasion, particularly from fishing boats. When a repeat trial was planned, the team, staffed and funded by the U.S. Department of Agriculture, had to take into account public disquiet about possible contamination with chemosterilant residues from released mosquitoes. Knowing from earlier studies that pupal irradiation of this species of mosquito reduced its sexual competitiveness, they used a technique of sterilizing chilled male adults at 24–48 hours old. After releasing 13,000 males daily (ratio 10:1 with wild males), they again eliminated the population from the island on a temporary basis.

Further experience of controlling this species with chemosterilized males was gained in continental India, a more demanding challenge. Techniques for larger-scale rearing were developed at a joint WHO/ICMR unit at Kilokri, New Delhi, and important refinements made to methods of marking and transporting mosquitoes in an ice-cooled vehicle to the release points. The relatively inexpensive technology was capable of handling and releasing 1 million mosquitoes per day. The village of Dhulsiras, near Delhi, chosen for a trial, was small enough that releases of 150,000–300,000 sterile males per day would achieve a ratio of sterilized to native males of 24:1. The village was surrounded by a specially prepared buffer zone, planned to be "breeding free." Despite this provision there was strong evidence of both inward and outward migration, with the disappointing result that the success of the released males was constantly being "diluted," and the village could not be protected, even after $5\frac{1}{2}$ months of daily releases. An analysis of the field data

indicated that a future program would have to be targeted at a much larger population, with correspondingly larger releases. Under such circumstances the data showed that the release rate at Dhulsiras was more than sufficient.

The most comprehensive and fully documented SIT trial against a mosquito vector was targeted against *Anopheles albimanus* in an area of 20 km² on the Pacific coast of El Salvador. This was another USDA project, in which chemosterilized male pupae were released at a rate of 1 million per day. Despite efforts made to isolate the release site by a buffer zone, it proved impossible, as in India, to prevent inward migration of fertile females. A temporary decline in the density of mosquitoes in the treated zone by 75% hardly compensated for the effort and expense. Either greater isolation had to be ensured or release rates needed to be much higher.

The evident value of SIT in other contexts has led to continuing hope that a way may still be found to adapt the technique to mosquitoes. As well as evaluating and improving various aspects of the technology, it is necessary to identify more suitable sites for trials. One proposal has centered on isolated towns in southern India in which the only malaria vector is *Anopheles stephensi*, which is totally absent from the surrounding countryside. Another possibility under consideration is *Anopheles arabiensis*, found in desert oases in Sudan.

Technical Problems

The major technical aspects of mass rearing to be evaluated and improved in mosquitoes are (1) automated mass rearing, (2) genetic sexing, (3) rapid sterilization, (4) quality control, (5) transportation and release, (6) trapping to evaluate progress. Aspects 1 and 2 probably require the most radical advances.

Mass Rearing

The extensive literature on large-scale rearing in mosquitoes records no example in which production exceeds 7 million per week. This compares with 500 million per week from a modern screwworm factory, and natural mosquito densities are usually much higher than those of screwworm flies. Clearly there is great scope for technical development in mass production if SIT for mosquitoes is to play a major part in disease control on an areawide basis.

The essential activities to be accommodated in the breeding factory are mating, feeding, resting, egg laying, larval development, and pupae harvesting, all at an unnaturally high density. Essential operations on adults (blood, sugar feeding, water provision, and egg collection) need to be controlled with zero escape. For the provision of blood, conveniently and hygienically, there is the example of the tsetse fly as a guide. Membrane feeding for other vectors has yet to be developed on a factory scale, although various laboratory systems have been designed for mosquitoes, one of which is now produced commercially. Electronically regulated for temperature from portable 12-volt DC power units, the feeders, in sets of up to six, are placed externally on the cages, where mosquitoes take their blood meals through the mesh.

In all its aspects, mass rearing must take account of each vector's particular requirement in terms of cage volume and dimensions, optimum population density, and sex ratio. In most cases a factory rearing system for a vector mosquito would have to deal with many more cages than in a screwworm fly or medfly factory of comparable output, which is bound to require radical rethinking in terms of cage handling. A major consideration is the collection of eggs without the escape of adults. A simple answer may lie in carrying out this operation, and certain other essential tasks, at reduced temperature. There is no reason for cages to be stationary and no reason why any "housekeeping" tasks, apart from blood feeding, need to be carried out under the "tropical" conditions, in which the insects are kept for most of the time. Larval rearing and pupal harvesting also offer scope for fresh technical rethinking. The starting point for future developments in all aspects of mass rearing has to be the experience of those who have already traveled some way along this path, particularly the teams active in earlier years in India and the United States, where progress was most extensive.

Genetic Sexing Techniques

The safe use of SIT for controlling vector-borne disease in humans is dependent on releasing only males. A way, therefore, has to be found either (1) to avoid the production of females or (2) to ensure that they are eliminated as soon as possible in the factory production line. A need is thus created in each species for a genetic sexing technique (GST) of absolute efficiency. For the most successful SIT trial against mosquitoes to date, against *Anopheles albimanus* in Costa Rica, sexing was achieved by exposure of the eggs to an insecticide. The method depended on the translocation of a propoxur resistance gene to the Y chromosome from an autosome so that only eggs destined to develop into males survived exposure to propoxur. The translocation was stabilized by an inversion induced between the resistance gene and the translocation junction (Fig. 44.2) to suppress recombination.

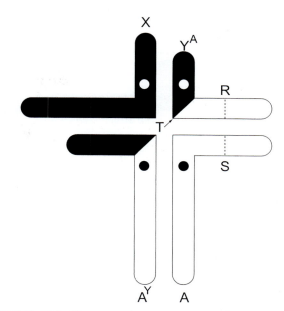

FIGURE 44.2 Diagrammatic representation of a Y autosome translocation in the male of a sexing strain in which an insecticide resistance gene R is segregating with its susceptible allele S. Such a male produces two kinds of gametes, XS and YR. A corresponding female produces only XS gametes, so the difference in resistance between the sexes is maintained in the next and following generations, *unless* recombination takes place across the translocation junction, T.

The system generated 99.9% males and reduced the cost of production by one-half.

Comparable translocation-based techniques, using different selectable genes, have been developed for various other species, and the value of an inversion to stabilize a translocation has been confirmed. There is, however, an intrinsic weakness in translocation-based sexing strains, arising from their vulnerability to outside contamination. Unless they can be absolutely isolated from chance contamination by a wild-type male, or a female mated to a wide-type male, or the eggs of such a female, they are in danger of losing their effectiveness rapidly, because the possession of a translocated Y chromosome reduces fertility due to the production of chromosomally unbalanced gametes. Untranslocated Y chromosomes are therefore at a selective advantage and spread rapidly through the population, as was demonstrated experimentally in a medfly sexing strain.

A completely different method of genetic sexing has been devised in *Drosophila melanogaster*, arising from recombinant DNA technology. This system works on the principle of dietary-dependent lethality, which two independent research studies have tested. A strain carries an autosomal, dominant lethal gene that is

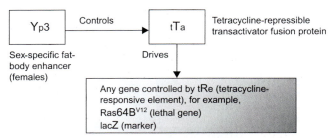

FIGURE 44.3 Schematic reopresention of the genetic basis of the RIDL system of male production in *Drosophila melanogaster*. From Thomas et al. 2000.

suppressed by tetracycline in the diet. When active, the gene is expressed only at the larval stage and only in females (Fig. 44.3). When males are required, a suitable proportion of stock is moved onto a diet without tetracycline. Absolutely no females are produced in the next generation, but male production is normal because the lethal acts in a tissue not present in males. When such males are released to mate with wild females, they produce only males, although their sons generate a proportion of females (see later).

Systems of this kind, referred to as RIDL (release of insects carrying a dominant lethal), developed in *D. melanogaster*, have the potential to be transferred, in modified form, into insects of medical importance, by microinjection of eggs. However, because the release of such a transgenic stock into a wild population is bound to meet public opposition, a more acceptable course of action is probably the development of similar systems isolated from the vector species' own genomes. This will require a greatly increased knowledge of vector genomics.

INHERITED STERILITY

Release of Insects Carrying a Dominant Lethal (RIDL)

The production of dominant lethals in mosquitoes, either by exposure to ionizing radiation or by chemical mutagenesis, is not always without problems. As an alternative for the future, the answer may lie in a modification of RIDL. The aim would be to isolate a genetic arrangement that would not only exclude females but produce males that would automatically become sterile when released. Preliminary experiments have demonstrated that dietary control of a dominant lethal gene active in *both* sexes is probably feasible.

It can be argued that sterility may not be necessary. Fertile RIDL males may provide more cost-effective control than sterile ones, because their inherited male-producing properties would be enhanced if the gene were to be inserted at several points in the genome. In the case of a single insertion, the second-generation progeny of such males is expected to be $1♀:2♂$, but with n insertions, independently assorting, it would be $1♀:2^n♂$. In the case of mosquitoes, with three pairs of chromosomes, the maximum number of independently assorting insertions is likely to be six, which predicts a maximum second-generation sex ratio of $1♀:64♂$. With such a degree of female depletion, constantly induced by repeated releases, the effect on population decline could well be greater than SIT at the same release rate. The effectiveness of multiple insertions has been explored in computer models, which indicate that with insertions at six loci, the release ratio necessary to reduce a population to $1/100$ of prerelease size in 10 generations would be remarkably favorable as compared with SIT.

MEIOTIC DRIVE AND CHROMOSOME REARRANGEMENTS

Vector control by male production is not a new idea. It arose early in relation to the discovery of sex-linked meiotic drive in the mosquitoes *Aedes aegypti* and *Cx. p. quinquefasciatus*. Meiotic drive is any mechanism causing inequalities in chromosome transmissions during meiosis, revealed by deviation from the 1:1 recovery of segregating alleles. In *Ae. aegypti*, a gene, *D* (Distortor), linked to the male-determining locus *M* on the Y chromosome has the potential to cause the X chromosome with which it is paired at meiosis to fragment. The degree to which female progeny are reduced depends on the sensitivity of the particular X chromosome, which is genetically controlled. In this widely distributed species, the *D* gene is restricted to parts of Africa, America, Sri Lanka, and Australia, where it is more or less completely suppressed. In other regions, where *D* is absent, the X chromosomes vary in sensitivity when tested experimentally against a *D* strain. The degree of sex ratio distortion observed can be extreme, as in some samples from the Pacific area. But more usually, as in Indian populations, where X chromosomes are polymorphic for sensitivity, the sex ratio lies around 30%♀ when tested experimentally. Sex ratio distortion can be enhanced by an associated chromosome translocation.

Translocations and some other chromosome rearrangements are associated with reduced fertility, because of the production of chromosomally unbalanced gametes, a property that led to research on their potential for practical control. By releasing them as a control agent in sufficient numbers, it was hoped to place an intolerable inherited genetic load on a population. In an experimental trial in India, a WHO/ICMR team compared three sets of releases of *Ae. aegypti* in walk-in field cages. Chemosterilized males were released in one cage, a double-translocation stock (T_1T_3) in a second cage and a double-translocation combined with meiotic drive stock (DT_1T_3) in a third. Into a fourth cage went an untreated control population. Releases were made daily at 10 times the initial daily rate of male emergence. Eradication of the population was achieved in all cages, but suppression took almost twice the time in T_1T_3 than in the other two, in which no further females were generated after 15 weeks. Clearly, the *D* gene was adding to the genetic load created by the two translocations. Although the sex ratio distortion was quite modest, because some of the X chromosomes in the population were resistant, the effect of successive daily releases was cumulative.

During the 1970s a great deal of enthusiastic effort went into open-field releases of translocation stocks (without *D*) against *Ae. aegypti* in trials in Kenya and Florida. Releases aimed at suppressing populations were carried out both with double-translocation heterozygous and translocation homozygotes. In several cases the strains proved difficult to maintain under culture and showed low competitiveness in the field. In Africa there was the added problem of infiltration of *Ae. aegypti* subsp. *formosus*, which did not mate in the field with the "type form" released. Experience suggested that if translocations and other chromosome arrangements had a part to play in vector population suppression, it would never be more than as an adjunct to other techniques, i.e., to add extra genetic load to an already weakened target. Although a degree of sterility was inherited for a few generations, the effect was always temporary.

NATURAL HYBRID STERILITY AND CYTOPLASMIC INCOMPATABILITY

Sterility in the hybrids of crosses between closely related species and subspecies exists in several medically important groups of insects. Hope that they might be useful for genetic control of the *Anopheles gambiae* complex did not survive the test of experience, although favorable results from certain nonvectors have kept the subject alive. Hybrid sterility in tsetse flies shows sufficient promise to attract continuing research.

Hope has also been placed in sterility between populations of the same species made incompatible by a microbial infection. The phenomenon is referred to as *cytoplasmic incompatibility* (CI), due to symbiotic bacteria of the genus *Wolbachia*. The bacterium is transmitted through the egg cytoplasm, which leaves an "imprint" on the sperm, thus preventing it from fertilizing any egg except one carrying the same kind of *Wolbachia*. In using CI as a substitute for SIT, the idea would be to release males of a strain that is uniformly incompatible with the local wild population. Such males would be functionally sterile without the expense of artificial sterilization.

The potential of CI for genetic control was first appreciated in the mosquito *Cx. p. quinquefasciatus*, which exists in a complex mosaic of cytoplasmic forms over its worldwide range. After the first test of CI against *Cx. p. quinquefasciatus* in the field in 1967, the outlook seemed optimistic. The trial took place in the village of Ockpo, 10 miles north of Rangoon, Burma, where *Cx. p. quinquefasciatus* is the major vector of lymphatic filariasis. Isolated in winter by dry paddy fields, the village population was eliminated in only 3 months (five to six generations) by the daily release of about 5,000 incompatible males with cytoplasm of a strain originating in Paris. Unfortunately, the initial promise was not maintained in more extensive trials, the most informative of which took place in two villages in the Delhi Union territory, India. Despite much larger releases than at Ockpo, population suppression was only partial and temporary. At the time this was attributed largely to population movement leading to reinfestation. But incompleteness of incompatibility due to male aging, when the imprinting of a sperm is less certain, is another possible explanation. The usefulness of the technique has been further brought into question by objections raised on the grounds of "genetic pollution," arising from the danger of creating "inadvertent population replacement," by cytoplasmically inherited elements (in mitochondria, for example) hitchhiking with the *Wolbachia*. So, the use of CI simply as a substitute for SIT is no longer favored.

POPULATION REPLACEMENT

Genetic control need not be directed toward *population suppression*. An alternative, so far unproven in the field, although with obvious attractions, is *population replacement*, in which the genetic structure of a pest population is modified in a way that favors human benefit. In the case of vectors of disease agents, the desired modification ("useful gene") would be one that prevents disease transmission, either by altering vector behavior or by producing some physiological change in parasite–vector interaction. Transforming malarial vectors into harmless insects that do not transmit disease to man has been a goal of WHO (TDR Molecular Entomology Initiative) since 1991. To bring about population replacement, a "transporting mechanism" is required, a genomic element linked to the useful gene, to facilitate its incorporation into the natural population in the face of selective forces acting against it.

Useful Genes

In earlier years, traditional selective breeding techniques were used to isolate genes controlling disease transmission, although most of the examples studied were not natural vectors of the parasites studied. However, these model experiments provided the basis for progress in understanding "vector competence," which has since benefited from outstanding advances in molecular genetics, in discovering key molecules in vector mosquitoes essential for parasitic growth (Chapter 30).

Transporting Mechanisms

The isolation of a gene critical to vector competence in the transmission of a major vector-borne disease will focus interest at once on the possible means of driving it into natural populations. Is there a reliable transport mechanism that can be guaranteed not to break down in the face of natural selection? Several possibilities have been researched, although none offers immediate prospects.

Negative Heterosis

When two crossed strains are less fertile than when bred singly or when their hybrids are sterile or partially sterile, they are said to exhibit *negative heterosis*. Under such circumstances, one of the strains is bound to eliminate the other, a phenomenon theoretically open to manipulation as a means of population replacement. As originally proposed, this concept was applied to the semisterility observed in reciprocal-translocation heterozygotes. Later, it was considered in relation to CI. But after tests in the field or in field cages, neither approach could be recommended as a practical proposition.

Meiotic Drive

Under circumstances in which one allele of a pair is driven at meiosis, any gene linked with it has a chance of increasing in frequency. The phenomenon was demonstrated experimentally with the Y-linked *D* gene in *Ae aegypti*, using an eye color marker (*re*). A black-eyed laboratory population (*n* = 1,000–5,000) was transformed within nine generations into one that was more than 60% red eyed. A complete transformation in color was not expected because of recombination, occurring naturally between *D* and *re*. Although successful as a laboratory demonstration, the experiment revealed several limitations: (1) The strain targeted was chosen for its extreme sensitivity to *D*; most wild populations are less sensitive; (2) the eye color marker was chosen for its closely sex-linked location; most genes are not closely sex linked; (3) a control experiment (without *D*) revealed that the *re* phenotype had a small selective advantage over wild type under laboratory conditions (although insufficient to deny the effect of *D*). Meiotic drive as a possible transporting mechanism remains a matter for research. Experience from other insects suggests that autosomal systems await discovery and will be open to close control after molecular study.

Transposable Elements

These are segments of DNA that can move and insert at multiple positions in the genome. They are used routinely as genomic vectors to transform a germ line with a DNA fragment of interest. In natural populations of *Drosophila melanogaster*, the *P* element moves spontaneously when a population containing it comes into contact with one that does not. The result is that the *P* element enters the new population and takes it over. A similar phenomenon might be exploitable in insect vectors: to transport closely associated genome with such an element, if research reveals the potentiality, and such releases receive public acceptance. Potential elements with trans-specific activity have been identified, among which *Piggybac* and *Hermes* are considered to have wide potential.

Wolbachia

In theory, infection of the mosquito *C. quinquefasciatus* with *Wolbachia pipientis* is favored by natural selection. This is because a fertilization by an imprinted sperm of an uninfected egg is sterile, while the fertilization by a nonimprinted sperm of an infected egg is fertile. On this basis, the use of *Wolbachia* itself has been considered a transporting mechanism, to carry a gene for vector incompetence. This would depend on (1) introducing such an "incompetence" gene into *Wolbachia*, (2) associating the *Wolbachia* with *Anopheles* (a symbiosis never yet observed), and (3) ensuring that the "incompetence" gene product, expressed in

Wolbachia, reaches the gut or salivary glands, which are sites occupied by the malarial parasite. Attempts to introduce *Wolbachia pipientis* into *Anopheles* populations by injecting eggs have so far proved unsuccessful.

Further discussion on potential methods of genetic transformation of vector populations in ways favorable to humankind is a topic bound to be speculative, and no firm recommendation can be made at this time, although the potential for molecular advance is high.

CONCLUSIONS

1. Among all potential methods of genetic control, SIT has the greatest chance of being effective and economically viable (although there remains the problem of how to pay for it in the case of vectors of disease).

2. SIT is at its most effective in combination with other control measures, and it probably extends the effective life of insecticides by slowing down the evolution of resistance.

3. To extend the scope for SIT, further research is required on safe and effective autosterilization.

4. If RIDL-type systems of dietary-conditioned lethality can be developed, the potential will be created for (1) an expansion of SIT at reduced cost and (2) 100% effective sexing techniques.

5. Control by male production from multiple RIDL-type insertions seems theoretically feasible but unacceptable in the present state of knowledge because of the release of fertile, man-biting insects transformed with recombinant DNA.

6. Population replacement remains a distant dream, although research on it provides a rich source of new information on vector physiology and genomics.

Readings

Asman, S. M., McDonald, P. T., and Prout, T. 1981. Field studies for genetic control systems for mosquitoes. *Annu. Rev. Entomol.* 26: 289–318.

Cosgrove, J. B., Wood, R. J., Petric, D., Evans, D. T., and Abbott, R. H. R. 1994. A convenient mosquito membrane feeding system. *J. Amer. Control Assoc.* 10: 434–436.

Curtis, C. F. 1985. Genetic control of insect pests: Growth industry or lead balloon? *Biol. J. Linnaean Soc.* 26: 359–374.

Curtis, C. F., and Sinkins, S. P. 1998. *Wolbachia* as a possible means of driving genes into populations. *Parasitology* 116: 371–374.

Curtis, C. F., and Townson, H. 1998. Malaria: Existing methods of vector control and molecular entomology. *Br. Med. Bull.* 54: 311–325.

Curtis, C. F., Grover, K. K., Suguna, S. G., et al. 1976. Comparative field cage tests of the population suppressing efficiency of three genetic control systems for *Aedes aegypti. Heredity.* 36: 11–29.

Heinrich, J. C., and Scott, M. J. 2000. A repressible female-specific lethal genetic system for making transgenic insect strains suitable for a sterile release program. *Proc. Natl. Acad. Sci. U.S.A.* 97: 8229–8232.

IAEA. 2001. STEP — Tsetse fly eradication in Ethiopia. *Insect and Pest Control Newsletter.* Joint FAO/IAEA Division, International Atomic Energy Agency, Vienna. 57 (July): 12–13.

Kidwell, M. G., and Ribeiro, J. M. C. 1992. Can transposable elements be used to drive refractiveness genes into vector populations? *Parasitol. Today* 8: 325–329.

Langley, P. A., and Hall, M. J. R. 1986. Tsetse control by autosterilization. *Parasitol. Today* 2: 125–126.

Schliekelman, P., and Gould, F. 2000. Pest control by the release of insects carrying a female killing allele on multiple loci. *J. Econ. Entomol.* 93: 1566–1579.

Seawright, J. A. 1988. Genetic methods for control of mosquitoes and biting flies. In *Modern Insect Control: Nuclear Techniques and Biotechnology.* International Atomic Energy Agency, Vienna.

Seawright, J. A., Keiser, P. E., Dame, D. A., and Lofgren, C. S. 1978. Genetic method for preferential elimination of females of *Anopheles albimanus. Science* 300: 1303–1304.

Thomas, D. D., Donnelly, C. A., Wood, R. J., and Alphey, L. S. 2000. Insect population control using a dominant, repressible lethal genetic system. *Science* 287: 2474–2476.

Vargas-Terán, M., Hursey, B. S., and Cunningham, E. P. 1994. Eradication of the screwworm from Libya using the sterile insect technique. *Parasitol. Today* 10: 119–122.

Vreysen, M. J. B., Saleh, K. M., Ali, M. Y., et al. 2000. *Glossina austeni* (Diptera: Glossinidae) eradicated on the island of Unguja, Zanzibar, using the sterile insect technique. *J. Econ. Entomol.* 93: 123–135.

Wood, R. J., and Newton, M. E. 1991. Sex-ratio distortion caused by meiotic drive in mosquitoes. *Am. Naturalist* 137: 379–391.

Wood, R. J., Cook, L. M., Hamilton, A., and Whitelaw, A. 1977. Transporting the marker gene *re* (red eye) into a laboratory cage population of *Aedes aegypti* (Diptera: Culicidae) using meiotic drive at the *MD* locus. *J. Med. Entomol.* 14: 461–464.

Yasuno, M., MacDonald, W. W., Curtis, C. F., et al. 1978. A control experiment with chemosterilized male *Culex pipiens fatigans* Wied. in a village near Delhi surrounded by a breeding-free zone. *Jpn. J. Sanitary Zool.* 29: 325–343.

45

Immunological Control of Vectors

STEPHEN K. WIKEL, FRANCISCO ALARCON-CHAIDEZ, AND
UWE MÜLLER-DOBLIES

INTRODUCTION

Blood-feeding insects and ticks have great medical and veterinary public health importance. Transmission of infectious agents to humans, wildlife, and domestic animals causes enormous suffering and economic loss. Factors that contribute to emergence and resurgence of vector-borne diseases include: population movements, urbanization, and social change; deforestation, irrigation, and changing climate; limited public health infrastructure; reliance on short-term solutions to suppress vector-borne diseases; shortage of trained specialists; altered pathogen virulence; and increased numbers of vectors. The key element in control of arthropod-borne infectious agents is vector suppression. In light of current control problems, novel strategies are needed to achieve effective protection against vector-borne diseases.

Development of antiarthropod vaccines reduces blood-feeding and/or pathogen transmission. Their advantages include specificity, safety, cost, ease of administration, long-term protection, and absence of residues in the environment. A number of other factors contribute to heightened interest in immunologic-based control. One factor is the success of the anti–*Boophilus microplus* vaccine, which is based on induction of immunity to a tick gut antigen; another factor is the promising studies identifying candidate immunogens associated with other blood-feeding arthropods. The practicality of vector-blocking vaccines has been given a significant boost by studies that established that immunity to the bite of the vector provides protection against pathogen transmission. Advances in immunobiology, genomics, proteomics, DNA immunization, and other vaccine-related technologies provide a foundation for new ways to control vectors and vector-borne diseases.

Host Immunity to Ectoparasitic Arthropods

Infestation by tissue-dwelling and blood-feeding arthropods stimulates a spectrum of host immune responses that can reduce feeding, impair development, and kill the arthropod. A detailed discussion of immunologic-based acquired resistance to arthropod infestation is beyond the scope of this chapter. Readers interested in this topic are referred to the review articles cited later.

Acquired resistance to infestation is not a universal phenomenon; it depends upon the species of arthropod and host. Tick–host immune interactions have been the most extensively studied. The immunology of host interactions has been the subject of reviews for the following ectoparasitic arthropods: sucking lice, fleas, and bugs (C. J. Jones 1996); flies, mosquitoes, and myiasis (Sandeman 1996); scabies mites (Arlian 1996); and mange mites and chiggers (Wrenn 1996).

Development of acquired resistance by hosts to blood-feeding and tissue-dwelling arthropods was the foundation for experiments exploring the feasibility of developing antiarthropod vaccines. Little is known about the molecules that elicit acquired resistance and are the targets of protective responses.

Pharmacology of Arthropod Saliva and Host Defenses

Blood-feeding and tissue-dwelling arthropods must counteract a number of host defenses: hemostasis, pain/itch responses, and immunity. A mosquito engorges completely within 10 minutes, while a triatomine bug may require an hour to obtain a blood meal. On the other hand, ticks feed for days, and larvae of the myiasis flies, *Hypoderma bovis* and *Hypoderma lineatum*, migrate through bovine tissues for approximately 8 months. Pharmacologically active molecules secreted in saliva of hematophagous arthropods and by tissue-dwelling larvae are the key to successful parasitism (Ribeiro 1995; Moire et al. 1997).

Blood feeders and myiasis larvae alike have developed multiple ways to counteract host hemostatic defenses, consisting of coagulation pathways, platelet aggregation, and vasoconstriction (Ribeiro 1995; Champagne and Valenzuela 1996; Stark and James 1996). Multiple biologically active molecules, such as bradykinin and histamine, can mediate pain and itch responses. Bradykinin is an important mediator of both itch (Alexander 1986) and pain (Clark 1979). Saliva of the tick, *Ixodes scapularis*, contains a kininase that blocks the action of bradykinin (Ribeiro and Mather 1998). In addition, histamine-binding proteins are found in saliva of the triatomid, *Rhodnius prolixus*, and ticks.

Although blood-feeding and tissue-dwelling arthropods can induce host immune responses that damage them, their ultimate survival in an immunocompetent host requires adaptations to those defenses. Not surprisingly, ectoparasitic arthropods have evolved multiple ways to modulate or deviate host innate and specific acquired immune defenses (Schoeler and Wikel 2001). In addition, it is important to realize that virtually all infectious agents themselves have evolved ways of modulating host immunity (Brodsky 1999).

Undeniably, only a fraction of the pharmacologically active molecules occurring in arthropod saliva have been identified and characterized. Not only do these molecules represent potential targets of anti-arthropod and vector-blocking vaccines, but many likely have utility as novel therapeutic agents.

Arthropod Saliva and Pathogen Transmission

Salivary glands of several insects and ticks produce molecules that enhance pathogen transmission and establishment (Wikel and Alarcon-Chaidez 2001; Schoeler and Wikel 2001; Chapter 28). Those molecules are increasingly viewed as candidate immunogens for vector-blocking vaccines, targeting molecules essential for pathogen transmission rather than each individual pathogen transmitted by a blood-feeding arthropod.

Leishmania major infection is enhanced by the saliva of both Old World, *Phlebotomus papatasi*, and New World, *Lutzomyia longipalpis*, sand flies (Titus and Ribeiro 1988; Belkaid et al. 1998). Maxadilan is a major immunomodulator produced by *L. longipalpis* (Gillespie et al. 2000). Although lacking maxadilan, saliva of *P. papatasi* suppresses macrophage expression of an inducible nitric oxide synthase gene and production of nitric oxide (Waitumbi and Warburg 1998). Likewise, *P. papatasi* salivary gland lysate enhances lymph node cell interleukin (IL)-4 and reduces interferon (IFN)-γ, IL-12, and nitric oxide synthase (Mbow et al. 1998).

Introduction of Cache Valley virus into the feeding sites of *Ae. aegypti*, *Ae. triseriatus*, or *Cx. pipiens* enhances infection (Edwards et al. 1998). Likewise, saliva of *Ae. triseriatus* potentiates transmission of vesicular stomatitis virus (Limesand et al. 2000). Tick saliva facilitates transmission of Thogoto virus (Jones et al. 1989), tick-borne encephalitis virus (Labuda et al. 1993), vesicular stomatitis virus (Hajnicka et al. 2000), and the apicomplexan, *Theileria parva* (Shaw et al. 1993). The importance of tick-induced host immunomodulation in pathogen transmission was confirmed by the observation that reversing tick suppression of host cytokines provided protection against transmission of *Borrelia burgdorferi* by *I. scapularis* nymphs (Zeidner et al. 1996).

In addition to providing a target for immunologic control of vector-borne diseases, the presence of molecules in arthropod saliva that facilitate pathogen transmission and establishment highlights the importance of studying these diseases by using vector transmission rather than by needle inoculation of the infectious agents.

Immunity to the Vector Blocks Pathogen Transmission

Exposure of hosts to bites of pathogen-free, blood-feeding arthropods can induce responses that subsequently prevent or reduce transmission of disease-causing agents by the same species of infected arthropod. Interestingly, these vector-blocking responses are not necessarily associated with host-acquired resistance to arthropod feeding. Rabbits infested with pathogen-free *Dermacentor andersoni* were significantly more resistant to *D. andersoni* transmission of *Francisella tularensis* than naive rabbits receiving an infestation with infected ticks (Bell et al. 1979). Four repeated infestations of BALB/c mice with pathogen-free *I. scapularis* nymphs induced resistance to transmission of *B. burgdorferi* by a subsequent infestation with infected nymphs (Wikel et al. 1997). No signs of acquired resistance to *I. scapularis* were evident. Likewise, repeated infestations of guinea pigs with pathogen-free *I. scapularis* nymphs induced acquired resistance and prevented subsequent tick transmission of *B. burgdorferi* (Nazario et al. 1998).

Prior exposure of mice to bites of the uninfected sand fly, *P. papatasi*, resulted in protection from fly-transmitted *L. major* (Kamhawi et al. 2000). Protection against *L. major* was attributed to an intense delayed-type hypersensitivity response and the production of IFN-γ at the bite site, where parasites were deposited with fly saliva. Induction of the protective response has been linked to a 15-kdal salivary protein of *P. papatasi* (Valenzuela et al. 2001). This protein induced a protective, delayed-type hypersensitivity response in B-cell-deficient mice. In addition, protection was induced by administration of a DNA vaccine containing the cDNA encoding the 15-kdal protein.

ANTI-ARTHROPOD VACCINES

Anti-arthropod vaccine research was initiated when guinea pig resistance to *Dermacentor variabilis* larvae was successfully induced by immunization with an extract prepared from whole larvae of the same tick species (Trager 1939a). Numerous studies followed, using whole arthropod extracts, homogenates of specific tissues, and, more recently, individual molecules important to the physiological integrity of a parasitic arthropod. During the past two decades, a central focus has been on the use of "concealed" antigens, those molecules not normally introduced into the host (Willadsen and Kemp 1988). Anti-ectoparasite vaccines have been the topic of reviews (Wikel 1982; Opdebeeck 1994; Kay and Kemp 1994; Barriga 1994; Jacobs-Lorena and Lemos 1995; Wikel et al. 1996; Willadsen 1997; Lee and Opdebeeck 1999).

Anti-insect Vaccines

Initially, ticks were the predominant focus of anti-arthropod vaccine research, and an effective vaccine was developed to aid in the control of *B. microplus* (Tellam et al. 1992; Wikel et al. 1996). Initial efforts to establish the feasibility of anti-insect vaccines often produced results that were contradictory (Jacobs-Lorena and Lemos 1995). Crude antigen extracts were used in many experiments, and variations in responses to immunization would be expected (Wikel et al. 1996). A similar situation was observed during the early attempts to induce anti-tick immunity (Wikel 1982). Crude extracts likely contain hundreds of potentially antigenic molecules, which could induce highly variable responses. Fortunately, interest in vaccines to limit blood-feeding insects and tissue-dwelling larvae has

increased, and results are promising, particularly when more well-characterized molecules are used as immunogens (Moire et al. 1994; Sandeman 1996; Tellam and Eisemann 1998; Wijffels et al. 1999; Sukarsih et al. 2000).

Sucking Lice

Immunization with a sonicate of the mouse louse, *Polyplax serrata*, induced resistance to infestation (Ratzlaff and Wikel 1990). Louse burden weight was reduced by 62% on immunized mice. Likewise, an extract of *Polyplax spinulosa* was used to induce resistance to infestation of rats (Volf and Grubhoffer 1991). An immunogenic 31-kdal protein was detected in the midut epithelia, partially digested gut contents, and feces of *P. spinulosa* (Volf 1994).

Initial studies with rodent-sucking lice led to efforts to develop a vaccine to limit infestation with the human body louse, *Pediculus humanus humanus*. Rabbits immunized with louse midgut developed immune responses that resulted in reduced blood meal size, arrested development, reduced ova production, and increased mortality (Ben-Yakir et al. 1994). Many of the lice that fed on immunized rabbits became red, indicating that their digestive tracts were disrupted. Protection-inducing antigens were immunolocalized to the microvilli of louse midgut cells (Mumcuoglu et al. 1996b). Louse feces also contain midgut antigens (Mumcuoglu et al. 1996a). Using immunoaffinity chromatography, antibodies induced by midgut immunization were used to isolate proteins with molecular masses of 17, 29, and 35 kdal (Mumcuoglu et al. 1996a). Cross-reactive immunogens were identified among *P. humanus humanus*, the cattle louse, *Haematopinus africanus*, and the goat louse, *Linognathus stenopsis* (Mumcuoglu et al. 1996a). Immunoblot analysis of a midgut extract of *P. humanus humanus* revealed antibodies reactive with bands of 12, 17, 29, 35, 40, 55, 63, 97, and 117 kdal (Ochanda et al. 1996). Feces of *P. humanus humanus* were used to isolate antigen for immunization of rabbits (Muncuoglu et al. 1997). Engorgement weights were reduced 29% for females that fed on immunized rabbits. Likewise, consumption of blood from an immunized rabbit resulted in a significant reduction in the mean number of ova produced; however, immunization did not reduce louse survival. These studies established the feasibility of developing an effective immunization regimen to prevent or reduce infestation with the human body louse.

Fleas

Fleas are important due to their ability to transmit the causative agent of plague, *Yersinia pestis*, and their bites can induce and elicit intense cutaneous hypersensitivity reactions. Two beneficial outcomes could arise from vaccination with flea immunogens. The first would be prevention of flea feeding and pathogen transmission, and the second would be desensitization against allergens in flea saliva.

In what appears to be the first report of vaccination against an insect, humans administered an extract of whole fleas were protected, or desensitized, against subsequent flea bites (Cherney et al. 1939). Individuals were immunized with an extract of the dog flea, *Ctenocephalides canis*. An immunization regimen was initiated similar to that used in desensitization with repeated immunizations with increasing doses of antigen. Protection was defined as reduced reactivity to flea bites.

Immunization of cats with gut membranes of the cat flea, *C. felis felis*, neither reduced the number of fleas recovered from immunized cats nor altered their production of oocytes. Possibly, the antigen profile on the midgut epithelium of the feeding flea differs from that of the unfed flea. An interesting aspect of this study was the impact that the choice of adjuvant made on the response of cats to immunization. Administration of flea midgut membranes in combination with Quil A caused depression, edema, fever, and inflammation. No adverse reacts were evident when the same immunogens were administered with Ribi adjuvant. Further studies are needed to identify molecules essential to the physiological integrity of the flea that would be accessible to immune effector elements in the blood meal.

Myiasis

Myiasis is the invasion of tissues or open body cavities by dipteran larvae. Vaccine-based control of myiasis is appealing, because fly larvae can reside in direct contact with elements of the host immune system for months. Not unexpectedly, natural infections do not generally induce inflammatory responses (Chabaudie and Boulard 1992) or host immunity to reinfestation (Sandeman et al. 1992). The most extensively studied causes of myiasis are the warble flies, *Hypoderma bovis* and *H. lineatum*, and a sheep blowfly, *Lucilia cuprina* (Sandeman 1996).

The feasibility of vaccination against *H. bovis* and *H. lineatum* was established when vaccinated bovines were capable of killing fly larvae *in vivo* (Khan et al. 1960). Cell-mediated immune responses to the enzymes hypodermin A, B, and C of *H. lineatum* were stimulated in vaccinated cattle, resulting in development of fewer larvae (Baron and Colwell 1991). Hypodermin A has been used to induce high levels of antibodies that provided protection (Pruett et al. 1987). Sequence and gene expression have been determined for hypodermins A, B, and C in larvae of *H. lineatum* (Moire et al. 1994). In addition, the three-dimensional structure of hypodermin C has been described (Broutin et al. 1996).

An important consideration regarding vaccination against hypodermatosis is the ability of hypodermins to modulate host immunity. Hypodermins A and B degrade complement component C3 (Boulard 1989), and hypodermin A also degrades IgG (Pruett 1993). Hypodermin A impairs in vitro proliferation of bovine T-cells, which is likely due to its ability to reduce IL-2 production by T-lymphocytes (Nicolas-Gaulard et al. 1995).

Myiasis caused by *L. cuprina* can progress rapidly and result in the death of an infested sheep. Infestations are most common in young sheep, suggesting that acquired resistance might occur. Significant progress has been made in developing an anti–*L. cuprina* vaccine (Tellam and Bowles 1997). Peritrophic membrane molecules, peritrophins, induced sheep antibodies that inhibited growth of *L. cuprina* larvae both in vitro and *in vivo*, but larval killing was not significant *in vivo* (East et al. 1993). Preliminary biochemical characterizations of the peritrophins have been completed (Wijffels et al. 1999). The cDNA and amino acid sequences were determined for a peritrophin, designated peritrophin 48 (Schorderet et al. 1998). In an effort to identify protective immunogens, first-instar larvae of *L. cuprina* were fractionated by preparative isoelectric focusing, and a fraction with a pH range of 5.9–6.7 contained proteins that induced an ovine immune response that reduced *in vitro* growth of larvae by a mean of $84 \pm 7\%$ (Tellam and Eisemann 1998). This effect was shown to be due to ingestion of antibodies, and they were immunolocalized to the larval peritrophic membrane, cuticle, and, in a lesser amount, the microvilli and basement membranes of digestive tract epithelial cells (Tellam and Eisemann 1998).

Like other ectoparasitic arthropods, *L. cuprina* modulates host immune defenses, which could affect vaccination-induced responses. Excretory/secretory products of larvae inhibit antibody production (Kerlin and East 1992). In addition, secreted larval enzymes degrade complement component C3 (O'Meara et al. 1992) as well as IgG (Sandeman et al. 1995).

Based on strategies developed for vaccination and assessment of immunity to *L. cuprina*, vaccination was assessed as a means of controlling the Old World screwworm fly, *Chrysoma bezziana* (Sukarsih et al. 2000). Immune responses that reduce the growth of *C. bezziana* larvae can be induced by administration of larval extracts, peritrophic membrane, and the cardia of the midgut, the organ that produces the peritrophic matrix. These studies supported the idea that vaccination could be an effective tool for the control of this important pest.

Biting Flies

Rabbits immunized with tissues of the stable fly, *Stomoxys calcitrans*, developed responses that damaged flies subsequently feeding on them (Schlein and Lewis 1976). Significant fly mortality was associated with obtaining a blood meal from rabbits immunized with fly thoracic muscle. Furthermore, flies feeding on those rabbits developed leg and wing paralysis as well as difficulty in probing. Immunization of rabbits with whole-fly, abdomen, or gut extract resulted in antibody responses that reacted with eight antigens from gut and 12 each from whole-fly and abdomen extracts (Webster et al. 1992). Flies fed on rabbits immunized with gut antigen had higher mortality and the lowest percentage of viable eggs, at 15.5%.

The horn fly, *Haematobia irritans irritans*, and the buffalo fly, *H. irritans exigua*, are serious pests of cattle. Their blood-feeding habits make these flies likely candidates for vaccination-based control (Baron and Lysyk 1995; East et al. 1995; Wijffels et al. 1999). Cattle exposed to horn flies in nature slowly developed an antibody response to salivary gland antigens (Baron and Lysyk 1995). The antibody response was weak during two 21-day experimental infestations that were separated by a 21-day fly-free period. However, antibody titers increased after the flies were removed. The hypothesis was made that horn fly saliva might have an immunomodulatory effect.

Buffalo flies that fed on cattle with high antibody titers to fly immunogens did not have a higher mortality rate than flies fed on control animals (Kerlin and Allingham 1992). Sheep vaccinated with a trypsin-like enzyme from the buffalo fly developed antibodies that inhibited in vitro enzyme activity but did not affect fly development in vitro (East et al. 1995). Adult buffalo flies were not damaged when fed bovine blood *in vitro* containing antibodies to peritrophin derived from adult buffalo flies (Wijffels et al. 1999).

Mortality of tsetse flies, *Glossina morsitans*, was significant when they obtained a blood meal from rabbits immunized with cuticle and adhering hypodermal cells and wing buds of *S. calcitrans* (Schlein and Lewis 1976).

Mosquitoes

Antimosquito vaccines could be designed to disrupt blood feeding, impair reproduction, and block devel-opment and transmission of mosquito-borne infectious agents. An important resource for anti-mosquito vaccines is being generated from mosquito EST and genome sequencing projects.

The first attempt to vaccinate against mosquitoes involved immunization of rabbits with extracts of whole female *An. quadrimaculatus* (Dubin et al. 1948). Obtaining a blood meal from immunized rabbits did not alter engorgement. The design of this experiment was not optimal for determining the impact of anti-mosquito antibodies. Mosquitoes were allowed to feed for 5 minutes. They were killed immediately upon removal from the host and examined microscopically. Subsequent studies have revealed that a much longer time postfeeding is needed to assess the full effects of host immunity on mosquitoes.

Several investigators have immunized animals with whole-mosquito extracts, inducing variable levels of immunity to subsequent mosquito feeding. Rabbits were divided into three groups and immunized with one of the following homogenates of *An. stephensi*: low-speed supernatant of whole mosquitoes, low-speed pellet of whole mosquitoes, and midgut (Alger and Cabrera 1972). Mortality was greatest for those mosquitoes that fed on midgut-immunized rabbits. This appears to be the first publication demonstrating the utility of arthropod digestive tract as a source of immunogens. Rabbits and guinea pigs were immunized with a homogenate of whole, sugar-fed, *Ae. aegypti* and challenged with *Ae. aegypti* and *Cx. tarsalis* (Sutherland and Ewen 1974). The fecundity of mosquitoes was reduced 24% to 31%, but no effect on mortality was observed. A slight increase in mortality occurred when *Ae. aegypti* were fed on mice immunized with whole-mosquito or midgut preparations (Hatfield 1988).

Mosquitoes might differentially express molecules as a result of blood feeding. Rabbits were immunized with one of the following extracts prepared from female *Ae. aegypti* that had blood fed 24 hours prior to extract preparation: head/thorax, midgut, and the remainder of the abdomen (Ramasamy et al. 1988). No significant differences in mortality were detected for mosquitoes fed on immunized or control rabbits. Fecundity was reduced for mosquitoes that fed on only some rabbits in each treatment group. Rabbits were immunized with homogenates of head/thorax, midgut, or the remainder of the abdomen of *An. tessellatus* (Ramasamy et al. 1992). Fecundity was reduced for *An. tessellatus* but not for *Cx. quinquefasciatus*, which fed on *An. tessellatus* homogenate–immunized rabbits. Increased mortality was observed only for *Cx. quinquefasciatus*. Essentially the same study was performed with *An. tessellatus*–immunized mice (Srikrishnaraj et al. 1993). Fecundity of *An. tessellatus* that consumed antibodies from immunized mice was reduced to a maximum of 29%, and antimidgut antibodies were found to inhibit peritrophic membrane formation by *An. tessellatus* (Ramasamy et al. 1996). *Anopheles stephensi* fed on mice immunized with midguts of the same species had significantly reduced longevity (Almeida and Billingsley 1998).

Antibodies to mosquito midgut influence development of infectious agents within the mosquito. *Aedes aegypti* were significantly less susceptible to infection with Murray Valley encephalitis and Ross River viruses when consuming blood that contained both antibodies to mosquito midgut and virus (Ramasamy et al. 1990). Likewise, fewer oocysts of *Plasmodium berghei* developed in *An. farauti* fed on mice immunized with midgut antigens (Ramasamy and Ramasamy 1990). Similar results were obtained when *An. stephensi* ingested anti-midgut antibodies along with infectious *P. berghei* (Lal et al. 1994) and when *An.*

tessellatus ingested *Plasmodium vivax*–infected erythrocytes in the presence of anti-mosquito midgut antibodies (Srikrishnaraj et al. 1995). Anti-mosquito midgut antibodies have recently been shown to block development of both *P. falciparum* and *P. vivax* in multiple species of *Anopheles* mosquitoes and also to reduce vector fecundity and survival (Lal et al. 2001).

An overview of anti-insect vaccinesis is provided in Table 45.1. The influence of anti-mosquito immunity on pathogen development is summarized in Table 45.2.

Anti-tick Vaccines

Ticks have been the major focus of anti-arthropod vaccine research since the induction of resistance to infestation with *D. variabilis* larvae by administration of a whole-larva extract (Trager 1939a). During the decades since that report, many attempts have been made to induce anti-tick immunity with whole-tick extracts, salivary gland extracts, and most recently

TABLE 45.1   Overview of Immunization Against Ectoparasitic Insects

Arthropod	Immunogen	Comments	Reference
Myiasis			
Hypoderma lineatum	Hypodermins A, B, C	Vaccinated cattle produced fewer larvae	Baron and Colwell (1991)
Lucilia cuprina	Peritrophins	Inhibits larval growth	East et al. (1993)
Lucilia cuprina	Fractionated first-instar larvae	Induced antibodies that reduced in vitro growth	Tellam and Eisemann (1998)
Chrysoma bezziana	Larval extracts, peritrophic matrix, cardia	Reduced larval growth	Sukarsih et al. (2000)
Lice			
Polyplax serrata	Whole-louse sonicate	Reduced louse burden	Ratzlaff and Wikel (1990)
Polyplax spinulosa	Whole-louse extract	Resistance to infestation	Volf and Grubhoffer (1991)
Polyplax spinulosa	Midgut, gut contents, feces	31-kdal immunogen	Volf (1994)
Pediculus humanus humanus	Louse midgut	Reduced blood meal, fewer ova, arrested development, and death	Ben-Yakir et al. (1994)
Pediculus humanus humanus	Protective immunogen on midgut microvilli		Muncuoglu et al. (1996b)
Biting flies			
Stomoxys calcitrans	Fly thoracic muscle	Increased mortality	Schlein and Lewis (1976)
Stomoxys calcitrans	Gut	Increased fly mortality, reduced viability of ova	Webster et al. (1992)
Glossina morsitans	*S. calcitrans* cuticle, hypodermal cells, and wing buds	Increased mortality	Schlein and Lewis (1976)
Mosquitoes			
Anopheles quadrimaculatus	Whole-female extract	No effect	Dubin et al. (1948)
Anopheles stephensi	Midgut	Increased mortality	Alger and Cabrera (1972)
Aedes aegypti and *Culex tarsalis*	Whole–*Aedes aegypti* homogenate	Slight reduction in fecundity	Sutherland and Ewen (1974)
Aedes aegypti	Whole mosquito or midgut	Slight increase in mortality	Hatfield (1988)
Anopheles tessellatus	Midgut	Inhibit peritrophic membrane development	Ramasamy et al. (1996)
Anopheles stephensi	Midgut	Reduced longevity	Almeida and Billingsley (1998)

TABLE 45.2   Immune Response to Mosquito Midgut Alters Development of Infectious Agents

Mosquito	Comments	Reference
Aedes aegypti	Less susceptible to Murray Valley encephalitis and Ross River viruses	Ramasamy et al. (1990)
Anopheles farauti	Fewer *Plasmodium berghei* oocysts	Ramasamy and Ramasamy (1990)
Anopheles stephensi	Fewer *Plasmodium berghei* oocysts	Lal et al. (1994)
Anopheles tessellatus	Fewer *Plasmodium vivax* oocysts	Srikrishnaraj et al. (1995)
Multiple *Anopheles* spp.	Blocked development of *Plasmodium falciparum* and *Plasmodium vivax*	Lal et al. (2001)

with defined immunogens. The commercial development of the anti–*B. microplus* vaccines TickGARDTM Plus in Australia and GAVACTM Plus in Latin America is proof of the feasibility and utility of antitick vaccines (Willadsen 1997; Willadsen and Jongejan 1999). This discussion of antitick vaccines focuses primarily on recent advances involving the use of "concealed" immunogens. Additional reviews address that topic as well as other aspects of antitick vaccine research (Tellam et al. 1992; Wikel et al. 1996; Willadsen 1997; Willadsen and Jongejan 1999; Lee and Opdebeeck 1999).

Whole-body extracts were administered, with variable success for inducing immunity to *Ixodes holocyclus* (Bagnall 1975), *Amblyomma americanum* (McGowan et al. 1981), *Rhipicephalus appendiculatus* (Mongi et al. 1986), and *A. hebraeum* and *A. marmoreum* (Tembo and Rechav 1992). Likewise, immunity induced to challenge infestations was variable for animals immunized with salivary gland extracts of *B. microplus* (Brossard 1976), *Dermacentor andersoni* (Wikel 1981), and *Hyalomma anatolicum anatolicum* (Banerjee et al. 1990). Whole-tick and whole-tissue extracts contain hundreds to thousands of different molecules at different concentrations. Immunization with those preparations likely induces antibodies to predominant as well as immunodominant components of extracts. Those antibodies or cell-mediated immune responses might be of little or no value in providing anti-tick immunity; however, minor components may be useful protection-inducing immunogens. The key to a successful vaccine is to fractionate and characterize molecules in tissues that would be suitable immune targets. That approach is exactly what resulted in the development of an effective anti–*B. microplus* vaccine.

Tick digestive tract cells would be in contact with host antibodies and effector cells in a blood meal obtained from an immunized animal. Integrity of the midgut is essential for successful feeding, molting, and reproduction. In addition, damage to the midgut should disrupt development and transmission of tick-borne infectious agents. Tick midgut was first used as an immunogen by Trager (1939b). Female *D. andersoni*

produced fewer ova and no viable larvae after feeding on guinea pigs immunized with female *D. andersoni* midgut combined with reproductive tract (Allen and Humphreys 1979). Resistance to *A. americanum* was induced by immunization with brush border of *A. americanum* digestive tract (Wikel 1988).

Ova production was reduced by 91% for *B. microplus* females that fed on Hereford cattle vaccinated with a $100,000 \times g$ pellet of *B. microplus* midgut (Opdebeeck et al. 1988). Protective immunogens were isolated from *B. microplus* gut cell membranes by nonionic detergents (Wong and Opdebeeck 1989). Furthermore, immunity was correlated directly with anti–*B. microplus* antibodies (Jackson and Opdebeeck 1990). A panel of monoclonal antibodies was generated against *B. microplus* gut immunogens, and the monoclonal QU13 immunoprecipitated a highly protective molecule (Lee and Opdebeeck 1991). The protective antigen reactive with monoclonal antibody QU13 was found in 10 different Australian isolates of *B. microplus*; however, the isolates did differ as much as fourfold in endpoint titers in their reactivity with the monoclonal antibody (Knowles and Opdebeeck 1996).

A series of well-crafted studies resulted in the development of the commercial vaccine for the control of *B. microplus* (Tellam et al. 1992; Willadsen 1997; Willadsen and Jongejan 1999). The initial studies showed that immunization of cattle with an extract of whole female *B. microplus* induced partial immunity (Johnston et al. 1986), and ticks collected from those cattle had damaged digestive tracts (Agbede and Kemp 1986). Starting with an estimated 50,000 partially fed ticks, approximately 100 micrograms of a minor midgut membrane component was isolated and used to induce immunity to *B. microplus* (Willadsen et al. 1988). The protection-inducing immunogen was purified by a combination of detergent solubilization of a membrane extract, lectin affinity chromatography, preparative isoelectric focusing, and HPLC gel filtration (Willadsen et al. 1989). This immunogen, designated as Bm86, is a glycoprotein with a molecular weight of 89 kdal and an isoelectric point of 5.1–5.6. Immunization of cattle with microgram quantities of Bm86 induced responses that reduced engorgement weights, egg laying, and survival of feeding ticks. Antibodies to Bm86 bind to the surface of tick digestive cells and inhibit their ability to endocytose the blood meal (Willadsen et al. 1989). Since this molecule would not be introduced into the host during feeding, it is referred to as a "concealed" antigen (Willadsen and Kemp 1988). An additional antigen, Bm91, is an 86-kdal glycoprotein with an isoelectric point of 4.8–5.2, which is present in both digestive tract and salivary gland (Riding et al. 1994). Bm91 is a carboxydipeptidase with homology to human angiotensin-converting enzyme (ACE) (Jarmey et al. 1995).

The processes involved in commercialization of the Bm86 vaccine have been reviewed (Tellam et al. 1992; Willadsen 1997; Lee and Opdebeeck 1999). Enhanced antitick immunity was achieved by combined administration of Bm86 and a mucin-like membrane glycoprotein derived from partially fed female *B. microplus* (McKenna et al. 1998). Additionally, immunization with Bm86 provided cross-protection against infestation with *Boophilus decoloratus*, *H. anatolicum anatolicm*, and *Hyalomma dromedarii*, but vaccination with Bm86 had no effect on infestation with *Amblyomma variega-tum* or *R. appendiculatus* (de Vos et al. 2001). The development of the Bm86 vaccine in Latin America has been reported, along with a potentially more broadly protective immunogen, Bm95 (de la Fuente et al. 2000). An overview of anti-tick vaccination research is provided in Table 45.3.

FUTURE CONSIDERATIONS

"Concealed" immunogens have proved to be an effective strategy for immunization against ticks. Likewise, other defined immunogens show promise for inducing protection against ectoparasitic insects. The initial concerns about the feasibility of anti-arthropod vaccines were due primarily to the variability of responses that were induced by whole-arthropod or whole-tissue extracts. The use of defined immunogens has been much more effective, and consideration is being given to unique adjuvant formulations to enhance responses.

A combination of genomic and proteomic strategies will likely provide further insights into the array of possible vaccine candidate immunogens; however,

TABLE 45.3   Overview of Immunization Against Ticks

Tick	Immunogen	Comments	References
Ixodes holocyclus	Extract of whole larvae	Larval reject of 38% to 68%	Bagnall (1975)
Amblyomma americanum	Whole-adult homogenate	Variable levels of reduced engorgement	McGowan et al. (1981)
Rhipicephalus appendiculatus	Antibodies complexed to whole-female extract	Delayed attachment, reduced feeding, and delayed drop-off	Mongi et al. (1986)
Amblyomma hebraeum *Amblyomma marmoreum*	Nymphal homogenate	Resistance to infection	Tembo and Rechav (1992)
Boophilus microplus	Salivary gland extract	Resistance to infestation	Brossard (1976)
Dermacentor andersoni	Salivary gland extract	Resistance to larvae	Wikel (1981)
Hyalomma anatolicum	Different supernatants and pellets of salivary gland	Variable results	Banerjee et al. (1990)
Dermacentor variabilis	Midgut	Variable results	Trager (1939b)
Dermacentor andersoni	Midgut and reproductive tract	Fewer ova, and none hatched	Allen and Humphreys (1979)
Amblyomma americanum	Digestive tract brush border	Reduced engorgement and increased mortality	Wikel (1988)
Boophilus microplus	100,000× g pellet of midgut	Ova production reduced 91%	Opdebeeck et al. (1988)
Boophilus microplus	Midgut membrane molecule	Highly protective	Lee and Opdebeeck (1991)
Boophilus microplus	Midgut membrane, Bm86	Highly protective Bm86 vaccine commercialization	Willadsen et al. (1989) Tellam et al. (1992) Lee and Opdebeeck (1999)
Boophilus microplus	Bm91 of salivary gland and midgut	Additional protective immunogen	Riding et al. (1994)
Boophilus microplus	Bm86 plus mucin-like membrane glycoprotein	Enhanced antitick immunity	McKenna et al. (1998)

these powerful tools should be combined with traditional biochemical purification schemes for identification of molecules of interest. Central to the success of these approaches are high-throughput screening strategies to identify those activities of interest. One particularly promising approach is the analysis of expressed sequence tags (ESTs) prepared from mRNA derived from the salivary glands of blood-feeding arthropods. An EST is a partial nucleic acid sequence derived from a cDNA generated from the mRNA within a tissue or cell (Adams et al. 1991). Combining genomics with the tools of proteomics, two-dimensional gel electrophoresis and tandem mass spectrometry, provides a powerful way to study gene expression at the protein level (Naaby-Hanseb et al. 2001). Again, a key element in all these analyses is development of high-throughput screening assays for the activities of interest.

The use of ESTs for arthropod tissues will result in the identification of many genes for which there are currently no homologues in the EST database, dbEST, which is the fastest-growing division of GenBank (Pandey and Lewitter 1999). Screening strategies need to be developed to assess the vaccine utility of those genes that currently encode proteins of unknown function. Periodic rescreening of those proteins in the databases will be essential as new information becomes available. One alternative approach is expression library immunization (ELI), which was first reported in 1995 (Barry et al. 1995). ELI is DNA vaccination with genomic libraries constructed in a eukaryotic expression vector. When using ELI, a knowledge of candidate immunogens is not required. Theoretically, every gene in an expression library can be screened for its ability to induce a protective immune response. Furthermore, DNA immunization induces both antibody and cell-mediated immune responses (Ellis et al. 2001). Adjuvants are being formulated to elicit specific types of immune responses with protein-, peptide-, virus-, and DNA-based vaccines (Moingeon et al. 2001; Scheerlinck 2001).

The likelihood of developing new anti-arthropod vaccines is significantly enhanced by the rapid advances in immunobiology, vector biology, genomics, proteomics, and vaccine technology. Antiarthropod and vector-blocking vaccines will become primary tools in efforts to control ectoparasitic arthropods and vector-borne infectious agents.

Acknowledgments

This work was supported in part by U.S. Public Health Service Grant AI46676 from the National Institute of Allergy and Infectious Diseases, National Institutes of Health, and Cooperative Agreement number U50/CCU614673 and U50/CCU119575 from the Centers for Disease Control and Prevention to S.K.W.

Readings

Adams, M. D., Kelley, J. M., Gocayne, J. D., Dubnick, M., Polymeropoulos, M. H., Xiao, H., Merril, C. R., Wu, A., Olde, B., Morens, D. M., Kerlavage, A. R., McConnell, D. J., and Venter, J. C. 1991. Complementary DNA sequencing: Expressed sequence tags and human genome project. *Science* 252: 1651–1656.

Agbede, R. I. S., and Kemp, D. H. 1986. Immunization of cattle against *Boophilus microplus* using extracts derived from adult female ticks: Histopathology of tick feeding on vaccinated cattle. *Int. J. Parasitol.* 16: 35–41.

Alexander, J. O'D. 1986. The physiology of itch. *Parasitol. Today* 2: 345–351.

Alger, N. E., and Cabrera, E. J. 1972. An increase in death rate of *Anopheles stephensi* fed on rabbits immunized with mosquito antigen. *J. Econ. Entomol.* 65: 165–168.

Allen, J. R., and Humphreys, S. J. 1979. Immunization of guinea pigs and cattle against ticks. *Nature* 280: 491–493.

Almeida, A. P., and Billingsley, P. F. 1998. Induced immunity against the mosquito *Anopheles stephensi* Liston (Diptera: Culicidae): Effects on mosquito survival and fecundity. *Int. J. Parasitol.* 28: 721–1731.

Arlian, L. G. Immunology of scabies. 1996. In: *The Immunology of Host–Ectoparasitic Arthropod Relationships* (S. K. Wikel, ed.), CAB International, Wallingford, England, pp. 232–258.

Bagnall, B. G. 1975. Cutaneous immunity to the tick *Ixodes holocyclus*. Unpublished Ph. D. thesis, University of Sydney.

Banerjee, D. P., Momin, R. R., and Samantaray, S. 1990. Immunization of cattle (*Bos indicus × Bos taurus*) against *Hyalomma anatolicum anatolicum* using antigens derived from tick salivary gland extracts. *Int. J. Parasitol.* 20: 969–972.

Baron, R. W., and Colwell, D. D. 1991. Enhanced resistance to cattle grub infestation (*Hypoderma lineatum* de Vill) in calves immunized with purified hypodermin A, B and C plus monophosphoryl lipid A (MPL). *Vet. Parasitol.* 38: 185–198.

Baron, R. W., and Lysyk, T. J. 1995. Antibody responses in cattle infested with *Haematobia irritans irritans* (Diptera: Muscidae). *J. Med. Entomol.* 32: 630–635.

Barriga, O. O. 1994. A review on vaccination against protozoa and arthropods of veterinary importance. *Vet. Parasitol.* 55: 29–55.

Barry, M. A., Lai, W. C., and Johnston, S. A. 1995. Protection against mycoplasma infection using expression-library immunization. *Nature* 377: 632–635.

Bell, J. F., Stewart, S. J., and Wikel, S. K. 1979. Resistance to tick-borne *Francisella tularensis* by tick-sensitized rabbits: Allergic klendusity. *Am. J. Trop. Med. Hyg.* 28: 876–880.

Belkaid, Y., Kamhawi, S., Modi, G., Valenzuela, J., Noben-Trauth, N., Rowton, E., Ribeiro, J., and Sacks, D. L. 1998. Development of a natural model of cutaneous leishmaniasis: Powerful effects of vector saliva and saliva preexposure on the long-term outcome of *Leishmania major* infection in the mouse ear dermis. *J. Exp. Med.* 188: 1941–1953.

Ben-Yakir, D., Mumcuoglu, K. Y., Manor, O., Ochanda, J., and Galun, R. 1994. Immunization of rabbits with a midgut extract of the human body louse *Pediculus humanus humanus*: The effect of induced resistance on the louse population. *Med. Vet. Entomol.* 8: 114–118.

Boulard, C. 1989. Degradation of bovine C3 by serine proteases from parasites *Hypoderma lineatum* (Diptera, Oestridae). *Vet. Immunol. Immunopathol.* 20: 387–398.

Brodsky, F. M. 1999. Stealth, sabotage and exploitation. *Immunol. Rev.* 168: 5–11.

Brossard, M. 1976. Relations immunologiques entre Bovins et Tiques, plus particulierement entre Bovins et *Boophilus microplus*. *Acta Tropica* 33: 15–36.

Brossard, M., and Wikel, S. K. 1997. Immunology of interactions between ticks and hosts. *Med. Vet. Entomol.* 11: 270–276.

Broutin, I., Arnoux, B., Riche, C., Lecroisey, A., Keil, B., Pascard, C., and Ducruix, A. 1996. A structure of *Hypoderma lineatum* collagenase: A member of the serine proteinase family. *Acta Cristalographica* D52: 380–392.

Chabaudie, N., and Boulard, C. 1992. Effect of hypodermin A, an enzyme secreted by *Hypoderma lineatum* (Insect: Oestridae), on the bovine immune system. *Vet. Immunol. Immunopathol.* 31: 167–177.

Champagne, D. E., and Valenzuela, J. G. 1966. Pharmacology of haemathophagous arthropod saliva. In: *The Immunology of Host–Ectoparasitic Arthropod Relationships* (S. K. Wikel, ed.), CAB International, Wallingford, England, pp. 85–106.

Cherney, L. S., Wheeler, C. M., and Reed, A. C. 1939. Flea-antigen in prevention of flea bites. *Am. J. Trop. Med.* 19: 327–332.

Clark, W. G. 1979. Kinins and the peripheral and central nervous systems. *Handb. Exp. Pharmacol.* 25: 311–356.

de la Fuente, J., Rodriguez, M., and Garcia-Garcia, J. C. 2000. Immunological control of ticks through vaccination with *Boophilus microplus* gut antigens. *Ann. N.Y. Acad. Sci.* 916: 617–621.

Dubin, I. N., Reese, J. D., and Seamans, L. A. 1948. Attempt to produce protection against mosquitoes by active immunization. *J. Immunol.* 58: 293–297.

East, I. J., Fitzgerald, C. J., Pearson, R. D., Donaldson, R. A., Vuocolo, T., Cadogan, L. C., Tellam, R. C., and Eisemann, C. H. 1993. *Lucila cuprina*: Inhibition of larval growth induced by immunization of host sheep with extracts of larval peritrophic membrane. *Int. J. Parasitol.* 23: 221–229.

East, I. J., Allingham, P. G., Bunch, R. J., and Matheson, J. 1995. Isolation and characterization of a trypsin-like enzyme from the buffalo fly, *Haematobia irritans exigua*. *Med. Vet. Entomol.* 9: 120–126.

Edwards, J. F., Higgs, S., and Beaty, B. J. 1998. Mosquito feeding–induced enhancement of Cache Valley virus (Bunyaviridae) infection in mice. *J. Med. Entomol.* 35: 261–265.

Ellis, R. 2001. Technologies for design, discovery, formulation and administration of vaccines. *Vaccine* 19: 2681–2687.

Gillespie, R. D., Mbow, M. L., and Titus, R. G. 2000. The immunomodulatory factors of blood-feeding arthropod saliva. *Parasite Immunol.* 22: 319–331.

Gratz, N. G. 1999. Emerging and resurging vector-borne diseases. *Annu. Rev. Entomol.* 44: 51–75.

Gubler, D. J. 1998. Resurgent vector-borne diseases as a global health problem. *Emerg. Infect. Dis.* 4: 442–450.

Hajnicka, V., Kocakova, P., Slovak, M., Labuda, M., Fuchsberger, N., and Nuttall, P. A. 2000. Inhibition of the antiviral action of interferon by tick salivary gland extract. *Parasite Immunol.* 22: 201–206.

Hatfield, P. R. 1988. Antimosquito antibodies and their effects on feeding, fecundity and mortality of *Aedes aegypti*. *Med. Vet. Entomol.* 2: 331–338.

Jackson, L. A., and Opdebeeck, J. P. 1990. Humoral immune responses of Hereford cattle vaccinated with midgut antigens of the cattle tick, *Boophilus microplus*. *Parasite Immunol.* 12: 141–151.

Jacobs-Lorena, M., and Lemos, F. J. A. 1995. Immunological strategies for control of insect disease vectors: A critical assessment. *Parasitol. Today* 11: 144–147.

Jarmey, J., Riding, G. A., Pearson, R. D., McKenna, R. V., and Willadsen, P. 1995. Carboxydipeptidase from *Boophilus microplus*: A "concealed" antigen with similarity to angiotensin-converting enzyme. *Insect Biochem. Mol. Biol.* 25: 969–974.

Johnston, L. A. Y., Kemp, D. H., and Pearson, R. D. 1986. Immunization of cattle against *Boophilus microplus* using extracts derived from adult female ticks: Effects of induced immunity on tick populations. *Int. J. Parasitol.* 16: 27–34.

Jones, C. J. 1996. Immune responses to fleas, bugs, and sucking lice. In: *The Immunology of Host–Ectoparasitic Arthropod Relationships* (S. K. Wikel, ed.), CAB International, Wallingford, England, pp. 150–174.

Jones, L. D., Hodgson, E., and Nuttall, P. A. 1989. Enhancement of virus transmission by tick salivary glands. *J. Gen. Virol.* 70: 1895–1898.

Kamhawi, S., Belkaid, Y., Modi, G., Rowton, E., and Sacks, D. 2000. Protection against cutaneous leishmaniasis resulting from bites of uninfected sand flies. *Science* 290: 1351–1354.

Kay, B. H., and Kemp, D. H. 1994. Vaccines against arthropods. *Am. J. Trop. Med. Hyg.* 50: 87–96.

Kerlin, R. L., and Allingham, P. G. 1992. Acquired immune response of cattle exposed to buffalo fly (*Haematobia irritans exigua*). *Vet. Parasitol.* 43: 115–129.

Kerlin, R. L., and East, I. J. 1992. Potent immunosuppression by secretory/excretory products of larvae from the sheep blowfly, *Lucillia cuprina*. *Parasite Immunol.* 14: 595–604.

Khan, M. A., Connell, R., and Darcel, C. leQ. 1960. Immunization and parenteral chemotherapy for the control of cattle grubs *Hypoderma lineatum* (DeVill) and *H. bovis* (L.) in cattle. *Can. J. Comp. Med.* 24: 177–180.

Knowles, A. G., and Opdebeeck, J. P. 1996. Uniformity of protective antigens among isolates of the cattle tick, *Boophilus microplus*. *Med. Vet. Entomol.* 10: 301–304.

Labuda, M., Jones, L. D., Williams, T., and Nuttall, P. A. 1993. Enhancement of tick-borne encephalitis virus transmission by tick salivary gland extracts. *Med. Vet. Entomol.* 7: 193–196.

Lal, A. A., Schriefer, M. E., Sacci, J. B., Goldman, I. F., Louis-Wileman, V., Collins, W. E., and Azad, A. F. 1994. Inhibition of malaria parasite development in mosquitoes by antimosquito midgut antibodies. *Infect. Immun.* 62: 36–318.

Lal, A. A., Patterson, P. S., Sacci, J. B., Vaughan, J. A., Paul, C., Collins, W. E., Wirtz, R. A., and Azad, A. F. 2001. Antimosquito midgut antibodies block development of *Plasmodium falciparum* and *Plasmodium vivax* in multiple species of *Anopheles* mosquitoes and reduce vector fecundity and survivorship. *Proc. Natl. Acad. Sci. U.S.A.* 98: 5228–5233.

Lee, R. P., and Opdebeeck, J. P. 1991. Isolation of protective antigens from the gut of *Boophilus microplus* using monoclonal antibodies. *Immunology* 72: 121–126.

Lee, R. P., and Opdebeeck, J. P. 1999. Arthropod vaccines. *Infect. Dis. Clinics North Amer.* 13: 209–226.

Limesand, K. H., Higgs, S., Pearson, L. D., and Beaty, B. J. 2000. Potentiation of vesicular stomatitis New Jersey virus infection in mice by mosquito saliva. *Parasite Immunol.* 22: 461–467.

Mbow, M. L., Bleyenberg, J. A., Hall, L. R., and Titus, R. G. 1998. *Phlebotomus papatasi* sand fly salivary gland lysate down-regulates a Th1, but up-regulates a Th2, response in mice infected with *Leishmania major*. *J. Immunol.* 161: 5571–5577.

McGowan, M. J., Barker, R. W., Homer, J. T., McNew, R. W., and Holscher, K. H. 1981. Success of tick feeding on calves immunized with *Amblyomma americanum* (Acari: Ixodidae) extract. *J. Med. Entomol.* 18: 328–332.

McKenna, R. V., Riding, G. A., Jarmey, J. M., Pearson, R. D., and Willadsen, P. 1998. Vaccination of cattle against *Boophilus microplus* using a mucin-like membrane glycoprotein. *Parasite Immunol.* 20: 325–336.

Moingeon, P., Haensler, J., and Lindberg, A. 2001. Towards the rational design of Th1 adjuvants. *Vaccine* 19: 4363–4372.

Moire, N., Nicolas-Gaulard, I., LeVern, Y., and Boulard, C. 1997. Enzymatic effects of hypodermin A, a parasite protease, on bovine lymphocyte membrane antigens. *Parasite Immunol.* 19: 21–27.

Mongi, A. O., Shapiro, S. Z., Doyle, J. J., and Cunningham, M. P. 1986. Immunization of rabbits with *Rhipicephalus appendiculatus* antigen–antibody complexes. *Insect Sci. Appl.* 7: 471–477.

Mumcuoglu, K. Y., Ben-Yakir, D., Gunzberg, S., Ochanda, J. O., and Galun, R. 1996a. Immunogenic proteins in the body and faecal material of the human body louse, *Pediculus humanus*, and their homology to antigens of other lice species. *Med. Vet. Entomol.* 10: 105–107.

Mumcuoglu, K. Y., Rahamim, E., Ben-Yakir, D., Ochanda, J. O., and Galun, R. 1996b. Localization of immunogenic antigens on midgut of the human body louse *Pediculus humanus humanus* (Anoplura: Pediculidae). *J. Med. Entomol.* 33: 74–77.

Mumcuoglu, K. Y., Ben-Yakir, D., Ochanda, J. O., Miller, J., and Galun, R. 1997. Immunization of rabbits with faecal extract of *Pediculus humanus*, the human body louse: Effects on louse development and reproduction. *Med. Vet. Entomol.* 11: 15–318.

Naaby-Hansen, S., Waterfield, M. D., and Cramer, R. 2001. Proteomics — postgenomic cartography to understand gene function. *Trends Pharmacol. Sci.* 22: 376–384.

Nicolas-Gaulard, I., Moire, N., and Boulard, C. 1995. Effect of the parasite enzyme, hypodermin A, on bovine lymphocyte proliferation and interleukin-2 production via the prostaglandin pathway. *Immunology* 84: 160–165.

Ochanda, J. O., Mumcuoglu, K. Y., Ben-Yakir, D., Okuru, J. K., Oduol, V. O., and Galun, R. 1996. Characterization of body louse midgut proteins recognized by resistant hosts. *Med. Vet. Entomol.* 10: 35–38.

O'Meara, T. J., Nesa, M., Raadsma, H. W., Saville, D. G., and Sandeman, R. M. 1992. Variation in skin inflammatory responses between sheep bred for resistance of susceptibility to fleece rot and blowfly strike. *Res. Vet. Sci.* 52: 205–210.

Opdebeeck, J. P. 1994. Vaccines against blood-sucking arthropods. *Vet. Parasitol.* 54: 205–222.

Opdebeeck, J. P., Wong, J. Y. M., Jackson, L. A., and Dobson, C. 1988. Hereford cattle immunized and protected against *Boophilus microplus* with soluble and membrane-associated antigens from the midgut of ticks. *Parasite Immunol.* 10: 405–410.

Pandey, A., and Lewitter, F. 1999. Nucleotide sequence databases: A gold mine for biologists. *Trends Biochem. Sci.* 24: 276–280.

Pruett, J. H. 1993. Proteolytic cleavage of bovine IgG by hypodermin A, a serine protease of *Hypoderma lineatum* (Diptera: Oestridae). *J. Parasitol.* 79: 829–833.

Pruett, J. H., Barrett, C. C., and Fisher, W. F. 1987. Kinetic development of serum antibody to purified *Hypoderma lineatum* proteins in vaccinated and nonvaccinated cattle. *Southwest Entomol.* 12: 9–88.

Ramasamy, M. S., and Ramasamy, R. 1990. Effect of antimosquito antibodies on the infectivity of the rodent malaria parasite *Plasmodium berghei* to *Anapholes farauti*. *Med. Vet. Entomol.* 4: 161–166.

Ramasamy, M. S., Ramasamy, R., Kay, B. H., and Kidson, C. 1988. Antimosquito antibodies decrease the reproductive capacity of *Aedes aegypti*. *Med. Vet. Entomol.* 2: 87–93.

Ramasamy, M. S., Sands, M., Kay, B. H., Fanning, I. D., Lawrence, G. W., and Ramasamy, R. 1990. Anti-mosquito antibodies reduce the susceptibility of *Aedes aegypti* to arbovirus infection. *Med. Vet. Entomol.* 4: 49–55.

Ramasamy, M. S., Srikrishnaraj, K. A., Wijekoone, S., Jesuthasan, L. S. B., and Ramasamy, R. 1992. Host immunity to mosquitoes: Effect of antimosquito antibodies. *Med. Entomol.* 29: 34–938.

Ramasamy, M. S., Raschid, L., Srikrishnaraj, K. A., and Ramasamy, R. 1996. Antimidgut antibodies inhibit peritrophic membrane formation in the posterior midgut of *Anopheles tessellatus* (Diptera: Culicidae). *J. Med. Entomol.* 33: 162–164.

Ratzlaff, R. E., and Wikel, S. K. 1990. Murine immune responses and immunization against *Polyplax serrata* (Anoplura: Polyplacidae). *J. Med. Entomol.* 27: 1002–1007.

Ribeiro, J. M. C. 1987. Role of saliva in blood-feeding by arthropods. *Annu. Rev. Entomol.* 32: 463–478.

Ribeiro, J. M. C. 1989. Role of saliva in tick/host interactions. *Exp. Appl. Acarol.* 7: 15–20.

Ribeiro, J. M. C., and Mather, T. N. 1998. *Ixodes scapularis*: Salivary kininase activity is a metallo dipeptidyl carboxypeptidase. *Exp. Parasitol.* 89: 213–221.

Riding, G. A., Jarmey, J., McKenna, R. V., Pearson, R., Cobon, G. S., and Willadsen, P. 1994. A protective "concealed" antigen from *Boophilus microplus*. Purification, localization and possible function. *J. Immunol.* 153: 158–166.

Sandeman, R. M. 1996. Immune responses to mosquitoes and flies. In: *The Immunology of Host–Ectoparasitic Arthropod Relationships* (S. K. Wikel, ed.), CAB International, Wallingford, England, pp. 5–203.

Sandeman, R. M., Bowles, V. M., Stacy, I. N., and Carnegie, P. R. 1986. Acquired resistance in sheep to infection with larvae of the blowfly, *Lucilia cuprina*. *Int. J. Parasitol.* 16: 69–75.

Sandeman, R. M., Chandler, R. A., and Seaton, D. S. 1995. Antibody degradation in blowfly strike. *Int. J. Parasitol.* 25: 621–628.

Scheerlinck, J-P. Y. 2001. Genetic adjuvants for DNA vaccines. *Vaccine* 1: 2647–2656.

Schlein, Y., and Lewis, C. T. 1976 Lesions in haematophagous flies after feeding on rabbits immunized with fly tissues. *Physiol. Entomol.* 1: 55–59.

Schoeler, G. B., and Wikel, S. K. 2001. Modulation of host immunity by haematophagous arthropods. *Ann. Trop. Med. Parasitol.* (In press).

Schorderet, S., Pearson, R. D., Vuocolo, T., Eisemann, C., Riding, G. A., and Tellam, R. 1998. cDNA and deduced amino acid sequences of a peritrophic membrane glycoprotein, "peritrophin 48," from larvae of *Lucilia cuprina*. *Insect Biochem. Mol. Biol.* 28: 99–111.

Shaw, M. K., Tilney, L. G., and McKeever, D. J. 1993. Tick salivary gland extract and interleukin-2 stimulation enhance susceptibility of lymphocytes to infection by *Theileria parva* sporozoites. *Infect. Immunity* 61: 1486–1495.

Srikrishnaraj, K. A., Ramasamy, R., and Ramasamy, M. S. 1993. Fecundity of *Anopheles tessellatus* reduced by the ingestion of murine antimosquito antibodies. *Med. Vet. Entomol.* 7: 66–68.

Srikrishnaraj, K. A., Ramasamy, R., and Ramasamy, M. S. 1995. Antibodies to *Anopheles* midgut reduce vector competence for *Plasmodium vivax* malaria. *Med. Vet. Entomol.* 9: 353–357.

Stark, K. R., and James, A. A. 1996. The salivary glands of disease vectors. In: *The Biology of Disease Vectors* (B. J. Beaty, and W. C. Marquardt, eds.), University Press of Colorado, Niwot, pp. 333–348.

Sukarsih, Partoutomo, S., Satria, E., Wijffels, G., Riding, G., Eisemann, C., and Willadsen, P. 2000. Vaccination against the Old World screwworm fly (*Chrysomya bezziana*). *Parasite Immunol.* 22: 545–552.

Sutherland, G. B., and Ewen, A. B. 1974. Fecundity decrease in mosquitoes ingesting blood from specifically sensitized mammals. *J. Insect Physiol.* 20: 655–660.

Tellam, R. L., and Bowles, V. M. 1997. Control of blowfly strike in sheep: Current strategies and future prospects. *Int. J. Parasitol.* 27: 261–273.

Tellam, R. L., and Eisemann, C. H. 1998. Inhibition of growth of *Lucilia cuprina* larvae using serum from sheep vaccinated with first-instar larval antigens. *Int. J. Parasitol.* 28: 439–450.

Tellam, R. L., Smith, D., Kemp, D. H., and Willadsen, P. 1992. Vaccination against ticks. In: *Animal Parasite Control Utilizing Biotechnology* (W. K. Yong, ed.), CRC Press, Boca Raton, FL, pp. 303–331.

Tembo, S. D., and Rechav, Y. 1992. Immunization of rabbits against nymphs of *Amblyomma hebraeum* and *A. marmoreum* (Acari: Ixodidae). *J. Med. Entomol.* 29: 757–760.

Titus, R. G., and Ribeiro, J. M. C. 1988. Salivary gland lysates from the sand fly, *Lutzomyia longipalpis*, enhance *Leishmania* infectivity. *Science* 239: 1306–1308.

Trager, W. 1939a. Acquired immunity to ticks. *J. Parasitol.* 25: 57–81.

Trager, W. 1939b. Further observations on acquired immunity to the tick *Dermacentor variabilis* Say. *J. Parasitol.* 25: 137–139.

de Vos, S., Zeinstra, L., Taoufik, O., Willadsen, P., and Jongejan, F. 2001. Evidence for the utility of Bm86 antigen from *Boophilus microplus* in vaccination against other tick species. *Exp. Appl. Acarol.* 25: 245–261.

Volf, P. 1994. Localization of the major immunogen and other glycoproteins of the louse *Polyplax spinuloa. Int. J. Parasitol.* 24: 1005–1010.

Volf, P., and Grubhoffer, L. 1991. Isolation and characterization of an immunogen from the louse *Polyplax spinulosa. Vet. Parasitol.* 38: 225–234.

Waitumbi, J., and Warburg, A. 1998. *Phlebotomus papatasi* saliva inhibits protein phosphatase activity and nitric oxide production by murine macrophages. *Infec. Immun.* 66: 1534–1537.

Webster, K. A., Rankin, M., Goddard, N., Tarry, D. W., and Coles, G. C. 1992. Immunological and feeding studies on antigens derived from the biting fly, *Stomoxys calcitrans. Vet. Parasitol.* 44: 143–150.

Wijffels, G., Hughes, S., Gough, J., Allen, J., Don, A., Marshall, K., Kay, B., and Kemp, D. 1999. Peritrophins of adult ectoparasites and their evaluation as vaccine antigens. *Int. J. Parasitol.* 29: 1363–1377.

Wikel, S. K. 1981. The induction of host resistance to tick infestation with a salivary gland antigen. *Am. J. Trop. Med. Hyg.* 30: 284–288.

Wikel, S. K. 1982. Immune responses to arthropods and their products. *Annu. Rev. Entomol.* 27: 21–48.

Wikel, S. K. 1988. Immunological control of hematophagous arthropod vectors: Utilization of novel antigens. *Vet. Parasitol.* 29: 235–264.

Wikel, S. K. 1996. Host immunity to ticks. *Annu. Rev. Entomol.* 41: 1–22.

Wikel, S. K. 1999. Tick modulation of host immunity: An important factor in pathogen transmission. *Int. J. Parasitol.* 29: 851–859.

Wikel, S. K., and Alarcon-Chaidez, F. J. 2001. Progress toward molecular characterization of ectoparasite modulation of host immunity. *Vet. Parasitol.* (In press).

Wikel, S. K., and Bergman, D. K. 1997. Tick-host immunology: Significant advances and challenging opportunities. *Parasitol. Today* 13: 383–389.

Wikel, S. K., Bergman, D. K., and Ramachandra, R. N. 1996. Immunological-based control of blood-feeding arthropods. In: *The Immunology of Host–Ectoparasitic Arthropod Relationships.* (S. K. Wikel, ed.), CAB International, Wallingford, England, pp. 290–315.

Wikel, S. K., Ramachandra, R. N., Bergman, D. K., Burkot, T. R., and Piesman, J. 1997. Infestation with pathogen-free nymphs of the tick *Ixodes scapularis* induces host resistance to transmission of *Borrelia burgdorferi* by ticks. *Infect. Immun.* 65: 335–338.

Willadsen, P. 1997. Novel vaccines for ectoparasites. *Vet. Parasitol.* 71: 209–222.

Willadsen, P., and Jongejan, F. 1999. Immunology of the tick–host interaction and the control of ticks and tick-borne diseases. *Parasitol. Today* 15: 258–262.

Willadsen, P., and Kemp, D. H. 1988. Vaccination with "concealed" antigens for tick control. *Parasitol. Today* 4: 196–198.

Willadsen, P., McKenna, R. V., and Riding, G. A. 1988. Isolation from the cattle tick, *Boophilus microplus*, of antigenic material capable of eliciting a protective immunological response in the bovine host. *Int. J. Parasitol.* 18: 183–189.

Willadsen, P., Riding, G. A., McKenna, R. V., Kemp, D. H., Tellam, R. L., Nielsen, J. N., Lahnstein, J., Cobon, G. S., and Gough, J. M. 1989. Immunologic control of a parasitic arthropod. Identification of a protective antigen from *Boophilus microplus. J. Immunol.* 143: 1346–1351.

Wong, J. Y. M., and Opdebeeck, J. P. 1990. Larval membrane antigens protect Hereford cattle against infestation with *Boophilus microplus. Parasite Immunol.* 12: 75–83.

Wrenn, W. J. 1996. Immune responses to mange mites and chiggers. In: *The Immunology of Host–Ectoparasitic Arthropod Relationships* (S. K. Wikel, ed.), CAB International, Wallingford, England, pp. 259–289.

Zeidner, N., Dreitz, M., Belasso, D., and Fish, D. 1996. Suppression of acute *Ixodes scapularis*–induced *Borrelia burgdorferi* infection using tumor necrosis factor-alpha, interleukin-2, and interferon-gamma. *J. Infect. Dis.* 173: 187–195.

SPECIAL METHODS AS APPLIED TO VECTORS OF DISEASE AGENTS

STEPHEN HIGGS

46

Transgenic Mosquitoes and DNA Research Safeguards

PETER W. ATKINSON

INTRODUCTION

The years since the late 1990s have witnessed significant progress in mosquito transgenic technology. At least five mosquito species, *Aedes aegypti* (Jasinskiene et al. 1998), *Anopheles gambiae* (Grossman et al. 2001), *Anopheles stephensi* (Catteruccia et al. 2000) *Anopheles albimanus* (A. M. Handler, personal communication), and *Culex quinquefasciatus* (Allen et al. 2001), can now be genetically transformed using transposable elements. It is anticipated that transformation will also be extended into other mosquito species, since the paradigm used to transform mosquitoes appears to be of general use across insect orders. There are now four transposable elements available that have been successfully used to transform mosquitoes, and each of these possesses a wide host range. While not as robust as *P*-element transformation of *Drosophila melanogaster*, mosquito transformation has been robust enough to be applied by a number of laboratories recently (Chapters 39 and 40). As a consequence, transgenic lines of mosquitoes now exist in several locations in the United States and Europe, and the number can be expected to increase.

At present all of these transgenic lines have been created in order to demonstrate either the process of transformation itself (Jasinskiene et al. 1998; Coates et al. 1998) or the effectiveness of genetic markers used to identify transgenic mosquitoes (Pinkerton et al. 2000) or to examine whether so-called "effector" genes can alter the phenotype of the mosquito in such a way

that its ability to vector pathogens is reduced (Kokoza et al. 2001). These constitute laboratory-based applications of this technology, since the mosquitoes produced are not intended to be released into the field. The principal application of mosquito transformation technology is used to validate the function of genes identified by computational means from genomics projects to answer questions about mosquito molecular biology and pathogen transmission. Another application of mosquito transformation technology is the generation and release of mosquito strains designed to achieve a desired environmental effect. The most commonly cited example is that of malaria control, in which it has been proposed that wild-type infectious mosquitoes can be replaced with a population of mosquitoes genetically engineered to prevent transmission of *Plasmodium falciparum* (James et al. 1999). The same scenario can be applied to other mosquito-borne diseases, such as dengue and encephalitis, that are resurgent viruses and require new approaches to control. Population replacement or inundative release of genetically engineered mosquitoes is one approach to this global problem. This will not likely be the principal application of this technology (but it may attract the most public comment), since the number of strains released into the field will probably be small and will be the outcome of comprehensive lab and field cage testing of prototype genetic strains.

In a climate of increasing concern over the inappropriate, and sometimes malicious, use of biological material, it is important to consider the containment and accidental or deliberate release of genetically engineered mosquitoes. The two applications of transgenic technologies applied to mosquitoes present different risk assessment questions. For laboratory-reared strains, the question of risk concerns predicting what the consequences of accidental escape may be. For strains intended for field release, questions of risk are concerned with what the unintended consequences of field release on the environment may be. Here, I discuss, from the perspective of DNA research safeguards, some of the parameters that should be considered when contemplating the production, generation, and release of genetically engineered mosquitoes.

RISKS ARISING FROM THE REARING OF GENETICALLY ENGINEERED MOSQUITOES

Preliminary advisory guidelines for the physical containment of genetically engineered insects have been included in *Arthropod Containment Guidelines*, Draft version 2.3, prepared by the American Committee of Medical Entomology (ACME) and the American Society for Tropical Medicine and Hygiene (ASTMH). These guidelines are quite specific and cover only the indirect effects arising from the culturing of insects infected with pathogens. *Indirect* effects are defined as those arising from infection by the pathogen vectored by the transgenic insect. *Direct* effects, such as the likelihood of biting and infestation following accidental release, are not covered in this document, yet they could be of concern to the public if an accidental release of transgenic mosquitoes occurred. These guidelines do address important parameters that may change as a consequence of genetically altering the insect vector, an associated endosymbiont, or the pathogen the insect is carrying. These include:

1. Altering the vectoral capacity of the arthropod for pathogens it is known to transmit
2. Decreasing the ability to control the arthropod following accidental or deliberate release
3. Altering the range or seasonal abundance of the arthropod and by doing so increasing the likelihood of its encountering and then vectoring new pathogens
4. Altering the vector so that it is unlikely to survive outside the laboratory
5. Increasing the reproductive capacity of the vector transmitting the pathogen
6. The stability of the genetic modification within the host genome
7. The ability of the transgene to be mobilized in natural populations, either within or between species
8. Altering the host range or antigenicity of endosymbionts that have been genetically modified
9. The DNA sequence of the transgene and the insertion site

The pace of progress in the production of transgenic arthropods, and specifically mosquitoes, has been rapid. Questions relating to any risks arising from generating them have yet to be investigated at the experimental level. This is due, in part, to the fact that strains designed for field release, for example, to limit the spread of a pathogen vectored by the arthropod, have yet to be generated, so testing them for any changes in the parameters just listed is neither possible nor practical. The gene transfer vectors and genetic markers used to generate transgenic mosquitoes have been developed and may well be used to generate transgenic strains for field release; risks associated from using these can now be addressed at a preliminary level. Information gained will be important for designing field-release strategies and for developing proper assessments of risk.

The level of containment of mosquito laboratory cultures has been dictated both by the type of pathogen that these strains harbor and by the need for good laboratory practice, i.e., the separation and hygienic maintenance of multiple strains within and between insectaries. Risks arising from maintaining mosquitoes containing pathogens are known, and the necessary precautions to guard against accidental infection of laboratory staff and against accidental release are well documented (*Biosafety in Microbiological and Biomedical Laboratories*, HHS Publication, 4th edition, April 1999). At present these risks are thought to be greater than those arising from generating and rearing transgenic mosquitoes, so draft containment guidelines for transgenic mosquitoes are based on those already established for mosquitoes harboring pathogens. Nevertheless, transgenic mosquitoes present new challenges in assessing what the intended and unintended consequences of transgenesis might be, both on the transgenic individual and on natural mosquito populations it might enter. It is difficult to predict what unintended consequences of transgenesis might be for an individual, let alone an entire population. Discussions of pleiotropic effects arising from the insertion and expression of a transgene have been limited to the inactivation of genes at, or close to, the point of insertion. Clearly, these effects cannot be catastrophic, or else the transgenic individual would not survive. The phenomenon of *position effect*, in which promoters, enhancers, or heterochromatic sequences adjacent to the point of insertion influence the expression of transgenes, is yet to be studied in mosquitoes. An attempt to minimize position-effect variegation using insulator sequences in mosquitoes is currently in progress and may well result in reducing the variability in expression arising from this phenomenon in transgenic mosquitoes (C. J. Coates, personal communication).

Discussion of several of the nine points raised by ACME/ASTMH draft guidelines at this point in time is strictly conjectural, since few transgenic strains of mosquitoes have been generated and none has as yet been examined with respect to these nine criteria. Alteration of vectoral capacity, seasonal abundance, and reproductive capacity by the process of transgenesis are all theoretically possible. These can be achieved either directly through the genotype conferred by the transgene or indirectly through unforeseen modification of the genome, for example, by inactivation of a gene or genes that affect these traits. Each of these traits can be measured in the laboratory. Analyses of transgenic lines measured against the appropriate wild-type strain would produce information concerning fertility, fecundity, daily survivorship, duration of the extrinsic incubation period, longevity, host preference, and vector competence. Variation from the wild-type strain might indicate that the transgenic line possesses altered properties with respect to vectoral capacity, and so the level of containment conditions should then be adjusted accordingly. Similarly, life table analyses conducted on transgenic mosquitoes in which temperature, humidity, and oviposition sites are varied may, along with the results of host preference studies, indicate that the range and seasonal abundance may be altered in the transgenic strains. Due to the recent development of transgenic techniques in mosquitoes, there is an absence in the literature of these studies' being applied to transgenic mosquitoes. In the future these studies should be applied to transgenic strains that could survive outside of the laboratory so that proper risk assessments of their effect on the environment could be made. At that point, several of the nine points identified by the ACME/ASTMH committee could be addressed with more certainty.

Decreasing the Ability to Control the Arthropod Following Accidental or Deliberate Release

Problems arising from the introduction of transgenic insects into the environment are similar to those encountered in studies involving transgenic fish: The organisms are mobile in the medium they inhabit and are difficult or almost impossible to recall. In the case of mosquitoes, which have an aquatic juvenile stage and an air-borne adult stage, the problem of dispersal is doubly compounded. One important consideration is the composition of the local mosquito population surrounding the laboratory. Historical surveys will show which species are most likely to be present and which are exotic, but these surveys can be dated. It is prudent for the investigator to gain "real-time" knowledge of the local mosquito population. This can be quickly achieved by placing the appropriate traps at varying distances from the laboratory and identifying these trapped mosquitoes. Checked on a regular basis, traps can also serve to monitor laboratory containment effectiveness. Should transgenics appear in the traps, then there has been an escape and the necessary notification and eradication protocols can be implemented.

Current trapping surveys along with previous surveys will show if the laboratory species is exotic or native to the local environment. Species that are exotic may not survive outside of the laboratory, and that may increase the chances of controlling any accidental release of transgenic mosquitoes. Species that are local to the region present a more difficult problem. Although potentially an infrequent event, a transgene may enter the local population through interbreeding between the transgenic population and the wild population. The effects of this interbreeding would be exacerbated if the transgene is an autonomous transposable element (see later), since these elements are capable of supporting their own transposition through genomes and populations. Particular care should therefore be taken with the containment of transgenic mosquitoes in regions in which the same species is present in the local environment.

Another issue affecting the ability to contain transgenic mosquitoes following accidental release concerns the nature of the transgenic sequences and particularly the genetic markers used. It would clearly be imprudent to use genes that confer insecticide resistance to transgenic mosquitoes, since this would lead to the possible spread of these genes into field populations. The development of the fluorescent protein genes as genetic markers has eliminated the need for dominant selectable markers such as insecticide-resistance genes. It would be wise, however, if any gene that is known to provide a clear selective advantage to transgenic mosquitoes is not routinely used other than in mature vector control strategies in which it is desired that this gene, and others tightly linked to it, spread through the target population.

Altering the Mosquito Vector So That It Is Unlikely to Survive Outside the Laboratory

Transgenic insects generated in the laboratory typically have reduced survivorship in the environment. *Drosophila* strains have usually contained multiple genetic markers, each of which reduces fitness outside the laboratory should any accidentally escape. The same range and accessibility of recessive mosquito mutants is not available to mosquito geneticists, and the utilization of dominant selectable markers such as the gene encoding the green fluorescent protein has led to the transformation of wild-type strains that, in principle, could survive and propagate outside the laboratory. The fitness costs to mosquitoes (or, for that matter, any insect) of expressing fluorescent proteins in all, or some, tissues has not been investigated. Life table analyses of representatives of these transgenic strains are required so that proper risk assessment measures can be developed. Perhaps the major factor leading to a reduction in fitness of any transgenic strain is that the recipient strain and the first few generations of transgenic strains will inevitably be lab-reared insects. The inability of lab-reared mosquitoes to mate competitively in the wild has frustrated attempts at using genetic control against mosquito-borne disease. Modern techniques, such as transformation, reduce the number of generations that need to be reared in the laboratory in order to produce a desired genotype. If the genes are dominant, they can be out-crossed into a wild strain quite rapidly, but there is still a requirement to rear these strains in captivity for at least a few generations.

In the near term, effort should be directed at developing laboratory strains of mosquitoes that are used as recipients of transgenes and that cannot sustain their numbers should they accidentally escape. One option currently available is to use white-eyed recipients for transformation experiments that use the Pax6 promoter to drive expression of the several fluorescent protein genes now commonly available. This promoter directs expression to the anal papillae of larvae, and the ommatidia of adults and transgenics are most easily detected in nonpigmented eyes (Horn and Wimmer 2000). While mating competition and life table analyses have not been undertaken on Pax6:fluorescent protein transgenic strains, it seems unlikely that these mosquitoes would enjoy an advantage over wild-type strains should they be accidentally released. Other options available to develop laboratory strains that could not survive outside the laboratory include using transgenesis itself combined with RNA interference technology to selectively knock out genes

required for dispersal, such as genes homologous to the vestigial gene of *Drosophila*. These transgenics would be unable to fly and so would be ideal as recipient strains. Furthermore, the several transposable elements now available for mosquito transgenesis mean that the recipient strain could be created using one transposable-element system, leaving the other systems available for further genetic manipulations.

The Stability of the Genetic Modification Within the Host Genome

Transgenic gene stability in mosquitoes has become increasingly important as the transformation technology has developed. Stability affects risk assessment questions. If a transgene is known to be stable, then risk assessments based on its persistence can be developed. Alternatively, if the transgene is known to be unstable after several generations, particularly in the absence of selection either for the transgene or a closely linked genetic marker, then concerns arising from accidental release may be alleviated by the likelihood that the transgene would become dysfunctional and even eliminated from the population.

Focused analyses of transgene stability in mosquitoes have yet to be performed. Many of the transgenic lines generated have been able to be maintained for many generations with intense selection for the transgenic genetic marker. These observations do not, however, necessarily address stability in terms of what might be required for mosquitoes reared in the absence of selection. Indeed, if there is selective pressure against some or all of the transgenes, then these would be rapidly lost from the population. Stability in the absence of selection can be easily addressed in the laboratory. The questions that can also now be addressed experimentally relate to the sequence motifs of a transposable element that make it susceptible to movement or recombination within the mosquito genome. These motifs would be the sequences of the transposable element that are recognized by its own transposase, related transposases that are endogenous to the mosquito host, and other host-encoded factors that may interact with and destabilize the transposable element. While such sequences are yet to be identified for any of the four transposable elements currently used to genetically transform mosquitoes, it is clear from the structure–function analyses of other transposable elements, such as the *Tag1* element of *Arabidopsis thaliana*, that these motifs exist and are small, sometimes being as short as 6 bp (Mack and Crawford 2001). Technologies to remove these following insertion of the transposable element exist. The ability of the yeast FLP recombinase system to work in mosquitoes has been known for over 10 years and could be used to selectively remove these motifs, rendering the transposable element immobile (Morris et al. 1991). Similarly, recent developments in homologous gene replacement in *Drosophila* (Rong and Golic 2000) might soon be extended into mosquitoes, leading to gene replacement without the insertion of transposable-element sequences that could destabilize the transgene in subsequent generations.

Transgene Stability and the Ability of the Transgene to be Mobilized in Natural Populations, Either Within or Between Species

Transposable elements are used to genetically transform mosquitoes. It is the mobility properties of these elements that enabled them to be used as genetic tools in mosquitoes. These properties also cause concerns over their ability to be subsequently and uncontrollably mobilized both within and between species. We know very little about the mobility properties of the four transposable elements used to transform mosquitoes. None of the four elements is from mosquitoes, they transform mosquitoes at about the same frequency, usually less than 10%, and they have broad host ranges. All four transpose by cut-and-paste transposition, with the exception of the *Hermes* element which, in mosquito germ-line nuclei, appears to insert via an as-yet-unknown, transposase-dependent mechanism (Jasinskiene et al. 2000; Allen et al. 2001).

How could subsequent remobilization of any of these transposable elements occur once they have been inserted into a mosquito genome? Most of these are nonautonomous and therefore incapable of movement in the absence of the corresponding transposase elements (autonomous transposable elements are discussed later). Unless this is also integrated into the genome during transformation, which is unlikely given the absence of inverted terminal repeats and other sequences required for transposition, this transposase gene is unlikely to be present in subsequent generations. If remobilization is to take place, the enzymatic source must be novel. The most likely source is the functional transposase of an endogenous related transposable element. Such interactions might be common given that transposable elements can spread through and between species. We know little about the distribution of *hAT*-like elements, of which *Hermes* is one member, and *piggyBac* elements through insects except that several insect *hAT* elements have been discovered (for a review, see Atkinson and O'Brochta 2000). While highly conserved, *piggyBac* elements have been found in such diverse species as *T. ni* and the

fruitfly *Bactrocera dorsalis* (Handler and McCombs 2000). Two *hAT* element–like sequences, *huni* (D. A. O'Brochta, P. W. Atkinson, and F. H. Collins, unpublished data) and an unnamed *hobo*-like element discovered during the French National Genomics Center —Genoscope *Anopheles* sequencing project, have been found in *An. gambiae*, while a similar sequence has been found in *Ae. aegypti* (Stark and James 1998). In all three cases, no inverted terminal repeats have been located, and it is not known if the transposase encoded by these sequences is transcribed and, if so, if it is functional. Nonetheless, the presence of these sequences indicates that at least DNA sequences very similar to the transposable elements now used for mosquito transformation are most likely all present in these species and that the completion of whole genome sequencing projects for *An. gambiae* and *Ae. aegypti* will enable a more comprehensive estimate of their abundance to be determined. The distribution of members of the *mariner-Tc1* superfamily of transposable elements in arthropods was undertaken by Robertson and MacLeod (1993); these sequences were found in *An. gambiae*. As for the vast majority of these elements discovered in insects, these are not functional, based on the absence of open reading frames encoding the transposase gene.

The single study illustrating cross-mobilization between transposable elements in transgenic insects was performed by Sundararajan et al. (1999), who examined cross-mobilization, as measured by excision of the transposable element between the *hobo* and *Hermes* elements in *D. melanogaster*. They reported an asymmetry between the ability of these two elements to cross-mobilize each other, with hobo transposase being able to cross-mobilize both the *Hermes* and *hobo* elements, while the *Hermes* transposase could mobilize the *Hermes* element but not the *hobo* element. The outcomes of this study have been overinterpreted to indicate that cross-mobilization between *hAT* elements will present a problem when using these elements to transform any insect species. However, such a conclusion remains speculative. In the system used by Sundararajan et al. (1999), both elements were newly introduced into the E (empty of *hobo* elements) *Drosophila* strains, and so no repression system for suppressing the movement of *hobo* elements had developed. Furthermore, it is presumptuous to assume that the behavior of transposable elements in *Drosophila* necessarily indicates how they will behave in other insect species. The example of *Hermes* is particularly pertinent because in the germ line of mosquitoes, it integrates into the chromosomes in a different fashion than it does in *Drosophila*. How this might affect subsequent remobilization in mosquitoes in the presence of *Hermes* transposase or of related transposases has not been determined.

The ability of each of the four elements now used to transform mosquitoes to be subsequently cross-mobilized in each transgenic mosquito strain needs to be determined empirically. This information will be extremely vital in determining long-term strain stability and the viability of genetic control strategies based on the expression of transgenes linked to transposable elements.

The assays for cross-mobility are relatively simple. Interactions between elements in somatic nuclei can be observed using plasmid-based excision or transposition assays conducted in developing embryos. Evidence of germ-line interaction requires the construction of two strains: one containing the transposase gene, typically placed under the control of an inducible heat shock promoter, and a second containing the target transposable element. The latter can contain an indicator gene or, alternatively, be placed inside an indicator gene, for example, a gene encoding a fluorescent protein placed under the control of a tissue-specific promoter, so that excision can, in some cases, lead to the restoration of the function of the indicator gene. Progeny resulting from crossing these strains would be examined for fluorescent protein expression in clones of cells in the appropriate tissue.

A separate question arises from the transformation of mosquito strains with autonomous transposable elements. These elements are capable of self-mobilization, and so, in principle, they could spread through a caged population of mosquitoes, rendering all the mosquitoes, by definition, transgenic, and, if accidentally released into the environment, could spread through wild populations of the same species. Indeed, even if the laboratory strain had severely reduced fitness relative to the field population, a chance, low-frequency mating that resulted in the transfer of the autonomous element into the field population could act as a springboard for the subsequent rapid and irreversible spread of the element, since it would now be free of its original, fitness-compromised host. In perhaps the worst-case scenario, the element might be as successful in invading a mosquito species as the *P* element has been in invading the global *Drosophila melnaogaster* population in the past 50–70 years. Clearly, such a situation is to be avoided, and transgenic autonomous strains of mosquitoes should perhaps be contained at higher levels of containment than strains transformed with nonautonomous elements until the mobility properties of the autonomous element in the new host species have been determined. The desire to determine if these transposable elements can function as genetic drive agents in mosquitoes increases the likelihood that autonomous forms of the four elements now used to transform mosquitoes will be created and placed into laboratory populations of mosquitoes. Indeed, this has already occurred with the *Hermes* element (P. W. Atkinson and D. A. O'Brochta, unpublished data).

The success of transposable elements as gene vectors in mosquitoes and other insects has led to concerns that these elements might be horizontally transferred to other organisms. Transposable elements are subject to two competing selection pressures. One maintains their ability to move within and between genomes, while the second limits their ability to move within a genome so that the fitness consequences on the host are minimized. The use of a transposable element as a gene vector, particularly an autonomous element, carries with it the risk that it might move between species.

Experiments to detect horizontal transfer are difficult to design, and the horizontal transfer of transposable elements between different eukaryotic species has not been detected in the laboratory. Horizontal transfer may occur at an infinitesimal frequency, yet, once it has taken place, the element may then move quickly through a new host species. How should the risk arising from horizontal transfer be measured? One approach, as described by O'Brochta (D. A. O'Brochta, personal communication), is to examine the mobility properties of the transposable element gene vector used to transform the mosquitoes in those organisms likely to come into intimate and routine contact with the insect. For mosquitoes these would include humans and the animal host used for blood feeding. Excision and transposition assays could be performed in cell culture using human and rodent cell culture lines. While the outcomes of these assays would not shed light on the likelihood that horizontal transfer will occur, they would allow its impacts to be estimated should it occur into likely host species.

BIOTERRORISM AND TRANSGENIC MOSQUITOES

The exploitation and misuse of biological material as agents of terror or warfare has recently attracted much public concern, and some of this has extended into questions concerning the likelihood that mosquitoes will be engineered to spread harmful pathogens through human populations. Even though progress in mosquito genetic engineering has been rapid, deliberate misuse of this technology to create new, harmful strains of mosquitoes is highly unlikely. Mosquito transformation is not a trivial technology to undertake.

Laboratories that utilize it require constant supplies of electricity and clean water in order to maintain wild-type colonies at levels that will consistently produce the large numbers of eggs needed to undertake transformation experiments. Identical requirements are needed for the maintenance of transgenic colonies once they have been created. Great attention needs to be paid to the shape of the fine needles required for embryo microinjection, and postinjection rearing needs to be performed under clean conditions. Insectary space is needed for the simultaneous multiple matings, and access to fluorescence microscopy is needed for the detection of transgenics if fluorescent protein genes are used as genetic markers. With the exception of *Ae. aegypti*, in which eggs can be stored for several months, continued attention must be paid to the maintenance of the colonies and, in the case of transgenic colonies, particular care must be directed toward them since they often have reduced fecundity and so can be difficult to maintain. To date, this technology has been used only in modern laboratories in the United States and Europe, where mosquito strains have been maintained for several years.

Even if a harmful biological agent were introduced into a mosquito using modern genetic technology, the question remains as to how to spread it through a human population. This could be achieved in two ways. One is by inundative release: massrearing millions of modified mosquitoes and then releasing them en masse into the environment. The physical and financial requirements for the mass rearing of insects are even more demanding than those required for generating transgenic lines, and there are only a handful of facilities worldwide that are devoted to rearing massive quantities of insects, these being tephritid fruit flies (such as the Mediterranean fruit fly) and moths. In all cases, specialist skills are required for the maintenance of these large numbers of insects since, as for any massive population reared in confined conditions, opportunity for disease that could debilitate the colony is rife.

The alternate approach would be to develop genetic driving agents that would spread the pathogen or genes required for the maintenance of the pathogen through wild mosquito populations. This technology does not, at present, exist. Indeed, the question of genetic drive through insect populations has been a longstanding one, with no clear answer in sight despite our recent advances in the identification of potential driving agents, such as transposable elements and *Wolbachia* species. For a comprehensive review of the spread of genetic constructs through insect populations, the reader is referred to Braig and Yan (2002). The likelihood of harnessing insect transgenic technologies for malicious purposes therefore appears to be very small.

CONCLUSION

In a recent review article, Spielman et al. (2002) listed several criteria that should be satisfied before a transgenic arthropod that serves as a vector for human pathogens is released. Simply put, these criteria can be summarized by stating that Spielman et al. (2002) concluded that it behooves those who embark on the genetic manipulation of mosquitoes to make sure that we are improving the health and welfare of the human population rather than the reverse. This requires us to establish, by experimentation, the link between the new genotypes we seek to confer on the mosquito and the phenotype resulting from these manipulations. Many of the parameters affecting vectorial capacity and competence, transgene stability, and the likely abundance and distribution of these mosquitoes can be addressed in the laboratory. From these studies, the role that the environment plays in determining phenotype can be estimated, leading to rational estimates of the risk arising from rearing, and perhaps releasing, these mosquitoes. The rapidity with which mosquito transgenesis has developed increases the need for these experiments to be undertaken on appropriately modified strains of mosquitoes and, in fact, increases the level of enthusiasm that has permeated this endeavor for more than a decade.

Readings

Allen, M. L., O'Brochta, D. A., Atkinson, P. W., and Levesque, C. S. 2001. Stable, germ-line transformation of *Culex quinquefasciatus* (Diptera: Culicidae). *J. Med. Entomol.* 38: 701–710.

Atkinson, P. W., and O'Brochta, D. A. 2000. Hermes and other hAT elements as gene vectors in insects. In: *Insect Transgenesis—Methods and Applications* (A. M. Handler, and A. A. James, eds.), CRC Press, Boca Raton, FL, pp. 219–236.

Braig, H. R., and Yan, G. 2002. The spread of genetic constructs in natural insect populations. In: *Genetically Engineered Organisms* (D. K. Letourneau, and B. E. Burrows, eds.), CRC Press, Boca Raton, FL, pp. 251–314.

Catteruccia, F., Nolan, T., Loukeris, T. G., Blass, C., Savakis, C., Kafatos, F. C., and Crisanti, A. 2000. Stable germ-line transformation of the malaria mosquito *Anopheles stephensi*. *Nature* (London) 405: 959–962.

Coates, C. J., Jasinskiene, N., Miyashiro, L., and James, A. A. 1998. *Mariner* transposition and transformation of the yellow fever mosquito, *Aedes aegypti*. *Proc. Natl. Acad. Sci. U.S.A.* 95: 3748–3752.

Grossman, G. L., Rafferty, C. S., Clayton, J. R., Stevens, T. K., Mukabayire, O., and Benedict, M. Q. 2001. Germ-line transformation of the malaria vector, *Anopheles gambiae*, with the *piggyBac* transposable element. *Insect Mol. Biol.* 10: 597–604.

Handler, A. M., and McCombs, S. M. 2000. The *piggyBac* transposon mediates germ-line transformation in the Oriental fruit fly and closely related elements exist in its genome. *Insect Mol. Biol.* 9: 605–612.

Horn, C., and Wimmer, E. A. 2000. A versatile vector set for animal transgenesis. *Dev. Genes Evol.* 210: 630–637.

James, A. A., Beernsten, B. T., de Lara Capurro, M., Coates, C. J., Coleman, J., Jasinskiene, N., and Krettli, A. U. 1999. Controlling malaria transmission with genetically engineered, *Plasmodium*-resistant Mosquitoes: Milestones in a model system. *Parassitologia* 41: 461–471.

Jasinskiene, N., Coates, C. J., Benedict, M. Q., Cornel, A. J., Rafferty, C. S., James, A. A., and Collins, F. H. 1998. Stable transformation of the yellow fever mosquito, *Aedes aegypti*, with the *Hermes* element from the housefly. *Proc. Natl. Acad. Sci. U.S.A.* 95: 3743–3747.

Jasinskiene, N., Coates, C. J, and James, A. A. 2000. Structure of *Hermes* integrations in the germ line of the yellow fever mosquito, *Aedes aegypti. Insect Mol. Biol.* 9: 11–18.

Kokoza, V., Ahmed, A., Cho, Wimmer, E. A., and Raikhel, A. S. 2001. Efficient transformation of the yellow fever mosquito, *Aedes aegypti*, using the *piggyBac* transposable element vector pBac[3xP3-EGFP afm]. *Insect Biochem. Molec. Biol.* 31: 1137–1144.

Mack, A. M., and Crawford, N. M. 2001. The *Arabidopsis* TAG1 transposase has an N-terminal zinc finger DNA–binding domain that recognizes distinct subterminal repeats. *Plant Cell* 13: 2319–2331.

Morris, A. C., Schaub, T. L., and James, A. A. 1991. FLP-mediated recombination in the vector mosquito, *Aedes aegypti. Nucleic Acids Res.* 19: 5895–5900.

Pinkerton, A. C., Michel, K., O'Brochta, D. A., and Atkinson, P. W. 2000. Green fluorescent protein as a genetic marker in transgenic *Aedes aegypti. Insect Mol. Biol.* 9: 1–10.

Robertson, H. M., and MacLeod, E. G. 1993. Five major subfamilies of *mariner* transposable elements in insects, including the Mediterranean fruit fly, and related arthropods. *Insect Mol. Biol.* 2: 125–139.

Rong, Y. S., and Golic, K. G. 2000. Gene targeting by homologous recombination in *Drosophila. Science* 288: 2013–2018.

Spielman, A., Beier, J. C., and Kiszewski, A. E. 2002. Ecological and community consideration in engineering arthropods to suppress vector-borne disease. In: *Genetically Engineered Organisms* (D. K. Letourneau and B. E. Burrows, eds.), CRC Press, Boca Raton, FL, pp. 315–329.

Stark, K. R., and James, A. A. 1998. Isolation and characterization of the gene encoding a novel factor Xa–directed anticoagulent from the yellow fever mosquito, *Aedes aegypti. J. Biol. Chem.* 273: 20802–20809.

Sundararajan, P., Atkinson, P. W., and O'Brochta, D. A. 1999. Transposable element interactions in insects: Cross-mobilization of *hobo* and *Hermes. Insect Mol. Biol.* 8: 359–368.

47

Gene Bank Data for Hematophagous Arthropods

CHRISTOS LOUIS

INTRODUCTION

Although high school has taught us that asking *"What if . . ."* is not particularly helpful in the understanding of history, I wonder what molecular biology would look like today had research in life sciences not been accompanied by an equally gigantic progress in both computer technology and the science of informatics. Three years after its establishment as the first nucleic acid sequence database, Genbank contained fewer than 6 million bases in close to 6 thousand entries. In contrast, by March 2002, more than 18 billion bases distributed in more than 16 million entries had been deposited to the EMBL nucleic acid sequence database. It is obvious that if this logarithmically increasing mass of information were to be used meaningfully by the biological research community, it had to be accompanied by the development of more powerful hardware, in terms of storage capacity, data transfer, and data handling. Furthermore, more sophisticated software was needed for searching, retrieving, and analyzing the stored sequences as well as the other biological information that is being accumulated. In January 1984, *Nucleic Acids Research* published for the first time an issue entirely dedicated to what is known today as bioinformatics. This issue contained, among others, one of the most cited papers in life sciences, which described "GCG," the most successful package of DNA analysis. In addition, one could find rarities such as making use of a programmable pocket calculator and, perhaps unexpected for some of the younger readers, an article coauthored by one of the contributors of this volume. Today, 20 years later, the one-issue-per-year forum for the publication of papers in bioinformatics has evolved to several international journals covering one of the most rapidly expanding fields of science.

This chapter cannot review bioinformatics in a few pages; it concentrates only on "gene banks." Since these are the biological databases that are most commonly visited by researchers, it attempts to provide a brief introduction to their contents and their handling. The term *gene bank* is used here to describe two kinds of biological databases: the strict sequence databases, both nucleic acid and protein, such as Genbank, EMBL, and SWISSPROT, and genomic/genetic databases that may be all-inclusive, such as FlyBase, or more restricted to actual genome data. In addition, this chapter provides a brief, nontechnical primer to the methods available for accessing and analyzing these data. A list of URLs at the end of this chapter provides links to the databases and to most of the tools mentioned. The reader is urged to visit these sites, to "surf" through all available biological Internet resources, and to "play" with the different programs. Hands-on experience is the only way to go!

THE SEQUENCE DATABASES

Nucleic Acid Sequence Databases and Data Retrieving

There are three major nucleic acid sequence databases, each developed and maintained in large institutions in different continents. These are (1) Genbank, a responsibility of the U.S. National Center for Biotechnology Information (NCBI), (2) EMBL, established at the European Molecular Biology Laboratory and maintained at its outstation, the European Bioinformatics Institute (EBI), and (3) DDBJ, found at the Japanese National Institute of Genetics. Although independent of one another, these three databases aim at providing the best possible service for the worldwide research community, being in close collaboration in order to avoid problems such as redundancies. All three databases use common accession numbers for their entries, and the quasi-constant exchange of data between them leads to an almost complete overlap of the stored data. Thus, the choice of which database to use is, to a large extent, either "geographical" (e.g., American scientists tend to submit their sequences to Genbank, while Europeans prefer EMBL) or a matter of taste. The "taste" in this case, then, refers to the software platform used by the databases to submit data, on one hand, or to retrieve them, on the other.

The major databases have developed user-friendly software that makes sequence submission extremely easy. Genbank offers the choice of Sequin and Bankit, for complicated and simple submissions, respectively, while WEBIN, EMBL's tool, is the method of choice for the European database. DDBJ, finally, also offers the possibility to use the SAKURA program, which allows the submission of data in Japanese. Although these tools may seem somewhat complicated when used for the first time, they are easy to learn, especially since the learning process can be assisted by tutorials that are offered online.

Retrieving data is naturally much more complicated than submission. Genbank, for example, is accessible through a set of search-and-retrieval software called ENTREZ. This package not only allows searching of the actual sequence databases, but it is also used to access other, additional NCBI datasets, such as the bibliographical database PubMed, a resource of the National Library of Medicine. It should be stressed at this point that the nucleic acid sequence databases are only part of the individual databases developed and curated by the bioinformatics centers. DDBJ, EBI, and NCBI all offer access to a large number of databases, some of them not discussed in this chapter, that include, among others, protein data (from sequences to structures), genomes, and taxonomy.

ENTREZ also provides a user interface to perform BLAST analysis (see later), a resource that is also available through SRS, the package developed to perform searches using EMBL and other databases handled by the EBI. DDBJ uses SRS because it does not have its own set of tools for data retrieval. Irrespective of whether one uses ENTREZ or SRS for searching and retrieval, the output, whose format depends on the database searched, consists of one entry, or more, depending on what the search strings were, which contains a large variety of information. This information includes general material such as author, addresses, reference, and organism and specific facts, called *features*, that relate directly to the respective sequence. The catalogue of features has grown longer over the years, since it now refers to actual structures, such as repeats, miscellaneous signals, such as binding sites and polyadenylation sites, gene description, and information on the source of the sequence or mapping data. Moreover, the entries contain links to the PubMed database or other databases, such as FlyBase (see later). This way, allied data can be accessed directly without the need of further searches.

Protein Sequence Databases and Data Retrieving

So far only nucleic acid sequences have been addressed. Since the vast majority of genes encode proteins, many of the analyses in molecular biology are, or should be, performed directly in protein databases rather than in nucleic acid databases. In contrast to the latter, which contain primary data only, protein sequence databases often rely on the translation of experimentally deduced coding regions for their entries. Naturally, experimentally established sequences are also banked, and they can be searched, together with databases for conceptually translated putative proteins, using either ENTREZ or SRS, as described earlier.

SWISSPROT is perhaps the best known of all protein databases. Established in 1986, it is now maintained collaboratively by the EBI and the Swiss Institute of Bioinformatics (SIB). In March 2002 this database contained more than 700,000 entries, of which about one-seventh was found in the actual SWISSPROT, a fully annotated form, while the rest were in TrEMBL, a "satellite" database containing sequences translated from the EMBL nucleic acid database. The entries of TrEMBL are only computer annotated. In contrast to the nucleic acid databases mentioned earlier, these two databases are not all-inclusive; they have now been established as the ones most often used when homology searches are concerned. The individual entries contain the sequences and general data as described for the nucleic acid sequences, but this time the features refer to protein structures (e.g., domains, variants). Similar to the nucleic acid counterparts, cross-links to other databases are provided as well as comments by the authors.

Sequence Analysis

Sequence analysis is a complicated procedure that is briefly described later. The software available varies tremendously, depending on the computer system/platform used, price, level of sophistication, etc. As mentioned earlier, "GCG" is the most successful and most comprehensive software package. It is, however, not unique. Since sequence analysis programs are often bought as software packages, individual programs designed for specific tasks are frequently found in more than one package. Although a package can be made out of excellent components, an individual program may be of rather low efficiency or be problematic, while its counterpart in a package of lower quality may, in contrast, be much superior. The complexity of the software packages, the many tasks they are called to help solve, and the "taste" of the individual researcher make it difficult to evaluate the available tools. It is therefore not possible even to attempt to recommend a product.

The only software that will be mentioned here is the family of programs commonly known as BLAST. This suite of programs, which was developed originally at NCBI, allows the comparison of sequences whereby a sequence database is searched/queried with any given nucleic acid or protein sequence. Additional BLAST programs offering more capabilities are now available (e.g., WU-BLAST) based on the original code publicly available by NCBI. BLAST searches are used to find similarities and homologies between genes of the same or different organisms. The queries can involve nucleic acid databases searched with a nucleic acid sequence, protein databases searched with a polypeptide sequence, and also cross-platform scans: TBLAST, for example, translates a nucleic acid sequence in all six frames and searches a protein database. BLAST programs are found as tools in the software packages mentioned earlier, and they can be also used from within ENTREZ or SRS. BLAST servers can be found at different locations in the World Wide Web, which allow searches of smaller datasets faster than if the complete sequence databases were searched at, say, Genbank. Finally, one should say once more that these programs could only be learned on a hands-on basis: Experience is required not only to run the program (setting the parameters), but also, more importantly, to evaluate the significance of the results obtained in a search.

Vectors and Sequence Databases

Vectors are obviously treated by the sequences' databases in the same way as any other organism.

What is interesting, though, is to look at the number of entries found for individual vector species and to use this information as a "fossil record" to describe the history of research in vector biology. Table 47.1 illustrates this point. It becomes immediately apparent that *Anopheles gambiae*, whose genome is in the finishing stages of sequence acquisition, is, by far, the vector arthropod with the highest number of entries. As a matter of fact, by the time of publication of this book, the number will likely have more than doubled, and it will be within the same range as that of the fruit fly,

TABLE 47.1 Number of Entries in EMBL Nucleic Acid Database for Different Arthropod Vectors as Well as Two Beneficial Insects and One Genetic Model Organism

Species	Number of Entries		
	March 2002	August 1998	June 1995
Drosophila melanogaster	329,469	98,157	5,200
Bombyx mori	15,767		252
Apis mellifera	15,528		
Musca domestica	162		30
Sarcophaga bullata	13		
Cochliomyia hominivorax	3		
Aedes spp.	2,383		
Aedes aegypti	2,067		31
Ae. triseriatus	16		
Ae. albopictus	73		7
Anopheles spp.; non-*gambiae*	1,303		
Anopheles gambiae	155,343	579	81
An. stephensi	101	15	8
An. minimus	85	0	0
An. darlingi	23	8	5
An. albimanus	55	44	28
An. funestus	70	2	0
Culex spp.	493		
Glossinidae	128		5
Simuliidae	191		11
Culicoididae	89		
Lutzomyia spp.	90		
Phlebotomus spp.	261		
Fleas (Siphonaptera)	188		
Rhodnius spp.	72		1
Ixodida	3,720		
Ixodes spp.	713		

A blank entry means data for the respective date were not available.

whose genome was sequenced in the year 2000. Although an upward trend can be seen for most of the vectors listed here, one should say that many of the entries represent different versions of the same DNA fragment (e.g., rDNA or mtDNA) sequenced for population biology studies or molecular markers such as microsatellites. In contrast, two beneficial insects listed in this table have a relatively high number of entries, indicating the existence of small-scale genome projects and, more importantly, increased funding of this research. It is hoped that the number of entries relating to vectors will greatly increase in the future.

GENOMIC AND GENETIC DATABASES

The last 20 years have seen not only an increase of sequence data but also an increase in biological knowledge in general. For example, a summer 1998 search of the PubMed database with the keyword *Drosophila* yielded 28,409 hits, while a 35% increase to 38,362 hits was achieved by March 2002. Similarly, increases of 22% and 42% could also be seen for *Anopheles* and *Anopheles gambiae*, respectively (from 4,494 to 5,493 and from 598 to 849 hits, respectively). This mass of information can obviously best be presented to the research community in the form of species-specific databases. This need is naturally enhanced by the different genome projects that must have the data not only stored in sequence databases, but also presented in a more workable form.

FlyBase

In the eyes of many scientists, the most successful database presenting these kind of data is FlyBase, the *Drosophila* database. It contains a variety of data dealing with the fruit fly, ranging from information on genes, stocks, and addresses of researchers to *Drosophila*-related literature and the annotated genome, the last through GadFly. FlyBase is a relational database, that is, a project the fruit fly research community has wholeheartedly supported; this explains its success to a large extent. Naturally, the field of *Drosophila* has had a tradition of open communication and free exchange of information. It is apparent that this policy has led to many success stories, making the fruit fly the best-studied higher animal. The contents of FlyBase, as mentioned, cover a broad spectrum of data, which are all cross-referenced. Thus, if one looks up a given gene, for example, *neuralized* (*neur*), FlyBase provides information on alleles, expression and phenotypes, maps, proteins and transcripts,

constructs, etc. This information, in this particular example, has a length of more than 15,000 words, and, considering that *neur* is a gene for which "only" 56 alleles exist, one can imagine the wealth of the overall information stored in FlyBase.

In addition to this kind of information, FlyBase gives access to GadFly, the Genome Annotation Database. Here, among other searches, one can browse annotations, look up information and maps concerning the full genome sequence, and use the BLAST server. Although GadFly is an independent database developed specifically for the annotation of the genome, its potential becomes more evident though its linkage to FlyBase. Again, it should be stressed that the best way of learning everything about both FlyBase and GadFly is to visit and browse through them.

Vector Databases

Only a few genomic/genetic databases exist for arthropod disease vectors, reflecting the relative scarcity of information available as compared to the fruit fly. The Mosquito Genomics Server at Fort Collins (Colorado), "housing" information on a variety of mosquitoes, is an example of an ambitious database that uses the ACeDB format originally developed for the genome project of the nematode *Caenorhabditis elegans*. This database format, an excellent tool for genomic research and analysis, is used here to store a variety of data on more than one mosquito species. In addition, the Mosquito Genomics Server provides several links to interesting sites as well as other information relating to mosquitoes.

AnoDB, the *Anopheles* database, on the other hand, is a simple text database that is responsible for data on all anophelines. An ACeDB format is also available at AnoDB, as is a BLAST server, a tool that is equally available at the Fort Collins database. With the completion of the *Anopheles gambiae* genome project, AnoDB is in the process of "upgrading" to a relational format that is fully compatible with FlyBase. This work is in close collaboration with the fruit fly team. In addition to the upgraded database, the completion of the genome project will also lead to the availability of annotation databases, similar to GadFly, that at the time of writing this chapter were not yet available.

SUMMARY

One can no longer consider molecular biological research without the help of bioinformatics. The existence of extensive databases and the tools developed to handle the data stored prove this. The term *in silico* has now become as widely used as *in vitro*; it is a common belief that dependence on computers will become even greater in the future. Indeed, the whole field of postgenomic research, from microarray profiling to proteomics, is absolutely unthinkable without the sophisticated software that makes this research feasible. The completed genome of *Anopheles gambiae* and other genomes that will certainly follow will soon make this a reality also for vector biology. The only advice one can give to facilitate the use of computer programs in biology is "to jump in and swim." Never has anybody drowned in a sea of data.

Some Useful URLs

The following URLs give direct access to the databases and most of the tools mentioned in this chapter.

http://www.ncbi.nih.gov/	Home page of the NCBI
http://www.ncbi.nih.gov/Entrez/	Direct access to NCBI's ENTREZ
http://www.ncbi.nih.gov/BLAST/	"Home page" of the BLAST programs
http://blast.wustl.edu/	"Home page" of the WU-BLAST programs
http://www.ncbi.nlm.nih.gov/PubMed/	PubMed, provides access to bibliographic citations
http://www.ebi.ac.uk/	Home page of the EBI
http://srs.ebi.ac.uk/	SRS at the EBI
http://www.ddbj.nig.ac.jp	Home page of the DDBJ
http://www.expasy.org/sprot/	Swiss ExPaSy server in Geneva
http://flybase.bio.indiana.edu/	Home page of FlyBase
http://www.fruitfly.org/annot/	GadFly, the Genome Annotation Database of *Drosophila*
http://www.anobase.org/	AnoBase, the *Anopheles* database
http://klab.agsci.colostate.edu/	Mosquito Genomics Server at Fort Collins
http://www.ensembl.org/Anopheles	The *Anopheles* genome pages

Acknowledgments

I would like to thank Inga Sidén-Kiamos, Christos Delidakis, and Pantelis Topalis, for critical reading of the manuscript, and Pantelis Topalis and Tassos Koutsos, for being such excellent collaborators on the AnoDB project. Work on AnoDB, the *Anopheles* database, has been supported by the John D. and Catherine T. MacArthur Foundation, the UNDP/World Bank/World Health Organization Special Programme for Research and Training in Tropical Diseases (TDR), and the Hellenic Secretariat General for Research and Technology.

Readings

Altschul, S. F., Gish, W., Miller, W., Myers, E. W., and Lipman, D. J. 1990. Basic local alignment search tool. *J. Mol. Biol.* 215: 403–410.

Bairoch, A., and Apweiler, R. 2000. The SWISS-PROT protein sequence database and its supplement TrEMBL in 2000. *Nucleic Acids Res.* 28: 45–48.

Baxevanis, A., and Ouellete, F. 2001. *Bioinformatics: A Practical Guide to the Analysis of Genes and Proteins*. Wiley-Liss, New York.

Benson, D. A., Karsch-Mizrachi, I., Lipman, D. J., Ostell, J., Rapp, B. A., and Wheeler, D. L. 2002. GenBank. *Nucleic Acids Res.* 30: 17–20.

Burks, C., Fickett, J. W., Goad, W. B., Kanehisa, M., Lewitter, F. I., Rindone, W. P., Swindell, C. D., Tung, C. S., and Bilofsky, H. S. 1985. The GenBank nucleic acid sequence database. *Comput. Appl. Biosci.* 1: 225–233.

Devereux, J., Haeberli, P., and Smithies O. 1984. A comprehensive set of sequence analysis programs for the VAX. *Nucleic Acids Res.* 12: 387–395.

Eeckman, F. H., and Durbin, R. 1995. ACeDB and macace. *Methods Cell. Biol.* 48: 583–605.

The FlyBase Consortium. 2002. The FlyBase database of the *Drosophila* genome projects and community literature. *Nucleic Acids Res.* 30: 106–108. http://flybase.bio.indiana.edu/

Mount, D. W. 2001. *Bioinformatics: Sequence and Genome Analysis*. Cold Spring Harbor Press, Cold Spring Harbor, New York.

Stoesser, G., Baker, W., van den Broek, A., Camon, E., Garcia-Pastor, M., Kanz, C., Kulikova, T., Leinonen, R., Lin Q., Lombard V., Lopez, R., Redaschi, N., Stoehr, P., Tuli, M. A., Tzouvara, K., and Vaughan, R. 2002. The EMBL Nucleotide Sequence Database. *Nucleic Acids Res.* 30: 21–26.

Tateno, Y., Imanishi, T., Miyazaki, S., Fukami-Kobayashim, K., Saitou, N., Sugawara, H., and Gojobori, T. 2002. DNA Data Bank of Japan (DDBJ) for genome-scale research in life science. *Nucleic Acids Res.* 30: 27–30.

Wheeler, D. L., Church, D. M., Lash, A. E., Leipe, D. D., Madden, T. L., Pontius, J. U., Schuler, G. D., Schriml, L. M., Tatusova, T. A., Wagner, L., and Rapp B. A. 2001. Database resources of the National Center for Biotechnology Information. *Nucleic Acids Res.* 29: 11–16.

Zdobnov, E. M., Lopez, R., Apweiler, R., and Etzold, T. 2002. The EBI SRS server—recent developments. *Bioinformatics* 18: 368–373.

48

The Containment of Arthropod Vectors

STEPHEN HIGGS

INTRODUCTION

The safe containment of arthropods that are vectors for pathogenic organisms is a critical issue that must be considered by all who work with these organisms regardless of the vector's infection status. It is the researchers' responsibility to maintain the vectors in appropriate facilities and to use protocols that prevent escape. To prevent the introduction of an exotic species to a new area where it could establish itself, it is important to contain uninfected vectors as well as infected ones. The immediate and long-term impacts of such an introduction may be difficult to assess but are potentially serious. In recent years at least two species, *Aedes albopictus* and *Ae. japonicus*, have been accidentally introduced into the United States. Both species have become established and implicated in the transmission of arboviruses, including West Nile and La Crosse viruses. Had these mosquitoes been released from a laboratory rather than as a result of international trade, the consequences for research on vectors could have been severe.

The USDHHS publication *Biosafety in Microbiological and Biomedical Laboratories* (the BMBL), published in 1984 and revised in 1999, provides guidelines for the handling and containment of pathogens and establishes the biosafety level (BSL) designations for most human pathogens. A four-tiered system is prescribed, depending upon the relative hazards associated with the pathogens. The U.S. government has recently implemented new restrictions on certain pathogens designated as *select agents*. These current and proposed U.S. federal regulations place legal restrictions on the possession, use, and transfer of the select agents that

limit who can work with these agents and what facilities are secure enough to contain them.

There is little available literature that considers the issue of vector containment. Technically, the handling and containment requirements when working with arthropods that are infected with agents covered in the BMBL are relatively easy to understand. For example, an arthropod infected with a level 2 agent should be contained using the procedures and facilities for BSL-2 research. With regard to uninfected "normal" vectors, in 1980 the Subcommittee on Arbovirus Laboratory Safety (SALS) of the American Committee on Arthropod-Borne Viruses stated, "Studies with live arthropods, either normal or infected, using viruses at Level 1 should be conducted using arthropod practices and containment described for Containment Level 2." This does not necessarily set precedence for increasing containment requirements of a pathogen when in a vector. For example, it is unnecessary to use BSL-3 handling practices for a BSL-2 agent in a vector.

From 1999 to 2003, a subcommittee of the American Committee of Medical Entomology (ACME) formulated guidelines for the containment of arthropod vectors. Drafts were posted on web sites, discussion meetings were held, and input was solicited from researchers (Aultman *et al.*, 2000; Higgs, 2003; Higgs et al., 2003; Hunt, 2001). In 2002, the guidelines were endorsed by ACME. They were published in 2003 (Benedict *et al.*, 2003) and are also available at http://www.liebertpub.com/ and http://thesius.ingentaselect.com/vl=15259108/cl=22/nw=1/rpsv/cw/mal/15303667/v3n2/contp1-1.htm. The guidelines include a consideration of risk assessment in which arthropods are assigned into one of four categories: 1) arthropods known to be free of specific pathogens; 2) arthropods known to contain specific pathogens; 3) arthropods containing unknown agents or whose status is uncertain; or 4) arthropods containing recombinant DNA molecules. By using these broad categories, researchers are guided to estimate the relative risk associated with an escape and thereby decide the level at which the arthropods should be contained. Arthropod containment levels (ACLs) range from one to four. Although the ACLs parallel the BSL system used for pathogens, unique characteristics of arthropods—for example active dispersal—must be considered. Whilst useful, a pathogen-based approach is overly simplistic. It is recommended that most uninfected vectors be housed at ACL-1, but uninfected vectors that are exotic and could become established in a new area should be contained in at least an ACL-2 insectary. Guidelines published in 1999 by the DHHS for handling organisms that contain recombinant DNA recommend that genetically engineered arthropods should be contained using BSL-2 standards. Issues relating to the rearing and containment of genetically engineered vectors are discussed in Chapter 46. Transportation and transfer of vectors were also discussed in the new guidelines.

The safe containment of arthropod vectors depends upon many factors, such as physical barriers, appropriate protocols, and a fastidious, well-trained staff. Many factors must be considered prior to working with vectors, since the potential for escape is a function of multiple vector characteristics. For example, size, mobility, appearance, behavior, reproductive rate, longevity, and numbers being used all influence the likelihood of escape. Work must therefore be performed within an appropriate insectary facility. It should be realized that design ideals for microbiological laboratories and insectaries are not identical. Since neither safety nor containment can be compromised, we discuss next the relative importance of design features and how one can minimize the risks associated with research using arthropod vectors.

KNOW YOUR VECTOR

Vectors are represented in many forms and have evolved to thrive in a broad range of environments. A thorough knowledge of the vector life cycle and ecology is essential in order to minimize the risk of escape. Some differences in escape potential are obvious; for example, adult mosquitoes fly, ticks crawl, and fleas jump. The potential for escape can be both an active and a passive process. Insectary design and maintenance protocols should be tailored to target these differences. Chapters in this section on vector-

specific maintenance discuss these characteristics. Variation in size is important when selecting materials, such as mesh for wall and container construction. When "insect proofing" a facility, one must consider all life stages and not just the adult. Mosquito eggs, for example, are small and, once embryonated, are frequently highly resistant to desiccation. This is the least visible stage and probably the most likely to escape by passive transfer into drains or by simply drying one's hands on a paper towel and throwing it into a trash can. Plumbing design that incorporates holding tanks to heat-treat water prior to release into the sewer system minimizes such an escape, but staff should be trained not to tip wastewater from rearing pans into general-use sinks. Eggs and even some larvae seem to be quite resistant to disinfectant chemicals, including sodium hypochlorite, and many will survive prolonged freezing; however, heating to 60°C for just a few minutes is generally lethal (Christophers 1960). Autoclaving at 121°C guarantees that no viable stages are discarded; however, a less expensive solution for dealing with rearing water and surplus immatures is to pour rearing water into a large-capacity coffee percolator to heat-kill all stages.

THE INSECTARY

A key component of any containment facility is limiting access to suitably trained and qualified personnel. The insectary should be locked when unoccupied, and appropriate signs should be posted to notify personnel of specific hazards and requirements such as vaccines (Fig. 48.1). The insectary design features recommended by SALS (1980) should be regarded as the basic minimum for vector research. These state, "Normal arthropods for vector competence studies are reared within an insectary that is physically separated from rooms holding infected laboratory animals by at least a solid wall and four screened or solid doors opening inwards and closing automatically. Space between doors must be sufficient for a person entering so that each door is closed before another is opened. If effectiveness is documented with the arthropod being maintained, an air curtain or cloth can be substituted for one door." (See Fig. 48.2.) Subdivision of the insectary into rooms of designated purposes (e.g., rearing species A, rearing species B, holding infected vectors, etc.) provides additional barriers to escape.

Although many workers and institutional committees assume that uninfected and infected vectors must not be located in the same room, this is not mandated. Suitable subdivision of a room is acceptable. SALS (1980) states, "Arthropods are infected and held within a separate screened or solid-walled section or room of

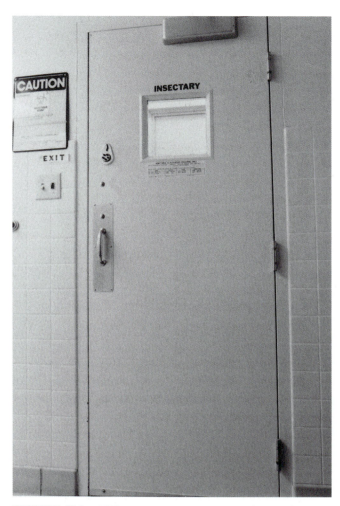

FIGURE 48.1 AIDL insectary. Main entrance door with appropriate signs.

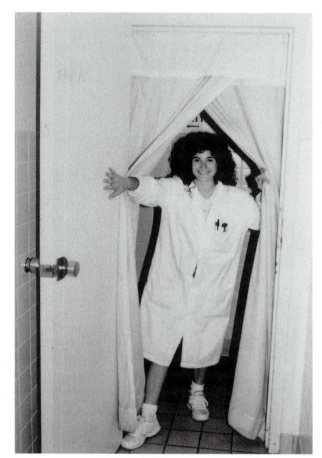

FIGURE 48.2 AIDL insectary. Overlapping cloth sheets stapled across internal door to dislodge mosquito escapees from clothing.

the insectary, which is separated from the remainder of the insectary by at least two screened or solid doors that open into the work area and close automatically (Fig. 48.3). If effectiveness is documented with the arthropod being maintained, an air curtain or cloth can be substituted for one door." A low ceiling aids in capturing loose arthropods and can be achieved by screening if retrofitting an existing facility (Fig. 48.4).

An example of a working insectary that fulfills these design criterions was described by Higgs and Beaty (1996) and is shown in Figure 48.5. Recently, insectary-specific design issues were discussed from an engineering viewpoint to enable institutions to develop facilities that both meet the need for research on vectors and satisfy the biocontainment requirements to work with pathogens (Duthu et al. 2001). It is important that engineers understand that the environmental requirements of many arthropods may be difficult to maintain in the containment laboratory. For example, high humidity and temperature can be compromised

by directional and constant airflow. Optimization of environmental conditions by balancing humidification, airflow diffusion, and lighting systems must therefore be considered as well as the selection of the laboratory location, construction materials, and finishes (Duthu et al. 2001). When Level 3 pathogens are used, additional design elements are required; for example, HEPA filtration of exhaust air. There are good guidelines for laboratory design (Richmond 1997; BMBL) and some features actually assist in preventing vector escape. The directional airflow in many microbiological laboratories to produce pressure gradients can also function to pull an escaped flying vector back into a room.

ACCOUNTING AND MONITORING

When working with infected vectors, one is usually dealing with relatively small numbers (a few hundred perhaps), and so containment and accounting of all

FIGURE 48.3   AIDL insectary. Partitioned area containing cages of manipulated mosquitoes.

FIGURE 48.4   AIDL Insectary. Suspended ceiling panels.

individuals is relatively straightforward and escape can effectively be prevented. Infected adult vectors should be counted into an unbreakable primary container, several of which may then be placed in a cage or environmental chamber for secondary containment, within a designated room of the insectary (tertiary containment). Throughout the experiment, for example, whenever samples are collected, the inventory of experimental arthropods must be updated so that the fate of all individuals can be tracked.

Routine large-scale rearing of arthropods for colony maintenance usually makes it impossible to know exactly how many individuals are present. Oviposition papers used to collect mosquito eggs, for example, may contain several thousand eggs, and it would be impractical, if not impossible, to count them. Although both physical barriers and procedures should minimize the risk of uninfected arthropod escape, it seems inevitable that some may escape into the general insec-

tary. Procedures must therefore be in place that guarantee that, although they may be loose in the insectary, they must not escape from it. Monitoring for and locating an escaped arthropod is vital. Monitoring may be achieved, for example, by using attractive oviposition cups, CO_2-baited light traps, or adhesive boards placed where an escapee is most likely to hide. As an aid to location, since most vectors are darkly colored, all surfaces in the insectary should be plain white. It should be emphasized that anyone seeing a loose arthropod should remain in the room until it is killed. In a unique risk assessment experiment, Hunt and Tabachnick (1996) released adult *Culicoides sonorensis* (syn. *C. variipennis sonorensis*) in a laboratory prior to its certification for BSL-3 research. The distribution of escapees was then determined to prove the effectiveness of their control and surveillance program in preventing loose insects from exiting the facility. Such an experiment is useful to identify areas that require additional precautions, although it must be conducted in a manner that

KEY TO INSECTARY ROOMS

A.	Autoclave
A.a.c.	*Aedes aegypti*, colony
A.a.e.	*Aedes aegypti*, experimental
A.t.c.	*Aedes triseriatus*, colony
A.t.e.	*Aedes triseriatus*, experimental
BL-3.a.	BL-3 containment area
BL-3.b.	BL-3 tissue culture/vector manipulation.
C.p.c.	*Culex pipiens*, colony
C.p.e.	*Culex pipiens*, experimental
Ent.	Entrance
Em.e.	Emergency exit
M.c.	Main corridor
V.c.	Vertebrate containment (BL-3)
V1.	Vestibule 1 (with shower)
V2.	Vestibule 2
V3.	Vestibule 3
W.C.	Toilet

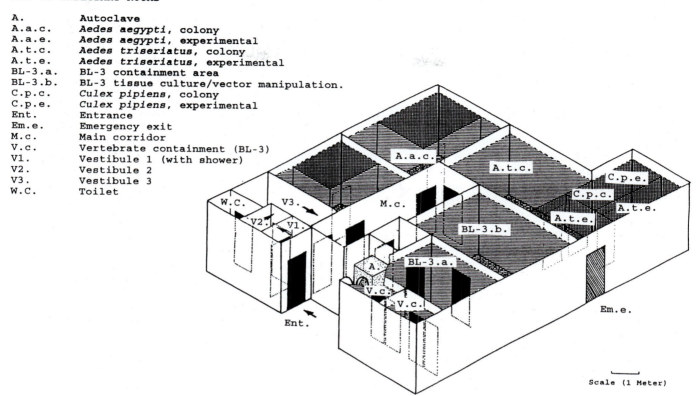

FIGURE 48.5 Plan of the insectary at the Arthropod-borne and Infectious Diseases Laboratory (AIDL), Colorado State University.

guarantees that insects do not actually escape from the facility. Having plain white surfaces throughout the insectary aids in the location of an escapee, but minimizing potential resting places is also important. Open shelving for storage of equipment should be avoided and superfluous apparatus should not be stored within the insectary.

HANDLING ARTHROPODS

At all times, personnel must wear appropriate personal protective equipment, taking into account the procedures used and the related risks of infection. White laboratory coats or scrubs with long pants are recommended. All stages of arthropods should be maintained or stored in containers with lids. Large numbers of dipteran vectors used as breeding colonies can be maintained in sturdy aluminum cages with fine-mesh screening that are available from various sources (e.g., BioQuip, 17803 LaSalle Ave., Gardena, CA 90248-3602). The integrity of cages should be checked weekly to ensure that no tears exist through which an arthropod could escape. Feeding procedures should be conducted in a manner that prevents arthropods from being inadvertently removed with a vertebrate host. Vertebrates should be carefully examined prior to returning them to their cages. Feeding arthropods on living vertebrates typically requires approval from institutional animal care and use committees, and protocols should also address the manner in which vertebrates are transferred and housed. Infected vertebrates should not be housed in the insectary and, if not euthanized on site, should be transferred alive in secure and nonbreakable containers (SALS 1980). Vectors may be similarly transferred for transmission experiments if the vertebrate cannot be brought into the insectary.

The greatest potential for escape occurs during the transfer of vectors from one container to another and during manipulative procedures. Chilling vectors prior to transfer can be used to immobilize them, but this is not always practical. An alternative to chilling is exposure to CO_2 gas. Where aspiration is necessary, constant suction mechanical aspirators (Fig. 53.1), for example, as made by Haushers Machine Works (1186

Old Freehold Rd, Toms River, NJ 08753) should be used in preference to mouth aspirators, the latter being forbidden when working in Level 2, 3, or 4 laboratories. A transparent transfer box that can house multiple containers can be used so that if an insect gets loose it is still effectively contained and can be destroyed. When infecting arthropods, the category of agent being used must be considered when designing protocols. Handling small arthropods in biosafety cabinets may actually increase the risk of escape due to the airflow. In these circumstances it may be permissible to work with the insects in an isolator with sleeve openings. However, when working with some agents, the use of biosafety cabinets is mandated and the option of working in an alternative manner is forbidden. SALS suggested that if working with arthropods infected with BSL-4 agents, the designs of experiments should be reviewed and approved individually by appropriate, knowledgeable, impartial consultants.

Inactive arthropods cannot actively escape, and so, whenever possible, prior to commencing infection procedures such as intrathoracic inoculation (Figs. 53.9 and 53.10), arthropods should be immobilized by refrigeration/immersion in ice and then be manipulated on a chill table (Fig. 53.2). This sensitivity to cold may be used as an additional precaution when working with vectors. In the event of an escape, air conditioning can be used to cool the facility and therefore slow down or even immobilize the escapee. Also, if an insectary is located in a region with a seasonal climate, then high-risk experiments can be performed during the winter. In the event of an escape, vectors would freeze to death. The area surrounding an insectary cannot be considered as just an escape zone but also as a potential source of wild vectors. Design, trapping, and monitoring should also be appropriate to preclude the intrusion of wild vectors and other pest species.

SUMMARY

Recent events have focused attention on research dealing with pathogenic organisms. New regulations and laws now apply to some of these agents, with an emphasis on safety, security, and accountability. The involvement of arthropod vectors in the transmission of many pathogens is therefore under scrutiny. As vector biologists, it is our responsibility to maintain the vectors in appropriate facilities and to use protocols that prevent escape. Safe containment of the arthropods is critical. It is imperative that no arthropod escape and no accidental transmission occur. The arthropod must be reared in secure insectary facilities, and strict practices are essential.

Readings

Aultman, K. S., Walker, E. D., Gifford, F., Severson, D. W., Beard, C. B., and Scott, T. W., 2000. *Managing risks of arthropod vector research. Science.* 288: 2321–2322.

Benedict, M. Q., Tabachnick, W. J., and Higgs, S. (key authors) 2003. Arthropod Containment Guidelines (Version 3.1). *Vector-Borne and Zoonotic Diseases.* 3: 59–98

Christophers, S. R. 1960. *Aedes aegypti* (L.) *The Yellow Fever Mosquito. Its Life History, Bionomics and Structure.* Cambridge University Press, New York.

Drolet, B., Campbell, C., and Mecham, J. 2003. Protect yourself and your sample: processing arbovirus-infected biting midges for viral detection assays and differential expression studies. In: *Anthology of Biosafety. VI. Arthropod Borne Diseases* (Richmond, J. Y., ed. 2003). American Biological Safety Association, Mundelein, IL. 53–62.

Duthu, D. B., Higgs, S., Beets, R. L., and McGlade, T. J. 2001. Design issues for insectaries. In: *Anthology of Biosafety. IV. Issues in Public Health* (J. Y. Richmond, ed.). pp. 227–244.

Escita, Y., Takasaki, T., Yamada, K-i., and Kurane, I. 2003. Isolation of arboviruses from field-collected mosquitoes. In: *Anthology of Biosafety. VI. Arthropod Borne Diseases* (Richmond, J. Y., ed. 2003). American Biological Safety Association, Mundelein, IL. 63–72.

Higgs, S. 2003 Editorial. *Vector-Borne and Zoonotic Diseases.* 3: 57–58.

Higgs, S., and Beaty. 1996. Rearing and containment of arthropod vectors. In: *Biology of Disease Vectors* (B. J. Beaty, and W. C. Marquardt, eds.). University of Colorado Press, Niwot, CO, pp. 595–605.

Higgs, S., Benedict, M. Q., and Tabachnick, W. J. 2003. Arthropod containment guidelines. In: *Anthology of Biosafety. VI. Arthropod Borne Diseases* (Richmond, J. Y., ed. 2003). American Biological Safety Association, Mundelein, IL. 73–84.

Hunt, D. L. 2001. Position paper—arthropod containment. *Appl. Biosafety.* 46–47.

Hunt, G. J., and Tabachnick, W. J. 1996. Handling small arbovirus vectors safely during biosafety level 3 containment: *Culicoides variipennis sonorensis* (Diptera: Ceratopogonidae) and exotic bluetongue viruses. *J. Med. Entomol.* 33: 271–277.

Hunt, G. J. and Schmidtmann, E. T. 2003. Safe and secure handling of virus-exposed biting midges within a BSL-3-AG containment facility. In: *Anthology of Biosafety. VI. Arthropod Borne Diseases* (Richmond, J. Y., ed. 2003). American Biological Safety Association, Mundelein, IL. 85–98.

Olson, K., Larson, R. E., and Ellis, R. P. 2003. Biosafety issues and solutions for working with infected mosquitoes. In: *Anthology of Biosafety. VI. Arthropod Borne Diseases* (Richmond, J. Y., ed. 2003). American Biological Safety Association, Mundelein, IL. 25–38.

Powers, A. M. and Olson, K. E. 2003. Working safely with recombinant viruses and vectors. In: *Anthology of Biosafety. VI. Arthropod Borne Diseases* (Richmond, J. Y., ed. 2003). American Biological Safety Association, Mundelein, IL. 39–52.

Richmond, J. Y., ed. 2003. *Anthology of Biosafety. VI. Arthropod Borne Diseases.* American Biological Safety Association, Mundelein, IL. 178 pp.

Richmond, J. Y. 1997. *Designing a Modern Microbiological/Biomedical Laboratory.* American Public Health Association, Washington, DC.

Spielman, A., Pollack, R. J., Kiszewski, A. E., and Telford, S. R., III. 2001. Issues in public health entomology. *Vector-borne Zoonotic Dis.* 1: 3–19.

Subcommittee on Arbovirus Laboratory Safety of the American Committee on Arthropod-Borne Viruses. 1980. Laboratory safety for arboviruses and certain other viruses of vertebrates. *Am. J. Trop. Med. Hyg.* 29: 1359–1381.

U.S. Department of Health and Human Services, Public Health Service. 1999. *Biosafety in Microbiological and Biomedical Laboratories. Guidelines for Research Involving Recombinant DNA Molecules.* 63 FR 25361. Washington, DC.

49

Care, Maintenance, and Experimental Infestation of Ticks in the Laboratory Setting

KEITH R. BOUCHARD AND STEPHEN K. WIKEL

INTRODUCTION

The global emergence and resurgence of arthropod-borne diseases continue to pose serious threats to world health. Since the discovery in 1982 of *Borrelia burgdorferi* as the causative agent of Lyme disease, 15 previously unrecognized bacterial pathogens transmitted by ixodid ticks have been described (Parola and Raoult 2001). Ticks are second only to mosquitoes as the leading vectors of infectious diseases to humans, transmitting an extensive array of viruses, rickettsia, bacteria, and protozoa. Ticks are also the most important vectors of disease-causing agents for both domestic animals and wildlife. The colonization of ticks in the laboratory setting has long been the goal of many researchers, in hopes of better understanding tick biology, host responses to ticks, and pathogen transmission (Gregson 1966; Sonenshine 1999).

There are many different approaches to tick colonization, based on tick species, hosts to be studied, and interests of a particular investigator. Here, we will describe basic methods used to rear large numbers of ticks in a standard laboratory setting. The techniques described are designed with both efficiency and economy in mind, and they can be implemented for most multihost hard-tick species.

TICK BIOLOGY AND MORPHOLOGY

Ticks are ectoparasitic, hematophagous arthropods that feed on birds, reptiles, mammals, and, less frequently, amphibians. They can be classified into three families: hard-bodied ticks (Ixodidae), soft-bodied ticks (Argasidae), and a unique third family with one representative species, the *Nuttalliellidae*. Our laboratory currently maintains seven species, including 11 different strains, of the most medically relevant hard-bodied ticks in North America.

There are approximately 670 species and 13 genera of Ixodidae worldwide. They are characterized and distinguished by a waxy, sclerotized cuticle dorsal region, referred to as the scutum. Their life cycle involves three distinct post-embryonic stages: larva, nymph, and adult. A blood meal at each stage in the life cycle is required for tick development and, in adults, for reproduction. Feeding can last anywhere from 3 to 12 days, depending on stage, species, and conditions. A molting period occurs following each blood meal, during which the tick undergoes its morphological development into the following stage.

Argasids, or soft ticks, account for an estimated 167 species throughout the world. They are distinguished by their soft, leathery bodies and lack of dorsal shield. Argasids undergo shorter feeding periods, with most species reaching engorgement within hours. Soft ticks also undergo a more complex life cycle, including up to seven instar nymphal growth stages. Argasid colony maintenance involves separate unique methods, as described by Endris et al. (1992).

INFESTATIONS—COLONY MAINTENANCE

Our experience has shown that even a complex, high-output, multispecies tick colony can be maintained conveniently using a combination of mouse (*Mus musculus*) and rabbit (*Oryctolagus cuniculus*) hosts. It is essential to work in collaboration with your institutional animal care and use committee, biosafety committee, and facility veterinary staff in the design of your protocols.

Larvae

Larval ticks of most hard-bodied species can be adapted to feed on murine hosts with minimal stress and without the need for anesthesia. The mouse must be gently restrained in a small wire-screen, cylindrically shaped cage with hinged doors on both ends (Fig. 49.1). The mouse is guided into the cage, and both doors are closed using cable ties. The cage is then placed into a Tupperware®-style plastic bowl with high sides and tape placed around the top edge, with the sticky side facing inward to prevent tick escape.

Larval ticks are essentially painted onto the neck and upper back of the mouse, using a small camel hair brush. Having a looped strip of masking tape handy will assist in capturing any runaway ticks. The ticks are allowed to attach for 1–2 hours before the mouse is released. The bowl containing the empty cage should be kept overnight at 60°C to kill any larvae that did not attach.

The mice must be housed individually in conventional shoebox-style mouse cages with stainless steel wire-rack bottoms (Fig. 49.2). Beneath the rack, a $\frac{1}{2}$-inch depth of distilled water allows ticks to fall into a secure drop-off environment following detaching from the host. A ring of tape surrounding the top of the cage will prevent tick escape. Cage water should be checked and changed daily so as to keep a clean environment for both the mouse and the ticks. Engorged larvae will fall off into the water and may be collected using forceps or a wire inoculating loop. They should be then placed onto filter paper to dry sufficiently before being placed into vials. Larvae may also be collected from the underside of the wire grating or from the tape surrounding the cage. Special care should be taken not to rupture the delicate bodies of the engorged ticks. One previously reported alternative larval infestation model involves the layering of white paper towels, instead of water, beneath the mice. Engorged larvae are then carefully collected using a vacuum-driven aspiration device (Sonenshine 1999).

Most larval ixodid ticks require an average of 3–5 days to feed until repletion. Variation in feeding duration may be observed when attempting to colonize

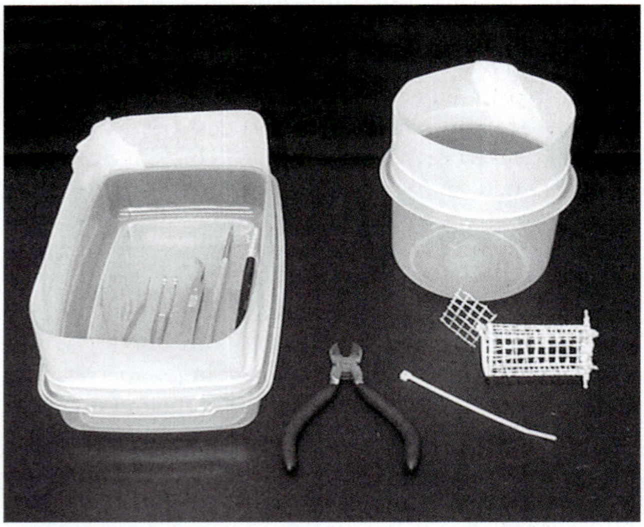

FIGURE 49.1 Instruments required for larval infestation. Clockwise, from left: forceps, camel hair brush housed within a taped container; taped infestation container with high sides; mouse infestation enclosure with hinged doors; cable tie; wire cutters. Photo courtesy Ken Bourell.

FIGURE 49.2 Conventional shoebox-style mouse cage with wire-rack bottom used for housing mice during infestation with larvae or nymphs. The cage must be taped along the junction between bottom and lid to prevent tick escape. Photo courtesy Ken Bourell.

field-collected material. Although previous literature has indicated some species' willlingness to feed more readily on alternative hosts, it has been our experience that larvae adaptation to murine hosts is generally possible.

Nymphs

Whole-body nymphal feedings for most ixodid species may be done in a manner similar to that described for larvae, only in fewer numbers. In our hands, up to 20 nymphs may be fed on an average mouse, though species and strain variation can be dramatic. Nymphs will generally require a slightly longer feeding duration, anywhere from 3 to 7 days, on average.

When defined feeding parameters are of interest, nymphs may be fed within a capsule fastened to the mouse's back. This capsule allows for a protected feeding area and facilitates controlling host exposure. Capsules can be fashioned from halved Eppendorf® 1.5-mL tubes (Fig. 49.3). It has been our experience that capsules fashioned from some alternative types of tubes may contain components that alter the ticks' willingness to feed. The lid is removed and punctured, and a small piece of mesh placed between the lid and tube allows for adequate air exchange. The hair on the upper back of the mouse should be trimmed close to the skin with a small standard electric razor. The capsules are attached using a melted 4:1 (w:w) composition of gum rosin (colophonium) and beeswax. This

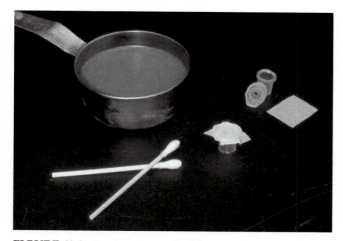

FIGURE 49.3 Instruments required for nymphal capsule infestation. Clockwise, from right: Eppendorf®-style capsule with removable lid and mesh barrier; assembled capsule; applicator swabs used to apply rosin:beeswax mixture; small metal pan containing a 4:1 (w:w) composition of gum rosin (colophonium) and beeswax used to affix capsule. Photo courtesy Ken Bourell.

mixture allows for secure adherence to the mouse's back throughout the duration of the feeding while causing minimal distress to the host. Following the infestation, the capsule can be removed easily by trimming away the regrown hair with a small pair of scissors.

Nymphs may be directly hand-applied into the capsule using forceps, or they may first be counted onto a small piece of gauze or into a gelatin capsule, which is then transferred. Recommended ixodid feeding numbers vary greatly depending on species, though with our *Ixodes scapularis* experiments, we feed 10 nymphs per capsule. The mice may be housed in groups of up to five per cage without compromising capsule security due to cross-grooming. The cage should also be taped as described for larval infestations to prevent the escape of any ticks.

Adults

Most adult ixodid species will adapt to feeding upon rabbits. Larger numbers of nymphs (100–200, depending on species) may also be fed on rabbits using a technique that involves feeding within a capsule secured to the animal's side. An approved anesthetic is needed to sedate the rabbit for this procedure. Special attention should always be given to animals under sedation. Anesthetized rabbits should always be kept on a heated pad in order to maintain body temperature, and the animal's eyes should be treated with a veterinary opthalmic lubricant to prevent drying.

The capsule is fashioned from a wide-mouth Nalgene® 500-mL bottle cut approximately 2 inches from the top (Fig. 49.4). We have found that this capsule, when fixed to a Duoderm GCF 4″ × 4″ wound dressing pad (ConvaTec) by using a glue gun, creates an optimal feeding microenvironment and provides a high degree of security when applied to the host. This method also causes no noticeable distress to the host during the feeding period. The capsule may also be constructed without the Duoderm pad, using only elastic tape and Elmer's® school glue and applying the capsule directly onto the rabbit's skin. The school glue is water soluble and will not cause irritation to the host's skin.

The hair on the rabbit's midsection should be clipped to the skin all the way around the body prior to capsule application. The DuoDerm pad is then attached to the upper side of the rabbit, using its adhesive backing. We have found that a supplemental ring of VetBond (3M) veterinary tissue adhesive around the inner perimeter of the feeding area provides additional security in confining ticks within the feeding area. The rabbit's midsection is then wrapped twice around in

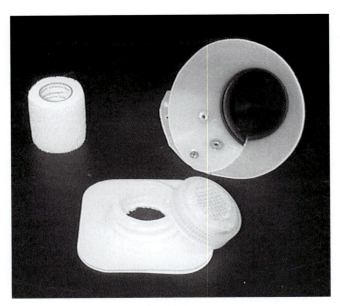

FIGURE 49.4 Clockwise, from bottom: Nalgene®-style infestation capsule fixed to a 4″×4″ DuoDerm wound dressing pad; hypoallergenic elastic tape used for securing capsule; Elizabethan collar to prevent grooming (note small bolts for opening and closing). Photo courtesy Ken Bourell.

expandable cloth tape to further secure the capsule. An additional outer layer of duct tape prevents the rabbit's nails from catching the fabric tape when grooming is attempted. An Elizabethan-style collar should be applied comfortably to the rabbit's neck in order to prevent grooming of the capsule area. Trimming a rabbit's nails before an infestation will also prevent damage to the infestation site.

The entire cage must be secured with masking tape (sticky side facing in) for security in the unlikely event of capsule breach. Adults should always be counted prior to feeding to ensure complete recovery, and equal numbers of males and females should be introduced into the capsule for most ixodid species. Gregson (1966) describes detailed species-specific data regarding the suggested number of feeding ticks per host without compromising the health of the animal.

Females who have fed to repletion and mated successfully will usually undergo a noticeable color change. Both the rabbit and the capsule should be checked at least several times daily. Detached females may be removed from the capsule using a pair of blunt live-insect forceps. The males of most ixodid species may be reused for several adult feedings, whereas females who feed to repletion will undergo oviposition, or egg laying, within a few weeks and die.

Following the infestation, the capsule can be conveniently trimmed away using surgical scissors. Capsule removal should not require the use of anesthesia. Some tick species will induce a moderate host

skin reaction, which should be treated as necessary. Rabbits may be reused as hosts, although resistance induction has been observed for exposure to some tick species.

An alternative method previously used in our laboratory involves feeding the ticks within stockinet bags affixed to the rabbit's ear (Sonenshine 1999). A critical aspect of this method is making sure that the bags are taped securely and that the rabbit cannot groom off the lightweight bags. The use of an Elizabethan collar for this technique, as described earlier, is essential. When larger animal facilities are available, mammals such as sheep and cattle may also be used as hosts for rearing larger numbers of ticks. These animals must be first stanchioned in an appropriate restraining device before capsules can be applied. Infestation "cells" can be fashioned from stockinet sleeves and glued to the host's shaved skin with an adhesive cement (Sonenshine 1999). Since cattle are the natural hosts for a number of tick species, this can be an effective method for rearing large numbers of adult ticks at a single time.

ARTIFICIAL FEEDING

The concept of an *in vitro* feeding system has long intrigued researchers in the field of tick research (Waladde et al. 1996). Such a model would eliminate the need for animal hosts. *In vitro* tick feeding on isolated host skin or artificial membranes has been accomplished, though with limited success. Ticks do not attach readily to synthetic membranes (Kemp et al. 1982), and they require constant stimulation to remain feeding. An *in vitro* model described by Voigt et al. (1993) involves the successful feeding of all life cycle stages of *Amblyomma variegatum* on rabbit and cattle skin membranes. Although this method required increased levels of carbon dioxide, blood heparinization, and both anti-fungal and antibiotic treatments, it was used to successfully transmit the cattle pathogens *T. mutans* and *C. ruminantium* to ticks. Levels of successful in vitro feeding upon skin membranes will vary greatly, depending on the feeding system and tick species being studied.

In vitro silicone membrane feeding often requires treatment of the membrane with the natural host skin, constant pheromonal stimulation, the presence of high levels of carbon dioxide, or even activation by prefeeding on a natural host. Such complicated procedures, coupled with the observation of decreased tick engorgement weights, molting success, and egg laying, have reduced the prospect of developing a convenient and successful in vitro model that could effectively be used for colony maintenance.

Chabaud (1950) originally described the use of glass capillary tubes to artificially feed ticks. This technique has been used to effectively transmit the Lyme disease spirochete, *Borrelia burgdorferi*, into *Ixodes scapularis* nymphs, as described by Broadwater et al. (2002). These nymphs took a smaller-than-average blood meal, perhaps due to the fact that the ticks' palps, which naturally remain outside a host when feeding, were positioned inside the capillary tubes. This procedure was proved effective in introducing infectious material into ticks at the experimental level, though it has yet to demonstrate convincing results in terms of maintaining a colony.

Most importantly, the use of in vitro feeding models eliminates the investigator's ability to concurrently monitor the host's responses during and following tick feeding. Such *in vitro* models are perhaps most useful only when specialized feeding needs must be addressed.

TICK HANDLING

One must exercise a great deal of caution when handling ticks. We have found that handling unfed nymphs and adults by the hind leg with fine forceps and minimal pressure works best. Engorged ticks of all life cycle stages should be handled with blunt live-insect forceps, so as to prevent rupturing.

Ticks are best handled when cooled prior to use. Placing the vial in ice or on a chill table for a few minutes will slow the ticks' activity. Caution must be used when working with ticks so as to not expose them to low temperatures for prolonged periods of time. Conversely, ticks become active in response to heat and carbon dioxide. Breathing into a tick vial warmed by the heat of one's palm will give an immediate indication of the ticks viability and possible willingness to feed.

Feeding larger numbers of larvae and nymphs and then selecting a diverse and reasonable number of adults for reproduction will best sustain the genetic background of a given species. It is also critical to occasionally introduce new genetic material into a colony. Continued inbreeding and the use of too few females may result in a "bottleneck effect," which may compromise the ticks' specific traits, including the ability to effectively transmit pathogens.

HOUSING

Following engorgement, ticks may be housed in sterile clear-glass sample vials. It is our experience that the Wheaton (Millville, NJ) 16-mL sample bottle with plastic snap cap works quite well. A dime-sized hole cut into the non-autoclavable lightweight plastic lid, coupled with a piece of fine mesh approximately $4\,cm^2$ in size, serves as a secure tick barrier while allowing sufficient air exchange (Fig. 49.5). Consistency in hole size can assist in estimating tick numbers, for most species will crawl up onto the mesh when ready to feed. A folded narrow strip of autoclaved filter paper inside the vial absorbs excess moisture and allows the ticks to crawl upward after molting.

The vials should be housed in glass dessicators containing saturated salt solutions in the dessicator basins (Fig. 49.6). Perhaps the most critical parameter in maintaining a laboratory colony is relative humidity. Most ixodid colonies require a relative humidity between 90% and 97%, depending upon species. It has been shown that saturated solutions of potassium nitrate (94%) and potassium sulfate (97%) produce a relative humidity (Winston and Bates 1960) within the ideal range for housing ticks. Other researchers use a mixture of plaster of Paris and charcoal at the vial base to help maintain humidity, but it has been our experience that this is unnecessary. The rim around the dessicator should be coated with a thin film of silicone vacuum grease, so as to ensure a tight seal and as added security against tick escape.

Vials should be grouped by species and labeled accordingly. We have found that computer-printed labels (GA International Cryo-Labels) that do not fade, collect mold, or wrinkle easily under humid conditions

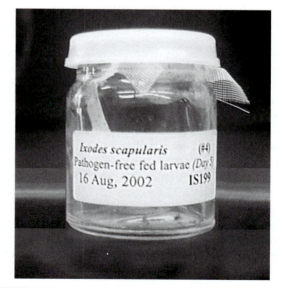

FIGURE 49.5 Glass vial with mesh barrier and plastic lid used to house ticks. The label clearly identifies species, date, and infection status as well as our numeric inventory system. Photo courtesy Ken Bourell.

FIGURE 49.6 Glass dessiccators within a temperature- and light/dark-cycle-controlled environmental chamber used to house tick colonies. The dessiccators basins should contain saturated salt solutions to maintain sufficient humidity, critical to the ticks' viability (see text). Photo courtesy Ken Bourell.

work best. Label information should include genus and species, stage, date of last feeding, whether or not the contents are infected with a pathogen, as well as any numeric inventory systems implemented in your laboratory. Keeping a computer database can assist in population inventory and scheduling of feeding, thereby simplifying management of complex multi-species colonies. Additional parameters, such as feeding times, engorged tick weights, and molting times, may also be useful in colony management.

The desiccators can be stored in environmental chambers with temperature and light/dark cycle controls. Cabinets with humidity controls are more costly and can be avoided by using dessiccators with saturated salt solutions, as described earlier. Most species can be continuously kept at 22°C, though when a colony is maintained at 15°C, the ticks' viability is prolonged. This phenomenon allows tick researchers the option of keeping a separate, long-term backup colony. It has been our experience that a constant 14-hour:10-hour light/dark cycle is sufficient for most species. Infected tick colonies should be stored separately and only in an approved chamber with clear markings.

It is recommended that a separate, clean work area in the laboratory be dedicated to tick rearing. This area should be free of clutter, and all work surfaces should be covered with white, disposable bench pads. Forceps and other handling material should be thoroughly washed and sterilized with 10% bleach between uses, to inhibit cross-contamination with any potential fungi in tick vials.

INFECTIOUS MATERIAL

Additional precautions within the laboratory are necessary when conducting research with ticks infected with pathogens. First and foremost, any work involving infectious material should be in strict compliance with your institution's biosafety committee guidelines. All persons working in direct contact with infectious material should be given training and made fully aware of the potential hazards of handling the pathogen, infected ticks, and infected animals.

When handling infected ticks, it is strongly recommended that white laboratory coats and gloves be worn and that the junction between the sleeve and the wrist of the glove be taped with masking tape. Splash-proof eyewear and masks are also suggested for both infected tick infestations and cage changing. A thorough inspection of one's hands, arms, and laboratory coat immediately following tick handling is strongly encouraged. Appropriate barriers should be placed on and around cages in order to trap any ticks that might egress from the host.

Stored infectious material should be clearly labeled as such and kept separate from pathogen-free material. Any instruments or cages coming in contact with pathogens should be placed within taped containers, treated with a 10% bleach solution following use, and then later washed and autoclaved. A sticky tape barrier along the perimeter of your work area will provide additional defense against tick escape.

Acknowledgments

The authors acknowledge the contribution made by Drs. Johnathan Pope, Doris Müller-Doblies, and Uwe Müller-Doblies in developing a rabbit capsule secured to the DuoDerm GCF wound dressing pad. The modification of the original "Nalgene® bottle" capsule used in this laboratory has proven to be a very useful change in capsule design.

Readings

Broadwater, A. H., Sonenshine, D. E., Hynes, W. L., Ceraul, S., and De Silva, A. M. 2002. Glass capillary tube feeding: A method for infecting nymphal *Ixodes scapularis* (Acari: Ixodidae) with the Lyme disease spirochete, *Borrelia burgdorferi*. *J. Med. Entomol.* 39(2): 285–292.

Chabaud, A. G. 1950. Sur la nutrition artificielle des tiques. *Ann. Parasitol.* 25: 42–47.

Endris, R. G., Hess, W. R., and Caiado, J. M. 1992. African swine fever virus infection in the Iberian soft tick, *Ornithodoros (Pavlovskyella) marocanus* (Acari: Argasidae). *J. Med. Entomol.* 29(5): 874–878.

Gratz, N. G. 1999. Emerging and resurging vector-borne diseases. *Annu. Rev. Entomol.* 44: 51–75.

Gregson, J. D. 1966. Ticks. In: *Insect Colonization and Mass Production* (C. N. Smith, ed.). Academic Press, New York, pp. 49–72.

Kemp, D. H., Stone, B. F., and Binnington, K. E. 1982. Tick attachment and feeding: Role of the mouthparts, feeding apparatus, salivary gland secretions, and the host response. In: *Physiology of Ticks* (F. D. Obenchain, and R. Galun, eds.). Pergamon Press, Oxford, UK, pp. 119–168.

Parola, P., and Raoult, D. 2001. Ticks and tick-borne bacterial diseases in humans: An emerging infectious threat. *Clin. Infec. Dis.* 32: 897–928.

Sonenshine, D. E. 1999. Maintenance of ticks in the laboratory. In: *Maintenance of Human, Animal, and Plant Pathogen Vectors* (K. Maramorosch, and F. Mahmood, eds.). Science Publishers, Enfield, NH, pp. 59–82.

Voigt, W. P., Young, A. S., Mwaura, S. N., Nyaga, S. G., Njihia, G. M., Mwakima, F. N., and Morzaria, S. P. 1993. *In vitro* feeding of instars of the ixodid tick *Amblyomma variegatum* on skin membranes and its application to the transmission of *Theileria mutans* and *Cowdria ruminantium. Parasitology* 107: 257–263.

Waladde, S. M., Young, A. S., and Morzaria, S. P. 1996. Artificial feeding of ixodid ticks. *Parasitol. Today* 12(7): 272–278.

Winston, P. W., and Bates, D. H. 1960. Saturated solutions for the control of humidity in biological research. *Ecology* 41(1): 232–237.

50

Care, Maintenance, and Experimental Infection of Human Body Lice

PIERRE-EDOUARD FOURNIER AND DIDIER RAOULT

INTRODUCTION

Human body lice, *Pediculus humanus corporis*, or *Pediculus humanus humanus* (Fig. 50.1), are obligate hematophagous ectoparasites strictly associated with humans. These insects without wings belong to the superorder *Psocodea*, order *Phthiraptera*, group *Anoplura*, family *Pediculidae*. They have a short and constricted head with two antennae that each have five segments. The thorax is compact, and the seven-segment abdomen is long and membranous with lateral paratergal plates. The louse's life cycle begins as an egg, laid in the folds of clothing. Eggs are held in place by an adhesive produced by the mother's accessory gland. The eggs hatch in 6–9 days after being laid. The emerging louse immediately moves onto the skin to feed before returning to the clothing, where it remains until feeding again. A louse typically feeds five times a day. The growing louse molts three times after hatching, normally at days 3, 5, and 10. After the final molt, the mature louse typically lives for another 20 days. Females lay about eight eggs per day. A pair of mating lice may generate up to 200 lice during their life spans. Blood is the only source of water for lice. Digestion of the blood meal is rapid; the erythrocytes are quickly hemolyzed and remain liquified.

Body lice have been recognized as human parasites for thousands of years. They are associated with lack of hygiene and promiscuity, conditions encountered worldwide either in poverty-stricken communities, during wartime, or in jails. Lice are especially prevalent during cold periods. The threat posed by body lice is not the louse itself but three bacteria causing human infections that it harbors: *Rickettsia prowazekii*, the agent of epidemic typhus, *Bartonella quintana*, respon-sible for trench fever, bacillary angiomatosis, chronic lymphadenitis, asymptomatic bacteremia, and endo-carditis, and *Borrelia recurrentis*, causing relapsing fever. The relationship of these bacteria and their host has been the subject of numerous studies, most of which were performed between WWI and WWII (Table 50.1).

Initially, laboratory colonies of body lice were used for the laboratory growth of *Rickettsia prowazekii*, the production of anti–*R. prowazekii* vaccine, the testing of antibiotic susceptibility of this bacterium, or the diag-nostic testing of *B. quintana* infection. However, lice were inoculated intrarectally, which does not reflect the physiological infection, or by feeding on artificially infected monkeys. The infection of lice with *R. prowazekii* was monitored by the survival and/or the color of the arthropods, with red lice being considered to be suffering from *R. prowazekii* infection. Later, a few colonies of lice were adapted to feeding once a day on humans and then on rabbits (Culpepper 1944, 1946; Smith and Eddy 1954). The Orlando colony, also named Culpepper colony, named after the city in Florida where it was bred, has formed the basis of many other colonies in other laboratories and is now considered the reference (Mumcuoglu et al. 1990). Another model of louse feeding on artificial mem-branes has been reported but will not be described in this chapter. Recently, we developed experimental models of body lice infection using rabbits artificially made bacteremic for *Bartonella quintana* (Fournier et al. 2001) and *Rickettsia prowazekii* (Houhamdi et al. 2002). We also demonstrated that the human body louse may play a role as an occasional vector of *Rickettsia typhi* (Houhamdi et al. 2003).

TABLE 50.1   Infections Transmitted by and Their Incidence on Body Lice

	Rickettsia prowazekii	Bartonella quintana	Borrelia recurrentis
Disease name	Epidemic typhus	Trench fever	Relapsing fever
Fatality rate (without antibiotics)	30%	<1%	10–40%
Chronic carriage	Yes	Yes	No
Relapse	One single, late (Brill–Zinsser)	Usually several (quintan fever)	Usually one
Reservoir	Human (+ flying squirrel in the United States)	Human	Human
Infection in lice	Stomach	Intestinal, generalized	Hemolymph
Disease in lice	Red louse syndrome	None	None
Human contamination	Contamination of scratches by louse feces or crushed lice	Contamination of scratches by louse feces or crushed lice	Contamination of scratches by hemolymph

TABLE 50.2   Primers Used to Detect *Bartonella quintana, Rickettsia prowazekii,*
and *Borrelia recurrentis*

Bacterium	Gene Amplified	Size of Expected Product (bp)	Primer Name (Ref.)	Primer Sequence (5'–3')
Bartonella quintana	16S–23S rRNA	704	QHVE1[F] (16) QHVE3[R] (16)	TTCAGATGATGATCCCAAGC AACATGTCTGAATATATCTTC
Rickettsia prowazekii	Citrate synthase (*gltA*)	342	CS877F[F] (18) CS1258R[R] (18)	GGGGGCCTGCTCACGGCGG ATTGCAAAAAGTACAGTGAACA
Borrelia recurrentis	16S rRNA	1,356	BF1[F] (14) BR1[R] (14)	GCTGGCAGTGCGTCTTAAGC GCTTCGGGTATCCTCAACTC
Pediculus humanus corporis	18S rRNA	513	SAIDG[F] SBIDG[R]	TCTGGTTGATCCTGCCAGTA ATTCCGATTGCAGAGCCTCG

[F] = forward; [R] = reverse.

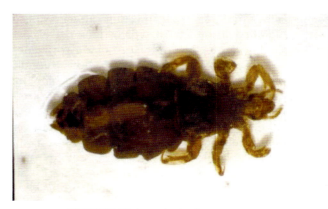

FIGURE 50.1   *Pediculus humanus corporis.*

BODY LICE

We use 15-day-old *Pediculus humanus corporis* from the Orlando strain. Lice are shown to be free from *R. prowazekii*, *B. quintana*, or *B. recurrentis* by periodic PCR-amplification (Table 50.2). Lice are kept on 4-cm² pieces of cotton tissue in 250-cm³ plastic pots. Ten to 15 lice live on each piece of tissue. Pinhead-sized holes in the lid of the pots let the air in. Lice are kept in an incubator at 30 ± 1°C and 70–80% humidity. Lice prefer a temperature between 29 and 32°C. Relative humidity (RH) is also a critical factor, for lice are susceptible to rapid dehydration. The optimal humidity for survival is in the range of 70–90% RH.

Rabbits

We use 3-month-old specific pathogen–free (SPF) New Zealand white rabbits. Usually, two rabbits, kept in individual cages in a controlled, arthropod-free university animal facility, are used alternately for the daily feeding of lice. Prior to louse feeding, one rabbit is anesthetized using an intramuscular injection of a combination of 67 mg of ketamine chlorhydrate and 17 mg of chlorpromazine. The rabbit is kept on its back, with the abdomen shaved. Lice on their cotton tissue pieces are deposited on the rabbit's abdomen and are allowed to feed for 20–30 minutes (Fig. 50.2). Meanwhile, feces and dead lice are removed from the plastic pot and thrown away. The degree of feeding is appreciated visually by the size and color of the abdomen of the lice. Once fed, the abdomen appears to be red.

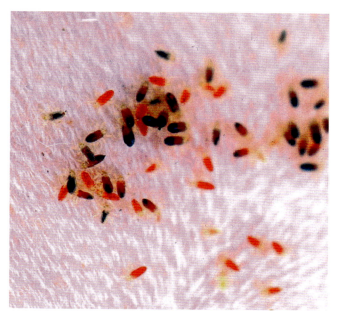

FIGURE 50.2 Human body lice on tissue pieces deposited on a rabbit's abdomen for feeding.

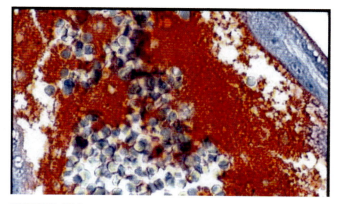

FIGURE 50.3 Immunohistological detection of *Bartonella quintana*. Bacteria appear as red clumps in the intestinal lumen and against the intestinal wall (blue). Note the numerous erythrocytes. Mag: ×660.

EXPERIMENTAL INFECTION OF BODY LICE

A suspension of 10^5–10^{10} bacteria is prepared extemporaneously in PBS or K36 buffer. One SPF rabbit, anesthetized as described earlier, is inoculated intravenously in the rostral vein of the ear with 20 mL of the bacterial suspension. During the injection of bacteria, a series of 200–800 lice is allowed to feed on the shaved abdomen of the rabbit and then kept at 30°C and 70% RH. The day of infection is referred to as *Day 1*. The following days, lice are fed daily alternately on two SPF rabbits. As a negative control, 200 SPF lice are fed on an uninfected rabbit on day 1 and thereafter alternately on two SPF rabbits.

EFFECTS OF LOUSE INFECTION

The number of dead lice is noted daily. Detection of bacteria may be performed using either polymerase chain reaction (PCR) or culture of lice, eggs, larvae, or feces. The location of bacteria in lice may be estimated using immunohistochemistry or electron microscopy. One live louse and any dead lice are taken every day for each test.

DNA is prepared from lice, eggs, larvae, or feces to be used for PCR. Specimens are placed in an Eppendorf tube and crushed with a sterile pipette tip. DNA

is extracted using the QIAamp Tissue kit (QIAGEN, Hilden, Germany) according to the manufacturer's instructions. As a negative control, DNA is extracted from an uninfected louse. As a positive control for DNA extraction, the louse's 18S rRNA is amplified (Table 50.2).

For culture, lice, eggs, and larvae are decontaminated by a 5-min immersion in 70% ethanol–0.2% iodine followed by a 5-min immersion in sterile distilled water. For feces, 0.2 μg/mL cotrimoxazole, 25 μg/mL gentamicin, and 10 μg/mL amphotericin B are added to the culture medium. Prior to inoculation, lice, eggs, and larvae are crushed with a sterile pipette tip. For *B. quintana* culture, specimens are inoculated onto 5% sheep-blood agar and incubated at 37°C under a 5% CO_2 atmosphere for up to 60 days. For *R. prowazekii* culture, specimens are inoculated on L929 cells using the shell vial technique (Marrero et al. 1990) and incubated at 35°C for 10 days. For *B. recurrentis* culture, specimens are inoculated in Barbour–Stoenner–Kelly (BSK) II medium (Sigma Chemical Co., St. Louis, MO) and incubated at 37°C for five days.

For immunochemistry, a leg of each louse is cut, and then the louse is fixed in 10% formalin for 24 hours. After lice were embedded in paraffin, 5-μm-thick sections were deparaffinized, treated with 0.3% hydrogen peroxide for 10-min, and then incubated with monclonal antibodies (Fig. 50.3).

Electron microscopy may be performed on the whole louse or on its intestine only. For the study of the louse, a leg is cut, and then the louse is fixed in 2.75% glutaraldehyde for 24 hours. The extraction of the gut from the abdomen is manual. After excision of the terminal abdominal segment, the abdominal content is squeezed out into 2.75% glutaraldehyde by immobilizing the thorax and compressing the abdomen from the thorax posteriorly with a needle under an enlarge-scope.

CONCLUSION

Pediculus humanus corporis, the human body louse, is highly specific to its host. This arthropod is the vector of three strictly human diseases: trench fever, epidemic typhus, and relapsing fever. It represents a good example of host–parasite coevolution. We have developed a laboratory model of louse infection and used it to study the relationships of *B. quintana*, *R. prowazekii*, and *B. recurrentis* with their vector. Body lice may be in contact with other bacteria due to feeding on bacteremic humans. Whether or not these other bacteria may persist in lice and be transmitted to other humans is not known and should be the subject of further studies.

Readings

Barbour, A. G. 1984. Isolation and cultivation of Lyme disease spirochetes. *Yale. J. Biol. Med.* 57: 521–525.

Barker, S. C. 1994. Phylogeny and classification, origins, and evolution of host associations of lice. *Int. J. Parasitol.* 24: 1285–1291.

Barker, S. C. 1996. Lice, cospeciation and parasitism. *Int. J. Parasitol.* 26: 219–222.

Birg, M. L., La Scola, B., Roux, V., Brouqui, P., and Raoult, D. 1999. Isolation of *Rickettsia prowazekii* from blood by shell vial cell culture. *J. Clin. Microbiol.* 37: 3722–3724.

Cadavid, D., and Barbour, A. G. 1998. Neuroborreliosis during relapsing fever: Review of the clinical manifestations, pathology, and treatment of infections in humans and experimental animals. *Clin. Infect. Dis.* 26: 151–164.

Culpepper, G. H. 1944. The rearing and maintenance of a laboratory colony of the body louse. *Am. J. Trop. Med.* 3: 327–329.

Culpepper, G. H. 1946. Factors influencing the rearing and maintenance of a laboratory colony of the body louse. *J. Econ. Entomol.* 39: 472–474.

Fournier, P. E., Minnick, M. F., Lepidi, H., Salvo, E., and Raoult, D. 2001. Experimental model of human body louse infection using green fluorescent protein–expressing *Bartonella quintana*. *Infect. Immun.* 69: 1876–1879.

Haddon, W., Jr. 1956. The maintenance of the human body louse *Pediculus humanus corporis* through complete cycles of growth by serial feeding through artificial membranes. *Am. J. Trop. Med. Hyg.* 5: 326–330.

Houhamdi, L., Fournier, P. E., Fang, R., Lepidi, H., and Raoult, D. 2002. An experimental model of human body louse infection with Rickettsia prowazekii. *J. Infect. Dis.* 186: 1639–1646.

Houhamdi, L., Fournier, P. E., Fang, R., and Raoult, D. 2003. An experimental model of human body louse infection with Rickettsia typhi. *Ann. N. Y. Acad. Sci.* 990: 617–627.

Lyal, C. H. C. 1985. Phylogeny and classification of the *Psocodea*, with particular reference to the lice (Psocodea: Phthiraptera). *Syst. Entomol.* 10: 145–165.

Marrero, M., and Raoult, D. 1989. Centrifugation-shell vial technique for rapid detection of Mediterranean spotted fever rickettsia in blood culture. *Am. J. Trop. Med. Hyg.* 40: 197–199.

Mumcuoglu, K., Miller, J., Rosen, L., and Galun, R. 1990. Systemic activity of ivermectin on the human body louse (Anoplura: Pediculidae). *J. Med. Entomol.* 27: 72–75.

Page, R. D. M., Clayton, D. H., and Paterson, A. M. 1996. Lice and cospeciation: A response to Barker. *Int. J. Parasitol.* 26: 213–218.

Raoult, D., Ndihokubwayo, J. B., Tissot-Dupont, H., Roux, V., Faugere, B., Abegbinni, R., and Birtles, R. J. 1998. Outbreak of epidemic typhus associated with trench fever in Burundi. *Lancet* 352: 353–358.

Raoult, D., and Roux, V. 1999. The body louse as a vector of reemerging human diseases. *Clin. Infect. Dis.* 29: 888–911.

Roux, V., and Raoult, D. 1995. The 16S-23S rRNA intergenic spacer region of *Bartonella* (*Rochalimaea*) species is longer than usually described in other bacteria. *Gene* 156: 107–111.

Roux, V., and Raoult, D. 1999. Body lice as tools for diagnosis and surveillance of reemerging diseases. *J. Clin. Microbiol.* 37: 596–599.

Roux, V., Rydkina, E., Eremeeva, M., and Raoult, D. 1997. Citrate synthase gene comparison, a new tool for phylogenetic analysis, and its application for the rickettsiae. *Int. J. Syst. Bact.* 47: 252–261.

Smith, C., and Eddy, G. 1954. Techniques for rearing and handling body lice, Oriental rat fleas and cat fleas. *Bull. WHO* 10: 127–137.

51

Care, Maintenance, and Handling of Infected Triatomines

PAMELA PENNINGTON, CHARLES B. BEARD, AND JENNIFER ANDERSON

INTRODUCTION

Triatomines are obligate hematophagous hemipterans. All five nymphal stages, or instars, and both sexes of the adults require regular blood meals for successful development and reproduction. In sylvatic and domestic habitats they are commonly associated with warm-blooded hosts, although some may feed on cold-blooded animals and birds in nature and in the laboratory. Triatomines are the vectors for *Trypanosoma cruzi*, the etiologic agent of Chagas disease. Sylvatic species maintain the zoonotic transmission cycle, and domestic or peridomestic species transmit the disease to humans. Of approximately 130 recognized species, fewer than 10 are considered to be of medical importance in the transmission of Chagas disease to humans, due to their ability to invade and colonize domestic habitats. The species of major medical importance include *Triatoma infestans*, *Triatoma brasiliensis*, *Triatoma dimidiata*, *Triatoma pallidipennis*, and *Rhodnius prolixus*.

Due to its fast developmental rate (3–4 months from egg to adult), *Rhodnius prolixus* has long been used in the laboratory to study insect physiology. In addition, colony establishment and maintenance of other medically important species has allowed studies of vector–parasite interactions, xenodiagnosis, bacterial symbiotic relationships, and, more recently, vector paratransgenesis. Our laboratory has experience in raising colonies of a number of species, including *Rhodnius prolixus*, *R. robustus*, *R. neglectus*, *Panstrongylus megistus*, *Paratriatoma hirsuta*, *Triatoma dimidiata*, *T. gerstaeckerii*, *T. infestans*, *T. pallidipennis*, and *T.*

sanguisuga. The following guidelines are based on data gathered by a number of investigators as well as through our own personal experience in rearing colonies from field-collected specimens.

The maintenance of triatomine colonies can be a prolonged and time-consuming task, especially when considering that full development from egg to adult can range from 3 to 12 months, depending mainly on feeding frequency, temperature, and the biology of the species. Carefully selecting the rearing conditions and recording the developmental status of the specimens are both integral to the long-term success of the colony.

REARING CONDITIONS

Temperature, Relative Humidity, and Circadian Cycle

The temperature for optimal triatomine development generally ranges between 27 and 30°C, depending on the species. Eggs will not hatch at temperatures higher than 34°C, and lower temperatures result in longer times to complete development from egg to adult. Humidity does not appear to be as critical a parameter as temperature. An experiment with *T. infestans* showed that there was no difference in the developmental rates of groups raised at a constant temperature (30°C) with varying relative humidities (RH) between 32% and 96%. For optimal colony development, it is recommended to have an incubator or a small room with controlled temperature and humidity

717

that can maintain a constant temperature of 30°C and 50–85% RH (Percival Scientific, Perry, IA). The first- and second-instar stages are highly susceptible to temperature fluctuations; therefore, if a controlled temperature room is not available, it is recommended to maintain them in an incubator. A less expensive way to control humidity is by placing a pan with a saturated salt solution inside the incubator. Combinations of different salt solutions (e.g., sodium chloride or other salt solutions) and different temperatures can be selected to achieve the desired humidity inside a closed incubator. A light source with a day/night timer is also recommended to maintain a circadian cycle of 12:12 light:dark. When starting a colony from field-collected specimens, it may be helpful to adjust the environmental conditions to resemble those of the collection site to decrease the stress imposed on the adults while they adapt to laboratory conditions.

Rearing Containers

The size of the container depends on the species, the instars, and the number of individuals to be kept (Figs. 51.1 and 51.2). Falcon brand conical tubes are practical containers to keep small numbers of individuals; approximately 10–20 nymphs of any species or up to two adults of the larger species can be accommodated (BD Biosciences, Bedford, MA). Wide-mouth pint jars (ca. 450 mL), such as transparent Nalgene brand containers or other similar plastic containers, can be used to keep 100–200 individuals, with no more than 50

adults of any species (Nalgene Nunc, Rochester, NY). A piece of paper (Whatman or brown packing paper), cut to the same height as the container, is inserted in the rearing jar to give the bugs a vertical surface to cling to during feeding and to provide a hiding place. The paper also absorbs the large quantities of feces and urine excreted during feeding. The paper may be folded accordion style, or two flat pieces may be cut to have a width identical to the diameter of the rearing jar and placed in the shape of an "X." The second method allows the handler to easily monitor the status of the individuals in a transparent plastic rearing jar without manipulating them.

Tubes and jars are covered with a fine mesh nylon cloth. Mesh greater than 40 threads per inch may interfere with feeding of the larger instars, and coarser mesh may allow first-instar nymphs to escape. The nylon is cut to cover the mouth of the container, leaving some material overhanging. This is then fastened in place with a rubber band and further secured with masking tape. A container with a screw cap lid may be cut to leave a hole on the top, and the nylon mesh can than be screwed in place. The paper should be flush against the nylon to allow first instars to reach the blood source when placed upright. For double containment, the plastic container can be placed in a gallon jar (4 L), such as a transparent plastic container or cardboard ice cream cartons (Fig. 51.1A).

Cleaning the rearing jars periodically is necessary because of the large quantities of urine and feces excreted after feeding. This is best accomplished by

FIGURE 51.1 Rearing jars, double-containment cartons, and storage rack in the insectary. (A) A plastic rearing jar containing paper folded accordion style and covered with nylon cloth (left). A gallon carton covered with nylon for double containment of the rearing jar (right). (B) A storage rack in the insectary at the Centers for Disease Control and Prevention in Atlanta, GA.

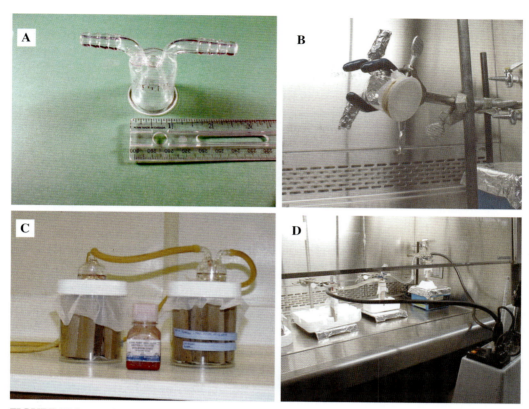

FIGURE 51.2 Artificial membrane feeding apparatus. (A) Feeding chamber (5-mL capacity). (B) Feeding membrane outfitted with a latex membrane. (C) Feeding chambers placed over rearing jars and rabbit blood container. (D) Feeding chambers clamped over conical tubes in a laminar flow hood.

transferring the paper from the old container to a clean rearing jar. Most individuals will cling to the paper, which minimizes handling that can be deleterious for development. In addition, contact with the fecal material from the older nymphs is necessary for optimal development of first instars (see later section on Bacterial Symbionts).

Containment and Manipulation

Although Triatomine adults are capable of flight, the greater risk of escape in the laboratory is due to their ability to quickly climb up the sides of containers. *Rhodnius* spp. have special climbing organs that allow them to climb efficiently even on smooth vertical surfaces. Therefore, it is best to anesthetize them in the jar with CO_2 gas or to lightly chill them in a freezer prior to manipulation. Sorting should be done by placing rearing jars inside one or two larger metal or plastic pans. The sides can be coated with a solution of Fluon® (Northern Products Inc., Woonsocket, RI) that will prevent them from crawling up the sides. It is recommended to have double-sided sticky tape along the edges of the table where they will be manipulated to further prevent escapes. The bugs should be gently picked up by the legs with long, blunt forceps. Special care should be taken when handling engorged bugs because they are more susceptible to injury. The handler should wear a white lab coat with the sleeves tucked inside latex gloves.

When working with *T. cruzi*–infected specimens, procedures should be carried out in accordance with practices described in the CDC-NIH Biosafety in Microbiological and Biomedical Laboratories guidelines for work involving biosafety level (BSL) 2 pathogens. Additional information and guidance can be obtained from the American Committee of Medical Entomology, Arthropod Containment Guidelines, which assign a specific risk-associated arthropod containment level (ACL) to various laboratory-based procedures involving arthropod vectors of human disease agents. At the present time, this document exists in draft form and can be accessed at the following URL: http://www.astmh.org/subgroup/archive/ACGv31.pdf. These guidelines can also be obtained by e-mail request directed to mbenedict@cdc.gov or cbeard@cdc.gov.

Only authorized personnel should be allowed in the room during these manipulations. Infected specimens should be double contained and kept in an incubator/room specifically designated for these rearing jars. All manipulations should be performed wearing protective clothing, gloves, goggles, and/or face shield. The use of sharps should be avoided when possible. Whenever possible, feeding and dissections of infected specimens should be performed in a laminar flow hood to protect the personnel from aerosolized fecal material (Fig. 51.2D). All material used to handle or contain infected bugs should be surface-sterilized with 10% bleach and autoclaved after each manipulation.

Insectary Maintenance

Access to the insectary should be restricted to trained personnel. Rearing jars should be stored on racks whose legs are immersed in a dish containing soapy water or oil. The racks should not make contact with walls. These measures will prevent ants, a common predator of triatomines, from crawling up on the racks. If control of ants or cockroaches is necessary, it is best to use a nonvolatile ingestible-bait insecticide. The working area should be cleaned with 10% bleach before and after every manipulation, and any garbage should be disposed of daily to maintain a clean environment. If mouse or rat infestations are a problem, rodent control should be initiated.

FEEDING TECHNIQUES

Live Animals

This is the technique of choice for general colony maintenance. The rabbit is most commonly used because it is large enough to feed a greater number of individual bugs. However, mice, chickens, or guinea pigs may also be used. In general, the animal should be restrained and anesthetized so that its movement is minimized during the procedure. Adult New Zealand rabbits can be anesthetized IM with a 1.5-mL dose of xylazine (5 mg/kg), Ketamine (35 mg/kg) and Acepromazine (0.75 mg/kg). This dose will be sufficient to anesthetize the animal for 90 min. It is important for all procedures to comply with local animal care and use committee guidelines. A rabbit can safely give 15 mL of blood every 15 days. It can be lethal to feed too many bugs on one rabbit. Based on blood uptake per instar, approximately 40 adults or 2500 first instars of *Triatoma* spp. and up to 75 adults or 3000 first instars of *Rhodnius* spp. may feed on one rabbit.

The fur may be clipped or shaved off to allow easy access to the skin, which should be wiped with ethanol (not soap) before and after feeding. The nylon

cloth–covered jar containing the bugs is placed flush against the shaved skin. In most cases, the jar is placed against the abdomen. If the animal is placed on its side, the jar may be strapped sideways with a rope. Ryckman (1952) described a feeding device consisting of a hollow stand with a hole on the top. The animal is restrained with leather straps on top of the hole, and the rearing jar is placed underneath the stand, against the rabbit's abdomen. Lop-eared rabbits may be used to avoid the use of harnesses and hardware cloth. The rabbit is comfortably placed in a box without other restraint, the ears are warmed, and the bugs are fed on the tip of the ears that hang over the side of the box. Feeding should be performed in a quiet room with dim lights to calm the animal and to emulate nighttime feeding for the bugs. Feeding should proceed for at least 30 minutes. Blowing gently across the mouth of the jar before presenting the meal alerts the bugs and hastens the feeding process.

It is difficult to synchronize colonies because not all individuals will feed the same day, and many will refuse to feed even after you repeatedly offer a blood meal. If most individuals have not fed during that period, it may be necessary to repeat several times during the day or to try again 1 week later. Difficulty may occur when individuals have recently hatched or molted; this is due to a lag period of 3–7 days during which they will not accept a blood meal. Individuals that refuse feeding after you offer several blood meals should be discarded. It is important to ensure complete feeding of the first instars because they are less resistant to starvation than later instars.

Artificial Feeding

In recent years, the use of living animals to support research activities has become increasingly more difficult and expensive. For these reasons, it may be more desirable to maintain bug colonies using an artificial feeding apparatus. Some experimental designs, such as the assessment of the efficacy of different blood meal sources and infection with *T. cruzi* or other parasites, may also require feeding through an artificial membrane. The general apparatus consists of a circulating water bath, connected through Tygon tubing to a feeding glass chamber (Lillie Glass, Smyrna, GA) (Fig. 51.2D). The glass chamber consists of an inner funnel-shaped cavity that holds the blood meal and an outer compartment through which warm water circulates (Fig. 51.2A). The wide mouth of the inner funnel is sealed with the membrane of choice by stretching it across the mouth and holding it in place with a rubber band (Fig. 51.2B). The blood to be used is then pipetted into the sealed funnel through the narrow end. The outer compartment is connected through the tubing to the circulating water bath, which heats the water to 37°C and circulates it through the outer compartment to maintain the blood at this temperature (Figs. 51.2C and D). Higher and lower temperatures may fail to stimulate the bugs to feed. Different membranes have been tried, such as Parafilm, condoms, latex gloves, and baudruche (dry intestine of cattle or sheep). The membrane should be washed with sterile water or PBS. The PBS should not contain magnesium, which may induce clotting when using anticoagulants such as citrate. We have used both sheepskin and latex condoms that are free of lubricants and spermicidal solutions. We have found that sheepskin condoms become brittle during prolonged feedings; therefore, we favor latex condoms. Latex condoms can be auto-claved in water to provide a sterile membrane. The glass feeding chamber is washed in bleach and auto-claved after each use.

Blood Sources and Additives

The blood used should be collected aseptically from animals that are not raised on feed containing antibi-otics. Trace antibiotics in blood may affect the bacter-ial endosymbionts found in the midguts of these bugs (see later section on Bacterial Symbionts). Some common blood sources are pig, rabbit, human, mouse, sheep, cow, horse, and chicken. Pig, rabbit, human, and mouse blood appear to be the best sources for optimal development. Sheep, cow, horse, and chicken blood result in retarded development, lower repro-ductive rates, or death. For convenience and economy, we use sheep blood for short-term experiments. The blood should be handled using BSL 2 standards. Fresh blood is preferred, but when refrigerated, blood may be used for up to 2 weeks after it is drawn. The blood may be defibrinated with glass beads (freshly drawn blood is vigorously shaken in a sterile Erlenmeyer flask with glass beads until the beads adhere to each other), or it may be collected with anticoagulants such as heparin (5–8 IU/mL blood) or sodium citrate (3.8 g sodium citrate/100 mL blood). Some phagostimulants include glucose (0.18 g/100 mL blood) and ATP (1 mM or 5.51 mg/10 mL blood). In all cases, the blood placed in the feeding chamber should be tested microbiolog-ically after feeding to verify that bacterial contamina-tion has not occurred during the feeding process. We typically spread an aliquot of the blood on a brain heart infusion plate and incubate it at 28°C for a week.

Feeding Intervals to Optimize Development

Based on average intervals between molts (Table 51.1), feeding every 3 weeks is sufficient to maintain a healthy colony. One large blood meal usually leads to molting (except with *Triatoma sordida*). The develop-mental time between molts depends on the species and the instar, and the time increases with each molt. The time between molts may be as little as 1 week and as long as 4 months, averaging 3–4 weeks.

T. infestans, *Panstongylus megistus*, and *R. prolixus* have the fastest developmental rates, from 3 to 6 months to develop from egg to adult. *T. dimidiata* is the slowest of all species tested by Szumlewicz (1975), taking 8.5 to 12 months to develop. The table compares data from different sources regarding different devel-opmental parameters of three medically important species, showing the two ends of the spectrum, from the fastest to the slowest development. These parame-ters are commonly used to monitor the developmental status of the colony. It may be necessary to adapt the feeding schedule to the developmental rate of the species of interest.

The first instars are the most susceptible to starva-tion and should be fed no later than 1 week after hatch-ing. It is noteworthy that all stages are somewhat resistant to starvation, ranging from 13 to 200 days, depending on the species and instar, with fourth and fifth instars being the most resistant. Some species refuse to feed during the first days after hatching and after molting.

Feeding Intervals to Optimize Reproduction

The life span of an adult female ranges from 9 months to 2.5 years, depending on the species. Ovipo-sition is optimal when the female is fed every 2 weeks and is maintained in contact with fed males. Each female may lay 200–1,000 eggs during its life, with regular egg-laying intervals interrupted by resting periods. It should be noted that adults from colonies maintained in the laboratory for several generations have shorter life spans and lower oviposition rates than field-collected individuals.

STARTING COLONIES FROM FIELD-COLLECTED SPECIMENS

Parasite Infections

Depending on the species and collection site, field-collected specimens are likely to be infected with *T. cruzi*. These specimens should be handled according to appropriate BSL and ACL guidelines. They should be kept in containers clearly labeled "Biohazard" and quarantined in a room separate from "clean" colonies. All manipulations should be performed wearing

TABLE 51.1 Developmental parameters for nymphal and adult stages of three triatomine species

	Szumlewicz, 1975[$] T. infestans	Schaub and Lösch, 1989[Y] T. infestans	Buxton, 1930[&] R. prolixus	Szumlewicz, 1975[$] R. prolixus	Lake and Friend, 1968[$] R. prolixus	Szumlewicz, 1975[$] T. dimidiata	Zeledón, 1981[#] T. dimidiata
Developmental periods (days)							
Egg	8–21	NA	10–37	11–24	NA	21–28	28–31
1st instar	7–25	NA	12–15	11–50	9–13	16–61	24–32
2nd instar	12–13	11–13	11	10–30	10–13	21–25	38–79
3rd instar	13–23	13–15	16–20	10–50	10–15	31–51	49–66
4th instar	22–32	16–32	30	30–40	12–19	44–124	73–90
5th instar*	35–66	NA	27	20–30	20–26	69–89	67–99
Male	NA	NA	NA	NA	NA	NA	157–1139
Female	488 ± 18	NA	NA	NA	NA	522 ± 31	355–994
Preoviposition time (days)	15 ± 26	NA	13	NA	NA	21 ± 11	17–47
# eggs laid/female	920 ± 35	NA	200–300	NA	NA	684 ± 22	446–2054
Eggs hatched (%)	92	NA	82	NA	NA	72	91

Values are given as minimum-maximum or mean ± standard deviation.

* Females usually ecdyse before males.

Maintained at room temperature (21°C–24°C).

& Maintained at 24°C.

Y Maintained at 26°C.

$ Rearing conditions unavailable.

NA (Not available).

gloves, goggles, and protective clothing, as described previously under Containment and Manipulation. To test specimens for infection with *T. cruzi*, the bugs should be fed preferably through an artificial membrane or on an animal designated for this purpose only. Some field isolates do not feed readily through membranes. The jar should be placed in an upright position to ensure that no fecal material is deposited on the animal's skin. Once fed, each specimen to be tested is placed upright inside a conical tube ranging in size from 1.5 to 15 mL, depending on the size of the individual, and allowed to defecate. Defecation will occur minutes to hours following feeding. The dark feces or clear urine collected in the tube can then be observed microscopically for metacyclic trypanosomes. All material to come in contact with the bugs must be surface sterilized with 10% bleach and autoclaved before discarding. To prevent cross-contamination, colony and field specimens should be sorted in separate containers. It may be desirable for personnel working with potentially infected colonies to be serologically tested periodically for exposure to *T. cruzi*. A baseline serum sample should also be maintained for all personnel working in the laboratory.

It is not uncommon for field-collected triatomines to be infected with other trypanosomatids. For example, *Rhodnius* spp. may be infected with *T. rangeli*, and *Triatoma* spp may be infected with *Blastocrithidia triatomae*. These entomopathogenic parasites can affect development and ultimately kill their triatomine hosts. The parasites are transmitted horizontally by coprophagy or by cannibalism. In the case of unexplained deaths during the maintenance of field-collected specimens, diagnosis may be performed by microscopically analyzing the hemolymph or salivary glands of *R. prolixus* for *T. rangeli* and the feces of *T. infestans* for *B. triatomae* cysts.

Bacterial Symbionts

Bacterial endosymbionts of *R. prolixus* were described in the early 1900s. These actinomycete bacteria, *Rhodococcus rhodnii*, were found in pure cultures in the midgut contents and in the feces of *R. prolixus* from laboratory colonies. Newly hatched instars were found to acquire their bacterial flora through coprophagy. This behavior allows the bacteria to be efficiently spread within a colony. It has been suggested that these bacteria provide essential vitamins that are lacking in the blood meal and are required for development. Cleaning the eggs and raising them in sterile containers separate from the older nymphs leads to an aposymbiotic state (i.e., triatomines lacking bacterial flora), which results in developmental arrest

of fourth-instar nymphs. This symbiotic relationship is specific; different triatomine species harbor different bacterial species. For this reason, it is important to maintain eggs and early-instar nymphs in rearing jars containing older nymphs and adults or with the fresh feces from older individuals. If the rearing jar is to be cleaned or if the colony must be split up, feces-contaminated papers from the old container should always be transferred to the new rearing jar. When starting colonies from eggs collected in the field, the best results will be obtained with eggs that are contaminated with fecal material.

Parasitoids and Ectoparasites

Field-collected specimens can also introduce parasitoids and ectoparasites, such as wasps and mites. The parasitoid wasp, *Telenomus fariai*, lays its eggs inside the triatomine eggs. Once it hatches, it can survive for 10 days without a sugar source. All egg-laying females should be quarantined inside a second jar with extra fine mesh, and all eggs should be stored in closed containers until hatching to ensure that all parasitoid wasps that hatch die before they can feed and reproduce. This is important in insectaries where mosquitoes are also maintained because the emerging wasps may feed on the sugar water used to feed the adults.

Mites of the genus *Pimeliaphilus* are commonly found in field-collected specimens and can be devastating if introduced into a colony. Even though they show a preference for their natural host, they will colonize other species in the laboratory. All individuals should be inspected under a stereoscopic microscope before initiating a colony. If there are mites, it may be possible to physically remove some by anesthetizing the bug with ether or CO_2 and picking the mites off with a cotton applicator. If there are too many to remove, the individual should be discarded before introducing it into the colony. Eggs laid by infested females may be collected and allowed to hatch in a separate container to prevent cross-infestations, always ensuring that they are in contact with fecal material.

CONCLUSION

Initiating and maintaining triatomine colonies requires knowledge of the intricate biology of these hematophagous insects. Factors such as the differences in the developmental rates of the different species and their unique bacterial symbionts are important in establishing routine feeding and cleaning procedures. In addition, special methodologies must be established when working with field-collected specimens because these can transmit *T. cruzi* to humans. Additionally, they can be naturally infected with entomopathogenic parasitoids or ectoparasites, which can greatly decrease the fitness of the colony and/or spread to other, healthy colonies that are maintained in the same facility.

Materials and Suppliers

- Plastic sorting container with the sides coated with Fluon® [$110/quart, Northern Products Inc., 152 Hamlet Ave., P.O. Box 1175, Woonsocket, RI 02895; (401) 766-2240]
- Nalgene containers with the lids cut open or other plastic containers (Nalgene Nunc International Corp., 75 Panorama Creek Drive, P.O. Box 20365, Rochester, NY 14602-0365)
- Falcon tubes (BD Biosciences, Two Oak Park Drive, Bedford, MA 01730)
- Gallon "ice cream" cartons
- Net material to cover both Nalgene container and cartons; small mesh: nylon organdy style 65200, ivory; large mesh: nylon white toule 106 × 108
- Soft-tip forceps
- Scissors
- Condoms (plain)
- Rubber bands
- Glass membrane feeder ($75 per feeder, listed as "membrane feeders" in the ordering catalog, Lillie Glass, Attn. Don Lillie, 3431 Lake Drive, Smyrna, GA 30082)
- Brown paper, cut and folded accordion style
- 10% bleach solution and kaydrys
- Circulating water bath
- Whole young rabbit blood, aseptically collected, Na citrate, 100 mL, code #31134-1 ($65.00/100 mL; Pel-Freeze Biologicals, P.O. Box 68, Rogers, AR 72757; (800) 643-3426)
- Biological incubator (Percival Scientific Inc., 505 Research Drive, Perry, IA 50220; (515) 465-9363)
- Double-sided sticky tape

Acknowledgments

The authors are grateful to Lic. Celia Cordón-Rosales (Medical Entomology Research and Training Unit/Centers for Disease Control and Prevention/Guatemala City, Guatemala) for her continued support in the establishment and maintenance of colonies of *T. dimidiata* and *R. prolixus*.

Readings

Biosafety in microbiological and biomedical laboratories. 1999. U.S. Government Printing Office, Washington, DC.

Azambuja, P., and Garcia, E. S. 1997. Care and maintenance of triatomine colonies. In *The Molecular Biology of Insect Disease Vectors* (J. M. Crampton, C. B. Beard, and C. Louis, eds.), Chapman and Hall, New York, pp. 56–64.

Beard, C. B., O'Neill, S. L., Tesh, R. B., Richards, F. F., and Aksoy, S. 1992. Transformation of an insect symbiont and expression of a foreign gene in the Chagas' disease vector *Rhodnius prolixus. Am. J. Trop. Med. Hyg.* 46: 195–200.

Brecher, G., and Wigglesworth, V. B. 1944. The transmission of *Actinomyces rhondii* Erikson in *Rhodnius prolixus* Stal Hemiptera and its influence on the growth of the host. *Parasitology* 35: 220–224.

Buxton, P. A. 1930. The biology of a blood-sucking bug, *Rhodnius prolixus. Trans. Entomol. Soc. London* 78: 227–236.

Durvasula, R. V., Gumbs, A., Panackal, A., Kruglov, O., Aksoy, S., Merrifield, R. B., Richards, F. F., and Beard, C. B. 1997. Prevention of insect-borne disease, an approach using transgenic symbiotic bacteria. *Proc. Nat. Acad. Sci. U.S.A.* 94: 3274–3278.

Eichler, S., Reintjes, N., Jung, M., Yassin, A. F., Schaal, K. P., Junqueira, A. C. V., Coura, J. R., and Schaub, G. A. 1996. Identification of bacterial isolates and symbionts from wild populations of *Triatoma infestans* and *T. sordida. Memorias do Instituto Oswaldo Cruz* 91 (Suppl.) 125.

Garcia, E. S., and Azambuja, P. 1997. Infection of triatomines with *Trypanosoma cruzi*. In *The Molecular Biology of Insect Disease Vectors* (J. M. Crampton, C. B. Beard, and C. Louis, eds.), Chapman and Hall, New York, pp. 146–155.

Goodchild, A. J. P. 1955. The bacteria associated with *Triatoma infestans* and some other species of Reduviidae. *Parasitology* 45: 441–448.

Lake, P., and Friend, W. G. 1967. A monoxenic relationship, *Nocardia rhodnii* Erikson in the gut of *Rhodnius prolixus* Stal (Hemiptera, Reduviidae). *Proc. Entomol. Soc. Ontario* 98: 53–57.

Lake, P., and Friend, W. G. 1968. The use of artificial diets to determine some of the effects of *Nocardia rhodnii* on the development of *Rhodnius prolixus. J. Insect Physiol.* 14: 543–562.

Perlowagora-Szumlewicz, A., Muller, C. A., and Moreira, C. J. 1990. Studies in search of a suitable experimental insect model for xenodiagnosis of hosts with Chagas' disease 4—The reflection of parasite stock in the responsiveness of different vector species to chronic infection with different *Trypanosoma cruzi* stocks. *Revista Saude Pub.* 243: 165–177.

Rabinovich, J. E., Leal, J. A., and Feliciangeli de Pinero, D. 1979. Domiciliary biting frequency and blood ingestion of the Chagas's disease vector *Rhodnius prolixus* Stahl (Hemiptera, Reduviidae), in Venezuela. *Trans. Royal Soc. Trop. Med. Hyg.* 73: 272–283.

Ribeiro, J. M., Hazzard, J. M., Nussenzveig, R. H., Champagne, D. E., and Walker, F. A. 1993. Reversible binding of nitric oxide by a salivary heme protein from a bloodsucking insect. *Science* 260: 539–541.

Ryckman, R. E. 1952. Laboratory culture of Triatominae with observations on behavior and a new feeding device. *J. Parasitol.* 38: 210–214.

Schaub, G. A. 1988. Direct transmission of *Trypanosoma cruzi* between vectors of Chagas' disease. *Acta Trop.* 45: 11–19.

Schaub, G. A. 1990. Membrane feeding for infection of the reduviid bug *Triatoma infestans* with *Blatocrithidia triatomae* Trypanosomatidae and pathogenic effects of the flagellate. *Parasitol. Res.* 76: 306–310.

Schaub, G. A. 1992. The effects of Trypanosomatids on insects. *Adv. Parasitol.* 31: 255–304.

Schaub, G. A., Böker, C. A., Jensen, C., and Reduth, D. 1989. Cannibalism and coprophagy are modes of transmission of *Blastocrithidia triatomae* Trypanosomatidae between triatomines. *J. Protozool.* 36: 171–175.

Schaub, G. A., and Lösch, P. 1989. Parasite/host interrelationships of the trypanosomatids *Trypanosoma cruzi* and *Blastocrithidia triatomae* and the reduviid bug *Triatoma infestans*, influence of starvation of the bug. *Ann. Trop. Med. Parasitol.* 83: 215–223.

Schofield, C. J. 1994. *Triatominae Biología y Control*. Eurocommunica Publications, West Sussex, UK, 76 pp.

Szumlewicz, A. P. 1975. Laboratory colonies of Triatominae, biology, and population dynamics. In *American Trypanosomiasis Research Conference*. Brazil, Pan American Health Organization Sc. Pub. 318, pp. 63–82.

Wigglesworth, V. B. 1936. Symbiotic bacteria in a blood-sucking insect, *Rhodnius prolixus* Stahl (Hemiptera, Triatomidae). *Parasitology* 28: 284–289.

Wigglesworth, V. B. 1972. *The Principles of Insect Physiology*. Wiley, New York.

Winston, P. W., and Bates, D. H. 1960. Saturated solutions for the control of humidity in biological research. *Ecology* 41: 232–237.

Zeledón, R. 1981. El *Triatoma dimidiata* y su relación con la enfermedad de Chagas. INCIENSA, EUNED, Costa Rica.

52

Care, Maintenance, and Experimental Infection of Fleas (Siphonaptera)

REX E. THOMAS, KEVIN R. MACALUSO, AND ABDU F. AZAD

INTRODUCTION

Flea colonies have been maintained in laboratories for nearly 100 years to support research efforts in anatomy, physiology, behavior, and pathogen transmission. For the most part, laboratory rearing and infection of fleas has focused on establishing host/flea associations close to those found in nature. This has often involved maintaining colonies of vertebrate hosts to support the reproductive cycle of the fleas or flea-borne pathogens. Some flea species can be reared successfully on a wide variety of host species, including laboratory mice or rats (e.g., *Ctenocephalides felis*, *Xenopsylla cheopis*, *X. astia*, and *Leptopsylla segnis*). Other species tolerate alternative hosts with reduced reproductive capacity (e.g., *Pulex irritans*, *Stenoponia* sp.) In contrast to cosmopolitan species of fleas (e.g., *C. felis*, *C. canis*, *X. cheopis*, and *L. segnis*), many species are host specific and are found only within the home ranges of their mammalian hosts. Laboratory rearing of fleas that are host specific has proven difficult.

Within the past 50 years much work has gone into developing methods for in vitro feeding and ultimately mass rearing of medically and economically important species, especially *C. felis*. The laboratory rearing of fleas, particularly the plague flea, *X. cheopis*, has been carried out successfully using suckling mice or rats. Although *C. felis* has been maintained on glass membrane feeders containing rat or cow blood and membranes such as Baudruche, parafilm, and shaved rat skin, the flea yields are normally small. Nevertheless, such a system has been used successfully to infect

fleas with flea-borne pathogens. In the late 1980s an in vitro rearing system became available commercially. It is commonly referred to as the "artificial dog." The system provides a convenient means of maintaining colonies and/or infecting fleas with various flea-borne pathogens (e.g., *Rickettsia typhi*, *Yersinia pestis*, and *Bartonella henselae*). Because *C. felis* has become such an economically important species worldwide and because many academic, corporate, government, commercial production, and consulting laboratories now use this in vitro system, the remainder of this chapter describes its use as well as its limitations.

REARING CONDITIONS

The artificial dog, in essence, is a heated Plexiglas box from which chambers of fleas are suspended and in which a blood source is heated for the fleas (Fig. 52.1). There have been many design modifications to accommodate the needs of various users over the years. Generally, it comprises round chambers with a fine screen on the bottom to contain adult fleas, their feces and eggs, and a coarser recessed screen on the top through which fleas feed. Metal or glass cylinders containing blood fit into the recessed top of each chamber (Fig. 52.2A). The cylinder is covered with a lid or stopper to prevent blood from drying out in the heated Plexiglas box. The bottom of the cylinder is covered with thinly stretched Parafilm to contain blood and provide a membrane through which fleas probe.

FIGURE 52.1 Flea-rearing apparatus. The plexiglas unit consists of a heated (top) chamber to keep the blood meal warm, while fleas are suspended in cages below at room temperature.

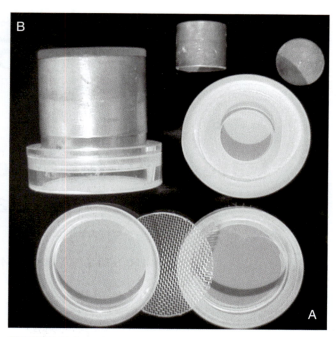

FIGURE 52.2 Individual components of flea-rearing cage.

Stretching Parafilm to an appropriate thickness is the more difficult part of using the system for most new users. Stretched too thinly, the membrane splits and the flea chamber floods with blood; too thick, and the fleas are not be able to probe through it. The technique for correctly stretching Parafilm for this system is best learned from someone skilled in the art. Once assembled (Fig. 52.2B) the chamber is suspended from the bottom of the Plexiglas box. The cylinder containing blood fits through a hole into the heated box, while the flea chamber rests underneath. The maximum number of chambers that can be used depends on their diameter. This also defines the maximum number of adult fleas that can be placed in each chamber.

Adult cat fleas can be obtained from a number of commercial sources in the United States. Some vendors rear them on cats; others use the artificial dog. Use of the system first involves placing adult fleas in the feeding chambers. Generally, this is done with a vacuum loading system available from the manufacturer. Adult fleas are released into a containment vessel, such as a pickle bucket or aquarium, and then vacuumed through a flexible tube into the bottom half of the chamber. The top is placed on the chamber before it is removed from the vacuum. If precise numbers of fleas and/or an exact sex ratio are desired, fleas may be anesthetized with carbon dioxide (CO_2), counted, and/or sorted before being put into chambers. After exposing fleas to CO_2, placing them in a dish or tube on ice will prolong anesthesia. Use of CO_2 for anesthesia in other arthropod groups has been shown to alter their innate susceptibility to some vector-borne pathogens, and this should be considered if it is used.

Once fleas are in the chambers, blood cylinders are placed on top and the combination is positioned under the heated chamber. The amount and type of blood used depends on the number of fleas in the chamber and the investigator's needs. Less blood may be used for short-term experiments; however, when rearing *C. felis*, excess blood should be used so that it will not run out before being replenished. Because *C. felis* is a persistent feeder, once female fleas have begun to feed in this system, they will not survive long in the absence of blood. Changing the blood daily helps to reduce problems with bacterial and fungal contamination. Many laboratories use cattle blood for mass rearing *C. felis*, because it is readily available from commercial sources or from slaughterhouses and is relatively inexpensive and because, whether it is defibrinated, heparinized, or citrated, cat fleas feed on it. Blood from other species (e.g., sheep) may work as well or better, but availability and cost may be problems.

The optimal number of fleas per cage depends on cage size. Smaller chambers generally are ca. 2.5 cm in diameter and are appropriate for pairs to tens of fleas. Chambers of 6- to 7.5-cm diameter will support tens to thousands of fleas, depending on the objectives of the program. Several hundred fleas per larger cage are appropriate for colony maintenance purposes. The optimal number of fleas per rearing chamber can be assessed by comparative egg counts. Normally, fleas begin feeding within minutes to an hour of being placed in the system. Feeding is evident by the presence of adult flea feces (dried blood) in the bottom of the chamber. Female fleas typically lay no eggs within the first 24 h of feeding on the artificial dog. During the second 24 h they lay approximately one-half of their eventual daily egg output. Maximal fecundity should be reached by day 3; 10–20 eggs per female per day is normal, depending on the type and quality of the blood source and other laboratory conditions.

Eggs hatch approximately 72 h after being laid; therefore, chambers are opened every 48 h starting on day 3. Adult fleas are transferred to clean cages and back onto the artificial dog. The remaining cage contents (eggs and adult flea feces) are placed onto a larval-rearing medium, either in a large containment vessel such as a crock or in smaller individual containers (e.g., 100-mm petri dish). There are nearly as many recipes for larval-growth medium as there have been laboratories raising fleas. Generally, they contain 90% sand and/or sawdust as a base, a small amount of dried blood (in addition to the adult flea feces), a ground meal such as dog, mouse, or rabbit pellets, and often a portion of brewer's yeast. Larvae can survive on adult flea feces alone; however, each additional component appears to enhance the ultimate survival of the larval population. This may be due to a reduction in cannibalism within the rearing dish. Contamination of larval-rearing containers by mites, psocids, and/or fungus is common. This can be delayed by sterilizing the medium before use, keeping the humidity of the rearing environment to a minimum (approximately 75% RH for *C. felis*), and using the medium for a limited time (one larval generation). Cat flea larvae have fewer requirements for temperature and humidity than other species. They can be maintained near room temperature (25–27°C) and at lower relative humidity (approximately 75%) than many other species. In contrast, *Xenopsylla cheopis* requires humidity around 90% for survival of immature stages.

After three larval molts (ca. 7–8 d after eggs are collected), third-instar larvae spin silken cocoons into which they incorporate detritus from the growth medium. Two to 3 days later adults begin to emerge. Prior to pupation, third-instar larvae shed their gut contents and become solid white. At this stage they are known as *prepupae*. As prepupae begin to spin a cocoon they fold themselves in half. This appears to facilitate use of their mandibles to draw silk from glands at the posterior end of the body. Some investigators differentiate this stage as "U" or "V" larvae, for their shape.

If large vessels are used to rear larvae, pupae are sieved from the medium every day to facilitate adult collection away from the rearing medium. When smaller containers are used, they are placed in a larger vessel (e.g., an aquarium) into which adults are released and then collected. Blowing gently on the dish will cause emerged adults to jump from the dish. Small containers must be sealed to prevent escape of larval or adult fleas. This requires that such containers have perforations or mesh in the lid for gas exchange. Larval fleas produce large amounts of ammonia that may reach levels lethal to the larvae if it is contained. Female fleas emerge from cocoons 2–3 days earlier than males. Maximal female emergence occurs 14–15 days after eggs are collected. Male emergence peaks on or about day 18. These time periods vary with rearing temperature. If individual cohorts of eggs are reared to the adult stage, daily collection of adults results in sex-biased collections during early and late emergence. Overall sex ratio for any individual cohort approximates 50:50. Adult cat fleas survive days to weeks if left unfed. However, as mentioned previously, once *C. felis* adults adapted to the artificial dog have been fed blood they require a nearly constant blood supply for survival.

The artificial dog system has substantially enhanced research into flea biology, control, and flea-borne diseases; however, along with the benefits of this system, there are potential limitations. One drawback is the lack of its suitability to rear wild-caught fleas. A period of selection and adaptation is required to rear fleas on the artificial dog; indeed, a flea colony that has been adapted to this system is characteristically different from the wild-caught cat fleas as a result of this selection process. Differences may include smaller size, infection rates with symbionts, and potentially a decreased ability to feed successfully on animal hosts. It is of interest that cat flea infection rates with *Rickettsia felis* range between 1 and 5% in wild caught fleas as compared to ~70% in some cat-reared laboratory colonies and 0% in some colonies maintained on the artificial dog. Whether infection with *R. felis* in colonized fleas is a factor that affects sex ratio, as described for other insect–bacteria relationships, is yet to be determined. Differences in size between wild-caught and colonized fleas may be due to the feeding chamber design. Smaller fleas may have a feeding advantage as a result of better positioning on a screen within the chambers (trampoline). Some laboratories have discontinued use of this internal screen successfully. While it has been our experience that *C. felis* will feed on animal hosts after membrane feeding, the effects of adaptation to in vitro feeding are an important consideration for pathogen transmission studies. Likewise, potentially altered behavior and/or physiology in a cat-flea colony maintained by an in vitro system means that their obvious utility in insecticide testing must be validated in vivo.

Although the artificial dog has its limitations, it provides researchers with uniformly consistent parameters for experimental design. The relative ease in mass-rearing fleas makes this system a useful tool for producing large numbers of fleas required for laboratory studies while reducing animal welfare concerns.

MARGINALIA

Blood-fed female *C. felis* lay nearly the same number of eggs per day in the absence of males, but they do not hatch. If adults are collected from cohorts where emergence has just begun (primarily female) and from cohorts that have been emerging for 7–9 days (primarily male) and combined, the resulting sex ratio approximates 50:50. Female flea fecundity has been shown to be nearly the same with sex ratios as low as 20% male.

Samples of clean eggs can be collected by removing adult fleas (living and dead) from feeding chambers, resealing the chambers, and then submerging them in water to dissolve the adult flea feces. Once done, the chamber contains clean flea eggs. This also is a convenient technique for counting eggs to assess fecundity.

Placing washed eggs in a petri dish at appropriate humidity and collecting larvae as they hatch results in a pool of unfed first-instar larvae. Alternatively, placing eggs and adult flea feces into a clean petri dish is a convenient means to collect fed first-instar larvae.

Not all third-instar larvae spin cocoons. Close inspection of rearing dishes will result in observation of "naked" pupae. Alternatively, placing third-instar larvae in the wells of round-bottom ELISA plates to pupate without making cocoons is a published technique. Either method is relatively convenient for collection of clean pupae.

LABORATORY INFECTION OF FLEAS

Methods for in vitro infection of fleas with various pathogens, primarily bacteria, have been published. The artificial dog can be used as well. Any work with pathogens preferably should be done in the secure BSL-2 or BSL-3 to ensure the safety of laboratory personnel. Glass blood cylinders are recommended to facilitate sterilization. Plexiglas feeding chambers for adult fleas do not withstand autoclaving, but they can be bleach-sterilized. Over time this may weaken the glue that holds the external screens.

Pathogenic microbes can be titrated in a known volume of buffered saline or blood and added to measured quantities of blood to produce desired concentrations for infecting fleas. The advantage of using the artificial dog is that one can control the number of pathogens and thus quantitatively determine pathogen intake by fleas. This is important in establishing minimum infectious doses. It is our experience that cat fleas can be readily infected with *Rickettsia* sp. and *Bartonella* sp. via addition of pathogens to defibrinated blood. Because cat fleas are persistent feeders, infected blood should be removed 3–4h after the addition of pathogens and replaced with fresh blood after removal of unfed fleas.

Containment must be a priority consideration when using this system to infect fleas. The upper portion of the artificial dog contains a fan-driven heater that drives air through the feeding chambers. The potential to aerosolize dried adult flea feces is significant. Placement of the entire unit within a laminar-flow hood may be necessary. Careful inspection of the screens on both sides of feeding chambers before adding fleas for infection is important. Extremely small tears at glued seams or holes in the mesh allow adults to escape. A modified artificial dog that has a sealed lower portion allows for containment of potentially infected fleas in the event of escape during handling of the chambers. Other means of containment include use of a pan filled with water, containing a detergent, placed underneath the flea cages. Daily cleaning of laboratory floors with bleach to kill fleas that may have escaped is recommended. Some laboratories also use commercially available light traps on laboratory floors. Careful planning and attention to detail prevents escape of fleas from the artificial dog.

Collection of potentially infected adult fleas from feeding chambers can be done by several means, depending upon the needs of the investigator. Flea feces may be contaminated with the pathogen. Chambers can be submerged in several changes of clean water, mild detergent, or bleach solution to remove external contamination from adult fleas. If the fleas are to be dissected, to confirm infection of internal organs, they can be anesthetized with CO_2 or frozen in the chamber prior to collection, with or without prior washing.

Readings

Akin, D. E. 1984. Relationship between feeding and reproduction in the cat flea, *Ctenocephalides felis* (Bouche) (Siphonaptera Pulicidae). M.S. Thesis, University of Florida, ••.

Amin, O. M., Jun, L., Shangjun, L., Yumei, Z., and Lianzhi, S. 1993. Development and longevity of *Nosopsyllus laeviceps kuzenkovi* (Siphonaptera) from Inner Mongolia under laboratory conditions. *J. Parasitol.* 79: 193–197.

Bacot, A. 1914. A study of the bionomics of the common rat fleas and other species associated with human habitations, with special reference to the influence of temperature and humidity at various periods of the life history of the insect. *J. Hygiene* (Plague Suppl. III): 447–453.

Bar-Zeev, M., and Sternberg, S. 1962. Factors affecting the feeding of fleas (*Xenopsylla cheopis* Rothsch.) through a membrane. *Entomol. Exp. Appl.* 5: 60–68.

Benton, A. H., Surman, M., and Krinsky, W. L. 1979. Observations of the feeding habits of some larval fleas (Siphonaptera). *J. Parasitol.* 65(4): 671–672.

Bruce, W. N. 1948. Studies on the biological requirements of the cat flea. *Ann. Entomol. Soc. Am.* 41: 346–352.

Cavanaugh, D. C., Stark, H. E., Marshall, Jr., J. D., and Rust, Jr., J. H. 1972. A simple method for rearing fleas for insecticide testing in the field. *J. Med. Entomol.* 9: 113–114.

Cerwonka, R. H., and Castillo, R. A. 1958. An apparatus for artificial feeding of Siphonaptera. *J. Parasitol.* 44: 565–566.

Dean, S. R., and Meola, R. W. 2002. Effect of diet composition on weight gain, sperm transfer and insemination in the cat flea (Siphonaptera: Pulicidae). *J. Med. Entomol.* 39: 370–375.

Dodson, B. L., and Robinson, W. H. 1986. The economic impact of fleas. *Pest Control Tech.* 14(7): 58.

Dryden, M. W. 1989a. Biology of the cat flea, *Ctenocephalides felis felis. Comp. Anim. Pract.* 19(3): 23–27.

Dryden, M. W. 1989b. Host association, on-host longevity and egg production of *Ctenocephalides felis felis. Vet. Parasitology* 34: 117–122.

Dryden, M. W., and Gaafar, S. M. 1991. Blood consumption by the cat flea, *Ctenocephalides felis* (Siphonaptera: Pulicidae). *J. Med. Entomol.* 28: 394–400.

Dryden, M. W., and Smith, V. 1994. Cat flea (Siphonaptera: Pulicidae) cocoon formation and development of naked pupae. *J. Med. Entomol.* 31: 272–277.

Edney, E. B. 1947. Laboratory studies of the bionomics of the rat fleas, *Xenopsylla brasiliensis*, Baker and *X. cheopis* Roths. 3. Factors affecting adult longevity. *Bull. Entomol. Res.* 38: 389–404.

Elbel, R. E. 1951. Comparative studies on the larvae of certain species of fleas (Siphonaptera). *J. Parasitol.* 37: 119–128.

Galun, R. 1966. Feeding stimulants of the rat flea, *Xenopsylla cheopis. Life Sciences.* 5: 1335–1342.

Hastriter, M. W., and Cavanaugh, D. C. 1981. An apparatus for colonizing fleas (Siphonaptera) and collecting pupal cocoons. *J. Med. Entomol.* 18: 251–252.

Hastriter, M. W., Robinson, D. M., and Cavanaugh, D. C. 1980. An improved apparatus for safely feeding fleas (Siphonaptera) in plague studies. *J. Med. Entomol.* 17: 387–388.

Higgins, J. A., Sacci Jr., J. B., Schriefer, M. E., Endris, R. G., and Azad, A. F. 1994. Molecular identification of rickettsia-like microorganism associated with colonized cat fleas (*Ctenocephalides felis*). *Insect Molec. Biol.* 3(1): 27–33.

Higgins, J. A., Radulovic, S., Jaworski, D. C., and Azad, A. A. 1996. Acquisition of the cat scratch disease agent *Bartonella henselae* by cat fleas (Siphonaptera: Pulicidae). *J. Med. Entomol.* 33(3): 490–495.

Hinkle, N. C., Koehler, P. G., and Patterson, R. S. 1992. Flea rearing in vivo and in vitro for basic and applied research. In: *Advances in Insect Rearing for Research and Pest Management* (T. E. Anderson, and N. C. Leppla eds.), Westview Press, San Francisco, pp. 119–131.

Hudson, B. W., and Prince, F. M. 1958. A method for large-scale rearing of the cat flea, *Ctenocephalides felis felis* (Bouche). *Bull. Wrld. Hlth. Org.* 19(6): 1126–1129.

Joseph, S. A. 1981. Studies on the bionomics of *Ctenocephalides felis orientalis* (Jordan). *Cheiron* 10: 275–280.

Kamala Bai, M., and Prasad, R. S. 1976. Studies on the host-flea relationship. III. Nutritional efficacy of blood meal of rat fleas *Xenopsylla cheopis* and *X. astia. Ann. Trop. Med. Parasitol.* 70: 467–472.

Karandikar, K. R., and Munshi, D. M. 1950. Life-history and bionomics of the cat flea *Ctenocephalides felis* (Bouche). *J. Bombay Nat. Hist. Soc.* 49: 169–177.

Kern, W. H., Jr. 1991. Cat flea larvae in the unknown life stage. *Pest Mgmt.* 6: 20–22.

Kern, W. H., Jr., Koehler, P. G., and Patterson, R. S. 1992. Diel patterns of cat flea (Siphonaptera: Pulicidae) egg and fecal deposition. *J. Med. Entomol.* 29: 203–206.

Larsen, K. S. 1995. Laboratory rearing of the squirrel flea *Ceratophyllus sciurorum sciurorum* with notes on its biology. *Entomol. Exp. Appl.* 76: 241–245.

Lavoipierre, M. M. J., and Hamachi, M. 1969. An apparatus for observations on the feeding mechanism of the flea. *Nature London* 192: 998–999.

Leeson, H. S. 1932. Methods of rearing and maintaining large stocks of fleas and mosquitoes for experimental purposes. *Bull. Ent. Res.* 33: 25–31.

Linardi, P. M., de Maria, M., and Botelho, J. R. 1997. Effects of larval nutrition on the postembryonic development of *Ctenocephalides felis felis* (Siphonaptera: Pulicidae). *J. Med. Entomol.* 34(4): 494–497.

Mead-Briggs, A. R. 1964. The reproductive biology of the rabbit flea, *Spilopsyllus cuniculi*, and the dependence of this species upon the breeding of its host. *J. Exp. Biol.* 41: 371–402.

Mellanby, K. 1933. The influence of temperature and humidity on pupation of *Xenopsylla cheopis. Bull. Entomol. Res.* 24: 197–203.

Meola, R., Meier, K., Dean, S., and Bhaskaran, G. 2000. Effect of pyriproxifen in the blood diet of cat fleas on adult survival, egg viability and larval development. *J. Med. Entomol.* 37: 503–506.

Metzger, M. E., and Rust, M. K. 1996. Egg production and emergence of adult cat fleas (Siphonaptera: Pulicidae) exposed to different photoperiods. *J. Med. Entomol.* 33(4): 651–655.

Metzger, M. E., and Rust, M. K. 2001. Laboratory techniques for rearing the fleas (Siphonaptera: Ceratophyllidae and Pulicidae) of the California ground squirrels (Rodentia: Sciuridae) using a novel nest box. *J. Med. Entomol.* 38: 465–467.

Molyneux, D. H. 1967. Feeding behavior of the larval rat flea *Nosopsyllus faciatus* Bosc. *Nature* 215: 779.

Moser, B. A., Koehler, P. G., and Patterson, R. S. 1991. Effect of larval diet on cat flea (Siphonaptera: Pulicidae) development times and adult emergence. *J. Econ. Entomol.* 84: 1257–1261.

Osbrink, W. L. A., and Rust, M. K. 1984. Fecundity and longevity of the adult cat flea, *Ctenocephalides felis felis* (Siphonaptera: Pulicidae). *J. Med. Entomol.* 21(6): 727–731.

Panchenko, G. M., and Fraenkel, G. 1966. The development of rat fleas in Siberia and the Far East, and its dependence on conditions of maintenance and feeding of larvae. *Rev. Appl. Entomol. B* 65(4): 281–282.

Pausch, R. D., and Fraenkel, G. 1966. The nutrition of the larva of the oriental rat flea, *Xenopsylla cheopis* (Rothschild). *Physiol. Zool.* 39: 202–222.

Prasad, R. S. 1969. Influence of host on fecundity of the Indian rat flea *Xenopsylla cheopis*. *J. Med. Entomol.* 6: 443–447.

Pullen, S. R., and Meola, R. W. 1995. Survival and reproduction of the cat flea (Siphonaptera: Pulicidae) fed human blood on an artificial membrane system. *J. Med. Entomol.* 32: 467–470.

Reitblat, A. G., and Belokopytova, A. M. 1974. Cannabilism and predatism in flea larvae. *Zoologicheskii Zhurnal* 53: 135–137.

Sharif, M. 1937. On the life history and the biology of the rat flea, *Nosopsyllus fasciatus* (Bosc.). *Parasitology* 29: 225–238.

Sharif, M. 1948. Nutritional requirements of flea larvae, and their bearing on the specific distribution and host preferences of the three Indian species of *Xenopsylla* (Siphonaptera). *Parasitology* 38: 253–263.

Sikes, E. K. 1931. Notes on breeding fleas, with reference to humidity and feeding. *Parasitology* 23: 243–249.

Silverman, J. M., and Appel, A. G. 1994. Adult cat flea (Siphonaptera: Pulicidae) excretion of host blood proteins in relation to larval nutrition. *J. Med. Entomol.* 31: 265–271.

Silverman, J., and Rust, M. K. 1985. Extended longevity of the pre-emerged cat flea (Siphonaptera: Pulicidae) and factors stimulating emergence from the pupal cocoon. *Ann. Entomol. Soc. Am.* 78: 763–768.

Silverman, J., Rust, M. K., and Reierson, D. A. 1981. Influence of temperature and humidity on survival and development of the cat flea, *Ctenocephalides felis*. *J. Med. Entomol.* 18: 78–83.

Smith, C. N., and Eddy, G. W. 1954. Techniques for rearing and handling body lice, oriental rat fleas, and cat fleas. *Bull. Wrld. Hlth. Org.* 10: 127–137.

Stark, H. E., Campos, E. G., and Elbel, R. E. 1976. Description of the third instar larva of *Monopsyllus wagneri* (Baker) (Siphonaptera: Ceratophyllidae). *J. Med. Entomol.* 13: 107–111.

Strenger, A. 1973. Zur ernaehrungsbiologie der larve von *Ctenocephalides felis felis*. (B.) *Zool. J. Syst. Bd.* 100: 64–80.

Thomas, R. E., Wallenfels, L., and Popiel, I. 1996. On-host viability and fecundity of *Ctenocephalides felis* (Siphonaptera: Pulicidae), using a novel chambered flea technique. *J. Med. Entomol.* 33: 250–256.

Thomas, R. E., Ozols, V. V., Hausser, N., and Silver, G. M. 1997. The biology of cat fleas, *Ctenocephalides felis*, reared on an in vitro feeding system. p. 25. In: *Proceeding, 4th International Symposium on Ectoparasites of Pets* (N. Hinkle, ed.), April 1997, University of California, Riverside.

Vashchenok, V. S., Karandina, R. S., and Briukhanova, L. V. 1988. Blood amount engorged by fleas of different species under experiment. *Parazitologia* 22: 312–320.

Vaughan, J. A., and Coombs, M. E. 1979. Laboratory breeding of the European rabbit flea, *Spilopsyllus cuniculi*. *J. Hygiene.* 83: 521–529.

Vaughan, J. A., Thomas, R. E., Silver, G. M., Wisnewski, N., and Azad, A. F. 1998. Quantitation of cat immunoglobulins in the hemolymph of cat fleas (Siphonaptera: Pulicidae) after feeding on blood. *J. Med. Entomol.* 35: 404–409.

Wade, S. E., and Georgi, J. R. 1988. Survival and reproduction of artificially fed cat fleas, *Ctenocephalides felis* Bouche (Siphonaptera: Pulicidae). *J. Med. Entomol.* 25: 186–190.

Yasumatsu, K. 1949. Rearing of flea larvae on various diets. *J. Fac. Agric.* 9: 121–126.

Yurgensov, I. A., and Ilíina, S. V. 1972. The use of pelleted food concentrate for standard rearing of fleas. *Rev. Appl. Entomol. B* 60(9): 342.

Care, Maintenance, and Experimental Infection of Mosquitoes

STEPHEN HIGGS

INTRODUCTION

Mosquitoes are the most frequently studied group of arthropod vectors. Various aspects of the mosquito life cycle have been known since ancient times, but perhaps the first published description of larva, pupa, and adult mosquitoes was given in 1669, by Jan Swammerdam. The idea that mosquitoes might be involved in disease transmission may date to 500 BC, when the Brahmin priest, Susruta, described malarial fever. In 1807 the *Baltimore Observer* published an article, "Mosquital Origin of Malaria," by Dr. John Crawford. And sometime in the 1870s, Dr. Carlos Juan Finlay of Cuba developed the mosquito theory for transmission of the etiologic agent causing yellow fever, which he formally announced in 1881. Finlay performed transmission experiments between 1883 and 1890. Mosquitoes were allowed to feed on yellow fever patients and later fed on human volunteers, none of whom developed disease. Infection of an arthropod with a parasite was first demonstrated in 1876 by Patrick Manson, who observed filarial worms in mosquitoes after they had fed upon his gardener. In 1893 Smith and Kilbourne published the first report of transmission of a pathogen by an arthropod—*Babesia bigemina* by ticks. Exactly when mosquitoes were first bred in captivity specifically for experiments is uncertain. In 1896, Ronald Ross, using adult *Anopheles* mosquitoes reared from larvae, observed malaria cysts attached to the mosquito's stomach 4 days after the mosquito had fed on an infected human. Two years later he demonstrated transmission of avian malaria

from infected to uninfected sparrows and described the different life stages. Laboratory-bred mosquitoes were raised in 1900 in Havana, Cuba, by Jesse W. Lazear for the transmission experiments that were to cost him his life. Major Walter Reed and colleagues finally proved the transmission of yellow fever virus by mosquitoes in 1901. Perhaps the first facility specifically designed for mosquito transmission experiments was used in 1900 by Patrick Manson's team working on malaria in Ostia, Italy, who built a mosquito-proof hut, with wire gauze covering the doors and windows.

There are approximately 3,450 named species and subspecies of mosquitoes, assigned to 30 genera. The variety of habitats in which these species exist and the species-specific host preferences, mating behaviors, and environmental requirements make it impossible to provide universal recommendations on rearing. Many species, including some significant vectors, have never been successfully maintained in the laboratory. This chapter aims to provide some generalized guidelines that will enable the reader to gain experience in working with some relatively easy species in the most common genera of vectors: *Aedes*, *Culex*, and *Anopheles*. The reader can then use the references provided here to obtain further information pertinent to specific species.

Planning

Prior to embarking upon the often-labor-intensive work of using mosquitoes in the laboratory, it is essential to know exactly what question you are addressing and what information is already available. For example, if you are interested in a particular pathogen, you should know which species of mosquito are the principle vectors in the field. If colonization fails, then use of another species that is closely related to the principal vector might be one's only option. It cannot be assumed that closely related species have similar vector characteristics, but it is at least a logical start. A thorough search of literature and Web sites as listed at chapter's end can save the researcher much time and effort.

Obtaining Mosquitoes

Numerous species and strains of mosquito have been reared in laboratories around the world. These are often well characterized in terms of vector competence, genetics, etc. Since the community of vector biologists is relatively small, a search for sources of a particular species can be completed quickly. The publication *World Directory of Arthropod Vector Research and Control Specialists* (Gerberg 1998) is useful but quickly outdated. An increasing number of mosquito-related Web sites exist (see later). Although some are narrowly focused, they can provide useful leads to groups working with a particular species.

Techniques for developing a colony from field-collected material are described by Service (1993). Various methods can be used to collect larvae or adults. This could include dipping larvae from water and collecting adults in mechanical traps. Since many species are poorly attracted to conventional insect traps, using human subjects as bait may be effective, especially when working with vectors of human pathogens. When working in areas where pathogens are being actively transmitted, the risks and ethics of using humans as bait should be considered.

Facilities

The features that should be considered when designing and building an insectary depend upon its intended use. As described in Chapter 48, information on insectary design, construction, and operation for mosquito research is available from several sources (Duthu et al. 2001; Gerberg et al. 1994; Higgs and Beaty 1995). An insectary can be a simple room with shelves, but it is often critical to carefully control the humidity, temperature, and light regimen within ranges that are appropriate for the mosquitoes that are being maintained. Smooth, white surfaces, bright, even illumination, and a low ceiling are recommended and aid in locating mosquitoes if they escape. When working with nonindigenous species or infected vectors, loose mosquitoes are unacceptable. Vector containment is discussed in Chapter 48. When planning to rear and conduct research with mosquito vectors, institutional biosafety committees and perhaps national authorities, such as the Department of Agriculture, should be consulted prior to commencing the work. Guidelines for the containment of arthropods have recently been published (Benedict *et al.*, 2003; http://www.liebertpub.com/VBZ/default1.asp) and are discussed in Chapter 48.

ESTABLISHING THE LIFE CYCLE

The basic mosquito life cycle consists of the egg, four aquatic larval instars, the pupa, and the adult. Eggs may be laid singly or in batches close to water, for example, on the water surface (*Culex* spp.), at the water/terrestrial interface (*Aedes* and *Anopheles* spp.), in moist soil (*Psorophora* spp.), or in vegetation (*Mansonia* spp.). Following embryonation, the eggs of some species, such as *Aedes*, may remain viable even when desiccated for many months. Others, such as the egg rafts produced by *Culex*, must remain wet and be hatched quickly. Rearing the former is therefore easier, since eggs can be stored, whereas the latter must be continuously maintained. Immersion in water is the key stimulus to induce egg hatching. Installment hatching often occurs when some eggs require multiple exposures in order to hatch. Synchronized rearing may accelerate hatching and can be accomplished by placing eggs in tap water and then reducing the oxygen tension by placing under a vacuum or by bubbling nitrogen into the water for 10 minutes. Once eggs have hatched, larvae should be provided with food immediately. For *Aedes aegypti* we use a 1:1 mixture of Tetramin fish flakes and powdered laboratory rodent diet. For anopheline larvae, which feed on the surface, we use finely powdered liver or brewers yeast and TetraMin Baby-E fish food. The required standards of water cleanliness are species dependent. Some *Culex* species are highly tolerant of water contamination, whereas *Anopheles gambiae* is fastidious. Each day the surface film of the water should be "skimmed" with a piece of absorbent paper towel to remove the old food and dust. Growth rate, final size, and survival are influenced by food and larval density. A main cause of mortality is overcrowding. Optimal density may be species specific, and so it is common practice to begin at high density in one rearing pan and, as larvae grow, to separate them out into more pans.

As an example, *Ae. aegypti* can easily be reared in the plastic "shoebox"-size pans with lids that are available from many department stores. Approximately 100 larvae should be placed in 500 mL of water. Development from hatching to pupation takes 7–10 days, depending upon rearing temperature. Pupae are transferred from the rearing pan into a plastic cup, either by individually sucking them up using a wide-bore pipette with bulb (e.g., an inverted Pasteur pipette) or by pouring the water through a tea strainer. Synchronized pupation and eclosion are more likely if larvae are reared in noncrowded conditions with adequate food. A cup is placed in a rearing cage for eclosion. Mortality is normally low at this stage for *Ae. aegypti*; but if adults are seen to die on the surface, pieces of floating material can be added to the cup or a paper towel can be wrapped around the inside so that the adults can pull themselves free.

The key to establishing the mosquito colony is the ability to have the adults mate, feed, and lay eggs. Some species have never been successfully reared; others, such as *Ae. aegypti*, are considered to be easy, and even single-pair matings in small cartons can be achieved. Many factors may influence the successful colonization of vectors. Accurately mimicking the natural environment can promote captive breeding; for example, species that "swarm" mate may require large cages or whole rooms. Day length and humidity can be critical, with some species requiring gradual light changes to simulate the sunset and sunrise. If natural mating cannot be achieved, forced mating, although labor intensive, is an effective procedure to maintain some species. Anautogenous species will require a blood meal within a few days of eclosion. Some species are extremely host specific and may feed only during a specific time period, for example, in the dimming light of evening. Others seem to feed on a wide variety of vertebrates at almost any time. *Aedes aegypti* feed readily on most laboratory animals and then rest for 2–3 days while producing an egg batch. The females then readily lay eggs on a moist paper towel wrapped around the inside of a cup half-filled with water. For tree-hole-breeding species, such as *Ae. triseriatus*, the water is more attractive if it has had oak leaves soaking in it. *Anopheles gambiae* lay their eggs on filter paper floating on the water in a cup, and *Culex* spp. lay their egg rafts directly on the water surface. Female *Psorophora* spp. may have to be fed every day and may lay a few eggs each day on moist soil. Eggs melanize soon after being laid but must remain moist during the embryonation period. The eggs of *Aedes* spp. can then be stored dry for several months. Many other species cannot withstand dessiccation, and those of *Culex* spp. must be left in the water, where they hatch within a few days of being laid. Since *An. gambiae* is prone to infection with microsporidian parasites, it may occasionally be necessary to surface-sterilize the eggs by immersion for 30 seconds in 1% sodium hypochlorite.

Sucrose or sugar cubes and water or a 10% sugar solution in water presented in a bottle with a cotton wick can be used as a general food source for most mosquitoes, but raisins and sliced apple may also be used. As a blood source, anaesthetized or restrained laboratory animals (mice, hamsters, chickens, etc.) can be placed on or in the cages for an hour. Institutional animal care and use committees must approve appropriate protocols. Alternatives to animals are available (see later) but are not suitable in all circumstances. Gerberg (1994) describes rearing methods for many species. Detailed protocols for rearing commonly used species have been published by Christophers (1960) and Munstermann (1997) for *Aedes aegypti*, by Benedict (1997) for *Anopheles gambiae*, and by Vinogradova (2000) for *Culex pipiens*.

MOSQUITO MANIPULATION AND PATHOGEN TRANSMISSION EXPERIMENTS

Any manipulation procedure should be designed to cause minimal handling and physical stress to the vector and yet satisfy the appropriate containment requirements. When working with BSL-3 pathogens, all procedures, including mosquito inoculations and oral infections, must be performed in a suitable biosafety cabinet, and infected mosquitoes must be housed under BSL-3 containment conditions. As described earlier, vectors should be collected using mechanical aspirators (Fig. 53.1) and manipulated using forceps or other instruments that can be sterilized. Throughout any manipulation procedure, a

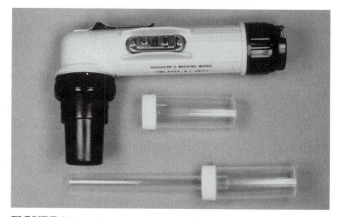

FIGURE 53.1 Mechanical aspirator (Hausherr's Machine Works).

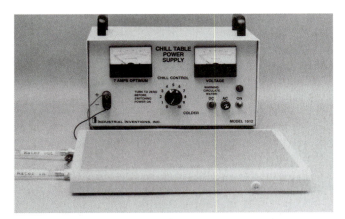

FIGURE 53.2 Chill table (Industrial Inventions Inc).

FIGURE 53.3 Mosquito salivating into capillary tube.

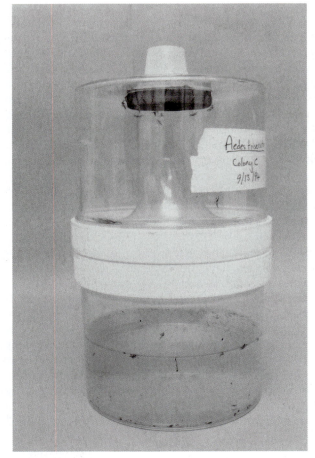

FIGURE 53.4 Mosquito breeder (BioQuip Products). Mosquitoes emerge from pupae held in the lower chamber and fly upward for easy isolation in the upper chamber.

constant tally of the numbers being used must be maintained. Only small numbers (10–15) of arthropods should be out of their container at a given time. Accurate records must be kept to determine that all manipulated vectors are accounted for throughout the experimental period.

Vector manipulation frequently requires that the arthropods be anaesthetized. This can be achieved most simply by chilling vectors in a refrigerator. Differences in susceptibility to cold must be determined to avoid killing the mosquitoes. *Aedes aegypti* is relatively susceptible to cold, whereas *Ae. triseriatus* and most temperate zone species of *Culex* are not. Carbon dioxide gas is a useful alternative method of immobilization, but it must also be used with caution. Once anaesthetized, vectors can be maintained in a torpid state by performing manipulations on a chill table (Fig. 53.2). A glass plate placed on ice can be used, but melting ice and condensation can be problems. Larvae

can be anaesthetized by immersion in ice-cold water but only for short periods, because low temperatures stop respiration and larvae may drown. Legs and wings can be removed from females prior to saliva collection (Fig. 53.3), and even the head can be removed from males used for forced mating with anaesthetized females. Additional preparation to maximize success of the procedure may be necessary; e.g., mosquitoes may have to be starved for 24–48 hours to promote feeding. If adult mosquitoes of a particular age are required, pupae can be held in mosquito breeder containers (Fig. 53.4) that allow emerged adults to fly into the upper chamber for easy separation. Behavioral manipulation, for example, induction of feeding or egg laying at a particular time, may be achieved by shielding colonies from ambient conditions and by controlling light cycles and day length.

Infecting mosquitoes by allowing them to feed on a viremic/parasitemic host is often problematic due to

FIGURE 53.5 Various designs of glass feeders used to present blood meals to mosquitoes.

FIGURE 53.6 Detail of glass feeder. Water is pumped into the feeder's water jacket via the left inlet tube and exits via the right outlet tube.

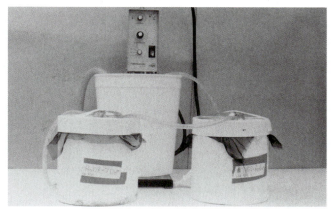

FIGURE 53.7 Apparatus used for orally infecting mosquitoes. Water at 37°C is circulated through the feeder, which is held on top of cartons of mosquitoes.

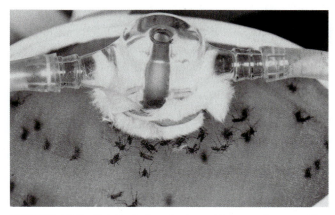

FIGURE 53.8 Detail of glass feeder, with mosquitoes probing through a mouse-skin membrane to reach the blood meal.

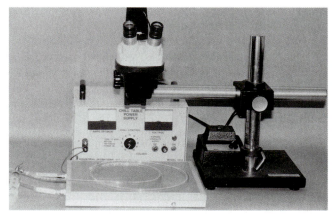

FIGURE 53.9 Apparatus for intrathoracic inoculation.

the variation and unpredictability associated with biological systems. An alternative to infected animals is the use of tissue-culture-derived pathogens. These can often be reliably propagated *in vitro*, quantified, mixed with blood, and then presented to the mosquitoes as drops or soaked cotton pledgets placed on the top of the cage or in artificial membrane feeders as designed by Rutledge (1964) (Figs. 53.5–53.8). Several feeding systems have been designed that are attractive to mosquitoes, since they mimic the vertebrate host by operating at 37°C and providing warm blood via a skin or skinlike membrane. To enhance feeding, the mosquitoes can be deprived of food overnight. Adenosine triphosphate can be added to the blood (final concentration 0.02 M) as a phago stimulant to induce contraction of the pharyngeal pump. Gently breathing into the carton can provide additional stimulation.

If it is not critical that a truly natural infection process be followed, then an efficient method of infection is intrathoracic (parenteral) inoculation (Figs. 53.9

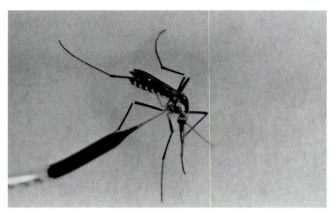

FIGURE 53.10   Intrathoracically inoculated mosquito.

and 53.10). The apparatus is simple and inexpensive, consisting of glass capillary tubes drawn to a fine point over a flame, a tube holder, rubber tubing, and a syringe with a three-way valve. An aqueous suspension of pathogens is drawn into the needle and inoculated into the thorax of anaesthetized mosquitoes on a chill table.

Following infection, mosquitoes are held in secure unbreakable cartons in cages within an appropriate-containment-level insectary. Cardboard ice cream containers are inexpensive and can be adapted for use by replacing the card top with fine mesh (Fig. 53.11). Food is supplied either by placing sugar cubes and water in an inverted cup on the mesh or by using absorbent cotton soaked in a 10% sucrose solution.

The establishment and distribution of the pathogen in the mosquito can be determined both qualitatively by organ dissection and appropriate microscopic observation via, for example, immunofluorescence assay, and quantitatively by, for example, trituration and titration *in vitro* (Higgs *et al.* 1997). To demonstrate transmission, either mosquitoes can be fed upon susceptible vertebrates that are then observed for signs of infection, or saliva may be collected directly for pathogen isolation. Following the conclusion of infection experiments, all mosquitoes and potentially infectious materials must be appropriately discarded.

Specialized Insectary Suppliers

There are numerous suppliers of equipment that is suitable for use in insectaries and for rearing various species of arthropods. Here are some suppliers of specialized equipment. Mention of a trade name, proprietary product, or specific equipment does not constitute a guarantee or warranty by the authors and does not imply its approval to the exclusion of other products that can be suitable.

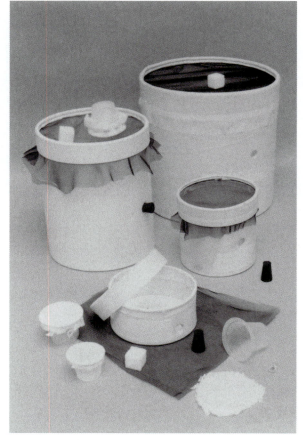

FIGURE 53.11   Cardboard cartons and accessories used to house small groups of mosquitoes.

BioQuip Products, 17803 LaSalle Avenue, Gardena, CA 90248-3602; tel: (310) 324-0620; fax (310) 324-7931

Hausherr's Machine Works, 1186 Old Freehold Road, Toms River, NJ 08753; tel: (908) 349-1319; fax (908) 286-4919

Industrial Inventions Inc., 694 Village Road West, Lawrenceville, NJ 08648; tel: (609) 275-1500

World Wide Web Sites for Mosquito Research

http://www.aphis.usda.gov/biotech/arthropod/
http://klab.agsci.colostate.edu/index.html
http://konops.imbb.forth.gr/AnoDB/
http://www.cmosquito.com/
http://www.mosquito.org/
http://www.astmh.org/subgroup/acme.html
http://www.malaria.mr4.org/mr4pages/index.html
http://www.bioquip.com/

http://www.watdon.com/
http://www.entsoc.org/
http://www.floridamosquito.org/
http://www.royensoc.co.uk/
http://www.liebertpub.com/
http://thesius.ingentaselect.com/vl=15259108/cl=22/
nw=1/rpsv/cw/mal/15303667/v3n2/contp1-1.htm

Readings

Benedict, M. Q. 1997. Care and maintenance of anopheline mosquito colonies. In: *The Molecular Biology of Insect Disease Vectors* (J. M. Crampton, C. B. Beard, and C. Louis, eds.). Chapman and Hall, London, pp. 3–12.

Benedict, M. Q., Tabachnick, W. J., and Higgs, S. (key authors) (2003). Arthropod Containment Guidelines (Version 3.1). *Vector-Borne and Zoonotic Diseases*. 3: 59–98.

Collins, W. E. 1997. Infection of mosquitoes with primate malaria. In: *The Molecular Biology of Insect Disease Vectors* (J. M. Crampton, C. B. Beard, and C. Louis, eds.). Chapman and Hall, London, pp. 92–100.

Cosgrove, J. B., Wood, R. J., Petric, D., Evans, D. T., and Abbott, H. R. 1994. A convenient mosquito membrane feeding system. *J. Am. Mosquito Cont. Assoc.* 10: 434–436.

Duthu, D. B., Higgs, S., Beets, R. L., and McGlade, T. J. 2001. Design issues for insectaries. In: *Anthology of Biosafety. IV. Issues in Public Health* (J. Y. Richmond, ed.). ••, pp. 227–244.

Gerberg, E. J. 1998. *World Directory of Arthropod Vector Research and Control Specialists*, 4th ed. American Mosquito Control Association, Lake Charles, LA.

Gerberg, E. J., Barnard, D. R., and Ward, R. A. 1994. *Manual for Mosquito Rearing and Experimental Techniques*, Bulletin 5 (revised). American Mosquito Control Association, Lake Charles, LA.

Higgs, S., and Beaty, B. J. 1996. Rearing and containment of mosquito vectors. In: *Biology of Disease Vectors* (B. J. Beaty, and W. C. Marquardt, eds.). University of Colorado Press, Niwot, CO, pp. 595–605.

Higgs, S., Olson, K. E., Kamrud, K. I., Powers, A. M., and Beaty, B. J. 1997. Viral expression systems and viral infections in insects. In: *The Molecular Biology of Insect Disease Vectors* (J. M. Crampton, C. B. Beard, and C. Louis, eds.). Chapman and Hall, London, pp. 459–482.

Maramorosch, K., and Mahmood, F. 1999. *Maintenance of Human, Animal, and Plant Pathogen Vectors*. Science Publishers, Enfield, NC.

Munstermann, L. E. 1997. Care and maintenance of *Aedes* mosquito colonies. In: *The Molecular Biology of Insect Disease Vectors* (J. M. Crampton, C. B. Beard, and C. Louis, eds.). Chapman and Hall, London, pp. 13–20.

Service, M. W. 1993. *Mosquito Ecology: Field Sampling Methods*, 2nd ed. Chapman and Hall, London.

Sinden, R. E. 1997. Infection of mosquitoes with rodent malaria. In: *The Molecular Biology of Insect Disease Vectors* (J. M. Crampton, C. B. Beard, and C. Louis, eds.). Chapman and Hall, London, pp. 67–91.

Townson, H. 1997. Infection of mosquitoes with filaria. In: *The Molecular Biology of Insect Disease Vectors* (J. M. Crampton, C. B. Beard, and C. Louis, eds.). Chapman and Hall, London, pp. 101–111.

Vinogradova, E. B. 2000. Culex pipiens pipiens *Mosquitoes: Taxonomy, Distribution, Ecology, Physiology, Genetics, Applied Importance and Control*. Russian Academy of Sciences Zoological Institute, Pensoft, Sofia, Moscow.

54

Care, Maintenance, and Experimental Infection of Biting Midges

GREGG J. HUNT AND EDWARD T. SCHMIDTMANN

INTRODUCTION

Understanding pathogen–vector–host interactions is essential to control arthropod-borne diseases. Many hematophagous arthropods, including *Culicoides* (Diptera: Ceratopogonidae) biting midges, represent confirmed or suspected vectors of various animal and human pathogens. Vector competence studies of biting midges have been hampered because most of these small insects are regarded as being difficult to colonize or maintain in the laboratory. Nevertheless, various species of *Culicoides* have been incriminated as intermediate hosts or as vectors of animal and human etiological agents such as helminths, protozoa, and viruses.

The laboratory colonization and rearing of *Culicoides sonorensis* (syn. *C. variipennis sonorensis*) Wirth & Jones conducted at the Arthropod-borne Animal Diseases Research Laboratory (ABADRL) are essential in the study of this vector or potential vectors of several viral pathogens of veterinary importance in the United States, such as the bluetongue (BLU) viruses, epizootic hemorrhagic disease (EHD) viruses, vesicular stomatitis (VS) viruses, and West Nile (WN) virus. *Culicoides sonorensis* is distributed throughout Mexico, the southern and western regions of the United States, and the southwestern region of Canada. This biting midge is the primary vector of the BLU viruses in sheep, cattle, and wild ruminants in North America. Bluetongue, an Office of International Epizooties List A animal disease, is a serious economic threat to the U.S. livestock industry that causes morbidity and mortality in

sheep, reproductive impairment in sheep and cattle, and restrictions in the international movement of breeding livestock and germplasm. Bluetongue disease is one of the most economically important arthropod-borne animal diseases in the United States.

REARING OF *CULICOIDES* *Sonorensis*

A program was initiated in 1955 to colonize *C. sonorensis* so that adequate studies could be conducted on the transmission of BLU viruses among sheep and on vector control. Since 1956, numerous colonies of *C. sonorensis* have been started from field-collected larvae and pupae in which various investigators have made improvements in the large-scale production to standardize the rearing techniques. Current production is about 2.5 million adults annually, in which this ceratopogonid is the only species of *Culicoides* that is maintained continuously in prominent numbers at an insectary.

Field and Laboratory Bionomics

The process of establishing and maintaining insect colonies is based upon insectary design, environmental controls, rearing procedures, and nutrition, with an understanding of the field bionomics, such as insect biology, behavior, habitat, and diet. The successful rearing of numerous short- and long-term colonies of

C. sonorensis at the ABADRL has been attributed to the reproduction of various field conditions in our laboratory.

Immatures

Natural breeding sites of *C. sonorensis* consist of shallow soft mud with standing or slow-moving water, high concentrations of livestock manure, moderate to high levels of salt-forming ions and salinity, direct sunlight, and minimal vegetation. Larval habitats usually are associated where livestock feed and drink. Manure-polluted water sources, such as dairy waste-water ponds, may generate large populations of *C. sonorensis*. Aquatic microorganisms, mostly bacterial contaminants of polluted water, are the primary food source of *C. sonorensis* larvae in the field. In temperate regions, larvae are the overwintering stage.

At the ABADRL, the laboratory environment for *C. sonorensis* eggs, four larval instars, and pupae consists of plastic pans in which deionized (DI) water serves as the rearing medium and inert dacron fibrous matting represents the substrate for the aquatic immatures. A photoperiod of 13-hour light/11-hour dark, water temperature of $27.0° ± 1°C$, aeration of the rearing medium, and dispersal of the bacterial scum in the rearing medium are used for both larval and pupal development. Bacteria along with other aquatic microorganisms and particulate detritus constitute the larval nourishment. This nutrition is supplemented with an artificial diet containing carbohydrates, lipids, minerals, proteins, sterols, and vitamins for optimal larval growth.

Adults

Resting sites are obscure for *C. sonorensis* in the field. Females imbibe vertebrate blood using biting and sucking mouthparts to obtain protein for ovarian development; autogeny does not occur. This biting midge has been observed to feed on cattle, deer, hares, horses, humans, mice, mules, rabbits, sheep, and swine in the field (and the laboratory). Males and females require plant sugars as an energy source. Aerial mating is common, and it is often accompanied by swarming or aggregations composed of males near an object (e.g., vegetation or an animal) in which females enter the swarms to mate. The natural longevity of adults is unknown; however, at least 2.5 weeks has been estimated. A generation time of 2–7 weeks, depending on seasonal daily temperatures, has been observed for field populations.

At the ABADRL, mixtures of male and female *C. sonorensis* are maintained in wax-lined cardboard containers held at $27.0° ± 1°C$ and 40–50% relative humidity (RH), with a 13-hour light/11-hour dark photoperiod. An artificial blood-feeding apparatus is used to feed females on defibrinated sheep blood through a reinforced silicone membrane. Adults are provided with both 10% sucrose solution and DI water. Colonized males and females and some field populations mate repeatedly within a confined space, such as an adult-holding cage or blood-feeding cone. The typical life span of females is about 2 weeks, whereas males live for just a few days. Larger females have been observed to survive 1.5 times longer than smaller females. The duration of the life cycle is at least 18 days: 2 days for egg; at least 10 days for larva; 3 days for pupa; 1 day for adult; and at least 2 days for blood digestion and preoviposition.

Large-Scale Production

The procedures and equipment used to produce and maintain *C. sonorensis* at the ABADRL have been described in detail. A nonspecific version of these protocols is given herein. The protocols can be modified for the small-scale rearing of *C. sonorensis*.

Eggs

An artificial blood-feeding apparatus (Fig. 54.1) is used to feed females. The apparatus consists of a water-jacketed glass cylinder with a reinforced silicone feeding membrane attached to the central well, where up to 10 mL of sheep blood is mechanically stirred and maintained at 36.5°C (skin temperature of sheep).

FIGURE 54.1 Female *Culicoides sonorensis* (located inside insect-secured blood-feeding cones) feeding on sheep blood contained inside an artificial blood-feeding apparatus. (*Note:* Insects feeding on BSL-3 infectious bloods would require the feeding apparatus to be operated within a biological safety cabinet.)

Insectary personnel exhale occasionally on the cones so that the carbon dioxide encourages blood-feeding activity. After blood feeding, the engorged females are maintained up to 4 days in adult-holding cages. A cage consists of a 3.8-liter cardboard container and lid, fine mesh polyester organdy, clear plastic film, plastic container with an oviposition substrate (filter paper positioned on top of moistened sterile cotton), and vials containing 10% sucrose solution and DI water. Thousands of eggs are collected from each cage during the 2- to 4-day oviposition period. For short-term storage, the oviposition substrates are stored at 4°C for up to 30 days; however, the viability of the eggs decreases significantly (<50%) after 30 days. Cryopreservation may be used for the long-term storage of eggs.

Larvae

Plastic 1,239-cm^2 rearing pans (Fig. 54.2) consist of a dacron island, a plastic paddle used to circulate the rearing medium slowly and to minimize the formation of bacterial scum, and two metal bars used to anchor the island during the movement of the medium. About 3.2 liters of DI water and varying amounts of albumin, alfalfa, algicide, bacterial inoculum, brain heart infusion medium, high-protein supplement, nutrient broth fluid concentrate, and yeast extract medium are added to each pan. The islands must be located at the water surface so that the larvae can maneuver easily on and off these substrates. Oviposition papers containing a few thousand eggs are placed on each island. Large and robust adults are produced when the optimal level of ~3,500 larvae (and pupae) is achieved for each pan.

FIGURE 54.2 Insect-rearing pans (located inside an insect-rearing rack) used to rear the eggs, larvae, and pupae of *Culicoides sonorensis*.

For maintenance of the rearing pans, nutrient broth fluid concentrate is added every other day for 24 days.

Pupae

Those islands containing pupae are submerged so that the pupae congregate primarily along the water edges. The dislodged buoyant pupae, along with some free-moving larvae, are aspirated using a vacuum pump device, and then the larvae are separated from the pupae using a sieve in which these larvae are transferred back to an established rearing pan. Up to 1,500 pupae are prepared for each cage, which includes moistened sterile cotton as the emergence substrate inside a plastic container. Up to 60 rearing pans containing varying ages of eggs, larvae, and pupae are maintained for each rearing rack.

Adults

Males and females typically emerge during the next 3 days.

Problems with Colonization

The success of our large-scale colonies of *C. sonorensis* has relied on the detailed procedures and insectary equipment that have been developed during the previous 4^1/$_2$ decades. However, the corresponding small-scale rearing procedures have proved unsuitable for initiating colonies or maintaining many field populations collected throughout the United States. Fundamental problems consist of insufficient blood-feeding rates and unsuccessful mating. Additional problems include field-collected mud samples containing various Diptera immatures, naturally occurring insect pathogens, temperature variation causing mortality, excessive handling inflicting mortality, changes in environmental and/or genetic variability, and changes in oral susceptibility to the BLU viruses.

Insect Security

During the colonization and routine maintenance of *C. sonorensis*, precautions must be taken to prevent the escape of biting midges and to prevent the introduction of exogenous insects into the ABADRL insect-rearing facility. Physical restrictions, such as a double-screened vestibule at each rearing room, the sealing of all openings and room penetrations, and numerous strategically placed insect light traps that serve as monitoring and abatement devices, are used to contain and prohibit the movement of escaped adults throughout the facility. Insectary personnel

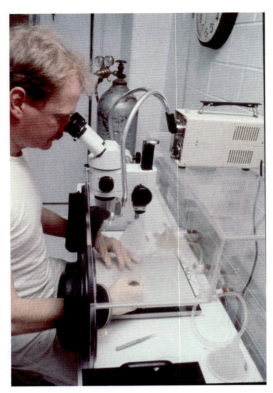

FIGURE 54.3 Anesthetized virus-exposed female *Culicoides sonorensis* being manipulated on a chill table inside an insect-secured glove box within the ABADRL BSL-3 agricultural containment facility.

adhere to security protocols, such as keeping the screened doorway zippers closed, wearing laboratory coats assigned to each rearing room, using hot water or bleach solution, autoclaving all disposable items, prohibiting the return of unused living insects, and inspecting for the presence of intruding arthropods.

Equipment and protocols used to handle virus-infected *C. sonorensis* safely and securely during vector competence studies have been reported. Four levels of barriers are used for biting midges within the ABADRL biosafety level 3 (BSL-3) agricultural containment facility. Primary containment consists of insect-secured adult-holding cages or blood-feeding cones. Secondary containment consists of insect-secured glove boxes (Fig. 54.3), artificial blood-feeding apparatus within a biological safety cabinet, or insect-secured storage boxes inside environmental incubators. Multiple zones of screened doorways and standard or nitrogen-inflatable gasketed doors along with insect light traps serve as tertiary containment. Quaternary containment is produced by the BSL-3

facility. Insect security requires accountings for all biting midges that are experimentally exposed to infectious pathogens.

SUMMARY

The laboratory colonization and maintenance of *C. sonorensis* and other veterinary- or medically-important arthropods requires knowledge of their biology and diligence in evaluating and conducting insectary protocols. The importance of different arthropods and their association with animal or human disease pathogens forces continual improvements in rearing so that investigators are provided with an essential component, the vector, to study the pathogen-vector-host interrelationships of arthropod-borne diseases.

Acknowledgments

The authors thank G. Letchworth, R. Nunamaker, and D. White for reviewing this chapter. J. Kempert, F. Stanek, and W. Yarnell provided technical assistance.

Readings

Boorman, J. 1993. Biting midges. In: *Medical Insects and Arachnids* (R. P. Lane and R. W. Crosskey, eds.). Chapman and Hall, London, pp. 289–309.

Holbrook, F. R., Tabachnick, W. J., Schmidtmann, E. T., McKinnon, C. N., Bobian, R. J., and Grogan, W. L. 2000. Sympatry in the *Culicoides variipennis* complex (Diptera: Ceratopogonidae): A taxonomic reassessment. *J. Med. Entomol.* 37: 65–76.

Holbrook, F. R. 1996. Biting midges and the agents they transmit. In: *The Biology of Disease Vectors* (B. J. Beaty and W. C. Marquardt, eds.). University Press of Colorado, Niwot, CO, pp. 110–116.

Hunt, G. J. 1994. A procedural manual for the large-scale rearing of the biting midge, *Culicoides variipennis* (Diptera: Ceratopogonidae). ARS-121. U.S. Department of Agricultural, Agricultural Research Service, Washington, DC.

Hunt, G. J., and McKinnon, C. N. 1990. Evaluation of membranes for feeding *Culicoides variipennis* (Diptera: Ceratopogonidae) with an improved artificial blood-feeding apparatus. *J. Med. Entomol.* 27: 934–937.

Hunt, G. J., and Tabachnick, W. J. 1995. Cold storage effects on egg hatch in laboratory-reared *Culicoides variipennis sonorensis* (Diptera: Ceratopogonidae). *J. Am. Mosq. Control Assoc.* 11: 335–338.

Hunt, G. J., and Tabachnick, W. J. 1996. Handling small arbovirus vectors safely during biosafety level 3 containment: *Culicoides variipennis sonorensis* (Diptera: Ceratopogonidae) and exotic bluetongue viruses. *J. Med. Entomol.* 33: 271–277.

Hunt, G. J., Tabachnick, W. J., and McKinnon, C. N. 1989. Environmental factors affecting mortality of adult *Culicoides variipennis* (Diptera: Ceratopogonidae) in the laboratory. *J. Am. Mosq. Control Assoc.* 5: 387–391.

Hunt, G. J., Mullens, B. A., and Tabachnick, W. J. 1999. Colonization and maintenance of species of *Culicoides*. In: *Maintenance of Human, Animal, and Plant Pathogen Vectors* (K. Maramorosch and F. Mahmood, eds.). Science Publishers, Enfield, NH, pp. 33–55.

Mellor, P. S., Boorman, J., and Baylis, M. 2000. *Culicoides* biting midges: Their role as arbovirus vectors. *Annu. Rev. Entomol.* 45: 307–340.

Nunamaker, R. A., and Lockwood, J. A. 2001. Cryopreservation of embryos of *Culicoides sonorensis* (Diptera: Ceratopogonidae). *J. Med. Entomol.* 38: 55–58.

Schmidtmann, E. T., Bobian, R., and Belden, K. 2000. Soil chemistries define aquatic habitats of immature populations of the *Culicoides variipennis* complex (Diptera: Ceratopogonidae). *J. Med. Entomol.* 37: 58–64.

Tabachnick, W. J. 1996. *Culicoides variipennis* and bluetongue virus in the United States. *Annu. Rev. Entomol.* 41: 23–43.

Walton, T. E., and Osburn, B. I. 1991. *Bluetongue, African Horse Sickness, and Related Orbiviruses.* CRC Press, Boca Raton, FL.

55

Care, Maintenance, and Experimental Infection of Black Flies

JAMES B. LOK

INTRODUCTION

Prior to the mid-1970s studies of vector–parasite interactions, potential control measures, and advanced biochemical, pharmacological, and molecular biological aspects of black flies lagged behind similar studies of other vector species due to the lack of well-established laboratory rearing methods. This chapter is a brief discussion of focused efforts, primarily in the period 1970–1990, to address this deficiency. For more comprehensive reviews and protocols, the reader may consult several recent publications (Edman and Simmons 1987; Cupp and Ramberg 1997; Gray and Noblet 1999).

Historically, the development of rearing methods for black flies has stressed attempts to simulate the natural environments of the species in question. The general biology of simuliids is reviewed in Chapter 11. From that and the following discussion, it can be gleaned that successful rearing methods have taken into account the rheophilic or stream-loving nature of larval black flies and the facts that mating occurs under conditions where there is a preponderance of male flies, favoring a high frequency of male-to-female contacts, that female black flies seeking a host for blood feeding orient along an ascending temperature gradient, and that females of many species deposit eggs on the moist splash areas of emergent objects in their stream habitats. New methods for physiological and vector biological study followed closely on the incorporation of these biological attributes into reliable rearing techniques.

COLLECTING STARTING MATERIAL

Immature black flies at various stages of development may be collected from their stream sites and transferred to the laboratory for rearing. Eggs of many species may be found in dense clusters above the waterline of partially submerged objects in stream habitats. In central New York State it is possible to collect gram quantities of *Simulium decorum* and *S. vittatum* eggs (Brenner et al. 1980; Tarrant et al. 1987) from the margins of spillways at the outflow from both natural and artificial lakes or ponds. Other species oviposit on emergent vegetation, rocks, or debris in the stream flow. Frequently, these objects may be dislodged from a stream site, returned to the lab under cool temperatures, and then transferred into a larval rearing device, where eggs will hatch and larvae will migrate into the flowing water. Similarly, vegetation, smaller rocks, and twigs to which black fly larvae are attached may be removed from the stream and transferred in insulated containers or portable aquaria to a laboratory rearing system, where attached larvae will frequently migrate from them and into areas of optimal flow. Larvae of some species, such as the nearctic *S. pictipes*, attach to exposed bedrock or other immovable objects. In these cases, late-instar larvae may be gently scraped from the substrate using a Teflon spatula and captured in a hand-held net positioned 30–50 cm downstream (Bernardo et al. 1986).

As an efficient alternative to collecting natural objects colonized by black fly immatures, artificial

747

substrates may be placed in natural habitats known to support a species of interest and then recovered at some interval and returned to the laboratory. For example, narrow strips of plastic sheeting may be secured to the streambeds of shallow creeks and streams, where, perhaps because they resemble some of the physical characteristics of trailing vegetation, they become attractive substrates for oviposition and larval development of species such as *Simulium venustum*. These objects may be collected periodically and replaced with fresh substrates. Such plastic or fabric substrates have the advantages of being quickly located and recovered from the field and then easily attached to the troughs or rearing chambers of laboratory systems.

Alternative sources of eggs for laboratory rearing are gravid female black flies. In some instances, natural oviposition sites, such as those of the autogenous *S. decorum* and *S. vittatum*, are well characterized and easily recognized. Large numbers of ovipositing females may be collected by sweep netting above these sites at twilight. Blood-engorged females of anautogenous species such as *S. damnosum* s.l. may be acquired from biting collections and held until gravid. Eggs may be collected from gravid flies using oviposition chambers of various designs (see later discussion). The advantage of this approach is that field-collected adults are relatively easy to identify and thus establish a population of conspecific eggs. Naturally occurring egg masses, on the other hand, may contain individuals of more than one species.

REARING LARVAL BLACK FLIES

Stirred Systems and Aerated Aquaria

The primary challenge in rearing larval black flies is the production of a well-aerated moving-water system in the laboratory. The most basic approach to this problem has been the modification of simple aerated aquaria with vertical substrates to which field collected larvae may attach. This approach has been used to rear larvae of a number of temperate species, including *S. decorum* and *S. ornatum*, but it has shown limited applicability to African species. Aerated aquaria do not accurately simulate the natural moving-water habitats of most black flies and generally support development to the adult only when cultures are initiated with well-nourished advanced-stage larvae from the field. These systems are generally not adequate for rearing large numbers of vigorous flies from eggs or early-instar larvae. The technique is valuable, however, as a method of transporting field-collected larvae to the laboratory, especially given the availability of battery-operated aeration pumps and for limited rearing of field-collected larvae to adults for taxonomic purposes.

Similarly, electromagnetically stirred containers have been used with some success as closed circulation systems for rearing larvae of temperate species such as *S. erythrocephalum*, for which Grunewald (1973) developed a highly sophisticated stirred apparatus capable of controlling temperature and various hydrochemical parameters (Fig. 55.1). Stirred systems

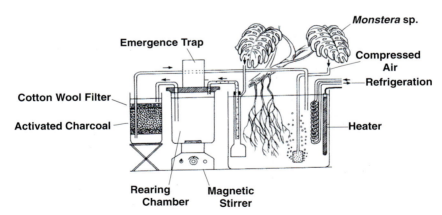

FIGURE 55.1 A stirred apparatus capable of controlling temperature and dissolved oxygen and of filtering out organic pollutants. Black fly larvae attach to the sides of the rearing chamber, and water flow is maintained by the magnetic stirrer. Aerated, temperature-controlled water from the aquarium at the right is delivered into the rearing chamber by a pump. Outflow from the rearing chamber passes through the activated charcoal filter on the left and then back to the aquarium, where it is further cleansed by exposure to roots of the aquatic plant *Monstera* sp. Redrawn from Grunewald (1973); used with permission.

offer the advantages of simplicity of construction and the ability to conduct cohort rearings in small containers, but the devices reported to date do not allow creation of turbulence microhabitats and have not proven applicable to African onchocerciasis vectors.

Gravity-Flow Troughs— "Artificial Streams"

To date, the most practical and successful laboratory systems have employed a gravity-flow trough to create an "artificial stream" more closely simulating the natural habitats of larval black flies. Some gravity-flow systems in field laboratories adjacent to black fly habitats have involved diversion of natural stream water into laboratory troughs. Such systems obviate the regulation of hydrochemical parameters and artificial feeding, but they are obviously limited to use in close proximity to black fly habitats and defeat the purpose of making black flies broadly available to researchers outside of these foci.

Closed gravity-flow systems constitute the best opportunities among the current methodologies of rearing vigorous adult black flies from eggs. They provide ready access to individual larvae for observation and sampling, and they make it possible to create turbulence microhabitats by placing baffles or other impediments in the stream flow. In addition to such baffles, gravity-flow systems frequently include a system of screens at the base of the trough to catch drifting larvae and remove large particulate debris (Fig. 55.2B). Closed circulation systems generally involve electrically operated pumps, and some have devised an automatic flooding system to protect larvae in the event of power failure.

Closed circulation systems allow for control of water temperature and concentration of food particles, but they also require the investigator to regulate critical hydrochemical parameters, particularly the concentration of oxygen and organic wastes, such as ammonia. Black flies are highly sensitive to both organic and inorganic pollutants; therefore, nitrogenous metabolites arising from bacterial growth and decay of suspended food particles, as well as the excreta of the fly larvae themselves, must be prevented from reaching toxic levels in the system. Filtration may be necessary to remove ammonia from systems in which the reservoir capacity is small relative to the surface area of the trough or stirred container. Biological filters incorporating activated charcoal, gravel, or even the submerged roots of the aquatic plant *Monstera* sp. (Grunewald 1973) have been used to cleanse small-capacity closed systems (Fig. 55.1). Ammonia and other nitrogenous metabolites appear to remain at safe levels in larger-capacity systems without biological filtration. Brenner and Cupp (1980) attribute this to the relatively large volume of the reservoir and the evaporative capacity of the larger surface area of the gravity trough or raceway. This design concept has formed the basis of several successful rearing systems (Fig. 55.2) (Simmons and Edman 1982; Brenner and Cupp 1980; Ham and Bianco 1984; Tarrant et al. 1987).

Feeding Systems

Most larval black flies are filter feeders, straining suspended food particles from stream water by means of their cephalic fans. Others graze on algae attached to the surrounding substrate (Chapter 11). Beyond simulating the physicochemical attributes of natural black fly habitats, provision of an adequate level of particulate food is another major challenge in rearing larval black flies to vigorous adults. Sufficient quantities of food particles, sized to conform to the cephalic fans of filter-feeding species, obviously serve to provide a critical nutritional resource. Less obvious is the fact that many species of larval black fly release their hold on the substrate and drift downstream in response to declining levels of suspended particulates. In nature, this response is adaptive, in that it allows filter-feeding species to relocate to points in the stream with more abundant food. In an artificial laboratory stream, this behavior can result in the loss of an entire population of reared larvae; therefore, maintenance of a continuous supply of suspended particulate food is necessary for optimum results.

Although suspensions of natural plankton have been used as food for laboratory-reared black fly larvae in a few instances, modifications of various commercial animal foods, such as those formulated for aquarium fish (Ham and Bianco 1984) or laboratory rabbits and guinea pigs (Brenner and Cupp 1980; Brenner et al. 1980), have proven to be more widely useful for this purpose. When appropriately modified, these provide a relatively standardized nutrient content and are generally available to investigators who do not have ready access to natural black fly habitats. Correct sizing of suspended particles gives the most efficient conversion of food and decreases the buildup of toxic metabolites due to decay of undigested material. Suspended particles measuring 50–60 μm in diameter are readily captured and ingested by filter-feeding black fly larvae. Proprietary fish and rabbit diets may be pulverized and sieved to remove larger particles and added to the rearing system as a concentrated slurry. Automated delivery systems incorporating continuously operating peristaltic pumps (Ham and Bianco 1984) or electrically timed

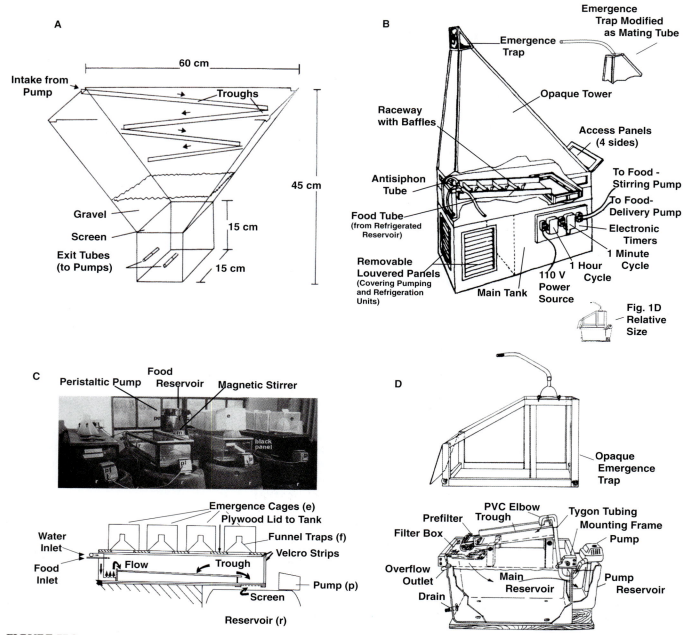

FIGURE 55.2 Four variations on the closed gravity-flow trough, or "artificial stream," concept. (A) Diagram of the inner tank assembly of a Plexiglas unit designed and used by Simmons and Edman (1982) to establish a continuous colony of *Simulium soubrense/sanctipauli* natural hybrids. The unit is fed with water from an outer tank by means of centrifugal pumps. A gravel filter is situated downstream of the series of four alternating troughs. Redrawn and used with permission. (B) A system designed by Brenner and Cupp (1980) for rearing temperate black flies and later used for complete generation rearing of *S. damnosum s.s.* This apparatus, later dubbed the Cornell system, is based on a commercially available unit used for maintaining marine and freshwater fish and crustacea (Dayno AquaLab®) with modifications, including the inclined trough with baffles, electronic timers comprising the automated food delivery system, and an opaque tower with an emergence trap later modified as a mating tube. The AquaLab® is capable of delivering temperature-controlled water from a 643-liter reservoir at a predetermined flow rate. The inset at the lower right shows the relative size of the benchtop rearing unit in panel D. Redrawn from Brenner and Cupp (1980) and used with permission. (C) A gravity-flow rearing system designed by Ham and Bianco (1984) for colonization of *S. erythrocephalum* and *S. lineatum*. The system, depicted in diagrammatic form below and in various configurations in the photograph above, includes a high-capacity reservoir of ambient-temperature water fashioned from an ordinary plastic barrel, an inclined trough fed by a pump through an anti-turbulence diffuser box, and a system of emergence traps consisting of muslin cages surmounting wide-stemmed powder funnels. At the time of pupation, black panels are affixed to the transparent walls of the trough enclosure to promote attraction of emerging flies to the translucent funnel traps. Slurries of pulverized fish food at various concentrations are delivered continuously from the food reservoir (an Erlenmeyer flask) into the trough by the peristaltic pump. (Redrawn and used with permission.) (D) A benchtop gravity-flow system, designed by Tarrant et al. (1987) for small-scale cohort or sibling rearings, includes many features of the larger "Cornell" system. The 10-L reservoir is a polypropylene cage bottom for laboratory rats modified to include a head tank, a pump, and an inclined trough with screens. A miniaturized emergence canopy with mating tube is lowered at the time of pupation. An inset in panel B illustrates the relative sizes of the benchtop apparatus and the Aqua Lab–based system.

bursts of food slurry (Brenner and Cupp 1980) have been devised to ensure a consistent level of suspended particles in closed gravity-flow trough systems. These automated delivery systems decrease maintenance and allow for periodic adjustments in the rate of food delivery to accommodate the increased requirements of growing larvae as rearings progress.

REARING AND MAINTENANCE OF ADULT BLACK FLIES

Capture of Newly Emerged Adults

Schemes for collecting adults arising from laboratory-reared larvae generally exploit the pronounced phototaxis of these insects. Most systems involve draping the trough enclosure with dark fabric or lowering an opaque emergence canopy at the time of pupation. These darkened enclosures are surmounted by transparent or translucent funnels or other collecting channels that admit light and attract emerging adults into some type of cage or collecting tube. The dimensions of the emergence enclosures are often dictated by the need to collect the flies into a confined space for mating. In this case a length of plastic tubing may substitute for an emergence cage (Figs. 55.2B–D).

Mating Systems

Beyond the need to create an artificial moving-water environment, inducement of adult black flies to mate in the laboratory has proven to be the greatest impediment to continuous colonization. Manual, or "forced," copulation techniques allowed the colonization of certain mosquitoes that exhibit swarm mating behavior in nature and consequently fail to mate in laboratory cages. Unfortunately, despite numerous attempts, these methods have not proven transferable to black flies as yet.

Many species of black flies exhibit protandry, or emergence of males prior to females. This adaptation ensures that a preponderance of males is present to inseminate females virtually as soon as they eclose, and it follows that mating occurs in a limited zone proximal to the site of pupation. In general, strategies for inducing laboratory-reared black flies to mate have involved a combination of attempts to simulate these natural conditions and to select strains genetically predisposed to mate in confinement. As in the case of emergence traps, their marked positive phototaxis has been exploited to form aggregations of males and newly emerged females in which mating can occur. A chamber for mating of the Kibweze form of *S.*

damnosum was designed upon this principle by Simmons and Edman (1982) and later modified for field use by Raybould and colleagues (Boakye and Raybould 1985).

Colonization of temperate-climate black flies such as *S. decorum*, *S. vittatum*, and *S. erythrocephalum* using high-capacity rearing systems involved selection of strains capable of copulating in confined spaces. Selection for such mating behavior in these species was rapid, occurring within one or two generations of selective pressure, and was probably pivotal in the relative ease with which they could be colonized. The mating response of *S. decorum* and *S. vittatum* could be enhanced by confinement in tubes of diameter 1–3 cm. The emergence trap of one automated gravity-flow system was subsequently modified by substituting a 3-cm-diameter Tygon tube for the original emergence cage (Fig. 55.2B).

Critical behaviors such as eclosion, mating, blood feeding, and oviposition in black flies frequently occur on a distinct diurnal cycle, and therefore control of both photoperiod and light intensity are often keys to stimulating these behaviors in the laboratory. Simulation of crepuscular lighting conditions was necessary for optimum mating and oviposition of *S. decorum* in the laboratory, and various electronic devices (Brenner et al. 1980; Simmons and Edman 1981) have been used to achieve controlled, gradual increases and decreases in the intensity of artificial lighting.

Adult Maintenance

Studies of vector–pathogen interactions, production of black fly-associated parasite stages, and basic physiological and behavioral study require maintenance of both laboratory-reared and field-collected adult black flies for significant periods of time. This requirement has represented a significant technical hurdle, since standard techniques for maintaining other biting flies in the laboratory generally do not suffice to sustain adult black flies in significant numbers for more than a few days. In general, careful control of temperature and humidity and availability of food is required. Among these, humidity is probably most critical. At one extreme, black flies are extremely susceptible to desiccation at low humidities; at extremely high humidities, adults readily succumb to infection with saprophytic fungi and bacteria and are prone to becoming trapped in moisture films. In addition, measures that retard caloric expenditures, such as limiting flight and interactions with other flies, tend to promote longer survival among adult flies.

The latter concern has been addressed by schemes that involve maintaining flies individually in small

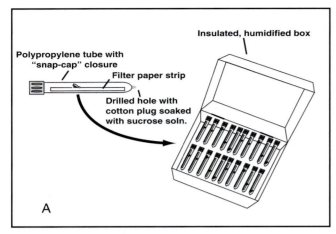

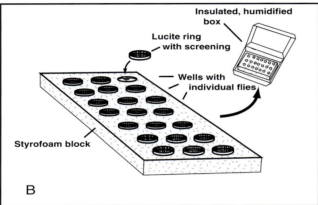

FIGURE 55.3 Two systems for individual holding of adult black flies. (A) Technique of Horacio-Figueroa et al. (1977), in which flies are captured during biting and collected in specialized tubes containing a filter paper strip to reduce condensation and outfitted with a cotton wick soaked with sugar solution and inserted into a hole drilled in the end of the tube. Groups of tubes may be stored in an insulated container humidified with moist paper. (B) System devised by Raybould and Mhiddin (1974) and used by this author to house individual flies infected with *Onchocerca* spp. The individual wells may be made by drilling or punching holes in commercial insulation material with a laboratory cork borer. The wells are enclosed by lucite rings screened with nylon mesh. This unit may be stored in a humidified box or incubator as indicated in panel A.

enclosures. One such method (Fig. 55.3A) involves housing flies individually in 5-mL polypropylene culture tubes modified to accept cotton or paper wicks soaked in sucrose solution and containing filter paper strips to absorb condensation (Horacio-Figueroa et al. 1977). This method serves especially well for flies captured in biting collections, since the same modified tubes can be used for capture and maintenance. Another approach (Fig. 55.3B) involves placing adult flies in screened circular chambers created by boring holes in blocks of styrofoam with a laboratory cork borer and affixing nylon netting to the openings by means of tight-fitting plastic rings (Raybould and Mhiddin 1974). Individual tubes or batteries of styro-

foam chambers are maintained within humidified boxes. Insulated coolers serve to maintain temperature under field conditions.

Housing flies individually tends to limit mortality due to excessive activity or to the spread of pathogenic microorganisms, but maintaining adults in this manner is labor intensive, especially where large numbers are involved. If significant numbers of flies must be maintained over relatively long periods of time, such as production operations for parasites, some type of cohort housing is required. Generally, cohorts of adult black flies survive longer when housed in small cages, fashioned from materials, such as styrofoam and cardboard, that resist the formation of moisture films from condensation. Pint-sized commercial food containers outfitted with screening of nylon mesh serve this purpose well. Wicks soaked with sugar solution that resist dripping or pooling provide a carbohydrate source, and these feeding solutions may be supplemented with antibiotic or antifungal compounds to protect against infection with adventitious pathogens. Such cages may be housed in groups within humidified chambers with flow-through ventilation to further enhance survival. The system of Bianco et al. (1989) pictured in Fig. 55.4 exemplifies these features and has been used extensively by this author to maintain black flies infected with *Onchocerca* spp.

Blood Feeding

Inducing anautogenous black flies to blood feed under laboratory conditions represents one of the chief obstacles in colonization of these species. Early attempts to feed laboratory-reared black flies by blood feeding on human volunteers or on animals such as rabbits, mice, or guinea pigs generally met with limited success, resulting in only 10–33% of the females feeding. In many, but not all, cases, more consistent results have been obtained with artificial membrane feeding systems.

Bernardo and Cupp (1986) review the literature in this area and give a detailed account of a membrane feeding system that they used for complete generation rearing of the nearctic species *S. pictipes*. This system incorporates an inverted, water-jacketed funnel with a thin artificial membrane affixed to the large opening. Materials used successfully for membrane feeding blackflies include mouse or chick skin, Badruch membrane, and thinly stretched laboratory paraffin film. Blood with anticoagulant is added through the upturned stem of the funnel, and the screened opening of a small cage containing the flies designated for feeding is brought in contact with the membrane (Fig. 55.5). In this and similar systems, warm water circulating through the jacket is used to maintain the

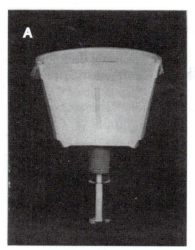

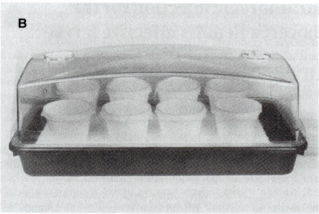

FIGURE 55.4 Elements of a successful cohort housing system for black flies as designed by Bianco et al. (1989; reproduced with permission). (A) Individual cages consist of a styrofoam food container with tight-fitting lid modified as a ring to affix a sheet of nylon screening. An absorbent wick conducts sugar solution, supplemented in some cases with antibiotic and antifungal agents, into the cage from the reservoir bottle below. (B) The cages are maintained in groups in a humidified, aerated chamber.

FIGURE 55.5 Typical membrane feeding apparatus from the author's laboratory and similar to that used by Ham and Bianco (1984) for colonization of *S. lineatum* and *S. erythrocephalum* and for complete generation rearing of *S. pictipes* and infection of both *S. pictipes* and *S. vittatum* with several black fly-associated pathogens (Bernardo and Cupp 1986; Bernardo et al. 1986). Membranes consisting of chick skin or Badruch material are stretched over the broad opening of the water-jacketed funnel, and heparinized blood (bovine, porcine, or ovine) is introduced through the top stem. Host body temperature is simulated by circulating water at 37–39°C through the water jacket. The feeding response is enhanced when a significant differential (ca. 22°C) between air and membrane temperatures is maintained. Significant quantities of black fly saliva may be collected by inducing flies to feed on nonnutritive solutions such as Tyrode's saline via such membrane feeding systems.

feeding funnel and its contents at a temperature approximating that of a mammalian or avian host (37–39°C). The best feeding responses were obtained when the temperature of the surroundings was maintained at 16–17°C to create a significant differential between the air and the membrane feeder and blood. A similar system was used for complete generation rearing of *S. lineatum* and for continuous colonization of *S. erythrocephalum* (Ham and Bianco 1984).

In addition to colony maintenance, membrane feeding systems have found experimental applications in studies of the physiology, vector biology, and control of black flies. Such systems have been used to experimentally infect black flies with filarial, protozoan, and viral pathogens (Ham and Gale 1984; Bernardo and Cupp 1986). They have also been used in efficacy studies of chemical repellents (Bernardo

and Cupp 1986). Finally, membrane feeding systems incorporating nonnutritive solutions such as Tyrode's saline or even distilled water have served as a means of collecting significant quantities of black fly saliva for immunopharmacologic study (Cupp and Cupp 1997).

OVIPOSITION AND EGG STORAGE

As with larval rearing, inducing black flies to oviposit under laboratory conditions has proven to be, in part, a matter of simulating the natural physical environment in which these insects lay their eggs. With some exceptions, gravid female black flies tend to place their eggs in the moist splash zone of emergent rocks, twigs, or vegetation in the stream habitat; in many cases this activity is concentrated at twilight. Accordingly, reliable systems for inducing oviposition by black flies in the laboratory have incorporated some provisions for an oviposition substrate within a moist spray or splash zone and for simulating crepuscular lighting conditions. Two such successful systems, one for *S. damnosum* (Simmons and Edman 1982) and the other for *S. decorum* (Brenner et al. 1980), are illustrated in Figs. 55.6A and B, respectively. Both systems present the ovipositing flies with a choice of flat and three-

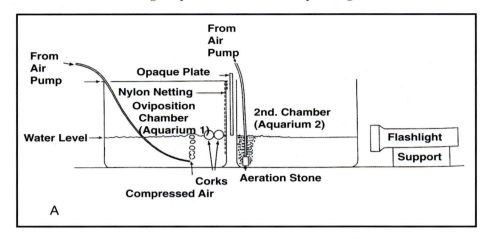

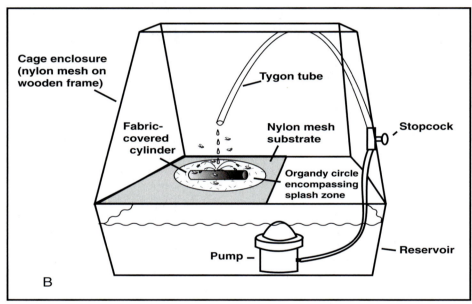

FIGURE 55.6 Oviposition chambers. (A) Design based on that of Simmons and Edman (1982) for colonization of *S. soubrense/sanctipauli* hybrids and modified for field use by Boakye and Raybould (1985). Flies, in some cases individual flies, are introduced into the covered aquarium 1 at the left. Water in this aquarium is kept in motion around floating corks by bubbling compressed air through a large-bore tube. Dark-adapted flies were induced to oviposit by exposure to simulated twilight created in the original construction by an automated crepuscular lighting system and in the field modification by simply shining a flashlight through aerated water in aquarium 2. Flies oviposited on the floating corks and/or the water-soaked nylon netting draped over the side of aquarium 1 adjacent to the opaque plate separating it from aquarium 2. (B) Chamber designed to stimulate oviposition by *Simulium decorum* (Brenner et al. 1980) and also used for *S. vittatum*. Recirculated water is dripped slowly on the cylinder fashioned by wrapping a 5-cm length of Tygon tubing in brown fabric. The cylinder lies on a circle of organdy, which in turn is supported by a wide-mesh nylon screen. The entire assembly is enclosed in a cage into which mated, gravid flies are introduced at simulated twilight. Flies typically oviposit in the splash zones on either the brown cloth cylinder or the underlying organdy. These pieces of fabric are easily affixed to the trough of the larval rearing system (Fig. 55.2B) after eggs are embryonated.

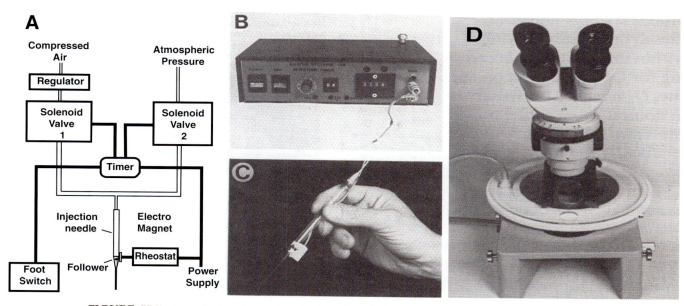

FIGURE 55.7 A semiautomated system for intrathoracic injection of black flies with microfilariae of *Onchocerca spp* developed by Bianco et al. (1989). (A) Diagram of the solenoid-controlled pressure regulator and agitated micropipette system. (B) Actual electronic control unit connected to the handheld micropipette (C) containing the microfilariae maintained in uniform suspension by means of a miniature electromagnetic agitator. (D) Flies for injection are positioned by means of the turntable stage of this microscope and maintained under CO_2 anesthesia by a slow stream of the gas from the inlet tube on the left. In some cases the electromagnetic agitator system may be replaced by smaller capillary micropipettes containing the inocula for single flies. These are repeatedly refilled from a stock culture of microfilariae by a second operator and transferred onto a pipette holder pressurized by the solenoid-driven regulator.

dimensional substrates within a splash area created by bubbling or gently falling water. Crepuscular lighting is simulated either electronically or by filtering dim light through an aerated, water-filled chamber, the latter arrangement being more practical under field conditions in the West African habitats of *S. damnosum*.

Black fly eggs hatch when embryonation is complete, a fact that necessitates virtually continuous rearing of laboratory strains. Nevertheless, as a practical matter, embryonation may be delayed by low-temperature storage to extend the life span of a given generation of flies and to reduce the frequency of active rearings somewhat. Generally, black fly eggs may be stored for many months on a wet substrate at temperatures ranging from 1° to 4°C (Brenner et al. 1980, Simmons and Edman 1981), with decreased O_2 levels extending storage life even further (Goll et al. 1989).

EXPERIMENTAL INFECTION WITH BLACK FLY-ASSOCIATED PARASITES AND PATHOGENS

Black flies may be infected per os with microfilariae, protozoan, and viral agents using artificial membrane feeding systems (Ham and Gale 1984; Bernardo and Cupp 1986). Membrane feeding has the advantage of relative ease of exposing large numbers of flies, but the numbers of parasites ingested by individual flies may vary greatly due to nonuniform suspensions in the membrane feeding device. This is particularly troublesome in the case of the filariae, where intensity of infection with developing larvae is closely correlated with the concentration of microfilariae in the infecting blood meal. For this reason, intrathoracic injection has been used extensively to infect black flies with *Onchocerca* spp. (Lok et al. 1980; Bianco et al. 1989), and a semiautomated system (Fig. 55.7; Bianco et al. 1989) has been designed that greatly speeds the inoculation process, allowing production of large numbers of infective larvae of *O. lienalis*. To date, the most important black fly-associated pathogen, *Onchocerca volvulus*, has not been adapted to any conventional laboratory animal host and consequently is virtually unavailable outside of endemic areas. For this reason, a systematic program of collecting wild *S. damnosum s.l.* after blood feeding on naturally infected human volunteers and maintaining these flies to infectivity has been used to amass large inventories of cryopreserved *O. volvulus* third-stage larvae (Cupp et al. 1988).

Readings

Bernardo, M. J., and Cupp, E. W. 1986. Rearing black flies (Diptera: Simuliidae) in the laboratory: Mass-scale in vitro membrane feeding and its application to collection of saliva and to parasitological and repellent studies. *J. Med. Entomol.* 23: 666–679.

Bernardo, M. J., Cupp, E. W., and Kiszewski, A. E. 1986. Rearing black flies (Diptera: Simuliidae) in the laboratory: Bionomics and life table statistics for *Simulium pictipes*. *J. Med. Entomol.* 23: 680–684.

Bianco, A. E., Ham, P. J., Townson, S., Mustafa, M. B., and Nelson, G. S. 1989. A semiautomated system of intrathoracic injection for the large-scale production of *Onchocerca lienalis* infective larvae. *Trop. Med. Parasitol.* 40: 57–64.

Boakye, D. A., and Raybould, J. N. 1985. The effect of different methods of oviposition inducement on egg fertility rates in a *Simulium damnosum* Theobald complex species (Diptera: Simuliidae). *J. Am. Mosq. Control Assoc.* 1: 535–537.

Brenner, R. J., and Cupp, E. W. 1980. Rearing black flies (Diptera: Simuliidae) in a closed system of water circulation. *Trop. Med. Parasitol.* 31: 247–258.

Brenner, R. J., Cupp, E. W., and Bernardo, M. J. 1980. Laboratory colonization and life table statistics for geographic strains of *Simulium decorum* (Diptera: Simuliidae). *Trop. Med. Parasitol.* 31: 487–497.

Cupp, E. W., and Cupp, M. S. 1997. Black fly (Diptera: Simuliidae) salivary secretions: Importance in vector competence and disease. *J. Med. Entomol.* 34: 87–94.

Cupp, E. W., and Ramberg, F. B. 1997. Care and maintenance of blackfly colonies. In: *Molecular Biology of Insect Disease Vectors: A Methods Manual* (J. M. Crampton, C. B. Beard, and C. Louis, ed.). Chapman and Hall, London, pp. 31–40.

Cupp, E. W., Bernardo, M. J., Kiszewski, A. E., Trpis, M., and Taylor, H. R. 1988. Large-scale production of the vertebrate infective stage (L3) of *Onchocerca volvulus* (Filarioidea: Onchocercidae). *Am. J. Trop. Med. Hyg.* 38: 596–600.

Edman, J. D., and Simmons, K. R. 1987. Maintaining black flies in the laboratory. In: *Black Flies — Ecology, Population Management and Annotated World List* (K. C. Kim and R. W. Merritt, eds.). Pennsylvania State University, University Park, PA, pp. 305–314.

Goll, P. H., Duncan, J., and Brown, N. 1989. Long-term storage of eggs of *Simulium ornatum*. *Med. Vet. Entomol.* 3: 67–75.

Gray, E. W., and Noblet, R. 1999. Large-scale laboratory rearing of black flies. In: *Maintenance of Human, Animal and Plant Pathogen Vectors* (K. Maramorosch and F. Mahmood, eds.). Science Publishers, Enfield, NH, pp. 85–105.

Grunewald, J. 1973. Die hydrochemischen Lebensbedingungen der praimaginalen Stadien von *Boophthora erythrocephala* De Geer (Diptera, Simuliidae) 2. Die Entwicklung einer Zucht unter experimentellen Bedingungen. *Z. Tropenmed. Parasit.* 24: 232–249.

Ham, P. J., and Bianco, A. E. 1984. Maintenance of *Simulium Wilhelmia lineatum* Meigen and *Simulium erythrocephalum* de Geer through successive generations in the laboratory. *Can. J. Zool.* 62: 870–877.

Ham, P. J., and Gale, C. L. 1984. Blood-meal-enhanced Onchocerca development and its correlation with fecundity in laboratory reared blackflies (Diptera, Simuliidae). *Trop. Med. Parasitol.* 35: 212–216.

Horacio-Figueroa, M., Collins, R. C., and Kozek, W. J. 1977. Postprandial transportation and maintenance of *Simulium ochraceum* infected with *Onchocerca volvulus*. *Am. J. Trop. Med. Hyg.* 26: 75–79.

Lok, J. B., Cupp, E. W., and Bernardo, M. J. 1980. The development of *Onchocerca* spp. in *Simulium decorum* Walker and *Simulium pictipes* Hagen. *Trop. Med. Parasitol.* 31: 498–506.

Raybould, J. N., and Mhiddin, H. K. 1974. A simple technique for maintaining *Simulium* adults including onchocerciasis vectors under artificial conditions. *Bull. Wld. Hlth. Org.* 51: 309–310.

Simmons, K. R., and Edman, J. D. 1981. Sustained colonization of the black fly *Simulium decorum* Walker (Diptera: Simuliidae). *Can. J. Zool.* 59: 1–7.

Simmons, K. R., and Edman, J. D. 1982. Laboratory colonization of the human onchocerciasis vector *Simulium damnosum* complex (Diptera: Simuliidae), using an enclosed, gravity-trough rearing system. *J. Med. Entomol.* 19: 117–126.

Tarrant, C. A., Scoles, G., and Cupp, E. W. 1987. Techniques for inducing oviposition in *Simulium vittatum* (Diptera: Simuliidae) and for rearing sibling cohorts of simuliids. *J. Med. Entomol.* 24: 694–695.

56

Care, Maintenance, and Experimental Infection of Phlebotomine Sand Flies

LEONARD E. MUNSTERMANN

INTRODUCTION

Phlebotomines are small flies associated with the transmission of *Leishmania* protozoans, *Bartonella* bacteria, and several *Phlebovirus* viruses. Because of their medical importance, attempts to develop efficient laboratory rearing methods for phlebotomine "sand flies" have a long history. In common parlance and even in scientific literature, the term *sand fly* does not distinguish between phlebotomine sand flies of the family Psychodidae and the beach sand flies of the family Ceratopogonidae (the biting midges, or no-see-ums). The following treatment refers to phlebotomines only; the rearing of Ceratopogonidae is treated in Chapter 54, and they are referred to as *biting midges*. Laboratory rearing and maintenance of phlebotomines is labor intensive and complicated compared with some other dipterous vectors. In technically skilled and experienced hands, however, colonies have been maintained for decades.

Psychodidae are separated as two groups, the free-living Psychodinae and the blood-feeding Phlebotominae. A representative of the former, *Clogmia albipunctata*, is extremely easy to rear and maintain, but its potential as a genetic and physiological model for the family Psychodidae (Troiano 1988) has not been exploited rigorously. In contrast, the Phlebotominae have had a 75-year history of laboratory rearing, focused largely on such representative species as *Phlebotomus papatasi* (Scopoli) and *Lutzomyia longipalpis* (Lutz and Neiva). More recent incremental improvements have come with plaster of Paris pots and suspended cloth cages (Hertig and Johnson 1961), larval diet (Endris et al. 1982; Young et al. 1981), immature rearing cages (Modi and Tesh 1983), mass rearing (Lawyer et al. 1991), initial colony establishment practices (M. Killick-Kendrick and Killick-Kendrick 1991), and oviposition substrate (Santamaría et al. 2002).

The subfamily Phlebotominae includes approximately 380 species of the New World genus *Lutzomyia*, 100 species of Eurasian *Phlebotomus*, 260 species of sub-Saharan and Asian *Sergentomyia*, and 100 species allocated to several additional genera. The habitats and behaviors of species within these genera are highly diverse, and, as a consequence, a single rearing–maintenance method cannot suit all; some of the improvements indicated earlier are applicable only to select species subsets. R. Killick-Kendrick et al. (1991) listed 44 species with some colonization effort: 23 *Lutzomyia*, 16 *Phlebotomus*, and 5 *Sergentomyia*. The subsequent decade saw several additional colonization reports (Table 56.1), most concerned with species implicated or incriminated in transmission of leishmaniasis.

If sand fly rearing is considered difficult, the difficulty resides mostly with the larval stages. A lengthy, 5- to 8-week period is required for development from egg to pupa, with four intervening larval molts (Chapter 12). Since the larval habitat and natural food requirements have been observed for few species, devising an optimum larval diet has been a process of trial and error. For most species, moisture is critical, and optimal humidity and temperature also promote growth of fungi and mites. If these contaminants are

TABLE 56.1 Phlebotomine Sand Fly Species That Have Been Subjected to Colonization

Genus Species	Subgenus/Group*	Colony Dates	Reference
Lutzomyia (n = 29)	(After Young and Duncan 1994)		
L. vespertilionis	*Coromyia*	1971	R. Killick-Kendrick et al. (1991)
L. anthophora	*Dampfomyia*	1945, 1981, 1982	R. Killick-Kendrick et al. (1991)
L. sanguinaria	*Helcocyrtomyia*	1961	R. Killick-Kendrick et al. (1991)
L. stewarti	*Helcocyrtomyia*	1967	R. Killick-Kendrick et al. (1991)
L. vexator	*Helcocyrtomyia*	1964, 1967, 1981, 1982	R. Killick-Kendrick et al. (1991)
L. diabolica	*Lutzomyia*	1936, 1982	R. Killick-Kendrick et al. (1991)
L. cruciata	*Lutzomyia*	1981, 1982	R. Killick-Kendrick et al. (1991)
L. gomezi	*Lutzomyia*	1961	R. Killick-Kendrick et al. (1991)
L. longipalpis	*Lutzomyia*	1940–1986, 1998	R. Killick-Kendrick et al. (1991); Ferro and Morales (1998)
L. californica	*Micropygomyia*	1967	R. Killick-Kendrick et al. (1991)
L. cayennensis hispanolae	*Micropygomyia*	1982	R. Killick-Kendrick et al. (1991)
L. migonei	*Migonei* group	1997	Nieves et al. (1997)
L. walkeri	*Migonei* group	1984, 1998	R. Killick-Kendrick et al. (1991); Ferro and Morales (1998)
L. flaviscutellata	*Nyssomyia*	1977, 1984	R. Killick-Kendrick et al. (1991)
L. intermedia	*Nyssomyia*	1940, 1985, 1986	R. Killick-Kendrick et al. (1991)
L. trapidoi	*Nyssomyia*	1974, 1975, 1981, 1986	R. Killick-Kendrick et al. (1991)
L. whitmani	*Nyssomyia*	1941	R. Killick-Kendrick et al. (1991)
L. ylephiletor	*Nyssomyia*	1961	R. Killick-Kendrick et al. (1991)
L. trinidadensis	*Oswaldoi* group	1972	R. Killick-Kendrick et al. (1991)
L. shannoni	*Psathyromyia*	1981, 1982, 1998	R. Killick-Kendrick et al. (1991); Ferro et al. (1998)
L. panamensis	*Psychodopygus*	1961	R. Killick-Kendrick et al. (1991)
L. evansi	*Verrucarum* group	1995, 1998	Olviedo et al. (1995); Montoya-Lerma et al. (1998)
L. longiflocosa	*Verrucarum* group	1998	Ferro and Morales (1998)
L. ovallesi	*Verrucarum* group	1999	Cabrera et al. (1999)
L. quasitownsendi	*Verrucarum* group	1998	Ferro and Morales (1998)
L. serrana	*Verrucarum* group	2002	Santamaría et al. (2002)
L. spinicrassa	*Verrucarum* group	1998	Ferro and Morales (1998)
L. youngi	*Verrucarum* group	1977, 1984, 1985, 1997	R. Killick-Kendrick et al. (1991); M. Killick-Kendrick et al. (1997)
L. furcata	*Viannamyia*	1984	R. Killick-Kendrick et al. (1991)
Phlebotomus (n = 18)	(After Lewis 1982)		
P. chinensis	*Adlerius*	1983	R. Killick-Kendrick et al. (1991)
P. halepensis	*Adlerius*	1997	Modi (1997)
P. sichuanensis	*Adlerius*	1990	R. Killick-Kendrick et al. (1991)
P. argentipes	*Euphlebotomus*	1925–1989, 1997	R. Killick-Kendrick et al. (1991); Modi (1997)
P. kiangsuensis	*Euphlebotomus*	1985	R. Killick-Kendrick et al. (1991)
P. ariasi	*Larroussius*	1940–1987	R. Killick-Kendrick et al. (1991)
P. langeroni	*Larroussius*	1987	R. Killick-Kendrick et al. (1991)
P. longipes	*Larroussius*	1970, 1976	R. Killick-Kendrick et al. (1991)
P. neglectus	*Larroussius*	1989	R. Killick-Kendrick et al. (1991)
P. orientalis	*Larroussius*	1964	R. Killick-Kendrick et al. (1991)
P. pedifer	*Larroussius*	1977, 1986	R. Killick-Kendrick et al. (1991)
P. perfiliewi	*Larroussius*	1983, 1989	R. Killick-Kendrick et al. (1991)
P. perniciosus	*Larroussius*	1928–1987	R. Killick-Kendrick et al. (1991)
P. tobbi	*Larroussius*	1989	R. Killick-Kendrick et al. (1991)
P. sergenti	*Paraphlebotomus*	1997	Modi (1997)
P. papatasi	*Phlebotomus*	1910–1988, 1997	R. Killick-Kendrick et al. (1991); Modi (1997)
P. duboscqi	*Phlebotomus*	1986, 1987, 1997	R. Killick-Kendrick et al. (1991); Modi (1997)
P. martini	*Synphlebotomus*	1982, 1983	R. Killick-Kendrick et al. (1991)
Sergentomyia (n = 6)	(After Ashford 1991; Lewis 1978)		
S. squamipleuris	*Grassomyia*	1942	R. Killick-Kendrick et al. (1991)
S. bailyi	*Nic-nic* group	1985	R. Killick-Kendrick et al. (1991)
S. africana	*Parratomyia*	1942, 1982, 1983	R. Killick-Kendrick et al. (1991)
S. dentata	*Sergentomyia*	1999	Naucke (1999)
S. minuta	*Sergentomyia*	1999	Naucke (1999)
S. schwetzi	*Sergentomyia*	1942, 1968, 1982, 1983, 1997	R. Killick-Kendrick et al. (1991); Modi (1997)

* Subgeneric classifications were designated after authors at column head.

present at a high density, both can have negative effects on larval development and survival.

STAGE-BY-STAGE MAINTENANCE

Laboratory Environment

The complexity of the rearing environment depends to some extent on the species of fly, the numbers of fly colonies, and the numbers of flies that are anticipated. Larger-scale operations may require an insectary dedicated to sand flies, with preset temperature, humidity, and photoperiod controls. For more typical operations with targeted, research-oriented goals, refrigerator-sized, reach-in incubators are available with light and temperature controls (Modi 1997). However, even in ambient laboratory spaces (fluctuating temperatures, no humidity, or no photoperiod controls), inexpensive styrofoam enclosures are adequate and nearly equal to an incubator (Cárdenas et al. 1999). Typical rearing temperatures may be 22–26°C, but the optimal temperature varies with species. Regardless of the general laboratory environment, an enclosure must be provided for all stages for maintaining high humidity in the microenvironment.

The Egg Stage Microenvironment

A substrate for oviposition acceptable to many species is the porous surface provided by plaster of Paris (Endris et al. 1982). Plastic containers are prepared in advance by layering the bottoms with plaster to a depth of 75–150 mm (Fig. 56.1). For some species, adding a stone protruding from the plaster provides added incentive for oviposition (Santamaría et al. 2002). Sufficient water is added to the plaster to maintain 90% RH, but not so much that beads of water form on container surfaces. Mated and blood-fed females are introduced through an opening in the mesh top of the container. A recent evaluation of carrying capacity indicated that in a 100-cm^3 (height = diameter) container, 22 *Lutzomyia serrana* females was the optimum number for maximum production of the next generation (Santamaría et al. 2002). After oviposition, living and dead flies are removed and a light sprinkling of larval food is added. In the 2–3 days during egg embryonation, temperature is held constant, and humidity is maintained by adding drops of water to the plaster substrate.

The Larval Stages Microenvironment

Larvae are retained in the oviposition container until pupation. Humidity is controlled, first, by the

FIGURE 56.1 Container for oviposition, larval rearing, and adult emergence of phlebotomine sand flies. The plastic container is 12 cm in diameter, with holes drilled in its bottom, and layered with 1.5 cm of plaster of Paris. Fine mesh over the top is held in place tightly with a rubber band. A cotton plug in the mesh opening permits admission or removal of adults with the aspirator shown in Figure 56.2. Photo by L. E. Munstermann.

moist plaster of Paris in the container and, second, by placing the oviposition/larval rearing containers in a larger plastic or styrofoam chamber. A wet cloth or sponge provides moisture for the outer chamber. The larvae undergo four molts over a 5- to 8-week period. Throughout this time, small amounts of larval food are added on the surface of the plaster of Paris container bottom as needed. The growth must be monitored by a thrice-per-week inspection for three factors: (1) sufficient food and moisture, (2) fungal growth, and (3) mite densities. Food and moisture must be regulated carefully; too much promotes growth of mites and fungus, and too little retards larval development. Agitating container contents (food, larvae, feces) by shaking at each inspection breaks fungal hyphae and thereby retards fungal growth. It may be necessary to crush individual mites in order to control their numbers.

Preparation of larval food has proven a critical step in successful colonies. The formulation commonly used has evolved from that of Young et al. (1981). The following is a précis of a description by Lawyer et al. (1991) (or see Modi 1997). Dried rabbit feces and rabbit chow in equal parts are ground together and spread in a 1.5-cm layer on a large tray. This is saturated with water and stored in enclosed cabinets at 90–100% humidity to encourage growth of mold. The mixture is stirred and dehydrated once a week until spontaneous mold growth ceases, usually in 1–2 months; beef liver powder is stirred into the mixture in the final stages of the aging process. After drying, the transformed mixture is pulverized and kept frozen until use. Alternate food recipes have been used for several *Lutzomyia* species, usually involving aquarium fish food (containing dried and pulverized meat, fish, shrimp, wheat germ, and algal components) in combination with sand particles or other substrate (e.g., Rangel et al. 1986).

The Pupal Stage Microenvironment

Before pupation, the fourth-stage larvae often migrate to the edges and sides of the container. They remain there in a quiescent state and can be readily manipulated with a small camel hair brush. Pupae can be sexed by moistening and removing the larval exuvium that remains glued to the abdominal terminus of the pupa (males have a bulbous terminus, corresponding to the external genitalia). Pupae are best moved to adult cages prior to emergence, because the adults are fragile.

The Adult Microenvironment

Adults may be removed by aspirator (Fig. 56.2) from larval rearing chambers to suspended cloth cages (Hertig and Johnson 1961). High humidity is essential for the New World *Lutzomyia* sand flies, although this requirement may be less stringent for species associated with arid zones of the Middle East or western Asia. Humidity is maintained by enclosing the entire cage in a plastic bag with a wet sponge placed inside (Modi and Tesh 1983). A pad of cellulose cotton moistened with a 10% sucrose solution provides the energy nutrient; apple slices or other sugar sources are sometimes substituted.

Blood Feeding of Adult Females

Blood feeding is required for the development of the eggs in the ovaries. Blood meals are provided by white mice or hamsters, first anesthetized with kedamine hydrochloride (for hamsters, 100–200 mg/kg) and placed on the floor of the cloth cage (Fig. 56.3). If the blood-feeding rate is low, offering additional blood meals at 1- to 2-day internals may be necessary. Feeding can be stimulated by covering the cage with a black cloth to simulate evening or by removing the sugar source several hours previously. Blood-fed females are removed gently by aspirator from the cloth cage and deposited in the larval rearing container. Several males are also retained with the females to increase probability of mating. The presence of males may also stimulate oviposition. After 5–7 days, oviposition is completed. The females usually die after oviposition and are regularly removed from the container.

Infecting Sand Flies

Reliable infections of sand flies with *Leishmania* protozoans are best obtained by means of a membrane feeder. The following protocol is summarized from Tesh and Modi (1984). The infected blood medium is prepared by mixing packed human red blood cells, Schneider's medium, and infected macrophages in a 1:1:1 v/v ratio. The macrophage cell line was infected 2–3 days previously by adding cultured promastigote

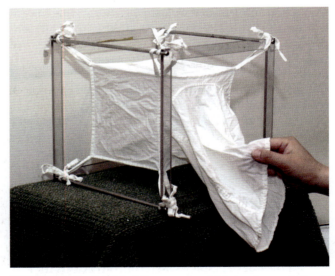

FIGURE 56.3 Adult holding cage for phlebotomine sand flies. It consists of a fine mesh bag with an open sleeve suspended on a Plexiglas–metal rod frame. Photo by L. E. Munstermann.

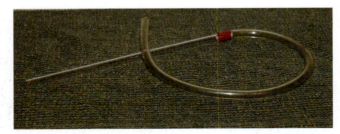

FIGURE 56.2 Hand aspirator for collecting and transferring phlebotomine sand flies. The 40-cm-long, straight tube consists of 1.5-cm-diameter Plexiglas with a fine mesh-and-cotton plug taped at one apex; the Nalgene tubing is attached over the plug. Photo by L. E. Munstermann.

Leishmania; the parasites were transformed into the amastigote form. (Only this form is infectious to the sand fly.) Skin for the membrane feeder is prepared from defeathered, newly hatched chicks; it is stretched across the end of a glass chamber into which the warmed blood mixture is placed. The membrane face of the apparatus is offered to the flies in the hanging cloth cage; after feeding, flies are moved to the chambers appropriate to the intended study.

CAUTIONARY CONSIDERATIONS

In the foregoing, the goal was to provide an introduction to the process, to the literature, and to the problems of sand fly rearing. Three facets of sand fly rearing require additional emphasis.

1. The procedures provided herein have proved to be successful for several of the species listed in Table 56.1. However, the procedures are not sufficient to establish colonies of every sand fly species. The best approach is to locate the most recent literature for the species in question in order to identify a rearing framework best suited to that species and the local laboratory setting. Special requirements for specific sand flies may be obtained from the references listed in Table 56.1. For more explicit details concerning cage construction, timed procedures, handling specimens, and sources of materials, the following references provide important guidelines: Endris et al. (1982), Lawyer et al. (1991), and Modi (1997).

2. The sand fly species selected for laboratory colonization is not random. Undoubtedly, the criteria for choice of species has been (a) accessibility, (b) ease of colonization, and (c) economic/medical importance. Species of flies found in close association with humans or domestic animals in high numbers are the first and most frequently colonized—*Lutzomyia longipalpis*, *Phlebotomus papatasi*, *P. argentipes*, and *P. perniciosus* best represent this association. Furthermore, each of the four is a major vector of leishmaniasis in its respective geographic area. As Table 56.1 indicates, however, the spectrum of subgeneric groupings is poorly represented. For the genus *Lutzomyia* (~380 species, 29 colonized), for example, representatives of only 11 of the 23 subgenera/groups have been colonized. For *Sergentomyia* (~260 species), a genus seldom implicated in disease transmission to humans, only six species have been listed.

3. The colony represents a small subset of the genetic variation that occurs in field populations. Consequently, inferences about behavior, genetics, physiology, or systematics of field populations must be circumspect when the inferences are based on examination or manipulation of laboratory colonies. Selection of the genetic subset occurs at two levels. (1) The initial collection from the field eliminates many rare variants from the colony by sampling alone, effectively, in population genetic terms, a founder event. (2) Subsequently, within the colony, passage from one generation to the next exerts continuous genetic bottlenecks. Often, enhancement of heterozygosity due to diallelic, balanced polymorphisms occurs; this appears to be associated with the effects of selection in the adaptation to laboratory conditions. The result, an overall lowering of allelic variants, has been documented in mosquito vectors (Munstermann 1994) as well as in the comparison of several colonies of *Lutzomyia longipalpis* to their field congeners (Morrison et al. 1995; Mukhopadhyay et al. 1997); these trends are clear even with methods of lowered resolving power (Lanzaro et al. 1998). For other species, e.g., *L. shannoni*, no overall reduction in allele number may be discerned (Mukhopadhyay et al. 2001), but highly distorted allelic associations are documented at particular loci (Cárdenas et al. 2001). In this case, the distortion took the form of 100% heterozygosity for the GPD locus in the males and 0% in females.

Nonetheless, these cautionary remarks do not contradict the fact that most of the knowledge concerning sand fly behavior, physiology, and vector competence have come from laboratory colonies. The particular value of colonies is that sand flies are small and secretive in habit, making field observations difficult. Furthermore, the inability to locate larvae in the field makes colonization essential for even the most basic biological observations of immature stages. The current epidemic increases in phlebotomine-transmitted diseases guarantee that increasing value will accrue to institutions where sand fly colonies are maintained and to the individuals with the skills to maintain those colonies.

Acknowledgments

Appreciation is offered to those who provided experience with rearing sand flies over the past two decades: C. Ferro, K. Ghosh, H. Guzman, P. Lawyer, J. Mukhopadhyay, E. D. Rowton, and R. B. Tesh. Sand fly research at Yale was supported by NIH grant AI-56254.

Readings

Ashford, R. W. 1991. A new morphological character to distinguish. *Sergentomyia* and *Phlebotomus. Parassitologia* 33(suppl): 79–83.
Cabrera, O. L., Neira, M., Bello, F., and Ferro, C. 1999. Ciclo de vida y colonización de *Lutzomyia ovallesi* (Diptera: Psychodidae),

vector de *Leishmania* spp. en América Latina. *Biomédica* 19: 223–229.

Cárdenas, E., Ferro, C., Corredor, D., Martinez, O., and Munstermann, L. E. 1999. Reproductive biology of *Lutzomyia shannoni* (Dyar) (Diptera: Psychodidae) under experimental conditions. *J. Vector Ecol.* 24: 158–170.

Cárdenas, E., Munstermann, L. E., Martínez, O., Corredor, D., and Ferro, C. 2001. Genetic variability among populations of *Lutzomyia (Psathyromyia) shannoni* (Dyar 1929) (Diptera: Psychodidae: Phlebotominae) in Colombia. *Mem. Inst. Oswaldo Cruz* 96: 189–196.

Endris, R. G., Perkins, P. V., Young, D. G., and Johnson, R. N. 1982. Techniques for laboratory rearing of sand flies (Diptera: Psychodidae). *Mosquito News* 42: 400–407.

Ferro, C., and Morales, A. 1998. Flebótomos de Colombia: Estudios realizados por el Laboratorio de Entomología 1965–1997. In *Instituto National de Salud 1917–1997: Una Historia, un Compromiso.* (G. Toro, C. A. Hernández, and J. Raad, eds.), Santa Fe de Bogotá, Colombia, pp. 219–233.

Ferro, C., Cárdenas, E., Corredor, D., Morales, A., and Munstermann, L. E. 1998. Life cycle and fecundity of *Lutzomyia shannoni* (Dyar) (Diptera: Psychodidae). *Mem. Inst. Oswaldo Cruz* 93: 195–199.

Hertig, M., and Johnson, P. T. 1961. The rearing of Phlebotomus sand flies (Diptera: Psychodidae). I. Technique. *Ann. Entomol. Soc. Am.* 54: 753–764.

Killick-Kendrick, M., and Killick-Kendrick, R. 1991. The initial establishment of sand fly colonies. *Parassitologia* 33(suppl): 315–320.

Killick-Kendrick, M., Killick-Kendrick, R., Añez, N., Nieves, E., Scorza, J. V., and Tang, Y. 1997. The colonization of *Lutzomyia youngi* and the putative role of free-living nematodes in the biology of phlebotomine sand fly larvae. *Parasite* 4: 269–271.

Killick-Kendrick, R., Maroli, M., and Killick-Kendrick, M. 1991. Bibliography on the colonization of phlebotomine sand flies. *Parassitologia* 33(suppl): 321–333.

Lanzaro, G. C., Alexander, B., Mutebi, J. P., Montoyo-Lerma, J., and Warburg, A. 1998. Genetic variation among natural and laboratory colony populations of *Lutzomyia longipalpis* (Lutz and Neiva, 1912) (Diptera: Psychodidiae) from Colombia. *Mem. Inst. Oswaldo Cruz* 93: 65–69.

Lawyer, P. G., Rowton, E. D., Perkins, P. V., and Johnson, R. N. 1991. Recent advances in laboratory mass rearing of phlebotomine sand flies. *Parassitologia* 33(suppl): 361–364.

Lewis, D. J. 1978. The phlebotomine sand flies (Diptera: Psychodidae) of the Oriental Region. *Bull. Brit. Mus. (Nat. Hist.)* 37: 217–343.

Lewis, D. J. 1982. A taxonomic review of the genus *Phlebotomus* (Diptera: Psychodidae). *Bull. Brit. Mus. (Nat. Hist.)* 45: 121–209.

Modi, G. 1997. Care and maintenance of phlebotomine sand fly colonies. In *The Molecular Biology of Insect Disease Vectors: A Methods Manual.* (J. M. Crampton, C. B. Beard, and C. Louis, eds.), Chapman and Hall, Cambridge, UK, pp. 21–30.

Modi, G., and Tesh, R. B. 1983. A simple technique for mass rearing *Lutzomyia longipalpis* and *Phlebotomus papatasi* (Diptera: Psychodidae) in the laboratory. *J. Med. Entomol.* 20: 568–569.

Montoya-Lerma, J., Cadena-Peña, H., and Jaramillo-Salazar, C. 1998. Rearing and colonization of *Lutzomyia evansi* (Diptera: Psychodidae), a vector of visceral leishmaniasis in Colombia. *Mem. Inst. Oswaldo Cruz* 93: 263–268.

Morrison, A. C., Munstermann, L. E., Ferro, C., Pardo, R., and Torres, M. 1995. Ecological and genetic studies of *Lutzomyia longipalpis* in a central Colombia focus of visceral leishmaniasis. *Bol. Dir. Malariol. Saneamiento Ambiental* 35(suppl): 235–248.

Mukhopadhyay, J., Rangel, E., Ghosh, K., and Munstermann, L. E. 1997. Patterns of genetic variability in colonized strains of *Lutzomyia longipalpis* (Diptera: Psychodidae) and its consequences. *Am. J. Trop. Med. Hyg.* 57: 216–221.

Mukhopadhyay, J., Ghosh, K. N., Ferro, C., and Munstermann, L. E. 2001. Distribution of phlebotomine sand fly genotypes (*Lutzomyia shannoni*, Diptera: Psychodidae) across a highly heterogeneous landscape. *J. Med. Entomol.* 38: 260–267.

Munstermann, L. E. 1994. Unexpected genetic consequences of colonization and inbreeding: Allozyme tracking in Culicidae. *Ann. Entomol. Soc. Am.* 87: 157–164.

Naucke, T. J. 1999. Colonization of the sand fly species *Sergentomyia dentata* and *S. minuta* (Diptera, Psychodidae) in filter crucibles filled with glass beads. *Third Int. Symp. Phlebotomine Sandflies,* Montpellier, France, p. O-51.

Nieves, E., Ribeiro, A., and Brazil, R. 1997. Physical factors influencing the oviposition of *Lutzomyia migonei* (Diptera: Psychodidae) in laboratory conditions. *Mem. Inst. Oswaldo Cruz* 92: 733–737.

Olviedo, M., Moreno, G., and Graterol, D. 1995. Bionomia de los vectores de leishmaniasis visceral in el estado Trujillo, Venezuela. III. Colonización de *Lutzomyia evansi. Bol. Dir. Malariol. Saneamiento Ambiental* 35(suppl): 269–276.

Rangel, E. F., Souza, N. A., Wermelinger, E. D., Barbosa, A. F., and Andrade, C. A. 1986. Biologia de *Lutzomyia intermedia* Lutz and Neiva, 1912 e *Lutzomyia longipalpis* Lutz and Neiva, 1912 (Diptera, Psychodidae), em condicões experimentais. I. Aspectos da alimentação de larvas e adultos. *Mem. Inst. Oswaldo Cruz* 81: 431–438.

Santamaría, E., Munstermann, L. E., and Ferro, C. 2002. Estimating carrying capacity in a newly colonized sand fly *Lutzomyia serrana* (Diptera: Psychodidae). *J. Econ. Entomol.* 95: 149–154.

Tesh, R. B., and Modi, G. 1984. A simple method for experimental infection of phlebotomine sand flies with *Leishmania. Am. J. Trop. Med. Hyg.* 33: 41–46.

Troiano, G. 1988. Heterozygous heterochromatin in Giemsa C-banded chromosomes of *Clogmia albipunctata (Telmatoscopus albipunctatus)* (Diptera: Psychodidae). *Caryologia* 41: 201–208.

Young, D. G., and Duncan, M. A. 1994. Guide to the identification and geographic distribution of *Lutzomyia* sand flies in Mexico, the West Indies, Central and South America (Diptera: Psychodidae). *Mem. Am. Entomol. Inst. No. 54,* Associated Publ., Gainesville, FL.

Young, D. G., Perkins, P. V., and Endris, R. G. 1981. A larval diet for rearing phlebotomine sand flies (Diptera: Psychodidae). *J. Med. Entomol.* 18: 446.

57

Care, Maintenance, and Experimental Infection of Tsetse Flies

SERAP AKSOY

INTRODUCTION

Establishment of tsetse colonies involves collecting puparia, sexing and sorting the emerging adults, setting up matings between males and females, separating the mated females into new cages, and maintaining the breeding females by providing a blood meal every 40–50 hours. Each female deposits a single larva about every 9 days, resulting in a total of six to eight offspring over a 3- to 4-month life span. This chapter describes the conditions required for maintaining a small colony that can be used for research purposes.

LIFE HISTORY OF TSETSE

Female tsetses are viviparous, ovulation proceeds alternately from the two polytrophic ovarioles, and the egg is fertilized by sperm stored in the paired spermathecae. An inseminated female deposits her first fully developed larva at about 18 days of age and continues to deposit a single larva every 9 days. Young females usually mate once and utilize the sperm stored in the spermathecae. The tsetse life cycle is detailed in Chapter 13.

Tsetse colonies can be initiated either by field-collected puparia or, alternatively, from puparia deposited by field-collected adult females. The field-caught females can be brought into the insectary and maintained long enough to deposit puparia and initiate the colony. The field-collected puparia have the advantage that the emerging adults are free from pathogens, since the field-collected flies may be infected with trypanosomes and require special handling. The difficulty with obtaining puparia from the field is that it may be a challenge to locate deposition sites for tsetse when large numbers of flies are needed to begin the colony. Since the deposited larvae quickly burrow their way into the soil and pupate, soil from suitable habitats needs to be collected and sifted to obtain the puparia. It is important, however, that sufficient material be obtained to ensure ample sampling of the gene pool for success of colony.

PUPARIA COLLECTION AND EMERGENCE

In colonies where fewer than 1,000 breeding females are maintained, the deposited puparia can be collected weekly. The puparia are stored in tubes or petri dishes, with the container open to air, making sure that the thickness of the puparia in each container is less than 2.5 cm. Cages made from metal frames or from PVC tubing covered with a black Terylene netting through which the flies may feed and through which the larvae, but not the adults, may pass are placed directly over storage tubes to allow the adults to emerge into the cages. Once the adults begin to emerge, the cages may be replaced daily by placing a cork or rubber stopper at the opening, and the emerged adults can then be sorted, fed, and mated. The cages are kept on metal, plastic, or Plexiglas trays, with two appropriately placed ridges or rods in the tray on which the cages rest. This is especially needed so that larva deposited by the breeding females can escape from the cage, squeeze through the netting, and fall onto the tray. The puparia can be collected from the trays, and having some coarse-grained sand in the trays makes collection easier.

One side of the cage should be reserved for feeding, and the opposite side should be placed face down on the tray. The cages should be dated and the number of flies housed in each cage marked; as flies die they should be removed and recorded. The cages should be handled with care; cages housing the breeding females should not be disturbed, for this often results in aborted larvae or eggs. To assess the quality of the colony in large insectaries, the size and/or the weight of the collected puparia is determined. It is recommended that in a healthy colony no more than 10% of the puparia be below the lower value of the mean. A significant increase in the number of smaller puparia indicates that the health of the colony is being compromised, and it is important at this point to evaluate the nutritional value of the blood meal source and the environmental conditions in the facility to prevent significant loss of fecundity. In addition to suboptimal environmental conditions, excessive crowding in cages or bacterial or chemical contamination may contribute to reduced fecundity. Both puparia and adults of most species are maintained at 24°C and about 60% humidity for the *morsitans* group, and 70–80% humidity for the *palpalis* group flies is recommended.

FEEDING AND SOURCE OF BLOODMEAL

Flies should be fed under environmental conditions that are as close to those used for maintenance as just described. In recent years, most insectaries have switched to using an in vitro feeding system where fresh or freshly frozen defibrinated blood is provided to the flies through a silicone membrane. It is important to maintain the trays and the membranes sterile. The feeding trays are washed and autoclaved after each use, and the membranes are washed and kept in an oven at 120°C overnight before use.

Blood is maintained at ambient temperature by placing the trays on a suitable heating apparatus that has a rheostat for controlling its temperature. A thin layer of blood is poured onto the tray, and the membrane is placed on the blood gently so as not to trap air bubbles. Blood is uniformly distributed under the membrane by use of a sterile plastic rod. Cages are placed with the feeding surface of the cage next to the membrane, and flies are allowed to feed for about 10 minutes. If flies are found dead with indigested blood in their guts shortly after feeding, it may be due either to contaminated blood or to a temperature of the blood that was too high above the ambient temperature.

BREEDING TSETSE

In order to obtain the maximum insemination rates in females, it is important to separate newly emerged females and males. This is done by chilling the emerged new adults and sexing flies manually. In males, the posterior end of the abdomen is pointed, the superior claspers or external genitalia are heavily sclerotized, and the fifth abdominal sternum is largely sclerotized and carries many setae. In contrast, the female's abdomen is truncated, and its midline region is not sclerotized and has a pale appearance. Although some important variations are noted among species, males (*Glossina morsitans morsitans*) reach full sexual capability within a week, and females are sexually receptive when they are 2–4 days old. At the appropriate times, a 1:2 ratio of males to females is used to establish cage matings. Males are separated from the females after 24–48 hours, and they can be used for two or three additional matings. The pregnant females are maintained for 12–16 weeks, during which time they deposit one pupa every 9–10 days.

EXPERIMENTAL INFECTION OF FLIES WITH TRYPANOSOMES

Flies can be infected with trypanosomes either by feeding on infected animals or by feeding on trypanosome-containing infectious blood meals via the artificial membrane system. Infective feeds are prepared from frozen stabilates of parasites suspended in the defibrinated blood at about $1–3 \times 10^5$ trypanosomes/mL.

Frozen stabilates are prepared from trypanosome-infected mice or rat blood collected at peak parasitaemia. For preparation of the frozen stabilates, an equal volume of phophate-buffered saline glucose (PSG: for a final volume of 1 liter use 8.09 g Na_2HPO_4, 0.47 g $NaH_2PO_2 \cdot 2H_2O$, 14.95 g glucose, and 2.63 g NaCl) 6:4 with 14% glycerol is added slowly to the infected blood pooled from a group of animals; the mixture is dispensed into cryotubes, frozen overnight at −70°C, and then transferred to liquid nitrogen. When desired, the cryotubes can be defrosted at room temperature, diluted fourfold with PSG, centrifuged at 1500 g for 10 minutes at 4°C. The parasites in the pellet are microscopically checked for viability before being added to the blood meal. Alternatively, culture form procyclic parasites can be used to infect the flies, but a higher dose is needed in the infectious blood meal ($1–5 \times 10^6$ cells/ml). For higher yields of infection, newly emerged (teneral) flies are used for infections, and male flies typically result in higher salivary gland infections. To artificially increase the infection rates, lectin inhibitory carbohydrates such as D+glucosamine can be included in the infectious meal at a concentration of 0.015–0.03 M. Parasite infections in the midgut can be detected microscopically about 10 days postinfection, while the development of mature infections in the salivary glands requires about 25–30 days.

Readings

Aksoy, S. 1999. Establishment and maintenance of small scale tsetse colonies. In *Maintenance of Human, Animal and Plant Pathogen Vectors* (K. Maramorosch and F. Mahmood, eds.). Science Publishers, NH, pp. 123–136.

Bauer, B., and Wetzel, H. 1976. A new membrane for feeding *Glossina morsitans* (Westwood) (Diptera:Glossinidae). *Bull. Entomol. Res.* 65: 319–326.

Gooding, R., Feldman, U., and Robinson, A. 1997. Care and maintenance of tsetse colonies. In *The Molecular Biology of Insect Disease Vectors: A Methods Manual* (J. Crampton, C. Beard, and C. Louis, eds.). Chapman and Hall, Cambridge, UK, pp. 41–55.

Moloo, S. K. 1971. An artificial feeding technique for *Glossina*. *Parasitology* 63: 507–512.

Moloo, S. K., and Kutuza, S. B. 1988. Large-scale rearing of *Glossina brevipalpis* in the laboratory. *Med. Vet. Entomol.* 2: 201–202.

Welburn, S. C., and Maudlin, I. 1987. A simple in vitro method for infecting tsetse with trypanosomes. *Ann. Trop. Med. Parasitol.* 81: 453–455.

Index